4th International Conference on Mathematical Modeling in Physical Sciences (IC-MSQUARE 2015)

Journal of Physics: Conference Series Volume 633

Mykonos, Greece
5 – 8 June 2015

Editors:

Elias C. Vagenas
Dimitrios S. Vlachos

ISBN: 978-1-5108-1314-4
ISSN: 1742-6588

Printed from e-media with permission by:

Curran Associates, Inc.
57 Morehouse Lane
Red Hook, NY 12571

Some format issues inherent in the e-media version may also appear in this print version.

Printed by Curran Associates, Inc. (2016)

For permission requests, please contact the Institute of Physics
at the address below.

Institute of Physics
Dirac House, Temple Back
Bristol BS1 6BE UK

Phone: 44 1 17 929 7481
Fax: 44 1 17 920 0979

techtracking@iop.org

Additional copies of this publication are available from:

Curran Associates, Inc.
57 Morehouse Lane
Red Hook, NY 12571 USA
Phone: 845-758-0400
Fax: 845-758-2634
Email: curran@proceedings.com
Web: www.proceedings.com

Table of contents

Volume 633

4th International Conference on Mathematical Modeling in Physical Sciences (IC-MSquare2015)
5–8 June 2015, Mykonos, Greece

Accepted papers received: 3 July 2015
Published online: 21 September 2015

Preface

Papers

012144

Relaxed elastic lines of second kind in semi-dual spaces OPEN ACCESS Gülşah Aydin Şekerci and A Ceylan Çöken pg. 733

Editors:

Elias C. Vagenas, Kuwait University, Kuwait
Dimitrios S. Vlachos, University of Peloponnese, Greece

Co-Editors:

Eric Suraud, Universite Paul Sabatier, France.
Andrey V. Solov'yov, MBN Research Center, Germany.
Irene Paola De Padova, Consiglio Nazionale delle Ricerche-Istituto di Struttura della Materia, Italy.
Guy Le Lay, Aix-Marseille University, France.
Kalman Varga, Vanderbilt, United States.

Journal of Physics: Conference Series **633** (2015) 011001 doi:10.1088/1742-6596/633/1/011001

Preface

The 4th International Conference on Mathematical Modeling in Physical Sciences (IC-MSQUARE) took place in Mykonos, Greece, from Friday 5th June to Monday 8th June 2015.

The Conference was attended by more than 150 participants and hosted about 200 oral, poster, and virtual presentations. There were more than 600 pre-registered authors. The 4th IC-MSQUARE consisted of different and diverging workshops and thus covered various research fields where Mathematical Modeling is used, such as Theoretical/Mathematical Physics, Neutrino Physics, Non-Integrable Systems, Dynamical Systems, Computational Nanoscience, Biological Physics, Computational Biomechanics, Complex Networks, Stochastic Modeling, Fractional Statistics, DNA Dynamics, Macroeconomics etc.

The scientific program was rather intense as after the Keynote and Invited Talks in the morning, three parallel oral and one poster session were running every day. However, according to all attendees, the program was excellent with a high quality of talks creating an innovative and productive scientific environment for all attendees.

We would like to thank the Keynote Speaker and the Invited Speakers for their significant contribution to IC-MSQUARE. We also would like to thank the Members of the International Advisory and Scientific Committees as well as the Members of the Organizing Committee.

The Conference Chairmen

Elias Vagenas
Kuwait University

Dimitrios Vlachos
University of Peloponnese

International Advisory Committee

- **Professor Constantino Tsallis,** Centro Brasileiro de Perquisas Fisicas, Brazil
- **Professor Athanasios S. Fokas,** University of Cambridge, UK
- **Professor Henrik J. Jensen**, Imperial College London, UK
- **Professor George E. Karniadakis,** Brown University, USA
- **Professor Nikolaos E. Mavromatos**, King's College London, UK
- **Professor John E. Straub**, Boston University, USA
- **Professor Jorge Zanelli**, Centro de Estudios Científicos (CECs), Chile

International Scientific Committee

- **Bari Ferenc,** University of Szeged, Hungary
- **Mark Cross,** Swansea University, UK
- **Morten Hjorth-Jensen,** University of Oslo, Norway
- **Nathan A Baker,** Pacific Northwest National Laboratory, Richland, WA, USA
- **Nikolas Markatos,** National Technical University of Athens, Greece
- **Shiyi Chen,** Peking University, People's Republic of China
- **Kostas Kokkotas,** Eberhard Karls University of Tübingen, Germany
- **Sergey Nazarenko,** University of Warwick, UK
- **Rho Shin Myong,** Gyeongsang National University, Republic of Korea
- **Michel A. Van Hove,** City University of Hong Kong, Hong Kong
- **Jacob Barhen,** Oak Ridge Natl. Lab, Oak Ridge, USA
- **Javier Hernández-Andrés,** University of Granada, Spain
- **Maia Martcheva,** University of Florida, USA
- **Emilio Elizalde,** CSIC, Barcelona, Spain
- **Angelo Tartaglia,** Politecnico di Torino, Italy
- **Yoshiharu Yamamoto,** The University of Tokyo, Japan
- **Amaury Mouchet,** Université de Tours, France
- **Mairi Sakellariadou,** King's College of London, UK
- **Kostas Skenderis,** University of Amsterdam, The Netherlands
- **Orfeu Bertolami,** University of Porto, Portugal
- **Saurya Das,** University of Lethbridge, Canada
- **Theocharis Kosmas,** University of Ioannina, Greece
- **Georgios Loukes-Gerakopoulos,** TPI, FSU Jena, Germany

Journal of Physics: Conference Series **633** (2015) 011001 doi:10.1088/1742-6596/633/1/011001

Keynote Speaker

- **Professor Henrik J. Jensen,** Imperial College of London, Faculty of Natural Sciences, Department of Mathematics, UK

Invited Speakers

- **Professor Panos Argyrakis,** University of Thessaloniki, Department of Physics, Greece
- **Professor Kalman Varga,** Vanderbilt University, Department of Physics and Astronomy, USA
- **Professor Jorge Pullin,** Louisiana State University, Department of Physics and Astronomy, USA
- **Professor Tassos Bountis,** University of Patras, Department of Mathematics, Greece
- **Professor Gabriela Gonzalez**, Louisiana State University, Department of Physics and Astronomy, USA

Peer review statement

All papers published in this volume of *Journal of Physics: Conference Series* have been peer reviewed through processes administered by the proceedings Editors. Reviews were conducted by expert referees to the professional and scientific standards expected of a proceedings journal published by IOP Publishing.

Forecasting systemic transitions in high dimensional stochastic complex systems

Duccio Piovani[1], Jelena Grujic[2] and Henrik Jeldtoft Jensen[1]

[1]Centre for Complexity Science and Department of Mathematics, Imperial College London, South Kensington Campus, SW7 2AZ, UK

E-mail: duccio.piovani@gmail.com,jelena.grujic@gmail.com,h.jensen@imperial.ac.uk

Abstract. We briefly describe a new procedure to monitor and forecast transitions occurring in high dimensional complex systems. The method is illustrated by applications to two model systems: firstly to the Tangled Nature Model of evolutionary ecology and secondly to a stochastic replicator system. The quasi-stable configurations of the stochastic dynamics are taken as input for a stability analysis of the deterministic mean field approximation of the dynamics. We demonstrate that the largest overlap between the observed configuration and the unstable eigendirections serves as a precursor that allows us to forecast transitions with an efficiency of about 80% even if we only know the couplings matrix describing the dynamics to with 10% accuracy.

1. Introduction

Many types of complex systems are prone to sudden transitions that are challenging to forecast[1]. High dimensional complex systems both physical and biological exhibit intermittent dynamical evolution consisting of stretches of relatively little change interrupted by often sudden and dramatic transitions to a new meta-stable configurations[1, 2].

These transitions may change the state of the affairs in dramatic ways, think for instance of ecological catastrophes or crashes in financial markets. Forecasting could help to alleviate the consequences. Different types of complex systems will have different types of dynamics and each type may very well need specific types of analysis. Scheffer and collaborators have developed a method pertinent to systems in which the transition takes the form as a bifurcation captured by a robust macroscopic variable[2].

We describe an alternative scenario, suggested recently in [3], in which the transitions are induced by intrinsic fluctuations at the microscopic level of the individual components. Our approach is relevant to systems in which the dynamics creates new hitherto components that can perturb the existing order. One may think of a new and more virulent virus being created through a mutation of an existing strain (e.g. the SARS virus in 2003), or a new economic agent arriving in the market (e.g. the dot-com bubble in 1997-2000).

We exemplify the methodology by applications to to two models: (1) The Tangled Nature (TaNa) model of evolutionary ecology[4]. This model has had considerable success in reproducing both macro-evolutionary aspects such as the intermittent mode of extinctions[5] and ecological aspects such as species abundance distributions[6] and species area laws[7].(2) A high dimensional replicator with a stochastic element of mutation[8, 9].

Journal of Physics: Conference Series **633** (2015) 012001 doi:10.1088/1742-6596/633/1/012001

We recall that the stochastic replicator system is an example of the generic Langevin approach is a generic way to including a stochastic element to otherwise deterministic dynamics, see e.g. [10] and that the deterministic version of the replicator dynamics used routinely in a large variety of applications, not least because of its relation to game theory[11, 12]. This makes us hopeful that successful application of our method to the stochastic replicator indicates a broad relevance of the way we identify and analyse precursors of endogenous transitions.

2. Methodology

The procedure consists in a stability analysis of a deterministic approximation of the dynamics. If the time evolution consists of a set of stochastic processes (as they do for the Tangled Nature model, see below) a mean filed dynamics will have to be established. In the case of Langevin dynamics one simply needs to eliminate the stochastic part of the dynamics (this is the situation for the stochastic replicator system).

We define a state vector $\vec{n}(t) = (n_1(t), \ldots, n_d(t))$ and consider deterministic dynamics of the form

$$\vec{n}(t+1) - \vec{n}(t) \simeq \mathbb{T}(\vec{n}(t))\vec{n}(t) \tag{1}$$

where the matrix $\mathbb{T}$ is given by the mean field evolution matrix of derived from the stochastic dynamics.

For intermittent dynamics Eq. (10) will support periods during which little change occur from time step to time step. I.e. $\vec{n}(t+1) \simeq \vec{n}(t) \simeq \vec{n}^*$. This corresponds to configurations for which the right hand side of eq.(10) will essentially be zero. These are to a good approximation the fixed point configurations of our dynamical system. Since we are interested in very high dimensional systems, it is typically impossible to solve $\mathbb{T}(\vec{n}(t))\vec{n}(t) = 0$. Instead we use as an approximation the observed metastable state during which $\vec{n}(t+1) \simeq \vec{n}(t)$. We do a local time average of the observed $\vec{n}(t)$ and denote this average by $\vec{n}^*$. We then linearise the dynamics around this observed state by expressing the configuration vector as $\vec{n}(t) = \vec{n}^* + \delta\vec{n}(t)$

$$\begin{aligned}\delta\vec{n}(t+1) - \delta\vec{n}(t) &= \left(\mathbb{T}(\vec{n}^*) + \partial_n \mathbb{T}(\vec{n}^*)\vec{n}^*\right)\delta\vec{n}(t) \\ &= \mathbb{M}(\vec{n}^*) \cdot \delta\vec{n}\end{aligned} \tag{2}$$

and extract the stability matrix $\mathbb{M}(\vec{n}^*)$

$$\mathbb{M}(\vec{n}^*) = \left(\mathbb{T}(\vec{n}^*) + \partial_n \mathbb{T}(\vec{n}^*)\vec{n}^*\right) \tag{3}$$

As usual the stability of the dynamics near $\vec{n}^*$ is determined to linear order by the nature of the eigenvalues of $\mathbb{M}(\vec{n}^*)$. We know that the eigenvectors or *generalised* eigenvectors (in case of a non diagonalizable Jacobian) $\vec{e}_+$ associated with λ with $\mathrm{Re}(\lambda) > 0$ indicate unstable directions. These can be identified with *toxic* components n_t of the configuration vector.

A sudden growth of these components would indicate the approach of a transition. We now use as a quantitative precursor of transitions

$$Q(t) = \max_{\lambda \in Sp^+(\mathbb{M}[\bar{\vec{n}}^*])} |\langle \delta n(t), \vec{e}_\lambda \rangle| \,. \tag{4}$$

Here $\langle \delta\vec{n}(t)), \vec{e}_\lambda \rangle$ estimates the time evolution (over one time unit) of the overlap between the deviation away from the metastable state and the eigendirection $\vec{e}_\lambda$. $Q(t)$ is then the maximal overlap between this forward propagated deviation and the unstable eigendirections.

3. Applications to two models

3.1. The Tangled Nature Model

The dynamical entities of the TaNa model consist of agents represented by a sequence of binary variables with fixed length L [13], denoted as, where. We denote by $n(\mathbf{S}^a, t)$ the number of agents of type $\mathbf{S}^a$ at time t and the total population is $N(t) = \sum_{a=1}^{2^L} n(\mathbf{S^a}, \mathbf{t})$. A time step is defined as a succession of one annihilation and of one reproduction attempt. Annihilation consists of choosing an agent at random with uniform probability and remove the agent with probability p_{kill}, taken to be constant in time and independent on the type. Reproduction: choose with uniform probability an agent, $\mathbf{S^a}$, at random and duplicate the agent (and remove the mother) with probability: $p_{off}(\mathbf{S^a}, \mathbf{t}) = \frac{\exp(\mathbf{H}(\mathbf{S^a},\mathbf{t}))}{\mathbf{1}+\exp(\mathbf{H}(\mathbf{S^a},\mathbf{t}))}$ which depends on the occupancy distribution of all the types at time t through the weight function:

$$H(\mathbf{S^a}, \mathbf{t}) = \frac{\mathbf{k}}{\mathbf{N(t)}} \sum_{\mathbf{b}} \mathbf{J}(\mathbf{S^a}, \mathbf{S^b}) \mathbf{n_b(t)} - \mu \mathbf{N(t)} \tag{5}$$

In Eq. (5), the first term couples the agent $\mathbf{S^a}$ to one of type $\mathbf{S^b}$ by introducing the interaction strength $\mathbf{J}(\mathbf{S^a}, \mathbf{S^b})$, whose values are randomly distributed in the interval $[-1, +1]$. For simplification and to emphasize interactions we here assume: $\mathbf{J}(\mathbf{S^a}, \mathbf{S^a}) = \mathbf{0}$. The parameter k scales the interactions strength and μ can be thought of as the carrying capacity of the environment. An increase (decrease) in μ corresponds to harsher (more favourable) external conditions.

Mutations occur in the following way: For each of the two copies $\mathbf{S^{a_1}}$ and $\mathbf{S^{a_2}}$, a single mutation changes the sign of one of the genes: $S_i^{a1} \to -S_i^{a1}$, $S_i^{a2} \to -S_i^{a2}$ with probability p^{mut}. We define a generation to consist of $N(t)/p^{kill}$ time steps, i.e. the average time needed to kill all the individuals at time t. These microscopic rules generate intermittent macro dynamics[5] as showen in Fig. 1.

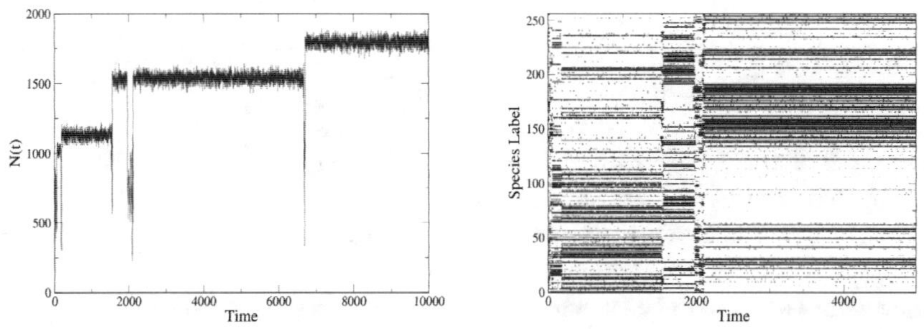

Figure 1. Left Panel: total population as a function of time (in generations) for a single realization of the TaNa. The punctuated dynamics is clearly visible: quasi-stable periods alternate with periods of hectic transitions, during which $N(t)$ exhibits large amplitude fluctuations. Right panel: occupancy distribution of the types. The genotypes are labelled arbitrarily and a blue (whilte) dot indicates a type which is (non) occupied at the time t. The lines therefore represent fit core species that fom the qESS, while the dots are the cloud of unfit species that are populated by mutations but rapidly go extinct.

3.1.1. Mean Field Description In this model the state vector $\vec{n}(t) = (n_1(t), \cdots, n_d(t))$ tells us how many individuals of each species are present at time t, therefore $n_i(t) \in \mathbb{N}$. The microscopic dynamics of the TaNa model is fully stochastic: reproduction, mutations and deaths. The average of this dynamics leads [3, 14] to a mean field evolution matrix given by

$$T_{ij} = \left(\alpha p_j^{\rm off}(t) - p^{kill}\right) \cdot \delta_{ij} + p_j^{\rm off} \cdot p_{j\to i}^{\rm mut} \cdot (1 - \delta_{ij}) . \quad (6)$$

where $\alpha = (2p_o - 1)$, the corresponding stability matrix of the linearized dynamics is

$$\mathbb{M}_{ij} = (\alpha p_j^{\rm off} - p^{kill})\delta_{ij} + 2(1 - \delta_{ij}) p_j^{\rm off} p_{j\to i}^{\rm mut} \quad (7)$$

$$+ \sum_k \left[\alpha\delta_{ik} + (1 - \delta_{ik}) \cdot p_{k\to i}^{mut}\right] \frac{\partial p_k^{\rm off}}{\partial n_j} n_k^*,$$

from which we can determine the stability spectrum and compute the indicator $Q(t)$ by use of Eq. (4).

3.2. The Stochastic Replicator System

Describes a large number of agents involved in games and applying different strategies [12]. We consider a version in which strategies are removed from the system if the frequency with which appear are too low. We also include a stochastic element, which describes players that change their strategy[9].

The instantaneous state of the system is given by a configuration vector $\vec{n}$, which contains the frequency (hence $n_i \in [0, 1]$) of each possible strategy. There a total of d possible strategies. The outcome of the games are described by a dxd payoff matrix J. This matrix is constant. Each strategy is related to the other strategies by a set of payoffs, or interactions.

The deterministic part of the time evolution is done by first computing the *fitness*, $h_i(t) = \sum_j J_{ij} n_j(t)$ of each active strategy and compare it with the average fitness $\bar{h}(t) = \sum_{ij} J_{ij} n_i(t) n_j(t)$. Each frequency is updated according to

$$n_i(t+1) = n_i(t) + \left(\sum_j J_{ij} n_j(t) - \sum_{ij} J_{ij} n_i(t) n_j(t)\right) \cdot n_i(t) \quad (8)$$

The stochastic part of the dynamics is implemented in the following way. With probability $p^{\rm mut}$ a strategy number i mutates into another one. In this process a fraction $\alpha^{\rm mut}$ of the frequency of that strategy is moved to a strategy with a label chosen from the normal distribution $N(n_i(t), \sigma)$ centred on label i with variance σ. The closer the labels of two strategies the more likely it is for one to mutate into the other.

Should a frequency become smaller than an extinction threshold $n_i(t) < n^{\rm ext}$ it is remove $n_i(t+1) = 0$. After an extinction event the system is renormalized in order to respect the condition $\sum_i n_i(t) = 1$.

We show in Fig.(2) and example of the complex macroscopic dynamics generated by these dynamical rules.

3.2.1. Mean Field Description

The mean field evolution equation obtained by averaging over the stochastic mutations and keeping in mind that strategies may go extinct takes the form

$$n_i(t+1) \simeq n_i(t) + \left(\sum_j J_{ij} n_j(t) - \sum_{ij} J_{ij} n_i(t) n_j(t)\right) \cdot n_i(t)$$

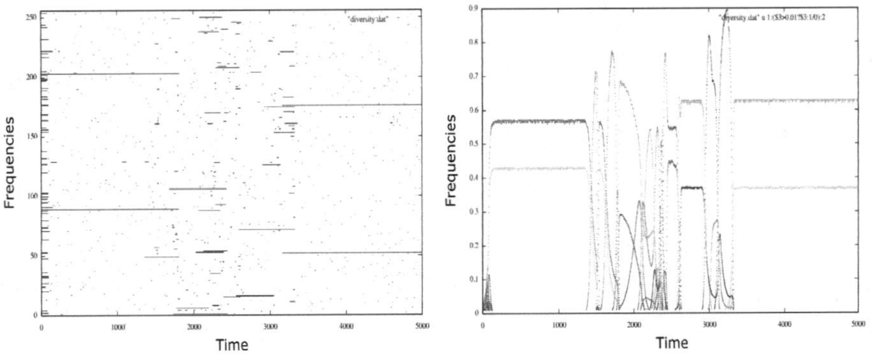

Figure 2. Left panel: We present once again the occupation plot. This is done exactly like in the Tangled Nature's case. The blue lines represent strategies, typically very fit, that are consistently in the system. The blue dots are strategies populated by mutations, that quickly go extinct. Populating a new fit strategy usually indicates the arrival of a transition (label 50 around $t = 1500$) Right panel: Each colour is associated with a different strategy, and indicates the frequency of players that are using it. We can clearly see how the system jumps from one configuration to the other.

$$-\alpha^{\text{mut}}p^{\text{mut}} \cdot n_i(t) + \frac{p^{\text{mut}}}{N_a(t)} \sum_j \alpha^{\text{mut}} n_j(t) p_{j\to i} \tag{9}$$

where $p_{j\to i} = \frac{\exp(-\frac{\|i-j\|}{2\sigma^2})}{\sqrt{2\pi\sigma^2}}$ follows the normal distribution. We can express Eq.(9), in compact form as

$$n_i(t+1) - n_i(t) \simeq \sum_j \mathbb{T}_{\text{ij}} n_j(t) \tag{10}$$

where

$$\mathbb{T}_{ij} = \left(\sum_j J_{ij} n_j(t) - \sum_{ij} J_{ij} n_i(t) n_j(t) - \alpha_{\text{mut}} \right) \cdot \delta_{ij} \tag{11}$$

$$-\frac{\alpha_{\text{mut}}}{2^L - 1} \cdot (1 - \delta_{ij}) \, .$$

The corresponding linear stability matrix takes the form

$$\mathbb{M}_{ij} = \mathbb{T}_{ij}(\vec{n}) + \left[J_{ij} - \sum_k (J_{ik} + J_{ki}) \vec{n}_k^* \right] \vec{n}_i^* \tag{12}$$

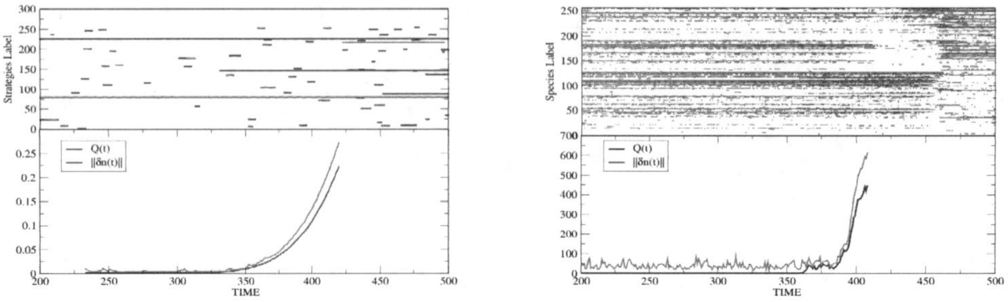

Figure 3. We show the behaviour of our stability indicator $Q(t)$, in both the Replicator Model (Left Panel) and the Tangled Nature (Right Panel). We compare Q's behaviour (blue curve) both to the displacement's $\delta n(t)$ (red curve) behaviour and the occupation plot. We can see how in both models the $Q(t)$ peaks only before the transition, while it doesn't feel the previous fluctuations.

4. Results

We now present how the indicator defined in Eq. (4) behaves as function of time for the two models. A comparison between the temporal evolution of $Q(t)$ and the deviation between the meta-stable configuration and the instantaneous state

$$\|\delta\vec{n}(t)\| = \|\vec{n}(t) - \bar{\vec{n}}_{\text{stoc}}\| \tag{13}$$

is given in Fig.(3). We see that $Q(t)$ is not sensible to the stochastic fluctuations of that happen during the qESS, but peaks only just before the transition actually occurs. This suggests that the $Q(t)$ correctly distinguishes between harmless fluctuations, that happen in the stable subspace, and the dangerous ones that are capable to push the system out of its configuration. As one can see from Fig.(3) the $Q(t)$ starts peaking while the $\delta n(t)$ is stil in its characteristic values. This happens several steps before the transition occurs, proving the forecasting power of the variable.

A more systematic analysis is shown in Fig.(4) where we investigate the behaviour of the the indicator $Q(t)$ and $\delta n(t)$ indicator just before the transition, on 10000 transitions for each model. As we can see far from the transition (t=20) most of the Q(t) are close to zero, but approaching the transition (t = 0) the average values of $Q(t)$ start increasing while the $\delta n(t)$ stays more or less constant.

As one can clearly see in Fig.(4) some of the transitions are not predicted by our method. This means, in these *missed transitions*, that the system is driven out of its configuration by a perturbation parallel to what we consider the the stable subspace; for details see [3].

5. Conclusion

We have presented a procedure for forecasting the approach of systemic transitions in which stochastic dynamics is combined with mean field stability analysis about the "observed" configuration at a given time. The crux of the method is to include in the analysis new mutants or new strategies which originates from the existing agents but are distinct from them. We believe that our method can be applied to the dynamics of many real complex systems such as ecosystems and the financial market. The crucial point will be to carefully monitoring what happens at the perimeter of the extant agents.

6. Acknowledgments

This work is supported by the European project CONGAS (Grant FP7-ICT-2011-8-317672)

7. References

[1] P. Sibani and H.J. Jensen. *Stochastic Dynamics of Complex Systems*. Imperial College Press, 2013.

[2] Marten Scheffer, Jordi Bascompte, William A Brock, Victor Brovkin, Stephen R Carpenter, Vasilis Dakos, Hermann Held, Egbert H Van Nes, Max Rietkerk, and George Sugihara. Early-warning signals for critical transitions. *Nature*, 461(7260):53–59, 2009.

[3] A. Caroli, D. Piovani, and H.J. Jensen. Forecasting transitions in systems with high dimensional stochastic complex dynamics: A linear stability analysis of the tangled nature model. *Phys. Rev. Lett.*, 113:264102, 2014.

[4] K. Christensen, S. A. di Collobiano, M. Hall, and H. J. Jensen. Tangled nature model: A model of evolutionary ecology. *J. Theor. Biol.*, 216:73–84, 2002.

[5] M. Hall, K. Christensen, S. A. di Collobiano, and H. J. Jensen. Time-dependent extinction rate and species abundance in a tangled-nature model of biological evolution. *Phys. Rev. E*, 66, 2002.

[6] E. P. Anderson and H. J. Jensen. Network properties, species abundance and evolution in a model of evolutionary ecology. *J. Theor. Biol.*, 232:551–558, 2005.

[7] D. Lawson and H. J. Jensen. The species-area relationship and evolution. *J. Theor. Biol.*, 241:590–600, 2006.

[8] Diederich Sigurd and Manfred Opper. Replicators with random interactions: A solvable model. *Physical Review A*, 39:4333, 1989.

[9] K.and Yasutomi A. Tokita. Emergence of a complex and stable network in a model ecosystem with extinction and mutation. *Theoretical population biology*, 63:131–146, 2003.

[10] G. A. Pavliotis. *Stochastic Processes and Applications Diffusion Processes, the Fokker-Planck and Langevin Equations*. Springer, 2014.

[11] R. Veltri A. Kianercy and K.J. Pienta. Critical transitions in a game theoretic model of tumour metabolism. *Interface focus*, 4(4):20140014, 2014.

[12] Josef Hofbauer and Karl Sigmund. *Evolutionary Games and Population Dynamics*. Cambridge University Press, 1998.

[13] P.G. Higgs and B. Derrida. Genetic distance and species formation in evolving populations. *J. Mol. Evolution*, 35(5):454–465, 1992.

[14] Simone Avogadro di Collobiano, Kim Christensen, and Henrik Jeldtoft Jensen. The tangled nature model as an evolving quasi-species model. *J. Phys A*, 36:883–891, 2003.

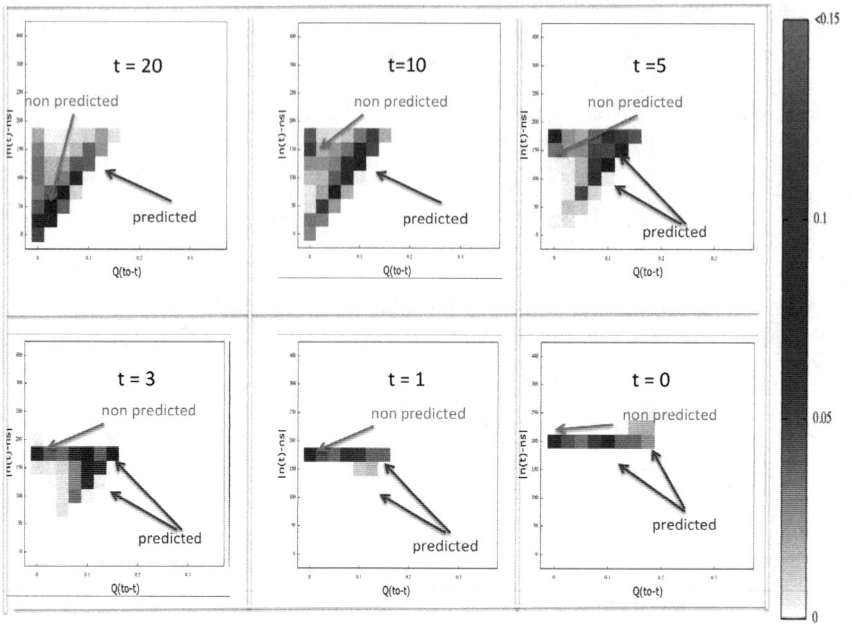

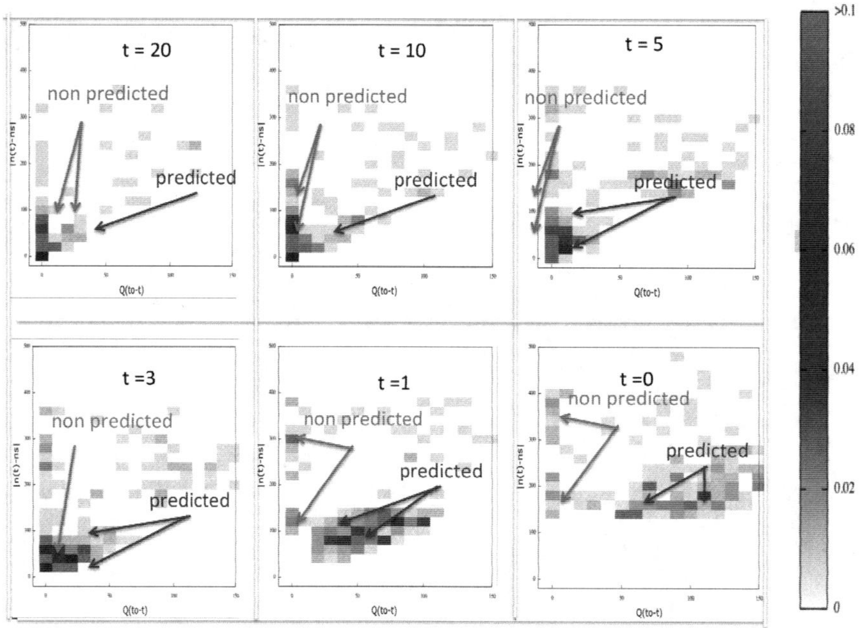

Figure 4. Top panel Replicator Model and bottom panel Tangled Nature model,. The figures compare the indicator $Q(t^*-t)$ (x-axis) and $\delta n(t^*-t)$ (y-axis) where t^* is the time the transitions begins. As we can see for $t = 20$, meaning 20 steps before the transitions most of the transitions are not predicted ($Q(t^* - t) \simeq 0$), but as the system approaches the transition the vast majority of transitions (80/90%) are predicted by an increase of $Q(t)$.

Investigating ionisation cluster size distribution due to sub-1 keV electrons in view of Heisenberg's Uncertainty

B Li[1,2,*], H Palmans[1], L Hao[1] and A Nisbet[2,3]
[1] National Physical Laboratory, Teddington, Middlesex TW11 0LW, U.K.
[2] The Department of Physics, University of Surrey, Guildford GU2 7XH, U.K.
[3] Royal Surrey County Hospital, Guildford, Surrey GU2 7XX, U.K.

E-mail: bo.li@surrey.ac.uk

Abstract. As the wavelengths of low energy electrons become comparable with the length scale of the mean ionisation step size, each event particle should be treated with care as the condition outlined in Heisenberg's uncertainty principle (HUP) should also be satisfied. Within this quantum-classical regime, spatial delocalisations of individual ionisation event sites that are generated outside the target region are calculated, and particular attention is given to the validity of using classical transport methods in simulations of nanodosimetric parameters such as mean cluster size, first and second moments, variance and cumulative frequency of ionisation cluster-size probability distributions. This paper presents the comparison between conventionally calculated nanodosimetric quantities and the ones where interacting particles are treated semi-classically with spatial uncertainties satisfied by HUP. The simulated primary charged particles are electrons of energies between 100 eV and 1 keV in DNA equivalent target aqueous water volumes using GEANT4-DNA.

1. Introduction

High levels of cellular damages upon exposure to irradiation are thought to be initiated by low energy secondary electrons, typically of energies of a few hundreds eV [5, 13]. In view of radiation induced initial damage to biologically sensitive targets, typical studied volumes are of biological equivalence to segments of DNA of 10 base pairs [3, 7] has shown that electrons of a few hundreds eV yield the highest number of double-strand breaks in the DNA, which can be explained as the probability of electron-induced ionisation clusters greater than 2 also peaks for initial electron energies of a few hundred eV [6].

In the frame of radiation transport modelling, methods that are generally used for macroscopic target sizes such as condensed theory methods are no longer applicable for target sizes approaching the biological sensitive volumes, typically of a few nanometres. When the target volume of interest is reduced, the occurrences of particle interactions are considered as stochastic. Most computational studies of radiation track structure are achieved by event-driven simulations based on Monte Carlo methods, such as GEANT4-DNA, where each particle is treated as a classical object with precisely known position and energy. In contrary to the multiple scattering theories used in condensed history methods, each particle in the track-structure simulations is treated individually with discretisation of interaction events. However, care should be taken in the use of these 'classical' track-structure codes

* To whom any correspondence should be addressed.

as dimensions of the target volume approach nanoscale, as well as energies of interest (i.e. ionisation energies of liquid water) enter the sub-1keV range [1,15]. This is due to spatial delocalisation in the location of the interaction sites, which is dependent on the velocity and the energy loss of the interacting particle [10].

2. Background

2.1. Quantum limit of classical mechanics

In the quantum limit of classical mechanics, the product of $\Delta \mathbf{x}$ and $\Delta \mathbf{p}$ will approach the quantum limit at $\hbar$ [4], where quantum quantities $\Delta \mathbf{x}$ and $\Delta \mathbf{p}$ are the delocalisation in position and uncertainty in momentum respectively. For singular events such as the individually calculated interaction event by the Monte Carlo methods, the associated position uncertainty depends on the interacting particle's initial velocity $\boldsymbol{v}_x$ and the energy deposition ΔE of the associated interaction event. This relation is shown by Kaplan and Miterev [10] as, $\Delta \mathbf{x} \geq (\hbar \boldsymbol{v}_x) \cdot (\Delta E)^{-1}$. By introducing the lower bound of the spatial uncertainty $\Delta \boldsymbol{x}$ as the critical value $\Delta \boldsymbol{x}_c$, the inequality sign is dropped, as shown in equation 1.

$$\Delta \boldsymbol{x} \geq \frac{\hbar \boldsymbol{v}_x}{\Delta E} \tag{1}$$

Two conditions must be simultaneously satisfied for the classical treatment of particles to remain as valid. First, the particle's associated spatial delocalisation must be greater or equal to Δx_c. And second, $\Delta \boldsymbol{x}_c$ must be small with respect to the mean free ionisation (the interaction of interest for radiobiological studies) path length $\lambda_{ion}(T)$ of electrons in aqueous water, at a given energy T of interest. I.e. $\Delta \boldsymbol{x}_c < \lambda_{ion}(T)$. Typical electron ionisation path lengths in liquid water $\lambda_{ion}(T)$ increase steadily from ~1.5 nm at electron energy of 100 eV to ~3.8nm for 1 keV electrons. A small local minimum occurs at electron energy of ~150 eV at step length of ~1.4 nm.

2.2. Nanodosimetric parameters of track structure

The propagation of a charged particle at initial energy T through a target volume of given geometry dimension and composition G, the ionisation cluster-size probability distribution can be written as $P_\upsilon(T;G)$. This describes the probability that exactly υ number of ionisations is produced within a pre-defined target volume, where υ is defined as the number of ionisation event counts per single track of a primary charged particle. The mean ionisation cluster number of the distribution $P_\upsilon(T;G)$ is known as the first moment $M_1(T;G)$, and will be referred to as the mean cluster size hereforth. This can be calculated according to the generic equation, where ζ denotes the order of moment, as shown in equation 2.

$$M_\zeta(T;G) = \sum_{\upsilon=0}^{\infty} \upsilon^\zeta P_\upsilon(T;G) \tag{2}$$

The stochastic fluctuations on the number of ionisation formation within microscopic target volumes can be characterised by the variance $V_{rel}(T;G)$ of the distribution $P_\upsilon(T;G)$. This is proposed by Fano as the quotient of the variance and the mean of the cluster-size distribution, shown in equation 4.

$$V_{\mathrm{rel}}(T;G) = \frac{M_2(T;G) - M_1^2(T;G)}{M_1(T;G)} \tag{3}$$

Another nanodosimetric parameter that is examined in this paper is the cumulative probability $F_2(T)$. This is defined as the probability for the primary electron to undergo at least two ionisation interactions within the target volume. It is thought that this quantity could be related to the probability of the formation of double strand breaks within DNA segments [8,14].

$$F_2(T) = \sum_{\upsilon=2}^{\infty} P(\upsilon|T;G) \tag{4}$$

3. Methods

3.1. Simulation of ionisation cluster formation in biologically equivalent DNA target volumes

Low electromagnetic package GEANT4-DNA version 10.0.0 is used to obtain ionisation cluster distributions in this paper. Detailed description of the simulation packages can be found in the literature [9,11].

The simulated environment consisted of a cube of liquid-filled world volume. The side length of the cube is chosen to be greater than the electron total interactions range at given primary particle energy levels. A target cylindrical volume consisted of liquid water, which is biologically equivalent to a DNA segment of 10 base pairs, has dimensions of 2.3 nm diameter and 3.4 nm height. The target volume is placed in the center of the world volume. The use of nanodosimetric cylindrical liquid water environment to study nanodosimetric parameters has been a common practice [5,12,13]. Primary particles are initiated from the center of the target volume with momentum towards the positive z-direction (along the principal axis of the target cylinder). The center is defined as the point of half height of the target cylinder and at the origin with respect to its base plane.

Ionisation cluster-sizes are recorded per primary particle within the target volume in two ways. First, without the consideration of the spatial delocalisation constraint due to HUP, a bench mark ionisation cluster-size distribution is collected with well-defined target volume boundaries. In a second case, ionisation events that occur outside the defined target volume are also considered if their delocalised spatial range overlaps or enters the target region. The boundaries of the delocalised position are determined according to equation 1 given the coordinates of the interaction event and the initial velocity of the interacting particle. This will be referred to as HUP-cluster-size distributions.

4. Results and discussions

The average spatial delocalisations associated with ionisation events within the DNA-equivalent target region are calculated from equation 1 as 0.2615 nm and 0.726 nm for 100 eV and 1 keV electron primaries respectively. Despite the average increases in the ionisation energy deposition within the DNA equivalent target volume: from 12.55 eV to 13.73 eV for primary electrons of 100 eV and 1 keV respectively, the average velocity of the interacting particles at the ionisation sites has increased by a much larger proportion: from 4.81 mm/ns (for 100 eV primaries) to 14.33 mm/ns (for 1 keV primaries). This has resulted in the average increase in spatial uncertainties in ionisation interaction sites as electron primary energies increase from 100 eV to 1 keV, within the predefined nano-target volume. Mean free ionisation path length of electrons in liquid water is ~1.5 nm and ~3.8 nm for electron energy of 100 eV and 1 keV respectively. The spatial uncertainties associated with both energy levels are smaller, however, and are comparable, with respect to the mean ionisation free path length. This suggests that the simulated environment is in the quantum-classical regime and the constraint posed by HUP should also be satisfied for the classical treatments of particles to hold.

Figure 1 shows the cluster-size distribution of primary electrons due ionisation events in liquid water at two different initial energies T: 100 eV and 1 keV. In the case when T=100 eV, only 5% of the electron primaries as simulated in the HUP-cluster-size distribution received additional counts of ionisations from outside of the target volume. Within the total 5% of the affected primaries, the majority at 4.76% had an increase shift in cluster-size by one. The remaining 0.2% and 0.01% were contributed from two and three additional counts respectively. As seen from figure 1 the shape of the two cluster size distributions coincides very almost perfectly. The percentage difference between the two distributions with respect to the well-defined target boundaries (i.e. the conventional method of cluster-size definition) is also represented. It can be seen that their differences over most cluster sizes are below 0.1%. The slight increase in the tail at higher cluster size number is due to the skewed amplification of the difference from small sample sizes as larger cluster sizes become less probable. Similar results are seen for electron primaries of 1 keV. Here, the general shape of the probability distribution becomes logarithmic. This can be explained by the increase in ionisation mean free path for increase in electron energy in liquid water. As the ionisation mean free path becomes larger than the half height (primary electron is initiated from the half height) of the target volume, it is expected that most of the primary electrons do not undergo ionisation events and thus energy is transported out of the target region via primaries. The percentage discrepancies between the conventionally defined target volume simulation and the HUP-cluster-size distribution are also shown in figure 1. Again, it is seen that the two distributions overlay almost perfectly. Percentage differences between the two are even smaller than the case with 100 eV primary electrons.

Figure 2 shows the mean cluster sizes for the two methods in energy ranges from 100 eV to 1 keV. Discrepancies between the two distributions are below 3.5%. Furthermore, it is seen that these differences are not significant to change the general shape of the distributions. This can be seen in figure 3 as the percentage of variance difference fluctuates at ~1.5%. Similarly, there are no significant differences found in other nanodosimetric parameters: M_2, and cumulative frequency (figure 4). All of the above nanodosimetric parameters, including M_1 and ionisation cluster-size distributions show good agreement with other published results using the classical treatments [7,12].

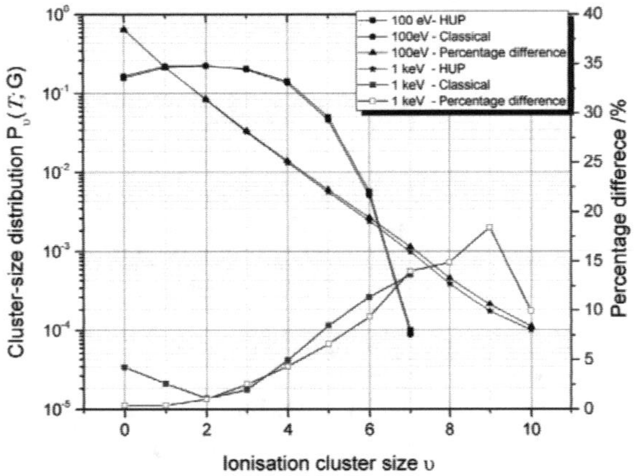

Figure 1. This is the ionisation cluster-size probability distribution $P_υ(T;G)$ for primary electrons at initial energy of 100 eV and 1 keV for both cases: conventional cluster-size distribution and HUP-cluster-size distributions.

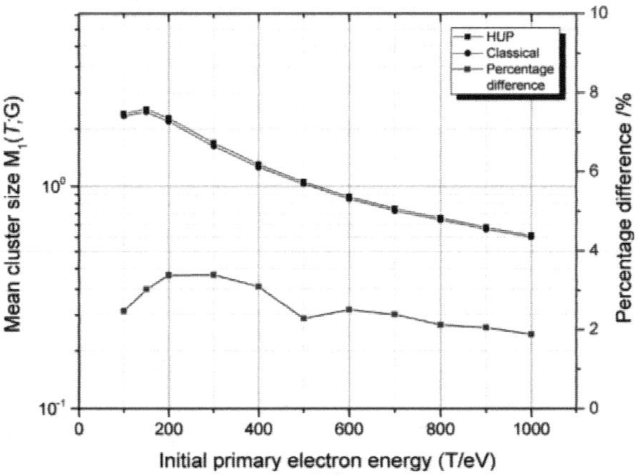

Figure 2. shows the mean ionisation cluster size of primary electrons with energy T from 100 eV to 1 keV.

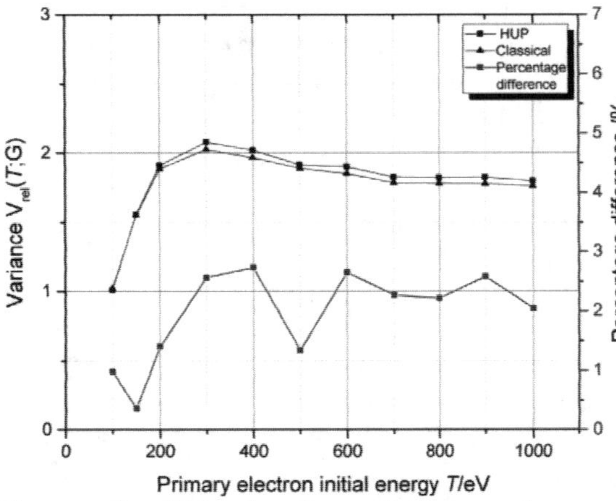

Figure 3. This is the relative variance as calculated by equation 3 for primary electron energies from 100 eV to 1 keV.

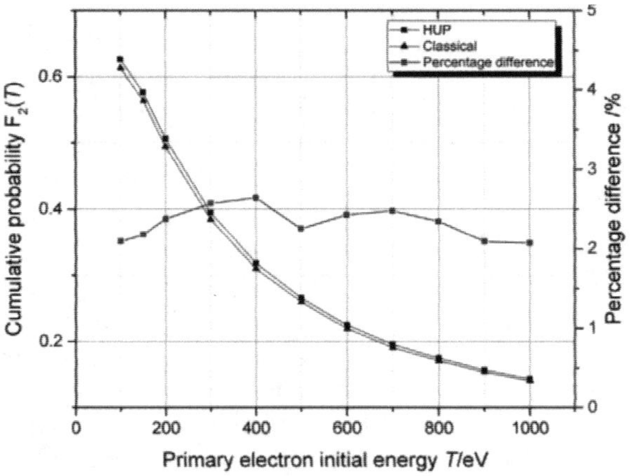

Figure 4. The cumulative frequency $F_2(T)$ is shown (calculated from equation 4) for primary electron energies from 100 eV to 1 keV.

5. Results and discussions

It has been demonstrated in this paper that for electron primaries of energy between 100 eV -1 keV special care needs to be taken when studying their track structure at nanometer scale, using classical mechanics. This can be seen as the mean free ionisation path length is comparable with the spatial uncertainties of the particles that are involved with ionisation interactions in the defined DNA-equivalent target regions. Given the non-violation of Heisenberg's uncertainty principle (HUP), which proposes constraints on the spatial and momentum uncertainties, nanodosimetric quantities are simulated using the Monte Carlo codes GEANT4-DNA. In particular, the studied nanodosimetric quantities are mean ionisation cluster-size distribution, first and second ionisation moment, and the ionisation variance as well as the cumulative frequency of the ionisation cluster-size distribution. The effect on the ionisation cluster-size based on the determination of each interaction's associated spatial delocalisation (as given by HUP), as opposed to the conventional consideration of well-localised position is examined in this paper. It has been found that the ionisation cluster-size distributions for electrons of energy 100-1k eV do not show significant variation due to the inclusion of HUP, as their significant (discarding larger, improbable cluster sizes) percentage differences are only within 5%. It also follows that the other studied nanodosimetric parameters are relatively unaffected by the inclusion of HUP.

6. Acknowledgement

This work is funded by Biologically weighted Quantities in RadioTherapy (BioQuaRT), a Joint Research Project within the European Metrology Research Programme EMRP, Engineering and Physical Sciences Research Council U.K. (EPSRC) and NHS Invention for Innovation U.K. (i4i).

7. References

[1] Emfietzoglou D, Papamichael G, Kostarelos K, & Moscovitch M. 2000 *Physics in Medicine and Biology,* **45**(11), 3171
[2] Fano U. 1947 *Physical Review*, **72**(1), 26.
[3] Friedland W, Jacob P, Paretzke H G and Stork T. 1998 Radiation research, **150**(2), 170-182.
[4] Gottfried K and Yan T M 2003. Springer Science & Business Media.
[5] Grosswendt B 2002. *Radiation and environmental biophysics*, **41**(2), 103-112.
[6] Grosswendt B 2005. *Radiation protection dosimetry*, **115**(1-4), 1-9.
[7] Grosswendt B 2006. Radiation protection dosimetry, **122**(1-4), 437-445.
[8] Grosswendt B, Pszona S and Bantsar A. 2007 *Radiation protection dosimetry*, **126**(1-4), 432-444.
[9] Incerti S, Baldacchino G, Bernal M, Capra R, Champion C, Francis Z, and Zacharatou C 2010 *International Journal of Modeling, Simulation, and Scientific Computing*, **1**(02), 157-178.
[10] Kaplan I G and Miterev A M. 1985 *Radiation Physics and Chemistry* (1977), 26(1), 53-56.
[11] Ivanchenko V N, Incerti S, Francis Z, Tran H N, Karamitros M, Bernal M A, and Guèye P. 2012 *Nuclear Instruments and Methods in Physics Research Section B: Beam Interactions with Materials and Atoms*, **273**, 95-97.
[12] Lazarakis P, Bug M U, Gargioni E, Guatelli S, Rabus H and Rosenfeld A B. 2012 *Physics in medicine and biology*, **57**(5), 1231.
[13] Nikjoo H and Goodhead D T. 1991 *Physics in medicine and biology*, **36**(2), 229.
[14] Rabus H and Nettelbeck H. 2011 *Radiation Measurements*, **46**(12), 1522-1528.
[15] Thomson, R. M., & Kawrakow, I. 2011 *Medical physics*, **38**(8), 4531-4534.

Journal of Physics: Conference Series **633** (2015) 012003 doi:10.1088/1742-6596/633/1/012003

Modelling irradiation by EM waves of multifunctionalized iron oxide nanoparticles and subsequent drug release

Feng Wang, Florent Calvayrac, Véronique Montembault, and Laurent Fontaine

Université du Maine (le Mans), Institut des Molécules et Matériaux du Mans, France

E-mail: `florent.calvayrac@univ-lemans.fr`

Abstract. Thermal transport in the environment close to the periphery of the nanoparticle, from a few angstroms to less than a nanometer scale, is becoming increasingly important with the advent of several biomedical applications of multifunctional magnetic nanoparticles, including drug delivery, magnetic resonance imaging, and hyperthermia therapy. We present a multiscale and multiphysics model of the irradiation by electromagnetic waves of radiofrequency of iron oxide nanoparticles functionalized by drug-releasing polymers used as new multifunctional therapeutic compounds against tumors. We compute ab initio the thermal conductivity of the polymer chains as a function of the length, model the unfolding of the polymer after heat transfer from the nanoparticle by molecular mechanics, and develop a multiscale thermodynamic and heat transfer model including the surrounding medium (water) in order to model the drug release.

1. Introduction

Magnetic nanoparticles can chemically anchor or physically adsorb specific polymers for further functionalization, depending on the desired application[1]. Functionalized magnetic nanoparticles with certain polymers as drug carriers are very promising for applications in magnetic-field-directed drug targeting and magnetic-field-controlled drug delivery[2, 3, 4]. The magnetic-field-directed drug targeting is to guide functional magnetic nanoparticles to a chosen site under the localized magnetic field gradients, while magnetic-field-controlled drug-releasing is based on the fact that remote release of drugs is initiated by the heat generated upon alternating magnetic field exposure. The controlled placement and drug delivery of these functional magnetic nanoparticles by means of an external magnetic field enables their biomedical application of having the potential to achieve high local concentration, and avoiding toxicity and other adverse side effects arising from high drug doses in other parts of the organism. Readers who are interested in a more detailed understanding of the physical properties and behavior of these magnetic nanoparticles, or biomedical and biotechnology applications, are referred to specific reviews[5, 6, 7, 8, 9, 10, 11, 12, 13, 14]. In the following, after briefly showing a diagrammatic representation of research ideas in the context of functional iron oxide magnetic nanoparticles (IONPs) with hyperthermia-induced drug-releasing of recent interest, we focus mainly on how heat is transferred through the polymer chains between the drug linker and the nanoparticle surface in such functional IONPs.

Journal of Physics: Conference Series **633** (2015) 012003 doi:10.1088/1742-6596/633/1/012003

2. Diagrammatic Representation of Research Ideas

To build an suitable computational model of functional IONPs, we employ a reported design scheme[15] that contains a phosphonic acid group for the binding to the iron oxide surface of IONPs, and two orthogonal clickable groups: an alkyne moiety for installing an azide end-functionalized hydrophilic polymer through 1,3-dipolar cycloaddition to impart aqueous dispersability/stability, antifouling property, and biocompatibility to the IONPs; and a furan ring that serves as a drug linker for a biologically active molecule through reversible Diels-Alder (DA) chemistry (as shown in Figure 1 and Figure 2).

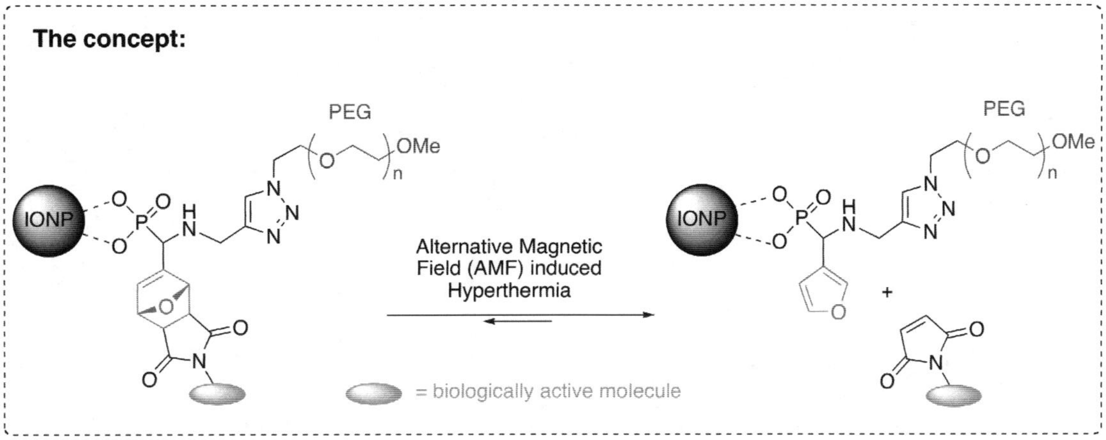

Figure 1. A reported design scheme of functional iron oxide magnetic nanoparticles with hyperthermia-induced drug-releasing.

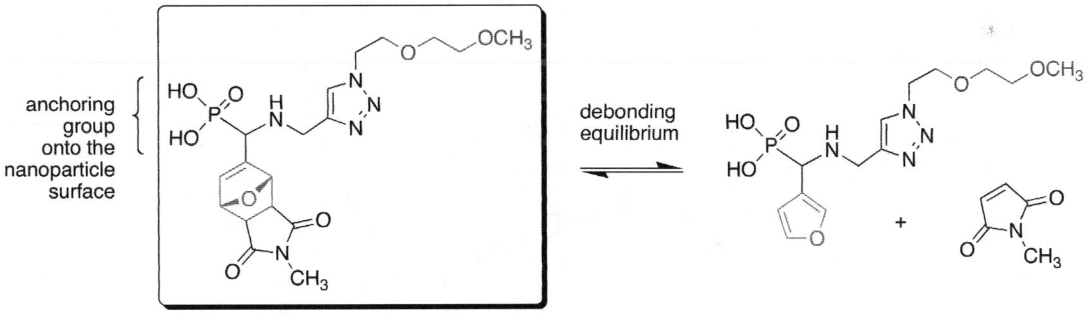

Figure 2. The structure to be considered for modeling.

It is well known that the microscopic reverse of the DA reaction, namely the retro-Diels-Alder (retro-DA) reaction, can be accomplished spontaneously with a supply of activation of heating. Consequently, one can speculate that when such functional IONPs are exposed to a varying magnetic field, the particles would become powerful heat sources, and sufficient heat could be brought in close proximity to the furan ring to initiate the retro-DA reaction. Indeed, recent experiments have shown that localized heating is observed without any significant heating of the medium[15, 16, 17] when IONPs are exposed to alternating magnetic field, forming very high temperature gradient from the nanoparticle surface. Obviously, for the given magnetic field frequency and the size of the IONPs, the chain length between the drug linker and the nanoparticle surface should be precisely designed/optimized to bring enough heat to engage the retro-DA reaction, (as shown in Figure 3). Theoretical studies of the possible heat transfer mechanisms have been initiated, but to date obtaining an atomic- and microscale-level

Journal of Physics: Conference Series **633** (2015) 012003 doi:10.1088/1742-6596/633/1/012003

understanding of how heat is transferred through the polymer chains between the furan ring and the nanoparticle surface remains the greatest challenge that must be overcome in order to realize the full potential of this new class of functional magnetic nanoparticles.

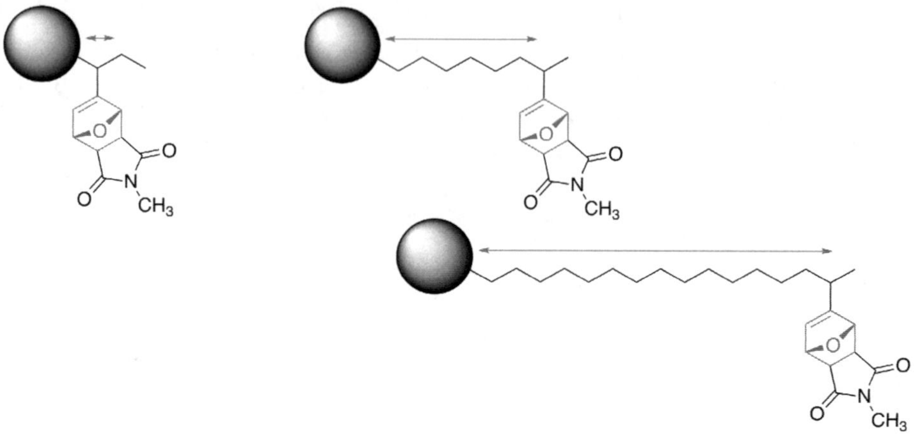

Figure 3. To look at the influence of the chain length between the drug linker and the nanoparticle onto heat transfer.

3. Atomic-level Modeling of Heat Transfer of Polymer Chains

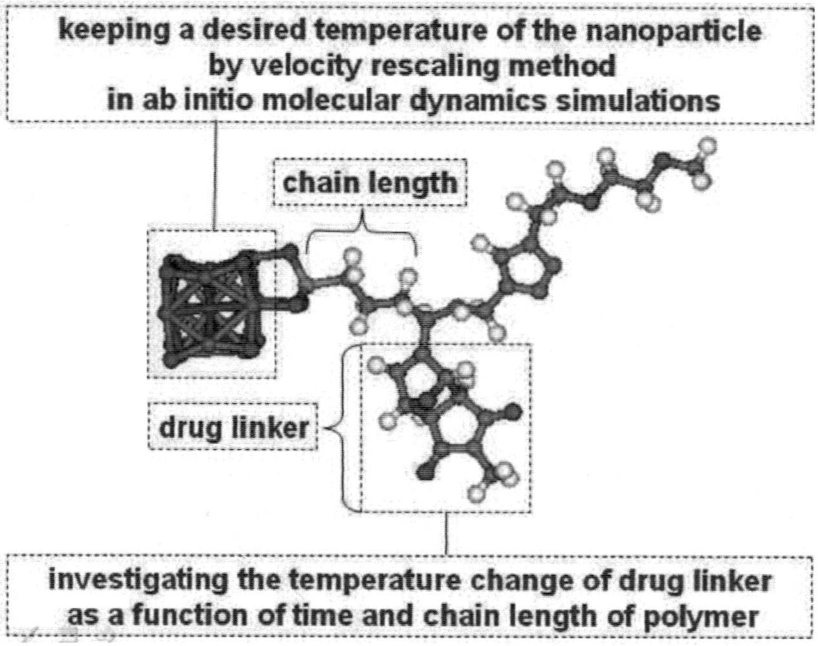

Figure 4. Atomic-level modeling of heat transfer of polymer chains from density-functional-theory-based ab initio molecular dynamics simulations.

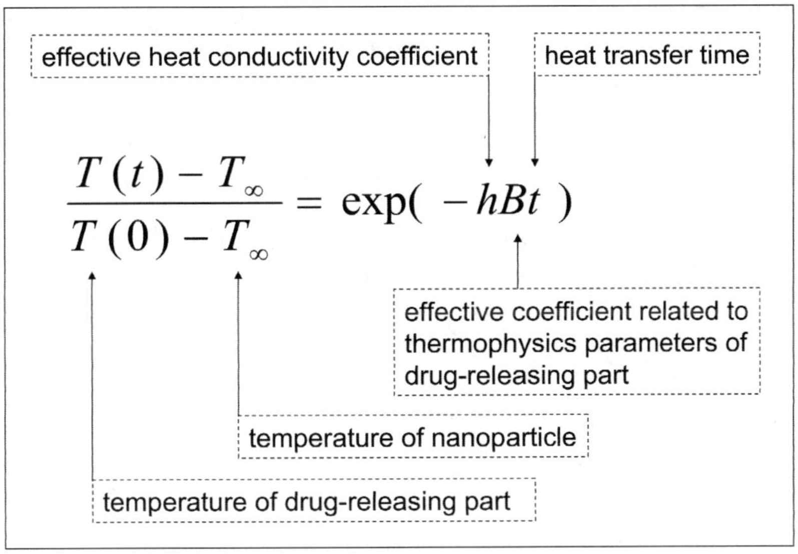

Figure 5. Heat conduction model with negligible internal resistance.

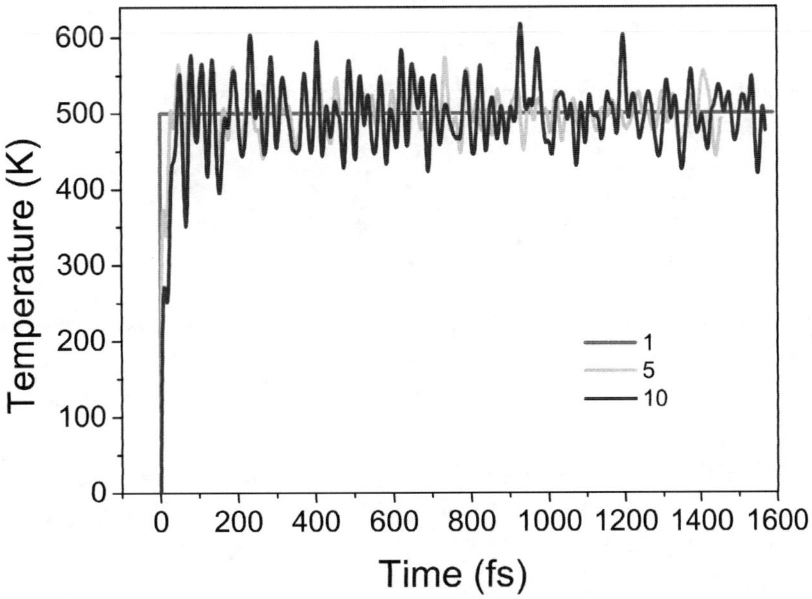

Figure 6. Checking velocity rescaling method with different relaxation time.

This study develops one of the most comprehensive series of molecular dynamics simulations to gain insight into thermal transport between the thermoreversible linker and the nanoparticle surface for functional IONPs in more realistic atomic-level modeling. Our approach is to hold the nanoparticle of the simulation structure at some desired temperature by rescaling velocities of nanoparticle atoms as a heat source while investigating temperature rising process of the drug linker as a function of time and chain length (as shown in Figure 4).

An approach is described by Figure 5 for determining thermal conductivity of polymer chains. The approach idea, which is directly analogous to the lumped thermal capacity model[18], is that the resistance to heat transfer within the drug-releasing part is negligible when compared with the resistance to heat transfer with the polymer chains.

The velocity rescaling method has been checked with different 'relaxation time' (as shown in Figure 6), which gives a rough estimates of the time needed to achieve the given target temperature, and controls the global temperature fluctuation intensity around the given target temperature.

4. Results and Discussion

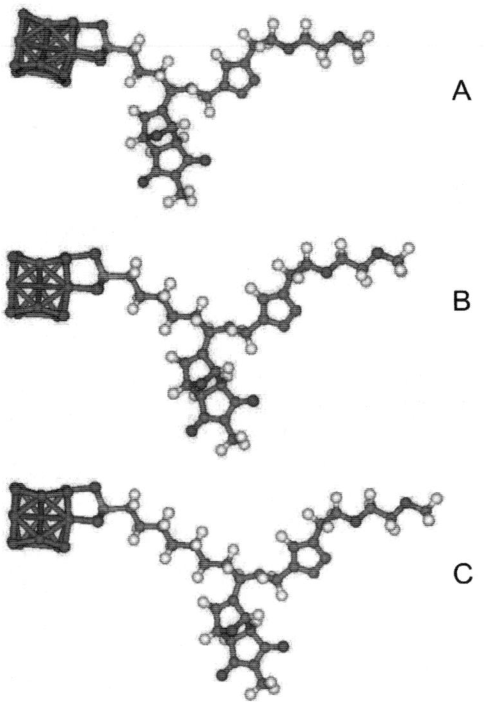

Figure 7. Optimized geometries of iron oxide nanoparticles functionalized by polymers with different chain length.

To investigating temperature rising process of the drug linker as a function of chain length, we have optimized three geometries of iron oxide nanoparticles functionalized by polymers with different chain length (as shown in Figure 7).

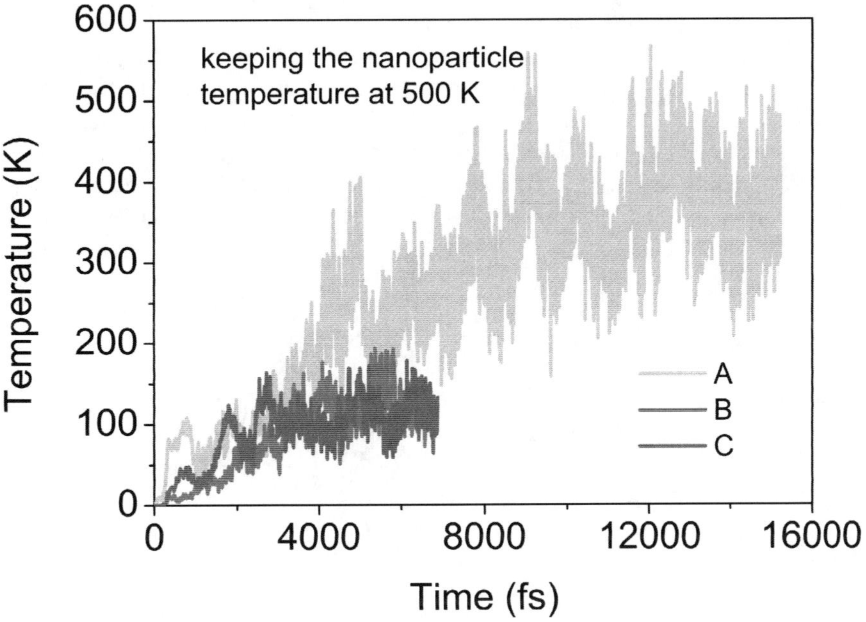

Figure 8. The temperature rising process of the drug linker in cases of A ,B and C, keeping the nanoparticle temperature at 500 K.

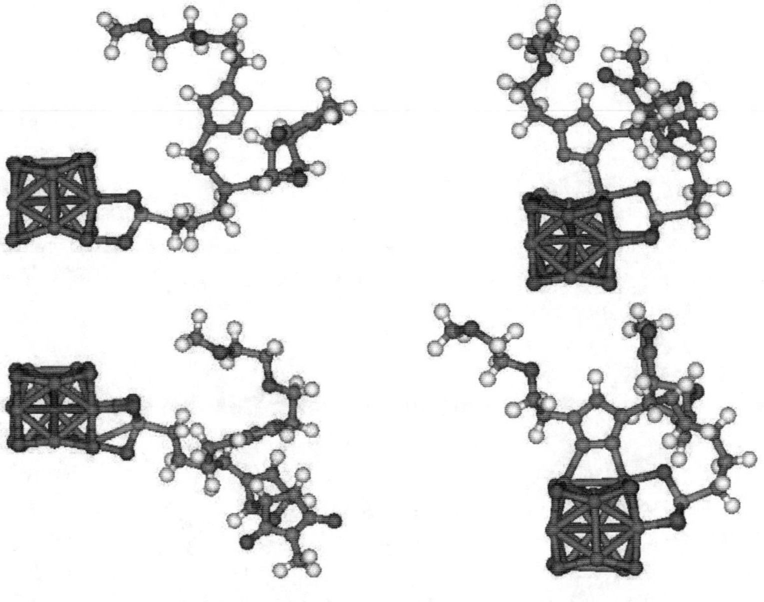

Figure 9. The polymer folding occurs with increasing temperature in cases of A, keeping the nanoparticle temperature at 500 K.

Keeping the nanoparticle temperature at 500 K, the temperature rising process of the drug linker of case A is obviously different from cases of B and C (as shown in Figure 8). This shows that the chain length between the drug linker and the IONP can influence the drug delivery deeply. After checking the time evolution of geometric structure, we find that the polymer folding occurs with increasing temperature in the case A (as shown in Figure 9).

The polymer folding can enhance the heat transfer rate between the IONP and drug linker by adding heat transfer pathway and decreasing heat transfer distance. One of the key findings of this study is that there is a significant thermal transport dependence on the polymer chain folding, invoked by heat accumulation in the polymer. Our calculations suggest that a new/existing heat transfer pathway is activated/inactivated by an increased/decreased link between the nanoparticle and the folded polymer chain, as indicated by the qualitative change of temperature rising process of the polymer drug-releasing part (as shown in Figure 10).

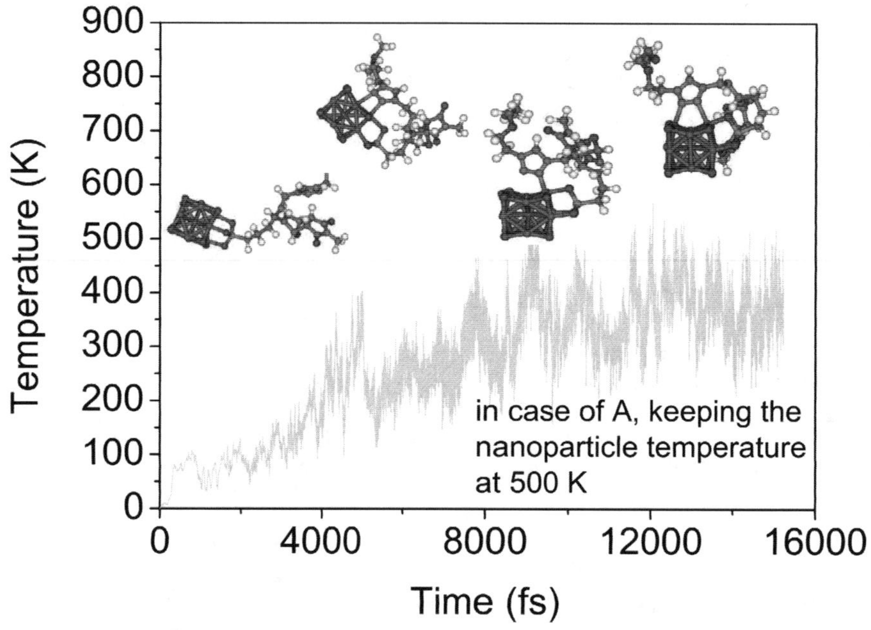

Figure 10. The qualitative change in temperature rising process of the drug linker might indicate opening/closing of a new/existing heat transfer pathway, in case of A, keeping the nanoparticle temperature at 500 K.

5. Summary and Conclusions

In summary, thermal transport in the environment close to the periphery of the nanoparticle, from a few angstroms to less than a nanometer scale, is becoming increasingly important with the advent of several biomedical applications of multifunctional magnetic nanoparticles, including drug delivery, magnetic resonance imaging, and hyperthermia therapy. Can we look at the influcence of the distance between the drug linker and the IONP onto the debonding? How heat is transferred from the IONP to the drug linker through polymer chain? Stimulated by above interesting and unanswered questions, we have built a model method for the heat transfer investigation in novel functional IONPs showing unprecedented, active control over drug release by retro-DA reaction by using hyperthermia effects, and achieved following conclusions: 1) The polymer folding can enhance the heat transfer rate between the IONP and drug linker by adding heat transfer pathway and decreasing heat transfer distance; 2) The chain length between the drug linker and the IONP can influence the drug delivery deeply.

References

[1] Lu A H, Salabas E L and Schüth F 2007 *Angew. Chem. Int. Ed.* **46** 1222
[2] Shubayev V I, Pisanic T R, Jin S H 2009 *Adv. Drug Delivery Rev.* **61** 467
[3] Ho D, Sun X L and Sun S H 2011 *Acc. Chem. Res.* **44** 875
[4] Mura S and Couvreur P 2012 *Adv. Drug Delivery Rev.* **64** 1394
[5] N'Guyen T K T 2012 *Magnetic Nanoparticles: From Fabrication to Clinical Applications* (Boca Raton, FL: CRC)
[6] Cheng Z L, Al Zaki A, Hui J Z, Muzykantov V R and Tsourkas A 2012 *Science* **338** 903
[7] Reddy L H, Arias J L, Nicolas J and Couvreur P 2012 *Chem. Rev.* **112** 5818
[8] Amstad E, Textor M and Reimhult E 2011 *Nanoscale* **3** 2819
[9] Hao R, Xing R J, Xu Z C, Hou Y L, Gao S and Sun S H 2010 *Adv. Mater.* **22** 2729
[10] Boyer C, Whittaker M R, Bulmus V, Liu J Q and Davis T P 2010 *NPG Asia Mater.* **2** 23
[11] Gao J H, Gu H W and Xu B 2009 *Acc. Chem. Res.* **42** 1097
[12] Frey N A, Peng S, Cheng K and Sun S H 2009 *Chem. Soc. Rev.* **38** 2532
[13] Laurent S, Forge D, Port M, Roch A, Robic C, Elst L V and Muller R N 2008 *Chem. Rev.* **108** 2064
[14] Lu A H, Salabas E L, Schüth F 2007 *Angew. Chem.* **119** 1242
[15] N'Guyen T T T, Duong H T T, Basuki J, Montembault V, Pascual S, Guibert C, Fresnais J, Boyer C, Whittaker M R, Davis T P and Fontaine L 2013 *Angew. Chem. Int. Ed.* **52** 14152
[16] Huang H, Delikanli S, Zeng H, Ferkey D M and Pralle A 2010 *Nat. Nanotechnol.* **5** 602
[17] Creixell M, Bohorquez A C, Torres-Lugo M and Rinaldi C 2011 *ACS Nano* **5** 7124
[18] Mellor P H, Roberts D and Turner D R 1991 *IEE Proceedings B (Electric Power Applications)* **138(5)** 205

Radio Resource Allocation on Complex 4G Wireless Cellular Networks

Kostas E. Psannis[1]
Department of Applied Informatics, University of Macedonia,156 Egnatia Street, Thessaloniki 54006

E-mail: kpsannis@uom.gr

Abstract. In this article we consider the heuristic algorithm which improves step by step wireless data delivery over LTE cellular networks by using the total transmit power with the constraint on users' data rates, and the total throughput with the constraints on the total transmit power as well as users' data rates, which are jointly integrated into a hybrid-layer design framework to perform radio resource allocation for multiple users, and to effectively decide the optimal system parameter such as modulation and coding scheme (MCS) in order to adapt to the varying channel quality. We propose new heuristic algorithm which balances the accessible data rate, the initial data rates of each user allocated by LTE scheduler, the priority indicator which signals delay- throughput- packet loss awareness of the user, and the buffer fullness by achieving maximization of radio resource allocation for multiple users. It is noted that the overall performance is improved with the increase in the number of users, due to multiuser diversity. Experimental results illustrate and validate the accuracy of the proposed methodology.

1. Introduction

Nowadays users are demanding continuous delivery of increasingly higher data over the Internet, in both wired and wireless networks. Due to its real-time nature, wireless data delivery typically has bandwidth, delay, and loss requirements. Moreover, the 3GPP Long Term Evolution (LTE) is the new standard developed to cope with future mobile data and emerging media applications. The generic characteristics of wireless networks are time-varying and their performance is generally inferior to those of wired networks. Therefore, it is still a challenging problem to efficiently provide data delivery service of high quality over 4G LTE Wireless Networks. However, for wireless data delivery in LTE, higher data rate could lead to higher packet loss rate, thus degrading the user's Quality of Experience (QoE). Generally, wireless communication systems should support a large number of users with flexibility in their quality of services (QoS), and Quality of Experience (QoE). The challenges to ensure the fulfillment of these requirements arise from the limited availability of frequency spectrum, and the nature of the wireless channel. To solve this issue, intelligent radio resource algorithms interacting in both the physical and the application layers are critical. Radio resource allocation are categorized into two major classes with different objectives. The objective of the first class is to

[1] To whom any correspondence should be addressed.

minimize the total transmit power with the constraint on users' data rates whereas in the second class, the objective is to maximize the total throughput with the constraints on the total transmit power as well as users' data rates [1-4].

2. Radio Resource Allocation

The main task of the scheduler is to allocate radio resources to all users equipments (UEs) in a multi-user scenario. Radio resources are elements from the time, frequency, and spatial domain of the LTE system [1,2,4]. Consider a wireless 4G-system with K-users and N-subCarriers. The data rate of the kth user R_k in bits/s is given by:

$$R_k = \frac{B}{N} \sum_{k=1}^{K} \sum_{n=1}^{N} subCarrier_{k,n} \, log_2(1 + SNR_{k,n}) \tag{1}$$

where B is the total bandwidth of the system. The effective Signal-to-Noise-Ratio (SNR) has to be adjusted according to the modulation and coding scheme (MCS) for a desired Bit Error Rate (BER). Hence, knowing the MCS, the effective $SNR_{k,n}$ is adjusted accordingly to meet the BER requirements. Moreover, in a wireless 4G-system system with K users and N subcarriers, each of N subcarriers is to be allocated to one of K users. In addition, the power allocated to each of K users should be optimized.

The total transmit power of the system P_T is given by

$$P_T = \sum_{n=1}^{N} \sum_{k=1}^{K} subCarrier_{k,n} \, p_{k,n} \tag{2}$$

where where $p_{k,n}$ is the power allocated for user k in sub-channel n

The resource allocation problem is formulated as follows

$$max_{k,n} \frac{B}{N} \sum_{k=1}^{K} \sum_{n=1}^{N} subCarrier_{k,n} \, log_2(1 + SNR_{k,n}) \tag{3}$$

$$min_{k,n} \sum_{n=1}^{N} \sum_{k=1}^{K} subCarrier_{k,n} \, p_{k,n} \tag{4}$$

subject to :

Constraint 1:

$$\sum_{n=1}^{N} \sum_{k=1}^{K} subCarrier_{k,n} \, p_{k,n} \le \Delta_{k,n}$$

where $\Delta_{k,n}$ denotes the priority indicator which signals transmission delay- throughput- packet loss rate (PLR) awareness of the kth user. It should be noted that PLR is affected by the varying channel quality and the adopted MCS on the resource blocks. On the other hand, the optimal MCS can effectively decrease the transmission delay while satisfying the constraint of packet loss rate (PLR). At the PHY layer, a set of possible modulation and coding schemes (MCS) are available for transmission. Although the formulation considers all combinations of modulation and coding schemes, the link adaptation algorithm implicitly drops combinations that provide lower throughput and higher PLR than their counterparts.

Constraint 2:

$$\sum_{k=1}^{K} R_k \geq R_{k,min}$$

where R_k and $R_{k,min}$ are the accessible, inside the available rate region C, and the minimum required data rate for the kth user, respectively. The maximum rate could be still achieved by assigning each subcarrier to the user with the largest channel gain on it. The available rate region C defined by the LTE scheduler. Hence, $R_k \leq C$

Constraint 3:

$$\sum_{k=1}^{K} R_k \leq B_k(i)$$

where $B_k(i) = B_1(1), B_1(2), B_2(1),....,$ denotes the buffer fullness at each time instant (i) and based on predefined thresholds, which depend on each User Equipment (UE)-capabilities. Therefore the limitation in buffer storage capacity is considered, which means that clients will continue access data as long as the channel characteristics and buffer fullness allows for it. Hence, the buffer at the receiver is monitored and the state is fed back to the transmitter to be used in source adaptation. The playback buffer starvation constraint is intended to avoid aggressive attempts to maximize quality that could potentially result in data freezes and re-buffering, specifically, in a bandwidth-limited environment. LTE's scheduler can make use of information defined in the proposed method to optimize radio resource allocation among the different users' equipments (UEs).
The objective is to maximize the total rate within the total power constraint of the system while maintaining rate proportionality among the users indicated in various constraints (Constraint 1-Constraint 3).

3. Experiments

We consider a scenario with k = 200 active users per cell (5MHz), where most of the users are far from the Base Station (BS) with resulting average SNR ranging from 2 to 19 dB. Without losing generality, we assume that there are K= 200 active users with variable delays constraints 16.66 ms, 20 ms, 33.33 ms, and 41.66 ms [1-3]. The available modulation and coding schemes (MCSs) are chosen from QPSK 1/2, QPSK 2/3, QPSK 4/5, 16-QAM 1/2, 16-QAM 2/, 16-QAM 4/5, 64-QAM 2/3, 64-QAM 3/4 and 64-QAM 4/5. We use nine types of MCSs as in the LTE standard [1]. Table I lists the MCSs we use in our simulations as well as the SNR minimum requirements for each MCS and the average spectrum efficiency (bps/Hz) for channel model based on 3GPP [2]. Average spectrum efficiency is defined as the aggregate throughput of all users (the number of correctly received bits) over a certain period of time (bps/Hz) [1,2]. Fig. 1 indicates the percentage of users that can support each MCS in our simulations. It is worth pointing out that the channel conditions of the users vary in a wide range. We adopted the LTE system model defined in [2]. Given an average SNR, the instantaneous link quality can be randomly produced from Rayleigh distribution [1-3]. User Equipment (UE) is the device that a consumer uses for 4G wireless communication. We assume that the initial pre-buffer at the UEs is set to 1ms. It should be noted that the buffer size can be used to tune the trade-off between quality and the mean playback buffer size. From a user quality of experience (QoE) viewpoint, playback buffer starvation is unwanted because it results re-buffering.

The data rate is calculated considering equal minimum power on all the subcarriers. Adaptive minimum power allocation has a very significant role in maximizing the total accessible data rate by using multiuser diversity. Specifically the optimal set of subcarriers is selected and the total power is equally distributed among those subcarriers involved in data transmission. Fig. 2 shows the comparison of proposed algorithm and the two conventional algorithms in terms of the total accessible data rate versus different numbers of users. In the simulation results, the two conventional algorithms referred to as “rate adaptive”, and “margin adaptive (MA)”, respectively. The optimization problem in MA allocation algorithm is formulated with the objective of minimizing the total transmit power while providing each user with its required quality of service in terms of data rate and BER [6]. On the other hand, the RA objective is to maximize the total data rate of the system with the constraint on the total transmit power [5].

As shown in Fig. 2, significant performance enhancement is achieved by using the proposed algorithm compared to conventional approaches, the RA and MA, respectively. We compare the proposed algorithm, where the design goal is to maximize the total rate within the total power constraint of the system while maintaining rate proportionality among the users indicated in various constraints. The graph with the highest spectral efficiency in Fig. 2 belongs to the proposed algorithm. In this algorithm, the data rate is calculated considering equal minimum power on all the subcarriers. Based on this result, one may conclude that in a system with K-users and N sub-channels, adaptive minimum power allocation has a very significant role in maximizing the total accessible data rate by using multiuser diversity, and consequently would always give close to optimum performance under the variable given- application delays constraints.

Table 1. MCS with SNR minimum requirements and Average efficiency

	Code	Aver. Efficiency (bps/Hz)	SNR[dB]
QPSK	1/2	0.15	[2]
QPSK	2/3	0.41	[4.3]
QPSK	4/5	1.05	[6.2]
QAM16	1/2	1.48	[7.9]
QAM16	2/4	1.89	[11.3]
QAM16	4/5	2.38	[12.8]
QAM64	2/3	3.37	[15.3]
QAM64	3/4	4.14	[17.5]
QAM64	4/5	5.05	[18.6]

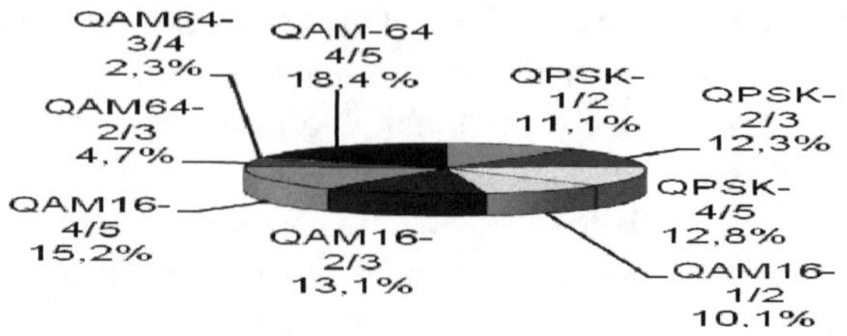

Figure 1. MCS and the percentage of users supporting them.

4. Conclusion

We proposed a new heuristic algorithm which balances the accessible data rate, the initial data rates of each user allocated by LTE scheduler, the priority indicator which signals delay- throughput- packet loss awareness of the user, and the buffer fullness by achieving maximization of radio resource allocation for multiple users. It is noted that the overall performance is improved with the increase in the number of users, due to multiuser diversity. Future work will estimate Quality of Experience (QoE) parameters from Quality of Service (QoS) parameters with a high degree of accuracy for LTE and LTE-Advanced radio access networks. It will also investigate the performance of the proposed methodology over hybrid LTE and Digital Video Broadcasting (DVB) channels. This is expected to be an active area of research in years to come.

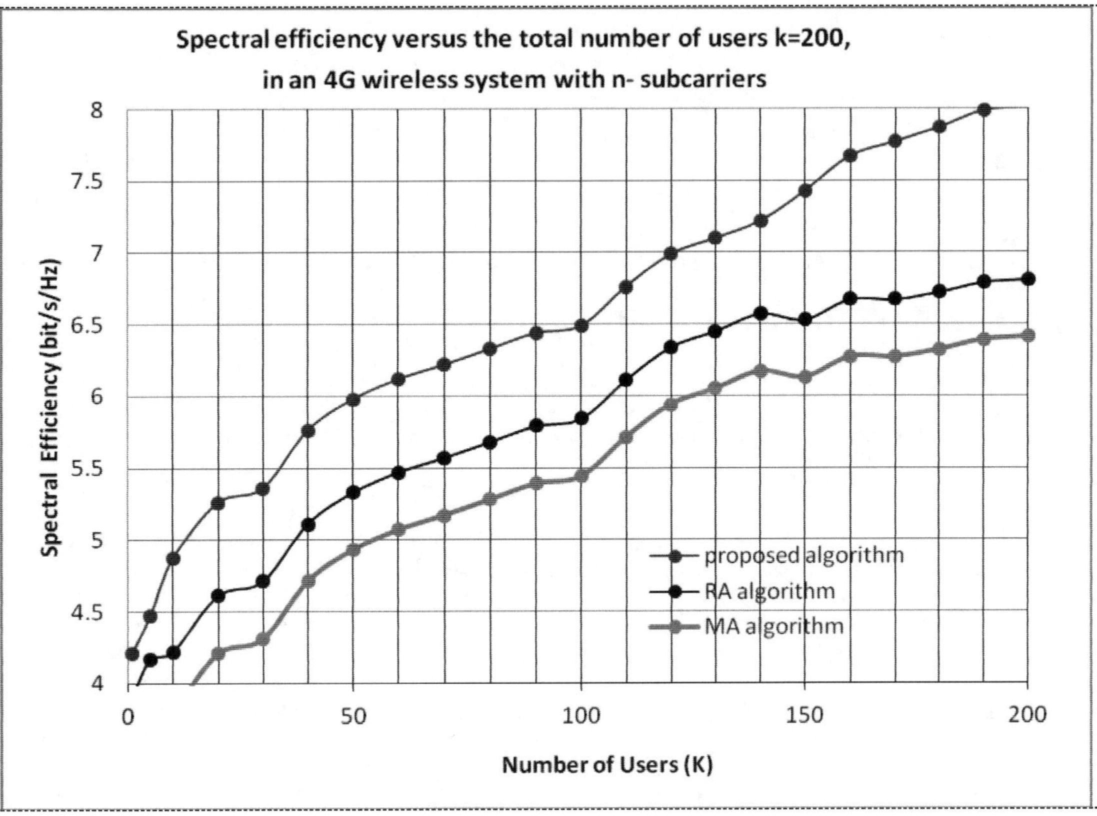

Figure 2. Spectral efficiency versus the total number of users k=200, for the proposed algorithm, the RA algorithm and the MA algorithm.

Acknowledgements
The author would like to thank the anonymous reviewers for their valuable comments and feedback which was extremely helpful in improving the quality of the paper.

References

[1] 3rd Generation Partnership Project, Technical Specification Group Radio Access Network, Requirements for further advancements for Evolved Universal Terrestrial Radio Access (E-UTRA) (LTE-Advanced), Release 11, 3GPP TR 36.913 V11.0.0 (2012-09)

[2] Universal Mobile Telecommunications System (UMTS), Spatial channel model for Multiple Input Multiple Output (MIMO) simulations, 3GPP TR 25.996 version 11.0.0, Release 11.

[3] H. Schulze and C. Lueders, Theory and Applications of OFDM and CDMA Wideband Wireless Communications. John Wiley, 2005

[4] S. Sadr, A. Anpalagan, and K. Raahemifar, Radio resource allocation algorithms for the downlink of multiuser OFDM communication systems, IEEE Communications Surveys and Tutorials, vol. 11, no. 3, pp. 92–106, 2009.

[5] J. Jang and K. B. Lee, Transmit power adaptation for multiuser OFDM systems, *IEEE J. Select. Areas Commun.*, vol. 21, pp. 171–178, February 2003.

[6] C. Y. Wong, R. S. Cheng, K. B. Letaief, and R. D. Murch, Multiuser OFDM with adaptive subcarrier, bit and power allocation, *IEEE J. Select. Areas Commun.*, vol. 17, pp. 1747–1758, October 1999.

MUSIC-type imaging of small perfectly conducting cracks with an unknown frequency

Won-Kwang Park

Department of Mathematics, Kookmin University, Seoul, 136-702, Korea

E-mail: parkwk@kookmin.ac.kr

Abstract. MUltiple SIgnal Classification (MUSIC) is a famous non-iterative detection algorithm in inverse scattering problems. However, when the applied frequency is unknown, inaccurate locations are identified via MUSIC. This fact has been confirmed through numerical simulations. However, the reason behind this phenomenon has not been investigated theoretically. Motivated by this fact, we identify the structure of MUSIC-type imaging functionals with unknown frequency, by establishing a relationship with Bessel functions of order zero of the first kind. Through this, we can explain why inaccurate results appear.

1. Introduction

The inverse scattering problem for the reconstruction of a single, perfectly conducting crack in $\mathbb{R}^2$, satisfying a Dirichlet boundary condition, has been studied in [1]. In this remarkable research, a Newton-type iterative reconstruction algorithm is proposed. In general, for a successful application, a good initial guess close to the unknown crack is required in order to guarantee convergence. Furthermore, the algorithm generally requires a large amount of computational time, and it is difficult to extend to the reconstruction of multiple cracks.

As an alternative, non-iterative algorithms have also been developed. Among them, MUltiple SIgnal Classification (MUSIC)-type algorithms are applied to various problems for the imaging of small electromagnetic scatterers, cracks, and extended targets. For more details, one can refer to [2, 3, 4, 5, 6, 7, 8, 9, 10, 11, 12, 13, 14] and references therein. However, to obtain a good result, the value of the applied frequency must be known. If not, inaccurate results are extracted via MUSIC. Until now, this fact has only been investigated through the results of numerical simulations. For this reason, a rigorous mathematical theory about the structure of MUSIC-type imaging functionals must be established, in order to diagnose this phenomenon.

In this contribution, we identify the structure of MUSIC-type imaging functionals of small, perfectly conducting cracks with an unknown frequency, by finding a relationship with Bessel functions of order zero of the first kind. This is based on the fact that the far-field pattern can be presented as an asymptotic expansion formula in the presence of a small crack. The derived structure illuminates the theoretical reason for the appearance of inaccurate locations.

The paper is organized as follows. In section 2, two-dimensional direct scattering problems and MUSIC-type imaging algorithms are briefly introduced. In section 3, the structure of MUSIC-type imaging functionals without information about the applied frequency is identified, and the corresponding results of a numerical simulation are exhibited. A short conclusion is given in Section 4.

2. Direct scattering problems and MUSIC-type imaging algorithm

Let Γ_m, $m = 1, 2, \cdots, M$, be a linear crack with a small length $2h$ and center $\mathbf{z}_m$. We denote Γ be the collection of all of the Γ_m and assume that the Γ_m are sufficiently separated from each other. In this paper, we consider the case of so-called Transverse Magnetic (TM) polarization. Let $u(\mathbf{x}, \boldsymbol{\theta})$ be the time-harmonic total field, satisfying the Helmholtz equation

$$\nabla^2 u(\mathbf{x}, \boldsymbol{\theta}) + k^2 u(\mathbf{x}, \boldsymbol{\theta}) = 0 \quad \text{in} \quad \mathbb{R}^2 \backslash \Gamma, \tag{1}$$

with a Dirichlet boundary condition $u(\mathbf{x}, \boldsymbol{\theta}) = 0$ on Γ. Here, $\boldsymbol{\theta}$ is an incident direction on the unit circle $\mathbb{S}^1$, and $k = 2\pi/\lambda$ denotes a strictly positive wavenumber, with wavelength λ. We assume that k^2 is not an eigenvalue of (1) and $h \ll \lambda$.

Note that $u(\mathbf{x}, \boldsymbol{\theta})$ be decomposed as $u(\mathbf{x}, \boldsymbol{\theta}) = u_{\text{inc}}(\mathbf{x}, \boldsymbol{\theta}) + u_{\text{scat}}(\mathbf{x}, \boldsymbol{\theta})$, where $u_{\text{inc}}(\mathbf{x}, \boldsymbol{\theta}) = \mathrm{e}^{ik\boldsymbol{\theta}\cdot\mathbf{x}}$ is the given incident field, and $u_{\text{scat}}(\mathbf{x}, \boldsymbol{\theta})$ is the unknown scattered field satisfying the Sommerfeld radiation condition uniformly in all directions $\hat{\mathbf{x}} = \mathbf{x}/|\mathbf{x}|$. Let $u_\infty(\hat{\mathbf{x}}, \boldsymbol{\theta})$ denote the far-field pattern of $u_{\text{scat}}(\mathbf{x}, \boldsymbol{\theta})$. Based on [5], the $u_\infty(\hat{\mathbf{x}}, \boldsymbol{\theta})$ can be represented as

$$u_\infty(\hat{\mathbf{x}}, \boldsymbol{\theta}) = -\frac{2\pi}{\ln(h/2)} \sum_{m=1}^{M} u_{\text{inc}}(\mathbf{z}_m, \boldsymbol{\theta}) \overline{u_{\text{inc}}(\mathbf{z}_m, \hat{\mathbf{x}})} + O\left(\frac{1}{|\ln h|^2}\right) \approx -\frac{2\pi}{\ln(h/2)} \sum_{m=1}^{M} \mathrm{e}^{ik(\boldsymbol{\theta} - \hat{\mathbf{x}})\cdot\mathbf{z}_m}. \tag{2}$$

We apply (2) in order to establish a MUSIC-type imaging functional. For this, one uses the eigenvalue structure of the MSR matrix $\mathbb{K} = [u_\infty(\hat{\mathbf{x}}_j, \boldsymbol{\theta}_l)]_{j,l=1}^{N}$. Suppose that $\boldsymbol{\vartheta}_j = -\boldsymbol{\theta}_j$ for all j. Then, $\mathbb{K}$ is a complex symmetric matrix, but is not Hermitian. Therefore, instead of Eigenvalue Decomposition, we perform a Singular Value Decomposition (SVD) of $\mathbb{K}$ (see [7]), with

$$\mathbb{K} = \sum_{m=1}^{M} \sigma_m \mathbf{U}_m \mathbf{V}_m^*, \tag{3}$$

where the superscript $*$ denotes the Hermitian conjugate. Then, $\{\mathbf{U}_1, \mathbf{U}_2, \cdots, \mathbf{U}_M\}$ is the basis for the signal space of $\mathbb{K}$. Therefore, one can define the projection operator onto the null (or noise) subspace, $\mathbf{P}_{\text{noise}} : \mathbb{C}^{N\times 1} \longrightarrow \mathbb{C}^{N\times 1}$. This projection is given explicitly by

$$\mathbf{P}_{\text{noise}} := \mathbb{I}_N - \sum_{m=1}^{M} \mathbf{U}_m \mathbf{U}_m^*, \tag{4}$$

where $\mathbb{I}_N$ denotes the $N \times N$ identity matrix. For any point $\mathbf{x} \in \mathbb{R}^2$, we define a test vector $\mathbf{f}(\mathbf{x}; k) \in \mathbb{C}^{N\times 1}$, as $\mathbf{f}(\mathbf{x}; k) = \frac{1}{N}\left[\mathrm{e}^{ik\boldsymbol{\theta}_1\cdot\mathbf{x}}, \mathrm{e}^{ik\boldsymbol{\theta}_2\cdot\mathbf{x}}, \cdots, \mathrm{e}^{ik\boldsymbol{\theta}_N\cdot\mathbf{x}}\right]^T$. With this, a MUSIC-type imaging functional $\mathcal{I}(\mathbf{x}; k)$ can be constructed as follows:

$$\mathcal{I}(\mathbf{x}; k) = |\mathbf{P}_{\text{noise}}(\mathbf{f}(\mathbf{x}; k))|^{-1}. \tag{5}$$

Then, the map of $\mathcal{I}(\mathbf{x}; k)$ will have peaks of large magnitudes at $\mathbf{z}_m \in \Gamma$.

3. Structure of MUSIC-type imaging functional and numerical results

Based on the above, one must know the exact value of k. Without this, one cannot detect the locations of the Γ_m exactly. This fact has been identified in various research, but has so far relied on the results of numerical simulations. Motivated by this, we explore the structure of (5) with an unknown frequency. Since the exact value of k is unknown, we choose a fixed value η, and apply the corresponding test vector $\mathbf{f}(\mathbf{x}; \eta) = \frac{1}{N}\left[\mathrm{e}^{i\eta\boldsymbol{\theta}_1\cdot\mathbf{x}}, \mathrm{e}^{i\eta\boldsymbol{\theta}_2\cdot\mathbf{x}}, \cdots, \mathrm{e}^{i\eta\boldsymbol{\theta}_N\cdot\mathbf{x}}\right]^T$ to (5), such that

$$\mathcal{I}(\mathbf{x}; \eta) = |\mathbf{P}_{\text{noise}}(\mathbf{f}(\mathbf{x}; \eta))|^{-1}. \tag{6}$$

Then, we can obtain the following result. A detailed derivation is to appear in an extended version of this work.

Journal of Physics: Conference Series **633** (2015) 012005 doi:10.1088/1742-6596/633/1/012005

Theorem 3.1. *For sufficiently large N, ω, and η, $\mathcal{I}(\mathbf{x};\eta)$ is of the form:*

$$\mathcal{I}(\mathbf{x};\eta) \approx \left(1 - \sum_{m=1}^{M} J_0(|\eta\mathbf{x} - k\mathbf{z}_m|)\right)^{-1/2},$$

where J_0 denotes the Bessel function of order zero and of the first kind.

Proof. Based on [12], $\mathbf{f}(\mathbf{z}_m k) \approx \mathbf{U}_m$ for all $m = 1, 2, \cdots, M$. It follows that

$$\mathbf{P}_{\text{noise}}(\mathbf{f}(\mathbf{x};\eta)) = \frac{1}{\sqrt{N}}\begin{bmatrix} e^{i\eta\boldsymbol{\theta}_1\cdot\mathbf{x}} \\ \vdots \\ e^{i\eta\boldsymbol{\theta}_N\cdot\mathbf{x}} \end{bmatrix} - \frac{1}{N\sqrt{N}}\sum_{m=1}^{M}\begin{bmatrix} e^{ik\boldsymbol{\theta}_1\cdot\mathbf{z}_m}\sum_{n=1}^{N} e^{i\boldsymbol{\theta}_n\cdot(\eta\mathbf{x}-k\mathbf{z}_m)} \\ \vdots \\ e^{ik\boldsymbol{\theta}_N\cdot\mathbf{z}_m}\sum_{n=1}^{N} e^{i\boldsymbol{\theta}_n\cdot(\eta\mathbf{x}-k\mathbf{z}_m)} \end{bmatrix}.$$

Since N is sufficiently large, the following relationship holds for $\boldsymbol{\theta}_n, \boldsymbol{\theta} \in \mathbb{S}^1$, and $\mathbf{x} \in \mathbb{R}^2$:

$$\frac{1}{N}\sum_{n=1}^{N} e^{ik\boldsymbol{\theta}_n\cdot\mathbf{x}} = \frac{1}{2\pi}\int_{\mathbb{S}^1} e^{ik\boldsymbol{\theta}\cdot\mathbf{x}} d\boldsymbol{\theta} = J_0(k|\mathbf{x}|),$$

where J_0 denotes the Bessel function of order zero of the first kind (see [15]). Therefore,

$$\mathbf{P}_{\text{noise}}(\mathbf{f}(\mathbf{x};\eta)) = \frac{1}{\sqrt{N}}\begin{bmatrix} e^{i\eta\boldsymbol{\theta}_1\cdot\mathbf{x}} - \sum_{m=1}^{M} e^{ik\boldsymbol{\theta}_1\cdot\mathbf{z}_m} J_0(|\eta\mathbf{x} - k\mathbf{z}_m|) \\ \vdots \\ e^{i\eta\boldsymbol{\theta}_N\cdot\mathbf{x}} - \sum_{m=1}^{M} e^{ik\boldsymbol{\theta}_N\cdot\mathbf{z}_m} J_0(|\eta\mathbf{x} - k\mathbf{z}_m|) \end{bmatrix}.$$

With this, we arrive at

$$|\mathbf{P}_{\text{noise}}(\mathbf{f}(\mathbf{x};\eta))|^2 = \frac{1}{N}\sum_{n=1}^{N}\left\{1 - \sum_{m=1}^{M}(\Phi_1 + \overline{\Phi}_1) + \left(\sum_{m=1}^{M}\Phi_2\right)\left(\sum_{m=1}^{M}\overline{\Phi}_2\right)\right\},$$

where $\Phi_1 := e^{i\boldsymbol{\theta}_n\cdot(\eta\mathbf{x}-k\mathbf{z}_m)} J_0(|\eta\mathbf{x} - k\mathbf{z}_m|)$ and $\Phi_2 := e^{ik\boldsymbol{\theta}_n\cdot\mathbf{z}_m} J_0(|\eta\mathbf{x} - k\mathbf{z}_m|)$. Since ω and η are sufficiently large we can obtain the following through a tedious calculation:

$$\frac{1}{N}\sum_{n=1}^{N}\Phi_1 = \frac{1}{N}\sum_{n=1}^{N}\sum_{m=1}^{M} e^{i\boldsymbol{\theta}_n\cdot(\eta\mathbf{x}-k\mathbf{z}_m)} J_0(|\eta\mathbf{x} - k\mathbf{z}_m|) = J_0(|\eta\mathbf{x} - k\mathbf{z}_m|)^2$$

$$\frac{1}{N}\sum_{n=1}^{N}\Phi_2\overline{\Phi}_2 = \frac{1}{N}\sum_{n=1}^{N}\sum_{m=1}^{M}\sum_{m'=1}^{M} e^{ik\boldsymbol{\theta}_n\cdot(\mathbf{z}_m - \mathbf{z}_{m'})} J_0(|\eta\mathbf{x} - k\mathbf{z}_m|) J_0(|\eta\mathbf{x} - k\mathbf{z}_{m'}|) = J_0(|\eta\mathbf{x} - k\mathbf{z}_m|)^2.$$

Therefore, $|\mathbf{P}_{\text{noise}}(\mathbf{f}(\mathbf{x};\eta))|$ can be represented as

$$|\mathbf{P}_{\text{noise}}(\mathbf{f}(\mathbf{x};\eta))| = \left(1 - \sum_{m=1}^{M} J_0(|\eta\mathbf{x} - k\mathbf{z}_m|)\right)^{1/2}.$$

This completes the proof. □

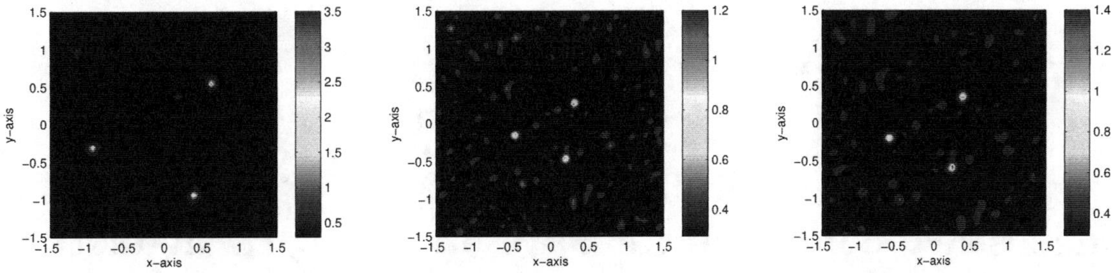

Figure 1. Maps of $\mathcal{I}(\mathbf{z}; 10)$ (left), $\mathcal{I}(\mathbf{z}; 20)$ (center), and $\mathcal{I}(\mathbf{z}; k)$ (right).

Note that $J_0(x)$ has a maximum value of 1 at $x = 0$. Therefore, the map of $\mathcal{I}(\mathbf{x}; \eta)$ will have plots of the magnitude (theoretically $+\infty$) at $\mathbf{x} = (k/\eta)\mathbf{z}_m$, instead of at the true location $\mathbf{z}_m$. This provides the theoretical reason for why inexact locations of cracks are extracted.

Now, let us consider the results of a numerical simulation, with the true value of $k = 2\pi/0.4$. Figure 1 displays maps of $\mathcal{I}(\mathbf{x}; \eta)$ for three small cracks. Following the result from Theorem 3.1, locations of $(k/\eta)\mathbf{z}_m$ are identified, instead of $\mathbf{z}_m$. This means that we can identify the existence of cracks, but it is impossible to find the true locations without *a priori* information about k.

4. Conclusion

Based on an asymptotic expansion formula, the structure of MUSIC-type imaging functionals with unknown frequencies is derived, by establishing a relationship with Bessel functions of order zero. On the basis of the identified structure, we were able to confirm the reason for the appearance of inaccurate locations of cracks.

Based on the result of the current contribution, we expect it is possible to develop an algorithm to detect the exact locations of cracks. The current contribution deals with cracks with a Dirichlet boundary condition. Along the same lines, an extension to cracks with a Neumann boundary condition would make an interesting topic for future research.

Acknowledgments

This research was supported by Basic Science Research Program through the National Research Foundation of Korea (NRF) funded by the Ministry of Education(No. NRF-2014R1A1A2055225) and the research program of Kookmin University in Korea.

References

[1] Kress R 1995 *Math. Meth. Appl. Sci.* **18** 267–293
[2] Ammari H, Garnier J, Kang H, Park W K and Sølna K 2011 *SIAM J. Appl. Math.* **71** 68–91
[3] Ammari H, Iakovleva E and Lesselier D 2005 *Multiscale Model. Simul.* **3** 597–628
[4] Ammari H, Iakovleva E, Lesselier D and Perrusson G 2007 *SIAM J. Sci. Comput.* **29** 674–709
[5] Ammari H, Kang H, Lee H and Park W K 2010 *SIAM J. Sci. Comput.* **32** 894–922
[6] Chen X and Zhong Y 2009 *Inverse Problems* **25** 015008
[7] Cheney M 2001 *Inverse Problems* **17** 591–595
[8] Hou S, Sølna K and Zhao H 2006 *Inverse Problems* **22** 1151–1178
[9] Iakovleva E, Gdoura S, Lesselier D and Perrusson G 2007 *IEEE Trans. Antennas Propag.* **55** 2598–2609
[10] Joh Y D, Kwon Y M and Park W K 2014 *Appl. Math. Comput.* **240** 273–280
[11] Joh Y D and Park W K 2013 *Prog. Electromagn. Res.* **138** 211–226
[12] Park W K 2015 *SIAM J. Appl. Math.* **75** 209–228
[13] Park W K and Lesselier D 2009 *J. Comput. Phys.* **228** 8093–8111
[14] Park W K and Lesselier D 2009 *Inverse Problems* **25** 075002
[15] Park W K 2015 *J. Comput. Phys.* **283** 52–80

Journal of Physics: Conference Series **633** (2015) 012006 doi:10.1088/1742-6596/633/1/012006

A necessary condition for applying MUSIC algorithm in limited-view inverse scattering problem

Taehoon Park and Won-Kwang Park

Department of Mathematics, Kookmin University, Seoul, 136-702, Korea

E-mail: thpark@kookmin.ac.kr and parkwk@kookmin.ac.kr

Abstract. Throughout various results of numerical simulations, it is well-known that MUltiple SIgnal Classification (MUSIC) algorithm can be applied in the limited-view inverse scattering problems. However, the application is somehow heuristic. In this contribution, we identify a necessary condition of MUSIC for imaging of collection of small, perfectly conducting cracks. This is based on the fact that MUSIC imaging functional can be represented as an infinite series of Bessel function of integer order of the first kind. Numerical experiments from noisy synthetic data supports our investigation.

1. Introduction

From the pioneering work of MUltiple SIgnal Classification (MUSIC) algorithm for estimating the locations of a number of point-like scatterers, it has been remarkably developed and successfully applied in various inverse scattering problems for finding locations of small scatterers or shapes of crack-like inhomogeneities/extended targets, refer to [1, 2, 3, 4, 5, 6, 7, 8, 9, 10, 11, 12, 13].

Almost every works are focused on the application of MUSIC in full-view inverse scattering problems. In these results, it is confirmed that MUSIC produces very good results in full-view problems but, theoretical reason of its effectiveness for imaging of curve-like perfectly conducting crack is recently developed in [14, 15]. In these works, a relationship between MUSIC-type imaging functional and Bessel function of integer order of the first kind is investigated and the effectiveness of MUSIC and some undiscovered phenomenons are clearly identified. However, in the limited-view inverse scattering problem, MUSIC produces poor results. This is recognized as a general property of MUSIC, and this fact has been identified via various numerical examples, refer to [2].

Motivated this, theoretical reason of this limitation of MUSIC has been identified and examined that MUSIC can be applicable for detecting locations of point-like scatterers or cracks of small length and extended thin electromagnetic inhomogeneities if the range of incident and observation directions is not too narrow, refer to [16, 17]. However, when one tries to apply MUSIC for imaging extended perfectly conducting crack, the selection of range of incident and observation directions significantly affects the imaging performance. However, this phenomenon is heuristically discovered so that theoretical treatment of MUSIC in limited-view problem should be considered carefully.

In this contribution, we explore a necessary condition of MUSIC in limited-view inverse scattering problem for imaging of arc-like perfectly conducting crack in two-dimensional space.

This is based on the relationship with MUSIC-type imaging function and infinite series of Bessel functions of integer order.

This paper is organized as follows. In Section 2, we briefly introduce the two-dimensional direct scattering problem and the structure of MUSIC-type imaging function. In Section 3, we discover a necessary condition for applying MUSIC in limited-view problem. In Section 4 results of numerical simulation are exhibited in order to support our discover.

2. Direct scattering problem and structure of MUSIC

Let Γ be a smooth curve describes crack and we denote $u(\mathbf{x})$ be the time-harmonic total field which satisfies

$$\Delta u(\mathbf{x}) + k^2 u(\mathbf{x}) = 0 \quad \text{in} \quad \mathbb{R}^2 \backslash \Gamma$$

(with positive wave number k) and a Dirichlet boundary condition $u(\mathbf{x}) = 0$ on Γ. Note that $u(\mathbf{x})$ can be decomposed as $u(\mathbf{x}) = u_{\text{inc}}(\mathbf{x}) + u_{\text{scat}}(\mathbf{x})$, where $u_{\text{inc}}(\mathbf{x}) = e^{ik\boldsymbol{\theta}\cdot\mathbf{x}}$ is given incident field with incident direction $\boldsymbol{\theta} \in \mathbb{S}^1_\star$ and $u_{\text{scat}}(\mathbf{x})$ is scattered field which satisfies the Sommerfeld radiation condition. Here, $\mathbb{S}^1_\star$ is a connected, proper subset of two-dimensional unit circle $\mathbb{S}^1$. Let $u_\infty(\boldsymbol{\vartheta}, \boldsymbol{\theta})$ be the far-field pattern of $u_{\text{scat}}(\mathbf{x})$. Then, it satisfies

$$u_{\text{scat}}(\mathbf{x}) = \frac{e^{ik|\mathbf{x}|}}{\sqrt{|\mathbf{x}|}} \left\{ u_\infty(\boldsymbol{\vartheta}, \boldsymbol{\theta}) + \mathcal{O}\left(\frac{1}{|\mathbf{x}|}\right) \right\}$$

uniformly in all directions $\boldsymbol{\vartheta} = \mathbf{x}/|\mathbf{x}|$ and $|\mathbf{x}| \longrightarrow \infty$.

Now, let us introduce MUSIC-type imaging algorithm. Before starting, we assume that Γ is divided into M different segments of size of order half the wavelength $\lambda/2$. Then, based on the Rayleigh resolution limit from far-field data, any detail less than one-half of the wavelength cannot be seen, and only one point, say $\mathbf{x}_m$, at each segment is expected to contribute at the image space of the response matrix $\mathbb{K}$ (see [11, 12] for instance). Having in mind this, we consider the collected Multi-Static Response (MSR) matrix such that

$$\mathbb{K} := [K_{jl}(\boldsymbol{\vartheta}_j, \boldsymbol{\theta}_l)]_{j,l=1}^N = [u_\infty(-\boldsymbol{\theta}_j, \boldsymbol{\theta}_l)]_{j,l=1}^N,$$

where $\{\boldsymbol{\vartheta}_j \in \mathbb{S}^1_\star : j = 1, 2, \cdots, N\}$ and $\{\boldsymbol{\theta}_l \in \mathbb{S}^1_\star : l = 1, 2, \cdots, N\}$ are set of observation and incident directions, respectively. Then, the singular value decomposition of $\mathbb{K}$ can be written as

$$\mathbb{K} = \mathbb{UDV}^* \approx \sum_{m=1}^{M} \sigma_m \mathbf{U}_m \mathbf{V}_m^*.$$

Note that the first M left-singular vectors, $\{\mathbf{U}_1, \mathbf{U}_2, \cdots, \mathbf{U}_M\}$, provide an orthonormal basis for $\mathbb{K}$. Hence, one can define a projection operator onto the null (or noise) subspace:

$$\mathbf{P}_{\text{noise}} := \mathbb{I}_N - \sum_{m=1}^{M} \mathbf{U}_m \mathbf{U}_m^*.$$

For any point $\mathbf{z} \in \mathbb{R}^2$, define a vector $\mathbf{f}(\mathbf{z}) \in \mathbb{C}^{N\times 1}$ such that $\mathbf{f}(\mathbf{z}) = [e^{ik\boldsymbol{\theta}_1\cdot\mathbf{z}}, e^{ik\boldsymbol{\theta}_2\cdot\mathbf{z}}, \cdots, e^{ik\boldsymbol{\theta}_N\cdot\mathbf{z}}]^T$. Then, an image of $\mathbf{z} \in \Gamma$, follows from computing

$$\mathcal{I}(\mathbf{z}) = |\mathbf{P}_{\text{noise}}(\mathbf{f}(\mathbf{z}))|^{-1}.$$

Based on [1, 4, 11, 12], map of $\mathcal{I}(\mathbf{z})$ will exhibit large peaks at $\mathbf{z} = \mathbf{x}_m \in \Gamma$.

Unfortunately, throughout above description, it is very hard to answer that why MUSIC yields unexpected results in limited-view inverse scattering problem. For this, the structure of MUSIC-type imaging functional $\mathcal{I}(\mathbf{z})$ has been derived in [17] as follows.

Lemma 2.1 *Assume that N and k are sufficiently large. Then, $\mathcal{I}(\mathbf{z})$ cam be written*

$$\mathcal{I}(\mathbf{z}) = \left\{ 1 - \sum_{m=1}^{M} \left| J_0(k|\mathbf{z}-\mathbf{x}_m|) + \frac{\mathcal{D}(k|\mathbf{z}-\mathbf{x}_m|)}{\theta_N - \theta_1} \right|^2 \right\}^{-1},$$

where $\mathbf{z} - \mathbf{x}_m = r_m[\cos(\phi_m), \sin(\phi_m)]$ and

$$\mathcal{D}(k|\mathbf{z}-\mathbf{x}_m|) := 4\sum_{n=1}^{\infty} \frac{\mathrm{i}^n}{n} J_n(k|\mathbf{z}-\mathbf{x}_m|) \sin\left(\frac{n(\theta_N - \theta_1)}{2}\right) \cos\left(\frac{n(\theta_N + \theta_1 - 2\phi_m)}{2}\right). \quad (1)$$

Based on the property of Bessel functions, the terms $J_0(k|\mathbf{z}-\mathbf{x}_m|)$ and $\mathcal{D}(k|\mathbf{z}-\mathbf{x}_m|)$ will contribute to and disturb the imaging performance, respectively. Therefore, we can conclude that $\mathcal{I}(\mathbf{z})$ is highly influenced by the range of incident directions.

3. A necessary condition for applying MUSIC in limited-view problems

Based on the result in Lemma 2.1, it is expected that if the term $\mathcal{D}(k|\mathbf{z}-\mathbf{x}_m|)$ is disappear, very good results can be obtained via MUSIC. Note that $\mathbf{z}$ is arbitrary and $\mathbf{x}_m$ is unknown. This means that we cannot make $J_n(k|\mathbf{z}-\mathbf{x}_m|) = 0$ in (1) so, we must find a condition of the range of incident directions such that

$$\sin\left(\frac{n(\theta_N - \theta_1)}{2}\right) \cos\left(\frac{n(\theta_N + \theta_1 - 2\phi_m)}{2}\right) \equiv 0.$$

Note that if $\theta_N - \theta_1 = \pi$ and $\theta_N + \theta_1 - 2\phi_m = \pi$ then, $\mathcal{D}(k|\mathbf{z}-\mathbf{x}_m|)$ is identically zero. Hence, if one can set the range $\theta_1 = \phi_m$ and $\theta_N = \phi_m + \pi$, good results can be obtained via MUSIC.

Based on above observation, a necessary condition for applying MUSIC in limited-view inverse scattering problem: let $\psi(\mathbf{x}_m)$ denotes the slope of tangential line at $\mathbf{x}_m \in \Gamma$. Then, for obtaining good results, the range of incident directions θ_1 and θ_N must satisfy

$$\theta_1 = \min\{\psi(\mathbf{x}_m)\} \quad \text{and} \quad \theta_N = \pi + \max\{\psi(\mathbf{x}_m)\} \quad \text{for} \quad \mathbf{x}_m \in \Gamma,\ m = 1, 2, \cdots, M. \quad (2)$$

A detailed description is to appear in an extended version of this contribution.

4. Results of numerical simulation

In this section, we exhibit some numerical results. For this, $\lambda = 0.4$ is applied and two Γ_s are chosen

$$\Gamma_1 = \left\{[s - 0.2, -0.5s^2 + 0.5] : -0.5 \le s \le 0.5\right\}$$
$$\Gamma_2 = \left\{[s + 0.2, s^3 + s^2 - 0.3] : -0.5 \le s \le 0.5\right\}.$$

Figure 1 shows maps of $\mathcal{I}(\mathbf{z})$ with various range of directions when the cracks are Γ_1 and Γ_2. This result shows that, an approximate shape of Γ_1 can be identified when $\theta_N - \theta_1 = \pi$ (see Figure 1(a)), and almost complete shape of crack can be identified when θ_1 and θ_N satisfy (2), refer to Figure 1(b). On the basis of the result in [15], complete shape of Γ_1 can be imaged via the map of $\mathcal{I}(\mathbf{z})$ (see Figure 1(c)) when $\theta_N - \theta_1 = 2\pi$.

Now, let us consider the imaging of Γ_2. Opposite to the previous result, we cannot recognize the shape of Γ_2 via the map of $\mathcal{I}(\mathbf{z})$ with $\theta_1 = 0$ and $\theta_N = \pi$ (see Figure 1(d)). On the basis of the result in Figure 1(e), we can observe that if we select θ_1 and θ_N so as to satisfy (2), the shape of Γ_2 can be recognized via the map of $\mathcal{I}(\mathbf{z})$. Same as the result in Figure 1(c), complete shape of Γ_2 can be imaged via the map of $\mathcal{I}(\mathbf{z})$ when $\theta_0 = 0$ and $\theta_N = 2\pi$, refer to Figure 1(f).

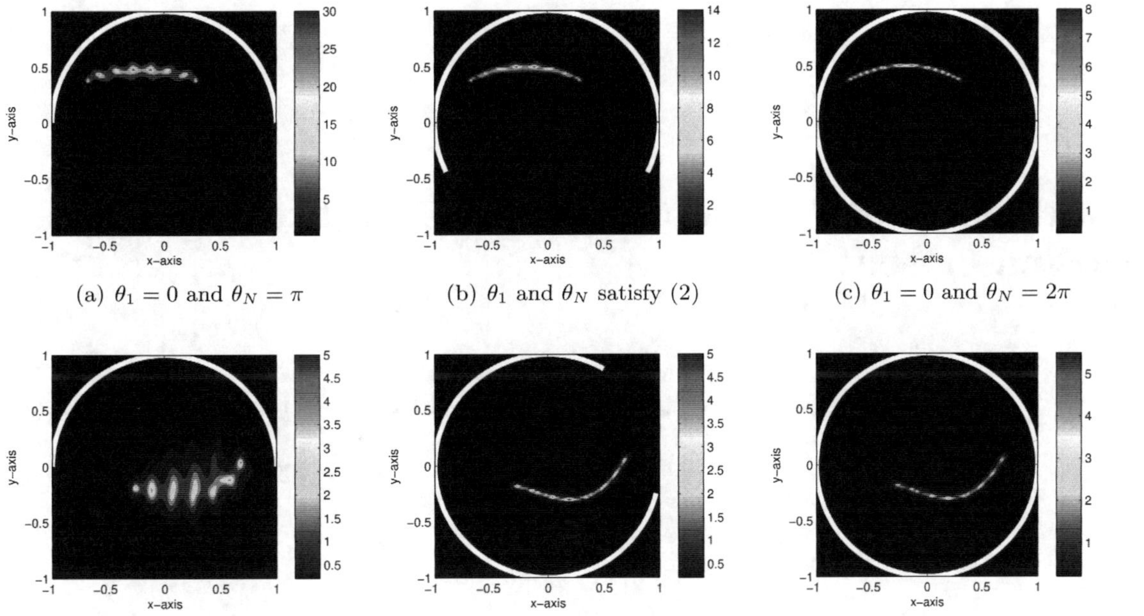

(a) $\theta_1 = 0$ and $\theta_N = \pi$ (b) θ_1 and θ_N satisfy (2) (c) $\theta_1 = 0$ and $\theta_N = 2\pi$

(d) $\theta_1 = 0$ and $\theta_N = \pi$ (e) θ_1 and θ_N satisfy (2) (f) $\theta_1 = 0$ and $\theta_N = 2\pi$

Figure 1. Maps of $\mathcal{I}(\mathbf{z})$ with various range of directions for Γ_1 (top row) and Γ_2 (bottom row).

Acknowledgments
This research was supported by Basic Science Research Program through the National Research Foundation of Korea (NRF) funded by the Ministry of Education(No. NRF-2014R1A1A2055225) and the research program of Kookmin University in Korea.

References
[1] Ammari H, Garnier J, Kang H, Park W K and Sølna K 2011 *SIAM J. Appl. Math.* **71** 68–91
[2] Ammari H, Iakovleva E and Lesselier D 2005 *Multiscale Model. Simul.* **3** 597–628
[3] Ammari H, Iakovleva E, Lesselier D and Perrusson G 2007 *SIAM J. Sci. Comput.* **29** 674–709
[4] Ammari H, Kang H, Lee H and Park W K 2010 *SIAM J. Sci. Comput.* **32** 894–922
[5] Chen X and Zhong Y 2009 *Inverse Problems* **25** 015008
[6] Cheney M 2001 *Inverse Problems* **17** 591–595
[7] Hou S, Sølna K and Zhao H 2006 *Inverse Problems* **22** 1151–1178
[8] Hou S, Sølna K and Zhao H 2007 *Inverse Problems* **23** 1533–1546
[9] Iakovleva E, Gdoura S, Lesselier D and Perrusson G 2007 *IEEE Trans. Antennas Propag.* **55** 2598–2609
[10] Kirsch A 2002 *Inverse Problems* **18** 1025–1040
[11] Park W K and Lesselier D 2009 *J. Comput. Phys.* **228** 8093–8111
[12] Park W K and Lesselier D 2009 *Inverse Problems* **25** 075002
[13] Zhong Y and Chen X 2007 *IEEE Trans. Antennas Propag.* **55** 3542–3549
[14] Joh Y D and Park W K 2013 *Prog. Electromagn. Res.* **138** 211–226
[15] Park W K 2015 *SIAM J. Appl. Math.* **75** 209–228
[16] Ahn C Y, Jeon K and Park W K 2015 *J. Comput. Phys.* **291** 198–217
[17] Joh Y D, Kwon Y M and Park W K 2014 *Appl. Math. Comput.* **240** 273–280

Study of deformed quasi-periodic Fibonacci two dimensional photonic crystals

K. Ben Abdelaziz[1], Y.Bouazzi [1] et M. Kanzari [1,2]
[1] Laboratoire de Photovoltaïques et Matériaux de Semi-conducteurs-ENIT-Université de Tunis el Manar, BP 37,le belvedère, 1002 Tunis, Tunisie
[2] Laboratoire de Photovoltaïques et Matériaux de Semi-conducteurs-ENIT-IPEITunis Montfleury-Université de Tunis, Tunisie

E mail:bab_khdadija@yahoo.fr

Abstract. Quasi-periodic photonic crystals are not periodic structures. These structures are generally obtained by the arrangement of layers according to a recursive rule. Properties of these structures make more attention the researchers especially in the case when applying defects. So, photonic crystals with defects present localized modes in the band gap leading to many potential applications such light localization.
The objective of this work is to study by simulation the effect of the global deformation introduced in 2D quasiperiodic photonic crystals. Deformation was introduced by applying a power law, so that the coordinates y of the deformed object were determined through the coordinates x of the non-deformed structure in accordance with the following rule: $y = x^{1+k}$. Here k is the coefficient defining the deformation. Therefore, the objective is to study the effect of this deformation on the optical properties of 2D quasiperiodic photonic crystals, constructed by Fibonacci generation. An omnidirectional mirror was obtained for optimization Fibonacci iteration in a part of visible spectra.

1. Introduction:

Photonic crystals can be periodic and/or quasi periodic structures which are characterized by photonic band gaps (PBG). These structures can be in 1D, 2D or 3D dimensions. One more attractive property of photonic crystals is 2D configuration due to the inherent simplicity of calculation, visualization, and understanding in 2D. [1]
For example, two-dimensional (2D) photonic crystals can be integrated with existing planar optical waveguide technology [2]. So, two classes of two-dimensional photonic crystal are considered. The first is a photonic crystal formed by a square lattice of dielectric columns of dielectric constant ε_a in a different dielectric background of dielectric constant ε_b. The second is a photonic crystal made by arranging the dielectric columns in a triangular lattice. One particularly interesting aspect of these systems is the possibility of creating crystal defects that confine light in localized modes [1]
Quasi-periodic crystals named quasi crystals were discovered by D. Shechtman et al [3], and exhibit unique electronic properties. It has been noticed that one advantage of these aperiodic structures is to present many inequivalent sites, and consequently many possible different defects [4, 5]. A significant property of photonic quasi crystals is that total band gaps can be obtained in the presence of sufficiently high statistical symmetry [6, 7], even if the value of the minimal index contrast has been subject of controversy.[3]

The most important and well-known quasi periodic structure is the Fibonacci sequence (FS) [8, 9].
In this paper we study photonic band gap (PBG) characteristics of deformed 2 D quasi-periodic crystal which is constructed according to the Fibonacci sequence so that the coordinate y of the deformed object were determined through the coordinate x of the Fibonacci stack in accordance with the following rule: $y=x^{k+1}$ .k is the deformation degree. We assume all the regions to be linear, homogenous, no absorbing and we will show that the PBG depends on the Fibonacci iteration, the degree of deformation and the reference wavelength.

2. Model Description:

HFSS (High Frequency Simulator Structure) is a simulator of high-performance which models and feigns in 3D the global fields shone by the structures. The program is based on Finite Element Method which consists to solving the Maxwell equations.
Global defect is done by a power law $y=x^{k+1}$. K is the coefficient defining the deformation or the asymmetry degree into the structure. So that 3 states can be occurred: k>0, k<0 and k=0.
Fibonacci structures are the paradigm of quasi-periodic order [9]. In two spatial dimensions, they can be generated by combining two different materials from a seed letter A or B by applying two elementary one-dimensional Fibonacci sequence maps:

$$f_A: A \rightarrow AB,\ B \rightarrow A$$
$$f_B: A \rightarrow B,\ B \rightarrow BA$$

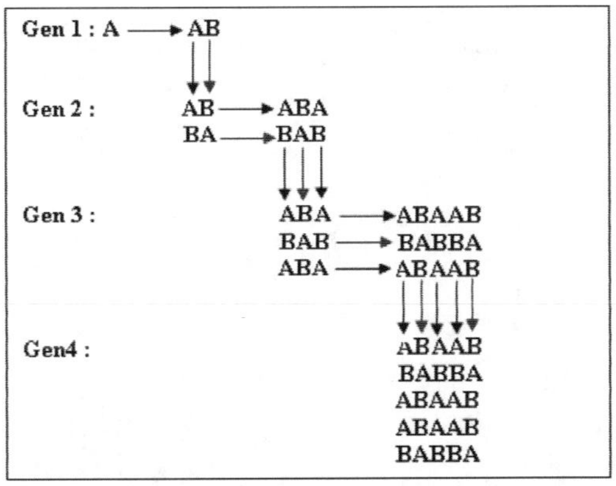

Figure 1. First three generations of 2D Fibonacci sequence shown along with the inflation rules,

In our study, A and B are two elementary materials, with high (TiO_2) and low refractive index (air). The refractive index of the study structure are taken as n_H=5.29 and n_L=1.

Figure 3 shows 2D quasi-periodic system for the 3rd, 4th, 5th and 6th generation according to Fibonacci model.

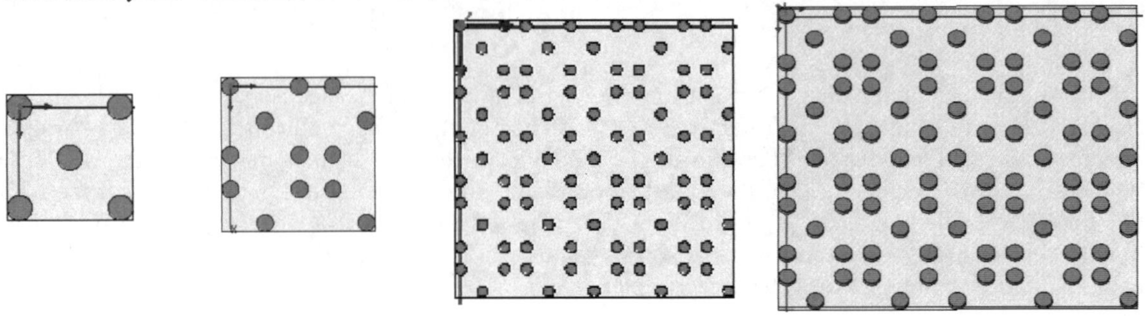

(a) (b) (c) (d)

Figure 2. 2D quasi-periodic structure for the 3rd (a), 4th (b) 5th (c) and 6th (d) generation according to Fibonacci model

3. Results and discussions:

To study the effect of deformation degree k on the optical response of our system we start by modeling the 2D structure and then we launch simulation on the computer.

3.1 Effect of generation number of Fibonacci sequence:

To see the effect of generation number of Fibonacci sequence on the Photonic Band Gap, We proceed by increasing the iteration order (N) of this structure for the asymmetry degree k=0. So that we determined the reflectance for N=4,5 and 6. The figure 3 shows the evolution of the structure response according to the number of Fibonacci generation.

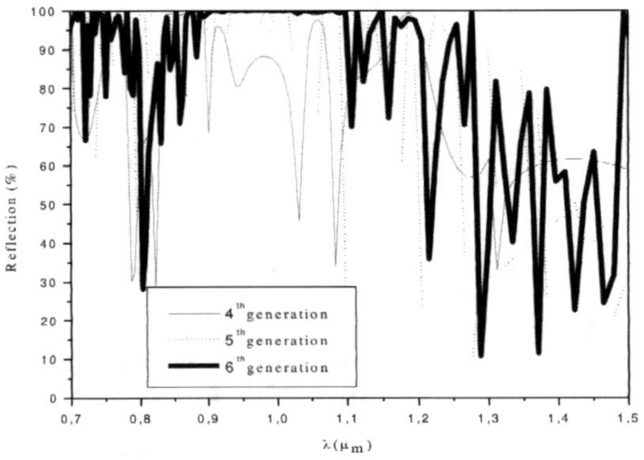

Figure3. Variation of the reflectance spectrum according to the number of iteration of Fibonacci sequence for asymmetry degree k=0.

From the figure 3 we can see that by increasing the number of Fibonacci sequence (N) an improvement in optical properties was observed. In fact we note that when N<5 there is no BIP. But from N=5 a Photonic Band Gap appears and an increase in width was observed. For N=5 the photonic band Gap cover the domain (0.93…1.05) μm for 99% of reflection. This domain becomes (0.89…1.09) μm for the 6th Fibonacci generation. The PBG width passes from 0.12μm to 0.29 μm. This number can be improved if we increase the number of Fibonacci sequence but it requires high computer performance, the reason why we will stop our research at the 6th Fibonacci generation.

3.2 Effect of asymmetry degree:

As it is mentioned, we use a 2D quasi-periodic structure obtained by applying the Fibonacci sequence for the 6th generation. The effect of deformation degree (k) on the BIP is giving by modifying the values of k. we will take negative and positive values of deformation degree and we will note the effect on the forbidden gap behavior with reference wavelength λ_0=0.6μm.

The Figure 4 shows reflection spectra according to wavelength for different values of k.

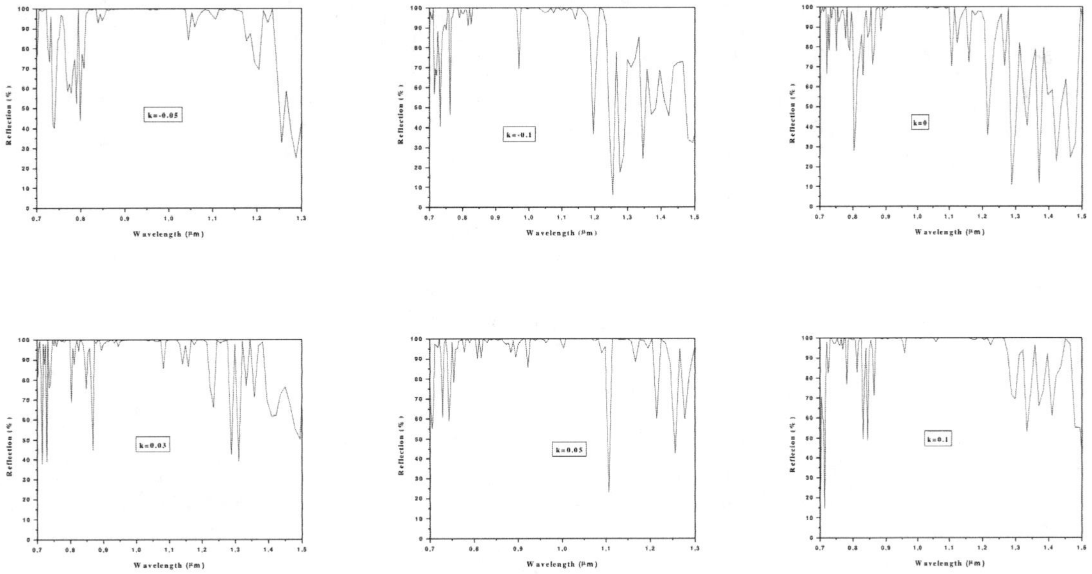

Figure 4. Reflection spectra according to wavelength while varying deformation degree k for the 6th generation of Fibonacci sequence applied on 2D structure of TiO_2 on air for reference wavelength λ_0=0.6(μm)

The figure above shows that the photonic Band Gap widens gradually according to the deformation degree. When k=0 the total bandwidth ($\Delta\lambda=\lambda_{long}-\lambda_{short}$) is 0.209 μm this value is increasing according to k and it reach 0.386 μm

The table 1 confirms the results above.

K=	-0.1	-0.05	0	0.03	0.05	0.1
$\Delta\lambda$(μm)	0.289	0.232	0.209	0.204	0.264	0.386

Table1. Improvement of the PBG width for the 2D quasi-periodic structure according to deformation degree (k)

3.3 Effect of reference wavelength:

For studying the impact of the reference wavelength λ_0 on the forbidden band we choose different values and we observe the response.

The results are reported in figure 5.

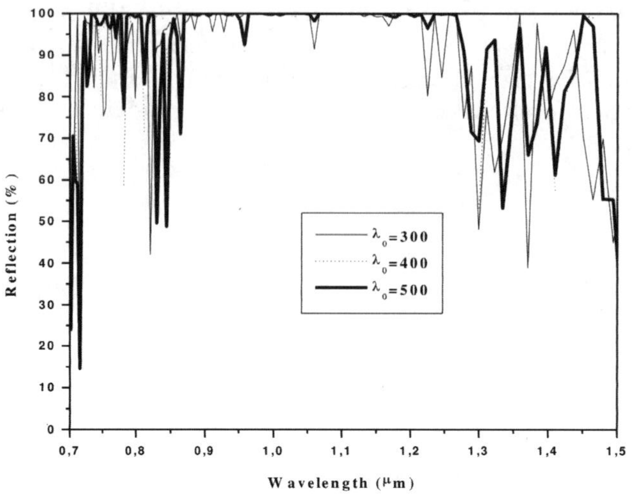

Figure 5. Reflection response according to reference wavelength λ_0 for a deformed 2D quasi-periodic structure k=0.1

Figure 5 shows that there is amelioration of the PBG behavior when we raise the reference wavelength. Indeed for λ_0=300μm the the PBG shows some transmission peaks whose decreases when λ_0=500 μm.

4. Conclusion

We have shown that the use of quasi-periodic photonic structures in the 2D configuration, and in the case where these structures are deformed as described above shows an improvement in their optical responses. Indeed, photonic band gap has been demonstrated dependent on the rank of the iteration, the reference wavelength, and the degree of deformation k. We varied these parameters in order that the BIP cover a wide band in the optical domain. The model that is based on the method of finite elements seems effective for extracting optical properties of 2D photonic crystals.

References:

[1] M. L. Povinelli, Steven G. Johnson, Shanhui Fan, and J. D. Joannopoulos, physical reviews B, volume 64, 075313 (2001)

[2] M.J.A. DE DOOD,E. SNOEKS ,A. MOROZ AND A. POLMAN, 2002, Optical and Quantum Electronics 34: 145–159,

[3] A. Della Villa, S. Enoch, G. Tayeb, V. Pierro, V. Galdi, and F. Capolino, 2005, *Band Gap Formation and Multiple Scattering in Photonic Quasicrystals with a Penrose-Type Lattice*,PHYSICAL REVIEW LETTERS,

[4] Y. S. Chan, C. T. Chan, and Z.Y. Liu, Phys. Rev. Lett. 80, 956 (1998).

[5] [13] M. Bayindir et al., Phys. Rev. B 63, R161104 (2001).

[6] P. Han, H. Z. Wang, *Effect of invariant transformation in one-dimensional randomly-perturbed photonic crystal*, Chinese Phys. Lett., Vol. 20, 2003, pp. 15-20.

[7] A. R.Merlin, K. Bajema, R. Clarke, F. Y. Juang, P. K. Bhattacharya, *Quasiperiodic GaAs-AlAs hetero structures*, Phys. Rev. Lett.,Vol. 55, No. 17,1985, pp. 1768-1770.

[8] Y. Trabelsi, N. Ben Ali, Y. Bouazzi, M. Kanzari, *Reflectance spectra using one-dimensional deformed quasi-periodic Band Gap structure in high frequency*, The Mediterranean Journal of Electronics and Communications, Vol. 4, No. 3, 2008.

[10] Ramona Dallapiccola1, Ashwin Gopinath2, Francesco Stellacci1, and Luca Dal Negro2, Quasi-periodic distribution of plasmon modes in two-dimensional Fibonacci arrays of metal nanoparticles, 14 April 2008 / Vol. 16, No. 8 / OPTICS EXPRESS 5544

Journal of Physics: Conference Series **633** (2015) 012008 doi:10.1088/1742-6596/633/1/012008

Simulations of a epidemic model with parameters variation analysis for the dengue fever

C. L. T. F. Jardim[1], D. B. Prates[2], J. M. Silva[3], L. A. F. Ferreira[4] and M. V. Kritz[5]

Universidade Federal dos Vales do Jequitinhonha e Mucuri, Minas Gerais - Brazil

E-mail: caiodt4@hotmail.com[1], derekbomfim@hotmail.com[2], jaqueline.silva@ufvjm.edu.br[3], leticiaalifig@hotmail.com[4] and kritz@lncc.br[5]

Abstract. Mathematical models can be widely found in the literature for describing and analyzing epidemics. The models that use differential equations to represent mathematically such description are specially sensible to parameters involved in the modelling. In this work, an already developed model, called SIR, is analyzed when applied to a scenario of a dengue fever epidemic. Such choice is powered by the existence of useful tools presented by a variation of this original model, which allow an inclusion of different aspects of the dengue fever disease, as its seasonal characteristics, the presence of more than one strain of the vector and of the biological factor of cross-immunity. The analysis and results interpretation are performed through numerical solutions of the model in question, and a special attention is given to the different solutions generated by the use of different values for the parameters present in this model. Slight variations are performed either dynamically or statically in those parameters, mimicking hypothesized changes in the biological scenario of this simulation and providing a source of evaluation of how those changes would affect the outcomes of the epidemic in a population.

1. Introduction

Mathematical models are the major tool for scientific and methodic analysis of most natural phenomena. Its versatility is used in the branch of applied mathematics, since this analysis allows the achievement of several data to the study and interpretation of the studied area.

Mathematical epidemiology aims to use tools to describe, simulate and predict the behavior of a population affected by an epidemic, the environment where it takes place or the pathogenic agents involved. Such analysis can be modeled either generalizing the study to a broad range of epidemics or specifying models to each type of epidemic. For the latter one, there are mathematical tools that can be chosen in order to better describe each situation.

Kermack and McKendrick proposed in 1927 the SIR model, precursor in mathematical epidemiology. It proposes to divide a population into three classes of individuals: susceptible, infected and recovered (KERMACK, 1927). The model uses a differential equations system fed with parameters that characterize the dynamics of those classes. Such parameters have a great importance in mathematical epidemiology because they represent intrinsic characteristics of either the population or the epidemic.

The dengue fever is a viral disease transmitted by the vectors Aedes aegypti and Aedes albopictus mosquitoes and presents the peculiarity of being caused by four viral strains. This makes possible an infection of the same population by more than a viral strain simultaneously; thus, there is a situation where cross-immunity among subtypes virus is observed (YANG and FERREIRA, 2008). The dissemination of the dengue fever is directly related to environmental factors, such as temperature and volumetric precipitation that may affect directly the epidemic dynamics. The dengue fever is characterized by recurrent epidemics, mainly in large cities, where control actions have been shown inefficient facing the disorganized population growth and basic sanitation problems. This recurrence is the motivational factor for using a model that includes seasonal parameters.

Among its symptoms, the dengue may cause fever, retro orbital pain and skin patches, from five to seven days after the infection, characterizing a classic dengue fever. More severe symptoms, as multiple hemorrhage, decrease in the number of platelets, increase in the vascular permeability and signals of circulatory insufficiency characterize hemorrhagic dengue (RICO-HESSE, 2007). Studies indicate that secondary infections caused by the different virus strains are associated with hemorrhagic dengue fever (YANG, 2003).

Modifications in the traditional epidemiological models by inserting parameters and classes that simulate real conditions of the biological epidemical scenario in question have been created over time. Mathematical tools that represent the use of seasonal forces, cross-immunity factors, age-structured systems and medicine insertion in the population were developed to better describe some specific situations. In this study, some of those tools are used to analyze the overall behavior of a population subjected to a dengue epidemic situation. In order to do so, a mathematical model that supports the presence of two disease strains, the possibility of temporary cross-immunity and that considers a seasonal influence is used. Finally, a huge attention is given to a study about the influence of the parameters associated with the epidemic in the epidemiological behavior of the population in question. Results of those studies are generated, illustrated, interpreted and discussed in their mathematical and biological sense.

2. The Model

In order to do the already proposed analysis, a model studied by (AGUIAR, 2012) is used. In this system, the population is divided in 10 classes: those susceptible to the two strains, S; those infected for the first time with the strain i, I_i; the recovered people from the first infection i, R_i; those susceptible to the infection j but with immunity to the already suffered case of infection i, S_i; there are also the classes of people that were infected with the strain i for the first time and then infected with the strain j, I_{ij}, and those recovered from the two infections, R. The model is mathematically represented by the following differential equation system:

$$\frac{dS}{dt} = -\frac{\beta(t)}{N}S(I_1 + \rho N + \phi I_{21}) - \frac{\beta(t)}{N}S(I_2 + \rho N + \phi I_{12}) + \mu(N - S)$$

$$\frac{dI_i}{dt} = \frac{\beta(t)}{N}S(I_i + \rho N + \phi I_i j) - (\gamma + \mu)I_i$$

$$\frac{dI_2}{dt} = \frac{\beta(t)}{N}S(I_2 + \rho N + \phi I_1 2) - (\gamma + \mu)I_2$$

$$\frac{dR_1}{dt} = \gamma I_1 - (\alpha + \mu)R_1$$

$$\frac{dR_2}{dt} = \gamma I_2 - (\alpha + \mu)R_2$$

$$\frac{dS_1}{dt} = -\frac{\beta(t)}{N}S_1(I_2 + \rho N + \phi I_{12}) + \alpha R_1 - \mu S_1$$

$$\frac{dS_2}{dt} = -\frac{\beta(t)}{N}S_2(I_1 + \rho N + \phi I_{21}) + \alpha R_1 - \mu S_1$$

$$\frac{dI_{12}}{dt} = \frac{\beta(t)}{N}S_1(I_2 + \rho N + \phi I_{12}) - (\gamma + \mu)I_{12}$$

$$\frac{dI_{21}}{dt} = \frac{\beta(t)}{N}S_{21}(I_1 + \rho N + \phi I_{21}) - (\gamma + \mu)I_{21}$$

$$\frac{dR}{dt} = \gamma(I_{21} + I_{12}) - \mu R.$$

Where:

$\beta(t) = \beta_0(1 + \cos(\omega(t + \psi)))$.

The parameters in the system are described as following: β is the infection rate between susceptible and infected classes; ϕ is the correction factor related with the second infection in the same individual; γ is the recovery rate; μ represents the mortality (both natural and caused by the infection) and birth rate; α is the temporary cross-immunity factor; ρ is the import rate parameter of the population, which says how many people coming from other places are included in the population; N is the population size, chosen 100 in these simulations so they can be represented as percentages. Finally, in the seasonality function, β_0 is the value associated with the infection rate in the beginning of the simulation, η is the degree of seasonality and ψ is the phase (considered null in these simulations).

The study of the model stability is presented in [1]. Intervals for parameters values and their consequence for the model stability and existence of solutions are also thoroughly presented in those references, being omitted here due to this work purpose. Nevertheless, such intervals are respected in the parameters variation done to prevent chaotic and indeterministic behavior.

The parameter variation is performed either statically or dynamically. This means that a value can be attributed to a parameter and being hold constant, or a function can be used to describe how this parameter value changes in the time. Whether this happens (and, if it does, which type of function is used) is specified in each scenario analysis, as long with a basic interpretation about the meaning of holding it constant or as a variable.

By modifying such parameter numerical values it is possible to simulate certain hypothesis in the biological scenario. This work proposes to analyze the variation in the birth and death rate, in the infection force β_0 and in the cross-immunity and recovery rates. The diseases intrinsic parameters are only slightly changed, in order to preserve the dengue fever parameters set identity. If it were changed significantly, this study would not be effective to represent a dengue fever scenario.

The simulations generates enough data to describe the temporal variation of the classes and are presented in the next section. This data appears when the differential equation system is numerically solved using a general Runge-Kutta method.

3. Results and Discussions

3.1. Original Simulation

The first simulation is done with the parameters presented in (AGUIAR, 2012). The following table presents the parameters:

Paramater	Description	Value
N	Population size	100
μ	Birth and death rate	$\frac{1}{65}y$
γ	Recovery rate	$52y^{-1}$
β_0	Infection rate	$104y^{-1}$
η	Degree of seasonality	0.3
ρ	Import parameter	10^{-10}
α	Temporary cross-immunity rate	$2y^{-1}$
Φ	Ratio of secondary infections contributing to force of infection	0.7

Table 1: Parameters used in the First Simulation

The Figure 1 show an irregular cyclic behavior in the class of individuals infected, where each cycle contains several outbreaks followed by a period with low disease levels, due to both seasonal force and cross-immunity parameters. The curves describing I_{12} and I_{21} present a lower number of infected individuals when compared to I_1 and I_2 due to the concept of cross-immunity, whereby an individual already recovered of the strain i acquires partial immunity to the strain j. The data obtained in this first simulation is presented as graphics in the following Figure 1 and Figure 2.

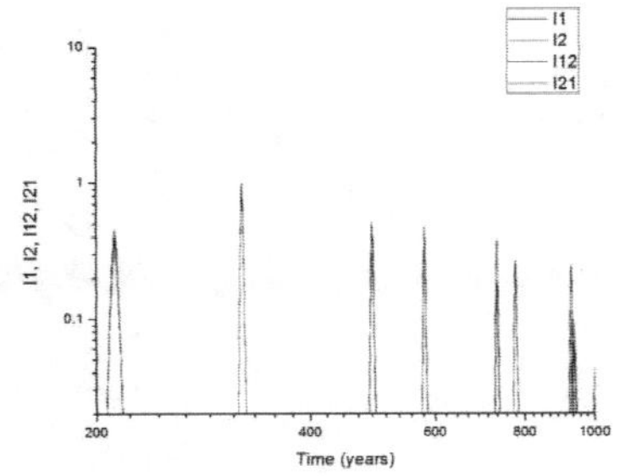

Figure 1: Infected population behavior.

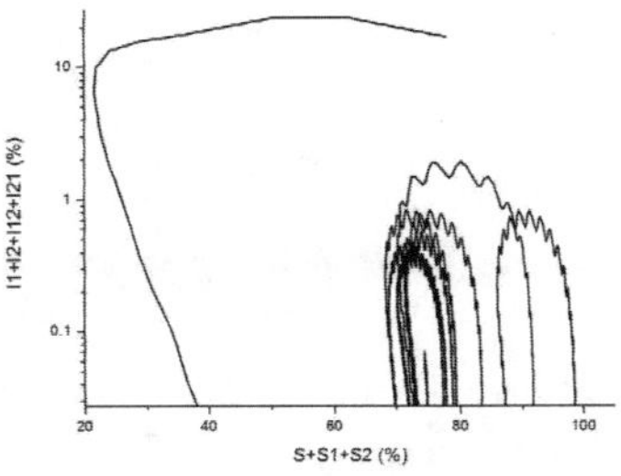

Figure 2: Dynamic between infected and susceptible populations.

The Figure 2 has a representative character, since it is possible to extract the dialogue between infected populations and susceptible populations throughout all the simulation. It is noticed the influence of the seasonality parameter in the regular trace acquired after some time

Journal of Physics: Conference Series **633** (2015) 012008 doi:10.1088/1742-6596/633/1/012008

3.2. Variation in the birth and death rate

These simulations present the epidemic behavior after variation in the birth and death rate (μ) - other parameters keep values seen in Table 1. The alteration made in this parameter was $\frac{4}{65}$ of the original value described in (AGUIAR, 2012). It is proposed an analysis of the Law of Mass Action, which basically says that the infection rate is directly proportional to the the number of infected and the number of susceptible individuals. This assumption shows consistence in many epidemiological scenarios and it is an important tool of mathematical modeling study. The changes made in the parameters achieved the expected results without compromising the reliability of the simulations. The value was chosen carefully, so that there was not a big distortion of its original value, which would cause the loss of its biological sense.

The Figure 3 show the same irregular behavior presented in the class of infected individuals in Section 3.1, but with more outbreaks in each cycle due to the stronger relation between the population size and the infection rate after the variation. Both graphs still representing infected individuals behavior influenced by seasonal force and cross-immunity parameters. By comparing them with Figure 1 and Figure 2 it is possible to see how affected the dynamic was after a slight change in its birth and death rate.

The Figure 4 shows a result that strengthens this discussion. The more condensed and frequent dynamic of this simulation is given by a quick inspection in the phase plan and its comparison with Figure 2. This considerable increase in the number of outbreaks was achieved in a very slight change in one of the parameters, showing how true and meaningful is the Law of Mass Action in real simulations.

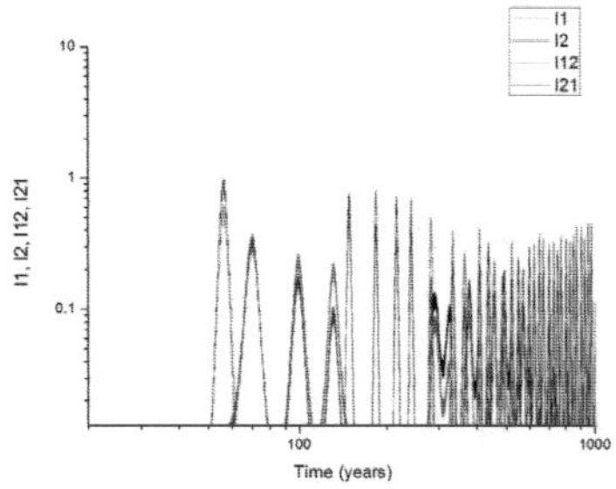

Figure 3: Infected population behavior.

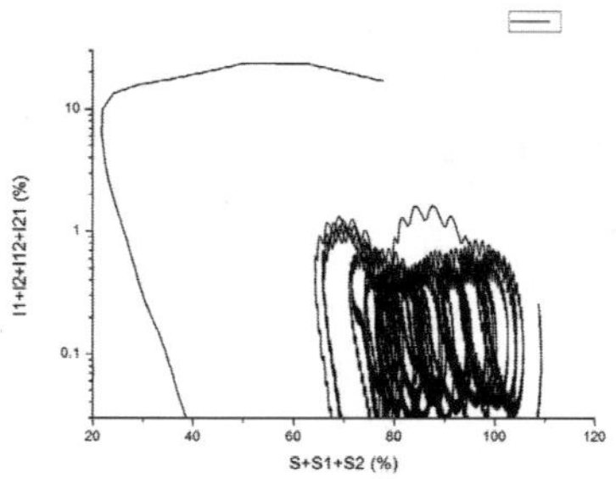

Figure 4: Dynamic between infected and susceptible populations.

3.3. Variation in the initial infection rate

Simulations are presented to describe the epidemic behavior after an first variation in the infect rate β and all other parameters keep values seen in Table 1. Now, the infect rate is 1.5 of its original value; thus, β= 156. The Figure 5 is used to present the infected population behavior. It is possible to notice the same behavior previous described in Figure 3, but with the curve presenting less outbreaks due to the decrease in the infect rate.

The phase diagram presented in Figure 6 indicate a more intense epidemic dynamics, thereby having more spirals and an increase in the number of dead individuals. This behavior happens during the same period used in simulations presented in the section 3.1.

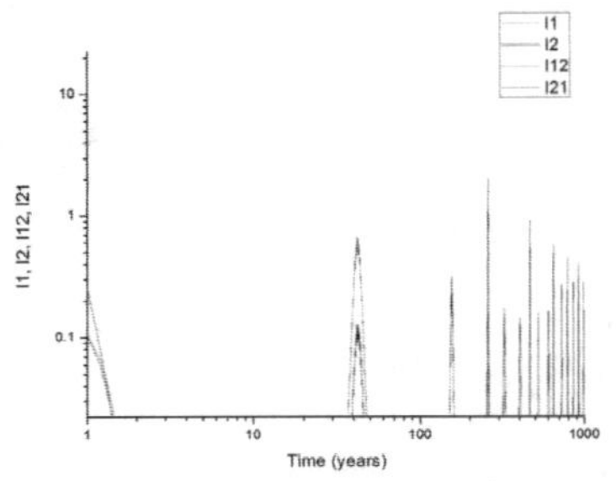

Figure 5: Infected population behavior.

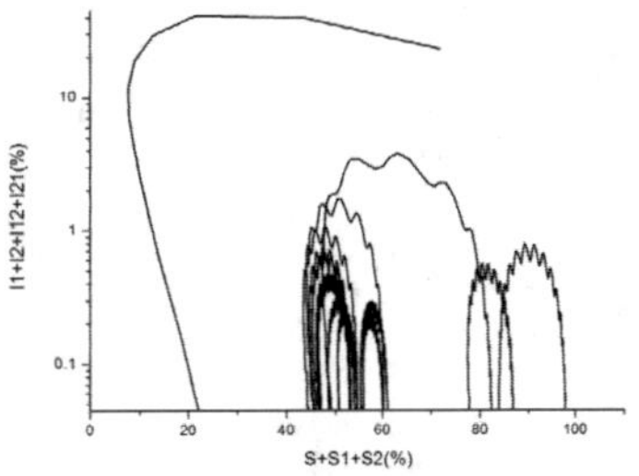

Figure 6: Dynamic between infected and susceptible populations.

4. Conclusion

Throughout the simulations performed in this study, it is noticed the strong effect that even slight variations in parameters can cause in the model and in the population behavior in an epidemic scenario. Such sensibility is so expressive in the mathematical modeling that researches give as much attention to the correct choice of a set of parameters as they give to the right model to be used for describing a certain epidemic. Furthermore, in field works, it is always sought to better mimic the real population characteristics and the more sensible and accurate way to do that is through the correct estimation of parameters from previous collected data, especially if a differential equations-based model is being used.

The importance of epidemiological modeling is increased when it is possible to verify, through those simulations, that changes in parameters (and therefore in biological conditions) actually promote real changes in the solutions and in the classes conduct. Such change when the epidemic environment is altered is a fact in biological phenomena, thus it is also expected from an effective model to reproduce mathematically those changes, bilaterally.

Furthermore, it is actually possible to extend that discussion for all the model with similar structure to the one used in this work, and for all the epidemics described in this way, not only the dengue fever. Such conclusion comes from the fact that this bilateral relation between a reasonable effective model (feed by parameters) and the biological scenario described needs to exist in order to the model be really effective. Thus, respecting the intervals in what the solutions exist and are not chaotically oriented, the expected response in a variation of parameters is a variation in the solution itself, even a minimal one. The absence of solution changes can be an evidence that either the chosen model does not represent properly the epidemic in question or it has a poor mathematical representation or accuracy.

The ability of better describe an epidemic using those models and their respective parameters goes far beyond the mathematical achievement itself and it is crucial for the prevention and cure actions taken by the responsible health organizations. Only understanding, dominating and improving the tools we have against those natural disasters it is possible to eradicate them and avoid the death or illness of uncountable populations.

Acknowledgments
We greatly thank FAPEMIG for the financial support, which makes possible the participation in this event. We also thank CNPq for the scientific scholarship given to the undergraduate student Derek B. Prates, who participates in this work.

References

[1] Adams, B.; Boots, M. (2006). Modelling the relationship between antibodydependent enhancement and immunological distance with application to dengue. Jounal of Theoretical Biology, 242, 337–46.

[2] Aguiar, M.; Ballesteros, S.; Kooi, B. W.; Stollenwerk, N. 2011 *The role of seasonality and import in a minimalistic multi-strain dengue model capturing differences between primary and secondary infections: Complex dynamics and its implications for data analysis.*. Journal of Theoretical Biology **268** 2011.

[3] Aguiar, M., Stollenwerk, N., Kooi, B.W., 2009. *Torus bifurcations, isolas and chaotic attractors in a simple dengue model with ADE and temporary cross immunity.* Int. J. Comput. Math. **86**, 1867–1877.

[4] Aguiar, M. (2012). Rich dynamics in multi-strain models: non-linear dynamics and deterministic chaos in dengue fever epidemiology. PhD Thesis, ISBN 978-989-20-2882-8, Edited by Maira Aguiar, Portugal.

[5] Aguiar, M.; Kooi, B.W.; Stollenwerk, N. (2008). Epidemiology of Dengue Fever: A Model with Temporary Cross-Immunity and Possible Secondary Infection Shows Bifurcations and Chaotic Behaviour in Wide Parameter Regions (see Chapter 2). Math. Model. Nat. Phenom., 4, 48-70.

[6] Yang, H. M.; Ferreira, C. P. 2008 *Assessing the effects of vector control on dengue transmission.* Applied Mathematics and Computation. **198** 400-413.

[7] Rico-Hesse, R. 2007 *Assessing the effects of vector control on dengue transmission.* Clinical Infectious Diseases. **44** 1462-1466.

[8] Yang, H. M. 2003 *Epidemiologia da Transmissão da Dengue.* Tendência sem Matemática Aplicada e Computacional. **4** 387-396.

[9] Kermack, W.; McKendrick, A. 1927 *Contributions to the mathematical theory of diseases.*Proceedings of the Royal Society, **141** 1933.

[10] Barros, A. M. R. 2007 *Modelos Matemáticos de equações diferenciais ordinárias aplicados à epidemiologia* UFLA.

[11] Bassanezi, R. C. 2006 *Ensino-aprendizagem com modelagem matemática* (São Paulo: COntexto).

[12] Castillo-Chavez, C.; Hethcote, H. W.; Andreasen, V.; Levin, S. A.; Liu, W. M. 1989 *Epidemiological models with age structure, proportionate mixing, and cross-immunity,* J. Math. Bioi. **27** 233-258.

[13] Dietz, K. 1975 *Transmission and control of arbovirus diseases.* In: Cooke, K. L. (ed.) Epidemiology, pp. 104-121. Philadelphia: Society for Industrial and Applied mathematics.

[14] Esteva, L.; Vargas, C. (2000) Influence of vertical and mechanical transmission on the dynamics of dengue disease. Mathematical Biosciences, 167, 51-64.

[15] Oliveira, I.M. (2008), Modelos Epidemiológicos SEIR. Dissertação de Mestrado. Departamento de Matemática Pura. Universidade do Porto. Portugal.

[16] World Health Organization, 2009. *Dengue: Guidelines for diagnosis, treatment, prevention and control.*

Virtual Laboratories in Physics with Autogenerated Parameters

M A Maksimov, V V Monakhov and A V Kozhedub
Department of Computational Physics, Faculty of Physics, St Petersburg State University, St Petersburg, 198504, Russia

E-mail: 2xmax@mail.ru

Abstract. The paper is devoted to a virtual laboratory system, which in particular can be used to test knowledge through research. The participant can prefer which tools to operate and what actions should be taken. For the most of the tasks, there are copious ways to obtain the correct solution. One of the most important features of the system that distinguish this one among other simulation packages and educational systems is the pseudo-random physical parameter generation technique. The technique supports constraints and relationships between variables. As a result, it provides correctness and equal complexity of the generated task. The system can be very complex and is highly customizable by internal script system executed on server-side. The system is used as a part of distolymp Learning Management System with about 40 thousand participants per year.

1. Introduction

Online Competition in Physics [1] provides a unique opportunity not only to test theoretical knowledge, but also the ability to apply this knowledge in practice. Checking the ability to conduct a physical experiment is achieved by means of virtual laboratories. This virtual laboratory is a computer simulation with virtual physical system and virtual measuring devices. The contest participants are given a set of tools which is necessary to perform the research. For the majority of tasks, there are various ways to obtain a correct solution. In order to solve the problem, an attendee of the Online Competition chooses the tools and makes decisions by himself. Moreover, those assessments check the attendee's ability to plan action that leads to a correct result among redundant conditions and obscurity of the system.

2. Distolymp Learning Management System

The developed system provides three types of assessments: tests, theoretical tasks and virtual experiments. Thus, tests check student's basic knowledge of the subject. The virtual experiments allow us to identify the most talented students, as they assume creative research approach. The software can be used both for simple tasks and for hard problems with complexity level which can be compared with the tasks of the final tour of the Online Competition in Physics. In such a way, once the Organizing Committee of the competition constructed quite a challenging task which was completely solved by only one of 930 participants of the final stage of the competition. The detailed description of the task and the solution are available here [2].

As it often happens in physical experiment (especially in a complex environment), it is not always easy to get the correct result from the first attempt. Thereupon, the attendee receives a feedback about whether he completed the task correctly or not. Then he can retry to solve the task again with some penalty. As a result, this feature tests the candidate's ability to correct his errors on the results of his activity. This is one of the most important features which distinguish distolymp virtual laboratories from some other Open Online Courses.

3. The generation algorithm

3.1. Motivation

First, all participants receive similar tasks, models and equipment. Thereupon the server generates some unique pseudorandom state of the system. Obviously, the generated problem should be correct and solvable. Another important criterion for the generation algorithm is the rule that the complexity of the generated assessments between various attendees should be equal. To accomplish such result, authors of the Learning Management System have developed a server-side system of physical parameter generation.

The parameter generation algorithm generates a pseudorandom set of input parameters according to given input constraints, calculates the constraints got other physical parameters corresponding to the generated set and repeats the process until all the constraints are satisfied.

The constraints can be divided into a number of categories:

- A pseudorandom value from a uniform grid
- A pseudorandom value from given set
- A result of function evaluation with already specified variables calculated by substitution into a formula

In order to optimize server load and extend capabilities of the LMS, some values are processed at the server side, while the other is processed at the client side.

3.2. Accuracy measurements

In the first version of the system we employed a fixed absolute error for checking the results. This approach, however, turned out to be problematic for our pseudorandom system if the generated dependent parameters differ by some magnitude, the calculated absolute error may be very inconsistent with the specified answer. As a consequence, there were a few times when the system rejected the nearly correct answers, and accepted solutions submitted by unfair participants.

Afterwards, the technique was improved with an algorithm which could specify the error for each interval of values. After all, it was decided to use the relative error. The part of source code and details of the algorithm are described in [3].

4. Client-side features

BARSIC programming language [4] is used to develop client-side software of virtual laboratories. It is a Domain-Specific Language with some unique features. Client-side technologies are evolving so fast, so it is quite important to be always ready to a new frontend technology. For this and other reasons, source code of the virtual laboratories is reused by implementing an extensible syntax-directed translation system for different platforms such as J2SE, Android, JavaScript. To put it another way, it is not necessary to rewrite the entire implementation of client library, it is possible to rewrite only the user interface controls and graphical primitives, the behaviour will remain the same.

Conclusion

Relying on our experience of organizing the Online Competitions in Physics for 10 years, it can be concluded that the implementation of virtual laboratories in Physics with server-side algorithm of generation of parameters contains a lot of advantages to test knowledge through research.

References

[1] Monakhov V *et al.* 2013 *Distolymp program complex for online competitions and distance learning* (Petrozavodsk: Proc. of XII International Conference "PHYSICS IN THE SYSTEM OF MODERN EDUCATION" vol 2) pp 221-223

[2] Monakhov V, Kozhedub A and Utkin A 2011 *Distinctive features of Online Competition in Physics* (St Petersburg: Computer Instruments in School № 6) pp 30–38.

[3] Monakhova E 2014 *Enhancement of problem generation and result processing systems in Distolymp program complex* (St Petersburg: St Petersburg University; BSc thesis) pp 4–11

[4] Monakhov V *et al.* 2005 *BARSIC: A Programming System for Physicists* (St Petersburg: Programming and Computer Software vol 31 № 3) pp 157–165.

Computational models of an inductive power transfer system for electric vehicle battery charge

AO Anele[1,2], Y Hamam[1], L Chassagne[2], J Linares[3], Y Alayli[2], K Djouani[1]

[1] Dept. of Electrical Engineering, Tshwane University of Technology, Pretoria, South Africa
[2] LISV Laboratory, Université de Versailles Saint-Quentin-en-Yvelines, Versailles, France
[3] GEMaC Laboratory, Université de Versailles Saint-Quentin-en-Yvelines, Versailles, France

E-mail: aneleamos@gmail.com

Abstract. One of the issues to be solved for electric vehicles (EVs) to become a success is the technical solution of its charging system. In this paper, computational models of an inductive power transfer (IPT) system for EV battery charge are presented. Based on the fundamental principles behind IPT systems, 3 kW single phase and 22 kW three phase IPT systems for Renault ZOE are designed in MATLAB/Simulink. The results obtained based on the technical specifications of the lithium-ion battery and charger type of Renault ZOE show that the models are able to provide the total voltage required by the battery. Also, considering the charging time for each IPT model, they are capable of delivering the electricity needed to power the ZOE. In conclusion, this study shows that the designed computational IPT models may be employed as a support structure needed to effectively power any viable EV.

1. Introduction

Taking into account high oil prices and environmental awareness, the development of electric vehicles (EVs) is considered as a healthier mode of transportation. This is because the electricity they consume can be generated from a wide range of sources which include fossil fuel, nuclear power and renewable sources (e.g. hydroelectricity, solar power etcetera) [1, 2].

According to [3], it was stated that inductive power transfer (IPT) systems for EV battery charge can be successful if issues such as interoperability between systems of different manufacturers, vehicle design towards effective integration of the inductive charging system, parking guidance and fundamental evaluation of the environmental impacts of the electromagnetic fields are solved.

This paper focuses on the aspect of improving the efficiency of the electric power transferred to the on-board battery storage system of EVs. Regarding this aspect, several studies have been presented in [3–7]. Nonetheless, to achieve the aim of this paper, 3 kW single phase and 22 kW three phase computational IPT models for EV battery charge are designed in MATLAB/Simulink and computed based on the technical specifications of the lithium-ion battery and charger type of Renault ZOE.

2. Analysis of IPT Transformer and Effects of Capacitive Compensation

The most significant part of IPT systems is its air-cored transformer. Its aim is to provide maximum electric power to the on-board battery storage system of EVs. However, it is faced with the problem of weak coupling (i.e., suffers from poor efficiency) due to the relatively large leakage reactances which is associated with its primary and secondary coils. Before presenting the computational models of 3 kW single phase and 22 kW three phase IPT systems, the behaviour of IPT transformer and the effects of applying capacitive compensation are studied based on Figs. 1 and 2.

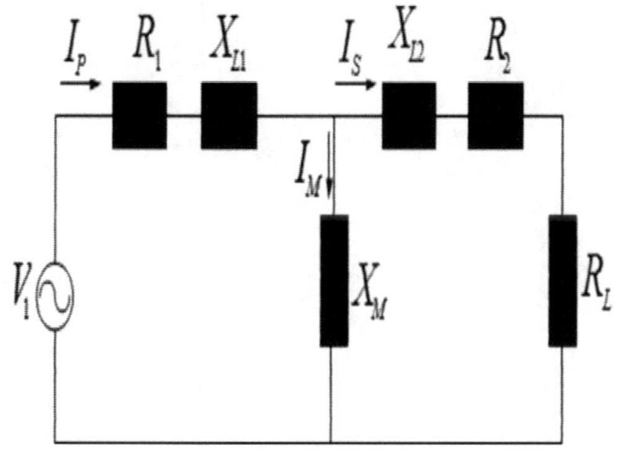

Figure 1. Equivalent circuit of a single-phase IPT transformer.

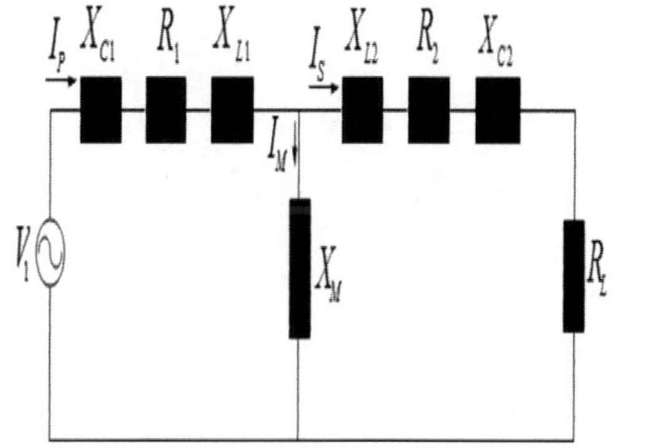

Figure 2. Compensated equivalent circuit.

where I_P, I_M and I_S are the primary, mutual and secondary currents respectively, R_1, R_2 and R_L are the primary, secondary and load resistances respectively, X_{L1} and X_{L2} are the series leakage reactances, X_{C1} and X_{C2} are the capacitive reactances and V_1 is the root mean square (RMS) AC voltage. In addition, X_M is the mutual inductance between the coils. It is given by $2\pi fM$, where $M{=}k\sqrt{L_1L_2}$; k is the coupling coefficient, f is the operating frequency while L_1 and L_2 are the series leakage inductances.

Based on Fig. 1, the equivalent impedance of the circuit as seen by the AC voltage source is given by

$$Z_{eq} = R_1 + jX_{L1} + \frac{jX_M R_2 + jX_M R_L - X_M X_{L2}}{jX_M + jX_{L2} + R_2 + R_L} \tag{1}$$

By applying current divider rule, the current division ratio is given by

$$K_c = \frac{I_S}{I_P} = \frac{jX_M}{R_2 + R_L + j(X_{L2} + X_M)} \tag{2}$$

Also, based on Fig. 1, the power transfer efficiency is given by

$$\eta = \frac{R_L I_S^2}{R_L I_S^2 + R_2 I_S^2 + R_1 I_P^2} = \frac{R_L}{(R_L + R_2) + R_1(\frac{(R_2+R_L)+j(X_{L2}+X_M)}{jX_M})} \tag{3}$$

Amongst other capacitive compensation techniques proposed in [3–5], series-series (SS) technique is employed because it helps to improve the power transfer capability of the transmission line due to its partial compensation of the series leakage reactances.

Based on Fig. 2, the equivalent impedance of the circuit as seen by the voltage source is given by

$$Z_{eq} = jX_{C1} + R_1 + jX_{L1} + \frac{jX_M(jX_{L2} + R_2 + jX_{C2} + R_L)}{jX_{L2} + R_2 + jX_{C2} + R_L + jX_M} \tag{4}$$

For an exactly compensated transmission link of the IPT transformer, the following reactance conditions are vital [4]:

$$\begin{aligned} X_{C1} + X_{L1} + X_M &= 0 \\ X_{C2} + X_{L2} + X_M &= 0 \end{aligned} \tag{5}$$

Thus, substituting (5) into (4) gives the compensated equivalent impedance of the circuit as

$$Z_{eqC} = \frac{R_1R_2 + R_1R_L + X_M^2}{(R_2 + R_L)} \tag{6}$$

Based on Fig. 2, the compensated current division ratio is given by

$$K_{cc} = \frac{I_S}{I_P} = \frac{jX_M}{R_2 + R_L} \tag{7}$$

Also, the compensated power transfer efficiency is given by

$$\eta_C = \frac{R_L I_S^2}{R_L I_S^2 + R_2 I_S^2 + R_1 I_P^2} = \frac{R_L}{(R_L + R_2) + R_1(\frac{R_2+R_L}{jX_M})} \tag{8}$$

3. Discussion of Results for IPT Transformer

Figure 3 shows the variation of K_c and K_{cc} as a function of frequency f for different values of mutual inductance M and resistive load R_L. It is obtained by computing equations (2) and (7) in MATLAB M-File. In the case without compensation, the values of K_c increase as f increases. However, at higher frequencies, they become constant in spite of the change in R_L but, different values of K_c are obtained for different values of M. In the case with capacitive compensation, the values of K_{cc} kept increasing as f increases but, different values are obtained for different values of M and R_L.

The result obtained in Fig. 3 shows that SS capacitive compensation helps to improve the power transfer capability of the IPT transformer (see values obtained for K_c and K_{cc}). In addition, based on the IPT parameters used for MATLAB simulation (see Table 1), the results obtained show that the IPT system considered in [9] must be operated above 100 kHz.

Figure 4 shows the power transfer efficiency of the IPT transformer with and without SS capacitive compensation. It is obtained by computing equations (3) and (8) in MATLAB. Without compensation, a maximum power transfer efficiency of 78.14% is obtained while with capacitive compensation, 93.18% is obtained. Thus, indicating that with SS capacitive compensation, the power transfer efficiency is increased by 15.04%.

Table 1. Parameters for IPT Transformer [8]

Physical Parameters	Description	Values
L_1	Leakage inductance of primary coil	103.4 μH
L_2	Leakage inductance of secondary coil	12.67 μH
R_1	Resistance of primary coil	0.1530 Ω
R_2	Resistance of secondary coil	0.0660 Ω

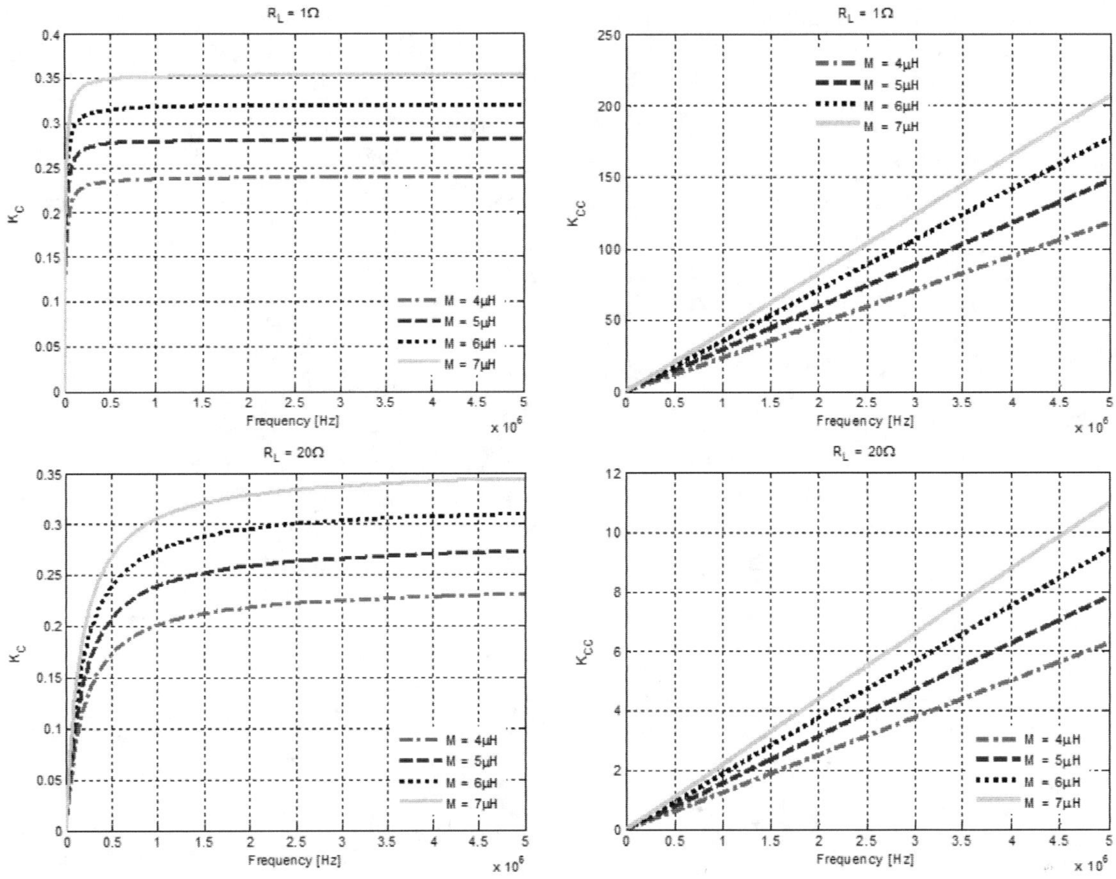

Figure 3. Variation of K_c and K_{cc} as a function of frequency f for different mutual inductances M and resistive loads R_L.

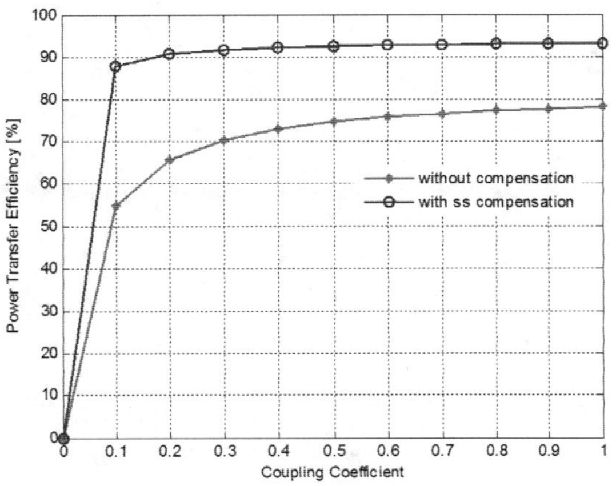

Figure 4. Power transfer efficiency vs coupling coeffiecient.

4. Computational IPT Models for Electric Vehicle Battery Charge

Figures 5 and 6 show the computational models of 3 kW single phase and 22 kW three phase IPT systems for EV battery charge designed using MATLAB/Simulink.

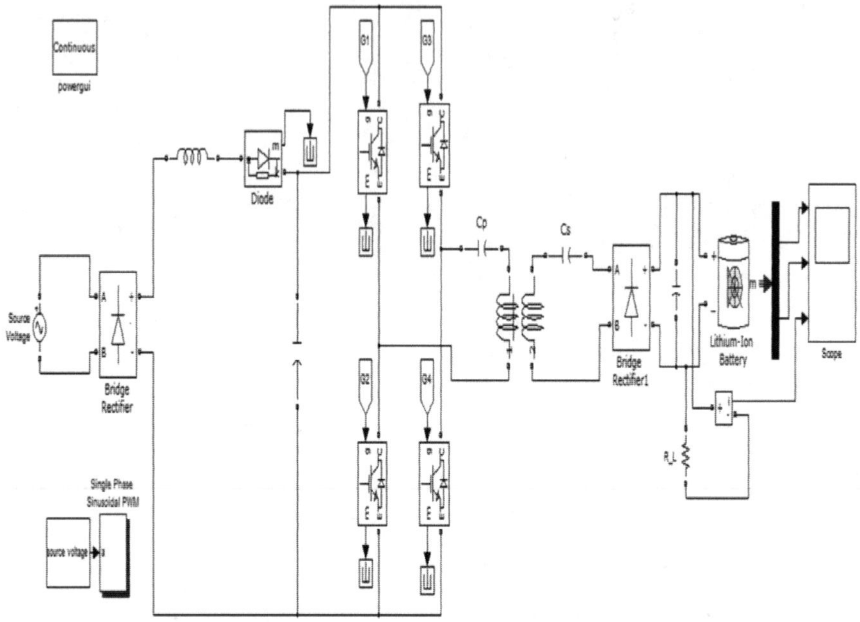

Figure 5. 3 kW single phase IPT model.

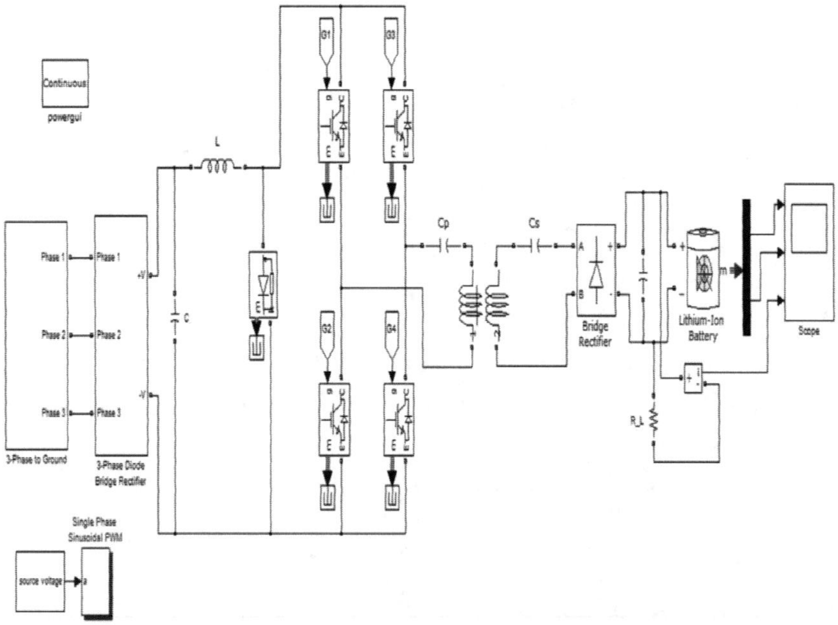

Figure 6. 22 kW three phase IPT model.

The waveforms of 3 kW single phase AC source voltage, its full wave varying DC voltage and stepped up DC voltage as well as the 22 kW three phase AC source voltage, its full wave constant DC voltage and stepped down voltage are shown in Fig. 7. With a single-phase uncontrolled bridge rectifier (UBR), the AC source voltage (i.e., 220 V) is transformed to a full wave varying DC voltage. In order to meet the demands of the lithium-ion battery, boost converter is used to produce a stepped up voltage of 400 V as input for the primary full bridge inverter "IGBT". Also, with a three-phase UBR, the AC source voltage (i.e., 380 V) is transformed to a full constant DC voltage and a buck converter is used to provide stepped down voltage of 400 V.

Figure 8 shows the waveform of the voltage supplied to the primary coil while Fig. 9 provides a better visibility of the waveform. The single phase pulse width modulation (PWM) (see Figs. 5 and 6) is used to control the switching of the gates of IGBT.

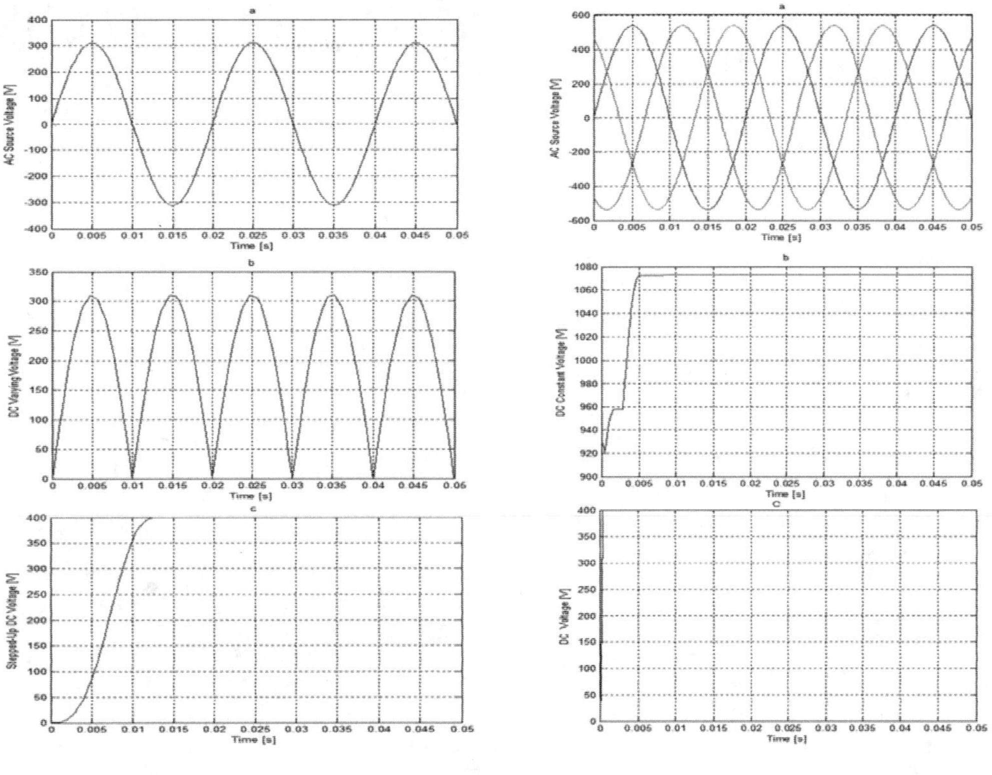

Figure 7. 3 kW (a) single phase AC source voltage, (b) full-wave varying DC voltage and (c) stepped up DC voltage. Also, 22 kW (a) three phase AC source voltage, (b) full-wave constant DC voltage and (c) stepped down voltage.

Based on the technical specifications given in [9], the IGBT is able to convert the direct voltage of 400 V to a higher and fixed frequency of 40 kHz AC voltage. Thus, the primary coil is able to generate a higher frequency electromagnetic field which is then coupled with the secondary coil. Furthermore, in order to guarantee maximum power transfer to the lithium-ion battery, the useful information presented in sections 2 and 3 regarding the application of SS capacitive compensation is incorporated in the designed computational IPT model. Furthermore, the battery charger unit consists of UBR and capacitor filter. This unit is responsible for creating the connection between the IPT system and the lithium-ion battery to be charged. The UBR is used to convert the high frequency AC voltage into a direct voltage while the capacitor filter is added to its output to reduce voltage ripples.

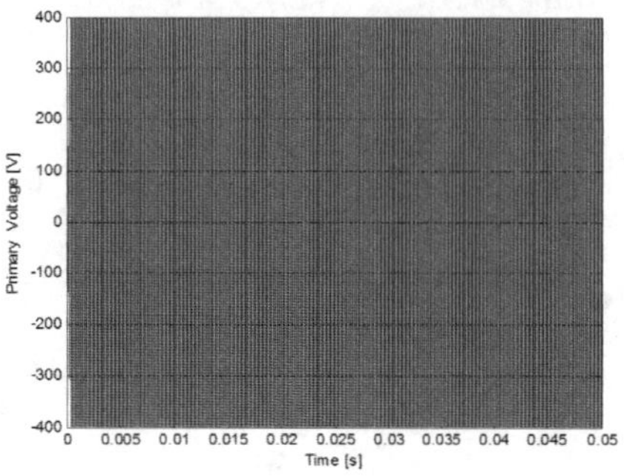

Figure 8. Voltage supplied to the primary coil.

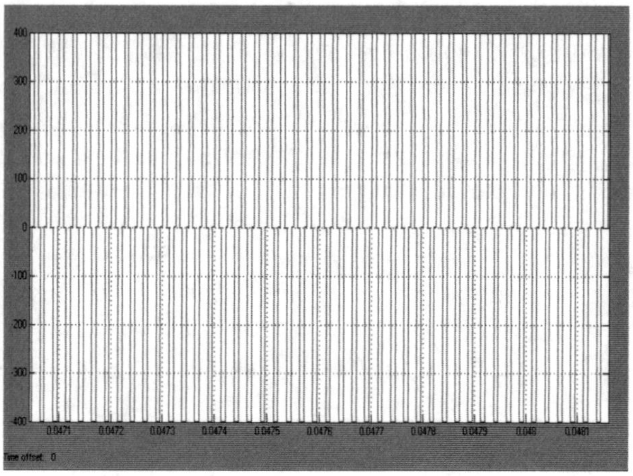

Figure 9. Better visibility of the primary voltage.

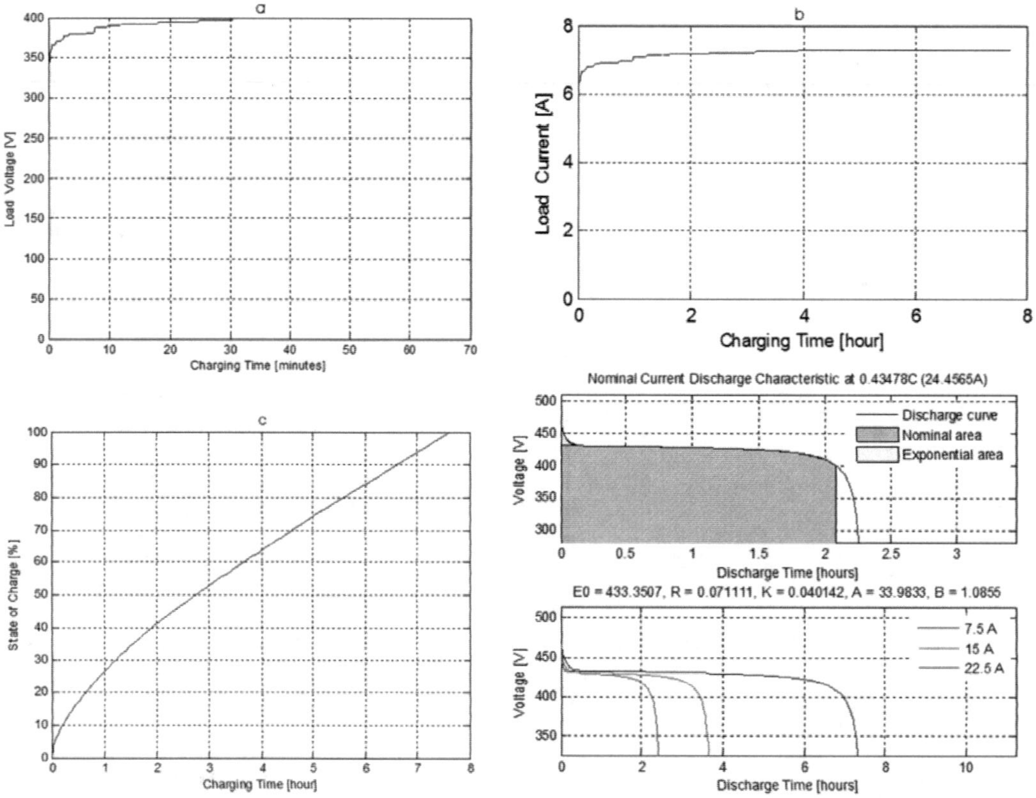

Figure 10. 3 kW IPT model (a) load voltage, (b) load current, (c) state of charge and (d) discharge characteristics of the battery.

The computation of the IPT models given in Figs. 5 and 6 is achieved based on the technical specifications of the lithium-ion battery and charger type of Renault ZOE as presented in [9]. The results (see Figs. 10 and 11) obtained for the load voltage, load current, state of charge of the battery and its discharge characteristic show that the computational models are able to provide the total voltage required by the battery. Also, considering the charging time for each IPT model, they are capable of delivering the electricity needed to power the battery of the Renault ZOE.

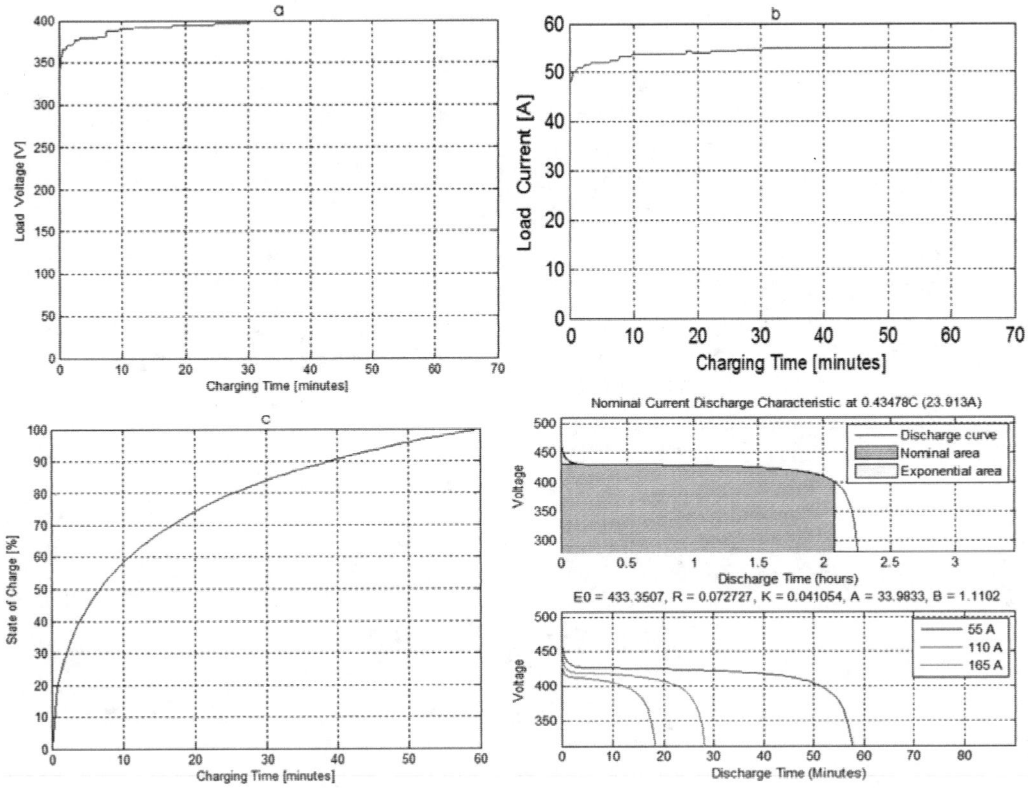

Figure 11. 22 kW IPT model (a) load voltage, (b) load current, (c) state of charge and (d) discharge characteristics of the battery.

5. Conclusion

In this paper, computational models of an inductive power transfer (IPT) system for EV battery charge are presented. To achieve this task, 3 kW single phase and 22 kW three phase IPT systems for Renault ZOE are designed in MATLAB/Simulink. The results obtained based on the technical specifications of Renault ZOE show that the IPT models may be employed as a support structure needed to effectively power any viable EV.

Reference

[1] AO Anele, Y Hamam, Y Alayli and K Djouani 2014 JMMC, 1, 83-104
[2] AO Anele, Y Hamam, Y Alayli and K Djouani 2014 JMMC, 1, 31-50
[3] H Barth, M Jung, M Braun, B Schmuelling and U Reker 2011 3rd European Conference Smart Grids, 1-14
[4] B Schmuelling, SG Cimen, T Vosshagen and F Turki 2012, EPE-PEMC, 110-115.
[5] A Neves, DM Sousa, A Rogue and JM Terras 2011, Power Electronics and Applications Conference, 1-10
[6] M Fareq, M Fitra, M Irwanto, Syafruddin Hasan and M Arinal 2014, J. Phys.: Conf. Series 495 1-7
[7] J Lawson, M Pinuela, DC Yates, S Lucyszyn and PD Mitcheson 2013, J. Phys.: Conf. Series 476 1-6
[8] S Chopra and P Bauer 2011, 33rd INTELEC,
[9] http://www.media.renault.com/download/media/specialfile/45115_1_5.aspx

Computation of the Mutual Inductance between Air-Cored Coils of Wireless Power Transformer

AO Anele[1,2], Y Hamam[1], L Chassagne[2], J Linares[3], Y Alayli[2], K Djouani[1]

[1] Dept. of Electrical Engineering, Tshwane University of Technology, Pretoria, South Africa
[2] LISV Laboratory, Université de Versailles Saint-Quentin-en-Yvelines, Versailles, France
[3] GEMaC Laboratory, Université de Versailles Saint-Quentin-en-Yvelines, Versailles, France

E-mail: aneleamos@gmail.com

Abstract. Wireless power transfer system is a modern technology which allows the transfer of electric power between the air-cored coils of its transformer via high frequency magnetic fields. However, due to its coil separation distance and misalignment, maximum power transfer is not guaranteed. Based on a more efficient and general model available in the literature, rederived mathematical models for evaluating the mutual inductance between circular coils with and without lateral and angular misalignment are presented. Rather than presenting results numerically, the computed results are graphically implemented using MATLAB codes. The results are compared with the published ones and clarification regarding the errors made are presented. In conclusion, this study shows that power transfer efficiency of the system can be improved if a higher frequency alternating current is supplied to the primary coil, the reactive parts of the coils are compensated with capacitors and ferrite cores are added to the coils.

1. Introduction

Wireless power transfer (WPT) systems allow the transfer of electric power between the air-cored coils of its transformer via high frequency magnetic fields to a consuming device (e.g. electric vehicle) [1–3]. However, due to its coil separation distance and misalignment, maximum power transfer is not guaranteed.

The computation of the mutual inductance between circular coils with and without misalignment is of fundamental practical interests to electrical engineers and physicists [1–6]. Several mutual inductance models based on the application of Maxwell's formula, Neumann's formula, Biot-Savart law and magnetic vector potential approach have been presented in [3–7]. These models are developed based on analytical or semi-analytical forms expressed over elliptical integrals of the first, second and third kinds, Heuman's Lambda function, Bessel functions and Legendre functions [6–10]. Nonetheless, a more advanced and relevant model for evaluating the mutual inductance between circular coils with misalignment have been formulated in [6, 7].

The aim of this paper is to rederive the mathematical models for evaluating the mutual inductance between circular coils with and without lateral and angular misalignment based on the more efficient and general model given in [6, 7]. Also, rather than presenting the computed results numerically as given in [7, 10], the results in this paper are graphically implemented using MATLAB codes. Lastly, the computed results are compared with the results of [7, 10] and clarification regarding the errors made are presented.

2. Advanced and Relevant Model for Mutual Inductance Computation

In [6], Grover formulated a more efficient and general model for computing the mutual inductance between two filamentatary circular coils with misalignment (see Eq. (1) and Fig. 1). The model given in Eq. (1) was achieved based on the application of Neumann integral approach. In order to confirm the validity of the model, Kim used the magnetic vector potential approach to obtain another mutual inductance model (see Eq. (20) of [10]). However, using the same approach employed by Kim, Babic retrieved Grover's formula through well detailed derivations [7].

$$M = \frac{2\mu_0}{\pi}\sqrt{R_P R_S}\int_0^{\pi}\frac{[cos\theta - \frac{d}{R_S}cos\phi]\Psi(k)}{k\sqrt{V^3}}d\phi \tag{1}$$

where

$\alpha = \frac{R_S}{R_P}$, $\beta = \frac{c}{R_P}$, $\xi = \beta - \alpha cos\phi sin\theta$, $k^2 = \frac{4\alpha V}{(1+\alpha V)^2+\xi^2}$, $\Psi(k) = (1-\frac{k^2}{2})K(k) - E(k)$

$V = \sqrt{1 - cos^2\phi sin^2\theta - 2\frac{d}{R_S}cos\phi cos\theta + \frac{d^2}{R_S^2}}$

$K(k) = \int_0^{\frac{\pi}{2}} \frac{1}{\sqrt{1-k^2 sin^2\theta}}d\theta$ and $E(k) = \int_0^{\frac{\pi}{2}} \sqrt{1-k^2 sin^2\theta}d\theta$.

where μ_0 is the magnetic permeability of free space, R_P and R_S are the radii of the primary and secondary coils respectively, d is the lateral misalignment, α is the shape factor of the coils' physical geometry, c is the separation distance between the coils' centres, θ is the angular misalignment, k is a variable, parameter $\Psi(k)$ is function of k, ϕ is the angle of integration at any point of the secondary coil, β, V and ξ are dimensionless parameters, $K(k)$ and $E(k)$ are the complete elleptic integral of the first and second kinds respectively .

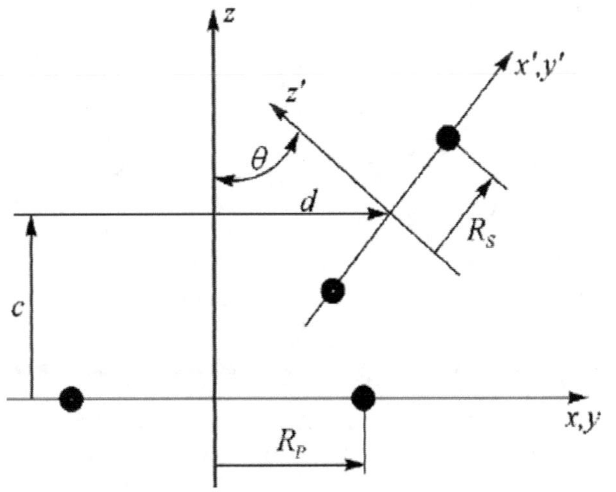

Figure 1. Filamentary circular coils with angular and lateral misalignment [6].

3. Rederived Mathematical Models for Mutual Inductance Computation

It is important to note that in Eq. (1), the number of turns (i.e., N_P and N_S) for the primary and secondary coils are not expressed. Therefore, taking into account N_P and N_S based on the example considered in [7], the rederived models are presented in the following subsections (see Figs. 2-4):

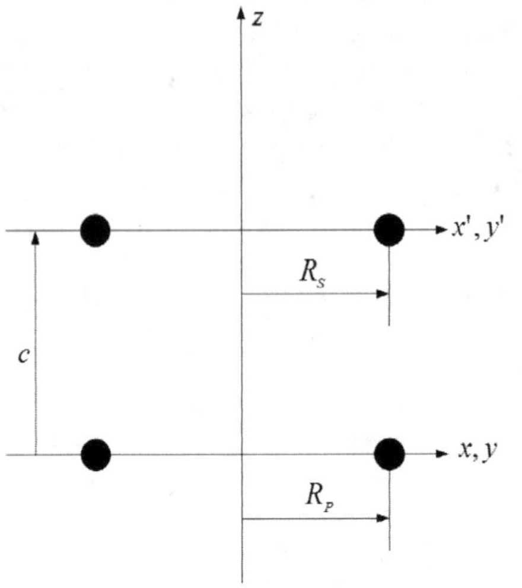

Figure 2. Case 1: without misalignment

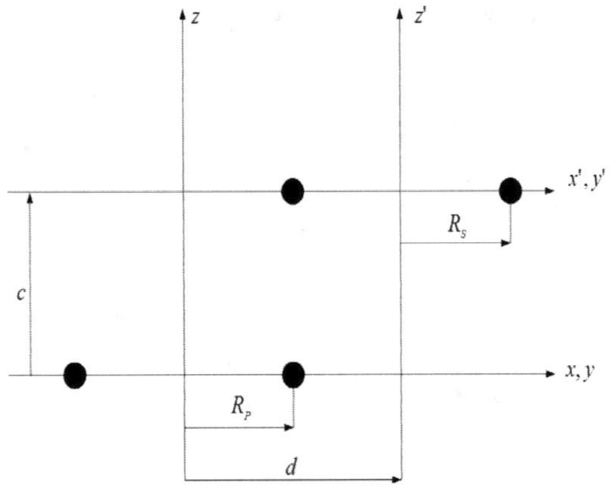

Figure 3. Case 2: with lateral misalignment.

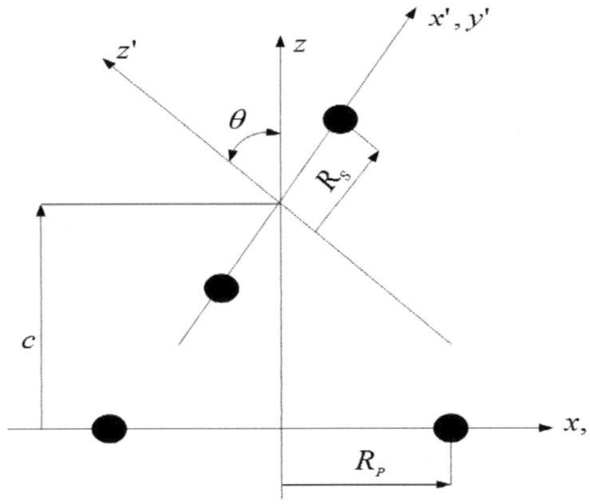

Figure 4. Case 3: with angular misalignment.

3.1. Case 1: without Misalignment

Figure 2 shows the case without misalignment (i.e., $\theta = 0$ and $d = 0$). Thus, the mutual inductance model between air-cored filamentary circular coils without misalignment is given by

$$M = \frac{2N_P N_S \mu_0}{\pi} \sqrt{R_P R_S} \int_0^{\pi} \frac{\Psi(k)}{k} d\phi \tag{2}$$

where
$\alpha = \frac{R_S}{R_P}$, $\beta = \frac{c}{R_P}$, $\xi = \beta$, $k^2 = \frac{4\alpha}{(1+\alpha)^2+\xi^2}$, $\Psi(k) = (1 - \frac{k^2}{2})K(k) - E(k)$, $K(k) = \int_0^{\frac{\pi}{2}} \frac{1}{\sqrt{1-k^2 sin^2\theta}} d\theta$ and $E(k) = \int_0^{\frac{\pi}{2}} \sqrt{1 - k^2 sin^2\theta} d\theta$.

3.2. Case 2: with Lateral Misalignment only
Figure 3 shows the case with lateral misalignment only (i.e., $\theta = 0$). Thus, the mutual inductance model between air-cored filamentary circular coils with only lateral misalignment is given by

$$M = \frac{2N_P N_S \mu_0}{\pi} \sqrt{R_P R_S} \int_0^{\pi} \frac{[1 - \frac{d}{R_S} cos\phi]\Psi(k)}{k\sqrt{V^3}} d\phi \qquad (3)$$

where
$\alpha = \frac{R_S}{R_P}$, $\beta = \frac{c}{R_P}$, $\xi = \beta$, $k^2 = \frac{4\alpha V}{(1+\alpha V)^2+\xi^2}$, $\Psi(k) = (1 - \frac{k^2}{2})K(k) - E(k)$
$K(k) = \int_0^{\frac{\pi}{2}} \frac{1}{\sqrt{1-k^2 sin^2\theta}} d\theta$, $E(k) = \int_0^{\frac{\pi}{2}} \sqrt{1 - k^2 sin^2\theta} d\theta$ and $V = \sqrt{1 - 2\frac{d}{R_S} cos\phi + \frac{d^2}{R_S^2}}$

3.3. Case 3: with Angular Misalignment only
Figure 4 shows the case with angular misalignment only (i.e., $d = 0$). Thus, the mutual inductance model between air-cored filamentary circular coils with only angular misalignment is given by

$$M = \frac{2N_P N_S \mu_0}{\pi} \sqrt{R_P R_S}\, cos\theta \int_0^{\pi} \frac{\Psi(k)}{k\sqrt{V^3}} d\phi \qquad (4)$$

where
$\alpha = \frac{R_S}{R_P}$, $\beta = \frac{c}{R_P}$, $\xi = \beta - \alpha cos\phi sin\theta$, $k^2 = \frac{4\alpha V}{(1+\alpha V)^2+\xi^2}$, $\Psi(k) = (1 - \frac{k^2}{2})K(k) - E(k)$
$V = \sqrt{1 - cos^2\phi sin^2\theta}$, $K(k) = \int_0^{\frac{\pi}{2}} \frac{1}{\sqrt{1-k^2 sin^2\theta}} d\theta$ and $E(k) = \int_0^{\frac{\pi}{2}} \sqrt{1 - k^2 sin^2\theta} d\theta$

3.4. Case 4: with Lateral and Angular Misalignment
Figure 1 shows the case with both misalignment. Thus, taking into account N_P and N_S, the mutual inductance model between air-cored filamentary circular coils with lateral and angular misalignment is given by

$$M = \frac{2N_P N_S \mu_0}{\pi} \sqrt{R_P R_S} \int_0^{\pi} \frac{[cos\theta - \frac{d}{R_S} cos\phi]\Psi(k)}{k\sqrt{V^3}} d\phi \qquad (5)$$

4. Presentation and Discussion of Results
The results obtained for the computation of the mutual inductance between air-cored circular coils with and without lateral, d and angular, θ misalignment are shown in Figs. 5-9. MATLAB software is used to compute the rederived models given in (2), (3), (4) and (5). In addition, the results are presented graphically (see Figs. 5-9) and numerically tabulated (see Table 1).

The result shown in Fig. 5 is compared with the numerical results obtained in [7, 10] (see Table 1). This is so because the example studied in [7, 10] considered the case with only lateral misalignment. Although different coil separation distances (i.e., $c = 0$, $0.05\ m$ and $0.1\ m$) are studied in Fig. 5, the authors of [7] and [10] focused only on $c = 0$.

The numerical results obtained in [7] showed that from $d = 0$ to $15.5\ mm$ (i.e., for the cases where the secondary coil is located inside the primary coil), the value of the mutual inductance M increases from $0.152875989\ mH$ to $0.181647101\ mH$. This outcome is in agreement with the results obtained in this paper (see Fig. 5 and Table 1). However, there are discrepancies compared to the results obtained in [10] (see Table 1). Furthermore, from $d = 69.5\ mm$ to $1000\ mm$ (i.e., the region where the secondary coil is located outside the primary coil), the results of [7] showed that negative values are obtained for M and for larger d (i.e., from $d = 200\ mm$ to

1000 mm), its value approached zero. This outcome is in agreement with the results obtained in this paper (see Fig. 5 and Table 1). This behaviour corresponds with the theory, as the flux lines linked by the secondary coil change their orientation outside the primary coil [6–8]. However, in [10], M slowly approached zero when the smaller coil is outside the bigger coil and never reached negative values, which is not correct according to the theory. Thus, confirm with the authors of [7] that Eq. (20) of [10] is erroneous.

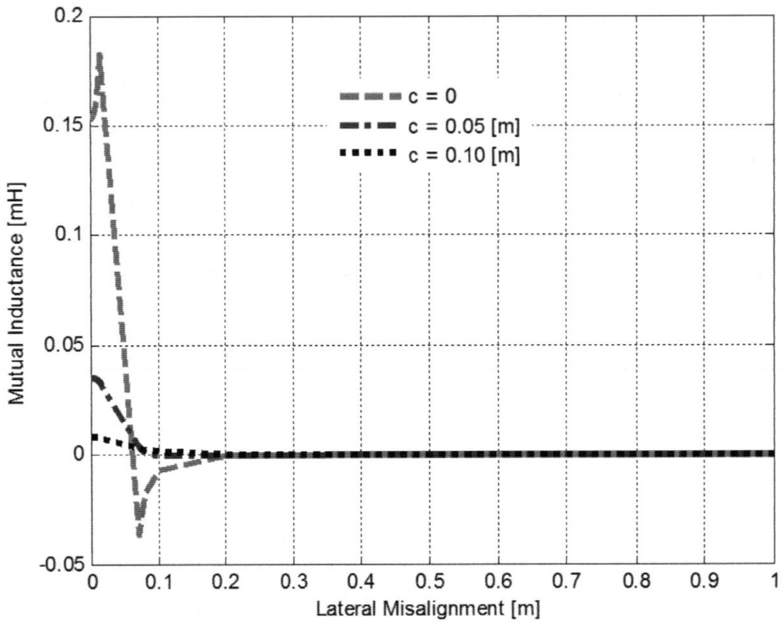

Figure 5. Case 2: with lateral misalignment

Table 1. Example studied for $R_P = 42.5\ mm, R_S = 20.0\ mm, N_P = 150, N_S = 50$

d (mm)	c (mm)	M ($10^{-4}H$) [7]	M (mH): This paper	M ($10^{-4}H$) [10], Eq. (20)	Discrepancy [%], Eq. (3) vs. Eq. (20)
0.0	0	1.52875989	0.152875989	1.52875989	0.00
1.0	0	1.52967999	0.152967999	1.53063181	0.062
2.5	0	1.53453878	0.153453878	1.54049796	0.388
5.0	0	1.55229043	0.155229043	1.57627671	1.545
7.0	0	1.57597265	0.157597265	1.62337093	3.001
10.0	0	1.63029226	0.163029226	1.72877701	6.041
15.0	0	1.79299153	0.179299153	2.02559287	12.973
15.5	0	1.81647101	0.181647101	2.06644759	13.762
69.5	0	−0.36397730	−0.036397730	1.57888396	–
80.0	0	−0.17767980	−0.017767980	1.05074054	–
100.0	0	−0.07231342	−0.007231342	0.60979570	–
200.0	0	−0.00718415	−0.000718415	0.13810829	–
500.0	0	−0.00043535	−0.000043535	0.02158082	–
1000.0	0	−0.00005401	−0.000005401	0.00537760	–

This study shows that as the coil separation distance and misalignment increase, the mutual inductance between the coils decreases (see Figs. 5-9). Thus, an increased value of mutual inductance can be achieved when a higher frequency AC current is supplied to the primary coil, the reactive parts of the coils are compensated with capacitors and ferrite cores are incorporated around the coils.

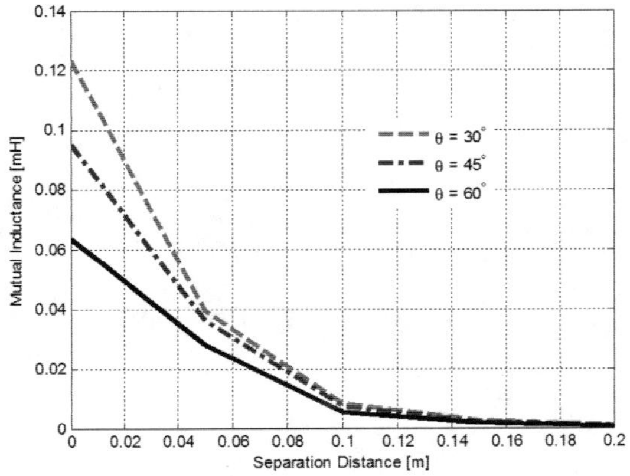

Figure 6. Case 3: with angular misalignment.

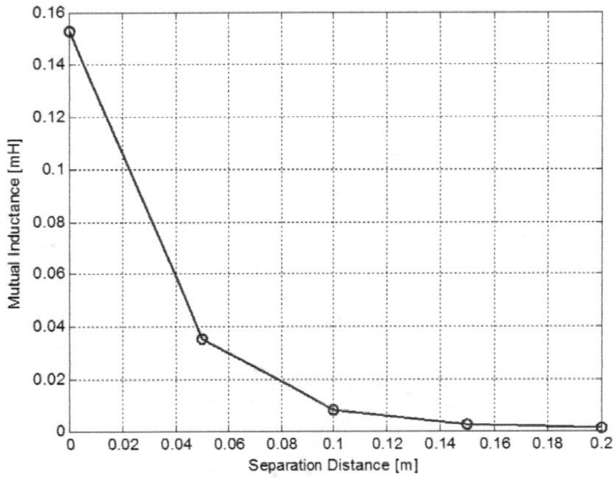

Figure 7. Case 1: without misalignment.

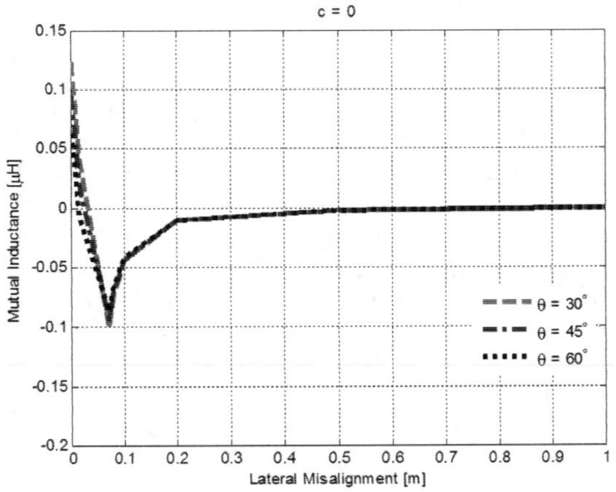

Figure 8. Case 4: with misalignments.

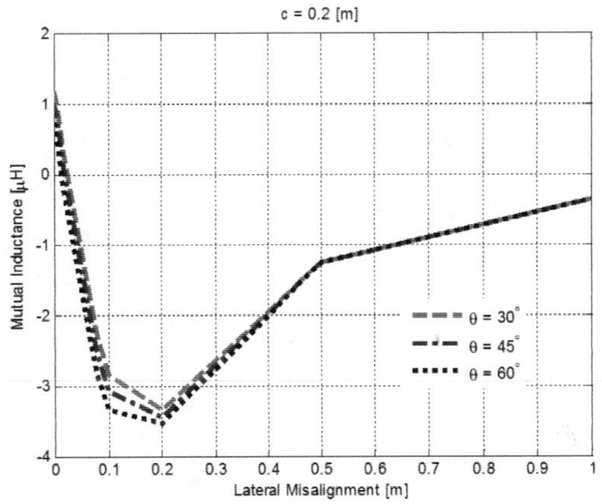

Figure 9. Case 4: with misalignments.

5. Conclusion

In this paper, rederived models for the mutual inductance between circular coils with and without misalignment are expressed and computed using MATLAB codes. The results obtained are compared with the published ones and clarification regarding the errors made are presented.

Reference

[1] AO Anele, Y Hamam, Y Alayli and K Djouani 2013, PACT, 200-205
[2] AO Anele, Y Hamam, Y Alayli and K Djouani 2013, PACT, 110-115
[3] AO Anele, Y Hamam, Y Alayli and K Djouani 2013, IEEE AFRICON, 1-5
[4] SI Babic and C Akyel 2008, IEEE Trans. Magn., Vol. 44, No. 7 1743-1750
[5] JC Maxwell 1954, A Treatise on Electricity and Magnetism, 1954
[6] FW Grover 1944, Proceedings of the I.R.E., 620-629
[7] SI Babic, F Sirois and C Akyel 2009 PIER M, Vol. 8, 15-26
[8] JT Conway 2007, IEEE Trans. Magn., Vol. 43, No. 3, 1023-1034
[9] AO Anele, Y Hamam, Y Alayli and K Djouani 2014, JMMC, Vol. 1, No. 1, 83-104
[10] KB Kim, E Levi, Z Zabar and L Birenbaum 1997, IEEE Trans. Mag., Vol. 33, No. 5, 3916-3921

Evaluation of the Magnetic Fields and Mutual Inductance between Circular Coils Arbitrarily Positioned in Space

AO Anele[1,2], Y Hamam[1], L Chassagne[2], J Linares[3], Y Alayli[2], K Djouani[1]

[1] Dept. of Electrical Engineering, Tshwane University of Technology, Pretoria, South Africa
[2] LISV Laboratory, Université de Versailles Saint-Quentin-en-Yvelines, Versailles, France
[3] GEMaC Laboratory, Université de Versailles Saint-Quentin-en-Yvelines, Versailles, France

E-mail: aneleamos@gmail.com

Abstract. This paper presents the evaluation of the magnetic fields and mutual inductance between circular coils arbitrarily positioned in space. Firstly, based on an advanced and relevant model available in the literature, MATLAB code is implemented to evaluate the mutual inductance between circular coils arbitrarily positioned with respect to each other. The computed results are compared with the numerical results previously published in the literature and a detailed clarification regarding the huge computational errors made are presented. In the second part, a complex and relevant model available in the literature for evaluating the magnetic fields due to a circular coil is presented. Based on the useful information, the model for computing the magnetic fields between two circular coils is formulated. The computed results are validated with experimental measurements. The comparison of the results shows that the developed model and the experimental measurements conducted are accurate and effective.

1. Introduction

Air-cored coils are widely used in various electromagnetic applications. They are preferred to the iron-cored type due to their design objectives of controllability and capability of a high power transfer [1]. Nonetheless, the measure of the magnetic fields and mutual inductance between them are of importance to electrical engineers and physicists because within certain limits of coil separation and misalignment, a minimal amount of power is guaranteed [2–5].

Several contributions have been made in the literature concerning the computation of the magnetic fields and mutual inductance between air-cored coils. In this view, mathematical modelling based on the application of Maxwell's formula, Neumann's formula, Biot-Savart law, Lorentz and magnetic vector potential have been presented in [2–9].

In the first part of this paper, MATLAB coding is implemented to compute the mutual inductance between two circular coils arbitrarily positioned with respect to each other. This computation is achieved based on the advanced and relevant model given in [4]. The computed results are presented graphically and numerically and then, compared with the ones presented in [4, 5]. Also, based on a more efficient and relevant model given in [7], the formula for computing the magnetic fields between two air-cored circular coils with and without misalignment is presented. Lastly, experimental measurements are used to validate the computed results.

2. Mutual Inductance Model

According to [4], the formula for calculating the mutual inductance between inclined circular filaments placed in any position (most general case see Fig. 1a) as a function of the: primary and secondary coils radii, R_P and R_S, parameters a, b and c defining the centre of the secondary coil and coordinates (x_C, y_C, z_C) defining the centre of the secondary coil is given by

$$M = \frac{\mu_0 R_S}{\pi} \int_0^{2\pi} \frac{[p_1 cos\varphi + p_2 sin\varphi + p_3]\Psi(k)}{k\sqrt{V_0^3}} d\varphi \qquad (1)$$

where
$\alpha = \frac{R_S}{R_P}$, $\beta = \frac{x_C}{R_P}$, $\gamma = \frac{y_C}{R_P}$, $\delta = \frac{z_C}{R_P}$, $\ell = \sqrt{a^2+c^2}$, $L = \sqrt{a^2+b^2+c^2}$, $p_1 = \pm\frac{\gamma c}{\ell}$, $p_2 = \mp\frac{\beta\ell^2+\gamma ab}{\ell L}$, $p_3 = \frac{\alpha c}{L}$, $p_4 = \mp\frac{\beta ab-\gamma\ell^2+\delta bc}{\ell L}$, $p_5 = \mp\frac{\beta c-\delta a}{\ell}$, $k = \sqrt{\frac{4V_0}{A_0+2V_0}}$, $\Psi(k) = (1-\frac{k^2}{2})K(k) - E(k)$, $A_0 = 1+\alpha^2+\beta^2+\gamma^2+\delta^2+2\alpha(p_4 cos\varphi + p_5 sin\varphi)$, $V_0^2 = \alpha^2[(1-\frac{b^2c^2}{\ell^2L^2})cos^2\varphi + \frac{c^2}{\ell^2}sin^2\varphi + \frac{abc}{\ell^2 L}sin2\varphi] + \beta^2+\gamma^2 \mp 2\alpha\frac{\beta ab-\gamma\ell^2}{\ell L}cos\varphi \mp \frac{2\alpha\beta c}{\ell}sin\varphi$, $K(k) = \int_0^{\frac{\pi}{2}} \frac{1}{\sqrt{1-k^2sin^2\theta}}d\theta$ and $E(k) = \int_0^{\frac{\pi}{2}} \sqrt{1-k^2sin^2\theta}d\theta$.

where μ_0 is the magnetic permeability of free space, k is a variable, parametr $\Psi(k)$ is function of k, α, β, γ, ℓ, L, p_1, p_2, p_3, p_4, p_5, A_0 and V_0 are dimensionless parameters, φ is the angle of integration at any point of the secondary coil, $K(k)$ and $E(k)$ are the complete elleptic integral of the first and second kinds respectively.

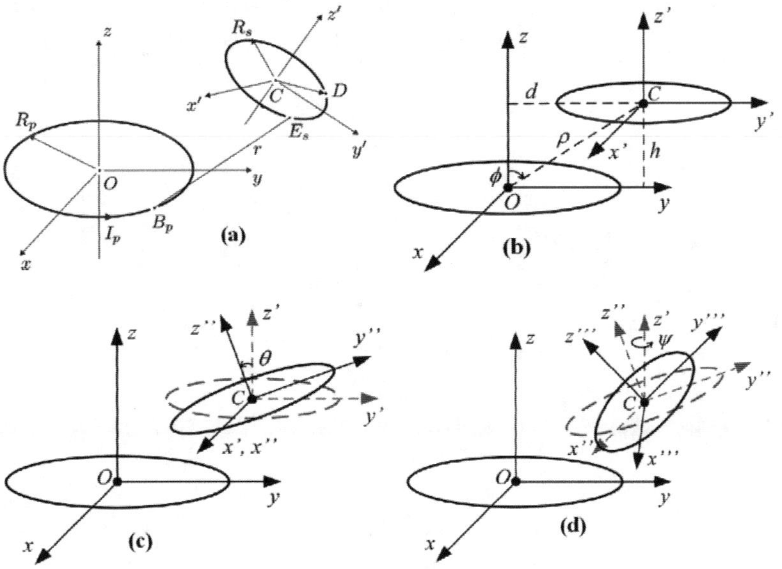

Figure 1. (a) Filamentary circular coils with arbitrary misalignment. Also, the geometric configurations and common notation used in the examples studied in [4] correspond to the following cases: (b) lateral misalignment only ($\theta = 0$, $\psi = 0$, axes $y-z$ and $y'-z'$ are coplanar), (c) lateral and angular misalignment ($\psi = 0$, axes $y-z$ and $y'-z'$ are coplanar) and (d) arbitrary lateral and angular misalignment (no coplanar axes anymore).

The geometric configurations considered in the examples studied in [4] are shown in Figs. 1b to 1d. In order to have easier link with Grover's formula, the geometric configurations are restated in terms of a common notation namely: $h = z_C$, $d = \sqrt{x_C^2 + y_C^2}$, $\rho = \sqrt{h^2 + d^2}$, $cos\phi = \frac{h}{\rho}$ and in all cases, the primary coil is located in plane XOY, with its centre at the origin $O(0,0,0)$. The horizontal and vertical distances between the centres of the coils are d and h respectively and ρ is the lateral misalignment.

The first part of this paper is to compute the mutual inductance between filamentary circular coils arbitrarily positioned with respect to each other. In this view, example 12 was studied in [4] and for such case, the following common notations are applied to Eq. (1):

- The centre of the secondary coil must be taken at point $C(x_C = 0, y_C = d, z_C = h)$. where $d = x_2 sin\theta$, $h = x_1 - x_2 sin\theta$ and $\rho = \sqrt{h^2 + d^2}$. It is vital to note that x_1 is the vertical distance when axis z'' intersects with axis z from origin O, x_2 is the distance between the intersection and the centre of the secondary coil and θ is the angular misalignment.
- The parameters defining the positioning of the secondary coil plane is that of a spherical Cartesian system of coordinates which is given by: $a = sin\psi sin\theta$, $b = -cos\psi sin\theta$ and $c = cos\theta$. where ψ is the rotation angle around axis z', which gives the complete positioning of the secondary coil.

3. Presentation and Discussion of Results for Mutual Inductance Computation

The results obtained for the mutual inductance between filamentary circular coils with arbitrary lateral ρ and angular θ misalignment are shown in Fig. 2 and Table 1. Based on the parameters given in example 12 (i.e., $R_P = 16\ cm, R_S = 10\ cm$, $x_1 = 20\ cm$ and $x_2 = 5\ cm$), MATLAB software is used to achieve the coding and computation of Eq. (1). Although the authors of [4] studied only for the case with $\theta = 60°$ and $\rho = 0.1803\ m$, the authors of this paper also studied for the cases with $\theta = 30°, \rho = 0.1587\ m$ and $\theta = 45°, \rho = 0.1684\ m$. The results obtained in this paper are graphically presented (see Fig. 1) and also numerically tabulated (see Table 1).

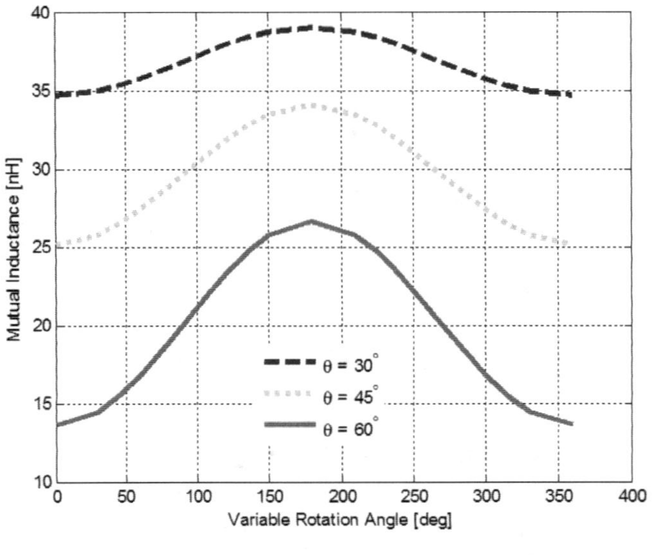

Figure 2. Results for mutual inductance between circular filaments with arbitrary misalignment

In the case with $\theta = 60°$ and $\rho = 0.1803\ m$, Fig. 2 shows that the value of the mutual inductance increases from 13.6113 nH to 26.6433 nH when the variable rotation angle ψ ranges from 0 to π and decreases from 26.6433 nH to 13.6113 nH when ψ ranges from π to 2π. This outcome is in agreement with the results obtained in [4] and [5] (see Table 1).

However, it is vital to note that the authors of [4] obtained very wrong values for the parameters which define the centre of the secondary coil (i.e., a, b and c see Table 1). With such outcome, it is never possible for the authors to obtain the correct mutual inductance values. Nonetheless, the correct values are obtained in this paper (see Table 1) based on the given common notations, which are $a = sin\psi sin\theta$, $b = -cos\psi sin\theta$ and $c = cos\theta$.

Table 1. Computed Values in Example 12, with Variable Rotation Angle ψ and $\theta = 60°$

ψ	(a, b, c) [4]	$M\ (nH)$ [4]	$M\ (nH)$ [5] Eq. (179)	(a, b, c): This paper	$M\ (nH)$: This paper
0	$(0; -0.0866; 0.05)$	13.6113	13.6113	$(0; -0.866; 0.5)$	13.6113
$\pi/6$	$(0.0433; -0.075; 0.05)$	14.4688	14.4688	$(0.433; -0.75; 0.5)$	14.4688
$\pi/4$	$(0.075; -0.0433; 0.05)$	15.4877	15.4877	$(0.6124; -0.6124; 0.5)$	15.4877
$\pi/3$	$(0.06124; -0.06124; 0.05)$	16.8189	16.8189	$(0.75; -0.433; 0.5)$	16.8189
$\pi/2$	$(0.0866; 0; 0.05)$	20.0534	20.0534	$(0.866; 0; 0.5)$	20.0534
$2\pi/3$	$(0.075; 0.0433; 0.05)$	23.3253	23.3253	$(0.75; 0.433; 0.5)$	23.3253
$3\pi/4$	$(0.06124; 0.06124; 0.05)$	24.6936	24.6936	$(0.6124; 0.6124; 0.5)$	24.6936
$5\pi/6$	$(0.0433; 0.075; 0.05)$	25.7493	25.7493	$(0.433; 0.75; 0.5)$	25.7493
π	$(0; 0.0866; 0.05)$	26.6433	26.6433	$(0; 0.866; 0.5)$	26.6433
$7\pi/6$	$(-0.0433; 0.075; 0.05)$	25.7493	25.7493	$(0.433; 0.75; 0.5)$	25.7493
$5\pi/4$	$(-0.06124; 0.06124; 0.05)$	24.6936	24.6936	$(0.6124; 0.6124; 0.5)$	24.6936
$4\pi/3$	$(-0.075; 0.0433; 0.05)$	23.3253	23.3253	$(0.75; 0.433; 0.5)$	23.3253
$3\pi/2$	$(-0.0866; 0; 0.05)$	20.0534	20.0534	$(0.866; 0; 0.5)$	20.0534
$5\pi/3$	$(-0.075; -0.0433; 0.05)$	16.8189	16.8189	$(0.75; -0.433; 0.5)$	16.8189
$7\pi/4$	$(-0.06124; -0.06124; 0.05)$	15.4877	15.4877	$(0.6124; -0.6124; 0.5)$	15.4877
$11\pi/6$	$(-0.0433; -0.075; 0.05)$	14.4688	14.4688	$(0.433; -0.75; 0.5)$	14.4688
2π	$(0; -0.0866; 0.05)$	13.6113	13.6113	$(0; -0.866; 0.5)$	13.6113

4. Formulation of the Magnetic Fields Model for two Filamentary Circular Coils

According to [7], the magnetic fields in an arbitrary point $Es(x_S, y_S, z_S)$ produced by the primary coil of the radius R_P carrying the current I_P (see Fig. 1a) is given by

$$B_x(x_S, y_S, z_S) = -\frac{\mu_0 I_P z_S x_S k}{8\pi\sqrt{R_P}(x_S^2 + y_S^2)^{5/4}} L_0$$
$$B_y(x_S, y_S, z_S) = -\frac{\mu_0 I_P z_S y_S k}{8\pi\sqrt{R_P}(x_S^2 + y_S^2)^{5/4}} L_0$$
$$B_z(x_S, y_S, z_S) = -\frac{\mu_0 I_P k}{8\pi\sqrt{R_P}(x_S^2 + y_S^2)^{3/4}} S_0 \tag{2}$$

where

$x_S = x_C + R_S u_x cos\varphi + R_S v_x sin\varphi$, $y_S = y_C + R_S u_y cos\varphi + R_S v_y sin\varphi$, $z_S = z_C + R_S u_z cos\varphi + R_S v_z sin\varphi$, $\overrightarrow{u} = [u_x, u_y, u_z] = [-\frac{ab}{\ell L}, \frac{\ell}{L}, -\frac{bc}{\ell L}]$, $\overrightarrow{v} = [v_x, v_y, v_z] = [-\frac{c}{\ell}, 0, \frac{a}{\ell}]$, $L_0 = 2K(k) - \frac{2-k^2}{1-k^2}E(k)$, $S_0 = 2\sqrt{x_S^2 + y_S^2}K(k) - \frac{2\sqrt{x_S^2+y_S^2}-(R_P+\sqrt{x_S^2+y_S^2})k^2}{1-k^2}E(k)$, $k^2 = \frac{4R_P\sqrt{x_S^2+y_S^2}}{(R_P+\sqrt{x_S^2+y_S^2})^2+z_S^2}$

where

x_S, y_S and z_S are the parametric coordinates of the secondary coil, $\overrightarrow{u}$ is the unit vector lying in the secondary coil plane between points C and D, $\overrightarrow{v}$ is the cross product of the unit vector of the axis z' and $\overrightarrow{u}$, S_0 and L_0 are dimensionless parameters while $K(k)$ and $E(k)$ are the complete elliptic integrals of the first and second kinds respectively.

Thus, for two coils, the total magnetic fields is the sum of the magnetic fields from each of the coils, which is formulated by the authors of this paper as

$$\begin{aligned} B_x = B_{xP} + B_{xS} &= [-\frac{\mu_0\mu_r I_P N_P (z_P + \frac{l}{2}) x_P k_P}{8\pi\sqrt{R_P}(x_P^2+y_P^2)^{5/4}} L_P] + [-\frac{\mu_0\mu_r I_S N_S (z_S - \frac{l}{2}) x_S k_S}{8\pi\sqrt{R_S}(x_S^2+y_S^2)^{5/4}} L_S] \\ B_y = B_{yP} + B_{yS} &= [-\frac{\mu_0\mu_r I_P N_P (z_P + \frac{l}{2}) y_P k_P}{8\pi\sqrt{R_P}(x_P^2+y_P^2)^{5/4}} L_P] + [-\frac{\mu_0\mu_r I_S N_S (z_S - \frac{l}{2}) y_S k_S}{8\pi\sqrt{R_S}(x_S^2+y_S^2)^{5/4}} L_S] \\ B_z = B_{zP} + B_{zS} &= [\frac{\mu_0\mu_r I_P N_P k_P}{8\pi\sqrt{R_P}(x_P^2+y_P^2)^{3/4}} S_P] + [\frac{\mu_0\mu_r I_S N_S k_S}{8\pi\sqrt{R_S}(x_S^2+y_S^2)^{3/4}} S_S] \end{aligned} \quad (3)$$

where
$x_P = x_C + R_P u_x cos\phi + R_P v_x sin\phi$, $x_S = x_C + R_S u_x cos\varphi + R_S v_x sin\varphi$, $y_P = y_C + R_P u_y cos\phi + R_P v_y sin\phi$, $y_S = y_C + R_S u_y cos\varphi + R_S v_y sin\varphi$, $z_P = z_C + R_P u_z cos\phi + R_P v_z sin\phi$, $z_S = z_C + R_S u_z cos\varphi + R_S v_z sin\varphi$, $L_P = 2K_P(k_P) - \frac{2-k_P^2}{1-k_P^2}E_P(k_P)$, $L_S = 2K_S(k_S) - \frac{2-k_S^2}{1-k_S^2}E_S(k_S)$, $S_P = 2\sqrt{x_P^2+y_P^2}K_P(k_P) - \frac{2\sqrt{x_P^2+y_P^2}-(R_P+\sqrt{x_P^2+y_P^2})k_P^2}{1-k_P^2}E_P(k_P)$, $S_S = 2\sqrt{x_S^2+y_S^2}K_S(k_S) - \frac{2\sqrt{x_S^2+y_S^2}-(R_S+\sqrt{x_S^2+y_S^2})k_S^2}{1-k_S^2}E_S(k_S)$, $k_P^2 = \frac{4R_P\sqrt{x_P^2+y_P^2}}{(R_P+\sqrt{x_P^2+y_P^2})^2+z_P^2}$, $k_S^2 = \frac{4R_S\sqrt{x_S^2+y_S^2}}{(R_S+\sqrt{x_S^2+y_S^2})^2+z_S^2}$

5. Computed and Experimentally Validated Results for Magnetic Fields

The results obtained for the computation and experimental validation of magnetic fields between two filamentary circular coils without misalignment (i.e., $\theta = 0$ and $\rho = 0$) is shown in Fig. 3. Figures 3a–3c are obtained by applying the following to Eq. (3):

- In the case without misalignment, the parameters determinig the positioning of the secondary coil plane are defined as $a = 0$, $b = 0$ and $c = 1$.
- The centre points for the two coils become $(x_C = 0, y_C = 0, z_C = h)$. where h is the varying distance along the symmetry axis of the coils (see Fig. 3d).
- The parameters used for the computation and experimental measurements are: $\mu_r = 1$, $\mu_0 = 4\pi \times 10^{-7}\,\frac{H}{m}$, $R_P = R_S = 0.065\ cm$, $N_P = N_S = 95$, $I_P = I_S = 1\ A$.

The magnetic field results are obtained when the separation l between the two circular coils are $l = 0.01\ m$ (see Fig. 3a), $l = 0.035\ m$ (see Fig. 3b) and $l = 0.065\ m$ (see Fig. 3c). The measurement of the magnetic fields is conducted on the symmetry axis of the two coils by using Teslameter (see Fig. 3d. For each l, Teslameter is operated to measure the magnetic fields from the mid-point between the two coils to the point outside the coil (i.e., $z_C = h$, which ranges from $-0.09\ m$ to $0.09\ m$).

Figure 3a shows that uniform magnetic fields is obtained when the separation l between the two circular coils is almost zero (i.e., $z_P + \frac{l}{2}$ and $z_S - \frac{l}{2}$ where $l = 0.01\ m$). Figures 3b and 3c show that as l increases, the values of the magnetic fields reduces and a different waveform is obtained. Furthermore, Figs. 3a–3c show that the values and waveforms for B_x and B_y along the $z-$ axis are symmetrical which is always true whenever the magnetic fields is computed on the symmetry axis of two circular coils.

The magnetic fields model formulated in Eq. (3) can be used for the cases with and without misalignment. However, the case without misalignment is considered in this section. The comparison of the results shows that the developed model and experimental measurements conducted are accurate and effective.

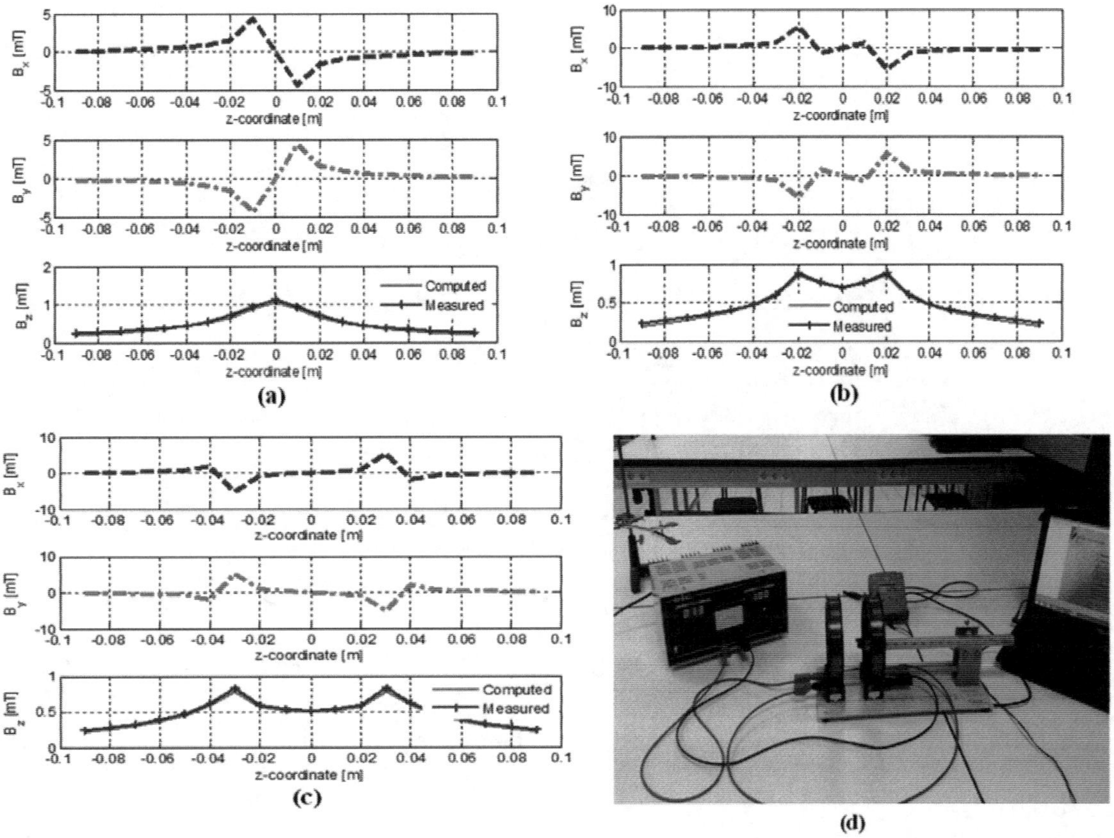

Figure 3. Magnetic fields results and experimental set up.

6. Conclusion

The evaluation of the magnetic fields and mutual inductance between circular coils arbitrarily positioned in space is presented in this paper. In the first part of the paper, MATLAB coding is implemented to compute the mutual inductance between two circular coils arbitrarily positioned with respect to each other. This computation is achieved based on the advanced and relevant model given in [4]. The computed results are presented graphically and numerically and then, compared with the ones presented in [4, 5]. Also, the formula for computing the magnetic fields between two air-cored filamentary circular coils with and without misalignment is presented and validation of the results shows that the developed model (see Eq. 3) and the experimental measurements conducted are accurate and effective.

Reference

[1] CS Wang, OH Stielau and GA Covic 2000, PowerCon, 1053-1058
[2] AO Anele, Y Hamam, Y Alayli and K Djouani 2014, JMMC, Vol. 1, No. 1, 83-104
[3] AO Anele, Y Hamam, Y Alayli and K Djouani 2014, JMMC, Vol. 1, No. 1, 31-50
[4] S Babic, F Sirois, C Akyel and C Girardi 2010, IEEE Trans. Magn., Vol. 46, No. 9, 3591-3600
[5] FW Grover 1944, Proc. IRE, 620-629
[6] JT Conway 2007, IEEE Trans. Magn., Vol. 43, No. 3, 1023-1034
[7] S Babic, and C Akyel 2012, PIER B, Vol. 38, 333-349
[8] JC Maxwell 1954, A Treatise on Electricity and Magnetism
[9] AO Anele, Y Hamam, Y Alayli and K Djouani 2013, IEEE AFRICON, 1-5

From infinitesimal symmetries to deformed symmetries of Lax-type equations

Jean-Pierre Magnot

Lycée Jeanne d'Arc - Avenue de Grand Bratagne - F-63000 Clermont-Ferrand

E-mail: `jean-pierr.magnot@ac-clermont.fr`

Abstract. Using the procedure initiated in [14], we deform Lax-type equations though a scaling of the time parameter. This gives an equivalent (deformed) equation which is integrable in terms of power series of the scaling parameter. We then describe a regular Frölicher Lie group of symmetries of this deformed equation

MSC(2010): 22E65, 22E66, 58B25, 70G65.
Keywords: Lax equations; Frölicher spaces; exponential of Lie groups

Introduction

In [14], one of the main advances of this work is to get a (non formal) principal bundle where the concept of holonomy makes sense rigorously. The geometric objects under consideration are diffeological or Frölicher groups, which are regular in the sense that the exponential map exists and is smooth. Diffeological spaces, developped by Souriau and his coworkers [2, 7, 10, 16] are generalizations of manifolds. Independently, Frölicher spaces give a more rigid framework, that also generalize the notion of manifolds [1, 3, 8]. The comparison of the two frameworks has been made independently in [12] and in [18], see e.g. [14]. The aim of this paper is to show how this framework can apply to the theory of Lax equations. A Lax equation [9] is a formally integrable equation of the type : $\partial_t L = [P, L]$ where P, L are in most cases differential, pseudo-differential, or difference operators. This equation integrates heuristically as a classical equation on a group of matrices: there should have an unique solution, up to the initial value $L(0)$, given by $L(t) = Ad_{ExpP(t)}L(0)$. Unfortunately, very often, the operator $ExpP(t)$ exists only at a formal level. We propose to apply a scaling $t \mapsto qt$ to the time variable. The operator $P(t)$ is changed into an operator $P_q(t)$ which is a monomial of order 1 in the $q-$variable, adapting the ideas of [14]. This allows the machinery of $q-$deformed operators: the algebras considered are now Lie algebras of (smooth) regular Lie groups. As a simple consequence, we get smoothness of the unique solution $L_q(t)$ with respect to $P(t)$ and $L(0)$; another consequence is that the full space of symmetries is a diffeological or Frölicher group, and that a class of symmetries of the $q-$deformed Lax equation obey also a Lax-type equation $\partial_t S_q = [ad_{P_q}, S_q]$. This equation is here interpreted as a holonomy equation, which integrates by virtue of the results of [14]. These symmetries S_q are rigorously constructed when L_q is a q-deformed formal pseudo-differential operator.

Journal of Physics: Conference Series **633** (2015) 012013 doi:10.1088/1742-6596/633/1/012013

1. Preliminaries: Frölicher Lie groups of formal series

We now turn to key results from [14]:

Theorem 1.1 *Let $(A_n)_{n\in\mathbb{N}^*}$ be a sequence of complete locally convex (Frölicher) vector spaces which are regular, equipped with a graded smooth multiplication operation on $\bigoplus_{n\in\mathbb{N}^*} A_n$, i.e. a multiplication such that $A_n.A_m \subset A_{n+m}$, smooth with respect to the corresponding Frölicher structures. Then, the set $1+\mathcal{A} = \left\{1 + \sum_{n\in\mathbb{N}^*} a_n | \forall n \in \mathbb{N}^*, a_n \in A_n\right\}$ is a Frölicher Lie group, with regular Frölicher Lie algebra $\mathcal{A} = \left\{\sum_{n\in\mathbb{N}^*} a_n | \forall n \in \mathbb{N}^*, a_n \in A_n\right\}$. Moreover, the exponential map defines a bijection $\mathcal{A} \to 1+\mathcal{A}$.*

We mimick and extend the procedure used in [14].

Theorem 1.2 *Let $\mathcal{A} = \bigoplus_{i\in I} A_i$ be a Frölicher $I-$graded regular algebra. Let G be a regular Frölicher Lie group, acting on $\mathcal{A}$ componentwise. Then, $G \oplus A$ is a regular Frölicher Lie group.*

Proof. Considering the exact sequence $0 \to 1+\mathcal{A} \to G \oplus \mathcal{A} \to G \to 0$; there is a (global) slice $G \to G \oplus \{0_{\mathcal{A}}\}$.

In our work of Lax-type equations, we use the following group from [14]:Let M be a compact manifold without boundary. We denote by $\mathcal{F}Cl$ be the space of formal classical pseudo-differential operators acting on $C^\infty(M,\mathbb{R})$. We denote by $\mathcal{F}Cl^*$ the groups of the units of the algebras $\mathcal{F}Cl$. Let q be a formal parameter. We define the algebra of formal series $\mathcal{F}Cl_q = \left\{\sum_{t\in\mathbb{N}^*} q^k a_k | \forall k \in \mathbb{N}^*, a_k \in \mathcal{F}Cl\right\}$. This is obviously an algebra, graded by the order (the valuation) into the variable q. Thus, setting $\mathcal{A}_n = \{q^n a_n | a_n \in \mathcal{F}Cl\}$, we can set $\mathcal{A} = Cl_q(M,E)$ and state the following consequence of Theorem 1.1: Let $\mathcal{F}Cl^{0,*}$ be the Lie group of invertible pseudo-differential operators of order 0. This group is known to be a regular Lie group since Omori, but the most efficient proof is actually in [5], to our knowledge. We remark a short exact sequence of Frölicher Lie groups: $0 \to 1+\mathcal{F}Cl_q \to \mathcal{F}Cl^{0,*} + \mathcal{F}Cl_q \to \mathcal{F}Cl^{0,*} \to 0$, which satisfies the conditions of Theorem 1.2. Thus, we have the following:

Theorem 1.3 *The group $1+\mathcal{F}Cl_q$ is a regular Frölicher Lie group with regular Frölicher Lie algebra $\mathcal{F}Cl_q$, and $\mathcal{F}Cl^{0,*} + \mathcal{F}Cl_q$ is a regular Frölicher Lie group with Lie algebra $\mathcal{F}Cl^0 + \mathcal{F}Cl_q$.*

2. On Lax equations and their symmetries

A PDE is of Lax type if there is a representation of the solutions $u(t,x) \in C^\infty(\mathbb{R}\times M, \mathbb{C})$ in terms of Lax operators, i.e. a smooth map $u \mapsto L(u) \in \mathcal{F}PDO$ (formal pseudo-differential operators) (very often, L is a differential operator), and another smooth map $u \mapsto P(u) \in \mathcal{F}PDO$ which satisfy a **Lax equation** such that u is a solution of the initial PDE if and only if the following equation is fulfilled:

$$\begin{cases} \partial_t L(t) &= [P(t), L(t)] \\ L(0) &= \text{fixed operator (initial value)} \end{cases} \tag{2.1}$$

(here and in the sequel, we write L and P instead of $L(u)$ and $P(u)$ when it carries no ambiguity) The couple (L,P) is called a **Lax pair.** If the path P is a smooth path of the Lie algebra $\mathfrak{g}$ of a regular Lie group G, if G acts on a Fréchet algebra of operators $\mathcal{B}$ that contains $L(0)$, the path

$$L(t) = Ad_{Exp_G P(t)} L(0) \tag{2.2}$$

is a solution of equation 2.1, yet very often a formal solution.

Example: the KdV equation. The KdV equation reads as $\partial_t u = 6u\partial_x u - \partial_x^3 u$ where $u(t,x) \in C^\infty(\mathbb{R}^2, \mathbb{R})$ ad has a Lax pair $L = -\partial_x^2 + u$ and $P = -4\partial_x^3 + 3(\partial_x u + u\partial_x)$. the operator P is of order 4, so that there is no Lie group G such that $Exp_G P(t)$ exists. Moreover, P depends on u (essentially because the KdV equation is non linear).

Let us note $\mathcal{S}$ the set of solutions of the initial PDE, which is assumed non empty and equipped with the diffeology spanned by (see [11] for the link between diffeological spaces and Fréchet manifolds):

- the trace diffeology as a subset of $C^\infty(\mathbb{R} \times M, \mathbb{C})$,
- and the pull-back of the diffeology of $\mathcal{F}PDO$ from the maps $u \mapsto L(u)$ and $u \mapsto P(u)$.

The **total set of symmetries** of the initial PDE is the group $Diff(\mathcal{S})$, which is a diffeological group. This space of symmetries is actually, to our knowledge, not studied. Instead of working with $\mathcal{S}$, we work with $L(\mathcal{S})$. If $\mathcal{S}$ and $L(\mathcal{S})$ are (diffeologically) isomorphic, then $Diff(\mathcal{S})$ and $Diff(L(s))$ are also isomorphic. We now restrict ourselves to smooth linear maps acting on the vector space spanned by $L(\mathcal{S})$ and $P(\mathcal{S})$ in $\mathcal{F}PDO$. Let us write formally the action of such a symmetry S: if $L(t)$ is a solution of 2.1, $S(t).L(t)$ is also a solution, from which we get

$$(\partial_t S).L(t) = [ad_P, S]\,.L(t) \tag{2.3}$$

Here, the map S is a smooth map $S : L(\mathcal{S}) \to \mathcal{L}(Span(L(\mathcal{S})))$. The map $\phi : S \mapsto (\partial_t S).L(t) - [ad_P, S]\,.L(t)$ is linear and the (restricted groups of) symmetries of (2.3) are the zeros of ϕ. This relation is now linear in S which allows to pass to infinitesimal symmetries, if the algebra of symmetries under consideration is equipped with the functional diffeology. We get here symmetries that are not in general exactly the ones described in [15], where *projectable* symmetries are the symmetries coming from changing of coordinates, i.e. infinitesimal symmetries in $Vect(\mathbb{R} \times M)$.

3. Integration, symmetries and time scaling

We only assume that both P and L are in a fixed Fréchet algebra $\mathcal{A}$ with unit element, or in a $c^\infty-$algebra if one prefers to work in the convenient setting [8]. Let us now build a corresponding Lax equation in $\mathcal{A}[[q]]$. We consider the paths $P(qt)$ and $L(qt)$ obtained by time scaling $t \mapsto qt$. Then, $\partial_t L(qt) = q(\partial_t L)(qt) = [qP(qt), L(qt)]$ for a fixed parameter q. We note by $L_q(t) = L(qt)$ and by $P_q(t) = qP(qt)$. We get the following equation:

$$\left\{ \begin{array}{lcl} \partial_t L_q(t) & = & [P_q(t), L_q(t)] \\ L(0) & = & \text{fixed operator (initial value) in } \mathcal{A} \end{array} \right. \tag{3.1}$$

Let val_q be the valuation of formal series in $\mathcal{A}[[q]]$ with respect to the q variable. We remark that $val_q L_q = 0$ and $val_q P_q = 1$. We note by $\mathcal{A}[[q]]_{>0}$ the ideal made of formal series S such that $val_q S > 0$.

Theorem 3.1 *The solutions of equation 3.1 in* $\mathcal{A}[[q]]$ *are such that:* $L_q(t) = exp(P_q)(t).L(0).\left(exp(P_q)(t)\right)^{-1}$ *where the the map exp is the group exponential* $\mathcal{A}[[q]]_{>0} \to Id + \mathcal{A}[[q]]_{>0}$.

The proof is a straightforward consequence of basic results on Lie groups. The serie $exp(P_q)(t)$, read as $exp(P_q)(t) = \sum_{i=0}^{\infty} a_i(q)$ where $a_i(q) = \int_{t \geq s_1 \geq \ldots \geq s_i \geq 0} \left[\prod_{j=1}^{i} P_q(s_j)\right] (ds)^i$

Let us now look for symmetries of a Lax equation. A symmetry is a path S of linear invertible operators on $\mathcal{A}$ such that, Assuming smoothness, we shall quickly go into more restricted classes of symmetries along the lines of the last section. Applying the time scaling, we get, with the obvious notations:

$$(\partial_t S_q).L_q(t) = \left[ad_{P_q}, S_q\right].L_q(t) \tag{3.2}$$

The map $S \to S_q$ is an homomorphism from the group of symmetries of (2.1) to the group of symmetries of (3.1), and it appears to us that there should exist symmetries of (3.1) that are not induced from symmetries of (2.1). The map $\phi_q : S_q \mapsto (\partial_t S_q).L_q(t) - \left[ad_{P_q}, S_q\right].L_q(t)$ is linear and the symmetries of (3.2) are the zeros of ϕ_q. Such a problem appears non relevant to the methods of resolution of this paper, and we leave the question of solving these two equations open. Let us now turn to a special class of solutions.

Journal of Physics: Conference Series **633** (2015) 012013 doi:10.1088/1742-6596/633/1/012013

4. Symmetries as holonomy elements

Let us now simplify this equation, avoiding the L_q- term. Then, we get another Lax-type equation

$$\partial_t S_q = \left[ad_{P_q}, S_q\right] \tag{4.1}$$

and we can remark that the operator ad_{P_q} is an inner derivation of A, which is of order 1 in q since P_q is of order 1. Let $In(A)$ be the Lie algebra of inner derivations of A. Let $In_q(A)$ be the $q-$graded algebra of operators spanned by qIn_A, endowed with the push-forward Frölicher structure from A.

We have to check:

(i) In_q is a smooth regular algebra

(ii) $Id_A + In_q(A)$ is a regular Frölicher Lie group with Lie algebra $In_q(A)$.

Let us remark that (ii) is a straightforward consequence of (i) and Theorem 1.1. Now, we recall that smoothness in $In(A)$ is induced by smoothness in A. Moreover, the inclusion $In(A) \to C^\infty(A, A)$ is smooth in the Frölicher sense [8]. So that $In_q(A)$ is a smooth algebra, where the composition is smooth and bilinear. Finally the only checkpoint is that the paths $ad_{a(t)}ad_{b(t)}...$ are integrable . This is in particular true in algebras of formal pseudo-differential operators, using the rules of composition of formal symbols. We can now apply the procedure that we used for equations (3.1): the exponential $exp_{Id_A+In_q(A)}$ exists and

$$S_q(t) = Exp_{Id_A+In_q(A)}(ad_{P_q}).S_q(0).\left(Exp_{Id_A+In_q(A)}(ad_{P_q})\right)^{-1}$$

is the unique solution to equation (3.2) with initial value $S_q(0)$. We now analyze equation 4.1: it extends to the path space of $C^\infty(M, \mathbb{R})[[q]] \times A_q$ which can be viewed as a trivial vector bundle. Setting $\nabla = d + ad_{P_q}$, we get a smooth connection on this fiber bundle. By the last discussion, ad_{P_q} takes values in a regular Frölicher group, and hence any path on $C^\infty(M)[[q]]$ lifts to a smooth path on A_q by holonomy theorem [14], and for any linear map S(0) that transforms an initial value $L(0)$ into another initial solution $S(0).L(0)$, we get a smooth path of operators $t \mapsto S_q(t)$ such that, if $t \mapsto L_q(t)$ is a smooth path which is solution of 3.1, the path $t \mapsto S_q(t).L_q(t)$ is also a solution of 3.1.

[1] Cherenack, P.; Ntumba, P. 2001 *Demonstratio Math.* **34** no 1 161-180

[2] Donato, P. 1984 *Revtements de groupes diffrentiels* Thse de doctorat d'état, Université de Provence, Marseille

[3] Frölicher, A; Kriegl, A 1988 *Linear spaces and differentiation theory* Wiley series in Pure and Applied Mathematics, Wiley Interscience

[4] Gilkey, P 1984 *Invariance theory, the heat equation and the Atiyah-Singer index theorem* Publish or Perish

[5] Glöckner, H 2002 *Studia Math.* **153**, no2 147-177

[6] Hector, G., and E. Macias-Virgos, Diffeological groups, Research Exp. Math. 25 (2002), 247260

[7] Iglesias-Zemmour, P. 2013 *Diffeology* Mathematical Surveys and Monographs, **185**, AMS

[8] Kriegl, A.; Michor, P.W. 2000 *The convenient setting for global analysis* Math. surveys and monographs **53**, AMS

[9] Lax, P. 1968 *Comm. Pure Appl. Math.* **21** no5 467490

[10] Leslie, J. 2003 *J. Lie Theory* **13**, 427-442

[11] Losik, M. V. 1992 *Russ. Math.* **36**, 3137.

[12] Magnot, J-P. 2006 *C. R. Math. Acad. Sci., Soc. R. Can.* **28**, no. 4, 121127

[13] Magnot, J-P. 2008 *Bull. Sci. Math.* **132** no. 2, 112-127

[14] Magnot, J-P. 2013 *Int. J. Geom. Meth. Mod. Phys.* **10**, No. 9, Article ID 1350043

[15] Olver, P. 1993 *Applications of Lie groups to differential equations* 2nd edition, GTM **107**, Springer (1993)

[16] Souriau, J.M. 1985 *Astérisque*, Hors Série, 341-399

[17] Robart, T. 1997 *Can. J. Math.* **49** (4) 820-839

[18] Watts, J.; *Diffeologies, differentiable spaces and symplectic geometry* PhD thesis arXiv:1208.3634v1

A new approach for modelling lattice energy in finite crystal domains

Y Bilotsky and M Gasik

Department of Material Science and Engineering, Aalto University Foundation School of Chemical Technology, P.O. Box 16200, FIN-00076 AALTO, Finland

E-mail: yevgen.bilotsky@aalto.fi

Abstract. Evaluation of internal energy in a crystal lattice requires precise calculation of lattice sums. Such evaluation is a problem in the case of small (nano) particles because the traditional methods are usually effective only for infinite lattices and are adapted to certain specific potentials. In this work, a new method has been developed for calculation of lattice energy. The method is a generalisation of conventional geometric probability techniques for arbitrary fixed lattices in a finite crystal domain. In our model, the lattice energy for wide range of two-body central interaction potentials (including long-range Coulomb potential) has been constructed using absolutely convergent sums. No artificial cut-off potential or periodical extension of the domain (which usually involved for such calculations) have been made for calculation of the lattice energy under this approach. To exemplify the applications of these techniques, the energy of Coulomb potential has been plotted as the function of the domain size.

1. Introduction

We recently developed a new method for evaluation of the lattice energy in finite crystals [1]. This method initially was created for the medium with continuous distributed mass. The central idea of this method is an adaptation of the geometric probability approach [2] for the lattices in finite domain. We applied the above mentioned method for calculation of long-range Coulomb interaction between lattice nodes (point charges) and continually distributed charge density (jellium). The Coulomb interaction, which has been intensively discussed in the literature from the end of nineteenth century, plays a central role in condensed matter crystals systems [3]. We assumed that the domain is neutral (the number of positive and negative charges is equal) and it has zero dipole moment. Our final result made under this assumption has been expressed in terms of unconditionally convergent sums. In real crystals negative and positive charges on the surface usually separated, and they form the surface dipole moment. For example, the electron gas above the surface of the metals together with positive charges of the ions on the metal surface creates the dipole moment. In presence of the dipole moment some sums do not converge absolutely for Coulomb potential. The problem has been solved by considering the finite size of the lattice nodes (ions or molecules). This leads to the expressions of the lattice energy in the form of absolutely convergent sums. We derived the expression for interaction energy in lattices domains by using this method. The interaction energy included node-node, jellium-jellium and node-jellium components. This method is applicable for any two-body central potential. Essentially, that we do not use periodic boundary conditions or vanishing exponential method [4] in our approach. Moreover, all kinds of interactions: lattice-lattice, lattice-jellium and jellium-jellium can be evaluated separately.

2. Geometric probability techniques for crystals lattice

The common expression for the self-energy of two-body potential

$$\langle U(R)\rangle \equiv \int\int d\vec{r}_1 d\vec{r}_2 \rho(\vec{r}_1) U(|\vec{r}_1 - \vec{r}_2|)\rho(\vec{r}_2), \tag{2.1}$$

where $\rho(\vec{r})$ is a density of the particles, has been written by Schleef et al. [2] in very useful form

$$\langle U(R)\rangle = \int_{r_0}^{2R} G(r_{12};\rho_1,\rho_2) U(|\vec{r}_{12}|) dr_{12}, \tag{2.2}$$

with r_0 being the minimum possible distance between the points. Inside the jellium, we assume that $r_0 = 0$ due to contiunuum. The radial density distribution function (RDF) $G(r_{12};\rho_1,\rho_2)$ for the number of points of the pairs inside the sphere can be written [2] as

$$G(r_{12},R) = r_{12}^2 \int_0^R 4\pi r_1 \rho(r_1)\left[\int_0^\pi 2\pi \sin\theta_{12}\rho\left(|\vec{r}_{12}+\vec{r}_1|\right)d\theta_{12}\right]dr_1 = 16\pi^2 r_{12} \int_{r_{12}/2}^{R} r_1\rho(r_1)\left(\int_{|r_{12}-r_1|}^{r_1} r_2\rho(r_2)dr_2\right)dr_1. \tag{2.3}$$

The normalized probability density for finding two points in the sphere to be a distance r_{12} apart, is simply normalized RDF

$$P(r_{12}) = \frac{G(r_{12};\rho_1,\rho_2)}{N},\ N = \int_{r_0}^{2R} G(r_{12};\rho_1,\rho_2)dr_{12}\,. \tag{2.4}$$

Therefore, using these expressions, the problem of self-energy of two-body potential can be considered with the probability theory. The advantage of this approach is in separation of purely geometrical factor (in the function $G(r_{12},R)$), from the interaction potential $U(|\vec{r}_{12}|)$. For that reason, when RDF solution was obtained once, it can be used for calculation of the self-energy for the different central two-body potentials by simple integration in (2.2). In our recent publication1 [1] this technique has been extended for crystals systems, by using the set of Dirac delta-functions as a microscopic density distribution function.

The interaction energy (self-energy) of N particles set

$$H = \frac{1}{2}\sum_{i=1}^{N}\sum_{j=1}^{N} U(|\vec{r}_i - \vec{r}_j|) \tag{2.5}$$

may be written in the continuous form as

$$H = \frac{1}{2}\int_V\int_V \rho(\vec{r}_1) U(|\vec{r}_1 - \vec{r}_2|)\rho(\vec{r}_2) d\vec{r}_1 d\vec{r}_2 \tag{2.6}$$

with the microscopic density distribution function [5]

$$\rho(\vec{r}) = \sum_{\vec{n}\in \inf} \delta(\vec{r} - n_1\cdot\vec{a}_1 - n_2\cdot\vec{a}_2 - n_3\cdot\vec{a}_3), \tag{2.7}$$

where δ being the Dirac delta-function, $\vec{a}_i$ is the basis vectors of the 3-D lattice, n_i - integers. The Eq. (2.3) is applicable only for a spherical-symmetric density $\rho(r)$, which depends only on the distance $|\vec{r}|$ (scalar). In our case the density function in (2.7) depends on the vector $\vec{r}$. Therefore, the approach has to be modified. By repeating steps leading to (2.3), but without initial integrations over azimuthal angles φ_1 and φ_2, we came to the expression

$$G(r_{12},R) = 2r_{12}^2 \int_{\frac{r_{12}}{2}}^{R} r_1 \int_{\theta_1=0}^{\pi}\left(\int_{\varphi_1=0}^{2\pi}\rho(\vec{r}_1)d\varphi_1\right)d\theta_1\left\{\int_{|r_1-r_{12}|}^{r_1}\left(\int_{\varphi_2=0}^{2\pi}\rho(\vec{r}_2)d\varphi_2\right)r_2 dr_2\right\}dr_1. \tag{2.8}$$

3. A new model

In the conventional model of lattice crystal described above, the real charges (which occupy some finite volume) are substituted by the infinite small charged nodes (points). The advantage of this discretization method is its relative simplicity, but the main disadvantage is the problem with convergence of the lattice sums, especially for spherical lattice domain [3]. Many publications, starting from early Ewald's article [5] have considered the solution to this problem for infinite lattice domain by using a lattice assembled from finite size objects rather than point nodes. Under this approach, atoms or ions in the lattice are represented as smeared-out 3D density distribution of lattice nodes [7]. This method also describes lattice vibrations.

In our new model, explicitly distribution has been used due to the specifics of the geometric probability method. The integrals in (2.8) for point charges (atoms) of the lattice model now are generalized for smeared-out spatially-distributed objects with the distribution function $f_d(\chi)$. It means that while the integration over r_1 gives the contribution for counting the number of the centers of the charges on the spherical surface, the integration over χ corrects this expression for smeared-out objects by integration over the spherical surface layer. The distribution function can also involve effects of thermal vibrations, lattice imperfections and the uncertainty principle. Here we are using 1D Gaussian

$$f_g(\chi, r_g) = \frac{1}{r_g\sqrt{\pi}} e^{-\frac{\chi^2}{r_g^2}} \tag{3.1}$$

as the distribution function. Generalizing the Eq. (2.8) with this distribution function (3.1) we receive RDF as

$$G(r_{12}, R) = G_{jj}(r_{12}, R) + G_{jl}(r_{12}, R) + G_{ll}(r_{12}, R) \tag{3.2}$$

with jellium-like component

$$G_{jj}(r_{12}, R) = \frac{16\pi^2 r_{12}}{v^2 N} \int_{r_{12}/2}^{R} r_1 dr_1 \int_{|r_{12}-r_1|}^{r_1} r_2 dr_2, \tag{3.3}$$

jellium-lattice correction

$$G_{jl}(r_{12}, R) \simeq \frac{16\pi^2 r_{12}}{v^2 N} \left[\int_{r_{12}/2}^{R} r_1 dr_1 \sum_{\vec{\gamma} \in \inf}{}' \frac{1}{2\pi\gamma} \int_{|r_{12}-r_1|}^{r_1} \int_{-\infty}^{\infty} \frac{1}{r_g\sqrt{\pi}} e^{-\frac{\chi^2}{r_g^2}} \cdot \sin(2\pi\gamma \cdot r_2 + 2\pi\gamma \cdot \chi) dr_2 + \right.$$
$$\left. + \sum_{\vec{\gamma} \in \inf}{}' \frac{1}{2\pi\gamma} \int_{r_{12}/2}^{R} \int_{-\infty}^{\infty} \frac{1}{r_g\sqrt{\pi}} e^{-\frac{\chi^2}{r_g^2}} \cdot \sin(2\pi\gamma \cdot r_1 + 2\pi\gamma \cdot \chi) dr_1 \int_{|r_{12}-r_1|}^{r_1} r_2 dr_2 \right] \tag{3.4}$$

and lattice-lattice correction

$$G_{ll}(r_{12}, R) \simeq \frac{16\pi^2 r_{12}}{v^2 N} \left[\begin{array}{l} \sum_{\vec{\gamma} \in \inf}{}' \frac{1}{2\pi\gamma} \int_{r_{12}/2}^{R} \int_{-\infty}^{\infty} \frac{1}{r_g\sqrt{\pi}} e^{-\frac{\chi_1^2}{r_g^2}} \cdot \sin(2\pi\gamma \cdot r_1 + 2\pi\gamma \cdot \chi_1) d\chi_1 dr_1 \times \\ \times \sum_{\vec{\gamma} \in \inf}{}' \frac{1}{2\pi\gamma} \int_{|r_{12}-r_1|}^{r_1} \int_{-\infty}^{\infty} \frac{1}{r_g\sqrt{\pi}} e^{-\frac{\chi_2^2}{r_g^2}} \cdot \sin(2\pi\gamma \cdot r_2 + 2\pi\gamma \cdot \chi_2) d\chi_2 dr_2 \end{array} \right] \tag{3.5}$$

components. Here the Poisson summation and the Rayleigh expansion formulas (up to second order) similar to Ref. [1] have been used.

4. Discussion

The part of the whole lattice energy, which corresponds the uniform charge distribution, is contained within the term:

$$H_{jj}(R,r_g)=\frac{1}{2}\int_{r_0}^{2R}G_{jj}(r_{12},R)U(|\vec{r}_{12}|)r_{12}dr_{12}. \quad (4.1)$$

This energy is compensated by the background energy of the opposite charge in the Wigner model. Such compensation eliminates the lattice Coulomb energy from divergence in infinite crystal. Other parts of lattice energy contain only absolutely convergent sums, even for Coulomb energy. RDF $G_{jl}(r_{12},R)$ can be interpreted as the number pairs of points with one point belongs to continuum distributed matter and another - to the lattice nodes. Therefore, the corresponding $H_{jl}(R,r_g)$ contribution to the whole lattice energy is equal to background-lattice interaction energy. Last energy component H_{ll} (with RDF $G_{ll}(r_{12},R)$) has exclusively discrete lattice origin.

As an example of application of this theory we calculate jellium-lattice contribution $H_{jl}(R)$

$$H_{jl}(R,r_g)=\frac{e^2}{v}\sum_{\vec{\gamma}\in \mathrm{inf}}{}' \left\{R^{-1}\frac{\sin(2\pi\gamma R)}{2\pi^2\gamma^3}-R^{-2}\frac{3}{4\pi^3}\frac{\cos(2\pi\gamma R)}{\gamma^4}-R^{-3}\frac{3\sin^2(2\pi\gamma R)}{8\pi^4\gamma^5}\right\}\exp(-\pi\gamma r_g) \quad (4.2)$$

for Coulomb potential in the spherical domain with zero total charge. The result is shown in the Figure 1.

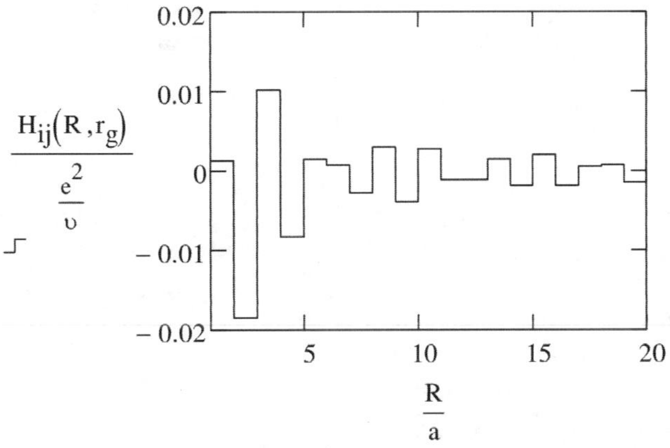

Figure 1. Variations of the jellium-lattice component (Eq. 4.2) per one charge, for Coulomb potential in primitive cubic lattice vs. sphere radius *R*, expressed in numbers of the minimal distance between nodes a and $r_g = 0.3\,a$.

This new approach allows calculations of lattice energy in different finite systems with various interaction potentials and lattice structures, which does not suffer from convergence problems commonly seen in traditional lattice sums methods.

References

[1] Y. Bilotsky: Adv. Mater. Sci. Appl. Vol. 2 (2013), p. 127.
[2] D. Schleef, M. Parry, Tu Shu-Ju, B. Woodahl and E. Fischbach: J. Math. Phys. Vol. 40 (1999), p. 1103.
[3] L.P. Buhler, R. E. Crandal: J. Phys. A: Math. Gen. Vol. 23 (1990), p. 2523.
[4] L.N. Kantorovich and I.I. Tupitsyn, Coulomb: J. Phys.: Condens. Matter Vol.11 (1999), p. 6159.
[5] P. P. Ewald: Ann. Phys. Vol. 64 (1921), p.253.
[6] T.R.S. Prasanna: Philos. Mag. Lett. Vol. 92 (2012), p. 29.
[7] D. M. Heyes and A. C. Branka: J. Chem. Phys. Vol. 138 (2013), p. 034504.

The singularity of the anharmonic oscillator

Y. Z. Pchelkina[1,2], **I. V. Alimenkov**[1]
[1] Samara State Aerospace University (SSAU), 34, Moskovskoye shosse Street, Samara, 443086, Russia
[2] 148-32, Unyh Pionerov Prospect Street, Samara, 443105, Russia

E-mail: musina@yandex.ru

Abstract. We have shown that a general solution to an anharmonic oscillation equation of a physical pendulum considered as a nonlinearity parameter function has a singularity at zero. This implies that applying different methods of regular decompositions in accordance with this parameter is impossible. We have developed a perturbation theory by using a new parameter of smallness in transformed canonical variables.

1. Introduction.

Anharmonic oscillator is a physical model which is widely used in various fields of physics, e.g. in classical nonlinear optics [1]-[3] anharmonic oscillator is the basis of the phenomenological theory of medium nonlinear susceptibility.

In this paper we have considered anharmonic oscillations of a physical pendulum. This model is quite common and convenient because there is no need in zero dimension of the original equation.

As it is known, anharmonic oscillations of a physical pendulum are described by the equation:

$$\frac{1}{\omega^2}\frac{d^2\varphi}{dt^2} = -\varphi + \varphi^3/6, \qquad (1)$$

were ω - is a cyclic frequency, which is determined by the given length of the pendulum. The right part is the expansion of function $-\sin\varphi$, in which the first two members are retained.

Now in (1) we pass on to dimensionless time $\tau = \omega t$ and we receive:

$$\ddot{\varphi} = -\varphi + \lambda\varphi^3, \qquad (2)$$

where $\lambda = 1/6$ is an anharmonicity parameter. The point denotes derivation on dimensionless time. Value $\lambda = 1/6$ is the parameter of infinitesimal in regular expansions various methods of [4] - [6].

It will be shown that the general solution of equation (2) singularity when $\lambda = 0$. This deperturbation of solution means that it is impossible to obtain the solution of the unperturbed oscillator in the form of harmonic oscillations with the passage to the limit $\lambda \to 0$.

Therefore, the general solution of equation (2) cannot be obtained by perturbation theory starting with linear equations. The right side of equation (2) satisfies the Cauchy theorem, so there exists a general solution. As a preliminary in (2) we pass on to a new dependent variable $q = \varphi^2/2$. So we get:

$$2q\ddot{q} - \dot{q}^2 = -4q^2 + 8\lambda q^3. \qquad (3)$$

Let us demonstrate the validity of theorem.

Theorem. The general solution of equation (3), considered as a parameter function λ has a pole of the first order at zero.

Proof of theorem. Equation (3) is equivalent to the following system of two equations of the first order:

$$\dot{q} = 2pq, \tag{4}$$

$$\dot{p} = -p^2 - 1 + 2\lambda q. \tag{5}$$

Indeed, by differentiating (4) and substituting (4) and (5) into equation, we find $\ddot{q} = -2q + 4\lambda q^2 + 2p^2 q$. Substituting $p = \dot{q}/2q$ from (4) we get (3). For the proof of theorem from (5) we express :

$$q = \left(\dot{p} + p^2 + 1\right)/2\lambda. \tag{6}$$

Now, by substituting (6) into (4), we get an equation of second order for p:

$$\ddot{p} = 2p\left(p^2 + 1\right). \tag{7}$$

This equation does not contain λ, so its general solution $p = f(\tau, C_1, C_2)$, after being substituted in (6) proves the theorem. The obvious solution of the last equation is $p = tg\,\tau$, then from (6) it follows that $q = 1/\lambda\cos^2\tau$ is a simple illustration of the theorem.

2. The perturbation theory

A consequence of the above theorem is to substitute $\varphi \to \varphi/\sqrt{\lambda}$ in the original equation (2) which leads (2) to the form:

$$\ddot{\varphi} = -\varphi + \varphi^3, \tag{8}$$

which does not contain the parameter of infinitesimal. Considering (8) as the Newton's equation, we find the potential energy:

$$W = \varphi^2/2 - \varphi^4/4, \tag{9}$$

which has a maximum value of $E = 1/4$ and it may be the parameter of infinitesimal in the new perturbation theory. Now in (9) we pass on to the new dependent variable $q = \varphi^2/2$. We obtain the equation:

$$2q\ddot{q} - \dot{q}^2 = -4q^2 + 8q^3, \tag{10}$$

which is equivalent to the system of two equations of the first order:

$$\dot{q} = 2pq, \tag{11}$$

$$\dot{p} = -p^2 - 1 + 2q, \tag{12}$$

which is Hamilton system with a Hamiltonian:

$$H = (p^2 + 1)q - q^2, \tag{13}$$

which allows to apply a well developed theory of canonical transformations with its powerful generating function. It should be noted that we have previously discussed the perturbation theory in the equation of Hamilton-Jacobi [7] in the action-angle variables with the old parameter of infinitesimal λ. As a result, we can find corrections to the unperturbed energy quite fast. But these corrections change the coordinates of the action-angle hamper a cumbersome solution of equations with significant coefficients in the Fourier series. The results of this perturbation theory are suitable for quantum anharmonic oscillator with its quasi-classical consideration. We are primarily interested in the law of anharmonic oscillations, which is important, for example, in nonlinear optics for determination of the nonlinear electric susceptibility.

Now in (11), (12), (13) we pass on to the interaction representation, considering $H_0 = (p^2 + 1)q$ to be a free Hamiltonian. For this case we use the theory of canonical transformations. As it follows from Jacobi-Poincare theorem, if a twice differentiable function $S(q, \overline{p}, \tau)$ exists, the one that $\left|\partial^2 S/\partial q \partial \overline{p}\right|$ is not identical to zero, then transformation $(q, p) \leftrightarrow (\overline{q}, \overline{p})$ generated by this function is:

$$p = \partial S/\partial q; \quad \overline{q} = \partial S/\partial \overline{p} \tag{14}$$

is a canonical function. And the new Hamiltonian function has the form:

$$\overline{H}(\overline{q},\overline{p},\tau)=H\big(q(\overline{q},\overline{p},\tau),\ p(\overline{q},\overline{p},\tau),\ \tau\big)+\frac{\partial S}{\partial \tau}\big(q(\overline{q},\overline{p},\tau),\ \overline{p},\ \tau\big). \tag{15}$$

In the theory of canonical transformations for two-dimensional phase space we will consider four types of generating functions that depend on one new and one old variable [8]. For further purposes, the most appropriate is the function that depends on the old momentum and new coordinates: $F=R_2(p,\overline{q},\tau)$.Then the explicit form of the transformations will be the resolution from equations:

$$q=\frac{\partial F}{\partial p};\quad \overline{p}=\frac{\partial F}{\partial \overline{q}}, \tag{16}$$

and the new Hamiltonian function:

$$\overline{H}(\overline{q},\overline{p},\tau)=H\big(q(\overline{q},\overline{p},\tau),\ p(\overline{q},\overline{p},\tau),\ \tau\big)-\frac{\partial F}{\partial \tau}\big(p(\overline{q},\overline{p},\tau),\ \overline{q},\ \tau\big). \tag{17}$$

Now we pass on from the dynamic system $\{q,p,H\}$ to the system $\{\overline{q},\overline{p},\overline{H}\}$ in the interaction representation with the help of the generating function $F(p,\overline{q},\tau)$, the one that the new Hamiltonian function is only determined by the perturbation in the new variables. This is done by using generating functions of the form:

$$F=\overline{q}Arctgp+\overline{q}\tau, \tag{18}$$

which is a complete integral of the unperturbed equation of Hamilton-Jacobi: $\partial F/\partial\tau=\left(p^2+1\right)\partial F/\partial p$. The clear form of transformations follows from equations (16):

$$q=\overline{q}\cos^2(\overline{p}-\tau);\quad p=tg(\overline{p}-\tau), \tag{19}$$

The form of the new Hamiltonian function follows from equations (17):

$$\overline{H}=-\overline{q}^2\cos^4(p-\tau), \tag{20}$$

New Hamiltonian equations:

$$\dot{\overline{q}}=\partial\overline{H}/\partial\overline{p};\quad \dot{\overline{p}}=-\partial\overline{H}/\partial\overline{q}, \tag{21}$$

determine the dynamics of the new canonical variables $\overline{q}$ и $\overline{p}$:

$$\dot{\overline{q}}=4\overline{q}^2\cos^3(\overline{p}-\tau)\sin(\overline{p}-\tau),\quad \dot{\overline{p}}=2\overline{q}\cos^4(p-\tau). \tag{22}$$

As in the interaction representation the Hamiltonian depends on the time in combination $(p-\tau)$, the first integral of the system (22) is easy to find:

$$\overline{q}-\overline{q}^2\cos^4(\overline{p}-\tau)=E. \tag{23}$$

Indeed, having taken the total time derivative of this expression we receive an identical expression with respect to (20) and (21). From (23) we get:

$$\overline{q}=\frac{1-\sqrt{1-4E\cos^4(\overline{p}-\tau)}}{2\cos^4(\overline{p}-\tau)}. \tag{24}$$

For the final solution we will find $\overline{p}$ as a function of time. For this let us take a second transformation from $\{\overline{q},\overline{p},\overline{H}\}$ to $\{Q,P,\tilde{H}\equiv 0\}$, which implies that $Q=E=const; P=const$, and the generating function of the transformation $\overline{F}(\overline{p},Q,\tau)$ satisfies the equation $\overline{H}(\partial\overline{F}/\partial\overline{p},\ \overline{p})=\partial\overline{F}/\partial\tau$.

Now expressions

$$\overline{q}=\partial\overline{F}/\partial\overline{p},\quad P=\partial\overline{F}/\partial Q \tag{25}$$

in implicit form set the first integrals of the Hamiltonian system (21). The equation for the generating function based on (20) has the form:

$$\frac{\partial\overline{F}}{\partial\tau}=-\left(\frac{\partial\overline{F}}{\partial\overline{p}}\right)^2\cos^4(\overline{p}-\tau). \tag{26}$$

Let us factor the full integral $\overline{F}(\overline{p},Q,\tau)$ in the series to the powers $Q=E$:

$$\overline{F}=\overline{F}_o+Q\overline{F}_1+Q^2\overline{F}_2+\ldots \tag{27}$$

Now let us substitute this expansion into equation (26). By equating expressions with equal powers Q on the left and on the right, we get the system of equations:

$$\frac{\partial \overline{F}_o}{\partial \tau} = -\left(\frac{\partial \overline{F}_o}{\partial \overline{p}}\right)^2 \cos^4(\overline{p}-\tau), \quad \frac{\partial \overline{F}_1}{\partial \tau} = -2\left(\frac{\partial \overline{F}_o}{\partial p}\right)\left(\frac{\partial \overline{F}_1}{\partial \overline{p}}\right)\cos^4(\overline{p}-\tau), \quad \frac{\partial \overline{F}_2}{\partial \tau} = -2\left(\frac{\partial \overline{F}_o}{\partial p}\right)\left(\frac{\partial \overline{F}_2}{\partial \overline{p}}\right)\cos^4(\overline{p}-\tau) + \left(\frac{\partial \overline{F}_1}{\partial \overline{p}}\right)^2,$$

etc.

As the original equation (8) was obtained in the first order decomposition of the sinus, then in this system it is suffice to use the second order. The solution of this system is not difficult: from the first equation we find $\frac{\partial \overline{F}_o}{\partial \overline{p}} = 1/\cos^4(\overline{p}-\tau)$, by substituting this into the last two equations and by solving them, we get $\overline{F}_1 = \overline{p} - 2\tau$, $\overline{F}_2 = -\int \cos^4(\overline{p}-\tau) d(\overline{p}-\tau) = -\frac{3}{8}(\overline{p}-\tau) - \frac{1}{4}\sin 2(\overline{p}-\tau) - \frac{1}{32}\sin 4(\overline{p}-\tau)$.

According to the second equation (25) $P = \partial \overline{F} / \partial Q$, taking into account (27), we get:

$$\overline{p} = P + 2\tau + \frac{3Q}{4}(P+\tau) + \frac{Q}{2}\sin 2(P+\tau) + \frac{Q}{16}\sin 4(P+\tau). \quad (28)$$

The problem is solved. Arbitrary constants Q и P have a simple physical meaning: $Q = E \leq 1/4$; $P = -\tau_o$, where τ_o - the initial moment of time. Now we have to go back to the original variable φ from the equation (2):

$$\varphi = \sqrt{\frac{1-\sqrt{1-4E\cos^4(\overline{p}-\tau)}}{\lambda \cos^2(\overline{p}-\tau)}}. \quad (29)$$

Expression (28) should be substituted here, but this would make the final result extremely cumbersome.

3. Conclusions

Thus, the problem of finding the general solution of equation (2) is reduced by using the theory of Hamiltonian systems to the two expressions (28) and (29), the first of which is approximate, and the second is exact. Given the known internal decomposition of the radical in a power series we come to the conclusion that the right-hand side of the formula (29) is a continuous function of infinitesimal time τ and of the parameter E and it has a singularity at zero on the anharmonicity parameter λ.

4. References

[1] Shubert, M. *Introduction in Nonlinear Optics* / M. Shubert, B. Wilgelmi. – Moscow: “Mir”, 1973. – 246p. – (in Russian).
[2] Bloembergen N., *Nonlinear optics*, Benjamin, New York, 1965.
[3] Zernike F., Midwinter J.E., *Applied Nonlinear Optics*, John Wiley & Sons, London, 1973.
[4] Giacaglia, G.E.O. *Perturbation Methods in Nonlinear Systems*. – Springer-Verlag, New York, 1972.
[5] Nayfeh, A. *Introduction to Perturbation Techniques*. – WILEY-VCH, 1993. – 543p.
[6] J. D. Cole, *Perturbation Methods in Applied Mathematics*. –Blaisdell, Waltham, Massachusetts, 1968 , - 276p.
[7] D. ter Haar. *Elements of Hamiltonian Mechanics*. – Pergamon Press, Oxford , 1971, - 201 p.
[8] Schmutzer, E. Grundprinzipien *der klassischen Mechanik und der klassischen Fieldtheorie (kanonischer Apparat)*, - Berlin:VEB Deutscher Verlag der Wissenschaften, 1973.- 160p.

Acknowledgements

This work was supported by the Ministry of education and science of the Russian Federation in the framework of the implementation of the Program of increasing the competitiveness of SSAU among the world's leading scientific and educational centers for 2013-2020 years.

Modeling of the size effect on the hysteresis characteristics of core-shell nanoparticles Co-Au

M A Chrenova[1*], **I G Iliushin**[1], **L L Afremov**[1]

[1]Department of Theoretical and Experimental Physics, School of Natural Sciences FEFU, 8 Syhanova st., Vladivostok, Russia

*corresponding author's e-mail: cmariaa@mail.ru

Abstract. Study of the dependence of hysteretic characteristics of core-shell Co-Au nanoparticles on size was conducted within the model of two-phase nanoparticles. It was shown, that coercive field, residual saturation magnetization and saturation magnetization decreasing with sizes of Co-Au particles due to the increasing role of thermal fluctuations. The simulation results of the size effect on the hysteresis characteristics of the core-shell nanoparticles are in good agreement with the experimental data.

Keywords: Core/shell nanoparticles, magnetic characteristics, model hysteresis calculation.

Introduction.

There are new opportunities for modeling complex processes due to the increase of computing power, this includes the simulation of physical processes and different types of materials that are relevant today. One of such materials are magnetic core-shell nanoparticles with unique properties, which allows their use in various fields [1-7]. In this paper we will present modeling of the size effect on the coercive force, residual saturation magnetization, saturation magnetization of the Co-Au core-shell nanoparticles, which due to their unique properties are widely used in biomedicine. [8] Necessity of Co coating with noble metals such as gold [9], due to their high sensitivity to oxidation, which significantly increases with decreasing size of the nanoparticles. Gold has good biocompatibility with living organisms and, effectively reducing the oxidation process increases the hysteresis characteristics of core-shell nanoparticles compared with cobalt nanoparticles. Core-shell nanoparticles used in practice are generally small in size so that a part of them may be in a superparamagnetic state. That is why such magnetic characteristics as coercivity, remanence and saturation magnetization essentially depend on their size [10].

Model used:

According to the two-phase nanoparticles' theory[11,12] following model was proposed:
1) ellipsoid-shaped Co-Au core-shell nanoparticle with volume V contains uniformly magnetized ellipsoidal Co core with volume $v = \varepsilon V$, and long axis coincides with the long axis of the core-shell nanoparticle oriented along an axis Oz (fig 1)
2) it is believed that the axes of crystallographic anisotropy of cobalt and gold are parallel to axis Oz;
3) spontaneous magnetization vectors of phases $\boldsymbol{I}_s^{(Au)}$ and $\boldsymbol{I}_s^{(Co)}$ are located in a plane yOz and make angles $\theta^{(Au)}$ and $\theta^{(Co)}$ with axis Oz, respectively;
4) external magnetic field H is applied along xis Oz;
5) it is assumed that the total energy of nanoparticles composed of the anisotropy energy E_A, magnetostatic interaction cnergy E_m, the energy of the interfacial exchange interaction E_ex and magnetic moment of nanoparticle in an external magnetic field E_H: $E = E_A + E_m + E_{ex} + E_H$.

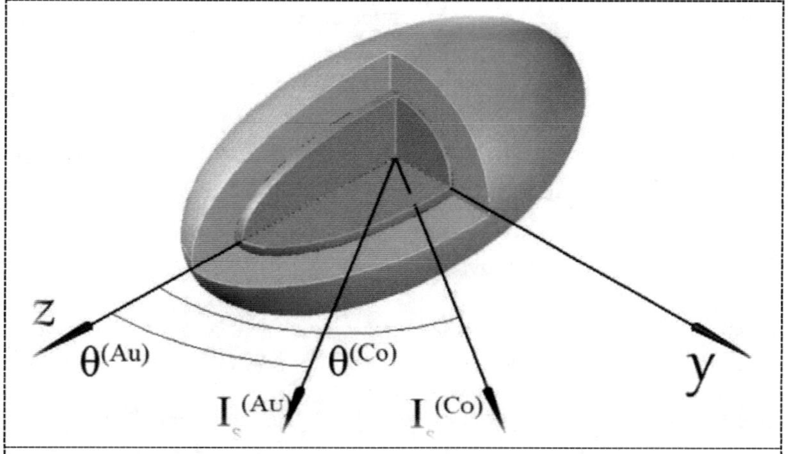

Figure 1. Model of core-shell nanoparticles

In the article [12] analysis of the full energy has shown that neglecting thermal fluctuations (T = 0) and absence of an external magnetic field lead to the situation where core-shell nanoparticle may be in one of four states:

- *first state* "(↑↑)-state" magnetic moments of both phases are parallel and oriented along axis OZ;
- *second state* "(↑↓)-state" magnetic moment of second phase is antiparallel to axis OZ;
- *third state* "(↓↓)-state" magnetic moments of both phases are antiparallel to axis OZ;
- *fourth state* "(↓↑)-state" magnetic moment of first phase is antiparallel to axis OZ.

If ($T \neq 0$), then system of core-shell nanoparticles which is in a nonequilibrium state over time t must get into a state with a population as determined by the vector $\boldsymbol{n}(t) = \{n_1(t), n_2(t), n_3(t),\ n_4(t)\}$, which components $n_i(t)$ can be regarded as the probability of finding nanoparticles in the i state. Transition probabilities between i and k states is determined by the height of the potential barrier $E_{ik}V$:

$$W_{ik} = f_0 exp\left(-\frac{E_{ik}V}{k_B T}\right) \tag{1}$$

where $f_0 \sim 10^7 \div 10^{10} s^{-1}$ – frequency of "attempts" to overcome potential barrier, which is assumed 10^{10} s^{-1} [13], k_B - boltzmann constant. If the initial state of the ensemble of two-phase particles $\mathbf{n}(t = 0) = \{\mathrm{n}_{01}, \mathrm{n}_{02}, \mathrm{n}_{03}, \mathrm{n}_{04}\}$ non equilibrium, then the transition to equilibrium is described by a system of four equations [11]:

$$\frac{dn_i(t)}{dt} = \sum_{k \neq i}^{4} \left(-W_{ik} n_i(t) + W_{ki} n_k(t)\right),\ \ n_i(t = 0) = \mathrm{n}_{0i}. \tag{2}$$

Using the normalization condition $\mathrm{n}_1(t) + \mathrm{n}_2(t) + \mathrm{n}_3(t) + \mathrm{n}_4(t) = 1$ and excluding $\mathrm{n}_4(t)$ from (2), we can rewrite system (2) in matrix form:

$$\frac{d\boldsymbol{N}(t)}{dt} = \widetilde{\boldsymbol{W}} \cdot \boldsymbol{N}(t) + \boldsymbol{V}, \tag{3}$$

$$\widetilde{W}_{ik} = \begin{cases} -\sum_{j \neq i}^{4} W_{ij} - W_{4i}, & i = k, \\ W_{ki} - W_{4i}, & i \neq k, \end{cases}, \quad \boldsymbol{N} = \begin{pmatrix} \mathrm{n}_1(t) \\ \mathrm{n}_2(t) \\ \mathrm{n}_3(t) \end{pmatrix}, \quad \boldsymbol{V} = \begin{pmatrix} \mathrm{W}_{41} \\ \mathrm{W}_{42} \\ \mathrm{W}_{43} \end{pmatrix}, \boldsymbol{N_0} = \begin{pmatrix} \mathrm{n}_{01} \\ \mathrm{n}_{02} \\ \mathrm{n}_{03} \end{pmatrix}$$

Solution of equation (3) can be conveniently written using the matrix exponential:

$$\boldsymbol{N}(t) = \exp\left(\widetilde{\boldsymbol{W}} t\right) \cdot \boldsymbol{N}_0 + \int_0^t exp\left(\widetilde{\boldsymbol{W}}(t - \tau)\right) d\tau \cdot \boldsymbol{V}. \tag{4}$$

Using equation (4) we can determine the magnetization of the system of core-shell nanoparticles consisting of particles of various sizes

$$I(t) = \int \Big[\big(n_1(t,b) - n_3(t,b)\big)\left((1-\varepsilon)I_S^{(Au)} + \varepsilon I_S^{(Co)}\right) + \\ +\big(n_2(t,b) - n_4(t,b)\big)\left((1-\varepsilon)I_S^{(Au)} - \varepsilon I_S^{(Co)}\right)\Big] F(b)db, \quad (5)$$

where $n_4(t,b) = 1 - n_1(t,b) - n_2(t,b) - n_3(t,b)$, $F(b)$ – distribution function of the size b of nanoparticles.

Hysteretic characteristics of Co-Au nanoparticles

Simulation algorithm of the size effect on the hysteresis characteristics of Co-Au nanoparticles involves the solution of two problems: the calculation of the population vector $\boldsymbol{n}(t)$ and formation of the hysteresis loops for nanoparticles of different sizes. The first task is divided into two parts:

a) calculation of the spectrum of potential barriers $E_{ik}V = ((E_{ikmax})_{min} - E_{imin})V$ and the transition probability matrix W_{ik}, which reduces to finding the equilibrium values of the energy of magnetic moment of core-shell nanoparticles E_{imin} and minimum of maximum values of several energy barriers separating i and k equilibrium states $(E_{ikmax})_{min}$;

b) calculation of the matrix exponent $\exp(\widetilde{\boldsymbol{W}}t)$ and it's integral $\int_0^t exp\left(\widetilde{\boldsymbol{W}}(t-\tau)\right) d\tau$.

Depending on the relationship between the eigenvalues $s_i$of the matrix $\widetilde{\boldsymbol{W}}$, matrix exponent can be represented as follows:

If eigenvalues of the matrix $\widetilde{\boldsymbol{W}}$ are not equal ($s_1 \neq s_2 \neq s_3$), then

$$\exp(\widetilde{\boldsymbol{W}}t) = \sum_{i=1}^{3} \exp(s_i t)\left[\prod_{k\neq i=1}^{3}(\widetilde{\boldsymbol{W}} - s_k\boldsymbol{E}) \Big/ \prod_{k\neq i=1}^{3}(s_i - s_k)\right] \quad (6)$$

$$\int_0^t exp\left(\widetilde{\boldsymbol{W}}(t-\tau)\right) d\tau = \sum_{i=1}^{3} \varphi_1(s_i,t)\left[\prod_{k\neq i=1}^{3}(\widetilde{\boldsymbol{W}} - s_k\boldsymbol{E}) \Big/ \prod_{k\neq i=1}^{3}(s_i - s_k)\right] \quad (7)$$

where $\varphi_1(s_i,t) = [\exp(s_i t) - 1]/s_i$. If there are two multiple eigenvalues of the $\widetilde{\boldsymbol{W}}$, **e.g.**, $s_i = s_k \neq s_m$:

$$\exp(\widetilde{\boldsymbol{W}}t) = \frac{\left(\widetilde{\boldsymbol{W}} - s_k\boldsymbol{E}\right)^2}{(s_m - s_k)^2}\{\exp(s_m t) - \exp(s_k t)[1 - (s_m - s_k)t]\} + \\ +(\widetilde{\boldsymbol{W}} - s_k\boldsymbol{E})t\exp(s_k t) + \boldsymbol{E}\exp(s_k t), \quad (8)$$

$$\int_0^t exp\left(\widetilde{\boldsymbol{W}}(t-\tau)\right) d\tau = \frac{\left(\widetilde{\boldsymbol{W}} - s_k\boldsymbol{E}\right)^2}{(s_m - s_k)^2}\{\varphi_1(s_m,t) - \varphi_1(s_k,t) - \varphi_2(s_k,t)(s_m - s_k)\} + \\ +(\widetilde{\boldsymbol{W}} - s_k\boldsymbol{E})\varphi_2(s_k,t) + \boldsymbol{E}\varphi_1(s_k,t). \quad (9)$$

In the case of three multiple eigenvalues of the $\widetilde{\boldsymbol{W}}$ ($s_1 = s_2 = s_3$):

$$\exp(\widetilde{\boldsymbol{W}}t) = \left(\frac{1}{2}\left(\widetilde{\boldsymbol{W}} - s_1\boldsymbol{E}\right)^2 t^2 + (\widetilde{\boldsymbol{W}} - s_1\boldsymbol{E})t + \boldsymbol{E}\right)\exp(s_1 t), \quad (10)$$

$$\int_0^t exp\left(\widetilde{\boldsymbol{W}}(t-\tau)\right) d\tau = \frac{1}{2}\left(\widetilde{\boldsymbol{W}} - s_1\boldsymbol{E}\right)^2\varphi_3(s_1,t) + (\widetilde{\boldsymbol{W}} - s_1\boldsymbol{E})\varphi_2(s_1,t) + \boldsymbol{E}\varphi_1(s_1,t). \quad (11)$$

For comparison with the experimental results [10] simulation was carried out for the core-shell Co-Au nanoparticles with sizes $0 \leq b \leq 45$nm. In integrating the expression (5) the normal distribution law was used:

$$F(b) = \frac{1}{\sqrt{2\pi\sigma^2}}\exp\left(-\frac{(b-\langle b\rangle)^2}{2\sigma^2}\right) \quad (12)$$

Average values of the nanoparticle size $\langle b\rangle$ and dispersion σ calculated from given in [10] experimental diagrams of particle sizes are presented in Table 1

Table 1. Experimental and calculated values of coercive field H_c, saturation magnetization I_H, residual saturation magnetization I_{rs} for Co-Au nanoparticles of different sizes Co-Au. $\langle b\rangle$ – average size, σ – dispersion of the distribution of nanoparticles.

$\langle b\rangle$, nm	σ, nm	H_c, Oe		I_H,G		I_{rs},G	
		Exp. [10]	Calc.	Exp. [10]	Calc.	Exp. [10]	Calc.
9	6.00	32.2	29.01	1.74	1.81	0.03	0,025
12	8.00	83.1	86.05	6.40	6.25	0.07	0.14
14	8.57	250.0	257.3	7.04	7.07	0.10	0.46
19	10.92	199.0	188.3	40.80	38.29	0.17	0.37

Calculations have shown (see table 1.) that increasing size of the nanoparticles leads to increase of the coercivity, residual and saturation magnetization, described hysteretic characteristics behavior is associated with decreasing (with the increasing of average size) portion of core-sell nanoparticles moved into the superparamagnetic state.

Conclusion

Within our model the basic calculation algorithms are described and dependence of the hysteresis characteristics of Co-Au nanoparticles on their size are simulated. It has been shown that reducing the size of the nanoparticles leads to a decrease in coercive force, saturation magnetization, residual saturation magnetization. This behavior of the hysteretic characteristics associated with an increased proportion of superparamagnetic particles. The obtained simulation results are in good agreement with experimental studies.

Acknowledgement
Work supported by Ministry of Education and Science, project № 559.2014.

References

[1] F. Lei, B. Yan, H.-H. Chen, Q. Zhang, J.-T. Zhao, *Crystal Growth and Design* **9** (2009) 3730–3736.
[2] Y. Hu, W. Zhuang, H. Ye, D. Wang, S. Zhang, X. Huang, *Journal of Alloys and Compounds* **390** (2005) 226–229.
[3] V.B. Mikhailik, H. Kraus, G. Miller, M. S. Mykhaylyk, D. Wahl, *Journal of Applied Physics* **97** (2005) 083523.
[4] M. Mai, C. Feldmann, *Journal of Materials Science*, **47**, 1427–1435.
[5] Motl, N. E.; Bondi, J. F.; Schaak, R. E., *Chem. Mater.* 2012, **24,** 1552−1554.
[6] Kim, Y.; Park, K. Y.; Jang, D. M.; Song, Y. M.; Kim, H. S.; Cho, Y. J.; Myung, Y.; Park, J. J. *Phys. Chem. C* 2010, **114**, 22141−22146.
[7] Ghosh Chaudhuri, R.; Paria, S. *Chem. Rev.* 2011, **112**, 2373−2433.
[8] Q A Pankhurst, J Connolly, S K Jones and J Dobson, *J. Phys, D: Appl. Phys.*, **36**, 2003, R167–R181
[9] G. Cheng, A. R. Hight Walker, Journal of Magnetism and Magnetic Materials, **31** (2007) p.311
[10] Gh. Bahmanrokh, M. Hashim, N. Soltani, I. Ismail, P. Vaziri,M. Navaseri, R. Sabbaghizadeh, *Digest Journal of Nanomaterials and Biostructures* **Vol. 7**, No. 4, p. 1799 – 1810
[11] Afremov L.L., Panov A.V. *The Physics of Metals and Metallography*. 1996. V. **82**. No. 5. pp. 439-444; // *The Physics of Metals and Metallography*. 1996. **V. 82**. No. 5. pp. 445-448.
[12] Afremov. L.L.,Ilyushin I.G., *Journal of Nanomaterials,* **Vol. 2013**, 15.
[13] BrownW.F., *Phys.Rev.*, 1963, **130**, p. 1677

Modelling study of phase transitions in amorphous ultrathin films

Afremov Leonid, Petrov Aleksandr

Departament of Theoretical and Experimental Physics, The School of Natural Sciences, Far Eastem Federal University, 8, Sukhanova St., 690950, Vladivostok, Russian Federation

e-mail: afremov.ll @ dvfu.ru, aleksandr.al.petrov@gmail.com

Abstract. The phase transition's critical temperature dependence of the thickness and critical magnetic concentration are studied using the method of configuration averaging over the interaction fields in the framework of the Ising model in amorphous ultrathin film to define the transition from ferromagnetic state to paramagnetic one. It is shown that the Curie temperature increases quickly with a rise in number of monolayers and critical magnetic concentration decreases to bulk material value.

1. Introduction

The amorphous materials' phase magnetic transitions have been studied theoretically for a long time (see for example [1, 2]) but it has been still timely [3]. The magnetic phase transition was studied in amorphous material by the method of random field in the framework of the Ising model in [1]. There was the phase transition's temperature dependence of the ferromagnetic's bulk concentration to study. Besides the critical magnetic concentration is calculated whereby the amorphous ferromagnetic becomes the paramagnetic. The Curie temperature dependence of the intensity of exchange interaction, the transverse and longitudinal susceptibilities and magnetization dependence of the temperature, the pyromagnetic coefficient dependence of the temperature was studied in the framework of the effective field theory in [2,3]. Particular attention is paid to the influence of the amorphization on the ferromagnetic's hysteresis characteristics [3]. Kaneyoshi was made similar study much earlier [4, 5] where he studied the influence of the amorphization on the magnetic characteristics.
It should be noted that there are many experimental and theoretical studies [6-15] about size-effect's influence on the crystalline ultrathin film's magnetic characteristics near phase transition point. Experimental studies have been shown that phase transition's relative change of the temperature decreases proportionally $L^{-1/\nu}$ with film thickness increasing where ν-critical exponent of spin-spin correlations [14, 15].

Magnetic [16] and concentration [17] phase transitions' theoretical studies have been shown that the dependence of the relative change in the Curie temperature of the thickness of an ultrathin film obeys a power law with an exponent equal to reverse critical exponent of spin-spin correlations $\nu = 0{,}69$. Moreover, the critical concentration of the transition from the ferromagnetic to the paramagnetic state decreases with increasing film thickness up to the percolation threshold [17].

It is assumed by the random interaction fields [1, 12, 13, 17] to carry out the theoretical study of the size-effect's influence on the magnetic and concentration phase transition.

2. The model

- Ultrathin film consists of N monolayers of diluted amorphous magnetic material with the bulk concentration of "magnetic atoms" c;
- the direct exchange interaction occurs only between the nearest neighbors; moreover, the interaction fields h between the spin magnetic moments of the atoms are distributed in a random manner;
- spin magnetic moments are oriented along an axis Oz (approximation of the Ising model).

According to gived a detailed account of formalism [1, 12, 13, 17] the functions of random fields distribution $h = H/m\,J_{1,1}$ of each $N-th$ moloyaer are as follows:

$$W_n(h)dh = \frac{1}{\sqrt{2\pi b_n}} Exp\left[-\frac{(h-h_n)^2}{2b_n}\right], \qquad 1 \le n \le N, \tag{1}$$

where m – atom's magnetic moment, $J_{1,1}$ – is the exchange constant between atoms in the first layer and , and distribution parameters (1) may be represented as follows:

$$h_1 = c\,(\mu_1 + i_{12}\mu_2), \quad b_1 = c(1 + i_{1,2}^2); \quad h_n = c(i_{n,n-1}\,\mu_{n-1} + i_{n,n}\,\mu_n + i_{n,n+1}\,\mu_{n+1}),$$

$$b_n = c(i_{n,n-1}^2 + i_{n,n}^2 + i_{n,n+1}^2); \; h_N = c\big(i_{N,N-1}\,\mu_{N-1} + i_{N,N}\,\mu_N\big), \; b_N = c(i_{N,N-1}^2 + i_{N,N}^2). \tag{2}$$

In relations (2) $i_{n,k} = J_{n,k}/J_{1,1}$, $J_{n,k}$ – the constant of the exchange interaction between atoms located in n –th and k –th layers. The average magnetic moment $\mu_n = \langle\alpha_n\rangle - \langle\beta_n\rangle$ is defined as the difference between the average probalities orientation of the magnetic moment of the atom at ($\langle\alpha_n\rangle$) and against ($\langle\beta_n\rangle$) Oz-axis respectively:

$$\langle\alpha_n\rangle = \int \alpha_n(h) W_n(h)dh = \int \frac{e^{mh/k_BT}}{e^{mh/k_BT} + e^{-mh/k_BT}} W_n(h)dh, \tag{3}$$

$$\langle\beta_n\rangle = \int \beta_n(h) W_n(h)dh = \int \frac{e^{-mh/k_BT}}{e^{mh/k_BT} + e^{-mh/k_BT}} W_n(h)dh, \tag{4}$$

where m – atom's magnetic moment, T – temperature, k_B – Boltzmann constant. We obtain a system of self-consistent equations according to the definition of the average magnetic moment quoted above:

$$\mu_1 = \int \tan\left(\frac{h}{t}\right) W_n(h,\mu_1,\mu_2)dh\,, \qquad \mu_n = \int \tan\left(\frac{h}{t}\right) W_n(h,\mu_{n-1},\mu_n,\mu_{n+1})dh\,,$$

$$\mu_N = \int \tan\left(\frac{h}{t}\right) W_n(h,\mu_{N-1},\mu_N)dh, \tag{5}$$

where $t = k_BT/m$.

Equations (5) and (7) enable us to find the relative magnetic moment and the phase transition's critical temperature dependence of the amorphous magnetic film's number of layers and «magnetic» atom's concentration.

3. The Curie temperature dependence of the film monolayers' number and «magnetic» atoms' concentration

The relative Curie temperature $t_c = k_BT_c/J_{1,1}m$ dependence of the ultrathin amorphous film's number of monolayers N is shown in figure 1. The Curie temperatures differ essentially for films less than 5 monolayers and practically don't depend of the thickness for $N > 8$.

Journal of Physics: Conference Series **633** (2015) 012017 doi:10.1088/1742-6596/633/1/012017

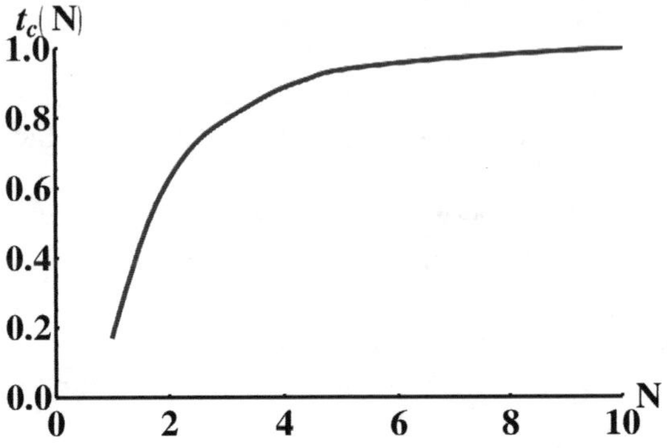

Fig 1. The relative Curie temperature $t_c = k_B T_c / J_{1,1} m$ dependence of the amorphous ultrathin film's number of monolayers N with «magnetic» atoms' concentration $c = 1$.

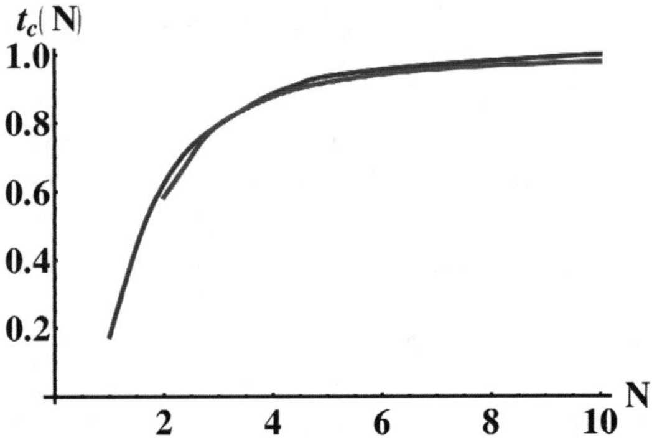

Fig 2. The relative Curie temperature dependence $t_c(N) = T_c(N)/T_c(N \to \infty)$ of the amorphous (blue curve 1) and crystalline (purple curve 2) [12] ultrathin films' number of monolayers N with «magnetic» atoms' concentration $c = 1$.

The calculation's results of the relative Curie temperature $t_c(N)$ dependence of the crystalline film's thickness with simple cubic lattice are shown in figure 2 for comparison [12]. It is evident that phase transition's critical temperature of the amorphous and crystalline films varies similarly with increasing thickness $t_c(N) \to 1$ but the amorphous film's Curie temperature varies faster. This is attributable to the large number of nearest neighbors z in amorphous film than in a crystalline one ($z > 6$).

The «magnetic» atoms' concentration markedly affects on the magnetic ordering as in the case of the crystalline ultrathin films [16, 17] (see for example figure 3).

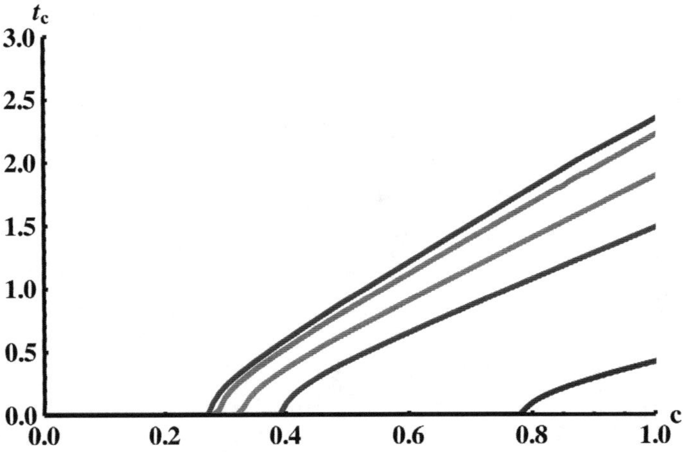

Fig 3. The relative Curie temperature $t_c = k_B T_c / J_{1,1} m$ dependence of the «magnetic» atoms' concentration c for the films with different thickness N. From left to right: blue curve-$N = 10$, green curve- $N = 5$, light brown curve- $N = 3$, purple curve- $N = 2$, blue curve- $N = 1$

Moreover the ferromagnetic's critical concentration c_{cr} decreases with with increasing film thickness up whereby the ferromagnetic becomes the paramagnetic. The amorphous film's critical concentration c_{cr} varies through bigger range in comparison with crystalline films [16, 17].

4. Conclusion

It is shown as the thoretical study's result of the influence of size-effect on the magnetic and concentration phase transition in ultrathin amorphous films:

1) the Curie temperature increases with increasing monolayers' number of the ultrathin amorphous films where the Curie temperature's type of the dependence is similar in behavior of crystalline one;
2) magnetic atoms' critical concentration decreases with increasing film thickness up whereby the magnetic phase transition occurs.

5. Acknowledgements

The work is supported by grant of Ministry of Education and Science. Reference number is Government Order № 559.2014.

References

[1] Belokon V and Semkin S 1992 *Journal of Experimental and Theoretical Physics* **101** (10) 1254-1258
[2] Ainame A, El-Atri A and Saber M 1995 *Journal of Magnetism and Magnetic Materials* **145** 139-146
[3] Essaoudi I, Ainame A, Saber M and J J de Miguel 2009 *Journal of Magnetism and Magnetic Materials* **321** 38-42
[4] Kaneyoshi T 1985 *Journal of the Physical Society of Japan* **54** 3514-3525
[5] Kaneyoshi T 1992 *Introduction to Amorphous Magnets* (Singapore: World Scientific) p 87
[6] Ambrose T and Chien C L 1996 *Physical Review Letters* **76** (10) 1743-1746
[7] Ambrose T and Chien C L 1996 *Journal of Applied Physics* **79** 5920-5922
[8] Bramfeld T S and Willis R F 2008 *Journal of Applied Physics* **103** 07C718
[9] Willis R F, Bramfeld T S, Podolak K R 2007 *Journal of Allpied Physics* **101** 09G119
[10] Gradman U 2001 *Encyclopedia of Materials: Science and Technology* 5832-5837
[11] Marques M I and Gonzalo J A 2000 *Acta Physica Polonica A* **97** 1033-1038
[12] Afremov L L and Kirienko Y V 2013 *Journal of Physics:Conference Series* **410** 012017
[13] Afremov L L and Petrov A A 2014 *Journal of Physics:Conference Series* **490** 012021
[14] Huang F, Mankey G J, Kief M T and Willis R F 1993 *Journal of Applied Physics* **73(10)** 6760-6762
[15] Huang F, Mankey G J, Kief M T and Willis R F 1994 *Journal of Applied Physics* **49(6)** 3962-3971
[16] Afremov L L, Kirienko Y V 2013 *Applied Mechanics and Materials* **328** 823-826
[17] Afremov L L, Il'yushin I G 2013 *Advanced Materials Research* **683** 69-72

Simulation of deformations in magnetic media by the movable cellular automata method

Victor V. Usachev, Petr D. Andriushchenko, Leonid L. Afremov

The School of Natural Sciences, Far Eastern Federal University,
8, Sukhanova St., 690950, Vladivostok, Russian Federation

e-mail: usachev.vv@dvfu.ru, pitandmind@gmail.com, afremov.ll@dvfu.ru

Abstract. A deformable model of the magnetic medium is considered in the research paper. Simulating algorithms of the impact of the external magnetic field on the deformation of the magnetic medium have been developed on the basis of the movable cellular automata (MCA) method.

1. The deformable model of the magnetic medium

Model of the medium is presented as ensemble of N interacting discrete elements (automata). Each element simulates a spherical particle of the medium and has a small size compared to the linear dimensions of the system. This element describes a set of physical parameters, which are typical for elastic media, such as density or bulk modulus. As a first step let's consider the problem in two-dimensional system. Each element can belong to one or another material and differ in their physical properties. In addition, certain groups of particles may be subject to special conditions, for example, the surface layer can be firmly fixed in the starting position during the simulation and the other group of particles may be subjected to an external force. This approach can help simplify description of heterogeneous and composite materials with different geometries.

The movable cellular automata method (MCA) is applied in the research paper. The feature of this method is the ability to simulate the deformation of solids over a wide range, up to the formation of cracks and failure of the material [1, 2]. Each element of the system is a movable cellular automaton. Rules of interaction and the possible states are defined for automaton. In the MCA method the concept of state is applied to a pair of automata; that's is the difference between the MCA and the classical cellular automata. The concept of the environment is introduced, according to which neighbors can be dynamically changed during evolution of the system: automata are capable of forming pair bonds with each other and break them. Thus, each pair can be in one of two states - connected or unconnected one. Bound pair of automata is a solid piece of material, and the unconnected automata belong to different fragments. This interaction mechanism allows to simplify entering the short-range forces and allows the destruction of the sample with strong deformations. It will also assume that N_m of N automata are magnetic, which requires the introduction of long-range forces.

To determine the status of the connection between pairs of automata must enter the criterion for connection and disconnection. In the simplest case, the criterion may be the distance between the centers of mass of the particle: automata i and j are connected on condition $r_{ij} \leq \alpha r_{ij}^0$, and if $r_{ij} > \alpha r_{ij}^0$ - automata are not connected. Here $r_{ij} = |\boldsymbol{r}_{ij}|$ - distance between centers of automata i and j, $r_{ij}^0 = \frac{(d_i + d_j)}{2}$ - equilibrium distance, d_i - diameter of the automaton i, $\alpha = 1.2$ - coupling parameter, chosen in this paper empirically. Figure 1 schematically illustrates a pair of connected automata, which are at a maximum (threshold) distance between automata $r_{ij} = \alpha r_{ij}^0$. Connection of any pair of automata has an indirect impact on the state of the entire system through the connection of neighboring elements. The state of each of the automaton system at any given time is determined by its previous state, and also depends on the condition of each adjacent pair of the previous time step. These automata form the active medium.

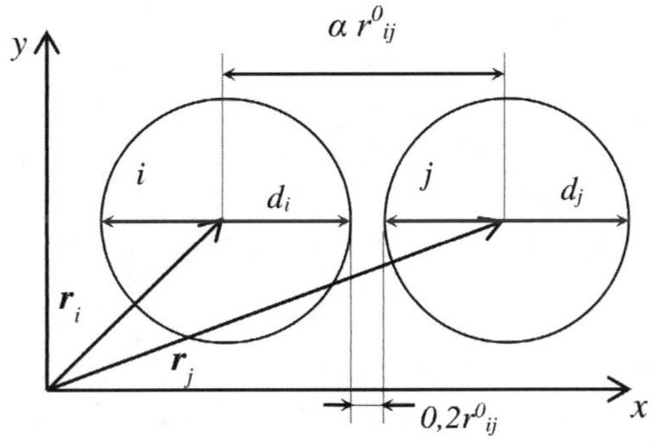

Fig 1. Existence of a link depends on the distance between the automata.

2. Algorithms for the calculation of deformations

To find the displacement of of the automaton from its position at the previous step scheme Werle is applied. At each time step of integration *dt* the forces acting on each automata from the system or from the outside is calculated, and the resulting speed and new coordinates of automata is determined. According to [3, 4], the displacement of the *i*-th automaton can be identified by its speed using the equations of translational and rotational motion:

$$m_i \frac{d^2 \boldsymbol{r}_{ij}}{dt^2} = \boldsymbol{F}_i, \quad J_i \frac{d^2 \boldsymbol{\theta}_i}{dt^2} = [\boldsymbol{r}_{ij}, \boldsymbol{F}_i], \tag{1}$$

Where $m_i, \boldsymbol{J}_i, \boldsymbol{r}_{ij}, \boldsymbol{\theta}_i$ – mass, moment of inertia, the linear and angular coordinates of the *i*-th automaton, $\boldsymbol{F}_i$ – the resultant force acting on the *i*-th automaton. The resulting force is the sum of the elastic and viscous forces: $\boldsymbol{F}_i = \boldsymbol{F}_i^{\Omega} + \boldsymbol{F}_{ji}^{\mathrm{norm}} + \boldsymbol{F}_{ij}^{\mathrm{tang}} + \boldsymbol{F}_i^{(m)}$. Here $\boldsymbol{F}_i^{\Omega} = \sum_{j=1}^{z} P_j S_{ij} \boldsymbol{n}_{ij}$ – volume-dependent force acting on automaton *i* and due to pressure from the surrounding automata, $P_j = -K_j\,(\Omega_j - \Omega_j^0)/\Omega_j^0 = -K_j D \bar{\varepsilon}_j$ – pressure of the *j*-th neighboring automaton, Ω_j^0 – initial equilibrium volume automaton, Ω_j – current volume of automaton, K_j – bulk modulus compression of the material of this automaton, D - coefficient depending on the dimension of the model (D = 1,2,3), $\bar{\varepsilon}_j = \sum_{j=1}^{z} \varepsilon_{ij}/z$ - average deformation, S_{ij} - contact area of *i*-th automaton with the *j*-th, $\mathbf{n}_{ij} = (\boldsymbol{r}_i - \boldsymbol{r}_j)/r_{ij}$ - normal vector between the *i*-th and *j*-th automata, z - number of nearest neighbors. $\boldsymbol{F}_{ji}^{\mathrm{norm}} = -(2G(\varepsilon_{ij} - \bar{\varepsilon}_i) + \eta\,(\boldsymbol{v}_j - \boldsymbol{v}_i, \mathbf{n}_{ij})/r_{ij})S_{ij}\mathbf{n}_{ij}$ - pair force of the normal interaction between the *i*-th and *j*-th automata which is determined by the superposition of the pair forces of their interaction, G - shear modulus, η - coefficient of viscosity , $\boldsymbol{v}_i$ - speed of the *i*-th Automaton . $\boldsymbol{F}_{ij}^{\mathrm{tang}} = -2G[\boldsymbol{\gamma}_{ij} + \boldsymbol{\gamma}_{ji}, \mathbf{n}_{ij}] - \eta[(\boldsymbol{\omega}_{ij} r^{ij} - \boldsymbol{\omega}_i q_{ij} - \boldsymbol{\omega}_j q_{ji}), \mathbf{n}_{ij}]/r_{ij}$ - the pair force of tangential interaction, $\boldsymbol{\gamma}_{ij} = \sum_{\mathrm{k}} [\mathbf{n}_{ij}{}^{\mathrm{k-1}}, \mathbf{n}_{ij}{}^{\mathrm{k}}] - \boldsymbol{\theta}_i$, $\boldsymbol{\gamma}_{ji} = \sum_{\mathrm{k}} [\mathbf{n}_{ji}{}^{\mathrm{k-1}}, \mathbf{n}_{ji}{}^{\mathrm{k}}] - \boldsymbol{\theta}_j$ - the angle of rotation of each automaton, $\boldsymbol{\omega}_{ij} = [\mathbf{n}_{ij}\,, (\boldsymbol{v}_j - \boldsymbol{v}_i)]/r^{ij}$, $\boldsymbol{\omega}_i$ - angular speed of rotation of pairs and *i*-th automaton, respectively, q_{ij} - distance from the center of *i*-th automaton to automata contact surface *i* and *j*, m_i and J_i- mass and moment of inertia *i*-th automaton, respectively. $\boldsymbol{F}_i^{(m)} = -\sum_{j\neq i,j=1}^{N_m} \{3((\boldsymbol{m}_i, \boldsymbol{r}_{ij})\boldsymbol{m}_j + (\boldsymbol{m}_j, \boldsymbol{r}_{ij})\boldsymbol{m}_i + (\boldsymbol{m}_i, \boldsymbol{m}_j)\boldsymbol{r}_{ij})/r_{ij}^3 - 15(\boldsymbol{m}_i, \boldsymbol{r}_{ij})(\boldsymbol{m}_j, \boldsymbol{r}_{ij})\boldsymbol{r}_{ij}/r_{ij}^5\}$ - magnetic force acting on the automaton *i*, $\boldsymbol{m}_i$ - magnetic moment of *i*-th automaton.

To estimate the critical step dt integration over time, affecting the accuracy of the simulation, we use the expression for the velocity of sound in the medium: $v_{env} = \sqrt{(K + 4G/3)/\rho}$.

$$dt_i = d_i\sqrt{\frac{\rho_i}{(K_i + {}^4/_3 G_i)}}, \qquad dt = 0.25\min(dt_i) \tag{2}$$

where ρ_i - density, K_i - bulk modulus, G_i - shear modul of the medium.

With the transition to the next time step, automata synchronously change state depending on its state and the states of neighboring automata in the previous step. This may be accompanied by a change in connections, orientation automata in space under the influence of external and internal forces, as well as a change in the orientation of the magnetic moments of automata depending on the distribution of the resultant magnetic field in the sample. At each time step the interaction field is recalculated and select automaton with maximum resulting field. In the case where the field value in automaton has reached a critical value and direction of the vectors field and magnetic moment do not coincide, the orientation of the magnetic moment changes according to the minimum energy.

3. The magnetic interaction automata

For the calculation of the deformation of the magnetic medium, we assume that N_m automata ($N_m \leq N$) is uniformly magnetized and interact with each other like a magnetic dipole. Mechanical movement of magnetic automata is defined as short-range and long-range forces, and the orientation of the magnetic moment of automaton depends only on the resulting magnetic field H_i, which is composed of the magnetic fields of the dipoles h_i and automata external field H. To account for the interaction of the magnetic automata at each time step recalculated interaction field. Distribution of interaction fields in the particle system has an effect on its mechanical behavior. However, the real physical picture of the magnetization reversal is very diverse, and writing of algorithms for the calculation of the magnetic moments in the result field is a non-trivial task in this paper is not considered. Therefore allowed the following simplification: the magnetic moments of automata can be oriented only along the positive and negative directions of the axis Oy, i.e. the magnetic moment of automaton is completely determined by a single component: $\boldsymbol{m}^{i(j)} = \{0, m_y^{i(j)}, 0\}$. Then the magnetic field interactions $\boldsymbol{h}^i$ produced in the i-th magnetic particle by all other automata may be expressed by the two respective projection:

$$\begin{cases} h_x^i = \sum_{j=1, j\neq i}^{N_m} \left(\frac{3m_y^j (y^j - y^i)(x^j - x^i)}{r^{ij^5}} \right), \\ h_y^i = \sum_{j=1, j\neq i}^{N_m} \left(-\frac{m_y^j}{r^{ij^3}} + \frac{3m_y^j (y^j - y^i)^2}{r^{ij^5}} \right), \end{cases} \tag{3}$$

here h_x^i, h_y^i - the projection of the interaction field generated in the i-th particle by all other magnetic particles. The whole procedure of the magnetization reversal can be reduced to a simple scheme:

1) calculated the interaction field created by all other particles in the magnetic automata;
2) selected one automaton from the entire set of magnetic particles, with maximum absolute value resulting field H_i (taking into account the external magnetic field H);
3) The condition of flip of the magnetic moment of the selected automaton is checked. If magnetic moment of automaton is flip, the whole loop is executed for all the particles again from the step 1; otherwise repeat the search of the particle with the largest field (step 2), except for the automata, whose moments are not overturned.

The flip of the magnetic moment of automaton is determined by the magnitude and direction of the field on this automation. Simplifications of the flip of magnetic moment for a given condition, can be expressed as follows:

1) the value of the field $\boldsymbol{H}^i$ on automaton i must reach the set value of the critical field h_{crit}^i,
2) directions of the vectors of the magnetic moment m^i and the resulting field $\boldsymbol{H}^i$ on the automation i have to be opposed.

Journal of Physics: Conference Series **633** (2015) 012018 doi:10.1088/1742-6596/633/1/012018

The given conditions can be written as follows:

$$if\ \left(H_y^i \geq h_{crit}^i\ and\ m_y^i H_y^i < 0\right)\ then\ m_y^i = -m_y^i, \tag{4}$$

where *if* - checking the condition, enclosed in parentheses, *and* - logic operation «And», *then* - the beginning of the instructions to be executed if the condition is satisfied.

Simulating the effect of magnetic forces applied to all magnetic automata in system, regardless of the connections between them, because the magnetic forces are long-range. The magnetic force $\boldsymbol{F}_{(m)}^{ij}$, acting on the i-th particle by particle j, is determined by its two components:

$$\begin{cases} f_{ij\,x}^{(m)} = -3\left(\frac{m_y^i m_y^j (x^i - x^j)}{r^{ij^5}} - \frac{5 m_y^i m_y^j (y^i - y^j)^2 (x^i - x^j)}{r^{ij^7}}\right), \\ f_{ij\,y}^{(m)} = -3\left(\frac{m_y^i m_y^j (y^i - y^j)}{r^{ij^5}} - \frac{5 m_y^i m_y^j (y^i - y^j)^3}{r^{ij^7}}\right). \end{cases} \tag{5}$$

Here $f_{ij\,x}^{(m)}, f_{ij\,y}^{(m)}$ – projection magnetic interaction force $\boldsymbol{F}_{ij}^{(m)}$, between *i*-th and *j*-th magnetic automaton, acting by *i*-th particle at *j*-th particle, $x^{i(j)}$ and $y^{i(j)}$ - coordinates of the center of mass *i*-th (*j*-th) automation. Illustration of the results of modeling the deformation of the magnetic medium is shown in figure 2.

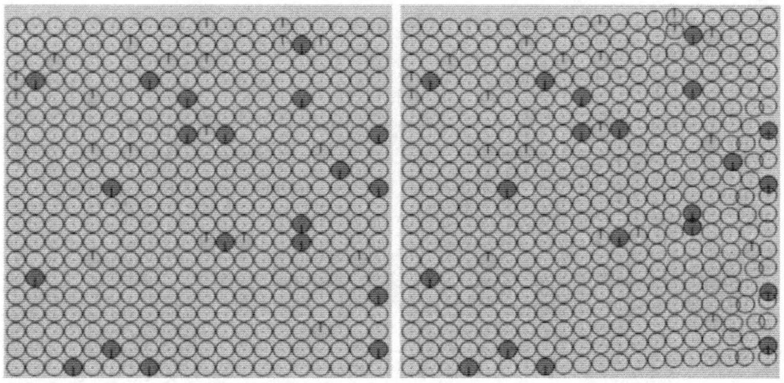

Fig 2. Magnetic properties of automata have an impact on the result of their interaction.

Obviously, the algorithms developed within the framework of the model described above allow to investigate the influence of an external magnetic field on the deformation of the magnetic medium.

4. Acknowledgements

The research work has been financed by Ministry of Education & Science of Russian Federation. Reference number is Government Order № 559.2014.

References

[1] Psakhie S G, Ostermeyer G P, Dmitriev A I, Shilko E V, Smolin A Yu, Korostelev S Yu. 2000 Method of movable cellular automata as a new trend of discrete computational mechanics. I. Theoretical description. *Physical Mesomechanics* vol 3, **2** 5–13.

[2] Psakhie S G, Horie Y, Ostermeyer G P, etc. 2001 Movable cellular automata method for simulating materials with mesostructure. *Theoretical and Applied Fracture Mechanics* **37** (1-3) 311-333.

[3] Psakhie S G, Horie Y, Korostelev S Yu, Smolin A Yu, Dmitriev A I, Shilko E V, and Alekseev S V 1995 Method of movable cellular automata as a tool for simulation within the framework of mesomechanics *Russian Physics Journal* **11** 1157–1168.

[4] Hockney R W, Eastwood J W 1989 Computer Simulation Using Particles. – Adam Hilger, IOP Publishing Ltd - 523.

Canonical quantisation via conditional symmetries of the closed FLRW model coupled to a scalar field

Adamantia Zampeli

Nuclear and Particle Physics Section, Physics Department, University of Athens, GR 15771 Athens, Greece

E-mail: azampeli@phys.uoa.gr

Abstract. We study the classical, quantum and semiclassical solutions of a Robertson-Walker spacetime coupled to a massless scalar field. The Lagrangian of these minisuperspace models is singular and the application of the theory of Noether symmetries is modified to include the conditional symmetries of the corresponding (weakly vanishing) Hamiltonian. These are found to be the simultaneous symmetries of the supermetric and the superpotential. The quantisation is performed adopting the Dirac proposal for constrained systems. The innovation in the approach we use is that the integrals of motion related to the conditional symmetries are promoted to operators together with the Hamiltonian and momentum constraints. These additional conditions imposed on the wave function render the system integrable and it is possible to obtain solutions of the Wheeler-DeWitt equation. Finally, we use the wave function to perform a semiclassical analysis following Bohm and make contact with the classical solution. The analysis starts with a modified Hamilton-Jacobi equation from which the semiclassical momenta are defined. The solutions of the semiclassical equations are then studied and compared to the classical ones in order to understand the nature and behaviour of the classical singularities.

1. Introduction

In this paper, we examine the classical, quantum and semiclassical behaviour of the FLRW spacetime metric coupled to a massless scalar field with the method of conditional symmetries which was first introduced in [1, 2]. The theory of variational symmetries has been extended to include cases in which the lapse function is not gauge fixed. This supplies the superspace with an additional degree of freedom and the theory with a gauge freedom, which leads to the appearance of additional symmetries. At the classical level, the importance of the method consists in solving first-order differential equations instead of the second-order Einstein equations. This method is presented in section 2. At the quantum level, it can be considered as an approach to the problem of time of quantum gravity in which one first quantises and then selects time variable [3]. However, we adopt a more ambitious analysis and exploit the quantum solution to perform a semiclassical analysis in order to gain insight on the behaviour of the classical singularities. This analysis is performed in section 3. Finally, we discuss the results in section 4.

2. Classical analysis for the massless field $\lambda = 0$ in FLRW with non flat $k \neq 0$ spatial metric

The model we study is the FLRW closed universe coupled to a massless scalar field with the following spacetime metric

$$ds^2 = -N^2(t)dt^2 + a^2(t)\Big[\frac{dr^2}{1-kr^2} + r^2\sin\theta^2 + r^2\sin^2\theta d\varphi^2\Big] \tag{1}$$

where $N(t)$ is the lapse function which is not gauge fixed, $a(t)$ is the scale factor and k is the spatial curvature, which in our case takes the value $k = 1$. The total action of the gravitational plus matter system is $S_{tot} = S_{grav} + S_{mat} = \int d^4x\sqrt{-g}[R - \frac{1}{2}g^{\mu\nu}\nabla_\mu\phi\nabla_\nu\phi]$ from which we obtain the total Lagrangian of the system after discarding a term of total derivative is equal to

$$L = 6Nka - \frac{6a\dot{a}^2}{N} + \frac{a^3\dot{\phi}^2}{2N} \tag{2}$$

This Lagrangian has the singular form

$$L = \frac{1}{2N}G_{\alpha\beta}(q)\dot{q}^\alpha\dot{q}^\beta - N\mathcal{V}(q) \tag{3}$$

where $G_{\alpha\beta}(q)$ is the supermetric defined on the configuration space of the dependent variables $q^\alpha(t)$ and $\mathcal{V}(q)$ the superpotential. The presence of the lapse function makes this Lagrangian explicitly invariant under the transformation $t = f(\tilde{t})$ which in turn leads to the existence of additional symmetries on the superspace called conditional symmetries [1]. In the constant superpotential parametrisation of the Lagrangian which can be obtained by the change $N \to n = \frac{N}{V(a,\phi)}$, the conditional symmetries become the usual symmetries of the supermetric plus the scaling symmetry. This leads to the existence of conserved quantities corresponding to the Killing fields of the supermetric and a rheonomic integral, $Q_i = \kappa_i, Q_h = \kappa_h + \int n(t)dt$.

The supermetric of our model in this parametrisation becomes

$$\tilde{G}_{\alpha\beta} = \begin{pmatrix} -72ka & 0 \\ 0 & 6ka^4 \end{pmatrix} \tag{4}$$

and it has three Killing vector fields generating the symmetries of the configuration space (a, ϕ)

$$\xi_1 = (\frac{e^{\phi/\sqrt{3}}}{a}, -\frac{2\sqrt{3}e^{\phi/\sqrt{3}}}{a^2}), \quad \xi_2 = (\frac{e^{-\phi/\sqrt{3}}}{a}, \frac{2\sqrt{3}e^{-\phi/\sqrt{3}}}{a^2}), \quad \xi_3 = (0, 1) \tag{5}$$

plus one homothetic field

$$\xi_h = (\frac{a}{4}, 0) \tag{6}$$

The system of equations of the conserved quantities becomes

$$-\frac{12e^{\frac{\phi}{\sqrt{3}}}ka\left(6\dot{a} + \sqrt{3}a\dot{\phi}\right)}{n} = \kappa_1, \quad \frac{12e^{-\frac{\phi}{\sqrt{3}}}ka\left(-6\dot{a} + \sqrt{3}a\dot{\phi}\right)}{n} = \kappa_2 \tag{7}$$

$$\frac{6ka^4\dot{\phi}}{n} = \kappa_3, \quad -\frac{18ka^3\dot{a}}{n} = \kappa_h + \int n(t)dt \tag{8}$$

The solution of the system cannot be fully specified because it contains a gauge freedom resulting from the freedom of the lapse function. In order to find the full solution, we choose to gauge fix the matter field $\phi = \ln t$. After performing suitable coordinate transformations to absorb the redundant constants, the final form of the spacetime metric is

$$ds^2 = -\frac{\lambda}{4\sqrt{t}(1+t)^3}dt^2 + \frac{\lambda\sqrt{t}}{(1-r^2)(1+t)}dr^2 + \frac{\lambda\sqrt{t}r^2}{1+t}d\theta^2 + \frac{\lambda\sqrt{t}r^2\sin^2\theta}{1+t}d\varphi^2 \tag{9}$$

where we have set $\lambda = -\frac{\kappa_3}{\sqrt{3}k^{3/2}}$. This solution satisfies the Euler-Lagrange equations of the Lagrangian (2) and therefore it is valid. This metric has a singularity for early and late times, that is when $t \to 0$ and $t \to \infty$. We will canonically quantise the system and then perform the semiclassical analysis in order to study whether the singularity persists for the semiclassical metric as well.

3. Canonical quantisation and semiclassical analysis

For the quantisation of our system we follow the scheme of canonical quantisation of constrained systems [4] extended in [1] to include the quantisation of the conserved quantities Q_i according to the relation $\hat{Q}_i \equiv -\frac{i}{2\mu}(\mu\xi_i^\alpha\partial_\alpha + \partial_\alpha\mu\xi_i^\alpha)$, with eigenvalues equal to their classical values κ_i and measure on the Hilbert space of states equal to $\mu(a,\phi) = 6\sqrt{3}a^3k$. The quantum operators $\hat{Q}_i$ cannot be imposed simultaneously because of the condition $c^n_{ij}\kappa_n = 0$. Therefore the compatible algebras are the two-dimensional $(\hat{Q}_1, \hat{Q}_2)$ and the one-dimensional $\hat{Q}_3$. The case of the one-dimensional algebra $\hat{Q}_2$ is not considered since it already belongs to a higher dimensional subalgebra and gives a trivial result. Following the above mentioned steps and solving the quantum equations

$$\hat{Q}_i\Psi(\alpha,\phi) = \kappa_i\Psi(\alpha,\phi), \quad \hat{H}\Psi(\alpha,\phi) = 0 \tag{10}$$

where $i = 1, 2, 3$ we find that the wave functions are respectively

$$\Psi_{12}(a,\phi) = A_{12}e^{i\frac{1}{4}a^2e^{-\frac{\phi}{\sqrt{3}}}(\kappa_1+\kappa_2e^{\frac{2\phi}{\sqrt{3}}})} \tag{11}$$

$$\Psi_3(a,\phi) = e^{i\phi\kappa_3-\frac{\sqrt{3}}{2}\pi\kappa_3}(c_1e^{\sqrt{3}\pi\kappa_3}I_{-i\sqrt{3}\kappa_3}(6a^2\sqrt{k})\Gamma(1-i\sqrt{3}\kappa_3) + c_2I_{i\sqrt{3}\kappa_3}(6a^2\sqrt{k})\Gamma(1+i\sqrt{3}\kappa_3)) \tag{12}$$

In order to make contact with the classical solutions, we perform a semiclassical analysis based on the Bohmian approach [5] first presented in the context of minisuperspace models in [2, 6]. To this end, the wave function is written in the polar form $\Psi(q_i) = \Omega(q_i)e^{iS(q_i)}$ where $\Omega(q_i)$ is the amplitude and $S(q_i)$ is the phase of the wave function, which satisfies a modified Hamilton-Jacobi equation with an additional quantum potential term [5] of the form $U(x) \equiv -\frac{1}{2}\Box\Omega$. When this quantum term vanishes we obtain the classical solution. The Hamilton-Jacobi equation also defines the quantum canonical momenta defined as $p_i = \frac{\partial S}{\partial q_i}$ from which the semiclassical equations are formed by equating them with the classical canonical momenta defined by the Lagrangian, thus having $\frac{\partial S}{\partial q_i} = \frac{\partial L}{\partial \dot{q}_i}$. For the case of the two-dimensional algebra (Q_1, Q_2), the wavefunction is $S_{12} = \frac{1}{4}a^2e^{-\frac{\phi}{\sqrt{3}}}(\kappa_1 + \kappa_2e^{\frac{2\phi}{\sqrt{3}}})$ and the coefficient A_{12} constant, thus leading to the vanishing of the quantum potential. Therefore, it is expected that the solution will not differ from the classical one. This indeed turns out to be the case when we solve the system of semiclassical equations $\frac{\partial L}{\partial \dot{q}_i} = \frac{\partial S_{12}}{\partial q_i}$ and the classical singularity is not resolved.

In the case of the one-dimensional algebra however things turn out to be different. First, the wave function does not have the assumed polar form. In order to write it in this form, two approximation limits are considered, one for small values of the scale factor and one for large. This represents early and late times of the universe repsectively. For early times, the wave function is approximated by (up to constants)

$$\Psi_{3_{sm}} \approx e^{i\kappa_3\phi} \cos\left(\sqrt{3}\kappa_3 \ln\left(3a^2\sqrt{k}\right)\right) \tag{13}$$

while for late ones

$$\Psi_{3_{la}} \approx e^{i\kappa_3\phi} \frac{\cosh\left(6a^2\sqrt{k}\right)}{\sqrt{a^2\sqrt{k}\sqrt{3\pi}}} \tag{14}$$

The quantum potential in both cases is non-zero and equal to $V_{sm} = \frac{\kappa_3^2}{12ka^4}$ for early times and $V_{la} = -1 - \frac{1}{144ka^4}$ for late times. Even though the quantum potential is not the same in the two limits, the phase function is common, $S = \kappa_3\phi$ and the solution will be the same at the two limits. The semiclassical equations $-\frac{72ka^2\dot{a}}{n} = 0$, $\frac{6ka^4\dot{\phi}}{n} = \kappa_3$ have a gauge freedom for the scalar field which we select to be such that the lapse function $N(t)$ of the semiclassical element is the same as the classical one, that is

$$\phi(t) = \alpha - \frac{8 \times 3^{3/4} t^{\sqrt{3}/2} \sqrt{-\frac{48kt^{2/\sqrt{3}}}{\kappa_1} - \frac{\kappa_1}{3}} \kappa_1 \kappa_3^{3/2} \left(-3 + \sqrt{1 + \frac{144kt^{2/\sqrt{3}}}{\kappa_1^2}}\, {}_2F_1\left(\frac{1}{2}, \frac{3}{4}; \frac{7}{4}; -\frac{144kt^{\frac{2}{\sqrt{3}}}}{\kappa_1^2}\right)\right)}{c^3(144kt^{2/\sqrt{3}}\kappa_1 + \kappa_1^3)} \tag{15}$$

The semiclassical line element will then be

$$ds^2 = -\frac{\lambda}{4\sqrt{t}(1+t)^3}dt^2 + \frac{1}{1 - (-\frac{4}{\lambda})^{2/3} r^2} dr^2 + r^2 d\theta^2 + r^2 \sin^2\theta d\varphi^2 \tag{16}$$

This spacetime metric does not contain any singularities since invariants do not diverge for $t \to 0$ and $t \to \infty$. Therefore, for this case the singularities of the classical metric vanish.

4. Conclusions

The FLRW cosmological model filled with a massless scalar field was studied at a classical and quantum level, exploiting the presence of additional symmetries. The innovation in this method is that the lapse function is not gauge fixed. Under the presence of the conditional symmetries the classical solution can be obtained by a first-order system, which is much easier to be integrated. This method also has advantages on the quantisation, since the integrals of motion are imposed as conditions on the wave function and the system of constraints is integrable. The final new element in our viewpoint first introduced in [6] is the semiclassical analysis performed in the context of a minisuperspace model following Bohm. The result was that, for the one-dimensional subalgebra the spacetime is not the same as the classical one and the singularities dissapear for early and late times, exactly at the limits where the classical singularities existed. This is encouraging for obtaining interesting results for other more complicated cosmological models such as the Bianchi cosmologies.

References

[1] Christodoulakis T, Dimakis N, Terzis P A, Doulis G, Grammenos T *et al.* 2013 *J.Geom.Phys.* **71** 127–138 (*Preprint* `1208.0462`)

[2] Christodoulakis T, Dimakis N, Terzis P A and Doulis G 2014 (*Preprint* `1405.0363`)

[3] Isham C 1992 (*Preprint* `gr-qc/9210011`)

[4] Dirac P A M 1964 *Lectures on Quantum Mechanics* Belfer Graduate School Sci. Mono. (New York, NY: Belfer Graduate School of Science)

[5] Bohm D 1952 *Phys.Rev.* **85** 166–179

[6] Christodoulakis T, Dimakis N, Terzis P A, Vakili B, Melas E *et al.* 2014 *Phys.Rev.* **D89** 044031 (*Preprint* `1309.6106`)

Journal of Physics: Conference Series **633** (2015) 012020 doi:10.1088/1742-6596/633/1/012020

Wave function of the Universe, preferred reference frame effects and metric signature transition

HOSSEIN GHAFFARNEJAD

Faculty of Physics, Semnan University, Semnan, IRAN, Zip Code: 35131-19111

E-mail: hghafarnejad@yahoo.com

Abstract. Gravitational model of non-minimally coupled Brans Dicke (BD) scalar field ϕ with dynamical unit time-like four vector field is used to study flat Robertson Walker (RW) cosmology in the presence of variable cosmological parameter $V(\phi) = \Lambda\phi$. Aim of the paper is to seek cosmological models which exhibit metric signature transition. The problem is studied in both classical and quantum cosmological approach with large values of BD parameter $\omega >> 1$. Scale factor of RW metric is obtained as $R(t) = 6\sqrt{\frac{3}{\Lambda}}\cosh\left(\frac{t}{4}\sqrt{\frac{\Lambda}{3}}\right)$ which describes nonsingular inflationary universe in Lorentzian signature sector. Euclidean signature sector of our solution describes a re-collapsing universe and is obtained from analytic continuation of the Lorentzian sector by exchanging $t \to it$ as $R(t) = 6\sqrt{\frac{3}{\Lambda}}\cos\left(\frac{t}{4}\sqrt{\frac{\Lambda}{3}}\right)$. Dynamical vector field together with the BD scalar field are treated as fluid with time dependent barotropic index. They have regular (dark) matter dominance in the Euclidean (Lorentzian) sector. We solved Wheeler De Witt (WD) quantum wave equation of the cosmological system. Assuming a discrete non-zero ADM mass $M_j = 4\sqrt{2}(2j+1)\sqrt{\frac{\Lambda}{3}}$ with $j = 0, 1, 2, \cdots$, we obtained solutions of the WD equation as simple harmonic quantum Oscillator eigen functionals described by Hermite polynomials. Absolute values of these eigen functionals have nonzero values on the hypersurface $R = 6\sqrt{\frac{3}{\Lambda}}$ in which metric field has signature degeneracy. Our eigen functionals describe nonzero probability of the space time with Lorentzian (Euclidean) signature for $R > 6\sqrt{\frac{3}{\Lambda}}$ ($R < 6\sqrt{\frac{3}{\Lambda}}$). Maximal probability corresponds to the ground state $j = 0$.

1. Introduction

Lorentz invariance is a fundamental requirement of the standard model of particle physics, verified to high precision by many tests [1], whereas, string theory predicts that we may live in a universe with non-commutative coordinates [2] which leads to a violation of Lorentz invariance [3]. Furthermore, astrophysical observations suggest the presence of high energy cosmic rays above the Greisen-Zatsepin-Kuzmin cutoff [4] which may be explained by a breaking down of Lorentz invariance. More generally, Lorentz invariance violation seems to be related to unknown physical effects of space time at the Planck scale ($l_p = (16\pi G)^{-1/2}$ with $c = \hbar = 1$). A straightforward method of implementing local Lorentz violation in a gravitational setting is to introduce a tensor field with a non-vanishing expectation value, and then couple this tensor to gravity or matter fields. The simplest proposal of this approach is to consider a single time-like vector field N_μ with a fixed norm [5]. Physical interpretation of such a vector field denotes to gravitational effects of preferred reference frames. Regarding general covariance principle, N_μ should be taken as a dynamical field. There have been also elsewhere, cosmological studies in the presence of vector fields without the fixed norm. Lorentz invariance violation directly causes the change of the metric signature of space time. The motivation for the metric signature transition

is usually caused by the extension of real functions to corresponding functions with complex variables via trivial Wick rotations in a complex plane. According to the work presented by Barbero et al [5], the authors attempt was previously to generalize the BD gravity action [6,7] by transforming its background metric as $g_{\mu\nu} \rightarrow g_{\mu\nu} + 2N_\mu N_\nu$ where N_μ is assumed to be a dynamical unit time like vector field. Further, dust, radiation and inflationary cosmological models were obtained.
In this lecture we first present our scalar-vector-tensor gravity [6,7] in the presence of variable cosmological parameter. Then we seek flat RW cosmological models which exhibit metric signature transition from Lorentzian (-,+,+,+) to Euclidean (+,+,+,+) sectors in both classical and quantum approaches. Last section contains the concluding remark.

2. Brans-Dicke scalar-vector-tensor Robertson Walker cosmology

Let us take the scalar-vector-tensor-gravity action $I_{total} = I_{BD} + I_\Lambda + I_N$ [6,7] where the first term is well known BD scalar tensor gravity [8] as $I_{BD} = \frac{1}{16\pi}\int dx^4\sqrt{g}\left\{\phi R - \frac{\omega}{\phi}g^{\mu\nu}\nabla_\mu\phi\nabla_\nu\phi\right\}$, the second term denotes the action of self interacting BD field (called as 'variable cosmological parameter') such that $I_\Lambda = \frac{\Lambda}{16\pi}\int dx^4\sqrt{g}\phi^n$, in which Λ, n are real constant parameters and the last term given by $I_N = \frac{1}{16\pi}\int dx^4\sqrt{g}\{\zeta(x^\nu)(g^{\mu\nu}N_\mu N_\nu + 1) + 2\phi F_{\mu\nu}F^{\mu\nu} - \phi N_\mu N^\nu(2F^{\mu\lambda}\Omega_{\nu\lambda} + F^{\mu\lambda}F_{\nu\lambda} + \Omega^{\mu\lambda}\Omega_{\nu\lambda} - 2R^\nu_\mu + \frac{2\omega}{\phi^2}\nabla_\mu\phi\nabla^\nu\phi)\}$ with $F_{\mu\nu} = 2(\nabla_\mu N_\nu - \nabla_\nu N_\mu)$, and $\Omega_{\mu\nu} = 2(\nabla_\mu N_\nu + \nabla_\nu N_\mu)$, describes action of unit time like dynamical four vector field $N_\mu(x^\nu)$. Non-minimally vector field N_μ is called usually as four velocity of preferred reference frame. The action I_{total} is written in units $c = \hbar = 1$ with Lorentzian signature (-,+,+,+). The undetermined Lagrange multiplier $\zeta(x^\nu)$ controls that N_μ to be an unit time-like vector field. Self-interacting coupling constant Λ has dimension as $(length)^{2n-4}$. Absolute value of determinant of the metric $g_{\mu\nu}$ is defined by g. BD field ϕ describes inverse of variable Newton's gravitational coupling parameter and its dimension is $(length)^{-2}$ in units $c = \hbar = 1$. Present limits of ω based on time-delay experiments [9,10,11] require $\omega \geq 4 \times 10^4$ and general relativistic approach I_{BD} is obtained with $\omega \rightarrow \infty$ [12,13,14]. As an example we set RW line element which with Lorentzian signature (-,+,+,+) is given from point of view of free falling comoving observer as $ds^2 = -dt^2 + R^2(t)\{dx^2 + dy^2 + dz^2\}$ where $R(t)$ is spatially part scale factor of space time. In this case one can obtain explicitly time dependent fields $\zeta(t)$, $\phi(t)$ and $R(t)$ such as follows: $\phi(t) = \phi_0\exp[8/(2\omega - 5)(\Omega t)]$ and $R(t) = R_0\cosh(\Omega t)$ (details of the mathematical calculations are given in ref. [15]), where $\phi_0 = \phi(0), R_0 = R(0)$ and we defined $\Omega = \frac{3(2\omega-5)}{4\sqrt{(6\omega-3-2\sqrt{6})(6\omega-3+2\sqrt{6})}}\sqrt{\frac{\Lambda}{3}}$. The parameter Ω has real value under the conditions $\frac{1}{2} - \frac{\sqrt{6}}{3} < \omega < \frac{1}{2} + \frac{\sqrt{6}}{3}$ with $\Lambda < 0$ or $\frac{1}{2} - \frac{\sqrt{6}}{3} > \omega > \frac{1}{2} + \frac{\sqrt{6}}{3}$ with $\Lambda > 0$. Our obtained scale factor describes non-singular inflationary expanding flat universe. With the above obtained solutions the undetermined Lagrange multiplier $\zeta(t)$, fluid density $\rho(t)$ and corresponding pressure $p(t)$ become respectively $\zeta(t) = \left\{\frac{(6\omega-31-4\sqrt{46})(6\omega-31+4\sqrt{46})}{(2\omega-5)(6\omega-3-2\sqrt{6})(6\omega-3+2\sqrt{6})}\right\}\Lambda\phi_0\exp(\Omega t)$, $\rho(t) = 3\Omega^2\tanh^2\Omega t$, and $p(t) = -\Omega^2\{2 + \tanh^2\Omega t\}$ where we set initial condition as $\rho(0) = 0$. One can obtain equation of state as $\eta(t) = \frac{p(t)}{\rho(t)} = -\frac{1}{3} - \frac{2}{3}\coth^2\Omega t$ which describes dark matter dominant [16] of the fluid from point of view of free falling comoving frame. It is applicable to choose a preferred reference frame with its corresponding time coordinate T related to the cosmological comoving time t as $T(t) = (\Omega t)^{\frac{2}{3}}$. From the point of view of the latter preferred reference frame the RW line element can be rewritten as $ds^2 = R_0^2 d\bar{s}^2$ in which $d\bar{s}^2 = -TdT^2 + \cosh^2(T^{3/2})\{dx^2 + dy^2 + dz^2\}$ and we set $R_0 = \frac{3}{2\Omega}$. We see that hypersurfaces $t = constant$ with $T > 0$ correspond to Lorentzian signature (-,+,+,+) of the space time metric. Its analytic continuation is obtained with hypersurfaces $it = constant$ which correspond to Euclidean signature (+,+,+,+) with negative times $T < 0$. Hence metric components have degeneracy on the signature transition

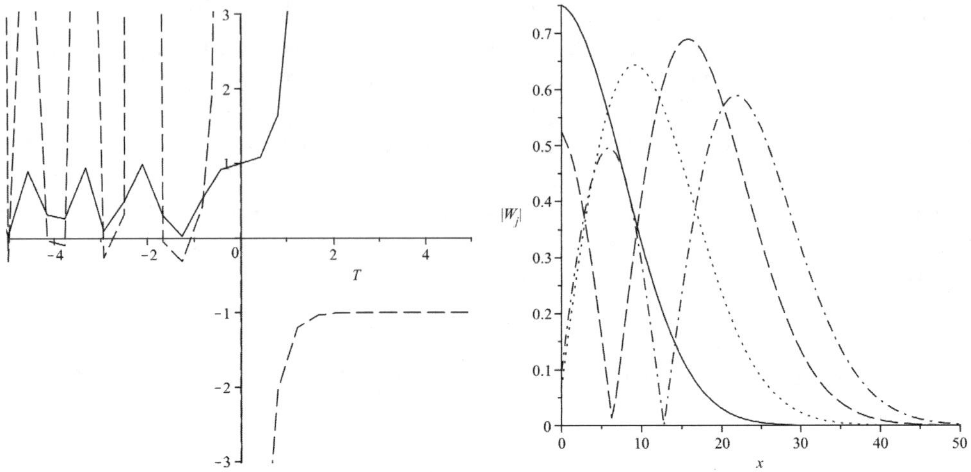

Figure 1. Solid line in left diagram describes Lorentzian (Euclidean) sector scale factor of RW space time for times $T > 0 (T < 0)$. Dash line in left diagram describes barotropic index $\eta(T)$ of Lorentzian (Euclidean) sector for times $T > 0 (T < 0)$. Right diagram describes variation of absolute value of the WD eigen functionals $|W_j(x)|$ against scale factor x for $j = 0, 1, 2, 3$ with solid, dot, dash and dash-dot lines respectively.

hypersurface $T = 0$ (see LHS in the figure 1). Diagram of the barotropic index $\eta(T)$ is plotted against T in LHS of the figure 1 with dash line. It shows that the scalar-vector matter source treats as regular (dark) matter in the Euclidean (Lorentzian) sector of the space time. In the quantum level we obtain below that the Lorenzian sector of space times may be tunneled quantum mechanically to the corresponding Euclidean sector and vice versa. Using minisuperspace approach of canonical quantum cosmology one can obtain corresponding Wheeler DeWitt probability wave equation of the system as $\{\frac{(6\omega^2-81\omega-27)}{4(2\omega+3)^2}\frac{\delta^2}{\delta R^2} + \frac{3(7\omega+12)}{(2\omega+3)^2}\frac{\phi}{R}\frac{\delta}{\delta R}\frac{\delta}{\delta\phi} - \frac{(2\omega+21)}{2(2\omega+3)^2}\frac{\phi^2}{R^2}\frac{\delta^2}{\delta\phi^2} - \phi R(M + \Lambda\phi R^3)\}W(R, \phi) = 0$ (details of mathematical calculations are given in ref. [15]). In general relativistic approach $\omega \to +\infty$ our classical solution leads to $\lim_{\omega\to+\infty}\phi(t) = \phi_0 = constant$ and $\lim_{\omega\to+\infty} R(t) = R_0 \cosh\left(\frac{t}{4}\sqrt{\frac{\Lambda}{3}}\right)$ where $R_0 = \frac{3}{2\Omega} = 6\sqrt{\frac{3}{\Lambda}}$ and $\phi_0 = \frac{1}{16\pi G}$. The above WD wave equation has not a simple solution with small ω, but we can obtain some physical interpretation from solution of its general relativistic approach $\omega \to +\infty$. In the latter approach the WD wave equation becomes $W''_\infty + U(x)W_\infty(x) \cong 0$ where we defined $U(x) = -x(a + bx^3)$, $a = \frac{108M}{\pi G\Lambda}\sqrt{\frac{3}{\Lambda}}$, $b = \frac{1944}{\pi^2 G^2\Lambda^2}$, $x = \frac{R}{R_0}$ and over-prime $'$ denotes differentiation with respect to the minisuperspace variable x. Subscript of the WD wave solution W_∞ denotes to $\omega \to +\infty$. We point that R_0 is scale factor of space time where the metric has signature degeneracy. Solving $U'(x) = 0$ we obtain minimum point of the potential $U(x)$ as $x_m = -\frac{1}{2}\left(\frac{\pi GM}{3}\sqrt{\frac{\Lambda}{3}}\right)^{1/3}$ for which corresponding Taylor series expansion becomes $U(x) \cong \frac{81}{2}\left(\frac{M}{\Lambda\sqrt{\pi G}}\right)^{\frac{4}{3}} - \frac{1944}{\Lambda}\left(\frac{M}{\pi^2 G^2\Lambda}\right)^{\frac{2}{3}}(x-x_m)^2+\cdots$. Defining $x - x_m = \sigma y$ where $\sigma = \left(\frac{\Lambda}{1944}\right)^{\frac{1}{4}}\left(\frac{\pi^2 G^2\Lambda}{M}\right)^{\frac{1}{6}}$ the WD wave equation becomes $W''(y) + (\varepsilon - y^2)W(y) = 0$ where we eliminated subscript ∞ and defined $\varepsilon = \frac{M}{4\sqrt{2}}\sqrt{\frac{3}{\Lambda}}$. This equation describes simple harmonic quantum Oscillator and its solutions are described in terms of Hermite polynomials $H_j(y) = (-1)^j e^{y^2}\frac{d^j}{dy^j}e^{-y^2}$ as

$W_j(y) = \left(\frac{1}{2^j j!\sqrt{\pi}}\right)^{\frac{1}{2}} \exp(-\frac{y^2}{2})H_j(y)$ with eigen values $\varepsilon_j = (2j+1)$, $j = 0, 1, 2, \cdots$. In terms x our obtained WD eigen functions become $W_j(x) = \left(\frac{1}{2^j j!\sqrt{\pi}}\right)^{\frac{1}{2}} \exp\{-\frac{\sigma_j^2(x-x_m^j)^2}{2}\}H_j(\sigma_j(x-x_m^j))$ in which $x_m^j = -\left(\frac{\pi\sqrt{2}(2j+1)\Lambda G}{18}\right)^{\frac{1}{3}}$, $\sigma_j = \left(\frac{1}{1944}\right)^{\frac{1}{4}}\left(\frac{\pi^2\Lambda^2 G^2}{4(2j+1)}\sqrt{\frac{3}{2}}\right)^{\frac{1}{6}}$ and discrete ADM mass becomes $M_j = 4\sqrt{2}(2j+1)\sqrt{\frac{\Lambda}{3}}$. Their superposition leads to general wave solution of quantum universe of our model as $W(x) = \sum_{j=0}^{\infty} P_j W_j(x)$ where P_j is probability amplitude of the universe which should stay on the eigen state j with ADM eigen mass (energy) M_j. However our knowledge about the universe is not enough and so we can not determine the coefficients P_j, because LHS in the above summation '$W(x)$' is unknown. But we can obtain some physical statements about the quantum tunneling of metric signature via eigen states $W_j(x)$: Our classical solution $x(T) = \frac{R}{R_0} = \cosh(T^{3/2})$ predicts a metric signature transition on the hypersurface $T = 0$ corresponding to $x = 1$. Lorentzian sector of space time corresponds to $x > 1$ but its Euclidean sector corresponds to $0 < x < 1$ (see LHS of the figure 1). The metric equation has signature degeneracy on the hypersurface $x = 1$ corresponding to the time $T = 0$ in the classical regime. $W_j(0 < x < 1)$ ($W_j(x > 1)$) describe the probability of Euclidean (Lorentzian) signature of the space time. Nonzero values $W_j(x = 1) \neq 0$ describe the probability of metric signature quantum tunneling which takes maximal value when the universe exhibits its ground state $j = 0$ (see RHS of the figure 1 where we set $\Lambda G\pi = 1$).

3. Summary and concluding remarks

Flat RW space time is studied by using the BD scalar-vector tensor gravity and variable cosmological parameter. A nonsingular inflationary cosmological model is obtained which exhibits metric signature transition from Lorentzian to Euclidean topology. Unit time like dynamical vector field and BD scalar field are treated as regular (dark) fluid in Euclidean (Lorentzian) sector of flat RW space time with time dependent barotropic index. WD wave equation of the cosmological system is solved by assuming nonzero ADM mass. Its solutions are obtained as simple harmonic quantum Oscillator eigen functionals. Maximal probability of signature quantum tunneling is obtained when the quantum Oscillating RW universe lives in its ground state. As future work the author intends to investigate the dynamical effects of preferred reference frames on black holes and anisotropic-inhomogeneous Bianchi cosmological metrics .

References

[1] Kostelecky A 2001 *Preprint* hep-ph/0104227
[2] Connes A, Douglas M R, and Schwarz A 1998 *JHEP.* **02** 003 *Preprint* hep-th/9711162
[3] Carroll S M, Harvey J A, Kostelecky A, Lane C D and Okamoto T, 2001 *Phys. Rev. Lett.* **87** 141601 *Preprint* hep-th/0105082
[4] Zatsepin G T, Kuzmin V A and Prima Zh Eksp 1966 *Teor.Fiz* **4** 114
[5] J F Barbero G and Villaseñor E J S2003 *Phys. Rev* D **68** 087501 *Preprint* gr-qc/0307066
[6] Ghaffarnejad H 2008 *Gen. Relativ. Gravit.* **40** 2229
[7] Ghaffarnejad H 2009 *Gen. Relativ. Gravit.* **41** 2941 (E)
[8] Brans C and Dicke R 1961 *Phys. Rev.* **124** 925
[9] Will C M 1993 *Theory and Experiment in Gravitational Physics* (Cambridge University Press) *Preprint* gr-qc/9811036.
[10] Gaztanaga E and Lobo J A 2001 *Astrophys. J.* **548** 47
[11] Reasenberg et al 1961 *Astrophys. J.* **234** 925
[12] Will C M 2006 *Living Rev. Rel.9* http://WWW.livingreviews.org/lrr-2006-3
[13] Cho Y M 1997 *Class. Quantum. Grav.* **10** 2963
[14] Garay L J and Bellido L G 1993 *Nucl.Phys.* B **400** 416-434 *Preprint* gr-qc/9209015 (1992).
[15] H. Ghaffarnejad 2013 *Preprint* gr-qc/1306.1827v5
[16] Overduin J M and Wesson P S 2008 *The Light Dark Universe* (World Scientific, Publisher)

4th International Conference on Mathematical Modeling in Physical Sciences (IC-MSquare2015) IOP Publishing
Journal of Physics: Conference Series **633** (2015) 012021 doi:10.1088/1742-6596/633/1/012021

TensorPack: a Maple-based software package for the manipulation of algebraic expressions of tensors in general relativity

P A Huf [a*] and J Carminati [a]

[a]School of Information Technology, Deakin University, Waurn Ponds, Victoria, AUSTRALIA 3216 *Corresponding author: Tel 61 429380524, PO Box 6091, Highton, Victoria, AUSTRALIA 3216

E-mail: peterhuf@deakin.edu.au, peter@bach2roq.com (Peter Huf), john.carminati@deakin.edu.au (John Carminati)

Abstract. In this paper we: (1) introduce TensorPack, a software package for the algebraic manipulation of tensors in covariant index format in Maple; (2) briefly demonstrate the use of the package with an orthonormal tensor proof of the shearfree conjecture for dust. TensorPack is based on the Riemann and Canon tensor software packages and uses their functions to express tensors in an indexed covariant format. TensorPack uses a string representation as input and provides functions for output in index form. It extends the functionality to basic algebra of tensors, substitution, covariant differentiation, contraction, raising/lowering indices, symmetry functions and other accessory functions. The output can be merged with text in the Maple environment to create a full working document with embedded dynamic functionality. The package offers potential for manipulation of indexed algebraic tensor expressions in a flexible software environment.

1. Introduction

Since the 1970s there has been a steady increase in the use of algebraic computing for manipulating tensor equations, expressions and components. A summary some common packages is shown at [1]. In general many of these systems compute components and tensors related to general relativity (GR); fewer packages provide the facility to develop fluent algebraic arguments. In this context we introduce TensorPack, an indexed string-based system that generates tensor expressions in covariant formalism consistent with the major literature and texts [2-5].

The main features of the new TensorPack package include:

- an indexed-string based format for tensorial algebra
- presentation of tensors in standard 1+3 covariant format
- functions for many common algebraic operations on tensors
- mapping algebraic expressions to tensor components
- an interface with the functionality of the Riemann package (in addition to compatibility with other Maple packages eg. physics)

- flexibility to be used as both a research and an instructional tool
- a structure that can be used in text or computational styles, or as a combination

We discuss the main features in this paper (section 2) and demonstrate its use in the proof of the theorem σ =0 =>ω Θ =0 for dust, based on [6] (section 3). Full downloads of the TensorPack package, and proofs of the examples of the shearfree theorem can be found at [1]. The Riemann and Canon packages are described in [7,8].

2. Summary of the package

2.1. General

The current set of functions available in TensorPack are shown using the 'with' command:
> *with* (*TensorPack*);

[*Absorbd*, *Absorbg*, *CDF*, *CDS*, *PrintArray*, *PrintSubArray*, *T*, *TEDS*, *TELS*, *TPset*, *anglesymm*, *antisymm*, *cod*, *contract*, *dotT*, *fpcodiff*, *raise*, *symm*]

2.2. The programming format of TensorPack - the T() function

The essence of the package is that tensor expressions are stored as strings using an indexed structure. The T() processes the string for visual and computational output. For example, the input (>) line below followed by T(%) results in:
>u[a, -B] = (1/3)*theta*P[a, -b]+omega[a, -b]+sigma[a, -b]:T(%);

$$u^{a}{}_{;b} = \frac{1}{3}\,\theta\,P^{a}{}_{b} + \omega^{a}{}_{b} + \sigma^{a}{}_{b}$$

The above example indicates that mixed tensors are represented by the combination of positive (contravariant) and negative (covariant) indices as required. In addition, a covariant derivative is formed by the use of an uppercase index. Furthermore, in the above and following examples, it will be seen that the T() function obeys the basic linear algebra of tensors.

2.3. Index operations: raising, lowering, contraction

Changing the type of indices simply involves the Maple substitution function. The sign of the index to be raised or lowered is reversed. The following example displays raising the index b, then contracting b with a:
>expr:=X[-a, -b] = Y[-a,-b] + W[-a]*Z[-b]: T(%); expr2 := subs(b = -b, expr):T(%);expr3 := contract(expr2, a, b);T(%)

$$X_{a\,b} = Y_{a\,b} + W_{a}\,Z_{b}$$
$$X_{a}{}^{b} = Y_{a}{}^{b} + W_{a}\,Z^{b}$$
$$X_{a}{}^{a} = Y_{a}{}^{a} + W_{a}\,Z^{a}$$

2.4. Substitution of a tensor equation into a tensor expression

A tensor expression can be substituted as a string into the string representing another tensor, using the TEDS() function. For example, replacing the contracted tensor with a scalar in the previous equation:
>TEDS(X[-a, a] = X, expr3); T(%);

$$X = Y_{a}{}^{a} + W_{a}\,Z^{a}$$

2.5. Covariant and time differentiation

The format of the covariant derivative of a tensor was introduced in section 2.2. An equation can be differentiated (in this case with respect to b, according to the Liebniz rule) using the cod() function, as follows:

>expr:=X[-a, -b] = Y[-a,-b] + W[-a]*Z[-b]: cod(expr, -b); T(%);

$$X^{a\ b}{}_{;b} = Y^{a\ b}{}_{;b} + W^{a}{}_{;b} Z^{b} + W^{a} Z^{b}{}_{;b}$$

Time differentiation is of significance in GR. The time derivative of a tensor A[a,b] is expressed in TensorPack as dotA[a,b]. It is obtained by the function dotT(). For an equation (using the Liebniz rule) we have the resulting time derivative of the previous expression:
>dotT(expr): T(%);

$$dotX^{ab} = dotY^{ab} + dotW^{a} Z^{b} + W^{a} dotZ^{b}$$

2.6. *Symmetry functions:symm(),antisymm(),anglesymm()*

Symmetric and antisymmteric tensors can be fully expressed for tensors with 4 indices or less. For example, the symmetric expression is found using symm():
>symm(X[a, b, c], a, c); T(%);

$$(1/6)X^{abc} + (1/6)X^{bca} + (1/6)X^{cab} + (1/6)X^{acb} + (1/6)X^{cba} + (1/6)X^{bac}$$

2.7. *Absorbing the metric and kronecker delta*

Tensor terms containing the metric or kronecker delta can usually be simplified:
>expr := delta[-a, b]*X[-b]+g[-b, -a]*X[b];T(%);

$$\delta_{a}{}^{b} X_{b} + g_{b\ a} X^{b}$$

>Absorbg(Absorbd(expr)); T(%);

$$2\ X_{a}$$

2.8. *General functions & interfaces*

It is possible to interface TensorPack with general functions in many ways:

a. The output of a list of expressions. eg. outputting a sublist of equations, uses the PrintSubArray command:

b. The general maple file and interface functions can be applied, including save, read, printing, copying and pasting code etc...

c. Displaying components of tensor expressions, usually after setting the dimension, defining coordinates and the metric

For a complete description of these functions, see [1].

3. Demonstration of a theorem proof using TensorPack: shearfree conjecture for dust

In this section we present a brief example of the use of the package by discussing a sample proof of the shearfree conjecture σ =0 =>ω Θ =0 for dust as given by Senovilla et al. [6]. Our version of the proof follows the same sequential order of proofs as for the original authors. The complete set of proofs using TensorPack is available at [1] but is too long to show in complete form in this document. Instead we demonstrate the approach, and the use of the software, in the following short example.

Example. Proof of equation 11b of [6] (Vorticity vector and tensor are orthogonal)
We proceed to give a proof of the identity $\omega_{a\ b}\,\omega^{b} = 0$
Consider the left side of the identity:
>eq1 := omega[-a, -b]*omega[b]; T(%); (1)

$$\omega_{a\ b}\,\omega^{b}$$

We commence by substituting:

Journal of Physics: Conference Series **633** (2015) 012021 doi:10.1088/1742-6596/633/1/012021

>eq2 := omega[a] = (1/2)*eta[a, b, c, d]*u[-b]*omega[-c, -d]:T(%); (2)

$$\omega^{a} = \frac{1}{2}\eta^{abcd} u_{b} \omega_{cd}$$

an identity for the vorticity vector into expression (1). We first re-arrange the indices in (2) to facilitate the substitution:

>eq3 := subs(b = e, c = f, d = g, a = b, eq2);T(%); (3)

$$\omega^{b} = \frac{1}{2}\eta^{befg} u_{e} \omega_{fg}$$

Completing the substitution, using the TEDS command in TensorPack, we find the expression:

>eq4 := expand(TEDS(eq3, eq1)):T(%); (4)

$$\frac{1}{2}\omega_{ab}\eta^{befg} u_{e} \omega_{fg}$$

In a similar way we substitute the following expression for the vorticity tensor into expression (1) :

>eq5 := omega[-a, -b] = eta[-a, -b, -c, -d]*u[d]*omega[c]:T(%); (5)

$$\omega_{ab} = \eta_{abcd} u^{d} \omega^{c}$$

to obtain:

>eq6 := expand(TEDS(eq5, eq4)):T(%); (6)

$$\frac{1}{2}\eta^{befg} u_{e} \omega_{fg} \eta_{abcd} u^{d} \omega^{c}$$

The skew pseudotensor is fully antisymmetric, so the expression (6) is equivalent to:

>eq7 := expand(TEDS(eta[-a, -b, -c, -d] = -eta[-b, -a, -c, -d], eq6)): T(%); (7)

$$-\frac{1}{2}\eta^{befg} u_{e} \omega_{fg} u^{d} \omega^{c} \eta_{bacd}$$

and hence we have, equating expressions (1) and (7), we obtain equation (8):

>eq8 := eq1 = eq7:T(%); (8)

$$\omega_{ab}\omega^{b} = -\frac{1}{2}\eta^{befg} u_{e} \omega_{fg} u^{d} \omega^{c} \eta_{bacd}$$

Now we substitute an identity for the skew pseudotensor into (8) which then yields:

>eq9 := expand(TEDS(eta[b, e, f, g]*eta[-b, -a, -c, -d] = -6*antisymm(delta[e, -a]*delta[f, -c]*delta[g, -d], e, g), eq8)):T(%); (9)

$$\omega_{ab}\omega^{b} = -\frac{1}{2} u_{e}\omega_{fg} u^{d}\omega^{c}\delta^{f}{}_{a}\delta^{e}{}_{c}\delta^{g}{}_{d} + \frac{1}{2} u_{e}\omega_{fg} u^{d}\omega^{c}\delta^{e}{}_{a}\delta^{f}{}_{c}\delta^{g}{}_{d} - \frac{1}{2} u_{e}\omega_{fg} u^{d}\omega^{c}\delta^{e}{}_{a}\delta^{g}{}_{c}\delta^{f}{}_{d} + \frac{1}{2} u_{e}\omega_{fg} u^{d}\omega^{c}\delta^{f}{}_{a}\delta^{g}{}_{c}\delta^{e}{}_{d} + \frac{1}{2} u_{e}\omega_{fg} u^{d}\omega^{c}\delta^{g}{}_{a}\delta^{e}{}_{c}\delta^{f}{}_{d} - \frac{1}{2} u_{e}\omega_{fg} u^{d}\omega^{c}\delta^{g}{}_{a}\delta^{f}{}_{c}\delta^{e}{}_{d}$$

Next we use the Absorbd function of TensorPack to simplify terms with kronecker delta:

>eq10 := Absorbd(Absorbd(Absorbd(eq9))): T(%); (10)

$$\omega_{ab}\omega^{b} = -\frac{1}{2} u_{c}\omega_{ad} u^{d}\omega^{c} + \frac{1}{2} u_{a}\omega_{cd} u^{d}\omega^{c} - \frac{1}{2} u_{a}\omega_{dc} u^{d}\omega^{c} + \frac{1}{2} u_{d}\omega_{ac} u^{d}\omega^{c} + \frac{1}{2} u_{c}\omega_{da} u^{d}\omega^{c} - \frac{1}{2} u_{d}\omega_{ca} u^{d}\omega^{c}$$

Using $u^{d} u_{d} = -1$ we obtain:

>eq11 := expand(TEDS(u[d]*u[-d] = -1, eq10)): T(%); (11)

$$\omega_{ab}\omega^{b} = \frac{1}{2} u_{a}\omega_{cd} u^{d}\omega^{c} - \frac{1}{2} u_{a}\omega_{dc} u^{d}\omega^{c} + \frac{1}{2} u_{c}\omega_{da} u^{d}\omega^{c} - \frac{1}{2} u_{c}\omega_{ad} u^{d}\omega^{c} - \frac{1}{2}\omega_{ac}\omega^{c} + \frac{1}{2}\omega_{ca}\omega^{c}$$

and the repeated use of the orthogonality of the velocity and vorticity (in three sets involving the TEDS command) yields:
>eq12a := expand(TEDS(u[d]*omega[-a, -d] = 0, eq11)); -1; eq12b := expand(TEDS(u[d]*omega[-c, -d] = 0, eq12a)); -1; eq12c := expand(TEDS(u[d]*omega[-d, -a] = 0, eq12b)); -1; eq12d := expand(TEDS(u[d]*omega[-d, -c] = 0, eq12c)): T(%); (12)

$$\omega_{ab}\,\omega^{b} = -\frac{1}{2}\,\omega_{ac}\,\omega^{c} + \frac{1}{2}\,\omega_{ca}\,\omega^{c}$$

The asymmetry of the vorticity tensor gives
>eq13 := expand(TEDS(omega[-c, -a] = -omega[-a, -c], eq12d)):T(%); (13)

$$\omega_{ab}\,\omega^{b} = -\omega_{ac}\,\omega^{c}$$

Finally, switching dummy indices yields:
>eq14 := subs(c = b, eq13):T(%); (14)

$$\omega_{ab}\,\omega^{b} = -\omega_{ab}\,\omega^{b}$$

and
>eq15 := eq14-omega[-a, -b]*omega[b]: T(%); (15)

$$0 = -2\,\omega_{ab}\,\omega^{b}$$

which completes the proof of the example (equation 11b of [6]).

Acknowledgments

We are grateful to J. Senovilla and C. Sopuerta for providing details of some of the more difficult sections of their proofs [6]. We are also grateful to the authors of the Riemann and Canon packages.

References

[1] Huf P A and Carminati J 2014 TensorPack *Preprint* http://www.deakin.edu.au/˜peterhuf http://www.bach2roq.com/science/maths/GR/TensorPack.html
[2] Ellis G F R, Maartens R and MacCallum M A H 2012 *Relativistic Cosmology* (Cambridge University Press, Cambridge, UK)
[3] Ehlers J 1993 Contributions to the Relativistic Mechanics of Continuous Media, *Gen.Rel.Grav.* **25** (12) 1225-66
[4] Misner C W, Thorne K S and Wheeler J A 1973 *Gravitation* (Freeman, San Francisco)
[5] Aris R 1960 *Vectors, Tensors and the Basic Equations of Fluid Mechanics* (Dover, New York)
[6] Senovilla J M M, Sopuerta C F and Szekeres P 1998 Theorems on shear-free perfect fluids with their Newtonian analogues, *Gen.Rel.Grav.* **30** 389-411
[7] Portugal R, Sautú S L 1997 Applications of Maple to General Relativity, *Comput. Phys. Commun.* **105** 233-53
[8] Martn-García J M, Portugal R and Manssur L R U 2007 The Invar Tensor Package, *Comput. Phys. Commun.* **177** 640-648

Numerical Analysis of Thirring Model under White Noise

Fatma Aydogmus
Istanbul University, Department of Physics, Istanbul, TURKEY
E-mail: fatma.aydogmus@gmail.com, fatmaa@istanbul.edu.tr

Eren Tosyali
Istanbul Bilgi University, School of Advanced Vocational Studies, Istanbul, TURKEY
E-mail: eren.tosyali@bilgi.edu.tr

Abstract. Today, the effects of noise on dynamical systems are an attractive area of research. The noise acts as a driving term in the equations of motion in nonlinear systems. In this work, we present conformally invariant pure spinor nonlinear Thirring model. Thirring model describes Dirac fermions in (1+1) space-time dimensions with local current-current interaction. This model has rich dynamic of the quantization of relativistic quantum field theories. We investigate the response of Thirring oscillator to white noise by constructing phase space displays.

1. Introduction

It is well known that the Korteweg-de Vries (KdV) equation derived by Korteweg and de Vries in 1895 which describes weakly nonlinear shallow water waves [1]. Zabusky and Kruskal found that the stable pulse-like waves could exist in a system described by the KdV equation. These solitary waves could collide with each other and preserve their shapes and speeds after the collision [2]. They were named as solitons by Zabusky and Kruskal. There are four leading solitonic characters: instanton, monopole, vortex, and kink ones. Instantons are localized, classical topological solutions in quantum field theories. They have finite action with zero energy so they have been considered as configurations of quantum fields that provide a tunnelling effect between the vacuums [2, 3]. Thirring model which describes Dirac fermions in (1+1) space-time dimensions with the non-linear $\left(\bar{\psi}\psi\right)^2$ self-interaction term [4].

Thirring model is also conformally invariant and with no mass term. It played a very fruitful role in the progress of quantum field theory [4]. The spinor-type instanton solutions of Thirring model were found via breaking of conformal symmetry and were shown that these solutions were stable [5].

In this paper, we present the conformal invariant pure spinor Thirring model, which admit instanton-type solutions. We focus the effects of white noise in the long time behaviors of excited 2D Thirring instantons. For this purpose, we consider the general dynamics of Thirring nonlinear differential equations system formed by the use of Heisenberg ansatz with the addition of a white noise term.

2. The Model

The wave equation of model as below

$$i\sigma_\mu \partial_\mu \psi + g\left(\bar{\psi}\psi\right)\psi = 0 \tag{1}$$

follows from the conformal invariant Lagrangian

$$\mathcal{L} = i\bar{\psi}\partial\psi + \frac{g}{2}\left(\bar{\psi}\psi\right)^2 \tag{2}$$

Where g is the positive coupling constant and the fermion field $\psi(x)$ has scale dimension $\frac{1}{2}$ [4]. The complex form of the Euclidian configuration of Heisenberg ansatz [6] is

$$\psi = [ix_\mu \gamma_\mu \chi(s) + \varphi(s)]c \tag{3}$$

with an arbitrary spinor constant, c. $\chi(s)$ and $\varphi(s)$ are real functions of $s = x^2 + t^2\,(x_1 \equiv x, x_2 \equiv t)$. Inserting ansatz into wave equation of model we get

$$\chi(s) + s\frac{d\chi(s)}{ds} + \alpha[s\chi(s)^2 + \varphi(s)^2]\varphi(s) = 0 \tag{4a}$$

$$\frac{d\varphi(s)}{ds} - \alpha[s\chi(s)^2 + \varphi(s)^2]\chi(s) = 0 \tag{4b}$$

where $\alpha = g\left(\bar{C}C\right)$. By writing $\chi = As^{-\sigma}F(u)$ and $\varphi = Bs^{-\tau}G(u)$ with $u \equiv \ln s$ and $\sigma = \tau + \frac{1}{2}$, $\tau = \frac{1}{4}$ and $A^2 = B^2$ [7], we achieved the dimensionless form of the non-linear ordinary coupled differential equation system (4) as

$$2\frac{dF(u)}{du} + \frac{1}{2}F(u) - \alpha AB\left(F(u)^2 + G(u)^2\right)G(u) = 0 \tag{5a}$$

$$2\frac{dG(u)}{du} - \frac{1}{2}G(u) + \alpha AB\left(F(u)^2 + G(u)^2\right)F(u) = 0 \tag{5b}$$

Here F and G are dimensionless functions of u and A, B are constants [4, 7, 5]. In Ref. [5], the two types of classical solution were found in the case of the two-dimensional conformal invariant Lagrangian exploiting the same symmetry arguments as for gauge and scalar fields. Either of them leads to the vanishing of energy-momentum tensor and finite action for $\alpha(AB) = 1$. It is called instanton solution (Thirring Instantons) [4, 5]. Recently, the role of the coupling constant in the evolution of Thirring instantons in phase space has been investigated [8]. Also common behaviours of spinor type 2D Thirring instantons and 4D Gursey instantons have been investigated by constructing their vector field [9].

In this paper we focus on effects of white noise on the dynamics of 2D excited Thirring instantons. For this purpose we perturbe this fermionic model by an external force under white noise as

$$2\frac{dF(u)}{u} + \frac{1}{2}F(u) - \beta\left(F(u)^2 + G(u)^2\right)G(u) = 0 \tag{6a}$$

$$2\frac{dG(u)}{u} - \frac{1}{2}G(u) + \beta\left(F(u)^2 + G(u)^2\right)F(u) - ACos(\omega H(u)) + \eta(u) = 0 \tag{6b}$$

$$\frac{H(u)}{du} = \Omega \tag{6c}$$

with a constant Ω by adding an extra dimension for numerical calculations. Here A is the amplitude of the external forcing, ω is its frequency and $\eta(u)$ is white noise term.

3. Simulation Results

Today, noise interference is a long-term problem for researchers in detection technology. In this paper we investigate the response of 2D excited Thirring instantons to white noise, is a random signal with a constant power spectral density, to provide a better understanding of the spinor type instanton dynamics in phase space. Firstly, we show the phase space dynamics of Thirring instantons for the unperturbed case without and with the addition of noise respectively in Figures 1 (a) and (b). They exhibit stability behaviours in phase space as it is realized in Figure 1.

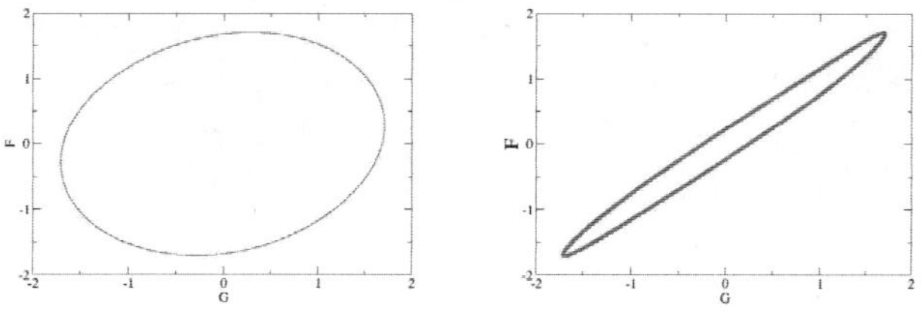

(a) (b)

Figure 1. Phase space displays of Thirring instantons with the initial conditions $F(0) = G(0) = 1.29904$ and $\beta = 1$ (a) without noise (b) with noise.

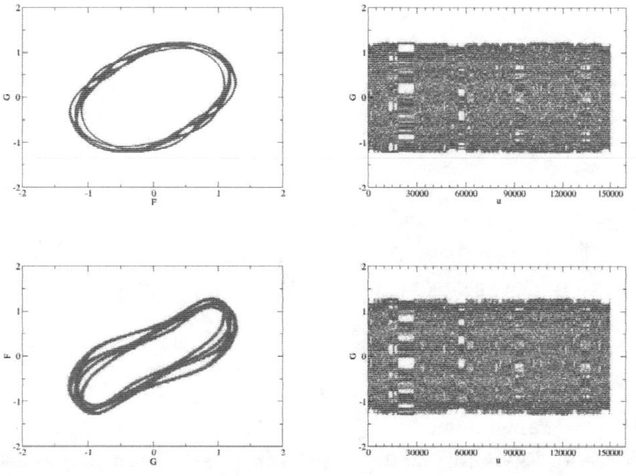

Figure 2. Phase space displays and time evolutions of the excited Thirring instantons without (first row) and with noise (second row) at $A = 0.6$.

In Figure 2-3, phase space displays of excited Thirring instantons with the initial conditions $F(0) = 1.1$,$G(0) = 0.1$, for $\beta = 1$, $\omega = \pi$ are seen. It is known that the effect of the external force on the spinor-type instantons sensitively depend on the initial conditions [10]. From the obtained time evolution and phase space displays, we can say that the system exhibits similar behaviours to forced Duffing oscillator [11] type characterization without noise in phase space. Also the system show small changes in phase space for not large values of noise (noise ratio r=0.015). It is interesting to note that 2D excited Thirring instantons remains forced Duffing-type characterization under white noise. This state can be interpreted as excited Thirring instantons have noise immunity in statistical sense.

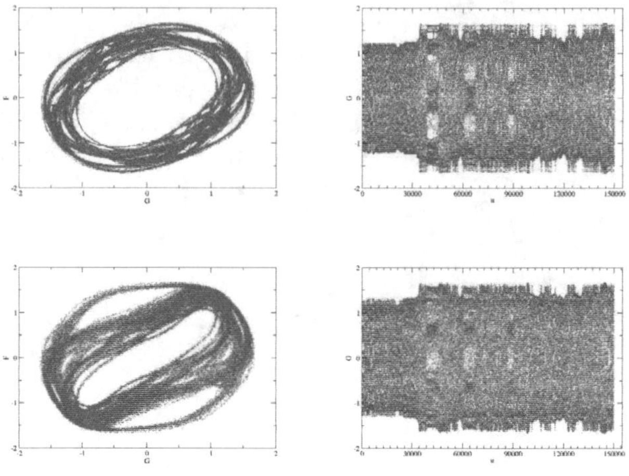

Figure 3. Phase space displays and time evolutions of the excited Thirring instantons without (first row) and with noise (second row) at $A = 0.7$.

Acknowledgement
This work was supported by the Scientific Research Projects Coordination Unit of Istanbul University; Project no: 35737.

References

[1] Korteweg, D.J., de Vries, G.: On the Change of Form of Long Waves Advancing in a Rectangular Canal, and on a New Type of Long Stationary Wave s; *Phil. Mag.*, **39**, pp. 422443, 1895.
[2] R. Rajaman, Solitons and Instantons, Volume 15: An Introduction to Solitons and Instantons in Quantum Field Theory, Elsevier, 1992
[3] A.A. Belavin, A.M. Polyakof, A. S. Schwartz and Yu.S. Tyupkin, Pseudo-Particle Solutions of the Yang-Mills Equations, *Phys. Lett.* **B59**, 85-87, 1975.
[4] W. E. Thirring, A Soluble Relativistic Field Theory, *Anal Physics*, **3**, 91-112, 1958.
[5] Akdeniz K.G. and Smailagic A., Classical Solitons for Fermionic Models, *Il Nuovo Cimento*, **51A**, 345-357, 1979.
[6] W. Heisenberg, "Zur Quantentheorie nichtrenormierbarer Wellengleichungen", *Zs. Naturforsch.*, **9a**, 292, 1954.
[7] Kortel F. ,On some solutions of Gurseys Conformal-invariant Spinor Wave Equation, *Il Nuovo Cim.* **4**, 210, 1956.
[8] B. Canbaz, C. Onem, F. Aydogmus and K. G. Akdeniz, From Heisenberg Ansatz to Attractor of Thirring Instanton, *Chaos Solitons & Fractals*, **45**, 188-191, 2012.
[9] F. Aydogmus and E. Tosyali, Common Behaviours of Spinor Type Instantons in 2-D Thirring and 4-D Gursey Fermionic Models, *Advances in High Energy Physics*, 148375, 2014.
[10] F. Aydogmus, Dynamics of Excited Instantons in the Forced Gursey Nonlinear Differential Equations System, *Journal of Experimental and Theoretical Physics*, **147**, 240-247, 2015.
[11] G. Duffing, Erzwungene Schwingungen bei Vernderlicher Eigenfrequenz., F. Vieweg u. Sohn, Braunschweig, 1918.

Nonergodicity in binary alloys.

Leonid Son[1] , Valery Sidorov[2] , Pjotr Popel[2] , Dmitry Shulgin[1]

[1] Ural Federal University, 19 Mira st., 620002, Ekaterinburg, Russia
[2] Ural State Pedagogical University, 18 Cosmonavtov ave., 620219, Ekaterinburg, Russia

E-mail: ldson@yandex.ru

Abstract. For binary liquids with limited miscibility of the components, we provide the corrections to the equation of state which arise from the nonergogic diffusivity. It is shown that these corrections result in lowering of critical miscibility point. In some cases, it may result in a bifurcation of miscibility curve: the mixtures near 50% concentration which are homogeneous at the microscopic level, occur to be too stable to provide a quasi - eutectic triple point. These features provide a new look on the phase diagrams of some binary systems. In present work, we discuss Ga-Pb, Fe-Cu, and Cu-Zr alloys. Our investigation corresponds their complex behavior in liquid state to the shapes of their phase diagrams.

1. Introduction

Traditional approach to the statistics and dynamics of binary mixture consisting of atoms of A and B type is based on a coarsened lattice gas model. Namely, it is supposed that possible atomic positions correspond to the sites of spatial lattice. Possible configurations of the mixture are described in terms of filling numbers $n(r)$, where r is the coordinate of the site, and $n(r) = 1$ if the site is occupied by the atom of A type, and $n(r) = 0$ for the atom of B type. Then, one introduces the effective Hamiltonian for certain spatial configuration of variables $\{n(r)\}$:

$$H\{n\} = -\frac{1}{2}\sum_{r,r'} n(r)J(r-r')n(r') - \sum_{r} \mu(r)n(r), \tag{1}$$

Hamiltonian (1) serves to provide Hibbs statistics (but, in contrast with "real" Hamiltonian, does not determine the dynamics!). Then, the statistics and thermodynamics are described by partition Z and thermodynamic potential G:

$$Z = \sum_{\{n\}} \exp\left[-\frac{H\{n\}}{T}\right], \quad G = -T\ln Z \tag{2}$$

To calculate the partition, one uses the mean field (MF) approximation, which replaces full statistics with statistics of single site variable $n(r)$, embedded into surrounding "field" of neighboring sites. For these sites, one suggests mean value of filling number $< n(r) >= x(r)$. For this single site, the partition includes only two terms

$$Z_{MF} = \sum_{n=0,1} \exp\left[n\frac{J}{T}x + n\frac{\mu}{T}\right] = 1 + \exp\left[\frac{J}{T}x + \frac{\mu}{T}\right], \tag{3}$$

where

$$J = \sum_{r'} J(r - r'). \tag{4}$$

Using (3), one easily calculates the mean value $< n(r) >$. Obvious self consistency $x =< n(r) >$ gives the well known mean - field equation on the mean concentration x:

$$\frac{T}{J}x = \frac{\exp(x+\nu)}{1+\exp(x+\nu)}. \tag{5}$$

The dimensionless chemical potential ν serves as a Lagrange multiplier to fix out the concentration. This equation at $T > J/4$ has only one solution for all ν, but at $T \le J/4$ in the vicinity of $x = 0.5$ it splits up in two solutions, and this split results in miscibility gap. Thus, the concentration - temperature phase diagram looks as a cupola dividing two - phase and single - phase domains. This is statistical base of regular solutions theory.

2. Nonergodic behavior of binary mixtures.

Hibbs statistics is based on the ergodic hypothesis, which implies that states with equal energies have equal probabilities. At the same time, binary system has obvious dynamical limitations, arising from the principle of local equilibrium: atom cannot disappear at one spatial point and appear at another at once, so starting from some certain configuation, system can not reach some configurations per observation time, in spite of the fact that those have big Hibbs probability. In [1, 2] we accounted these limitations. It was shown, that below $T = J/4$, equation (5) should be rewritten as

$$\frac{T}{J}x = \frac{e^\beta}{1+e^\beta} + C\,\frac{e^\beta(1-e^\beta)}{(1+e^\beta)^3}, \quad \beta = x + \nu \tag{6}$$

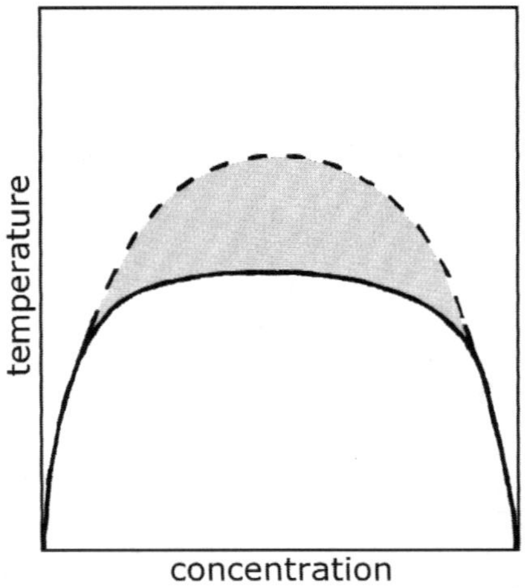

Figure 1. Phase diagram according to (5) is drawn by dashed line. Solid line corresponds to (6)

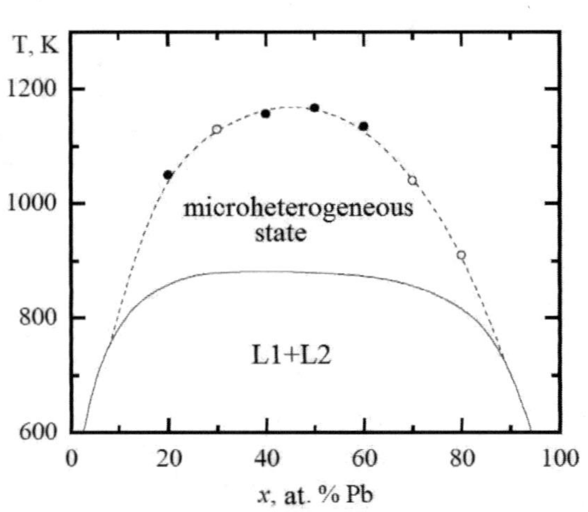

Figure 2. Miscibility cupola in Ga-Pb system and microheterogeneous area [3], obtained by ultrasonic technique. Dark and open circles correspond to the samples obtained by different smelting regimes

In [1, 2] equation (6) was obtained by strict accounting of dynamical limitations. It may be derived via mathematically not rigorous but simple considerations. Below $T = J/4$, the tendency to the phase separation occurs, and system becomes microheterogeneous: some local volumes become enriched with component A, while others - with component B. This microheterogeneity may be described by random spatial fluctuations of chemical potential:

$$\nu(r) = \nu + \Delta\nu(r). \tag{7}$$

Thus, one gets a set of equations (5) in different spatial points with different $\nu(r)$. Suggesting smallness of $\Delta\nu$, one can expand (5) with respect to this value up to second order terms. Then, one averages the obtained equation over spatial ensemble. While averaging, the first order term vanishes due to the conservation of the mean concentration. Denoting spatial mean value $<\Delta\nu^2(r)>= 2C$, one gets equation (6). In mean field approximation, constant C should be considered as a phenomenological parameter, which characterizes microheterogeneity. At small C, the phase diagram based on (6), does not change drastically, only lowering of the critical point occurs - see fig.1. The lowering of the critical temperature (difference between tops of cupolas in fig.1) is $\Delta T = JC/8$. Shadowed area corresponds to the microheterogeneous state. Note, that this state is nonergodic: ergodic behavior leads to the phase separation according to the dashed curve.

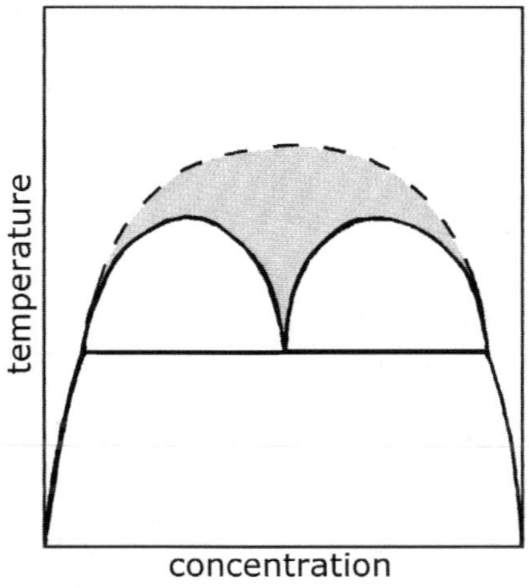

Figure 3. Bifurcation of the miscibility cupola at $C > 1/2$. Nonergodic microheterogeneous area is shadowed.

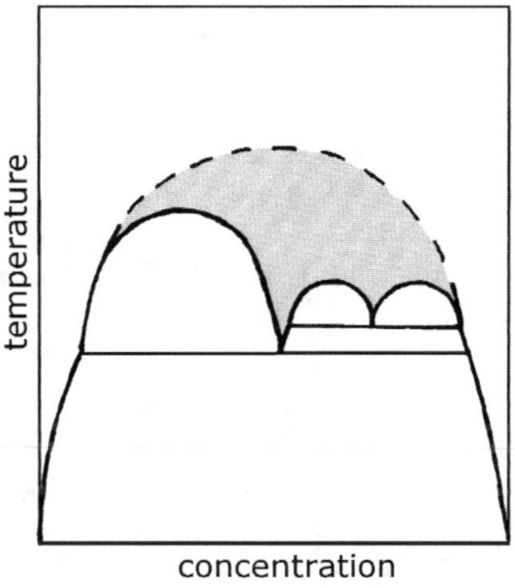

Figure 4. The small cupola corresponds to the quasi – binary system, and thus has a possibility to further splitting.

If $C > 1/2$, the shape of the phase diagram changes drastically. It splits into two cupolas, and triple point corresponding to the horizontal line in fig.3 occurs. Temperature of the tops of new cupolas T_{max} and triple point T_3 may be obtained from the analysis of (6):

$$T_3 = J\frac{E_1 - E_2}{\ln E_1 - \ln E_2} \tag{8}$$

$$T_{max} = J\left(\frac{E_{1,2}}{(1+E_{1,2})^2} + C\frac{E_{1,2} - 4E_{1,2}^2 + E_{1,2}^3}{(1+E_{1,2})^4}\right), \tag{9}$$

where

$$E_{1,2} = \frac{5C-1}{C+1} \pm \sqrt{\left(\frac{5C-1}{C+1}\right)^2 - 1}. \tag{10}$$

Here, plus corresponds to the lower index 1, mines – to the index 2. In (9), substitution of any of two values (10) leads to the one and the same result. Above T_{max}, equation (6) has only one solution while changing ν, between T_{max} and T_3 – three stable solutions, separated by two unstable, and below T_3 – two stable, separated by one unstable. Note, that each of the two cupolas in fig.3, corresponds to the quasi – binary situation: on some intermediate space scale, the separated phases behave like two pure components, so all considerations leading to the bifurcation of the miscibility cupola may be repeated again. Thus, each of cupolas has the possibility to further splitting (fig.4)

3. Experimental status of nonergodic microheterogeneity

As an obvious experimental justification of the presented approach, we can refer to the Ga-Pb system [3], fig.2. Its behavior near the miscibility cupola in liquid state justify the theory very well. Bifurcation of the cupola and triple point are well extended also, but following considerations should be realized. In binary mixtures with limited miscibility, crystal phases may correspond to the pure components, or to several very special compositions, which have very narrow concentration intervals of existence. In fig.5, possible conjugations of diagrams in figs.1,3,4 with melting points of the pure components are presented. Fragment a) shows conjugation with the high – temperature cupola, and fragment b) – two possible variants of conjugation with low - temperature two – phase domains. Intermediate phases may be conjugated with the miscibility diagrams in the vicinity of the top of the cupola (fig.6 a), or in the vicinity of the triple point (fig.6 b).

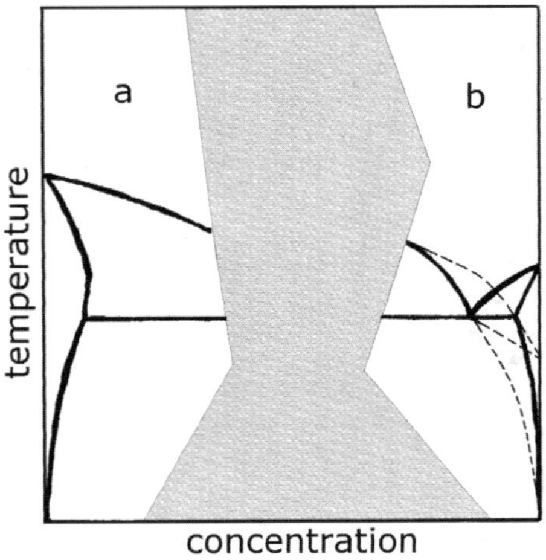

Figure 5. Possible conjugations of the miscibility curve with the melting points of pure components.

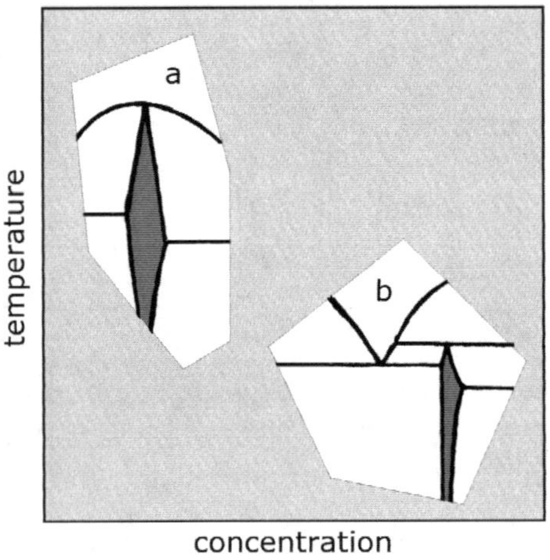

Figure 6. Possible conjugations of the miscibility curve with intermediate phase domains.

Thus, figs.1,3–6 provide possible fragments, from which phase diagrams of binary systems with limited miscibility of the components may be constructed. This provides some new look at the mentioned diagrams. In particular, our approach implies that triple point arises due to the inherent slowing down of diffusion dynamics in the liquid, and the crystallization of the products of phase separation is a supplying phenomena. Thus, slightly above eutectic points, nonergodic microheterogeneity must exist. The first example is Cu-Fe system (fig.7). Its phase diagram is constructed from elements presented in figs.1,5. The central wide two – phase domain arises from the miscibility in liquid state. This may be justified by the data on magnetic susceptibility

[5], which suggest liquid - liquid phase separation in undercooled alloy (open circles correspond to cooling, while solid – to heating). For this system, long – living microheterogeneity above the *liquidus* line was established by viscosity measurments [6]

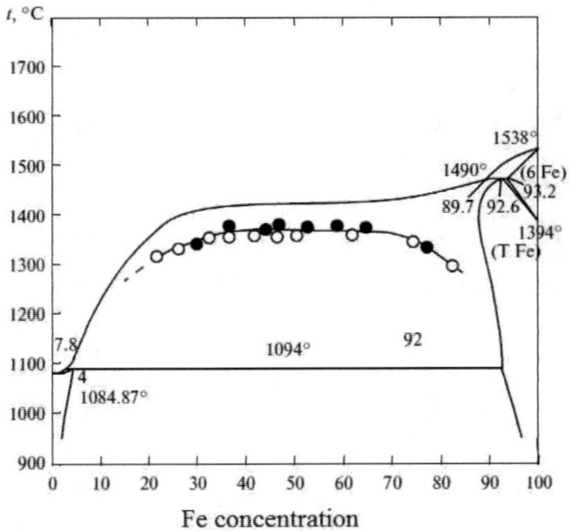

Figure 7. Cu-Fe phase diagram [4]. Undercooled cupola from magnetic susceptibility measurments is taken from [5]

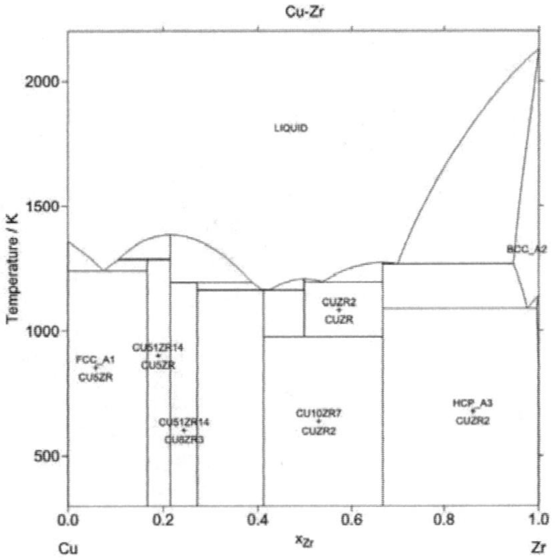

Figure 8. Cu-Zr phase diagram taken from [7]

Cu-Zr phase diagram is presented in fig.8. It may be suggested, that *Liquidus* line corresponds to four cupola splitting, afterward it was conjugated with existing crystal phases by patterns of figs.5,6. Note, that this system is strongly nonergodic, so that it can form a bulk metallic glass [8]. According to our approach, maximum of dynamic slowing down corresponds to 50% concentration, and this really the alloy with maximal glass forming ability.

Acknowledgments
The work was supported by RFBR (projects 13-03-96055, 13-03-00598, 14-02-00359) and Russian Ministry of Science and Education (Government job 2014/392 projects 2391 and 1177) One of us (L. Son) is grateful to RSCF (project 14-12-01185)

References
[1] Son L 2014 *Journal of Non - Crystalline Solids* **401** 213
[2] Son L D 2014 *JETP Letters* **99** 689
[3] Filippov V and Popel P 2007 *J. Non-Cryst. Solids* **235** 3269
[4] Turchanin M A, Agraval P G, Nikolaenko I V 2003 *J. Phase Equilibria* **24** 307
[5] Chikova O A, Vitjunin M A, Chentsov V P, Sakun G V 2010 *Colloid Journal* **72** 2 1
[6] Guschin V S, Shulgin D B, Baum B A, Buhler T P 1989 *Metallofizika* **11** 6 33
[7] http://resource.npl.co.uk/mtdata/phdiagrams/png/cuzr.png
[8] Inoue A, Takeuchi A 2011 *Acta Materialia* **59** 2243-2267

Journal of Physics: Conference Series **633** (2015) 012024 doi:10.1088/1742-6596/633/1/012024

Stability of de Sitter solution in mimetic $f(R)$ gravity

Nurgissa Myrzakulov

Eurasian International Center for Theoretical Physics, Eurasian National University, Astana, 010008, Kazakhstan

E-mail: nmyrzakulov@gmail.com

Abstract. The mimetic $f(R)$ gravity is revisited from the stability of de Sitter point of view. We show that how de Sitter evolves stable in the cosmological era. Our investigation can be used as a potentially interesting idea for early inflation as well as late time acceleration expansion.

1. Introduction

Recently, observational dates such as supernovae Ia (SN Ia)[1], large-scale structure (LSS)[2], baryon acoustic oscillations (BAO) [3] and cosmic microwave background (CMB) [4] indicated accelerating expansion of our universe. Observational data shows that the universe has undergone two phases of cosmic acceleration. One of phase which is believed to have occurred prior to the radiation domination is called *inflation*. This phase can explain a flat spectrum of temperature anisotropies which is observed in CMB. The another phase of accelerating is unknown phenomena *dark energy* which has started after the matter domination.

The generalization of general relativity is considered as alternative for a unified description of the inflation and dark energy. The properties and structure of modified gravities such as $f(R)$ gravity, $f(T)$ gravity, scalar-tensor theory, Gauss - Bonnet gravity, non-local gravity, string inspired, non-minimally coupled models, dilaton gravity are considered. The cosmological reconstruction of these modified theories is investigated in great detail.

One of simplest approach in scalar field models of dark energy and inflation correspond to a modification of the energy-momentum tensor. Second approach to explain the acceleration of the universe is modification of gravitational part. Generally, to modify gravitational part, adds higher-order corrections to the Einstein-Hilbert action. One of simplest model that is $f(R)$ gravity. Its perform by replace Ricci scalar R by an arbitrary function $f(R)$ in the action [5]. The modified $f(R)$ gravity studied also in Ref. [6, 7, 8, 9, 10, 11, 12, 13].

Recently has been proposed a new type modified gravities as titled mimetic model [14]. Modification of mimetic model was investigated in [15, 16, 17]. The idea of model proposed scalar theory of gravitation, conformally-invariant in which ghosts does not cause any problem with scalar degree of freedom. The approach is to parametrize metric of an auxiliary metric as a conformal transformation. Unification $f(R)$ gravity with mimetic gravity as mimetic $f(R)$ gravity proposed in [17]. This new modified mimetic theory interesting, because of its has more physical solutions and complexity. The purpose of our work is study stability of de Sitter solution in mimetic $f(R)$ gravity.

The structure of paper is organized as following. In Sect. II we briefly review modified theory of gravity such as $f(R)$ gravity. In Sect. III we analyzed equations of motion for mimetic $f(R)$ gravity in FRW spacetime. Also investigated de-Sitter solution and stability.

2. Preliminaries of $f(R)$ gravity

We start with the action in $f(R)$ gravity

$$S = \frac{1}{2\kappa^2}\int d^4x\sqrt{-g}\, f(R) + \mathcal{L}_m\,, \tag{1}$$

where $\kappa^2 = 8\pi G$, g is the determinant of the metric $g_{\mu\nu}$ and $\mathcal{L}_m$ is a matter Lagrangian that depends on $g_{\mu\nu}$ and matter fields ϕ_m. The $f(R)$ is an arbitrary function f of Ricci scalar R. The Ricci scalar R is defined by

$$R = g^{\mu\nu}R_{\mu\nu} \tag{2}$$

where the Ricci tensor $R_{\mu\nu}$ is

$$R_{\mu\nu} = R^{\alpha}{}_{\mu\alpha\nu} = \partial_\lambda\Gamma^{\lambda}_{\mu\nu} - \partial_\mu\Gamma^{\lambda}_{\lambda\nu} + \Gamma^{\lambda}_{\mu\nu}\Gamma^{\rho}_{\rho\lambda} - \Gamma^{\lambda}_{\nu\rho}\Gamma^{\rho}_{\mu\lambda}\,. \tag{3}$$

In the metric formalism, the connections $\Gamma^{\alpha}_{\beta\gamma}$ are the usual metric connections defined by metric tensor $g_{\mu\nu}$ as

$$\Gamma^{\alpha}_{\beta\gamma} = \frac{1}{2}g^{\alpha\lambda}\left(\frac{\partial g_{\gamma\lambda}}{\partial x^\beta} + \frac{\partial g_{\lambda\beta}}{\partial x^\gamma} - \frac{\partial g_{\beta\gamma}}{\partial x^\lambda}\right)\,. \tag{4}$$

This can be obtained from the metricity relation

$$\nabla_\lambda g_{\mu\nu} = \partial g_{\mu\nu}/\partial x^\lambda - g_{\rho\nu}\Gamma^{\rho}_{\mu\lambda} - g_{\mu\rho}\Gamma^{\rho}_{\nu\lambda} = 0. \tag{5}$$

The field equation can be obtained by varying the action (1) with respect to $g_{\mu\nu}$:

$$f(R)R_{\mu\nu}(g) - \frac{1}{2}f(R)g_{\mu\nu} - \nabla_\mu\nabla_\nu f(R) + g_{\mu\nu}f(R) = \kappa^2 T_{\mu\nu} \tag{6}$$

where $f(R) \equiv \frac{\partial f}{\partial R}$.
Energy-momentum tensor $T^{(m)}_{\mu\nu}$ of the matter fields is the defined by

$$T^{(m)}_{\mu\nu} = -\frac{2}{\sqrt{-g}}\frac{\delta\mathcal{L}_M}{\delta g^{\mu\nu}}\,. \tag{7}$$

Continuity equation satisfied

$$\nabla^\mu T^{(m)}_{\mu\nu} = 0\,, \tag{8}$$

From equation (6) the trace gives

$$3\Box f(R) + f(R)R - 2f(R) = \kappa^2 T\,, \tag{9}$$

where

$$T = g^{\mu\nu}T^{(m)}_{\mu\nu} \tag{10}$$

and

$$\Box f = \left(\frac{1}{\sqrt{-g}}\right)\partial_\mu\left(\sqrt{-g}g^{\mu\nu}\partial_\nu f\right) \tag{11}$$

We now assume the Friedmann-Lemetre-Robertson-Walker (FLRW) space-time

$$ds^2 = -dt^2 + a(t)^2\sum_{i=1,2,3} dx^{i2}\,, \tag{12}$$

where t is cosmic time.
For FLRW metric the Ricci scalar R is given by

$$R = 6\frac{\ddot{a}}{a} + 6\frac{\dot{a}^2}{a^2} \tag{13}$$

Then, we can rewritten as

$$\frac{\ddot{a}}{a} = \frac{1}{6}(R - 6(\frac{\dot{a}}{a})^2) \tag{14}$$

where $H \equiv \frac{\dot{a}(t)}{a(t)}$ is the Hubble parameter.
The energy-momentum tensor of matter is given by

$$T^{\mu(m)}_{\ \nu} = \text{diag}\,(-\rho_m, p_m, p_m, p_m) \tag{15}$$

where ρ_m is the energy density and P_m is the pressure. The field equations (6) in the flat FLRW background give

$$3fH^2 = \frac{fR - f}{2} - 3H\dot{f} + \kappa^2\rho_m\,, \tag{16}$$

$$-2F\dot{H} = \ddot{f} - H\dot{f} + \kappa^2(\rho_m + p_m)\,, \tag{17}$$

where the perfect fluid satisfies the conservation equation

$$\dot{\rho}_m + 3H(\rho_m + p_m) = 0\,. \tag{18}$$

3. Mimetic $f(R)$ gravity

In mimetic theory, we will parametrize the metric tensor as the following

$$g_{\mu\nu} = -\hat{g}^{\rho\sigma}\partial_\rho\phi\partial_\sigma\phi\hat{g}_{\mu\nu}\,, \tag{19}$$

In this case we consider the variation with respect to $\hat{g}$ and ϕ. According (19), we can rewrite action (1) following form as

$$S = \int d^4x\sqrt{-g\left(\hat{g}_{\mu\nu},\phi\right)}\left[\frac{f\left(R\left(\hat{g}_{\mu\nu},\phi\right)\right)}{2\kappa^2} + \mathcal{L}_m\right]. \tag{20}$$

So, if we perform variation with respect to metric tensor $\hat{g}$, we obtain

$$\begin{aligned}
&\frac{1}{2}g_{\mu\nu}f\left(R\left(\hat{g}_{\mu\nu},\phi\right)\right) - R\left(\hat{g}_{\mu\nu},\phi\right)_{\mu\nu} f_R\left(R\left(\hat{g}_{\mu\nu},\phi\right)\right)\\
&+\nabla\left(g\left(\hat{g}_{\mu\nu},\phi\right)_{\mu\nu}\right)_\mu \nabla\left(g\left(\hat{g}_{\mu\nu},\phi\right)_{\mu\nu}\right)_\nu f_R\left(R\left(\hat{g}_{\mu\nu},\phi\right)\right)\\
&-g\left(\hat{g}_{\mu\nu},\phi\right)_{\mu\nu}\Box\left(\hat{g}_{\mu\nu},\phi\right) f_R\left(R\left(\hat{g}_{\mu\nu},\phi\right)\right) + \kappa^2 T_{\mu\nu}\\
&+\partial_\mu\phi\partial_\nu\phi\Big[2f\left(R\left(\hat{g}_{\mu\nu},\phi\right)\right) - R\left(\hat{g}_{\mu\nu},\phi\right) f_R\left(R\left(\hat{g}_{\mu\nu},\phi\right)\right)\\
&-3\Box\left(g\left(\hat{g}_{\mu\nu},\phi\right)_{\mu\nu}\right) f_R\left(R\left(\hat{g}_{\mu\nu},\phi\right)\right) + \kappa^2 T\Big] = 0\,,
\end{aligned} \tag{21}$$

Also, if we take variation respect to the scalar field ϕ, the field equation written as

$$\begin{aligned}
\nabla\left(g\left(\hat{g}_{\mu\nu},\phi\right)_{\mu\nu}\right)^\mu &\Big\{\partial_\mu\phi\Big[2f\left(R\left(\hat{g}_{\mu\nu},\phi\right)\right) - R\left(\hat{g}_{\mu\nu},\phi\right) f_R\left(R\left(\hat{g}_{\mu\nu},\phi\right)\right)\\
&-3\Box\left(g\left(\hat{g}_{\mu\nu},\phi\right)_{\mu\nu}\right) f_R\left(R\left(\hat{g}_{\mu\nu},\phi\right)\right) + \kappa^2 T\Big]\Big\} = 0\,,
\end{aligned} \tag{22}$$

Here, in our case we define

$$T = g\left(\hat{g}_{\mu\nu}, \phi\right)^{\mu\nu} T_{\mu\nu} \tag{23}$$

Then, from (19) we have

$$g\left(\hat{g}_{\mu\nu}, \phi\right)^{\mu\nu} \partial_\mu \phi \partial_\nu \phi = -1 \tag{24}$$

Now, we will slightly modify action (20) for mimetic $f(R)$ gravity by including a Lagrange multiplier λ and potential $V(\phi)$ as the following

$$S = \int d^4x \sqrt{-g\left(\hat{g}_{\mu\nu}, \phi\right)} \left[\frac{f\left(R\left(\hat{g}_{\mu\nu}, \phi\right)\right)}{2\kappa^2} - \lambda(\partial_\mu \phi \partial^\mu \phi + 1) - V(\phi) + \mathcal{L}_m\right]. \tag{25}$$

From action (25) find point-like Lagrangian $\mathcal{L}(a, R, \phi, \dot{a}, \dot{R}, \dot{\phi})$ as

$$\mathcal{L} = a^3 f - a^3 R f_R - 6\Big(a\dot{a}^2 f_R + a^2 \dot{a}\dot{R} f_{RR}\Big) - pa^3 - \lambda(1 - \dot{\phi}^2)a^3 - Va^3 \tag{26}$$

Note that p depends on a, $\dot{a}$ and ϕ. The Euler-Lagrange equations are

$$\frac{\partial \mathcal{L}}{\partial a} - \frac{d}{dt}\frac{\partial \mathcal{L}}{\partial \dot{a}} = 0 \tag{27}$$

$$\frac{\partial \mathcal{L}}{\partial \phi} - \frac{d}{dt}\frac{\partial \mathcal{L}}{\partial \dot{\phi}} = 0 \tag{28}$$

By using Euler-Lagrange equations in (27) and (28) we obtain equation of motion for a and ϕ

$$\begin{aligned} &6f_{RR}\ddot{R} + 6\dot{R}^2 f_{RRR} + a\ddot{a}p_{\dot{a}\dot{a}} + \dot{a}ap_{a\dot{a}} + ap_{\dot{a}t} + 3\dot{a}p_{\dot{a}} - ap_a \\ &+f_R\Big(-3R + 6(\frac{\dot{a}}{a})^2 + 12\frac{\ddot{a}}{a} - 3(f - \lambda + \lambda\dot{\phi}^2 - V(\phi) - p)\Big) + 12\frac{\dot{a}\dot{R}}{a} = 0 \end{aligned} \tag{29}$$

$$\ddot{\phi} + \dot{\phi}(\frac{\dot{\lambda}}{\lambda} + 3H) + \frac{V_\phi}{2\lambda} = 0. \tag{30}$$

If we set $p = p(a), \lambda = 1, \phi(t) = t, V(\phi) = 0$, we becomes simple equation of motion for $f(R)$ gravity.
Then, we consider de Sitter solution, where $H = H_0$ we find

$$R = R_0 = 12H_0^2 \tag{31}$$

$$\dot{R} = \ddot{R} = 0 \tag{32}$$

Now we perform small perturbations for H and R as

$$H = H_0 + \delta H_0, \quad R = R_0 + \delta R_0, \quad \delta \ll 1, \tag{33}$$

By substituding this perturbation into (29) and taking only linear terms of $\delta(H, R)$, we obtain

$$\begin{aligned} &\frac{6f_{RR}R_0\dot{a}}{a^2}\ddot{\delta} + \left(\frac{12H_0^2 R_0}{a} + 12H_0 f_R - 18H_0 f_R\right)\dot{\delta} + (36f_R + 18f_R + 3p_{\dot{a}})\,H_0^2 - \\ &-p_a H_0 - \frac{3ff_R\dot{a}}{a^2} + \frac{3\lambda f_R\dot{a}}{a^2} - \frac{3\lambda\dot{\phi}^2 f_R\dot{a}}{a^2} + \frac{3Vf_R\dot{a}}{a^2} + \frac{3pf_R\dot{a}}{a^2} = 0 \end{aligned} \tag{34}$$

We take form of $\delta(H,R) = \exp(\eta t)$, where η is a constant. In our case, if $\eta > 0$, then amplitude of value η grows as the cosmic time passes, the de Sitter solution is unstable. So that the universe can exit from the inflationary stage. By plugging $\delta = \exp(\eta t)$, we have

$$\frac{6f_{RR}\dot{a}}{a^2}\eta^2 + \left(\frac{12H_0}{a} + 12f_R - 18f_R\right)\eta + \tag{35}$$
$$(36f_R + 18f_R)\,H_0 - p_a -$$
$$-\frac{3ff_R\dot{a}}{a^2} + \frac{3\lambda f_R\dot{a}}{a^2} - \frac{3\lambda\dot{\phi}^2 f_R\dot{a}}{a^2} + \frac{3Vf_R\dot{a}}{a^2} + \frac{3pf_R\dot{a}}{a^2} = 0$$

Then, solve it approximately, we obtain

$$\eta_\pm = \frac{\left(-\frac{12H_0}{a} - 12f_R + 18f_R\right) \pm \sqrt{\mathcal{D}}}{\frac{12f_{RR}\dot{a}}{a^2}} \tag{36}$$

$$\mathcal{D} = \left(\frac{12H_0}{a} + 12f_R - 18f_R\right)^2 - 4\mathcal{M} \tag{37}$$

$$\mathcal{M} = \frac{6f_{RR}\dot{a}}{a^2} \times (36f_R + 18f_R)\,H_0 - p_a - \tag{38}$$
$$-\frac{3ff_R\dot{a}}{a^2} + \frac{3\lambda f_R\dot{a}}{a^2} - \frac{3\lambda\dot{\phi}^2 f_R\dot{a}}{a^2} + \frac{3Vf_R\dot{a}}{a^2} + \frac{3pf_R\dot{a}}{a^2}$$

The de Sitter solutions are stable only if the value of η_- is a real and positive number.

4. Conclusion

In this paper we investigated stability of de Sitter solution in mimetic $f(R)$ gravity. Also, analyzed our results for explanation early and late time acceleration expansion.

References

[1] Perlmutter S and etal 1999 *Astrophys. J.* **517** 565-586
[2] Tegmark M and et al 2004 *Phys. Rev. D* **69** 103501
[3] Eisenstein D J and etal 2005 *Astrophys. J.* **633**, 560-574
[4] Spergel D N and etal 2003 *Astrophys. J. Suppl.* **148** 175-194
[5] Nojiri S and Odintsov S D 2007 *Int. J. Geom. Methods Mod. Phys.* **04** 115
[6] Nojiri S and Odintsov S D 2011 *Phys. Rep* **505** 59-144
[7] Bamba K, Nojiri S, Odintsov S D and Saez-Gomez D 2014 *Phys. Rev. D* **90** 124061
[8] Nojiri S and Odintsov S D 2003 *Phys. Rev. D* **68** 123512
[9] Lopez-Revelles A J and Elizalde E 2012 *Gen. Rel. and Grav* **44** 751-770
[10] Elizalde E, Nojiri S, Odintsov S D, Sebastiani L and Zerbini S 2011 *Phys. Rev. D* **83** 086006
[11] Elizalde E, Nojiri S, Odintsov S D and Saez-Gomez D 2010 *Eur. Phys. Jour. C* **70** 351-361
[12] Astashenok A V, Capozziello S and Odintsov S D 2010 *Astroph. Space Sci* **355** 333-341
[13] Bamba K, Geng C Q, Nojiri S and Odintsov S D 2010 *Mod. Phys. Lett. A* **25** 900
[14] Chamseddine A H and Mukhanov V 2013 *Jour. High Energy Phys.* **1311** 135
[15] Matsumoto J, Odintsov S D and Sushkov S V *Preprint* 1501.02149
[16] Momeni D, Altaibayeva A and Myrzakulov R 2014 *Int. J. Geom. Methods Mod. Phys.* **11** 1450091
[17] Nojiri S and Odintsov S D 2014 *Mod. Phys. Lett. A* **29** 1450211

Mie-type potential from a class of multiparameter exponential-type potential: Bound state solutions in D dimensions

J. J. Peña[1], A. Menéndez[2], J. García-Ravelo[2] and J. Morales[1]

[1]Universidad Autónoma Metropolitana - Azcapotzalco, CBI - Area de Física Atómica Molecular Aplicada, Av. San Pablo 180, 02200 México, D.F., MEXICO.

[2]Escuela Superior de Física y Matemáticas, Instituto Politécnico Nacional - Zacatenco, Edificio 9, Unidad Profesional Adolfo López Mateos, 07738 México, D.F.

Corresponding author: jjpg@correo.azc.uam.mx

Abstract. The Mie potential is a model of molecular interaction, very useful in the study of diatomic molecules because allows one to describe the softness/hardness of the repulsive interactions as well as the range of attraction. As a consequence, the Mie potential and one of their particular cases, the Lennard-Jones potential, have been extensively used in many branches of physics and chemistry. In this work, the exact bound state solutions of the D-dimensional Schrödinger equation with the Mie-type potential are presented. These eigen-functions/values are obtained as a particular case of the exactly solvable Schrödinger equation for a class of multiparameter exponential-type potential. Furthermore our approach does not need any approximation to the centrifugal term. As an example of the usefulness of our proposition, we show how the bound state solutions of the Kratzer-Fues and Coulomb potentials in D-dimensions are particular cases from the proposal.

1. Introduction

The molecular vibrational and rotational spectroscopy is one of the most important fields of research due that it is a valuable tool in practically all the areas of scientific sciences. At this regard, numerous molecular interaction models have been proposed to describe the behavior of diatomic molecules as well as different approaches to the treatment of the centrifugal term. Specifically, the Mie (n,m) potential [1] has been advantageously used in many branches of physics and chemistry because this model of molecular interaction comprises a repulsive term at short distances and an attractive part for large distances. This feature has been exploited in the investigation of the effect of Mie-type potential range on the cohesive energy of metallic nanoparticles using the size-dependent potential parameters method [2]. Also, the thermodynamic properties of the Mie n-6 potential has been studied by means of a comparison between the statistical associating fluid theory of variable range approach and molecular dynamics results [3]. In the same way, the particular case (n=12, m=6), the Lennard-Jones potential [4], is one of the most celebrated interaction potential models used to get thermodynamic properties [5], phase diagrams of fluids [6], Molecular dynamics calculations [7] and so on. On the other hand, from the theoretical point of view different methods have been employed in the search of bound states solutions of Mie potential in the non relativistic and relativistic cases. For example, the Nikiforov-Uvarov method has been used in the search of the exact solution of Dirac equation for the Mie-type potential under pseudospin and spin symmetry limit [8-9] as well as for the Mie-type potential including a Coulomb-like tensor potential [10]. Similarly, the Laplace transform approach was used to find the exact solutions of the Schrödinger equation for the Mie-type potential [11-12] and for relativistic studies of spin-1/2 particles [13]. Also, the factorization method has been employed to obtain the exact solution of Schrödinger equation for the Mie–type potential [14] as well as the SU(1,1) dynamical algebra to built their corresponding ladder operators. [15]. However, as far as we know, all works dedicated to the Mie interaction potential model, have as common characteristic the selection of a specific method to solve the corresponding Schrödinger and/or Dirac equations. In this work, we are showing how the Mie-type potential can be solved as a particular case of a class of multiparameter exponential-type potential [16] in D-dimensions without using any approximation to the centrifugal term as will be seen next.

2. A class of multiparameter exponential-like potential.

Recently, Peña et al [17] have proposed a canonical transformation approach applied to the Schrödinger equation to identify those exactly solvable exponential-type potentials with hypergeometric wavefunctions. Accordingly to their results, the radial potential

$$V(r) = \frac{Gqe^{-r/k}+Hq^2e^{-2r/k}}{\left(1-qe^{r/k}\right)^2} \tag{1}$$

with

$$G = [4ab - 2c(a+b+1)]/4k^2 \quad \text{and} \quad H = [(a-b)^2 - (c-1)^2]/4k^2 \tag{2}$$

where a, b, and c are the parameters of the hypergeometric function, has physically acceptable wavefunctions

$$\psi_n = \left(e^{-r/k}\right)^{(c-1)/2}\left(1-e^{-r/k}\right)^{(h+1)/2} {}_2F_1\left(-n, b; c; qe^{-r/k}\right) \tag{3}$$

and energy spectra given by

$$E_n = -\frac{1}{4k^2}\left[\frac{\left(n+\frac{h+1}{2}\right)^2 - k^2 H}{n+\frac{h+1}{2}}\right]^2 \tag{4}$$

where

$$b = \frac{(h+1)(h-a)-2k^2G}{h+1-2a} \quad \text{and} \quad c = \frac{2a(h-a)-2k^2G}{h+1-2a} \tag{5}$$

such that

$$c = a + b - h \quad \text{with} \quad h = \sqrt{1+4k^2(G+H)}, \tag{6}$$

on condition that $a = -n$ where $n = 0, 1, 2, \ldots$

To obtain the D-dimensional form of the above potential [18], we consider $q = 1$ and the redefinition of parameters

$$G = k^{-2}\left(\boldsymbol{G} + L(L+1)\right) + k^{-1}A \quad \text{and} \quad H = k^{-2}\boldsymbol{H} - k^{-1}A \tag{7}$$

that allows rewrite

$$V(r) = \frac{1}{k^2}\left(\frac{\boldsymbol{G}e^{-r/k}+\boldsymbol{H}e^{-2r/k}}{\left(1-e^{r/k}\right)^2} + \frac{L(L+1)e^{-r/k}}{\left(1-e^{r/k}\right)^2}\right) \tag{8}$$

with

where $L = \ell + (D-3)/2$ such that $L(L+1) = \ell_D(\ell_D+1) - (D-1)(3-D)/4$ being $\ell_D = \sqrt{1/4 + \ell(\ell + D - 2)} - 1/2$ with ℓ the usual angular momentum. In this new situation the energy spectra is given by

$$E_n = -\frac{1}{4k^2}\left[\frac{\left(n+\frac{h+1}{2}\right)^2 - \boldsymbol{H} + kA}{n+\frac{h+1}{2}}\right]^2 \tag{9}$$

with eigenfunctions

$$\psi_{nL}(r) = \left(e^{-r/k}\right)^{(c_L-1)/2}\left(1-e^{-r/k}\right)^{(h_L+1)/2}\ {}_2F_1\left(-n, b_L; c_L; e^{-r/k}\right) \tag{10}$$

where

$$b_L = \frac{(h_L+1)(h_L+n) - 2(\boldsymbol{G}+kA+L(L+1))}{h_L+1+2n}, \tag{11}$$

$$c_L = \frac{-2n(h_L+n) - 2(\boldsymbol{G}+kA+L(L+1))}{h_L+1+2n} \tag{12}$$

and

$$h_L = \sqrt{(2L+1)^2 + 4(\boldsymbol{H}+\boldsymbol{G})}. \tag{13}$$

3. Bound state solutions of the Mie-Type potential in D-dimensions.

As well known, the Mie intermolecular pair potential has the standard form [1]

$$V_M(r) = \left(\frac{n}{n-m}\right)\left(\frac{n}{m}\right)^{\frac{m}{n-m}} \varepsilon\left[\left(\frac{\sigma}{r}\right)^n - \left(\frac{\sigma}{r}\right)^m\right] \tag{14}$$

where $r = |r_1 - r_2|$, σ is the value of r at $V(r)$ and ε is the well depth; we notice the particular case $(n = 12, m = 6)$ refers to the Lennard-Jones. With the aim to obtain the above $V_M(r)$form the D-dimensional multiparameter exponential-type potential given in Eq.(8), we consider the limit $k \to \mathbb{N}$leading to

$$V_{M-t}(r) = \frac{A}{r} + \frac{\boldsymbol{G}+\boldsymbol{H}}{r^2} + \frac{L(L+1)}{r^2} \tag{15}$$

with energy spectra

$$E_{nL} = -\frac{A^2}{(1+2n+h_L)^2} \tag{16}$$

which indicates that the new potential $V_{M-t}(r)$ is of Mie-type (M-t) for the case $n = 2, m = 1$. At this point it is important to notice that despite of the divergences in the parameters b_L and c_L when

$k \to \mathbb{N}$ the wavefunctions can be written without divergences. To show that, we begin by using Eq.(11) and Eq.(12) along with the value of h_L given in Eq.(13) to find

$$b_L = \frac{h_L+1}{2} + \nu + \xi\, k \quad \text{and} \quad c_L = -n - \frac{h_L-1}{2} + \nu + \xi\, k \tag{17}$$

where

$$\nu = \frac{2H}{2n+h_L+1} \quad \text{with} \quad \xi = -\frac{2A}{2n+h_L+1}. \tag{18}$$

With this new definitions, the corresponding wave functions for $V_{M-t}(r)$ are given by

$$\begin{aligned}\psi_{nL} &= e^{-\xi\, r}\left(e^{-r/k}\right)^{(\nu-n)/2}\left(e^{-r/k}\right)^{\frac{h_L+1}{4}}\left(1-e^{-r/k}\right)^{\frac{h_L+1}{2}}\, {}_2F_1\left(-n, b_L; c_L; e^{-r/k}\right) \\ &= e^{-\xi\, k}\left(e^{-r/k}\right)^{(\nu-n)/2}\left(e^{r/2k}-e^{-r/2k}\right)^{\frac{h_L+1}{2}}\, {}_2F_1\left(-n, b_L; c_L; e^{-r/k}\right) \\ &= \left(\frac{1}{2k}\right)^{\frac{h_L+1}{2}} e^{-\xi r}\left(e^{-r/k}\right)^{(\nu-n)/2} r^{(h_L+1)/2}\left[\frac{\sinh(r/2k)}{r/2k}\right]^{\frac{h_L+1}{2}}\, {}_2F_1\left(-n, b_L; c_L; e^{-r/k}\right).\end{aligned} \tag{19}$$

Furthermore, using the relationship between the hypergeometric function and Jacobi polynomials

$${}_2F_1\left(-n, \alpha+\beta+n+1, \alpha+1; \frac{1-z}{2}\right) = \frac{n!}{(\alpha+1)_n} P_n^{\alpha,\beta}(z) \tag{20}$$

along with

$$P_n^{\alpha,\beta}(z) = (-1)^n P_n^{\beta,\alpha}(-z), \tag{21}$$

the wavefunctions of Eq.(19) can be written as

$$\psi_{nL}(r) = C_{nk} e^{-\xi\, r}\left(e^{-\frac{r}{k}}\right)^{\frac{\nu-n}{2}} r^{\frac{h_L+1}{2}}\left[\frac{\sinh\left(\frac{r}{2k}\right)}{\frac{r}{2k}}\right]^{\frac{h_L+1}{2}} P_n^{h_L,\ \xi k+\nu-n-\frac{h_L-1}{2}}\left(2e^{-\frac{r}{k}}-1\right) \tag{22}$$

where

$$C_{nk} = \frac{\left(\frac{1}{2k}\right)^{\frac{h_L+1}{2}} n!(-1)^n}{\left(\xi k+\nu-n-\frac{h_L-1}{2}\right)_n}. \tag{23}$$

At this point, we want to mention that the function

$$\begin{aligned}\varphi_{nL}(r) &= (C_{nk})^{-1}\, \psi_{nL}(r) \\ &= e^{-\xi\, r}\left(e^{-\frac{r}{k}}\right)^{\frac{\nu-n}{2}} r^{\frac{h_L+1}{2}}\left[\frac{\sinh\left(\frac{r}{2k}\right)}{\frac{r}{2k}}\right]^{\frac{h_L+1}{2}} P_n^{h_L,\ \xi\, k+\nu-n-\frac{h_L-1}{2}}\left(2e^{-\frac{r}{k}}-1\right)\end{aligned} \tag{24}$$

is also solution of the Schrödinger equation for the potential given in Eq.(8). Consequently, in order to apply the limit $k \to \mathbb{N}$ to the function $\varphi_{nL}(r)$ we make use of the identities [19]

$$\mathrm{Lim}_{k\to\mathbb{N}}\ P_n^{\alpha,\ 2\beta\, k+\gamma}\left(1-\frac{z}{k}\right) = L_n^{\alpha}(\beta z) \tag{25}$$

$$P_n^{\alpha,\beta}(z) = (-1)^n P_n^{\beta,\alpha}(z) \tag{26}$$

to obtain

$$R_{nL}(r) = \mathrm{Lim}_{k\to N}\ \varphi_{nL}(r) = e^{-\xi r}\, r^{\frac{h_L+1}{2}}\, L_n^{h_L}(\xi\, r), \tag{27}$$

for the wavefunction of the potential $V_{M-t}(r)$ given in Eq.(15). At this regard, we notice that the calculation of the above limit was possible after using the expansion ($k \gg 1$)

$$1 - 2e^{-\frac{r}{k}} = 1 - \frac{2r}{k} + \frac{r^2}{k^2} - \frac{r^3}{3k^3} + \frac{r^4}{12k^4} \ldots.. \frown 1 - 2r/k \tag{28}$$

4. Applications

In this section, some useful examples of potentials derived directly from our proposal $V_{M-t}(r)$ are shown next.

a) Coulomb potential.

Straightforward, by choosing the values $A = -Ze^2$ and $\boldsymbol{G} + \boldsymbol{H} = 0$ in Eq.(15) one gets

$$V_C(r) = -\frac{Ze^2}{r^2} + \frac{L(L+1)}{r^2} \tag{29}$$

that is the D-dimensional Coulomb potential. Consequently, from Eq.(16) the corresponding energy spectra results as

$$E_{nL} = -\frac{Z^2e^4}{4(n+L+1)^2}, \tag{30}$$

in agreement with Eq.(18) of Abgoola [20]. Furthermore, accordingly to Eq.(27) the respective wavefunctions are given by

$$R_{nL}(r) = e^{-\xi\, r}\, r^{L+1} L_n^{2L+1}(\xi\, r) \tag{31}$$

where $\xi = \frac{Ze^2}{n+L+1}$.

b) Kratzer-Fues potential

Another particular case comes from the choice $A = -2aD_e$, $\boldsymbol{G} = 0$ and $\boldsymbol{H} = a^2D_e$ in Eq.(15) to obtain the potential

$$V_{MK}(r) = -\frac{2aD_e}{r^2} + \frac{a^2D_e}{r^2} + \frac{L(L+1)}{r^2} \tag{32}$$

that corresponds to the Kratzer potential with energy spectra given, through Eq.(16), by

$$\begin{aligned} E_{nL} &= -\frac{4a^2D_e}{\left[1+2n+\sqrt{(2L+1)^2+4a^2D_e}\right]^2} \\ &= -\frac{a^2D_e}{\left[n+1/2+\sqrt{(L+1/2)^2+a^2D_e}\right]^2} \end{aligned} \tag{33}$$

in agreement with the exact analytical solution obtained by Bayrak *et al* [21] using the asymptotic iteration method. Similarly, from Eq.(27) the corresponding wavefunction becomes

$$R_{nL}(r) = e^{-\xi\, r}\, r^{\frac{h_L+1}{2}}\, L_n^{h_L}(\xi\, r) \tag{34}$$

where

$$h_L = \sqrt{(2L+1)^2 + 4a^2 D_e} \tag{35}$$

c) A non-central potential

A class of non-central potential that can be solved with the approach given before, is

$$V(r,\theta) = V_1(r) + \frac{1}{r^2} V(\theta) \tag{36}$$

where

$$V_1(r) = \frac{A}{r} + \frac{B}{r^2} \qquad \text{and} \qquad V_2(\theta) = \frac{C + D\cos\theta}{\sin^2\theta}, \tag{37}$$

which, although has dependence on the polar angle θ, the D-dimensional Schrödinger equation can be solved via the separation variables method. In fact, following the procedure given by Ikhair and Sever [22] and using the notation of reference [18] for $(r,\Omega) = (r,\theta_1,\theta_2,\theta_3 \ldots \theta_{D-2},\phi)$, the radial equation is given by

$$\left(-\frac{d^2}{dr^2} + V_1(r) + \frac{L(L+1)}{r^2}\right) u(r) = Eu(r). \tag{38}$$

Similarly, the angular part becomes

$$\left(\frac{1}{sin^{D-2}\theta_1}\frac{d}{d\theta_1}\left(sin^{D-2}\theta_1\frac{d}{d\theta_1}\right) + \ell_D(\ell_D+1) - \frac{\Lambda_{D-2}}{sin^{D-2}\theta_1} - V_2(\theta_1)\right) H_1(\theta_1) = 0 \tag{39}$$

until

$$\left(\frac{1}{sin^{j-1}\theta_j}\frac{d}{d\theta_j}\left(sin^{j-1}\theta_j\frac{d}{d\theta_j}\right) + \Lambda_j - \frac{\Lambda_{j-1}}{sin^2\theta_j}\right) H_j(\theta_j) = 0, \tag{40}$$

$$\left(\frac{d^2}{d\phi^2} + \Lambda_D\right) H_\phi = 0 \tag{41}$$

where $j = 2,3,\ldots.D-2$ and Λ_j are separation constants. Since the angular part has been already solved [22], in the following we consider only the radial equation. Thus, by comparing the potential $V_1(r)$ with $V_{M-t}(r)$ given in Eq.(15), it becomes $\boldsymbol{G} = 0$ and $\boldsymbol{H} = B$ to have the corresponding energy spectra

$$E_{nL} = -\frac{A^2}{(2n + h_L + 1)^2} \tag{42}$$

as well as the wavefunctions

$$u_{nL}(r) = e^{-\xi r} r^{\frac{h_L+1}{2}} L_n^{h_L}(\xi r) \tag{43}$$

where $h_L = \sqrt{(2L+1)^2 + 4B}$, in good agreement with Eq.(16) of reference [20].

5. Concluding remarks

This work is devoted to obtain the bound state solutions of the Mie-type potential in D-dimensions. The proposal is direct because the corresponding solution is found as a particular case of the exactly solvable multiparameter exponential-type potential under the action of the limit $k \to \mathbb{N}$. With this approach, our proposal does not need any approximation of the centrifugal term as usually is done. In addition, the eigenvalues and wavefunctions of the Kratzer, Coulomb and a class of non-central potentials are derived as particular cases from our proposition. Advantageously, instead of solving the Schrödinger equation for a specific potential with a particular solution method, we are offering a procedure that consider the general solution of a class of multiparameter exponential-type potential from where particular or specific potentials can be derived. Beyond the case considered, by selecting different possibilities of the involved parameters A, B and C the proposed method can be also applied directly to other kinds of potentials.

Acknowledgments

This work was partially supported by the projects UAM-A-CBI-2232004 and 009. One of us (JGR) thanks to the Instituto Politécnico Nacional for the financial support given through the COFAA-IPN project SIP-20150019. A. Menéndez acknowledges the IPN-ESFM, for the hospitality during his undergraduate studies in Science and Technology. We are grateful with the SNI - Conacyt -México for the stipend received.

References

[1] G. Mie, Annalen der Physik, **11** (1903) 657.
[2] T. Barakat, O. M. Al-Dossary and A. A. Alharbi, Int. J. Nanosci. **06** (2007) 461.
[3] G. Galliero, T. Latte, D. Bessieres and C. Boned., J. Chem. Phys., **127** (2007) 184506.
[4] J. E. Lennard-Jones, Proceedings of the Royal Society of London. Series A,Containing Papers of a Mathematical and Physical Character, **106** pp. 441-462 and 463-477 (1924).
[5] J. Mairhofer, R. J. Sadus, J Chem Phys. **139** (2013) 154503.
[6] B. Smit, J. Chem. Phys. **96**, (1992) 8639,
[7] N. Tchouar, M. Benyettou, F Ould Kadour, Int. J. Mol. Sci., **4** (2003) 595.
[8] M. Hamzavi, H. Hassanabadi and A. A. Rajabi, Mod. Phys. Lett. A, **25** (2010) 2447.
[9] O. Aydogdu and R. Sever, Annals of Physics, **325** (2010) 373.
[10] M. Hamzavi, A. A. Rajabi, H. Hassanabadi, Few-Body Systems, **48** (2010) 171.
a. Arda and R. Sever, J. Math.Chem., **50** (2012) 971.
[11] T. Das, J. Math.Chem., **53** (2015) 618.
[12] M. Eshghi and S. M. Ikhdair, Chinese Phys. B, **23** (2014) 120304.
[13] B. Okon, E. E. Ituen, O. Popoola and A. D. Antia, Int. J. Recent Adv. Physics, **2** (2013) No.2
[14] D Martínez, J C Flores-Urbina, R D Mota and V D Granados, J. Phys. A:Math. Theor. **43** (2010) 135201.
[15] García-Martínez, J. García-Ravelo, J.Morales and J.J. Peña, Int. J. Quan. Chem., **112** (2012) 195.
[16] J.J. Peña, J. Morales, J. García-Martínez and J. García-Ravelo, Mol. Phys,**113** (2015) 260.
[17] J.J. Peña, J.García-Martínez, J.García-Ravelo and J.Morales, Int. J. Quant, Chem, **115** (2015) 158.
[18] Koekoek, R. Koekoek, J. Comp.App.Math. **126** (2000) 1-31.
[19] D. Agboola, Acta Phys. Polon. A., **120** (2011) 371.
[20] O. Bayrak, I. Boztosun and H. Ciftci, Int. J. Quan.Chem, **107** (2007) 540.
[21] S. M. Ikhdair and R. Sever., Cent. Eur. J. Phys., **6** (2008) 685.

4th International Conference on Mathematical Modeling in Physical Sciences (IC-MSquare2015) IOP Publishing
Journal of Physics: Conference Series **633** (2015) 012026 doi:10.1088/1742-6596/633/1/012026

Simple description of shear dynamics of glasses with crystalline local order.

Leonid Son

Ural Federal University, 19 Mira st., 620002, Ekaterinburg, Russia

E-mail: ldson@yandex.ru

Abstract. For those glassformers (especially metallic glasses), which structure may be presented as a set of linear topological defects embedded into a media with crystalline local order, we suggest a description of the shear dynamics in terms of kinks motion along the topological defect lines, as it is customary for crystalline materials. Locally, these defects are similar to dislocations and disclinations. For the motion of the kink, we write out the Fokker - Planck equation in a self - consistent potential. The glass transition occurs to be described as a localization of the kink.

1. Introduction

Near melting, the structure of condensed substance is often being described in terms of local order and topological defects. This approach arises from the articles [1, 2]. The base suggest of the approach is that in an overwhelming majority of local clusters of the substance (including first coordination shell) the local arrangement of the atoms is similar to the corresponding crystal arrangement, and such clusters form topologically connected area. Clusters, where the local order differs sharply from the crystalline order, are not too numerous and are organized into linear defects, which can not be terminated in the bulk of the substance. Locally, these defects are similar to dislocations and disclinations in crystal. The difference is that in globally disordered state (glass or liquid), the density of the defects is high enough to allow the disorientation of local crystalline axis at distances larger than orientational order correlation length. Since the loss of global orientation order is the main content of melting, than the approach considered implies melting being described in terms of statistics of topological defects. In two dimensions, the description of melting in terms of topological defects statistics is rather common approach [3, 4]. For three dimensions, the theory becomes mathematically complicated [5, 6], but still applicable. Melting occurs to be described as a jump rise of defect density. Recently, this approach was applied to the liquid - glass transition [7]. It is known, that metals may be transformed into nano structured (crystalline at nano scale but disordered at larger distances) state by large plastic deformation [8], i.e. by mechanical rising of topological defects density. This fact may be considered as a qualitative justification (at least for metals) of the approach considered. For the metals and alloys, the approach may be treated as a complementary to the traditional one, based on the formalism of correlation functions (for the example of the last, see [9, 10]).

In present article, we realize the following idea. In crystalline state, the shear deformation is provided by the motion of topological defects (dislocations). The last is due to the motion of point objects (kinks) along the defect line [11]. For the substance with crystalline local order, the same mechanism may be suggested in liquid and amorphous states also. The difference is that in the case of high defect density, the dislocation sliding plane curves into a sliding surface, and the dislocation motion is strongly retarded by the intersections with other defects. Nevertheless, the local picture of the kink motion is the same, and one can describe the shear relaxation in that terms.

2. Model of the kink motion.

Consider a certain dislocation and a certain kink moving along it. The kinks motion shifts the dislocation in its sliding surface, so the excess density of intersections with other defects occurs during the motion just behind the kink. This excess density disappears, if the kink returns conversely in a short time. Thus, if the excess intersection lifetime is infinite, the kink moves in a potential $U(x) = u_0 \mid x \mid$, where u_0 is the energy of excess intersections per unit length, and x is a coordinate along the defect line. However, the excess intersection contributes into the excess energy, so it relaxes in time, and we denote the probability to relax during time t as $m(t)$. With respect to the motion of the kink, all intersections may be divided into "excess" and "relaxed". Let us denote the probability of kink to have the coordinate below x as $F(x,t)$, and the kink probability density as $f(x,t) = F'(x,t)$. Here and below, we'll denote the derivative with respect to x by the bar, and the time derivative by the dot. It can be shown that

$$U'(x,t) = (1 - 2F(x,t))u_0(\omega_e - \omega_r), \tag{1}$$

where ω_e, ω_r are the probabilities of the intersections at (x,t) to be "excess" or "relaxed" respectively. Let us write out the equation on $\omega_{e,r}$. Within a small Δt, the probability of relaxed state at point x changes as follows:

$$\omega_r = \omega_e(x,t)m(\Delta t) + \omega_r(x,t)(1 - f(x,t)\Delta t). \tag{2}$$

Here, the first term corresponds to the relaxation from the excess to relaxed state, and the second describes possible excitation from relaxed state to the excess one by the kink motion across point x during the time interval $t \div t + \Delta t$. The obvious condition $\omega_e + \omega_r = 1$ gives closed equation

$$\dot{\omega}_e + \omega_e(m_0 + f(x,t)) = f(x,t) \ , \quad m_0 = \dot{m}(t) \mid_{t \to 0} . \tag{3}$$

This equation may be easily solved

$$\omega_e = e^{-m_0 t - G} \int_0^t f(x,\tau) e^{m_0 \tau + G(x,\tau)} d\tau, \quad G(x,t) = \int f(x,t) dt \tag{4}$$

The probability density function $f(x,t)$ obeys the Fokker - Planck equation [12]

$$\frac{T}{\gamma} f'' + \frac{1}{\gamma}(U'f)' = \dot{f}, \tag{5}$$

where T is the temperature in energy units, and γ - effective friction coefficient. Relations (1,4,5) together with the condition $\omega_e + \omega_r = 1$ form a closed set of equations on the kink's distribution function $F(x,t)$. The initial condition is that at $t = 0$ the kink is at point $x = 0$. Below, we'll consider its stationary solution.

3. Stationary solution

For stationary solution, $f(x,t) = f(x)$, $\dot{\omega}_e = 0$, so that from (3) one gets

$$\omega_e = \frac{f(x)}{m_0 + f(x)}. \tag{6}$$

Equation (5) may be once integrated with respect to x:

$$\frac{T}{\gamma}F'' + \frac{u_0}{\gamma}(1-2F)F'\frac{F'-m_0}{F'+m_0} = 0. \tag{7}$$

Using standard substitution $F'' = F'dF'/dF$, one arrives to the equation

$$F'\left[\frac{dF'}{dF} + \frac{u_0}{T}(1-2F)\frac{F'-m_0}{F'+m_0}\right] = 0, \tag{8}$$

which has trivial solution $F' = 0 = f(x)$. This solution corresponds to the vanishing homogeneous distribution of the kink along the defect line. Besides, there is nontrivial solution

$$F' + 2m_0 \ln\left(\frac{m_0 - F'}{m_0}\right) = \frac{u_0}{T}(F^2 - F). \tag{9}$$

Here, we choose the integration constant as to fulfil the condition $F' = 0$ when $F = 0$ $(x \to -\infty)$ or $F = 1$ $(x \to +\infty)$. Let us analyze the solution of (9) at small x, where the $F(x)$ may be expanded into power series:

$$F = \frac{1}{2} + \alpha x - \beta x^3. \tag{10}$$

Using (9), one gets equations on the α, β coefficients:

$$\alpha + \frac{u_0}{4T} = -2m_0 \ln\left(1 - \frac{\alpha}{m_0}\right), \quad \beta = \frac{\alpha^2 u_0(m_0 - \alpha)}{3T(m_0 + \alpha)}. \tag{11}$$

At large x, one can use the smallness of the F'. In that case, linear approximation of the logarithm in (9) gives

$$F = \frac{1}{2} + \frac{1}{2}\tanh\left(\frac{u_0}{2T}x\right). \tag{12}$$

Note, that nontrivial solution exists only if smooth stitching of approximations (11) and (12) at some intermediate x is possible. In the first case, the probability density function $f(x) = F'(x)$ is a parabolic cupola which width is

$$\Delta x_1 \simeq \sqrt{\frac{2T(m_0+\alpha)}{\alpha u_0(m_0-\alpha)}}. \tag{13}$$

In the second case, the probability density is described by the exponential cupola

$$f(x) = \frac{u_0}{4T}\cosh^{-2}\frac{u_0 x}{2T} \tag{14}$$

with the width

$$\Delta x_2 \simeq 4T/u_0. \tag{15}$$

The smooth stitching is possible if $\Delta x_1 \geq \Delta x_2$. Equations (11,13,15) give the parameters values, which correspond to the appearance of the nontrivial solution:

$$\frac{u_0}{Tm_0} = z, \quad \frac{\alpha}{m_0} = y, \tag{16}$$

where z, y are the solutions of the algebraic system

$$\begin{cases} 8y(1-y)/(1+y) = z \\ y + z/4 = -2/\ln(1-y) \end{cases} \tag{17}$$

4. Conclusion

Our reasoning may be supplemented by one natural consideration. The kink motion causes shifting of the dislocation in its sliding surface. Thus, the intersection with another defect, arising due to this shift (excess intersection), looks as a local shift of the line of this another defect according to the Burgers vector of initial dislocation, i.e. it is a kink on the another defect line. Thus, the process of its relaxation is a diffusion of this kink out from the place of intersection, which is the Poison random process

$$m(t) = 1 - e^{-\lambda t}, \tag{18}$$

where λ is the probability of the kink to move out from the initial position per unit time, which is the half of the diffusion coefficient in the Fokker - Planck equation (5):$\lambda = 2T/\gamma$. Thus, one gets

$$m_0 = \lambda = 2T/\gamma. \tag{19}$$

Relations (16), which correspond to the point of appearance (or disappearance) of non - trivial solution, give

$$T_c = \sqrt{\frac{u_0 \gamma}{2z}}. \tag{20}$$

Above T_c, the kink moves along the defect line fluently, and shear deformations may relax (liquid). Below T_c, the kink is localized at the length $l \simeq 4T/u_0$ counted along the defect, or at the three dimensional area of size

$$r_l \sim l^d \sim T^d, \tag{21}$$

where d is the fractal dimensionality of the defect line (glass with finite shear rigidity). To produce plastic flow, one has to provide relative shear deformation $\varepsilon \sim 1/r_l \sim T^{-d}$. Measuring temperature dependence of plastic deformation, one gets fractal dimensionality of defects.

T_c in (20), is the temperature of glass instability. The effective friction coefficient γ is proportional to the viscosity, while the density of intersections u_0 is proportional to the activation energy of viscose flow. Thus, the viscosimetry experiment in the liquid state gives the T_c.

Thus, we suggested relatively simple approach which allows one to get an interesting results concerning shear dynamics of glassformers.

Acknowledgments

The work was supported by RFBR (projects 13-03-96055, 13-03-00598, 14-02-00359) and Russian Ministry of Science and Education (Government job 2014/392 projects 2391 and 1177)

References

[1] Rivier N 1979 *Phil.Mag.* A **10** 859
[2] Nelson D R 1983 *Phys.Rev.* B **28** 5515
[3] Kosterlitz J M, Thouless D J 1973 *J. Phys.* C **6** 1181
[4] Halperin B I, Nelson D R 1979 *Phys. Rev.* B bf 19 2457
[5] Patashinski A Z, Shumilo B I 1985 *JETP* **62** 177
[6] Patashinski A Z, Son L D 1993 *JETP* **76** 534
[7] Vasin M G 2014 *J. Non-Cryst. Solids* **401** 78
[8] Valiev R 2009 *Int. J. Mat. Res.* **100** 757
[9] Dubinin N E, Son L D, Vatolin N A 2007 *Defect and Diffusion Forum* **263** 105
[10] Dubinin N E, Son L D, Vatolin N A 2008 *Journal of Physics: Condensed Matter* **20** 114111
[11] Nabarro F R N 1979 *Theory of crystal dislocations* (Oxford: Clarendon Press)
[12] Bazarov I P, Gevorkyan E V, Nikolaev P N 1989 *Neravnovesnaya termodinamika i fizicheskaya kinetika* (Moscow: Moscow University Press (in Russian))

Euclidean Complex Relativistic Mechanics: A New Special Relativity Theory

Spyridon Vossos[1, 3] and Elias Vossos[2]

[1] Chemist, [2]Physicist, School of Science, National and Kapodistrian University of Athens, Panepistimiopolis, GR15784 Zografos, Athens, Hellas

E-mail: spyrolia@hotmail.com

Keywords: electromagnetic tensor, Euclidean Request, fine structure of atomic spectrum, human senses, linear spacetime transformation, Lorentz Transformation, Relativistic Doppler Shift, Special relativity, Thomas Precession.

PACS: 02.10.Ud, 02.40.Dr, 03.30.+p, 03.50.De, 03.65.-w, 32.10.Fn

Abstract. Relativity Theory (RT) was fundamental for the development of Quantum Mechanics (QMs). Special Relativity (SR), as is applied until now, cancels the transitive attribute in parallelism, when three observers are related, because Lorentz Boost (LB) is not closed transformation. In this presentation, considering Linear Spacetime Transformation (LSTT), we demand the maintenance of Minkowski Spacetime Interval (S^2). In addition, we demand this LSTT to be closed, so there is no need for axes rotation. The solution is the Vossos Matrix (Λ_B) containing real and imaginary numbers. As a result, space becomes complex, but time remains real. Thus, the transitive attribute in parallelism, which is equivalent to the Euclidean Request (ER), is also valid for moving observers. Choosing real spacetime for the unmoved observer (O), all the natural sizes are real, too. Using Vossos Transformation (VT) for moving observers, the four-vectors' zeroth component (such as energy) is real, in contrast with spatial components that are complex, but their norm is real. It is proved that moving (relative to O) human O′ meter length, according to Lorentz Boost (LB). In addition, we find Rotation Matrix Vossos-Lorentz (R_{BL}) that turns natural sizes' complex components to real. We also prove that Speed of Light in Vacuum (c) is invariant, when complex components are used and VT is closed for three sequential observers. After, we find out the connection between two moving (relative to O) observers: $X'' = \Lambda_{LO''(O)} \Lambda_{LO(O')} X'$, using Lorentz Matrix ($\Lambda_L$). We applied this theory, finding relations between natural sizes, that are the same as these extracted by Classic Relativity (CR), when two observers are related (i.e. relativistic Doppler shift is the same). But, the results are different, when more than two observers are related. VT of Electromagnetic Tensor ($F^{\mu\nu}$), leads to Complex Electromagnetic Fields (CEMFs) for a moving observer. When the unmoved observer O and a moving observer O′ are related, O measures the same Real Electromagnetic Fields (REMFs) as those are given, using LB, but O′ measure CEMFs with the same formula. Complex Electromagnetic Tensor (F'_B) turns to Real Electromagnetic Tensor (F') using the formula $F' = \tilde{R}_{BLO'(O)} F'_B \tilde{R}_{BLO(O')}$. When there are two moving (relative to O) observers O′ and O′′, their real electromagnetic tensors are related, using the form $F'' = \Lambda_{LO''(O)} \Lambda_{LO(O')} F' [\Lambda_{LO''(O)} \Lambda_{LO(O')}]^T$. In addition, we prove that

[3] Spyridon Vossos, 39 Epidavrou Str., GR-15233 Halandri, Athens, Hellas, e-mail: spyrolia@hotmail.com.

the relation between two moving (relative to O) observers, when they use Real Coordinates (RCs), causes a real rotation between their frames. The turn is opposite to the turn of Thomas and has a little different measure, when the velocities are small. We apply these to the Uniform Circular Motion (UCM) and to the Hydrogen Atom (H), considering that the Proton (p) is the unmoved observer O and the laboratory observer O′ has infinitesimal velocity. Using Perturbation Theory (PT) we calculate the position of the fine structure peaks of the atomic spectrum. The result is better not only than this extracted using P. Dirac Theory, but also than that extracted using L. H. Thomas Method.

1. Introduction

SR as is applied until now uses LB. But it is known that LB is not closed transformation. In contrast, Lorentz Transformation (LT) (product of Rotation and LB) is closed [1]. Thus, if three observers O, O′ and O′′ are related, where O′ has parallel axes not only to O axes, but also to O′′ axes, then the axes of observers O and O′′ are not parallel. This cancels the transitive attribute in parallelism, when more than two observers are related. This option leads to successful results, such as Thomas Precession (TP) which explains the fine structure of atomic spectra. But this happens only if we take successive observers O′ and O′′ with the Thomas order [2]. If we reverse the order of this sequence, it yields a result with 200% relative error. In this presentation, we prove that there is closed LSTT which maintains the S^2 for every inertial observer. Thus, the transitive attribute in parallelism is valid and the axes rotation that happens in case that RC are used, when more than two observers are related, is only an equivalent phenomenon. The full proofs of the following equations there are at the referred papers in Greek [3,4,5] and in English [6].

2. The Matrix Λ_B

Initially, we find out Λ_B that expresses the LSTT which we postulate to a) be closed, b) be invariant to the rotation and c) maintain the spacetime interval S^2. For simplicity reasons, from now on wherever it is written i, is meant ±i. The result is VT: $X' = \Lambda_{B(\beta)} X$. So, the four-vectors of two whoever inertial observers O and O′ are related with *Vossos Matrix*:

$$\Lambda_{B(\beta)} = \gamma \begin{bmatrix} 1 & -\beta_x & -\beta_y & -\beta_z \\ -\beta_x & 1 & i\beta_z & -i\beta_y \\ -\beta_y & -i\beta_z & 1 & i\beta_x \\ -\beta_z & i\beta_y & -i\beta_x & 1 \end{bmatrix} \tag{1}$$

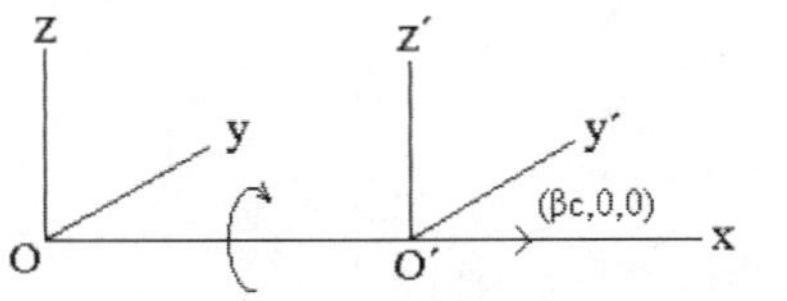

Figure 1. Two coordinate systems: O*xyz* and O′*x*′*y*′*z*′ which is moved with velocity (*β*c, 0, 0) relative to O*xyz*.

The typical Vossos Matrix along *x*-axis (Figure 1) is

$$\Lambda_{B\,\tau\upsilon\pi} = \gamma \begin{bmatrix} 1 & -\beta & 0 & 0 \\ -\beta & 1 & 0 & 0 \\ 0 & 0 & 1 & i\beta \\ 0 & 0 & -i\beta & 1 \end{bmatrix} \tag{2}$$

Using Minkowski four-vector, the corresponding typical *Vossos Matrix* in Minkowski space becomes

$$\Lambda^{M}_{B\,\tau\upsilon\pi} = \gamma \begin{bmatrix} 1 & -i\beta & 0 & 0 \\ i\beta & 1 & 0 & 0 \\ 0 & 0 & 1 & i\beta \\ 0 & 0 & -i\beta & 1 \end{bmatrix} \tag{3}$$

which is a rotation matrix. We observe that the same happens to (ct', x') and (z', y'). We observe that

$$\mathrm{A}_{(\beta)}\,\delta = \left[-\vec{\beta}\times\vec{\delta}\right] = \left[\vec{\delta}\times\vec{\beta}\right] \tag{4}$$

where

$$\beta = \begin{bmatrix} \beta_x \\ \beta_y \\ \beta_z \end{bmatrix}, \ \delta = \begin{bmatrix} \delta_x \\ \delta_y \\ \delta_z \end{bmatrix}, \ \mathrm{A}_{(\beta)} = \begin{bmatrix} 0 & \beta_z & -\beta_y \\ -\beta_z & 0 & \beta_x \\ \beta_y & -\beta_x & 0 \end{bmatrix} \tag{5}$$

Thus

$$\Lambda_{\mathrm{B}(\beta)} = \gamma \begin{bmatrix} 1 & -\beta^T \\ -\beta & \mathrm{I} + \mathrm{i}\,\mathrm{A}_{(\beta)} \end{bmatrix} \tag{6}$$

3. Measure of Time and Length

As Λ_B contains some elements which are imaginary numbers, we conclude that the space-time of one moving observer is complex. Thus, we put an index M or B in complex natural sizes. We have already made the option that O measures real space-time. This simplifies the problem, because the followings are proven: a) Time is real for every observer, b) The unmoved observer measures real velocity, c) Every observer measures square of velocity which is real and positive, d) Every observer measures γ-factor which is real and positive, for particles moved with $\upsilon < c$, e) The unmoved observer measures natural sizes that, can be suitably determined to be real. VT may be written as

$$X' = \Lambda_{\mathrm{B}(\beta)} X \tag{7}$$

Using vectors VT becomes

$$ct' = \gamma(ct - \vec{\beta}\cdot\vec{x}), \ \vec{x}'_M = \gamma\left(\vec{x} - \vec{\beta}ct\right) - \mathrm{i}\gamma\vec{\beta}\times\vec{x} \tag{8}$$

$$\Rightarrow c\,dt' = \gamma\,(c\,dt - \vec{\beta}\cdot d\vec{x}\,), \ d\vec{x}'_M = \gamma\left(d\vec{x} - \vec{\beta}c dt\right) - i\gamma\vec{\beta}\times d\vec{x} \tag{9}$$

3.1. Λ_B Properties – Inverse VT

$$\Lambda_{\mathrm{B}(0)} = \mathrm{I} \tag{10}$$

$$\Lambda^{-1}_{\mathrm{B}(\beta)} = \Lambda_{\mathrm{B}(-\beta)} \tag{11}$$

$$\det \Lambda_{\mathrm{B}(\beta)} = 1 \tag{12}$$

So, VT is closed subgroup of proper LT, under the operation

$$g_1 * g_2 = (\Lambda_{B2}\Lambda_{B1}, \Lambda_{B2} b_1 + b_2) \tag{13}$$

where b^{μ}_{1M} is μ-CC measured by O′, if O measures $x^{\nu} = 0$ and b^{μ}_{2M} is μ-CC measured by O′′, if O′ measures $x'^{\nu}_{M} = 0$ [7].

3.2. Lorentz Boost

Human realize real spacetime, through senses. In contrast, according to VT for a moving observer, time is real, but space is complex. This conflict is solved as following: Let have a rod O′Σ, perpendicular to x-axis moving along it, with velocity (βc, 0, 0) relative to Oxyz (Figure 2). Observer O project point Σ (edge or the rod), perpendicularly to y' and z'-axis. Observer O′ consider as RC the length of the resultant rods O′A και O′B. It emerges that for O′, RCs of point Σ are:

$$y' = \left|\left(\mathrm{O}'\vec{\mathrm{A}}\right)'\right| = y, \quad z' = \left|\left(O'\vec{B}\right)'\right| = z$$

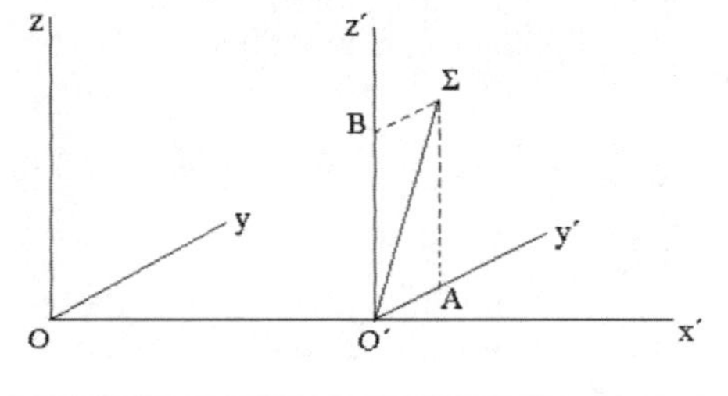

Figure 2. Moving rod O′Σ, perpendicular to x-axis moving along it, with velocity (βc, 0, 0) relative to Oxyz

We observe that using RCs, only the transformation from O to O′, happens according to Proper LT: $X' = \Lambda_{\mathrm{L}\upsilon\pi(\beta)} X$.

4. Electromagnetism, Vossos Precession and Fine Structure of Atomic Spectrums

We applied VT to electromagnetism and we find out that every classic physical law keep its form. After, we prove that VT is in order with QMs. Because QMs produce real eigenvalues, we work with three frames using RCs (Figure 3). It is proven that O′ measures for O′′ *Vossos Real Angular Velocity*

$$\vec{\omega}'_{\mathrm{B}} = -\frac{\gamma}{(\gamma+1)c^2}\vec{\upsilon}\times\vec{\alpha} \tag{14}$$

We observe that it emerges angular velocity γ times smaller value than L. H. Thomas Method.

According to QMs electron (e) motion around nucleus has not definite path. But, for a very short time interval the phenomenon may be approached with UCM. So, the above results may be applied, where particle is e and point K is the position of nucleus. In laboratory frame O′, the magnetic energy of an atom with reduced mass *m* is

$$U' = -\left(\frac{g''}{2}\frac{m}{m_e} - \frac{1}{\gamma(\gamma+1)}\right)\frac{e}{m^2c^2}\frac{1}{r}\frac{dV}{dr}\vec{L}\cdot\vec{S}'' \tag{15}$$

In case of H, the energy difference between 2 $^2P_{3/2}$ and 2 $^2P_{1/2}$ is

$$\Delta\overline{U}' = \frac{1}{256}\left(\frac{g''}{2}\frac{m}{m_e} - \frac{1}{\gamma(\gamma+1)}\right)\frac{me^8}{c^2h^4{\varepsilon_0}^4} \tag{16}$$

and is calculated [8]: $\Delta f = 10956.9\pm9.5$ MHz

The corresponding experimental data [9] is

$\Delta f_{\mathrm{exp}} = 10969.127\pm0.095$ MHz

The calculation has relative error $E_r = -0.00111(87) \approx -0.111\%$ less than P. Dirac and L. H. Thomas Methods.

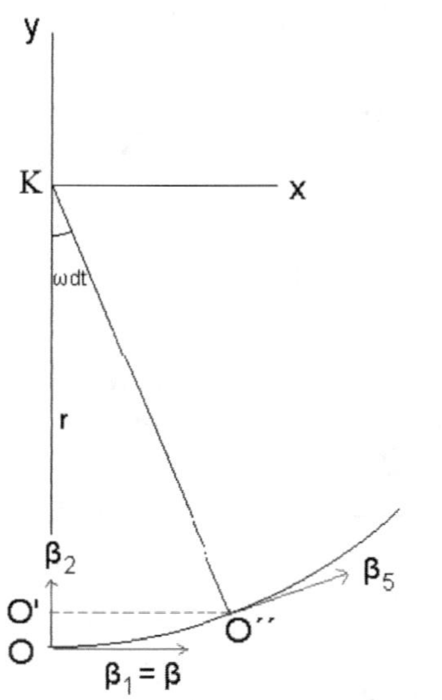

Figure 3. UCM with radius *r*. When t = 0, the particle has velocity β_1c = βc, while t = d*t*, the particle has velocity β_5c (relative to O). Inertial observer O′ is moved with velocity β_2c = α d*t* (relative to O).

References

[1] Tsamparlis M 2010 *Special relativity: An introduction with 200 problems and solutions* (Berlin Heidelberg: Springer-Verlag) pp 35, 41, eq 1.104 [ISBN: 978-3-642-03836-5]

[2] Thomas L H 1926 The Motion of the Spinning Electron *Nature (London)* **117** 514

[3] Vossos E and Vossos S 2013 Euclidean Complex Relativistic Mechanics I: The Matrix *Epitheor. Klin. Farmakol. Farmakokinet.* **31** 127-133 (http://pharmakonpress.gr/?p=464)

[4] Vossos E and Vossos S 2014 Euclidean Complex Relativistic Mechanics II: Measure of Time, Length, Velocity, Momentum, Energy and Doppler Effect *Epitheor. Klin. Farmakol. Farmakokinet.* **32** 45-59 (http://pharmakonpress.gr/?p=1335)

[5] Vossos E and Vossos S 2014 Euclidean Complex Relativistic Mechanics III: Electromagnetism, Vossos Precession and fine structure of atomic spectrums *Epitheor. Klin. Farmakol. Farmakokinet.* **32** 125-141 (http://pharmakonpress.gr/?p=1680)

[6] Vossos E and Vossos S 2015 Euclidean Complex Relativistic Mechanics IV: Special Relativity *Rev. Clin. Pharmacol. Pharmakokinet. Int. Ed.* **29**(1), in press

[7] Chistodoulakis T and Korfiatis E 2002 *Simeioseis Eidikis Theorias tis Sxetikotitas* (Athens: University of Athens, Department of Physics) pp 35-37 http://users.uoa.gr/~thapostol/SPREL/LectureNotes.pdf

[8] Lide D R ed 2009 *CRC Handbook of Chemistry and Physics* 89th Edition (Internet Version) (FL: CRC Press/Taylor and Francis, Boca Raton) pp 6-8

[9] Erickson G W 1977 Energy Levels of One Electron *Atoms. J. Phys. Chem. Ref. Data* **6** 849 (http://www.nist.gov/data/PDFfiles/jpcrd100.pdf)

Journal of Physics: Conference Series **633** (2015) 012028 doi:10.1088/1742-6596/633/1/012028

A novel method for determining the mean-field directly from the single particle matter density

S. Shlomo[1,2], M. R. Anders[1] and I. Talmi[2]
1Cyclotron Institute, Texas A&M University, College Station, Texas 77840
2The Weizmann Institute of Science, Rehovot, Israel

E-mail: s-shlomo@tamu.edu; manders14@tamu.edu; igal.talmi@weizmann.ac.il

Abstract. We present a novel method, using the single particle Schrodinger equation, to determine the central potential directly from the single particle matter density and its first and second derivatives. As an example, we consider the experimental data for the charge density difference between the isotones ^{206}Pb – ^{205}Tl that corresponds to the shell model $3s_{1/2}$ proton orbit, and determine the corresponding single particle potential (mean-field). We also present results of least-square fits to parametrized single particle potentials.

1. Introduction

The shell model, which is based on the assumption that nucleons in the atomic nucleus move independently in single particle orbits associated with a single particle potential, has been very successful in explaining many features of nuclei [1]. In this work we present a novel method, using the single particle Schrodinger equation for a wave function $\Psi(\vec{r})$ with eigenenergy E, to determine the central potential $V(\vec{r})$ directly from the measured single particle matter density, $\rho(\vec{r}) = [\Psi(\vec{r})]^2$ and its first and second derivatives, assuming known for all $\vec{r}$. We apply the method to the experimental data of the charge distribution of the proton $3s_{1/2}$ orbit given by the charge density difference, $\Delta\rho_c(r)$, between charge density distributions of the isotones ^{206}Pb – ^{205}Tl, determined by analysis of elastic electron scattering measurements [2].

We point out that the resulting single particle potential, if found, will provides a stringent limit on the effects of short correlation on the expected values of long-range operators, an important test for the shell model. The potential can also be used as an additional experimental constraint in determining a modern energy density functional (EDF) for more reliable prediction of properties of nuclei and nuclear matter [3,4]. In the next section we present the novel method and in section III we present some results with the conclusion given in section IV. Some results were presented in Ref. [5].

2. Formalism

Consider the single particle Schrodinger equation,

$$-\frac{\hbar^2}{2m}\Delta\Psi + V\Psi = E\Psi \,, \tag{1}$$

where $V(\vec{r})$ is a real local and non-singular potential and $\Delta = \vec{\nabla}\cdot\vec{\nabla}$, with $\vec{\nabla}$ being the gradient operator. Given a single particle wave function $\Psi(\vec{r})$, known for all $(\vec{r})$, with an eigenvalue E, the corresponding single particle potential V in (1) is uniquely determined [1] from

$$V(\vec{r}) = E + \frac{\hbar^2}{2m}S(\vec{r}) \,, \qquad S(\vec{r}) = \frac{\Delta\Psi(\vec{r})}{\Psi(\vec{r})} \,. \tag{2}$$

For a nonsingular V, $\Delta\Psi(\vec{r}) = 0$ when $\Psi(\vec{r}) = 0$. Operating with Δ on $[\Psi(\vec{r})]^b$, where b is positive and real, and using the relation $\vec{\nabla}\Psi^b = b\Psi^{b-1}\vec{\nabla}\Psi$ with the definition $S(\vec{r}) = \frac{\Delta\Psi(\vec{r})}{\Psi(\vec{r})}$ of Eq. (2), we obtain the general relation

$$S(\vec{r}) = \frac{\Delta\,[\Psi(\vec{r})]^b}{b\,[\Psi(\vec{r})]^b} - \frac{b-1}{b^2}\left[\frac{\vec{\nabla}\,[\Psi(\vec{r})]^b}{[\Psi(\vec{r})]^b}\right]^2 . \tag{3}$$

Note that Eq. (2) is a special case of Eq. (3) for b=1. For b > 2, $\vec{\nabla}[\Psi(\vec{r})]^b = 0$ and $\Delta[\Psi(\vec{r})]^b = 0$ when $[\Psi(\vec{r})]^b = 0$.

In the spherical case we have for the wave function of a nucleon

$$\Psi_{nlj}(\vec{r}) = \frac{R_{nlj}(r)}{r} Y_{lj} \,, \tag{4}$$

where $R_{nlj}(r)$ is the radial wave function for the orbit with principle number n, orbital angular momentum l and total angular momentum j and Y_{lj} is the known spin harmonic wave function.
Using Eqs. (2) and (4) we have that nuclear central potential is given by,

$$V_{cen}(r) = E + \frac{\hbar^2}{2m}S(r) - \frac{\hbar^2}{2m}\frac{l(l+1)}{r^2} - \frac{1}{2}(1-\tau_z)V^{coul}(r) \; - c_{ls}\, V_{s.o.}(r) \,, \tag{5}$$

$$S(r) = \frac{d^2R_{nlj}}{dr^2}\frac{1}{R_{nlj}(r)} \,.$$

The last three terms in (5) are the centrifugal, Coulomb and spin-orbit potentials. The single particle radial density $\rho_{nlj}(r)$ is related to the square of the single particle radial wave function R^2_{nlj} by

$$R^2_{nlj}(r) = 4\pi r^2 \rho_{nlj}(r) \,. \tag{6}$$

Adopting Eq. (1) for the radial wave-function $R_{nlj}(r)$ one can determine the central potential using Eq. (2). But this leads to numerical complication, particularly in the vicinity of the nodes of the wave function. From Eq. (1) for the radial wave-function $R_{nlj}(r)$ and Eq. (3) for b = 2 we have that the corresponding single particle potential V can be obtained from (5) by using the simple relation

$$S(r) = \frac{1}{2R^2_{nlj}}\left[\frac{d^2(R^2_{nlj})}{dr^2} - \frac{1}{2}\frac{1}{R^2_{nlj}}\left[\frac{d(R^2_{nlj})}{dr}\right]^2\right]. \tag{7}$$

From Eqs. (6) and (7) we find the relation

$$S(r) = \frac{1}{2\rho_{nlj}}\left[\frac{d^2\rho_{nlj}}{dr^2} + \frac{2}{r}\frac{d\rho_{nlj}}{dr} - \frac{1}{2\rho_{nlj}}\left(\frac{d\rho_{nlj}}{dr}\right)^2\right]. \tag{8}$$

Eqs. (8) can also be derived from Eq. (3) with b = 2 using the (real) three-dimensional wave function and the operators Δ and $\vec{\nabla}$ in spherical coordinates.
We note that for the central nuclear potential it is common to use the Woods Saxon (WS) form,

$$V(r) = V_0/[1 + exp((r - R_1)/a_0)] \,, \tag{9}$$

where, V_0, R_1 and a_0 are the depth, half radius and diffuseness parameters, respectively. Furthermore, for the Coulomb potential we adopt the form,

$$V^{coul}(r) = Ze^2\begin{cases}(3 - r^2/R^2_{ch})/2R_{ch} & r < R_{ch} \\ 1/r & r > R_{ch}\end{cases} \,, \tag{10}$$

In Eqs. (10) R_{ch} is the equivalent uniform charge distribution radius, obtained from the charge root-mean-square radius, $\langle r^2\rangle_{ch}$, by

$$R^2_{ch} = (5/3)\langle r^2\rangle_{ch} \,. \tag{11}$$

In elastic electron-nucleus scattering measurements the charge density distribution, $\rho_c(\vec{r})$, is determined by carrying out a phase shift analysis of the cross section [6]. In theoretical models the point proton density distribution, $\rho_p(\vec{r})$ is calculated. They are related by the convolution relation

$$\rho_c(\vec{r}) = \int \rho_p(\vec{r'})\,\rho_{pfs}(\vec{r} - \vec{r'})d^3\vec{r'} \,, \tag{12}$$

where $\rho_{pfs}(\vec{r})$ is the charge density distribution of the free proton, commonly parametrized as

$$\rho_{pfs}(\vec{r}) = \frac{1}{8\pi a^3}e^{-r/a} \,, \tag{13}$$

where $a^2 = \frac{1}{12} r_{pfs}^2$ with $r_{pfs} = 0.85$ fm being the corresponding charge root mean square (rms) radius [5]. The Fourier transform for $\rho_c(r)$, Eq. (12), is given by

$$F_c(q) = F_{pfs}(q) F_p(q) \,, \tag{14}$$

where $F_c(q)$, $F_{pfs}(q)$ and $F_p(q)$, are the Fourier transforms of $\rho_c(\vec{r})$, $\rho_{pfs}(\vec{r})$ and $\rho_p(\vec{r})$, respectively. Eq. (14) can be used to determine the form factor $F_p(q)$. Then $\rho_p(r)$ can be obtained from $F_p(q)$ by the inverse Fourier transform and compared with theoretical predictions.

3. Results

Starting In Figure 1a we present (solid line) the experimental data [2] for the charge density difference,

$$\Delta\rho_c(r) = \rho_c(r;\, {}^{206}\text{Pb}) - \rho_c(r;\, {}^{205}\text{Tl}), \tag{15}$$

between the isotones ^{206}Pb – ^{205}Tl. It is normalized to a total charge of one proton. The dotted lines indicate the experimental uncertainty. The two nodes associated with the proton $3s_{1/2}$ orbit are clearly seen in the figure. The experimental values of the charge rms radii of ^{206}Pb and ^{205}Tl are 5.4897 and 5.4792 fm, respectively. To assess the possible rearrangement effect (from ^{205}Tl to ^{206}Pb) on the charge rms radius of the 81 protons core in ^{206}Pb, we assume that it increases by 0.005 fm, similar to the change between nuclei in this region [7]. The rearrangement effect is approximated (see Ref. [7]) by scaling the charge distribution of ^{205}Tl . We thus obtain

$$\Delta\rho_{Rc}(r) = \rho_c(r;\, {}^{206}\text{Pb}) - \alpha^3 \rho_c(\alpha r;\, {}^{205}\text{Tl}), \tag{16}$$

where the scaling parameter $\alpha = 5.4792/(5.4792 + 0.005) = 0.9990$ is the ratio between the charge rms radius of ^{205}Tl to that of the 81 core protons in ^{206}Pb. The results for $\Delta\rho_{Rc}(r)$ is shown in Figure1a (dashed line). Using the relation (6) we determined the values of $R_p^2(r) = 4\pi r^2 \Delta\rho_p(r)$ and $R_{Rp}^2(r) = 4\pi r^2 \Delta\rho_{Rp}(r)$, deduced from the results of Figure 1a using Eqs. (12) and (14), shown as solid and dashed lines in Figure 1b. The dotted lines indicate the experimental uncertainty. The magnitude of the difference between $R_p^2(r)$ and $R_{Rp}^2(r)$ is similar to that of the experimental uncertainty.

The corresponding potentials, for $R_p^2(r)$ and $R_{Rp}^2(r)$ of Figure 1b, obtained by employing Eqs. (5) and (6), are shown in Figure 2a by the solid and dashed lines, respectively. The dotted lines (constant potentials) are extracted from fits of the corresponding wave-function in the vicinity of the minima to $\Psi = A\sin(kr+\phi) + C$. The Coulomb potential of Eq. (10), with $R_{ch} = 7.1$ fm, was adopted in the calculations. For the $3s_{1/2}$ orbit, there is no contribution from the centrifugal and spin-orbit potentials. Note the large uncertainties in the extracted potential in the vicinity of the nodes (at 2.6 and 4.9 fm).

We have therefore considered several nuclear central potentials with parameters obtained by fits to the corresponding experimental data, taking into account the Coulomb potential, Eq. (10). In Figure 2b the resulting potentials fitted to $R_p^2(r)$ of Figure 1b. Solid and dashed lines are fits to multipole lines potential to the solid and dashed lines of Figure 1b, respectively. We obtained good agreement with data with $\chi^2/N = 1.15$ and 1.81, respectively. The dashed-double dotted and dashed-dotted potential are for the fitted and the conventional Wood-Saxon potential (9) with $\chi^2/N = 3.28$ and 8.85 respectively.

4. Conclusions

Starting from the single particle Schrodinger equation for the function $\Psi(\vec{r})$, with eigen-energy E, we have derived a novel method for determining the corresponding single particle potential V from $[\Psi(\vec{r})]^b$, where b is a real number, assuming that $[\Psi(\vec{r})]^b$, $\vec{\nabla}\Psi$ and $\Delta[\Psi(\vec{r})]^b$ are known for all positions $(\vec{r})$, see Eqs. (2) and (3). It is clear from the Schrodinger equation that for a nonsingular V, $\Delta\Psi(\vec{r}) = 0$ when $\Psi(\vec{r}) = 0$. For b > 2 this condition is extended to the requirements that, $\vec{\nabla}[\Psi(\vec{r})]^b = 0$ and $\Delta[\Psi(\vec{r})]^b = 0$ when $[\Psi(\vec{r})]^b = 0$. For $b = 2$, we have from Eqs. (2) and (3) that the potential

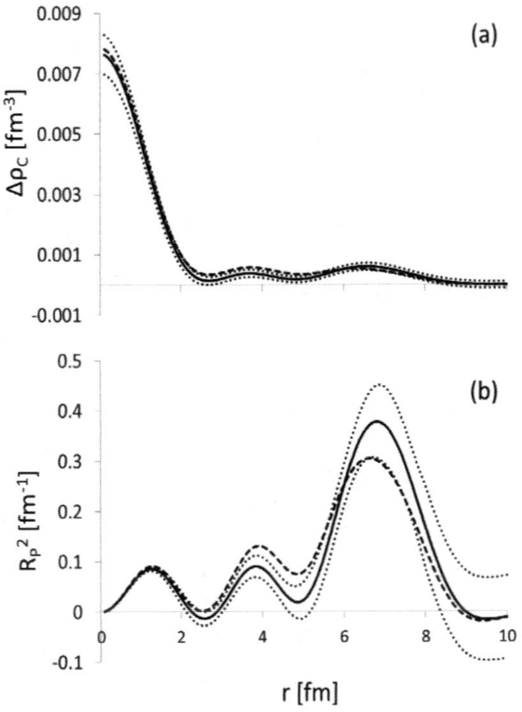

Figure 1. See text.

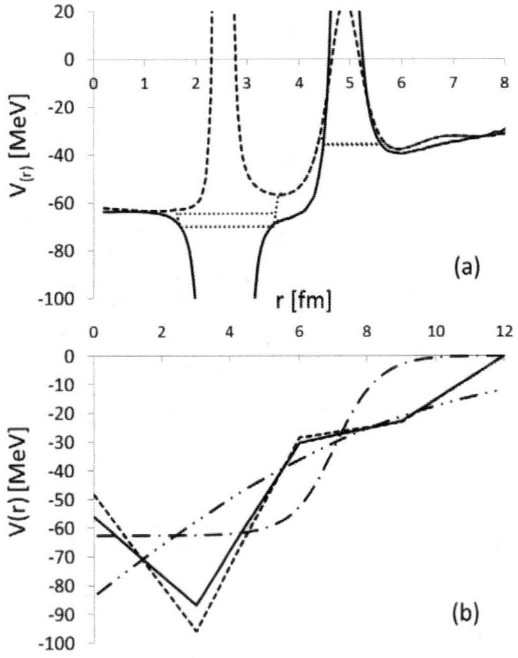

Figure 2. See text.

$V(\vec{r})$ is given in term of the corresponding single particle matter density $\rho(\vec{r})$ (for real $\Psi(\vec{r})$) and its first and second derivatives.

As an example, we considered the experimental data for the charge density difference between the isotones ^{206}Pb – ^{205}Tl that corresponds to the shell model $3s_{1/2}$ proton orbit, and determined the corresponding single particle potential (mean-field). We also presented results of least-square fits to parametrized single particle potentials. The $3s_{1/2}$ wave functions of the determined potentials reproduce fairly well the experimental data within the quoted errors. The fair agreement with fitted potentials may be an indication that effects of short range correlations on charge distributions are not significant. More accurate experimental data, with uncertainty smaller by a factor of two or more, may answer the question how well can the data be reproduced by a calculated $3s_{1/2}$ wave function.

5. References

[1] A. de Shalit and I. Talmi, *Nuclear Shell Model*, Academic Press, New York, N.Y., 1963.
[2] B. Frois, et al, Nucl. Phys. **A396**, 409c (1983).
[3] B. K. Agrawal, S. Shlomo and V. Kim Au, Phys. Rev. C **72**, 014310-1-13 (2005).
[4] S. Shlomo, in *The Universe Evolution: Astrophysical and Nuclear Aspects,"* Eds., L. Blokhintsev and I. Strakovsky, Nova Science Publishers, 2013.
[5] M. R. Anders, S. Shlomo and I. Talmi, ArXiv:1504.05507 [nucl-th]; Sub. to Phys. Lett B.
[6] L. R. B. Elton, *Nuclear Sizes*, Oxford University Press, London, 1961.
[7] H. Euteneuer, J. Friedrich and N. Voegler, Nucl. Phys. **A298**, 452 (1978).

Acknowledgments

S. S. thanks The Weizmann Institute of Science for the kind hospitality. This work was supported in part by US Department of Energy under Grant No. DOE-FG03-93ER40773.

Post-quantum attacks on key distribution schemes in the presence of weakly stochastic sources

S W Al–Safi and C M Wilmott

Department of Physics and Mathematics, Nottingham Trent University, Clifton Campus, Nottingham NG11 8NS, UK

E-mail: sabri.alsafi@ntu.ac.uk, colin.wilmott@ntu.ac.uk

Abstract. It has been established that the security of quantum key distribution protocols can be severely compromised were one to permit an eavesdropper to possess a very limited knowledge of the random sources used between the communicating parties. While such knowledge should always be expected in realistic experimental conditions, the result itself opened a new line of research to fully account for real–world weak randomness threats to quantum cryptography. Here we expand of this novel idea by describing a key distribution scheme that is provably secure against general attacks by a post-quantum adversary. We then discuss possible security consequences for such schemes under the assumption of weak randomness.

1. Introduction

Considerable efforts have been advanced in recent years which have yielded good insights into the relative structure of quantum theory. These efforts have been important from the view of better understanding the quantum framework, but, interestingly, they have also provided speculative arguments regarding the existence and structure of physical theories which may well supersede quantum theory. These alternative physical theories have been formulated under a set of mathematically intuitive operational representations, and are referred to as general probabilistic theories. Within this class of general probabilistic theories, a possible alternative to quantum theory called *Boxworld* has emerged. That aspect of Boxworld which has identified itself from other general theories relates to the set of correlations it produces. While a defining feature of quantum theory has been the non–local correlations it produces, these being correlations that cannot be produced by any local hidden variable theory, Boxworld, on the other hand, generates non–signalling correlations that are beyond those admitted by quantum theory. These non–signalling correlations have extremely powerful features, and were first studied by Popescu and Rohrlich, and accordingly called *PR–boxes* (Popescu and Rohrlich 1994). Indeed, to understand the potential of Boxworld, and general probabilistic theories is to appreciate the significance of PR–boxes. Amongst others, PR–boxes offer a non-local resource that would render communication complexity a trivial task, and they can also be used to violate information casuality.

While Boxworld finds itself on the upside of advantage, the same certainly remains true for quantum theory. Examples abound, but in relation to security and cryptography, there are a number of pertinent results. Shor's quantum algorithm holds profound implications for the security of classical cryptography (Shor 1994), and, yet, despite this outstanding individual result, the concept of cryptography has managed to evade redundancy as quantum theory comes equipped with its own style of cryptography called *quantum key distribution (QKD)*. QKD protocols enable pairs of communicating parties to produce shared random secret keys, which in turn, can be used to implement an unconditionally secure encryptions. Assuming the correctness of quantum theory, usual QKD protocols have been shown to be provably secure against eavesdropping attacks. Against the backdrop of general probabilistic theories, QKD has been found to be quite resilient. In particular, Barrett *et al.* (2005) have described a quantum key distribution scheme that is provably secure against arbitrary attacks by a post-quantum eavesdropper. In this post-quantum scenario, the eavesdropper is permitted to break the laws of quantum theory, but is limited solely by the impossibility of superluminal signalling. This is a remarkable result because, while the implication is that current QKD protocols may no longer be appropriate, Barrett *et al.* (2005) have, simultaneously, provided a revised quantum scheme that is secure against a non–signalling, post–quantum eavesdropper.

Quantum theory via key distribution has provided us with a provably secure way to communicate, but were quantum theory to ever fail, Barrett *et al.* (2005) have shown that we can rely on a quantum scheme that is secure against general attacks by a post-quantum eavesdropper. So, a big hurrah for QKD! Or, is it? Recent work by Bouda *et al.* (2012) examined QKD security issues arising when an adversary, aside from having full control of the various quantum and classical channels, has very limited access to the random sources employed by the communicating parties. They showed that with an increasing key length, only a negligible control of the randomness was necessary to render QKD insecure. This result illustrated that a vulnerability of quantum technology lies not with its science but in its practice of assuming perfectly uniform random processes. Indeed, both academia and industries specializing in commercial quantum technologies have blindly overlooked this fact by making assumptions that are fine in theory but not in practice. In this paper, we will outline an approach to examine the setup of a key distribution scheme robust against post-quantum attacks in light of results presented by Bouda *et al.* (2012). In section 2, we will outline some some basic facts relating to generalized probabilistic theories; in section 3, we present a quantum scheme based on Barrett *et al.*'s construction; and in section 4, we discuss various quantities relating to weak sources of randomness that may affect the real–world practical implementation of post-quantum cryptographic setups.

2. Preliminaries

2.1. Non-signalling constraints

We shall now motivate an operational setup for defining general probabilistic theories. As mentioned earlier, general probabilistic theories are compared and contrasted against the correlations they produce. These correlations are best described in terms of system probabilities that are ascribed to an outcome as a result of local measurements. Let $p(a,b|x,y)$ denote the probability of outcomes a and b given respective measurement choices, x and y. The value $p(a,b|x,y)$ respects the usual probability constraints. Thus, we have

$$0 \leq p(a,b|x,y) \leq 1 \tag{1}$$

for all measurement and outcome choices, and, for any fixed choice of local measurements,

$$\sum_{x,y} p(a,b|x,y) = 1. \tag{2}$$

We assume that correlations occur independently of the spatial separation between the systems, and demand that information cannot be communicated between the systems. This latter condition forces us to exclude those correlations that would permit the superluminal transmission of information, and, instead, we ascribe to position that outcome probabilities, for any subset of systems, must not be affected by local measurement choices of other systems. In mathematical terms, we, therefore, require that:

$$p(a|x) = \sum_b p(a,b|x,y) \quad \forall y;$$
$$p(b|y) = \sum_a p(a,b|x,y) \quad \forall x. \tag{3}$$

We refer to this requirement as the non-signalling principle.

2.2. Generalized probabilistic theories

Post-quantum correlations are described in terms of a set of conditional probability distributions. However, it is often convenient to represent the associated generalized probabilistic theory by a real vector space $\mathbb{R}$ whereby states s and effects e are specified by vectors such that $\langle e, s\rangle$ defines the probability of the effect e occurring in the state s. Therefore, analogous to quantum theory, we can prepare the usual combination of states s_i with probabilities p_i in the mixed state $s = \sum_i p_i s_i$ with $p_i > 0$ and $\sum_i p_i = 1$. The set of allowed states is given by the convex state space $\mathcal{S} \subset \mathbb{R}$, for which the set of pure states are represented by the extremal points of $\mathcal{S}$.

Next, we correspond a measurement outcome to an effect. This is a function that maps every state to a probability in the interval [0,1], and is well–defined for probabilistic mixtures of states. Consequently, the set of allowed effects is the set of vectors $e_i \in \mathbb{R}$ such that $0 \leq \langle e_i, s\rangle \leq 1$ and $\sum_i \langle e_i, s\rangle = 1$ for all $s \in \mathcal{S}$. The unit effect u is the unique effect that corresponds to the measurement with only one outcome that is certain to occur; $\langle u, s\rangle = 1$ for all $s \in \mathcal{S}$. Finally, we shall make use of unnormalized states $\tilde{s} = \lambda s$, $\lambda \geq 0$, and note that the set of unnormalized states yields the positive cone $\mathcal{S}_+$.

By way of example, we can demonstrate the versatility of generalized probabilistic theories and recover both classical and quantum theories. In the case of classical theory, every pure state s_i corresponds to an effect e_i via $\langle e_i, s_j\rangle = \delta_{ij}$, and mixed state representations have a unique decomposition into pure states. The dimension of the associated system is given by the maximal number of linearly independent vector that form a basis, and the system itself is given by a simplex. In terms of quantum theory, states and effects are given as positive hermitian matrices. Here, the action of an effect on a state is defined by the Hilbert-Schmidt inner product. The unit effect is the identity matrix and normalized states have trace one.

3. Post-quantum cryptography

The motivation for wanting to consider theoretical frameworks beyond that of quantum theory is entirely legitimate. If for no other reason than the fact that Shor's algorithm revealed fundamental implications for classical notions of security, we can be justified in wanting to ensure nothing is lurking around the quantum corner (Shor 1994). While quantum theory has been validated time and again, it is conceivable that some future experiment may expose some limitation. This practical concern provides a compelling argument to investigate post–quantum theories and their effects on quantum theory. To this end, the formulation of general probabilistic theories have helped to uncover natural alternatives to quantum theory. For instance, in Boxworld we have a theory which admits all non–local, non–signalling correlations. We caution that theories of this type may be superseded by superluminal signalling theories.

Standard quantum key distribution schemes have been proven robust within realistic environments. Interestingly, this robustness has only been demonstrated with respect to possible attacks on the quantum information exchanged during the protocol. On the other hand, there is an assumption that the eavesdropper possesses all knowledge of classical information exchanged.

Journal of Physics: Conference Series **633** (2015) 012029 doi:10.1088/1742-6596/633/1/012029

Furthermore, sources of classical randomness used in standard QKD protocols are implicitly assumed to be unbiased and completely inaccessible to the eavesdropper. Bouda *et al.* 2012 argued that perfect randomness is, in fact, unrealistic and that information eventually leaks via side channels, rendering it potentially vulnerable to the eavesdropper. As a result, Bouda *et al.* 2012 showed that the admission of weak randomness negatively influences the security of typical QKD protocols. It is precisely against this result, and the subsequent role of weakly random stochastic processes, that we question the security levels of more recent key distribution schemes. We shall now present a version of Barrett *et al.*'s construction, before going on to outline how this approach may allow an analysis of weak stochastic sources on the robustness of the construction, thereby taking into account all real-world practical threats.

Let us assume access to a d–dimensional irreducible state cone $\mathcal{S}_+$. Given a set of d linearly independent pure states $\{s_1, \ldots, s_d\}$, any measurement on this set will either disturb at least one of the states or provide zero information in which case, we have a non–disturbing measurement.

Protocol 1: A post-quantum key distribution scheme

Let us suppose that a communicating party wishes to discuss in a manner that is secure from arbitrary attacks by a post–quantum eavesdropper. The various stages of the corresponding protocol are as follows:

1. The first half of the communicating party, Alice, prepares a set of k *key* systems, each of which is randomly assigned to a state from the set $\{s_1, s_2\}$. Following this, Alice prepares a set of n *test* systems, each of which is randomly assigned to a state from the set $\{s_1, \ldots, s_d\}$. Alice then sends all $n + k$ prepared states in some random order to the second half of the communicating party, Bob;
2. On successful receipt of the communicated set of $n + k$ states, Alice announces those states that were originally assigned as key states and those that were assigned as test states;
3. On the set of test systems, Bob performs the measurement $\mathcal{M} = \{e_1, \ldots, e_d\}$ with $e_i, i = 1, \ldots d$ a linearly independent set. Alice and Bob abort the protocol on the occasion that any disturbance is revealed;
4. On the set of key systems, Bob performs the measurement $\mathcal{M} = \{e_1, e_2\}$ yielding

$$\langle e_1|s_2\rangle = 0, \qquad \langle e_1|s_1\rangle > 0$$
$$\langle e_2|s_1\rangle = 0, \qquad \langle e_2|s_2\rangle > 0$$

5. As the protocol has proceeded to stage 4, the resulting outcomes can be used to obtain the secret key, whereupon s_1 can be taken as 0 and s_2 can be taken as 1.

4. Modelling weak stochastic processes

Within quantum key distribution, research has demonstrated the absolute necessity of perfectly random processes to theoretically guarantee the secrecy of quantum transmissions – a secrecy that, remarkably, is not vulnerable to technological progress. Within entanglement theory, meanwhile, findings have illustrated how the manipulation of perfect randomness implies computational and cryptographic tasks for which no classical analogues exit. Perfect randomness is ubiquitous and, thus, quantum technologies have been dominated by this strict and expensive assumption, despite that fact that devices may show a significant bias in their outputs. We shall now outline a technique by which we can use this fact as a basis to model the sources of randomness employed in the protocol above. By taking this line of argument, we hope to reveal an understanding of how random processes possibly impact the protocol.

Sources of randomness are described by probability distributions, and the ideal source of randomness for almost all protocols is the perfectly random source. However, perfect randomness is very difficult to obtain in practice, and so we argue that a more appropriate model of

Journal of Physics: Conference Series **633** (2015) 012029 doi:10.1088/1742-6596/633/1/012029

randomness is that of weak randomness. Weak randomness has been extensively studied and is well understood in classical information processing, see, for example, Dodis *et al.* 2004, Maurer and Wolf 1997, and Renner and Wolf 2003. In light of this, we assume that a source exhibits randomness to some degree. Thus, we allow the outputs of such a source to be distributed according to any probability distribution displaying an adequate level of randomness. For what follows, we shall quantify the amount of randomness of a distribution by the *min–entropy* of its source. The min–entropy of the random variable $\mathbf{X}$ is defined by

$$H_\infty(\mathbf{X}) = \min_{x \in \mathbf{X}} \left(-\log_2 Pr\left(\mathbf{X} = x\right)\right). \tag{4}$$

Min–entropy is a widely accepted measure of weak randomness, and possesses a varied list of highly desirable properties (McInnes and Pinkas 1991). Importantly, min–entropy is sufficiently general which means that many real–world stochastic sources can be defined in terms of a min–entropy source. A noteworthy property of this source is that it readily models the most general of information leaks. In particular, a fall in min–entropy directly relates to the the amount of information learned by an eavesdropper, and in fact, a relatively small min–entropy decrease may well yield a sizeable amount of information. Another aspect of min–entropy relates to expected sequences. Here, an eavesdropper may be able exclude some sequences completely, which may aid with the design of future attacks.

Some important min–entropy quantities that may be used to assess the relative security merits of protocol above are as follows. A weak source of randomness is said to be an (n, b)–*source* if it outputs n–bit strings that are drawn according to a probability distribution with a min–entropy of at least b bits. Therefore, n–bit sequences have associated probabilities less than or equal to 2^{-b}. A perfect source occurs in the instance $b = n$. The *min–entropy loss* describes the bias of the source and is denoted $c = n - b$. We have an (n, b)*–flat distribution* if we have an (n, b)–source that is uniform on a subset of 2^b strings. The *min–entropy rate* is given by the quantity b/n, which achieves unity for perfectly random sources that deliver one bit of entropy per bit produced. Finally, particular interest lies with the *min–entropy loss rate*, which is denoted by quantity c/n. All of these quantities could prove to be very fruit tools when considering the real–world implementation key distributions protocols, including those known to be robust against general post-quantum eavesdroppers.

5. Future work and conclusion

In this paper, we have discussed a new topic of study called generalized probabilistic theories, which paves the way for a more robust discussion of physical systems should limitations to quantum theory be ever revealed. An interesting aspect of this discourse relates to cryptography and general security matters. Here, we outlined a novel approach for assessing the security parameters of cryptographic key distribution protocols that are known to be robust against general post–quantum attacks.

References

Barrett J, Hardy L and Kent A 2005 *Phys. Rev. Lett.* **95** 010503
Bouda J, Pivoluska M, Plesch M and Wilmott C 2012 *Phys. Rev. A* **86** 062308
Dodis Y, Jin Ong Shien, Prabhakaran M and Sahai A 2004 *Proc. 45th Ann. IEEE Symp. Found. Comp. Sci.* (IEEE Computer Society, Washington, USA) p 196.
Maurer U and Wolf S 1997 *Lect. Notes Comput. Sci. 1294, 307*
McInnes J L and Pinkas B 1991 *Proc. CRYPTO '90*
Popescu S and Rohrlich D 1994 *Foundations of Physics* **24** 3
Renner R and Wolf S 2003 *Lect. Notes Comput. Sci. 2729, 78*
Shor P 1994 *Proc. 35th Ann. Symp. on Foundations of Computer Science* (Santa Fe, NM)

Journal of Physics: Conference Series **633** (2015) 012030 doi:10.1088/1742-6596/633/1/012030

The Geometrical Meaning of Time in the Presence of Matter

Asher Yahalom
Ariel University, Ariel 40700, Israel
E-mail: asya@ariel.ac.il

Abstract. It is stated in many text books that the any metric appearing in general relativity should be locally Lorentzian i.e. of the type $g_{\mu\nu} = \text{diag}\ (1, -1, -1, -1)$ this is usually presented as an independent axiom of the theory, which cannot be deduced from other assumptions. The meaning of this assertion is that a specific coordinate (the temporal coordinate) is given a unique significance with respect to the other spatial coordinates. It was shown that the above assertion is a consequence of requirement that the metric of empty space should be linearly stable and need not be assumed. In this work we remove the empty space assumption and investigate the consequences of matter on the stability of Lorentzian space-time. It will be shown that in the presence of matter an upper limit scale to the size of a locally Lorentzian universe exists which incidentally is about the size of the observable universe.

1. Introduction

It is well known that our daily space-time is approximately of Lorentz (Minkowski) type that is, it possesses the metric $\eta_{\mu\nu} = \text{diag}\ (1, -1, -1, -1)$. The above statement is taken as one of the central assumptions of the theory of special relativity and has been supported by numerous experiments. But why should it be so?

Many textbooks [3, 4] state that in the general theory of relativity any space-time is locally of the type $\eta_{\mu\nu} = \text{diag}\ (1, -1, -1, -1)$, although it can not be presented so globally due to the effect of matter. This is a part of the demands dictated by the well known equivalence principle. The above principle is taken to be one of the assumptions of general relativity other assumption such as diffeomorphism invariance, and the requirement that theory reduce to Newtonian gravity in the proper regime lead to the Einstein equations[1]:

$$G_{\mu\nu} = -\frac{8\pi G}{c^4} T_{\mu\nu} \tag{1}$$

in which $G_{\mu\nu}$ is the Einstein tensor, $T_{\mu\nu}$ is the stress-energy tensor, G is the gravitational constant and c is the velocity of light.

Generic solution of Einstein equations for empty (and non-empty) spaces usually show different signatures as was shown by Eddington [5, page 25]. Among those possible solutions it was shown that for empty space-times [1, 2] only the Minkowskian solutions are stable.

Thus one need not assume that space-time is locally Minkowskian based on an empirical (unexplained) fact, rather one can derive this property from the field equations themselves based

[1] This specific form of Einstein equations is taken from Narlikar [3].

on the stability of the Minkowskian solutions. The number of assumptions needed to obtain the celebrated results of general relativity is reduced, by making the theory more compact we enhance its predictive strength. Moreover, it is shown that the existence of a temporal coordinate is a necessary consequence of the geometrical structure of the four dimensional space and not a separate ad-hoc assumption.

A different approach is due to Mukohyama & Uzan [6]. Those authors have derived a Lorentzian metric assuming a that at the microscopic level the metric is Riemannian, i.e., locally Euclidean, and that the Lorentzian structure, that we usually consider as fundamental, is in fact an effective property that emerges in some regions of a four dimensional space with a positive definite metric. In such a model, there is no dynamics nor signature flip across some hypersurface; instead, all the fields develop a Lorentzian dynamics in these regions because they propagate in an effective metric. They have shown that one can construct a decent classical field theory for scalars, vectors, and (Dirac) spinors in flat spacetime. And that gravity can be included but that the theory for the effective Lorentzian metric is not general relativity but of the covariant Galilean type. The constraints arising from stability, the equivalence principle, and the constancy of fundamental constants are detailed and a phenomenological picture of the emergence of the Lorentzian metric is also given. This construction, while restricted to classical fields, offers a new view on the notion of time. Unfortunately this approach is not consistent with Occam's razor as one is forced to introduce several scalar fields (one of them a "clock field") with fine tuned coupling in order to obtain the effective sign change.

What is deficient in the above approach (and similar approaches) is that additional theoretical structures & assumptions are needed in order to justify what appears to be a fundamental property of space-time. In previous works [1, 2, 7] it was claimed otherwise. It was shown that General relativistic equations and some "old fashioned" linear stability analysis will lead to a unique choice of the Lorentzian metric being the only one which is linearly stable in empty space-time. Other metrics are allowed but are unstable and thus can exist in only a limited region of empty space-time. The implications of this are discussed elsewhere [8, 9, 10]. It should be mentioned that the choice of coordinates in the Fisher approach to physics can be also be justified using the stability approach [7]. The nonlinear stability question of the Lorentzian metric for empty space-times was settled by D. Christodoulou & S. Klainerman [12].

This paper assumes that space-time must have four dimensions, it does not explain why this is so. For a possible explanation derived from string theory one can consult a paper by S. K. Rama [13]. The analysis discussed in this work is classical, for quantum aspects of space time difference one can consult a book by S. Hawking & R. Penrose [14]. Quantum field considerations made by A. White, S. Weinfurtner and M. Visser [15] have shown that regardless of the underlying classical theory, there are severe problems associated with any quantum field theory residing on a signature-changing background which does not respect the space-time difference (Such as the production of what is naively an infinite number of particles, with an infinite energy density). Those authors raises the question as to whether signature change transitions could be fully understood and dynamically generated within (modified) classical general relativity, or whether they require the knowledge of a theory of quantum gravity.

The plan of this paper is as follows: in the first section we outline the stability analysis of empty space-times, the next is dedicated to the modifications of those stability equations in the presence of matter. Finally we deduce an upper scale for a stable Minkowskian space-time and compare this scale with the size of the observable universe.

2. Stability analysis for empty space-times

In what follows I give an outline of the proof of the linear stability of Lorentzian space-time and the linear instability of other possible flat space-times. Unfortunately the details cannot be given here for the lack of space, but can be found elsewhere [1].

Consider a flat space-time, in this case one can find a set of coordinates such that the metric will be constant everywhere. A constant metric is nothing but a 4 by 4 symmetric matrix, therefore by a suitable choice of coordinates it can be made diagonal with eigenvalues all real. A final step would be to choose the scaling of the coordinates such that the metric will have the form:

$$\eta = \text{diag } (\pm 1, \pm 1, \pm 1, \pm 1). \tag{2}$$

Notice that no zero eigen values are allowed by virtue of the four dimensionality of space-time.

Once the family of possible canonical flat metrics is established we make a small perturbation $h_{\mu\nu}$ of the metric and consider a new metric $g_{\mu\nu} = \eta_{\mu\nu} + h_{\mu\nu}$ in which $h_{\mu\nu} << 1$. Inserting $g_{\mu\nu}$ into equation (1) and keeping only first order terms in $h_{\mu\nu}$ while taking $T_{\mu\nu} = 0$ we arrive at homogenous linear equations for $h_{\mu\nu}$. Solving those equations with **local** but otherwise arbitrary boundary conditions we see that only in the case of a Lorentzian metric the perturbation does not exponentially grow while in any other case it does. This explains why space-time is Lorentzian and why time is unique among the coordinates of space-time.

3. Stability analysis for a Lorentzian non empty space-time

Let us now assume that space-time is locally Lorentzian and contains matter. We choose following Narlikar [3] the non-perturbed metric to be:

$$\eta = \text{diag } (1, -1, -1, -1). \tag{3}$$

this metric is stable in empty space. We assume a fluid energy momentum tensor of the form [3]:

$$T_{\mu\nu} = (p + \rho c^2) u_\mu u_\nu - p g_{\mu\nu} \tag{4}$$

In the above p is the pressure, ρ is the density and $u_\mu = \frac{dx_\mu}{ds}$ in which the interval ds is defined as:

$$ds^2 = g_{\mu\nu} dx^\mu dx^\nu \tag{5}$$

We notice that $T_{\mu\nu}$ depend on metric perturbations both directly through the term $-p g_{\mu\nu}$ and through the ds term. Taking only first order terms and assuming that only h_{00} is non zero we arrive at the equation:

$$\eta^{00} \frac{\partial^2 h_{00}}{\partial x^{0 2}} + \eta^{ij} \frac{\partial^2 h_{00}}{\partial x^i \partial x^j} + 4\kappa \rho c^2 h_{00} = -2\kappa \rho c^2 \tag{6}$$

in which $\kappa = -\frac{8\pi G}{c^4}$. Taking a Fourier transform:

$$h_{00} = \frac{1}{(2\pi)^{\frac{3}{2}}} \int_{-\infty}^{\infty} A(x_0, \vec{k}) e^{i\vec{k}\cdot\vec{x}} d^3k, \quad \vec{k} = (k^1, k^2, k^3), \quad \vec{x} = (x^1, x^2, x^3) \tag{7}$$

We obtain the equation for A:

$$\frac{\partial^2 A}{\partial x^{0 2}} + (k^2 + 4\kappa \rho c^2) A = -2\kappa \rho c^2 (2\pi)^{\frac{3}{2}} \delta^3(\vec{k}), \qquad \vec{k}^2 \equiv k^2 \tag{8}$$

In the above $\delta^3(\vec{k})$ is a three dimensional Dirac delta function. Since a solution of an inhomogeneous differential equation is a sum of a specific solution to the inhomogeneous differential equation plus a general solution to the homogeneous equation. For stability analysis it will suffice to look at the homogeneous equation:

$$\frac{\partial^2 A}{\partial x^{0 2}} + (k^2 + 4\kappa \rho c^2) A = 0 \tag{9}$$

Journal of Physics: Conference Series **633** (2015) 012030 doi:10.1088/1742-6596/633/1/012030

This shows that in the presence of matter stability can not be maintained unless:

$$k > k_{crit} = \frac{4}{c}\sqrt{2\pi G\rho} \tag{10}$$

In terms of wavelengths we see that an upper scale for stable perturbations is:

$$\lambda < \lambda_{max} = \frac{2\pi}{k_{crit}} = \frac{c}{2}\frac{\sqrt{\pi}}{\sqrt{2G\rho}} \tag{11}$$

4. The scale of the observable universe

The a density of the universe is estimated to be [16]: $\rho \simeq 4.5 \cdot 10^{-28}\frac{Kilogram}{Meter^3}$ which leads to a $\lambda_{max} \simeq 1.08 \cdot 10^{27} Meter$. This is slightly larger than the radius of the observable universe: [17] $14 \cdot 10^9 Parsec \simeq 4.32 \cdot 10^{26} Meter$. Hence in "surprising" coincidence the diameter of the observable universe is about the size of the largest scale stable perturbation. At this time the size of the universe is not known [18] but it is suspected [9, 10] that above the stability scale the metric of space-time and hence physics may be quite different.

5. Conclusion

Mathematically speaking one of the main differences between time and space is encapsulated in the flat metric of our space-time which is **locally** of the Lorentzian type $\eta_{\mu\nu} =$ diag (1,-$-1, -1, -1$). But this is an empirical fact or a mathematical postulate thus unexplained. One can imagine also other flat metrics such as the Euclidian metric: diag $(1, 1, 1, 1)$. In Euclidian metrics there is no restriction on the speed of any moving body as the speed of light restricts the speed of propagation only in the presence of a Lorentzian metric. Why is our space-time Lorentzian and not Euclidean? The answer is that only the Lorentzian metric is stable [1] for an (almost) empty space. But space-time is not empty and thus the notion of time always progressing forward with the increase of entropy is probably just a consequence of the scales of reality that we are exposed to. In huge cosmological scale the Friedman-Lemaitre-Robertson-Walker universe losses its Lorentzian character [9, 10]. The horizon problem related to the homogeneity of cosmic microwave background can be solved if one takes into account the superluminal motion of particles for $r > r_c$ and the same particles moving into the $r < r_c$ Lorentzian domain. What is shown here is that a **locally** Lorentzian type universe although stable for all scales in an empty universe has a size limitation for a non empty universe and hence the suggested solution to the horizon problem is not only plausible but in fact necessary.

It should be noted that the stability analysis given here is rather restricted as we only allowed h_{00} perturbations. A full stability analysis is expected to involve the roots of a sixth order polynomial and hence we speculate the existence of a lower (as well as an upper scale) [9] for a Lorentzian Space-Time. This may have some consequences for elementary particle physics.

References

[1] Yahalom, A. (2008). The Geometrical Meaning of Time, Foundations of Physics, Volume 38, Number 6, Pages 489-497.

[2] Yahalom, A. (2009). The Gravitational Origin of the Distinction between Space and Time, International Journal of Modern Physics D, Vol. 18, Issue: 14, pp. 2155-2158. Gravity Research Foundation - 2009 Honorable Mention.

[3] Narlikar, J. V. (1993). Introduction to Cosmology, Second Edition. Cambridge University Press.

[4] Weinberg, S. (1995). The Quantum Theory of Fields, Cambridge University Press.

[5] Eddington, A. S. (1923). The mathematical theory of relativity, Cambridge University Press.

[6] Mukohyama, S. & Uzan, J. P. (2013). From configuration to dynamics: Emergence of Lorentz signature in classical field theory. Physical Review D, 87(6), 065020.

[7] Yahalom, A. (2011). Advances in Classical Field Theory, Chapter 6, Bentham eBooks eISBN: 978-1-60805-195-3.

[8] Yahalom A. (2013). Gravity and Faster than Light Particles. Journal of Modern Physics (JMP), Vol. 4 No. 10 PP. 1412-1416. DOI: 10.4236/jmp.2013.410169.
[9] Yahalom, A. (2014). On the Difference between Time and Space. Cosmology Vol. 18. 466-483. Cosmology.com.
[10] Yahalom, A. (2015). The Geometrical Meaning of Time - Some Cosmological Implications. Proceedings of 3rd International Conference on Mathematical Modelling in Physical Sciences (IC-MSQUARE 2014), 28-31 August 2014, Madrid, Spain. Journal of Physics: Conference Series (IOP Publishing), Volume 574, 012061. doi:10.1088/1742-6596/574/1/012061.
[11] Yahalom, A. (2010). Gravity and the Complexity of Coordinates in Fisher Information, International Journal of Modern Physics D, Vol. 19, No. 14 1-5, World Scientific Publishing Company.
[12] Christodoulou, D. & Klainerman, S. (1989-1990). The global nonlinear stability of the Minkowski space, Seminaire Equations aux derivees partielles (dit "Goulaouic-Schwartz"), Exp. No. 13, p. 29.
[13] Kalyana Rama, S. (2006). Los Alamos Archive hep-th/0610071 v2.
[14] Hawking, S. & Penrose, R. (2010). The Nature of Space and Time, Princeton University Press.
[15] White, A. Weinfurtner, S. and Visser, M. (2010). Signature change events: A challenge for quantum gravity?. Classical and Quantum Gravity 27.4 : 045007.
[16] "Universe 101: What is the Universe Made Of?". NASA: WMAP's Universe. Jan 24, 2014. Retrieved 2015-02-17.
[17] Bars, I. & Terning, J. (2009). Extra Dimensions in Space and Time. Springer. pp. 27. ISBN 978-0-387-77637-8. Retrieved 2011-05-01.
[18] Greene, B. (2011). The Hidden Reality. Alfred A. Knopf.

On the integrability of evolving membranes

Adrián Sotomayor[1], **Luis Cortés Vega**[1], **Alvaro Restuccia**[2]

[1] Departament of Matemathics , Antofagasta University, Antofagasta, Chile

[2] Physics Departament, Antofagasta University, Antofagasta, Chile

E-mail: adrian.sotomayor@uantof.cl,luis.cortes@uantof.cl,alvaro.restu@uantof.cl

Abstract. We analyze an integrable sector of the space of solutions of the four dimensional (super) membrane theory, where the target space is locally Minkowskian and hence a solution of the Supergravity equations. By performing a duality transformation we relate it to the 1+1 Born-Infeld integrable equation.

1. Introduction

The membrane field equations are a nonlinear coupled system which describes the evolution of the membrane on a target space. The equations are very interesting, they have the same dynamical structure as the Einstein equations of General Relativity. The initial data is restricted by first class contraints which are preserved by the membrane field equations. The constraints are the generators of diffeomorphisms on the base manifold, a three dimensional pseudo riemannian manifold. The evolution equations are not directly hyperbolic but they can be rewritten as an equivalent hyperbolic system which has a well-posed initial value problem and a blow up criterion [1]. From a physical point of view the membrane theory or more generally the supermembrane theory has the remarkable property of interacting with supergravity theories. In particular, the 11-dimensional supermembrane [6] is the unique geometrical object which interacts with the unique 11-dimensional supergravity. It is believed that the non-perturbative quantization of supermembranes is a fundamental step in the understanding of string theory. In this work we study the integrable properties of a class of solutions of the membrane field equations.

2. The duality transformation

In this section we analyze an integrable sector of Born-Infeld theory and of the membrane theory in a 4-dimensional target space. The duality among the theories allows to relate the $1+1$ integrable Born-Infeld (BI) equation:

$$\left[1+\phi'^2\right]\ddot{\phi}+\left[-1+\dot{\phi}^2\right]\phi''-2\dot{\phi}\phi'\dot{\phi}'=0 \tag{1}$$

where $\phi'=\partial_\sigma\phi$ and $\dot{\phi}=\partial_\tau\phi$, σ and τ are local coordinates on the $1+1$ formulation of Born-Infeld theory, to an integrable sector on the space of solutions of the membrane field equations. The transformation between the BI action and the membrane action in the case we will consider here, that is, in which the target space is a three dimensional Minkowski space times S^1 is a S-duality transformation in the string terminology.

In [2] the Cauchy problem of the 1+1 BI equation was solved and the quantization of the system was analyzed. In [3, 5, 4], for the same equation, an integrable criterion was shown (it has an infinite number of conserved quantities). In [5] the existence of a multi-Hamiltonian structure associated to it was proven.

The membrane theory in 4-dimensions is defined in terms of an embedding

$$X = (X^\mu) : M_3 \to T$$

from the base manifold M_3 to the target space T which we will take as a product of a Minkowski space times S^1. M_3 is assumed to be a product of a compact closed Riemann surface Σ times $\mathbb{R}$, the one dimensional vector space on which we define a time coordinate ξ^0.

The induced metric

$$g_{ij} = \partial_i X^\mu \partial_j X^\nu \eta_{\mu\nu} \tag{2}$$

defines the induced volume element $d^3\xi\sqrt{-g}$ where g is the determinant of the induced metric g_{ij}.

The action of the theory is defined by the volume of M_3 with the induced metric (2):

$$S(X) = \int_{M_3} \sqrt{-g}. \tag{3}$$

An equivalent formulation of the membrane theory expressed in terms of a polynomial lagrangian density is given by

$$S(\gamma, X) = -\frac{1}{2}\int d^3\xi\sqrt{-\gamma}\left(\gamma^{ij}\partial_i X^\mu \partial_j X^\nu \eta_{\mu\nu} - 1\right) \tag{4}$$

where the metric γ_{ij} on the 3-dimensional manifold M_3 is now taken as an independent field and not to be the induced metric. The -1 follows from the equivalence of this Polyakov type of lagrangian with the Nambu-Goto type of lagrangian (3).

If some of the embedding maps X^μ, say X, takes value on S^1 then we impose

$$\int_c dX = 2\pi q_c \tag{5}$$

where C is a basis of homology of dimension 1 elements on the base manifold M_3, while q_c is an integral number associated to each element of the basis. The set of q_c defines the winding of the map over S^1.

The S-dual formulation to (4) may be obtained by considering the master action

$$S(\gamma, X, L, A) = -\frac{1}{2}\int_{M_3} d^3\xi\sqrt{-\gamma}\left(\gamma^{ij}L_iL_j + \gamma^{ij}\partial_i X^m \partial_j X^n \eta_{mn} - 1\right) - \int_{M_3} d^3\xi \partial_j A_k L_i \tag{6}$$

where the closed two form with components $F_{ij} = \partial_i A_j - \partial_j A_i$ is restricted by

$$\oint_\Sigma F(A) = 2\pi p. \tag{7}$$

Here L_i is a vector field defined on the M_3 base manifold or in geometrical terms a section on a real vector bundle over M_3, $A_k d\xi^k$ is a $U(1)$ connection one-form over M_3 which arises from a connection on a $U(1)$ principal bundle with base manifold M_3 acting on an associated vector bundle and F_{ij} are the components of the curvature two-form associated to the connection one form $A_k d\xi^k$. The connection one form on M_3, valued on $U(1)$, is obtained by choosing a (local)

Journal of Physics: Conference Series **633** (2015) 012031 doi:10.1088/1742-6596/633/1/012031

section on the principal bundle over M_3 and taking a pullback of the connection on the principal bundle.

If we take variations of $S(\gamma, X, L, A)$ with respect to A_k we obtain locally

$$L_i = \partial_i X$$

for some scalar field X. If we replace the solution for L_i into the action (6) we obtain (4) or equivalently (3).

If instead we take variations of $S(\gamma, X, L, A)$ with respect to L_i, we obtain L_i in terms of γ^{ij} and A_k and after we replace it in $S(\gamma, X, L, A)$, we get the D-brane action [7, 8, 9]

$$S(\gamma, X, A) = -\frac{1}{2}\int_{M_3} d^3\xi\sqrt{-\gamma}\left(\gamma^{ij}\partial_i X^m \partial_j X^n \eta_{mn} + \frac{1}{2}\gamma^{ij}\gamma^{kl}F^r_{ik}F^r_{jl} - 1\right) \tag{8}$$

which can be proven to be equivalent to the following Born-Infeld action in terms of the induced metric $g_{ij} = \partial_i X^m \cdot \partial_j X^n \eta_{mn}$

$$S(X, A) = \int_{M_3}\left[-\sqrt{-\det\left(g_{ij} + F_{ij}\right)}\right].$$

For a global argument showing the equivalence of the action of the membrane theory compactified on a circle and the Born-Infeld action, in terms of the induced metric g_{ij}, see [10, 11, 12]. The condition (5) ensures that the closed two-form F is the curvature of a one-form connection on a principle bundle over the base manifold and it is a necessary condition to prove the equivalence.

We now consider a class of configurations which leads to an integrable system: the 1+1 Born-Infeld equation (1). The manipulation that we are going to do should be performed directly on the field equations. However the resulting action from them is the same as if we perform the manipulation directly on the action.

We consider

$$X^0 = \xi^0, X^1 = \xi^1, X^2 = \xi^2$$
$$A_0 = A_1 = 0, \partial_{\xi^2} A_2 = 0,$$

then

$$\det\left(g_{ij} + F_{ij}\right) = -1 + \left(\dot{\phi}\right)^2 - \left(\phi'\right)^2$$

where we have denoted $A_2 \equiv \phi$.

We now perform a dimensional reduction of the Born-Infeld action on M_3, it reduces to a two-dimensional formulation

$$\int_{S^1\times\mathbb{R}} d^2\xi\left(-\sqrt{1 - \left(\dot{\phi}\right)^2 + \left(\phi'\right)^2}\right)$$

which under variations with respect to ϕ yields the 1+1 Born-Infeld integrable equation (1) [13].

3. Solutions to the membrane field equations

We may now analyze the dual configurations in terms of the solutions of the membrane field equations.

They arise as the critical points of the action (3). The field equations obtained by taking variations of (3) with respect to X^μ give rise to submanifolds with vanishing mean curvature [1]:

$$\partial_i\left[\sqrt{-g}g^{ij}\partial_j X^\mu\right] = 0$$

where g_{ij} is the induced metric constructed from the maps X^μ, see (2). When $\sqrt{-g} \neq 0$ these equations can be expressed as

$$\left(\delta_{\mu\nu} - g^{ij}\partial_i X_\mu \partial_j X_\nu\right) g^{kl}\partial_k\partial_l X^\nu = 0. \tag{9}$$

If we consider the class of solutions such that

$$\begin{aligned} X^0 &= \xi^0 \\ X^1 &= \xi^1 \\ X^2 &= \xi^2 \\ X^3 &= \phi \end{aligned}$$

it follows that any solution of

$$g^{kl}\partial_k\partial_l\phi = 0$$

is a solution of the field equations (9). If we evaluate g^{kl} on the above configurations we obtain exactly the 1+1 Born-Infeld equation (1), as expected since the theories are related by a duality transformation. Although this equation resembles a laplacian operator acting of ϕ it isn't. In fact, g^{kl} itself depends on ϕ.

4. Conclusions

We have then characterized a class of solutions of the membrane field equations which describe an integrable equation. By performing a suitable change of variables this second order (in time) partial differential equation (1) reduces to a coupled system [5]

$$\begin{aligned} u_t + uu_x - ww_x &= 0 \\ w_t + uw_x - wu_x &= 0 \end{aligned}$$

describing a Chaplygin gas. The polynomial expression of this integrable system and its generalization

$$\begin{aligned} u_t + u^p u_x - ww_x &= 0 \\ w_t + uw_x - wu_x &= 0 \end{aligned} \tag{10}$$

(p an integer) can be compared to the known parametric coupled Korteweg-de Vries system [14, 15, 16].

The analysis of system (10) and generalizations of it in the context of membrane theory will be considered elsewhere.

Acknowledgments

A. S. , L. C. V. and A. R. are partially supported by Project Fondecyt 1121103, Chile.

References

[1] P. T. Allen, L. Andersson, A. Restuccia, Commun. Math. Phys. (2011) 301:383-410.
[2] B. M. Barbashov, N. A. Chernikov, Sov. Phys. JETP 23(1966) 861.
[3] B. M. Barbashov, N. A. Chernikov, Sov. Phys. JETP 24 (1966) 93.
[4] D. B. Fairlie and J.A. Mulvey, J. Phys. A: Math. Gen. 27 (1994) 1317.
[5] M. Arik, F. Neyzi, Y. Nutku, P. Olver and J. M. Verosky, J. Math. Phys. 30 (1989) 1338.
[6] E. Bergshoeff, E. Sezgin and P. K. Townsend, Phys. Lett. 189B (1987) 75.
[7] P. K. Townsend, Phys. Lett. B350 (1995) 184, hep-th/9501068, Phys. Lett. B373 (1996) 68, hep-th/9512062.
[8] R.G. Leigh, Mod. Phys. Lett. A4 (1989) 2767.
[9] C. Schmidhuber, D-brane Actions, hep-th/9601003.
[10] I. Martin, J. Ovalle, A. Restuccia, Phys. Lett. B472 (2000) 77-82.

[11] E. Witten, Selecta Math. (1995) 1:383, hep-th/9505186.
[12] M. I. Caicedo, I. Martin, A. Restuccia, Annals Phys. 300 (2002) 32-53.
[13] J. C. Brunelli, A. Das, Phys. Lett. B 426 (1998) 57-63.
[14] S. Yu. Sakovich, J. Nonlin. Math. Phys. (1999) 6, N 3 255-262 .
[15] P. Casati and G. Ortenzi, J. Geom. and Phys. (2006) 56, 418-449.
[16] L. C. Vega, A. Restuccia, A. Sotomayor, arXiv:1407.7743 v3.

The filamentation of the laser beam as a *labyrinth* instability

Florin Spineanu and Madalina Vlad

National Institute of Laser, Plasma and Radiation Physics, Magurele, 077125 Bucharest, Romania

E-mail: florin.spineanu@free.fr

Abstract. At incident powers much higher than the threshold for filamentation a pulse from a high-power laser generates in the transversal plane a complex structure. It consists of randomly meandering stripes defining connected regions where the field intensity is high; and, the complementary regions dominated by diffusive plasma with defocusing property. The pattern is similar to an ensemble of clusters of various extensions. We provide evidence that there is a correlation between this filamentation and the *labyrinth* instability in reaction-diffusion systems. Besides the similarity of the spatial organization in the two cases, we show that the differential equations that describe these two dynamical processes lead to effects that can be mutually mapped. For the laser beam at high power the Non-linear Schrodinger Equation in a regime of strong self-focusing and ionization of the air leads to multiple filamentation and the structure of clusters. Under the effect of the *labyrinth* instability a model of activator-inhibitor leads to a similar pattern. The origin of this connection must be found in the fact that both optical turbulence and the activator-inhibitor dynamics have the nature of competition between two phases of the same system.

1. Introduction

The propagation of the beam of a high power laser is an interaction with the atmospheric medium that involves the local activation of the polarization and of other processes that depends nonlinearly on the incident field. The first consequence is the tendency of self-suppression of the propagation, caused by the self-focusing of the beam. In an analytic description this is manifested by the existance of a singularity of the electric field of the wave, which arises in finite time [1]. Effectively, the characteristic trajectories of the propagation converge toward a caustic. The propagation (as a beam) has no meaning beyond this point.

In reality, the dynamics is more complex, including wave diffraction and dispersion of the group velocity, which act to prevent the singularity [2], [3]. More important, the self-focusing increases the intensity ($\sim |\mathbf{E}|^2$) of the incident field to a level that affects the gaseous medium: ionization of the neutral atoms generates a plasma which in turn has an effect of defocalization. The process toward singularity is stopped and a balance between the two opposite tendencies become possible. A regime of quasi-equilibrium is established with the structure of the beam in the transversal plane becoming inhomogeneous: symmetrically centered on the axis there is a channel of high beam intensity, having a diameter of the order of 100 μ, surrounded by a much larger (diameter of the order of several milimeters) cylindrical region where the intensity is much lower. In this latter region most of the energy is located. This structure is a filament. The

Journal of Physics: Conference Series **633** (2015) 012032 doi:10.1088/1742-6596/633/1/012032

fluctuation around the equilibrium, for powers in the beam higher than a threshold $P_{in} > P_{cr}$ allows the propagation on large distances, that can reach kilometers.

For powers that are a large factor (of the order of tens of units) greater than the critical power P_{cr} the picture changes. Essentially the symmetrical structure of the propagation, mentioned above, becomes azimuthally unstable [4]. The azimuthal perturbations evolve under weak mutual interaction and become sources of local filamentation, which dispose of sufficient power ($> P_{cr}$) to propagate individually. In this multi-filament structure the self-focalization followed by multiple ionization and defocalization produced by the plasma take place for each filament. The interaction leads to the random nucleation of filaments [5], then self-focusing followed by defocusing, everything leading to a random alternation of filaments with short time of existence in a field of small amplitude where the zones of plasma are dynamically redistributed in space [6], [7]. This regime has been called *optical turbulence* [8]. From measurements and numerical simulations it results that the plane transversal to the beam direction is organized in randomly meandering connected stripes (channels) where the field has higher intensity (in this region there are the filaments too) alternating and limited by, - similarly connected regions of plasma. This picture is clearly seen in the Figures of the work by Ettoumi et al. [9].

2. Hypothesis

The qualitative aspect of the pictures suggests the following association:

the shape and the dynamics of the meandering structures (connected ramnifications of two types) in the transversal plane to the direction of propagation are identical with structures that result from the reaction-diffusion dynamics of 2D media characterized by the competition of two components: activator and inhibitor.

This suggests to examine the possible common nature of the dynamics associated with

- optical turbulence of the regime of multiple filamentation
- the activator-inhibitor competition in nonlinear media

The simple qualitative comparison of the image of the transversal plane, and respectively of the two-dimensional domain of a system activator-inhibitor suggests that they may have a common nature (compare Figs. 1 and 2). Indeed , the activator-inhibitor is universal and there would be no surprise to be found in particular circumstances, as optical turbulence. The essential content of the two phenomena is common: it is a competition between two components, with one having auto-catalytic development and the other acting to limit and eventually to suppress this effect.

3. The equation of the envelope

The propagation of the laser beam exhibits various regimes. For a power greater than the critical threshold $P_{in} > P_{cr}$ the filament (central channel and the surrounding energy bath) self-focuses up to the limit that activates the opposing reaction of ionization followed by energization of the electrons: multiphoton ionization and inverse brehmstrahlung energy transfer. At much higher powers the same regime only persists for a finite time (or length of propagation) followed by multi-filamentation and the random dynamics produced by the modulational instability [2], [3].

The electric field is represented in a multiple space-time analysis by separating the slow evolution of the envelope $A(x, y, z, t)$, as $E = A(x, y, z, t)\chi(t)$ where $\chi(t) = \exp(-i\omega_0 t)$ is the fast wave factor. The nonlinear polarization of the air, together with processes of interaction with the plasma created by ionization, lead to the equation

$$\begin{aligned}\frac{\partial A}{\partial z} &= \frac{i}{2k_0}\Delta_\perp A - \frac{i}{2}\left(\frac{d^2k}{d\omega^2}\right)\frac{\partial^2 A}{\partial t^2} + i\omega n_2 |A|^2 A \\ &\quad -\frac{\sigma}{2}\rho A - \frac{i}{2}\sigma\omega\tau\rho A - \frac{\beta^{(K)}}{2\sqrt{K}}|A|^{2K-2} A\end{aligned}$$

where k_0 is the central wavenumber of the beam and the terms represent: the diffraction, the group-velocity dispersion (GDV), the Kerr nonlinearity of the polarization, the transfer of energy from the beam to the electrons of the plasma (σ is the cross section of the inverse brehmstrahlung effect) the rate of generation of the electrons (τ is the inverse of the collision frequency) and the multi-photon ionization. A separate equation, for the density of electrons ρ is explained below. The equation above, which is a modified Non-linear Schrodinger equation, is integrated numerically in various regimes of beam power. For $P_{in} \gg P_{cr}$ the result shows that from a rather homogeneous transversal state it occurs along propagation the nucleation of filaments. The first to be lost is the azimuthal symmetry followed by the quasi-independent evolution of the perturbations to definite filaments, by concentration of the photon energy from the surrounding medium. The self-focalization leads to episodic extinction and further re-nucleation of filaments [10]. This is the regime of *optical turbulence.* The equilibrium becomes a dynamical state placed at marginal stability. There is a competition between two opposite tendencies, and this is manifested as random fluctuations in close proximity of a statistical equilibrium. For part of the propagation, the fluctuations of the spatial distributions of the two fields A and ρ have correlations that do not exhibit any intrinsic scale, a situation that is characteristic to criticality. At longer distances from the source, a sharp transition occur and the transversal structure is broken into clusters of finite size [9].

4. The common analytical structure of the optical turbulence and of the *activator-inhibitor* dynamics

The transition from quasi-homogeneity in the transversal plane to a structure consisting of connected stripes where the fluence is high separated by similar branched channels of low fluence is similar to the fingering instability in a reaction-diffusion system. In the latter case the interface separates two distinct, competing, phases. In the high fluence region there is nucleation of filaments. The analogous effect in the activator-inhibitor system is the formation of "spotty-spiky" solutions [11].

The evolution of the envelope amplitude A is associated with that of the density ρ of the electrons of the plasma generated by ionization at focalization

$$\frac{\partial \rho}{\partial t} = D_\rho \Delta_\perp \rho + \frac{\beta^{(K)}}{K\hbar\omega_0}|A|^{2K}\left(1 - \frac{\rho}{\rho_{at}}\right)$$

Besides the last term that describes multi-photon ionization, we have introduced a new term, absent in the standard treatments, of diffusion of the electron density. We simplify the writing of the two equation (introducing coefficients a, b, ζ and ξ)

$$\begin{aligned}\frac{2k_0}{i}\frac{\partial \psi}{\partial z} &= \Delta_\perp \psi - a|\psi|^2\psi + b\rho\psi \\ \frac{\partial \rho}{\partial t} &= D_\rho \Delta_\perp \rho - \zeta\rho|\psi|^{2K} + \xi|\psi|^{2K}\end{aligned}$$

We first note that in the first equation above the term in the left and the first term in the right side, if alone, would give a set of multiple (Gaussian-like) bumps, disposed periodically on a line in the transversal plane. This may actually be seen as a local limit of a circular contour where spots of high intensity (core of filaments) exists (as confirmed by experiments [12]). This is very similar to what is found for activator-inhibitor systems, where the spot solutions are also periodic [13].

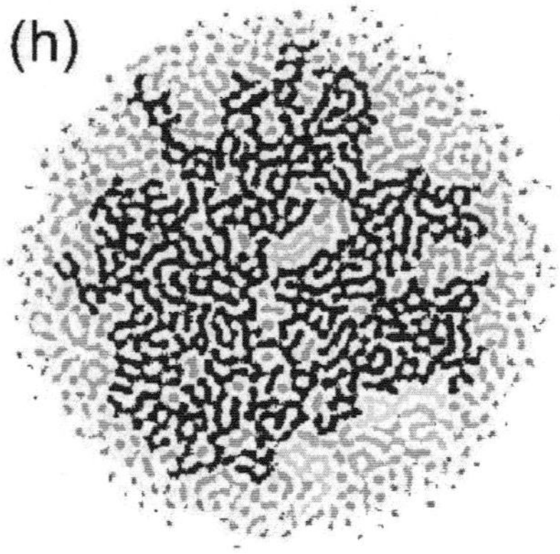

Figure 1. Structure of clusters in the transversal plane. This is subfigure (h) of Fig.1 from Ref [9]. (Courtesy W. Ettoumi)

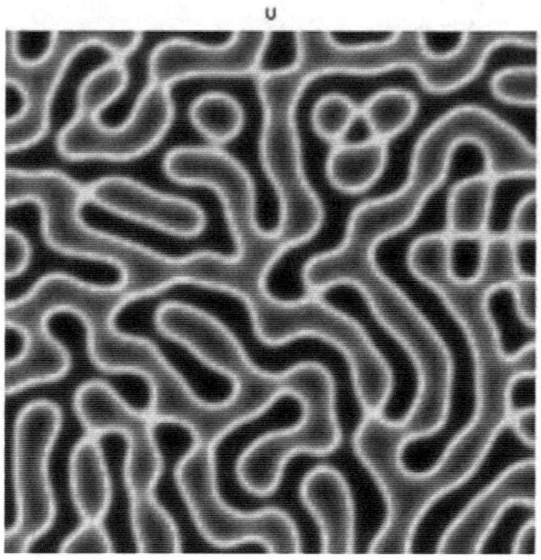

Figure 2. Labyrinth pattern for the Cahn-Hilliard activator-inhibitor system.

We note the similarity between the system of equation from where it results the *optical turbulence* and the analytical structure of the model FitzHugh-Nagumo, which exhibits the *labyrinth instability* [14]

$$\begin{aligned} \frac{\partial u}{\partial t} &= \varepsilon^2 \Delta u + f(u) - v \\ \frac{\partial v}{\partial t} &= \Delta v - \delta\gamma v + \delta u \end{aligned}$$

where $f(u) = u(1-u)(u-u_\kappa)$ for $u_\kappa \in (1, 1/2)$. The regime in which this dynamics consists of competing phases that occupy labyrinthic, mutually excluded, connected channels (like clusters) needs a fast inhibitor (v) diffusion.

The analogous behavior of the fields (ψ, ρ) and respectively (u, v) can be found to other models of the type *activator-inhibitor*, like Gierer-Meinhardt, Gray-Scott, Cahn-Hilliard [15]. The picture of meandering ramnifications in plane is similar and has at the origin the competition between the two physical fields [14], [16].

The activator-inhibitor systems have been shown to have a dynamics that is constructed on a deep level of order, the "shadow system" [17]. The existence of this system is a common property of several reaction-diffusion models and explains why they have similar behavior: labyrinthic interfaces, spot-like solutions, their spatially regular distribution (polygonal), their attraction and coalescence.

5. Results that become accessible by the mapping between optical turbulence and the *activator-inhibitor* dynamics

The analogy between the two systems may be useful. This is because the class of *activator-inhibitor* systems has been investigated mathematically and disposes of precise description of its various solutions. We expect to transfer some of these results to the model of optical turbulence, especially for the regimes where it has been examined experimentally or by numerical simulation

(i) The solutions of the Gierer-Meinhardt (GM) system can be, for the case of rapid diffusion of the *inhibitor*, localized ("spot-like") [11]; this corresponds to the filaments observed in the high-fluence region.

(ii) in some conditions, the GM solutions form groups (clusters) with regular spatial disposition, eventually polygonal. Correspondingly, in a laser beam it has already been found regular spacing of the multiple filaments. It seems supported theoretically by the application of the notion of *Chaplygin gas with anomalous polytropic* [18].

(iii) the solutions of GM, found to be grouped in clusters, have been proved to be unstable and a number of "spots" disappear after an evolution in time, being replaced by a single central bump solution. This looks to be the analogous case to the coalescence of filaments and re-formation of a single central filament.

A problem raised by this mapping: which of the diffusion-reaction systems that have a behavior of the type *activator-inhibitor* can be identified as the equivalent of the modified NSEq in the regime of multiple-filamentation? The response appears to not be constraining, because at fast inhibitor they have the property of being manifestation of a *shadow system* which means a common type of behavior. However we must confine therefore to those characteristics that are common and can be made to correspond to the multi-filamentation.

A specific property is the proliferation of interfaces caused in the activator-inhibitor system, by the *labyrinth* instability and in the laser field, by the competition between high fluence clustered (branched) spatial regions (where filaments can nucleate as spots) and zones of plasma with defocalization effect which keeps control on the local expansion of the first phase.

An interesting aspect that can result from a comparative investigation performed on the two systems is the effective interaction between filaments. This is based on the connection between the NSEq (in its extended form for beam propagation) and the Complex Ginzburg-Landau (CGL) equation [19]. The exact solution of the CGL equation is a soliton with an oscillating tail. If there would be no interaction then the sum of two such functions would also be a solution too. Replacing this sum in the expression of the energy, it should result a sum of two times the individual energy of a single solution. Or, this is not so, showing that besides the individual energies we have a term of interaction. This depends parametrically on the positions of the centers of the two solitons. When the relative distance between the centers is varied, the supplementary term decreases (if there is attraction) or increases (if there is repulsion). The method is identical with the one used to find the interaction between vortices of the Abelian-Higgs superconduction model [20] at non-self-duality. However the energy of interaction between solitons of the CGL equation is found to be exponentially small, which means that the coalescence of filaments is slow. Or, the mapping to an *activator-inhibitor* system helps to reformulate the problem: indeed there are solutions consisting of several localized bumps with a space distribution which is regularly periodic. This solution is unstable and the final state consists of a central spike [21].

6. Conclusion

The filamentation generated during the propagation of the pulse of a high power laser has many regimes and in particular the optical turbulence. It is the formation in the transversal plane of a system of randomly meandering ramnifications where the incident field is high, separated, and limited by -, a similar region dominated by defocusing plasma. This structure is dynamical and in addition in the high intensity zone new filaments nucleate. They are transient and end up by coalescing into a single chanel of propagation. This regime can be mapped onto the activator-inhibitor dynamics of a nonlinear reaction-diffusion system. Reformulated in the new framework, some problems of beam propagation can be simpler.

An important objective of further investigation is how is reflected in the activator-inhibitor model the inverse phase transition that suppresses progressively the long range correlations in the beam field, *i.e.* breaks up the large scale clusters.

Acknowledgments

The authors thank to W. Ettoumi for permitting to reproduce a figure from the paper [9].

This work is partially supported by the Contract PN 09 39 01 01.

References

[1] Marburger J 1975 *Progress in Quantum Electronics* **4, Part 1** 35 – 110 ISSN 0079-6727 URL `http://www.sciencedirect.com/science/article/pii/0079672775900038`
[2] Couairon A and Mysyrowicz A 2007 *Physics Reports* **441** 47 – 189 ISSN 0370-1573 URL `http://www.sciencedirect.com/science/article/pii/S037015730700021X`
[3] Bergé L 1998 *Physics Reports* **303** 259 – 370 ISSN 0370-1573 URL `http://www.sciencedirect.com/science/article/pii/S0370157397000926`
[4] Skupin S, Peschel U, Etrich C, Leine L, Michaelis D and Lederer F 2002 *Opt. Lett.* **27** 1812–1814 URL `http://ol.osa.org/abstract.cfm?URI=ol-27-20-1812`
[5] Kandidov V P, Kosareva O G, Tamarov M P, Brodeur A and Chin S L 1999 *Quantum Electronics* **29**(10) 911–915
[6] Mlejnek M, Wright E M and Moloney J V 1998 *Opt. Lett.* **23** 382–384 URL `http://ol.osa.org/abstract.cfm?URI=ol-23-5-382`
[7] Méchain G, Couairon A, Franco M, Prade B and Mysyrowicz A 2004 *Phys. Rev. Lett.* **93**(3) 035003 URL `http://link.aps.org/doi/10.1103/PhysRevLett.93.035003`
[8] Mlejnek M, Kolesik M, Moloney J V and Wright E M 1999 *Phys. Rev. Lett.* **83**(15) 2938–2941 URL `http://link.aps.org/doi/10.1103/PhysRevLett.83.2938`
[9] Ettoumi W, Kasparian J and Wolf J P 2015 *Phys. Rev. Lett.* **114**(6) 063903 URL `http://link.aps.org/doi/10.1103/PhysRevLett.114.063903`
[10] Tzortzakis S, Bergé L, Couairon A, Franco M, Prade B and Mysyrowicz A 2001 *Phys. Rev. Lett.* **86**(24) 5470–5473 URL `http://link.aps.org/doi/10.1103/PhysRevLett.86.5470`
[11] del Pino M, Kowalczyk M and Wei J 2003 *Ann. I. H. Poincaré* **20** 53–85
[12] Méchain G, CDAmico, André Y B, Tzortzakis S, Franco M, Prade B, Mysyrowicz A, Couairon A, Salmon E and Sauerbrey R 2005 *Optics Communications* **247** 171 – 180 ISSN 0030-4018 URL `http://www.sciencedirect.com/science/article/pii/S0030401804011812`
[13] Wei J and Winter M 2008 *Journal of Mathematical Biology* **57** 53–89 ISSN 0303-6812 URL `http://dx.doi.org/10.1007/s00285-007-0146-y`
[14] Goldstein R E, Muraki D J and Petrich D M 1996 *Phys. Rev. E* **53**(4) 3933–3957 URL `http://link.aps.org/doi/10.1103/PhysRevE.53.3933`
[15] Desai R C and Kapral R 2009 *Dynamics of self-organized and self-assembled structures* (Cambridge University Press) ISBN 978-0-511-71966-0
[16] Roland C and Desai R C 1990 *Phys. Rev. B* **42**(10) 6658–6669 URL `http://link.aps.org/doi/10.1103/PhysRevB.42.6658`
[17] Li F and Ni W M 2009 *Journal of Differential Equations* **247** 1762 – 1776 ISSN 0022-0396 URL `http://www.sciencedirect.com/science/article/pii/S0022039609001806`
[18] Trubnikov B and Zhdanov S 1987 *Physics Reports* **155** 137 – 230 ISSN 0370-1573 URL `http://www.sciencedirect.com/science/article/pii/0370157387901232`
[19] Malomed B A 1991 *Phys. Rev. A* **44**(10) 6954–6957 URL `http://link.aps.org/doi/10.1103/PhysRevA.44.6954`
[20] Jacobs L and Rebbi C 1979 *Phys. Rev. B* **19**(9) 4486–4494 URL `http://link.aps.org/doi/10.1103/PhysRevB.19.4486`
[21] Alikakos N D and Fusco G 1998 *Archive for Rational Mechanics and Analysis* **141** 1–61 ISSN 0003-9527 URL `http://dx.doi.org/10.1007/s002050050072`

Gravitational mass and Newton's universal gravitational law under relativistic conditions

Constantinos G. Vayenas[1,2,*], Athanasios Fokas[3,4] and Dimitrios Grigoriou[1]

[1]School of Engineering, Caratheodory 1, St., University of Patras, Patras GR 26504, Greece
[2]Division of Natural Sciences, Academy of Athens and [3]Centre of Mathematics, Academy of Athens, 28 Panepistimiou Ave.,GR-10679 Athens, Greece
[4]Department of Applied Mathematics and Theoretical Physics, University of Cambridge, Cambridge, CB3 0WA, UK

E-mail: cgvayenas@upatras.gr

Abstract. We discuss the predictions of Newton's universal gravitational law when using the gravitational, m_g, rather than the rest masses, m_o, of the attracting particles. According to the equivalence principle, the gravitational mass equals the inertial mass, m_i, and the latter which can be directly computed from special relativity, is an increasing function of the Lorentz factor, γ, and thus of the particle velocity. We consider gravitationally bound rotating composite states, and we show that the ratio of the gravitational force for gravitationally bound rotational states to the force corresponding to low ($\gamma \approx 1$) particle velocities is of the order of $(m_{Pl}/m_o)^2$ where m_{Pl} is the Planck mass $(\hbar c/G)^{1/2}$. We also obtain a similar result, within a factor of two, by employing the derivative of the effective potential of the Schwarzschild geodesics of GR. Finally, we show that for certain macroscopic systems, such as the perihelion precession of planets, the predictions of this relativistic Newtonian gravitational law differ again by only a factor of two from the predictions of GR.

1. Introduction

Since the formulation [1] and frequent experimental confirmation [2] of the theory of General Relativity it is generally accepted that Newton's universal gravitational law is not accurate enough under relativistic conditions [2]. For example it cannot describe the mercury perihelion precession, whereas the theory of General Relativity (GR) is "not inconsistent" with it [3, 4]. Also the deflection of light passing near massive objects predicted by Newton's gravitational law is half the value predicted by GR which is in good agreement with experiment [3].

In testing Newton's gravitational law little attention has being paid in the differences between the rest, relativistic, inertial and gravitational masses of the particles involved.

Inertial and gravitational masses are equal to each other, according to the equivalence principle [5], but the ratios of inertial or relativistic mass to the rest mass are unbounded as the particle velocity approaches the speed of light [6].

Consequently the following quite important question arises when using Newton's universal gravitational law under conditions where the Lorentz factor, γ, is significantly larger than unity: Should one use the particle rest mass, m_o, the relativistic mass, γm_o, or the gravitational mass, m_g, which is equal to the inertial mass, m_i? This is the question discussed in the present paper and although it appears to be obvious that one should use the gravitational mass, m_g, rather than the rest mass, m_o, or the relativistic mass, γm_o, one may wonder why this point has been discussed only recently [7]. There appear to be two reasons for this, first that it has been discussed for years that Newton's gravitational law fails under relativistic conditions, which is certainly true when using rest or relativistic masses [2, 7], as also shown here, and second that the velocities for the planets of our planetary system are of the order $10^{-4}c$, so that the corresponding Lorentz factors are of the order 10^{-8} and thus the differences between relativistic and gravitational masses are very small. This, however, is not the case for many other astronomical systems [8, 9].

2. Rest, relativistic, inertial and gravitational mass

The inertial, m_i, and thus the gravitational mass, m_g, is related to the rest mass, m_o, via

$$m_g = m_i = \gamma^3 m_o, \tag{1}$$

where $\gamma(= (1 - \mathrm{v}^2/c^2)^{-1/2})$ is the Lorentz factor. This result was first obtained in A. Einstein's pioneering special relativity paper for linear particle motion [6]. It has been shown recently to remain valid for arbitrary particle motion, including circular motion [7]. Consequently, Newton's universal gravitational law for the gravitational force between two particles of rest mass m_o, each, and velocities v_1, and v_2 relative to a laboratory observer, is

$$F = \frac{G m_{g,1} m_{g,2}}{r^2} = \frac{G m_o^2 \gamma_1^3 \gamma_2^3}{r^2}, \tag{2}$$

and for equal particle velocities v_1 and v_2 reduces to

$$F = \frac{G m_o^2 \gamma^6}{r^2}. \tag{3}$$

This equation has been found to predict the existence of gravitationally confined states with many of the properties of hadrons [7, 10] and to show that the rest masses of quarks are very small, in the mass range of neutrinos [7, 13]. The ratio F/F_o of the force keeping two or three-particle systems in orbit, divided by the same force computed for $\gamma \approx 1$, is $(1/2)(m_{Pl}/m_o)^2$, exactly half of the value predicted by GR [7].

3. SR treatment of the perihelion precession

According to Newton's 2nd law the equation of motion of a body or particle of rest mass m_o is given by

$$F = m_o a, \tag{4}$$

where a is the acceleration. For example, for linear motion we have [11]:

$$F = m_o a = m_o \frac{d^2 y}{dt^2} = m_o \frac{\gamma^{-1} d^2 y_o}{\gamma^2 dt_o^2} = m_o \gamma^{-3} \frac{d^2 y_o}{dt_o^2} = \gamma^{-3} m_o a_o. \tag{5}$$

If the force is a gravitational one, then using the equivalence principle of inertial and gravitational mass, m_i and m_g respectively, and accounting for $m_i = \gamma^3 m_o$ [6, 7, 11, 12] one obtains

$$F = \frac{G M m_o \gamma^3}{r^2}. \tag{6}$$

From (5) and (6) it follows

$$a_o = \gamma^6 G M / r^2. \tag{7}$$

Journal of Physics: Conference Series **633** (2015) 012033 doi:10.1088/1742-6596/633/1/012033

The proper acceleration, i.e. the one perceived on the rotating planet, e.g. on mercury, is

$$a_o = GM/r^2. \tag{8}$$

It thus follows that one can perform the analysis as if moving on mercury ($\gamma = 1$) and simply replace GM by GMγ^6.

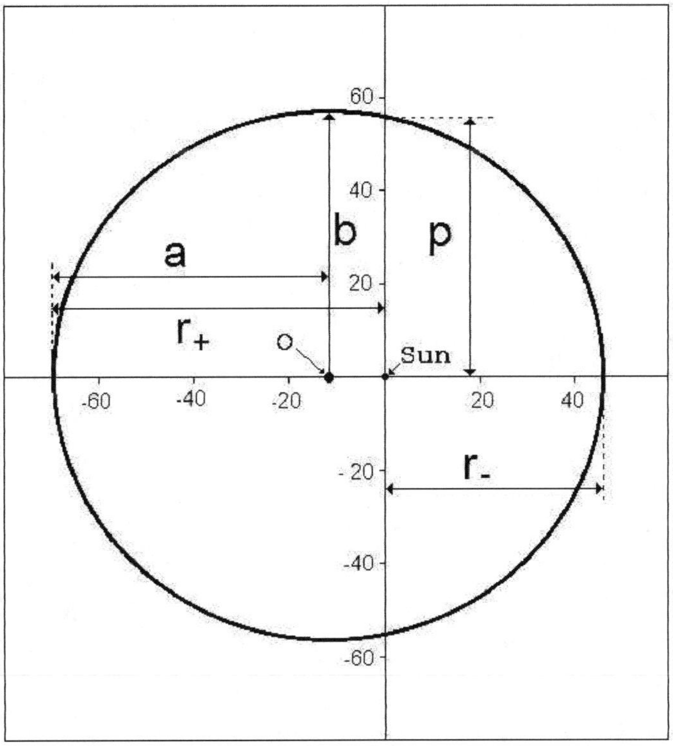

Figure 1. Schematic of the mercury elliptical orbit (e=0.2054), showing the definitions of the elliptical parameter (semilatus rectum) p, of the semimajor axis a, of the semiminor axis b, of the perihelion r_- and aphelion r_+; numbers indicate distances in 10^9 m.

The final expression obtained by GR for the perihelion precession of mercury or any other planet moving on an elliptical orbit is [2, 4, 14, 15, 16, 17]

$$\Delta\varphi = \frac{6\pi GM}{a(1-e^2)c^2} = \frac{6\pi GM}{pc^2}, \tag{9}$$

where a is the semimajor axis of the ellipse, p is the elliptic parameter, also known as semi-latus rectum, and e is the eccentricity (it is $e = 0$ for a cycle, $0 < e < 1$ for an ellipse, $e = 1$ for a parabola and $e > 1$ for a hyperbola) (Figure 1); p and a are related via

$$p = (1 - e^2)a. \tag{10}$$

One observes in the second equation (9) that when considering the set of all ellipses, including the cycle, of given ellipse parameter p, then $\Delta\varphi$ does not depend on the eccentricity e. Consequently in the SR treatment given here we simplify the computation by considering a cycle of radius $r = p = a$. Actually, in order to keep the definition of perihelion uniquely defined, we consider an ellipse of eccentricity $e = \epsilon$, where ϵ is an arbitrarily small positive number, e.g. $\epsilon < 10^{-12}$, which is sufficient for the present analysis.

For $r = p = a$ the equation of motion of the rotating planet or, more generally object, is

$$\frac{m_o \mathrm{v}^2}{p} = \frac{GMm_o\gamma^6}{p^2}, \tag{11}$$

obtained from the non-relativistic proper equation of motion observed on the planet, i.e.

$$\frac{m_o \mathrm{v}^2}{p} = \frac{GMm_o}{p^2}, \tag{12}$$

by replacing GM with GMγ^6. Equation (11) yields

$$\frac{\mathrm{v}^2}{\gamma^6} = \frac{GM}{p}, \tag{13}$$

which for $\mathrm{v} << c$, as is the case here, thus $\gamma \approx 1$, gives

$$\mathrm{v}^2 \approx GM/p, \tag{14}$$

and thus from the definition of γ it follows

$$\gamma \approx \frac{1}{(1 - GM/c^2p)^{1/2}}, \tag{15}$$

thus

$$\gamma^3 \approx \frac{1}{(1 - GM/c^2p)^{3/2}} \approx 1 + \frac{3GM}{2pc^2}, \tag{16}$$

and

$$\gamma - 1 \approx \frac{GM}{2pc^2}. \tag{17}$$

On the other hand, the rotational period, T, is given by

$$T = \frac{2\pi p}{\mathrm{v}}, \tag{18}$$

and using eq. (14) one obtains

$$T = 2\pi\left(\frac{p^3}{GM\gamma^6}\right)^{1/2} = 2\pi\left(\frac{p^3}{GM}\right)^{1/2}\frac{1}{\gamma^3} = T_o\left[1 - 3(\gamma - 1)\right], \tag{19}$$

where T_o is the non-relativistic period corresponding to $\gamma = 1$. It thus follows from (19)

$$\Delta T = T - T_o = -3(\gamma - 1)T_o, \tag{20}$$

thus

$$\frac{\Delta T}{T_o} = -3(\gamma - 1), \tag{21}$$

and using (17)

$$\frac{\Delta T}{T_o} = -\frac{3GM}{2pc^2}. \tag{22}$$

Accounting for the constant angular velocity, if follows

$$\frac{\Delta T}{T_o} = \frac{\Delta \varphi}{2\pi}, \tag{23}$$

and thus it follows

$$\Delta \varphi = \frac{3\pi GM}{pc^2}, \tag{24}$$

which is exactly 1/2 of the value predicted by GR [14, 15, 16, 17].

Consequently, when using gravitational masses in the universal Newtonian gravitational law, the resulting eq. (3) provides very good agreement (within a factor of 2) with the results of general relativity (GR). It is interesting that the SR treatment leads to 1/2 of the GR treatment results both for fm size systems and for macroscopic astronomical systems.

References

[1] Einstein A (1916) Die Grundlagen der allgemeinen Relativitätstheorie. Ann der Physik Bd. XLIX, S. 769-822

[2] Misner CW, Thorne KS, Zurek WH (2009) John Wheeler, relativity, and quantum information. Phys Today 62(4):40-46

[3] Das A (2011) Lectures on Gravitation. World Scientific Publishing Co. Pte. Ltd, Singapore

[4] Pireaux S, Rozelot J-P, Godier S (2001) Solar quadrupole moment and purely relativistic gravitation contributions to Mercury's perihelion Advance, arXiv:0109032v1 [astro-ph]

[5] Roll PG, Krotkov R, Dicke RG (1964) The equivalence of inertial and passive gravitational mass. Annals of Physics 26(3):442-517

[6] Einstein A (1905) Zür Elektrodynamik bewegter Körper. Ann. der Physik., Bd. XVII, S. 17:891-921; English translation On the Electrodynamics of Moving Bodies (*http : //fourmilab.ch/etexts/einstein/specrel/www/*) by G.B. Jeffery and W. Perrett (1923)

[7] Vayenas CG, Souentie S, Fokas A (2014) A Bohr-type model of a composite particle using gravity as the attractive force. Physica A 405:360-379.

[8] Bertone G (2010) Particle dark matter: Observations, models and searches. p. 762. Cambridge University Press, Cambridge. ISBN 9780521763684

[9] Das S & Vagenas EC (2009) Phenomenological Implications of the Generalized Uncertainty Principle. Can. J. Phys. 87:233-240

[10] Vayenas CG, Grigoriou D (2015) Microscopic black hole stabilization via the uncertainty principle. Journal of Physics 574:012059

[11] French AP (1968) Special relativity. W.W. Norton and Co., New York

[12] Freund J (2008) Special relativity for beginners. World Scientific Publishing, Singapore

[13] Vayenas CG, Souentie S, Fokas A (2014) A Bohr-type model of a composite particle using gravity as the attractive force. arXiv:1306.5979v4 [physics.gen-ph]

[14] Einstein A. Erklarung der Perihelbewegung des Merkur aus der allgemeinen Relativitatstheorie. Sitzungsberichte der Königlich Preussischen Akademie der Wissenschaften (Berlin), Seite 831-839 translated by Vankov AA (2011) Einstein's Paper: "Explanation of the Perihelion Motion of Mercury from General Relativity Theory" http://www.wbabin.net/eeuro/vankov.pdf.

[15] Magnan C (2007) Complete calculations of the perihelion precession of Mercury and the deflection of light by the Sun in General Relativity. arXiv:0712.3709v1 [gr-qc]

[16] Lemmon TJ, Mondragon AR (2010) First-Order Special Relativistic Corrections to Kepler's Orbits. arXiv:1012.5438v1 [astro-ph.EP]

[17] Rydin RA (2011) The Theory of Mercury's Anomalous Precession. Proceedings of the NPA 8: 501-506

Neutrinos in an expanding Universe

Richard Wigmans
Department of Physics
Texas Tech University
Lubbock, TX 79409-1051, U.S.A.

E-mail: wigmans@ttu.edu

Abstract. The Universe contains several billion neutrinos for each nucleon. In this paper, we follow the history of these relic neutrinos as the Universe expanded. At present, their typical velocity is a few hundred km/s and, therefore, their spectra are affected by gravitational forces. This may have led to a phenomenon that could explain two of todays great mysteries: The large-scale structure of the Universe and the increasing rate at which it expands.

1. Introduction

According to the Big Bang model, large numbers of neutrinos and antineutrinos have been around since the earliest stages of the evolving Universe. During the Leptonic Era, the density was so large that the (anti-)neutrinos were in thermal equilibrium with the other particles that made up the primordial soup: photons, electrons, positrons and nucleons. When the Universe was about 1 second old, the neutrinos decoupled and thus turned into relics of the Big Bang. Since that time, they have behaved like a gas of free particles. They are now the most abundant species, outnumbering baryons by more than 9 orders of magnitude. Yet, despite their enormous abundance (estimated at $\mathcal{O}(10^{87})$ for the observable Universe), relic neutrinos have until now escaped direct detection. Their extremely small kinetic energy makes it all but impossible to find a process through which they might reveal themselves as individual particles.

In this paper, we follow the relic neutrinos throughout the history of the Universe. In Section 2, their properties are briefly reviewed. In Section 3, the effects of gravitational acceleration and deceleration on the spectra and local densities of massive neutrinos are studied. It turns out that the density may exceed the maximum density allowed by the Pauli principle in the field-free region around the center-of-mass of galaxy clusters. In Section 4, we discuss the mechanism that prevents this from happening and analyze some of the consequences of this mechanism. Conclusions are presented in Section 5.

2. Properties of relic neutrinos

2.1. Spectra and densities

During the Leptonic Era, neutrinos and antineutrinos were in thermal equilibrium with the other constituents of the Universe. They converted protons into neutrons and vice versa. However, when the Universe was ≈ 1 second old and the temperature (or rather kT) had dropped to ~ 1 MeV, this thermal equilibrium ceased to exist, for two reasons. First, the production of neutrons through the reactions $\bar{\nu_e} + p \rightarrow n + e^+$ and $e^- + p \rightarrow \nu_e + n$ became energetically impossible, whereas the reverse reactions still occurred. This led to the proton/neutron asymmetry. Second, the density dropped to the point where the average lifetime of the neutrinos and antineutrinos became longer than the age of the Universe (*i.e.*, 1 second at that time).

Journal of Physics: Conference Series **633** (2015) 012034 doi:10.1088/1742-6596/633/1/012034

The decoupling of the neutrinos[1] marked the end of the Leptonic Era. At that time, the neutrino spectrum was described by a relativistic Fermi-Dirac distribution function. The number of particles per unit volume in the momentum bin $(p, p + dp)$ is given by

$$N(p)dp = \frac{\pi p^2 dp}{h^3[\exp(pc/kT_\nu) + 1]}(\frac{g_\nu}{2}) \quad (1)$$

where $kT_\nu \approx 1$ MeV, $p_\nu \approx 1$ MeV/c and g_ν denotes the number of neutrino helicity states: $g_\nu = 2$ for Majorana neutrinos, $g_\nu = 4$ for Dirac neutrinos [1].

The neutrino density, which can be found by integrating Equation 1 over the entire spectrum, was during the Leptonic Era related to the photon density as:

$$\frac{n_\nu}{n_\gamma} = \frac{3}{4} \text{ (Majorana)}, \ \frac{6}{4} \text{ (Dirac)}$$

About 10 seconds later, when kT had dropped below 0.511 MeV, the photon spectrum was reheated because of the electron-positron annihilation, and the photon density increased by a factor of 11/4. As a result, the neutrino and photon temperatures were no longer the same either:

$$T_\nu = \sqrt[3]{4/11}\, T_\gamma \quad (2)$$

After the decoupling, the de Broglie wavelengths ($\lambda = h/p$) of these relic neutrinos have been expanding in proportion to the size of the Universe. Even though the particles may well be non-relativistic at the present time, their momentum spectrum is still described by the same relativistic equation (1), albeit that the momenta have been redshifted by 10 orders of magnitude and the temperature T_ν has decreased correspondingly. Their present spectrum is thus believed to be a momentum-redshifted relativistic Fermi–Dirac distribution. Since the present value of T_γ amounts to 2.725 ± 0.001 K [2], this neutrino spectrum is characterized by a temperature $T_\nu = 1.95$ K. In the following we will define the Fermi momentum[2] p_F as kT_ν/c. For a neutrino temperature of 1.95 K, this momentum is $1.68 \cdot 10^{-4}$ eV/c, or $9.0 \cdot 10^{-32}$ kg.m.s^{-1}.

Since the current photon density has been measured to be 411 cm^{-3}, the neutrino density may be estimated to be $\sim$ 220 cm^{-3}, for each flavor, if neutrinos and antineutrinos are fundamentally different (Dirac) particles, as in the case of neutral kaons. If they are indistinguishable (Majorana particles), such as π^0s, then the densities are a factor of two smaller.

2.2. *Neutrino masses and velocities*

At these densities, there are thus $\mathcal{O}(10^{87})$ relic neutrinos in the observable Universe. Given the measured baryon/photon ratio of $(4.5 \pm 1.9) \cdot 10^{-10}$ [3], and the fact that there are three generations of neutrinos, there are thus $\approx$ 5 billion neutrinos for every baryon. Even a very small neutrino rest mass may thus have very significant consequences for the total mass of the Universe.

The experimental upper limit on the mass of the electron antineutrino[3] was determined at 2.2 eV (95% C.L.), from a study of the electron spectrum of ^{3}H decay [4]. The experimental results on atmospheric neutrinos obtained by the Superkamiokande [5] and SNO [6] Collaborations suggest that at least one of the neutrino mass eigenstates has a value in excess of 0.04 eV. Interpretations of other experimental results suggest mass values of the electron neutrino in the 0.2 - 0.6 eV range [7, 8]. There is no experimental information that rules out a neutrino rest mass of the order of 0.1 – 1 eV.

It is important to realize that, depending on their mass, the relic neutrinos might be very *nonrelativistic* at the current temperature. Since they decoupled, their momenta have been stretched by a factor 10^{10}, from $\sim$ 1 MeV/c to $\sim 10^{-4}$ eV/c. If their rest mass were 1 eV/c^2, their maximum velocity would thus be $\sim 10^{-4}c$.

[1] In the following, statements about "neutrinos" are intended to mean "neutrinos and antineutrinos", unless stated otherwise.
[2] This definition deviates from the one used in solid state physics, in which a non-zero chemical potential is used.
[3] In the following, we express masses in energy units, omitting the c^{-2} factor.

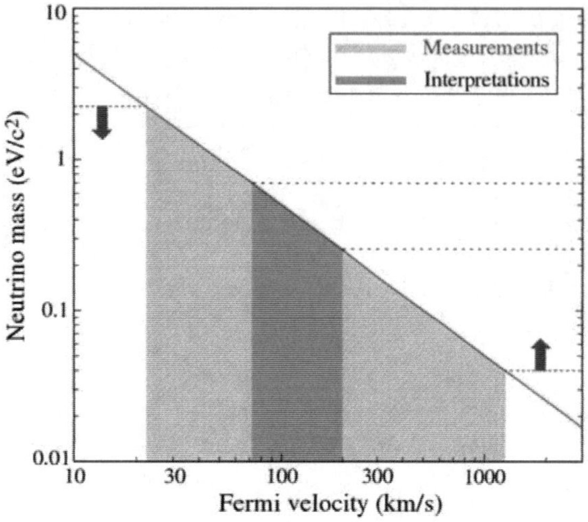

Figure 1. Relationship between the neutrino mass and the Fermi velocity at the present temperature ($T_\nu = 1.95$ K).

The relationship between the neutrino mass and the Fermi velocity ($v_F = kT_\nu/cm_\nu$) at the current neutrino temperature ($T_\nu = 1.95$ K) is shown in Figure 1, together with the limits on the neutrino mass derived from the aforementioned direct measurements and interpretations. The latter suggest v_F values between 70 and 200 km/s, while direct measurements put the velocity interval between 25 and 1200 km/s.

2.3. Degeneracy and expansion

In a Fermi-Dirac gas, the fermions are stacked into progressively higher energy states, beginning with the ground state. When all available states are filled, *i.e.*, when the gas is said to be *degenerate*, one cannot squeeze any more particles with momenta in the range $0 < p < p_{\rm max}$ into the volume ocupied by the gas without violating the Pauli principle. We may estimate the density at which violation of the Pauli principle becomes an issue with the aid of the Heisenberg uncertainty relation:

$$\rho_f(\max) = \frac{8\pi}{3}\left[\frac{p_{\rm max}}{h}\right]^3 \qquad (3)$$

In non-relativistic Fermi-Dirac gases, the maximum fermion density at a given temperature is thus proportional to the fermion mass cubed.

This density cannot be exceeded. Attempts to squeeze more fermions into the available volume than allowed by this maximum will trigger the *fermion degeneracy pressure*, which prevents this from happening. The outside pressure forces the fermions into higher and higher quantum states, up to the point where the pressure exerted as a result of their thermal motion balances the external force. This is the mechanism that prevents the gravitational collapse of White Dwarfs and neutron stars. In the case of White Dwarfs, the fermions are electrons, and the momenta are typically of the order of MeV/c. In neutron stars, the fermions are neutrons and the momenta are measured in units of GeV/c.

The Pauli principle applies also to relic neutrinos. Since the momenta of these particles are of the order of 10^{-4} eV/c, Eq. 3 implies that their current maximum density is 30 orders of magnitude smaller than that of electrons in a White Dwarf. Using Equations 1 and 3, this maximum density is calculated at a temperature $T_\nu = 1.95$K to be ~ 225 cm^{-3}, for each (Dirac) flavor. This result is in excellent agreement with the estimated present density of relic neutrinos, $n_\gamma \times 6/11 \sim 220$ cm^{-3} (Section 2.1).

The relic neutrinos thus form a degenerate fermion gas, at $T_\nu = 1.95$ K. It is important to note that the expansion of the Universe has not changed the degeneracy. Since the neutrino momenta, $p_\nu(t)$, are inversely proportional to the scale, $R(t)$, of the Universe, and since the maximum allowed density evolves in proportion to p^3 (Equation 3), the ratio of the actual neutrino density and the maximum allowed density is **independent of** R! The present situation is thus not new. The Universe has *always* been filled with a degenerate neutrino gas. All available quantum states with $p/p_F \ll 1$ are, and always have been, almost completely occupied, regardless of the neutrino temperature. Since the neutrino momenta also scale with the neutrino temperature of the expanding Universe (*e.g.*, $cp \propto kT_\nu$, see Equation 1), we conclude that this temperature must have evolved inversely proportional to the distance scale of the Universe:

$$R(t)T_\nu(t) = \text{constant} \qquad (4)$$

As we will see in the next section, this has important consequences.

Journal of Physics: Conference Series **633** (2015) 012034 doi:10.1088/1742-6596/633/1/012034

3. Massive neutrinos in the Matter Era

3.1. Spectral effects

During the radiation era, the neutrino spectrum was only affected by the gradual expansion, with $p_\nu(t)$ and $T_\nu(t)$ inversely proportional to the evolving distance scale. However, when gravity became the dominant force in the Universe, important changes were about to take place. Initially, during the period following the photon decoupling (at a redshift $z \sim 1100$), matter was still very uniformly distributed. The main effect of the newly dominant gravitational force was a general slowdown of the expansion rate. However, when galaxy formation started, around $z \sim 10$, matter started to cluster in certain areas, leaving other areas devoid. As a result, gravitation was no longer a uniform force. Figure 2 gives a conceptual impression of this phenomenon. It shows the gravitational field lines in a region of space where this process is taking place. The different galaxies manifest themselves as gravitational wells.

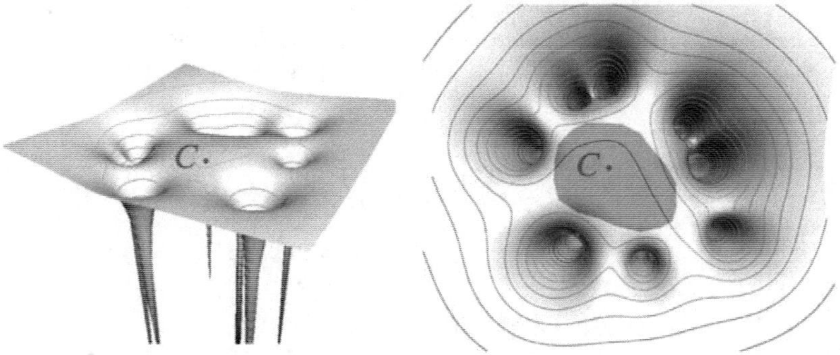

Figure 2. Gravitational field lines in a region of the Universe where several galaxies are forming.

Since neutrinos are subject to gravitational forces, their spectra were affected by this developing non-uniformity. The neutrino velocities either increased or decreased as a result of gravitational acceleration or deceleration, depending on the direction of motion of the particles with respect to the (dominant) source of gravitational fields. The effects of this extended over intergalactic distances, to regions far away from these sources. The velocity distribution of the relic neutrinos has thus gradually changed in the non-uniform gravitational fields that resulted from baryon clustering, *i.e.,* galaxy formation.

A well known consequence of this phenomenon is the gravitational clustering of relic neutrinos inside and around galaxies. The properties of such neutrino halos were extensively studied by Ringwald and Wong [9]. Their calculations, which concern neutrinos with masses ranging from 0.15 - 0.6 eV, show that the neutrino density in these halos may increase by as much as a factor of 20 (for $m_\nu = 0.6$ eV) with respect to the average universal density.

This is precisely what should be expected as a result of the neutrino "heating" that has occurred as a result of the process described above. Neutrinos roll into the gravitational wells (depicted in Figure 2) that are forming. They are trapped inside these wells if the kinetic energy they acquire in this process is insufficient to escape from the (deepened) well by the time they reach the other side. Because the neutrinos gain momentum in this process, their effective temperature increases, and the number of quantum states available to the higher-temperature Fermi-Dirac gas becomes larger as well.

In this paper, we concentrate on the neutrinos that experience the opposite effect. Neutrinos climbing out of the gravitational wells are *decelerated* in that process. As a result, the neutrino gas in the shaded area surrounding the center-of-gravity (C) of the galaxy cluster in Figure 2 will gradually be "cooled", *i.e.,* the effective temperature of the neutrinos in this area *decreases*. Just as for the formation of neutrino halos around galaxies, the magnitude of the gravitational cooling effect is very sensitive to the velocity distribution, and thus to the rest mass, of the neutrinos.

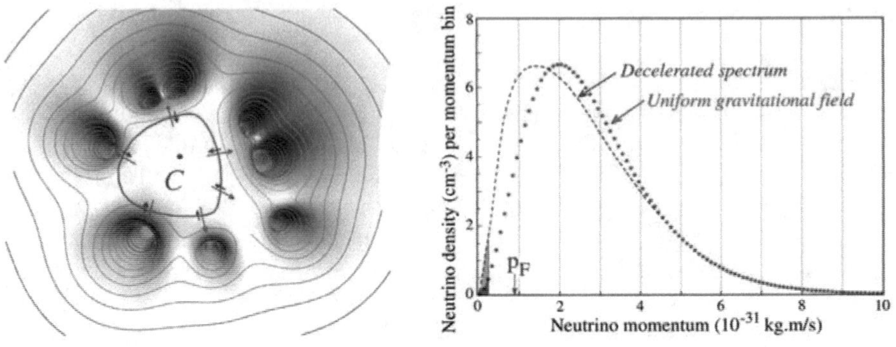

Figure 3. Gravitational field lines in a region of the Universe where several galaxies are forming (*left*). The neutrino spectrum inside the box surrounding the center-of-gravity of this galaxy cluster is affected by this process (*right*), since neutrinos leaving this box are replaced by neutrinos that are, on average, slower. See text for details.

We have simulated the effects of galaxy formation on the neutrino spectra and densities with a computer program that was especially developed for this purpose. Our simulations show that the low-field region around the center-of-mass of this galaxy cluster is gradually populated by neutrinos with a decreasing average velocity. This is schematically illustrated in Figure 3. The neutrinos leaving the bounded region around the center-of-mass are replaced by neutrinos which have, on average, a lower velocity, as a result of the gravitational deceleration they undergo when climbing out of the gravitational wells. In other words, the neutrino gas in this region is *"gravitationally cooled"* in this process, in the same way as the neutrino gas trapped inside the gravitational wells is gradually heated.

However, if sufficiently large numbers of neutrinos are sufficiently decelerated in the described process, then there may not be enough quantum states in this region to accommodate them. In that case, the Pauli principle is in jeopardy of being violated. Our calculations indicate that violation of the Pauli principle starts to play a role 600 - 800 million years after the galaxies start to form, depending on the time scale of the galaxy formation process and the neutrino mass.

3.2. The Pauli principle

The Pauli Exclusion Principle is one of the cornerstones of Quantum Mechanics. It is responsible for the atomic shell structure and its many implications. It also determines the properties of degenerate matter, and is thus responsible for astrophysical phenomena such as the helium flash in stellar evolution and the mass-volume relationship for White Dwarfs. Combined with Heisenberg's Uncertainty Principle, it implies that there is a maximum to the number of fermions that can be squeezed into a given volume. The impossibility to exceed this maximum is a source of pressure, which may counteract efforts to achieve otherwise, and thus prevents the gravitational collapse of objects such as White Dwarfs. Despite many attempts, violation of the Pauli principle has never been experimentally observed.

How does nature prevent violation of the Pauli principle in degenerate matter? In shrinking White Dwarfs, the electrons are forced into increasingly higher quantum states. The higher the quantum state, the higher the velocity, and thus the pressure that can be exerted by these electrons on the outside world, in this case the outside layers of the collapsing star. When the mass of the star is less than the Chandrasekhar limit, an equilibrium is found in which the inward gravitational pressure is balanced by the described outward *"fermion degeneracy pressure"*. Prevention of violation of the Pauli principle is thus achieved through the action of the forces to which the fermions in question are subjected, in this case the electromagnetic force.

However, the relic neutrinos are not affected by the electromagnetic force and are thus incapable of exerting pressure on their environment. Therefore, the question arises what happens when gravitationally decelerated neutrinos cannot find a quantum state to fit into? How is violation of the Pauli principle prevented in this case? Whatever the answer to these questions, the Pauli principle must be safeguarded through the action of a fundamental force to which neutrinos are sensitive. Neutrinos are only subject to the weak force and to gravity. And since the range of the weak force is many orders of magnitude smaller than the typical distance between the neutrinos in the degenerate conditions considered here, the weak force can be ruled out as a candidate. Violation of the Pauli principle in intergalactic space can thus only be prevented through the action of gravity, the same force that was responsible for the problem in the first place.

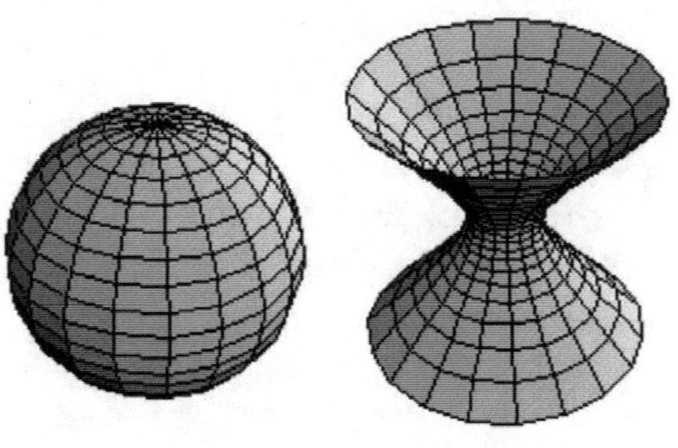

Figure 4. Schematic depiction of the curvature of space-time as a result of gravity. A positive curvature implies that the geodesics are located on the surface of a sphere, for a negative curvature they follow the surface of a hyperboloid.

In the general-relativistic description, gravity curves space-time in the vicinity of its source. When 10^{57} protons cluster together and form a star, space-time around the newly formed massive object is modified as if it were a converging gravitational lens. Massive particles, and even photons traveling in the vicinity follow trajectories that correspond to the geodesics of this curved space.

Similarly, we envision that when $\mathcal{O}(10^{77})$ relic neutrinos congregate in a region of space where there are insufficient quantum states to accommodate all of them, space-time is modified such as to prevent violation of the Pauli principle. In the example described in Figure 3, this means that the center-of-gravity C starts to act as a *diverging gravitational lens.* This is illustrated in Figure 4. Particles approaching point C will follow a trajectory that corresponds to a geodesic on the surface of a hyperboloid. The net effect is the same as if a *negative mass* were placed in point C: All neutrinos approaching point C experience a repulsive force that prevents them from reaching an area that cannot accommodate them. In the following section, we examine some of the consequences of what we we will call the *Pauli acceleration model.*

4. The Pauli acceleration model

4.1. Qualitative considerations

In the previous section, we saw that violation of the Pauli principle may occur (for proper values of the neutrino rest mass) in the low-field region near the center-of-mass of a newly forming galaxy cluster. Preventing such a violation could be achieved if space-time around this center-of-mass were negatively curved. Such a curvature would not only prevent relic neutrinos from reaching areas where quantum mechanics does not allow them to be, but would have other consequences as well.

The Equivalence Principle implies that any massive object in the gravitational field created this way would experience the same consequences as if a negative mass were placed in the center-of-mass of the galaxy cluster. The galaxies constituting the cluster would thus experience a repulsive force driving them away from their common center-of-mass. The void around the center-of-mass of the cluster would thus tend to grow over time. This *local* expansion of space is superimposed on the overall Hubble expansion.

If the galaxies are also part of another cluster, they may feel a repulsive force from the center-of-mass of that cluster as well. Both forces work in opposite directions and thus (partially) cancel each other. This may explain the Large Scale structure of the Universe, as observed in the Sloan Digital Sky Survey (Figure 5). Strings of galaxies are separated by large voids. This situation can be compared with that in soap foam above hot water. Temperature differences creates pressure differences, which causes the simultaneous growth of all soap bubbles in the relatively cold surrounding air, and the molecules arrange themselves in patterns very similar to the one shown in the figure.

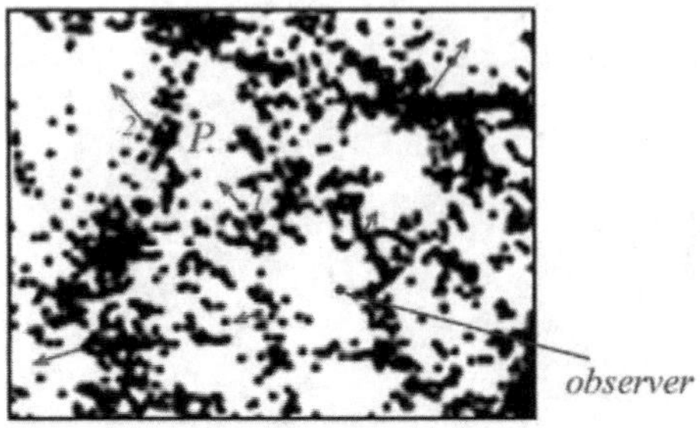

Figure 5. Small region of the Sky, as mapped by the Sloan Digital Sky Survey. Each dot represents a galaxy. The arrows represent the apparent velocity vectors of some representative galaxies with respect to the position of the observer.

It is important to note that the negative curvature around the center-of-mass is not a temporary feature, which disappears as soon as the local expansion of space caused by it has created the necessary quantum states for the relic neutrinos. Since the de Broglie wavelengths of the neutrinos expand in proportion, the degree of neutrino degeneracy is *not affected* by this local expansion (Section 2.3). Therefore, the phenomenon that necessitated the negative curvature is permanent and the void around the center-of-mass of the galaxy cluster will thus continue to expand over time. The same will be true for any cluster of galaxies.

A second consequence of this model is that to an observer located at any point in space it will seem as if all distant galaxies move away, *in an accelerated fashion*. This is illustrated in Figure 5, which shows an enlarged detail of the sky mapped by the Sloan Digital Sky Survey. The arrows represent the velocity vectors measured by the observer located at the indicated position. All galaxies in the ring surrounding the void in which the observer is located move away, with velocities proportional to the distance from the observer. Since this motion is governed by a repulsive force, it is accelerated. Had the observer measured these velocities in the (distant) past, smaller values would have been found. The motion of galaxies located outside the inner ring is determined by a superposition of their motion with respect to galaxies closer to the observer. For example, galaxies 1 and 2 both experience a repulsive force originating from point P. Therefore, galaxy 2 undergoes an accelerated motion with respect to galaxy 1. Its motion with respect to the position of the observer is governed by the vector sum of the acceleration vectors of galaxy 1 with respect to the observer and galaxy 2 with respect to galaxy 1. The result is also here an accelerated motion with a component along the direction away from the observer, and velocities measured today are larger than they were in the distant past.

The apparent accelerated motion of distant galaxies, *i.e.,* the fact that the current value of the Hubble constant is larger than it was in the past, has been concluded from careful analyses of the properties of Supernova Type Ia explosions in distant galaxies [10, 11]. This phenomenon has given rise to many speculations as to its origin. If our model were correct, it is in essence a quantum mechanical effect.

4.2. Quantitative considerations

We have tried to estimate the effects of the described Pauli acceleration quantitatively. Parameters that are of crucial importance in this respect are the galaxy masses and the distance separating the galaxies at the time when they started to form, as well as of course the value of the neutrino mass. Therefore, we have also investigated the sensitivity of our results to the choice of the values of these parameters.

Since it was necessitated by gravitationally decelerated relic neutrinos, the Pauli acceleration is a relatively recent phenomenon. It only started to play a role after the first galaxies had formed and the resulting non-uniform gravitational fields had sufficiently reduced the speed of a significant fraction of the relic neutrinos. Based on the current distance typically separating galaxies ($\approx$ 1Mpc), typical galaxy masses ($10^{10} - 10^{11} M_{\odot}$) and the measured value of the Hubble constant (73 km.s^{-1}.Mpc^{-1}), we find that the turning point, at which the accelerating effects became larger than the deceleration resulting from the gravitational attraction between the galaxies occurred at a redshift $z \sim 5$. Assuming a flat Universe, this corresponds to a time when $(1+z)^{-3/2} = 7\%$ of the total history of the Universe had unfolded, *i.e.,* when the Universe was $\sim$ 1 billion years old.

Our model implies that the expansion of the Universe will continue forever, albeit with a gradually decreasing acceleration. In Section 2.4, we saw that the local expansion of the intergalactic voids does *not* change the degree of degeneracy of the relic neutrinos, since the wavelengths of these neutrinos increase in proportion to the distance scale. Therefore, the mechanism that drives the expansion does not go away. Interestingly, this statement is *independent of the mass content* of the Universe. Our model implies that there is *no relationship* between the geometry (k) and the energy density (Ω) of the Universe.

The scenario that leads to the local expansion of intergalactic voids hinges of course critically on the value of the neutrino mass. If this mass were too small, then the neutrino velocities would be so large that gravitational effects on these velocities would be insignificant. Based on the assumptions listed above, we conclude that the rest masses of neutrinos should not be smaller than ~ 0.5 eV/c^2. This limit implies that direct measurements of the neutrino mass are possibly within reach. Planned efforts to improve the neutrino mass sensitivity in studies of ^{3}H decay to the 0.2 eV/c^2 level [4] may well pay off in the form of a precise measurement of this important parameter.

5. Conclusions

The relic neutrinos that fill the Universe form a degenerate Fermi-Dirac gas, with typical velocities of the order of 100 km/s. Some fraction of these neutrinos were significantly affected by the non-uniform gravitational fields that developed when galaxy formation started. We have shown that gravitationally decelerated neutrinos accumulating in the low-field region around the center of a developing galaxy cluster may have approached the density limits set by the Pauli Exclusion Principle. The only way to avoid a violation of the Pauli principle was a local expansion of space in that region. Such an expansion could lead to and would provide a natural explanation for the Large-Scale Structure observed in the Sloan Digital Sky Survey and other studies. It could also explain the apparent accelerated expansion suggested by the high-z Supernova data, without the need for "dark energy" or other unknown physics.

Fermion degeneracy pressure is a phenomenon that, until now, has been exclusively associated with the extremely high temperatures and densities that characterize the interior of degenerate stellar objects such as White Dwarfs and neutron stars. It is remarkable that the same phenomenon may also play a crucial role in the extremely cold and empty conditions of intergalactic space.

Of course, the model proposed in this paper has speculative elements. If the neutrino mass is indeed of the order of 0.5 eV/c^2, then violation of the Pauli principle must have been a serious issue when galaxies started to form. Nature might of course have used other ways to prevent such a violation, instead of the one proposed in this paper, although it is unclear what these other ways could be. However, if the model proposed in this paper turned out to be correct, it means that the expansion of the Universe and the observed peculiarities in this process are essentially a consequence of Quantum Mechanics. It would also be the first observation of the effects of gravity at the quantum level. The intimate link between the structure of matter in its smallest details and on its largest imaginable scales is further illustrated by the crucial role played by the value of the neutrino mass in the expansion of the Universe. If our model is correct, this value must be about 0.5 eV/c^2, which might make it accessible to direct measurements planned for the near future.

References

[1] S. Tremaine and J.E. Gunn, Phys. Rev. Lett. **42**, 407 (1979).
[2] S. Eidelman *et al.*, Phys. Lett. **B592**, 1 (2004).
[3] B.D. Fields and S. Sarkar, in K. Hagiwara *et al.*(Particle Data Group), Phys. Rev. **D66** (2002) 010001-162.
[4] A. Osipowicz *et al.*, *KATRIN: A next generation tritium β-decay experiment with sub-eV sensitivity for the ν_e mass*, e-print archive hep-ex/0109033 (2001).
[5] Y. Fukuda *et al.*, Phys. Rev. Lett. **81**, 1562 (1998).
[6] Q.R. Ahmad *et al.*, Phys. Rev. Lett. **87**, 071301 (2001).
[7] H.V. Klapdor-Kleingrothaus *et al.*, e-print archive hep-ph/0404088 (2004); *ibid.* Nucl. Instr. and Meth. in Phys. Res. **A510** (2003) 281; *ibid.* Phys. Lett. **B578** (2004) 54.
[8] R. Wigmans, Astroparticle Physics **19** (2003) 379.
[9] A. Ringwald and Y.Y.Y. Wong, e-print archive hep-ph/0408241 (2004).
[10] A.G. Riess *et al.*(High-z Supernova Search), Astron. J. **116** (1998) 1009.
[11] S. Perlmutter *et al.*(Supernova Cosmology Project), Astrophys. J. **517** (1999) 565.
[12] S. Perlmutter, Physics Today, April 2003, 53.

Testing General Relativity on cosmological scales

Athina S. Pouri, Spyros Basilakos

Research Center for Astronomy and Applied Mathematics of the Academy of Athens, Greece

E-mail: ath.pouri@gmail.com, svasil@academyofathens.gr

Abstract.
In order to test the validity of General Relativity (GR) on cosmological scales, it has been proposed that measuring the so called growth index, γ, could provide an efficient way to discriminate between scalar field dark energy models which admit to general relativity and modified gravity models. Using the clustering properties of the Luminus Red Galaxies and the growth data we perform a standard χ^2-minimization procedure between theoretical expectations and data, followed by a joint likelihood analysis and we find a value of $\gamma = 0.56 \pm 0.05$, perfectly consistent with the expectations of the ΛCDM model, and $\Omega_{m0} = 0.29 \pm 0.02$, in very good agreement with the latest *Planck* results. Our analysis provides significantly more stringent growth index constraints with respect to previous studies, as indicated by the fact that the corresponding uncertainty is only $\sim 0.09\gamma$.

1. Introduction

Recent cosmological observations support that the universe is experiencing an accelerated expansion at late times. To explain such a behavior one has to view Dark Energy (DE) either as a new field in nature or as a modification of General Relativity (see for review [1], [2], [3]). An interesting approach to discriminate between scalar field DE models and modified gravity is to use the evolution of the linear growth of matter perturbations. Specifically, a useful tool in this kind of studies is the so called growth rate of clustering, which is defined as $f(a) = \frac{d\ln D}{d\ln a} \simeq \Omega_m^{\gamma}(a)$, where $a(z) = (1+z)^{-1}$ is the scale factor of the universe, $\Omega_m(a)$ is the dimensionless matter density parameter, γ is the growth index and $D(a) = \delta_m(a)/\delta_m(a=1)$ is the linear growth factor usually scaled to unity at the present time [4], [5]. Note that γ can be seen as a function of redshift: $\gamma \equiv \gamma(z)$. Theoretically speaking, it has been shown that for those DE models which are within the framework of GR and have a constant Equation of State parameter, the growth index γ is equal to $\gamma \simeq 6/11$ for the ΛCDM model.

Due to the fact that gravity reflects, via gravitational instability, on the nature of clustering [4] it has been proposed to use the clustering/biasing properties of the mass tracers in order to constrain cosmological models (see [6], [7], [8], [9]) as well as to test the validity of GR on extragalactic scales ([10], for a recent review see [11]). Based on the aforementioned, the aim of the present study is to place constraints on the (Ω_m, γ) parameter space using the joint analysis of the measured two-point angular correlation function (ACF) of LRGs [12] and the growth rate of clustering data as collected by Basilakos et al. (2013) (see Table 1 of [13]) which are based on the PSCz, 2dF, VVDS, SDSS, 6dF, 2MASS, BOSS and *WiggleZ* galaxy surveys, for which their combination parameter of the growth rate of structure, $f(z)$, and the redshift-dependent rms fluctuations of the linear density field, $\sigma_8(z)$, is available as a function of redshift, $f(z)\sigma_8(z)$.

Journal of Physics: Conference Series **633** (2015) 012035 doi:10.1088/1742-6596/633/1/012035

2. Angular Correlation Function

Limber's inversion equation relates the ACF, $w(\theta)$, with the two point spatial correlation function $\xi(r, z)$:

$$w(\theta) = 2\frac{H_0}{c} \int_0^{\infty} \left(\frac{1}{N}\frac{dN}{dz}\right)^2 E(z)dz \int_0^{\infty} \xi(r, z)du \quad (1)$$

where $1/N\ dN/dz$ is the normalized redshift distribution of the LRGs. The redshift distribution for the case of the LRGs is given by the relation:

$$\frac{dN}{dz} \propto \left(\frac{z}{z_\star}\right)^{(a+2)} e^{-\left(\frac{z}{z_\star}\right)^\beta} \quad (2)$$

where $(a, \beta, z_\star) = (-15.53, -8.03, 0.55)$ and $z_\star$ is the characteristic depth of the subsample studied. The spatial correlation function of the mass tracers is given by:

$$\xi(r, z) = b^2(z)\xi_{DM}(r, z) \quad (3)$$

where $b(z)$ is the evolution of the linear bias, and ξ_{DM} is the corresponding correlation function of the underlying mass distribution which is written as:

$$\xi_{DM}(r, z) = \frac{1}{2\pi^2} \int_0^{\infty} k^2 P(k, z) \frac{\sin(kr/a)}{(kr/a)} dk \quad (4)$$

with $P(k, z) = D^2(z)P(k)$, and $P(k) = P_0 k^n T^2(k)$ denoting the CDM power spectrum of the matter fluctuations where $T(k)$ is the CDM transfer function of Eisenstein & Hu (1998) or this of Bardeen et al. (1986) and $n \simeq 0.9671$ following the recent reanalysis of the *Planck* data by Spergel et al. (2013). The variable r corresponds to the physical separation between two sources having an angular separation, θ (in steradians) and $E(z) = H(z)/H_0$, is the normalized Hubble parameter. Non linear effects have also been taken into account through the slope of the power spectrum at the relevant scales: $n_{\text{eff}} = d\ln P/d\ln k$ ([17], [18], [19]).

3. The evolution of linear bias, $b(z)$

Here we briefly present the model used in this analysis in order to trace the evolution of the linear bias factor, which reflects the relation between the overdensities of luminous and of dark matter [15]. We remind the reader that biasing is considered to be statistical in nature with galaxies and clusters being identified as high peaks of an underlying, initially Gaussian, random density field. The usual paradigm is of a linear and scale-independent bias and it is defined as the ratio of density perturbations in the mass-tracer field to those of the underling total matter field: $b = \delta_{tr}/\delta_m$. We use the bias evolution model of [20] and [10] which is based on the linear perturbation theory and the Friedmann-Lemaitre solutions of the cosmological field equations. It is valid for any DE model (scalar or geometrical) and it is given by:

$$b(z) = 1 + \frac{b_0 - 1}{D(z)} + C_2 \frac{\mathcal{J}(z)}{D(z)} \quad (5)$$

with $\mathcal{J}(z) = \int_0^z \frac{(1+y)}{E(y)} dy$. The constants b_0 (the bias at the present time) and C_2 depend on the host dark matter halo mass M_h (see [10]).

Table 1. Results in the $(\Omega_{m0}, \gamma, M_h, n_{\rm eff})$ parameter space for the different $T(k)$ and σ_8.

$T(k)$	Ω_{m0}	γ	$M_h/10^{13}M_\odot$	$n_{\rm eff}$	$\chi^2_{t,\min}/df$
		$\sigma_8 = 0.797\,(0.30/\Omega_{m0})^{0.26}$ [21]			
Hu	0.29 ± 0.02	0.56 ± 0.05	1.90 ± 0.10	0.10 ± 0.20	16.36/23
Bardeen	0.29 ± 0.01	0.56 ± 0.10	1.80 ± 0.30	$-0.10^{+0.30}_{-0.10}$	16.56/23
		$\sigma_8 = 0.818\,(0.30/\Omega_{m0})^{0.26}$ [16]			
Hu	$0.29^{+0.03}_{-0.02}$	$0.58^{+0.02}_{-0.06}$	1.70 ± 0.20	0.30 ± 0.20	15.90/23
Bardeen	$0.29^{+0.02}_{-0.03}$	0.56 ± 0.10	1.60 ± 0.4	$0.0^{+0.10}_{-0.20}$	16.13/23

4. Fitting Theoretical Models to the data

A standard χ^2 minimization statistical analysis is implemented in order to provide constraints in the (Ω_{m0}, γ) parameter space. It is defined as follows:

$$\chi^2_t = \sum_{i=1}^{n} \left[\frac{X_{obs} - X_{th}}{\sigma_i^2} \right]^2$$

where the index t stands for the clustering data of LRGs or the growth data of clustering, X is the angular correlation function or the $f\sigma_8$ respectively and σ_i is the corresponding uncertainty. We perform a joint likelihood analysis of the two cosmological probes using:

$$\chi^2 = \chi^2_{LRGs} + \chi^2_{gr}$$

The result of the analysis for the case of [14] transfer function is $(\Omega_{m0}, \gamma) = (0.29 \pm 0.02, 0.56 \pm 0.05)$, which is the strongest (to our knowledge) joint constraint appearing in the literature. The results are shown in Table 2. For a more detailed analysis, as well as for the case of a varying growth index with redshift $\gamma = \gamma(z)$, see [22].

References

[1] Copeland, E. J., Sami, M., & Tsujikawa, S. 2006, *Int. J. of Mod. Phys. D.*, 15, 1753
[2] Caldwell, R. R., & Kamionkowski, M. 2009, *Ann. Rev. Nucl. Part. Sci.*, 59, 397
[3] Amendola, L., & Tsujikawa, S. 2010, *Dark Energy: Theory and Observations, Cambridge University Press, Cambridge UK*
[4] Peebles, P. J. E. 1993, *"Principles of Physical Cosmology", Princeton University Press, Princeton New Jersey (1993)*
[5] Wang, L., & Steinhardt, P. J. 1998, *ApJ*, 508, 483
[6] Matsubara, T. 2004, *ApJ*, 615, 573
[7] Basilakos, S., & Plionis, M. 2005, *MNRAS*, 360, L35
[8] Basilakos, S., & Plionis, M. 2006, *ApJ*, 650, L1
[9] Krumpe, M., Miyaji, T., & Coil, A. L. 2013, *arXiv:1308.5976*, 650, L1
[10] Basilakos, S., Dent, J. B., Dutta, S., Perivolaropoulos, L. & Plionis, M. 2012, *Phys. Rev. D.*, 85, 123501
[11] Bean, R., et al. 2013, *arXiv:1309.5385*
[12] Sawangwit, U., et al. 2011, *MNRAS*, 416, 3033
[13] Basilakos, S., Nesseris, S., & Perivolaropoulos, L. 2013, *Phys. Rev. D.*, 87, 123529
[14] Eisenstein, D. J., & Hu, W. 1998, *ApJ*, 496, 605
[15] Bardeen, J. M., Bond, J. R., Kaiser, N., & Szalay A. S. 1986, *ApJ*, 304, 15
[16] Spergel, D., Flauger, R., & Hlozek, R. 2013, *arXiv:1312.3313*
[17] Peacock, J. A., & Dodds, S. J. 1994, *MNRAS*, 267, 1020

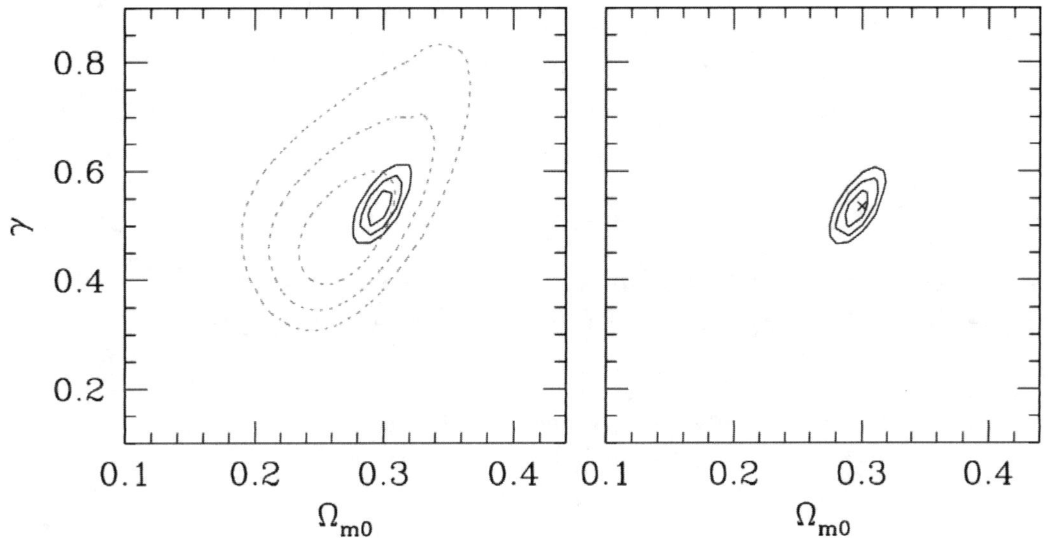

Figure 1. Likelihood contours for $\Delta\chi^2 = \chi_t^2 - \chi_{t,\rm min}^2$, corresponding to 1σ, 2σ and 3σ confidence levels, in the (Ω_{m0}, γ) plane using the [14] transfer function and the [20], [10] bias model. *Left Panel:* The joint likelihood contours correspond to the [21] power spectrum normalization. *Right Panel:* The contours here correspond to σ_8 provided by [16]. The best fit solutions are represented by the crosses. Note that using the [15] transfer function we find almost the same results within 1σ erros. Finally, the red dotted curves are the SNIa/BAOs/CMB$_{\rm shift}$/$f\sigma_8$ joint likelihood contours provided by [23].

[18] Smith, R. E. et al. 2003, *MNRAS*, 341, 1311
[19] Widrow, L. M., Elahi, P. J., Thacker, R. J., Richardson M., & Scannapieco, E. 2009, *MNRAS*, 397, 1275
[20] Basilakos, S., Plionis, M., & Pouri, A. 2011, *Phys. Rev. D.*, 83, 123525
[21] Hajian, A., Battaglia, N., Spergel, D. N., Bond, J. R., Pfrommer, C., & Sievers, J. L. 2013, *JCAP*, 11, 64
[22] Pouri, A., Basilakos, S., & Plionis, M. 2014, *arXiv:1409.0964*
[23] Nesseris, S., Basilakos, S., Saridakis, E. N., & Perivolaropoulos, L. 2013, *Phys. Rev. D.*, 88, 103010

New exact quasi-classical asymptotic beyond WKB approximation and beyond Maslov formal expansion

J Foukzon[1], E Men'kova[2], A Potapov[3], S Podosenov[4]
[1]Department of mathematics, Israel Institute of Technology, Haifa, Israel

E-mail: jaykovfoukzon@list.ru

[2]All-Russian Research Institute for Optical and Physical Measurements, Moscow, Russia

E-mail: E_Menkova@mail.ru

[3]Kotel'nikov Institute of Radioengineering and Electronics of the Russian Academy of Sciences, Moscow, Russia

E-mail: potapov@cplire.ru

[4]All-Russian Research Institute for Optical and Physical Measurements, Moscow, Russia

E-mail: podosenov@mail.ru

Abstract. New exact quasi-classical asymptotic of solutions to the n-dimensional Schrodinger equation beyond WKB-theory and beyond Maslov's canonical operator theory is presented. Quantum jump nature is considered successfully. We pointed out that an explanation of quantum jumps can be found to result from Colombeau solutions of the Schrödinger equation alone without additional postulates. Such jumps, does not depend on any single measurement of particle position $q(t)$ at instant t and completely obtained only from Schrodinger equation, without reference to any phenomenological master-equation of Lindblad form.

1.Introduction

A number of experiments on trapped single ions or atoms have been performed in recent years [1]-[5]. Monitoring the intensity of scattered laser light off of such systems has shown abrupt changes that have been cited as evidence of "quantum jumps" between states of the scattered ion or atom. The existence of such jumps was required by Bohr in his theory of the atom. Bohr's quantum jumps between atomic states [6] were the first form of quantum dynamics to be postulated. He assumed that an atom remained in an atomic eigenstate until it made an instantaneous jump to another state with the emission or absorption of a photon. Since these jumps do not appear to occur in solutions of the

[1] To whom any correspondence should be addressed.

Schrodinger equation, something similar to Bohr's idea has been added as an extra postulate in modern quantum mechanics. Stochastic quantum jump equations [7] were introduced as a tool for simulating the dynamics of a dissipative system with a large Hilbert space and their links with quantum measurement. The question arises whether an explanation of these jumps can be found to result from a Colombeau solution $(\Psi_\varepsilon(x,t;\hbar))_\varepsilon$ of the Schrödinger equation [8] alone without additional postulates. We found exact quasi-classical asymptotic of the quantum averages with position variable with well localized initial data. Note that an axiom of quantum measurement is: if the particle is in some state $|\Psi_{t,\varepsilon}\rangle$ that the probability $P(\boldsymbol{x},\delta\boldsymbol{x}), \boldsymbol{x}\in\mathbb{R}^d$ of getting a result $\boldsymbol{x}$ at instant t with an accuracy of $\|\delta\boldsymbol{x}\|\ll 1$ will be given by

$$P_t(\boldsymbol{x},\delta\boldsymbol{x},\boldsymbol{x}_0;\hbar,\varepsilon)=\int_{\|x-x'\|\le\|\delta x\|}\left|\langle x'|\Psi_{t,\varepsilon}\rangle\right|^2 d\boldsymbol{x}'=\int_{\|x-x'\|\le\|\delta x\|}|\Psi_\varepsilon(\boldsymbol{x}',t,\boldsymbol{x}_0;\hbar)|^2 d\boldsymbol{x}'. \quad (1.1)$$

We rewrite now Eq. (1.1) in the form

$$P_t(\boldsymbol{x},\delta\boldsymbol{x},\boldsymbol{x}_0;\hbar,\varepsilon)=\frac{1}{(2\pi)^{d/2}\|\delta\boldsymbol{x}\|^d}\int_{\mathbb{R}^d}|\Psi_\varepsilon(\boldsymbol{x}',t,\boldsymbol{x}_0;\hbar)|^2\exp\left[-\frac{\|x-x'\|^2}{\|\delta x\|^2}\right]d\boldsymbol{x}'. \quad (1.2)$$

Definition 1.1. [8]. We define well localized Colombeau limiting quantum trajectories $\boldsymbol{q}(t)=\boldsymbol{q}(t,\boldsymbol{x}_0)=\{q_1(t,\boldsymbol{x}_0),\dots,q_d(t,\boldsymbol{x}_0)\}\in\mathbb{R}^d$, $\boldsymbol{q}(0,\boldsymbol{x}_0)=\boldsymbol{x}_0, t\in[0,T]$ by $\left(\underline{\lim}_{\|\delta\boldsymbol{q}(t)\|\to 0}\underline{\lim}_{\hbar\to 0}P_t(\boldsymbol{q}(t),\delta\boldsymbol{q}(t),\boldsymbol{x}_0;\hbar,\varepsilon)\right)_\varepsilon=1$ and well localized limiting quantum trajectories $\boldsymbol{q}(t)=\boldsymbol{q}(t,x_0)=\{q_1(t,x_0),\dots,q_d(t,x_0)\}\in\mathbb{R}^d$, $\boldsymbol{q}(0,\boldsymbol{x}_0)=\boldsymbol{x}_0, t\in[0,T]$ by

$$\lim_{\varepsilon\to 0}\lim_{\|\delta\boldsymbol{q}(t)\|\to 0}\lim_{\hbar\to 0}P_t(\boldsymbol{q}(t),\delta\boldsymbol{q}(t),\boldsymbol{x}_0;\hbar,\varepsilon)=1. \quad (1.3)$$

2.Colombeau solutions of the Schrödinger equation and corresponding path integral representation

Let **H** be a complex infinite dimensional separable Hilbert space, with inner product $<\cdot,\cdot>$ and norm $|\cdot|$.

Let us consider Schrödinger equation:

$$-i\hbar\left(\frac{\partial\Psi(t)}{\partial t}\right)+\widehat{H}(t)\Psi(t)=0, \Psi(0)=\Psi_0(x),\ H(t)=-\left(\frac{\hbar^2}{2m}\right)\Delta+V(x,t). \quad (2.1)$$

Here operator $H(t)\colon\mathbb{R}\times\mathbf{H}\to\mathbf{H}$ is essentially self-adjoint, $\widehat{H}(t)$ is the closure of $H(t)$.

Theorem 2.1. [9]. Assume that: (1) $\Psi_0(x)\in L_2(\mathbb{R}^d)$,(2)$V(x,t)$ is continuous and $\sup_{x\in\mathbb{R}^d,t\in[0,T]}|V(x,t)|)<+\infty$. Then corresponding solution of the Schrödinger equation (2.1) exists and can be represented via formula

$$\Psi(t,x)=\lim_{n\to\infty}\left(\frac{nm}{4\pi it\hbar}\right)^{d(n+1)/2}\int_{\mathbb{R}^d}\dots\int_{\mathbb{R}^d}dx_0dx_1\dots dx_n\Psi_0(x_0)\exp\left[\frac{i}{\hbar}S(x_0,x_1,\dots,x_n,x_{n+1};t)\right], \quad (2.2)$$

where we have set $x_{n+1}=x$ and $S(x_0,x_1,\dots,x_n,x_{n+1};t)=\sum_{i=1}^n\left[\frac{m}{4}\frac{|x_{i+1}-x_i|^2}{(t/n)^2}-V(x_{i+1},t_i)\right]$, where $t_i=\frac{it}{n}$. Let $q_n(t)$ be a trajectory; that is, a function from $[0,t]$ to $\mathbb{R}^d$ with $q_n(0)=x_0$ and set $q_n(t_i)=x_i, i=1,\dots,n+1$.

Trotter and Kato well known classical results give a precise meaning to the Feynman integral when the potential $V(x,t)$ is sufficiently regular [9]. However if potential $V(x,t)$ is a non-regular this is well known problem to represent solution of the Schrödinger equation (2.1)-(2.2) via formula (2.3), see [9]. We avoided this difficulty using contemporary Colombeau framework. Using replacement $x_i\to\frac{x_i}{1+\varepsilon^{2k}|x|^{2k}}, \varepsilon\in(0,1], k\ge 1$ we obtain from potential $V(x,t)$ regularized potential $V_\varepsilon(x,t), \varepsilon\in(0,1]$, such that $V_{\varepsilon=0}(x,t)=V(x,t)$ and

(i) $\left(V_\varepsilon(x,t)\right)_\varepsilon\in G(\mathbb{R}^d)$, (ii) $\sup_{x\in\mathbb{R}^d,t\in[0,T]}|V_\varepsilon(x,t)|)<+\infty, \varepsilon\in(0,1]$.

Here $G(\mathbb{R}^d)$ is Colombeau algebra of Colombeau generalized functions [8]. Finally we obtain regularized Schrödinger equation of Colombeau form [8]

$$-i\hbar\left(\frac{\partial\Psi_\varepsilon(t)}{\partial t}\right)_\varepsilon + \left(\widehat{H}_\varepsilon(t)\Psi_\varepsilon(t)\right)_\varepsilon = 0, \left(\Psi_\varepsilon(0)\right)_\varepsilon = \Psi_0(x), H_\varepsilon(t) = -\left(\frac{\hbar^2}{2m}\right)\Delta + V_\varepsilon(x,t). \quad (2.3)$$

Using the inequality Theorem 2.1 asserts again that corresponding solution of the Schrödinger equation exists and can be represented via formula [8]:

$$\left(\Psi_\varepsilon(t,x)\right)_\varepsilon = \left(\lim_{n\to\infty}\left(\frac{nm}{4\pi it\hbar}\right)^{d(n+1)/2}\int_{\mathbb{R}^d}\dots\int_{\mathbb{R}^d} dx_0 dx_1 \dots dx_n \Psi_0(x_0)\exp\left[\frac{i}{\hbar}S_\varepsilon(x_0,x_1,\dots,x_n,x_{n+1};t)\right]\right)_\varepsilon, \quad (2.4)$$

where we have set $x_{n+1} = x$ and $S_\varepsilon(x_0,x_1,\dots,x_n,x_{n+1};t) = \sum_{i=1}^{n}\left[\frac{m}{4}\frac{|x_{i+1}-x_i|^2}{(t/n)^2} - V_\varepsilon(x_{i+1},t_i)\right]$,

where we have set $t_i = \frac{it}{n}$.

3.Exact quasi-classical asymptotic beyond Maslov canonical operator

Theorem 3.1. Let us consider Cauchy problem with initial data $\Psi_0(x)$ is given via formula

$$\Psi_0(x) = \frac{\eta^{d/4}}{(2\pi)^{d/4}\hbar^{d/4}}\exp\left[-\frac{\eta(x-x_0)^2}{2\hbar}\right], \text{ where } 0<\hbar\ll\eta\ll 1 \text{ and } x^2 = \langle x,x\rangle.$$

(1) We assume now that: (i) $\left(V_\varepsilon(x,t)\right)_\varepsilon \in G(\mathbb{R}^d)$, (ii) $V_{\varepsilon=0}(x,t) = V(x,t)\colon \mathbb{R}_+\times\mathbb{R}^d \to \mathbb{R}$ and (iii) $\forall t\in\mathbb{R}_+$ function $V(x,t)$ is a polynomial on variable $x = (x_1,\dots,x_d)$, i.e.

$$V(x,t) = \sum_{\|\alpha\|\le m} g_\alpha(t)x^\alpha, \alpha = (i_1,\dots,i_d), x^\alpha = x_1^{i_1}\times\dots\times x_d^{i_d}, \|\alpha\| = \sum_{r=1}^{d} i_r.$$

(2) Let $u(\tau,t,\lambda,x,y) = \left(u_1(\tau,t,\lambda,x,y),\dots,u_d(\tau,t,\lambda,x,y)\right)$ be the solution of the boundary problem

$$\frac{\partial^2 u^{\mathrm{T}}(\tau,t,\lambda,x,y)}{\partial\tau^2} = \mathrm{Hess}[V(\lambda,\tau)]u^{\mathrm{T}}(\tau,t,\lambda,x,y) + [V'(\lambda,\tau)]^{\mathrm{T}}, u(0,t,\lambda,x,y) = y, u(t,t,\lambda,x,y) = x. \quad (3.1)$$

Here $\lambda = (\lambda_1,\dots,\lambda_d)\in\mathbb{R}^d, u^{\mathrm{T}}(\tau,t,\lambda,x,y) = \left(u_1(\tau,t,\lambda,x,y),\dots,u_d(\tau,t,\lambda,x,y)\right)^{\mathrm{T}}$,

$$V'(\lambda,\tau) = \left(\left[\frac{\partial V(x,t)}{\partial x_1}\right]_{x=\lambda},\dots,\left[\frac{\partial V(x,t)}{\partial x_d}\right]_{x=\lambda}\right) \text{ and } \mathrm{Hess}[V(\lambda,\tau)] = \left[\frac{\partial^2 V(x,t)}{\partial x_i\partial x_j}\right]_{x=\lambda}.$$

(3) Let $S(t,\lambda,x,y)$ be the master action given by formula

$$S(t,\lambda,x,y) = \int_0^t L(\dot{u}(\tau,t,\lambda,x,y),u(\tau,t,\lambda,x,y),\tau)\,d\tau, \quad (3.2)$$

where master Lagrangian $L(\dot{u},u,\tau)$ are

$$L(\dot{u},u,\tau) = \frac{m}{2}\dot{u}^2(\tau,t,\lambda,x,y) - \hat{V}(u(\tau,t,\lambda,x,y),\tau), \dot{u} = \left(\frac{\partial u_1}{\partial\tau},\dots,\frac{\partial u_d}{\partial\tau}\right), \dot{u}^2 = \langle\dot{u},\dot{u}\rangle,$$

$$\hat{V}((\tau,t,\lambda,x,y),\tau) = u(\tau,t,\lambda,x,y)\mathrm{Hess}[V(\lambda,\tau)]u^{\mathrm{T}}(\tau,t,\lambda,x,y) + V'(\lambda,\tau)u^{\mathrm{T}}(\tau,t,\lambda,x,y). \quad (3.3)$$

Let $y_{cr} = y_{cr}(t,\lambda,x)\in\mathbb{R}^d$ be the solution of the linear system of algebraic equations

$$\left[\frac{\partial S(t,\lambda,x,y)}{\partial y_i}\right]_{y=y_{cr}} = 0, i = 1,\dots,d. \quad (3.4)$$

(4) Let $\hat{x} = \hat{x}(t,\lambda,x_0)\in\mathbb{R}^d$ be the solution of the linear system of algebraic equations

$$y_{cr}(t,\lambda,\hat{x}) + \lambda - x_0 = 0. \quad (3.5)$$

Assume that for a given values of parameters t,λ,x_0 the point $\hat{x} = \hat{x}(t,\lambda,x_0)$ is not a focal point on a corresponding trajectory is given by corresponding solution of the boundary problem (3.1). Then for the limiting quantum average given via formulae (1.1) the inequality is satisfied:

$$\underline{\lim}_{\substack{\hbar\to 0\\ \varepsilon\to 0}}|\langle i,t,x_0;\hbar\rangle - \lambda_i| \le 2\left[\left|\det S_{y_{cr}y_{cr}}\left(t,\lambda,\hat{x}(t,\lambda,x_0),y_{cr}(t,\lambda,\hat{x}(t,\lambda,x_0))\right)\right|\right]^{-1}|\hat{x}_i(t,\lambda,x_0)|, \quad (3.6)$$

$i = 1, \ldots, d$. Thus one can to calculate the limiting quantum trajectory corresponding to potential $V(x,t)$ by using *transcendental master equation*

$$\hat{x}_i(t, \lambda, x_0) = 0, i = 1, \ldots, d. \tag{3.7}$$

4.Quantum anharmonic oscillator with a cubic potential supplemented by additive sinusoidal driving

In this subsection we calculate exact quasi-classical asymptotic for quantum anharmonic oscillator with a cubic potential supplemented by additive sinusoidal driving. Using theorem 3.1 we obtain corresponding limiting quantum trajectories given by Eq. (1.3). Let us consider quantum anharmonic oscillator with a cubic potential $V(x) = (m\omega^2/2)x^2 - ax^3 + bx, x \in \mathbb{R}, a, b > 0$ supplemented by an additive sinusoidal driving, i.e. $V(x,t) = (m\omega^2/2)x^2 - ax^3 + bx - [A\sin(\Omega t)]x$. The corresponding master Lagrangian is given by (3.3), is

$$L(\dot{u}, u, \tau) = (m/2)\dot{u}^2 - m(\omega^2/2 + 3a\lambda/m)u^2 - (m\omega^2\lambda + 3a\lambda^2 - b - A\sin(\Omega t))u.$$

We assume now that $\omega^2/2 + 3a\lambda/m \geq 0$ and rewrite $L(\dot{u}, u, \tau)$ in the form $L(\dot{u}, u, \tau) = (m/2)\dot{u}^2 - (m\varpi^2\lambda/2)u^2 + g(\lambda, t)u$, where $\varpi(\lambda) = \sqrt{2|\omega^2/2 + 3a\lambda/m|}$ and $g(\lambda, t) = -[m\omega^2\lambda + 3a\lambda^2 - b - A\sin(\Omega \cdot t)]$.

Therefore corresponding master action $S(t, \lambda, x, y)$ given by Eq. (3.2) is

$$S(t, \lambda, x, y) = \frac{m\varpi}{2\sin\varpi t}[(\cos\varpi t)(y^2 + x^2) - 2xy + \frac{2x}{m\varpi}\int_0^t g(\lambda, \tau)\sin(\varpi\tau)\, d\tau +$$

$$+ \frac{2y}{m\varpi}\int_0^t g(\lambda, \tau)\sin(\varpi(t - \tau))\, d\tau - \frac{2}{m^2\varpi^2}\int_0^t \int_0^\tau g(\lambda, \tau)g(\lambda, s)\sin\varpi(t - \tau)\sin(\varpi s)\, ds d\tau]. \tag{4.1}$$

The linear system of algebraic equations (3.4) is

$$\frac{\partial S(t, \lambda, x, y)}{\partial y} = 2y\cos\varpi t - 2x + (2/m\varpi)\int_0^t g(\lambda, t)\sin((\varpi(t - \tau))d\tau = 0.$$

Therefore

$$y_{cr}(t, \lambda, x) = \frac{x}{\cos\varpi t} - \frac{1}{m\varpi\cos\varpi t}\int_0^t g(\lambda, t)\sin\big((\varpi(t - \tau)\big)d\tau. \tag{4.2}$$

The linear system of algebraic equations (3.5) is

$$\frac{x}{\cos\varpi t} - \frac{1}{m\varpi\cos\varpi t}\int_0^t g(\lambda, t)\sin\big((\varpi(t - \tau)\big)d\tau + \lambda - x_0 = 0. \tag{4.3}$$

Then the solution of the linear system of algebraic equations (3.5) is

$$\hat{x}(t, \lambda, x_0) = \frac{1}{m\varpi}\int_0^t g(\lambda, t)\sin\big((\varpi(t - \tau)\big)d\tau + (\lambda(t) - x_0)\cos\varpi t. \tag{4.4}$$

Transcendental master equation (3.7) is

$$\frac{1}{m\varpi}\int_0^t g(\lambda(t), t)\sin\big((\varpi(t - \tau)\big)d\tau + (\lambda(t) - x_0)\cos\varpi t = 0. \tag{4.5}$$

Finally from Eq. (4.5) one obtains

$$d\big(\lambda(t)\big)\left(\frac{\cos(\varpi t)}{\varpi} - \frac{1}{\varpi}\right) + \frac{A(\varpi\sin(\Omega t) - \Omega\sin(\varpi t))}{\varpi^2 - \Omega^2} - (\lambda(t) - x_0)m\varpi\cos(\varpi t) = 0, \tag{4.6}$$

where $d(\lambda) = m\omega^2\lambda + 3a\lambda^2 - b$.

Numerical Example. $x_0 = 0, m = 1, \Omega = 0, \omega = 9, a = 3, b = 10, A = 0$.

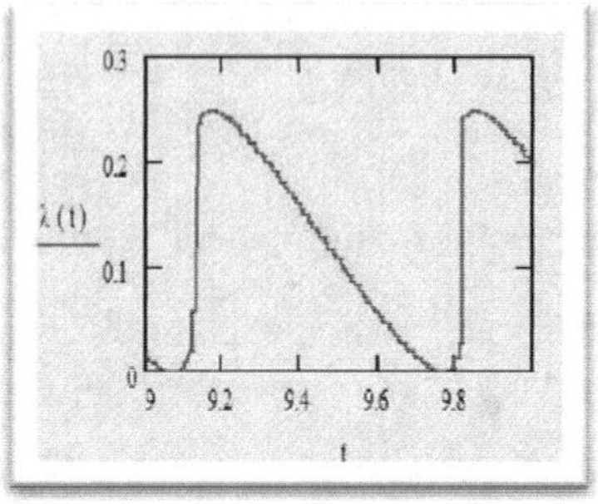

Figure 1. Limiting quantum trajectory $\lambda(t)$ with a jump.

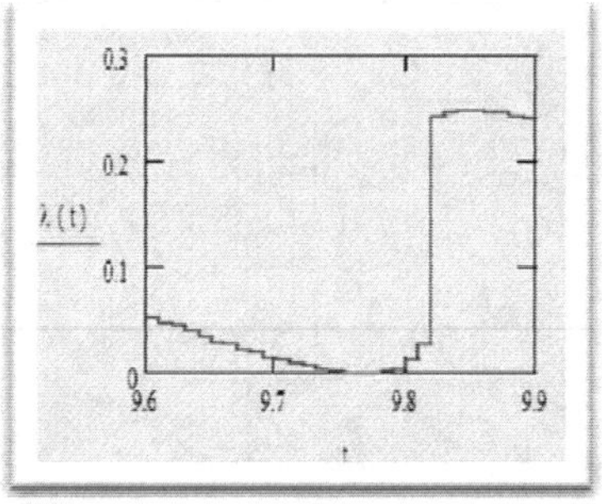

Figure 2. Limiting quantum trajectory $\lambda(t)$ with a jump.

5.Conclusion

We pointed out that there exist limiting quantum trajectories given via Eq. (1.3) with a jump. Such jumps does not depend on any single measurement of particle position $q(t)$ at instant t and obtained without any reference to a phenomenological master-equation of Lindblad form. An axiom of quantum mechanics is that we cannot predict the result of any single measurement of an observable of a quantum mechanical system in a superposition of eigenstates. However we can predict the result of any single measurement of particle position $q(t)$ at instant t with a probability $P(t, q, \delta q) \cong 1$ if the condition is valid: $\lim_{\varepsilon\to 0}\lim_{\|\delta q(t)\|\to 0}\lim_{\hbar\to 0} P_t(q(t), \delta q(t), x_0; \hbar, \varepsilon) = 1$, where $P_t(q(t), \delta q(t), x_0; \hbar, \varepsilon)$ given via Eq. (1.2). Complete explanation is given in our preprint [8].

6.References

[1] Vijay R, Slichter D H & Siddiqi I 2011 Observation of quantum jumps in a superconducting artificial atom *Phys. Rev. Lett.* **106** 110502

[2] Peil S & Gabrielse G 1999 Observing the quantum limit of an electron cyclotron: QND measurements of quantum jumps between Fock states *Phys. Rev. Lett.* **83** 1287–1290

[3] Nagourney W, Sandberg J & Dehmelt H 1986 Shelved optical electron amplifier: observation of quantum jumps *Phys. Rev. Lett.* **56** 2797–2799

[4] Sauter T, Neuhauser W, Blatt R & Toschek P E 1986 Observation of quantum jumps *Phys. Rev. Lett.* **57** 1696–1698

[5] Bergquist J C, Hulet R G, Itano W M & Wineland D J 1986 Observation of quantum jumps in a single atom *Phys. Rev. Lett.* **57** 1699–1702

[6] Bohr N 1913 *Philos. Mag.* **26** 1

[7] Dum R, Zoller P and Ritsch H 1992 *Phys. Rev. A* **45** 4879

[8] Foukzon J, Potapov A A, Podosenov S A Exact quasiclassical asymptotics beyond Maslov canonical operator. http://arxiv.org/abs/1110.0098

[9] Nelson E 1964 Feynman integrals and the Schrödinger equation *J. Math. Phys.* **5** 332-343

Effect of β^-–charged eradiation and its calculation in the nuclear electrodynamics theory

V Yu Tertychny-Dauri

Saint-Petersburg National Research University of Information Technologies, Mechanics & Optics
Department of Physics and Engineering
49 Kronverkskiy Prospect, Saint-Petersburg 197101, Russia

E-mail: tertychny-dauri@mail.ru

Abstract. The study of own fields and charged particles motion and also charged fission splinters of a heavy nucleuses into nonrelativistic approximation is the subject of this paper research. The main efforts are concentrated in quest of charged share components by the radioactive β^-–disintegration. The corresponding field equations and equations of motion in the nuclear electrodynamics processes are obtained and their solutions are found. Analysis of the microscopic equations is generalized to the level of the macroscopic description of continuous medium electrodynamics and is accompanied by quantumomechanical additions.

1. Introduction

The important peculiarity of presented work is a definition and solution of fields equations and motion equations for fission splinters in consideration of charged share of radioactive inside nuclon β^-–disintegration effect, i. e. in consideration of emited cascaded electrons but without consideration of γ–eradiation [1], [2], [3]. The obtained theoretical results can be a basis for the further mathematical modelling of generation phenomenon of a virulent directional electromagnetical fields on the toroid (the nuclear electrogenerator) in the process of nuclear chain fission [4], [5], [6]. We propose the toroid is the solenoid with the external winding. Moreover inside of this toroid all fragments of the nuclear disintegration remain in force. That is why light charges don't may escape the system and therefore enter the formalism.

2. The fields and motion of charged fission particles

We are interested by the calculation recipe of electrical $E_*(R,t) \in R^3$ and magnetical $B_*(R,t) \in R^3$ fields in the point with coordinate $R \in R^3$ at time moment t which are produced by the particles with charges Z_i, coordinates $R_i(t)$ and velocities $\dot{R}_i(t) = v_i(t)$, $i = 1, 2, \dots$. It is known that given fields into Gaussian units system (taking into account of values in order $1/c$, where c is velocity of light, and of multiplicator 4π in the expression for potentials) can be written with the aid of microscopic Maxwell-Lorenz equations system in the form

$$\nabla E_* = \sum_i Z_i\,\delta\,(R_i - R), \quad -E_*' + \nabla \times B_* = \frac{1}{c}\sum_i Z_i \dot{R}_i\,\delta\,(R_i - R),$$

$$\nabla B_* = 0, \qquad B_*' + \nabla \times E_* = 0. \tag{1}$$

We denote here: for the vector ∇ (the differential Hamiltonian operator) is the differentiation on coordinate R, the feature on top is the differentiation on ct; $\delta\,(R_i - R)$ is the delta-function from $R_i - R$ included in the field sources.

Taking into account of the vector potential $a(R,t)\,:\;\; B_* = \nabla \times a$, the scalar potential $\varphi(R,t)\,:\; E_* = -\,\nabla\varphi - a'$ and the Lorenzian calibrated condition: $\nabla a + \varphi' = 0$ also, the solution for system (1) in the form of nonrelativistic electromagnetical field can be found

$$E_* = \sum_i E_i, \qquad E_i = -\,\nabla\,\frac{Z_i}{4\pi\,\sigma_i}, \qquad B_* = \sum_i B_i, \qquad B_i = \frac{1}{c}\,\nabla \times \frac{Z_i\,\dot{R}_i}{4\pi\,\sigma_i},$$

where through $\sigma_i = |\,R_i - R\,|$ is denoted the distance between points with coordinates R_i and R in the space R^3.

The equation of i–th particle motion with the mass m_i under the influence Lorenzian force has the form

$$m_i\,\ddot{R}_i = Z_i\,[\,E(R_i,t) + \frac{1}{c}\,\dot{R}_i \times B(R_i,t)\,], \tag{2}$$

where $E(R_i,t) = \sum_j E_j + E_0$, $B(R_i,t) = \sum_j B_j + B_0$; (E, B) is full electromagnetical field in the point R_i at time moment t; (E_0, B_0) is external field, (E_*, B_*) is internal field of the charged particles in the given point, moreover $i \neq j$.

For equation (2) can been attached the canonical form: $\partial H/\partial P_i = \dot{R}_i$, $\partial H/\partial R_i = -\,\dot{P}_i$ with the aid of the Hamiltonian

$$H = \sum_i \frac{P_i^2}{2m_i} + \sum_i \sum_j \frac{Z_i\,Z_j}{8\pi\,\sigma_{ij}}$$

$$+ \sum_i Z_i \left[\varphi_0(R_i,t) - \frac{P_i}{cm_i}\,A_0(R_i,t)\right], \qquad \sigma_{ij} = |\,R_i - R_j\,|,$$

where $i \neq j$, into the the terms of impulse variables P_i, coordinate variables R_i and potentials (scalar φ_0 and vector A_0) of external field.

3. The fields and motion of charged fission splinters

Let us set the problem about conclusion of field equations and motion equations of the charged splinters (the united particles with internal nuclear structure) which are made in consequence of the chain nuclear fission reaction. These compound united particles can be considered as many time ionizated positive ions for the reason electrons upsetting of the atom outer skin of divided substance.

We proceed from the field equation of individual particles (1). Add to the index i the index k. Then instead of vector R_i we take the vector R_{ki}, where $R_{ki} = R_k + r_{ki}$; where k is the splinter index, i is the index of the particle of given splinter, r_{ki} is the internal coordinate of ki-th particle concerning the fixed point of k-th splinter.

The charged fission splinters are not "the stable complexes" but are the powerfully nonsteady particles groups are exposed the pronounced instantaneous radioactive fission (the β^-–fission accompanied by the γ–eradiation). We have in the k-th fission splinter the individual charged particles in the point R_{ki} with the charges Z_{ki} and the charged nucleus in the point R_k is exposed β^-–fission with the charge $Z_k\,:\;\; Z_k = X_k + Y_k$, where X_k is the proton charge of nucleus (i. e. the original proton charge of splinter + the proton charge of β^-–fission products), Y_k is the electron charge of β^-–fission products.

The solutions E_* and B_* of field equations can be approximated as converged rows on parameter $|\,r_{ki}\,|/|\,R_k - R\,|$, i. e. the dimensions of a splinter R_{ki} smaller a distance σ_k from the observation point R to fixed point (nucleus) R_k of k-th splinter. Transform equations (1) taking into account all expansions, included the δ–function expansion into the Taylor's row on r_{ki} in the locality of the point $(R_k - R)$

$$\nabla E_* = \sum_k \sum_i Z_{ki} \sum_{n=0}^{\infty} \frac{(-1)^n}{n!} (r_{ki}\nabla)^n \delta(R_k - R) + \sum_k X_k \delta(R_k - R) + \sum_k Y_k \delta(R_k - R),$$

$$-E_*' + \nabla \times B_* = \frac{1}{c} \sum_k \sum_i Z_{ki} (\dot{R}_k + \dot{r}_{ki}) \sum_{n=0}^{\infty} \frac{(-1)^n}{n!} (r_{ki}\nabla)^n \delta(R_k - R)$$

$$+ \frac{1}{c} \sum_k X_k \dot{R}_k \delta(R_k - R) + \frac{1}{c} \sum_k Y_k \dot{R}_k \delta(R_k - R), \qquad \nabla B_* = 0, \qquad B_*' + \nabla \times E_* = 0.$$

The motion equation of ki-th particle into k-th splinter with the charge Z_{ki}, the mass m_{ki} and the coordinate R_{ki} at the time moment t into full electromagnetical field $(E(R_{ki}, B(R_{ki}, t))$ has a form of equation (2), where $R_i \to R_{ki}$, i. e.

$$m_{ki}\ddot{R}_{ki} = Z_{ki} [\, E(R_{ki}, t) + \frac{1}{c} \dot{R}_{ki} \times B(R_{ki},)\,] \qquad (3)$$

and

$$E(R_{ki}, t) = -\sum_j \nabla_{ki} \frac{Z_{ki}}{4\pi\,\sigma_{k,ij}} - \nabla_{ki} \frac{(X_k + Y_k)}{4\pi\,\sigma_{k,ik}} - \sum_l \sum_j \nabla_{ki} \frac{Z_{lj}}{4\pi\,\sigma_{kl,ij}}$$

$$-\sum_l \nabla_{ki} \frac{(X_l + Y_l)}{4\pi\,\sigma_{kl,il}} + E_0(R_{ki}, t), \qquad B(R_{ki}, t) = B_0(R_{ki}, t), \qquad (4)$$

where $j \neq i$, $l \neq k$; ∇_{ki} is the gradient vector on the elements of vector R_{ki}. In the system (4) are denoted

$$\sigma_{k,ij} = |\, R_{ki} - R_{kj} \,|, \qquad \sigma_{k,ik} = |\, R_{ki} - R_{kk} \,| = |\, R_{ki} - R_k \,|,$$

$$\sigma_{kl,ij} = |\, R_{ki} - R_{lj} \,|, \qquad \sigma_{kl,il} = |\, R_{ki} - R_{ll} \,| = |\, R_{ki} - R_l \,|,$$

and moreover $|\, R_{ki} - R_k \,| = |\, r_{ki} \,|$.

On the analogy of (2) $(i \to k)$ can be written the motion equation of the nucleus of k-th splinter with the charge $Z_k = X_k + Y_k$, the mass m_k, the coordinate R_k into full electromagnetical field $(E(R_k, t),\ B(R_k, t))$

$$m_k \ddot{R}_k = Z_k [\, E(R_k, t) + \frac{1}{c} \dot{R}_k \times B(R_k, t)\,], \qquad (5)$$

where

$$E(R_k, t) = -\sum_j \nabla_k \frac{Z_{kj}}{4\pi\,\sigma_{k,kj}} - \sum_l \sum_j \nabla_k \frac{Z_{lj}}{4\pi\,\sigma_{kl,kj}}$$

$$-\sum_l \nabla_k \frac{(X_l + Y_l)}{4\pi\,\sigma_{kl,kl}} + E_0(R_k, t), \qquad B(R_k, t) = B_0(R_k, t), \quad l \neq k. \qquad (6)$$

After the summarizing of equations (3) on i and the adding with equation (5) for description of the motion of k-th splinter with the mass $\bar{m}_k$ and the coordinate of masses center $\bar{R}_k$

$$\bar{m}_k = \sum_i m_{ki} + m_k, \qquad \bar{R}_k = \frac{\sum_i m_{ki} R_{ki} + m_k R_k}{\bar{m}_k}$$

we obtain the equation

$$\bar{m}_k \ddot{R}_k = \sum_i Z_{ki} \,[\, E(R_{ki}, t) + \frac{1}{c}\, \dot{R}_{ki} \times B(R_{ki}, t) \,] + Z_k \,[\, E(R_k, t) + \frac{1}{c}\, \dot{R}_k \times B(R_k, t) \,]. \qquad (7)$$

Substitute the expressions (4) and (6) into the equation (7). Then the resultant force corresponding the intrasplinter field for the set of central forces is equal zero. Consequently the equation (7) acquires the form

$$\begin{aligned} \bar{m}_k \ddot{R}_k = &- \sum_l \sum_i \sum_j \nabla_{ki} \frac{Z_{ki} Z_{lj}}{4\pi\, \sigma_{kl,ij}} - \sum_l \sum_i \nabla_{ki} \frac{Z_{ki}\,(X_l + Y_l)}{4\pi\, \sigma_{kl,il}} - \sum_l \sum_j \nabla_k \frac{Z_{lj}\,(X_k + Y_k)}{4\pi\, \sigma_{kl,kj}} \\ &- \sum_l \nabla_k \frac{(X_l + Y_l)(X_k + Y_k)}{4\pi\, \sigma_{kl,kl}} + \sum_i Z_{ki} \,[\, E_0(R_{ki}, t) + \frac{1}{c}\, \dot{R}_{ki} \times B_0(R_{ki},) \,] \\ &+ (X_k + Y_k) \,[\, E_0(R_k, t) + \frac{1}{c}\, \dot{R}_k \times B_0(R_k, t) \,], \quad l \neq k. \end{aligned} \qquad (8)$$

For all that external field (E_0, B_0) satisfies the homogeneous equations

$$\nabla B_0 = 0, \qquad B_0' + \nabla \times E_0 = 0.$$

Thus, the equation (8) is the motion equation of k-th splinter into electromagnetical field of other splinters and external sources.

It is important to note the received equations on a level with the traditional items in the right part of the Lorenzian force contain also the items are made a force connected with radioactive charged eradiation of nucleuses into fission splinters (the accounting of β^-–charged eradiation by the radioactive electrodynamical effect).

4. Conclusions

Chain fission of heavy nuclei is one of many physical phenomena which proceed in an avalanche scheme. In fission there is an impetuous increase in the numbers of neutrons, charged particles and fission splinters with enormous kinetic energy. The present theory aims to describe the above process quantitatively. In the future it could find an application in powerful reactors operating exclusively on the basis of electronuclear conversions.

References

[1] de Groot S R, Suttorp L G 1972 *Foundations of Electrodynamics* (Amsterdam: North-Holland Publishing Company) 560 p

[2] Artsimovich L A, Lukyanov S Yu 1979 *Motion of Charged Particles into Electric and Magnetic Fields* (Moscow: Nauka) 224 p

[3] Tertychny-Dauri V Yu 2014 *Foundations of Nuclear Electrodynamics* (Saarbrucken: Lambert Academic Publishing) 304 p

[4] Keepin G R 1965 *Physics of Nuclear Kinetics* (Massachusetts, London: Addison-Wesley Publishing Company) 428 p

[5] Tertychny-Dauri V Yu 2006 Optimal stabilization in the problems of adaptive nuclear kinetics *Different. Equat.* **42** 374-384

[6] Tertychny-Dauri V Yu 2008 *Galamech* Vol 3 *Hyperreactive Mechanics* (Moscow: Fizmatlit) 576 p

Coexistence of pure and mixed states in nonlinear maps

Yehuda Roth
Oranim College
E-mail: yudroth@gmail.com

Abstract. Coherence and interaction are important concepts in physics. While interaction describes a relation between individual objects such as forces acting between distinguishable particles, coherent objects exist with the sole purpose of describing a single object. For example, each component of a vector provides us with only partial information. The whole picture is revealed only when the components are coherently related to their generating vector. Another example is a singlet of two spin $\frac{1}{2}$- particles. The true nature of these two coherent particles is described by a spin-less single particle. Apparently it seems that objects can be either coherent or non-coherent but they cannot be both simultaneously. This is almost true. We show that a system can be described simultaneously as coherent and non-coherent but an observer can distinguish only one concept at a time.

1. Introduction

Although coherence and interaction seem to describe similar concepts, in fact, they are referred to as entirely different concepts. While interaction describes relations between individual objects, coherent objects exists with the sole purpose of describing a single object. This substantial difference is stressed by the inconceivable quantum measurement problem (the hard problem) [1]-[6]: Quantum entities can compose a coherent state, such as the $|+\rangle = \frac{1}{\sqrt{2}}(|1\rangle + |2\rangle)$ state in two-level systems. Although the $|1\rangle$, $|2\rangle$-states are only the components of the $|+\rangle$-vector, the essence of the states $|1\rangle, |2\rangle$ measurement is the ability to detect them separately as distinguishable quantities. This apparent paradox in the measurement procedure requires a drastic solution reflected through the collapse procedure [1]-[6].

Thus, at first glance it seems that defining a system as both coherent and incoherent is inconceivable unless the brutal collapse procedure is implemented. Nevertheless, in this paper we present a system for which both descriptions are simultaneously valid. For that purpose we define an observer who can interpret reality in various ways.

2. Primary Definitions

Following previous works [7, 8] we define nonlinear maps to describe the system evolution. Such a system can be related to a negative feedback system where the term negative feedback describes an evolving system that is fed back in a manner tending to balance the changes [9]. In our model the system is represented through a nonlinear regular map while this system is in contact with an environment that balanced its evolution. The environment is represented by a chaotic map. We represent the observed system map (s-map) with the iterating equation

$$x_{n+1} = f_R(x_n), \quad \forall n \;\; x_n \in [0,1] \tag{1}$$

Journal of Physics: Conference Series **633** (2015) 012038 doi:10.1088/1742-6596/633/1/012038

where all values of x are within the interval $[0,1]$ and R is a control parameter that determines the map nature.
We also introduce an environment through a chaotic map [7, 8]

$$\chi_{\nu+1} = \mathcal{F}_{\mathcal{R}}\left(\chi_\nu\right), \quad \forall\nu \;\; \chi_\nu \in [0,1], \tag{2}$$

where now $\mathcal{R}$ is the environment map control parameter.
Many nonlinear maps behaves as follows:
For relatively low value of the control parameter, a typical map converges into a single stable point, $x^{[1]}$, calculated as

$$x_{n+1} = x_n \equiv x^{[1]} \Rightarrow x^{[1]} = f_R^{[1]}\left(x^{[1]}\right) \tag{3}$$

where the superscript in the square brackets [1] indicates that the map converges into a single value.
In general a higher rank map experiences the Andronov Hopf bifurcation[11]. However, for m-splits, it is possible to redefine a map that retrieves a single stable value simply by using an m-step iteration using the map

$$x_{n+m} = x_n \equiv x^{[m]} \Rightarrow x^{[m]} = f_R^{[m]}\left(x^{[m]}\right) \Rightarrow f_R^{[m]} = \underbrace{f_R^{[1]} \circ f_R^{[1]} \circ \ldots \circ f_R^{[1]}}_{\longleftarrow \; m-times \; \longrightarrow} = f_R^{[m/2]} \circ f_R^{[m/2]}. \tag{4}$$

Let us recall that this map of rank $[m]$ possesses an m- dimension [7, 8, 10].

3. Batch of Maps Definition as the Representative of a Physical Phenomenon

We define the concept **batch of maps** where a batch defines the system in hand while the participating maps stand for a possible variety of descriptions.
We consider a basic map $f_R^{[1]}$ that establishes a **batch** of maps containing the basic map and maps of higher ranks. For example a map $f_R^{[1]}$ that converges into two final values establishes the $\left\{f_R^{[1]}, f_R^{[2]}\right\}$-batch of maps. Each batch is associated with a rank-number $\{b\}$ (b for batch) defined as *the number of the final values reached by the basic map* $f_R^{[1]}$.
We associate a batch of rank $\{b\}$ with some physical phenomenon while $\{b\}$ - the batch rank - defines the alternative ways an observer can describe this physical phenomenon. Let us elaborate using the following example. Suppose we have an observer who explores a batch of rank $\{2\}$. He finds two alternative methods to describe the system:
The first order map, $f_R^{[1]}$, describes a $1-D$ system that alternates between two values. The second rank map, $f_R^{[2]}$, leads to a categorization method by cataloging the physical phenomenon into two sections with sets terminating at x_α or x_β. Increasing a map rank beyond the double rank batch provides us with no further descriptions. Furthermore, the time step enlargement of the second- rank map eliminates some of the early-stage iterations. In this present example, maps with rank greater than 2 (the maps $f_R^{[4]}$, $f_R^{[8]}$, $f_R^{[16]}$,...) behave basically as a map of rank two.
Following this logic, in the general case, we limit the number of maps in a $\{b\}$-rank batch to maps with rank m smaller or equal to the batch rank b, that is, the batch $\left\{f_R^{[1]}, f_R^{[2]}, \ldots, f_R^{[b]}\right\}$.
States-Model Construction. We define the following state:

$$|n\rangle = \left(x_{n+1} - x_n\right)|\emptyset\rangle + e^{i\theta_\nu}\sqrt{1-\left(x_{n+1}-x_n\right)^2}\,|1\rangle, \quad \theta_n = \chi_{n+1} - \chi_n \tag{5}$$

where a set of relative phases θ_ν allows us to define an ensemble of states that we consider to describe the same f_R-map-iteration.

Notation Conventions A state composed of a (k)-set (selected out of the m-dimensions) with rank $[m]$, embedded in a batch of rank $\{b\}$ appears as

$$|\{b\}, [m], (k), n\rangle . \tag{6}$$

Adding the environment $\{\beta \to \infty\}$-batch, the notation becomes

$$|\{b\}, [m], (k), n; \{\beta \to \infty\}, [\mu], (\kappa), \nu\rangle . \tag{7}$$

4. The Map Density Matrix

An observer who observes the world is subject to the influence of two types of batches: The system $\{b\}$-batch, as was defined in eq. 1, and the environment $\{\beta\}$-batch $(\beta \to \infty)$ composed by chaotic maps according to eq. 2. All types of phases in these states representation are determined by the same map-phase relations where the chaotic phases $\theta^{[\mu]}_{\nu,\kappa}$ are defined as

$$\sin\left(\theta^{[\mu]}_{\nu,(\kappa)}\right) = \left(\chi^{[\mu]}_{\nu+1,(\kappa)} - \chi^{[\mu]}_{\nu,(\kappa)}\right), \quad \kappa = 0, 1, 2, ...\mu - 1, \quad \mu = 1, 2, ..., \infty, \tag{8}$$

while in the same manner the regular phases $\phi^{[m]}_{n,(k)}$ are associated with the x-map by

$$\sin\left(\phi^{[m]}_{n,(k)}\right) = \left(x^{[m]}_{n+1,(k)} - x^{[m]}_{n,(k)}\right), \quad k = 0, 1, 2, ..., m - 1, \quad m = 1, 2, ..., b \tag{9}$$

where the subscripts k and κ count the system and environment-independent sets, respectively. We now consider a specific set (k) picked arbitrarily. We assume that the iterations act on the $|\emptyset_{(k)}\rangle$ and $|1_{(k)}\rangle$ states as a unitary operator to form the rotated state

$$|\psi\rangle = |[m, \mu], (k, \kappa), n, \nu\rangle = \sin\left(\phi^{[m]}_{n,(k)}\right)|\emptyset_{(k)}\rangle + Exp\left\{i\theta^{[\mu]}_{\nu,(\kappa)}\right\}\cos\left(\phi^{[m]}_{n,i}\right)|1_{(k)}\rangle . \tag{10}$$

An observer who tracks the vector rotation can use the projective operators $P = |\psi\rangle\langle\psi|$ of eq. 10. Selecting some k-series and implementing the Birkhoff Von-Neumann ergodic theorem[12] over the time- dependent recursive maps, we assume ν - the chaotic time interval - to be much shorter than n - the system time interval. Alternatively, we can use the ergodic representation and assume that the chaotic map possesses an enormous degree of freedom compared to the small regular system size.Thus, either by using the ergodic representation or through the time dependent representation we obtain:

$$\rho^{[m]}_{n,(k)} = \underbrace{\left\langle P^{[m,\mu]}_{n,(k),(\kappa\approx 1)}\right\rangle_n}_{TimeAverage} = \underbrace{\left\langle P^{[m,\mu]}_{n,\nu\approx 1,(k),(\kappa)}\right\rangle_\kappa}_{ErgodicAverage} = \begin{pmatrix} \sin^2\left(\phi^{[m]}_{n,(k)}\right) & 0 \\ 0 & \cos^2\left(\phi^{[m]}_{n,(k)}\right) \end{pmatrix}. \tag{11}$$

In the general scenario a multi-dimensional projective operator always possesses the chaotic phase terms in the non-diagonal parts. Thus, after averaging over the chaotic phases we are left only with the density matrix diagonal terms.

5. The $\{b = 1\}$-batch Converges toward a Single Pure Density Matrix

A $\{b = 1\}$-batch possesses only the basic map $f^{[1]}_R$ that converges into a single value. According to eq. 9 the corresponding phase ϕ vanishes, thereby generating a pure density matrix. We can say that regardless of the random θ-phase a $\{b = 1\}$-batch always corresponds with pure states density matrices. That is $\rho^{[1]}_{n\to\infty} \to |1\rangle\langle 1|$. (see eqs 11).

6. The $\{b = 2\}$-batch converges toward Pure and Mixed Density Matrices

A batch of the second rank consists of the maps $\left\{f_R^{[1]}, f_R^{[2]}\right\}$.

The $m = 1$ map embedded in a $b = 2$-Batch

The first map $f_R^{[1]}$ induces the following density matrices

$$\rho^{[1]}_{\{2\},(k),n} = \begin{pmatrix} \sin^2\left(\phi^{[1]}_{n,(k)}\right) & 0 \\ 0 & \cos^2\left(\phi^{[1]}_{n,(k)}\right) \end{pmatrix} \qquad k = 0, 1 \tag{12}$$

where $\sin\left(\phi^{[m]}_{n,(k)}\right) = \left(x^{[m]}_{n+1,(k)} - x^{[m]}_{n,(k)}\right)$ while $k = 0$ and $k = 1$ denote the two distinguishable sets. After sufficiently long iterations the trigonometric functions converge to

$$\lim_{n\to\infty} \sin\left(\phi^{[1]}_{n,(\alpha)}\right) = -\lim_{n\to\infty} \sin\left(\phi^{[1]}_{n,\beta}\right) = (x_\alpha - x_\beta)\,. \tag{13}$$

We then obtain the mixed-state density matrix

$$\begin{matrix}\rho^{[1]}_{\{2\},(k),n} \\ n\to\infty\end{matrix} = \begin{pmatrix} (x_\alpha - x_\beta)^2 & 0 \\ 0 & 1 - (x_\alpha - x_\beta)^2 \end{pmatrix}. \tag{14}$$

Thus, a map of rank one embedded in a batch of rank two generates a density matrix of mixed state.

The $m = 2$ - map embedded in a $b = 2$-batch

In the the second-rank map representation ($f_R^{[2]}$) the space splits into two parts: The α part containing all sets terminating at the x_α-value and the β part, the other sets ending at the x_β-value. Now there are four states divided into two categories: correlated and uncorrelated states. The uncorrelated bases of states are

$$\underbrace{\left|1_{(\alpha)}\right\rangle,\ \left|\emptyset_{(\alpha)}\right\rangle}_{\alpha\text{-Basis}},\quad \underbrace{\left|1_{(\beta)}\right\rangle,\ \left|\emptyset_{(\beta)}\right\rangle}_{\beta\text{- Basis}} \tag{15}$$

with the two independent density matrices

$$\rho^{[2]}_{\{2\},(\alpha),n} = \begin{pmatrix} \sin^2\left(\phi^{[2]}_{n,(\alpha)}\right) & 0 \\ 0 & \cos^2\left(\phi^{[2]}_{n,(\alpha)}\right) \end{pmatrix},\ \rho^{[2]}_{\{2\},(\beta),n} = \begin{pmatrix} \sin^2\left(\phi^{[2]}_{n,(\beta)}\right) & 0 \\ 0 & \cos^2\left(\phi^{[2]}_{n,(\beta)}\right) \end{pmatrix} \tag{16}$$

with

$$\sin\left(\phi^{[2]}_{n,(\alpha)}\right) = \left(x^{[2]}_{2n+2,(\alpha)} - x^{[2]}_{2n,(\alpha)}\right),\quad \sin\left(\phi^{[2]}_{n,(\beta)}\right) = \left(x^{[2]}_{2n+3,(\beta)} - x^{[2]}_{2n+1,(\beta)}\right). \tag{17}$$

For $n \to \infty$, both matrices converge into the double pure-state matrices $\left|1_{(\alpha)}\right\rangle\left\langle 1_{(\alpha)}\right|$ and $\left|1_{(\beta)}\right\rangle\left\langle 1_{(\beta)}\right|$.

The correlated-states density matrix

Correlated states are the basis of states that span a space obtained from a direct product between the α and β-spaces

$$\begin{matrix} \left|\emptyset_{(\beta)}, \emptyset_{(\alpha)}\right\rangle = \left|\emptyset_\beta\right\rangle\left|\emptyset_{(\alpha)}\right\rangle, & \left|1_{(\beta)}, \emptyset_{(\alpha)}\right\rangle = \left|1_\beta\right\rangle\left|\emptyset_{(\alpha)}\right\rangle, \\ \left|\emptyset_{(\beta)}, 1_{(\alpha)}\right\rangle = \left|\emptyset_\beta\right\rangle\left|1_{(\alpha)}\right\rangle, & \left|1_{(\beta)}, 1_{(\alpha)}\right\rangle = \left|1_\beta\right\rangle\left|1_{(\alpha)}\right\rangle. \end{matrix} \tag{18}$$

The density of states is composed as follows. We associate each $|\emptyset\rangle$- state with its compatible *sin*-term as a coefficient (with the appropriate α or β phases) while the coefficient of a $|1\rangle$-type-state is of the appropriate *cos*-form. As an example observe the term "$\sin^2\left(\phi^{[2]}_{n,(\beta)}\right)\left|\emptyset_{(\beta)}\right\rangle$". Thus, assuming only diagonal terms we obtain *the correlated-states density operator*

$$\rho^{[2]}_{\{2\}n} = \sin^2\left(\phi^{[2]}_{n,(\beta)}\right)\sin^2\left(\phi^{[2]}_{n,(\alpha)}\right)\left|\emptyset_{(\beta)},\emptyset_{(\alpha)}\right\rangle\left\langle\emptyset_{(\alpha)},\emptyset_{(\beta)}\right| + \cos^2\left(\phi^{[2]}_{n,(\beta)}\right)\sin^2\left(\phi^{[2]}_{n,(\alpha)}\right)\left|1_{(\beta)},\emptyset_{(\alpha)}\right\rangle\left\langle\emptyset_{(\alpha)},1_{(\beta)}\right| + \sin^2\left(\phi^{[2]}_{n,(\beta)}\right)\cos^2\left(\phi^{[2]}_{n,(\alpha)}\right)\left|\emptyset_{(\beta)},1_{(\alpha)}\right\rangle\left\langle 1_{(\alpha)},\emptyset_{(\beta)}\right| + \cos^2\left(\phi^{[2]}_{n,(\beta)}\right)\cos^2\left(\phi^{[2]}_{n,(\alpha)}\right)\left|1_{(\beta)},1_{(\alpha)}\right\rangle\left\langle 1_{(\alpha)},1_{(\beta)}\right|. \tag{19}$$

Unlike the uncorrelated density matrices defined in eqs. 16 that converged into the two pure-states density matrices $\left|1_{(\alpha)}\right\rangle\left\langle 1_{(\alpha)}\right|$ and $\left|1_{(\beta)}\right\rangle\left\langle 1_{(\beta)}\right|$, now the correlated-states density matrix converges only to the single pure-states density matrix $\left|1_{(\alpha)},1_{(\beta)}\right\rangle\left\langle 1_{(\beta)},1_{(\alpha)}\right|$.

7. Summary-The General Scenario

Suppose we have a batch of rank $\{b\}$. All maps below the batch rank generate only mixed-state density matrices. However, when the batch and a map share the same rank number, we obtain pure states density matrix. Thus, we obtained a system which can be interpreted as a mixed or pure system.

[1] Y. Roth (2014), *J. Phys.: Conf. Ser.* **490** 012055.
[2] R. Penrose, , *The Road to Reality: A Complete Guide to the Laws of the Universe*, Ch. 2, (Vintage Books, 2004)
[3] A. Bassi, *J. Phys.: Conf. Series* **67**, 012013 (2007)
[4] A. Bassi and D.G.M. Salvetti:, *Journ. Phys. A: Math. Theor*, **40**, 9859 (2007).
[5] A. Bassi, *Physics Today*, **44**, 36 (1991)
[6] S. L. Adler et al *J. Phys. A: Math. Theor.*, **40** 15083-15098 (2007)
[7] Y. G. Roth, **Advances in Pure Mathematics**, **4**, (2014)
[8] Y. G. Roth, *J. Phys.: Conf. Ser.* **574**, (2015)
[9] C.I. Byrnes, A. Isidori, J.C Willems, IEEE *Transactions on automatic control* **36, II** (1991)
[10] J. Awrejcewicz and P. Hagedorn, *Nonlinearity, Bifurcation and Chaos - Theory and Applications*, (InTech, 2012).
[11] J. E. Marsden, M. McCracken, *The Hopf Bifurcation and its Applications*, (Springer-Verlag, 1976) http://mina4-49.mc2.chalmers.se/ gojo71/KvantInfo/ DensityMatrix.pdf
[12] G. D. Birkhoff, *Proc Natl Acad Sci USA*, **17(12)**, (1931)

Pattern formation in non-equilibrium spinor polariton condensates

Florian Pinsker

E-mail: florian.pinsker@gmail.com
Department of Atomic and Laser Physics, University of Oxford, United Kingdom.

Abstract. Non-equilibrium polariton condensates entangle properties of lasers, atomic Bose-Einstein condensates (BEC) and semiconductor physics. They provide a great variety of physical phenomena while maintaining a simple theoretical description. Among those phenomena are nonlinear excitations such as solitons or spontaneous spin bifurcations. In this paper I first present a short overview on the theoretical basis of polaritons. Then starting from a scenario of excitation generation in equilibrium BEC I turn to corresponding phenomenona in polariton condensates such as dark soliton formation. Later the spin sensitive phenomena such as non-equilibrium bright solitons and half-bright solitons in semiconductor microcavities are discussed. Theoretically all the considered scenarios are described by partial differential equations (PDEs) and coupled systems thereof. The system of PDEs defines a so called condensate wave function, which completely describes the experimental relevant aspects of the physical system in a certain parameter regime where condensation occurs. The developed theories enables us particularly to make a variety of statements about excitations such as solitons and half-solitons in spinor systems forming within a non-equilibrium condensate. It turns out that by those means we can elucidate in particular the experimental implementation of a coherent superposition analog in a spin sensitive setting forming a macroscopic QUBIT within the semiconductor microcavity at temperatures in the Kelvin range.

1. Introduction

Coherent quantum matter has intriguing properties such as BEC, a finite speed of sound and nonlinear excitations [1] like dark solitons [2]. Polaritons are quasiparticles consisting of excitons and cavity photons within semiconductor micro-cavities which obey Bose-Einstein statistics [3, 4] and thus the potential to condense into a single particle mode [5]. Excitons are coupled electron-hole pairs of oppositely charged spin-half particles in a semiconductor held together by an effective Coulomb force between them [4] as the energy to form a pair is lower than a free electron and a free hole. States of excitons interact with light fields [6] and can form polaritons in the so called strong coupling regime when confined to a micro-cavity [7]. These polaritons posses integer spin and for dilute systems condense and because polaritons are 10^9 times lighter than rubidium atoms [5], this condensation is observed in CdTe/CdMgTe micro-cavities [5, 3] and even at room temperatures in flexible polymer-filled micro-cavities [8]. This stands in contrast to atomic BEC observed below 200nK [9, 10, 11]. The Hamiltonian describing the interaction between the light modes and excitons is [4, 3]

$$H = \sum_{\mathbf{k}} E_c(\mathbf{k})\phi^\dagger_{\mathbf{k}}\phi_{\mathbf{k}} + \sum_{\mathbf{k}} E_x(\mathbf{k})\xi^\dagger_{\mathbf{k}}\xi_{\mathbf{k}} + \hbar\Omega_R \sum_{\mathbf{k}} E_c(\mathbf{k}) \left(\phi^\dagger_{\mathbf{k}}\xi_{\mathbf{k}} + \xi^\dagger_{\mathbf{k}}\phi_{\mathbf{k}}\right). \tag{1}$$

Here $\phi_{\mathbf{k}}^{\dagger}$ is the creation and $\phi_{\mathbf{k}}$ the annihilation operator of a photon, while $\xi_{\mathbf{k}}^{\dagger}$ and $\xi_{\mathbf{k}}$ the corresponding operators for the exciton. By diagonalizing the Hamiltonian one gets two energy eigenvalues (lower and upper branch of the polariton dispersion) $E_{\mathrm{L,U}}(\mathbf{k}) = \frac{1}{2}\Big(E_c(\mathbf{k}) + E_x(\mathbf{k}) \mp \sqrt{(E_x(\mathbf{k}) - E_c(\mathbf{k}))^2 + 4\hbar^2\Omega_R^2}\Big)$ that are functions of the in-plane wave vector $\mathbf{k}$ and corresponding two basis vectors $\psi(\mathbf{k})_{\mathrm{LP}} = (h_{2,\mathbf{k}}, -h_{1,\mathbf{k}})$ and $\psi(\mathbf{k})_{\mathrm{UP}} = (h_{1,\mathbf{k}}, h_{2,\mathbf{k}})$, where the coefficients, the so called Hopfield factors [6], are given by $h_{1,\mathbf{k}} = \sqrt{\frac{1}{2}\left(1 - \frac{\delta(\mathbf{k})}{\sqrt{\delta(\mathbf{k})^2+4\hbar^2\Omega_R^2}}\right)}$ and $h_{2,\mathbf{k}} = \sqrt{\frac{1}{2}\left(1 + \frac{\delta(\mathbf{k})}{\sqrt{\delta(\mathbf{k})^2+4\hbar^2\Omega_R^2}}\right)}$. Generally the quantities $|h_{1,\mathbf{k}}|^2$ and $|h_{2,\mathbf{k}}|^2$ correspond to the fraction of photon and exciton within a given polariton. The dispersion of the cavity photon is $E_c(\mathbf{k}) = \frac{\hbar c}{n_c}\sqrt{\mathbf{k}_{\perp,c}^2 + \mathbf{k}^2}$ and the exciton energy $E_x(p) = \varepsilon + p^2/(2m_x)$ where m_x is the exciton mass and ε the energy offset. n_c is the refraction index between the cavity mirrors, c the speed of light. Although not considered by the stated Hamiltonian polaritons can be generated at will and decay usually between $1-100$ ps afterwards depending on the quality of confinement within the micro-cavity. When macroscopic coherent occupation of a single mode has been established the finite lifetime makes it essentially a *non-equilibrium* BEC with intriguing novel physical aspects and phenomena. The following section summarizes findings of the research paper [12], which discusses excitation generation in equilibrium BEC.

2. Quantum Piston

First, let us consider a scenario of equilibrium BEC. The many-body system can be described by a single giant matter wave when considering a dilute and weakly interacting atom cloud below the critical temperature [13]. In contrast to open systems the condensate wave function ψ is normalized to a fixed value for all times, $\|\psi\|_{L^2(\mathbb{R}^d)} = N$. Here d is the spatial dimension of the condensate. This models the scenario where particle numbers N remain the same and we assume the particles to be stable, i.e. non-decaying and there is no exchange of particles. The time evolution of the condensate wave function is given by the so-called Gross-Pitaevskii equation [14, 15],

$$i\hbar\partial_t\psi(\vec{r},t) = \left(-\frac{\hbar^2\vec{\nabla}^2}{2m} + V(\vec{r},t) + g(z)|\psi|^2\right)\psi(\vec{r},t). \tag{2}$$

To generate excitations within the condensate wave we utilize the possibility of changing interactions between the condensed particles (e.g. atoms) locally via Feshbach resonances. The basic idea is presented in Fig. 1. Here we assume a cold atom cloud trapped by the potential $V(\vec{r},t)$ to obey a 3D cigar shaped geometry. Once the self-interactions g are changed at one half of the atom cloud, particles move from highly interacting areas to less interactive areas of the cloud. Mathematically this change of self-interactions can be modeled by letting the initially constant interactions g become a step like function $g(z)$. This flow might carry topologically stable excitations within the superfluid. The excitations particularly depend on the dimension of the BEC. In Fig. 2 we see numerical results for 3D and 1D settings based on (2). The 1D setting becomes feasible for very tight trapping in the radial directions. The numerical results were generated by a 4th order Runge Kutta and 4th order finite differences method. All plots in Fig. 2 show the density of the superfluid. In a 3D scenario (Fig. 2 (a)) we observe the emergence of vortex rings propagating from the self-interaction step. In 1D we see the generation of so called dark soliton trains (Fig. 2 (b)). Those are regular arrays of dark solitons moving from

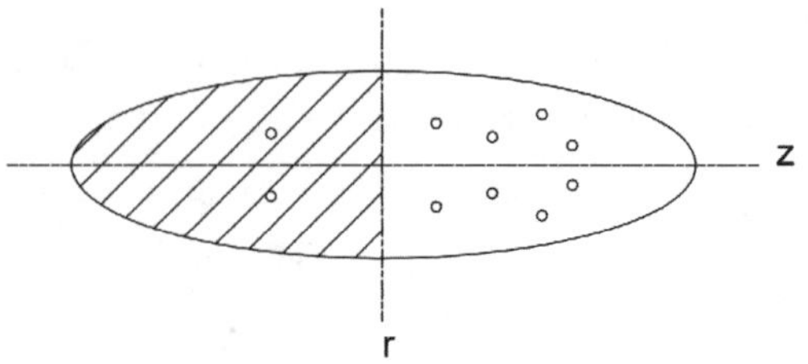

Figure 1. Schematic illustration of the quantum piston concept [12]. The area of change in self-interactions is illustrated by the hatched lines. Dots represent the emerging vortex rings as the self-interactions $g \to g(z)$ become a step function.

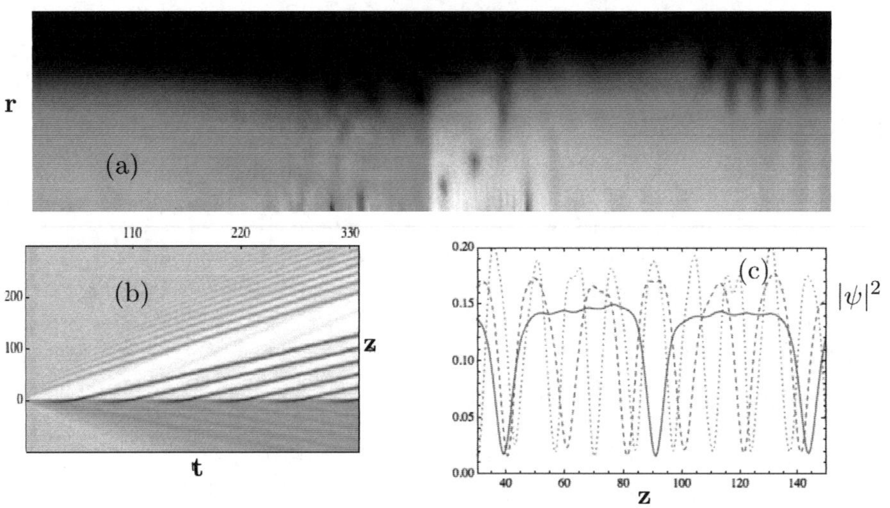

Figure 2. BEC density $n = |\psi|^2$ for 3D showing vortex rings (a) and in 1D dark-solitons (b). (c) shows dependence of the frequency on $g(z)$.

the self-interaction step. One observes that the frequency of excitation generation increases for higher changes in self-interaction aka higher superfluid velocities as shown in Fig. 2 (c).

3. Controlled soliton generation in polariton BEC

So far we have considered the scenario of excitation generation in equilibrium BEC (atoms, molecules, closed systems), where particle numbers are conserved. Now let us turn to non-equilibrium BEC of polaritons observed in semiconductor microcavities and pattern formation within the generalized condensate wave function as discussed in the research paper [17]. Polaritons are quasiparticles formed as superpositions of electrons and light fields that can be created at a laser pump spot and decay after 1-100ps. To theoretically model many observed phenomena usually a complex Ginzburg-Landau (cGL)-type equation incorporating energy relaxation and a stationary reservoir is utilized [3, 16]

$$i\partial_t \psi = (1 - i\eta) \left[-\frac{\hbar \nabla^2}{2m} + \alpha_1 |\psi|^2 + V(\vec{x}, t) + \frac{i}{2} \left(\delta_R N(\vec{x}, t) - \gamma_C \right) \right] \psi \tag{3}$$

with $\partial_t N = -(\gamma_R + \beta |\psi|^2) N + P$, where N is the magnitude of pre condensed polaritons. This set of equations models the scenario of incoherent pumping [3]. η denotes energy relaxation, α_1

the self-interaction strength, δ_R the scattering rate into the condensate, γ_C the condensate decay rate, γ_R is the constant reservoir decay rate, β the magnitude of scattering into the condensate and P the spatially dependent pumping function. Physically hot excitons are formed by external illumination, which by a variety of scattering processes then macroscopically occupy the lowest energy state on the polariton dispersion, i.e. a giant matter wave similar as in the atomic BEC case discussed before. To generate excitations within the corresponding condensate wave function we proceed as follows and as discussed in [17]. First we pump polaritons at a spatially localized area, which approximately obeys the form of a Gaussian. Subsequently we apply a potential step, which is feasible by adding a local electrical field. As new polaritons are formed at the pump spot the potential step induces a perturbation to the matter wave. The main physical quantities for soliton generation are the local speed of sound $c_s(x) = \sqrt{\mu(x)/m}$ of the superfluid, where $\mu(x) = \alpha_1 n(x)$ is the chemical potential, and the superfluid velocity, which is proportional to the gradient of the phase of the condensate wave function. Once the speed of sound is exceeded by the superfluid dark solitons emerge regularly. By changing the step size or the pumping strength, i.e. the amount of new polaritons entering the condensate, the period of dark solitons can be changed on demand. Results are presented in Fig. 3 for different pumping strengths. In (d) we see a Sinusoidal pumping signal.

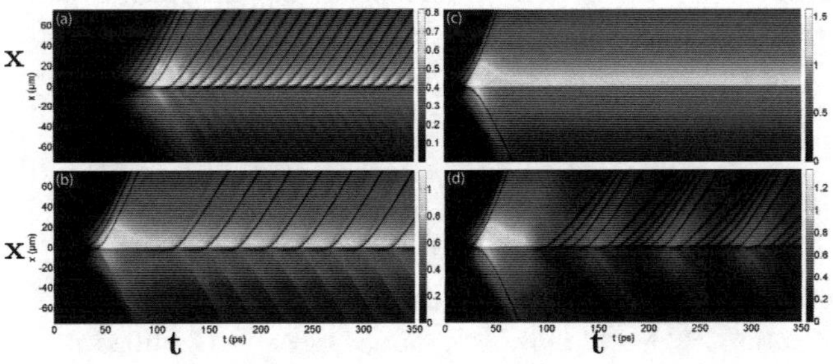

Figure 3. BEC density $n = |\psi|^2$ for 1D showing dark-solitons.

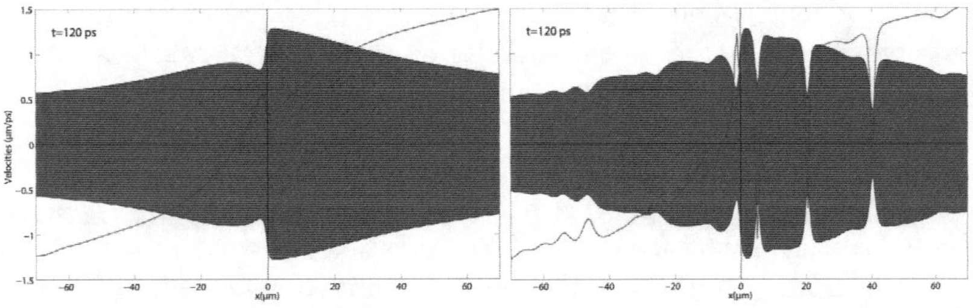

Figure 4. Snapshots of velocity related quantities taken at t_s = 120 ps. The blue envelope displays $\pm c_s(x, t_s)$ (speed of sound) and the red line shows the condensate velocity $v(x, t_s)$. The panel (a) shows results from a pump amplitude $600\Gamma_R$ where no soliton develop. The panel (b) shows the development of the train for a lower amplitude of $375\Gamma_R$ due to the local supersonic regions at the step.

The essential criterion for dark soliton generation is the breaking the speed of sound as Fig. 4 shows. Once the speed of sound is surpassed solitons are generated within the superfluid's stream. We observe that higher pump rates imply higher speed of sound, because the superfluid density increases. Therefore the number of generated dark solitons becomes smaller. On the other hand lower pump rates imply the emergence of more dark solitons in the same interval of time. In this setting one can further change the emergence of dark solitons on the fly by changing the step size or the pumping strength during the experiment.

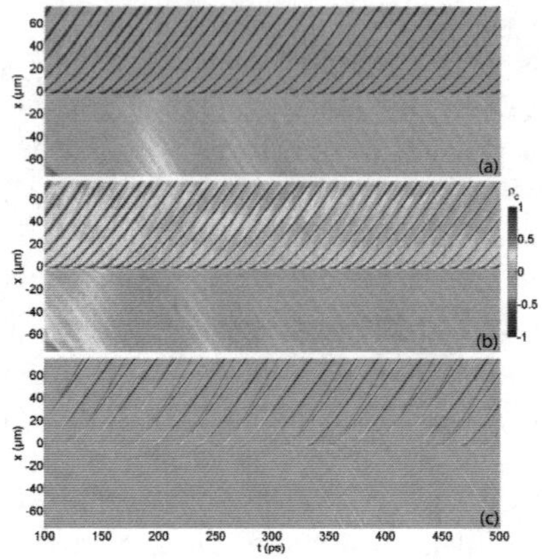

Figure 5. (Color online) Control over the polarization of the trains. The colormap shows the degree of circular polarization $\rho_c = (n_+ - n_-)/(n_+ + n_-)$. (a) $\delta_R^2/\delta_R^1 = 0.90$, (b) $\delta_R^2/\delta_R^1 = 1.11$ and (c) $\delta_R^2/\delta_R^1 = 0.99$.

Polaritons and polariton condensates have the property of spin, this is due the coupling of excitons of spin ± 1 with circularly polarized light. By this coupling either spin up or spin down polaritons and condensates thereof form. The physical system can be described mathematically by two coupled complex Ginzburg-Landau equations each representing one spin component of the spinor. The coupling between the two components depends on the particular regime, but for condensates away from $k = 0$ the so called longitudinal transversal splitting (TMTE splitting) becomes relevant, which introduces an energy separation of the two spin component wave functions. Physically the concept to generate spin sensitive phenomena can be to generate different populations of spin up and spin down polaritons by slightly generating more polaritons in one of the spin components. Then via the mechanism of TMTE splitting the two components distinguish themselves even further. Once a dark soliton train emerges in the spinor wave function, i.e. when the speed of sound is broken in a 1D scenario the dark soliton trains separate for differently populated components within each component. Interestingly each train is generated at the same time on the same breaking point of the potential step, but both evolve differently particularly due to the different speed of sounds in each component as Fig. 5 shows. Here the circular polarization is plotted for different ratios of growth rates δ_R^1 and δ_R^2, for spin up and spin down polaritons respectively, modeling a slight imbalance in pumping strengths into each spin component.

4. Spontaneous spin bifurcation

Next we turn to the system of spinor polariton condensates far away from the pumping spots as discussed in [19]. In Fig 6 (a) and (b) the schematics of the four pump spot setting is provided. The results are based on the experimental finding that the condensate forms away from the areas where they are created by external illumination through laser beams that act as repulsive potentials [18]. This is possible as polaritons have a relative long lifetime compared to excitons. As polaritons move away from the 4 pump spots generating them they get trapped in-between and the bulk of polaritons forms mainly there. In contrast to all previous experiments we observe a spontaneous emergence of either a spin up or spin down component, which mathematically corresponds to vanishing L^2 norm for one of the spin components, while having a fully populated complementary spin component. This holds although there is no pumping imbalance corresponding to a linearly polarized laser, i.e. both components are equally fed and initially should have the same population up to very small perturbations. The spinor polariton

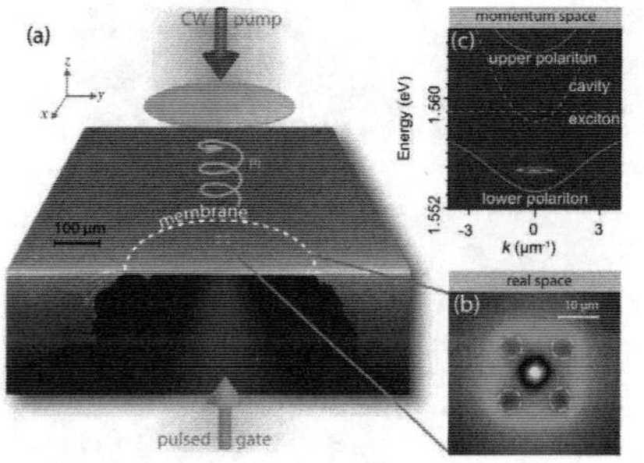

Figure 6. This Figure has been taken from [19].

condensate wave function's (i.e. $\Psi = (\psi_+, \psi_-)$) time evolution for this particular setting is given by: $i\partial_t\psi_\sigma = (1 - i\eta)\frac{\mathcal{H}_\sigma}{\hbar}\psi_\sigma$ with $\sigma \in \{+,-\}$ and

$$\frac{\mathcal{H}_\pm}{\hbar}\psi_\pm = \left(-\frac{\hbar\nabla^2}{2m^*} + aN_\pm + \alpha_1|\psi_\pm|^2 + \alpha_2|\psi_\mp|^2\right)\psi_\pm + \frac{i}{2}\left(bN_\pm - \gamma_C\right)\psi_\pm + iJ\psi_\mp \quad (4)$$

with complex Joesphson coupling $J = \varepsilon + i\gamma \in \mathbb{C}$ between the two spin components. The radiative dissipative part γ is due to interference of the light from different components and real Josephson coupling models tunneling between states ε. Here α_1 denotes self-interaction strength and $\alpha_2 = -0.1\alpha_1$ the cross-interaction strength. m^* is the effective mass of the polaritons of spin σ. The reservoir is approximately given by $N_\pm = P/\gamma_R(1 - \beta/\gamma_R(|\psi_\pm|^2 + |\psi_\mp|^2))$ and pumping of polaritons into the reservoir is $P(\vec{r}) = \sum_i^4 A \cdot \exp\left[-\theta\left((x - x_i)^2 + (y - y_i)^2\right)\right]$.

Fig 6 (b) shows the geometry in real space, while (c) shows the energy dispersions of the polariton, exciton and cavity photon as discussed in the introduction. Experimentally condensation happens at $k = 0$ and the width of the condensate peak is $\delta k = 0.4\mu m^{-1}$. Interestingly we note that such a setting can be described by a mean-field theory over several minutes [18] without taking into account the emergence of decoherence. Mathematically in addition to the usual terms for incoherently pumped polaritons with a spin up and down component we add a complex Josepshon coupling modeling the interference between both spin components and the tunneling i.e. interconversion between spin up and down particles, which can be understood as a transformation from spin up and down particles and vice versa. It turns out that this term is essential for the observed pattern formation. Figure 7 (a) shows the spontaneous realization of either a spin up or a spin down state. (b) clearly shows the stability of the spontaneously selected spin state as measured over time. Remember that the polariton lifetimes are between $1 - 100$ps in current settings and this figure shows stability over seconds. The degree of polarization is about 0.95 measured in experiments. Once a spin has been selected spontaneously, applying a constant gain of polaritons with opposite spin generates Bloch oscillations as indicated experimentally in second row of the Fig. 7 (a) and predicted theoretically in (b) in the same line. These findings indicate the existence of a macroscopic superposition between spin up and spin down states.

Acknowledgements.– I acknowledge support through my Schrödinger Fellowship.

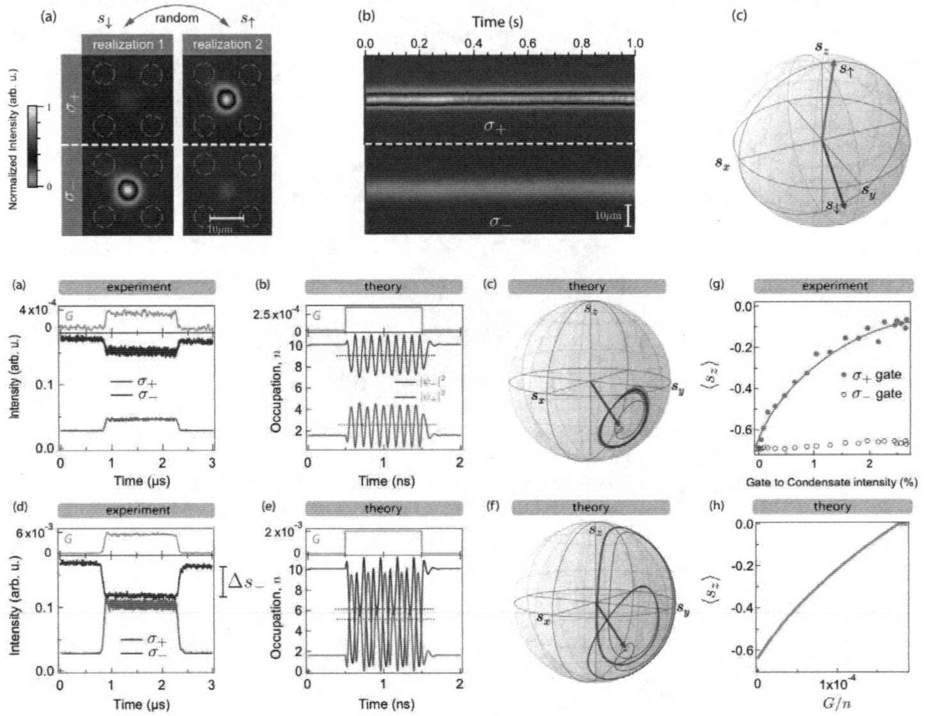

Figure 7. This Figure has been taken from [19].

5. References

[1] A. J. Leggett, Rev. Mod. Phys. **73**, 307 (2001).
[2] T. Tsuzuki, Journal of Low Temperature Physics **4**, 441 (1971).
[3] I. Carusotto and C. Ciuti, Rev. Mod. Phys. **85**, 299 (2013).
[4] H. Deng, H. Haug and Y. Yamamoto, Rev. Mod. Phys. **82**, 1489 (2010).
[5] J. Kasprzak *et al.*, Nature **443**, 409-414 (2006).
[6] J. J. Hopfield, Phys. Rev., **112**, 15551567 (1958).
[7] C. Weisbuch *et al.*, Phys. Rev. Lett. **69**, 33143317 (1992).
[8] J. D. Plumhof, *et al.*, Nature Materials, doi:10.1038/nmat3825 (2013).
[9] M. H. Anderson *et al.*, Science **269**, 198 (1995).
[10] K. B. Davis *et al.*, Phys. Rev. Lett. **75**, 3969 (1995).
[11] C. C. Bradley *et al.*, Phys. Rev. Lett. **75**, 1687 (1995).
[12] F. Pinsker, N. G. Berloff and V. M. Pérez-García, Phys. Rev. A **87**, 053624 (2013); arXiv:1305.4097 (2013).
[13] E.H. Lieb, *et al.* , Oberwolfach Seminars, **34**, Birkhäuser, Basel, 184pp. (2001).
[14] L. P. Pitaevskii, Zh. Eksp. Teor. Fiz. **40**, 646 (1961); Sov. Phys. JETP **13**, 451 (1961).
[15] E. P. Gross, Nuovo Cimento **20** (1961) 451; J. Math. Phys. **4** 195 (1963).
[16] J. Keeling and N.G. Berloff, Phys. Rev. Lett. **100**, 250401 (2008).
[17] F. Pinsker and H.Flayac, Phys. Rev. Lett. **112**, 140405 (2014); arXiv:1310.7500 (2013).
[18] A. Dreismann *et al.*, Proceedings of the National academy of Sciences of the United States, **111**, No. 24, 8770-8775 doi: 10.1073/pnas.1401988111 (2014).
[19] H. Ohadi *et al.*, *Spontaneous spin bifurcations and ferromagnetic phase transitions in a spinor exciton-polariton condensate*, to appear in PRX (2015).

Contact symmetries of constrained systems and the associated integrals of motion

N Dimakis[1], **Petros A Terzis**[2] **and T Christodoulakis**[2]

[1] Instituto de Ciencias Fisicas y Matematicas,
Universidad Austral de Chile, Valdivia, Chile
[2] Nuclear and Particle Physics Section, Physics Department,
University of Athens, GR 157–71 Athens

E-mail: nsdimakis@gmail.com, pterzis@phys.uoa.gr, tchris@phys.uoa.gr

Abstract. The conditions for the existence of (polynomial in the velocities) contact symmetries of constrained systems that are described by quadratic Lagrangians is presented. These Lagrangians mainly appear in mini-superspace reductions of gravitational plus matter actions. In the literature, one usually adopts a gauge condition (mostly for the lapse N) prior to searching for symmetries. This, however, is an unnecessary restriction which may lead to a loss of symmetries and consequently to the respective integrals of motion. A generalization of the usual procedure rests in the identification of the lapse function N as an equivalent degree of freedom and the according extension of the infinitesimal generator. As a result, conformal Killing tensors (with appropriate conformal factors) can define integrals of motion (instead of just Killing tensors used in the regular gauge fixed case). An example of a relativistic particle in a pp-wave space-time and under the influence of a quadratic potential is illustrated.

1. Introduction

Symmetries have always played a fundamental role in physical theories. In recent years there exists an increased interest on the subject of Noether point and/or generalized symmetries of physical systems and their geometric nature [1]-[7].

In general, one can discern two approaches regarding the treatment of symmetries in mini-superspace systems: The first entails an a priori fixation of the gauge (usualy $N = 1$) at the Lagrangian level and the treatment of the ensuing system as if it were regular ([8]-[12]). The second method ([6], [7], [13], [14]) utilizes the gauge invariance of parametrization invariant systems, resulting in the emergence of larger symmetry groups.

The general form of a mini-superspace Lagrangian is

$$L = \frac{1}{2N} G_{\mu\nu}(q)\dot{q}^{\mu}\dot{q}^{\nu} - NV(q), \quad \mu,\nu = 1,...,d, \quad \det(G_{\mu\nu}) \neq 0 \tag{1.1}$$

where $q^{\mu}(t)$ and $N(t)$ are the $d+1$ degrees of freedom of the system while the dot symbolizes the total time derivative $\frac{d}{dt}$. The Lagrangian denoted by (1.1) is singular, due to the quadratic constraint equation for N ($\frac{\partial L}{\partial N} = 0$). This constraint can be used to generalize what is considered as an integral of motion, since the time derivative of such a quantity need not be strictly zero, but just a multiple of the constraint.

Let us first review some basic facts from the theory of symmetries of the action for regular systems [16]. A contact transformation whose infinitesimal generator is $X = \Xi^\kappa(t,q,\dot{q})\frac{\partial}{\partial q^\kappa}$, constitutes a symmetry of the action if the condition

$$pr^{(1)}X(L) = \frac{dF}{dt} \tag{1.2}$$

is satisfied, where $pr^{(1)}X$ is the first prolongation of the generator (that expresses the change induced by X in the velocities $\dot{q}$'s) and $F(t,q,\dot{q})$ a gauge function. Given the form of the generator, a conserved quantity can be constructed as $\Xi^\kappa \frac{\partial L}{\partial q^\kappa} - F$.

For singular systems it has been shown [6] that the form of X should include dependence in N, since the latter is also a degree of freedom for the system. Thus, one is led to consider

$$X = \Xi^\kappa(t,N,q,\dot{q})\frac{\partial}{\partial q^\kappa} + \Omega(t,N,q,\dot{q})\frac{\partial}{\partial N}. \tag{1.3}$$

Notice that when the additional term $\Omega(t,N,q,\dot{q})\frac{\partial}{\partial N}$ is acting on the Lagrangian through (1.2), creates a multiple of the quadratic constraint. Thus, (1.2) becomes an equation that holds modulo the constraint.

2. General results for polynomial in the velocities contact symmetries

In our case we shall restrict ourselves to the consideration of symmetries with corresponding generators polynomial in the velocities. This means that Ξ^κ in (1.3) is taken as

$$\Xi^\kappa = \xi^\kappa(t,N,q) + S^\kappa{}_{\alpha_1}(t,N,q)\dot{q}^{\alpha_1} + \dots + S^\kappa{}_{\alpha_1\dots\alpha_n}(t,N,q)\dot{q}^{\alpha_1}\dots\dot{q}^{\alpha}_n. \tag{2.1}$$

This automatically restricts the forms of $\Omega(t,N,q,\dot{q})$ and $F(t,N,q,\dot{q})$. The first has to be polynomial of first order in $\dot{N}$ and of n order in $\dot{q}$'s, while the second of zero order in $\dot{N}$ and of $n+1$ order in $\dot{q}$'s. However, the existence of the gauge function can be proven to be trivial [7], so we won't refer to it any longer.

The procedure in order to derive the symmetry generator in these cases is the following: After the application of the symmetry criterion (1.2), the coefficients of terms involving velocities $\dot{N}$ and $\dot{q}^\mu$ of various orders are gathered and set to zero. By solving these partial differential equations, the general form for the symmetry generator is obtained:

$$X = \left(\xi^\kappa(q) + \frac{1}{N}S^\kappa_{\alpha_1}(q)\dot{q}^{\alpha_1} + \frac{1}{N^2}S^\kappa_{\alpha_1\alpha_2}(q)\dot{q}^{\alpha_1}\dot{q}^{\alpha_2} + \dots + \frac{1}{N^n}S^\kappa_{\alpha_1\dots\alpha_n}(q)\dot{q}^{\alpha_1}\dots\dot{q}^{\alpha_n}\right)\frac{\partial}{\partial q^\kappa}$$
$$- \left(\frac{N}{V}\tilde{\xi}^\sigma V_{,\sigma} + \frac{2V^{,\sigma}}{V}S_{\sigma\alpha_1}\dot{q}^{\alpha_1} + \frac{n+1}{N^{n-1}}\frac{V^{,\sigma}}{V}S_{\sigma\alpha_1\dots\alpha_n}\dot{q}^{\alpha_1}\dots\dot{q}^{\alpha_n}\right)\frac{\partial}{\partial N}$$

together with the subsequent conditions:

$$\mathcal{L}_\xi G_{\mu\nu} = -\frac{\mathcal{L}_\xi V}{V}G_{\mu\nu} \tag{2.2a}$$

$$S_{(\mu\alpha_1;\nu)} = -\frac{V^{,\sigma}}{V}S_{\sigma(\alpha_1}G_{\mu\nu)}, \; \dots \; , S_{(\mu\alpha_1\dots\alpha_n;\nu)} = -\frac{n+1}{2}\frac{V^{,\sigma}}{V}S_{\sigma(\alpha_1\dots\alpha_n}G_{\mu\nu)}. \tag{2.2b}$$

From the derived symmetry generator, one can construct integrals of motion of several orders in the momenta. It is clear that equations (2.2) already contain the possibility of ξ (the S's) being a Killing field (Killing tensors) of the metric and the potential ($S^{\sigma}\alpha_1 ... \alpha_n V_{,\sigma} = 0$), which are of course the known cases of symmetries for regular systems. It can also be easily verified that for systems described by (1.1), rheonomic integrals of motion can also be constructed by any conformal Killing field $\xi(q)$ [6]; for such a vector, the quantity

$$\tilde{Q} = \xi^{\alpha} p_{\alpha} + \int N(t) \left(\omega(q(t)) + f(q(t))\right) V(q(t)) dt, \tag{2.3}$$

where $f = \frac{\mathcal{L}_{\xi} V}{V}$ and ω is defined by $\mathcal{L}_{\xi} G_{\mu\nu} = \omega(q) G_{\mu\nu}$, is conserved on mass shell. We now proceed to consider a simple example.

3. Example: A particle in a curved space-time under the influence of a quadratic potential

We have already mentioned that Lagrangians of the form (1.1) are mainly encountered in cosmology. However, here we shall consider an example of a relativistic particle moving in a type Biv pp-wave space-time described by the line element

$$ds^2 = -2dvdu - \frac{2}{z^2} du^2 + dx^2 + dy^2 \tag{3.1}$$

under the influence of a quadratic potential of the form $V(u) = \frac{\mu}{2} u^2$. In the case where $\mu > 0$ the latter can be considered to represent an oscillator in the u direction. The space-time described by (3.1) has been examined in [17] with respect to its conformal algebra and its Killing tensors.

By using (1.1), the Lagrangian of the system reads

$$L = \frac{1}{N}\left(-\dot{u}\,\dot{v} - \frac{\dot{u}^2}{z^2} + \frac{1}{2}\dot{y}^2 + \frac{1}{2}\dot{z}^2\right) - N\frac{\mu\, u^2}{2} \tag{3.2}$$

and, as we can observe, it possesses five degrees of freedom. However, due to its constrained nature, four of them can be expressed as functions of the fifth on the solution space.

There exist three vector fields satisfying condition (2.2a) (which in this case happen also to be Killing fields of both the metric $G_{\mu\nu}$ and the potential $V(u)$). At the same time, there also exists a homothetic vector of both the metric and the potential. With their help the following four linear integrals of motion can be written:

$$Q_1 = -\frac{\dot{u}}{N}, \quad Q_2 = \frac{\dot{y}}{N}, \quad Q_3 = \frac{u\,\dot{y}}{N} - \frac{y\,\dot{u}}{N} \tag{3.3a}$$

$$Q_h = -\frac{2\,u\,\dot{u}}{N\,z^2} - \frac{u\,\dot{v}}{N} + \frac{y\,\dot{y}}{2\,N} + \frac{z\,\dot{z}}{2\,N} + \frac{3}{2}\mu \int N\,u^2\,dt. \tag{3.3b}$$

that consequently define the equations $Q_i = \kappa_i$ and $Q_h = \kappa_h$.

It can also be verified that there exist seven second rank non reducible conformal Killing tensors satisfying criterion (2.2b), all of which, albeit one, are also Killing tensors for the space-time under consideration. In our example we are going to make use of the following two

$$S_1^{\mu\nu} = \begin{pmatrix} -\frac{8}{u^2 z^2 \mu^2} & \frac{4}{u^2\mu^2} & 0 & 0 \\ \frac{4}{u^2\mu^2} & 0 & 0 & 0 \\ 0 & 0 & -\frac{4}{u^2\mu^2} & 0 \\ 0 & 0 & 0 & -\frac{4}{u^2\mu^2} \end{pmatrix}, \quad S_2^{\mu\nu} = \begin{pmatrix} \frac{8}{z^2\mu^2} & 0 & 0 & 0 \\ 0 & 0 & 0 & 0 \\ 0 & 0 & 0 & 0 \\ 0 & 0 & 0 & \frac{4}{\mu^2} \end{pmatrix},$$

Journal of Physics: Conference Series **633** (2015) 012040 doi:10.1088/1742-6596/633/1/012040

with $S_1^{\mu\nu}$ being a proper CKT. Thus, we are supplemented with two additional quadratic in the momenta integrals of motion $Q_4 = S_1^{\mu\nu} p_\mu p_\nu = \kappa_4$ and $Q_5 = S_2^{\mu\nu} p_\mu p_\nu = \kappa_5$. We can now proceed with the derivation of the solution.

3.1. Case $\kappa_1 \neq 0$ and $\kappa_5 \neq 0$

In order to completely integrate the equations, we choose to write $N(t) = \dot{f}(t)$, where $f(t)$ is considered to be a non-constant (scalar under time re-parameterizations) function. We first integrate equations $Q_1 = \kappa_1$ and $Q_2 = \kappa_2$ with respect to $u(t)$ and $y(t)$ respectively, and get

$$u(t) = c_u - \kappa_1 f(t) \quad \text{and} \quad y(t) = c_y + \kappa_2 f(t) \tag{3.4}$$

where c_u and c_y are constants of integration. With the help of the first, equation $Q_5 = \kappa_5$, yields the following quartet of solutions

$$z(t) = z_{\pm\pm}(t) = \pm \frac{\sqrt{\pm 4 c_z \kappa_5^2 \mu^4 f(t) + 4 c_z^2 \kappa_5^2 \mu^4 + \kappa_5^2 \mu^4 f(t)^2 + 32 \kappa_1^2}}{2\sqrt{\kappa_5}\,\mu}, \tag{3.5}$$

where c_z is a constant of integration; of course we must also require $\kappa_5 \neq 0$ (the case $\kappa_5 = 0$ leads to a complex solution). At this point, equation $Q_4 = \kappa_4$ can be integrated to give

$$\begin{aligned} v(t) = v_\pm(t) = & c_v - \frac{f(t)\left(3\left(\mu^2\left(c_u^2\kappa_4 + \kappa_5\right) + 4\kappa_2^2\right) - 3c_u\kappa_1\kappa_4\mu^2 f(t) + \kappa_1^2\kappa_4\mu^2 f(t)^2\right)}{24\kappa_1} \\ & + \sqrt{2}\tan^{-1}\left(\frac{\kappa_5\mu^2(f(t) \pm 2c_z)}{4\sqrt{2}\kappa_1}\right), \end{aligned} \tag{3.6}$$

with c_v being constant and $\kappa_1 \neq 0$; the plus sign $v_+(t)$ corresponds to solutions $z = z_{++}(t)$ and $z = z_{-+}(t)$, while the minus $v_-(t)$ is for the cases $z = z_{--}(t)$ and $z = z_{+-}(t)$.

By using the solutions (3.4), (3.5) and (3.6) in equation $Q_h = \kappa_h$ we see that, in order for Q_h to be constant for any function $f(t)$, one needs to set $\kappa_4 = \frac{4}{\mu}$. It is easy to check that substitution of the same solutions in the equations of motion together with the constraint $\kappa_4 = \frac{4}{\mu}$ satisfies the system. It is noteworthy that, we were able to obtain the general solution without even turning to the Euler-Lagrange equations of the system.

3.2. Case $\kappa_1 = 0$

Equations $Q_1 = 0$, $Q_2 = \kappa_2$ and $Q_5 = \kappa_5$ lead to the same result with (3.4) and (3.5) by simply setting $\kappa_1 = 0$. One can check that all the other equations regarding autonomous integrals of motion reduce to relations between constants. However, the one defined by the rheonomous integral of motion $Q_h = \kappa_h$ yields, upon integration,

$$v(t) = v_\pm(t) = c_v + \frac{\frac{1}{2} f(t)^2\left(12c_u^2\mu + 4\kappa_2^2 + |\kappa_5|\mu^2\right) + 2f(t)\left(2c_y\kappa_2 \pm c_z\sqrt{|\kappa_5|}\mu - 4\kappa_h\right)}{8c_u}, \tag{3.7}$$

where the $v_+(t)$ corresponds to $z_+(t)$, while $v_-(t)$ to $z_-(t)$ and of course in each case $c_u \neq 0$ (the subcase where $c_u = 0$ will be explored later on).

At this point we have used all equations available by the integrals of motion, substitution of their solutions into the Euler - Lagrange equations yields the constraint relation between constants $\kappa_5 = -\frac{4(c_u^2\mu + \kappa_2^2)}{\mu^2}$.

As we previously remarked, apart from this solution we must also check the case where $c_u = 0$. The application of the relations (3.4) and (3.5) under the condition $\kappa_1 = 0$, leads to the observation that Q_h is constant for every $f(t)$, if and only if $\kappa_5 = -\frac{4\kappa_2^2}{\mu^2}$. Substitution of all these results into the equations of motion leads to a final ordinary differential equation for $v(t)$ that by integration yields

$$v(t) = c_1 f(t) + c_2 \tag{3.8}$$

and thus, we have obtained the complete solution space.

4. Discussion

In this work we have considered, polynomial in the velocities, contact variational symmetries of the action. We derived that, for constrained systems described by (1.1), the infinitesimal criterion (1.2) allows for more symmetries than in the case of regular systems. The main reason for this result is the consideration of gauge invariance not as a redundant degree of freedom that must be fixed out, but rather as an essential part of the system under consideration.

The existence of the constraint gives rise to larger symmetry groups. Of course, this does not mean that one must never choose a gauge. The optimum procedure would be to first calculate all symmetries and then apply any gauge fixing, if necessary, in order to simplify the equations. The important thing is not to do so prior to the determination of the integrals of motion, since their existence may depend on the constraint itself.

In order to exhibit the importance of the method, we examined an example of motion of a relativistic particle in a curved space-time. Due to the consideration of conformal Killing tensors instead of just Killing, we were able to obtain the general solution with no need of gauge fixing and without even using the equations of motion themselves.

Acknowledgments

ND acknowledges financial support by FONDECYT postdoctoral grant no. 3150016.

References

[1] Sarlet W and Cantrijn F 1981 *J. Phys. A: Math. Gen.* **14** 479-92
[2] Kalotas T M and Wybourne B G 1982 *J. Phys. A: Math. Gen.* **15** 2077
[3] Rosquist K and Uggla C 1991 *J. Math. Phys* **32** 3412
[4] Tsamparlis M and Paliathanasis A 2010 *Gen. Rel. Grav.* **42** 2957-80
[5] Tsamparlis M and Paliathanasis A 2011 *J. Phys. A: Math. Theor.* **44** 175202
[6] Christodoulakis T, Dimakis N and Terzis P A 2014 *J. Phys. A: Math. Theor.* **47** 095292
[7] Terzis P A, Dimakis N, Christodoulakis T, Paliathanasis A and Tsamparlis M 2015 *Preprint gr-qc/1503.00932*
[8] Vakili B 2008 *Phys. Lett.* B **664** 16-20
[9] Capozziello S, Piedipalumbo E, Rubano c and Scudellaro P 2009 *Phys. Rev.* D **80** 104030
[10] Basilakos S, Tsamparlis M and Paliathanasis A 2011 *Phys. Rev.* D **83** 103512
[11] Paliathanasis A, Tsamparlis M and Basilakos S 2011 *Phys. Rev.* D **84** 123514
[12] Paliathanasis A, Basilakos S, Saridakis E N, Capozziello S, Atazadeh K, Darabi F and Tsamparlis M 2014 *Phys. Rev.* D **89** 104042
[13] Dimakis N, Christodoulakis T and Terzis P A 2014 *J. Geom. and Phys.* **77** 97-112
[14] Terzis P A, Dimakis N and Christodoulakis T 2015 *Phys. Rev.* D **90** 123543
[15] Dirac P A M 1964 *Lectures on Quantum Mechanics* (Yeshiva University, Academic Press, New York)
[16] Olver P J 2000 *Applications of Lie Groups to Differential Equations, Graduate Texts in Mathematics* vol 107 (Springer-Verlag)
[17] Keane A J and Tupper B O J 2010 *Clas. Quant. Grav.* **27** 245011

Canonical equations of Hamilton for the nonlinear Schrödinger equation

Guo Liang, Qi Guo*, and Zhanmei Ren
Guangdong Provincial Key Laboratory of Nanophotonic Functional Materials and Devices, South China Normal University, Guangzhou 510631, P. R. China
E-mail: guoq@scnu.edu.cn

Abstract. We define two different systems of mathematical physics: the second order differential system (SODS) and the first order differential system (FODS). The Newton's second law of motion and the nonlinear Schrödinger equation (NLSE) are the exemplary SODS and FODS, respectively. We obtain a new kind of canonical equations of Hamilton (CEH), which exhibit some kind of symmetry in form and are formally different from the conventional CEH without symmetry [H. Goldstein, C. Poole, J. Safko, Classical Mechanics, third ed., Addison-Wesley, 2001]. We also prove that the number of the CEHs is equal to the number of the generalized coordinates for the FODS, but twice the number of the generalized coordinates for the SODS. We show that the FODS can only be expressed by the new CEH, but not introduced by the conventional CEH, while the SODS can be done by both the new and the conventional CEHs. As an example, we prove that the nonlinear Schrödinger equation can be expressed with the new CEH in a consistent way.

1. Introduction

The Hamiltonian viewpoint provides a framework for theoretical extensions in many areas of physics. In classical mechanics it forms the basis for further developments, such as Hamilton-Jacobi theory, perturbation approaches and chaos [1, 2]. The canonical equations of Hamilton (CEH) for the continuous system in classical mechanics are expressed as [1]

$$\dot{q}_s = \frac{\delta h}{\delta p_s}, \quad (s = 1, \cdots, N), \tag{1}$$

$$-\dot{p}_s = \frac{\delta h}{\delta q_s}, \quad (s = 1, \cdots, N), \tag{2}$$

where the subscript s, which in all cases denotes $1, \cdots, N$ throughout the paper, represents the components of the quantity of the continuous system [1], $\frac{\delta h}{\delta q_s} = \frac{\partial h}{\partial q_s} - \frac{\partial}{\partial x}\frac{\partial h}{\partial q_{s,x}}$ and $\frac{\delta h}{\delta p_s} = \frac{\partial h}{\partial p_s} - \frac{\partial}{\partial x}\frac{\partial h}{\partial p_{s,x}}$ denote the functional derivatives of h with respect to q_s and p_s with $q_{s,x} = \frac{\partial q_s}{\partial x}$ and $p_{s,x} = \frac{\partial p_s}{\partial x}$, q_s and p_s are the generalized coordinate and the generalized momentum, respectively, and h is the Hamiltonian density of the continuous system. The generalized momentum p_s for the continuous system is defined as $p_s = \frac{\partial l}{\partial \dot{q}_s}$, and the Hamiltonian density h for the continuous system is obtained by the Legendre transformation as

$$h = \sum_{s=1}^{N} \dot{q}_s p_s - l, \tag{3}$$

where l is the Lagrangian density. It should be noted that for the continuous system both q_s and p_s are not only functions of time t, but also the spatial coordinate x, where the spatial coordinate x is not the generalized coordinate. To distinguish t from the coordinate x, we refer to time t as the evolution coordinate. p_s and q_s define the infinite-dimensional phase space. h is a function of q_s, p_s and $q_{s,x}$ but not $p_{s,x}$ [1], so $\frac{\delta h}{\delta p_s} = \frac{\partial h}{\partial p_s}$, then Eq.(1) can be also expressed as $\dot{q}_s = \frac{\partial h}{\partial p_s}$.

Up to now, to our knowledge, the CEH appearing in all the literatures are of the form (1) and (2), which are established based on the second order differential system (SODS) (the definition of the SODS will be given in the following). Besides the SODS, there are a number of the first order differential systems (FODS) (the definition of the FODS will also be given in the following) to model physical phenomena. For example, the nonlinear Schrödinger equation (NLSE) is a universal FODS. We will show that the conventional CEH, Eqs. (1) and (2), are not valid for the NLSE, from which it is impossible to obtain the NLSE or its complex-conjugate equation. Attempt was made to deal with the difficulty in Ref.[3], but the CEH for the NLSE they obtained were inconsistent, as will be shown later. In this paper, we obtain a new kind of CEH valid for the FODS, with which the NLSE can be expressed in a consistent way. We prove that the new CEH and the conventional CEH are equivalent for the description of the SODS. But the conventional CEH can not express the FODS.

2. First-order differential system and its canonical equations of Hamilton

The Newton's second law of motion in classical mechanics, based on which the Hamiltonian formulation is established, is the second order differential equation about the evolution coordinate (here the evolution coordinate is time). In this paper we define the system described by the second order partial differential equation about the evolution coordinate as the second order differential system (SODS). Similarly, the first order differential system (FODS) is the system described by the first order partial differential equation about the evolution coordinate. For example, the NLSE [4]

$$i\frac{\partial\varphi}{\partial t} + \frac{1}{2}\frac{\partial^2\varphi}{\partial x^2} + |\varphi|^2\varphi = 0, \tag{4}$$

is an exemplary FODS, where the evolution coordinate t is the propagation direction coordinate.

For the FODS, the Lagrangian density must be the linear function of the generalized velocities $\dot{q}_s$ (see Ref. [5] for details), which can be expressed as

$$l = \sum_{s=1}^{N} R_s(q_s)\dot{q}_s + Q(q_s, q_{s,x}),$$

Consequently, the generalized momentum p_s, which is obtained by the definition $p_s = \frac{\partial l}{\partial \dot{q}_s}$, as

$$p_s = R_s(q_s), \tag{5}$$

is only a function of q_s. There are $2N$ variables, q_s and p_s, in Eqs. (5). The number of Eqs. (5) is N, which also means there exist N constraints between q_s and p_s. So the degree of freedom of the system given by Eqs. (5) is N. Without loss of generality, we take $q_1, \cdots, q_\nu$ and $p_1, \cdots, p_\mu$ as the independent variables, where $\nu + \mu = N$. The remaining generalized coordinates and generalized momenta can be expressed with these independent variables as $q_\alpha = f_\alpha(q_1, \cdots, q_\nu, p_1, \cdots, p_\mu)(\alpha = \nu+1, \cdots, N)$, and $p_\beta = g_\beta(q_1, \cdots, q_\nu, p_1, \cdots, p_\mu)(\beta = \mu+1, \cdots, N)$. We obtained the CEH for the FODS (see Ref. [5] for details), which are

$$\frac{\delta h}{\delta q_\lambda} = \sum_{s=1}^{N}\left(\dot{q}_s\frac{\partial p_s}{\partial q_\lambda} - \dot{p}_s\frac{\partial q_s}{\partial q_\lambda}\right) + \sum_{\alpha=\nu+1}^{N}\frac{\partial}{\partial x}\frac{\partial h}{\partial q_{\alpha,x}}\frac{\partial f_\alpha}{\partial q_\lambda}, \tag{6}$$

$$\frac{\delta h}{\delta p_\eta}=\sum_{s=1}^{N}\left(\dot{q}_s\frac{\partial p_s}{\partial p_\eta}-\dot{p}_s\frac{\partial q_s}{\partial p_\eta}\right)+\sum_{\alpha=\nu+1}^{N}\frac{\partial}{\partial x}\frac{\partial h}{\partial q_{\alpha,x}}\frac{\partial f_\alpha}{\partial p_\eta} \tag{7}$$

($\lambda = 1,\cdots,\nu$, $\eta = 1,\cdots,\mu$, and $\nu+\mu = N$). The CEH, Eqs. (6) and (7), can be easily extended to the discrete system, which can be expressed as $\frac{\partial H}{\partial q_\lambda} = \sum_{s=1}^{N}\left(\dot{q}_s\frac{\partial p_s}{\partial q_\lambda} - \dot{p}_s\frac{\partial q_s}{\partial q_\lambda}\right)$ and $\frac{\partial H}{\partial p_\eta} = \sum_{s=1}^{N}\left(\dot{q}_s\frac{\partial p_s}{\partial p_\eta} - \dot{p}_s\frac{\partial q_s}{\partial p_\eta}\right)$, where $\lambda = 1,\cdots,\nu$, $\eta = 1,\cdots,\mu$, and $\nu+\mu = N$.

3. Application of the new canonical equations of Hamilton with symmetry to the nonlinear Schrödinger equation

In this part, we will discuss the application of the new CEH with symmetry, Eqs. (6) and (7), to the NLSE. It is known that the Lagrangian density for the NLSE can be expressed as [6] $l = -\frac{i}{2}(\varphi^*\frac{\partial\varphi}{\partial t} - \varphi\frac{\partial\varphi^*}{\partial t}) + \frac{1}{2}|\frac{\partial\varphi}{\partial x}|^2 - \frac{1}{2}|\varphi|^4$. The NLSE is complex, and therefore it is an equation with two real functions, the real part of φ and its imaginary part. It is convenient to consider instead the fields φ and φ^* which are treated as independent from each other. Therefore, N (the components of the quantity) for the NLSE equals two, i.e., there are two generalized coordinates, $q_1 = \varphi^*$ and $q_2 = \varphi$, and two generalized momenta can be obtained as $p_1 = \frac{i}{2}\varphi$ and $p_2 = -\frac{i}{2}\varphi^*$. The Hamiltonian density for the NLSE can be obtained by use of Eq.(3) as [4]

$$h = -\frac{1}{2}\left|\frac{\partial\varphi}{\partial x}\right|^2 + \frac{1}{2}|\varphi|^4. \tag{8}$$

If the generalized coordinate q_1 and the generalized momentum p_1 are taken as the independent variables, the remaining generalized coordinate q_2 and the remaining generalized momentum p_2 can be expressed as $q_2 = -2ip_1$ and $p_2 = -\frac{i}{2}q_1$, respectively. We should also note that the Hamiltonian density h is also the function of $q_{s,x}$, which are independent from q_s and p_s. Then for the NLSE, the Hamiltonian density (8) should be expressed with the independent variables $q_1, p_1, q_{1,x}$ and $q_{2,x}$ as $h = -\frac{1}{2}q_{1,x}q_{2,x} - 2q_1^2p_1^2$. We should note that $\nu = \mu = 1$ and $N = 2$ for the NLSE, which means that the equations (6) have only one equation, so do Eqs.(7). Therefore, the CEH (6) and (7) will yield two equations for the NLSE. The left side of Eq.(6) is obtained as $\frac{\delta h}{\delta q_1} = \frac{1}{2}\frac{\partial^2\varphi}{\partial x^2} + |\varphi|^2\varphi$, and its right side is $-\dot{p}_1 + \dot{q}_2\frac{\partial p_2}{\partial q_1} = -i\dot{\varphi}$. Then the NLSE (4) can be obtained. Using the other CEH, the left side of Eq.(7) is obtained as $\frac{\delta h}{\delta p_2} = -4q_1^2p_1 = -2i|\varphi|^2\varphi^*$, and its right side is $\dot{q}_1 - \dot{p}_2\frac{\partial q_2}{\partial p_1} + \frac{\partial}{\partial x}\frac{\partial h}{\partial q_{2,x}}\frac{\partial q_2}{\partial p_1} = 2\dot{\varphi}^* + i\frac{\partial^2\varphi^*}{\partial x^2}$. Then the complex conjugate of the NLSE is also obtained. Therefore, the CEHs (6) and (7) are consistent in the sense that the NLSE can be expressed with one of the two CEHs, and its complex conjugate can be expressed with the other.

In Ref.[4], the CEH for the NLSE were considered to be the same as those for the SODS, that is, Eqs. (1) and (2). Then, according to the definition $p_\varphi = \partial l/\partial\dot{\varphi}$, p_φ must be $-i/2\varphi^*$ but not $-i\varphi^*$. It was artificially doubled in Ref.[4] so that $p_\varphi = -i\varphi^*$ in Eq.[5.1.29] (to avoid confusion, we replace the parentheses by the brackets to represent the formulas in the references) to make the NLSE derived from the CEH (2). In fact, substitution of the Hamiltonian density (8) into Eq.(2) only yields $\frac{i}{2}\frac{\partial\varphi}{\partial t} + \frac{1}{2}\nabla_\perp^2\varphi + |\varphi|^2\varphi = 0$, which in fact is not the NLSE (4). In Ref.[3], the CEH obtained by the authors are Eqs.[3.87] and [3.81], the latter is $\dot{q}_s = \frac{\partial h}{\partial p_s}$.. Although the NLSE can be derived from Eq.[3.87], its complex conjugation could not be obtained from the other, Eq. [3.81]. We now show this claim. Substituting the Hamiltonian density (8) into the equation $\dot{q}_s = \frac{\partial h}{\partial p_s}$, the left side of it is obtained as $\frac{\partial\varphi^*}{\partial t}$, and the right side is $\frac{\partial h}{\partial p_{\varphi^*}} = \frac{\partial h}{\partial\varphi}\frac{\partial\varphi}{\partial p_{\varphi^*}} = -2i|\varphi|^2\varphi^*$, where $p_{\varphi^*} = \frac{\partial l}{\partial\dot{\varphi}^*}$. Then the equation $-\frac{i}{2}\frac{\partial\varphi^*}{\partial t} + |\varphi|^2\varphi^* = 0$ can be obtained, which is absolutely not the complex conjugate of the NLSE. Therefore, the CEH for the NLSE obtained in Ref.[3] are inconsistent.

4. Some further discussions about the new canonical equations of Hamilton with symmetry

Now, we know that the CEH valid for the FODS are Eqs. (6) and (7), and the conventional CEH valid for the SODS are Eqs. (1) and (2). We will prove here that the new CEH with symmetry, Eqs. (6) and (7), can also be used to express the SODS. If all the N generalized coordinates q_s and N generalized momenta p_s in Eqs.(6) and (7) are independent, then the CEH, Eqs. (6) and (7), can be reduced to Eqs.(1) and (2). In fact, this is just the case of the SODS, where all the generalized coordinates and the generalized momenta are independent. Consequently, the new CEH with symmetry obtained in the paper can express not only the FODS but also the SODS. In the other word, the new CEH, Eqs. (6) and (7), and the conventional CEH, Eqs.(1) and (2), are equivalent for the description of the SODS, but the former are with some kind of symmetry in form and the latter are lack of such symmetry. The conventional CEH, Eqs.(1) and (2), can only expresses the SODS.

In addition, the number of Eqs. (6) and (7) is N (N is the number of the generalized coordinates) in the case for the FODS, and is half of that of Eqs.(1) and (2). This can be explained in the following way. For the SODS, the Euler-Lagrange equations are the second order partial differential equations about the evolution coordinate, the number of which is N. It is well known that one second-order differential equation can be reduced to two first-order differential equations [7]. Then $2N$ CEHs, which are the first order partial differential equations about the evolution coordinate, can be obtained from the N Euler-Lagrange equations. But it is significantly different for the FODS that the Euler-Lagrange equations are the first order partial differential equations about the evolution coordinate, from which only N CEHs are obtained.

5. Conclusion

We obtain a new kind of CEH, which are of some kind of symmetry in form. The new CEH with symmetry can express both the FODS and the SODS, while the conventional CEH can only express the SODS but impossibly express the FODS. The number of the CEH for the first order differential system is N rather than $2N$ like the case for the second order differential system, where N is the number of the generalized coordinates. The NLSE can be expressed with the new CEH in a consistent way, but can not be expressed with the conventional CEH. The CEH for the NLSE are two equations, which are consistent in the sense that the NLSE can be expressed with one of the CEH and its complex conjugate can be expressed with the other. The Hamiltonian formulation for the continuous system can also be extended to the discrete system.

Acknowledgment

This research was supported by the National Natural Science Foundation of China, Grant Nos. 11274125 and 11474109.

References

[1] H. Goldstein, C. Poole, J. Safko, Classical Mechanics, third ed., Addison-Wesley, 2001, pp.334-343,558-598,368-421,34-63.
[2] Paul A. M. Dirac, Lectures on quantum mechanics, Belfer Graduate School of Science Monographs Series 2, Belfer Graduate School of Science, New York, MR 2220894, 1964, Reprinted by Dover in 2001.
[3] T. Dauxois, M. Peyrard, Physics of Solitons, Cambridge university Press, 2006, pp.92-94.
[4] A. Hasegawa, Y. Kodama, Solitons in Optical Communications, Clarenoon Press, Oxford, 1995.
[5] G. Liang, Q. Guo,and Z. Ren, http://arxiv.org/abs/1311.0115.
[6] D. Anderson, Variational approach to nonlinear pulse propagation in optical fibers, Phys. Rev. A, 27, (1983) 3135.
[7] S. M. Lea, Mathematics for Physicists, Brooks/Cole-Thomson Learning, 2004, pp.204-205.

Band parameters of phosphorene

L. C. Lew Yan Voon[1], J. Wang[2], Y. Zhang[2], M. Willatzen[3]

[1]School of Science and Mathematics, The Citadel, Charleston, SC, USA, [2]Department of Electrical and Computer Engineering, University of North Carolina, Charlotte, NC, USA, [3]Department of Photonics Engineering, Technical University of Denmark, Lygnby, Denmark.

E-mail: llewyanv@citadel.edu

Abstract. Phosphorene is a two-dimensional nanomaterial with a direct band-gap at the Brillouin zone center. In this paper, we present a recently derived effective-mass theory of the band structure in the presence of strain and electric field, based upon group theory. Band parameters for this theory are computed using a first-principles theory based upon the generalized-gradient approximation to the density-functional theory. These parameters and Hamiltonian will be useful for modeling physical properties of phosphorene.

1. Introduction

Two-dimensional nanomaterials are expected to revolutionalize material science and the next generation devices. The first example was graphene [1] but there are now a few candidates beyond graphene, particularly MoS_2 [2], silicene [3] and phosphorene [4]. Phosphorene is the latest promising candidate for both electronics and photonics due to the presence of a band gap ($\sim 2\,\text{eV}$) and relatively high electron mobility.

For applications, such as in nanoelectronics, it is necessary to have a band structure model that is efficient in order to simulate the electrical properties. A widely-used band-structure model is the $k \cdot p$ model [5], often giving energy bands analytically in the vicinity of extrema in terms of meaningful parameters such as effective masses and optical matrix elements. For phosphorene, a few $k{\cdot}p$ models have recently appeared [6, 7, 8]. In our recent paper, we presented a model in the presence of external fields. Our present goal is to evaluate the parameters of the model.

2. Theory

2.1. Structure

Phospherene can be viewed as a single layer of black phosphorus. They have the same in-plane translational symmetry and the nonsymmorphic space group is base-centered orthorhombic [7] (Fig. 1).

2.2. Effective Hamiltonian

The effective Hamiltonian for phosphorene in the presence of strain and an external electric field has recently been derived [8]:

$$\mathcal{H} = \mathcal{H}_i + \mathcal{H}_e, \quad (1)$$

$$\mathcal{H}_e = \mathcal{H}^{\epsilon} + \mathcal{H}^{E}, \quad (2)$$

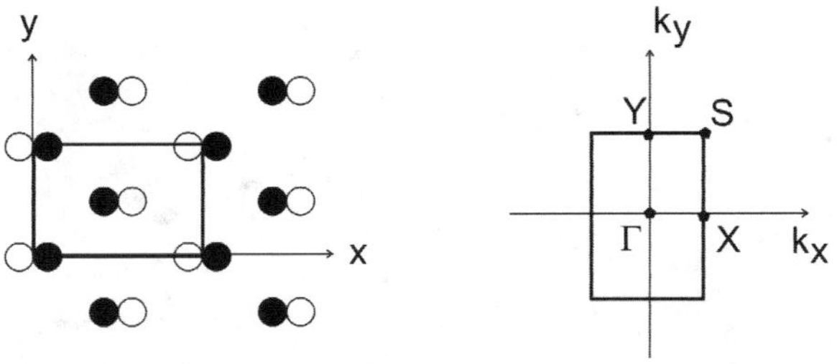

Figure 1. Real (left) and reciprocal (right) space drawings of phosphorene. The two atoms (solid and open circles) are at different vertical heights.

$$\begin{aligned}
\mathcal{H}_i &= a_1 k_x^2 + a_2 k_y^2 + \sum_{i\leq j} a_{ij} k_i^2 k_j^2 + \ldots, && (3)\\
\mathcal{H}^\epsilon &= \left(e_1 + e_2 k_x^2 + e_3 k_y^2\right)\epsilon_{xx} + \left(e_5 + e_6 k_x^2 + e_7 k_y^2\right)\epsilon_{yy} \\
&+ \left(e_9 + e_{10} k_x^2 + e_{11} k_y^2\right)\epsilon_{zz} + \ldots, && (4)\\
\mathcal{H}^E &= \left(c_4 + c_5 k_x^2 + c_6 k_y^2\right) E_x^2 + \left(c_8 + c_9 k_x^2 + c_{10} k_y^2\right) E_y^2 \\
&+ \left(c_{12} + c_{13} k_x^2 + c_{14} k_y^2\right) E_z^2 + \ldots, && (5)
\end{aligned}$$

where $\mathbf{k}$ is the two-dimensional wave vector, ϵ_{ij} is the strain tensor, and $\mathbf{E}$ is a homogeneous, external electric field. The above result is obtained in the absence of spin-orbit interaction.

2.3. Intrinsic band structure

It is known from group theory [8] that all the bands are nondegenerate; thus, the Hamiltonians give the electronic energies directly. Thus, from Eq. (3), the intrinsic band structure near the Brillouin zone center is given by

$$E_n(\mathbf{k}) = a_1 k_x^2 + a_2 k_y^2 + \sum_{i\leq j} a_{ij} k_i^2 k_j^2, \qquad (6)$$

for band n. It can be seen that only terms in even powers of k_i are allowed. Thus, the linear dispersion present in graphene and silicene is not present in the band structure of phosphorene in the neighborhood of the Brillouin zone center. a_1 and a_2 are band parameters related to the effective masses. We would expect an anisotropy in the effective masses due to the orthorhombic symmetry of the crystal lattice.

2.4. Strained band structure

At the Γ point, the band energies are

$$E_n(\epsilon) = E_n + e_1\epsilon_{xx} + e_5\epsilon_{yy} + e_9\epsilon_{zz}, \qquad (7)$$

where E_n are the band edges in the absence of strain. The e_i's are deformation potentials for a given band. Since the lattice is two-dimensional, a perpendicular strain from continuum elasticity theory is not defined (nor can it be described within the first-principles software since the latter uses an artificial superperiodic in the z direction). It can be defined as an internal deformation potential of the basis atoms inside the unit cell as some are outside the lattice plane. However, since it is not described by continuum elasticity theory, we will neglect in this work.

For a finite wave vector, the energies are

$$\begin{aligned} E_n(\epsilon) &= E_n(\epsilon) + (a_1 + e_2\epsilon_{xx} + e_6\epsilon_{yy} + e_{10}\epsilon_{zz})k_x^2 \\ &\quad (a_2 + e_3\epsilon_{xx} + e_7\epsilon_{yy} + e_{11}\epsilon_{zz})k_y^2. \end{aligned} \tag{8}$$

2.5. Electric field

Our prediction, from Eq. (5), is that the band gap changes quadratically with an externally-applied perpendicular electric field. This is in contrast to silicene where the dependence is linear, whereas there is no gap opening for graphene. At $k = 0$, the band gap as a function of a perpendicular field is

$$\Delta E_\Gamma(E_z) = E_\Gamma(0) + cE_z^2. \tag{9}$$

3. Results

Our density-functional theory (DFT)-based calculations were conducted using the projector augmented-wave method [9] and the PBE-GGA exchange-correlation functional [10] as implemented in the VASP code [11, 12, 13]. The wave functions were expanded using a plane-wave basis set with an energy cutoff of 1200 eV. Atomic positions were fully relaxed and k-point sampling was achieved by using a 9×9×1 special-points grid until residual forces were lower than 5 meV/Å. The interlayer separation in the z direction was chosen large ($\sim$ 30 Å) to minimize fictitious interactions. The optimized lattice constants obtained were 3.300 and 4.624 Å.

3.1. Intrinsic band parameters

The valence band structure near the Γ point is shown in Fig. 2. The gap is almost direct at the Γ point and the GGA value is about 0.912 eV; Tran *et al.* [14] have shown that a GW correction increases the gap to 2.0 eV. The strong anisotropy in the band structure is evident. We did a fit

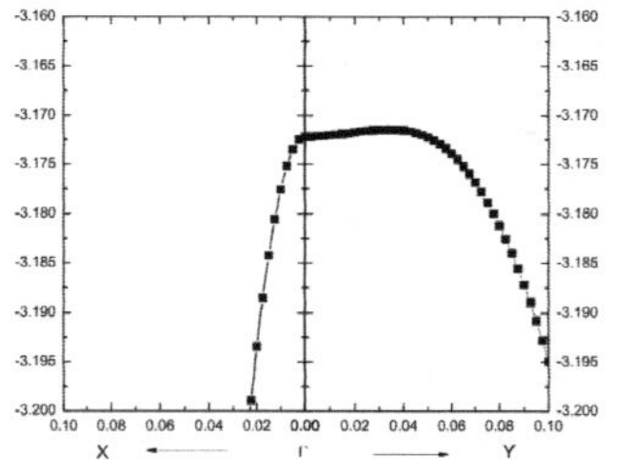

Figure 2. Highest valence band of phosphorene near the Γ point.

to our DFT calculations to obtain the intrinsic band parameters of the $k \cdot p$ theory (Table 1).

Table 1. Band parameters a_i (in units of $1/m_0$) and effective masses (in units of m_0) of the highest valence and lowest conduction bands of phosphorene at the Γ point. The effective mass of the VB along ΓY is very sensitive to the computation [8].

	a_1	$m^*(\Gamma Y)$	a_2	$m^*(\Gamma X)$
CB	0.808	1.24	5.804	0.173
		1.246 [15], 1.16 [16]		0.146 [15], 0.22 [16]
VB	0.106	7.20	−6.263	−0.160
		3.24 [16]		−0.19 [16]

3.2. Strain deformation potentials

We present here our results for the strain deformation potentials. Neglecting the e_9 deformation potential, we see that there are six deformation potentials for each band: e_1, e_2, e_3, e_5, e_6 and e_7. In order to compute the deformation potentials, strain is applied in one direction (e.g., x direction) while restricting the atomic displacements in the perpendicular direction (i.e., ϵ_{yy}) to be zero. Values obtained by fitting to our DFT calculations are given in Table 2 for the highest valence band and lowest conduction band. The anisotropy is again clear. Additionally, one of

Table 2. Strain deformation potentials (e_1 and e_5 in units of eV, all others in units of $1/m_0$) of the highest valence and lowest conduction bands of phosphorene.

	e_1	e_2	e_3	e_5	e_6	e_7
CB	5.801	-0.678	-27.241	1.584	-0.710	-1.497
VB	0.502	-6.286	28.850	-2.745	-0.435	6.291

the deformation potentials has the opposite sign which can lead to band crossing effects.

3.3. Electric Field

It is fairly straightforward to compute the band structure change in the presence of a perpendicular electric field E_z. Hence, this is the only electric deformation potential to be presented here. Using DFT and Eq. (9), for an electric field between 0 and 0.5 V/Å, we find that $c = -0.41\, e\text{Å}^2/\text{V}$.

4. Conclusion

The fundamental band parameters needed to simulate the electronic properties of phosphorene under the influence of an external in-plane strain and an external perpendicular electric field have been obtained by comparing a $k \cdot p$ theory to the band structures computed using a density-functional theory. Qualitative results are that the properties of phosphorene are expected to be hightly anisotropic in the plane, and that the electric field dependence of the energy gap is quadratic rather than the linear behaviour shown by silicene.

Acknowledgment

Lok Lew Yan Voon acknowledges the support of the Traubert Chair. Morten Willatzen would like to thank The Citadel for travel support. Yong Zhang acknowledges the support of the Bissell Distinguished Professorship.

5. References

[1] Neto A H C, Guinea F, Peres N M R, Novoselov K S and Geim A K 2009 *Reviews of Modern Physics* **81** 109 (pages 54) URL `http://link.aps.org/abstract/RMP/v81/p109`
[2] Radisavljevic B, Radenovic A, Brivio J, Giacometti V and Kis A 2011 *Nat. Nanotechnol.* **6** 147
[3] Guzmán-Verri G G and Lew Yan Voon L C 2007 *Physical Review B (Condensed Matter and Materials Physics)* **76** 075131 (pages 10) URL `http://link.aps.org/abstract/PRB/v76/e075131`
[4] Liu H, Neal A T, Zhu Z, Luo Z, Xu X, Tomnek D and Ye P D 2014 *ACS Nano* **8** 4033–4041 (*Preprint* `http://pubs.acs.org/doi/pdf/10.1021/nn501226z`) URL `http://pubs.acs.org/doi/abs/10.1021/nn501226z`
[5] Lew Yan Voon L C and Willatzen M 2009 *The k · p Method* (Heidelberg: Springer Verlag)
[6] Rodin A S, Carvalho A and Castro Neto A H 2014 *Phys. Rev. Lett.* **112**(17) 176801 URL `http://link.aps.org/doi/10.1103/PhysRevLett.112.176801`
[7] Li P and Appelbaum I 2014 *Phys. Rev. B* **90**(11) 115439 URL `http://link.aps.org/doi/10.1103/PhysRevB.90.115439`
[8] Lew Yan Voon L C, Lopez-Bezanilla A, Wang J, Zhang Y and Willatzen M 2015 *New Journal of Physics* **17** 025004 URL `http://stacks.iop.org/1367-2630/17/i=2/a=025004`

[9] Blöchl P E 1994 *Phys. Rev. B* **50**(24) 17953–17979 URL http://link.aps.org/doi/10.1103/PhysRevB.50.17953
[10] Perdew J P, Burke K and Ernzerhof M 1996 *Phys. Rev. Lett.* **77**(18) 3865–3868 URL http://link.aps.org/doi/10.1103/PhysRevLett.77.3865
[11] Kresse G and Hafner J 1993 *Phys. Rev. B* **48**(17) 13115–13118 URL http://link.aps.org/doi/10.1103/PhysRevB.48.13115
[12] Kresse G and Furthmüller J 1996 *Phys. Rev. B* **54**(16) 11169–11186 URL http://link.aps.org/doi/10.1103/PhysRevB.54.11169
[13] Kresse G and Joubert D 1999 *Phys. Rev. B* **59**(3) 1758–1775 URL http://link.aps.org/doi/10.1103/PhysRevB.59.1758
[14] Tran V, Soklaski R, Liang Y and Yang L 2014 *Phys. Rev. B* **89**(23) 235319 URL http://link.aps.org/doi/10.1103/PhysRevB.89.235319
[15] Fei R and Yang L 2014 *Nano Letters* **14** 2884–2889 pMID: 24779386 (*Preprint* http://dx.doi.org/10.1021/nl500935z) URL http://dx.doi.org/10.1021/nl500935z
[16] Li Y, Yang S and Li J 2014 *The Journal of Physical Chemistry C* **118** 23970–23976 (*Preprint* http://dx.doi.org/10.1021/jp506881v) URL http://dx.doi.org/10.1021/jp506881v

Electromagnetic Coherent Effects in Metamaterials with Randomly Rough Surfaces

Gerard Berginc
THALES, Elancourt, France
E-mail: gerard.berginc@fr.thalesgroup.com

Abstract.
In this paper, we present a theory of transport based on the Bethe-Salpeter equation for a three-dimensional disordered medium with randomly rough interfaces illuminated by an electromagnetic wave. The slab is defined by a random medium composed of nanospheres randomly distributed in a layer. We study the electromagnetic coherent effects of the wave scattering inside the random volume bounded by randomly rough interfaces. Coherent effects are related for example to backscattering enhancement due to the interaction of the wave with the randomly rough boundaries or the random volume. The phenomenon manifests itself at the level of configurationally averaged mean intensity. We examine the solution by moments of the Bethe-Salpeter equation and we introduce the ladder and most-crossed contributions of the transport theory including the randomly rough boundaries of the considered structure. In this paper we are concerned with the wave transport behavior beyond the effective medium picture in taking into account the multiply scattered behavior.

1. Introduction

Metamaterials are composite structures made of small metallic or dielectric particles. These particles play the same role as atoms or molecules in ordinary materials, which makes metamaterials considered as " artificial materials ". In this paper, we consider a three-dimensional disordered medium with randomly rough interfaces. For classical waves, the most interesting case is that of three-dimensional disorder. The quantity that obeys the disordered wave equation - the electric field - is a vector and not a scalar as in the Schrdinger equation. If many theoretical and numerical approaches have been developed to model transport of scalar light in random media, few or even no method at all nowadays can deal with ensembles of complex different nanoparticles in interaction between themselves and randomly rough boundaries. This aspect of the problem was never considered in detail up to now. In mesoscopic physics, multiple light scattering in disordered potentials gives rise to complex wave interference phenomena, like speckle correlations, backscattered enhancement or Anderson localization of light [1], and to applications like imaging through opaque biological tissues or photovoltaics. We present a theory of transport based on an equation for second statistical moment, i.e. the Bethe-Salpeter equation [1-6], in which the vectorial character of electromagnetic wave, the correlations between the scatterers are taken into account. We can derive the radiative transfer equation from the Maxwell equations. The problem is then reduced to find a good approximation to the solution of this equation. The Bethe-Salpeter equation is constructed in order that the medium and the boundaries are treated on the same footing [4-6]. This unified Bethe-Salpeter equation (for details see [4,6]) enables us to obtain a general expression, whatever the choice of the scattering operators used at the boundaries. In [7-10] we describe the application of the Small Perturbation Theory and Small Approximation Theory to the scattering of electromagnetic waves from two-dimensional randomly rough interfaces and three-dimensional film bounded by random surfaces, multiple-scattering effects on the randomly rough surfaces are incorporated in these theories, this can lead to describe new coherent phenomena due to the coupling on and between the random surfaces [11] . The Quasi-Crystalline Coherent Potential Approximation (QC-CPA) [2,4,12,13] is taken into account for the contribution of the random medium, which is made of spherical nanoparticles of given permittivity in a homogeneous background medium. The boundaries are described by random functions. Mathematically, the way to study the solution of the generalized Bethe-Salpeter equation is based on the use of diagrammatic perturbation theory [1-4]. The physical process of the backscattering enhancement is obtained by adding the contribution of the so-called most-crossed diagrams [4] to the diagrammatic expansions of the transfer equation. This paper deals with the coherent effects, as the backscattered enhancement called the weak localization effect [1], which occurs when the electromagnetic intensity is scattered back to the source, it is a coherent effect experienced by waves when propagating through a disordered system despite the randomness of the system. For three-dimensional slabs with nanoscale roughness and nanoparticles, the enhanced backscattering [14] is produced by different mechanisms, wave scattering by the same boundary, wave coupling by the two boundaries or wave scattering by the randomly distributed nanoparticles. Quantum structures as metallic or dielectric nanoparticle with a size of a few nanometers exhibit behaviors different than that of the bulk [15-19] due to confinement effects which change the electronic properties. Therefore the treatment of the permittivity of such nanoparticles can not be considered without taking into account quantum effects related to the behavior of electrons in the atom cloud. We use a calculation method in which conduction electrons are modeled as a gas of free electrons. In the first part of this paper we introduce the theory and the calculation of the total specific intensity scattered by the slab illuminated by a polarized electromagnetic plane wave. In the second part of the paper, we give some numerical examples of scattering properties of random films with discrete nanoparticles. These examples of simulation of light interaction with random structures are devoted to the study of the introduction of the multiple scattering to the scattering from random media with random boundaries.

2. Incoherent specific intensity

We consider harmonic waves with $e^{-\mathrm{i}\omega t}$ dependence.We are going to study a three-dimensional slab (see Fig.1), which is composed of an incident medium of permittivity ϵ_0, a first randomly rough boundary, a random medium defined by a permittivity ϵ_1 which contains scatterers, a second randomly rough boundary at the distance $z = -H$ and a semi-infinite homogeneous medium with a permittivity ϵ_2. The main characteristic we must study for the scattering from a random slab is the specific intensity $\boldsymbol{\mathcal{I}}(\boldsymbol{R},\boldsymbol{k})$. If we take into account the polarization of the wave, the specific intensity is defined as a Stokes vector or a tensor which gives the power flux per unit area and solid angle at the point $\boldsymbol{R}$ in the direction $\hat{\boldsymbol{k}}$. Writing an energy balance equation, it can be shown shown that the specific intensity satisfies a Boltzmann type equation called radiative transfer equation. If particles are inside a slab with a permittivity different from the outside medium, boundaries condition must be added to the radiative transfer equation in order to calculate the specific intensity. For rough surfaces, these boundaries condition are expressed in function of scattering operators, where several approximate analytical expressions exist depending on the roughness of the surface [7-10]. The specific intensity can be deduced from the Wigner function of the electrical field [4,6]. After having determined the relation between the electromagnetic field and the specific intensity, we can derive the radiative transfer equation from the Maxwell equations.

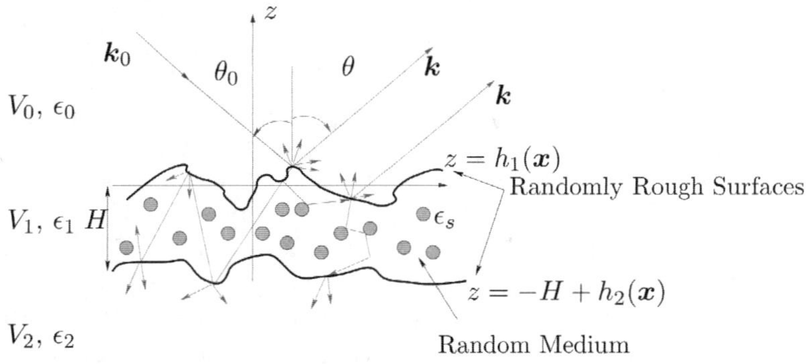

Figure 1. Random volume with rough boundaries.

The procedure is to write the Maxwell equations in an integral form with the help of Green functions and to apply the Wigner transform to the equation obtained. In our approach, we use scattering operators which are a unified way to describe how waves interact with the boundaries. To use these operators, we have introduced two kinds of Green functions [4]. The first one $\overline{\boldsymbol{G}}_{SV}$ describes the field scattered by the volume (V), which contains the scatterers, and by the rough surfaces (S). The second type of Green function is $\overline{\boldsymbol{G}}_S$, which gives the field scattered by a slab with rough boundaries where the scatterers have been replaced by an homogeneous medium described by an effective permittivity. With these Green functions, we can separate the contributions of the surfaces and the volume. The Quasi-Crystalline Coherent Potential Approximation (QC-CPA) [13] is taken into account for the contribution of the random medium, which is made of spherical nanoparticles of given permittivity in a homogeneous dielectric background medium. The main advantage of our approach is that the equation obtained are similar to the equations generally used to describe the wave scattered by an infinite random medium. For a random layer with rough interfaces, the incoherent specific intensity can be decomposed into four parts [4,6]:

$$\overline{\mathcal{I}}^{incoh} = \overline{\mathcal{I}}^{incoh}_{L=0} + \overline{\mathcal{I}}^{incoh}_{L=1} + \overline{\mathcal{I}}^{incoh}_{Ladder} + \overline{\mathcal{I}}^{incoh}_{Crossed}, \quad (1)$$

where the contributions are given by :

$$\begin{aligned} \mathcal{I}^{incoh}_{L=0}(\boldsymbol{R}, \boldsymbol{k}) = e_0 \int_{V_0} \mathrm{d}^3\boldsymbol{r} \mathrm{e}^{-\mathrm{i}\boldsymbol{k}\cdot\boldsymbol{r}} \Big[&< \boldsymbol{E}^{0\,s}_S(\boldsymbol{R} + \frac{\boldsymbol{r}}{2}) \otimes \boldsymbol{E}^{0\,s\,*}_S(\boldsymbol{R} - \frac{\boldsymbol{r}}{2}) >_S \\ &- < \boldsymbol{E}^{0\,s}_S(\boldsymbol{R} + \frac{\boldsymbol{r}}{2}) > \otimes < \boldsymbol{E}^{0\,s\,*}_S(\boldsymbol{R} - \frac{\boldsymbol{r}}{2}) >_S \Big] , \end{aligned} \quad (2)$$

$$\mathcal{I}^{incoh}_{L=1}(\boldsymbol{R}, \boldsymbol{k}) = e_0 \int_{V_0} \mathrm{d}^3\boldsymbol{r} \mathrm{e}^{-\mathrm{i}\boldsymbol{k}\cdot\boldsymbol{r}} \overline{\mathcal{G}}^{01}_{<S>} : \overline{\mathcal{P}}^{11} :< \boldsymbol{E}^{1\,t} \otimes \boldsymbol{E}^{1\,t} >_S (\boldsymbol{R} + \frac{\boldsymbol{r}}{2}, \boldsymbol{R} - \frac{\boldsymbol{r}}{2}), \quad (3)$$

$$\mathcal{I}^{incoh}_{Ladder}(\boldsymbol{R}, \boldsymbol{k}) = e_0 \int_{V_0} \mathrm{d}^3\boldsymbol{r} \mathrm{e}^{-\mathrm{i}\boldsymbol{k}\cdot\boldsymbol{r}} \; \overline{\mathcal{G}}^{01}_{<S>} : \overline{\mathcal{P}}^{11} : \overline{\mathcal{G}}^{11}_{\ll SV \gg} : \overline{\mathcal{P}}^{11} :< \boldsymbol{E}^{1\,t} \otimes \boldsymbol{E}^{1\,t} >_S (\boldsymbol{R} + \frac{\boldsymbol{r}}{2}, \boldsymbol{R} - \frac{\boldsymbol{r}}{2}), \quad (4)$$

$$\begin{aligned} \mathcal{I}^{incoh}_{Crossed}(\boldsymbol{R}, \boldsymbol{k}) = e_0 \int_{V_0} \mathrm{d}^3\boldsymbol{r} \mathrm{e}^{-\mathrm{i}\boldsymbol{k}\cdot\boldsymbol{r}} \; &\Big[\overline{\mathcal{G}}^{01}_{<S>} : \overline{\mathcal{P}}^{11} : \overline{\mathcal{G}}^{11}_{\ll SV \gg} \\ &: \overline{\mathcal{P}}^{11} :< \boldsymbol{E}^{1\,t} \otimes \boldsymbol{E}^{1\,t} >_S \Big]^{T_R} (\boldsymbol{R} + \frac{\boldsymbol{r}}{2}, \boldsymbol{R} - \frac{\boldsymbol{r}}{2}), \end{aligned} \quad (5)$$

with $e_0 = \epsilon_{vac}\, c_{vac}\, n_0/2$. ϵ_{vac} and c_{vac} are respectively the permittivity and the speed of light in vacuum. n_0 is the optical index of the medium 0. T_R defines the right transpose of a tensor. The superscripts of the operators and the Green functions correspond respectively to the receiver location and the source location (0 for the medium 0, 1 for the medium 1, 2 for the medium 2). The lower index $\ll SV \gg$ represents the average over the randomly rough surfaces and the disordered medium. The lower index $< S >$ represents the average over the randomly rough surfaces. We assume the ergodicity of the system and we look for a configurational average: i.e., the infinite-time average of the energy scattered by the slab is equivalent to its spatial average over many configurations. The fields $\boldsymbol{E}_S^{0s}(\boldsymbol{r})$, $\boldsymbol{E}_S^{1t}(\boldsymbol{r})$ represent respectively the field scattered by the interfaces without any interaction with the scatterers and the field transmitted in the layer before an interaction with the scatterers. The first term $\overline{\boldsymbol{\mathcal{I}}}_{L=0}^{incoh}$ gives the scattering by the slab with rough boundaries and defined by an effective permittivity ϵ_e given by the QC-CPA approximation [13], we will give an overview of the formulation in the second part of the paper. In order to take into account scattering effects, quantum multiple scattering theory has been transposed in the electromagnetic case , but as a rigorous analytical answer is unreachable, several approximation schemes have been developed. One of the most advanced is the Quasicrystalline Coherent Potential Approximation (QC-CPA) which takes into account the correlation between the particles to determine ϵ_e, we use the fact that the effective medium is not spatially dispersive, it can be shown, using a diagrammatic technique, that the coherent part of the field $< \boldsymbol{E} >_V$ which propagates inside an infinite random medium behaves as a wave in an homogeneous medium with a renormalized effective permittivity. The Quasi-Crystalline Approximation (QCA) is strictly valid when the particles have a fixed position, as in a crystal. The quasi-crystalline approximation is equivalent to neglect the fluctuation of the effective field on a particle located at r_j , due to a position deviation of a particle located at r_i from its average position. Under the (QC-CPA) approach, for Rayleigh scatterers, our approximate formula for ϵ_e [13] can generalize the usual Maxwell-Garnett formula and the Keller approximation. The second term $\overline{\boldsymbol{\mathcal{I}}}_{L=1}^{incoh}$ is related to the approximation where only one process of scattering by a particle is taken into account. The two last terms $\overline{\boldsymbol{\mathcal{I}}}_{Ladder}^{incoh}$ and $\overline{\boldsymbol{\mathcal{I}}}_{Crossed}^{incoh}$ correspond respectively to the ladder approximation and the most-crossed approximation (see [4,6] for the detailed expressions). Fig.2 gives the scattering processes for the ladder approximation. The most-crossed approximation gives the physical phenomenon producing a peak in the backscattering direction due to the interference of waves following the same path but in opposite directions. The most elegant way to introduce the enhanced backscattering in the theory is to use the reciprocity principle. Heuristically, it means that when we exchange the source and the detector position and also their polarization, the field measured is the same. In the previous formulas, we can follow from the right to the left, how the scattering process is defined by the combinations of the different operators. The introduction of tensor products of matrices is related to the fact we calculate an incoherent cross-section, which depends on the square modulus of a scattering amplitude. The tensor $\overline{\boldsymbol{\mathcal{P}}}^{11}$ describes the intensity scattered by the first particle the wave encounters. The expression of this tensor is given by the the Quasi-Crystalline Coherent Potential Approximation. This intensity operator is determined by using the energy conservation principle and takes into account the correlations with the other particles. Under the QC-CPA assumption, the effective permittivity satisfies a non-linear system of equations. Introducing these equation in a Ward identity, we obtain an expression, called the modified ladder approximation for the intensity operator $\overline{\boldsymbol{\mathcal{P}}}^{11}$ which satisfies the energy conservation [4,6]. The operator $\overline{\boldsymbol{\mathcal{G}}}_{\ll SV \gg}^{11}$ is the Green tensor which contains the interaction between the scatterers and the surfaces. As we have seen, the contributions for the most-crossed and the ladder diagrams are determined by the Green tensor $\overline{\boldsymbol{\mathcal{G}}}_{\ll SV \gg}^{11}$, which satisfies a Bethe-Salpeter equation. It contains the second order and the higher orders describing the scattering by particles. To calculate the previous expressions, we need to estimate the tensor $\overline{\boldsymbol{\mathcal{G}}}^{11}_{\ll SV \gg}$, as no exact solution exists for the expression of the Green tensor, we rely on a perturbative method. In this paper, we take into account the second-order approximation. The very first results for this second-order approximation was given in [4,14]. The Green tensor $\overline{\boldsymbol{\mathcal{G}}}^{11}_{\ll SV \gg}$ is given by the Bethe-Salpeter equation as follows:

$$\overline{\boldsymbol{\mathcal{G}}}^{11}_{\ll SV \gg} = \overline{\boldsymbol{\mathcal{G}}}^{11}_{\infty <S>} + \overline{\boldsymbol{\mathcal{G}}}^{11}_{\infty <S>} : \overline{\boldsymbol{\mathcal{P}}}^{11} : \overline{\boldsymbol{\mathcal{G}}}^{11}_{\ll SV \gg}, \tag{6}$$

where the previous tensors are defined by the following tensorial products of the Green functions:

$$\overline{\boldsymbol{\mathcal{G}}}^{11}_{\ll SV \gg} = \ll \overline{\boldsymbol{G}}^{11}_{SV} \otimes \overline{\boldsymbol{G}}^{11\,*}_{SV} \gg_{SV}, \tag{7}$$

$$\overline{\boldsymbol{\mathcal{G}}}^{11}_{\infty <S>} = < \overline{\boldsymbol{G}}^{11}_{S} \otimes \overline{\boldsymbol{G}}^{11\,*}_{S} >_{S}, \tag{8}$$

By iterating the Bethe-Salpeter equation (6) we obtain the following perturbative development:

$$\begin{aligned}\overline{\boldsymbol{\mathcal{G}}}^{11}_{\ll SV \gg} &= \overline{\boldsymbol{\mathcal{G}}}^{11}_{\infty <S>} + \overline{\boldsymbol{\mathcal{G}}}^{11}_{\infty <S>} : \overline{\boldsymbol{\mathcal{P}}}^{11} : \overline{\boldsymbol{\mathcal{G}}}^{11}_{\infty <S>} \\ &+ \overline{\boldsymbol{\mathcal{G}}}^{11}_{\infty <S>} : \overline{\boldsymbol{\mathcal{P}}}^{11} : \overline{\boldsymbol{\mathcal{G}}}^{11}_{\infty <S>} : \overline{\boldsymbol{\mathcal{P}}}^{11} : \overline{\boldsymbol{\mathcal{G}}}^{11}_{\infty <S>} + \dots .\end{aligned} \tag{9}$$

The first-order Green tensor expansion represents the double-scattering theory for the cross-section while the second-order corresponds to the third-order scattering for the cross-section. This second-order expansion is taken into account in the simulation we present in the next section.

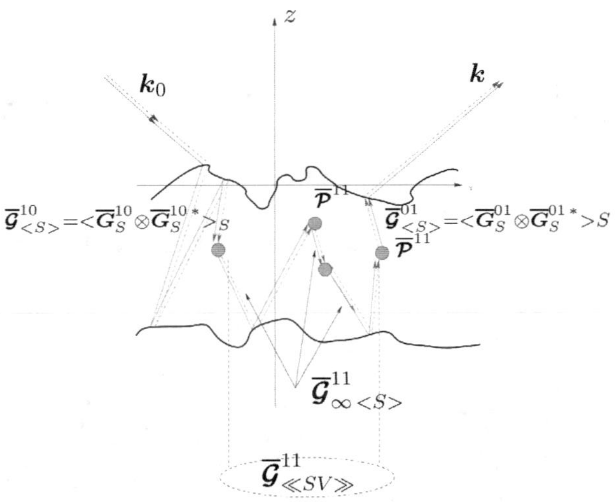

Figure 2. Scattering processes for the ladder approximation. The solid and dashed lines represent respectively the propagating wave on the right and left sides of the tensorial products $\overline{\boldsymbol{\mathcal{G}}}^{11}_{\ll SV \gg} = \ll \overline{\boldsymbol{G}}^{11}_{SV} \otimes \overline{\boldsymbol{G}}^{11\,*}_{SV} \gg_{SV}$.

3. Scattering from film with discrete nanoparticles

The term $\overline{\boldsymbol{\mathcal{I}}}^{incoh}_{L=0}$ contains the effective permittivity of the slab bounded by two randomly rough surfaces. In [12,13] we have introduced a formulation for the effective permittivity for different classes of spherical scatterers. In the following we take the example of two classes of scatterers a and b. The expression of the effective permittivity is given by:

$$\epsilon_e(\omega) = \epsilon_1 + \frac{1}{K^2_{vac}} \frac{n_a \overline{\boldsymbol{t}}^1_a(\omega) + n_b \overline{\boldsymbol{t}}^1_b(\omega)}{1 - \left(\frac{1}{3\,K_e^2} (n_a \overline{\boldsymbol{t}}^1_a(\omega) + n_b \overline{\boldsymbol{t}}^1_b(\omega)) + \frac{\mathrm{i}\, K_e}{6\,\pi} \left[(w_a - 1) \overline{\boldsymbol{t}}^1_a(\omega) + (w_b - 1) \overline{\boldsymbol{t}}^1_b(\omega) \right] \right)}, \tag{10}$$

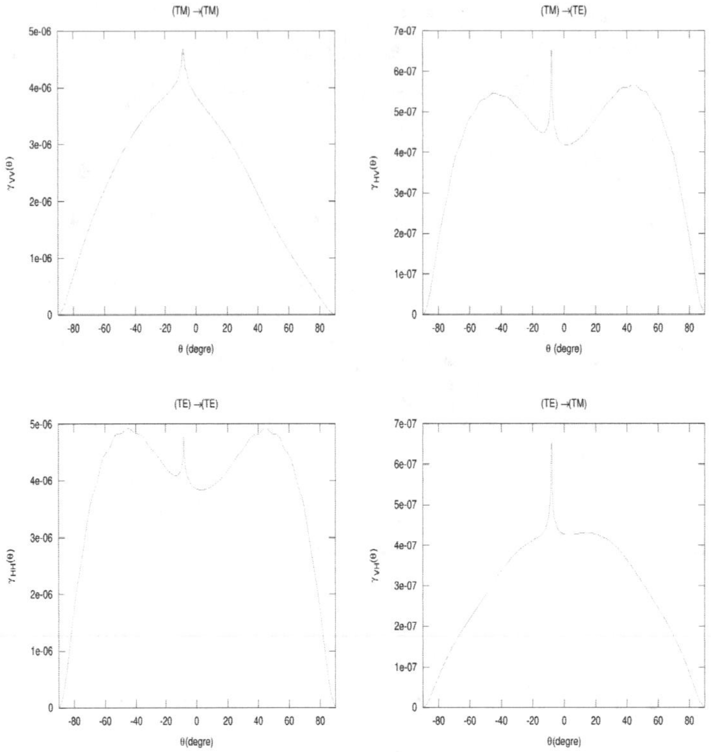

Figure 3. Incoherent cross-section in function of scattering angle, medium with nanoholes.

where the function $\overline{\boldsymbol{t}}_i^1(\omega)$ may be written in the form: $(i = a$ or $b)$

$$\overline{\boldsymbol{t}}_i^1(\omega) = \frac{3\,K_{vac}^2 \epsilon_e\, v_i\,(\epsilon_i - \epsilon_1)}{(\epsilon_i - \epsilon_1)[1 - \frac{\mathrm{i}3K_e K_{vac}^2}{6\,\pi}\epsilon_e v_i] + 3\epsilon_e}\,. \tag{11}$$

and

$$w_i = \frac{(1 - f_{vol}^i)^4}{(1 + 2\,f_{vol}^i)^2}\,. \tag{12}$$

In these equations, n_i is the number of particles of type i, v_i their volume and $f_{vol}^i = n_i\,v_i$ the volume fraction, $K_e^2 = \epsilon_e K_{vac}^2$. The equation (10) is a non linear equation connecting the effective permittivity ϵ_e and the permittivities ϵ_1, of the layer , and ϵ_a, ϵ_b of the scatterers. Its derivation is obtained with the assumptions (QC-CPA) and gives a tractable formula for the effective permittivity. The permittivity of a nanoparticle is given by a quantum expression where conduction electrons are modeled as a gas of free electrons. The relation between the specific intensity and the incoherent cross-section is given by a product of two tensors:

$$4\pi\cos\theta\,\boldsymbol{\mathcal{I}}^{incoh}(\boldsymbol{R},\hat{\boldsymbol{k}}) = \overline{\gamma}^{incoh}(\hat{\boldsymbol{k}}|\hat{\boldsymbol{k}}_0) : \boldsymbol{\mathcal{J}}^{0i}(\hat{\boldsymbol{k}}_0)\,, \tag{13}$$

where

$$\boldsymbol{\mathcal{J}}^{0i}(\hat{\boldsymbol{k}}_0) = e_0\,\boldsymbol{E}^{0i}(\hat{\boldsymbol{k}}_0) \otimes \boldsymbol{E}^{0i\,*}(\hat{\boldsymbol{k}}_0)\,. \tag{14}$$

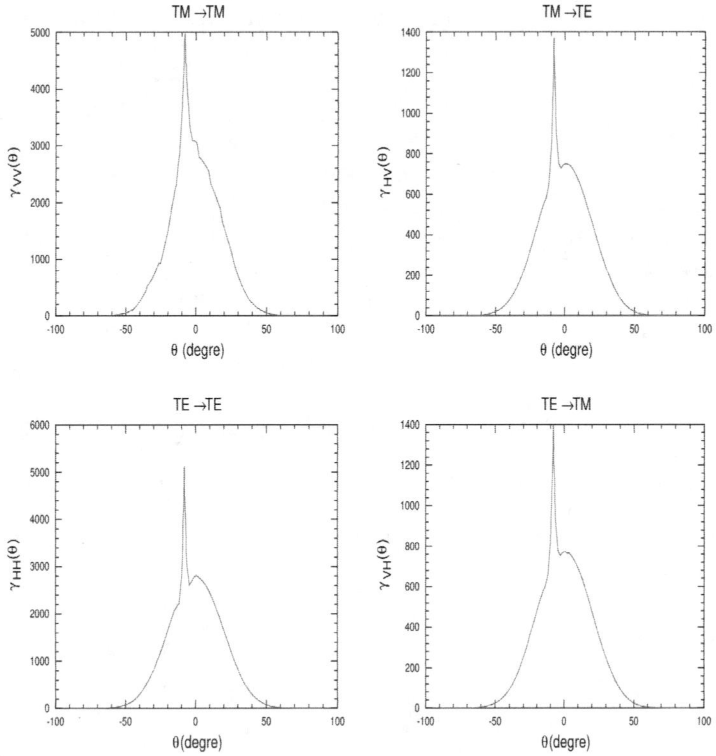

Figure 4. Incoherent cross-section in function of scattering angle, medium with metallic nanoparticles.

Eq.14 is a tensorial product of two vectors. We give two examples of calculation of the scattering from two thin films. The wavelength is given by $\lambda = 0.8\mu m$. The four terms of the incoherent cross-section are taken into account in these computations. We consider that the incident medium is vacuum ($\epsilon_0 = 1$). We compute the incoherent cross-section $\overline{\gamma}^{incoh}(\theta)$ for the four polarizations in the plane of incidence for a slab , which is defined as follows.The layer depth is $4.5\mu m$, the optical index of the layer is $n_1 = 1.49$. The upper layer boundary is defined as a slightly randomly rough surface defined by a root mean square height of 200nm and a correlation length of 462nm, we consider a Gaussian distribution for the distribution of surface heights and for the correlation function. The lower boundary is a flat surface, which separates the layer from a semi-infinite medium of optical index $n_2 = 1.5$. Inside the medium we consider nanoobjects. In the first example, we consider two types of spherical nanoholes with diameters of 10 and 100nm. The filling rate is 4% with a volume fraction of 1/2. The permittivity of the holes is that of the vacuum. In the second example, the medium contains two types of spherical metallic particles defined as Rayleigh spherical scatterers with the same parameters as for medium with nanoholes. We obtain the total incoherent cross-section for an incident angle $\theta = 8^{\circ}$, for the co and cross-polarization components in the case a medium with nanoholes (Fig.3) and a medium with metallic nanospheres (Fig.4). Despite the size of the metallic particles and the filling ratio, their influence on the scattering process is important as demonstrated by the patterns of the two curves.

Journal of Physics: Conference Series **633** (2015) 012043 doi:10.1088/1742-6596/633/1/012043

4. Conclusion

This paper was devoted to a theory of transport based on a Bethe-Salpeter equation. We presented a multiple scattering theory for a random slab with discrete nanoscale scatterers and bounded with randomly rough surfaces. By starting from the Maxwell equations, we are able to give an unambiguous definition of the specific intensity as a function of the electric field. Furthermore, we can take into account the correlations between the scatterers. We use a new theory of effective permittivity, which has been extended to random media with different types of particles and bounded with randomly rough surfaces. Finally, we have also incorporated the most-crossed contributions in our theory, with which we can estimate the enhanced backscattering phenomenon and a quantum expression of the permittivity of the metallic nanosphere. We gave two example of scattering calculation for spherical nanoholes and metallic nanospheres.

References

[1] Sheng P 1995 *Introduction to Wave Scattering, Localization, and Mesoscopic Phenomena* (New York: Academic Press)

[2] Tsang L and Kong J A 2001 *Scattering of Electromagnetics Waves: Advanced Topics* vol 3 (New York: Wiley-Interscience)

[3] Apresyan L A and Kravtsov Y A 1996 *Radiation Transfer: Statistical and Wave Aspects* (Amsterdam: Gordon and Breach)

[4] Berginc G and Maradudin A A 2013 Scattering properties of random structures in thin films, in *Optical Thin Films and Coatings* ed A Piegari and F Flory (Oxford: Woodhead Publishing Limited) pp 177-289

[5] Soubret A and Berginc G 2003 Electromagnetic wave scattering from a random layer with rough interfaces I: Coherent field arXiv:physics/0312133

[6] Soubret A and Berginc G 2003 Electromagnetic Wave Scattering from a Random Layer with Rough Interfaces II: Diffusive intensity arXiv:physics/0312136

[7] Berginc G 2002 Small-slope approximation method : a further study of vector wave scattering from two-dimensional surfaces and comparison with experimental data in *Progress in Electromagnetics Research PIER 37* ed Y Zhang and T M Grzegorczyk (Cambridge USA :EMW Publishing) pp 251-287

[8] Berginc G and Bourrely C 2007 The small-slope approximation method applied to a three-dimensional slab with rough boundaries in *Progress In Electromagnetics Research, PIER 73,* ed J A Kong (Cambridge USA: EMW Publishing) pp 131-211

[9] Soubret A Berginc G and Bourrely C 2001 Application of reduced Rayleigh equations to electromagnetic wave scattering by two-dimensional randomly rough surfaces *Phys. Rev. B* **63** pp 245411–245431

[10] Berginc G 2007 Small-amplitude perturbation theory for two-dimensionally rough surfaces Chapter 6 in *Light scattering and nanoscale surface roughness* ed Alexei Maradudin (New York: Springer-Verlag) pp 127-179

[11] Soubret A Berginc G and Bourrely C 2001 Backscattering enhancement of an electromagnetic wave scattered by two-dimensional rough layers *J. Opt. Soc. Am. A* **18** 11 pp 2778-2788

[12] Soubret A and Berginc G 2003 Effective dielectric constant for random medium arXiv:physics/0312117

[13] Berginc G 2013 Coherent light scattering of heterogeneous randomly rough films and effective medium in the theory of electromagnetic wave multiple scattering *Quantum Electronics* **43** 11 pp 1055-1064

[14] Berginc G and Bourrely C 2010 Light scattering from 3-D nanoscale disordered media *PIERS Online* **6** 8 pp 730-734

[15] Flory F Escoubas L and Berginc G 2011 Optical properties of nanostructured materials: a review *J. of Nanophoton.* **5** 1 052502

[16] Yi He and Zeng T 2010 First-principles study and model of dielectric functions of silver nanoparticles *J. Phys. Chem. C* **114**18023

[17] Ginzburg P and Zayats A 2013 Localized surface plasmon resonances in spatially dispersive nano-objects: phenomenological treatise *ACS Nano* **7** 4334

[18] Scholl J A Koh A L and Dionne J A 2012 Quantum plasmon resonances of individual metallic nanoparticles *Nature* **483** 421.

[19] Thierry F Le Rouzo J Flory F Berginc G Escoubas L 2015 Fast and reliable approach to calculate energy levels in semiconductor nanostructures *J. Nanophoton.* **9** 1 093080

Simulation of the adhesion contact at the interface between polymer and metal

EV Proutorov, OG Maksimova and AV Maksimov

Cherepovets State University, Pr. Lunacharskii 5, 162600, Cherepovets, Russian Federation

E-mail: Proutorov@gmail.com

Abstract. The dynamic Monte Carlo method for the modified lattice Langmuir's model of adsorption is developed. It is shown that there is an optimum temperature regime at which the greatest adhesion of monomer coating on the metal sheet surface occurs. The simulation results are confirmed by experimental data on the study of the adhesive strength of poly-vinyl-butyral primer layer on the metal surface.

1. Inrtoduction

In recent years, the study of the surface physical-chemical properties of different polymer structures has acquired not only theoretical but also practical meaning due to their unique engineering characteristics. On the one hand, the solution of the adsorption problem of macromolecules at different interfaces creates the basis for the development of surface physical chemistry of polymers [1, 2]. On the other hand, the development of polymer coverings (polyvinylidene fluoride, polyurethane and etc.) is necessary, e.g., for the effective protection of metal production from corrosion. As it is known, the corrosion control of metals is one of the most important technical and economic problems of today's society.

The quality of polymer coating is determined, first of all, by its durability, adhesion and resistance to aggressive external environment and also by strength, which mainly depends on the degree of orientational order of macromolecules in the primer layer [3]. These properties of polymer coating depend on the method of its formation and its chemical modification [4]. High efficiency of polymer coating is achieved only by proper choice of thermal and other conditions of formation of its multilayer "sandwich" structure. Therefore, it is necessary to study in detail adhesion and orientational ordering of macromolecules in different layers of the coating.

The process of formation of polymer coating on steel surface consists of the following stages:

- formation of the adhesion primer layer (the monolayer structure) after application of the monomer solution to the steel sheet before polymerization;
- polymerization of the monomer paint by special cross-linking agents;
- formation of the multilayer (macromolecular) structure during the evaporation of the solvent after polymerization.

Therefore, to predict the properties of polymer coating on the metal surface, both experimental and theoretical investigation of all the processes is necessary. In this paper, only the first stage of formation of polymer coating is under investigation.

In surface physics there is a whole class of problems associated with the study of adsorbed atom layers on a neutral substrate [1]. The adsorbent surface, which causes "sticking" of atoms, determines a periodic potential, and thus some "lattice" consisting of nodes possible for filling. In this case, the characteristics of the adsorbed layer on the investigated surface of three-dimensional solid must correspond to similar characteristics of two-dimensional system. Therefore, in the theory of monolayer adsorption of low-molecular substances on the surface of homogeneous neutral surface, the Langmuir's model of lattice gas [5] may be used, in which attraction forces between adsorbed molecules and the substrate and their mobility along its surface are neglected. Further development of the model was advanced in Brunauer, Emmett and Teller's theory of polymolecular adsorption (the BET method [6]). Hill and J. de Boer developed the theory [7], which takes into account interactions between the adsorbed molecules in Langmuir's model.

There is a formal analogy between the model of lattice gas and the two-dimensional Ising's model [8]. This analogy is widely used to study surface properties of ferromagnetic and ferro electric films [9, 10]. However, to study adhesion for various materials (metals, glasses and others), more complex models should be used, which take into account interactions of atoms of the adsorbed substance, both among themselves, and with the substrate surface, specified by periodic potential, determined by the type of the substrate. It becomes difficult to calculate surface properties of the coatings in such models using analytical methods of studying, and therefore it is necessary to use computer simulation methods, such as the Monte Carlo method [10].

The aim of this work is to study the first stage of formation of polymer coating on steel surface, namely the occurrence of adhesive contact at the interface between polymer and metal, which is necessary for creation of single-layer monomer structure before polymerization. For this purpose, the dynamic Monte – Carlo method for the modified version of Langmuir's model will be developed, which takes into account interactions between the monomers themselves and the specific form of the potential interaction with the metal surface.

2. Model

Coupling of surfaces of the lower (primer) layer of the coating and the steel sheet occurs due to adhesion process, which is determined by intermolecular interactions. This adhesive layer can be considered as a set of monomers interacting with metal surface. The dependence of potential energy of monomer on the distance to the surface [11] is schematically represented in Figure 1. By physical absorption, monomers interact with metal by means of the Van der Waals forces, characterized by mean energy E_p. In this case, there is no charge transfer from the substrate to monomers or vice versa. At higher temperatures monomers can reach metal surface (the bottom of the potential well with the depth E_c), overcoming some energy barrierE_b. Then, electron exchange between the monomers and the metal surface is possible; it leads to sufficiently strong chemical bonds between them (i.e. to chemical adsorption [11]). The energy sum determines the energy required for the transition of the monomer from the adsorbed state (on the metal surface) to the free state, i.e. $A = E_c + E_b$.

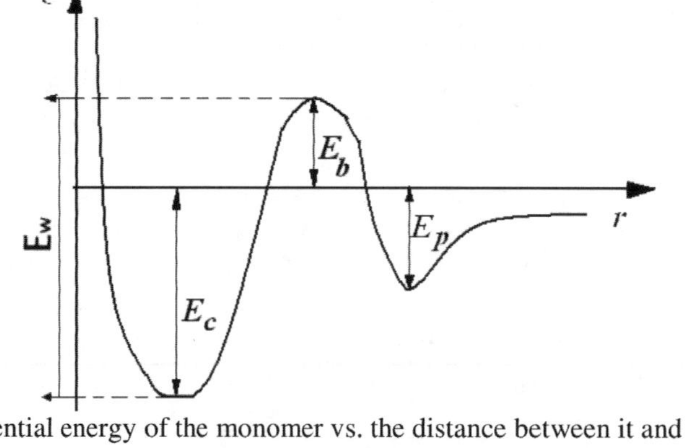

Figure 1. The potential energy of the monomer vs. the distance between it and the metal surface.

Journal of Physics: Conference Series **633** (2015) 012044 doi:10.1088/1742-6596/633/1/012044

In this paper, the degree of adhesion of monomers to metal is calculated by the dynamic Monte Carlo method for the modified Langmuir's model of adsorption. In this model, the steel sheet surface is represented as two-dimensional rectangular lattice consisting of N nodes (of adsorption) along the X axis and M nodes along the axis Y. The position of the node in this lattice is determined by two numbers i and j, and its occupation is characterized by number $n_{i,j}$, which are equal to zero (0) if the adsorption site is still free, and unity (1), otherwise. The model is based on the following assumptions:

- all adsorption lattice nodes are equivalent;
- only one particle (monomer) can occupy an adsorption node;
- the influence of the adsorption forces exists only within the thickness of one monolayer.

The interaction energy between the lattice sites (monomers) has the form as in Ising's model:

$$E = -\sum_{i,j=1}^{N,M} K n_{i,j} n_{i+1,j} - \sum_{i,j=1}^{N,M} K n_{i,j} n_{i,j+1}, \tag{1}$$

where K is the interaction constant. In the calculations by the formula (1), the interactions only between the nearest nodes are taken into account.

In this work at the initial moment ($t = 0$), it was assumed that the value was $n_{i,j} = 0$ for all lattice nodes, i.e. the adsorption hasn't occurred. The calculations were performed via the computer simulation by the dynamic Monte Carlo method [10, 12], in which a random process consisting of a set of consecutive configurations of the system was generated. At sufficiently large number of configurations in the ensemble of such system, the relative adhesion degree is defined as the ratio of number of the adsorbed monomers to total number of lattice nodes, i.e.

$$\theta = (1/NM) \sum_{i,j=1}^{N,M} n_{i,j}. \tag{2}$$

3. Results and discussion

Figure 2 shows the results of the calculation of time dependences of θ at equal heating rates of metal sheet $g = \Delta T/\Delta t$, but different values of constants of interaction of monomers ($K/k_Б T_0$) and barrier energy ($E_b/k_Б T_0$), where $T_0 = 300$ K. It can be seen, that with the rise of interaction constant of monomers, its adhesion degree to the metal surface increases because of the growth of the cooperative adhesion process (Fig. 2*a*). On the other hand, with the increase of the barrier height E_b the quantity θ reduces, that is, fewer and fewer monomers adhere to the metal surface (Figure 2*b*).

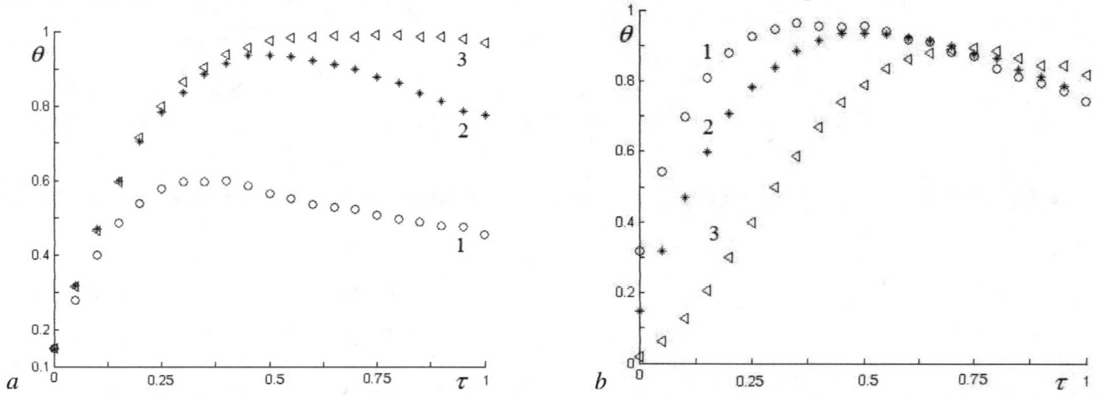

Figure 2. The adhesion degree of monomers θ at equal rates of heating of the metal sheet vs. the reduced time τ: *a* – at different values of the interaction constants of monomers ($K/k_Б T_0$= 0 (*1*), 1(2), 2(3)) for the given energy barrier ($E_b/k_Б T_0$=1); *b* – at different values of the energy barrier $E_b/k_Б T_0$=0.5(*1*), 1(*2*), 2(*3*) for the given value of the interaction constant of monomers ($K/k_Б T_0$=1), where T_0 = 300 ^{0}K.

Figure 3*a* represents the results of the calculation of time dependence of adhesion degree θ at different heating rates of the metal sheet for given values of the interaction constant of the monomers and energy barrier. It is clear, that at slow heating rate of the steel sheet, the degree of adhesion of monomers to the metal surface is small (curve *1*). At high heating rate and relatively high temperatures, there is some adhesion to the metal surface, but at its further heating the value of θ decreases (curve *3*). In this case (after the drying process) the primer layer may peel off from the metal surface that is one of the causes of various defects of polymer coatings. It can be seen, that the optimum temperature regime is determined by curve 2 in Figure 3*a*, when the adhesion is maximal.

Despite of well-known analogy of statistical properties of the lattice Langmuir's model and the two-dimensional Ising's model [9], the calculated temperature dependence of the adhesion degree, determining the adhesive strength of the polymer coating is non-monotonic function (Figures 2 and 3). Initial increase of the value θ in this model can be explained by the growth of kinetic energy of the monomers, it results in overcoming the energy barrier E_b (Figure 1), that prevents the formation of absorbed state of monomers on the metal surface. At further rise of temperature, adhesion degree of the monomers, conversely, decreases as well as magnetization in the two-dimensional Ising's model due to the growth of thermal fluctuations in the system. This result is consistent with the experimental data [13] from the study of temperature dependence of the adhesion strength of primer poly-vinyl-butiral adhesive layer on the steel surface (Figure 3*b*).

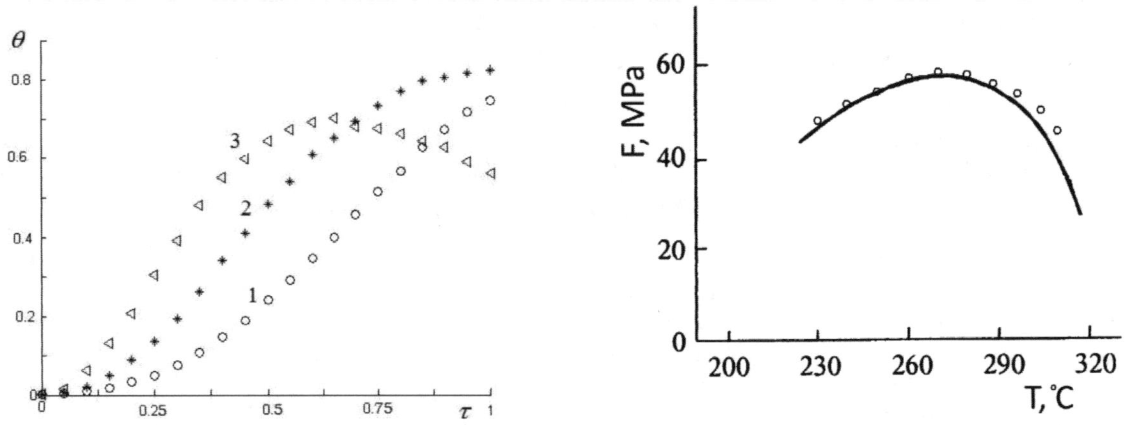

Figure 3. *a* – The adhesion degree of monomers θ vs. the reduced time τ at different heating rates of the metal sheet $g_0(1)$, $2g_0(2)$, $3g_0(3)$ for the given values of the interaction constant of monomers ($K/k_BT_0 = 1$) and the barrier energy ($E_b/k_BT_0 = 3$), where $T_0 = 300\ ^0$K. *b* – the dependence of the adhesion strength of polyvinyl butyral film formation temperature on the steel surface: solid line - the experiment [13], the symbols (o) - the simulation results by the Monte Carlo method.

4. Conclusion

Thus, in this paper we propose the model of absorption of monomers for the study of one of the most important stages in the process of forming the polymer coating, namely, the formation of single-layer structure on the surface of the metal sheet. It is shown that there is an optimum temperature regime at which there is the greatest adhesion of monomer coating to metal sheet surface.

Acknowledgements

This work was supported by the Ministry of Education and Science of the Russian Federation, project №1715 (contract № 2014/267) and by Foundation for Assistance to Small Innovative Enterprises in Science and Technology

References

[1] Karim A and Kumar S 2000 *Polymer Surfaces, Interfaces and Thin Films* (Singapore: World Scientific)
[2] Maksimov AV, Pavlov GM 2007 *Polymer Science.* A **49(7)** 828.
[3] Gotlib YuYa, Maximov AV 1992 *Polymer Science.* A **34(11)** 157
[4] Yilgor I 2013 *Abstracts 12-th Int.Conf. "Polymers in advanced technologies" (Berlin, Germany 27 September – 2 October 2014)* 34
[5] Langmuir I 1932 *J. Am. Chem. Soc.* **54(7)** 2798
[6] Brunauer S, Emmett PH, and Teller E 1938 *J. Am. Chem. Soc.* **60** 309
[7] Fridrihsberg DA 1984 *Colloid chemistry course (textbook for High Schools)* 2-nd ed. (Lenindrad: Chemistry)
[8] Lee T.D, Yang CN 1952 *Phys. Rev.* **87** 410
[9] Fisher ME and Ferdinand AE 1967 *Phys.Rev.Lett.* **19** 169
[10] *Monte Carlo and Molecular Dynamics Simulations in Polymer Science* 1995 ed by K Binder (Oxford: University Press).
[11] Prutton M 1994 *Introduction to Surface Physics.* (Oxford: Clarendon Press).
[12] Petrova TO, Maksimova OG, Gerasimov RA and Maksimov AV 2012 *Solid State Physics* **54(5)** 883.
[13] Yakovlev AD 2008 *Chemistry and technology of coatings* (Saint-Petersburg: Himizdat).

A Frontier orbital energy approach to redox potentials

Jeanet Conradie

Department of Chemistry, University of the Free State, 9300 Bloemfontein, Republic of South Africa.

conradj@ufs.ac.za

Abstract. The prediction of the oxidation and reduction potentials of molecules is important in many research areas. A review of relationships obtained between frontier orbital energies (eV), the calculated ionization potentials (IP in eV), or adiabatic electron affinities (EA in eV) with the experimental oxidation and reduction potentials is presented, for selected series of β-diketones, rhodium-β-diketonato complexes, as well as metal-tris-β-diketonato complexes, with the metal Fe or Mn. The good linear relationships obtained for related series of complexes show that the oxidation and reduction potentials of these complexes can be predicted by their DFT-calculated energies.

1. Introduction

The Koopmans' theorem, related to closed-shell Hartree–Fock theory, states that the first (vertical) ionization energy of a molecular system is equal to the negative of the Hartree–Fock orbital energy of the highest occupied molecular orbital (HOMO).[1] Although the original Koopmans' paper did not refer to the calculation of electron affinities as the energy of the lowest unoccupied molecular orbitals (LUMO), the theorem was extended to calculate electron affinities from LUMO energies.[2,3] The theorem was also extended to the Kohn-Sham orbital energies of density functional theory (DFT) calculations.[4,5] The energy involved in the oxidation (loss of electrons) or reduction (gain of electrons) of a complex, is closely related to the ionization energy (amount of energy required to remove an electron from the HOMO of a molecule in the gaseous state) and electron affinity (amount of energy released when an electron is added to the LUMO of a neutral molecule in the gaseous state) of the complex. In fact, the following approximate linear relationship between HOMO and LUMO energies and oxidation/reduction potentials can be deduced from the Marcus quadratic relationship for interfacial electron transfer: [6]

$E_{OX/RED} = a + b.E_{HOMO/LUMO}$; where $E_{OX/RED}$ are the experimental redox potentials; $E_{HOMO/LUMO}$ the calculated HOMO and LUMO energies; and a and b are constants.

We therefore here present a summary of empirical comparisons found between DFT calculated HOMO and LUMO energies and experimental oxidation and reduction values, obtained for the β-diketones and metal-β-diketonato complexes in Figure 1.

2. Discussion

Coordination of the enolated anion, $(RCOCHCOR')^-$ (Figure 1 (a)) of a β-diketone to a metal, produces a pseudo-aromatic ring with the metal. The electron donating properties of the R and R' groups on the β-diketonato ligand, will thus change the electron density on the metal through the conjugation and thus the ease of oxidation and/or reduction of the metal. The energy of the frontier orbitals of β-diketones and metal-β-diketonato complexes will also be determined by the electron donating properties of the R and R' groups. By choosing the R and R' groups to range from electron donating (like ferrocenyl, Fc) to electron withdrawing (like CF_3), the abovementioned relationships are obtained over a large HOMO/LUMO energy and oxidation/reduction potential range. The order of electronegativity of the different R and R' groups is as follows, from most to least electron withdrawing: $PhNO_2 > CF_3 > CCl_3 > CH_3 > C_4H_3O > Ph > C_4H_3S > Fc$.

The DFT calculated electronic energies of the neutral and oxidized/reduced complex with metal M were used to calculate the first ionization potential (IP) and the electron affinity (EA) values of the complexes:

$$IP(M) = E(M) - E(M^+) \quad \text{and} \quad EA(M) = E(M^-) - E(M)$$

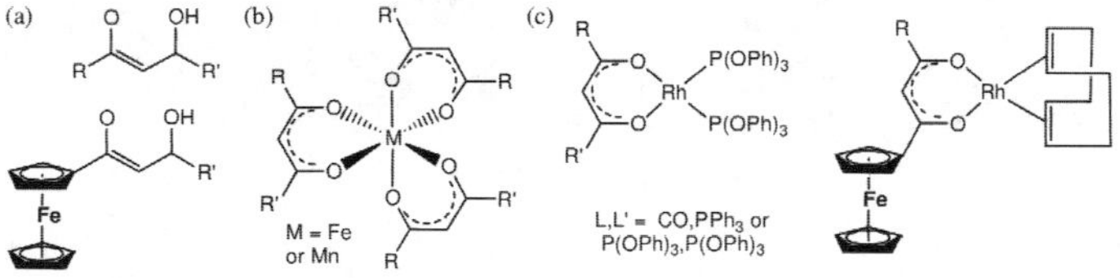

Figure 1. Structure of (a) β-diketones, (b) M(β-diketonato)$_3$ and (c) Rh-β-diketonato complexes.

2.1. *β-diketones*

The linear relations for the correlations found between the experimentally measured reduction potential of a series of β-diketones[7] and their calculated LUMO energies and EA, are shown in Figure 2 and Figure 3 respectively. In the electro-active centre, the LUMO has π-character that is distributed mainly over the conjugated β-diketone backbone, allowing for the delocalization of the electron density, see Figure 4. The reduction potentials vary over more than 1 eV, indicating excellent electronic communication *via* conjugation between the electro-active centre and the substituent R groups. The DFT calculations showed that β-diketones containing electron withdrawing groups, have lower more negative LUMO energies, higher EA and are reduced at a higher, less negative potential.

β-diketones: E_{pc} (V *vs.* Fc/Fc$^+$) = –0.63 E_{LUMO} – 3.63 $R^2 = 0.99$

E_{pc} (V *vs.* Fc/Fc$^+$) = 0.62 EA – 2.39 $R^2 = 0.84$

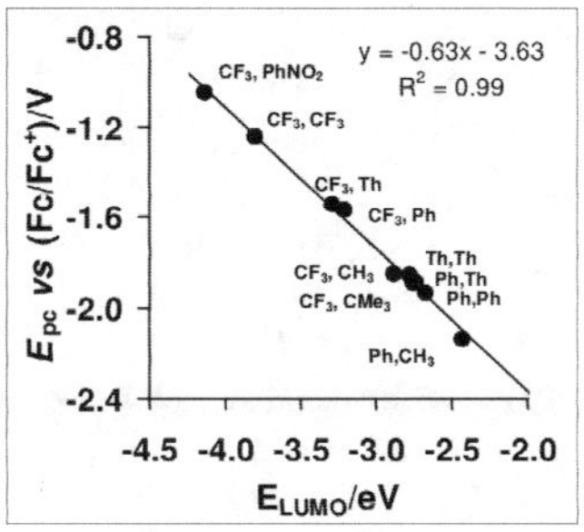

Figure 2. Linear correlation between the peak reduction potentials (E_{pc}) of 1,3-substituted β-diketones, RCOCHC(OH)R', and their OLYP/TZP calculated LUMO energies. R and R' are shown on the graph (1 eV = 96.48 kJ mol^{-1}). Data was obtained

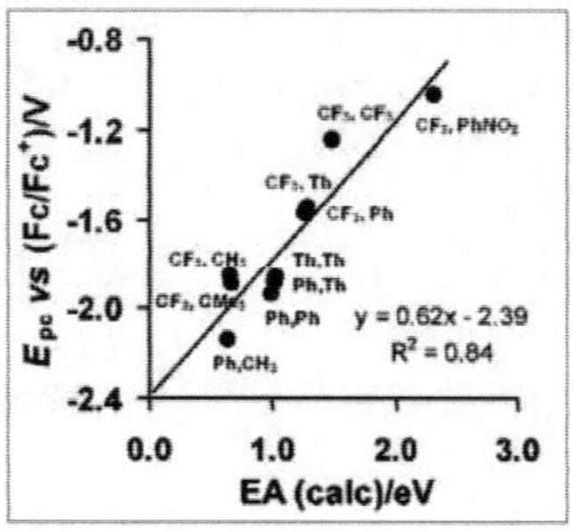

Figure 3. Linear correlation between the peak reduction potentials (E_{pc}) of 1,3-substituted β-diketones RCOCHC(OH)R', and their OLYP/TZP calculated electron affinity (EA). R and R' are shown on the graph. Data was obtained from reference 7.

Journal of Physics: Conference Series **633** (2015) 012045 doi:10.1088/1742-6596/633/1/012045

from reference 7.

The linear relations for oxidation of the metal (Fe) and their calculated HOMO energies and IP, in a series of ferrocenyl-containing β-diketones,[8] are shown in Figure 6 and Figure 7, respectively. The HOMO is Fe-based, confirming the Fe oxidation, see Figure 5. The relationships obtained are:

Fc-containing β-diketones: $E^{0'}$ (V *vs.* Fc/Fc$^+$) = –0.24 E_{HOMO} – 1.15 $R^2 = 0.92$

$E^{0'}$ (V *vs.* Fc/Fc$^+$) = 0.10 IP – 0.40 $R^2 = 0.86$

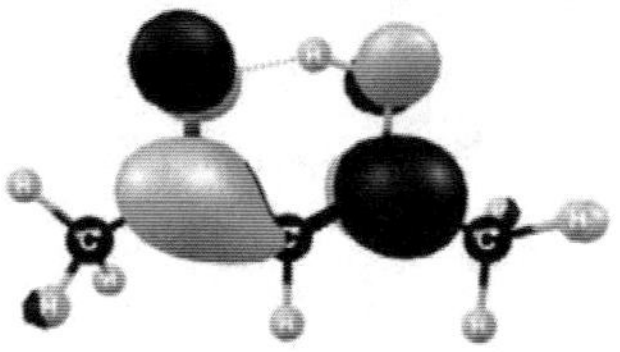

Figure 4. Kohn-Sham b3lyp/6-311G(d,p) calculated LUMO of the optimized geometry of $CH_3COCHC(OH)CH_3$.

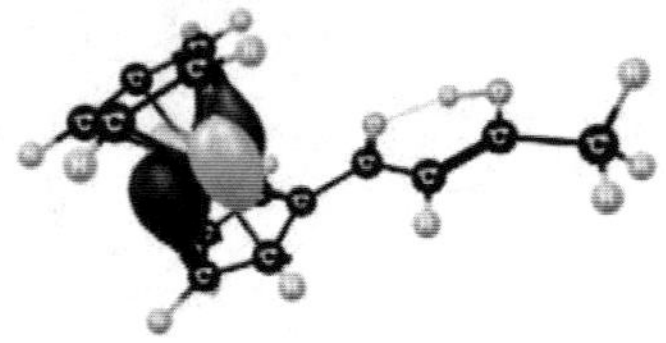

Figure 5. Kohn-Sham b3lyp/6-311G(d,p) calculated HOMO of the optimized geometry of $FcCOCHC(OH)CH_3$.

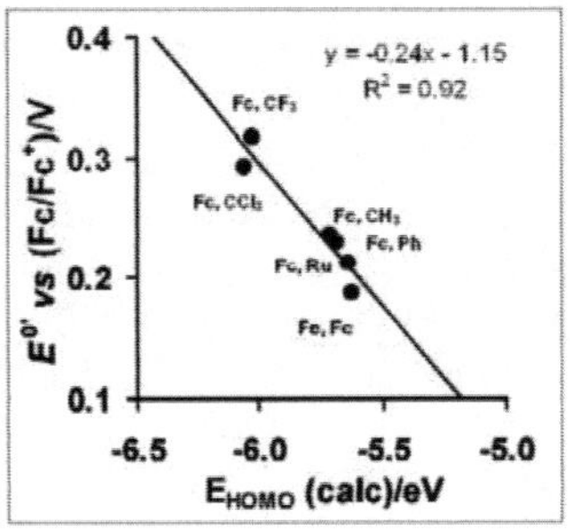

Figure 6. Linear correlation between the reversible oxidation potential ($E^{0'}$) of the ferrocenyl group of ferrocenyl-substituted β-diketones FcCOCHC(OH)R, and their b3lyp/6-311G(d,p) calculated HOMO energies. R is shown on the graph. Data was obtained from reference 8.

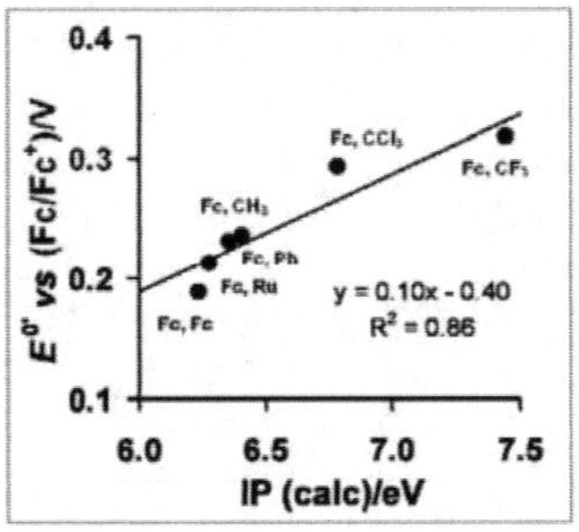

Figure 7. Linear correlation between the reversible oxidation potential ($E^{0'}$) of the ferrocenyl group of ferrocenyl-substituted β-diketones FcCOCHC(OH)R and their b3lyp/6-311G(d,p) calculated ionization potential IP. R is shown on the graph. Data was obtained from reference 8.

2.2. M(β-diketonato)$_3$ complexes, M = Fe or Mn

Reduction of a complex involves the addition of an electron to the LUMO. The LUMOs of the octahedral M(β-diketonato)$_3$ complexes with M = Fe[9] or Mn[10] are mainly metal based (Figure 8 and Figure 9), implying that the experimentally observed reduction during a cyclic voltammetry experiment, involve the reduction of the M^{III} to M^{II}. The influence of the electron donating/withdrawing power of the R and R' groups of the coordinated β-diketonato $(RCOCHCOR')^-$ complex on the metal reduction, is evident from the large experimental potential range. The relationship between the experimentally measured reduction potential and the calculated LUMO energies (Figure 10 and Figure 11) gave a better fit than the relationship between the experimentally measured reduction potential and the calculated EA (graphs not shown):

Fe(β-diketonato)$_3$:	$E^{0'}$ (V *vs* SCE) = –0.59 E_{LUMO} – 2.74	$R^2 = 0.96$
	$E^{0'}$ (V *vs* SCE) = 0.71 EA – 1.94	$R^2 = 0.94$
Mn(β-diketonato)$_3$:	$(E_{pc} + E_{pa})/2$ *vs* SCE/V = –0.76 E_{LUMO} – 2.75	$R^2 = 0.98$
	$(E_{pc} + E_{pa})/2$ *vs* SCE/V = 0.83 EA – 1.95	$R^2 = 0.91$

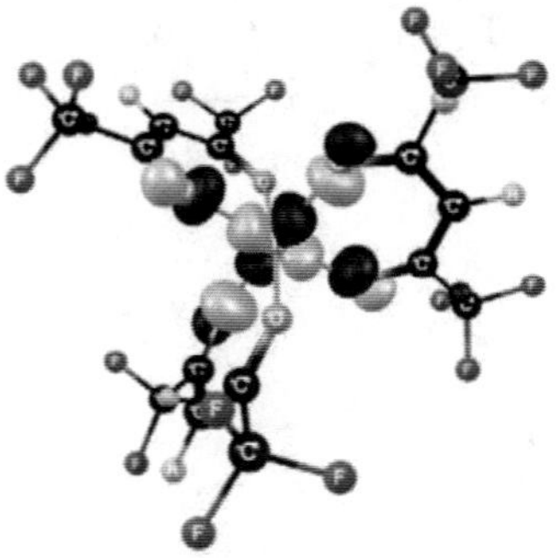

Figure 8. The Kohn–Sham type OLYP/STO-TZP calculated LUMO of d_{x2-y2} character, of complex $Mn(CF_3COCHCOCF_3)_3$.

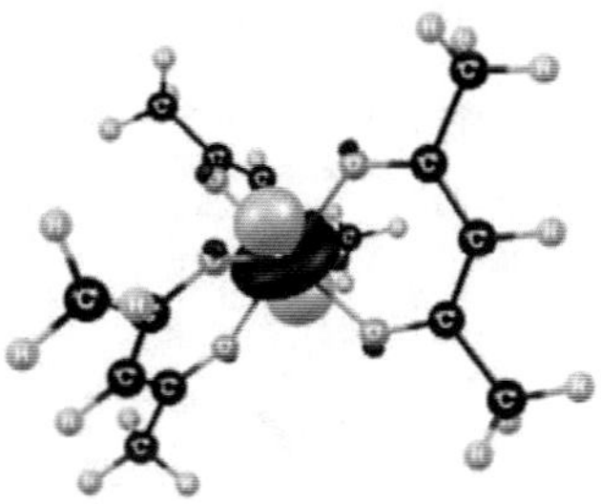

Figure 9. The Kohn–Sham OLYP/STO-TZP calculated LUMO of d_{z2} character, of complex $Fe(CF_3COCHCOCF_3)_3$.

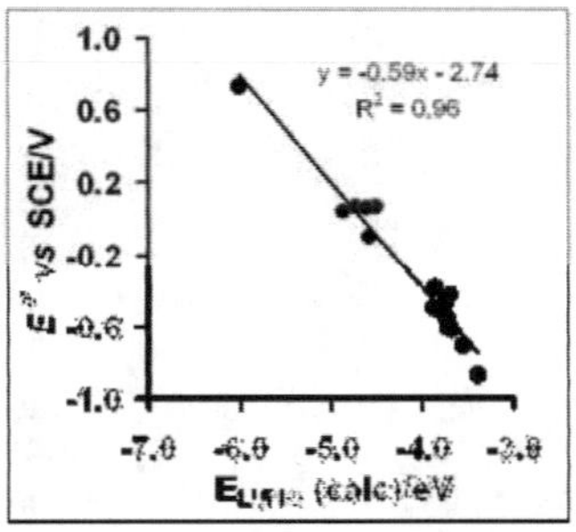

Figure 10. Linear correlation between the reversible $Fe^{III/II}$ reduction ($E^{0'}$) of Fe(β-diketonato)$_3$ complexes and their OLYP/TZP calculated LUMO energies. Data was obtained from reference 9.

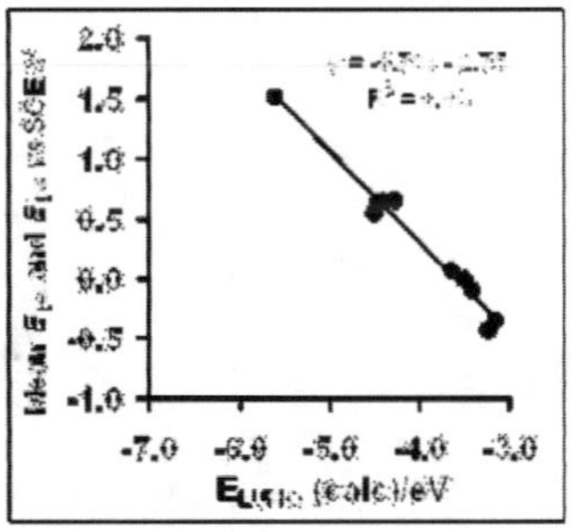

Figure 11. Linear correlation between the experimental mean of E_{pa} and E_{pc} of the $Mn^{III/II}$ reduction of Mn(β-diketonato)$_3$ complexes and their OLYP/TZP calculated LUMO energies. Data from reference 10.

2.3. Rh(β-diketonato)(L,L') complexes,[11] *(L,L') = (P(OPh)$_3$,P(OPh)$_3$); (cod) or (CO),(PPh$_3$)*

Rh^I(β-diketonato)(L,L′) complexes are experimentally oxidized to Rh^{III}(β-diketonato)(L,L′) in an irreversible two electron transfer process. The HOMO of the d^6 Rh^I(β-diketonato)(L,L′) complexes is mainly d_{z2} based (Figure 12). For the ferrocenyl-β-diketonato containing Rh(β-diketonato)(cod) complexes, the Rh^I to Rh^{III} oxidation, is followed by oxidation of the ferrocenyl group (Fe^{II} to Fe^{III}) at a slightly more positive potential. From the relationships presented in Figure 13 and given below we observe, as expected, that the higher the energy of the HOMO, the easier it is to remove the electrons during the oxidation process.

Rh(β-diketonato)(P(OPh)$_3$)$_2$:	E_{pa} (Rh in V *vs.* Fc/Fc$^+$) = −1.50 E_{HOMO} − 7.81	$R^2 = 0.85$
Rh(β-diketonato)(CO)(PPh$_3$):	E_{pa} (Rh in V *vs.* Fc/Fc$^+$) = −0.45 E_{HOMO} − 2.11	$R^2 = 0.89$
	E_{pa} (Rh in V *vs.* Fc/Fc$^+$) = 0.29 IP − 3.65	$R^2 = 0.97$
Rh(β-diketonato)(cod):	E_{pa} (Rh in V *vs.* Fc/Fc$^+$) = −0.35 E_{HOMO} − 1.31	$R^2 = 0.90$
	E_{pa} (Fc in V *vs.* Fc/Fc$^+$) = −0.39 E_{HOMO-1} − 1.52	$R^2 = 0.99$

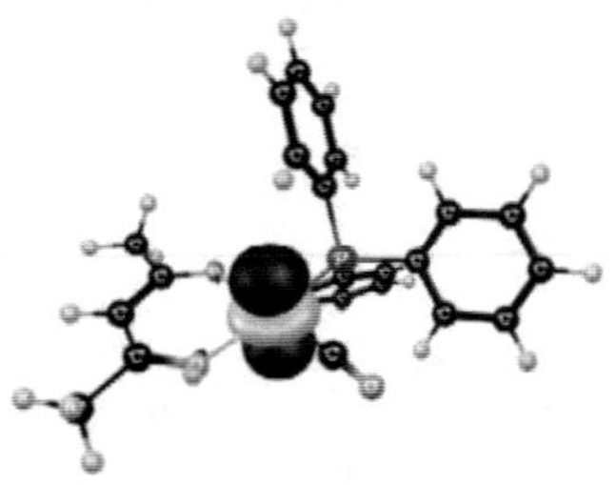

Figure 12. The Kohn–Sham b3lyp/6-311G(d,p) calculated d_{z2} HOMO of complex [Rh($CH_3COCHCOCH_3$)(CO)-(PPh_3)].

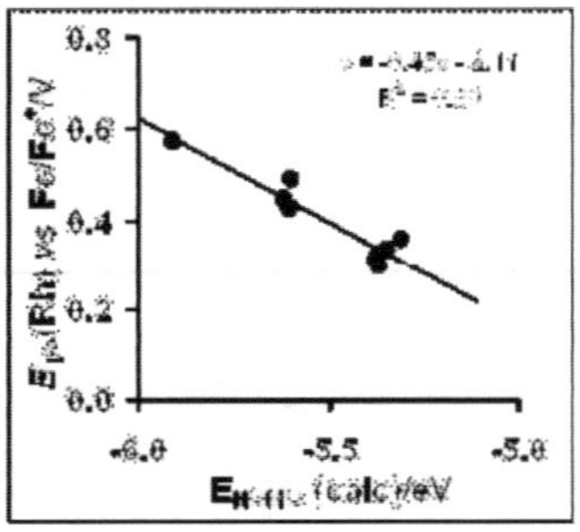

Figure 13. Linear relationship between b3lyp/6-311G(d,p) calculated HOMO energies of [Rh(β-diketonato)(CO)(PPh$_3$)] and the experimental oxidation potential of rhodium, E_{pa}(Rh). Data from reference 11 and 12.

3. Conclusion

We have illustrated that for a series of complexes with similar structures, a linear relationship exists between the molecular orbital energies and their reduction and oxidation potentials obtained from cyclic voltammetry. These linear correlations can be used to estimate the redox potentials of unknown complexes, using the calculated HOMO and LUMO energy levels.

Acknowledgements

This work has received support from the Norwegian Supercomputing Program (NOTUR) through a grant of computer time (Grant No. NN4654K), the South African National Research Foundation, HPC Warehouse Facility and the Central Research Fund of the University of the Free State, Bloemfontein.

References

[1] Koopmans T A 1934 *Physica* **1** 104-113
[2] Bozkaya U J 2014 *Chem. Theory Comput.* **10** 2041
[3] Heinrich N, Koch W and Frenking G 1986 *Chem. Phys. Lett.* **124**(1) 20
[4] Perdew J P and Levy M 1997 *Phys. Rev.* B **56** 16021

Casida M E 1999 *Phys. Rev.* B **59** 4694
[5] Hamel S, Duffy P, Casida M E and Salahub D R 2002 *J. Electron. Spectrosc. Relat. Phenom.* **123** 345
[6] Méndez-Hernández D D, Tarakeshwar P, Gust D, Moore T A, Moore A L and Mujica V 2013 *J. Mol. Model.* **19** 2845
[7] Kuhn A, von Eschwege K G and Conradie J 2011 *Electrochim. Acta* **56** 6211
[8] von Eschwege K G and Conradie J *S.* 2011 *Afr. J. Chem.* **64** 203
[9] Conradie M M and Conradie J 2015 *Electrochim. Acta* **152** 512
[10] Freitag R and Conradie J 2015 *Electrochim. Acta* **158** 418
[11] Conradie J 2015 *Dalton Trans.* **44** 1503
[12] Ferreira H, Conradie M M and Conradie J 2013 *Electrochim. Acta* **113** 519

A Finsler Geometry Modeling of the Liquid Crystal Elastomer

Hiroshi Koibuchi

Department of Mechanical and Systems Engineering, National Institute of Technology, Ibaraki College, Nakane 866, Hitachinaka, Ibaraki 312-8508, Japan

E-mail: koibuchi@mech.ibaraki-ct.ac.jp

Andrey Shobukhov

Faculty of Computational Mathematics and Cybernetics, Lomonosov Moscow State University, 119991, Moscow, Leninskiye Gory, MSU, 2-nd Educational Building, Russia

Liquid crystal elastomer (LCE) is a rubbery material composed of polymer chains and liquid crystals (LC). LCE is well known to undergo a shape transformation from the isotropic to the anisotropic phase. This shape transformation is caused by the nematic transition of the LC included in the LCE. However, the mechanism of this transformation is unknown because the interaction of LC with the bulk polymers is too complex. In this presentation, we extend the two-dimensional Finsler geometry model for membranes to a three-dimensional model for LCE. The Finsler geometry model for LCE is a coarse grained one: the Guassian bond potential S_1 is obtained by extending the one for membranes, which is originally obtained by a simple extension of the Guassian bond potential for the linear chain polymer model. The continuous Hamiltonian, which contains S_1 and the curvature energy S_2, is discretized using a three-dimensional rigid sphere composed of tetrahedrons. We study the shape transformation as a phase transition between the isotropic and anisotropic phases and report the results of the transition order, obtained by the Monte Carlo simulations.

Keywords: Liquid crystal elastomer, Finsler geometry, Anisotropy, Monte Carlo

1. Introduction

The linear chain model for polymer is defined by the Gaussian bond potential, which is a function of the sum of chain (or bond) length squares [1]. This model is extended to a model for membranes, where the extended Gaussian bond potential plays the role of surface tension energy [2]. Therefore it is natural to suppose that the Gaussian bond potential for membranes can also be extended to the one for the three-dimensional liquid crystal elastomer (LCE) [3].

The LCE is composed of polymer chains and liquid crystals (LC) and is well known because of its interesting properties such as the anisotropic shape transformation and the soft-elasticity [3]. At high temperature, the LC is in a random phase, and as a consequence the LCE has no preferred direction, and hence it is in the symmetric phase. On the contrary, as the temperature is lowered, the LCE starts to elongate in a spontaneously chosen direction. This phenomenon, which is the so-called anisotropic shape transformation, has its origin in the nematic transition of the constituent LC. However, the mechanism of the shape change is unknown. This is because the interaction of the LC and the bulk polymer is too complex. Therefore, a coarse grained model is helpful for better understanding of the LCE properties.

In this paper, we extend the $2D$ Finsler geometry model for membranes [4] to a $3D$ model for LCE. The discrete Hamiltonian is defined on the spherical body, which is discretized by tetrahedrons. The curvature energy is assumed on the surface of the sphere with the bending rigidity κ. By using the Monte Carlo (MC) technique, we study the anisotropic shape transformation and its dependence of the surface elasticity, and report the order of the transition of the anisotropic shape transformation.

2. Model

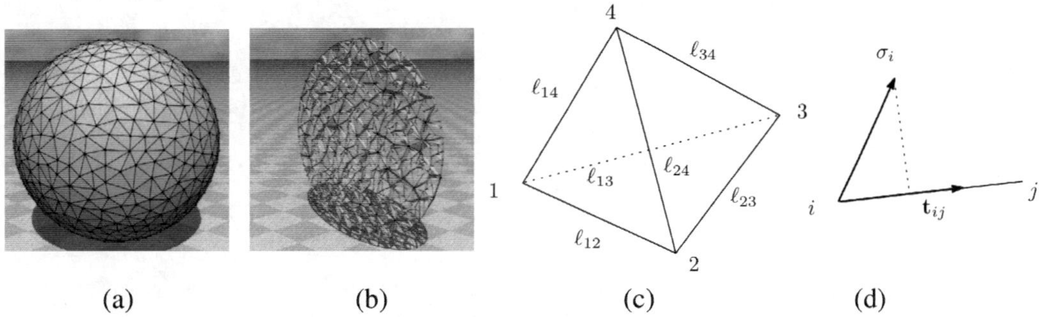

Figure 1. (a) A sphere of size $N = 2431$ and (b) its sectional slice. (c) A tetrahedron in which the discrete Gaussian bond potential S_1 is defined, and (d) the variable σ_i and the unit tangential vector $\mathbf{t}_{ij}$ along the edge vector $\vec{\ell}_{ij}$.

A sphere in $\mathbf{R}^3$ is randomly discretized by Voronoi tessellation (Fig. 1(a),(b)). In this sphere, the primitive object is the tetrahedron, which is composed of vertices, bonds, and triangles. The discrete model is defined by the Hamiltonian $S(\mathbf{r},\sigma)$ which is given by

$$\begin{aligned}
&S(\mathbf{r},\sigma) = \lambda S_0(\sigma) + S_1(\mathbf{r},\sigma) + \kappa S_2(\mathbf{r},\sigma) + U_{3D} + U_{2D},\\
&S_0(\sigma) = \sum_{ij}(1-\sigma_i\cdot\sigma_j),\\
&S_1(\mathbf{r},\sigma) = \frac{1}{4}\sum_{\rm tet}\left(\gamma_{12}\ell_{12}^2+\gamma_{13}\ell_{13}^2+\gamma_{14}\ell_{14}^2+\gamma_{23}\ell_{23}^2+\gamma_{24}\ell_{24}^2+\gamma_{34}\ell_{34}^2\right),\\
&S_2(\mathbf{r}) = \sum_{ij}(1-\mathbf{n}_i\cdot\mathbf{n}_j),\\
&U_{3D} = \sum_{\rm tet} U_{3D}({\rm tet}), \qquad U_{3D}({\rm tet}) = \begin{cases} 0 & ({\rm Vol(tet)}>0)\\ \infty & ({\rm otherwise})\end{cases},\\
&U_{2D} = \sum_{\Delta_i,\Delta_j} U_{2D}(i,j), \qquad U_{2D}(i,j) = \begin{cases} 0 & (i,j \ {\rm intersect})\\ \infty & ({\rm otherwise})\end{cases},
\end{aligned} \tag{1}$$

where $\mathbf{r}(\in \mathbf{R}^3)$ is the vertex position and $\sigma_i(\in S^2)$ (unit sphere) represents the LC defined at the vertex i. In S_1, ℓ_{ij}^2 denotes the bond length squares between the vertices i and j (Fig.1(c)), and γ_{ij} are defined by

$$\gamma_{12} = \frac{v_{12}}{v_{13}v_{14}} + \frac{v_{21}}{v_{23}v_{24}}, \quad \gamma_{13} = \frac{v_{13}}{v_{12}v_{14}} + \frac{v_{31}}{v_{32}v_{34}},$$

$$\gamma_{14} = \frac{v_{14}}{v_{12}v_{13}} + \frac{v_{41}}{v_{43}v_{42}}, \quad \gamma_{23} = \frac{v_{23}}{v_{21}v_{24}} + \frac{v_{32}}{v_{31}v_{34}},$$
$$\gamma_{24} = \frac{v_{24}}{v_{23}v_{21}} + \frac{v_{42}}{v_{41}v_{43}}, \quad \gamma_{34} = \frac{v_{34}}{v_{31}v_{32}} + \frac{v_{43}}{v_{41}v_{42}}. \tag{2}$$

The variable v_{ij} in Eq. (2) is the tangential component of σ_i along the bond ij (Fig.1(b))

$$v_{ij} = |\mathbf{t}_{ij} \cdot \sigma_i|, \quad \mathbf{t}_{ij} = \vec{\ell}_{ij}/\ell_{ij}. \tag{3}$$

Note that in general $v_{ij} \neq v_{ji}$. The $\sum_{\rm tet}$ in S_1 denotes the sum over tetrahedrons. The symbol $\mathbf{n}_i$ in S_2 denotes the unit normal vector of the triangle i on the surface, and κ is the bending rigidity. The potential U_{3D} protects the tetrahedron volume Vol(tet) from being negative. The potential U_{2D} also protects the surface (of a spherical body) from intersecting. We should note that the interaction of U_{3D} is local and the computation is not so time consuming, while that of U_{2D} is non-local and hence the computational time increases with increasing lattice size.

The partition function Z is given by

$$Z(\lambda, \kappa) = \sum_{\sigma} \int' \prod_{i=1}^{N} d\mathbf{r}_i \exp\left[-S(\mathbf{r}, \sigma)\right], \tag{4}$$

where $\int' \prod_{i=1}^{N} d\mathbf{r}_i$ are the multiple three-dimensional integrations, that are performed by fixing the center of mass to the origin of $\mathbf{R}^3$. $\sum_\sigma$ denotes the sum over the spin variables σ.

The continuous Gaussian bond potential S_1 is given by

$$S_1 = \int \sqrt{g} d^3x g^{ab} \partial_a \mathbf{r} \cdot \partial_b \mathbf{r}, \tag{5}$$

where $\mathbf{r}$ is a mapping from a three-dimensional space M to $\mathbf{R}^3$, g^{ab} is the inverse of the Finsler metric on M, and g is its determinant. The discrete form of g_{ab} is given by

$$g_{ab} = \begin{pmatrix} 1/v_{12}^2 & 0 & 0 \\ 0 & 1/v_{13}^2 & 0 \\ 0 & 0 & 1/v_{14}^2 \end{pmatrix}, \tag{6}$$

where v_{ij} are defined in Eq. (3).

The partial derivatives of $\mathbf{r}$ in S_1 are replaced by $\partial_1 \mathbf{r} \to \mathbf{r}_2 - \mathbf{r}_1$, $\partial_2 \mathbf{r} \to \mathbf{r}_3 - \mathbf{r}_1$, and $\partial_3 \mathbf{r} \to \mathbf{r}_4 - \mathbf{r}_1$ on the tetrahedron in Fig. 1(c), where the coordinate origin is at the vertex 1. Since there are four possible coordinate origins in a tetrahedron, the partial derivatives $\partial_a \mathbf{r}$ have several different discrete forms depending on the coordinate origin. Thus, summing over all possible forms of $\partial_a \mathbf{r}$, we have S_1 in Eq. (1).

3. Simulation Results

The model has a continuous transition of shape transformation between the spherical phase and the oblong phase shown in Figs. 2(a), (b), where the oblong direction in Fig. 2(b) is chosen spontaneously. The bending energy S_2/N_B' on the surface rapidly changes in the region of $\lambda \simeq 0.23$ in Fig.2(c), where N_B' is the total number of bonds on the surface. The specific heat for S_2 is defined by

$$C_{S_2} = (1/N_S)\langle (S_2 - \langle S_2 \rangle)^2 \rangle, \tag{7}$$

where the factor κ^2 is eliminated, and N_S denotes the total number of vertices on the surface. The peak value $C_{S_2}^{\rm max}$, growing with increasing N, indicates the existence of a phase transition.

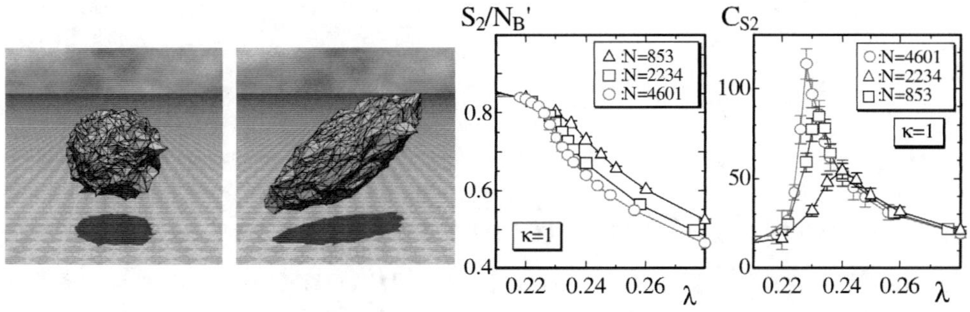

(a) (b) (c) (d)

Figure 2. Snapshots in (a) the isotropic phase at $\lambda=0.224$ and (b) anisotropic phase at $\lambda=0.54$, where $\kappa=1$ and $N=4601$. (c) The two-dimensional bending energy S_2/N_B' on the surface, and (d) the specific heat C_{S_2}.

The scaling coefficient for $C_{S_2}^{\max} \sim (N_S)^\alpha$ is $\alpha = 0.44 \pm 0.06$. This implies that the transition is of second order. This continuous transition resembles the one of the two-dimensional surface model [5]. Also the order of this transition is consistent with that of two-dimensional self-avoiding surface model in [6], although the model in [6] is fluid. The difference between the LCE surface and the membrane surface is that the former surface has neither the crumpled phase nor the branched polymer phase.

4. Summary

We have presented preliminary results of a Finsler geometry model for $3D$ liquid crystal elastomer (LCE). The anisotropic shape transformation, caused by the nematic transition of the liquid crystal, is understood as a continuous transition in the model. We expect that the dependence of the LCE length L on λ (or inverse temperature) and the soft elasticity [3] such as the strain vs. stress of a cylindrical LCE, the height of which is fixed, are obtained from the Finsler geometry model. Three dimensional energy for the rigidity of LCE can also be included in the Hamiltonian. The detailed information of the model and those MC data will be reported elsewhere.

This work is supported in part by the Grant-in-Aid for Scientific Research (C) Number 26390138. We acknowledge the support of the Promotion of Joint Research 2014, Toyohashi University of Technology. We are grateful to K. Osari and S. Usui for the computer analyses.

References

[1] M. Doi and F. Edwards, *The Theory of Polymer Dynamics*, (Oxford University Press, 1986).
[2] D. Nelson, in Statistical Mechanics of Membranes and Surfaces, Second Edition, edited by D. Nelson, T. Piran, and S. Weinberg, (World Scientific, 2004) p.1.
[3] V. Domenici, Prog. in Nucl. Mag. Res. Spec. **63** 1 (2012).
[4] H. Koibuchi and H. Sekino, Physica A **393** 37 (2014).
[5] M. Bowick, A. Cacciuto, G. Thorleifsson, and A. Travesset, Euro. Phys. J. E **5**, 149 (2001).
[6] H. Koibuchi and A. Shobukhov, Physica A **410**, 54 (2014).

Numerical analysis of the effect of the kind of activating agent and the impregnation ratio on the parameters of the microporous structure of the active carbons

Mirosław Kwiatkowski

AGH University of Science and Technology, Faculty of Energy and Fuels

al. A. Mickiewicza 30, 30-059 Krakow, Poland

kwiatkow@agh.edu.pl

Abstract. The paper presents the results of the research on the application of the *LBET* class adsorption models with the fast multivariant identification procedure as a tool for analysing the microporous structure of the active carbons obtained by chemical activation using potassium and sodium hydroxides as an activator. The proposed technique of the fast multivariant fitting of the *LBET* class models to the empirical adsorption data was employed particularly to evaluate the impact of the used activator and the impregnation ratio on the obtained microporous structure of the carbonaceous adsorbents.

1. Introduction

A range of activities have been undertaken in recent years with the aim of reducing the propagation of environmental pollution, while adsorptive processes and porous materials have been gaining in importance in the newly developed technologies. Carbonaceous adsorbents are among the most popular and commonly used groups of materials, mainly because their porous structure can be shaped in a number of ways, depending on the raw material chosen to make them and the production method applied as well as in terms of the control of the production process conditions and a substantial scope of modifying the chemical properties of their surface [1,2].

Carbonaceous adsorbents are usually produced by the carbonization and activation of materials of organic origin that are rich in carbon. Carbonization consists in the thermal processing of the raw material at about 500 to 800°C in the absence of air and without the involvement of chemical agents in the atmosphere of inert gases. As the process progresses, the organic matter is decomposed and the gaseous components are removed. As a result, an adsorptively inactive half-finished carbonization product is obtained, referred to as the carbonizate. The porous structure of the carbonizate is inadequate; in order to have it properly developed, the material must undergo activation [3].

Two types of activation are distinguished, namely physical activation and chemical activation. Physical activation consists in partially gasifying the remains of the carbonization process with oxidizing gases, namely carbon dioxide and water vapour, at increased temperatures [3-5]. The production of carbonaceous adsorbents by chemical activation consists in a one-off reaction of the raw material with any of the following activating agents: H_3PO_4, HNO_3, H_2SO_3, $ZnCl_2$, $MgCl_2$, KOH, or NaOH [6-9]. In that group, potassium hydroxide and sodium hydroxide are particularly noteworthy for their effect on the production of carbonaceous materials characterized by substantially developed porosity and a large real surface [10-12].

Chemical activation using potassium and sodium hydroxides proves to be highly beneficial as compared with physical activation with the use of steam or carbon dioxide, its major advantages being greater efficiency of the process and the fact that it is carried out in a single stage, lower temperatures and shorter time of the process, as well as a significantly enhanced development of the porous structure and the possibility to obtain a porous structure with a limited range of pore width. The disadvantages of chemical activation using potassium and sodium hydroxides include substantially higher process costs as compared with physical activation, the requirement to add the rinsing stage, and considerable corrosion of the hydroxides applied to the process.

The right choice of the activation process conditions for the given raw material which would ensure the production of microporous carbonaceous adsorbents suitable for the particular adsorptive process requires laborious tests and experiments. Still, for the experimental work to yield the anticipated results, mathematical tools have to be used with which the microporous structure of carbonaceous materials obtained under different conditions can be reliably verified. The tool in question should not only enable accurate evaluation of the structural parameters, but also permit an analysis of the microporous structure independently of the adsorbate used and an evaluation of the reliability of the parameters determined. Looking to meet these expectations, the author and his research team developed a group of mathematical adsorption models derived from *uniBET* theory, which is their antecedent universal adsorption theory [13-15]. Within the scope of the research presented in this paper, the said models were used to analyse the effect of the type of activating agent and the impregnation ratio on the parameters of the obtained porous structure of carbonaceous adsorbents.

2. The adsorption *LBET* class models

In the presented approach the adsorption is viewed as a clusterization of adsorbate molecules in pores of different shapes. Each cluster starts at a place (primary site) on a pore wall producing a local minimum of adhesion energy. Further molecules (the layers $n > 1$) are joined due to cohesive and adhesive forces [15-24]. The maximum number of layers k (referred to as the cluster size) is determined by a space preassigned to another cluster (competitive clusterization - 1st type) or by a geometry of pore (2nd type). Possible branching of clusters is taken into account that is expressed quantitatively by the pore shape factor β. It can significantly affect adsorption isotherms. The clusters have only one molecule in the first layer, i.e. only one original adsorption site on the surface can begin its formation, and each adsorbate molecule which belongs to the layer $n > 1$ can fill only one location on the layer n - 1. The areas assigned to the clusters are divided into classes in a way that the energy profiles within each class required to place a molecule on the successive layers n = 1, 2,..., are identical. Each place available on the layer n - 1 of given class can be occupied with the same probability, regardless of the condition of occupying the remaining places in the system - this way the interactions between particular cluster layers are taken into account.

In the proposed models, there exists a distinction between the two types of models. The first model type refers to the adsorption system in which the limitation in the number of layers results from competing physical adsorption. The second model type describes the systems in which limitations of cluster size result from pores size. Upon derivation of the *LBET* class models the following assumptions were taken into account [15-24]:

a) the layers $n > 1$ are homogeneous and energetic parameters B_C are the same for $n > 1$:

$$B_C = \exp\left(U_p(1 - 2 \cdot Z_{pp}) - Z_C \cdot Q_{cp}/RT\right), \tag{1}$$

where Q_{cp} is the molar adhesion energy in ideal adsorbent-adsorbate contacts, U_p is the molar cohesion energy of adsorbate and Z_{pp}, Z_C are the correcting factors.

The number of primary adsorption sites is expressed by the formula given below [15-24]:

$$m_{hAk} = m_{hA}(1 - \alpha)\alpha^{k-1}, \tag{2}$$

where α is the geometrical parameter of the porous structure, $\alpha \in \langle 0, 1)$, m_{hAk} is the amount of primary sites for the kth type clusters

c) the first layer physical adsorption energy is expressed in the following form:

$$Q_{A\kappa} = U_p - Z_{A\kappa} \cdot Q_{cp}, \tag{3}$$

where $Z_{A\kappa}$ is the factor uniformly distributed over a range depending on k, κ counts k^{th} type clusters of identical energy profile across the layers $n = 1, .., k$, $Q_{A\kappa}$ is the molar energy contributed by placing an adsorbate molecule at first layer of κth class clusters $Q_A = \min_{k=1}(Q_{A\kappa})$, Q_A is the the first layer physical adsorption energy and Z_A is the correcting factor of adsorbent-adsorbate contacts.

c) the distribution function related to the energetic parameter $B_{A\kappa} \in (B_{Ak}, B_{fk})$ is:

$$B_{Ak} \overset{def}{=} \exp\left(\frac{Q_A}{RT}\right)\exp\left(\frac{Q_{cp}}{RT}\right)^{Z_A(1-\zeta_{Ak})}, B_{fk} \overset{def}{=} \exp\left(\frac{Q_A}{RT}\right)\exp\left(\frac{Q_{cp}}{RT}\right)^{Z_A(1-\zeta_{fk})}. \tag{4}$$

The energetic parameters B_{Ak} and B_{fk} do not depend on k, at least for $k > 1 + d$, where $d = \{0,1\}$ is the binary variable where its value 1 provides for a more accurate variant of the model, with bimolecular clusters treated separately.

e) the coverage ratios θ_{kn} are the same for all $k > 1 + d$, $n > 1 + d$.

The heterogeneous adsorption model derived using the above assumptions assumes the following mathematical form, called the *LBET* class model [15-24]:

$$\frac{m_a}{m_{hA}} = (1-\alpha)\left\{1 - \frac{1}{\ln(B_A / B_{f1})} \cdot \ln\left(\frac{B_A + \pi}{B_{f1} + \pi}\right)\right\} +$$

$$+ d \cdot \alpha(1-\alpha)(1+\beta\theta_2)\left\{1 - \frac{1}{\ln(B_{A2} / B_{f2})} \ln\left(\frac{B_{A2}(1-\theta_2)^\beta + \pi}{B_{f2}(1-\theta_2)^\beta + \pi}\right)\right\} + \tag{5}$$

where: m_a is total adsorption [mmol/g], m_{hA} is total number of the adsorption primary sites, $m_{hA} = V_{hA}/V_a$, V_{hA} is the first adsorption layer capacity [mm^3/g], V_a is the molar volume of adsorbate [mm^3/mmol], θ_{kj} and θ are the coverage ratio of j-th - layer at k-th - type cluster and mean coverage ratio of layers $n > 1$ respectively, π is the relative pressure p/p_0, α and β are parameters of the geometrical and pore shape porous structure respectively. B_{Ak}, B_{fk} are both energetic parameters, Q_A is the first layer adsorption energy [J/mol] and B_C is the energetic parameter for adsorption layer $n > 1$ [15-24].

The *LBET* models involves five parameters: V_{hA}, Z_A, α, β, B_C which can be adjusted by fitting Eq. (5) to empirical adsorption isotherm, with a chosen variant of the surface energy distribution function [15-24].

4. Numerical calculations

The aim of the research was to analyse the possibilities of applying *LBET* class models and the fast multivariant procedure of adsorption system identification to perform a comparative evaluation of the effect of the activating agent and the impregnation ratio on the parameters of the microporous structure of carbonaceous adsorbents obtained by way of chemical activation. The calculations were carried out with the use of data taken from subject literature, namely the isotherms of adsorption of nitrogen on UA active carbons obtained from Spanish anthracites [12]. The fitting of the theoretical isotherm to the isotherms of adsorption of nitrogen on UA active carbons was performed using the method of non-linear optimization with limitations, which minimizes the smallest square fitting error. The presented analysis was carried out using the 30-variant version of the *LBET* class models in the calculations and the reliability of identification was assessed based on residual dispersion [15-24].

The results of the numerical analyses carried out for the isotherms of adsorption of nitrogen on the active carbons UA [12] were presented in a set of two figures. In the top right-hand figure, the grey dots '•' mark the spots of the fitted empirical adsorption isotherm, while the black solid line '-' marks the theoretical isotherm plotted according to the best-fitted variant of the *LBET* class model. The dotted line represents the theoretical coverage of the first adsorption layer. The heading of the top figure gives the name of the adsorption system, i.e. the name of the adsorbate (N_2) and the symbol of the adsorbent (UA); the subscript at the adsorbent symbol points to type of the activating agent used and the impregnation ratio. Next the name and number of the best-fitted *LBET* class model is given, preceded by a slash. The area of the top chart contains the parameters Q_A [J/mol], B_C, Z_A and the fitting error dispersion σ_e. The second figure presents adsorption energy distribution on the adsorbent surface; it contains the value of the volume of monolayer V_{hA} [mm^3/g], the values of the geometrical parameters α, β and the value of the parameter of heterogeneity h.

5 . Discussion of the obtained results

The results obtained during calculations carried out using computerized *LBET* class adsorption models for the isotherms of adsorption of nitrogen on active carbons produced from Spanish anthracites with sodium and potassium hydroxides as the activating agents [12] point to substantial discrepancies between the parameters of the microporous structure produced using the said activating agents at different impregnation ratios. As regards the analysis of active carbons obtained with the use of sodium hydroxide, it can be observed that an increase in the impregnation ratio results in an expansion of the volume of the first adsorption layer V_{hA} and a rise in the value of the geometric parameter α, which defines the height of the agglomerates of adsorbate particles formed in the micropores of the analysed material. Also, an increase in the impregnation ratio triggers a rise in the value of the energy parameter for the successive layers B_C.

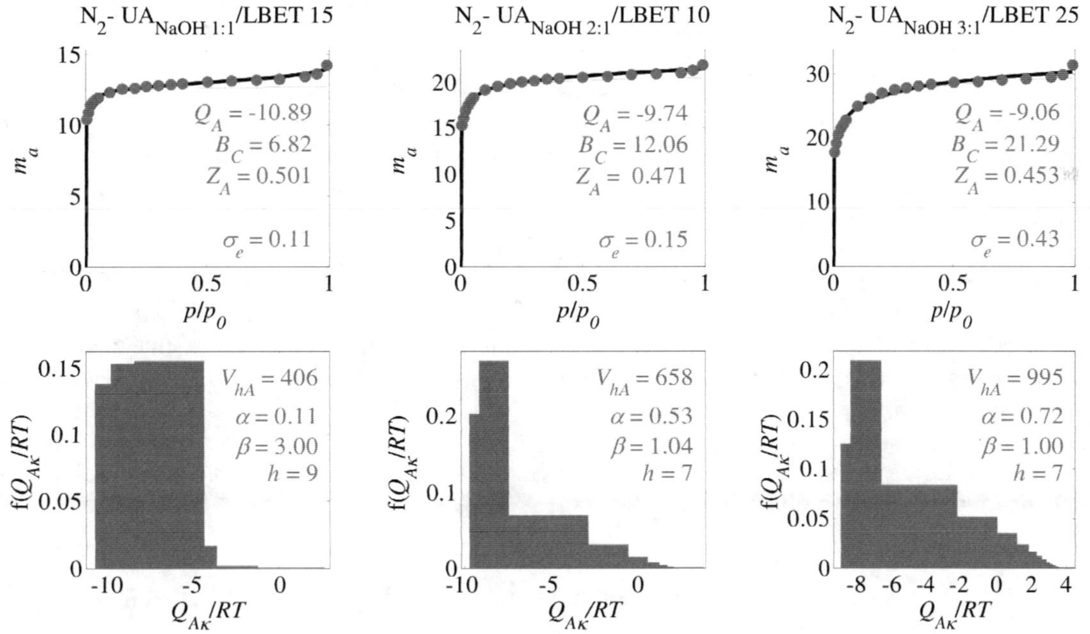

Fig 1. Multivariant identification of the isotherms of adsorption of nitrogen on UA activated carbon obtained from Spanish anthracite with varying NaOH to anthracite mass ratio [12].

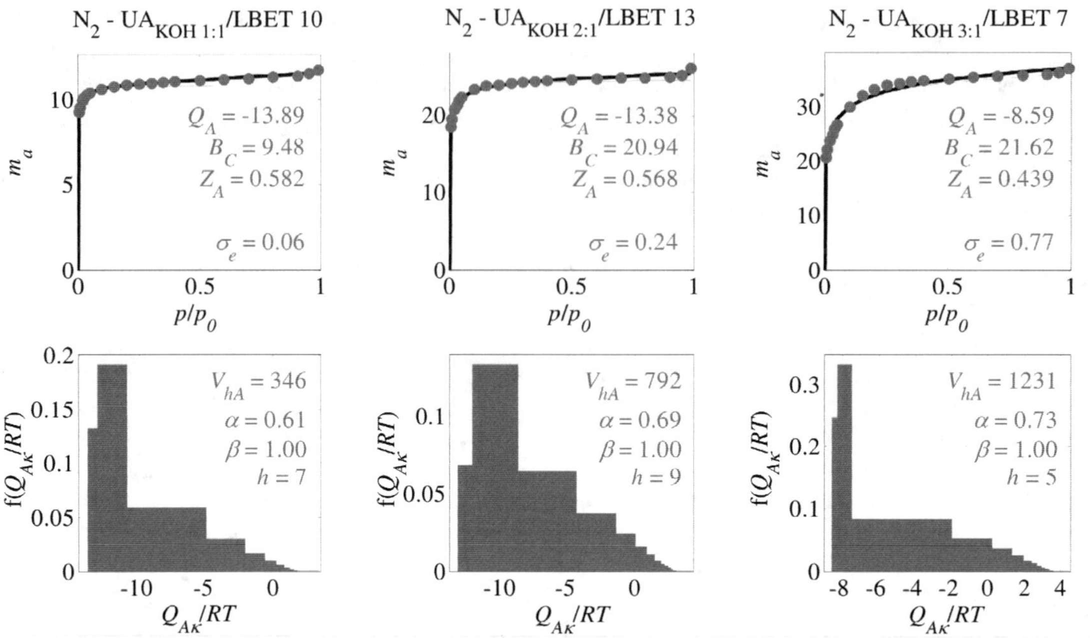

Fig 2. Multivariant identification of the isotherms of adsorption of nitrogen on UA activated carbon obtained from Spanish anthracite with varying KOH to anthracite mass ratio [12].

Moreover, it can be seen that a microporous structure with a lesser degree of heterogeneity, namely with the heterogeneity parameter $h = 7$, was obtained for the ratio of the mass of the activating agent to the mass of the activated substance, i.e. the impregnation ratio, exceeding 1:1. A material with the largest volume of the first adsorption layer was obtained for the NaOH impregnation ratio of 3:1. High, stack-like adsorbate particles are formed in the said material, which points to the occurrence of long and narrow micropores in it. When comparing the above results of analyses carried out for carbonaceous materials produced by way of chemical activation with sodium hydroxide (Figure 1) with the results obtained for active carbons produced by way of chemical activation with potassium hydroxide (Figure 2), one glance is enough to conclude that evidently larger volume parameters of the first adsorption layer were obtained for active carbons produced by way of chemical activation using potassium hydroxide. Also, for the active carbon obtained by way of chemical activation with the Spanish anthracite to potassium hydroxide ratio of 3:1 a microporous structure was formed that had the highest degree of homogeneity of all the analysed materials ($h = 5$) and the largest volume of the first adsorption layer, namely $V_{hA} = 1231$ [mm^3/g].

6. Conclusions

The work presents original results of a research into the use of *LBET* class models with an original method of fast multivariant fitting of theoretical models to empirical isotherms for evaluating the effect of the activating agent used and the impregnation ratio applied on the parameters of the microporous structure produced. On the basis of the research and analyses, a significant influence of the type of the activating agent used and the impregnation ratio applied on the formation of the microporous structure and, consequently, on the adsorptive properties of the produced adsorbate was observed. The author demonstrated that the effect of the conditions for producing carbonaceous adsorbents, namely in this case a substantial effect of the type of the activating agent and the impregnation ratio, can be analysed with success using the *LBET* class models along with the procedure of fast multivariant fitting of theoretical models to empirical adsorption isotherms.

The instruments presented in the article prove highly advantageous when compared with well-known and popular methods of microporous structure description by providing a wider range of information on the analysed microporous structure and offering unique possibilities of evaluating reliability of the obtained information on the analysed structure.

Acknowledgment. The research is led within the AGH University of Science and Technology in Krakow grant No. 11.11.210.217.

References:

[1] Jüntgen H 1977 *Carbon* **15** 273
[2] Scharff P 1998 *Carbon* **36** 481
[3] Kwiatkowski M 2008 *Microporous and Mesoporous Materials* **115** 314
[4] Abe I, Hitomi M, Ikuta N, Kawafune I and Kera Y 1996 *Carbon* **34** 1455
[5] Hayashi J, Kazehaya A, Muroyama K and Watkinson A P 2000 *Carbon* **32** 1873
[6] Kwiatkowski M and Broniek E 2012 *Przemysł Chemiczny* **91** 2355
[7] Evans M J B, Halliop E and MacDonald J A F 1999 *Carbon* **37** 269
[8] Benaddi H, Legras D, Rouzaud JN and Beguin F 1998 *Carbon* **36** 306
[9] Ahmadpour A and Do D D, 1997 *Carbon* **35** 1723
[10] Guo Y, Yu K, Wang Z and Xu H 2000 *Carbon* **41** 1645
[11] Lillo-Ródenas M A, Juan-Juan J, Cazorla-Amorós D and Linares-Solano A 2004 *Carbon* **42** 1371
[12] Lillo-Ródenas M A, Lozano-Castelló D, Cazorla-Amorós D and Linares-Solano A 2001 *Carbon* **39** 751
[13] Duda J T, Kwiatkowski M and Milewska-Duda J 2005 *Journal of Molecular Modeling* **11** 416
[14] Duda J T, Milewska-Duda J and Kwiatkowski M 2005 *Applied Surface Science* **252** 570
[15] Kwiatkowski M 2007 *Langmuir* **23** 2569
[16] Kwiatkowski M 2007 *Journal of Colloid And Interface Science* **313** 428
[17] Kwiatkowski M 2007 *Journal of Mathematical Chemistry* **42** 815
[18] Kwiatkowski M 2007 *Colloids and Surfaces A: Physicochem. Eng. Aspects* **294** 92
[19] Kwiatkowski M 2008 *Adsorption* **14** 37
[20] Kwiatkowski M 2008 *Journal of Molecular Modeling* **14** 183
[21] Kwiatkowski M 2008 *Przemysł Chemiczny* **87** 284
[22] Kwiatkowski M 2009 *Journal of Colloid and Interface Science* **330** 266
[23] Kwiatkowski M 2009 *Journal of Colloid and Interface Science* **340** 1
[24] Kwiatkowski M 2011 *Applied Surface Science* **257** 8912

Application of the clustering-based LBET class adsorption models to the analysis of the microporous structure of silica membranes

Mirosław Kwiatkowski[1], Magda Ziółkowska[2]

AGH University of Science and Technology, Faculty of Energy and Fuels

al. A. Mickiewicza 30, 30-059 Krakow, Poland

kwiatkow@agh.edu.pl

Abstract. This paper presents the application research results of the clustering-based *LBET* class adsorption models as a tool for the analysis of the porous structure of silica membranes. This research provides information on possible significant advantages that can be drawn while applying the new proposed method of microporous structure description in comparison with the conventional approaches. Furthermore, the obtained calculation results and additional information from adsorption isotherms analysis using *LBET* class models as well as their interpretation are shown. In particular, in the presented method it is possible to easily acquire not only information regarding the values of the obtained structure parameters, but also, more importantly, precise information on the reliability of these parameter calculations.

1. Introduction

A growing interest in gas phase separation with the application of ceramic membranes is mainly caused by the well-known stability of chemical and thermal parameters of this materials [1-2]. Furthermore, in comparison with polymeric membranes it turns out that ceramic membranes have higher gas flux and also high specific selectivity, which is related to pore size and occurs for the pores having less than 2 nm in diameter in accordance with the IUPAC definition [3-5]. Microporous silica, one of the commonly used ceramic materials, is chemically stable and does not expand while firing. Earlier studies concerning its catalytic properties also provide important additional information about the gas transport mechanism [6]. This makes silica an interesting material that can be used as a ceramic membrane.

Sol-gel processes are employed most commonly in order to obtain microporous silica. Two techniques can be distinguished within these processes: polymeric and particulate [6]. In the first technique, chainlike macromolecules are formed in the sol, which enables the preparation of low-cracking silica coatings having pore sizes within the range of micropore [4]. However, in order to better control pore volume and pore size distribution, the particle technique is commonly employed. In this technique the colloidal suspension of uniform particles is involved in membrane preparation containing uniformly distributed pores of a large volume. Moreover, in comparison with polymeric silica membranes, particulate membranes typically have higher gas flow and high surface cracking ratio. The formation of cracks is the result of tensile stress during liquid evaporation, which is difficult to avoid during the drying process. In many cases, the key step in avoiding these surface cracks in the course of silica membrane preparation is the application of coating process [7].

Practical application of silica membranes to the gas phase separation is related to their adsorption properties, which are directly resultant from their porous structure and chemical properties of the surface. In order to get information about these materials, porous structure adsorption techniques based on experimental adsorption isotherm measurements can be employed. However, the interpretation of the adsorption data requires an assumption of the mathematical model describing the adsorption equilibrium. A number of equations have already been proposed for this purpose. The ones used most commonly for the description of microporous materials include the *BET* equation [8] as well as *Dubinin-Raduskievitch* (*DR*) and *Dubinin-Astachow* (*DA*) equations [9,10]. Also, the *Density Functional Theory* (*DFT*) has recently become a strongly recommended tool for this purpose [11-14]. Nevertheless, our team developed an alternative approach to the analysis of microporous materials, employing the clustering-based adsorption theory (*uniBET*) with the *LBET* class models [15-26]. The *LBET* class models are applied to the examination of random porous structure adsorbents and can be viewed as an alternative to the classical techniques based on *BET* or *DR* equations and to the recently recommended *DFT*-based techniques [11-14].

2. Theoretical basis of the LBET class models

The *LBET* class adsorption models are based on the *uniBET* theory, described in our previous papers [15-19]. They are aimed at drawing information on pore structure and capacity by analysis of adsorption isotherms of small molecule adsorbates.

The heterogeneous adsorption *LBET* class models have the following form [20-25]:

$$\begin{aligned}\frac{m_a}{m_{hA}} &= (1-\alpha)\left\{1-\frac{1}{\ln(B_A/B_{f1})}\cdot\ln\left(\frac{B_A+\pi}{B_{f1}+\pi}\right)\right\}+\\ &+d\cdot\alpha(1-\alpha)(1+\beta\theta_2)\left\{1-\frac{1}{\ln(B_{A2}/B_{f2})}\ln\left(\frac{B_{A2}(1-\theta_2)^{\beta}+\pi}{B_{f2}(1-\theta_2)^{\beta}+\pi}\right)\right\}+\\ &+\alpha^{d+1}\left[d+(\beta\theta)^d\left(1+\frac{\beta\theta}{1-\alpha\beta\theta}\right)\right]\left\{1-\frac{1}{\ln(B_{A2+d}/B_{f\theta})}\ln\left(\frac{B_{A2+d}(1-\theta)^{\beta}+\pi}{B_{f\theta}(1-\theta)^{\beta}+\pi}\right)\right\},\end{aligned} \tag{1}$$

where: m_a is the total adsorption [mmol/g], m_{hA} is the total number of the adsorption primary sites, $m_{hA} = V_{hA}/V_a$, V_{hA} is the first adsorption layer capacity [mm^3/g], V_a is the molar volume of adsorbate [mm^3/mmol], θ_{kj} and θ are the coverage ratio of *j*-th - layer at *k*-th - type cluster and mean coverage ratio of layers $n > 1$ respectively, π is the relative pressure p/p_0, α and β are parameters of the geometrical and pore shape porous structure respectively. B_{Ak}, B_{fk} are both energetic parameters, Q_A is the first layer adsorption energy [J/mol] and B_C is the energetic parameter for adsorption layer $n > 1$ [20-25].

The *LBET* models involves five parameters: V_{hA} [mm^3/g], Z_A, α, β, B_C which can be adjusted by fitting Eq. (1) to empirical adsorption isotherm, with a chosen variant of the surface energy distribution function [20-25]. In the presented approach two types of adsorption at the top layer *k* are considered, corresponding to different nature of restrictions for the adsorbate clusters size represented by *k*. The first type adsorption, assuming the restrictions of formal nature, implies $\theta = \pi/(B_c(1-\theta)\beta-1)$ and may results in more compact or flat end large pores. For the second type adsorption (the restrictions are of geometrical nature, which is likely in small or narrow pores) the ratio θ averaged over the layers $n=2, \ldots, k$, $k=2+d, \ldots, \infty$ is taken [20-25].

The reliability of identification was assessed on the basis of residual dispersion σ_e and relative errors of parameters calculated for all 30 variants. To make possible a final choice of the best variant, the detailed results were presented for the three best fittings. Moreover, the 10 of well fitted variants from 30 ones were treated as acceptable ones. In order to get a measure of the identification uncertainty, the identifiability index w_{id} was defined [20-25]:

$$w_{id} = 1 - \frac{\sigma_{e\min}}{\frac{1}{10}\sum_{opt=1}^{10}\sigma_{e\,opt}} \tag{2}$$

where: σ_{emin} and σ_{eopt} denotes respectively dispersion error of the best fitting and the sequence of increasing dispersion errors of the ten well fitted models.

3. Numerical calculations

The purpose of the reported research was to evaluate the application of *LBET* class adsorption models to the porous structure analysis of crack-free unsupported silica membranes with different polyvinyl alcohol (PVA) and silica ratios on the basis of nitrogen isotherms at 77 K, obtained by Chu and Anderson [26]. The crack-free unsupported silica membranes with different PVA additive ratios were obtained by the sol-gel technique. Chu and Anderson employed the interfacial hydrolysis method to obtain silica sol by mixing tetraethylorthosilicate (TEOS) and tetramethylorthosilicate (TMOS) with ammonia solution (0.4 - 0.5 M silica concentration) [26]. After obtaining a homogenous mixture, sol was dialyzed and then diluted by adding nitric acid until the pH was 3, whereas PVA, used to improve support coating as an organic binder, was prepared by mixing PVA - water solution with previously obtained sol. Final unsupported silica membranes with different PVA/silica ratio by weight (10/100; 20/100; 30/100 respectively) were obtained after solvent evaporation and heating for 1 h at 500°C. The examination was concluded with the measurement of nitrogen adsorption isotherms at 77 K. Adsorbent preparation is described more precisely in paper [26]. The results of numerically calculated nitrogen isotherm for the best-fitted variants are illustrated in Figs. 1-3. Each figure presents the set of fitting results (subfigure in the first row) and adsorption energy distribution (subfigure in the second row) corresponding to the upper subfigure. The first row of subfigures presents the results of simultaneously fitted calculated isotherms (solid lines) to experimental data (marked with dots "∘"), and the theoretical coverage of the first adsorption layer is additionally presented (dotted lines). The headlines contain adsorbate and adsorbent name with PVA/silica ratio described in brackets, and the slash is followed by the best-fitted model name and the number of *LBET* class models. The values of identifiability index w_{id}, fitting error dispersion σ_e, fitted adsorption model type and variants (h, d, η) are also presented. The second row of subfigures presents adsorption energy distribution and the values of numerically calculated parameters Q_A, B_C, Z_A, V_{hA}, α and β.

4. Discussion of the results of numerical analysis

On the basis of the measurements of nitrogen adsorption isotherms on crack-free unsupported silica membranes with different PVA/silica ratios, it can be concluded that adsorption systems are well identifiable by way of the fitting of a *LBET* class model. It seems to confirm the abovementioned high applicability of the *LBET* class models with multivariant fitting procedure as a tool for the analysis of the microporous structure of silica membranes [26].

During the numerical calculations the dependence of preparation techniques of silica membranes on the obtained microstructure was considered. In particular, the PVA additive ratio influence on the formation of the microporous structure of silica membranes in sol-gel preparation techniques was considered. The obtained isotherm fitting results, presented in the subfigure in the first row (Fig. 1), show that silica membrane structure, with the PVA/silica ratio of 10/100 [26], is highly heterogeneous, as evidenced by the values of parameter $h = 7$. The porous structure consists of high stack-like adsorbate clusters ($\alpha = 1.00$, $\beta = 1.00$), whose enlargement is determined by a preassigned distance to another cluster – competing clusterisation, first type. Moreover, the subfigure showing adsorption energy distribution on the first layer (Fig. 1) indicates a wide spectrum of the primary adsorption sites.

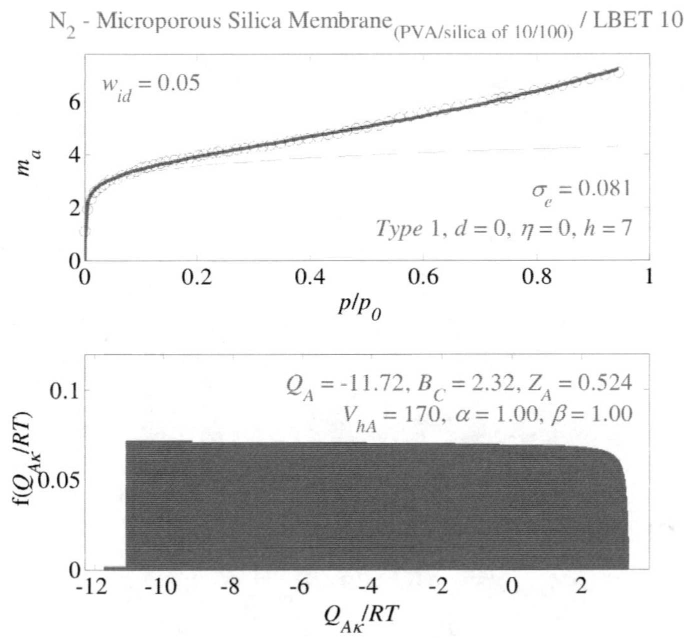

Fig. 1. The results of multivariant identification for nitrogen adsorption isotherms on crack-free unsupported silica membranes with PVA/silica ratio by weight of 10/100 [26]. The subfigure in the first row: fitted isotherm, the subfigure in the second row: adsorption energy distribution.

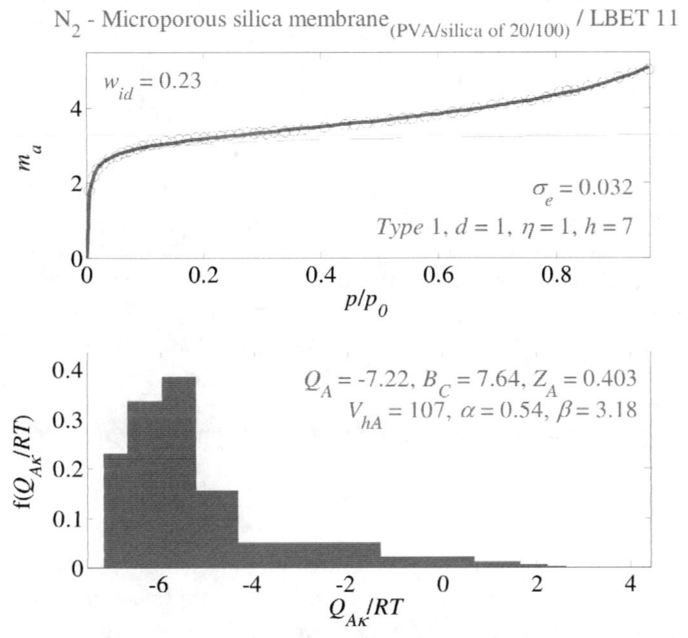

Fig. 2. The results of multivariant identification for nitrogen adsorption isotherms on crack-free unsupported silica membranes with PVA/silica ratio by weight of 20/100 [26].

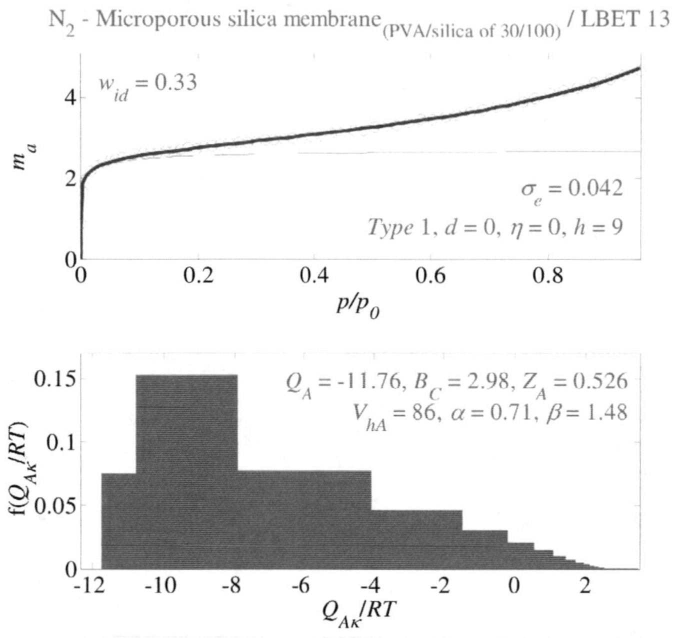

Fig. 3. The results of multivariant identification for nitrogen adsorption isotherms on crack-free unsupported silica membranes with PVA/silica ratio by weight of 30/100 [26].

An analysis of the results presented in the second figure, for silica membranes with the PVA/silica ratio of 20/100 [26], shows that – similarly to the case of the earlier results – silica membrane structure is highly heterogeneous (*h* parameter value is the same). However, further fitted parameter analysis indicates differences in the porous structure. The porous structure of the analyzed silica membranes consists of medium-sized branched adsorbate clusters, whose enlargement is determined by competing clusterisation ($\alpha = 0.54$, $\beta = 3.18$). It should be noticed that particular values of the parameter V_{hA} (first adsorption layer volume, $V_{hA} = 107$ [mm^3/g]) are smaller, as in the case of the previously analyzed sample. Also, the adsorption energy distribution on the first layer occurs with a narrower spectrum of the primary adsorption sites as compared with the first analyzed membrane (Fig. 2). Some interesting results were obtained for the third sample of silica membrane. It was found that higher PVA additive results in increasing surface heterogeneity ($h = 9$) and first adsorption layer volume reduction ($V_{hA} = 86$ [mm^3/g]). Particular values of the geometric parameters indicate the creation of high and branched adsorbate clusters ($\alpha = 0.71$, $\beta = 1.48$), in case of the silica membranes PVA/silica ratio of 30/100 [26]. First layer adsorption energy distribution subfigure (Fig. 3) indicates a wide spectrum of the primary adsorption sites, but with the dominant fraction of the highly energetic primary adsorption sites.

5. Conclusions

The results of the numerical analysis presented in this paper indicate complicated relationships between preparation techniques of crack-free unsupported silica membranes obtained at different PVA/silica ratios and microporous structure formation. Silica membranes structure identification requires an application of advanced numerical tools. Also, reliable identification was obtained while applying *LBET* class models. To sum up, the obtained results confirm the applicability of *LBET* class models to the silica membranes structure analysis as an efficient tool to evaluate surface capacity, monolayer adsorption energy, and outline pore geometry. Moreover, it gives an insight into the properties of the proposed identification technique presented in earlier papers by the author's team.

Acknowledgment: The research is led within the AGH University of Science and Technology grant No. 11.11.210.217.

References:

[1] Uhlhorn R J R, Keizer K and Burggraaf A J 1992 *J. of Membrane Sci.* **66** 271
[2] Bhave R R 1991 *Inorganic Membranes Synthesis, Characteristics and Applications* (New York Van Nostrand Reinhold)
[3] De Vos R M and Verweij H 1998 *J. Membrane Sci.* **143** 37
[4] Brinker C J, Ward T L, Sehgal R, Raman N K, Hietala S L, Smith D M, Hua D –W and Heedley T J 1993 *J. Membrane Sci.* **77** 165
[5] Sing K S W, Everett D H, Haul R A W, Moscou L, Pierotti R A, Rouquerol J and Siemieniewska T 1985, *Pure Appl. Chem.* **57** 603
[6] Brinker C J and Scherer G W 1990 *Sol-Gel Science: The physics and chemistry of sol-gel processing* (Boston Academic Press).
[7] Chu L, Tejedor-Tejedor M I and Anderson M A 1994 *Mat. Res. Soc. Symp. Proc.* **346** 855
[8] Brunauer S, Emmett P H and Teller E 1938 *J. Am. Chem. Soc.* **60** 309
[9] Dubinin M M 1960 *Chem. Rev.* **60** 235
[10] Dubinin M M and Astakhov V A, 1971 *Adv. Chem. Ser.* **102** 69
[11] Chmiel G, Lajtar L, Sokołowski S and Patrykiejew A 1994 *J. Chem. Soc., Faraday Trans.* **90** 1153
[12] Jagiełło J and Thommes M M 2004 *Carbon* **42** 1227
[13] Jagiełło J, Ansón A and Martínez M T 2006 *J. Phys. Chem. B* **110** 4531
[14] Duda J T, Jagiełło L, Jagiełło J and Milewska-Duda 2007 *J. Appl. Surf. Sci.* **253** 5616
[15] Milewska-Duda J and Duda J T 2002 *Appl. Surf. Sci.* **196** 115
[16] Duda J T and Milewska-Duda J 2002 *Langmuir* **18** 7503
[17] Duda J T and Milewska-Duda J 2005 *Langmuir* **21** 7243
[18] Duda J T, Kwiatkowski M and Milewska-Duda J 2005 *J. Mol. Model.* **11** 416
[19] Kwiatkowski M 2007 *Langmuir* **23** 2569
[20] Kwiatkowski M 2007 *J. Colloid Interface Sci.* **313** 428
[21] Kwiatkowski M 2007 *J. Math. Chem.* **42** 815
[22] Kwiatkowski M 2009 *J. Colloid Interface Sci.* **340** 1
[23] Duda J T, Kwiatkowski M and Milewska-Duda 2010 *J. Appl. Surf. Sci.* **256** 5243
[24] Kwiatkowski M 2011 *Appl. Surf. Sci.* **257** 8912
[25] Kwiatkowski M, Ziółkowska M, Milewska-Duda J and Duda J T 2014 *Applicability of the universal uniBET adsorption theory to silica gel structure description*: Science and Technology Conference "Mineral Sorbents" - Raw Materials, Energetics, Environmental Protection, Modern Technologies Proceedings pp 269
[26] Chu L and Anderson M A 1996 *J. Membrane Sci.* **110** 141

Journal of Physics: Conference Series **633** (2015) 012049 doi:10.1088/1742-6596/633/1/012049

Simulation of conductivity of polymer films on metal surface

OG Maksimova, AV Maksimov and AR Baidganov

Cherepovets State University, Pr. Lunacharskii 5, 162600, Cherepovets, Russian Federation

E-mail: og62@mail.ru

Abstract. In this paper, protective properties of polymer films are analyzed. The simulation is performed by means of the Monte-Carlo method on the basis of three-dimensional lattice model of polymer system with orientational interactions [1]. Initially, configuration of polymer system is calculated by the Metropolis algorithm taking into account the characteristics of the internal structure (constants of intermolecular interactions etc.), temperature regime and metal quality. Further, for the study of conductivity, the motion of charged particles within the proposed lattice model is investigated on the basis of the calculated configuration. The interaction energy of the oxygen atom with eight neighboring links of polymer chains and electric double layer on the metallic surface is accounted. The direction of movement of charged particles is calculated by the Monte-Carlo method according to the energy advantage of its position. This method allows to calculate the number of charged particles passing through the polymer film and reaching the metal sheet surface. The dependences of conductivity on temperature, film thickness, and distance between molecular layers are obtained. It is shown that there is an optimum density for the given film thickness at which it possesses protective properties. The adequacy of the developed mathematical models and calculated dependences are verified by comparison with laboratory data and production testing.

1. Introduction

The protective polymer film is necessary to eliminate interaction of metal surface with external oxygen ions. As a rule it is assumed that a thick polymer film is an insulator. However, in recent years, the interest to investigation of transport properties of thin films has arisen [2-4]. This interest is primarily associated with the synthesis of polymers having high conductivity. It is shown [5, 6] that the change in electrical conductivity of polymer films occurs due to change in thickness, pressure, density, etc. Therefore, the study of the influence of these characteristics of polymer coatings of metal is also the topical problem for creation of protection against corrosion.

2. Model

For the study of this problem, the Gotlib-Maksimov's models of ordered polymer systems are used [7], in which, on the one hand, an individual polymer chain with bending rigidity is chosen, and on the other hand, local interactions of the dipole type are taken into account. In these lattice models, the links of the polymer chains are localized in the nodes (*n, m, l*) of three-dimensional cubic "quasi-lattice". Figure 1 shows one of the planes (*n, m*) of the model. The position of a node is characterized by the set of three numbers *n, m* and *l*. Every node is related to a dipole (flexible or rigid kinetic unit of polymer chain); its orientation is determined by angles in the spherical coordinate system.

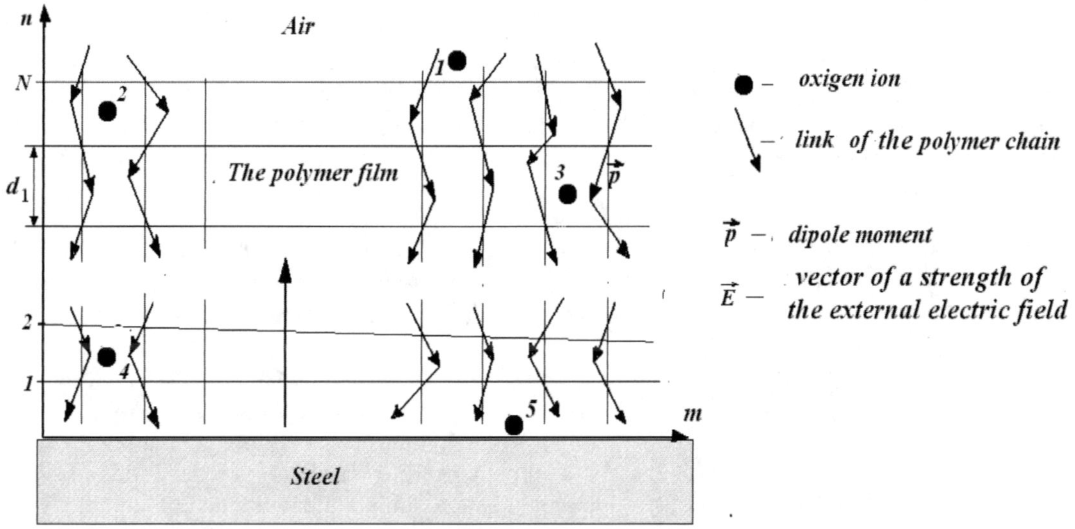

Figure 1. The movement of oxygen ion within the protective polymer film.

The potential energy of interaction of dipole type for the polymer chain system in the lattice has the form [8]:

$$V = -\sum_{n,m,l} K_1 \cos(\Phi_{n,m,l,n+1,m,l}) - \sum_{n,m,l} K_2 \cos(\Phi_{n,m,l,n,m+1,l}) - \sum_{n,m,l} K_2 \cos(\Phi_{n,m,l,n,m,l+1}) + pE \sum_{n,m,l} \cos(\phi_{n,m,l}), \quad (1)$$

where $\Phi_{n,m,l,n',m',l'}$ is the spatial angle between the dipoles located in the neighboring nodes n, m, l and n', m', l' of the lattice, p is the module of the dipole moment of the polymer chain link, E is the strength of the external electric field, that creates an electric double layer of the metal surface, $\varphi_{n,m,l}$ is the angle between the vectors $\vec{p}$ and $\vec{E}$.

The energetic constant K_1 along the longitudinal curved direction X of the "quasi-lattice" describes intrachain orientational interactions; it is related to the chain bending rigidity, which determines the persistence length and the mean cosine of the angle between neighboring units of the individual chain. The constant K_2 characterizes orientational interactions between neighboring chains in the "transverse" directions m and l of the "quasi-lattice". The constant K_2 is determined by the specific chemical structure of the interacting polymer chains. Its value can be estimated, for example, from the multipolar expansion of the interaction energy of two identical dipoles p in neighboring chains:

$$K_2 = \frac{1}{4\pi\varepsilon_0} \frac{p^2}{r^3}, \quad (2)$$

where r is the mean distance between neighboring chains, ε_0 is dielectric constant. The energetic constants are determined by the chemical structure of the polymer coating and its density.

In the model developed in this paper, the oxygen ions are initially at the interface "polymer – air" (ion 1 in Figure 1). It is assumed that the motion of the charged particles in this lattice model is carried by leaps in localized states in the centers of the cells (ions 2-4 in Figure 1). Similar approach has been used previously in the ref. [9]. For example, it has attributed the increase of conductivity of the layer by increasing the temperature and the frequency of the external electric field. The number of oxygen ions that pass through the polymer film and reach the metal surface (ion 5 in Figure 1) is calculated to determine the protective properties of the film. The interaction potential between the oxygen ion and the dipole is determined by the formula [10]:

$$V_{ion} = -\frac{qp \cdot \cos\alpha}{4\pi\varepsilon_0 r^2}, \tag{3}$$

where q is the charge of oxygen ion ($q = -2e$, where e is the charge of electron), p is the module of the dipole moment of the polymer chain link. The distance r and the angle α are shown in Fig. 2. The ion interactions with more than eight links of the polymer chains and the field of the electric double layer of metal (Figure 2) are taken into account in the simulation process of corrosion-resistance.

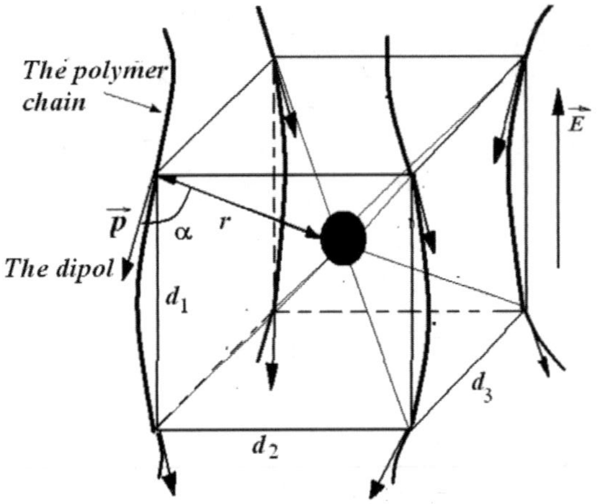

Figure 2. The location of the oxygen ion within the unit cell

3. Simulation method

The algorithm for the simulation of the full configuration of the polymer system on the metal surface and the motion of charged particles, takes into account all the features of occurrence of these phenomena and consists of two stages:

Stage1. The calculation of polymer chain links configuration depending on temperature, its inner structure (energetic constant of intermolecular interactions), and influence of the metal substrate is performed by the Monte Carlo method. The calculated values of the orientational order parameter ($\mu = \langle \cos(\phi_{i,j,k}) \rangle$) coincide with the experimental data on the investigation of thermal depolarization of the poly (vinylidene fluoride) [11]. Calculations of heat capacity, susceptibility and other macroscopic characteristics of polymer systems show that all of them have abnormal behavior at the same temperature, namely in the point of phase transition [1].

Stage 2. The study of the movement of oxygen ions within the simulated macromolecular (polymer) system is also carried also by the Monte Carlo method. The selection criterion of the direction of the ions motion is based on checking the energy advantage of its position in each of six neighboring lattice places (Figure 3). The new position of the ion is determined by the following algorithm:

1. The ion energy $V_{ion\,0}$ is determined if it is located at the point O.

2. The virtual displacement of the ion into each of the points O_s (s = 1, 2...6), i.e. in six neighboring cells, is produced (Figure 3).The energy of the ion is calculated in each of these positions by the formula (3).

3. The transition probability of the ion from the point O to each of the points O_s is calculated by the formula:

$$W_s = \exp\left(-\frac{V_{ion\,s} - U_{ion\,0}}{k_B T}\right). \tag{4}$$

4. A unit segment is divided into six parts. The length of each segment is equal to the value $W_s / \sum_{s=1}^{6} W_s$. Further, a random number R in the interval [0; 1] is selected by a uniform distribution. Its number determines the new position of the ion.

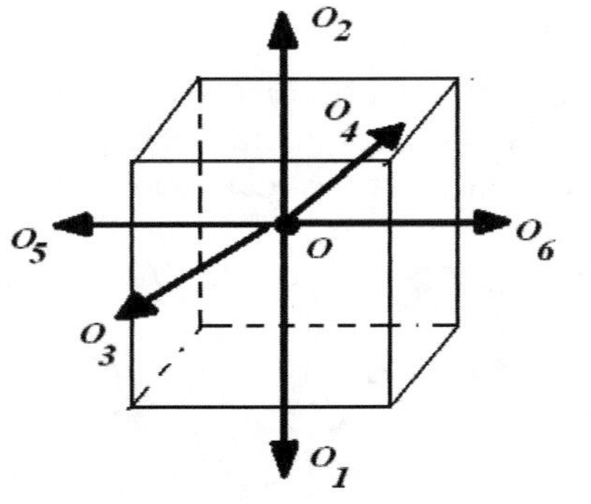

Figure 3. Possible motions of the oxygen ion

To characterize the anticorrosive properties of the coating, the corrosion parameter is introduced $\sigma = N_{cor} / N_{MC}$, where N_{cor} – the number of ions reached the metal surface and trapped by it, N_{MC} is the number of Monte Carlo steps. The calculations show that the value σ does not depend on the simulation parameters (initial conditions and the number of Monte Carlo steps). At sufficiently large number of steps, the corrosion parameter is determined only by the constants of the interaction potential and the value of the electrical field of double layer on the metal surface. Therefore, in this paper, the criterion $\sigma = 0$ (when corrosion is absent) is accepted as the main indicator of the polymer coating quality.

4. The simulation results

Figure 4a demonstrates the dependences of parameter of corrosion σ on the renormalized temperature (the value $k_B T / K_1$) at different values of the distance d_1 between the layers of the polymer coating. It is shown, that the number of the oxygen ions that have passed through the polymer film and have reached the metal surface, increases with the rise of temperature up to a certain critical one. The same behavior of the conductivity of polymers has also been observed experimentally, for example, in ref. [12]. The comparison of the simulation results with the data obtained in the ref. [1] show, that the temperature at which the value σ reaches its peak, is the point of structural phase transition. Beyond it all properties of the coating dramatically change (e.g., strength and hardness reduce). However, it should be noted, that this model has great restrictions for the study of macromolecular systems at temperatures above the phase transition point T_c , because the mobility of segments of sufficiently large chains cannot be considered in this model.

Figure 4a shows the dependences of the corrosion parameter of σ on the polymer film thickness N at different values of the distance d_1 between the layers of the polymer coating. The value of the parameter σ drops till zero with the increase of the polymer film thickness. At every value of the distance d_1, there exist the minimum number $N_{\min}$ of the coating layers, at which the oxygen ions stop to penetrate to the metal surface. In this case, the polymer film with thickness $N > N_{\min}$ becomes a corrosion protective barrier.

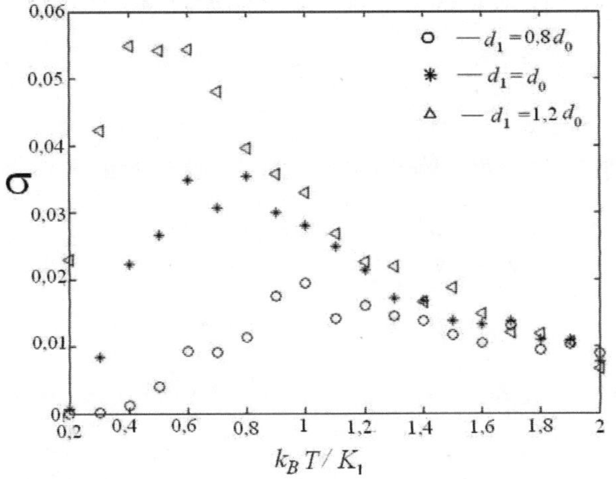

a

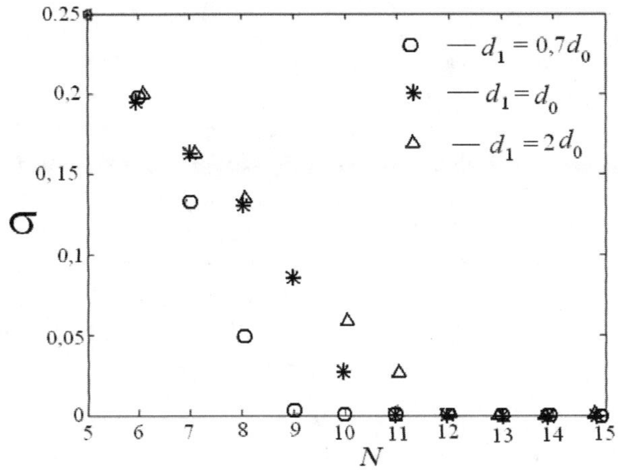

b

Figure 4. The parameter of corrosion σ vs. the renormalized temperature $k_B T / K_1$ (*a*) and the number of layers in the polymer film N (*b*) at different values of the distance d_1 between the layers.

5. Results of experimental study

The test samples were selected to investigate the corrosion properties of the coated metal (polyester enamel). Two incisions at 90 ° were made on each test sample. Further, the test samples were placed into saline mist chamber for a certain time. The results of corrosion resistance of the coating after 240, 360 and 500 hours have been obtained.

The test sample № 1. The coating thickness is 18.3 microns. Result: the polymer film is a protective coating.

The test sample № 2. The coating thickness also is 18.3 microns, but the density of the polymer film is decreased by 1.5% (the distance between the molecular layers is slightly increased). According to computer simulation, this density value of the polymer film is optimal. Result: at further decrease of the polymer film density it is not a protective cover.

The test sample № 3. The coatings thickness is 18.7 microns. The density is decreased by 3%, i.e. it is less than the optimal one. Therefore, though this coating is thicker, it is not a protective coating.

6. Conclusion

Simulation model is developed that allows to calculate the number of oxygen ions that penetrate through the protective film. Anti-corrosion properties of the polymer coating decrease if the temperature or the distance between the molecular layers increases. For the given value of intermolecular distance there is an optimal number of layers, by which the film becomes a protective coating. These results have not only theoretical but also practical importance for the production of polymer coating of metal, because its cost depends on the coating thickness and density.

Acknowledgements

This work was supported by the Ministry of Education and Science of the Russian Federation, project №1715 (contract № 2014/267)

References

[1] Petrova T, Maksimova O, Maksimov A, Nastuljavichus A, Baruzdina O, Baidganov A 2013 *Technical Proceedings of the NSTI Nanotechnology Conference and Expo, NSTI-Nanotech (Washington, USA, 12-16 May 2013)* pp. 611-614

[2] Shirakawa H 2002 *Synthetic Metals* **125(1)** 3-10

[3] Hill I G, Kahn A, Comil J, dos Santos D A, Bredas J L 2000 *Chem. Phys. Lett.* **317(3)** 444-450

[4] Hoppe H., Sariciftci N.S 2004 *J.Mater. Res.* **19(7)** 1921-1945.

[5] Yoshino K, Yun M S, Ozaki M, Kim S N, Inuishi Y 1983 *Japanese Journal of Applied Physics* **22(10)** 1510-1514.

[6] Naber R C G, Tanase K, Blom P W M, Geiinck G H, Marman A W, Touwslager F J, Setayesh S, Leeuw de D M 2005 *Nature Materials* **4(3)** 243-248.

[7] Gotlib Yu Ya, Maksimov A V 1992 *Polymer Science. Series A* **34(10)** 902

[8] Maksimova O G, Petrova T O, Maksimov A.V 2013 *Ferroelectrics* **444(1)** 111-115

[9] Kausch H H 1978 *Polymer fracture* (New York: Springer-Verlag)

[10] Kleman M and Lavrentovich O D 2003 *Soft matter physics* (Berlin: Springer)

[11] Petrova T O, Maksimova O G, Gerasimov R A and Maksimov A V 2012 *Physics of the Solid State* **54(5)** 937-939

[12] Ilyasov V Kh., Lachinov A N, Moshelev A V, Ponomarev A F 2008 *Physics of the Solid State* **50(3)** 568-572

Compactly supported kernels method of approximate particular solutions for solving elliptic problems

Marjan Uddin
Department of Basic Sciences and Islamiat, University of Engineering and Technology Peshawar, Pakistan
E-mail: marjankhan1@hotmail.com

Gul Shed
Department of Basic Sciences and Islamiat, University of Engineering and Technology Peshawar, Pakistan
E-mail: gul.shed@yahoo.com

Kamran
Department of Basic Sciences and Islamiat, University of Engineering and Technology Peshawar, Pakistan
E-mail: kamrank44@gmail.com

Amjad Ali
Department of Basic Sciences and Islamiat, University of Engineering and Technology Peshawar, Pakistan
E-mail: amjad_puet@yahoo.com

Abstract. In this work the method of approximate particular solutions using compactly supported kernels is investigated. In the work of [1] the globally supported radial kernels for the construction of solution are used, and it is observed that for large scaled PDEs the differentiation matrix is ill-conditioned. We extended the work of [1] for compactly supported kernels, in order to solve large-scaled PDEs engineering sciences. The numerical scheme of the present method of approximate particular solutions is very accurate and simple in implementation. Three benchmark problems are solved by the present numerical scheme and the results are compared to other methods in the literature.

1. Introduction

For the approximate solution of various types of PDEs, the method of approximate particular solution (MAPS) is recently developed in [1]. In this approach the approximate particular solution is obtained in such a way to satisfy the differential equation as well as the boundary conditions [2]. In this numerical approach very simple and accurate kernel based numerical scheme is obtained. In this approach, due to the radial symmetry of the special differential operator, the integration is performed in the radial direction. In the present method due the global structure the system matrix is dense and ill-conditioned, and is unable to solve large-scaled problems.

The finite element and finite difference methods are local in nature, and are very effective in solving large-scaled problems. Like these mesh-dependent methods there are also some kernel based methods which are local in their construction and are very effective for solving large system in irregular domains [3–7].

Apart from these methods there are also several methods to overcome this difficultly such as domain decomposition [8], the greedy algorithm [9], the extended precision arithmetic [10], the improved truncated singular valued decomposition [11], iterative methods [12], fast multipole expansion techniques [13].

Our purpose is to extend the MAPS for solving large-scaled problems. In our work we have adopted the idea of using locally supported kernel functions resulted a localized formulation in the construction of MAPS. In our procedure, we approximate the solution by a linear combination of locally supported kernels. Contrary to global method, the local method have sparse system matrix which can be solved easily and efficiently.

2. Description of the method

The idea to extend the method of approximate particular solution (MAPS) in [1] for locally supported radial kernel is similar to the construction of the DRBEM [14], in which the operator of Laplacian is retain as main operator on the left, and all the other terms are shifted to right side. We consider the following elliptic partial differential equation in 2D

$$\Delta u = f(\xi,\eta), (\xi,\eta) \in \Omega, \tag{1}$$

$$\mathcal{B}u = g(\xi,\eta), (\xi,\eta) \in \partial\Omega, \tag{2}$$

where Δ, is a linear differential operator and $\mathcal{B}$ is a boundary differential operator, and $f(\xi,\eta)$ and $g(\xi,\eta)$ are known functions. Using radial kernels, an approximate particular solution to (25)-(27) is given by

$$u = \sum_{i=1}^{n} \lambda_i \Phi(r_i), \tag{3}$$

where $r_i = \|(\xi,\eta)-(\xi_i,\eta_i)\|$ and $\{(\xi_i,\eta_i)\}_{i=1}^{n}$ are called the centers or trial points. In the method of approximate particular solution the equation defined by

$$\Delta\Phi = \phi, \tag{4}$$

can be solved analytically for kernel Φ in (19) for a given radial kernel ϕ [15]. Our aim is to solve this equation for locally supported radial kernels.

It was proved in ([16], Lemma 9.15, p. 130) that the transition $\Phi \to -\Delta\Phi$ allows to generate new kernels under certain circumstances. Here, the construction explicit and transparent for the standard families of radial kernels. In this construction the radial form of the Laplacian are apply it to any kernel to generate new kernels.

If we write a radial kernel K in f-form with $s = r^2/2 = \|\xi-\eta\|_2^2/2$, its d-variate Laplacian follows via

$$\frac{\partial}{\partial \xi_j} K(\xi-\eta) = \frac{\partial s}{\partial \xi_j}\frac{d}{ds} f(s) = f^{'}(s)(\xi_j-\eta_j) \tag{5}$$

$$\frac{\partial^2}{\partial \xi_j^2} K(\xi-\eta) = f^{''}(s)(\xi_j-\eta_j)^2 + f^{'}(s) \tag{6}$$

$$-\Delta K(\xi - \eta) = -2sf''(s) - 2df'(s) \tag{7}$$

the radial kernel defined by $-\Delta\Phi$ is positive definite, if Φ is positive definite. For conditionally positive definite Φ of order m, the same argument applies, but the order of conditional positive definiteness of of $\Delta\Phi$ then is $m-1$ because of the new factor in the Fourier transform.

Theorem 1. [17] *The transition on radial kernels generates a radial kernel consisting of a weighted sum (31)-(7) of two radial kernels, if f is the f-form of, and if the action of $-\Delta$ is valid on the kernel. If, furthermore, the class of kernels is invariant under taking pairs of forward and backward Fourier transforms in arbitrary dimensions. the resulting kernel is a weighted linear combination of two radial kernels of the same family.*

In the present study we choose the locally compactly kernel

$$\Phi(r, \varepsilon) = (1 - \varepsilon r)_+^8 (32(\varepsilon r)^3 + 25(\varepsilon r)^2 + 8\varepsilon r + 1) \tag{8}$$

then we recover generate the new kernel

$$\Delta\Phi(r, \varepsilon) = 44\varepsilon^2(1 - \varepsilon r)_+^6 (88(\varepsilon r)^3 + 3(\varepsilon r)^2 - 6\varepsilon r - 1) \tag{9}$$

From equation (30), we have

$$\Delta u = \sum_{i=1}^{n} \lambda_i \Delta\Phi(r_i) = \sum_{i=1}^{n} a_i \phi(r_i) \;\; in \;\; \Omega \tag{10}$$

putting these values in (25)-(27), we get

$$\sum_{i=1}^{n} \lambda_i \phi(r_i) = f(\xi, \eta) \;\; (\xi, \eta) \in \Omega \tag{11}$$

$$\sum_{i=1}^{n} \lambda_i \mathcal{B}\Phi(r_i) = g(\xi, \eta), (\xi, \eta) \in \partial\Omega \tag{12}$$

For the numerical implementation, we need to choose two sets of interpolation points as shown in Figure 1. We let n_i be the number of interior points , $\{(\xi_j, \eta_j)\}_1^n$ and n_b the number of boundary points, $\{(\xi_j, \eta_j)\}_{n_i+1}^{n_i+n_b}$,. Furthermore, n denotes the total number of points; i.e., $n = n_i + n_b$, By collocation method, we have

$$\sum_{i=1}^{n} \lambda_i \phi(r_{ij}) = f(\xi_j, \eta_j), \;\; 1 \le j \le n_i \tag{13}$$

$$\sum_{i=1}^{n} \lambda_i \mathcal{B}\Phi(r_{ij}) = g(\xi_j, \eta_j), \;\; n_i + 1 \le j \le n \tag{14}$$

where, $r_{ij} = \|(\xi_j, \eta_j) - (\xi_i, \eta_i)\|$. The above system of equations can be easily solved by standard matrix solver. Once $\{\lambda_i\}_1^n$ is determined, the approximate particular solution becomes the approximate solution u of (25)-(27); i.e.,

$$u(\xi, \eta) = \sum_{i=1}^{n} \lambda_i \Phi(r_i) \tag{15}$$

Problem 1. We consider the Poisson problem with the Dirichlet boundary condition:

Journal of Physics: Conference Series **633** (2015) 012050 doi:10.1088/1742-6596/633/1/012050

$$\Delta u(\xi, \eta) = -\pi^2 \left(\eta \sin(\pi\xi) + \xi \cos(\pi\eta)\right), \quad (\xi, \eta) \in \Omega. \tag{16}$$

$$u(\xi, \eta) = \eta \sin(\pi\xi) + \xi \cos(\pi\eta), \quad (\xi, \eta) \in \partial\Omega. \tag{17}$$

The parametric equation of the computational domain is given by

$$\Omega = \{(\xi, \eta) | \xi = \rho \cos\theta, \eta = \rho \sin\theta, 0 \leq \theta \leq 2\pi\}, \tag{18}$$

where

$$\rho = \left(\cos(3\theta) + \sqrt{2 - \sin^2(3\theta)}\right)^{1/3}. \tag{19}$$

The analytic solution is given by s

$$u(\xi, \eta) = \eta \sin(\pi\xi) + \xi \cos(\pi\eta) \tag{20}$$

We solved the problem by MAPS using compactly supported radial kernels (8)-(9). The computational domain considered for this problem and the analytic solution in the extended domain are shown in Figures 1-2 respectively. We compute the L_∞ and the RMS error norms using CS-MAPS for different values of interior and boundary nodes. It can be seen from Figures 3-4, that with the increase in c, the sparsity of the differentiation matrix increases and at the same time the accuracy decreases. We incorporate more and more interpolation nodes with a small support, but at the expanse of decrease in optimal accuracy. It is demonstrated from the table that these locally supported kernel-based MAPS is capable of solving large scaled Poisson's problems with reasonably acceptable accuracy.

Table 1. Comparison of the two method in terms of RMSE and L_∞ for compactly supported kernels.

n_b	20	40	80	120	1000
n_i	71	117	211	295	4832
CS-MAPS					
Support c	0.5	0.5	0.5	0.5	2
RMSE	6.349e-003	1.411e-003	2.171e-004	1.252e-004	1.026e-004
L_∞	3.124e-002	4.341e-003	1.801e-003	1.077e-003	8.920e-004

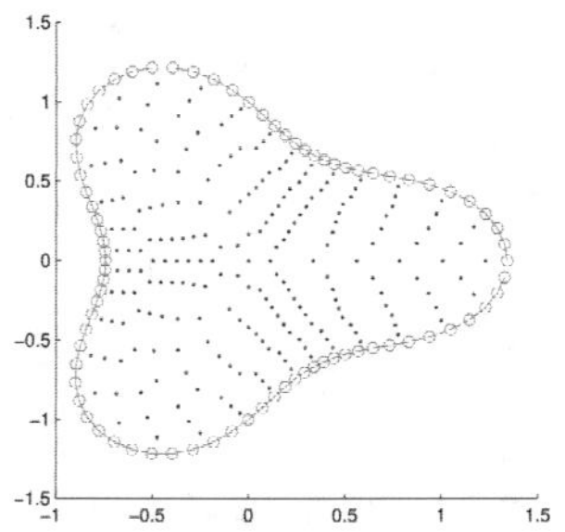

Figure 1. Plot of computational domain for problem 1.

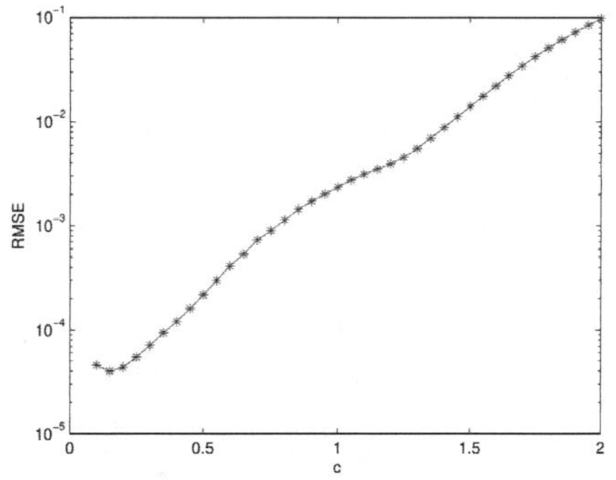

Figure 2. Plot of RMSE versus shape parameter for problem 1

Problem 2. In this problem we consider the inhomogeneous modified Helmholtz equation:

$$(\Delta - \lambda^2)u(\xi, \eta) = f(\xi, \eta), \quad (\xi, \eta) \in \Omega. \tag{21}$$

$$u(\xi, \eta) = g(\xi, \eta), \quad (\xi, \eta) \in \partial\Omega. \tag{22}$$

where $f(\xi, \eta)$ and $g(\xi, \eta)$ are chosen according to the following analytical solution The parametric equation of the computational domain is given by

$$\Omega = \{(\xi, \eta)|\xi = \cos\theta, \eta = \sin\theta, 0 \leq \theta \leq 2\pi\}, \tag{23}$$

where

The analytic solution is given by

$$u(\xi, \eta) = \sin(\pi\xi)\cosh(\eta) + \cos(\pi\xi)\sinh(\eta) \tag{24}$$

We solved this problem in a unit disk with CS-MAPS. For this problem the computational domain and the analytic solution in the extended domain are shown in Figures 5-6. We used the $\lambda = 10$, and different number of interior and boundary nodes. It shout be noted that Helmholtz-type equations with large value of λ are mostly difficult. But with CS-MAPS the solution accuracy is excellent for large values of λ. Further more we note that a very large as well as a very small number of nodes can be used to achieve reasonably good accuracy. Here we solve $(\Delta - \lambda^2)\Phi = \phi$, for the radial kernel given in equation (8).

Table 2. Comparison of the two method in terms of RMSE and L_∞ for compactly supported kernels.

n_b	50	100	150	200	500
n_i	170	295	436	933	5778
CS-MAPS					
Support c	1	1	1	1	4
RMSE	1.690e-003	7.285e-004	3.762e-004	9.741e-005	3.224e-003
L_∞	8.033e-003	3.939e-003	2.172e-003	5.904e-004	4.839e-004

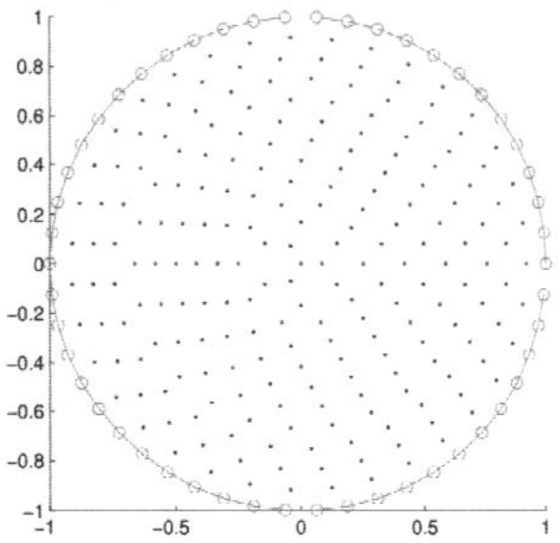

Figure 3. Plot of computational domain for problem 2.

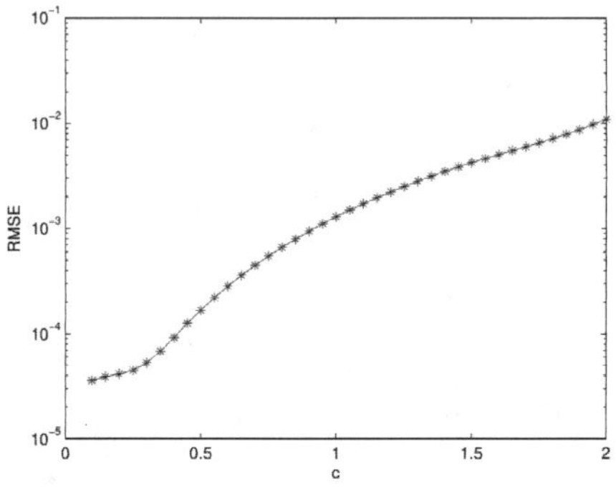

Figure 4. Plot of RMSE versus shape parameter for problem 2, $n_i = 211,\ n_b = 50$

Problem 3. In this problem we consider the equation:

$$\Delta u(\xi,\eta) = f(\xi,\eta), \quad (\xi,\eta) \in \Omega. \tag{25}$$

$$u(\xi,\eta) = \sin(\eta^2+\xi) - \cos(\eta-\xi^2), \quad (\xi,\eta) \in \partial\Omega_1. \tag{26}$$

$$\frac{\partial u(\xi,\eta)}{\partial \mathbf{n}} = (\nabla(\sin(\eta^2+\xi) - \cos(\eta-\xi^2)))\cdot\mathbf{n}, \quad (\xi,\eta) \in \partial\Omega_2. \tag{27}$$

where $f(\xi,\eta)$ is chosen according to the following analytical solution The parametric equation of the computational domain is given by

$$\Omega = [0,1]^2, \tag{28}$$

$$\partial\Omega_1 = \{(\xi,\eta) \mid \xi = 0,\ 0 \le \eta \le 1,\ \wedge\ \xi = 1,\ 0 \le \eta \le 1\} \tag{29}$$

$$\partial\Omega_2 = \{(\xi,\eta) \mid \eta = 0,\ 0 \le \xi \le 1,\ \wedge\ \eta = 1,\ 0 \le \xi \le 1\} \tag{30}$$

where The analytic solution is given by

$$u(\xi,\eta) = \sin(\eta^2+\xi) - \cos(\eta-\xi^2) \tag{31}$$

We solved this problem in a unit square $[0,1]^2$ with mixed boundary conditions. Along the boundary Ω_1 we applied Dirichlet boundary condition, while along the boundary Ω_2 we used Neumann boundary condition. We used here again the locally supported radial kernel (8). The results are shown in Table 3, and Figures 11-12, which demonstrate the capability of CS-MAPS with a very small as well as large number of nodes with mixed boundary conditions.

Table 3. Comparison of the two method in terms of RMSE and L_∞ for compactly supported kernels.

n_b	76	96	116	136	316
n_i	324	529	784	1089	6084
CS-MAPS					
Support c	0.6	0.6	0.6	0.6	2
RMSE	2.541e-004	1.106e-004	5.827e-005	2.646e-005	1.784e-004
L_∞	8.682e-004	5.183e-004	3.636e-004	1.760e-004	2.317e-003

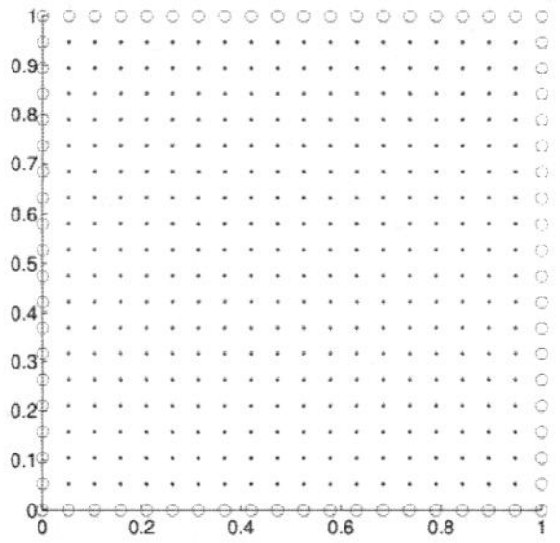

Figure 5. Plot of computational domain for problem 2.

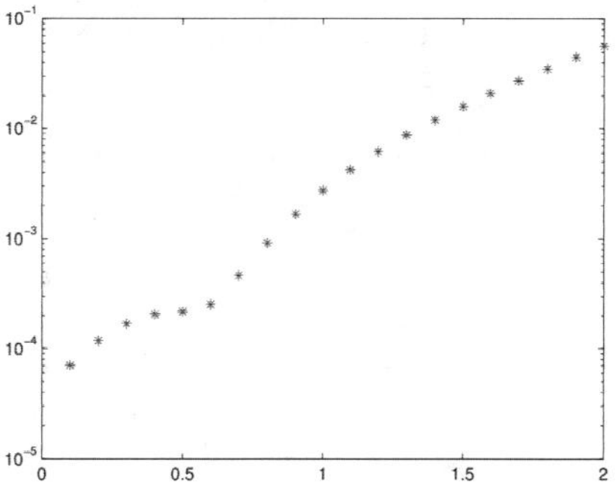

Figure 6. Plot of RMSE versus shape parameter for problem 2, $n_i = 324$, $n_b = 76$

Conclusions In this work, we extended the work of [1] for compactly supported radial kernels. We used the work of [?] to solve $\Delta\Phi = \phi$. The solution for the Φ was not very trivial for arbitrary radial kernel ϕ. It was not known why Φ be a radial kernel for a given arbitrary kernel ϕ. Theorem 1 show that Φ is again a radial kernel for a given arbitrary kernel ϕ. The present study demonstrated that the recently developed MAPS [1] may be very successful for large data points with locally supported kernels.

[1] Wen P H and Chen C S (2010), The method of particular solutions for solvinf scalar wave equations, *The International Journal for Numerical Methods in Biomedical Engineering*, **26** 1878-89.

[2] Chen C S, Fan C M, Wen P H, (2012), The method of particular solutions for solving certain partial differential equations, *Numerical Methods for Partial Differential Equations*, **28** 506-22.

[3] Lee C K, Liu X, Fan S C, (2003) Local multiquadric approximation for solving boundary value problems, *Computational Mechanics*, **30** 396409.

[4] Sarler B, Vertnik R, (2006) Meshfree explicit local radial basis function collocation method for diffusion problems, *Computers and Mathematics with Applications*, **21** 126982.

[5] Shu C, Ding H, Yeo K S, (2003) Local radial basis function-based differential quadrature method and its application to solve two dimensional incompressible navier-stoke equations, *Computer Methods and Applied Mechanics Engineering*, **192** 94154.

[6] Vertnik R, Sarler B, (2006) Meshless local radial basis function collocation method for convective-diffusive solid-liquid phase change problems, *International Journal of Numerical Methods for Heat and Fluid Flow* **16** 61740.

[7] Yao G, Kolibal J, Chen C S, (2011) A localized approach for the method of approximate particular solutions, *Computers and Mathematics with Applications*, **61** 2376-87.

[8] Kansa E J, Hon Y C, (2000) Circumventing the ill-conditioning problem with multiquadric radial basis functions: applications to elliptic partial differential equations, *Computcomputer and Mathematics with Applications* **39** 123-37.

[9] Hon Y C, Schaback, Zhou R X, (2003) An adaptive greedy algorithm for solving large RBF collocation problems , *Numerical Algorithms* **32** 13-25.

[10] Huang C S, Lee C F, Cheng A H D, Error estimate, optimal shape factor, and high precision computation of multiquadric collocation method, *Engineering Analysis with Boundary Elements*, **31** 614-23.

[11] Libre N A, Emdadi A, Kansa E J, Rahimian M, Shekarchi M, (2008) A stabilized rbf collocation scheme for neumann type boundary value problems, *Computer Modeling in Engineering and Science* **24** 61-80.

[12] Cheng A H D, Young D L, Tsai J J, (2000) Solution of Poisson's equation by iterative DRBEM using compactly supported, positive definite radial basis function, *Engineering Analysis with Boundary Elements* **24** 549-57.

[13] Beatson R K, Greengard L, (1997) A short course on fast multipole methods, in: M. Ainsworth, J. Levesley, W. Light, M. Marletta (Eds.),Wavelets, Multilevel Methods and Elliptic PDEs, *Oxford University Press*, 1-37.

[14] Partridge P W, Brebbia C A, and Wrobel L C, (1992) The Dual Reciprocity Boundary Element Method. CMP/Elsevier.

[15] Golberg M A, and Chen C S, The method of fundamental solutions for potential, Helmholtz and diffusion problems. In M. A. Golberg, editor, Boundary Integral Methods: Numerical and Mathematical Aspects, pages 103-176. WIT Press, 1998.

[16] Wendland H, (2005) Scattered Data Approximation, *Cambridge University Press.*

[17] Bozzini M, Rossini M, Schaback R, Volonte E, Radial kernels via scale derivatives, Advances in Computational Mathematics, DOI 10.1007/s 10444-014-9366-z. (2014)

A DFT and structural investigation of the conformations of Fischer carbene complexes

Marilé Landman

Department of Chemistry, University of Pretoria, Private Bag X20, Hatfield, 0028, South Africa

marile.landman@up.ac.za

Abstract. A set of different Fischer carbene complexes of group VI and VII metals, with varied heteroatom and heteroaromatic substituents on the carbene carbon atom, was studied. Density functional theory as well as single crystal diffraction techniques were employed to investigated the most stable conformation of these complexes. The complexes studied, $[M(CO)_4L\{C(X)Z\}]$, with L = PPh_3 or CO, X = ethoxy ($-OCH_2CH_3$) or amino ($-NH_2$ or NHCy) substituents as the heteroatom carbene substituents, Z = 2-furyl ($-C_4H_3O$), 2-thienyl ($-C_4H_3S$), 2-(N-methyl)pyrrolyl ($-C_4H_3NCH_3$) as the second carbene substituent had their substituents varied systematically to give all the possible conformations of these complexes. The conformations of the complexes, in particular the relative orientations of the heteroatoms in the molecule (*syn* vs. *anti*), *E*/*Z* isomerism in the aminocarbene complexes and *cis*/*trans* isomerism in the ligand substituted complexes as well as various combinations of these aspects, were studied. In general, it was found that the most stable conformation theoretically as well as in the solid state for most of the complexes preferred the *syn* conformation. The *Z*-isomer is generally preferred over the *E* isomer while the *cis* is more predominant than the *trans* isomer. Using DFT and NBO calculations, explanations for the preferred conformations were explored. It was concluded that both steric and electronic factors influence the conformations of the carbene complexes, with the extent of contribution of these two factors varying for each of the different carbene substituents.

1. Introduction

Since the discovery of the first transition metal carbene complex, $(CO)_5W{=}C(Ph)(OMe)$, in 1964 [1], Fischer carbene complexes have attracted attention as useful compounds in organic syntheses [2] and in catalytic processes [3]. Isomerism in Fischer carbene complexes has been shown to influence the reactivity and final product formation in reactions with different isomers [4].

Studies focusing on the synthesis, electronic (particularly of the $M\text{-}C_{carbene}$ bond) and spectroscopic properties of Fischer carbene complexes have been reported in the early 1970's[5]; however, little has been reported on the structure of these complexes, and more specifically on the conformations of these molecules. Selected mononuclear Fischer-type carbene complexes with heteroaromatic substituents of the form $[M(CO)_4L\{C(X)Z\}]$, with L = PPh_3 or CO, X = ethoxy ($-OCH_2CH_3$) or amino ($-NH_2$ or NHCy) substituents as the heteroatom carbene substituents, Z = 2-furyl ($-C_4H_3O$), 2-thienyl ($-C_4H_3S$), 2-(N-methyl)pyrrolyl ($-C_4H_3NCH_3$) of group VI metals have been synthesized and the different isomers/conformers formed in the reactions were studied using X-ray crystallography. In this study we report the experimental results compared with DFT calculated data for the different possible conformations.

Journal of Physics: Conference Series **633** (2015) 012051 doi:10.1088/1742-6596/633/1/012051

2. Results and discussion

2.1. Pentacarbonyl metal carbene complexes

For the pentacarbonyl metal carbene complexes with general formula [M(CO)$_5${C(X)Z}], with X = OEt or NH_2 and Z = 2-furyl, 2-thienyl or 2-(N-methyl)pyrrolyl, six model chromium complexes were investigated. Similar results were obtained with selected Mo and W complexes and therefore these six complexes can be seen as representative of group VI metal Fischer carbene complexes [6]. The complexes are listed in Table 1.

Table 1. Pentacarbonyl metal carbenes of this study

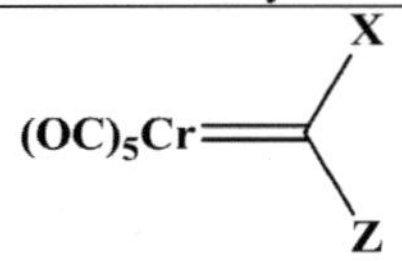

Complex	X Substituent	Z Substituent = C_4H_3Y
1	OEt	2-Furyl
2	OEt	2-Thienyl
3	OEt	2-(N-Methyl)pyrrolyl
1*	NH_2	2-Furyl
2*	NH_2	2-Thienyl
3*	NH_2	2-(N-Methyl)pyrrolyl

For these pentacarbonyl metal carbene complexes, only one conformational difference is possible. This difference lies in the orientation of the heteroatom (Y) of the heteroarene carbene substituent (Z = C_4H_3Y) relative to the heteroatom substituent (X), leading to *syn* or *anti* conformations (see figure 1).

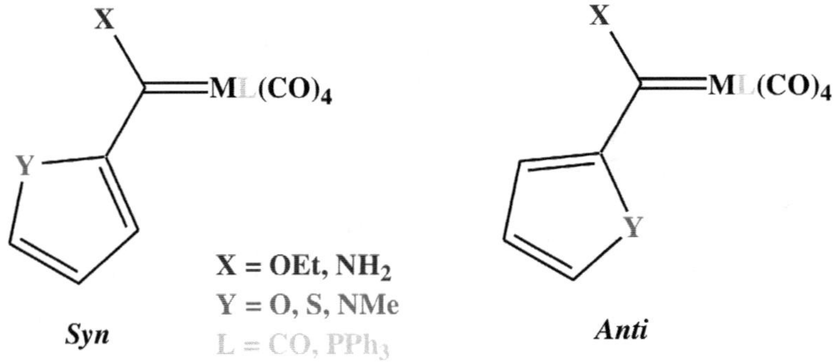

Figure 1. *Syn* versus *Anti* conformations

To determine the lowest energy conformation of the heteroaromatic substituent using DFT calculations, an energy scan around the dihedral angle of interest (X-$C_{carbene}$-C-Y) was performed to ascertain at which dihedral angle a minimum conformation of the complex can be expected. In this way the minimum energy *syn* and a *anti* conformations were obtained, as well as the transition between the two conformations. Default convergence criteria were used during the geometric optimizations. Vibrational frequencies were calculated at the optimized *syn* and *anti* geometries and no imaginary frequencies were observed for the E_{min} conformers. DFT calculated data is summarized in Table 2. From the data in the table it is clear that the predominantly favoured conformation for pentacarbonyl metal carbenes, is the *syn* conformation. The only complex for which the *anti* form (with the X-C-C-Y angle 180°) is most favoured is **1**, a 2-furyl complex. The DFT theoretical model predicts the geometries of the carbene complexes studies particularly well when compared to experimentally determined X-ray crystal structures.

Table 2. Results of *syn*- and *anti* conformation optimization of **1-3***

Complex	Dihedral Angle X-C-C-Y (°)	Conformation	Energy (kJ/mol)	Boltzmann Distribution
1	0.0	*Syn*	7.6	4.5%
	180.0	*Anti*	0.0	95.5%
	90.0	*TS*[a]	51.2	
2	0.0	*Syn*	0.0	99.4%
	177.3	*Anti*	12.8	0.6%
	90.0	*TS*	42.7	
3	4.2	*Syn*	0.0	100%
	153.1	*Anti*	23.9	0%
	90.0	*TS*	43.7	
1*	0.0	*Syn*	0.0	99.6%
	170.7	*Anti*	13.8	0.4%
	90.0	*TS*	48.6	
2*	23.5	*Syn*	0.0	93.9%
	148.9	*Anti*	6.7	6.1%
	90.0	*TS*	22.5	
3*	26.9	*Syn*	0.8	42.0%
	144.0	*Anti*	0.0	58.0%
	90.0	*TS*	19.1	

Data from ref [6]

To understand the preferred *anti* form of **1**, and NBO analysis was performed. The NBO calculations indicate a stabilising interaction in [Cr(CO)$_5${C(OEt)2-furyl}], **1**, between the lone pair of electrons on the furyl oxygen atom and an empty antibonding orbital of the neighbouring carbonyl groups when the molecule adopts an *anti* conformation (figure 2). This interaction is absent or < 0.2 kJmol^{-1} in the *anti*-conformation of each of the other cases.

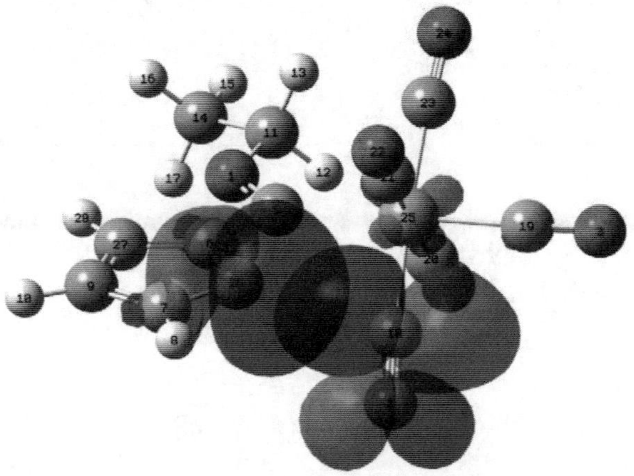

Figure 2. NBO-calculated donor-acceptor interaction in **1**

For the pentacarbonyl metal carbene complexes with general formula [M(CO)$_5${C(X)Z}], with X = NHCy and Z = 2-furyl or 2-thienyl, two model tungsten complexes were investigated. Due to the amino group containing two different substituents (H and Cy), one can distinguish between *E*- and *Z*-conformation of the Cy group relative to the metal M (figure 3).

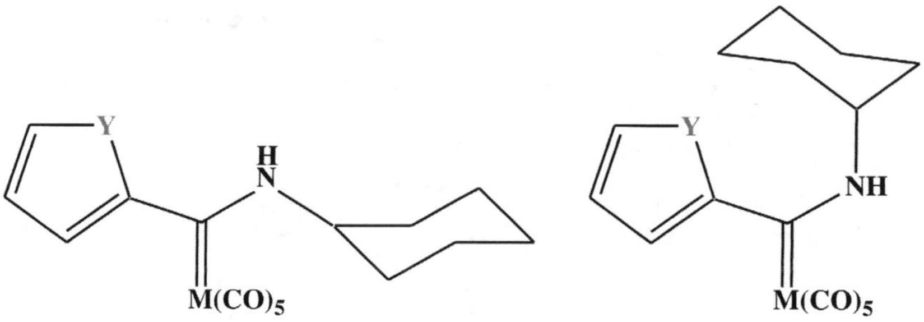

Z-isomer *E*-isomer

4: M = W, Y = O
5: M = W, Y = S

Figure 3. *Z*-isomer versus *E*-isomer

This isomerism in combination with the possible *syn* or *anti* conformations of the heteroaryl substituent, lead to four different possible conformations for each complex. These four conformations were optimized with DFT to determine the lowest energy conformation.

Table 2. DFT results of optimization of **4** and **5**

Complex	**Conformation**	**DFT calculated relative energy (eV)**		**% NMR**	**X-ray structure**	**DFT calculated rotation barrier of aryl ring (eV)**
		PW91	B3LYP			PW91
4	*syn,Z*	**0.00**	**0.00**	96.7	Crystal structure	0.38
	anti,Z	0.13	0.14			
	syn,E	0.16	0.15	3.3		0.25
	anti,E	0.21	0.22			
5	*syn,Z*	**0.00**	**0.00**	87.7	56 % disorder	0.18
	anti,Z	0.03	0.03		44 % disorder	
	syn,E	0.11	0.10	12.3		0.09
	anti,E	0.12	0.11			

Data from ref [6]

The lowest energy conformer in each case was the *syn,Z* conformer, the same conformer as was characterized by X-ray crystallography. The barrier of rotation of a furyl ring is on average 0.2 eV larger than that of the thienyl ring for related complexes. The rotation barrier of the *Z* isomers is generally higher than that of the *E* isomers.

2.2. Tetracarbonyl metal carbene complexes

For the tetracarbonyl carbene complexes, such as **6-8** in figure 4, one can distinguish between *cis* and *trans* isomers, with respect to substitution by PPh_3 relative to the carbene ligand. Three model molybdenum complexes of general formula [$M(CO)_4L\{C(X)Z\}$], with L = PPh_3, X = OEt, NH_2 or NHCy and Z = 2-furyl will be discussed. It is suggested on grounds of pi acceptor/donor arguments that the *cis* complex is formed by preference because the phosphine reduces the electrophilicity of the *trans*-carbonyl carbon due to the increased backbonding from the metal. The CO group trans to the M=carbene bond is therefore less available to react with the organolithium reagent, hence reducing the chance of the *trans* isomer forming. However, in a recent paper both *cis*- and *trans* isomers of

$[(CO)_4(PPh_3)W{=}C(OEt)(2\text{-furyl})]$ were reported and a crystal structure for the *trans* isomer was obtained [7].

For each of the Mo-carbene complexes shown in figure 4, a number of conformers of the different isomers are possible. Complexes **6-8** have PPh_3 ligands in the *cis* position relative to the carbene ligand, either nearest to the furyl ring or nearest to the heteroatom substituent X (OEt, NHCy or NH_2), leading to four, eight and four possible *cis*-PPh_3 conformers, respectively. If the PPh_3 is in the *trans* position, two, four and four *trans*-PPh_3 conformers are possible, respectively, in combination with *syn* or *anti* and *E* or *Z* configurations, where applicable (Table 4).

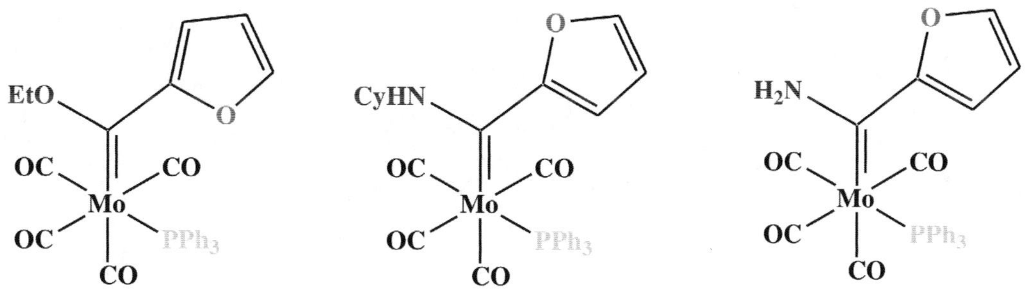

6 *cis* **7** *cis* **8** *cis*

Figure 4. Phosphine-substituted carbene complexes **6-8**

Table 4. DFT results of optimization of **6-8**

Complex	Conformation	Relative energy/ $kJmol^{-1}$
6	*cis-anti* (1)	**0.0**
	cis-syn (1)	7.5
	cis-anti (2)	1.1
	cis-syn (2)	8.1
	trans-anti	7.6
	trans-syn	17.1
7	*trans-anti,Z*	18.3
	trans-anti,E	24.8
	trans-syn,Z	7.4
	trans-syn,E	19.0
	cis-anti,Z (1)	8.2
	cis-anti,E (1)	14.1
	cis-syn,Z (1)	0.04
	cis-syn,E (1)	10.6
	cis-anti,Z (2)	13.7
	cis-anti,E (2)	11.2
	cis-syn,Z (2)	**0.0**
	cis-syn,E (2)	0.2
8	*cis-syn* (2)	**0.0**
	cis-anti (2)	15.3
	cis-anti (1)	20.5
	cis-syn (1)	6.4
	trans-anti	30.0
	trans-syn	16.9

The experimental solid state crystal structures of the complexes corresponded to the lowest DFT calculated gas phase electronic energy conformation determined by DFT calculations for **6-8**. However, it was possible to obtain also the *E*-isomer of **7** in the solid state. The corresponding *E*-isomer is only 0.2 kJ/mol higher in energy as the lowest energy *Z*-isomer. Therefore it is concluded that both are experimentally feasible. Further DFT calculations showed that the shorter experimental and DFT calculated intramolecular distance N–H· · ·O_{Fu} in *cis-syn,Z* (2) and the stronger conjugative LP1(O) → BD*(N–H) NBO calculated interaction associated with *cis-syn,Z* (2) compared to the related properties of the C–H· · ·O_{Fu} hydrogen bond in *cis-syn,E* (2), confirm the higher stabilization of the *Z* configuration of **7** compared to that of **7**-*E* (figure 5).

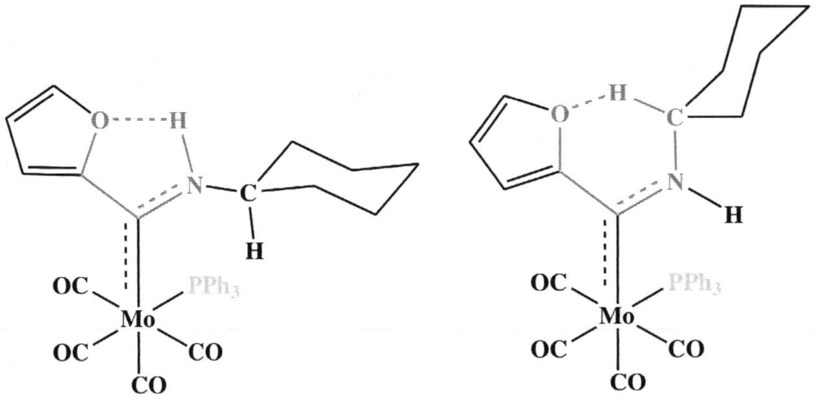

cis-syn,Z (2) *cis-syn,E (2)*

Figure 5. Two isomers of **7**

3. Conclusion

The experimental and DFT calculated results correlated very well. It was possible to explain the conformational preference of the metal carbonyl carbene systems, obtained experimentally, with DFT calculations.

References

[1] Fischer E O and Maasböl A 1964 *Angew. Chem.* **76** 645
[2] Dötz K H 1984 *Angew. Chem., Int. Ed. Engl.* **23** 587
[3] Koo B and McDonald F E 2007 *Org. Lett.* **9** 1737
[4] Waters M L, Brandvold T A, Isaacs, L and Wulff W D 1998 *Organometallics* **17** 4298
Sinha-Mahapatra D K, Hazra D, Puranik, V G and Sarkar A 2004 *J. Organomet. Chem.* **689** 3501
Vyklicky L, Dvořáková H and Dvořák D 2001 *Organometallics* **20** 5419
[5] Connor J A and Jones E M 1971 *J. Chem. Soc (A)* **12** 1974
[6] Thompson S, Wessels H R, Fraser R, van Rooyen P H, Liles D C and Landman M 2014 *J. Mol. Struct.* **1060** 111
Landman M, Fraser R, Twigge L and Conradie J 2015 *J. Coord. Chem.* in press http://dx.doi.org/10.1080/00958972.2015.1046852
Landman M, Levell T J, Conradie M M, van Rooyen P H and Conradie J 2015 *J. Mol. Struct.* **1086** 190
Landman M and Conradie J 2015 *J Mol. Struct.* **1094** 36
[7] Landman M, Levell T, van Rooyen P H and Conradie J 2014 *J. Mol. Struct.* **1065–1066** 29

Knowledge-Based Framework: its specification and new related discussions

Douglas Rodrigues, Rodrigo R Zaniolo and Kalinka R L J C Branco
Institute of Mathematics and Computer Science, University of São Paulo, São Carlos, Brazil
E-mail: {douglasr, kalinka}@icmc.usp.br, rrzaniolo@gmail.com

Abstract. Unmanned Aerial Vehicle is a common application of critical embedded systems. The heterogeneity prevalent in these vehicles in terms of services for avionics is particularly relevant to the elaboration of multi-application missions. Besides, this heterogeneity in UAV services is often manifested in the form of characteristics such as reliability, security and performance. Different service implementations typically offer different guarantees in terms of these characteristics and in terms of associated costs. Particularly, we explore the notion of Service-Oriented Architecture (SOA) in the context of UAVs as safety-critical embedded systems for the composition of services to fulfil application-specified performance and dependability guarantees. So, we propose a framework for the deployment of these services and their variants. This framework is called Knowledge-Based Framework for Dynamically Changing Applications (KBF) and we specify its services module, discussing all the related issues.

1. Introduction

Unmanned Aerial Vehicle (UAV) is a common application of critical embedded systems. The term UAV was adopted by FAA (Federal Aviation Administration) and by the international academic community to describe systems that include not only the aircraft, but also all the associated elements, like payload, ground control station and communication links. UAVs have been widely used in precision agriculture, remote sensing, mapping, traffic monitoring, air support for oil stations and pipelines, environmental monitoring, and national security. Several papers have been published in this area, demonstrating the feasibility of using these vehicles as important tools to perform those tasks, specially precision agriculture and environmental monitoring [1].

The heterogeneity and constraints of UAVs and the distinct nature of their interactions are challenges for their successful integration into an architecture for a shared exploitation of UAS's. The heterogeneity prevalent in UAVs in terms of services for avionics is particularly relevant to the elaboration of multi-application missions. Besides, this heterogeneity in UAV services is often manifested in the form of characteristics such as reliability, security and performance. Different service implementations typically offer different guarantees in terms of these characteristics and in terms of associated costs. The initial choice of a particular avionics service implementation can, therefore, become sub-optimal as new applications/services are deployed, needing a careful selection of services to fulfil particular performance and operational guarantees and, subsequently, to avoid compromising the mission.

Therefore, the main goal of this ongoing research is investigate the degree of heterogeneity present in UAVs in terms of services and propose architectural abstractions for the integration of these service-variants. Particularly, we explore the notion of Service-Oriented Architecture (SOA) in the context of UAVs as safety-critical embedded systems for the composition of services to fulfil application-specified performance and dependability guarantees. Based on a study of UAS's and their incumbent services [2], we propose a framework for the deployment of these services and their variants. This framework is called Knowledge-Based Framework for Dynamically Changing Applications (KBF) and we specify its services module, discussing all the related issues.

The rest of this paper is organized as follows. In Section 2, we describe related works in the field of SOA in embedded and safety-critical embedded systems. Section 3 specifies the KBF and its services module. Finally, Section 4 presents our conclusions regarding this paper and also some future prospects for this research.

2. Related works

Many complex embedded systems are coupled with a high-level information system. SOA can provide the integration of low-level embedded system services and high-level information system services. This integration is still an incomplete work, despite the many related works found in the literature [3] [4] [5] [6] [7] . In practice, the use of SOA in embedded systems can provide a lot of benefits, such as decoupling configuration from environment, improvement of reusability and maintainability, higher level of abstraction and interoperability, more interactive interface between devices and information systems, and easy use of resource-hungry services provided by more powerful internet servers.

The increasing use of SOA in critical applications demands for dependable and cost-effective techniques to ensure high security. SOA technology, underpinned by Web services (WS), are widely used for linking suppliers and clients in different sectors such as banking and financial services, transportation, manufacturing, to name a few. However, the problem of engineering secure Web services is a non-trivial task as several studies [8] [9] show that a large number of WS implementations are deployed with security flaws that range from code vulnerability to inadequate use of standards and protocols.

Most work related to UAVs in the literature and military roadmaps show implementations using traditional approaches [10] [11]. However, the USAF's roadmap [12] suggests the use of SOA in long term. This roadmap indicates that SOA can facilitate the modularization of the system components. Also, it is considered that the adoption and maintenance of standardized interfaces for UAVs can protect clients' investment in the development of new systems.

Our work has a different focus from the works reviewed, a large portion of which do not address critical systems. For this class of systems, the references on the use of SOA are poor and almost nonexistent. This is due to the specific requirements that standard SOA does not address.

3. KBF: Knowledge Based Framework for Dynamically Changing Applications

KBF was proposed in [13] and is currently under development. KBF extends the capability of a SOA broker's service discovery, adding knowledge about the application domain. In addition, KBF uses context and monitoring information to select or compose dynamically the best service to perform a specific mission. This selection or composition is based on a set of usage rules and quality attributes, like reliability, security, and performance. Figure 1 illustrates the KBF.

KBF uses a knowledge database to store all information and quality attributes defined by the user and the application. Another key factor is the assembly of reconfigurable matrix, a data structure that correlates the chosen service, its functionality, and the quality attribute to mission procedures.

3.1. Services module

Initially, a service provider develops its service, describes its interface (WSDL, for example), and publishes it in a service registry (UDDI, for example). Shortly after the publication, an automatic

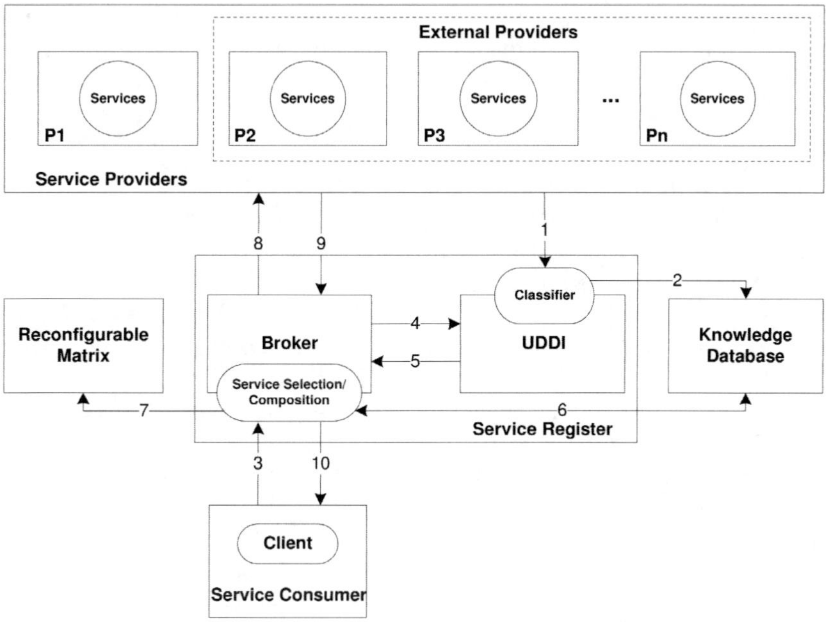

Figure 1. Knowledge Based Framework for Dynamically Changing Applications (KBF).

classifier rates the service according to the defined quality attribute (reliability, security and performance). This operation will be repeated for each new service being published.

Regarding to the quality attributes, the reliability of services is ensured using the WS-ReliableMessaging standard [14], responsible for ensuring the messages delivery. The security of services is ensured with the use of the following standards: WS-Security [15], XML Encryption [16], XML Signature [17], and WS-SecurityPolicy [18].

In the KBF's application domain, the published services use the basic information of the UAV (e.g., maximum cruise altitude, cruise speed, and others). Some of these services are illustrated in Figure 2.

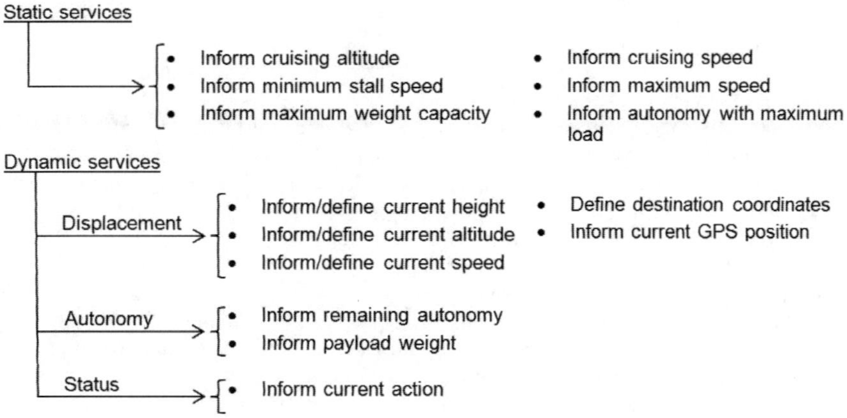

Figure 2. UAV's basic services.

Usually, static services are invoked only once, when the client is connected to the UAV, in order to the client gets specific information about the aircraft. Dynamic services may be invoked multiple times during the mission execution, since they can be very important to the mission success.

4. Conclusions

Several papers have been published in this area, demonstrating the feasibility of using UAVs in many different tasks. However, the references on the use of SOA in critical embedded systems and UAVs are poor and almost nonexistent. This is due to the specific requirements that standard SOA does not address. That is the reason we proposed the KBF.

KBF extends the capability of a SOA broker's service discovery, adding knowledge about the application domain. Moreover, KBF uses context and monitoring information to select or compose dynamically the best service to perform a specific mission.

Currently, the KBF is still in development, because the steps that compose the project continue to be implemented due to the high degree of difficulty. Moreover, most of the steps have been completed successfully. Therefore, the next papers will focus on the integration of KBF modules, making the framework fully operational.

When finished, the main result of this work will be the implementation and validation of KBF, with the establishment of the necessary resources to make the selection and composition of services more flexible and context-oriented, meeting the requirements established by the application in a more simple and efficient way.

References

[1] Branco K R L J C, Pelizzoni J M, Neris L O, Trindade O, Osório F S and Wolf D F 2011 *ICRA '11: IEEE International Conference on Robotics and Automation* pp 1–4 ISSN 1050-4729

[2] Rodrigues D, Chaves A A, Branco K R L J C, Ramdhany R and Coulson G 2014 *2014 11th International Conference on Information Technology: New Generations (ITNG)* pp 103–108

[3] Kakanakov N R and Spasov G 2005 *CompSysTech '05: Proceedings of the International Conference on Computer Systems and Technologies* pp 1–6

[4] Pastor E, Lopez J and Royo P 2007 *IEEE Aerospace and Electronic Systems Magazine* **22** 3–8 ISSN 0885-8985

[5] Thramboulidis K C, Doukas G and Koumoutsos G 2008 *EURASIP Journal on Embedded Systems - Embedded System Design in Intelligent Industrial Automation* **2008** 1–15 ISSN 1687-3955

[6] Lee M H, Yoo C J and Jang O B 2008 *International Journal of Advanced Science and Technology* **1** 55–64 ISSN 2005-4238

[7] Moritz G, Prüter S, Timmermann D and Golatowski F 2008 *ETFA '08: IEEE International Conference on Emerging Technologies and Factory Automation* pp 432–435

[8] Yau S S, Yin Y and An H G 2009 *ICWS '09: Proceedings of the 2009 IEEE International Conference on Web Services* (Washington, DC, USA: IEEE Computer Society) pp 287–294 ISBN 978-0-7695-3709-2

[9] Gruschka N, Jensen M, Iacono L and Luttenberger N 2011 *IEEE Transactions on Services Computing* **4** 272–285 ISSN 1939-1374

[10] Valavanis K P 2007 *Advances in Unmanned Aerial Vehicles: State of the Art and the Road to Autonomy* International Series on Intelligent Systems, Control, and Automation (Springer) ISBN 9781402061134

[11] DoD 2009 Unmanned systems integrated roadmap FY2009-2034 U.S. Department of Defense. Office of the Secretary of Defense

[12] USAF 2009 United states air force unmanned aircraft systems flight plan 2009-2047 Headquarters, United States Air Force

[13] Rodrigues D, Pires R M, Estrella J C, Vieira M, Correa M, Camargo J B, Branco K R L J C and Trindade O 2011 *Communications in Computer and Information Science* **206** 345–354

[14] OASIS 2009 Web services reliable messaging (ws-reliablemessaging) version 1.2 `http://docs.oasis-open.org/ws-rx/wsrm/v1.2/wsrm.html`

[15] OASIS 2012 Web services security: Soap message security version 1.1.1 `http://docs.oasis-open.org/wss-m/wss/v1.1.1/wss-SOAPMessageSecurity-v1.1.1.html`

[16] W3C 2002 Xml encryption syntax and processing `http://www.w3.org/TR/xmlenc-core/`

[17] W3C 2008 Xml signature syntax and processing (second edition) `http://www.w3.org/TR/xmldsig-core/`

[18] OASIS 2009 Ws-securitypolicy 1.3 `http://docs.oasis-open.org/ws-sx/ws-securitypolicy/v1.3/ws-securitypolicy.html`

Using Botnets to provide security for safety critical embedded systems - a case study focused on UAVs

Fernando Augusto Garcia Muzzi[1], Paulo Rogério de Mello Cardoso[2], Daniel Fernando Pigatto[3], Kalinka Regina Lucas Jaquie Castelo Branco[3]

[1]FATEC Lins – Faculdade de Tecnologia de Lins – FATEC.
[2]Centro Universitário Senac, SENAC/SP, Brasil.
[3]Universidade de São Paulo – Instituto de Ciências Matemáticas e de Computação, Av. Trabalhador São Carlense 400, Centro, CEP 13566-590, São Carlos, SP, BR.

E-mail: fagmuzzi@yahoo.com.br

Abstract. The use of unmanned aerial vehicles (UAVs) has been growing not only in military applications, but also in civilian. UAVs have enormous potential for use, which mostly still are unexplored. For the use of UAVs in the airspace, not only Brazilian new studies on methods of analysis and technologies should be incorporated into navigation systems, control among others, promoting security mechanisms for these aircraft. Implement security mechanisms using a platform with operating systems and botnet to simulate such attack Distributed Denial of Service (DDoS) in UAVs is an important task when it is aimed at containment and mitigation of attacks on this type of platform.

1. Introduction

Aerial vehicles have increasingly receiving more electronic components over the last years, which are connected through wired and wireless networks and running embedded software. This integration of dedicated computing devices, the physical environment and the network composes a Cyber-Physical System (CPS). CPS has thus become part of common vehicles, accessible to everyone, such as ground and aerial vehicles, including the autonomous ones. These vehicles have processing power increased and a very sophisticated embedded software, adding a big amount of sensors, becoming each day a more complex system. These complex embedded systems, like UAVs (Unmanned Aerial Vehicles), are able to perform lots of missions. If failure events occur, it may be caused losses of human lives and high-value assets, what makes such systems critical embedded systems.

The combination of high mobility and wireless communications in such systems has further increased their exposure to malicious threats and to faults deriving from uncertain connectivity or communication timeliness. Security and real-time operation, like other non-functional requirements, have become harder to fulfill, creating new challenges to such safety-critical embedded systems. This paper addresses the security problem in UAVs, focused in the wireless interface and in the Distributed Denial of Service attack by using BotNets.

2. Sphere: Safety and Security Platform under HAMSTER Architecture

Sphere is the platform for safety & security in HAMSTER architecture [7]. Although the platform may have centralized modules, it is not a centralized platform. Sphere is present in many parts of the unmanned vehicle (UV) according to its necessities and is responsible for information security (the way it is exchanged, stored, manipulated etc.) and also for health and safety of the overall vehicle and all subsystems that compose the unmanned system. Only a few parts of an UV are properly treated to ensure that all connected modules are authentic and have not been replaced or tampered with by a third party. The current policy adopted by most aircraft manufacturers uses a concept of "Accept all" which trusts in all components embedded in an aircraft. This proposal suggests the assumption of an "Almost Deny All" approach, which denies the authenticity of all mechanical components and peripherals attached to the vehicle until the opposite is proved, which may result in safer vehicles.

The categorization of every module is therefore crucial for such a new security model to be applied to UV. There are various peripheral devices embedded in an UV that require different levels of security, which leads to the necessity of a module categorization according to the criticality of their performed functions. The Sphere proposal suggests the modules categorization into primary, secondary, and so on, according to necessity.

Primary modules are those considered essential components for the UV to operate, to be aware of its location and to be able to perform an emergency abort operation safely, even when the mission was not entirely concluded. An autopilot (focus of this paper), a GPS receiver, and barometric/inertial units are examples of modules classified as primary, since they might cause a big lost if in failure state. In contrast, modules not considered as essential functions to the UV are classified as secondary modules.

Whether abnormal behaviors are detected in any secondary module, the operation of the primary components of the UV is not affected and the secondary module that presented the abnormality should be disabled or isolated. It implies that all primary modules must be authenticated before the operation begins. However, the secondary modules do not necessarily need an authentication before the mission execution. The creation of access profiles for modules is another concept associated with the proposal of authentication. As a mission is assigned to the UV, it must go through an authentication process, which assigns different access permissions to the UV modules. Such concept is similar to the one used in modern operating systems where an administrator user is allowed to install and uninstall software with no restrictions, unlike a visitor user who has access to the programs, but is not allowed to install/uninstall them. Applied to UV, such concept adds a layer of security that allows blocking the use of selected modules by specific users. Such specification is intended to prevent unauthorized access. Even if there is a single effective user of an UV, no other user (an attacker or not) will have privileged access to modules or their information.

Figure 1 presents the Sphere main modules. Central Security Unit (CSU) is responsible by modules authentication and "health" verification. The access policy assigned to each user must use cryptographic algorithms suitable for embedded or real-time sensitive environments. Several experiments have been carried out regarding security for critical embedded systems [4, 5].

To protect the UV against attacks coming from malicious components, Sphere implements strict security policies. It is necessary to ensure that all modules are authentic, so if one of them fails or presents an abnormal behavior, the others must not communicate with it. Furthermore, such policies must be applicable even during vehicle operations, considering that external factors may affect the components behavior e. g. climate or weather changes. During the vehicle start up, a mutual authentication phase should occur with CSU. It checks the database credentials of all modules, their criticality, and even if there is any access restriction. There is also the possibility of deciding whether a module should be initialized or not during the verification stage. The following steps will be authentication and exchange of encrypted messages to establish a secure channel for communication among modules and CSU. After such handshake, three situations are expected:

1. The module that is trying to authenticate and CSU have not been tampered with;
2. The module has been tampered with and therefore has not been authenticated:
    a. If it is a module of primary type, the UV must not operate;
    b. If it is a module of secondary type, communication with it must be interrupted and a notification be sent to a control station.
3. The module that is trying to authenticate may notice that CSU is not authentic, and must notify other components about it.

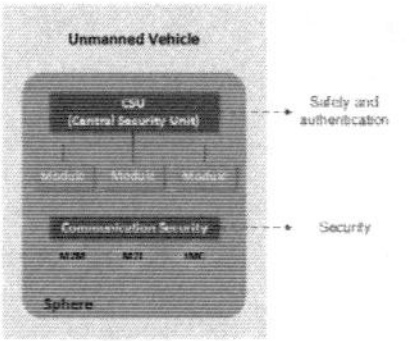

Figure 1. Composition of Sphere, the Safety & Security platform on HAMSTER architecture.

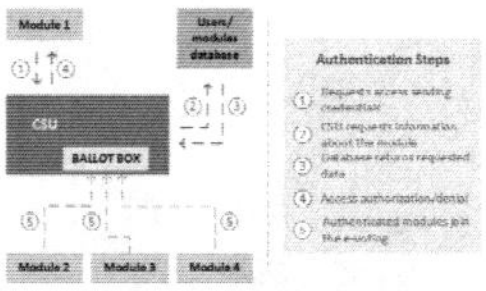

Figure 2. Modules and CSU authentication steps. The "Ballot box" is used by every other system modules for authenticating CSU through e-voting protocols.

From the point of view of communication security, an ideal situation would be if all modules could authenticate with others. However, this method would cause a system overload, since the increase of modules in the aircraft would cause an exponential increase in the number of exchanged messages. To solve such problem there exist the e-voting protocols [6]. Such model can be further expanded according to the needs of the UV, including a negotiation mediated by CSU to create a secure channel of communication among modules. A graphical representation of processes performed during the authentication module with CSU can be seen in Figure 2.

3. Threats to Unmanned Aerial Vehicles

In the scope of this paper three vulnerabilities related to vehicular wireless communication may be mentioned [1]: operating wireless network interfaces to DoS (Denial of Service) attacks and espionage (information theft); malicious code injection into embedded devices present in the vehicle; and induction to incorrect behavior when in cooperation with other vehicles.

Security is a big concern. In [2] the authors presented a UAV system cyber-security threat model, which can be seen in Figure 1. Note that there are several open issues related to communication. Open networks could be easily hacked by malicious entities. Studies show that using BotNets we can stop the communication, and using Botmaster to start the attack with bots to an UAV through vulnerabilities in existing Wi-Fi networks, we can identify vulnerabilities in communication and communication link, important to check for possible threats and stop attacks [2]. Using a real botnet is possible to attack an UAV autopilot and detect all the standard behavior. With the information will be possible to provide countermeasures to stop the attack. Based in Figure 3, we will focus on the attack presented in Figure 4. All these process is integrated with Sphere.

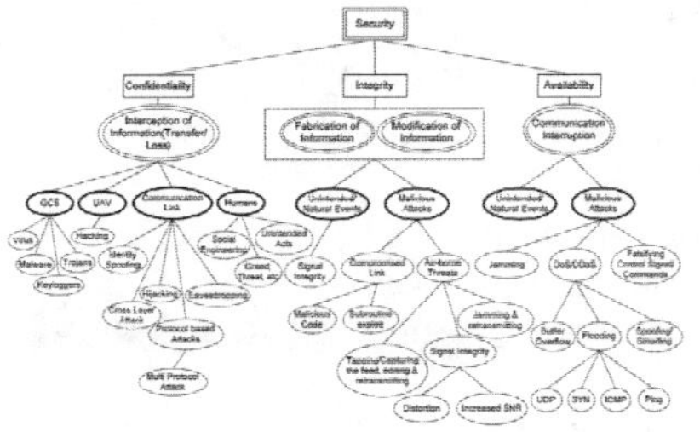

Figure 3. UAV system cyber-security threat model by [2].

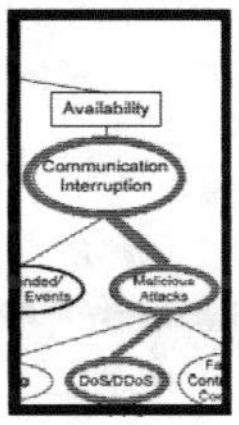

Figure 4. Attack to be analyzed.

4. Results and Discussions

The simulation platform is illustrated in Figure 5 and shows the network infrastructure that contains virtual machines and bots, IRC server for sending botmaster commands to the bots (hosts) to start the attack on a specific target in which is used an autopilot board to simulate DDoS attack on a UAV. We make an attack on the UAV and we detect patterns of behavior by the attack, in order to verify mistakes and to develop containment mechanisms and mitigation of attack.

The first step that we realize to model the DDoS attack was the specification using an extension tool called UML*intr* proposed by Hussein and Zulkernie [4]. But we note that the UML*intr* is not specified for this attack type making necessary the addition of new classes for the characters of DDoS attack can be represented with all its features. In Figure 6 we present a modeling of a DDoS attack using the UML*intr* extended to allow the DDoS specification (detailed in Figure 7).

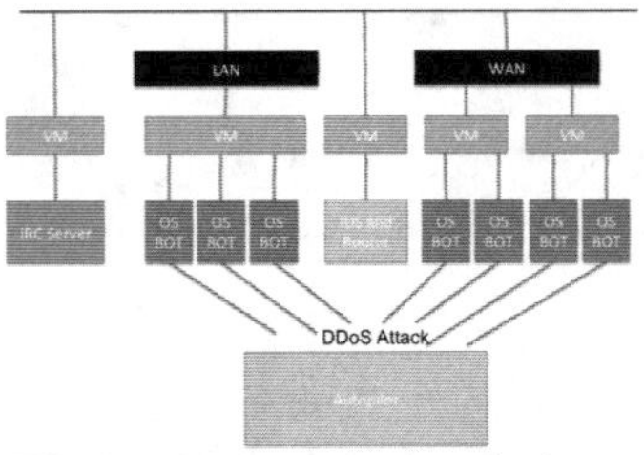

Figure 5. Attack simulation platform.

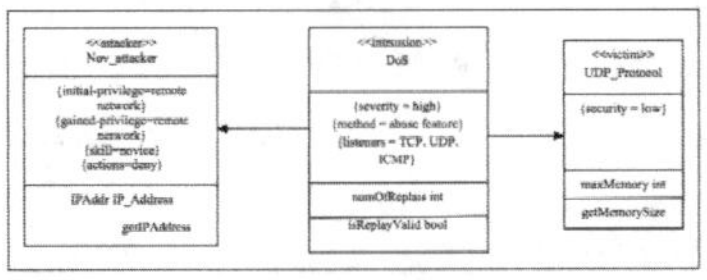

Figure 6. Modeling a DDoS attack.

Figure 7. Representation of a DDoS attack elements.

The processes had stalled instantly after the attack. Memory usage has not changed, but the processing was very high utilization rates. At different intervals were given three attacks on the "victim" and the CPU utilization, which left the Idle state (idle) 100 % and reached the level of 0% Idle. We measured the capture packets received by the network, which recorded a flood of UDP packets the type attack used for the DDoS. The percentage of UDP packets was very high during the attack period, representing 99.8% all traffic. Was possible see that when the attack was completed the network interface still responding to requests, but at a lower level of saturation, which proves that even after the effective attack, the victim continues to respond the actions of the attack. With these results, its also possible see the complexity involved in this type of attack and notes the efficiency of the same when struck the victim. Take these in account it is necessary to evaluate DDoS attack to create defense mechanisms.

5. Conclusion

The main objective of this paper was to present a platform to perform a DDoS attack to an UAV to find a way to mitigate this kind of attack. As first results we obtained a DDoS attack modeling and specification. We could also analyze the effects caused by this kind of attack. This is an important step towards to find a way of mitigating such attack. We aim to use the botnet platform to mitigate DDoS attacks in a future work.

Acknowledgment

The authors would like to thank the institutions where they conduct their research for supporting this project and also the financial support provided by CNPq, CAPES and FAPESP (2014/13713-8).

References

[1] Wyglinski, A. M., Huang, X., Padir, T., Lai, L., Eisenbarth, T. R., & Venkatasubramanian, K. (2013). Security of Autonomous Systems Employing Embedded Computing and Sensors. IEEE Micro, 33(1), 80–86.

[2] Javaid, A. Y.; Sun, W.; Devabhaktuni, V. K.; Alam, M.. Cyber Security Threat Analysis and Modeling of an Unmanned Aerial Vehicle System. Homeland Security (HST), 2012 IEEE Conf. on Technologies.

[3] Hussein, M.; Zulkernine, M. UMLintr: A UML for Specifying Intrusion. In: 13th ANNUAL IEEE International Symposium and Workshop on Engineering Based Systems, 27-30, 2006, Germany. p. 279-88.

[4] V. Schoaba, F. E. G. Sikansi, D. F. Pigatto, K. R. L. J. C. Branco, and L. C. Branco, "Digital Signature for Mobile Devices: A New Implementation and Evaluation," International Journal of Future Generation Communication and Networking, vol. 4, pp. 23–36, 2011.

[5] D. F. Pigatto, N. B. F. D. Silva, and K. R. L. J. C. Branco, "Performance Evaluation and Comparison of Algorithms for Elliptic Curve Cryptography with El-Gamal based on MIRACL and RELIC Libraries," Journal of Applied Computing Research, vol. 1, no. 2, pp. 95–103, Feb. 2012.

[6] H.-T. Liaw, "A secure electronic voting protocol for general elections," Computers & Security, vol. 23, no. 2, pp. 107–119, 2004.

[7] D. F. Pigatto, L. Goncalves, A. S. R. Pinto, G. F. Roberto, J. F. R. Filho, and K. R. L. J. C. Branco, "HAMSTER - Healthy, mobility and security-based data communication architecture for Unmanned Aircraft Systems," in 2014 Int. Conference on Unmanned Aircraft Systems (ICUAS). IEEE, May 2014, pp. 52–63.

Low frequency wave modes of liquid-filled flexible tubes

Yuan-Fang Chou[1] **and Tzu-Huan Peng**

Department of Mechanical Engineering, National Taiwan University,
Taipei, Taiwan 10617, Republic of China

E-mail: yfchou@ntu.edu.tw

Abstract. Many canals in the human body are liquid-filled thin wall flexible tubes. In general the P-wave and S-wave velocities of tube material are much slower than the sound velocity of the liquid. It is interested to study the dynamic deformation of the wall caused by pressure fluctuation of liquid. In the low frequency range, the liquid pressure is essentially axial symmetric. Therefore, axial symmetric wave propagation modes are investigated. The calculated spectrum shows there are two modes with zero frequency limit. Phase velocities of these two modes are much smaller than the sound velocity of the liquid. They are also slower than the P-wave velocity of the tube material. At very low wave number, radial displacements of both liquid particles and tube are very small compared to their axial counter parts. As the frequency goes higher, boundary waves are observed.

1. Introduction

Georg von Békésy measured the motion of basilar membrane and introduced the travelling wave theory for cochlea model [1]. He observed a slow wave with speed two orders of magnitude less than that of the water waves. Because he used a sound pressure level of 150 dB, it was questioned that there might be a contribution of broad tuning [2]. Olson developed a micro pressure sensor to measure pressure distribution inside cochlea [3, 4, 5]. She found slow and fast waves travelling in the cochlea in the sound pressure level range of 50 to 80 dB. Many researchers adopted lumped system to model acoustic waves in cochlear and confirmed that slow and fast waves do exist in the cochlea [6]. Since the lumped system model cannot completely describe pressure distribution, it is desired to employ the continuum model to study waves in cochlea. Del Grosso analysed multimode wave propagation in inviscid liquid-filled cylinders. He adopted continuum models with realistic boundary condition [7]. Lafleur and Shields employed the method developed by Del Grosso to study low-frequency mode waves in a liquid-filled elastic tube [8]. They found there are one subsonic and one supersonic modes existing in the liquid in the zero frequency limit for aluminium/water and PVC/water combinations. Baik, Jiang, and Leighton calculated waves in a PMMA tube filled with water [9]. They also found one subsonic and one supersonic modes existing in the zero frequency limit. In the studies above, the solid tube materials have P-wave and S-wave velocities higher than the acoustic wave velocity of filled-liquid. However, P-wave and S-wave velocities of the membrane materials in cochlea are slower than the liquid acoustic wave velocity. Therefore, it is interested in examining waves in liquid-filled flexible tubes to investigate dynamic behaviour of the system.

[1] To whom any correspondence should be addressed.

2. Formulation

2.1. Elastic tubes

In the absence of external body force, Newton's law gives the governing equation for the displacement $\mathbf{u}^t$ of an isotropic elastic solid as

$$(\lambda+\mu)\nabla(\nabla\cdot\mathbf{u}^t)+\mu\nabla^2\mathbf{u}^t=\rho^t\left(\partial^2\mathbf{u}^t/\partial t^2\right), \tag{1}$$

where and λ and μ are Lamé constants, ρ is the mass density, and the superscript "t" indicates "tube". The displacement can be decomposed into the scalar potential ϕ and vector potential $\boldsymbol{\psi}$ as

$$\mathbf{u}^t=\nabla\phi+\nabla\times\boldsymbol{\psi} \quad \text{with} \quad \nabla\cdot\boldsymbol{\psi}=0\,. \tag{2}$$

Substituting equation (2) into equation (1) obtains two wave equations

$$(\lambda+2\mu)\nabla^2\phi-\rho^t\left(\partial^2\phi/\partial t^2\right)=0 \quad \text{and} \tag{3}$$

$$\mu\nabla^2\boldsymbol{\psi}-\rho^t\left(\partial^2\boldsymbol{\psi}/\partial t^2\right)=0\,. \tag{4}$$

Equation (3) and (4) give velocities of dilatation and shear waves respectively as

$$c_p=\left[(\lambda+2\mu)/\rho^t\right]^{1/2} \quad \text{and} \quad c_s=\left(\mu/\rho^t\right)^{1/2}. \tag{5}$$

Axial symmetric modes are investigated here because they dominate most low frequency behaviour of canals in the human body. The axial symmetric displacement $\mathbf{u}^t$ of a circular tube can be expressed as

$$\mathbf{u}^t=u_r^t\mathbf{e}_r+u_z^t\mathbf{e}_z \tag{6}$$

where subscripts "r" and "z" denote respectively radial and axial directions in cylindrical coordinate. Equations (3) and (4) are separable and the solution for time harmonic motion with frequency ω can be written as

$$\phi=R_\phi(r)Z_\phi(z)e^{j\omega t} \quad \text{and} \tag{7}$$

$$\psi_r=R_{\psi_r}(r)Z_{\psi_r}(z)e^{j\omega t} \tag{8a}$$

$$\psi_\theta=R_{\psi_\theta}(r)Z_{\psi_\theta}(z)e^{j\omega t} \tag{8b}$$

$$\psi_z=R_{\psi_z}(r)Z_{\psi_z}(z)e^{j\omega t}. \tag{8c}$$

Therefore the Helmholtz equations corresponding to scalar potential ϕ are

$$\left(d^2Z_\phi/dz^2\right)+k_z^{\,2}Z_\phi=0 \quad \text{and} \tag{9}$$

$$\left(d^2R_\phi/dr^2\right)+\left(dR_\phi/dr\right)/r+P^2R_\phi=0 \tag{10}$$

where k_z is the wave number in the axial direction and $P^2=\omega^2/c_p^2-k_z^{\,2}$. Combining the general solutions of equations (9) and (10) to obtain

$$\phi=\left[C_1J_0(Pr)+C_2Y_0(Pr)\right]e^{j(\omega t-k_z z)} \tag{11}$$

where C_1 and C_2 are constants to be determined. By the same token, solutions of z-variable can be found as

$$Z_{\psi_r}(z)=Z_{\psi_\theta}(z)=Z_{\psi_z}(z)=e^{-jk_z z} \tag{12}$$

and the remaining Helmholtz equations for vector potential $\boldsymbol{\psi}$ are

$$\left(\partial^2R_{\psi_r}/\partial r^2\right)+\left(\partial R_{\psi_r}/\partial r\right)/r+\left(S^2r^2-1\right)R_{\psi_r}/r^2=0 \tag{13a}$$

$$\left(\partial^2R_{\psi_\theta}/\partial r^2\right)+\left(\partial R_{\psi_\theta}/\partial r\right)/r+\left(S^2r^2-1\right)R_{\psi_\theta}/r^2=0 \tag{13b}$$

$$\left(\partial^2R_{\psi_z}/\partial r^2\right)+\left(\partial R_{\psi_z}/\partial r\right)/r+S^2R_{\psi_z}=0 \tag{13c}$$

where $S^2=\left(\omega/c_s\right)^2-k_z^2$. The general solutions for components of vector potential are therefore

$$\psi_r = \left[C_3 J_1(Sr) + C_4 Y_1(Sr)\right] e^{j(\omega t - k_z z)} \tag{14a}$$

$$\psi_\theta = \left[C_5 J_1(Sr) + C_6 Y_1(Sr)\right] e^{j(\omega t - k_z z)} \tag{14b}$$

$$\psi_Z = \left[C_7 J_0(Sr) + C_8 Y_0(Sr)\right] e^{j(\omega t - k_z z)} \tag{14c}$$

where C_i's are constants. Substituting equations (11) and (14a, b, c) into equation (2) results in the displacement field of elastic tube

$$\begin{aligned}\mathbf{u}^t = &\left\{-P\left[AJ_1(Pr) + BY_1(Pr)\right] + jk_z\left[CJ_1(Sr) + DY_1(Sr)\right]\right\} e^{j(\omega t - k_z z)}\mathbf{e}_r \\ &+\left\{S\left[CJ_0(Sr) + DY_0(Sr)\right] - jk_z\left[AJ_0(Pr) + BY_0(Pr)\right]\right\} e^{j(\omega t - k_z z)}\mathbf{e}_z\end{aligned} \tag{15}$$

where A, B, C, and D are constants to be determined and the corresponding stress fields τ_{rr} and τ_{rz} can be found from the constitutive equations

$$\tau_{rr} = \lambda\left\{\left[\partial\left(ru_r^t\right)/\partial r\right]/r + \left(\partial u_z^t/\partial z\right)\right\} + 2\mu\left(\partial u_r^t/\partial r\right) \text{ and} \tag{16}$$

$$\tau_{rz} = \mu\left(\partial u_r^t/\partial z + \partial u_z^t/\partial r\right) \tag{17}$$

which give rise to

$$\begin{aligned}\tau_{rr}^t = &\left\{A\left[-\lambda\left(P^2 + k_z^2\right)J_0(Pr) - 2\mu P^2 J_0(Pr) + 2\mu P J_1(Pr)/r\right] + \right.\\ &B\left[-\lambda\left(P^2 + k_z^2\right)Y_0(Pr) - 2\mu P^2 Y_0(Pr) + 2\mu P Y_1(Pr)/r\right] + \\ &\left. j2\mu C\left[Sk_z J_0(Sr) - k_z J_1(Sr)/r\right] + j2\mu D\left[Sk_z Y_0(Sr) - k_z Y_1(Sr)/r\right]\right\} e^{j(\omega t - k_z z)}\end{aligned} \tag{18}$$

and

$$\tau_{rz}^t = \left\{j2\mu P k_z\left[AJ_1(Pr) + BY_1(Pr)\right] + \mu\left(k_z^2 - S^2\right)\left[CJ_1(Sr) + DY_1(Sr)\right]\right\} e^{j(\omega t - k_z z)}. \tag{19}$$

2.2. Fluid inside tube

The acoustic wave in fluid is the pressure fluctuation p^f caused by fluid compressibility that is measured by the bulk modulus B. The fluid is assumed to be lossless so there are no dissipative effects arising from viscosity or heat conduction. The analysis will be limited to waves of relatively small amplitude, so the density change of the medium will be small compared with its equilibrium value ρ^f. Therefore a corresponding linear, lossless displacement field $\mathbf{u}^f$ can be expressed with a wave equation

$$c_f^2 \nabla\left(\nabla \cdot \mathbf{u}^f\right) = \partial^2 \mathbf{u}^f / \partial t^2 \tag{20}$$

where c_f is the sound velocity. The particle velocity field corresponding to a small amplitude acoustic processes is irrotational. This means that the displacement field can be expressed as the gradient of a scalar potential χ

$$\mathbf{u}^f = \nabla \chi \tag{21}$$

and equation (26) becomes

$$c_f^2 \nabla^2 \chi = \partial^2 \chi / \partial t^2 \tag{22}$$

or

$$\partial\left[r\left(\partial\chi/\partial r\right)\right]/\partial r + r\left(\partial^2\chi/\partial z^2\right) - r\left(\partial^2\chi/\partial t^2\right)/c_f^2 = 0 \tag{23}$$

in cylindrical coordinate for axial symmetric case. The solution of equation (30) is similar to equation (11) and can be expressed as

$$\chi = \left[D_1 J_0(Fr) + D_2 Y_0(Fr)\right] e^{j(\omega t - k_z z)} \tag{24}$$

where D_1 and D_2 are constants and $F^2 = \left(\omega/c_f\right)^2 - k_z^2$. Due to the restriction of finite value at $r = 0$, D_2 must vanish and the solution becomes

$$\chi = D_1 J_0(Fr) e^{j(\omega t - k_z z)} \tag{25}$$

Substituting equation (25) into equation (21) results in the displacement field of fluid

$$\mathbf{u}^f = E\left[FJ_1(Fr)\mathbf{e_r} + jk_z J_0(Fr)\mathbf{e_z}\right]e^{j(\omega t - k_z z)}. \tag{26}$$

and the corresponding acoustic pressure

$$p^f = -E\rho^f \omega^2 J_0(Fr)e^{j(\omega t - k_z z)} \tag{27}$$

The constants A, B, C, D, and E in equations (15) and (27) are determined by satisfying tube's boundary conditions on the outer surface and the interface conditions on tube's inner surface.

2.3. Boundary and interface conditions

The illustration of a fluid filled elastic tube is shown in figure 1 where R_o and R_i are respectively the inner and outer radius of the tube. The outside surface of the tube is traction free as shown in figure 2. Therefore the corresponding stress components must vanish at $r=R_o$, i.e.,

$$\tau_{rr}^t(R_o) = 0 \text{ and} \tag{28}$$

$$\tau_{rz}^t(R_o) = 0. \tag{29}$$

The solid-fluid interface is located on the inner surface of the tube. Because the inviscid fluid cannot resist shear, the traction on the inner surface of the tube has no shear components. Therefore the stress components in the tube have to satisfy

$$\tau_{rz}^t(R_i) = 0 \text{ and} \tag{30}$$

$$\tau_{rr}^t(R_i) = -p^f(R_i). \tag{31}$$

Finally, the radial displacement component of fluid is equal to its counterpart of tube at the interface

$$u_r^t(R_i) = u_r^f(R_i). \tag{32}$$

These five conditions can be expressed with general solutions in tube and fluid as

$$\begin{aligned}&\left[QJ_0(PR_o) + PJ_1(PR_o)/R_i\right]A + \left[QY_0(PR_o) + PY_1(PR_o)/R_i\right]B \\ &+ jk_z\left[SJ_0(SR_o) - J_1(SR_o)/R_i\right]C + jk_z\left[SY_0(SR_o) - Y_1(SR_o)/R_i\right]D = 0\end{aligned} \tag{33}$$

$$jPk_zJ_1(PR_o)A + jPk_zY_1(PR_o)B + QJ_1(SR_o)C + QY_1(SR_o)D = 0 \tag{34}$$

$$jPk_zJ_1(PR_i)A + jPk_zY_1(PR_i)B + QJ_1(SR_i)C + QY_1(SR_i)D = 0 \tag{35}$$

$$\begin{aligned}&\left[QJ_0(PR_i) + PJ_1(PR_i)/R_i\right]A + \left[QY_0(PR_i) + PY_1(PR_i)/R_i\right]B \\ &+ jk_z\left[SJ_0(SR_i) - J_1(SR_i)/R_i\right]C + jk_z\left[SY_0(SR_i) - Y_1(SR_i)/R_i\right]D - \left(\rho^f\omega^2/2\mu\right)J_0(FR_i)E = 0\end{aligned} \tag{36}$$

$$-PJ_1(PR_i)A - PY_1(PR_i)B + jk_zJ_1(SR_i)C + jk_zY_1(SR_i)D - FJ_1(FR_i)E = 0 \tag{37}$$

The parameter Q in these equations are defined as $Q = k_z^{\,2} - \left(\omega^2\rho^t/2\mu\right)$.

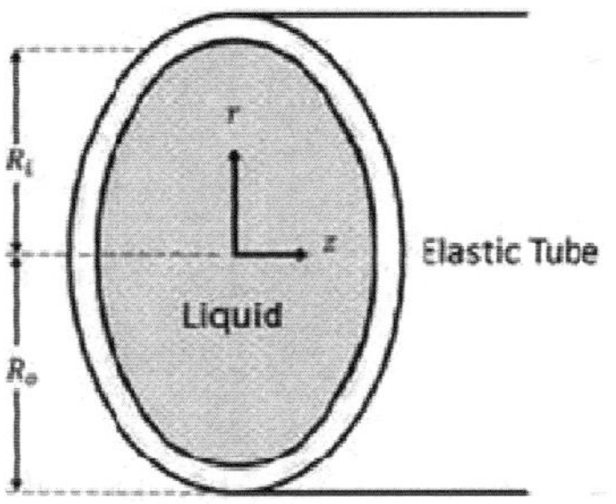

Figure 1. A fluid filled tube.

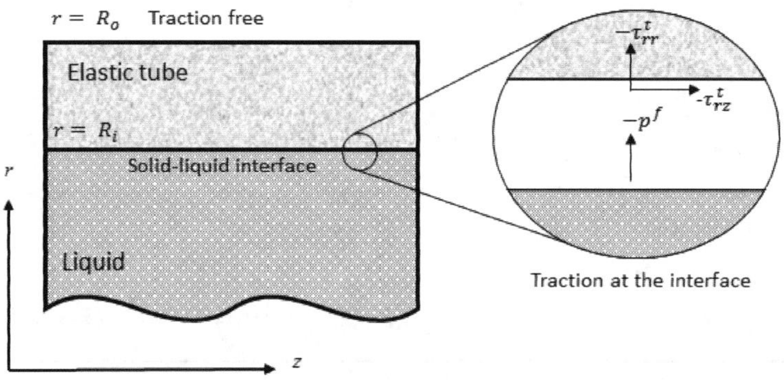

Figure 2. Boundary and interface conditions.

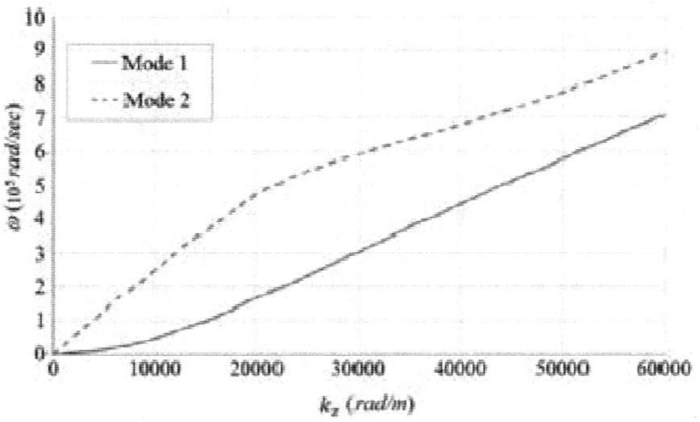

Figure 3. The lowest two dispersion curves.

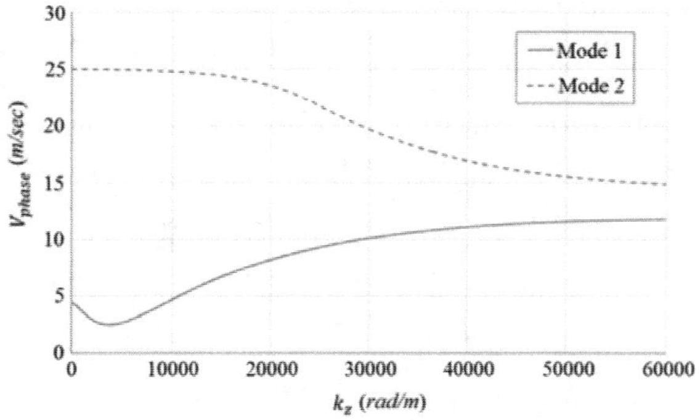

Figure 4. Phase velocities for lowest two modes.

The system of homogeneous equations (33), (34), (35), (36), and (37) forms an eigenvalue problem. Pairs of (k_z, ω) offering nontrivial solutions are dispersive relations and the corresponding eigenvector $\{A, B, C, D, E\}$ gives mode shape of waves.

3. Example and discussion

An elastic tube with inner radius R_i =1.5 *mm* and outer radius R_o =1.6 *mm* are chosen for study. Since the material properties of membranes in cochlea are similar to that of cartilage [10], Lamé constants λ=0.167 *MPa* and μ=0.25 *MPa* and the density ρ^t =1000 kg/m^3 are adopted. The filled fluid has a density ρ^f =1000 kg/m^3 and an acoustic wave velocity c_f =1500 *m/sec*. Dilatation and shear wave velocities are respectively 25.82 *m/sec* and 15.81 *m/sec* for tube material. These velocities are much slower than the acoustic wave velocity of the filled fluid. The system has two dispersion curves without cut-off frequency as shown in figure 3. Figure 4 gives phase velocities of these two low-frequency modes and they are respectively 4.366 *m/sec* and 24.98 *m/sec* as wave number approaches zero. Characteristics of waves that wavelength is much longer than tube's diameter, in the same order of tube's diameter, and much smaller than tube's diameter are studied as follows.

3.1. Low wave number ranges

Mode shapes corresponding to a specific point on dispersion curve is obtained from the eigenvector

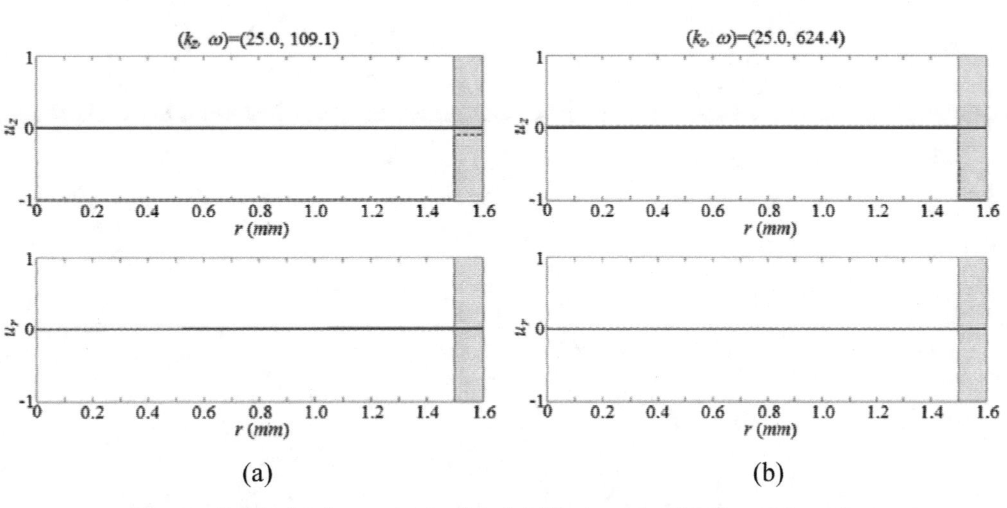

Figure 5. Mode shapes at k_z=25. (a) First mode. (b) Second mode.

{A, B, C, D, E} corresponding to the (k_z, ω) pair. Figure 5 shows the mode shape of these two modes at the wave number of 25 rad/m that has a wave length of 251.3 mm. The real parts of displacement fields are indicated with blue solid lines while the imaginary parts are drawn with green dashed lines. Figure 5(a) presents the displacement field of first mode at 17.36 Hz. The radial displacement is zero and the axial displacement field is mainly happened in fluid. This planelike wave has a phase velocity of 4.364 m/sec that is much slower than the acoustic velocity of 1500 m/sec for the fluid. The displacement field of second mode at 99.38 Hz is shown in figure 5(b). It is a longitudinal plane wave of the tube with axial component only. The phase velocity is 24.98 m/sec, very close to the solid P-wave velocity of 25.82 m/sec.

3.2. Mid wave number range

Mode shapes of the first and seond modes at the wave number of 2000 rad/m are shown in Figure 6. The wave length is 3.142 mm that is about the same as the tube diameter. Figure 6(a) presents the displacement field of first mode at 961.8 Hz. This wave has a phase velocity of 3.022 m/sec that is even slower than its low wavenumber counterparts. The radial and axial components of the displacement have the same order in amplitude but they hold a phase difference of $\pi/2$. Therefore the fluid particles move clockwise in elliptical loci while the radial component dominates the motion of tube particles. The displacement field of second mode at 7946 Hz is shown in figure 6(b). It is a longitudinal plane wave of the tube which particles move in axial direction. The phase velocity is 24.96 m/sec, very close to the P-wave velocity of 25.82 m/sec. The fluid is stay in rest.

3.3. High wave number range

Figure 7 shows the mode shape of these two modes at the wave number of 60000 rad/m that has a wave length of 0.1047 mm. It shows significant coupling between fluid and tube. Figure 7(a) presents the displacement field of first mode at 112554 Hz. This wave has a phase velocity of 11.79 m/sec. Most of the fluid particles do not move and only those close to the interface circle around clockwise in z-r plane. The displacement of solid particles on the tube surface is greater than that of interior particles. All solid particles move elliptically. The wave of second mode at 142143 Hz has a phase velocity of 14.89 m/sec. The displacement field is shown in figure 7(b) where maximum displacement occurred at the outer surface of the tube and particles move counterclockwise. Similar to the first mode, large clockwise motion is also observed for particles in the neighborhood of solid-fluid interface. The motion decays rapidly toward the center so that most of the fluid particles stay still.

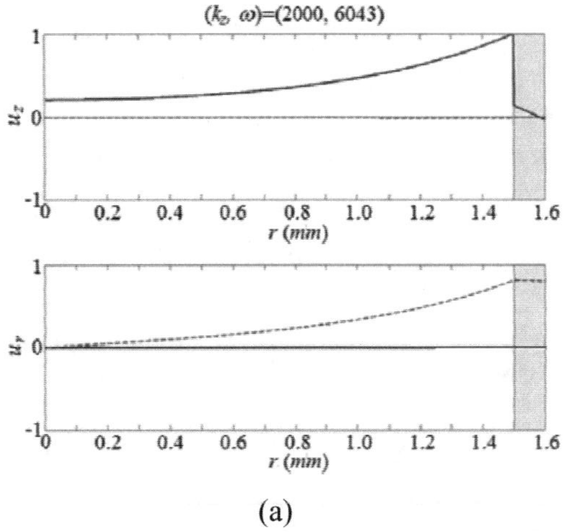

Figure 6. Mode shapes at k_z=2000. (a) First mode. (b) Second mode.

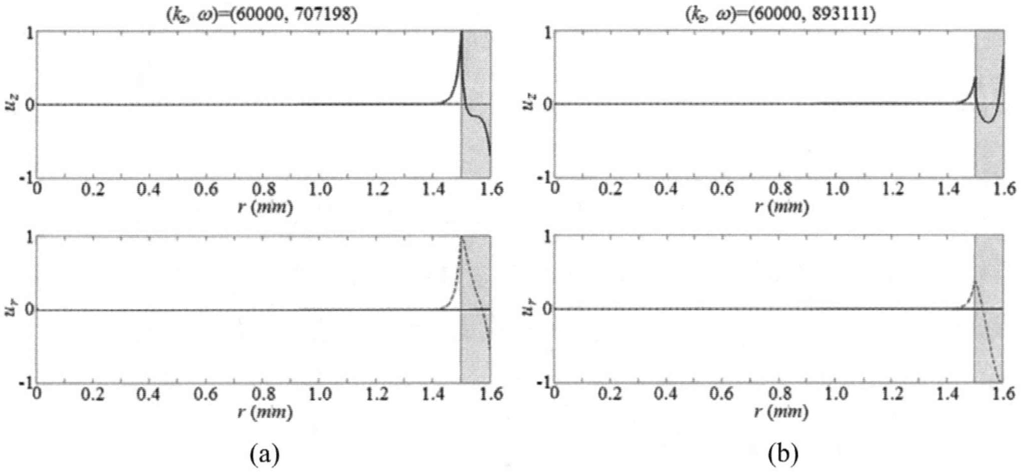

Figure 7. Mode shapes at k_z=60000. (a) First mode. (b) Second mode.

4. Conclusions

A liquid-filled thin wall flexible tube which P-wave and S-wave velocities are much slower than the sound velocity of the filled-liquid is studied in this paper. Two lowest axial symmetric modes with zero frequency limit are investigated. Their phase velocities are much smaller than both the sound velocity of the liquid. For very long wave length, the radial displacement fields in liquid and tube are much smaller than their axial counter parts. The first mode is a planelike wave of fluid and the second mode is a longitudinal plane wave of the tube. Displacement fields of these two waves are dominated by axial component. When the wave length is comparable to the tube diameter, the first mode has fluid particles move clockwise in elliptical loci while the radial component dominates the motion of tube particles. The displacement field of second mode is a longitudinal plane wave in the tube which particles move in axial direction and the fluid is stay in rest. As the frequency goes higher, boundary waves can be observed in very short wave length range. The slow wave observed by Von Békésy and Olsen is verified in this study.

References

[1] Von Békésy G 1960 *Experiments in Hearing* McGraw-Hill New York
[2] Johnstone B M, Patuzzi R, and Yates G K 1986 *Hearing Research*, **22** 147-53
[3] Olson E S 1998 *J. Acous. Soc. Am.* **103** 3445-63
[4] Olson E S 1999 *Nature* **402** 526-9
[5] Olson E S 2001 *J. Acous. Soc. Am.* **110** 349-67
[6] Duifhuis H 2012 *Cochlear Mechanics- Introduction to a Time Domain Analysis of the Nonlinear Cochlea*, Springer, New York
[7] Del Grosso V A 1971 *Acustica* **24** 299-311
[8] Lafleur L D and Shields F D 1995 *J. Acous. Soc. Am.* 97 1435-45
[9] Baik K, Jiang J, and Leighton T G 2009 *Theoretical Investigation of Phase Velocity, Group Velocity and Attenuation of Acoustic Waves in a Liquid-Filled Cylinder*, ISVR Technical Report 329, University of Southampton
[10] Mansour J M, 2003 *Biomechanics of cartilage, In Kinesiology: the Mechanics and Pathomechanics of Human Movement* Lippincott Williams & Wilkins, Philadelphia

Journal of Physics: Conference Series **633** (2015) 012055 doi:10.1088/1742-6596/633/1/012055

Development of a Galactic Magnetic Field Model and its application in identifying sources of Ultra-High-Energy Cosmic Rays in Northern Sky.

Pantea Davoudifar; Keihanak Rowshan Tabari

Research Institute for Astronomy and Astrophysics of Maragha (RIAAM), P.O.Box:55134-441, IRAN

E-mail: dfpantea@riaam.ac.ir

Abstract. New physical conditions were applied to our previous Galactic Magnetic Field model. The relative motion of a Galactic source were also considered. We simulated the propagation of Ultra-High-Energy particles under the influence of Galactic Magnetic Field. In this research the particles were originated from millisecond pulsars located in the northern sky. Considering the relative motion of Galactic sources for a proper time interval, sample test images of millisecond pulsars were produced using cosmic rays of energies ranged in $10^{18} - 10^{19}$, $10^{19} - 10^{20}$, and $10^{20} - 10^{21} eV$. The results were compared with our previous ones. For each part of the sky considering the structure of Galactic Magnetic Field, the source location and its relative motion to the observer, one may use these images as a guide to find possible sources of the Ultra-High-Energy Cosmic Ray' events. Consequently, a possible method of identifying the sources of these particles were introduced. Some physical limits were discussed.

1. Introduction

The deflection of Cosmic Ray (CR) charged particles in the Galactic and Extragalactic (G & EG) Magnetic Fields (MFs) has been the subject of many researches. At the highest energies, calculating these deflections plays a crucial role to find the origin of Ultra High Energy (UHE) CRs because in this case the deflection from straight-line paths is within few degrees [1]. As one of the most powerful G and EG sources, it is shown that pulsars are capable to accelerate particles up to $10^{19} eV$ for proton and up to $10^{20} eV$ for Iron nuclei [2, 3, 4, 5, 6]. So considering the deflections of these particles, one may construct hypothetical pulsar images in a CR sky.

Through studies based of physical processes such as Synchrotron Radiation, Polarization of electromagnetic waves in different wavelengths, Zeeman Effect and Zeeman splitting and Faraday Rotation Measure (FRM) [13, 14, 15, 16], G and EG MFs were mapped. These strength and correlation lengths of GMFs, may fitted to the putative models of GMF.

2. The specifications of the model

Essentially, three models were considered for GMF in the disk, included: Concentring Rings (CCR), Axisymmetric Spiral (ASS) and Bisymmetric Spiral (BSS) models of GMF [15]. Following our previous works [1, 7, 8], now an ASS GMF model is adopted for the Galactic disk in which 2 radial reversals [17] occurs. Such a model is in consistence with the turbulent dynamo theory [14].

In our previous model, we have been considered a constant magnetic field strength ($B_0 = 10\ nG$, Stanev et al., [18]) with a regular magnetic field $B = B_0 e^{\frac{-\mathcal{L}}{3000}}$ along the z axis [1, 7, 8]. Here, the model were developed using an ASS model of GMF in Galactic disk[9, 10, 11, 12], which is considered to have a cylindrical form with an average thickness of $0.3kpc$ [19]. The regular component considered in the direction of spiral arms. In the halo, the vertical distribution of total magnetic field is taken from Cox [20]. Here the only physical bond was the distribution of matter in the Galaxy, which is exerted through two essential parameters, i.e. B (strength of GMF) and L_c (MF coherence length). No other interactions were considered for UHECR' propagation in the Galaxy and the deflections from the straight line pathes were considered to be only due to different gyro-radiuses of particles. To consider a randomly oriented field we have used a coherence length, L_c, of order of $\sim 100pc$ and less in the vicinity of the source (i.e. pulsar). The total magnetic field were considered to have a regular component and a randomly oriented one. Far from the source, the coherence length were considered to have a value of around 2 kpc. A Random Walk Method (RWM) were used to calculate the actual path of the particle.

We also added a few other parameters, included the rigidity of the projected particle from proton to iron nuclei and also the relative motion of source during an observation epoch. GZK processes were off, as the propagation is just considered in the Galaxy. Now The time delay of observed particles is just due to their deflected paths in varying GMF:

$$\tau = \gamma\alpha(\mathcal{D} - d)years \tag{1}$$

where $\mathcal{D}$ and d are in parsecs and α is $\sim$3.26 and γ is 1 for particles with rigidity of 1.

2.1. Deflection Angles and the Picture of Source in a CR Sky

As what we followed in the previous works [1, 7, 8], for each possible source the time delay distributions for different energy ranges (10^{18}, 10^{19}, 10^{20} and 10^{21} eV) were calculated. Using the relation of the average time delay, τ, and deflection angle [21, 22, 23, 24]: $\tau(E, d) \approx \frac{\theta^2(E,d)d}{4}$, the values of θ were calculated for each energy interval.

In this research we constricted the picture of sources using particles of different energy intervals (figure 1).

3. The Results

As an example, our results were shown for two objects studied before: J1751-2857 ($l = 0.65, b = -1.12$), $d = 1.44\ kpc$, $p_0 = 3.9\ ms$ and J0621+25 ($l = 187.31, b = 4.93$), $d = 4.25\ kpc$, $p_0 = 2.7\ ms$ in $\sim$ anti-center direction [25, 26]. For J0621+25, we have tracked the deflected protons and calculated the resulted (l, b) pairs around the source. These results are shown in figure 1.

It is obvious that considering more details for GMF cause higher degrees of deflection up to 5 degrees for protons ejected from a Galactic source. Here it is also shown that adding a parameter of relative motion of an object may cause an elongation in the picture as seen here. But such an *oblateness* is observable in large epoches. For example to reach the above picture, we considered this source was ejecting particles for more than $10^3 years$.

Our results shows that considering the structure are GMF arms and disk, may cause large deflection angles in the directions of detected particles. But still more studies are needed to verify a Galactic origin of UHECRs. In fact the correlation lengths and magnetic field strengths between Galactic Arms and the direction of $B_{regular}$ are much more complicated than our simple models and the effect of adding more details is not clear.

An interesting case may occurs for a young pulsar, when the correlation length of random component of magnetic field around the pulsar and behind the shock is $\sim 100pc$ and less. In

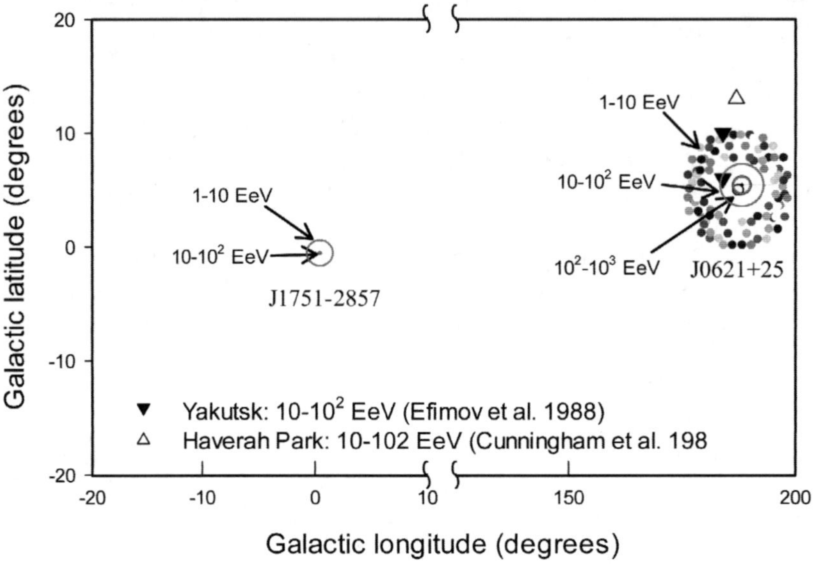

Figure 1. Deflected particles (protons) around J0621+25. The red circles are shown our previous results.

such a case our model predicts the highest degrees of deflection, i.e. $\sim 12degrees$ and more for protons up to a few $10EeV$, with a correlation length of $\sim 100pc$ and $\sim 15degrees$ and more for a correlation length of $\sim 50pc$. This result may shift the G to EG region to a few $10EeV$.

Acknowledgments

This work has been supported financially by Research Institute for Astronomy and Astrophysics of Maragha (RIAAM) under research projects No. 1/2352 and 1/2782-24.

References

[1] Davoudifar P., Fatemi S. J., Clay R., and Whelan B., 2011, *J. Sciences*, **22**(1), 75-84
[2] Berezhko E. G. 1991, *Proc. Int. Cosmic Ray Conf.*, **2**, 436-439
[3] Berezhko E. G., 1993, *Proc. Int. Cosmic Ray Conf.*, **2**, 348-351
[4] Berezhko E. G., 1994, *Astron. Lett.*, **20** (1), 75-79
[5] Wibig T. and Wolfendale A. W., 2011, *Open Astron. J*,**4**, 211-217
[6] Fang Ke. Kotera K. and Olinto A. V., 2012, *Astrophys. J.*, **750**, 118
[7] Davoudifar P., and Rowshan Tabari K., 2014, *J. Phys.: Conf. Series* **490** 012029-12034
[8] Davoudifar P., 2011, *Proc. Int. Cosmic Ray Conf.*, **2**, 230-233
[9] Vallee J. P., Simard-Normandin M., and Bignell R. C., 1988, *Astrophys. J.*, **331**, 321-324
[10] Vallee J. P., 1991, *Astrophys. J.*, **366**, 450-454
[11] Vallee J. P., 1995, *Astrophys. J.*, **454**, 119-124
[12] Vallee J. P., 1996, *Astron. Astrophys.*, **308**, 433-440
[13] Vallee J. P., 1997, *Fundamental of Cosmic PHysics*, **19**, 1-89
[14] Vallee J. P., 1998, *Fundamental of Cosmic PHysics*, **19**, 319-422
[15] Han J.L., 2003, *"The Magnetized Interstellar Medium"*, 3-12
[16] Andreasyan R.R., Makarov A.N., *Astrophysics*, **30**, 101-110
[17] Beck R., 1991, *Astron. Astrophys.*, **251**, 15-26
[18] Stanev T., Seckel D., and Engel R., 2003, *Phys. Rev. D*, **68**, 103004
[19] Jo Bovy Hans-Walter Rix, 2013, *Astron. Astrophys.*, **21**, 61(pp 58)
[20] Cox D.P., 2005, *Annu. Rev. of Astron. Astrophys.*, **43**, 337-385

[21] Waxman E., and Coppi P., 1996, *Astrophys. J.*, **464**, L75-L78
[22] Waxman E., and Miralda-Escude J., 1996, *Astrophys. J.*, **472**, L89-L92
[23] Miralda-Escude J., and Waxman E., 1996, *Astrophys. J.*, **462**, L59-L62
[24] Bhattacharjee P., 2000, *Phys. Rep.*, **327**, 109-247
[25] Manchester R. N., Hobbs G. B., Teoh A., and Hobbs M., 2005, *Astron. J.*, **129**, 1993-2006
[26] http://www.atnf.csiro.au/research/pulsar/psrcat

SEVIRI Cloud mask by Cumulative Discriminant Analysis

M G Blasi[1], C Serio[1,2], G Masiello[1,2], S Venafra[1] and G Liuzzi[1]

[1] School of Engineering, Università della Basilicata, Via Ateneo Lucano, 85100 Potenza, Italy
[2] CNISM, Unità di Ricerca Università della Basilicata, 85100 Potenza, Italy

E-mail: maria.blasi@unibas.it

Abstract. In the context of *cloud detection* for satellite observations we want to use the method of Cumulative Discriminant Analysis (CDA) as a tool to distinguish between clear and cloudy sky applied to Spinning Enhanced Visible and Infrared Imager (SEVIRI) data. The methodology is based on the choice of several statistics related to the cloud properties, whose correlation has been analyzed by Principal Component Analysis (PCA). Results have been compared with the SEVIRI reference cloud mask provided by the European Centre for the Exploitation of Meteorological Satellite (EUMETSAT), in order to find suitable thresholds able to discriminate between clear or cloudy conditions. We trained the statistics on a selected region, the Basilicata area, located in the south of Italy, in different periods of the year 2012, in order to take into account the seasonal variability. Moreover we separated land and sea surface and distinguished between day-time or night-time. The validation of thresholds, obtained through SEVIRI observations analysis, shows a good agreement with the reference cloud mask.

1. Introduction

At present EUMETSAT geostationary meteorological programme is preparing for Meteosat Third Generation (MTG) to give continuity to the scientific community with infrared and visible/ultraviolet observations through four imaging and two sounding satellites in a twin configuration. The imaging satellites, MTG-I, comprise the Flexible Combined Imager (FCI) that will continue the very successful operation of SEVIRI on Meteosat Second Generation (MSG) satellites. The new imager will be provided of additional channels with higher spatial, temporal and radiometric resolution compared to the current MSG instruments.
Nowadays SEVIRI is the operational imager on board of the Meteosat Second Generation (MSG) geostationary satellites, then we want to exploit its high temporal resolution (one acquisition every 15 min) in order to study and construct a very accurate cloud mask scheme.
Infrared data are affected by clouds, so we aim to develop a scene analysis methodology able to discriminate between clear and cloudy areas since this is a crucial operation for many remote sensing applications over continental or oceanic surfaces. There are different methodologies employed in the cloud detection context [2, 9, 10, 13] most of which make use of statistics sensitive to spectroscopic properties of clouds. At present it is preferred to develop new methodologies that consider multiwavelength statistics [1, 6], namely physical quantities that involve two or more wavelengths that are sensitive to the surface parameters and to the cloud phase.
In this context we want to evaluate the potentiality and the efficiency of Cumulative Discriminant Analysis (CDA) as a tool to develop a cloud mask, being able to classify clear and cloudy pixels

in several climatic conditions (different surface type and illumination condition), and also to analyze the functionality of the selected statistics used in this method. We will not focus on cloud phase classification. The structure of the paper is organized as follows.
In Section 2 we describe the characteristics of data sets used to train and validate CDA method, and the reference cloud mask. The statistics used to discriminate between clear or cloudy areas are defined in Section 3, while in Section 4 we present the basic theory of the methodology: PCA to analyze the statistics' inter-correlation and CDA as classification procedure for the cloud detection. Then we show the results in Section 5 and Conclusions are given in Section 6.

2. Instrument and data sets

The cloud detection scheme elaborated in this work was performed on SEVIRI data. SEVIRI is the imager on board of MSG geostationary satellites ([11]), characterized by very high temporal and spatial resolution, that are respectively one acquisition every 15 min and 3 km at nadir. It is provided of 12 spectral channels: 3 in the visible band (one of them has higher spatial resolution of 1 km), 1 in NIR band and 8 in the thermal IR band (from 3.9 μm to 13.4 μm).
The analysis was performed on four data sets: two were used to train the cloud detection method, while the other two data sets have been used to validate the cloud detection scheme. The region of interest was the same for all data sets: it is a rectangular area with latitude and longitude respectively ranging from 39.25° to 42° and from 13° to 18.75°, including the entire Basilicata region, located in the south of Italy.
Data sets for the training phase are SEVJAN_1 and SEVJUL_1, comprising SEVIRI data with the related SEVIRI cloud masks respectively from 1st to 10th of January 2012 and from 1st to 10th of July 2012, each one including 240 h observation and 960 acquisitions.
Data sets considered for validation are SEVJAN_2 and SEVJUL_2, that comprise SEVIRI data with the related SEVIRI cloud masks respectively from 11st to 20th of January 2012 and from 11st to 20th of July 2012, each one including 240 h observation and 960 acquisitions.
The two validation data sets are temporally located just ten days after the training ones, because validation should be done on periods with similar climatic conditions. Each SEVIRI acquisition on the area of interest corresponds to 9643 pixels, for a total of 9257280 radiances in each dataset. We took into account two main aspects to choose these data sets. On one side we want to analyze a statistically significant sample with two well and similarly populated classes. On the other side the statistics we choose to perform CDA are season dependent, and we aim to develop a method which is fairly accurate and sensitive with respect to seasonal variability.

2.1. SEVIRI reference cloud mask

The SEVIRI reference cloud mask we used has been developed at EUMETSAT by the Satellite Application Facility in Support to Nowcasting and Very Short Range Forecasting (NWC-SAF), and it describes the scene type identifying all cloud-free pixels. Each pixel is classified in one of the following four types: clear sky over water, clear sky over land, cloudy, or not processed. The algorithm is based on a multispectral threshold technique, consisting in a series of threshold tests applied to various channels combination for each pixel. For more details see e. g. [4, 5].

3. Statistics

In this section we provide a description of the statistics tested to discriminate between clear and cloudy pixels in the CDA, and to build up the estimated cloud mask. Each statistics refers to the Brightness Temperature (BT) spectrum, computed from the SEVIRI radiance inverting the Planck function. We considered ten statistics, computed from MSG Level 1.5 Image data: the BT from all 8 IR SEVIRI channels (from 3.9 μm to 13.4 μm), and two particular differences between these values. They are listed and defined in table 1.

Table 1. List of the statistics and their spectral characteristics.

Statistic	Spectral Range (μm)	Nominal Centre Wavelength (μm)	Method
T_1	3.48 - 4.36	3.92	Brightness Temperature
T_2	5.35 - 7.15	6.25	Brightness Temperature
T_3	6.85 - 7.85	7.35	Brightness Temperature
T_4	8.30 - 9.10	8.70	Brightness Temperature
T_5	9.38 - 9.94	9.66	Brightness Temperature
T_6	9.80 - 11.80	10.80	Brightness Temperature
T_7	11.00 - 13.00	12.00	Brightness Temperature
T_8	12.40 - 14.40	13.40	Brightness Temperature
W_1	-	-	Difference T_7-T_1
W_2	-	-	Difference T_7-T_6

Infrared SEVIRI channels provide information on clouds and other important geophysical parameters, such as surface temperature, water vapor and ozone, exploiting the thermal contrast between surface and clouds, that are usually colder than terrestrial surface. Moreover, a proper combination of spectral channels provides information on atmospheric instability.

SEVIRI IR channels at 3.9, 8.7, 10.8 and 12.0 μm are window channels, where the absorption is almost negligible, while IR channels at 6.2 and 7.3 μm are centred on the water vapor absorption bands, then they are used for monitoring the water vapor content in the troposphere, but also to observe winds. The IR channel at 9.7 μm is centered on ozone absorption band, so that the evolution of total ozone profile can be monitored. Finally the IR channel at 13.4 μm is centered on CO_2 absorption channel, but it also contributes to estimate static instability because it is able to give temperature information from the lower troposphere. The only remark is about the 3.9 μm IR channel, where the observed radiance is the sum of terrestrial and solar contribution so that for our purposes it can be exploited only at night-time.

The statistic W_1 exploits the fact that the cirrus and stratus cloud types have an higher reflectance than the most of surface features at 3.9 μm, and it has been widely used in the AVHRR cloud detection. The last statistic we considered consists in the split window difference $T_{12\mu m} - T_{10.8\mu m}$, that is similar to Inoue's slope test [7, 8], used to detect cirrus clouds.

4. Methodology

In this section we present the basic theory for both PCA and CDA, paying particular attention to define the categorical variables from the statistic point of view, in order to understand how the Cumulative Discriminant theory is applied to.

4.1. PCA decomposition

In order to obtain a single cloud mask applying CDA on one parameter we want to reduce the dimensionality of the problem, exploiting information coming from each single statistic and combining them into a single one. For this purpose we used PCA and the related Singular Value Decomposition (SVD), that are appropriate tools in the field of data reduction [3, 12].

PCA is a statistical procedure that estimates the inter-correlation between variables, projecting the initial variables in an orthogonal space and defining their Principal Components (PC).

It is necessary to build up the covariance matrix of the process, that PCA decomposes through the most suitable orthogonal basis (in order to maximize the variance of the first component), yielding orthogonal functions and their related eigenvalues.

Let's call $\mathbf{s}$ a vector of $n_s \times 1$ size, where n_s is the number of the statistics, $\mathbf{s} = (s_1, ..., s_{n_s})^T$. The representative matrix of the process, $\mathbf{S}$, is an $n_s \times N$ matrix, where N is the number of samples for a suitable data set. The covariance matrix of $\mathbf{S}$, namely $\mathbf{C} = \frac{1}{N}\, \mathbf{S}\,\mathbf{S}^T$, can be SVD decomposed, obtaining $\mathbf{C} = \mathbf{U}\Lambda\mathbf{V}^T$ where $\mathbf{U}$ and $\mathbf{V}$ are the two $n_s \times n_s$ orthogonal basis, so that $\mathbf{U}\mathbf{U}^T = \mathbf{I}$ and $\mathbf{V}\mathbf{V}^T = \mathbf{I}$, with $\mathbf{I}$ the identity matrix. It is worth to note that have to be $\mathbf{U}$=$\mathbf{V}$, because the covariance matrix is symmetric and definite positive; then Λ is a diagonal matrix whose elements are the singular values, namely the eigenvalues, positive definite of the covariance matrix: $\mathbf{\Lambda} = diag(\lambda_1^2, \ldots, \lambda_{n_s}^2)$. It shall be $\lambda_1^2 > \cdots > \lambda_{n_s}^2$, hence the PCs, are defined as $\mathbf{c} = \mathbf{U^T}\,\mathbf{s}$. The columns of matrix $\mathbf{U}$ define an orthogonal basis, hence an orthogonal space, characterized by the important property of orthogonality of PCs, that are the projection of the initial variables into this new orthogonal space. This orthogonal transformation is defined so that the first PC describes most of the variance of the initial data set and each eigenvalue is a measure of the PCs' inter-correlation. The first eigenvalue in our data sets is always several magnitude orders higher than following ones, that is the first PC has a far higher variance than that of the other PCs. Then we decided to apply CDA only on the first PC in the scene analysis.

4.2. Classification of categorical binary variable

We need to clarify the classification of a binary event: the *clear sky event* on one side and the *cloudy sky event* or *not-clear sky event* on the other side. From the probabilistic point of view the cloud detection process can be described by a 2×2 contingency table, which means to recognize four different categories comparing the prediction with the truth:

- *clear*: n. of occurrences a *clear sky* prediction agrees with the true *clear sky event*
- *I-type error*: n. of occurrences a *cloudy sky* prediction corresponds to a true *clear sky event*
- *II-type error*: n. of occurrences a *clear sky* prediction corresponds to a true *cloudy sky event*
- *cloudy*: n. of occurrences a *cloudy sky* prediction agrees with a true *cloudy sky event*

The first and last categories are also denoted as *True Positives* (TP) and *True Negatives* (TN). The efficiency of CDA in classifying the pixels of a certain scene can be evaluated computing the *score* $SC = (TP + TN)/(N_{clear} + N_{cloudy})$, where N_{clear} and N_{cloudy} are respectively the number of clear and cloudy sky events according to the reference cloud mask.

Many classification methods resort to estimate the probability density function, but in real applications they do not often fit very well real distributions. Anyway it is possible to recur to an estimator that is also convenient from the computational point of view: the Empirical Cumulative Density Function (ECDF), $F(x < \vartheta)$, that is the probability for the variable x to be below the threshold value, specified here with ϑ. An important result in asymptotic theory is that the estimator almost surely converges to the true cumulative function asymptotically [14]. The formulation related to the cloud detection problem is based on the individuation of a suitable variable, able to describe the cloudiness of the scene discriminating between clear or cloudy pixels, defining two populations. The best situation would be when the two populations do not overlap, but in practice the continuity of the variable involves a misclassifications for some pixels because there is no discontinuity in the passage from clear to cloudy scenes.

4.3. Cumulative Discriminant Analysis

In this section we describe CDA methodology in the univariate case, corresponding to a single statistic analysis. The basic assumption is that the predictor variable is of Discriminant Analysis type, where the estimator $\Gamma(\vartheta, \mathbf{x})$ is based on a threshold ϑ as follows:

$$\Gamma(x, \mathbf{x}) = \begin{cases} 1\ (Clear) & if\ x \geq \vartheta \\ 2\ (Cloudy) & if\ x < \vartheta \end{cases} \tag{1}$$

where $\mathbf{x} = x_1, \ldots, x_N$ is the statistic sample of size N defined by the data set, and each component x_i represents the predictor variable to be classified according to the threshold ϑ.

By CDA we are able to find the threshold ϑ, but we have to define a *cost function*, in order to minimize the error due to I-Type and II-Type misclassifications. Let E^I and E^{II} be respectively the I-type error and II-type error. According to the decision rule in Eq. 1 they are determined by: $E^I = F^{Clear}(\vartheta)$ and $E^{II} = 1 - F^{Cloudy}(\vartheta)$. Two possibilities of *cost function* can be:

$$A) \qquad C(\mathbf{x}, \vartheta) = E^I + E^{II} = F^{Clear}(\vartheta) + 1 - F^{Cloudy}(\vartheta) \tag{2}$$

$$B) \qquad C(\mathbf{x}, \vartheta) = \max(E^I, E^{II}) = \max(F^{Clear}(\vartheta), 1 - F^{Cloudy}(\vartheta)) \tag{3}$$

To find the best estimate of the threshold ϑ we have to minimize the *cost function*, which means to minimize both the I-type error and II-type error, and to obtain a balance between these two categories. If the training dataset is not balanced between Clear and Cloudy categories, the threshold overweights the most populated class, that will be better described from the estimated cloud mask. This is a common problem during the cloud detection process, because of the seasonal and meteorological variability of the scene, that could involve a quite significant difference in the relative fraction of Clear and Cloudy pixels.
The classification rule given in (1) is used to classify clear and cloudy pixels but CDA methodology does not assume the hypothesis of Gaussian probability density function, that does not occur in real conditions, conversely it computes the optimal estimate of the threshold ϑ making use of ECDF. We analyzed results for both *cost functions*, with the only difference that each error type is weighted with respect to the populousness of their own class, in order to take into account the different size of the two classes. Hereafter we indicate with A or B the results obtained considering respectively equations (2) and (3) in CDA.
CDA methodology can be generalized to the multidimensional case, that exploits simultaneously more than one statistic under the assumption they have independent density functions. In this way the multidimensional probability density function can be factorized as the product of each single probability density function.

5. Results

Here we show the main results keeping in mind this is a preliminary work to inspect and probe the potentiality of CDA method, applied to PCs of the considered statistics. We work out some sample test cases to obtain the optimal thresholds that we used for validation. The methodology has been applied for a small area, centred on the Basilicata region, located in the south of Italy, for two different periods, corresponding respectively to Mid-Latitude Winter and Mid-Latitude Summer climatic zones in the North hemisphere [12]. For each dataset we distinguished several cases according to the type of surface (land or sea) and illumination conditions (day-time or night-time). We selected the first PC for the cloud detection by CDA considering that the first eigenvalue is usually three or four orders of magnitude greater than the others from the SVD, which means that it alone is able to satisfactorily describe the scene cloudiness.

5.1. CDA for the training datasets

We computed the threshold ϑ applying CDA to SEVJAN_1 and SEVJUL_1 data sets for all the mentioned cases. In Table 2 we show the percentage of I-type error, II-type error and the performance of the cloud detection process computed according A or B method. We can see that the performance is usually higher for sea surface because of minor changes in temperature with respect to land surface, while in figure 1 there are two examples of ECDF.

5.2. CDA for the validation data sets

We used the thresholds obtained during the training to validate CDA methodology on SEVJAN_2 and SEVJUL_2 data sets. We evaluated the performance of the method comparing the estimated cloud mask with the SEVIRI reference cloud mask. Results are showed in Table 3.
In figure 2 there are two comparisons between the estimated cloudy scene and the reference cloud

Table 2. Training results applying CDA to the first PC of SEVJAN_1 and SEVJUL_1, for all cases.

Surface Type	Illumination condition	I-type error (%)		II-type error (%)		Total score (%)		Total of Soundings
		A	B	A	B	A	B	
SEVJAN_1								
Land	Day-time	9.14	9.33	9.65	9.48	81.21	81.20	1298142
	Night-time	5.24	13.14	19.40	13.25	75.36	73.61	2514978
Sea	Day-time	5.17	8.90	11.36	9.09	83.47	82.01	1866948
	Night-time	11.10	10.53	9.97	10.58	78.93	78.89	3577212
SEVJUL_1								
Land	Day-time	0.79	6.99	10.48	7.04	88.72	85.97	2202531
	Night-time	0.92	6.45	8.46	6.41	90.62	87.14	1610589
Sea	Day-time	0.45	8.04	10.65	8.44	88.90	83.52	3134058
	Night-time	0.44	3.11	4.01	3.27	95.55	93.62	2310102

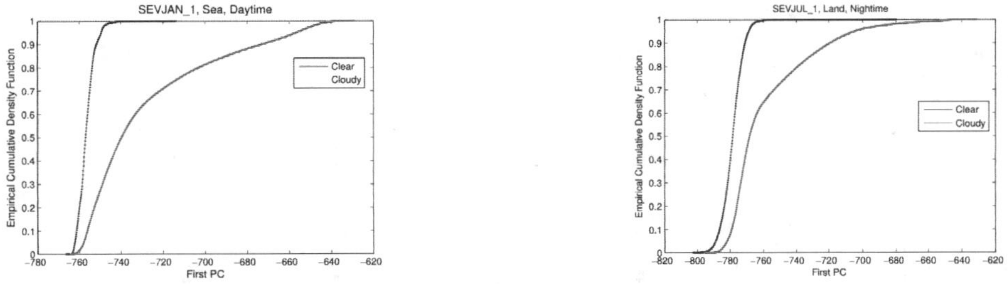

Figure 1. Two examples of ECDF: (left) ECDF applying CDA to sea surface in day-time for SEVJAN_1 dataset; (right) ECDF applying CDA to land surface in night-time for SEVJUL_1 dataset.

mask in two different cases. Comparing Table 2 and Table 3 we see that CDA methodology shows quite consistent scores both for winter and for summer data sets, meaning that the computed thresholds are suitable to differentiate between clear or cloudy areas.

Table 3. Validation results applied to the first PC, for SEVJAN_2 and SEVJUL_2 and for all cases.

Surface Type	Illumination condition	I-type error (%)		II-type error (%)		Total score (%)		Total of Soundings
		A	B	A	B	A	B	
SEVJAN_2								
Land	Day-time	10.20	10.02	11.32	11.53	78.48	78.45	1335666
	Night-time	16.54	9.85	12.65	23.57	70.81	66.59	2477454
Sea	Day-time	10.13	7.86	5.71	8.90	84.16	83.24	1918213
	Night-time	19.98	21.48	9.84	9.17	70.19	69.34	3525947
SEVJUL_2								
Land	Day-time	6.57	4.03	0.24	4.31	93.19	91.66	2172792
	Night-time	11.92	10.23	0.47	3.11	87.61	86.65	1640328
Sea	Day-time	7.12	6.31	0.05	0.81	92.82	92.88	3097970
	Night-time	9.75	9.66	0.03	0.14	90.22	90.20	2346190

6. Conclusions

In view of new missions, such as MTG, with improved satellite technologies, able to obtain enhanced quality data, we have experimented a new methodology in the field of cloud detection. We tested CDA potentiality as a new classification methodology to discriminate between clear

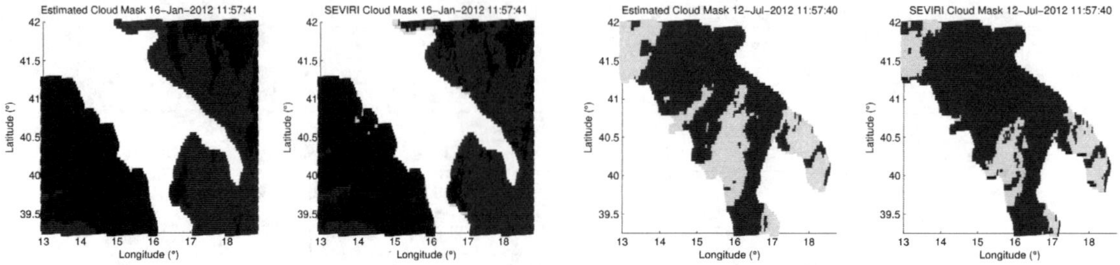

Figure 2. Comparison between CDA estimated cloudy scene and SEVIRI reference cloud mask for one SEVIRI acquisition in two cases. *Green* is for clear over land, *blue* for clear over sea and *dark red* for cloudy. (Left) Sea surface, Day-time for SEVJAN_2. (Right) Land surface, day-time for SEVJUL_2.

and cloudy pixels. We selected ten statistics, related to the cloud properties and computed from SEVIRI thermal IR data. PCA analysis has showed that most of information is contained in the first PC, that we analyzed by CDA. The method needs a reference cloud mask and a training data set in order to find the thresholds that allow us to classify clear and cloudy pixels.
We investigated the area focused on Basilicata region in two different periods aiming to evaluate the seasonal variability. For the validation we obtained quite good results: the methodology performance is higher for the summer data set, both for sea and land surface, with an agreement with respect to SEVIRI reference cloud mask that ranges from 87.6% to 93.2%, while the efficiency of the cloud detection process drops to about 70% when we applied CDA to the winter data set. We plan to improve this methodology analyzing other climatic zones and other statistics, that are descriptive of seasonal variability.

6.1. Acknowledgments
The research in this paper is a part of the corresponding author Ph.D. program, carried out in cooperation with the Italian Space Agency (ASI). The corresponding author also thanks ASI for supporting its Ph.D. scholarship.

References
[1] Ackerman S, Strabala K, Menzel W, Frey R A, Moeller C and Gumley L E, 1998 *J. Geophys. Res.*, **103**, 141
[2] Amato U, Antoniadis A, Cuomo V, Cutillo L, Franzese M, Murino L and Serio C, 2008 *Remote Sens. Envir.*, vol. **112**, 750
[3] Amato U, Lavanant L, Liuzzi G, Masiello G, Serio C, Stuhlmann R, and Tjemkes S A, 2014, *Atmospheric Measurement Techniques*, Vol. **7**, 3355
[4] Derrien M and Le Glau H, 2005, *Int. J. Remote Sens.*, **26**, 4707
[5] Derrien M, 2012, *Validation Report for Cloud Products* (CMa-PGE01 v3.2, CT PGE02 v2.2 and CTTH-PGE03 v2.2), EUMETSAT, Darmstadt.
[6] Heidinger A K, 2004, CLAVR-x Cloud Mask Algorithm Theoretical Basis Document, NOAA/NESDIS/Office of Research and Applications, Washington, D.C
[7] Inoue T, 1985 *J. Meteor. Soc. Japan*, **63**, 88
[8] Inoue T, Ackerman S A, 2002 *J. Meteor. Soc. Japan*, Vol. **80**, 138394
[9] Ricciardelli E, Romano F and Cuomo V, 2008 *Rem. Sens. Envir.*, **112**, 2741
[10] Rossow W B, 1989 *J. Climate*, **2**, 201213
[11] Schmetz J, Pili P, Tjemkes S, Just D, Kermann J, Rota S and Ratier A 2002 *Bull. Am. Met. Soc.*, **83**, 977
[12] Serio C, Masiello G, Venafra S, Amato U and De Feis I, 2013, Application Data for MTG-IRS Cloud Detection Method, *Final Report EUMETSAT*, EUMETSAT, Darmstadt
[13] Tapakis R and Charalambides A G, 2012 *Solar Energy*, **95**, 392−430
[14] Van der Vaart A W, 1998, *Asymptotic statistics*, Cambridge University Press, Cambridge

A multi-mode cantilever singular point detection using adaptive hypothesis testing

Jakub Dokoupil and Pavel Václavek

Central European Institute of Technology, Brno University of Technology, 616 00 Brno, Czech Republic

E-mail: `jakub.dokoupil@ceitec.vutbr.cz`, `pavel.vaclavek@ceitec.vutbr.cz`

Abstract. Fundamental analysis of a multi-mode model of the atomic force microscope cantilever shows that at some points; called here singular points, the mode is vanished. Consequently, the order of the input/output behavior is reduced. The singular points can be detected comparing possible candidates on the best model order. The detection is then naturally performed by applying the Bayesian model comparison. Since the exact position of the singular points is not available a priori, an explicit model of updating the probability of tested hypotheses in time is built. More specifically, a mechanism of suppressing absolute information is suggested based on the Bayesian decision problem where the Kullback-Leibler divergence is used.

1. Introduction

The problem of singular point detection of the atomic force microscope cantilever [1] can be viewed as the adaptive testing of hypotheses, each one matched to a certain model order. As the system model is consider to be stochastic, these hypotheses naturally represent the realizations of the mass function. To compensate the lack of knowledge about the true singular point positions, the idea of the posterior mass function flattening can be employed. The flattening is generally performed by exponential or linear forgetting [2]. In the present paper, we extend the approach of the exponential forgetting scheme. We attempt to give a stabilized forgetting strategy where an explicit model of updating the hypotheses in time is built. More specifically, the resulting formula is based on solving the statistical decision problem [3], [4], where the Kullback-Leibler divergence [5] is used to measure a distance between two mass functions.

Dealing with the problem of which from competing models is more suitable to explain the data generating process a subjective measure of confidence for each model is evaluated to choose the most probable one. The known solution satisfying the theorems of probability theory enables us to compare models with noninformative priors [6], [7] only over parameters which are common to all models. Hence, the given solution for the comparison of nonnested models produced by the Kalman based algorithms could not be put into practice. For this reason, the existing concept is reformulated preserving its Bayesian principle to be more beneficial in the case where on-line processing is needed.

2. Adaptive model comparison

Consider that the directly manipulated input u_k and the indirectly effected output y_k are both observed on the system at every discrete instant of time k. Let the sequence of input-output data pairs from time k_1 up to the time k be denoted by $D_{k_1}^k = \{u_i, y_i\}_{i=k_1}^k$. The j-mode cantilever model [1] can be described in its input/output form as follows:

$$ {}^{\lfloor j}\theta_{k+1} = {}^{\lfloor j}\theta_k, \tag{1} $$

$$ y_k = {}^{\lfloor j}h_k' \, {}^{\lfloor j}\theta_k + e_k, \quad e_k \sim \mathcal{N}(0, {}^{\lfloor j}r_k)\big|_{{}^{\lfloor j}r_{k+1} = {}^{\lfloor j}r_k}, \tag{2} $$

where ${}^{\lfloor j}h_k \in \mathbb{R}^{2j}$ represents the finite-dimensional regression vector containing observations $\{u_i, y_i\}_{i=k-j}^{k-1}$, h_k' stands for the transposition of h_k, ${}^{\lfloor j}\theta_k \in \mathbb{R}^{2j}$ is the vector of inner random variables. The unmeasurable variable e_k is presumed to be a discrete white noise, normally distributed with a zero mean and an unknown variance ${}^{\lfloor j}r_k$.

Solving the problem when more than one model order is considered to be possible, the selection of the best one can be performed by applying the Bayesian model comparison. Suppose that we have the finite set of different competing model orders $\{{}^{\lfloor j}\mathcal{M}_k\}_{j=1}^n$, which all seek to explain the system under study. To assess the degree of support for a particular model order ${}^{\lfloor j}\mathcal{M}_k$ corresponding to the parameter set ${}^{\lfloor j}\Theta_k = \{{}^{\lfloor j}\theta_k, {}^{\lfloor j}r_k\}$, the posterior model probability $f({}^{\lfloor j}\mathcal{M}_k | D_{1-n}^k)$ shall to be evaluated. The Bayes' rule gives

$$ f\left({}^{\lfloor j}\mathcal{M}_k \middle| D_{1-n}^k\right) \propto f\left(D_k^k \middle| D_{1-n}^{k-1}, {}^{\lfloor j}\mathcal{M}_k\right) f\left({}^{\lfloor j}\mathcal{M}_k \middle| D_{1-n}^{k-1}\right). \tag{3} $$

where $\propto$ stands for equal up to the normalization. Since the singular points are not known beforehand, an explicit model of order variations is not available. Therefore, we develop a forgetting scheme for the Bayesian model comparison that is linked to the following common idea. The probability of the realization ${}^{\lfloor j}\mathcal{M}_k$ given data D_{1-n}^k is described by the posterior probability $f({}^{\lfloor j}\mathcal{M}_k | D_{1-n}^k)$. The problem is to determine the predictive probability $f({}^{\lfloor j}\mathcal{M}_{k+1} | D_{1-n}^k)$ of the realization ${}^{\lfloor j}\mathcal{M}_{k+1}$ at the next time instant given the same data D_{1-n}^k. Let us introduce a prior information about the model orders through alternative mass functions. The problem then reads: given two alternatives on the mass function $f(\mathcal{M}_{k+1} | D_{1-n}^k)$ ($\mathcal{M}_{k+1}$ is the discrete random variable), namely

- the posterior mass function $f_0(\mathcal{M}_{k+1}) = f(\mathcal{M}_k | D_{1-n}^k)$ in the case of no order changes, and
- the alternative mass function $f_1(\mathcal{M}_{k+1}) \propto 1$ describing the maximal uncertainty about $\mathcal{M}_{k+1}$,

decide on an optimal mass function $\hat{f}(\mathcal{M}_{k+1})$. The task of choosing $\hat{f}(\mathcal{M}_{k+1})$ given $f_0(\mathcal{M}_{k+1})$ and $f_1(\mathcal{M}_{k+1})$ is formulated as a Bayesian decision problem. In particular, the probabilities λ and $1 - \lambda$ are assigned to the alternatives $f_0(\mathcal{M}_{k+1})$ and $f_1(\mathcal{M}_{k+1})$ and the best decision $\hat{f}(\mathcal{M}_{k+1})$ is defined as the mass function minimizing expectation of a distance between the true mass function $f(\mathcal{M}_{k+1})$ and the unknown mass function $f_A(\mathcal{M}_{k+1})$ restricted to the parameter space $f_A^*(\mathcal{M}_{k+1}) \equiv \{f_0(\mathcal{M}_{k+1}), f_1(\mathcal{M}_{k+1})\}$. In order to evaluate a proper distance between $f(\mathcal{M}_{k+1})$ and $f_A(\mathcal{M}_{k+1})$, the Kullback-Leibler divergence is used

$$ \mathcal{D}(f(\mathcal{M}_{k+1}) \| f_A(\mathcal{M}_{k+1})) = \sum_{j=1}^n f({}^{\lfloor j}\mathcal{M}_{k+1}) \ln \left(\frac{f({}^{\lfloor j}\mathcal{M}_{k+1})}{f_A({}^{\lfloor j}\mathcal{M}_{k+1})} \right). \tag{4} $$

Recall that ${}^{\lfloor j}\mathcal{M}_{k+1}$ is the realization of the random variable $\mathcal{M}_{k+1}$. Taking the expectation $\mathcal{E}[\cdot]$ of (4) results in

$$\begin{aligned}&\mathcal{E}\left[\mathcal{D}(f(\mathcal{M}_{k+1})\|f_A(\mathcal{M}_{k+1}))\right] \qquad (5)\\ =&\lambda\sum_{j=1}^{n} f({}^{\lfloor j}\mathcal{M}_{k+1})\ln\left(\frac{f({}^{\lfloor j}\mathcal{M}_{k+1})}{f_0({}^{\lfloor j}\mathcal{M}_{k+1})}\right)+(1-\lambda)\sum_{j=1}^{n} f({}^{\lfloor j}\mathcal{M}_{k+1})\ln\left(\frac{f({}^{\lfloor j}\mathcal{M}_{k+1})}{f_1({}^{\lfloor j}\mathcal{M}_{k+1})}\right)\\ =&\mathcal{D}\left(f(\mathcal{M}_{k+1})\middle\|\frac{f_0(\mathcal{M}_{k+1})^{\lambda}f_1(\mathcal{M}_{k+1})^{(1-\lambda)}}{\sum_{l=1}^{n} f_0({}^{\lfloor l}\mathcal{M}_{k+1})^{\lambda}f_1({}^{\lfloor l}\mathcal{M}_{k+1})^{(1-\lambda)}}\right)-\ln\left(\sum_{l=1}^{n} f_0({}^{\lfloor l}\mathcal{M}_{k+1})^{\lambda}f_1({}^{\lfloor l}\mathcal{M}_{k+1})^{(1-\lambda)}\right).\end{aligned}$$

The optimal prediction $\hat{f}(\mathcal{M}_{k+1})$ is found by solving the following decision problem:

$$\hat{f}(\mathcal{M}_{k+1}) = \arg\min_{f(\mathcal{M}_{k+1})} \mathcal{E}\left[\mathcal{D}(f(\mathcal{M}_{k+1})\|f_A(\mathcal{M}_{k+1}))\right]. \qquad (6)$$

The resulting form of (5) and the properties of the Kullback-Leibler divergence imply that for individual realizations generated by $\hat{f}(\mathcal{M}_{k+1})$ hold

$$f\left({}^{\lfloor j}\mathcal{M}_{k+1}\middle|D_{1-n}^{k}\right) = f^{\lambda}\left({}^{\lfloor j}\mathcal{M}_{k}\middle|D_{1-n}^{k}\right)/\sum_{l=1}^{n} f^{\lambda}\left({}^{\lfloor l}\mathcal{M}_{k}\middle|D_{1-n}^{k}\right). \qquad (7)$$

The probability $\lambda \in (0,1)$ can be viewed more prosaically as the forgetting factor with a lower value producing a higher degradation of less relevant information. At this point, equation (3) can be rewritten according to

$$f\left({}^{\lfloor j}\mathcal{M}_{k}\middle|D_{1-n}^{k}\right) \propto \int_{\Theta^*} f\left(y_k\middle|{}^{\lfloor j}h_k, {}^{\lfloor j}\Theta_k\right) f\left({}^{\lfloor j}\Theta_k\middle|D_{1-n}^{k-1}\right) d\,{}^{\lfloor j}\Theta_k f^{\lambda}\left({}^{\lfloor j}\mathcal{M}_{k-1}\middle|D_{1-n}^{k-1}\right), \qquad (8)$$

where Θ^* is the space of ${}^{\lfloor j}\Theta_k$. By invoking the system (1), (2), it allows us to algebraically quantify the functional form (8) as follows:

$$f({}^{\lfloor j}\mathcal{M}_k|D_{1-n}^{k}) = \frac{{}^{\lfloor j}c_k\,{}^{\lfloor j}\Sigma_{k-1}^{{}^{\lfloor j}\gamma_{k-1}/2}\,{}^{\lfloor j}\hat{r}_{y;k}^{-1/2}\,{}^{\lfloor j}\Sigma_{k}^{-{}^{\lfloor j}\gamma_k/2} f^{\lambda}\left({}^{\lfloor j}\mathcal{M}_{k-1}|D_{1-n}^{k-1}\right)}{\sum_{l=1}^{n}{}^{\lfloor l}c_k\,{}^{\lfloor l}\Sigma_{k-1}^{{}^{\lfloor l}\gamma_{k-1}/2}\,{}^{\lfloor l}\hat{r}_{y;k}^{-1/2}\,{}^{\lfloor l}\Sigma_{k}^{-{}^{\lfloor l}\gamma_k/2} f^{\lambda}\left({}^{\lfloor l}\mathcal{M}_{k-1}|D_{1-n}^{k-1}\right)}, \qquad (9)$$

where the following quantities have to be calculated for each model order:

$$\begin{aligned}
{}^{\lfloor j}\hat{r}_{y;k} &\equiv 1 + {}^{\lfloor j}h_k'\,{}^{\lfloor j}P_{k-1}\,{}^{\lfloor j}h_k, && (10)\\
{}^{\lfloor j}\hat{e}_k &\equiv y_k - {}^{\lfloor j}h_k'\,{}^{\lfloor j}\hat{\theta}_{k-1}, && (11)\\
{}^{\lfloor j}K_k &\equiv {}^{\lfloor j}P_{k-1}\,{}^{\lfloor j}h_k\,{}^{\lfloor j}\hat{r}_{y;k}^{-1}, && (12)\\
{}^{\lfloor j}\hat{\theta}_k &= {}^{\lfloor j}\hat{\theta}_{k-1} + {}^{\lfloor j}K_k\,{}^{\lfloor j}\hat{e}_k, && (13)\\
{}^{\lfloor j}P_k &= (I - {}^{\lfloor j}K_k\,{}^{\lfloor j}h_k')\,{}^{\lfloor j}P_{k-1}, && (14)\\
{}^{\lfloor j}\Sigma_k &= {}^{\lfloor j}\Sigma_{k-1} + {}^{\lfloor j}\hat{e}_k^2\,{}^{\lfloor j}\hat{r}_{y;k}^{-1}, && (15)\\
{}^{\lfloor j}\gamma_k &= {}^{\lfloor j}\gamma_{k-1} + 1 && (16)
\end{aligned}$$

and the auxiliary parameter ${}^{\lfloor j}c_k$ is defined as the ratio of two gamma functions ${}^{\lfloor j}c_k = \Gamma({}^{\lfloor j}\gamma_k/2)/\Gamma\left({}^{\lfloor j}\gamma_{k-1}/2\right)$. A detailed discussion about the particular quantities above and their prior settings can be found in [8].

From (9) it can be inferred that the Bayesian model comparison favors those models for which the addition of new observations will cause an increase in accuracy. The measure of the jth model order accuracy in terms of the model error rate in the output predictions is reflected by the sum of squared errors $\sum_{\kappa=j-n+1}^{k} (y_\kappa - {}^{\lfloor j}h'_\kappa\, {}^{\lfloor j}\hat{\theta}_k)^2$. These errors form the part of ${}^{\lfloor j}\Sigma_k$ with a lower value indicating a higher model accuracy. The model comparison also contains reward for coherency between the data and the prior information (via $({}^{\lfloor j}\hat{\theta}_k - {}^{\lfloor j}\hat{\theta}_{j-n})'({}^{\lfloor j}P^{-1}_{h;j-n})({}^{\lfloor j}\hat{\theta}_k - {}^{\lfloor j}\hat{\theta}_{j-n})$ appearing in ${}^{\lfloor j}\Sigma_k$) and parsimony.

Simulations produce excellent results and point out to the validity of our solution. However, due to space limitation, the simulation studies are not presented here.

3. Conclusion

In this paper a novel approach for detection of singular points of the multi-mode cantilever has been developed based on the Bayesian inference principle. More specifically, the algorithm which is capable of deciding which of the model orders is the most probable one was derived. To detect the singular points, which position is unknown, stabilized forgetting strategy that is close to the well-known technique of exponential forgetting was embed into the mechanism of time-evolution of the particular hypotheses. The algorithm generally provides a solution in the case where data are generated by the Gaussian probability density function through the unknown time-varying system parameters.

Acknowledgements

The first author was supported by the project CZ.1.07/2.3.00/30.0039 of Brno University of Technology. The work of the second author was realized at Central European Institute of Technology (CEITEC), with research infrastructure supported by the Project CZ.1.05/1.1.00/02.0068 financed from European Regional Development Fund. This support is gratefully acknowledged.

References

[1] Salapaka V M Bergh S H Lai J Majumdar A and McFarland E 1997 Multi-mode noise analysis of cantilevers for scanning probe microscopy *J. Appl. Phys.* **81** pp 2480-87
[2] Kulhavý R and Kraus J F 1996 On duality of regularized exponential and linear forgetting *Automatica* **32** pp 1403-15
[3] DeGroot H M 2004 *Optimal statistical decision* (Wiley-Interscience)
[4] Liese F and Miescke J K 2008 *Statistical decision theory: Estimation, testing, and selection* (Springer)
[5] Kullback S and Leibler R 1951 *On information and sufficiency* (Annals of Mathematical Statistics vol 22) pp 79-87
[6] Koop G 2003 *Bayesian econometrics* (Wiley-Interscience)
[7] Geweke J 2005 *Contemporary Bayesian econometrics and statistics* (Wiley-Interscience)
[8] Kárný M Bohm J Guy T Jirsa L Nagy I Nedoma P and Tesar L 2006 *Optimized Bayesian dynamic advising: Theory and algorithms* (Springer Verlag)

4th International Conference on Mathematical Modeling in Physical Sciences (IC-MSquare2015) IOP Publishing
Journal of Physics: Conference Series **633** (2015) 012058 doi:10.1088/1742-6596/633/1/012058

Numerical methods for solution of the stochastic differential equations equivalent to the non-stationary Parkers transport equation

A. Wawrzynczak[1], R. Modzelewska[2], M. Kluczek[2]
[1]Institute of Computer Sciences, Siedlce University, Poland,
[2]Institute of Mathematics and Physics, Siedlce University, Poland.
E-mail: awawrzynczak@uph.edu.pl, renatam@uph.edu.pl

Abstract. We derive the numerical schemes for the strong order integration of the set of the stochastic differential equations (SDEs) corresponding to the non-stationary Parker transport equation (PTE). PTE is 5-dimensional (3 spatial coordinates, particles energy and time) Fokker-Planck type equation describing the non-stationary galactic cosmic ray (GCR) particles transport in the heliosphere. We present the formulas for the numerical solution of the obtained set of SDEs driven by a Wiener process in the case of the full three-dimensional diffusion tensor. We introduce the solution applying the strong order Euler-Maruyama, Milstein and stochastic Runge-Kutta methods. We discuss the advantages and disadvantages of the presented numerical methods in the context of increasing the accuracy of the solution of the PTE.

1. Introduction

We employ the stochastic methodology to model the galactic cosmic rays (GCR) transport in the heliosphere. Modulation of the GCR is a result of an action of four primary processes: convection by the solar wind, diffusion on irregularities of the heliospheric magnetic field, particles drifts in the non-uniform magnetic field and adiabatic cooling. Transport of the GCR particles in the heliosphere can be described by the Parker transport equation (PTE) [1]:

$$\frac{\partial f}{\partial t} = \vec{\nabla} \cdot (K_{ij}^S \cdot \vec{\nabla} f) - (\vec{v}_d + \vec{U}) \cdot \vec{\nabla} f + \frac{R}{3}(\vec{\nabla} \cdot \vec{U}) \frac{\partial f}{\partial R}, \qquad (1)$$

where $f = f(\vec{r}, R, t)$ is an omnidirectional distribution function of three spatial coordinates $\vec{r} = (r, \theta, \varphi)$, particles rigidity R and time t; $\vec{U}$ is the solar wind velocity, $\vec{v}_d$ the drift velocity, and K_{ij}^S is the symmetric part of the diffusion tensor of the GCR particles.

This paper is an extension of our previous results presented in [2]. We have presented [2] that the GCR transport can be effectively modelled based on the solution of the set of stochastic differential equations (SDEs) corresponding to the PTE (1). Firstly, the PTE must be brought to the form of the backward Fokker-Planck equation (FPE) (e.g.[3]), and then the corresponding SDEs must be solved numerically. In [2, 4] the Euler-Maruyama method was applied. In this paper, we increase the accuracy of the SDEs solution by applying the higher order methods i.e. Milstein and stochastic Runge-Kutta.

2. Numerical methods for solving stochastic differential equations

The Euler-Maruyama method for solving the SDEs is unconditionally stable in the higher dimensions and doesn't rely on the numerical grid. All integration methods involve the statistical error, which can be reduced by increasing the number of simulated pseudoparticles and applying the higher order integration method e.g. Milstein or Runge-Kutta method e.g.[5]. These methods improve the approximation of the solution by applying higher order extension of the SDE solution in the Ito - Taylor series, e.g.[6]. The Euler-Maruyama method has an order of convergence $\gamma = 0.5$, by adding just one more term from the Ito - Taylor expansion the order of convergence increases for the Milstein method up to $\gamma = 1$. In turn, the stochastic Runge–Kutta method, apart from next term of the Ito-Taylor expansion, require to generate a new random variable $Z(t)$ resulting in a strong order of convergence with $\gamma = 1.5$.

Let's consider a basic SDE:

$$dX_t = f(X_t)dt + g(X_t)dW_t, \tag{2}$$

depending on the applied method the numerical approximation of the solution can be obtained by following formulas :

$$X_{j+1} = \underbrace{\underbrace{\underbrace{X_j + f \cdot dt + g \cdot dW_t}_{Euler-Maruyama} + \frac{1}{2} g \cdot g'(dW_j^2 - dt)}_{Milstein} + \Phi}_{Stochastic\ Runge-Kutta} \tag{3}$$

where $\Phi = f' \cdot g \cdot dZ_j + \frac{1}{2}(f \cdot f' + \frac{1}{2} g^2 \cdot f'')dt^2 + (f \cdot g' + \frac{1}{2} g^2 \cdot g'') \cdot (dW_j \cdot dt - dZ_j) + \frac{1}{2} g(g \cdot g'' + g'^2)(\frac{1}{3} dW_j^2 - dt)dW_j$ and $dZ_i = \frac{1}{2} dt(dW_i + \frac{dV_i}{\sqrt{3}})$ and dV_i is an additional Wiener process e.g.[6] .

3. Stochastic differential equation corresponding to the Parker transport equation

The PTE (Eq.1) in the 3-D heliocentric spherical coordinate system (r, θ, φ) written as time-backward FPE diffusion equation has the form:

$$\frac{\partial f}{\partial t} = A_1 \frac{\partial^2 f}{\partial r^2} + A_2 \frac{\partial^2 f}{\partial \theta^2} + A_3 \frac{\partial^2 f}{\partial \varphi^2} + A_4 \frac{\partial^2 f}{\partial r \partial \theta} + A_5 \frac{\partial^2 f}{\partial r \partial \varphi} + A_6 \frac{\partial^2 f}{\partial \theta \partial \varphi} + A_7 \frac{\partial f}{\partial r} + A_8 \frac{\partial f}{\partial \theta} + A_9 \frac{\partial f}{\partial \varphi} + A_{10} \frac{\partial f}{\partial R} \tag{4}$$

the coefficients A_1,...,A_{10} are presented in detail in [2]. Depending on the choice of the numerical approximation the corresponding to Eq.4 set of SDEs has a form:

$$\begin{aligned}
dr &= \underbrace{\underbrace{\underbrace{A_7 \cdot dt + B_{11} \cdot dW_r}_{Euler-Maruyama} + \frac{1}{2} B_{11} \frac{\partial B_{11}}{\partial r}(dW_r^2 - dt)}_{Milstein} + \Phi_1}_{Stochastic\ Runge-Kutta} \\
d\theta &= \underbrace{\underbrace{\underbrace{A_8 \cdot dt + B_{21} \cdot dW_r + B_{22} \cdot dW_\theta}_{Euler-Maruyama} + \frac{1}{2} B_{21} \frac{\partial B_{21}}{\partial \theta}(dW_r^2 - dt) + \frac{1}{2} B_{22} \frac{\partial B_{22}}{\partial \theta}(dW_\theta^2 - dt)}_{Milstein} + \Phi_2}_{Stochastic\ Runge-Kutta} \\
d\varphi &= \underbrace{\underbrace{\underbrace{A_9 \cdot dt + B_{31} \cdot dW_r + B_{32} \cdot dW_\theta + B_{33} \cdot dW_\varphi}_{Euler-Maruyama} + \Phi_3}_{Milstein} + \Phi_4}_{Stochastic\ Runge-Kutta} \\
dR &= A_{10} \cdot dt.
\end{aligned} \tag{5}$$

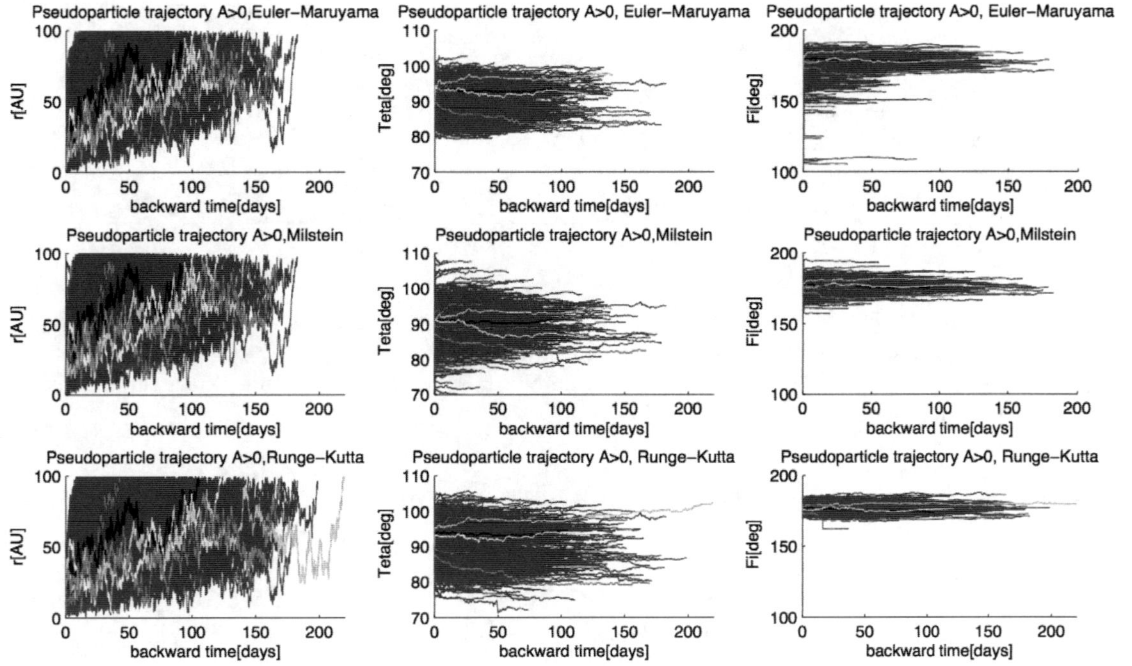

Figure 1. Trajectories of the pseudoparticles with rigidity 10 GV for the $A > 0$ solar magnetic cycle obtained applying the Euler-Maruyama, Milstein and stochastic Runge-Kutta methods. The specific colors highlight the trajectories of the sample pseudoparticles, based on the same Wiener process, traced backward in time from the position $r = 1AU$, $\theta = 90°$, $\varphi = 180°$ until they reach the heliosphere boundary.

The B_{ij}, $(i, j = r, \theta, \varphi)$, is the lower triangular matrix presented in [2], and the coefficients Φ_1, Φ_2, Φ_3 and Φ_4 have the form:

$$\Phi_1 = B_{11} \cdot dZ_r \frac{\partial A_7}{\partial r} + \frac{1}{2}(A_7 \frac{\partial A_7}{\partial r} + \frac{1}{2} B_{11}^2 \frac{\partial^2 A_7}{\partial r^2}) dt^2 + (A_7 \frac{\partial B_{11}}{\partial r} + \frac{1}{2} B_{11}^2 \frac{\partial^2 B_{11}}{\partial r^2})(dW_r \cdot dt - dZ_r) + \frac{1}{2} B_{11}(B_{11} \frac{\partial^2 B_{11}}{\partial r^2} + (\frac{\partial B_{11}}{\partial r})^2)(\frac{1}{3} dW_r^2 - dt) dW_r;$$

$$\Phi_2 = B_{21} \cdot dZ_r \frac{\partial A_8}{\partial \theta} + \frac{1}{2}(A_8 \frac{\partial A_8}{\partial \theta} + \frac{1}{2} B_{21}^2 \frac{\partial^2 A_8}{\partial \theta^2}) dt^2 + (A_8 \frac{\partial B_{21}}{\partial \theta} + \frac{1}{2} B_{21}^2 \frac{\partial^2 B_{21}}{\partial \theta^2})(dW_r \cdot dt - dZ_r) + \frac{1}{2} B_{21}(B_{21} \frac{\partial^2 B_{21}}{\partial \theta^2} + (\frac{\partial B_{21}}{\partial \theta})^2)(\frac{1}{3} dW_r^2 - dt) dW_r + B_{22} \cdot dZ_\theta \frac{\partial A_8}{\partial \theta} + \frac{1}{2}(A_8 \frac{\partial A_8}{\partial \theta} + \frac{1}{2} B_{22}^2 \frac{\partial^2 A_8}{\partial \theta^2}) dt^2 + (A_8 \frac{\partial B_{22}}{\partial \theta} + \frac{1}{2} B_{22}^2 \frac{\partial^2 B_{22}}{\partial \theta^2})(dW_\theta \cdot dt - dZ_\theta) + \frac{1}{2} B_{22}(B_{22} \frac{\partial^2 B_{22}}{\partial \theta^2} + (\frac{\partial B_{22}}{\partial \theta})^2)(\frac{1}{3} dW_\theta^2 - dt) dW_\theta;$$

$$\Phi_3 = \frac{1}{2} B_{31} \frac{\partial B_{31}}{\partial \varphi}(dW_r^2 - dt) + \frac{1}{2} B_{32} \frac{\partial B_{32}}{\partial \varphi}(dW_\theta^2 - dt) + \frac{1}{2} B_{33} \frac{\partial B_{33}}{\partial \varphi}(dW_\varphi^2 - dt);$$

$$\Phi_4 = B_{31} \cdot dZ_r \frac{\partial A_9}{\partial \varphi} + \frac{1}{2}(A_9 \frac{\partial A_9}{\partial \varphi} + \frac{1}{2} B_{31}^2 \frac{\partial^2 A_9}{\partial \varphi^2}) dt^2 + (A_9 \frac{\partial B_{31}}{\partial \varphi} + \frac{1}{2} B_{31}^2 \frac{\partial^2 B_{31}}{\partial \varphi^2})(dW_r \cdot dt - dZ_r) + \frac{1}{2} B_{31}(B_{31} \frac{\partial^2 B_{31}}{\partial \varphi^2} + (\frac{\partial B_{31}}{\partial \varphi})^2)(\frac{1}{3} dW_r^2 - dt) dW_r + B_{32} \cdot dZ_\theta \frac{\partial A_9}{\partial \varphi} + \frac{1}{2}(A_9 \frac{\partial A_9}{\partial \varphi} + \frac{1}{2} B_{32}^2 \frac{\partial^2 A_9}{\partial \varphi^2}) dt^2 + (A_9 \frac{\partial B_{32}}{\partial \varphi} + \frac{1}{2} B_{32}^2 \frac{\partial^2 B_{32}}{\partial \varphi^2})(dW_\theta \cdot dt - dZ_\theta) + \frac{1}{2} B_{32}(B_{32} \frac{\partial^2 B_{32}}{\partial \varphi^2} + (\frac{\partial B_{32}}{\partial \varphi})^2)(\frac{1}{3} dW_\theta^2 - dt) dW_\theta + B_{33} \cdot dZ_\varphi \frac{\partial A_9}{\partial \varphi} + \frac{1}{2}(A_9 \frac{\partial A_9}{\partial \varphi} + \frac{1}{2} B_{33}^2 \frac{\partial^2 A_9}{\partial \varphi^2}) dt^2 + (A_9 \frac{\partial B_{33}}{\partial \varphi} + \frac{1}{2} B_{33}^2 \frac{\partial^2 B_{33}}{\partial \varphi^2})(dW_\varphi \cdot dt - dZ_\varphi) + \frac{1}{2} B_{33}(B_{33} \frac{\partial^2 B_{33}}{\partial \varphi^2} + (\frac{\partial B_{33}}{\partial \varphi})^2)(\frac{1}{3} dW_\varphi^2 - dt) dW_\varphi.$$

4. Results and Summary

We performed the simulation applying all three methods given by Eqs.5. The trajectory of $n = 3000$ pseudoparticles was traced backward in time in the spherical heliocentric coordinate system. The pseudoparticles were initialized in the point representing the Earth's orbit (i.e.

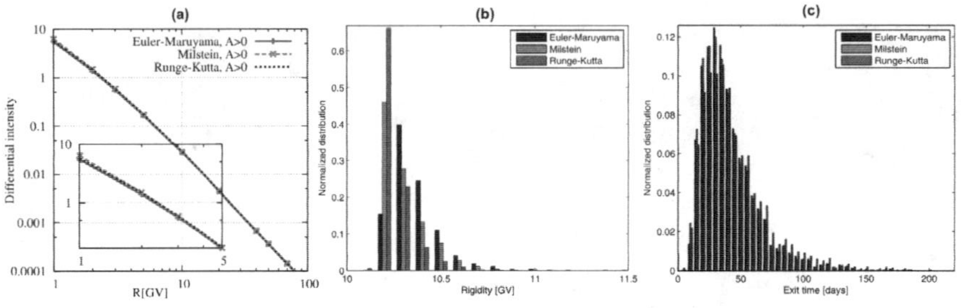

Figure 2. (a) Simulated galactic protons rigidity spectra and the histograms of (b) the particles rigidity and (c) the particles exit time for the pseudoparticles initialized with $R = 10GV$ from position $r = 1AU$, $\theta = 90°$, $\varphi = 180°$ for the $A > 0$ solar magnetic cycle obtained applying the Euler-Maruyama, Milstein and stochastic Runge-Kutta methods.

$r = 1AU$, $\theta = 90°$, $\varphi = 180°$) and traced backward in time until crossing the heliosphere boundary assumed at 100 AU (see Fig. 1 in [2]). The value of the particle distribution function $f(\vec{r}, R)$ for the starting point was found as an average of $f_{LIS}(R)$ value for pseudoparticles characteristics at the entry positions, i.e. $f(\vec{r}, R) = \frac{1}{N}\sum_{n=1}^{N} f_{LIS}(R)$, where $f_{LIS}(R)$ is the cosmic ray local interstellar spectrum taken as in [8] for rigidity R of the nth particle at the entrance point.

In the case when we know the analytical solution it is easy to compare the efficiency of each numerical method. However, here we do not have such a possibility. Thus, we have compared the simulated galactic protons rigidity spectra for $A > 0$ solar magnetic cycle. Fig. 2a presents that all applied numerical schemes resulted in the same spectra, slight differences are seen for the lower rigidity particles. The obtained spectrum is in agrement with results presented in [7].

Fig. 1 presents the trajectories of the pseudoparticles with rigidity 10 GV for the $A > 0$ solar magnetic cycle with respect to all coordinates. To allow the reliable comparison all simulations were based on the same Wiener process. One can see that there arises some subtle differences: the trajectories of pseudoparticles vs. the heliolongitude are the widest for Euler-Maruyama and narrow for the Runge-Kutta method. At the same time the opposite distribution is observed vs. the heliolatitude. Additionally, we have estimated the histograms of the particles rigidity and the particles exit time for all three methods. Fig. 2bc show that the time of the pseudoparticles travel throughout the heliosphere is the longest applying the stochastic Runge-Kutta method, at the same time particles loses less energy during travel.

The obtained results suggest that all applied methods give a reliable results. Performed tests proved that the Milstein and stochastic Runge-Kutta methods are more stable and return the same values of differential spectra when we decrease the number of simulated particles by factor of three. However, the slight differences between the pseudoparticles trajectories and theirs characteristics (rigidity and exit time) given by the tested numerical schemes must be analyzed in more detail.

5. Acknowledgments and References

This work is supported by The Polish National Science Centre grant awarded by decision number DEC-2012/07/D/ST6/02488.

[1] Parker E N 1965 *Planetary and Space Science* **13** 9-49
[2] Wawrzynczak A, Modzelewska R and Gil A 2015 *Journal of Physics: Conference Series* **574** 012078
[3] Gardiner C W 2009 *Handbook of stochastic methods. For physics, chemistry and the natural sciences* Springer Series in Synergetics
[4] Kopp A, Busching I, Strauss R D and Potgieter M S 2012 *Computer Physics Communications* **183** 530-542

[5] Riadh B, Ouerdiane H 2014 *Int. Journal of Math. Analysis* **8** 5 229-246
[6] Kloeden P E, Platen E, Schurz H 2003 *Numerical solution of SDE through computer experiments* 24, 140-146
[7] Zhang M 1999 *Astrophys. J.* **513** 409-420
[8] Webber W R and Lockwood J A 2001 *Journal of Geophysical Research* **106** 29323-29332

Hyper fast radiative transfer for the physical retrieval of surface parameters from SEVIRI observations

G Liuzzi[1], G Masiello[1,2], C Serio[1,2], M G Blasi[1] and S Venafra[1]

[1] School of Engineering, Università degli Studi della Basilicata, Via Ateneo Lucano 10, 85100 Potenza, Italy

[2] CNISM, Unità di Ricerca Università della Basilicata, 85100 Potenza, Italy

E-mail: giuliano.liuzzi@unibas.it

Abstract. This paper describes the theoretical aspects of a fast scheme for the physical retrieval of surface temperature and emissivity from SEVIRI data, their implementation and some sample results obtained. The scheme is based on a Kalman Filter approach, which effectively exploits the temporal continuity in the observations of the geostationary Meteosat Second Generation (MSG) platform, on which SEVIRI (Spinning Enhanced Visible and InfraRed Imager) operates. Such scheme embodies in its core a physical retrieval algorithm, which employs an hyper fast radiative transfer code highly customized for this retrieval task. Radiative transfer and its customizations are described in detail. Fastness, accuracy and stability of the code are fully documented for a variety of surface features, showing a peculiar application to the massive Greek forest fires in August 2007.

1. Introduction

Geostationary satellite platforms have the capability of monitoring large portions of our planet with the advantage of having temporal continuity between consecutive observations. Such continuity could enable to solve the diurnal cycle of many of geophysical parameters, and should be properly exploited when data are analyzed, in order to maximize the information content extracted by data themselves. In particular, the observational continuity can be useful to correctly constrain the physical retrieval of the geophysical parameters of interest, even with different time scales of variability. This job has been already done with success [1] on geostationary sensors data by adopting a Kalman Filter (KF) strategy [2].

Even though the observational continuity is essential to follow the evolution of the geophysical parameters of interest, the continuous data collection is very demanding from the computational point of view: an enormous amount of data must be processed in a very reduced time. This means that, besides a solid mathematical framework for inversion, one should also have a very fast and accurate way to perform radiative transfer calculations, which are the bases for physical inversion. In this paper, we will address this peculiar problem by describing the implementation of an hyper fast radiative transfer model for simulating the radiance as it is observed by the SEVIRI instrument, which is on board of the Meteosat Second Generation (MSG) platform, currently the operative core of the geostationary meteorological programme by EUMETSAT.

The aim of the KF methodology implementation, largely addressed in [1], is to develop a scheme in which the retrieval of the parameters of interest was driven exclusively by data, not combining a Numerical Weather Prediction scheme (NWP, see e.g. [3]) with this machinery.

Journal of Physics: Conference Series **633** (2015) 012059 doi:10.1088/1742-6596/633/1/012059

This points out why having a fast radiative transfer code is really a fundamental piece of the work: this approach relies only on data and the efficiency of Optimal Estimation in fitting observed radiance with radiative transfer calculations. Since the task we have dealt with is limited to surface temperature and emissivity retrieval, we have optimized the radiative transfer code paying particular attention to the accuracy of calculations in those spectral regions where atmospheric absorption is minimal. Section 2 will describe the model in its details; in section 3 we will show the results obtained with the whole procedure on the case study of the Greek fires occurred during August 2007. Finally, conclusions will be drawn in Section 4.

2. Fast radiative transfer and SEVIRI

2.1. The SEVIRI imager

The SEVIRI imager, on board of Meteosat-9 geostationary platform, produces a complete scan (from now on *Full Disk*, FD) of the visible hemisphere once every 15 minutes. The FD image is a cube with a spatial resolution of 3 km, and 12 broad spectral channels, 8 of which are in the thermal infrared band, that of our interest. These eight SEVIRI channels cover the spectral range [3.9 μm; 13.4 μm]. The spectral response for each channel is plotted in Fig. 1, where we indicate also the predominant information content therein.

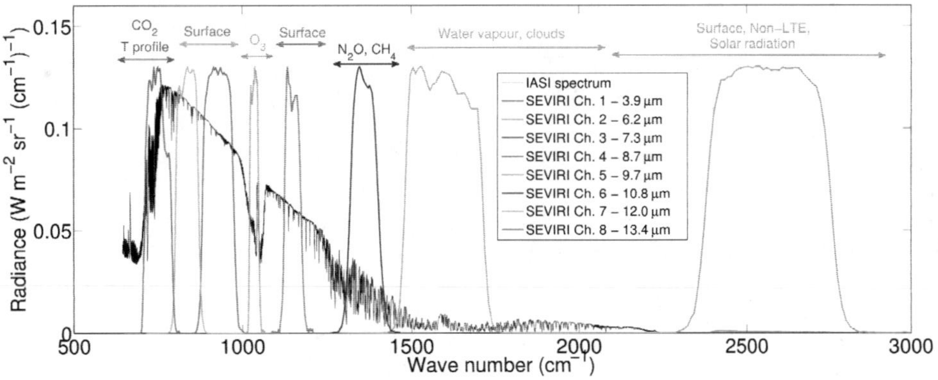

Figure 1. Spectral response of the eight thermal infrared SEVIRI channels. They are superposed to a typical IASI [4] spectrum at a resolution of 0.5 cm^{-1} to highlight the main spectral features that yield the radiance in each SEVIRI channel. SEVIRI spectral response functions are in arbitrary units.

It is clearly visible that four of the eight SEVIRI infrared channels (numbers 1, 4, 6 and 7 in Table 1) are centred on atmospheric windows, hence the corresponding radiance values are only marginally affected by gaseous absorption. It should also be noted that Channel 1, though it is entirely located in an atmospheric window, is also characterized by the lowest radiance value; what is more, the radiance observed in this channel is not easy to be modelled given the solar contamination during the day and other non-LTE effects (i.e. multiple scattering).

The way in which the radiative transfer code has been tailored is mainly driven by the fact that the problem of retrieving surface properties does not require to work with a very complicated (and consequently, slow) atmospheric model to obtain accurate values of radiance in the three SEVIRI channels we use for retrieval. In addition, SEVIRI spectral response function is far broader than any gas absorption line: this enables us to perform radiative transfer calculations at a relatively coarse resolution, which in turn steeply reduces computational burden and time.

2.2. Radiative transfer basic elements

The forward radiative transfer model we describe here is called σ-SEVIRI, and has been specifically developed for this instrument. The model's name is inspired by the σ-IASI [5] from which this model (and other similar, see e.g. [6]) has been derived. The radiative transfer scheme we adopt has been already extensively validated in many studies [7, 8, 9, 10] as far as both radiative transfer itself and retrieval applications are concerned. The σ-SEVIRI model solves the complete radiative transfer equation (RTE), which expresses the total radiance observed by the sensor, and has the following form:

$$R(\sigma) = R_{\uparrow}(\sigma) + (1-\epsilon_g)\tau_0 R_{\downarrow}(\sigma) + R_{\odot}(\sigma) =$$
$$\epsilon_g B(T_s)\tau_0 + \int_0^{+\infty} B(T)\frac{\partial \tau}{\partial z}\mathrm{d}z + (\epsilon_g - 1)\tau_0^2 \int_0^{+\infty} B(T)\frac{\partial}{\partial z}\left(\frac{1}{\tau}\right)\mathrm{d}z + \frac{1-\epsilon_g}{\pi}\tau' \mu_s I_s(\sigma) \quad (1)$$

where σ is the wave number, ϵ_g surface emissivity, T_s surface temperature, B the Planck function, τ the atmospheric transmittance, z the height on the surface, I_s the solar irradiance. The first term of the RTE is the radiance directly emitted from the surface, scaled with the atmospheric total transmittance τ_0; the second term represents the radiance emitted from the atmosphere integrated over the whole observation path; the third term is the down welling radiance, that is that emitted from the atmosphere, reflected from surface and then reaching the sensor. Both Lambertian and specular reflections can be modelled, in order to use the code both on land and sea surface. Finally, the last term is the solar irradiance which is reflected from surface; this last term becomes important only at wave numbers greater than $\approx$2000 cm^{-1}.

The RTE in the form (1) cannot be directly implemented, but it requires to be a bit manipulated. The key passage consists in its discretization, which is done by dividing the atmosphere in a certain number of layers, namely N_L. We choose to use a fixed pressure grid, where each layer is rigorously defined by its pressure boundaries; each layer is assumed to have an uniform temperature and composition. Equation (1) then becomes:

$$R(\sigma) = \epsilon_g B(T_s)\tau_0 + \sum_{j=1}^{N_L} B(T_j)(\tau_j - \tau_{j-1}) + (\epsilon_g - 1)\tau_0^2 \sum_{j=1}^{N_L} B(T_j)(\tau_j^{-1} - \tau_{j-1}^{-1}) + \frac{1-\epsilon_g}{\pi}\tau' \mu_s I_s(\sigma) \quad (2)$$

with the same notation as above. Equation (2) reduces the computed radiance to sums/products of transmittances τ_j from the bottom of the j-th layer to the atmospheric top. To accomplish calculations of transmittances, the σ-SEVIRI model uses a wide look-up table for the optical depth. This table is pre-calculated using the reference Line-By-Line Radiative Transfer Model (LBLRTM) most recent available version [11].

Before giving any detail about the two steps by which we have speeded up the σ-SEVIRI code, it is worth to explain how transmittances are expressed and how the look-up table is built. For simplicity, let us consider a single layer j in our pressure grid. At a given wave number σ, the partial transmittance of the layer is expressed through its relation with the optical depths of the S absorbing species in that layer:

$$\tau(\sigma) = \tau_{j-1}(\sigma)/\tau_j(\sigma) = \exp\left(-\sum_{i=1}^{S} \chi_{i,j}(\sigma)\right) \quad (3)$$

where $\chi_{i,j}(\sigma)$ is the optical depth of the i-th absorber in the j-th layer. In σ-SEVIRI, we take into account only gases as absorbing species, and the strategy we adopt is to exploit the dependence of the gases optical depths both on their concentration and the layer temperature T_j. In detail, the optical depth $\chi_{i,j}(\sigma)$ is expressed as a simple second-order polynomial of the temperature for carbon dioxide, ozone and all the other gases but water vapour treated all together as *mixed gases* (i=2,3,4). For water vapour (i=1), instead, we add a further coefficient in order to take into account effects depending on its abundance, such as self-broadening [12]. Hence

$$\chi_{1,j}(\sigma) = q_{1,j}\left(\sum_{n=0}^{2} c_{n,1,j}(\sigma)(\Delta T)^n + c_{3,1,j}(\sigma)q_{1,j}\right) \qquad \chi_{i,j}(\sigma) = q_{i,j}\sum_{n=0}^{2} c_{n,i,j}(\sigma)(\Delta T)^n \quad (4)$$

where $q_{i,j}$ is the molecule concentration, and ΔT is the difference between the actual layer temperature and a reference temperature, that is the central point of the parameterization. The $c_{\cdot,i,j}$ coefficients are stored in the look-up table used by the code. In principle, calculations should be performed at *infinite resolution* in such a way to take into account the single spectral line contribution. For Earth atmosphere, this would correspond to $\Delta\sigma$=10^{-4} cm^{-1}, making calculations extremely slow.

In the case of SEVIRI, thanks to its broad channels, we can work with a far coarser resolution, without losing in accuracy. The equivalent optical depths for water vapour and the other gases, can be written as follows

$$\chi_{1,j}^{\langle\sigma\rangle} = q_{1,j}\left(\sum_{n=0}^{2} c_{n,1,j}^{\langle\sigma\rangle}(\Delta T)^n + c_{3,1,j}^{\langle\sigma\rangle}q_{1,j}\right) \qquad \chi_{i,j}^{\langle\sigma\rangle} = q_{i,j}\sum_{n=0}^{2} c_{n,i,j}^{\langle\sigma\rangle}(\Delta T)^n \quad (5)$$

for all the other gases. The symbol $\langle\sigma\rangle$ indicates the average over wave number. The averaged coefficients are obtained via averaging transmittances and not optical depths, because of the form in which RTE is solved by the model, which deals with layers' transmittances. In the work described in [1], optical depths coefficients' database was generated at a resolution of 10^{-1} cm^{-1}, and with an atmosphere model made by N_L=60 layers. In this work, instead, the first step to speed up the code has been to enlarge the binning step until 5×10^{-1} cm^{-1}. Radiances calculated at this resolution are then convolved to SEVIRI resolution through its Instrument Spectral Response Function (ISRF). On the one hand, this has yielded a drastic reduction of the look-up table size; on the other hand, this has cut by a factor of 5 the time needed to compute a single spectrum and all its derivatives, that are used in the retrieval framework.

The second optimization step is related to the fact that we use only window channels, where atmospheric absorption is only marginal. Thus, there should be no or negligible bias in computed radiances if we reduce the number N_L of atmospheric layers used to perform radiative transfer calculations. Hence, we have chosen to further optimize the code choosing N_L=25. Such a choice is for convenience, since in the KF machinery we use the atmospheric profiles provided by ECMWF (European Centre for Medium-range Weather Forecast) whose analyses data are given on a 25-layer pressure grid. As a consequence, no preliminary interpolation is required to fit data into the pressure grid, further speeding calculations. The comparison between the two pressure grids (60 and 25 layers) is given in Fig. 2, while an inter-comparison between sample calculations performed with 60 and 25 layers is provided in Table 1. The tabulated radiances evidence how the coarsening of the resolution at which calculations are performed has introduced practically no bias in the results, while the reduction of N_L has an effect which is at most of the same order of magnitude of the SEVIRI radiometric noise, again introducing no significant bias in the radiance calculated in the three atmospheric window channels we use in KF retrieval.

With these modifications to the σ-SEVIRI model, the inversion scheme to retrieve surface parameters has been speeded up by a factor of two with respect to the original configuration.

Because of these results, this boosted version of σ-SEVIRI suits well in an application, like that we are going to show here, in which a very huge amount of data has to be analyzed in near-real time. σ-SEVIRI has been integrated in a physical retrieval framework, whose core is an iterative, non linear Optimal Estimation Gauss-Newton scheme, that has been already

Table 1. Sample radiance calculation results for the three window channels used in the KF scheme. The comparison points out that the differences between calculations made at 10^{-1} cm^{-1} (first row), those made at 5×10^{-1} cm^{-1} (second row), both with a 60-layer pressure grid, and those made at a resolution of 5×10^{-1} cm^{-1} and a 25-layer pressure grid (third row) are below SEVIRI radiometric noise (fourth row). Radiances are expressed in W m^{-2} sr^{-1} (cm^{-1})$^{-1}$ and are calculated for the U.S. standard climatology profile AFGL6 [13].

Sampling-layers	**Radiance @12.0 μm**	**@10.8 μm**	**@8.7 μm**
10^{-1} cm^{-1} - 60 layers	$1.0194\cdot10^{-1}$	$8.816\cdot10^{-2}$	$5.313\cdot10^{-2}$
5×10^{-1} cm^{-1} - 60 layers	$1.0193\cdot10^{-1}$	$8.816\cdot10^{-2}$	$5.306\cdot10^{-2}$
5×10^{-1} cm^{-1} - 25 layers	$1.0195\cdot10^{-1}$	$8.812\cdot10^{-2}$	$5.312\cdot10^{-2}$
SEVIRI NEdN	$2.63\cdot10^{-4}$	$1.86\cdot10^{-4}$	$1.35\cdot10^{-4}$

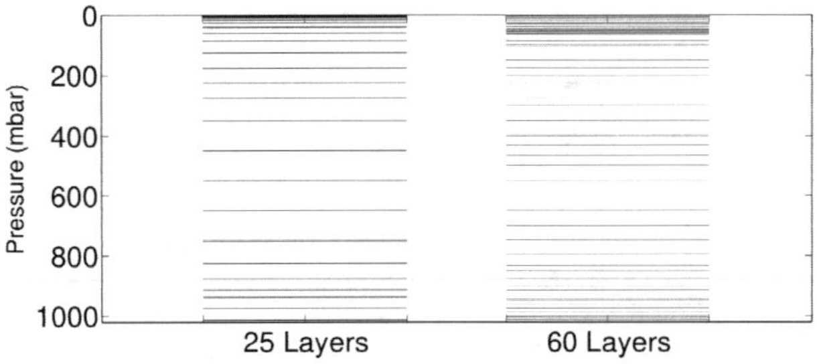

Figure 2. Comparison between the two pressure grids of 60 and 25 layers. In the 25-layer pressure grid most of the layers of the 60-layer grid in the upper atmosphere are suppressed.

implemented in many contexts. In order not to be too long-winded, here we will not recall any element neither of the inversion procedure nor of the Kalman Filter application to this case. The interested reader can refer to [1, 8, 14, 15] for details about these aspects.

3. Results on a case study: the Greek fires in August 2007

The case study we are going to show here proves the double potentiality of both the radiative transfer model σ-SEVIRI to calculate rapidly and accurately the radiance, and the capability of the Kalman Filter to follow the time cycle of skin temperature and emissivity independently of data voids and/or clouds, and without wasting computational resources. We examine SEVIRI data that have been acquired on the Greek peninsula along a month, from the 23rd of August to the 21st of September, 2007. Such a choice is due to the fact that, in the last decade of August 2007, many very extended fires affected the Peloponnese region. Such events are usually associated with sharp peaks in the observed radiance of the corresponding pixel, and to some kinds of discontinuities in the emissivity. Such phenomena, together with extended smoke clouds, are not only interesting from a scientific point of view, but they also offer the ideal terrain to test the robustness and accuracy of the KF procedure.

The target area we have investigated covers the whole Greek peninsula. We have selected 8619 MSG-9 pixels, 4047 over land and 4572 over sea for the period mentioned before. SEVIRI observations have undergone through a first selection, aimed to exclude from the KF analysis all the data corresponding to cloudy scenes, equalizing them to data voids. Then the KF has been run on each single SEVIRI pixel, without considering the spatial correlation between them. As said, we have adopted a simple persistence model to make the system strongly data-driven.

The parameters we want to retrieve are the three emissivity values in the SEVIRI channels 4, 6 and 7, and the skin temperature T_s. Hence, the parameters column vector $\mathbf{v}$ at a generic time t is $\mathbf{v}=(\epsilon_7, \epsilon_6, \epsilon_4, T_s)$. To begin with, Fig. 3 shows the averaged maps of the retrieved skin temperature (land and sea) and land emissivities in the three SEVIRI channels @8.7, @10.8 and @12.0 μm. Such results are useful to point out the capability of the KF approach, together with

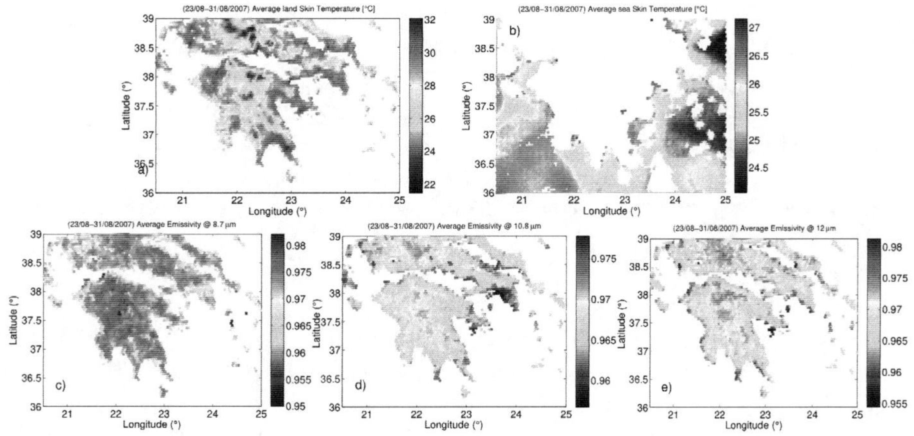

Figure 3. Panel a) Temporally averaged map of the skin temperature over the target area for land, in the period 23th-31th August, 2007; b) Same as a), but for sea surface; c) Same map, but for the averaged retrieved emissivity @8.7 μm, d) @10.8 μm, and e) @12.0 μm.

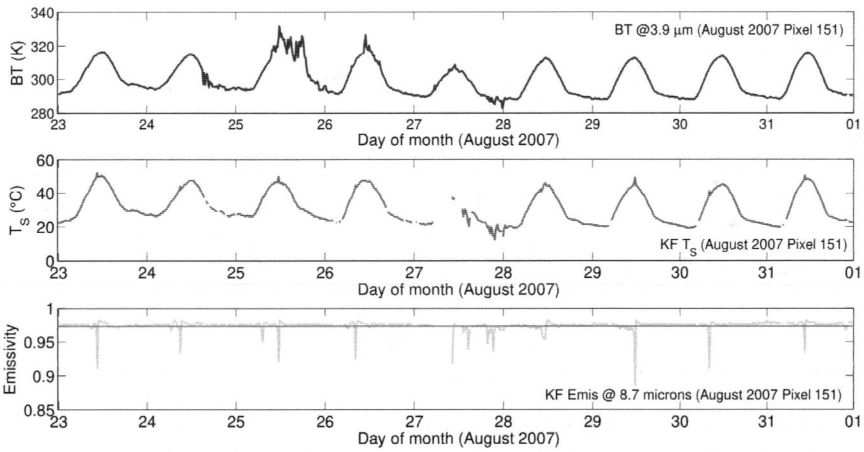

Figure 4. Top panel: observed radiance (in BT units) in the SEVIRI channels @3.9 μm (not used for retrieval) in one of the SEVIRI pixels interested by fires; for the same pixel, central panel shows the retrieved skin temperature; finally, bottom panel shows the retrieved emissivity @8.7 μm (green line), on which the background emissivity (black line) is overlapped.

a fast radiative transfer model, to process effectively and accurately a large volume of data. It is evident how the KF captures the most salient surface features as far as emissivity is concerned, while the orography is well visible from skin temperature map. In this case, one can evidence sea currents and/or the presence of deep water fluxes towards sea surface.

As already stated, this peculiar dataset was selected because of the occurrence of large fires. In order to verify the KF scheme behaviour in presence of such events, we have selected the SEVIRI pixels interested by fires, and plotted the corresponding retrieved skin temperature and emissivities time series. Results are shown in Fig. 4. Fire events are mostly associated to sharp, short peaks in both retrieved skin temperature and observed radiance, more visible in the SEVIRI channel @3.9 μm, not used for retrieval purposes. Such peaks in the skin temperature usually associate with to similar sharp variations in the retrieved emissivity. During the fire event, indeed, the emissivity @8.7 μm goes down, reaching a value of about 0.9, typical of a grey body [16]. It is interesting that, notwithstanding such anomalous events, the KF is stable with respect to them, as the temporal behaviour of the retrieved quantities shows.

Considering that, for a single SEVIRI pixel, the KF scheme requires no more than one minute for a one-month data series for a single pixel on a single modern CPU, one can perform a real-time data analysis for a region like that considered in this case, even using a reduced amount of computational resources.

4. Conclusions

We have presented a new version of the already existing σ-IASI model which is customized for the SEVIRI radiometer. The model, which is able to calculate the Earth radiance as it is observed by SEVIRI in a few hundredths of second, has been speeded-up preserving accuracy in those SEVIRI channels where the atmospheric absorption is negligible, making them useful for the retrieval of skin temperature and spectral emissivity.

The code has been used in a Kalman Filter framework to retrieve the temporal evolution of skin temperature and emissivity. The forward radiative transfer model, joint with the KF methodology, has proved to be able to retrieve such parameters despite possible data voids responsible for local temporal discontinuities in the retrieval. The whole procedure works efficiently both on restricted and on large spatial regions, where each pixel is processed individually. The machinery has been applied on a challenging case study, which involves SEVIRI observations acquired on the Greek peninsula during the huge fires occurred in August 2007. The KF has successfully processed such data, highlighting emissivity anomalies in the spatial pixels involved in fire events.

For its completeness and speed, the KF methodology is suitable to be applied on large regions, even on the SEVIRI Full Disk, in order to build a comprehensive skin temperature and emissivity atlas [17], with the aim to support other retrieval procedures, providing some background for surface, and to interface with Numerical Weather Prediction (NWP) models.

References

[1] Masiello G, Serio C, De Feis I, Amoroso M, Venafra S, Trigo I and Watts P 2013 *Atmos. Meas. Tech.* **6** 3613
[2] Kalman R E 1960 *J. Basic Eng.-ASME* **82** 35
[3] Nychka D W and Anderson J L 2010 *Data Assimilation. Handbook of Spatial Statistics* edited by: Gelfand A, Diggle P, Guttorp P and Fuentes M (Chapman & Hall/CRC, New York)
[4] Hilton F *et al.* 2012 *Bulletin of the American Meteorological Society* **93** 347
[5] Amato U, Masiello G, Serio C and Viggiano M 2002 *Env. Model. & Software* **17** 651
[6] Liuzzi G, Masiello G, Serio C, Fonti S, Mancarella F and Roush T L 2013 *Proceedings of SPIE* vol 8867 Strojnik Scholl M and Pàez G eds. (San Diego: SPIE) ID 88670B
[7] Saunders R *et al.* 2007 *J. Geophys. Res: Atmospheres* **112** D1
[8] Masiello G, Serio C, Carissimo A, Grieco G and Matricardi M 2009 *Atm. Chem. and Physics* **9(22)** 8771
[9] Masiello G and Serio C 2013 *Applied Optics* **51(11)** 2428
[10] Liuzzi G, Masiello G, Serio C, Fonti S, Mancarella F and Roush T L 2015 *Applied Optics* **54(9)** 2334
[11] Clough S A, Shephard M W, Mlawer E J, Delamere J S, Iacono M J, Cady-Pereira K, Boukabara S and Brown P D 2005 *J. Quant. Spectrosc. Radiat. Transfer* **91** 233
[12] Masiello G and Serio C 2003 *Proceedings of SPIE* vol 4882 Schaefer K P, Lado-Bordowsky O, Comeron A and Picard R H eds. (Crete, Greece: SPIE) 52
[13] Anderson G P, Clough S A, Kneizys F X, Chetwynd J H and Shettle E P 1986 *AFGL atmospheric constituent profiles (0-120 km)* (USA: Geophysics Laboratory, Hanscom Air Force Base)
[14] Carissimo A, De Feis I and Serio C 2005 *Env. Model. & Software* **20(9)** 1111
[15] Rodgers C D 2000 *Inverse methods for atmospheric sounding, Theory and Practice* (Singapore: World Scientific)
[16] Johnston J M, Wooster M J and Lynham T J 2014 *International Journal of Wildland Fire* **23(4)** 463
[17] Masiello G, Serio C, Venafra S, Liuzzi G, Göttsche F, Trigo I and Watts P 2015 *Atmos. Meas. Tech. Discuss.* **8** 4049

Journal of Physics: Conference Series **633** (2015) 012060 doi:10.1088/1742-6596/633/1/012060

A mathematical model for the secondary drying of a freeze-drying process

F. Font[1,2], **W. Lee**[1,2]

[1]Pharmaceutical Manufacturing Technology Centre, University of Limerick, Limerick, Ireland
[2]MACSI, Department of Mathematics and Statistics, University of Limerick, Limerick, Ireland

E-mail: francesc.font@ul.ie

Abstract. In this manuscript a mathematical model describing the secondary drying stage of a freeze-drying process is presented. The model consists in governing equations for the transport of an air-vapour mixture in a porous medium. The production of water vapour due to the desorption of bound water is accounted for by means of a source term in the equation for the water vapour concentration. We show how, in the limit of small Peclet numbers, the model can be solved analytically. In addition, we provide with an explicit expression for the total time for the secondary drying stage of the freeze-drying process amenable for real time control applications.

1. Introduction

Freeze-drying or lyophilisation is a dehydration process widely used in the food and pharmaceutical industries, mainly for product preservation and storage purposes. The process enables materials or slurry products, which have been previously frozen, to be dried under vacuum conditions. In the case of pharmaceuticals, the substance to be dried is distributed in vials and placed on trays in a freeze dryer. Then, the freeze-drying proceeds through three successive steps:

1. Freezing: The product in the vials is frozen to a very low temperature. Depending on the particular product involved this process must be fast or slow, as nucleation rates and the size of the ice crystals can affect the quality of the final product.
2. Primary drying: The pressure in the chamber is decreased to nearly vacuum conditions, resulting in the sublimation of the ice content of the product. Once all the ice is removed, a porous dried material is obtained.
3. Secondary drying: The temperature is increased to evaporate adsorbed water (desorption) that did not freeze during the first stage of the process. At the end of the secondary drying a dried porous material with typically less than 1% of water content is obtained.

There is a reasonable amount of literature dealing with mathematical models that describe the different stages of a freeze-drying processes. In general, these models consist in systems of coupled partial differential equations describing the heat and mass transfer in a vial where phase change and chemical reactions are typically involved [1, 2, 3]. Thus, solving this models is usually computationally expensive and not feasible for real time applications [4]. Soft sensors and other control systems require simple models that can be easily implemented in a software and provide information of a process in real time.

In the present study, we will focus on the description of the secondary drying step of the freeze-drying process. In section 2 we formulate a mathematical model describing the transport of water vapour in the porous medium inside a cylindrical vial. In section 3 we present some exact and approximate solutions and provide an explicit formula for the total time of the secondary drying. In section 4 we discuss the results and in section 5 we present the conclusions.

2. Mathematical Model

An illustration of the model is presented in figure 1. The cylindrical body on the left represents a vial containing a porous medium with adsorbed bound water at the beginning of the secondary drying stage. Note, vials are small glass containers with the upper section open. The right side of the image illustrates the interior of the porous medium. The white spots represent the non-porous dried parts of the material which are surrounded by a thin blue layer representing the adsorbed bound water. The gray area represents the porous space where the gas mixture circulates.

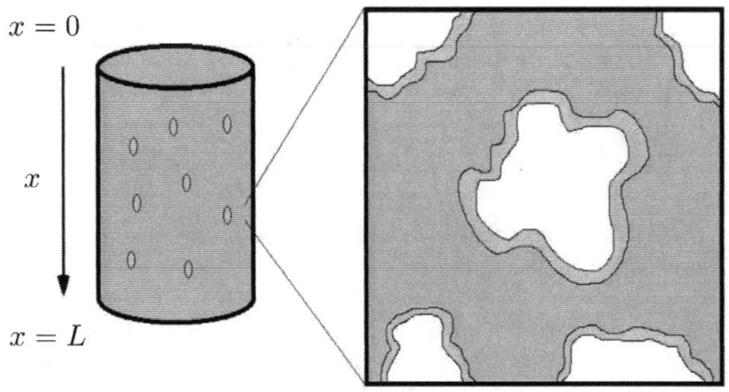

Figure 1. Illustration of the model. The cylinder on the left represents a vial with a porous material inside. The picture on the right represents the internal structure of the porous medium with a layer of adsorbed water (blue) on the surface of the pores. The area in gray represents the free space where the air-vapour mixture circulates.

In a porous medium the transport of a component in a gas mixture is determined by the local conservation of mass of each component in the mixture. One form of the mass conservation equations for an air-vapour mixture is

$$\epsilon \frac{\partial \rho_v}{\partial t} + \nabla \cdot (\mathbf{v} \rho_v) = -\nabla \cdot \mathbf{J}_v + \mathbf{q}, \tag{1}$$

$$\epsilon \frac{\partial \rho_a}{\partial t} + \nabla \cdot (\mathbf{v} \rho_a) = -\nabla \cdot \mathbf{J}_a, \tag{2}$$

where $\rho_{v,a}$ are mass concentrations of vapour and air, ϵ the porosity, $\mathbf{v}$ is a volume flux with units of velocity, $\mathbf{J}_{v,a}$ the diffusive fluxes and $\mathbf{q}$ the source term accounting for desorption. In the present study, the diffusive fluxes are given by the Fick's law, $\mathbf{J}_{v,a} = -\epsilon D \nabla \rho_{v,a}$. For more discussion about other forms of fluxes see [5, 6]. Addition of equations (1)-(2) gives the continuity equation for the mixture

$$\epsilon \frac{\partial \rho}{\partial t} = -\nabla \cdot (\mathbf{v} \rho) + \mathbf{q}, \tag{3}$$

where $\rho = \rho_v + \rho_a$. The source term $\mathbf{q}$ contains the information of how the desorption of water takes place in the porous medium. In the present study we take the approach followed in [2]: if

ρ_{bw}^0 is the initial amount of bound water and $\tilde{\rho}_v$ the vapour generated from bound water only, one approximation that models the desorption of water is

$$\frac{\partial \tilde{\rho}_v}{\partial t} = k\,\rho_{bw}^0\,e^{-k\,t}\,, \tag{4}$$

where k is the desorption rate.

As we are interested in the water vapour concentration we focus on equation (1). Noting that the gas circulation in the vial occurs mainly in the vertical direction, as the gas is only allowed to leave from the top of the vial, the radial and angular transport components can be dismissed. Then, equation (1) can be written as

$$\epsilon\,\frac{\partial \rho_v}{\partial t} + v\,\frac{\partial \rho_v}{\partial x} = \epsilon\,D\,\frac{\partial^2 \rho_v}{\partial x^2} + k\,\rho_{bw}^0\,e^{-\,k\,t} \qquad \text{on} \qquad 0 < x < L\,, \tag{5}$$

where $x = 0$ and $x = L$ represent the top and bottom of the vial, respectively. As a first approximation we assume the following boundary conditions

$$\rho_v(0,t) = \rho_v^0\,, \qquad \left.\frac{\partial \rho_v}{\partial x}\right|_{x=L} = 0\,, \tag{6}$$

accounting for a constant concentration of vapour in the chamber, and so at the top of the vial, and insulation at the bottom. We close the system with the initial condition $\rho_v(x,0) = \rho_v^0$.

The governing equation (5) can be nondimensionalised by introducing the following variables

$$\hat{t} = \frac{t}{\tau}\,, \qquad \hat{x} = \frac{x}{L}\,, \qquad \hat{\rho} = \frac{\rho_v - \rho_v^0}{\Lambda}\,, \qquad \hat{v} = \frac{v}{V}\,. \tag{7}$$

By looking at equation (5) it is clear that there are three different time scales in our model: the diffusion time scale, $\tau_{\text{dif}} = L^2/D$, the desorption time scale, $\tau_{\text{des}} = 1/k$, and the time scale of the convective flux, $\tau_{\text{con}} = L/V$. As we are interested in how long it takes for all the bound water to evaporate, we choose $\tau = \tau_{\text{des}} = 1/k$. The overall variation of vapour concentration in the vial will be mainly determined by the initial amount of bound water existent in the system, so we choose $\Lambda = k(\rho_{bw}^0 L)^2/\epsilon D$. The resulting nondimensional equation for the vapour concentration in the vial is

$$\delta\,\frac{\partial \hat{\rho}_v}{\partial \hat{t}} + \text{Pe}\,\hat{v}\,\frac{\partial \hat{\rho}_v}{\partial \hat{x}} = \frac{\partial^2 \hat{\rho}_v}{\partial \hat{x}^2} + e^{-\,\hat{t}} \qquad \text{on} \qquad 0 < \hat{x} < 1\,, \tag{8}$$

where $\text{Pe} = VL/D$ is the Peclet number and $\delta = L^2 k/D$.

3. Solution

The secondary drying step is a long process that can take several hours where the pressure in the freeze-dryer chamber is typically kept constant. In such conditions, it is believed that convection terms, usually accounted for by means of Darcy's law, are unimportant compared to diffusion terms, which translates in a small value of the Peclet number. Setting $\text{Pe} = 0$ in (8) and dropping the hat notation we obtain the boundary value problem

$$\delta\frac{\partial \rho_v}{\partial t} = \frac{\partial^2 \rho_v}{\partial x^2} + e^{-\,t}\,, \qquad \rho_v(0,t) = 0\,, \qquad \left.\frac{\partial \rho_v}{\partial x}\right|_{x=1} = 0\,, \qquad \rho_v(x,0) = 0\,. \tag{9}$$

The above problem can be solved analytically by means of the eigenfunction expansion method

$$\rho_v(x,t) = \sum_{n=1}^{\infty} \frac{2(e^{-t} - e^{-\lambda_n t/\delta})}{(\lambda_n - \delta)\sqrt{\lambda_n}}\,\sin\left(\sqrt{\lambda_n}\,x\right)\,, \qquad \lambda_n = \frac{(2n-1)^2\pi^2}{4}\,, \tag{10}$$

where λ_n are the eigenvalues and $\sin(\sqrt{\lambda_n}\, x)$ corresponding eigenfunctions.

The nondimensional parameter δ gives information of the relative importance between the diffusion and desorption time scales, $\delta = \tau_{\text{dif}}/\tau_{\text{des}}$. If the desorption time scale is much larger than the diffusion time scale, $\tau_{\text{des}} \gg \tau_{\text{dif}}$, then $\delta \ll 1$ and the boundary value problem (9) suggests an approximate series solution of the form $\rho_v = \rho_v^0 + \delta\, \rho_v^1 + \mathcal{O}(\delta^2)$. At leading order, we obtain

$$0 = \frac{\partial^2 \rho_v^0}{\partial x^2} + e^{-t} \qquad \Rightarrow \qquad \rho_v(x,t) \approx e^{-t}\left(1 - \frac{x}{2}\right) x \, . \tag{11}$$

Note, in (11) the initial condition $\rho(x,0) = 0$ is not satisfied as $\rho_v \neq 0$ for $t = 0$, indicating the presence of a boundary layer. In the next section we will compare the approximate solution (11) to the exact solution (10). We will also provide a numerical solution of (9), by means of backward Euler finite differences, to test that the exact solution found is correct. Applying such numerical scheme to this problem is straightforward and the details will not be provided in this manuscript.

In a freeze-drying process, a key factor to obtain a final product with the desired final concentration of water is the right adjustment of the primary and secondary drying times. An estimation of the total time for secondary drying, t_{2dry}, can be obtained by inverting expression (11). Noting that the highest concentration of water vapour occurs at $x = 1$ (bottom of the vial), this gives

$$t_{\text{2dry}} = -\ln\left(2\rho_v\right) \, . \tag{12}$$

By establishing a level of acceptance of water vapour concentration in the vial, e.g. $\rho_v = 0.01$, an explicit value for t_{2dry} can be found by means of (12).

4. Results

In figures 2 and 3 we show the concentration of water vapour inside the vial for $\delta = 0.1$ and $\delta = 0.5$, respectively, at two different times, $t = 0.29$ and $t = 1.29$. The solid lines represent the exact solution given by (10), the dash-dotted lines the approximate solution (11)b and the dashed lines the numerical solution. We observe that the numerical and the exact solution match in all cases, confirming that our exact solution is correct. By comparing figure 2 with figure 3, it can be seen how the approximate solution is much closer to the exact solution for $\delta = 0.1$ than for $\delta = 0.5$. Indeed, the approximate solution is based in the limit of $\delta \to 0$ and the behaviour observed is consistent with this fact.

As mentioned previously, the right adjustment of the primary and secondary drying times is of crucial importance in a freeze-drying process. Typically, the drying times, which are usually found experimentally by trial and error, strongly depend upon the physical and chemical properties of the product to be dried. In freeze-drying studies is a common practice to use skimmed milk for experiments and model testing. For skimmed milk the value of the desorption rate is $k = 7.8 \times 10^{-5}\,\text{s}^{-1}$ [2]. Then, using the diffusion coefficient for an air-vapour mixture $D = 0.28 \times 10^{-4}\,\text{m}^2\,\text{s}^{-1}$ and assuming a length scale of $L = 0.01\,\text{m}$ we obtain $\delta = 2.79 \times 10^{-4}$. In this case, where the desorption time scale strongly dominates the process, expressions (11) and (12) are very accurate. By substituting $\rho_v = 0.01$ in (12) we obtain $t_{\text{2dry}} = 3.91$. Multiplying the resulting time by $\tau_{\text{des}} = 1/k = 1.28 \times 10^4\,\text{s}$ leads to a total time for secondary drying of 13.93 h. This value shows good agreement with the values found in [2].

5. Conclusions

In this paper we have presented a mathematical model describing the secondary drying stage of a freeze-drying process. The model is based in the standard equations for mass transport of gas mixtures in porous medium with an additional term accounting for the production of vapour due to the desorption of bound water. After nondimensionalizing and making some consistent

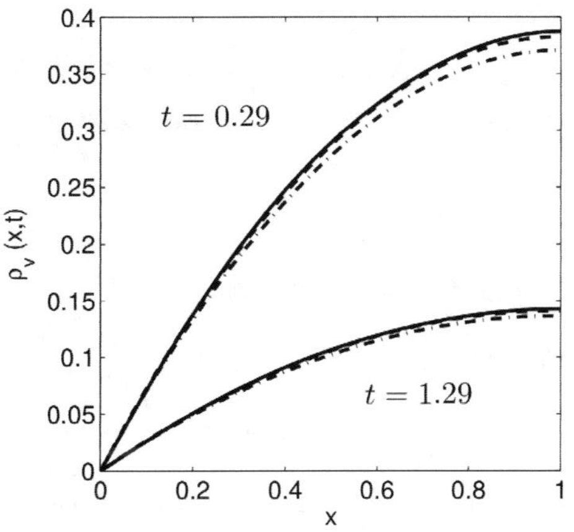

Figure 2. Concentration of water vapour in the vial for $\delta = 0.1$. Solid line is the exact solution, dashed line is a numerical solution and dash-dotted line is the approximate solution.

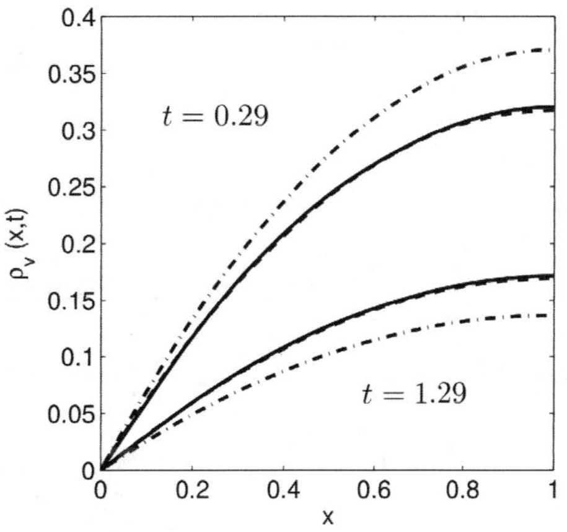

Figure 3. Concentration of water vapour in the vial for $\delta = 0.5$. Solid line is the exact solution, dashed line is a numerical solution and dash-dotted line is the approximate solution.

assumptions, we have reduced the model to a partial differential equation for the evolution of the water vapour concentration which has exact solution. In addition, we have provided with an explicit formula, valid in the limit of small ratio of diffusion to desorption time scales, for the total time of the secondary drying stage. The simplicity of the expression suggests an easy implementation for real time control applications.

[1] W.J. Mascarenhas, H.U. Akay, M.J. Pikal. A computational model for finite element analysis of the freeze-drying process. Computational Methods in Applied Mechanical Engineering, 148 (1997) 105–124.

[2] P. Sheehan, A.I. Liapis. Modeling of the primary and secondary drying stages of the freeze drying of pharmaceutical products in vials: numerical results obtained from the solution of a dynamic and spatially multi-dimensional lyophilization model for different operational policies. Biotechnology and Bioengineering, 60:6 (1998) 712–728

[3] E.A. Boss, R.M. Filho, E.C.V. de Toledo. Freeze drying process: real time model and optimization. Chemical Engineering and Processing, 43 (2004) 1475–1485.

[4] S.A. Velardi, A.A. Barresi. Development of simplified models for the freeze-drying process and investigation of the optimal operating conditions. Chemical Engineering Research and Design, 86 (2008) 9–22.

[5] C-S. Fen, L.M. Abriola. A comparison of mathematical model formulations for organic vapor transport in porous media. Advances in Water Resources, 27 (2004) 1005–1016.

[6] J.W. Veldsink, R.M.J. van Damme, G.F. Versteeg, W.P.M. van Swaaij. The use of the dusty-gas model for the description of mass transport with chemical reaction in porous media. Chemical Engineering Journal, 57 (1995) 115–125.

Model of transient cooperative phenomena triggered by THz-pulse irradiation

Kunio Ishida[1] **and Keiichiro Nasu**[2]

[1] Corporate Research and Development Center, Toshiba Corporation, Kawasaki 212-8582, Japan

[2] Institute of Materials Structure Science, KEK, Tsukuba 305-0801, Japan

E-mail: `ishida@arl.rdc.toshiba.co.jp`

Abstract. We studied the cooperative phenomena induced by injection of coherent phonons by THz optical pulses. Based on the analogy to the photoinduced phase transitions observed in various materials, we consider that cooperative interactions between electrons and coherent phonons will lead to the multiplication of excited electrons and/or growth of a transient phase, which is understood by bifurcation of quantum-mechanical wavepackets on adiabatic potential energy surfaces. Employing a model of localized electrons coupled with a quantized optical phonon mode, we discuss the dynamics of the cooperative phenomena by THz-pulse irradiation and, in particular, the role of the number and/or the initial distribution of phonons in the initial creation process of transient phases.

1. Introduction

Recent progress of intense THz-pulse generation technology has made it possible to inject coherent phonons in a macroscopic scale[1], and the lattice deformation induced by such a process will induce photoluminescence caused electronic transitions in coupled electron-phonon systems[2].

On the other hand, photoinduced cooperative phenomena have been observed in various materials, and the study of these phenomena has shed light on long-standing theoretical and experimental problems with nonequilibrium dynamics of excited states[3, 4, 5, 6]. In particular, they are concomitant with the change of crystal structure as well as the electronic transitions in coupled electron-phonon systems, which means that similar phenomena will be observed by excitation of phonons.

In this paper, we discuss the cooperative phonomena triggered by injection of coherent phonons by a model of localized electrons coupled with phonons[7]. In this case, adiabatic transition of electronic states is relevant to the cooperative phenomena, e.g., multiplication of excited electrons. In particular, nonadiabatic coupling of electrons is important in the initial relaxation dynamics as in the case of injection of electronically excited states, since nonadiabatic transition disturbs the generation of excited electrons. Thus, in considering the possibility of cooperative phenomena, we should carefully consider the dynamics of electrons where both adiabatic and nonadiabatic transition is possible.

2. Model and method

We employ a model of localized electrons coupled with an optical phonon mode which describes the general properties of the photoinduced structural change. In the present study we consider an array of molecules (unit cells) on a square lattice with two electronic levels and a single phonon mode interacting with each other, which is described by the following Hamiltonian:

$$\begin{aligned}\mathcal{H} &= \sum_{\vec{r}} \left\{ \frac{\hat{p}_{\vec{r}}^2}{2} + \frac{\omega^2 \hat{u}_{\vec{r}}^2}{2} \right. \\ &+ \left. (\sqrt{2\hbar\omega^3} s \hat{u}_{\vec{r}} + \varepsilon\hbar\omega + s^2\hbar\omega)\hat{n}_{\vec{r}} + \lambda\sigma_x^{\vec{r}} \right\} \\ &- \sum_{\langle \vec{r}, \vec{r'} \rangle} [\alpha\omega^2(\hat{u}_{\vec{r}} - \beta\hat{n}_{\vec{r}})(\hat{u}_{\vec{r'}} - \beta\hat{n}_{\vec{r'}}) \\ &- \{V - W(\hat{u}_{\vec{r}} + \hat{u}_{\vec{r'}})\}\hat{n}_{\vec{r}}\hat{n}_{\vec{r'}}], \end{aligned} \tag{1}$$

where $\hat{p}_{\vec{r}}$ and $\hat{u}_{\vec{r}}$ are the momentum and coordinate operators for the vibration mode of a molecule at site $\vec{r}$, respectively. Two electronic states are assigned to each site $\vec{r}$ which are denoted by $|\downarrow\rangle_{\vec{r}}$ (ground state) and $|\uparrow\rangle_{\vec{r}}$ (excited state). $\sigma_i^{\vec{r}}$ $(i = x, y, z)$ are the Pauli matrices which act only on the electronic states of the molecule at site $\vec{r}$. the nonadiabaticity in the dynamics is taken into account via "spin-flip" interaction between two electronic states as in typical organic molecules[8]. $\hat{n}_{\vec{r}}$ denotes the density of the electron in $|\uparrow\rangle_{\vec{r}}$, i.e., $\hat{n}_{\vec{r}} = \sigma_z^{\vec{r}} + 1/2$. The second sum which gives the intermolecular interaction is taken over all the pairs on nearest neighbor sites, and we take into account the dipole-dipole interaction between excited molecules as well as the elastic interactions. We note that this Ising-like model is similar to the one to study the thermodynamical properties of the Jahn-Teller effect[9], while our aim is to reveal the nonequilibrium dynamics of the excited states of the model.

The molecules that we consider have two diabatic potential energy surfaces (PESs) which cross each other, and the nonadiabatic coupling $\lambda\sigma_x^{\vec{r}}$ acts to reorganize them into two adiabatic PESs. We assume that the quantum mechanical nature of the nuclear wavepackets on the PESs plays an important role, which is the case particularly for nonadiabaticity of the dynamics is relevant. We chose the values of the parameters as: $\omega = 1$, $\varepsilon = 2.3$, $s = 1.4$, $V = 1.1$, $W = 0.2$, $\alpha = 0.1$, $\beta = 0.2$, and $\lambda = 0.2$, and the unit of time is taken to be $T = 2\pi/\omega$. We note that the values of the parameters do not correspond to any specific material but the order of magnitude for the parameters is estimated referring to those for typical organic materials. We numerically solved the time-dependent Schrödinger equation for the Hamiltonian (1) for systems with 128×48 sites in which the periodic boundary condition is taken only for the longer edge of the system, and the time-dependent wavefunction of the electron-phonon system $|\Phi(t)\rangle$ was obtained. Detail of the method of calculations is described in Ref. [7] and we do not mention it further to avoid redundancy.

3. Calculated results

Since the penetration of THz pulses is limited due to high absorption ratio of organic materials, we consider that the lattice deformation induced by those pulses is limited to the molecules in the vicinity of the surface of samples. Hence, as an initial condition of calculations, we consider that only five molecules closest to either of the longer edges of the system were excited, where the wavefunction of those molecules is given by a coherent state of phonons

$$|a\downarrow\rangle_{\vec{r}} = e^{-\frac{|a|^2}{2}} \sum_{n=0}^{\infty} \frac{a^n}{\sqrt{n!}} |n\downarrow\rangle_{\vec{r}}. \tag{2}$$

The other molecules are considered to stay in the ground state at $t = 0$.

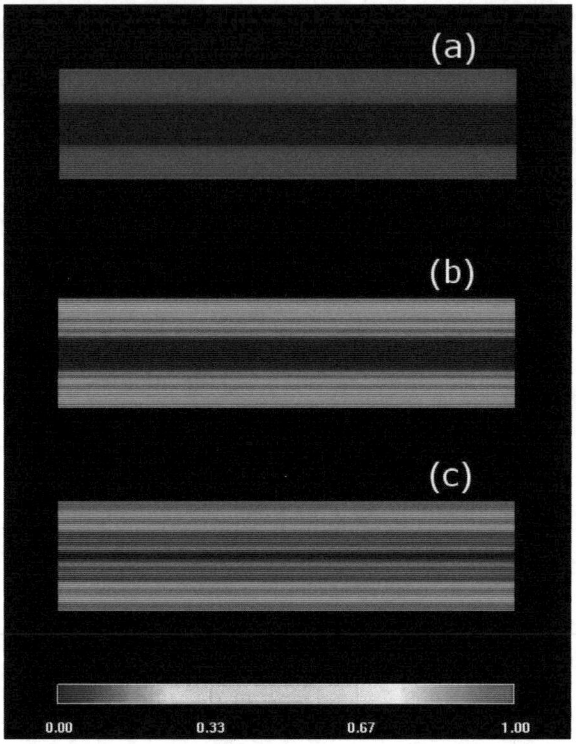

Figure 1. Spatial distribution of the population of excited electrons $\tilde{n}(\vec{r},t)$ for (a)$t = 10T$, (b)$t = 25T$, and (c)$t = 50T$.

Figures 1-(a)-(c) show the spatial distribution of the population of excited electrons $\tilde{n}(\vec{r},t) = \langle\Phi(t)|\hat{n}_{\vec{r}}|\Phi(t)\rangle_{\vec{r}}$ for $t = 10T$, $25T$, and $50T$ for $a = s$. When phonons are excited in a molecule, a wavepacket described by Eq. (2) starts to move on the ground state PES, and when it approaches the crossing point of the grounds state PES and the excited state PES, the electronic interaction $\lambda\sigma_x^{\vec{r}}$ induces electronic transition between $|\downarrow\rangle_{\vec{r}}$ and $|\uparrow\rangle_{\vec{r}}$. Hence, $\tilde{n}(\vec{r},t)$ increases only in the initially phonon-excited molecules as shown in Fig. 1-(a).

On the other hand, the electronic transition also affects the PES at the nearest neighbor molecules through the intermolecular interaction in Eq. (1), which causes the transition from $|\downarrow\rangle_{\vec{r}}$ to $|\uparrow\rangle_{\vec{r}}$ in those molecules. Thus, as the energy dissipation proceeds by the propagation of phonons in the system, the number of excited molecules increases, and all the molecules are in a superposition of $|\downarrow\rangle_{\vec{r}}$ and $|\uparrow\rangle_{\vec{r}}$ as shown in Figs. 1-(b) and (c).

In order to show that the increase of the excited state molecules corresponds to the multiplication of excited electrons, we calculated the sum of the population in the excited electronic state $N(t) = \sum_{\vec{r}}\tilde{n}(\vec{r},t)$ divided by the number of the initially phonon-excited molecules n_i. As shown in Fig. 2, $N(t)/n_i$ starts to increase at $t \sim 3$, which is reminiscent of the incubation period observed in photoinduced cooperative phenomena[7]. During this period, even the electrons at the initially phonon-excited molecules tend to stay in the ground state $|\downarrow\rangle_{\vec{r}}$ although the electronic interaction $\lambda\sigma_x^{\vec{r}}$ works. Thus, the origin of the incubation period in this case is the nonadiabaticity of the electron dynamics, which disturbs the adiabatic transition between $|\downarrow\rangle_{\vec{r}}$ and $|\uparrow\rangle_{\vec{r}}$.

Since, however, the electronic population gradually transfers to the excited electronic state, $\tilde{n}(\vec{r},t)$ also gradually increases, and the PES of the molecules at the ground state deforms.

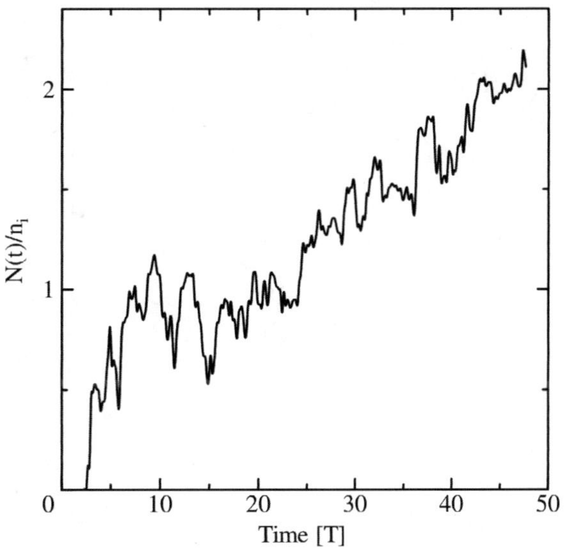

Figure 2. Sum of the population in the excited electronic state $N(t)$ divided by the number of the initially phonon-excited molecules n_i. When $N(t)/n_i > 1$, the multiplication of electrons in the excited states takes place.

Hence, the molecules in the vicinity of initially phonon-excited molecules starts to deform and the multiplication of excited electrons is triggered by this process, though the electronic transition is slower than that in the photoexcited case[7].

4. Summary

We calculated the dynamics of coupled electron-phonon system excited by THz optical pulses. We found that the electronic transition between the ground state and an excited state is induced by the propagation of phonons, and the region of excited molecules extends over the whole system, although initial deformation of the lattice is limited to the molecules in the vicinity of the surface of samples. As a result, the population of excited electrons increases, which shows that the multiplication of excited electrons is possible by the injection of phonons. Although these features are similar to the photoinduced cooperative phenomena, we also found a certain difference between "electron-excited" cases and "phonon-excited" cases, i.e., the rate of the increase of $N(t)$ is larger in former case. The origin of the difference is that the excitation energy given to the system is larger in the former case than in the latter, since the electronic excitation energy is larger than phonon energy in typical semiconductors and insulators. It is, however, still interesting that the irradiation of intense THz-pulses are effective to induce the multiplication of excited electrons in various systems, and thus it will be another method to reveal the dynamics of excited states.

Acknowledgments

This work was partly supported by JSPS Grant-in-Aid for Specially Promoted Research, Grant Number 24000006.

[1] Hase M *et al.* 2012 Nat. Photonics **6** 243.
[2] Shinokita K *et al.* 2013 Phys. Rev. Lett. **111** 067401.
[3] *Photoinduced phase transitions* ed Nasu K (World Scientific, Singapore).

[4] Yonemitsu K and Nasu K 2008 Phys. Rep. **465**, 1.
[5] Koshihara S *et al.* 1992 Phys. Rev. Lett. **68** 1148.
[6] Mino A *et al.* 1998 Mol. Cryst. Liq. Cryst. **314** 107.
[7] Ishida K and Nasu K 2008 Phys. Rev. **B77** 214303.
[8] Salem L 1976 Science **191** 822.
[9] Boukheddaden K 2004 Prog. Theor. Phys. **112** 205.

Turbulence modeling using test modes

M Vlad[1]**, F Spineanu**
National Institute of Laser, Plasma and Radiation Physics,
Atomistilor 409, Magurele 077125, Bucharest, Romania

E-mail: madalina.vlad@inflpr.ro

Abstract. We present an analytical self-consistent approach, which is able to analyze the processes that appear in the nonlinear stage of turbulence. The results concern drift turbulence in confined plasmas. The method is based on a combined study of test particle and test modes in turbulent plasmas. We show that the main cause of the nonlinear processes that appear beyond the quasilinear stage of turbulence is trajectory trapping or eddying. Trapping introduces quasi-coherent aspects in test trajectory statistics, which lead to large scale correlations, nonlinear damping of the drift modes and generation of the zonal flow modes.

1. Introduction

Turbulence is considered to be the last great problem of classical physics that is still unsolved. It is of fundamental importance in plasma physics, astrophysics, fluid mechanics and atmospheric and oceanic sciences. Direct numerical simulations, which have obtained important results in the last decades, largely dominate the actual research in this field. The analytical advance based on first principle description and mathematically justifiable approximations is very small. The reason is the stochastic advection process, which is the dominant nonlinearity in all turbulent system: the advected field depends on the velocity in plasmas or it is just the velocity field in fluids. Moreover, even the most simplified models that deal with passive fields have not been solved analytically because the basic problem of particle trajectories in stochastic velocity fields is strongly nonlinear.

We present an analytical modeling of the evolution of drift turbulence in magnetically confined plasmas, which consists of a coupled study of the test modes on turbulent plasmas and of the statistics of the particle trajectories. It is similar to the Lagrangian approach initiated by Dupree [1] and developed in the 70`s. Their assumption of random trajectories with Gaussian distribution has limited the application of this method to the quasilinear regime. We introduce a realistic description of the rather complex trajectories that are determined by the ExB drift in turbulence, which enables the description of the processes that appear in the nonlinear regime. Particle motion is not only random but it includes coherent aspects. A typical trajectory is a random sequence of trapping or eddying events and large jumps. We have developed semi-analytical statistical methods (the decorrelation trajectory method DTM [2] and the nested subensemble approach NSA [3]), which are able to describe the process of trapping and the related coherent aspects of trajectory statistics.

We show that trajectory trapping is the main cause of the nonstandard (anomalous) transport regimes and of the strong nonlinear effects in the evolution of drift turbulence (generation of large scale

[1] M. Vlad

Journal of Physics: Conference Series **633** (2015) 012062 doi:10.1088/1742-6596/633/1/012062

correlations and of the zonal flow modes). This is the first analytical result on this complex problem that is in agreement with numerical simulations.

2. Test modes on turbulent plasmas

Test modes on turbulent plasmas were studied for drift turbulence in constant magnetic field [4] starting from the basic description provided by the drift kinetic equations. Analytical expressions are derived, which approximate the growth rates γ and the frequencies ω of the test modes as functions of the characteristics of the background turbulence. They provide the tendencies in turbulence evolution. We consider the (universal) drift instability described by the drift kinetic equations in the collisionless limit. Electron kinetic effects produce the dissipation mechanism to release the energy and, combined with the ion polarization drift, make drift waves unstable. Beside this, the polarization drift

$$\mathbf{u}_p = \frac{m_i}{eB^2}\partial_t \mathbf{E}_\perp \tag{1}$$

has a more complex influence determined by its nonzero divergence, which produces compressibility effects in the background turbulence.

The dispersion relation of the test modes in turbulent plasma is shown to be formally the same as in quiescent plasma, except for a time dependent function $M(\tau)$, which embeds all the effects of the background turbulence:

$$-i\left(k_y V_{*e} - \omega \rho_s^2 k_\perp^2\right)\int_{-\infty}^{t} d\tau\ M(\tau)\exp\left[-i\omega(\tau - t)\right] = 1 + i\sqrt{\frac{\pi}{2}}\frac{\omega - k_y V_{*e}}{|k_z| v_{Te}} \tag{2}$$

$$M(\tau) \equiv \left\langle \exp\left[i\mathbf{k}\cdot\left(\mathbf{x}(\tau)-\mathbf{x}\right) - \int_\tau^t d\tau' \,\nabla\cdot\mathbf{u}_p\left(\mathbf{x}(\tau')\right)\right]\right\rangle. \tag{3}$$

$V_{*e} = T_e/(eBL_n) = \rho_s c_s / L_n$ is the diamagnetic velocity produced by the density gradient (taken along the x direction), $\rho_s = c_s/\Omega_i$, $c_s = \sqrt{T_e/mi}$, T_e is the electron temperature, m_i is the ion mass, e is the absolute value of electron and ion charge, $\Omega_i = eB/m_i$ is the cyclotron frequency and $k_\perp = \sqrt{k_x^2 + k_y^2}$ is the perpendicular wave number.

The function $M(\tau)$ depends on the background potential through the statistical average on the trajectories denoted by $\langle\ \rangle$, and also through the compressibility term determined by the polarization drift. The trajectories (the characteristics of the drift kinetic equation) are solutions of

$$\frac{d\mathbf{x}(\tau)}{d\tau} = -\frac{\nabla\phi(\mathbf{x}-\mathbf{V}_{*e}t)\times\mathbf{e}_z}{B}, \tag{4}$$

calculated backwards in time with the condition at $\tau = t$, $\mathbf{x}(t) = \mathbf{x}$. In the case of quiescent plasmas $M = 1$. The function $M(\tau)$ is estimated using the results of the test particle study.

3. Test particle statistics in stochastic velocity fields

The origin of trapping is the Hamiltonian structure of the equation of motion (4). The components x and y of particle position are conjugate variables and the potential $\phi(\mathbf{x},t)$ is the Hamiltonian function. The trajectories are periodic and they wind on the contour lines of ϕ when the potential is time independent. This defines the state of permanent trapping. If the motion is weakly perturbed (by time variation of the potential, particle collisions, etc.), the trapping is temporary and alternates with large jumps.

Original semi-analytical methods for determining the statistics of stochastically advected particles were developed (the DTM [2] and the NSA [3]), and they have been used in a series of studies of the transport in magnetically confined plasmas (see [5]-[7] and the references there in). This represents the first analytical approach that is able to go beyond the quasi-linear regime that corresponds to quasi-Gaussian transport. The general conclusion of these studies is that the existence of space correlations of the stochastic velocity can generate trajectory trapping or eddying, which leads to nonstandard statistics of trajectories: non-Gaussian distributions, long time Lagrangian correlations (memory), strongly modified transport coefficients and an increased degree of coherence. We have also shown that the trapped trajectories form quasi-coherent structures similar to fluid vortices. These effects appear in two-dimensional, zero divergence velocity fields as in Eq. (4).

Essentially, the DTM and the NSA reduce the problem of determining the statistics of the stochastic trajectories to the calculation of weighted averages of some smooth, deterministic trajectories determined from the Eulerian correlation (EC) of the stochastic potential. These methods are in agreement with the statistical consequences of the invariance of the potential so that they are able to describe trajectory trapping.

The distribution of displacements strongly depends on the ordering of the main characteristic times of the stochastic process: the diamagnetic time $\tau_* = \lambda_y / V_{*e}$ defined by the motion of the potential along y with the diamagnetic velocity, and the time of flight (or eddying time) $\tau_{fl} = \lambda_y / V_y$. The decorrelation process is determined by two mechanisms: the time variation of the stochastic potential and the parallel decorrelation due to ion motion along the magnetic field. The correlation time of the potential is larger than the parallel time $\tau_{II} = \lambda_z / v_i$, and thus the decorrelation parameter (Kubo number) is $K_d = \tau_{II} / \tau_{fl}$. Trapping is statistically relevant for drift turbulence if the time of flight is the smallest characteristic time. This condition corresponds to $K_d > 1$ and $K_* > 1$, where $K_* = \tau_* / \tau_{fl} = V_y / V_{*e}$ is the diamagnetic parameter. The fraction of trapped trajectories n_{tr} and the maximum size of trajectory structures are increasing functions of K_* and K_d (see Figure 1).

The function $M(\tau)$ that accounts for the background turbulence is evaluated using simplified analytical expressions for the probability of displacements with the characteristics of the results of the DTM.

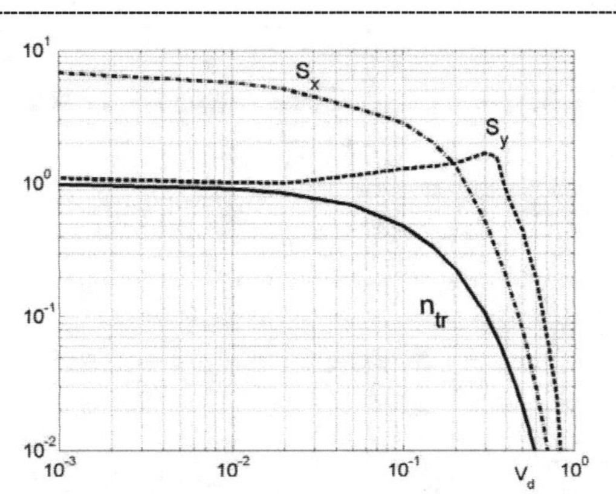

Figure 1. The fraction of trapped trajectories and the sizes of the trajectory structures as function of K_* for $K_d \to \infty$.

4. The evolution of the drift turbulence

The self-consistent modeling of the evolution of drift turbulence is obtained from these coupled studies of test modes and test particles. Although the input for both studies is the EC of the stochastic potential, it is possible to determine the evolution of this function. The difference between the characteristic time of turbulence evolution and the decorrelation time of ion trajectories (the parallel time) enables the evaluation of the EC evolution from the tendencies provided by the growth rates and the frequencies of the test modes.

Starting from an initial condition with a large spectrum of very small amplitude, the EC of the potential is determined (by the Fourier transform), and then the statistics of trajectories and the function $M(\tau)$ are calculated for this EC using the DTM. The dispersion relation (2) is solved and eventually the short time evolution of the test modes is determined from

$$\left\langle \left| \Phi^2(\mathbf{k},t+\tau) \right| \right\rangle = \left\langle \left| \Phi^2(\mathbf{k},t) \right| \right\rangle \exp\left(2\gamma(\mathbf{k};t)\tau\right) \tag{5}$$

for times τ of the order of the correlation time of the turbulence. We have taken the initial condition

$$\left\langle \left| \Phi^2(\mathbf{k},0) \right| \right\rangle = \Phi_0^2 \exp(-k^2\lambda_0^2/2), \tag{6}$$

where the initial amplitude Φ_0 is small and $\lambda_0 \in (0.4,\ 0.8)$. The EC of the potential at time $t+\tau$ is the Fourier transform

$$\begin{aligned} E(x,y',t+\tau) &\equiv \left\langle \phi(x_1,y_1,t+\tau)\,\phi(x_1+x,y_1+y,t+\tau) \right\rangle \\ &= \int d^2k \left\langle \left| \Phi^2(\mathbf{k},t+\tau) \right| \right\rangle \exp\left(ik_x x + ik_y y\right). \end{aligned} \tag{7}$$

This determines the statistics of the trajectories at time $t+\tau$, and the evolution of the EC is obtained by repeating these calculations.

4.1. The quasilinear regime

For small amplitude of the background potential trapping does not appear and the transport is Gaussian. One obtains

$$M(\tau) = \exp\left[-k_i^2 D_i (t-\tau) - 2ik_1 \frac{V^2}{\Omega_i V_*}\right] \tag{8}$$

and the solution of the dispersion relation is

$$\omega = \frac{k_y V_*}{1+k_\perp^2}, \tag{9}$$

$$\gamma = \gamma_0 \omega\left(k_y V_* - \omega\right) - k_i^2 D_i. \tag{10}$$

The effect of the background turbulence appears in the growth rate, while the frequency is the same as in quiescent plasma. It determines the diffusive damping (second term in Eq. (10)). The damping is produced by ion diffusion and it is important for the modes with large wave numbers.

The positive growth rate determines the increase of the amplitude of the modes accompanied by a strong modification of the shape of the EC (see Figure 2, dashed lines). The latter is determined by the fact that the growth rate is zero for $k_y = 0$, which leads to $\left\langle \left| \Phi^2(k_x, k_y=0,t) \right| \right\rangle = 0$ and to $\int dy E(x,y,t,\tau) = 0$. This change of the shape of the EC determines a strong decrease of D_y, which enables the development of the drift turbulence. The correlation lengths λ_i are of the order of ρ_s.

4.2. The weakly nonlinear regime

When the amplitude reaches values that make $K_* > 1,$ ion trajectory trapping appears and generates vortical structures of trapped ions. The fraction of trapped trajectories is small in this regime. The distribution of displacements is non-Gaussian, with an invariant maximum that is produced by trajectory structures and an expanding part that corresponds to the free ions and determines the diffusion. The effect of the background turbulence appears in the frequency

$$\omega = \frac{k_y F V_*}{1+Fk_\perp^2}, \quad F = \exp\left(-\frac{1}{2}k_i^2 S_i^2\right), \tag{11}$$

where S_i are the average sizes of the trajectory structures. Trajectory structures determine the decrease of the frequencies, which leads to the displacement of the unstable range of wave numbers to small values of the order $k_i \sim 1/S_i$. The solution for the growth rate remains as in the quasilinear regime (Eq. (10)), but γ decreases due to the decrease of the frequency.

Turbulence amplitude continues to increase in this stage, but with a smaller rate. The shape of the EC is not modified, but the correlation length increases in both directions (see Figure 2, the dashed-dotted line). An inverse cascade appears as the shift of the unstable wave-number range toward small wave numbers.

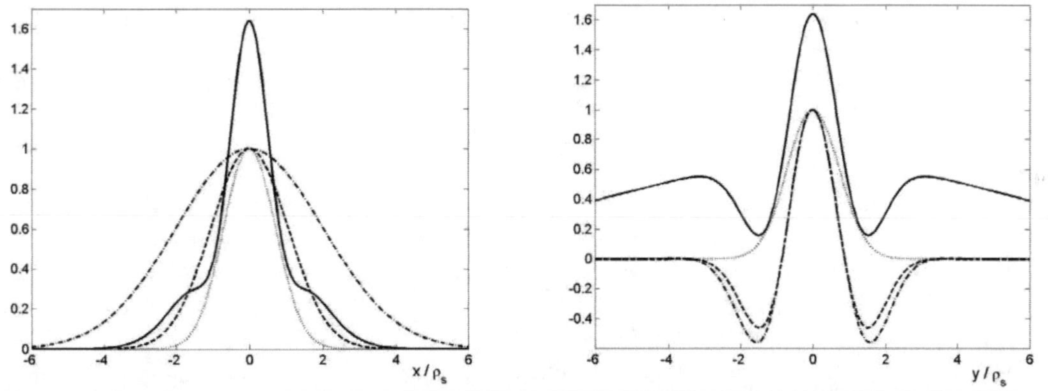

Figure 2. The Eulerian correlation of the potential normalized with the amplitude of the drift turbulence $E(\mathbf{x},t)/E(0,t)$ as function of x (left panel) and of y (right panel). The initial condition (dotted lines), the shape of the EC at the end of the quasilinear regime (dashed lines), in the weakly nonlinear regime (dashed-dotted lines) and in the strongly nonlinear regime where zonal flow modes are generated (continuous lines).

4.3. The strongly nonlinear regime

When the fraction of trapped ions n_{tr} becomes comparable with the fraction of free ions n_f, ion flows are generated by the trapping process. The trapped ions move with the potential, while the free ions move in the opposite direction with the velocity $V_f = -V_{*e} n_{tr} / n_f$, such that the total flux is zero. This determines the splitting of the distribution of ion displacement in two parts that depart from each other. A new term appears in the function $M(\tau)$ and the dispersion relation is strongly modified. It has two solutions (see [4] for the calculations).

One of the solutions corresponds to the drift modes. It is strongly modified by the ion flows, which determine a damping effect. It begins with the damping of small k_y modes and it extends to the entire spectrum as n_{tr} increases (the growth rates are negative for the entire range of $\mathbf{k}$ at $n_{tr} = 1/2.$). The other solution corresponds to a new type of modes, the zonal flow modes with $k_y = 0$ and very small frequencies. They are produced by the combined action of the ion flows and of the compressibility due to the polarization drift in the background turbulence. The growth rate of the zonal flow modes is proportional to n_{tr} and it increases with the decrease of the correlation length of the drift turbulence. The shape of the total potential is strongly changed due to the zonal flow mode component (Figure 2, continuous line). This determined the decrease of D_x and the strong increase of D_y.

The damping of the drift modes is not determined by the zonal flows mode. There is only an indirect contribution through the diffusive damping, which is increased by the zonal flow modes. The growth and the decay of turbulence are produced on different paths (hysteresis process). Large scales are generated at the increase of turbulence amplitude, when trapping is weak. Later in the nonlinear evolution, when trapping is stronger and produces ion flows, turbulence amplitude continues to increase, but it is accompanied by the decrease of the correlation length and by the generation of zonal flow modes. Then, the amplitude decays. A closed evolution curve in the (β, λ) space is described by the turbulence, which remains in the nonlinear stage characterized by trapping and oscillates between weak and strong trapping. The characteristic time Δt for turbulence and transport oscillations estimated as the inverse of the growth rates is in agreement with numerical simulations.
The predator-prey paradigm is not sustained by these results, although there is time correlation between the growth of zonal flow modes and the damping of the drift modes.

5. Conclusions

Ion trajectory trapping determines strong transitory changes in the turbulence. Drift turbulence does not saturate but has a complex oscillatory behavior. The trapped ions that form trajectory structures lead to longer correlations lengths. The potential motion combined with trapping stabilizes and damps the drift modes (first those with small k) and in the same time generates zonal flow modes, which eventually are damped when the fraction of trapped trajectories decreases.

In conclusion, this first principle semi-analytical approach is able to describe the complex evolution of drift turbulence and yields results in agreement with numerical simulations. A different perspective on important aspects of the physics of drift type turbulence in the strongly non-linear regime (as zonal flow mode generation) is deduced.

6. References

[1] Dupree T H 1972 *Phys. Fluids* **15** 334
[2] Vlad M, Spineanu F, Misguich J H and Balescu R 1998 *Phys. Rev. E* **58** 7359
[3] Vlad M and Spineanu F 2004 *Phys. Rev. E* **70** 056304
[4] Vlad M 2013 *Phys. Rev. E* **87** 053105
[5] Vlad M, Spineanu F, Misguich J H, Reuss J D, Balescu R, Itoh K and Itoh S I 2004 *Plasma Phys. Control. Fusion* **46** 1051
[6] Vlad M, Spineanu F and Benkadda S 2006 *Phys. Rev. Lett.* **96** 085001
[7] Vlad M and Spineanu F 2013 *Phys. Plasmas* **20** 122304

Journal of Physics: Conference Series **633** (2015) 012063 doi:10.1088/1742-6596/633/1/012063

Controlled Interaction of a Four-Level Quantum Emitter with a Plasmonic Nanostructure

Emmanuel Paspalakis[1], Efthymios Kallos[1] and Vassilios Yannopapas[2]

[1]Materials Science Department, School of Natural Sciences, University of Patras, Patras 265 04, Greece

[2]Department of Physics, National Technical University of Athens, Athens 157 80, Greece

E-mail: paspalak@upatras.gr, timaras@gmail.com, vyannop@mail.ntua.gr

Abstract. We present results on the controlled dynamics of an externally driven four-level quantum emitter coupled to a plasmonic metamaterial, specifically a periodic two-dimensional array of metal-coated dielectric nanospheres. For the study of the system's dynamics, we combine the density matrix approach for the quantum emitter with *ab initio* electromagnetic calculations for the plasmonic nanostructure. We then present results for the time evolution of the populations of the different levels of the quantum emitter in both the presence and the absence of the plasmonic nanostructure. The dependence of the population dynamics to different distances of the quantum emitter from the plasmonic nanostructure is also studied.

1. Introduction

Recently, there is increasing interest in the study of the interaction of quantum emitters (such as atoms, molecules and semiconductor quantum dots) with plasmonic nanostructures [1]. The large fields and the strong light confinement associated with the plasmonic resonances enable strong interaction between the electromagnetic field and the quantum emitters near plasmonic nanostructures. Also, the quantum emitter can be used for the controlled optical response of the hybrid quantum - plasmonic system.

A quantum emitter that has attracted particular attention is modeled by a four-level quantum system with two V-type transitions [2, 3, 4, 5, 6, 7, 8]. In the quantum system one V-type transition is influenced by the interaction with surface plasmons while the other V-type transition interacts with free space vacuum [see Fig. 1(a)]. This system has led to several interesting quantum coherence and interference phenomena, including optical transparency and slow light [2], transient gain without inversion [3], strongly modified Kerr nonlinearity [4], phase-dependent absorption and dispersion [5], strongly modified spontaneous emission spectrum [6, 7] and coherent population trapping [8].

In this work, we present new theoretical results on the controlled dynamics of the four-level quantum emitter coupled to a plasmonic metamaterial, namely a periodic two-dimensional array of metal-coated dielectric nanospheres. An external electromagnetic field is applied to the system and is used for the control of the dynamics of the quantum system. For the study of the system's dynamics, we combine the density matrix approach for the quantum emitter with *ab initio* electromagnetic calculations for the plasmonic nanostructure. We then present results for the time evolution of the population (occupation probability) of the different levels of the quantum emitter in both the presence and the absence of the plasmonic nanostructure. Also, in the presence of the plasmonic nanostructure, we consider the influence of different distances of the quantum emitter from the plasmonic nanostructure to the population dynamics.

2. Theoretical Model

The quantum system of interest is shown in Fig. 1(a). We consider a four-level system with two closely lying upper states $|2\rangle$ and $|3\rangle$, and two lower states $|0\rangle$ and $|1\rangle$. The quantum system is located in vacuum at distance D from the surface of the plasmonic nanostructure. We take states $|2\rangle$ and $|3\rangle$ to characterize two Zeeman sublevels. The dipole moment operator is taken as $\vec{\mu} = \mu^{'}(|2\rangle\langle 0|\hat{\varepsilon}_- + |3\rangle\langle 0|\hat{\varepsilon}_+) + \mu(|2\rangle\langle 1|\hat{\varepsilon}_- + |3\rangle\langle 1|\hat{\varepsilon}_+) + \text{H.c.}$, where $\hat{\varepsilon}_\pm = (\mathbf{e}_z \pm i\mathbf{e}_x)/\sqrt{2}$, describe the right ($\hat{\varepsilon}_+$) and left-rotating ($\hat{\varepsilon}_-$) unit vectors; and μ, $\mu^{'}$ are taken to be real.

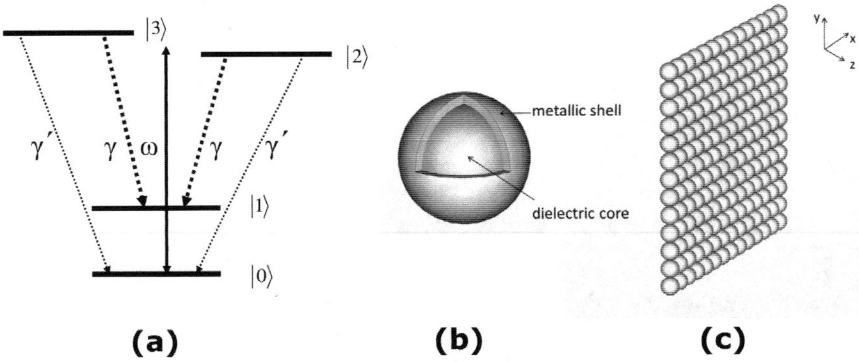

Figure 1. (a) The quantum system under study is a four-level system where two upper states $|2\rangle$ and $|3\rangle$ decay with spontaneous emission to the two lower states $|0\rangle$ and $|1\rangle$. The system interacts with an electromagnetic field that couples state $|0\rangle$ with states $|2\rangle$ and $|3\rangle$. (b) A metal-coated dielectric nanosphere and (c) the two-dimensional array of metal-coated dielectric nanospheres used in this work.

The quantum system interacts with a linearly polarized continuous wave laser field, with electric field $\vec{E}(t) = \hat{z}E_0\cos(\omega t)$, where E_0 is the electric field amplitude and ω the angular frequency of the electric field. The laser field couples state $|0\rangle$ with states $|2\rangle$ and $|3\rangle$.

Both excited states $|2\rangle$ and $|3\rangle$ decay spontaneously to state $|0\rangle$ with decay rate $2\gamma^{'}$ and to state $|1\rangle$ with decay rate 2γ. We assume that the transitions $|2\rangle$, $|3\rangle$ to $|1\rangle$ lie within the surface-plasmon bands of the plasmonic nanostructure, whereas the transitions $|2\rangle$, $|3\rangle$ to $|0\rangle$ are spectrally far from the surface-plasmon bands and are not influenced by the plasmonic nanostructure.

The Hamiltonian that describes the interaction of the electromagnetic field with the quantum system, in the dipole and rotating wave approximations, is given by

$$\begin{aligned} H &= \hbar\left(-\delta - \frac{\omega_{32}}{2}\right)|2\rangle\langle 2| + \hbar\left(-\delta + \frac{\omega_{32}}{2}\right)|3\rangle\langle 3| \\ &- \frac{\hbar\Omega}{2}\left(|0\rangle\langle 2| + |0\rangle\langle 3| + H.c.\right). \end{aligned} \quad (1)$$

Here, $\delta = \omega - \tilde{\omega}$ is the detuning from resonance with the average transition energies of states $|2\rangle$ and $|3\rangle$ from state $|0\rangle$, with $\tilde{\omega} = (\omega_3 + \omega_2)/2 - \omega_0$, $\omega_{32} = (\omega_3 - \omega_2)/2$, and Ω is the Rabi frequency defined as $\Omega = \mu^{'}E_0/(\sqrt{2}\hbar)$. Also, $\hbar\omega_n$ with $n = 0 - 3$, is the energy of state $|n\rangle$.

Using the Hamiltonian of Eq. (1) we obtain the following equations for the density matrix elements of the system, assuming a Markovian response:

$$\begin{aligned}
\dot{\rho}_{00}(t) &= 2\gamma^{'}\left[\rho_{22}(t)+\rho_{33}(t)\right] - i\frac{\Omega}{2}\left[\rho_{02}(t)-\rho_{20}(t)\right] \\
&- i\frac{\Omega}{2}\left[\rho_{03}(t)-\rho_{30}(t)\right], \qquad (2) \\
\dot{\rho}_{22}(t) &= -2(\gamma+\gamma^{'})\rho_{22}(t) + i\frac{\Omega}{2}\left[\rho_{02}(t)-\rho_{20}(t)\right] \\
&- \kappa\left[\rho_{23}(t)+\rho_{32}(t)\right], \qquad (3) \\
\dot{\rho}_{33}(t) &= -2(\gamma+\gamma^{'})\rho_{33}(t) + i\frac{\Omega}{2}\left[\rho_{03}(t)-\rho_{30}(t)\right] \\
&- \kappa\left[\rho_{23}(t)+\rho_{32}(t)\right], \qquad (4) \\
\dot{\rho}_{20}(t) &= (i\delta + i\frac{\omega_{32}}{2} - \gamma - \gamma^{'})\rho_{20}(t) - i\frac{\Omega}{2}\rho_{22}(t) \\
&- i\frac{\Omega}{2}\rho_{23}(t) + i\frac{\Omega}{2}\rho_{00}(t) - \kappa\rho_{30}(t), \qquad (5) \\
\dot{\rho}_{30}(t) &= (i\delta - i\frac{\omega_{32}}{2} - \gamma - \gamma^{'})\rho_{30}(t) - i\frac{\Omega}{2}\rho_{33}(t) \\
&- i\frac{\Omega}{2}\rho_{32}(t) + i\frac{\Omega}{2}\rho_{00}(t) - \kappa\rho_{20}(t), \qquad (6) \\
\dot{\rho}_{23}(t) &= (i\omega_{32} - 2\gamma - 2\gamma^{'})\rho_{23}(t) + i\frac{\Omega}{2}\rho_{03}(t) - i\frac{\Omega}{2}\rho_{20}(t) \\
&- \kappa\left[\rho_{22}(t)+\rho_{33}(t)\right], \qquad (7)
\end{aligned}$$

with $\rho_{00}(t)+\rho_{11}(t)+\rho_{22}(t)+\rho_{33}(t)=1$ and $\rho_{nm}(t)=\rho^{*}_{mn}(t)$. Here, κ is the coupling coefficient between states $|2\rangle$ and $|3\rangle$ due to spontaneous emission in a modified anisotropic vacuum [9]. This term is responsible for the appearance of quantum interference in spontaneous emission [10].

The values of γ and κ are obtained by [9, 11, 12]

$$\begin{aligned}
\gamma &= \frac{\mu_0\mu^2\bar{\omega}^2}{\hbar}\hat{\varepsilon}_{-}\cdot \mathrm{Im}\mathbf{G}(\mathbf{r},\mathbf{r};\bar{\omega})\cdot\hat{\varepsilon}_{+} = \frac{1}{2}\left(\Gamma_{\perp}+\Gamma_{\parallel}\right), \qquad (8) \\
\kappa &= \frac{\mu_0\mu^2\bar{\omega}^2}{\hbar}\hat{\varepsilon}_{+}\cdot \mathrm{Im}\mathbf{G}(\mathbf{r},\mathbf{r};\bar{\omega})\cdot\hat{\varepsilon}_{+} = \frac{1}{2}\left(\Gamma_{\perp}-\Gamma_{\parallel}\right). \qquad (9)
\end{aligned}$$

Here, $\mathbf{G}(\mathbf{r},\mathbf{r};\omega)$ is the dyadic electromagnetic Green's tensor, where $\mathbf{r}$ refers to the position of the quantum emitter, and μ_0 is the permeability of vacuum. Also, $\bar{\omega}=(\omega_3+\omega_2)/2-\omega_1$. In addition, we define the spontaneous emission rates normal and parallel to the surface as $\Gamma_{\perp,\parallel}=\mu_0\mu^2\bar{\omega}^2\mathrm{Im}[G_{\perp,\parallel}(\mathbf{r},\mathbf{r};\bar{\omega})]/\hbar$, where $G_{\perp}(\mathbf{r},\mathbf{r};\bar{\omega})=G_{zz}(\mathbf{r},\mathbf{r};\bar{\omega})$, $G_{\parallel}(\mathbf{r},\mathbf{r};\bar{\omega})=G_{xx}(\mathbf{r},\mathbf{r};\bar{\omega})$ denote components of the electromagnetic Green's tensor where the symbol $\perp$ ($\parallel$) refers to a dipole oriented normal - along the z-axis (parallel - along the x-axis) to the surface of the nanostructure.

The plasmonic nanostructure considered in this study is a two-dimensional array of touching silver-coated silica nanospheres [see Fig. 1(b) and (c)]. The dielectric function of the shell is provided by a Drude-type electric permittivity given by

$$\epsilon(\omega) = 1 - \frac{\omega_p^2}{\omega(\omega+i/\tau)}, \qquad (10)$$

where ω_p is the bulk plasma frequency and τ the relaxation time of the conduction-band electrons of the metal. The dielectric constant of SiO_2 is taken to be $\epsilon=2.1$. In the calculations we have

Table 1. The values of $\Gamma_\perp$ and $\Gamma_\parallel$ for different distances D from the surface of the plasmonic nanostructure for $\hbar\tilde{\omega} = 2.4$ eV. Γ_0 is the spontaneous decay rate in free space and in our work we take it $\Gamma_0 = 10^9$ ns^{-1}.

D (nm)	$\Gamma_\perp$ (Γ_0)	$\Gamma_\parallel$ (Γ_0)
10.4	27.081	0.105
20.8	6.417	0.038
31.2	1.774	0.021
41.6	0.559	0.021
52	0.196	0.026

taken $\tau^{-1} = 0.1\omega_p$ and $\hbar\omega_p = 3.8$ eV corresponding to silver. The lattice constant of the square lattice is $a = 104$ nm and the sphere radius $S = 52$ nm with core radius $S_c = 36.4$ nm.

3. Numerical Results

The corresponding electromagnetic Green's tensor which provides the corresponding spontaneous emission rates $\Gamma_\perp$ and $\Gamma_\parallel$ is given by [12, 13, 14]

$$G^{EE}_{ii'}(\mathbf{r},\mathbf{r}';\omega) = g^{EE}_{ii'}(\mathbf{r},\mathbf{r}';\omega) - \frac{i}{8\pi^2}\int\int_{SBZ} d^2\mathbf{k}_\parallel \sum_{\mathbf{g}} \frac{1}{c^2 K^+_{\mathbf{g};z}} \times v_{\mathbf{g}\mathbf{k}_\parallel;i}(\mathbf{r}) \exp(-\mathrm{i}\mathbf{K}^+_{\mathbf{g}}\cdot\mathbf{r})\hat{\mathbf{e}}_{i'}(\mathbf{K}^+_{\mathbf{g}}), \quad (11)$$

with

$$v_{\mathbf{g}\mathbf{k}_\parallel;i}(\mathbf{r}) = \sum_{\mathbf{g}'} R_{\mathbf{g}';\mathbf{g}}(\omega,\mathbf{k}_\parallel)\exp(-\mathrm{i}\mathbf{K}^-_{\mathbf{g}'}\cdot\mathbf{r})\hat{\mathbf{e}}_i(\mathbf{K}^-_{\mathbf{g}'}), \quad (12)$$

and

$$\mathbf{K}^\pm_{\mathbf{g}} = (\mathbf{k}_\parallel + \mathbf{g},\ \pm[q^2 - (\mathbf{k}_\parallel + \mathbf{g})^2]^{1/2}). \quad (13)$$

The vectors $\mathbf{g}$ denote the reciprocal-lattice vectors corresponding to the 2D periodic lattice of the plane of scatterers and $\mathbf{k}_\parallel$ is the reduced wavevector which lies within the surface Brillouin zone associated with the reciprocal lattice [15]. When $q^2 = \omega^2/c^2 < (\mathbf{k}_\parallel + \mathbf{g})^2$, $\mathbf{K}^\pm_{\mathbf{g}}$ defines an evanescent wave. The term $g^{EE}_{ii'}(\mathbf{r},\mathbf{r}';\omega)$ of Eq. (11) is the free-space Green's tensor and $\hat{\mathbf{e}}_i(\mathbf{K}^\pm_{\mathbf{g}})$ the polar unit vector normal to $\mathbf{K}^\pm_{\mathbf{g}}$. $R_{\mathbf{g}';\mathbf{g}}(\omega,\mathbf{k}_\parallel)$ is the reflection matrix which provides the sum (over $\mathbf{g}$'s) of reflected beams generated by the incidence of plane wave from the left of the plane of scatterers [15]. Also, in Eq. (11), the terms corresponding to s-polarized waves (those containing components with the azimuthial unit vector $\hat{\mathbf{e}}_i(\mathbf{K}^\pm_{\mathbf{g}})$ normal to $\mathbf{K}^\pm_{\mathbf{g}}$) have small contribution to the decay rates and have been, therefore, neglected.

We use the above methodology and calculate the spontaneous emission rates normal and parallel to the surface of the plasmonic nanostructure for different distances of the quantum emitter from the plasmonic nanostructure. The results of the calculations are shown in Table 1.

We assume that the quantum system is initially in state $|0\rangle$. This means $\rho_{00}(0) = 1$ and for all other elements $\rho_{nm}(0) = 0$. We are interested in the time evolution of the populations $\rho_{nn}(t)$, with $n = 0, 2, 3$, at $\delta = 0$ $(\omega = \tilde{\omega})$, because, at this frequency, coherent population trapping may occur in states $|0\rangle$, $|2\rangle$ and $|3\rangle$ if $\kappa = \gamma$ and $\gamma' = 0$ [8, 16, 17]. Here, κ and γ are never exactly equal to each other (see the values of Table 1); however, we expect an interesting behavior of the population dynamics at this frequency if $\gamma' = 0$. Therefore, we will present results obtained from a numerical solution of the density matrix equations (2)-(7) using a fourth-order Runge-Kutta method.

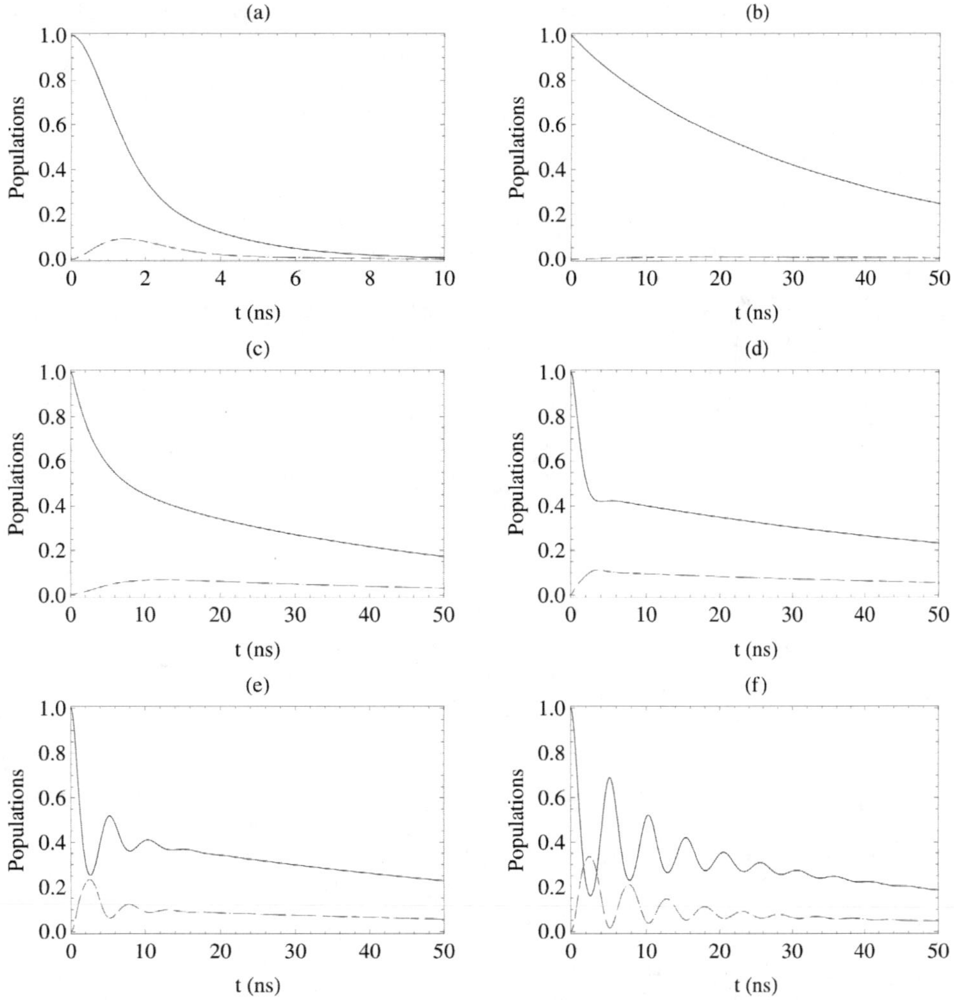

Figure 2. The time evolution of the population $\rho_{nn}(t)$ in states $|n\rangle$. With solid curve we present the population in state $|0\rangle$, with dashed curve the population in state $|2\rangle$ and with dot-dashed curve the population in state $|3\rangle$. Plot (a) shows results in the absence of the plasmonic nanostructure, while the other plots depict results in the presence of the plasmonic nanostructure. We take $\hbar\bar{\omega} = 2.4$ eV, $\Omega = 1$ ns^{-1}, $\omega_{32} = 2$ ns^{-1} and $\gamma' = 0$. In (b) $D = 10.4$ nm, (c) $D = 20.8$ nm, (d) $D = 31.2$ nm, (e) $D = 41.6$ nm and (f) $D = 52$ nm.

In Fig. 2 we present the population $\rho_{nn}(t)$, with $n = 0, 2, 3$ as a function of time for $\gamma' = 0$. We note that $\rho_{22}(t) = \rho_{33}(t)$. When the quantum system is placed in vacuum [Fig. 2(a)], i.e. in the absence of the plasmonic nanostructure, a transient weak excitation of the population to states $|2\rangle$ and $|3\rangle$ occurs and the whole population is transfer by spontaneous emission to state $|1\rangle$ in about 10 ns. In the presence of the plasmonic nanostructure [Figs. 2(b)-(f)] the evolution of the population is different and depends on the distance of the quantum emitter from the plasmonic nanostructure. For distances close to the plasmonic nanostructure, such those shown in Figs. 2(b) and (c), the population remains in states $|0\rangle$ (mainly), $|2\rangle$ and $|3\rangle$ for much longer times. For even longer distances, Fig. 2(d), a weak oscillation is obtained together with the slow population decay. Similar behavior is also found for distances close or equal to the nanoparticle radius, Figs. 2(e) and (f), with the addition that in this case the populations exhibit stronger damped oscillations. We note that similar behavior to the last figure is obtained for distances that are within a few nanoparticle radii. For much larger distances the influence of the plasmonic nanostructure becomes very weak, with no signature of the plasmonic nanostructure.

4. Summary

We have studied the controlled dynamics of the four-level quantum emitter coupled to a plasmonic metamaterial. The plasmonic system we examined is a periodic two-dimensional array of silver-coated silica nanospheres. An external electromagnetic field is applied to the system and is used for the control of the dynamics of the quantum emitter. For the study of the system's dynamics, we combined the density matrix approach for the quantum emitter with *ab initio* electromagnetic calculations for the plasmonic nanostructure. We then presented results for the time evolution of the population of the different levels of the quantum emitter in both the presence and the absence of the plasmonic nanostructure. We found that the population evolution depends strongly on the distance between the quantum emitter and the plasmonic nanostructure.

Acknowledgments

This work was implemented within the framework of the Action Supporting Postdoctoral Researchers of the Operational Program Education and Lifelong Learning (Actions Beneficiary: General Secretariat for Research and Technology), and was co-financed by the European Social Fund and the Greek State (Program Nanokallos PE3_26).

References

[1] Tame M S, McEnery K R, Ozdemir S K, Lee J, Maier S A and Kim M S 2013 *Nature Phys.* **9** 329.
[2] Evangelou S, Yannopapas V and Paspalakis E 2012 *Phys. Rev. A* **86** 053811.
[3] Evangelou S, Yannopapas V and Paspalakis E 2014 *Opt. Commun.* **314** 36.
[4] Evangelou S, Yannopapas V and Paspalakis E 2014 *J. Mod. Opt.* **61** 1458.
[5] Paspalakis E, Evangelou S, Yannopapas V and Terzis A F 2013 *Phys. Rev. A* **88** 053832.
[6] Evangelou S, Yannopapas V and Paspalakis E 2011 *Phys. Rev. A* **83** 023819.
[7] Gu Y, Wang L, Ren P, Zhang J-X, Zhang T-C, Martin O J F and Gong Q-H 2012 *Nano Lett.* **12** 2488.
[8] Gu Y, Wang L, Ren P, Zhang J-X, Zhang T-C, Xu J-P, Zhu S-Y and Gong Q-H 2012 *Plasmonics* **7** 33.
[9] Agarwal G S 2000 *Phys. Rev. Lett.* **84** 5500.
[10] Kiffner M, Macovei M, Evers J and Keitel C H 2010 *Prog. Opt.* **55** 85.
[11] Yang Y-P, Xu J-P, Chen H and Zhu S-Y 2008 *Phys. Rev. Lett.* **100** 043601.
[12] Yannopapas V, Paspalakis E and Vitanov N V 2009 *Phys. Rev. Lett.* **103** 063602.
[13] Sainidou R, Stefanou N and Modinos A 2004 *Phys. Rev. B* **69** 064301.
[14] Yannopapas V and Vitanov N V 2007 *Phys. Rev. B* **75** 115124.
[15] Stefanou N, Yannopapas V and Modinos A 1998 *Comput. Phys. Commun.* **113**, 49; 2000 *Comput. Phys. Commun.* **132** 189.
[16] Zhu S Y and Scully M O 1996 *Phys. Rev. Lett.* **76** 388.
[17] Paspalakis E and Knight P L 1998 *Phys. Rev. Lett.* **81** 293.

Influence of anisotropy on percolation and jamming of linear k-mers on square lattice with defects

Yu Yu Tarasevich[1], **V V Laptev**[1,2], **A S Burmistrov**[1], **T S Shinyaeva**[1]

[1]Astrakhan State University, 20a Tatishchev Street, Astrakhan, 414056, Russia
[2]Astrakhan State Technical University, 16 Tatishchev Street, Astrakhan, 414025, Russia

E-mail:
tarasevich@asu.edu.ru, laptev@ilabsltd.com, ksairen10@mail.ru, tae4ka19@mail.ru

Abstract. By means of the Monte Carlo simulation, we study the layers produced by the random sequential adsorption of the linear rigid objects (k-mers also known as rigid or stiff rods, sticks, needles) onto the square lattice with defects in the presence of an external field. The value of k varies from 2 to 32. The point defects randomly and uniformly placed on the substrate hinder adsorption of the elongated objects. The external field affects isotropic deposition of the particles, consequently the deposited layers are anisotropic. We study the influence of the defect concentration, the length of the objects, and the external field on the percolation threshold and the jamming concentration. Our main findings are (i) the critical defect concentration at which the percolation never occurs even at jammed state decreases for short k-mers ($k < 16$) and increases for long k-mers ($k > 16$) as anisotropy increases, (ii) the corresponding critical k-mer concentration decreases with anisotropy growth, (iii) the jamming concentration decreases drastically with growth of k-mer length for any anisotropy, (iv) for short k-mers, the percolation threshold is almost insensitive to the defect concentration for any anisotropy.

1. Introduction

Deposition of large particles such as colloids, polymers or nanotubes on substrates can be considered and studied as the random sequential adsorption (RSA) [1]. In RSA, objects randomly deposit on a substrate; this process is irreversible, and the newly placed objects cannot overlap or pass through the previously deposited ones. The substrate may be prepatterned (e.g., see [2]) or include some impurities (defects) (e.g., see [3]). The adsorbed objects may be identical or present a mixture of objects of different sizes and shapes (e.g., see [4, 5]). Moreover, anisotropy can be introduced by postulating unequal probabilities for deposition of elongated objects along different directions (e.g., see [5, 6]). The anisotropy of deposition can reflect the influence of the external fields, flows or anisotropy of the substrate. The adsorption of the elongated particles in the presence of the external fields produces the anisotropic layers (e.g., see [7, 8]).

Very often in RSA, a substrate is considered as a discrete space, e.g., regular or irregular lattice, a tree. The simplest, but the most commonly used instance of the discrete substrate is the square lattice. The linear segments are used very often as the objects adsorbed onto the square lattice. The linear segment (also denoted as needle, linear chain, stiff rod or k-mer) consists of k successive connected sites.

If the deposition of the objects goes infinitely long, a jamming state is reached. At the jamming state, there are still voids between the previously placed object on the substrate, but their size and shape are not sufficient to deposit even one additional object.

Definitions. For a perfect lattice (a lattice without any defects), the jamming concentration, p_j, is the fraction of the sites occupied by deposited objects. For a diluted (disordered, disturbed) lattice (a lattice with the defects or impurities), there are several ways to define the jamming concentration.

(i) The jamming concentration, ρ, is the fraction of all occupied sites, i.e. the fraction of the sites occupied by the deposited objects plus the fraction of the sites occupied by the impurities [9].

(ii) The pure object jamming limit, p_j, is defined as the fraction of total lattice sites occupied by the deposited objects [10]. This quantity equals $\rho - d$, where d means the fraction of the sites occupied by impurities.

(iii) The jamming concentration, c_j, in terms by Cornette et al. [11], is the fraction of the sites free of the impurities and occupied by the objects, i.e.,

$$c_j = \frac{p_j}{1-d} = \frac{\rho - d}{1-d}.$$

In the present research, we utilize everywhere the pure object jamming limit and denote it as the jamming concentration, p_j, for short.

Ben-Naim and Krapivsky obtained an important analytical result for the one-dimensional case [9]. Initially, a one-dimensional lattice is occupied by the point impurities with the uniform random distribution. Next, the linear k-mers randomly and sequentially adsorb on the lattice. In this case, the jamming concentration, ρ, is

$$\rho = d + k(1-d)^k \int_0^\infty \exp\left(-t - 2\sum_{i=1}^{k-1} \frac{1-\exp(-it)}{i}(1-d)^i\right) dt, \tag{1}$$

where d is the initial concentration of the point impurities, k is the length of the objects [9].

If the concentration of the deposited objects on the substrate is sufficiently large, there may be a path through the objects from one side of the system to its opposite side. Below this concentration, there is no spanning path through the system, while above this concentration, there is a connected component of the order of the system size. This concentration is called the percolation threshold. (See, e.g., [12] for the details; art of the state may be found, e.g., in [13].) For the some systems, the percolation never occurs even at the jamming concentration (e.g., the adsorbed layer produced by the deposition of the equal squares on a square lattice if the side of the square is greater than 3 sites [14]). In fact, the percolation threshold corresponds to a phase transition, e.g., insulator-conductor. Different definitions of the percolation threshold are utilized for the lattices with defects similar to the definitions of the jamming concentration. In the present research, we use the term 'the percolation threshold' (p_c) as the ration of the sites occupied by the objects to the whole number of the lattice sites.

Cornette et al. [11, 15] investigated numerically the percolation of the polyatomic species in the presence of impurities on a square lattice with periodic boundary conditions. Bond and site percolation problem have been taken in consideration. Linear k-mers as well so called SAW k-mers, i.e. segments of self avoiding walk, have been studied up to $k = 9$. Phase diagram where the critical concentration of impurities is plotted as a function of k has been offered.

The kinetics of the random sequential adsorption of line k-mers (values of k vary up to 64) has been studied on a disordered substrate occupied by the point impurities [10]. The coverage of the surface and the jamming limits are calculated by a Monte Carlo method. The coverage has an asymptotically exponential behaviour at low concentration of the impurities. The jamming limits depend on the concentration of the impurities d. At $d < d^*$ the jamming limits increase as d increases. At $d > d^*$ the jamming limits increase as d increases, where the value of d^* depends on k. In one-dimensional case, the results are in good agreement with published analytic results [9]. The coverage and the jamming limits on a two-dimensional disordered lattice are similar to the one-dimensional case. The jamming limits decrease monotonically as the length of line segments increases. The minimum locations of the jamming limits for both one and two dimensions are on the same values for a given length of the k-mer.

The effect of anisotropic deposition on the jamming [16] and the percolation [17] was intensively studied by means of the Monte Carlo simulation.

Recently, impact of defects on percolation in random sequential adsorption of linear k-mers on square lattices was investigated for rather wide range of k-mers (the length of the k-mers varies from 2 to 256) [18]. Two different case were studied: (1)it was assumed that the initial square lattice is nonideal and some fraction of sites d is occupied by nonconducting point defects (impurities); (2)it was assumed that some fraction of the sites in the k-mers d consists of defects, i.e., is nonconducting, whereas the initial square lattice is perfect.

Mixed site-bond percolation was studied for RSA of k-mers on heterogeneous lattices with of variable connectivity z [19]. The simulations were performed for $k = 1$–3 on a triangular lattice. The percolation phase diagrams in terms of the percolation threshold p_c versus lattice connectivity z were obtained. For RSA deposition of monomer onto triangle lattice with defects preliminary filled with k-mers, the percolation ($k = 3$–24) [20] and the jamming ($k \leq 50$) [21] were investigated by means of the Monte Carlo simulations. The nonmonotonicity of the percolation threshold as a function of the impurity concentration was observed [20].

RSA of polydisperse mixtures of k-mers was also intensively investigated [5, 22, 23]. A phase diagram separating a percolating from a non-percolating region for mixtures of monomers and k-mers on a square lattice ($k = 2$–7) was determined [22]. The jamming coverage for a mixture was greater than either of the jamming coverage of the components making the mixture [5]. From other hand the percolation threshold for a mixture was slightly greater than that of the longest k-mer [23].

Recently, continuum random sequential adsorption of polymer on a flat and homogeneous surface was studied [24].

Our goal is to study the influence of the anisotropy on the percolation threshold and the jamming concentration of the linear k-mers onto the square lattice with the point defects.

The rest of the paper is organized as follows. Section 2 describes our models of the k-mer deposition onto a square lattice. The results obtained using finite size scaling theory and dependencies of the percolation threshold, p_c, and the jamming concentration, p_j, vs order parameter, s, and the defect concentration, d, are examined and discussed in details in Section 3. We summarize the results and conclude the paper in Section 4.

2. Method

2.1. Anisotropy

To describe the anisotropy, we used the quantity

$$s = \frac{|N_| - N_-|}{N_| + N_-}, \tag{2}$$

where $N_|$ is the number of vertically oriented objects, N_- is the number of horizontally oriented objects.

2.2. Relaxation Random Sequential Adsorption

In our simulation, we utilized the modification of RSA, so-called relaxation random sequential adsorption (RRSA) [16]. In this model, the lattice site is randomly selected and an attempt of deposition of a k-mer with the orientation defined by the order parameter s is done. If the attempt is unsuccessful, a new lattice site is randomly selected until the object will be deposited. The objects may move all over the substrate searching for a sufficiently large empty space. The deposition terminates when a jamming state is reached along one direction. We considered the lattice with periodic (toroidal) boundary conditions to eliminate the effect of the border and treated the spiral clusters as wrapping (percolating) in contrast with [25]. We checked the percolation in two perpendicular directions and used two criteria: there is percolation in both directions (criterion AND), there is percolation at least along one direction (OR).

For each given anisotropy, s, and the concentration of the defects, d, we filled the lattice with the k-mers to the concentration p or to the jamming and checked whether the percolation. We repeated it 1000 times and found the probability, $R(p)$, that the percolation occurs at given concentration of the k-mers. The abscissa of the inflection point of the curve is treated as the estimation of the percolation threshold for the given lattice size. We used the lattices of the sizes $L = 100k, 200k, 400k$ and performed finite-size analysis to obtain the percolation threshold in the thermodynamic limit ($L \to \infty$) $p_c \propto L^{-1/\nu}$, where the critical exponent $\nu = 4/3$ [12]. All probability curves intersect each other in one point with the co-ordinates (p_c, R^*), where $R^* = R(p_c) \approx 0.859$. The value of R^* depends on how the percolation is defined [25]. The intersection point offers another way to estimate the percolation threshold. Examples of probability curves and scaling are shown in Fig. 1.

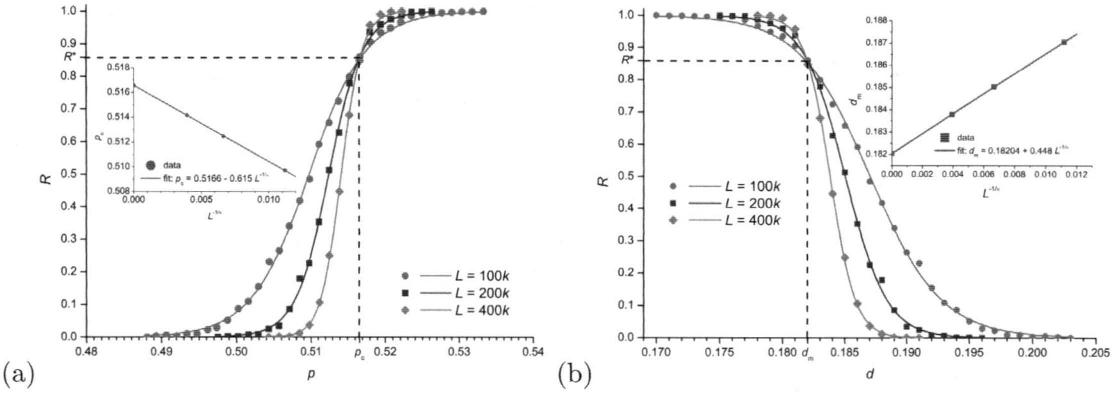

Figure 1. The probability curves for $k = 4$, $s = 0.5$. (a) $R(p)$. (b) $R(d)$. Insets: scaling. The statistical error is smaller than the marker size.

We compared our numerical simulation for completely aligned k-mers ($s = 1$) with the analytical results [9]. Figure 2 shows that there is not any visible difference between analytical and numerical results.

3. Results

We denote as d_m the critical defect concentration, i.e. the defect concentration above which the percolation never occurs even at jammed state. We denote as p_c^m the critical k-mer concentration for given concentration of defects, i.e. the object concentration at which the percolation occurs at jammed state only.

The critical defect concentration, d_m, decreases for short k-mers ($k < 16$) and increases for long k-mers ($k > 16$) as anisotropy, s, increases (see Figure 3a). The critical k-mer concentration,

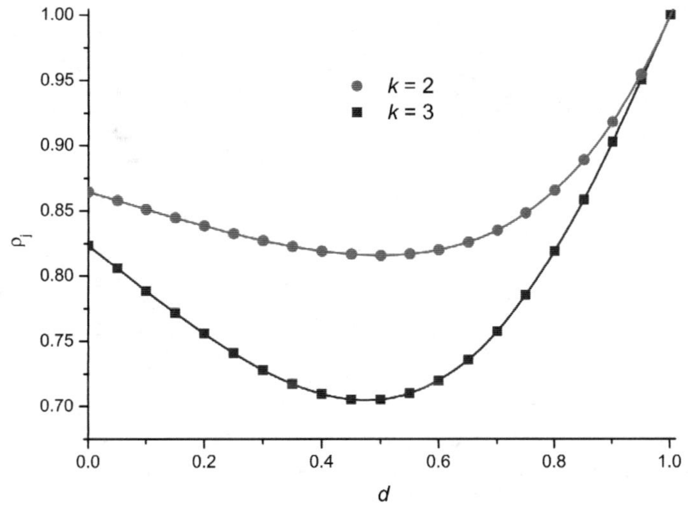

Figure 2. The total jamming coverage, ρ_j, for completely aligned ($s = 1$) dimers ($k = 2$) and trimers ($k = 3$) as a function of the concentration of point defects, d: theory (1) (lines) and our numerical simulations (symbols). The statistical error is smaller than the marker size.

p_c^m, decreases with anisotropy growth (see Figure 3b). It looks like, this dependence may be nonmonotonic and have a minimum somewhere about $s \approx 0.5$ for long objects ($k > 32$).

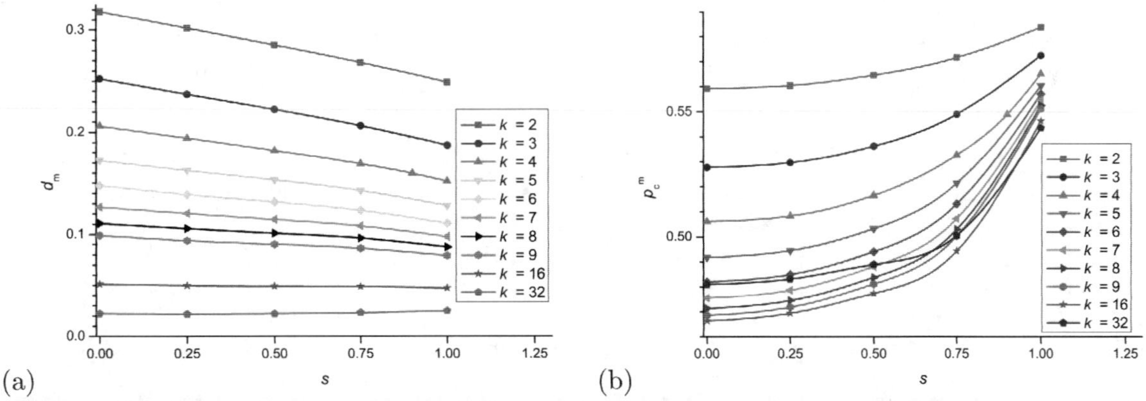

Figure 3. (a) The critical defect concentration, d_m, and (b) the critical k-mer concentration, p_c^m, as a function of anisotropy, s, for different length of the k-mers. The solid lines are a guide to the eye. The statistical error is smaller than the marker size.

The jamming concentration decreases drastically with growth of k-mer length for any anisotropy (see the samples for $s = 0.25$ and 0.75 in Figure 4).

For short k-mers, the percolation threshold, p_c, is almost insensitive to the defect concentration, d, for any anisotropy, s (see Figures 5).

4. Conclusion

In our work, the percolation and jamming behavior of partially ordered linear k-mers on a square lattice with periodic (toroidal) boundary conditions has been investigated by computer

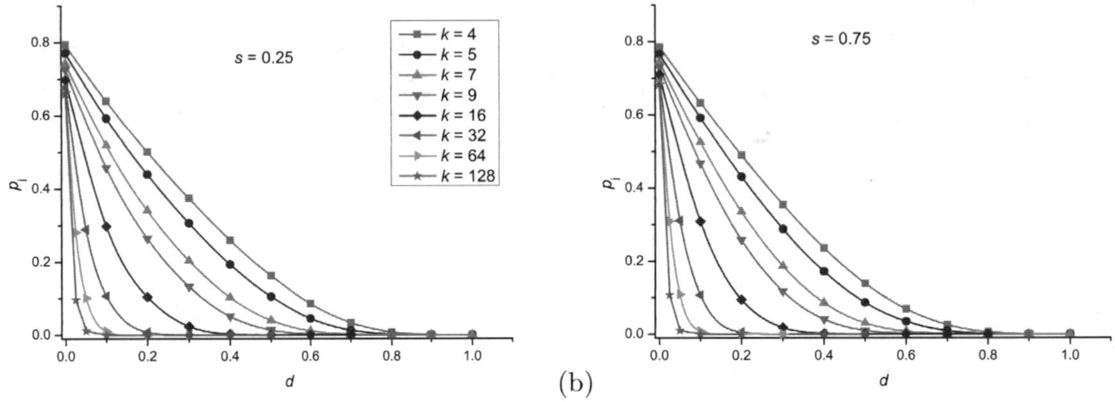

(a) (b)

Figure 4. The jamming concentration, p_j, as a function of the defect concentration, d, for different k and fixed anisotropy (a) $s = 0.25$, (b) $s = 0.75$. The solid lines are a guide to the eye. The statistical error is smaller than the marker size.

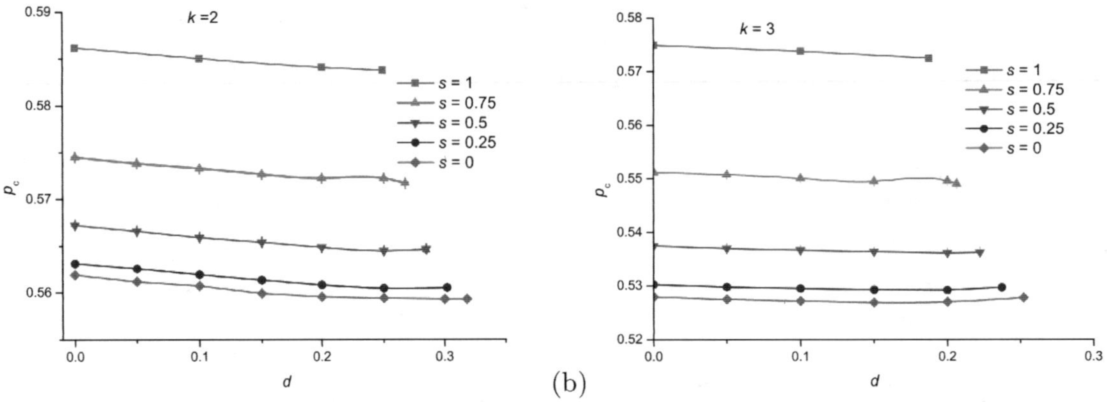

(a) (b)

Figure 5. The percolation threshold, p_c, as a function of the defect concentration, d, for different anisotropy, s, for (a) dimers ($k = 2$) and (b) trimers ($k = 3$). The solid lines are a guide to the eye. The statistical error is smaller than the marker size.

simulations. The lattice is imperfect, i.e. some fraction of the lattice sites is occupied by defects (impurities). The length of a k-mer varies from 1 to 32. The relaxation random sequential adsorption model [16] has been used to deposit the k-mers with the given anisotropy. The anisotropy is characterized by order parameter $s = 0 \dots 1$, where $s = 0$ corresponds to the isotropic system and $s = 1$ corresponds to the system with completely aligned objects. The defects hinder the adsorption of the elongated objects. If the concentration of the defect exceeds the critical value, d_m, the percolation cannot occur. The dependence of critical concentration on anisotropy, $d_m(s)$, is almost linear function for any value of k. Even a small concentration of the defects inhibits the deposition of long objects. Additional studies with long objects ($k > 32$) look promising because the real objects of practical interest (e.g., nanotubes), can have large aspect ratios.

Acknowledgments

The reported study was partially supported by Russian Foundation for Basic Research, research project No. 14-02-90402_Ukr_a, the Ministry of Education and Science of the Russian Federation, project no 266.

References

[1] Evans J W 1993 *Rev. Mod. Phys.* **65**(4) 1281–1329
[2] Cadilhe A, Araújo N A M and Privman V 2007 *Journal of Physics: Condensed Matter* **19** 065124
[3] Lee J W 1997 *Phys. Rev. E* **55**(3) 3731–3734
[4] Lee J W 2000 *Colloids and Surfaces A: Physicochemical and Engineering Aspects* **165** 363–372 ISSN 0927-7757
[5] Budinski-Petković L, Lončarević I, Jakšić Z M, Vrhovac S B and Švrakić N M 2011 *Phys. Rev. E* **84**(5) 051601
[6] de Oliveira M, Tomé T and Dickman R 1992 *Phys. Rev. A* **46**(10) 6294–6299
[7] Pagonabarraga I, Bafaluy J and RubíJ M 1995 *Phys. Rev. Lett.* **75**(3) 461–464
[8] Pagonabarraga I, Bafaluy and RubíJ M 1999 *Phys. Rev. E* **59**(4) 4285–4297
[9] Ben-Naim E and Krapivsky P L 1994 *Journal of Physics A: Mathematical and General* **27** 3575–3577
[10] Lee J W 1996 *Journal of Physics A: Mathematical and General* **29** 33–39
[11] Cornette V, Ramirez-Pastor A J and Nieto F 2006 *The Journal of Chemical Physics* **125** 204702
[12] Stauffer D and Aharony A 1992 *Introduction to Percolation Theory* (London: Taylor & Francis)
[13] Araújo N, Grassberger P, Kahng B, Schrenk K and Ziff R 2014 *The European Physical Journal Special Topics* **223** 2307–2321 ISSN 1951-6355
[14] Nakamura M 1987 *Phys. Rev. A* **36**(5) 2384–2388
[15] Cornette V, Ramirez-Pastor A and Nieto F 2011 *Physica A: Statistical Mechanics and its Applications* **390** 671–679 ISSN 0378-4371
[16] Lebovka N I, Karmazina N N, Tarasevich Y Y and Laptev V V 2011 *Phys. Rev. E* **84**(6) 061603
[17] Tarasevich Y Y, Lebovka N I and Laptev V V 2012 *Phys. Rev. E* **86**(6) 061116
[18] Tarasevich Y Y, Laptev V V, Vygornitskii N V and Lebovka N I 2015 *Phys. Rev. E* **91**(1) 012109
[19] Quintana M, Kornhauser I, López R, Ramirez-Pastor A and Zgrablich G 2006 *Physica A: Statistical Mechanics and its Applications* **361** 195–208 ISSN 0378-4371
[20] Kondrat G 2005 *The Journal of Chemical Physics* **122** 184718
[21] Kondrat G 2006 *The Journal of Chemical Physics* **124** 054713
[22] Dolz M, Nieto F and Ramirez-Pastor A 2007 *Physica A: Statistical Mechanics and its Applications* **374** 239–250 ISSN 0378-4371
[23] Budinski-Petković L, Lončarević I, Petković M, Jakšić Z M and Vrhovac S B 2012 *Phys. Rev. E* **85**(6) 061117
[24] Cieśla M 2013 *Phys. Rev. E* **87**(5) 052401
[25] Newman M E J and Ziff R M 2001 *Phys. Rev. E* **64**(1) 016706

4th International Conference on Mathematical Modeling in Physical Sciences (IC-MSquare2015) IOP Publishing
Journal of Physics: Conference Series **633** (2015) 012065 doi:10.1088/1742-6596/633/1/012065

Scaling laws of turbulence intermittency in the atmospheric boundary layer: the role of stability

Paolo Paradisi
Istituto di Scienza e Tecnologie dell'Informazione "A. Faedo" (ISTI-CNR), Via G. Moruzzi 1, 56124 Pisa, Italy
E-mail: paolo.paradisi@cnr.it

Rita Cesari
Istituto di Scienze dell'Atmosfera e del Clima (ISAC-CNR), Str. Prov. Lecce-Monteroni, Km 1.2, 73100 Lecce, Italy
E-mail: r.cesari@isac.cnr.it

Paolo Allegrini
Scuola Superiore Sant'Anna, P.zza Martiri della Libertà 7, 56127 Pisa, Italy
E-mail: allegrip@gmail.com

Abstract. Bursting and intermittent behavior is a fundamental feature of turbulence, especially in the vicinity of solid obstacles. This is associated with the dynamics of turbulent energy production and dissipation, which can be described in terms of coherent motion structures. These structures are generated at random times and remain stable for long times, after which they become suddenly unstable and undergo a rapid decay event. This intermittent behavior is described as a birth-death point process of self-organization, i.e., a sequence of critical events. The Inter-Event Time (IET) distribution, associated with intermittent self-organization, is typically a power-law decay, whose power exponent is known as complexity index and characterizes the complexity of the system, i.e., the ability to develop self-organized, metastable motion structures. We use a method, based on diffusion scaling, for the estimation of system's complexity. The method is applied to turbulence velocity data in the atmospheric boundary layer. A neutral condition is compared with a stable one, finding that the complexity index is lower in the neutral case with respect to the stable one. As a consequence, the crucial birth-death events are more rare in the stable case, and this could be associated with a less efficient transport dynamics.

1. Introduction

The atmospheric boundary layer (ABL) is a complex system characterized by time and spatial scales that can vary by different orders of magnitude. It is characterized by turbulent behavior [15] and by the presence of *coherent structures* ([13]). Following [7], a *coherent motion* for turbulent flow can be defined as a three-dimensional region of the flow over which at least one fundamental flow variable (velocity components, density, temperature, humidity, vorticity, other scalars or vector fields) exhibits significant correlation with itself or with another variable over a range of space and/or time that is significantly larger than the smallest local scales of the flow. This means that coherent structures are characterized by some kind of self-organization determined by the presence of cooperative dynamics. Coherent structures in the ABL are metastable, i.e., they are characterized by a relatively long life-time after which a rapid decay of memory and spatial coherence occurs. Near the ground, such fast transitions are typically associated with bursting events. In Ref. [3] the authors denote these quasi-istantaneous transitions as *critical events* and relate them to the intermittent behavior of the system, that is a fundamental feature of turbulent fluid flows [1]. Then, the dynamics of coherent structures is characterized by intermittent events or, in other words, a birth-death process of self-organization. However, the definition of intermittency varies between studies, depending on the examined scale and adopted methods [8]. In Ref. [9] the author distinguishes between small scale or microscale intermittency and global intermittency. In the former, the intermittent behavior arise from overall modulation of the turbulence by the main eddies or in connection with sharp edges of the main eddies. The dissipation of turbulent energy is then confined in small subregions of individual coherent eddies. On the contrary, global intermittency occurs at scales larger than the main coherent eddies [9], as in the case of episodic bursting of turbulence in a strongly stratified boundary layer. In Ref. [30] the terms *internal* and *external intermittency* are introduced to distinguish between intermittency originated from the mean turbulent kinetic energy dissipation rate and intermittency resulting from interaction between the flow and the boundary conditions. Being large and small scale strongly coupled, at finer scales external and internal intermittency interact.

Statistics dissimilarity between velocity and scalars within the inertial subrange are dependent on external intermittency. The authors of Refs. [30, 32] investigate the fine-scale intermittency analysing the clustering features of canopy turbulence. The intermittent behavior of coherent structures give a substantial contribution to turbulent transport. Many studies are devoted to investigate the role of coherent structures on turbulent transport of momentum and scalars. In Ref. [31] the authors analyze velocity, air temperature and scalar concentrations fluctuations above different surfaces in order to investigate the role of surface characteristic and atmospheric stratification on intermittency and clustering. Dissimilarity of turbulent transport of momentum and scalars and the relations with the topology and the role of turbulent coherent structures as sweep and ejections is also investigated ([11], [10],[12]) for different stability conditions.

The sequence of intermittent bursting events is mainly described by two statistical indicators: the Inter-Event Times (IETs) distribution and the correlation among evens and IETs. Regarding the correlation among events, a renewal condition is here assumed, i.e., both events and the IETs are statistically independent random variables. This is in agreement with several findings in literature. In Ref. [4] critical events in the ABL are introduced and the renewal condition is verified. Further, in Refs. [16] the authors developed the so-called *surface renewal model*. This is a model for the dynamical formation of turbulent structures near the surfaces, given as random intrusions of fluid alternated with a period (residence time) of unsteady Fickian diffusion. More recently, it has been recognized that the rate of surface renewals is not constant, a condition associated with a non-exponential IET distribution [17]. In summary, the renewal condition is a reasonable assumption in the ABL.

The IET distribution associated with self-organization is typically a power-law decay with an asymptotic power exponent, known as complexity index [3] and characterizing the complexity of the system, i.e., the ability to develop self-organized, metastable motion structures. In Ref. [14] the authors study the IET distribution associated with anomalous diffusion of gusts wind.

In the present work we investigate the time intermittency of turbulent transport associated with the birth-death of self-organized coherent structures in the atmospheric boundary layer. The intermittent behavior of ABL is investigated by means of a scaling analysis, the Event-Driven Diffusion Scaling (EDDiS) method, which is applied to the sequence of IETs extracted from a

set of wind data.
In Appendix A the EDDiS method is briefly recalled. In Section 2 we describe the results of the stastical analysis and show the main results about the estimated scaling exponents. In Section 3 we draw some conclusions.

2. Diffusion scaling analysis of turbulence data

Before applying the EDDiS method (see Appendix A), we need to extract the critical events from the turbulence data with an *event detection* algorithm. Wind velocity data observed in the Atmospheric Buondary Layer (ABL) were considered. The data were collected in a measurement campaign carried out in Lecce (Salentum Peninsula, Apulia region, South-East part of Italy) during March 2004 [33]. The instrumental set-up for the measurement of wind velocity was a ultrasonic anemometer Gill R3, with 100 Hz sampling frequency. This was mounted on a horizontal bar placed at the top of a telescopic mast, 9.6m above the ground. 8 hours of collected wind data were approximatively considered for the analysis.
Firstly, a sequence of events was extracted from the experimental time series. Our goal is to identify bursts of air motion in the vicinity of the ground, marking the birth or the death of a coherent structure of fluid motion. As bursting events are associated with a sudden fluid acceleration, the events are defined as abrupt changes in the moduli of the turbulent velocity increments:

$$|\Delta S(t)| = |S(t + \Delta t) - S(t)| > S_0 \ , \tag{1}$$

where $S(t)$ is the generic signal at time t, Δt the sampling time of the time series and S_0 a threshold defining the abrupt change. In our case, the signal $S(t)$ is a turbulent velocity fluctuation in the generic direction (x, y, z). The fluctuation is defined in such a way to minimize the non-stationary effects, which are related to large scale meteorological patterns. These long-time behaviors are mainly included in the direction and intensity of the mean wind and also in the turbulence intensity (i.e., kinetic energy). Following the common practice in meteorological studies, the turbulent velocity fluctuations are defined by the linear detrending of the air velocity, evaluated over time windows of 30 minutes [15]. The detrended velocity components were also normalized with the local standard deviation, again computed over time windows of 30 minutes, thus defining the fluctuating signal $S(t)$ in Eq. (1).
The turbulence features in the ABL depend mainly by the so-called *atmospheric stability*, which characterizes the local tendency of small air particles to remain in the same vertical position (quote) or not. This depends mainly by the temperature and humidity profiles, which can be combined to derive a stability parameter. In particular, a stable atmosphere means that an air particle, when perturbed, has the tendency to come back to the original quote (attractive bouyancy forces), while, in the unstable case, it escapes far from the starting quote (repulsive bouyancy forces). In the neutral case, there are no vertical forces and the wind dynamics is independent from temperature and humidity. Here, we compared two periods with different stability features: a neutral case and a stable one.
The detection of a bursting event formally depends on the threshold S_0, but threre's a range in which the scaling is not affected by this choice. This is easily seen in Fig. 1 (Left Panel), where different values of S_0 are compared.

Then, the events were defined as the 90th percentile in the statistical distribution of the increments, corresponding to a threshold of about 0.15: $\Delta S(t) > 0.15$. A sequence of occurrence times $\{t_n\}$ was derived in this way and used to define the discrete variable $\xi(t)$ for the three walking rules introduced in Appendix A. Then, the diffusion variable $X(t)$ was computed by applying Eq. (A.1). An example of the DFA function, evaluated from the turbulent bursting events, is reported in Fig. 1 (Right Panel) for the three walking rules. The DFA displays the typical power-law $F(t) \sim t^H$, valid for monoscaling signals. The SJ rule gives a normal scaling $H = 0.5$, which means that $\mu > 2$. The SV rule seems to give a normal scaling in the long-time

Journal of Physics: Conference Series **633** (2015) 012065 doi:10.1088/1742-6596/633/1/012065

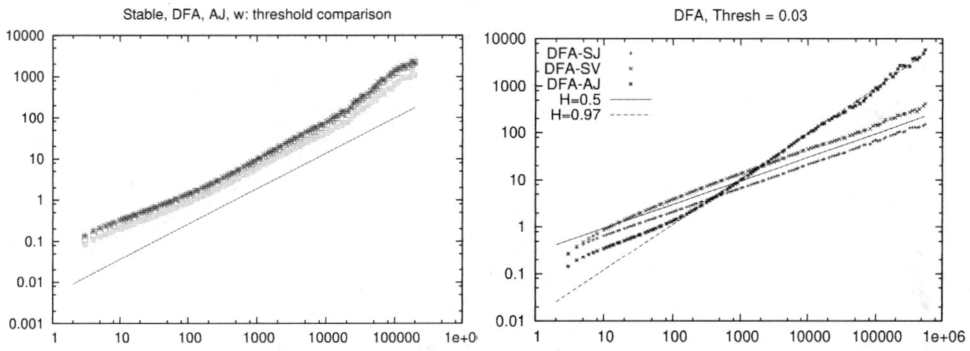

Figure 1. Left Panel: threshold dependence of the DFA function applied to the event-driven random walk with the AJ rule (vertical velocity). Right Panel: The DFA function applied to the three walking rules (AJ, SJ and SV). The diffusion scaling is also reported.

	H_{AJ}		μ	
	Neutral	Stable	Neutral	Stable
u	0.97	0.86	**2.06**	**2.28**
v	0.97	0.88	**2.06**	**2.24**
w	0.99	0.87	**2.02**	**2.26**

Table 1. Estimated values of H_{AJ}, from the AJ rule, and μ for the three turbulent velocity components.

limit. However, this rule is less accurate as it is affected by the presence of noise, which can be related to the experimental data and instrumentation, but also to the thresholding technique [5]. On the contrary, the AJ rule gives a well-defined anomalous scaling $H \sim 0.9$. The values of the diffusion scaling H and, under the renewal assumption, of the associated complexity index μ are reported in Table 2. For a given stability condition, all the fluctuacting, or turbulent, velocity components have the same H scaling and, consequently, the same complexity index. The diffusion scaling H is greater in the neutral case that in the stable one. In the renewal assumption, this gives a slower power-law decay of the IET distribution in the stable case. This means that the bursting, birth-death, events are more rare, so that the transport dynamics associated with these critical events is less efficient.

3. Conclusions

We have applied a scaling analysis, denoted as EDDiS method, based on the generation of three different random walks to a set of turbulence data in order to characterize the diffusive behavior associated with intermittent bursting events in the ABL turbulence. Two different stability conditions, one neutral and the other stable, have been compared, showing that scaling is affected by the atmospheric stability. We have found that the complexity index associated with intermittency is lower in the neutral case with respect to the stable one. As a consequence, the crucial birth-death events are more rare in the stable case, and this could be associated with a less efficient transport dynamics. Future investigations will carry out to better characterize the dissimilarity of different scalars and the consequences of the different scaling on the modeling of turbulent transport in the ABL.

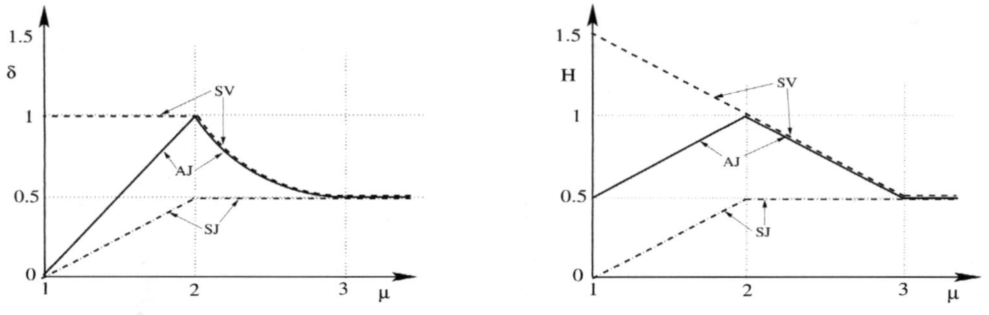

Figure 2. Scaling δ and H vs. complexity index μ for the three walking rules: AJ (continuous line), SJ (dotted-dashed line) and SV (dashed line). Left Panel: scaling δ. Right Panel: scaling H.

Appendix A. The diffusion scaling analysis of event-driven random walks

Given an experimental sequence of random events in time, this method is based on the generation of three different random walks and on the estimations of two different scaling exponents, the self-similarity index of the diffusion variable Probability Density Function (PDF) δ and the second moment scaling H. It is worth noting that the three random walks introduced here, and their scaling properties in the case of renewal events, are well known in literature by many years [21, 22, 26, 23, 24, 25]. The calculation of these scaling exponent is based on the analytical treatment of the Continuous Time Random Walk developed by Montroll and co-workers [18, 19, 20]. In particular, these walking rules have been extensively used as a tool of scaling analysis in experimental event sequences [6, 3, 5]. However, the joint use of all these random walks and scaling analyses was carried out in Ref. [6] for the first time as an application to brain data (see also [3] for a brief review). This scaling analysis, which is here denoted as Event-Driven Diffusion Scaling (EDDiS) method, allows to evaluate the diffusion scaling exponents δ and H, which are asymptotic features of the signal under consideration. In the renewal condition this allows also to derive a robust estimation of the complexity index μ. When a reliable estimation of all the different scaling exponents is carried out, the EDDiS also gives a reasonable justification of the renewal condition itself [6, 3, 5]. In fact, for renewal events, the relatiohships among the different scaling exponents and of these exponents with the complexity index are theoretically known. Then, it is possible to derive independent estimations of the complexity index. When these evaluations are compatible with each other, the renewal condition is reasonably proved and, further, a robust estimation of the complexity index is obtained.

Given the experimental sequence of event occurrence times t_0, t_1, t_2,... corresponding to the events 0, 1, 2,..., we have the following walking rules:

(a) Asymmetric Jump (AJ) rule:
the walker makes a positive jump ($\xi(t_n) = 1$) in correspondence of each event n, otherwise it stands ($\xi(t) = 0$). In other words, $\xi(t)$ is a sequence of unitary pulses

(b) Symmetric Jump (SJ) rule:
as in the AJ rule, but the walker can make positive or negative jumps in correspondence of an event: $\xi(t_n) = \pm 1$. The sign $\pm$ is randomly chosen.

(c) Symmetric Velocity (SV) rule:
the walker moves with constant velocity towards a given direction, in the time interval between two events, then a new random direction is chosen in correspondence of an event: $\xi(t) = \pm 1 \;;\; t_n < t \leq t_{n+1}$.

Each random walk is then defined through a signal $\xi(t)$, which is a kind of random discontinuous velocity, or with discontinuous derivative, and taking two or three different discrete values (two for AJ and SV and three for SJ). The diffusion variable $X(t)$ of the random walk is computed by integrating $\xi(t)$:

$$X(t) = X_0 + \int_0^t \xi(t')dt' \tag{A.1}$$

The scaling properties of these random walks were extensively investigated in several papers [21, 22, 26, 23, 24, 25]. (see [6, 3] for a brief account of the literature). The scaling exponents δ and H are defined in the following way:

$$P(x,t) = \frac{1}{t^\delta} F\left(\frac{x}{t^\delta}\right) , \tag{A.2}$$

$$\sigma^2(t) = \langle \left(X(t) - \overline{X}\right)^2 \rangle \sim t^{2H} , \tag{A.3}$$

where $\overline{X}$ is the mean value of $X(t)$. These exponents were analytically computed, in the asymptotic long-time limit, for random walks driven by renewal events with asymptotic power-law decay in the IET-PDF: $\psi(\tau) \sim \tau^\mu$ (this is also denoted as *fractal intermittency* [3]). The analytical expressions for the diffusion scaling exponents δ and H as a function of μ are given by:

$$\delta_{AJ}(\mu) = \begin{cases} \mu - 1 ; & 1 < \mu < 2 \\ 1/(\mu - 1) ; & 2 \leq \mu < 3 \\ 1/2 ; & \mu \geq 3 \end{cases} \tag{A.4}$$

$$H_{AJ}(\mu) = \begin{cases} \mu/2 ; & 1 < \mu < 2 \\ 2 - \mu/2 ; & 2 \leq \mu < 3 \\ 1/2 ; & \mu \geq 3 \end{cases} \tag{A.5}$$

$$\delta_{SV}(\mu) = \begin{cases} 1 ; & 1 < \mu < 2 \\ 1/(\mu - 1) ; & 2 \leq \mu < 3 \\ 1/2 ; & \mu \geq 3 \end{cases} \tag{A.6}$$

$$H_{SV}(\mu) = \begin{cases} 2 - \mu/2 ; & 1 < \mu < 3 \\ 1/2 ; & \mu \geq 3 \end{cases} \tag{A.7}$$

$$\delta_{SJ}(\mu) = H_{SJ}(\mu) = \begin{cases} (\mu - 1)/2 ; & 1 < \mu < 2 \\ 1/2 ; & \mu \geq 2 \end{cases} \tag{A.8}$$

These results are plotted in Fig. 2. All rules give a normal scaling $\delta = H = 1/2$ for $\mu \geq 3$, corresponding to normal (Gaussian) diffusion. This is a consequence of the generalized limit theorem for Lévy stable distribution [28]. For the SJ rule this is true also in the range $2 < \mu \leq 3$, while AJ and SV rules are super-diffusive ($H > 1/2$) in all the interval $1 < \mu < 3$. On the contrary, the SJ rule is sub-diffusive ($H < 1/2$) for $1 < \mu < 2$. For Poisson renewal processes, corresponding to the joint limit $\mu, T \to \infty$, the values of δ and H are again given by the normal scaling 1/2 and the diffusion is Gaussian.
The scalings δ and H are evaluated by means of the Diffusion Entropy (DE) method [23, 24] and of the Detrended Fluctuation Analysis (DFA) [27], respectively.

References

[1] U. Frisch, Turbulence: The legacy of A.N. Kolmogorov, Cambridge University Press (1995).
[2] D.R. Cox, Renewal Theory, Methuen, London (1962).
[3] P. Paradisi, R. Cesari, A. Donateo, D. Contini, P. Allegrini, Reports on Mathematical Physics **70**(2), 205-220 (2012).
[4] P. Paradisi, R. Cesari, D. Contini, A. Donateo and L. Palatella: Eur. Phys. J. Special Topics **174**, 207-218, 2009.
[5] P. Paradisi, R. Cesari, A. Donateo, D. Contini, P. Allegrini, Scaling laws of diffusion and time intermittency generated by coherent structures in atmospheric turbulence Nonlinear Proc. Geoph. 19 (2012) 113-126; P. Paradisi et al., Corrigendum, Nonlin. Processes Geophys. 19 (2012) 685.
[6] P. Allegrini, D. Menicucci, R. Bedini, L. Fronzoni, A. Gemignani, P. Grigolini, B.J. West, and P. Paradisi, Spontaneous brain activity as a source of ideal 1/f noise, Phys. Rev. E 80 (2009) 061914.
[7] Stephen K. Robinson, Annual Review of Fluid Mechanics, **23**, 601-639 (1991).
[8] L. Mahrt, Boundary Layer Meteorology, **135**, 1-18, (2010).
[9] L. Mahrt, J. Atmos. Sci., **46**, 7995, (1989).
[10] Linlin Wang, Dan Li, Zhiqiu Gao, Ting Sun, Xiaofeng Guo, Elie Bou-Zeid, Boundary-Layer Meteorology **150**, 485511, (2014).
[11] Dan Li, Elie Bou-Zeid, Boundary-Layer Meteorology **140**, 243262 (2011).
[12] D. Cava, G. G. Katul, A. M. Sempreviva, U. Giostra, A. Scrimieri, Boundary-Layer Meteorology **128**, 3357 (2008).
[13] K. Trumner, Th. Damian, Ch. Stawiarski, A. Wieser, Boundary-Layer Meteorology **154**, 125 (2015).
[14] Xueling Cheng, Fei Hu, Qingcun Zeng, Chin. Sci. Bull. **59**(34), 48904898 (2014).
[15] R.B. Stull: An introduction to boundary layer meteorology, Kluwer Academic Publishers, Dordrecht (The Netherlands), 1988.
[16] Higbie, Trans. Amer. Inst. Chem. Eng. **31**, 365 (1935); Danckwerts, Indust. Eng. Chem. **43**,1460 (1951); Perlmutter, Chem. Eng. Sci. **16**, 287 (1961).
[17] A. Kermani, H.R. Khakpour, L. Shen and T. Igusa, J. Fluid Mech. **678**, 379- 416 (2011).
[18] E.W. Montroll: Random walks on lattices, Proc. Symp. Appl. Math., Am. Math. Soc. **16**, 193-220, 1964.
[19] V.M. Kenkre, E.W. Montroll and M.F. Shlesinger: J. Stat. Phys. **9**(1), 45-50, 1973.
[20] G.H. Weiss and R.J. Rubin: Adv. Chem. Phys. **52**, 363-505, 1983.
[21] M. F. Shlesinger: J. Stat. Phys. **10**(5), 421-434, 1974.
[22] P. Allegrini, P. Grigolini and B.J. West: Phys. Rev. E **54**(5), 4760-4767, 1996.
[23] P. Grigolini, L. Palatella, G. Raffaelli: Fractals **9**(4), 439-449, 2001.
[24] P. Grigolini, D. Leddon, and N. Scafetta: Phys. Rev. E **65**(4), 046203, 2002.
[25] S.B. Lowen and M.C. Teich: Phys. Rev. E **47**(2), 992-1001, 1993.
[26] R. Metzler and J. Klafter: Phys. Rep. **339**(1), 1-77, 2000.
[27] C.K. Peng, S.V. Buldyrev, S. Havlin, M. Simons, H.E. Stanley, A.L. Goldberger: Phys. Rev. E **49**(2), 1685-1689, 1994.
[28] V.M. Zolotarev: One-dimensional Stable Distributions, American Mathematical Society, Providence, 1983; P. Lévy: Theorie de l'Addition des Variables Aleatoires, Gauthier-Villars, Paris, 1937.
[29] D. Belusic and L. Mahrt: J. Geophys. Res., **117**,D09115 (2012).
[30] G. Katul, A. Porporato, D. Cava, M. Siqueira, Physica D, **215**, 117126 (2006).
[31] D. Cava, G.G. Katul, A. Molini, C. Elefante, J. Geophys. Res., **117**, D01104 (2012).
[32] D. Cava, G.G. Katul, Bound. Layer Meteorol.,**130** (3), 307325 (2009).
[33] A. Donateo, D. Contini, F. Belosi, Atmospheric Environment **40** 1346-1360 (2006).

Journal of Physics: Conference Series **633** (2015) 012066 doi:10.1088/1742-6596/633/1/012066

Clustering of the points lying on monotonous curves as a partition into antichains

Eduard Lerner and Dmitry Voloskov

Kazan Federal University, Kremlevskaya str.18, Kazan 420018, Russia

E-mail: `eduard.lerner@gmail.com`, `voloskovdmitriy@gmail.com`

Abstract. Let us consider some set of points on the Cartesian plane. Each point is a part of one of few curves describing the dependency between abscissas and ordinates. In our case these are dependencies between the rock occurrence depth and the oil saturation described by Skelt-Harrison equation. In this work a problem of distributing these points into clusters corresponding to different curves is being investigated. Our original method based on presenting data points as elements of partial ordered sets with coordinate order is proposed. Thus to solve clustering problem one needs to find all the points which are parts of maximum length chains and to distribute them into corresponding antichains. One can propose obvious algorithm to solve the problem in quadratic time, based on Mirsky's theorem. In this work algorithm of $O(n \log n)$ complexity is proposed. The algorithm is based on the fact that Dushnik–Miller dimension of the partially ordered set is equal to 2 and can be applied to a wide class of dependencies.

The problem: Assume that P is a given set of distinct points (x_i, y_i), $i = 1, \ldots, n$, where n is sufficiently large (about a million or several millions). The main assumption is that there exists a small **unknown** set of continuous functions $f_j(x)$, $j = 1, \ldots, k$, whose graphs do not intersect in the considered domain, and each point (x_i, y_i) satisfies one of dependencies $y_i = f_j(x_i)$ for some j (the number k of these dependencies is also unknown). Denote the corresponding set of points by I_j, $j = 1, \ldots, k$. The simplest requirement to functions f_j is their increase. A weaker condition consists in the existence of a homeomorphism h of some domain D containing all points of P to a part of the plane D', under which all functions are increasing. This means that there exists a biunique continuous map h of the domain D to D' such that if $\{(x, y), (x'y')\} \in I_j$, then both coordinates of the difference $h(x', y') - h(x, y)$ have one and the same sign. The problem is to find the least value of k and to distribute as many points as possible over sets I_j. We propose an algorithm which solves this problem within the time of $O(n \log n + kn)$.

Note that the homeomorphism allows us to reduce to the case of increasing functions a wide class of functions such as the case when all functions satisfy the Lipschitz condition of the first order with one and the same constant L. It alows us to consider withour loss of generality only the case of increasing functions.

Let us describe one specific situation, when we needed to solve the stated problem. We were given a 70Mb array consisting of 12 million numbers representing oil saturation values in each cell of a 3D grid that described the underground part of an oil deposit. The data were provided as is, though we were not aware of the technique used for determining the oil saturation values. The problem consisted in describing the data in the shortest possible form (without loss of accuracy) in order to send them via Internet. We treated the oil saturation values in the given file as one of several (unknown to us) functions of the depth. Nonzero values started to appear at the depth of a little less than 2 489 meters, so by adding 2 489 to the (negative) value of the depth we have concluded that the depth that corresponded to nonzero values of the oil saturation ranged from 0.236 to 112.387 meters. All these points are shown in Fig. 1.

According to the Skelt–Harrison formula [1], the dependence of the oil saturation value (y) on the depth (x) takes the form $y = a\exp\{-b/(x-d)^c\}$, $x > d$. This function equals 0 with $x = d$ and increases with $x > d$. After the performed transformations we have got $d = 0$, but the rest parameters of the curves remained unknown.

In Fig. 2 we see 6 curves. First we had to pick out (to cluster) points of these curves and then to parameterize the curves themselves. Using the algorithm described below, we have obtained 6 sets of points $I_1, \ldots, I_6$, which had to be parameterized then (each set had to be parameterized by its own curve). The number of points in the set I_1 (located on the lowest curve) appeared to be much less than the number of points located on the rest curves. Note that $\sum_i |I_i| < n$ (here n is the total number of points) therefore, having determined values of parameters of curves by the Levenberg–Marquardt method (see Fig. 2), we had to refer each of points that initially remained nonclustered to the cluster whose functional formula gave the ordinate closest to that of the point under consideration. This approach has allowed us to describe the initial 70Mb array of 12 million oil saturation values with the help of subscripts 0–6 (i.e. 1–6 — numbers of functional dependencies and 0 corresponds to zero values) and thus to reduce the size of the archive file to 679KB.

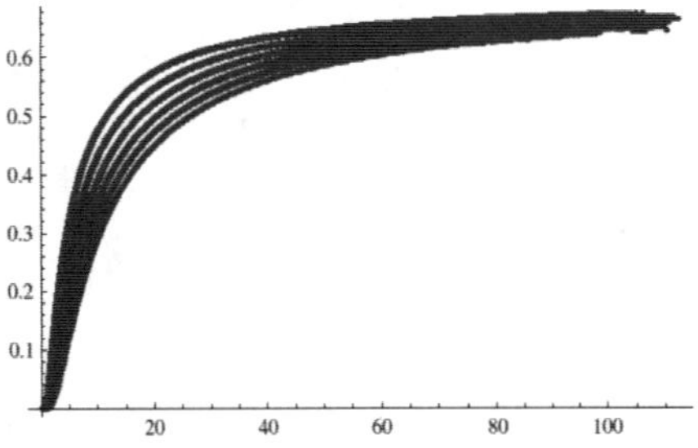

Figure 1. All points (depth, saturation) with nonzero saturation values.

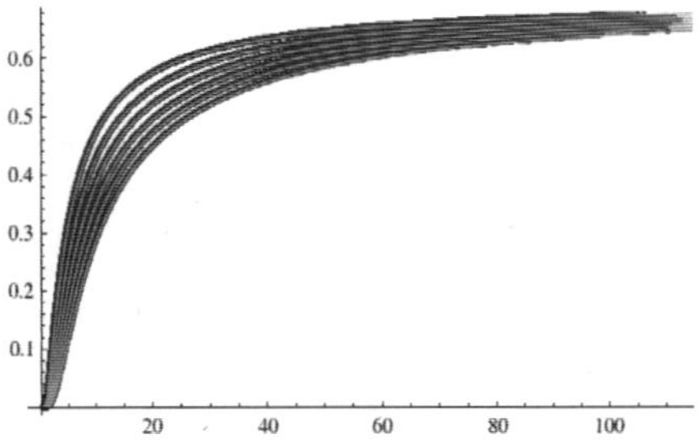

Figure 2. Approximating curves on the entire set of points (depth, oil saturation)

Before we describe the proposed method, let us consider other approaches. Note that the nearest neighbor method, which is a classical clustering technique, theoretically is applicable in our case. However, its time consumption equals $O(n^2)$, which is too much. Moreover, not everywhere in the domain under consideration the curves are separated enough.

We can simplify the problem by considering it in the characteristic part of curves (in our graph we can distinguish the domain, where the abscissa ranges from 4 to 25). The number of points in this domain is somewhat less. In view of the said above, it seems possible to solve the problem with a large number of clustered points by means of random sampling. However, according to results of our numerical experiments with the nearest neighbor method, in order to make the time consumption of the quadratic algorithm acceptable, it is necessary to reduce the sample size, at least, to several dozens of thousands of points. If the number of points on the lower curve (or the upper one) is not so much (in our specific example after the reduction of the sample size the number of such points appeared to be less than 100), then the clustering process leads to the merge of the lower curve with the upper one (because the distance between its points becomes comparable with the distance to it), or/and it falls into several parts (which has happened in our experiments).

Journal of Physics: Conference Series **633** (2015) 012066 doi:10.1088/1742-6596/633/1/012066

Thus, we did not succeed in solving even the considered specific problem without several tricks. Below we describe the algorithm which solves the clustering problem for a set of points satisfying various functional dependencies in the most general case and, which is essential, does it in reasonable time.

The key moment in the formal statement of the problem, in a general case, does not consist in imposing conditions under which points can belong to one curve. On the contrary, the key moment consists in imposing conditions that make this case impossible. Note that one can numerate curves "in a natural way"; we treat the first of them as the "lower" and do the last curve as the "upper" one. Let us now introduce a partial order on the mentioned set of points; namely, let us assume that for points a and b it holds $a \succ b$, if the point a belongs to the curve that is located "above" the curve that contains the point b. Note that for "increasing" polylines a partial order is introduced as follows: for $a = (x_1, y_1)$ and $b = (x_2, y_2)$ it holds

$$a \succ b \text{ iff } x_1 < x_2,\ y_1 \geq y_2 \text{ or } x1 = x2,\ y1 > y2. \tag{1}$$

Now we can easily construct a quadratic algorithm for our problem based on the use of a POSET (a partially ordered set). Recall [2] that a chain is defined as a set of pairwise comparable elements, while an antichain is a set, all whose elements are pairwise incomparable. Thus, the problem consists in finding the least possible number k of antichains, whose union represents the whole set of points, and in finding points which (under any partition) belong to the first, second, ..., kth antichain.

Really, it is evident that the number of curves (antichains) is not less than the number of elements in the maximal chain of a POSET. Moreover, according to the Mirsky theorem [2], such a partition on an antichain exists and can be obtained by two classical methods (evidently, it is nonunique). The first technique (the ascend) implies the determination of the least elements of the POSET as points of the first antichain and their consequent deletion. Then a similar process is performed for the second antichain, and so on, until there remain only incomparable points of one kth antichain. In the second technique (the descent) we, on the contrary, start with the maximal elements of the POSET which form the kth antichain and so on.

Recall that the problem under consideration consists in finding points which under any partition belong to the first, second, ..., kth antichain. Evidently, these points exactly coincide with points which belong to some chain of length k. It is clear that points, which under both partitions belong to an antichain with one and the same number, have the mentioned property. Therefore, the intersection of two techniques for constructing a partition on an antichain mentioned in the Mirsky theorem gives a solution to our problem.

As a result, we obtain a simple algorithm. However, it is quadratic and therefore inapplicable in the case of millions of points. Note that for an arbitrary POSET we cannot propose a better variant, because it is possible that a POSET contains only two comparable elements; in this case $k = 2$ and we can refer to the first and second curve (exactly) one element, namely, each of comparable ones. Evidently, in order to find a pair of comparable elements, we are forced to enumerate all possible pairs, so the time consumption of the program is quadratic. However, in our concrete POSET there exists a certain correlation between comparable elements, and the number of them is rather large.

Let us try to describe this correlation. Assume that along with the partial order there exists a strict order $>$ agreed with the partial order in the following sense: if $a \succ b$, then $a > b$. Evidently, this assumption is not restrictive, because one can introduce the indicated order in many ways. The next assumption imposes an essential constraint. Let $a \succ b$, then $a > b$; denote by (b, a) the set of elements c satisfying the inequality $b < c < a$. Our basic assumption is that

$$\text{for any } c \in (b, a) \text{ the element } c \text{ is comparable either with } a \text{ or with } b. \tag{2}$$

If the strict order is defined, then we can sort n points in the ascending order within the time of $O(n \log n)$. Let us now discuss a linear solution algorithm for the initial problem, where the list of points π is sorted in the strict order. More precisely, below we propose a linear ascend algorithm (the descent is performed by an analogous linear algorithm). Our algorithm divides points on an antichain within one pass through the list π; its time consumption is $O(kn)$.

Evidently, the first element in the list π belongs to the first (lower) antichain. We compare the next element a with the last point c referred to the lower antichain. If it is incomparable with c (in the sense of a partial order), then we refer it to the lower antichain; but if it exceeds c in the sense of the partial order (evidently, no other case is possible), then we immediately refer it to the second antichain, if the latter is empty, otherwise we compare it with the last element of the second antichain and so on.

Let us prove that the algorithm works correctly. It suffices to make sure that the set of points referred to the lower antichain contains no extra elements. Really, let a be the first point, for which there exists an element b such that $a \succ b$. Let us first make sure that b cannot belong to the set of elements referred to the lower antichain. In this case $c \in (b, a)$, but c (by the algorithm and by the induction hypothesis) is comparable neither with a nor with b, which contradicts the main assumption. Let us now assume that b does not belong to the set of elements referred to the lower antichain. This means that there exists b' referred to the antichain such that $b \succ b'$. But then in view of the transitivity of the partial order $a \succ b'$, which contradicts the proved assertion. Since, evidently, when referring points to the set of least elements in accordance with the algorithm, we cannot miss anyone (and, as was proved earlier, there are no extra elements), the algorithm works correctly.

It remains to verify the existence of a strict order satisfying property (2) in the case of POSET (1). We can easily make sure that the inverse lexicographical order $a = (x_1, y_1) > b = (x_2, y_2)$ iff either $x1 < x2$ or $x_1 = x_2$, $y_1 < y_2$, can serve as the desired strict order.

In conclusion, let us generalize the obtained results from the point of view of the POSET theory. Thus, let a certain nonstrict order be defined. It appears to be insufficient for solving the stated algorithmic problem (to determine all elements of the POSET that belong to all maximal chains and to divide them into the corresponding antichains within a linear time or close to it). Therefore we assume that it is possible to define one more construction, namely, a strict order which agrees with the partial one and, in addition, has property (2). If such an order exists, then, as was described above, after sorting elements in this strict order, we can solve the stated algorithmic problem within a linear time.

However, it remains to solve one more problem, namely, to find necessary and sufficient conditions for introducing order (2). It appears that all such POSETs are isomorphic to the POSET of points on the two-dimensional plane with the partial order generated by the coordinatewise comparison

$$a = (x_1, y_1) \succeq b = (x_2, y_2) \text{ iff } x_1 \geq x_2, \text{ and } y1 \geq y2. \tag{3}$$

Recall that the order dimension (or the Dushnik–Miller dimension) [2] of a POSET (P, $preceq$) is defined as the least number of linear orders (chains) $(P, \preceq_i), i \in \{1, \ldots, k\}$ on P such that their intersection gives $(P, \preceq)$: $(P, \preceq) = \bigcap_{i \in \{1,\ldots,k\}} (P, \preceq_i)$ (i.e., $a \preceq b$ iff $a \preceq_i b$ for all $i \in \{1, \ldots, k\}$.

In the Dushnik–Miller paper [3] one, in fact, proves that condition (2) is equivalent to the fact that the order dimension of a POSET does not exceed two (note that in modern papers condition (2) is not used so often). It remains only to note that if the order dimension of a POSET equals m, then we can associate each element of the POSET with a point in R^m so as to make the coordinatewise order of these points coincide with the order on the corresponding elements of the initial POSET. Therefore, all POSETs with property (2) are equivalent to POSET (3).

Evidently, if the direction of the abscissa changes to the opposite one, then the partial order (1) turns into order (3). Thus, the case of increasing functions, in fact, describes the most general case, which allows the application of the developed algorithm.

The authors are grateful to Vladislav Sudakov for the provided data and to Artur Aslanyan for the problem statement.

References

[1] Skelt C and Harrison B, 1995, *Transaction of the Society of Professional Well Log Analysts* 36th Annual Logging Symposium, NNN, 10
[2] Harzheim E, 2005, Ordered Sets, *Series: Advances in Mathematics* **7** Springer-Verlag, New York
[3] Dushnik Ben and Miller E.W. 1941, *American Journal of Mathematics* **63** (3) 600

Modeling of intense charged particle bunch dynamics in external magnetic fields

H Y Barminova, M S Saratovskyh
Department of Physics of Extreme Matter States, National Research Nuclear University MEPhI, Kashirskoye sh. 31, Moscow 115409, Russia

E-mail: EEBarminova@mephi.ru

Abstract. Program module CAMFT is developed to simulate the intense charged particle bunch dynamics in external magnetic fields. The program is based on the accurate solution of the motion equation for each particle of the intense bunch. The program is written on C++ language and uses standart OpenMP (version 2.0) for parallelization, so one can simulate the behavior of the bunch with intensity about 10^9-10^{10} particles. Visual C++ and library Qt 4.8.3 of qtcreator are used for the result visualization. Dynamics of the bunch with arbitrary phase distributions in magnetic fields of arbitrary geometry may be studied by means of the program. The actual CAMFT version is checked while simulating the accelerating structure with racetrack geometry. Modified CAMFT version is checked for ITEP Heavy-Ion Prototype charge-state separator.

1. Introduction.

There are numerous program codes for simulation of a charged particle beam dynamics in accelerator physics – TRACEWIN, PARMILLA, TRACE-3D, TRACK, ORBIT, DYNAMION, TRANSPORT, PATRIC, etc [1-7]. While developing a new accelerating structure or a facility, it is of a great importance to choose the suitable code which model approximations or restrictions of applicability correspond to the actual beamline requirements. In the case of intense beams the codes using PIC-method are mostly preferred to calculate the beam dynamics. But there are significant limitations for the application of the code based on a PIC-method, especially for simulation of a rather long-life intense particle ensemble, caused by the appearance of the artificial numerical effects. An attractive tool to avoid such the phenomena is to use the different schemes of parallel computing. As such a scheme the direct integration of a motion equation for each of the particles of the ensemble in physically correct situation of a bunched beam may be considered. The CAMFT program module based on a direct integration of the particle motion equation is discussed in present paper. The code is developed for intense bunch behaviour simulation in external magnetic fields. Program is tested while two real accelerating structure simulating.

2. Motivation and the basis of the program development

The external magnetic field changes the beam characteristics, especially the phase space configuration [8, 9], so the accurate beam dynamics description is necessary to avoid the possible problem of beam mismatching and particle losses, i.e. to avoid the problems of non-effective acceleration and radiation safety violation. The paper presented proposes such a method of accurate bunched beam dynamics description.

The accelerating structures which can be under consideration are the structures that can serve as a part of ERL (energy recovery linear accelerator) [10], new low energy rings (as example, USR, ELISA, CRYRING, LESR) [11-13], linear colliders (as example, ILC) [14], as well as the separated elements of traditional accelerator sections (chains of bending magnets). As a rule all the structures mentioned above use the dipole or quadrupole magnetic field to change the trajectory of the beam with the aims of decreasing of the facility dimensions, separating of the beams with specific values of the mass or of the charge-state of the particle, redirectoring the beams for the tasks of the experiment, input and output of the beams in circular accelerators and others. So the motion equation for the particle of the bunched beam has in its right part the constant Lorentz force or the force depending linearly from the particle coordinate. It means that one can't take into account the non-linear effects that can lead to strong violation of the Liouville's law. Let us consider the dipole magnet field as an element of accelerating facility with racetrack geometry.

The algotithm of the program module which describes the bunch behaviour in such a facility is shown at Figure 1.

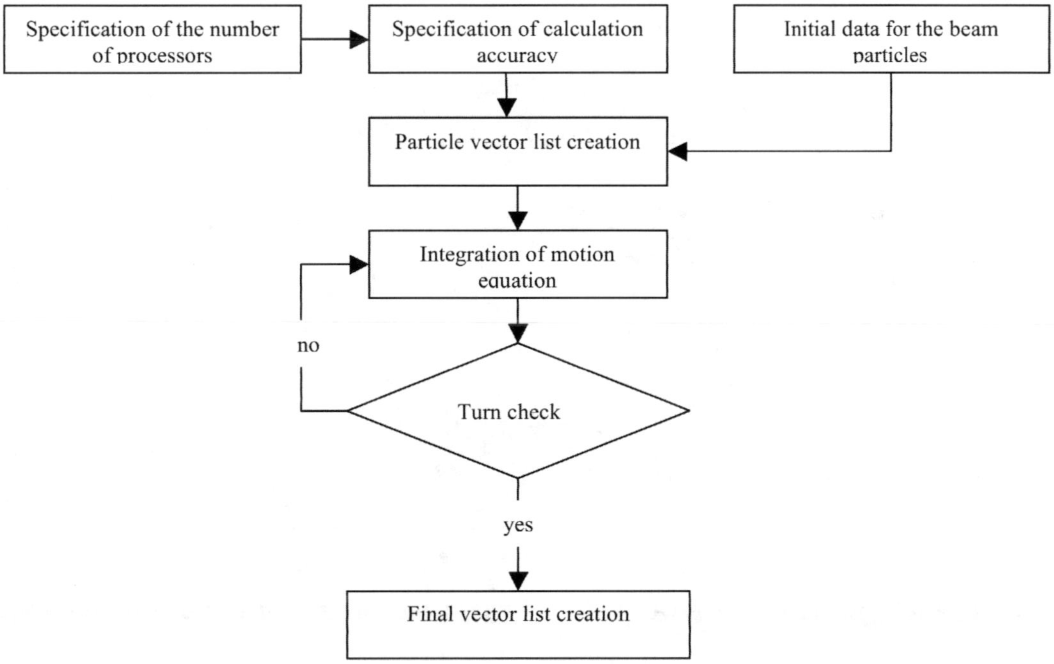

Figure 1. Short algorithm of CAMFT.

In CAMFT the particle-field interaction is treated by means of matrix description. The intrinsic space charge of intense beam should be taken into account while the motion equation integrating. It requires a significant time for the simulation because to calculate the quasistationary electrostatic field it is necessary to wait the completion of all the processes for the bunch particles. The calculation time may be decreased by means of the code optimization. The code efficiency is improved too by means of using MSVC 2013 with multiprocessor compilation (version 12.0) that results in the efficiency improvement of about 8% compared with previous version. Usage of MS Parallel Studio for post-treatment results in efficiency improvement equal approximately 15% .

3. Results of simulation.

The results of simulation of the charged particle bunch behavior in accelerating structure with racetrack geometry are presented in this Section. The racetrack geometry is characterized by the bending magnet with the total beam trajectory turn (total reflection). As a rule it is a dipole magnet. The calculation was performed using Intel Core i-5 4200M CPU 2.50 GHz with Hyper-V and for cache L3 (3 Mb) under operation of Open MP 2.0 at ITEP supercomputer. At Figure 2 the three-dimensional pictures of the particle spatial configurations are presented before the turn in the magnet and after the turn. The bunch was approximated as an arbitrary ellipsoid with Gaussian distribution of the own space charge (spatial density of the particles) and Maxwellian distribution of the particles in velocity space.

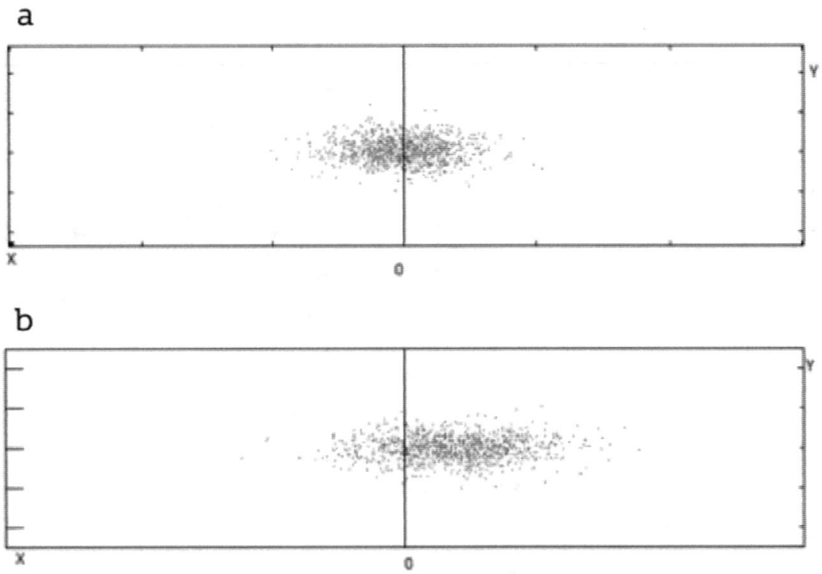

Figure 2. Bunch particle spatial distribution in XY-plate before the turn in magnet field (a) and after the turn (b).

The program allows represent the calculation results as 3D-images after the treatment by means of the wavelet transformation. At Figure 3 the example of such representation is shown.

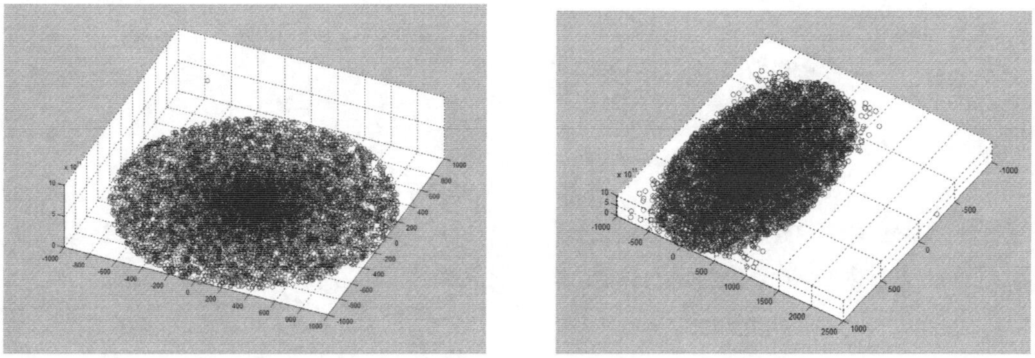

Figure 3. The 3D bunch in velocity space before the magnet and after the magnet.

4. Conclusions.

The program module CAMFT is developed allowing to accurate description of the bunched beam dynamics in external magnetic field. The program realized at C++ with Open MP for parallel calculations, and includes the different means of post-treatment such as MathLab and MS Parallel Studio. The scheme of parallel computing is used which allows to avoid the error cumulation during the field calculations and the motion equation integration. The program may be used for the bunch description with arbitrary phase and spatial distributions and for the arbitrary field configuration.

References.

[1] Boine-Frankenheim O and Kornilov V Simulations of transverse coherent effects in intense ion bunches: *Proc. ICAP 2012 (Rostock-Warnemunde, Germany, June 2012)*

[2] Kolomiets A, Pershin V, Vorobyov I, Yaramyshev S and Klabunde Ju DYNAMION – the code for beam dynamics simulations in high current ion linac: *Proc. EPAC-98 (Stockholm, Sweden, 1998)*

[3] Webb S, Bruhwiler D, Abell D, Danilov K, Nagaitsev S, Valishev A, Danilov V and Shishlo A Simulating high-intensity proton beams in nonlinear lattices with PyORBIT: *Proc. IPAC 2012 (New Orleans, Louisiana, USA, May 2012)*

[4] http://www.phy.anl.gov/atlas/TRACK/

[5] http://mcnp.lanl.gov

[6] PARMILA, LA-UR-96-4478, LANL, Los Alamos, New Mexico 87545, USA.

[7] Crandall K R and Rusthoi D P TRACE-3D Documentation, 3rd Edition. May 1997. LANL, Los Alamos, New Mexico 87545, USA.

[8] Barminova E E Model analysis of intense charged particle fluids to solve the task of the ion beam creation with high brightness. PhD Thesis, MEPhI, Moscow. 2001

[9] Barminova E E, Chikhachev A S 1991 *Radioph. and Quant. Electronics* **34** 818

[10] Herfurth F, Brauning-Demian A, Enders W, Danared H, et al The low energy storage ring CRYRING@ESR: *Proc. COOL2013 (Murren, Switzerland)*

[11] LSR FLAIR. Technical Design Report. Stockholm University. 6 May 2011

[12] Moncton D E (private communication). FLS Workshop, Hamburg, Germany, 2006

[13] Stiebing R E, et al. 2010 *Nucl. Instrum. Meth.* A**614** 10

[14] ILC. Technical Design Report. V.3. Part I, II. CERN, Switzerland. 12 June 2013.

Computational chemistry insights in the REDOX Behaviour of Cr and W Fischer carbene complexes

Marile Landman,[1] Jeanet Conradie[2] and Petrus H. van Rooyen[1]

[1] Department of Chemistry, University of Pretoria, Private Bag X20, Hatfield, 0028, Republic of South Africa
[2] Department of Chemistry, University of the Free State, 9300 Bloemfontein, Republic of South Africa.

phvr@up.ac.za

Abstract. An electrochemical study of a series of Fischer carbene complexes containing a hetero-aryl group showed that Cr and W carbenes exhibit different electrochemical behaviour. The Cr carbenes are oxidized in two one electron oxidation processes, namely Cr(0) to Cr(I) and Cr(I) to Cr(II). On the contrary, Fischer carbene complexes of tungsten are directly oxidized from W(0) to W(II). The first reduction process observed for both W- and Cr-carbenes, is a one electron process. A density functional theory (DFT) computational chemistry study of the electronic structure of the Cr- and W-carbenes, showed that the oxidation is metal based and the reduction is located on the carbene ligand. The DFT calculations further showed that the Cr(II) species is a triplet and the W(II) species a closed shell singlet. The DFT calculated energies of the HOMO and LUMO of the neutral carbenes relate linearly to the experimental oxidation and reduction potential, respectively. These mathematical relationships obtained can be used to predict experimentally measured potentials of related Fischer carbene complexes.

1. Introduction

A transition metal carbene complex is an organometallic compound having a metal-carbon double bond, often denoted by $M{=}CR_2$. The first example of such a complex was synthesized by Fischer and Maasböl in 1964, $(CO)_5W{=}C(Ph)(OCH_3)$ with the metal is in the 0 oxidation state [1]. Figure 1 shows how the structure of a typical Fischer carbene complex can be modified. For example, the carbene centre may be attacked by nucleophiles such as amines leading to the formation of aminocarbene complexes with liberation of alcohols. Hydrogens found on the α-carbon to the carbene are readily removed by a base, allowing the complex to be available for further reactions. Electrophiles, such as Lewis acids, may also coordinate to the heteroatom substituent through its lone pairs. Lastly, the ligand sphere of the metal may be altered in substitution reactions by replacing the carbonyl ligands with e.g. phosphines. The influence of the modifications on the electron density of the metal can theoretically be calculated utilizing computational chemistry techniques, and can experimentally be measured by cyclic voltammetry.

A cyclic voltammetry (CV) experiment [2] involves the application of linear potential scans to a stationary working electrode immersed in a solution containing the redox active species under investigation (analyte) while measuring the resulting current. The potential is scanned from a potential E_1 where no redox reaction takes place, to a potential where electron transfer is rapid as observed by an increasing current, till the current decays after the solution surrounding the working electrode is depleted of the analyte and then the potential is reversed and scanned back to E_1. When the material has oxidations (or reductions) in the experimental potential range possible for the solvent, an anodic (or cathodic) wave will appear on the positive forward (or negative backward) scan, and a corresponding cathodic (or anodic) wave will be detected on the reverse scan if the oxidized (reduced) analyte is stable in the timescale of the CV experiment. If the two oxidation states (Ox/Red) are stable in the time scale of the experiment, the maximum (peak) currents in these two coupled waves should be approximately the same and the redox process is called chemically reversible. If the redox couples exhibiting fast electron transfer and the two oxidation states are stable in the time scale of the experiment, the potential difference (ΔE_p) between the cathodic (E_{pc}) and anodic (E_{pa}) peaks potentials will be $\Delta E_p = E_{pc} - E_{pa} \approx (0.057 \text{ V}/ n)$ (n is the number of electrons transferred in the process) and the redox process is called electrochemically reversible. The potential midway between the two peak potentials is called formal electrode potential $E^{0'} = (E_{pc} - E_{pa})/2$ [3]. A slow exchange of redox species with the working electrode may lead to large peaks separations and the redox process is then considered electrochemical quasi-reversible (0.090 V < ΔE_p < 0.150 V) or electrochemical irreversible (ΔE_p > 0.150 V) [4]. A chemical irreversible system is one where only oxidation (or only reduction) is observed [5].

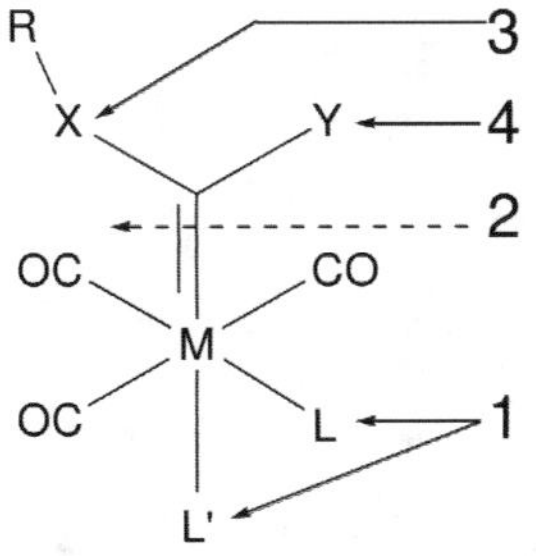

Figure 1. Potential sites of Fischer carbene modification with L and L' = Ligand(s); X = heteroatom; Y = Alkyl or Aryl; R = Alkyl

From a theoretical point of view, oxidation of the analyte in solution is the movement of an electron from the highest occupied molecular orbital (HOMO) of the analyte into the metal electrode and reduction is when the cell potential is sufficient to promote an electron into the lowest unoccupied molecular orbital (LUMO) of the analyte. The character of the HOMO (LUMO) of the analyte will thus tell where the oxidation (reduction) takes place. In addition, the ease of oxidation (reduction) should be related to the energy of the HOMO (LUMO). In fact, in many cases good linear relationships have been obtained between redox potentials and MO energy levels [6-8].

2. Discussion

2.1. Redox results

Fischer carbenes (Figure 1) have two electro-active centres, the metal centre and the carbene carbon [4],[9]. During cyclic voltammetry experiments of Fischer carbenes, generally one or more oxidation processes and one reduction process are observed; see Figure 2 [9] and Figure 3 [4]. The Cr carbenes are oxidized in two 1 e^- oxidation processes, namely Cr(0) to Cr(I) and Cr(I) to Cr(II). On the contrary, Fischer carbene complexes of tungsten are oxidized in two 2 e^- oxidation processes, namely from W(0) to W(II) and W(II) to W(IV). The first reduction process observed for both W- and Cr-carbenes, is a one electron process.

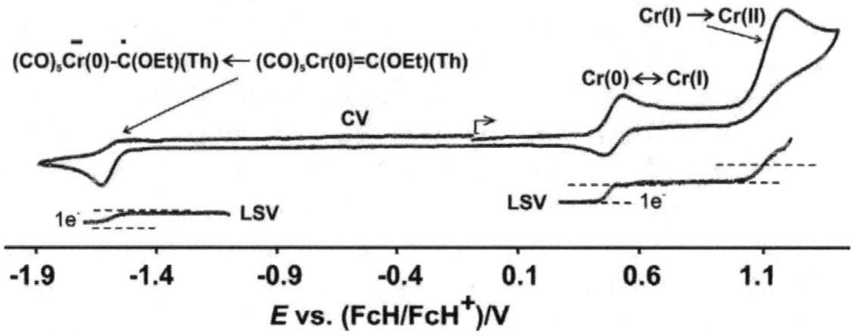

Figure 2. Cyclic voltammogram (CV) and linear sweep voltammogram (LSV) of *c.a.* 0.0005 mol dm^{-3} solutions of $[Cr(CO)_5\{C(OEt)Th\}]$ in CH_3CN / 0.1 mol dm^{-3} $[^n(Bu_4)N][PF_6]$ on a glassy carbon-working electrode at a scan rate of 0.100 V s^{-1}. Data from reference 9.

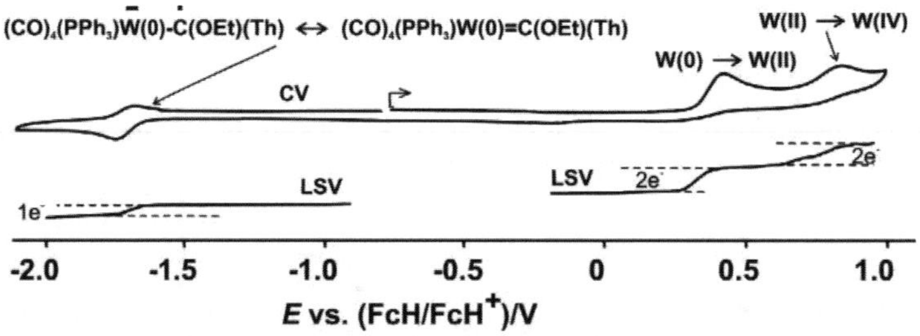

Figure 3. Cyclic voltammogram (CV) and linear sweep voltammogram (LSV) of *c.a.* 0.0005 mol dm^{-3} solutions of $[W(CO)_4(PPh_3)\{C(OEt)Th\}]$ *cis* in CH_3CN / 0.1 mol dm^{-3} $[^n(Bu_4)N][PF_6]$ on a glassy carbon-working electrode at a scan rate of 0.100 V s^{-1}. Data from reference 4.

2.2. Computational chemistry DFT results

The metal-d character of the HOMO and HOMO-1 of the density functional theory (DFT) optimized neutral carbene confirm the metal oxidation, while the LUMO clearly shows that during reduction the carbene carbon is reduced to form a carbene-radical with the added electron density found to be distributed over the carbene ligand. The delocalization of the added electron over the carbene ligand can also be visualized by the Mulliken spin density plot of the reduced carbene. The DFT calculations further showed that the oxidized Cr(II) species is a triplet [9] while the W(II) species a closed shell singlet [4].

Substitution of one of the CO groups in the penta-carbonyl $[(CO)_5W{=}C(OEt)(Th)]$ (L in Figure 1) with PPh_3, leads to tetra-carbonyl $[(CO)_4(PPh_3)W{=}C(OEt)(Th)]$ with the PPh_3 group coordinated in a *cis* position or a *trans* position relative to carbene ligand C(OEt)(Th). Experimentally, *trans*-$[(CO)_4(PPh_3)W{=}C(OEt)(Th)]$ is more easily oxidized at a lower potential (0.100 V lower) than the corresponding *cis* complex, see Figure 5(a). This is consistent with DFT calculations showing the *trans* complex to be more than 0.1 eV less stable than the corresponding *cis* complex, and that the calculated HOMO energy of the *trans* complex is 0.2 eV higher than that of the corresponding *cis* complex, implying that an electron will be more easily removed from the higher energy *trans* HOMO than from the *cis* HOMO. Experimentally it is found that the *trans* complexes isomerize to the more stable *cis* complexes, explaining the shoulder on the oxidation peak on the CV of *trans*-$[(CO)_4(PPh_3)W{=}C(OEt)(Th)]$, while the CV of the more stable *cis*-$[(CO)_4(PPh_3)W{=}C(OEt)(Th)]$

exhibits only one sharp oxidation peak corresponding to the oxidation of *cis*-$[(CO)_4(PPh_3)W{=}C(OEt)(Th)]$.

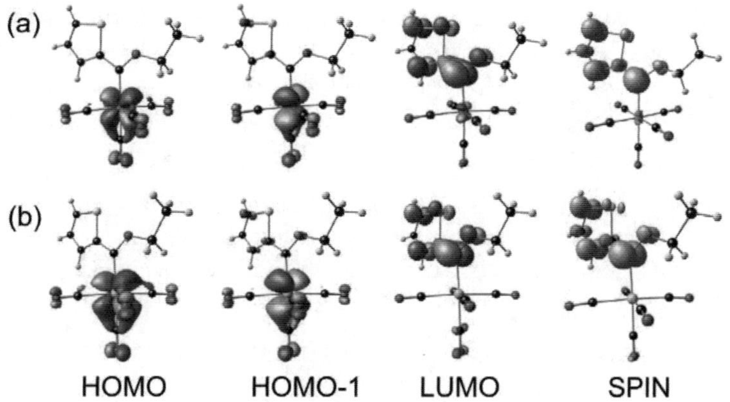

Figure 4. Selected frontier orbitals of the B3LYP/6-311G(d,p)/def2-TZVPP optimized structures of the neutral $[M(CO)_5\{C(OEt)Th\}]$ complexes. The spin density plot of the optimized reduced radical anion (q = -1, S = ½) is also shown. (a) M = Cr and (b) M = W. Colour code of atoms (online version): C (black), H (white), O (red), S (yellow),W (turquoise), Cr (blue).

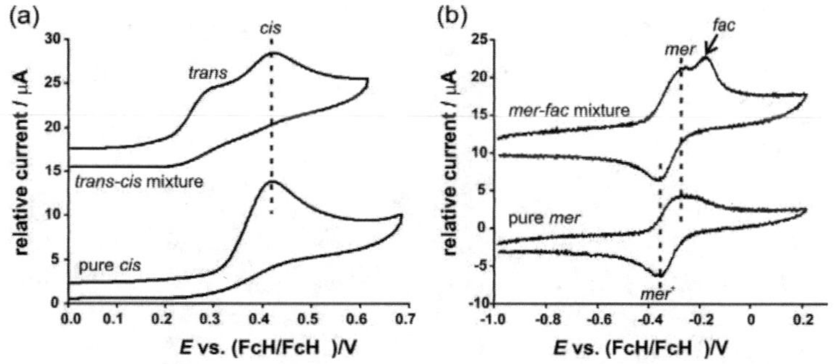

Figure 5. Cyclic voltammogram of the oxidation of *c.a.* 0.0005 mol dm^{-3} solutions of (a) $[W(CO)_4(PPh_3)\{C(OEt)Th\}]$ *cis* and *trans* and (b) $[Cr(CO)_3(dppe)\{C(OEt)Th\}]$ *mer* and *fac* in CH_3CN / 0.1 mol dm^{-3} $[^{n}(Bu_4)N][PF_6]$ on a glassy carbon-working electrode at a scan rate of 0.100 V s^{-1}.

Substitution of two of the CO groups in the penta-carbonyl $[(CO)_5Cr{=}C(OEt)(Th)]$ (L and L' in Figure 1) with 1,2-bis(diphenylphosphinoethane) (dppe), leads to tri-carbonyl $[(CO)_4(dppe)Cr{=}C(OEt)(Th)]$ with the dppe group coordinated in a *fac* or *mer* position relative to the carbene ligand C(OEt)(Th). DFT results on the relative energies of the *fac* and *mer* isomers indicated that the *mer* isomers are considerably more stable than the *fac* isomers, in agreement with the fact that the *mer* isomer is the main isomer observed experimentally. DFT calculations further showed that the energy of the HOMO of the *fac* isomer was lower (more negative), than that of the *mer* isomer. More energy will thus be needed to remove an electron from the more stable *fac* HOMO, implying that *fac* will be oxidized at a higher potential. This results agrees with experimental observation when comparing the CV of a pure *mer* complex with that of a *fac-mer* mixture, see Figure 5(b). DFT calculations also show that the oxidized *mer*-cation is at least 0.3 eV more stable than the *fac*-cation. Applying these energies to the Boltzmann distribution, results show the dominance of more than 99.9 % mer^+. Thus after oxidation fac^+ will convert to mer^+. This is also consistent with experimental observations that only one reduction peak coupled to the *mer* and *fac* oxidation peaks is observed on the CV that is ascribed to the reduction of the mer^+ cation [10].

2.3. Relationships

The close relationship between the energy of the frontier orbitals and the ease of removing (reduction) or adding (oxidation) an electron to it, is illustrated in the linear relationships obtained in Figure 6 for a series of Cr and W Fischer carbenes. These relationships can be utilized to design similar Cr or W Fischer carbenes with a specific stability towards oxidation and reduction.

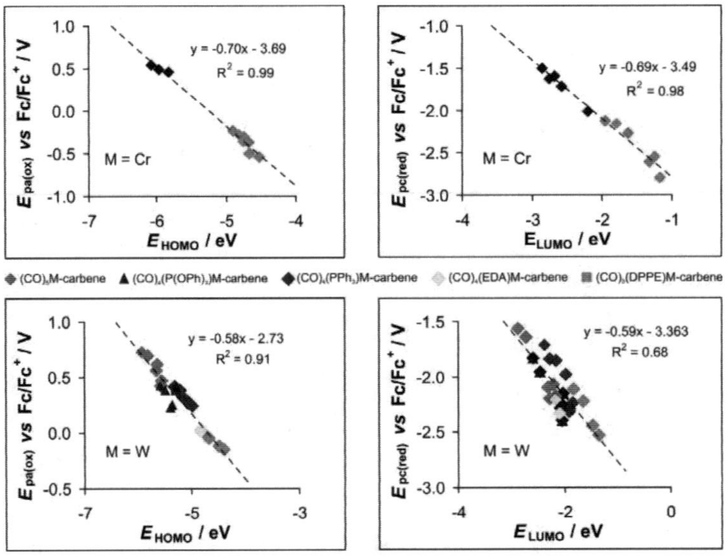

Figure 6. Linear relationships between the DFT calculated E_{HOMO} or E_{LUMO} and the experimental oxidation E_{pa} or reduction E_{pc} potential of a series of Cr and W Fischer carbenes. Data obtained from reference 4, 9,10,11.

3. Conclusion

DFT can be considered a very useful technique to investigate and understand experimental observations by calculation of molecular properties which are difficult to measure experimentally. Here we illustrated how DFT calculations can be used to understand the oxidation and reduction center of Cr and W Fischer carbenes, to predict the relative stability of *cis* and *trans* or *fac* and *mer* isomers, and to obtain linear relationships between experimental redox values and DFT calculated energies.

Acknowledgements
This work has received support from the Norwegian Supercomputing Program (NOTUR) through a grant of computer time (Grant No. NN4654K), the South African National Research Foundation, HPC Warehouse Facility, the University of Pretoria, Pretoria and the Central Research Fund of the University of the Free State, Bloemfontein.

References

[1] Fischer E O and Maasbol A 1964 *Angew. Chem.* **76** 645
[2] Sawyer D T and Roberts J L Jr. 1974 *Experimental Electrochemistry for Chemists* Wiley New York p. 54
[3] Bard A J and Faulkner L R 2001 *Electrochemical Methods* 2nd ed. John Wiley & Sons New York
[4] Landman M, Pretorius R, Fraser R, Buitendach B E, Conradie M M, van Rooyen, P H and Conradie J 2014 *Electrochim. Acta* **130** 104
[5] Christensen P A and Hamnett A 1994 *Techniques and Mechanisms in Electrochemistry* Blackie Academic & Professional London p.55 – 67, 170 – 175
[6] Suzuki T, Maruyama Y, Akasaka T, Ando W, Kobayashi K, Nagase S 1994 *J. Am. Chem. Soc.* **116** 1359
[7] Luo J, Xue Z Q, Liu W M, Wu J L and Qin Z 2006 *J. Phys. Chem. A* **110** 12005
[8] Conradie J 2015 *Dalton Trans.* **44** 1503
[9] Landman M, Lui R, van Rooyen P H and Conradie J 2013 *Electrochim. Acta* **114** 205
[10] Landman M, Liu R, Fraser R, van Rooyen P H and Conradie J 2014 *J. Organomet. Chem.* **752** 171
[11] Landman M, Pretorius R, Buitendach B E, van Rooyen P H and Conradie J 2013 *Organometallics* **32** 5491

Effect of Drying Temperature on the Chemical Properties and Diffusivity of *belimbi* (*averrhoa belimbi*)

N. Shahari[1], N. Jamil[1], KA. Rasmani[1] , Nursabrina[2]
[1]Faculty of Computer and Mathematical Sciences, [2]Faculty of Applied Sciences
Universiti Teknologi MARA Negeri Sembilan, Malaysia

Email:norazni@ns.uitm.edu.my

Abstract. In recent years, many dried fruit products have been developed in response to a strong demand by the customer. This type of fruit has a different composition and hence different moisture diffusivity (D). During drying, Fick's Law of diffusion, which describes the movement of liquid water was used to calculate this diffusivity. However diffusivity has strong effects on the material drying characteristics and these must be determined. In this paper, Fick's Law of diffusion with different kinds of boundary conditions was solve using separation of variable (SOV). In order to get the value of D, results obtained using SOV will be compared with the results from the drying of *belimbi* at temperature of 40°C, 50°C and 60°C. Although the results show that variation in the values of diffusivity for different temperatures is relatively small, but the variation in the total time required for drying is significantly bigger: between 3-7 hours. Its shown that diffusivity is an important measurement and should be considered in the modeling of the drying process. The chemical properties of *belimbi* slices in terms of vitamin C, total ash and antioxidant activity with different air temperatures and pre-treatment were also investigated. Higher drying temperatures gives less drying time, a lower vitamin C and antioxidant activity but a greater total of ash, whilst pre-treatment can increased vitamin C and antioxidant activity. The results show that pre-treatment and the drying temperature are important variables to improve mass and heat transfer, as well as the *belimbi* chemical properties.

1. Introduction

It is generally accepted that world food production is not increasing at the same rate as population growth. Therefore, the primary goal of food processing should be to conserve the food we have so that it can be distributed [1]. The way to make food available for later use is to convert it into a stable form because it is difficult to control chemical and biological activities while it is in a fresh state. Hence, drying can be described as the reduction of product moisture to the required dryness values as a definite process [2]. This research centres on the drying of fruit, especially the drying of *belimbi* (*averrhoe belimbi*), which have high levels of initial moisture. Currently, most dehydrated fruits are produced by the technique of hot air drying, which is the simplest and most economical of the various method [3]. However, for modelling purposes, in this study, a combination of osmotic dehydration (sucrose pre- treatment) with conventional hot air drying is assumed to be applied to the surface. The main feature of such a system is its ability to predict moisture and temperature inside the product, which is a very important way of providing structural knowledge about the quality of the new product.

Journal of Physics: Conference Series **633** (2015) 012069 doi:10.1088/1742-6596/633/1/012069

Even though experimentation is an important step for the enhancement of drying technology, with the aid of mathematical modeling and analytical solutions, fundamental research provides very useful information for investigating the complicated physics that evolve during the drying process. For example, diffusivity, *D*, is generally about transportation of the mass property of water and is an important parameter to be determined during the drying process. Most of the drying models used the 'method of slope' to find the diffusion coefficient, *D* (for example see [4, 5]). In this method, the Fick law of diffusion with specified boundary condition was solved analytically. Then, the solution is compared with the experiment drying curve by plotting the decay of moisture with drying time.

The basic requirement for drying, from an industrial point of view, is to achieve an amount of dried product in a reasonable time and to obtain acceptable product quality for minimal cost [6]. Hence, the optimisation of the processing variables, especially temperature and pre-treatment, is essential not only to achieve a precise control of the process but also to produce a high quality dried product. Thus, it is important to control this variable and generally achieve higher nutritional value.

Therefore, the aim of this study is to evaluate the behavior of the drying process of *belimbi* at different air temperatures with/without pre-treatment. Three different air temperatures of 40°C, 50°C and 60°C were used. In this study, a partial differential equation of mass transfer equation condition was solved analytically using separation of variable to simulate the drying process. The result from the analytical solution was compared with the experiment drying curve of *belimbi* to find the value of diffusivity, D. In addition, the effects of process treatment before drying and process temperature to chemical properties such as vitamin C, total ash, antioxidant capacity and total phenolic content were also evaluated.

2. Material and Methods

2.1. Mathematical Formulation

A mathematical model that used Fick's Law of diffusion, describing unsteady state mass transfer, is given as follows:

$$\frac{\partial M}{\partial t} = D\frac{\partial^2 M}{\partial x^2}, \qquad 0 < x < L \tag{1}$$

where D = effective diffusivity ($m^2\ sec^{-1}$), M = average local moisture content (kg water/ kg dry matter) and t = time.

2.1.1. Model 1

By assuming that the initial moisture content is uniform ($M=M_0$) throughout the solid, the surface of the solid is in equilibrium with the air for the time considered; the shape of the solid remains constant during the drying period(see [7]), thus giving the boundary and initial condition as below:

$$\text{at} \quad x=0, \quad \frac{\partial M(0,t)}{\partial x}=0, \quad \text{and} \quad x=L, \quad M(L,t)=M_e, \tag{2}$$

Decompose the equation and solve, using separation of variable. The solution is given by:

$$\frac{M(x,t)-M_e}{M_0-M_e} = \sum_{n=1}^{\infty}\frac{4(-1)^{n+1}}{(2n-1)\pi}e^{-Dt\left(\frac{(2n-1)\pi}{2L}\right)^2}\cos\left(\frac{(2n-1)\pi x}{2L}\right) \tag{3}$$

Integrate the solution between $0<x<L$ to find the moisture ratio (MR) and, for a long diffusion time (taken n=1), Equation (3) becomes

$$MR = \frac{8L}{\pi^2}e^{(-\frac{1}{4L^2}\pi^2 Dt)} \tag{4}$$

M_o shows the initial moisture, M_e is the moisture at equilibrium and L is the half thickness of the slice.

2.1.2. Model 2

Dincer [8] used another assumption on the boundary condition, which is given by

$$\text{at} \quad x=0, \quad \frac{\partial M(0,t)}{\partial x}=0, \quad \text{and} \quad x=L, \quad -D\frac{\partial M(0,t)}{\partial x}=h_m(M-M_e), \tag{5}$$

Using separation of variables, the solution for Equation (1) with boundary conditions Equation (5), is given by

$$\frac{M(x,t)-M_e}{M_0-M_e}=\sum_{n=1}^{\infty}\frac{2\sin\mu_n}{\mu_n+\sin\mu_n\cos\mu_n}e^{-Dt\left(\frac{\mu_n}{L}\right)^2}\cos\left(\frac{\mu_n x}{L}\right) \tag{6}$$

Taking (n=1) for long time diffusion, and the value of μ_1 is given by [8], Equation (6) becomes,

$$MR=Ge^{(-\frac{\mu_1^2}{L^2}Dt)} \tag{7}$$

On measuring the drying moisture content M with time, t, the drying curve of each experiment is obtained by plotting the decay of dimensionless moisture (MR) in the sample with the drying time. The basic decay equation that is normally used is an assumed exponential model for the moisture ratio, such as the Handerson and Pabis models (for example see [9]), given by,

$$MR=Ae^{-kt} \tag{8}$$

The data from the experiment were fitted with the Handerson and Pabis (Equation (8))for drying temperatures 40°C, 50°C, and 60°C to get the value of A and k; comparison was made with analytical solution (Equation (4) for Model 1 and Equation (7) for Model 2). From the comparison, the value of diffusivity, D, was determined at different drying temperatures.

2.2. Chemical Analysis and Antioxidant activity measurement

2.2.1. Drying of belimbi slices and moisture content determination

The *bilimbi* fruits were selected on the basis of their appearance and state of ripeness, which was evaluated for the total soluble solid content by using a digital hand held refractometer (N-1α,ATAGO, USA). The selected *bilimbi* were washed and sliced into 0.5cm thickness. Each slice was separated between control and pre-treatment. A pre-treatment was carried out using syrup blanching 70^0Brix sugar solutions for 10 minutes at room temperature. The fruit slices were removed from the syrup, rinsed, drained and the dehydration process continued. The slices were spread on stainless sieves in a convective drier with a horizontal air flow (0.2 m /s) at 40°C, 50^0C and 60^0C for about 8-15 hours. The *belimbi* were weighed every hour on an electronic balance with precision 10-3 g. The drying tests were terminated when 20% moisture content was reached.

The moisture content in the dried *bilimbi* slices was analyzed by using the moisture content determination method by AOAC [10]. The percentage of moisture (wt/wt) of the dried *bilimbi* slices calculated using the following formula:

$$\text{Percentage moisture (wt/wt)} = \frac{\text{Wt of wet sample} - \text{Wt of dried sample}}{\text{Wt of wet sample}} \times 100\%$$

2.2.2. Vitamin C Determination

The vitamin C or ascorbic acid content in the dried *belimbi* slices was determined by using the 2, 6-dichlorophenol indophenols titrimetric method. The 2, 6-dichlorophenol indophenol solutions is a solution that has a blue colour in alkaline conditions and pink colour in acidic conditions. Upon titration, these colours are reduced by ascorbic acid to become colourless. The vitamin C content was calculated as follows:

$$\text{mg ascorbic acid per 100g sample} = (X - B)\left(\frac{F}{W}\right)\left(\frac{V}{Y}\right),$$

where, X = Average volume, in ml, of dye used for sample titration, B= Average volume, in ml, of dye used for blank titration, F= mg ascorbic acid equivalent to 1.0 ml indophenols dye solution, V = Total volume, in ml, of the initial assay solution, Y= Total volume, in ml, of the sample aliquot titrated, W= Weight, in g, of sample

2.2.3. Total Ash determination

In the total ash determination for dried *bilimbi* slices, the method of AOAC [10] was followed. The total ash content of the dried *bilimbi* was calculated using the following formula:

$$\text{g ash per 100 g the total sample} = \frac{\text{Weight of sample after ashing}}{\text{Weight of fresh sample taken}} \times 100\%$$

2.2.4. Determination of Total Phenolic Content (TPC)

The total phenolic content (TPC) of the crude extracts was determined using Folin-Ciocalteu reagent following the method of Wolfe et al. [11], using gallic acid as a standard (1.0 mg of crude extract was diluted in 1.0 ml distilled water). The absorbance was measured at 760 nm using a UV-Vis Spectrophotometer (Lambda 35, Perkin Elmer, USA). The total phenolic content was determined as mg of gallic acid equivalent (GAE) per milligram of crude extract, using the equation obtained from the standard gallic acid calibration graph.

2.2.5. Ferric reducing antioxidant power (FRAP)

The FRAP assay was performed with some modifications [12]. FRAP reagent (10:1:1), 10 volumes of 300 mm acetate buffer, pH 3.6 were added with 1 volume of 10 mm 2,4,6-tripyridyl-s-triazine (TPTZ) in 40 mm HCl and 1 volume of 20 mM Fe_2Cl_3. The solvent mixture was incubated at 37^0C for 10 minutes. 0.1 ml sample then was added with 2.9 ml of the solvent mixture and incubated again for another hour. The absorbance was measured at 593 nm using UV/ Vis spectrophotometer Five different concentrations of trolox (100, 250, 500, 750, 1000 μmol/l) were used to construct a standard curve and the results for the samples were expressed as μmol/g (edible part).

2.2.6. DPPH Radical scavenging activity

The effect of plant extracts on 2,2 diphenyl-1-picrylhydrazyl (DPPH) radical is estimated according to the method of [13] with some modifications. 600 μl of the extracts (0.2 mg /1 ml distilled water =200 ppm) with increasing concentration were added to 4.5 ml of DPPH (1 mM in ethanolic solution) in a 10 ml bottle with a screw cap. The mixture was shaken and left to stand at room temperature for 20 minutes in a dark place. The absorbance of the resulting solution was measured spectrophotometrically at 517 nm. The radical scavenging activity was measured as a decrease in the absorbance of DPPH and was calculated by the following formula:

$$\text{Scavenging affect (\%)} = 1 - \frac{A_{sample(517nm)}}{A_{control(517nm)}} \times 100\%$$

3. Results and Discussions

3.1. Effective Diffusivity

With reference to the drying process of *belimbi* slices with length 4cm, thickness L=0.5cm, M_0=0.8 and M_e=0.1, the results from the experiment were fitted with 40°C, 50°C, and 60°C to get the value of A and k. The non-linear least square regression based on the Levenberg-Marquardt algorithm was used. The values of k and A and the overall statistical analysis results of Sums Square Error (SSE),

Root Square Mean Error (RMSE) and coefficient of determination (R^2) are shown in Table I. The values of SSE and RMSE are relatively small. From these data it can be seen that there is an adequate fit of the statistical analysis results. Figure I show the result from curve fitting of the pre-treatment and control *belimbi* data. The lines show the data that have been fitted on the basis of their SSE, RMSE and R^2.

Table 1. Result of statistical analysis of drying of Belimbi at 40°C, 50°C and 60°C

Temperature(°C)	Condition	A	k	SSE	RMSE	R^2
40	Control	1.0448	0.0774	0.0089	0.0235	0.9879
	Pre-treatment	1.0429	0.0716	0.0047	0.0191	0.9908
50	Control	1.0549	0.1014	0.0082	0.0252	0.9885
	Pre-treatment	1.0485	0.1140	0.0976	0.0312	0.9798
60	Control	1.0481	0.1428	0.0097	0.0348	0.9756
	Pre-treatment	1.0591	0.1208	0.0154	0.0393	0.9714

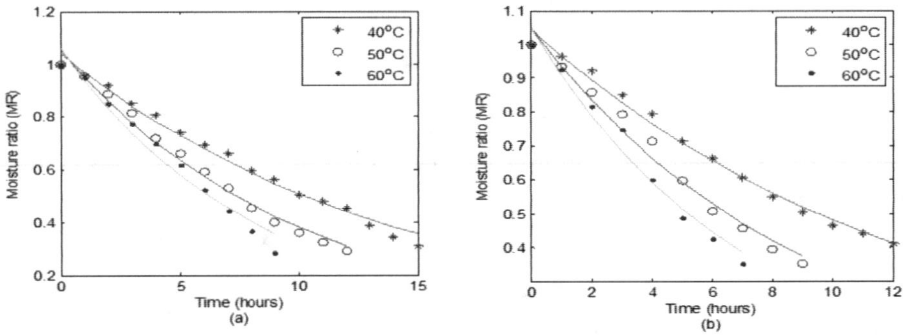

Figure 1.Predicted and experiment moisture ratio for control and pre-treatment, using three different temperatures.

Table 2. Diffusivity Value for control and pre-treatment of Belimbi using different air temperatures

Type of data	Temperature (°c)	Diffusivity(m^2/s) Model 1	Diffusivity(m^2/s) Model 2
Control	40	5.4516×10^{-11}	4.3326×10^{-10}
	50	7.315×10^{-11}	4.5467×10^{-10}
	60	8.4997×10^{-11}	4.9992×10^{-10}
Pre-treatment	40	5.0386×10^{-11}	2.2879×10^{-10}
	50	8.0209×10^{-11}	6.6808×10^{-10}
	60	1.1004×10^{-10}	7.3773×10^{-10}

Table 2 shows the value of diffusivity (D) of the control and pre-treatment of *belimbi*. The value of D from the results lies in the range between 1.1004×10^{-10} to 8.2112×10^{-11} (m^2/sec) for Model 1 and 2.28 10^{-10}-7.37×10^{-10} (m^2/sec) for Model 2 and is comparable with the data collected [14] for different kinds of food. These values depend on the experimental drying temperature: if the temperature is increased, the diffusivity will increase, which leads to a faster decrease in moisture content. The value of diffusivity for Model 1 is smaller than that of Model 2 and this value depends on the assumption at the boundary. Thus, it can be concluded that the diffusion process is a very slow process and it takes around 8-15 hours to dry *belimbi*, depending on the temperature.

3.2. Effect of temperature and pre-treatment on chemical properties of dried Belimbi

Many biochemical reactions can be induced during the drying of food. These reactions can be linked to product composition, temperature and water content (or chemical potential, or water activity), all of which may vary from one point to another point inside the food [15]. The results of the total ash, vitamin C and antioxidant properties of dried *belimbi* after drying at different temperatures are given in Table 3. Vitamin C (ascorbic acid) is an important nutrient, and is often taken as an index of the nutrient quality of processes. From the result, syrup blanching gives higher vitamin C compared to the control and this value decreased as the temperature increased. Ascorbic acid was mainly lost due to heat sensitive reactions, especially oxidation. This could explain the lower value of ascorbic acid in high temperature drying. A similar result obtained by [16] showed that higher osmotic temperature affects ascorbic acid content. [17] found out that a maximum loss of 98.2% of vitamin C occurred in a sample dried at 90°C. The amount of ash is an index of the mineral content in fruit. From the Table 3, the value of ash increases with the use of pre-treatment compared to the control; this value also decreased when the air temperature increased.

Table 3. Temperature and pre-treatment effect on chemical properties of dried *belimbi*

Temperature	Condition	Moisture Content	Ash (%)	Vitamin C(mg/g)	Antioxidant		
					Total Phenolic content(mg GAF/100g)	Scavenging effect (%)	FRAP Activiy
40	Control	69.26	70.20	5.95	15	56	0.72
	Pre-treatment	58.19	90.29	8.2	28	75	0.85
50	Control	71	60.17	5.5	14	56	0.58
	Pre-treatment	57.46	87.72	7.95	18	75	0.65
60	Control	71	65.47	4.5	13	56	0.4
	Pre-treatment	57.15	81.12	6.95	15	75	0.52

Antioxidant is a substance that has the ability to delay the oxidation of a substrate by inhibiting the initiation or propagation of oxidizing chain reactions caused by free radicals [18]. Estimation of the total phenolic content using the Folin-Ciocalteu reagent revealed that pre-treated *bilimbi* slices contained the highest amount of polyphenol, which is between 15-29 mg GAE/100 g extract, compared to control slices. A lower polyphenol amount was achieved by control slices, where the values ranged between 13-15 mg GAE/100 g extract. Drying using syrup blanching gives a higher total phenolic content compared to the control, and this value decreased as the temperature increased. This behavior could be related to the drying process at low temperatures, which implies that long drying times may promote a decrease in antioxidant capacity. Results by [19] showed that the highest antioxidant activity of dried pulp and peel oranges was determined when dehydration took place at 60°C. The total antioxidant activity was higher at a lower drying temperature in both unblanched and water blanched orange peel [20]. However, data on the effects of total phenolic content are conflicting due to several factors such as the drying method, type and extraction, pre-treatment and so on

From the results in Table 2, drying temperature has no effect on percentage of scavenging. However, the experiment by [17] shows that radical scavenging activity produces more antioxidant activity at high temperatures (80°C and 90°C) than at low temperatures (40°C-70°C). When drying using syrup pre-treatment, a higher scavenging effect is given, compared to control slices.

A ferric reducing antioxidant power (FRAP) assay was performed to measure the total reducing capacity of a compound, based on its ability to reduce $Fe3+$ or tripyridyltriazine complex to its blue-colored ferrous form. Table 2 also shows the relationship between FRAP activity of control and pre-treatment dried *bilimbi* slices with the increment of drying temperature. The results show that both the control and pre-treatment dried *bilimbi* slices have decreasing FRAP activity as the drying temperature increases. Pre-treated slices had higher FRAP activity when compared to the control slices. According to [12], a strong correlation between total phenolic content and FRAP assay existed as the phenolic compounds have redox properties, which allow them to act as reducing agents, hydrogen donators, and singlet oxygen quenchers. Thus, this redox potential of phenolic compounds can be used as the indicator to determine the antioxidant potential of foods

4. Conclusion

The quality of a dried product depends critically on the process parameters. Specifically, drying characteristics, such as temperature and pre-treatment, play an important role in the determination of physical and chemical properties. The optimization of drying temperature and pre-treatment concentration in order to get a desired product quality can be determined through mathematical modeling. Hence, the value of effective diffusivity has been investigated through the Fick Law of diffusion that is solved using separation of variable. We can conclude that, with the aid of an analytical solution, the parameter values involved in the fundamental physics equation can be determined. In this paper, different drying temperatures of 40°C, 50°C and 60°C and sucrose pre-treatment have been used. The major findings of this study are that the drying time is reduced when the air temperature increases and, furthermore, the use of pre-treatment can further reduce the drying time. A higher temperature can reduce the Vitamin C content in *belimbi* but the value of ash increased as the temperature increased. A lower temperature, which implies long drying times, may promote a decrease in antioxidant capacity, decreasing FRAP activity but this has no effect on the percentage of scavenging. There are close relationships between temperature and drying time with respect to the heat treatment of food. As a general rule, high drying temperatures tend to spoil the chemical properties of food products.

Acknowledgments

This work was supported by Ministry of Higher Education, Malaysia under Grant No 600-RMI/RAGS 5/3 (147/2014) and Universiti Teknologi MARA, Malaysia.

References

[1] Ramaswamy H and Marcotte M, 2006,*Food Processing: Principles and Applications*: Taylor and Francis,
[2] Ceylan I, Aktas M, and Dogan A, 2007, *Applied Thermal Engineering,* **27**. 1931.
[3] Fernandes F N, Rodrigues S, Law C, and Mujumdar A, 2011, *Food and Bioprocess Technology,* **4**. 163.
[4] Velic D, Planinic M, Tomas S, and Bilic S, 2004, *Journal of Food Engineering,* **64**. 97.
[5] Akpinar E, Midilli A, and Bicer Y, 2003, *Energy Conversion and Management,* **44**. 1689.
[6] Kiranoudis C T, Maroulis Z B, Marinos-Kouris D, and Tsamparlis M, 1997, *Journal of Food Engineering,* **32**. 269.
[7] Simal A, Femenia A, Garau M C, and Rossello C, 2005, *Journal of Food Engineering,* **66**. 323.
[8] Dincer I and Hussain M M, 2002, *International Journal of Heat and Mass Transfer* **45**. 3065.
[9] Shahari N, Hussein S M, Nursabrina M, and Hibberd S, 2014,"Mathematical modelling of cucumber (cucumis sativus) drying," in *AIP Conference Proceedings*. 307.
[10] AOAC., 2000,*Official methods of Analysis of AOAC International*, 17th Edition ed.: Association of Official Analytical Chemist, EAU.,
[11] Wolfe K, Wu X, and Liu R H, 2003, *Journal of Agricultural and Food Chemistry,* **51**. 609.
[12] Benzie I and Szeto Y T, 1999, *J Agric Food Chem,* **47**. 633.
[13] Yamaguchi T, Takamura H, Matoba T, and Terao J, 1998, *Bioscience, Biotechnology, and Biochemistry,* **62**. 1201.
[14] Pav´on-Melendez G, Hern´andez J A, Salgado M A, and Garc´ıa M A, 2002, *Journal of Food Engineering,* **51**. 347.
[15] Bonazzi C and Dumoulin E, 2011,*Quality Changes in Food Materials as Influenced by Drying Processes*. KGaA, Weinheim, Germany: Wiley-VCH Verlag GmbH & Co,
[16] Djendoubi Mrad N, Boudhrioua Milhoubi N, Kechaou N, Courtois F, and Bonazzi C, 2013, *Journal of Food Process Technology,* **4**. 1.
[17] Vega-Galvez A, Di Scala K, Rodriguez K, Lemus-Mondaca M, Miranda M, Lopez J, *et al.*, 2009, *Food Chemistry,* **1174**. 647.
[18] Azlim Almey A A, Ahmed Jalal Khan C, Syed Zahir I, Mustapha Suleiman K, Aisyah M R, and Kamarul Rahim K, 2010, *International Food Research Journal* **17**. 1077.
[19] Garau M C, Simal S, Rosselló C, and Femenia A, 2007, *Food Chemistry,* **104**. 1014.
[20] Akter M S, Ahmed M, and Eun J B, 2010, *Int J Food Sci Nutr.* , **61**. 702.

Computational neuroanatomy: mapping cell-type densities in the mouse brain, simulations from the Allen Brain Atlas

Pascal Grange

Xi'an Jiaotong-Liverpool University, Department of Mathematical Sciences, 111 Ren'ai Rd, Science Building SD557, Suzhou 215123, Jiangsu Province, China

E-mail: `Pascal.Grange@xjtlu.edu.cn`

Abstract. The Allen Brain Atlas of the adult mouse (ABA) consists of digitized expression profiles of thousands of genes in the mouse brain, co-registered to a common three-dimensional template (the Allen Reference Atlas).This brain-wide, genome-wide data set has triggered a renaissance in neuroanatomy. Its voxelized version (with cubic voxels of side 200 microns) is available for desktop computation in MATLAB. On the other hand, brain cells exhibit a great phenotypic diversity (in terms of size, shape and electrophysiological activity), which has inspired the names of some well-studied cell types, such as granule cells and medium spiny neurons. However, no exhaustive taxonomy of brain cell is available. A genetic classification of brain cells is being undertaken, and some cell types have been chracterized by their transcriptome profiles. However, given a cell type characterized by its transcriptome, it is not clear where else in the brain similar cells can be found. The ABA can been used to solve this region-specificity problem in a data-driven way: rewriting the brain-wide expression profiles of all genes in the atlas as a sum of cell-type-specific transcriptome profiles is equivalent to solving a quadratic optimization problem at each voxel in the brain. However, the estimated brain-wide densities of 64 cell types published recently were based on one series of co-registered coronal *in situ* hybridization (ISH) images per gene, whereas the online ABA contains several image series per gene, including sagittal ones. In the presented work, we simulate the variability of cell-type densities in a Monte Carlo way by repeatedly drawing a random image series for each gene and solving the optimization problem. This yields error bars on the region-specificity of cell types.

1. Introduction

The Allen Brain Atlas (ABA, [1, 2]) put neuroanatomy on a genetic basis by releasing voxelized, *in situ* hybridization data for the expression of the entire genome in the mouse brain (`www.mouse-brain.org`). These data were co-registered to the Allen Reference Atlas of the mouse brain (ARA, [3]). About 4,000 genes of special neurobiological interest were proritized. For these genes an entire brain was sliced coronally and processed (giving rise to the coronal ABA). For the rest of the genome the brain was sliced sagitally, and only the left hemisphere was processed (giving rise to the sagittal ABA).

From a computational viewpoint, gene-expression data from the the ABA can be studied collectively, thousands of genes at a time. Indeed the collective behaviour of gene-expression data is crucial for the analysis of [4], in which the brain-wide correlation between the ABA and cell-type-specific microarray data was studied. These microarray data characterize the transcriptome of

64 different cell types, microdissected from the mouse brain, and collated in [5]. However, for a given cell characterized in this way, it is not known where other cells of the same type are located in the brain. A linear model was proposed in [4], and used to estimate the region-specificity of cell types by linear regression with positivity constraint. The model was fitted using the coronal ABA only, which allowed to obtain brain-wide results. However, this restriction implies that only one ISH expression profile per gene was used to fit the model. This poses the problem of the error bars on the results of the model.

2. Spatial densities of cell types in the mouse mouse brain from the ABA and transcriptome profiles

Since all the ISH data in the ABA were co-registered to the voxelized ARA, so that data for the sagittal and coronal atlas can be treated computationally in the same way (results presented in this study where obtained using tha MATLAB toolbox Brain Gene Expression Analysis [6]). However, the ABA does not specify from which cell type(s) the expression of each gene comes.

Gene expression energies from the Allen Brain Atlas. In the ABA, the adult mouse brain is partitioned into $V = 49,742$ cubic voxels of side 200 microns, to which ISH data are registered [3, 1] for thousands of genes. For computational purposes, these gene-expression data can be arranged into a voxel-by-gene matrix. For a cubic labeled v, the *expression energy* [1, 2] of the gene g is a weighted sum of the greyscale-value intensities evaluated at the pixels intersecting the voxel:

$$E(v, g) = \text{expression energy of gene labeled } g \text{ in voxel labeled } v, \tag{1}$$

The analysis of [4] is restricted to numerized image series from the coronal ABA, for which the entire mouse brain was processed in the ABA pipeline (whereas only the left hemisphere was processed for the sagittal atlas).

Cell-type-specific transcriptomes and density profiles. On the other hand, the cell-type-specific microarray reads collated in [5] (for $T = 64$ different cell-type-specific samples) can be arranged in a type-by-gene matrix denoted by C, such that

$$C(t, g) = \text{expression of gene labeled } g \text{ in cell type labeled } t, \tag{2}$$

and the columns are arranged in the same order as in the matrix E of expression energies defined in Eq. 1.

We proposed the following linear model in [4] for a voxel-based gene-expression atlas in terms of the transcriptome profiles of individual cell types and their spatial densities:

$$E(v, g) = \sum_t \rho_t(v) C(t, g) + \text{Residual}(v, g), \tag{3}$$

where the index t denotes labels cell type, and $\rho_t(v)$ denotes its (unknown) density at voxel labeled v. The values of the cell-type-specific density profiles were computed in [4] by minimizing the value of the residual term over all the (positive) density profiles, which amounts to solving a quadratic optimization problem (with positivity constraint) at each voxel.

3. Monte Carlo simulation of variability of spatial densities of cell types

The optimization procedure in our model is deterministic. On the other hand, decomposing the density of a cell type into the sum of its mean and Gaussian noise is a difficult statistics problem (see [7]). Some error estimates on the value of $\rho_t(v)$ were obained in [4] using sub-sampling

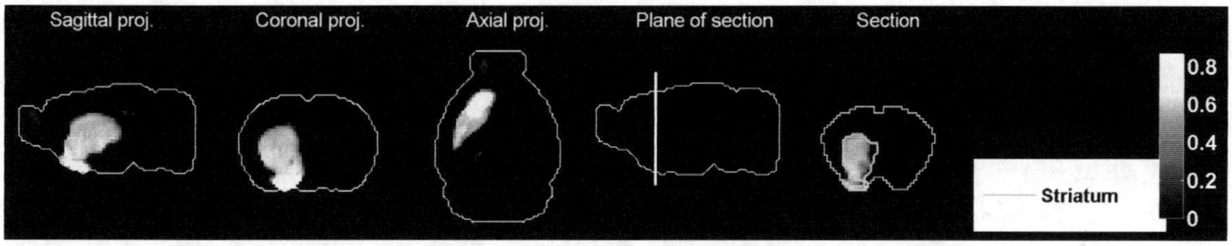

Figure 1. Heat map of the average density of cell types in the left hemisphere, $\langle \rho_t(v) \rangle$, defined in Eq. 4, for medium spiny neurons, labeled $t = 16$ in our data set. The restriction to the left hemisphere comes from the use we made of sagittal image series, which cover the left hemisphere only.

techniques (i.e. sub-sampling the data repeatedly by keeping only a random 10% of the coronal ABA). This induced a ranking of the cell types based on the stability of the results against sub-sampling. However, the 10 % fraction is arbitrary (even though it is close to the fraction of the genome covered by our coronal data set).

In the present work we simulated the variability of the spatial density of cell types by integrating the digitized sagittal image series into the data set. For gene labeled g, the ABA provides $N(g)$ expression profiles, where $N(g)$ is the number of image series in the ABA for this gene. Hence, instead of just one voxel-by-gene matrix, the ABA gives rise to a family of $\prod_{g=1}^{G} N(g)$ voxel-by-gene matrices, with voxels belonging to the left hemisphere. A quantity computed from the coronal ABA can be recomputed from any of these matrices, thereby inducing a distribution for this quantity. This is a finite but prohibitively large number of computations, so we took a Monte Carlo approach based on R random choices of images series, described by the following pseudo-code:

for all i in $[1..R]$
1. for all g in [1..G], choose an image series labeled by the integer $n_i(g)$ in $[1..N(g)]$;
2. construct the matrix $E_{[i]}$ with entries $E_{[i]}(v, g) = E^{(n_i(g))}(v, g)$,
3. estimate the density of cell type labeled t using this matrix, call the result $\rho_{t,[i]}$;
end

The larger R is, the more precise the estimates for the distribution of the spatial density of cell types will be. The only price we have to pay for th e integration of the sagittal ABA is the restriction of the results to the left hemisphere in step 2 of the pseudo-code.

4. Anatomical analysis of results

The average density across random draws of image series for cell type labeled t reads:

$$\langle \rho_t(v) \rangle = \frac{1}{R} \sum_{i=1}^{R} \rho_{t,[i]}(v). \tag{4}$$

A heat map of this average for medium spiny neurons (extracted from the striatum) is presented on Fig. 1. It is optically very similar to the (left) striatum, which allows the model to predict that medium spiny neurons are specific to the striatum (which confirms prior neurobiological knowledge and therefore serves as a proof of concept for the model).

To compare the results to classical neuroanatomy, we can group the voxels by region according to the ARA. Since the number of cells of a given type in an extensive quantity, we compute the fraction of the total density contributed by a given brain region denoted by V_r (see the legend of Fig. 2 for a list of possible values of V_r):

$$\phi_{r,[i]}(t) = \frac{1}{\sum_{v \in \text{left hemisphere}} \rho_{[i],t}(v)} \sum_{v \in V_r} \rho_{t,[i]}(v). \tag{5}$$

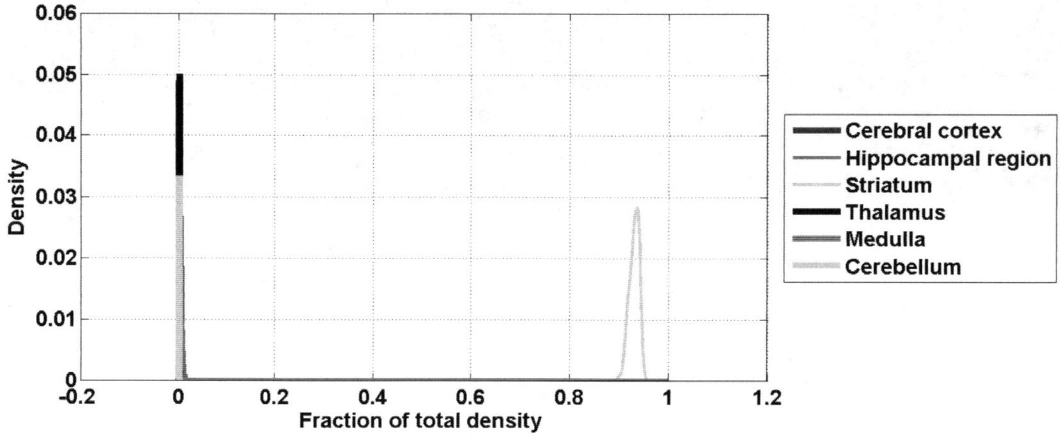

Figure 2. Estimated probability densities of fractions of density agglomerated in a few regions of the coarsest versions of the ARA (see Eq. 5), for medium spiny neurons, labeled $t = 16$, based on $R = 1000$ random draws. The right-most peak, corresponding to the striatum, is well-decoupled from the others, furthermore the other peaks are all centered close to zero (making most of them almost invisible). Medium spiny neurons have $93(\pm 3)$ percent of their densities supported in the striatum, without any region gathering more than 5 percent of the signal in any of the random draws.

We can plot the distribution of these R values for a given cell type and all brain regions (see Fig. 1 for medium spiny neurons, which gives rise to the best-decoupled right-most peak in the distribution of simulated densities). Moreover, we estimated the densities of the contribution of each region in the coarsest version of the ARA to the total density of each cell type in the data set. For most cell types, this confirms the ranking of cell types by stability obtained in [4], but based on error bars obtained from the same set of genes in every fitting of the model. The most stable results against sub-sampling tend to correspond to cell types for which the anatomical distribution of results is more peaked. The present analysis can be repeated when the panel of cell-type-specific microarray expands.

Acknowledgments

Microarray data were made available by Ken Sugino, Benjamin Okaty and Sacha B. Nelson. The Allen Atlas data were analysed under the guidance of Michael Hawrylycz and Lydia Ng. This work is supported by the Research Development Fund of Xi'an Jiaotong-Liverpool University.

References

[1] Lein ES, et al. (2007) Genome-wide atlas of gene expression in the adult mouse brain, *Nature* **445**, 168–176.
[2] Ng, L et al. (2009), An anatomic gene expression atlas of the adult mouse brain, *Nature Neuroscience* **12**, 356–362.
[3] Dong HW (2007), *The Allen reference atlas: a digital brain atlas of the C57BL/6J male mouse*, Wiley.
[4] Grange P, Bohland JW, Okaty BW, Sugino K, Bokil H, Nelson SB, Ng L, Hawrylycz M, Mitra PP, *Cell-type–based model explaining coexpression patterns of genes in the brain*, PNAS 2014 111 (14) 5397–5402.
[5] Okaty BW, Sugino K, Nelson SB (2011) A Quantitative Comparison of Cell-Type-Specific Microarray Gene Expression Profiling Methods in the Mouse Brain. *PLoS One* **6(1)**.
[6] Grange P, Hawrylycz M, Mitra PP (2013), Computational neuroanatomy and co-expression of genes in the adult mouse brain, analysis tools for the Allen Brain Atlas. *Quantitative Biology*, 1(1): 91–100. (DOI) 10.1007/s40484-013-0011-5.
[7] Meinshausen N (2013) Sign-constrained least squares estimation for high-dimensional regression. Electronic Journal of Statistics, 7, 1607–1631.

Theoretical study of a 0.22THz quasi-optical gyrotron

Xuesong Yuan, Bin Wang, Yang Yan
School of Physical Electronics,
University of Electronic Science and Technology of China, Chengdu 610054 China

E-mail: yuanxs@uestc.edu.cn

Abstract. The fully numerical calculation method for a quasi-optical gyrotron has been investigated in this paper. In this method, the analytical solution to different modes in the quasi-optical interaction cavity is replaced by the numerical solution based on electromagnetic simulation results. A 0.22THz quasi-optical gyrotron with HE06 mode has been investigated by using this method and simulation results show that this approach has a significant advantage of developing terahertz gyrotrons.

1. Introduction

Gyrotrons as so-called fast-wave devices on the basis of the stimulated radiation of electron cyclotron resonance are independent of the slow-wave system used extensively in traditional vacuum electronic devices. It is one of the most important sources which generate high power radiation in THz frequency region. For example, gyrotrons with the power of 2MW at 140/170GHz have been also developed for nuclear fusion plasma heating [1]. High frequency gyrotrons with 1 THz have been demonstrated and a very strong magnetic field (40Tesla) is required for operation at the fundamental [5]. Nevertheless, the magnetic field can significantly be decreased by using high harmonics. Thus high harmonic terahertz gyrotrons start to become one of the research hotspots in terahertz radiation sources and some special cross-section cavities have been proposed in order to develop high harmonic terahertz gyrotrons, such as the quasi-optical cavity. In a quasi-optical cavity gyrotron, mode competition in the high harmonic terahertz gyrotron can be solved well.

For a terahertz gyrotron with quasi-optical cavity we usually choose the traditional nonlinear theory of gyrotrons. Although the finite-difference time-domain (FDTD) and particle in cell (PIC) scheme, as the most accurate simulation method, can not be applied. Because terahertz gyrotrons are often designed to operate at the high order mode and high frequency, not only the spatial grid number is very large but also time step is very small, which results in very extensive computing resources and even simulation unavailable. However for the traditional nonlinear theory, three-dimensional vector equations of the beam-wave interaction have been simplified to one dimensional scalar equation and partial field components can be derived by analytical theory. The computational task and simulation time are greatly reduced. But in a terahertz gyrotron with quasi-optical cavity, it is difficult to obtain field components by using the analytical solution. Thus not only the traditional nonlinear theory calculation method but also the PIC simulation software is not available.

In order to solve problems above, a calculation method of a terahertz gyrotron with quasi-optical cavity is proposed. The terahertz gyrotron with special cross-section cavity can be better designed and studied through inserting the simulation results of the three-dimensional electromagnetic simulation software which combines the respective advantages of the simulation software and traditional nonlinear theory.

2. Formatting the title, authors and affiliations

For a terahertz gyrotron with quasi-optical cavity, we select the three-section waveguide cavity as an investigation mode. The structure diagram is shown in Figure.1. The first section L_1 is to ensure that no RF power can reach the gun. The second section L_2 is the beam-wave interaction section, and the third section L_3 permits the RF output to leave from the resonator.

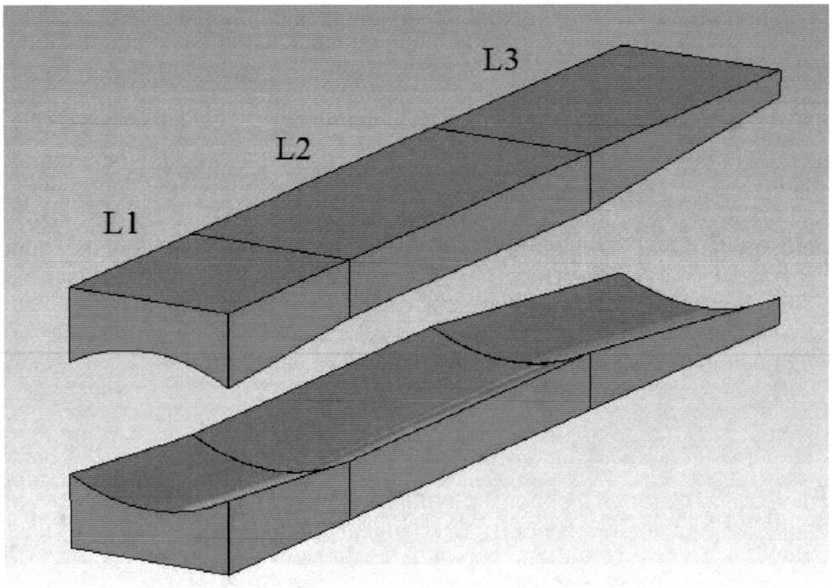

Figure 1. Structure diagram of three-section waveguide cavity.

The boundary of the cavity is the slowly varying function of the axial position z. Thus the electric field in the cavity is expressed in the form

$$\mathbf{E} = \mathbf{E_t} + \hat{z}E_z \tag{1}$$

For the HE mode in the quasi-optical cavity, we can obtain the transverse field based on simulation results,

$$\mathbf{E_t}(\mathbf{r}, z, t) = \sum\nolimits_{mp} f_{mp}(z,t)\mathbf{e}_{mp}(\mathbf{r}, z)\exp(-i\omega_0 t) \tag{2}$$

where $f_{mp}(z,t)$ is the profile function of TE_{mp} mode electric field amplitude, TM_{mp} mode electric field amplitude is very small and we can ignore it. If a steady state is assumed, f_{mp} is independent of time. $\mathbf{e}_{mp}$ is the eigenvector of the waveguide transverse field and defined by following equations:

$$\nabla_T(\nabla_T \cdot \mathbf{e_{mp}}) + k_{cmp}^2\mathbf{e_{mp}} = 0 \tag{3a}$$

$$\mathbf{e_{mq}} \cdot \hat{\mathbf{n}} \,|_{onS} = 0 \tag{3b}$$

Here $\hat{\mathbf{n}}$ is the normal unit vector of the cavity wall S. According to the wave equation and the transverse electric field, one obtains

$$\frac{d^2 f_{mp}(z)}{dz^2} + \left(\frac{\omega^2}{c^2} - k_{cmp}^2(z)\right) f_{mp}(z) = j\omega\mu_0 \frac{\int_s \mathbf{J}_t \cdot \mathbf{e}_{mp}^* ds}{\int_s \mathbf{e}_{mp} \cdot \mathbf{e}_{mp}^* ds} \tag{4}$$

where $\mathbf{J}_t$ is the transverse current density, and $\mathbf{J}_t$ is expressed in the form

$$\mathbf{J}_t(\mathbf{r},t) = \rho v_\perp = -\frac{I_0}{u_z} u_\perp \tag{5}$$

$k_{cmp}(z)$ is the cutoff wave number. If $k_{cmp}(z)$ is in inverse proportional to the axial distribution function of the cavity wall, one obtains

$$k_{cmp}(z) = k_{cmp}(z_0) * F_z(z_0) / F_z(z) \tag{6}$$

where z_0 is the middle point of the cavity axis and the $k_{cmp}(z_0)$ value can be obtained by using the waveguide port solver of the three-dimensional electromagnetic simulation software CST. If not, $k_{cmp}(z)$ can be obtained by the linear interpolation method. $k_{cmp}(z_i)$ ($i=1,2,3...n$) can be calculated based on CST, where $z_{i+1}-z_i=L/n$ (L is the length of cavity). And $k_{cmp}(z)$ is given by

$$k_{cmp}(z) = k_{cmp}(z_i) + \left[k_{cmp}(z_{i+1}) - k_{cmp}(z_i)\right](z - z_i)/(L/n) \tag{7}$$

where $z_i<z<z_{i+1}$. For the eigenvector of the waveguide transverse field $\mathbf{e}_{mp}$, we can adopt the interpolation method, too. The TE$_{mp}$ mode simulation results are shown in Figure 2. Where figure 2(b) is Ez patterns of TE$_{mp}$ mode in quasi-optical cavity, its maximum value is very small.

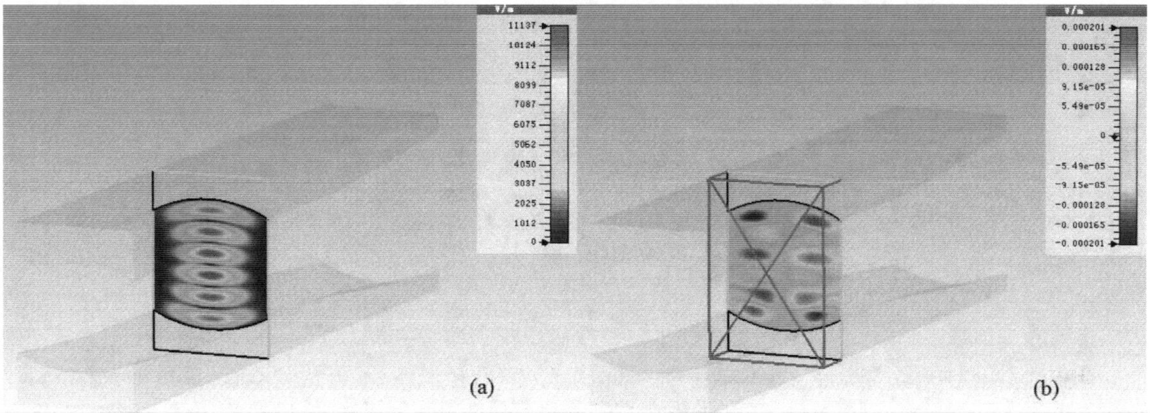

Figure 2. Field patterns of TE$_{mp}$ mode in quasi-optical cavity based on CST simulation (a), Ez field patterns of TE$_{mp}$ mode in quasi-optical cavity (b).

For cold cavity case there is no electron beam, we can get the cold cavity equation

$$\frac{d^2 f(z)}{dz} - [k^2 - k_c^2(z)] f(z) = 0 \tag{8}$$

Based on equation (7) and the condition of an evanescent wave and outgoing wave at the cavity input end and output end, respectively,

$$\begin{cases} \dfrac{df}{dz} - j\sqrt{k^2 - k_c^2} f = 0 & (z = z_{in}) \\ \dfrac{df}{dz} + j\sqrt{k^2 - k_c^2} f = 0 & (z = z_{out}) \end{cases} \tag{9}$$

the diffraction quality factor Q_d , resonant frequency and RF field profile $f(z)$ of the cold cavity can be obtained.

Based on equation (4),(5),(9) and the equations of electron motion, beam-wave self-consistent nonlinear interaction can be calculated by the shooting method. And the interaction efficiency can be expressed in the form

$$\eta = \frac{\langle \gamma_i \rangle - \langle \gamma_o \rangle}{\langle \gamma_i \rangle - 1} \tag{10}$$

where $\langle \gamma_i \rangle$ $\langle \gamma_o \rangle$ is the average value of electron beam γ at the entry and exit of the cavity, respectively.

According to above calculation method, we design a 0.4THz third harmonic TE33 mode gyrotron with corrugated waveguide cavity, and the main design parameters are shown in Table 1.

Table 1. The main parameters of 0.22TH HE_{06} mode gyrotron with quasi-optical cavity.

Cutoff section L1	5mm
Middle section L2	10mm
Ouput section L3	6mm
Radius of quasi-optical cavity R	2.11mm
the diffraction quality factor Q	1300
Cutoff frequency f_0	220.5GHz
The beam current I	2A
The beam voltage U	40kV
Magnetic field in cavity B	8.35T

3. Conclusion

In this paper, combining advantages of the traditional gyrotron nonlinear theory and electromagnetic simulation software we propose the fully numerical nonlinear theory for a gyrotron with quasi-optical cavity. And we design a 0.22THz HE06 mode gyrotron with quasi-optical cavity.

Acknowledgement

This work was supported by the National Basic Research Program of China (No.2013CB933603) and National Natural Science Foundation of China (No.U1134006, No.61101041).

[1] M. Yu. Glyavin, A. G. Luchinin, and G. Yu. Golubiatnikov, 2008, Phys. Rev. Lett., **100**, 015101.

[2] T. Notake, T. Saito, Y. Tatematsu, A. Fujii, S. Ogasawara, La Agusu, I. Ogawa, T. Idehara and V. N. Manuilov, 2009, Phys. Rev. Lett., 103, 225002

[3] XueSong Yuan, Ying Lan, Yu Han, and Yang Yan 2012 Physics of Plasmas 19, 053107

Calculating Hurst exponent and fractal dimension of neutron monitor data in a single parallel algorithm

A S Kussainov[1] and S G Kussainov[2]

[1] Physics and Technology Department, al-Farabi Kazakh National University, Almaty, Republic of Kazakhstan

[2] Insitute of Industrial Engineering, K.I. Satpaev Kazakh National Technical University, Almaty, Republic of Kazakhstan

E-mail: arman.kussainov@gmail.com

Abstract.
We implemented an algorithm for simultaneous parallel calculation of the Hurst exponent H and the fractal dimension D for the time series of interest. Parallel programming environment was provided by `OpenMPI` library installed on three machines networked in the virtual cluster and operated by `Debian Wheeze` operating system. We applied our program for a comparative analysis of week and a half long, one minute resolution, six channels data from neutron monitor. To ensure a faultless functioning of the written code we applied it to analysis of the random Gaussian noise signal and time series with manually introduced self-affinity features. Both of them have the well-known values of H and D. All results are in good correspondence with each other and supported by the modern theories on signal processing thus confirming the validity of the implemented algorithms. Our code could be used as a standalone tool for the different time series data analysis as well as for the further work on development and optimization of the parallel algorithms for the time series parameters calculations.

1. Introduction

The applications of the Hurst exponent H and fractal dimension D calculations are ranging from stock market analysis [1] to electron gas modeling [2] addressing data statistics and systems fractal properties [3]. Multiple studies have been done, including the studies of abundant data on cosmic rays variations. For example, Patra et al [4] analyzed 36 years long data series on cosmic rays density covering time period almost three solar cycles long and came to the conclusion that "the present data is anti-persistent in behavior and the process is a short memory process with the H value of 0.15". The data size does not always guarantee or necessary to perform an informative study of the subject. Flynn and Pereira, on the contrary, studied extra short, hundred points and less, data sequences [5] and extracted vital information from a data sample on population dynamics.

A fractal dimension D is a quantity reflecting how frequently and to what extent the self-affine patterns appear with the change of scale in picture. For self-affine processes we expect the local properties to be observable at the level of the global ones leading, in limited number of cases, to the relationship $D+H=n+1$ between D and H in n-dimensional space [6]. More often, in real experimental data, local and global behaviors are decoupled without any linear relation.

Journal of Physics: Conference Series **633** (2015) 012072 doi:10.1088/1742-6596/633/1/012072

Hurst exponent and fractal dimension estimates are mainly dependent on the nature of data, the length of data sample, the signal to noise ratio and the range of scales used for the analysis.

Our primary goal in this paper is to perform an effective parsing of original data text file, to identify similar steps in H and D calculations algorithms, and to provide reliability of results of the Hurst exponent H and fractal dimension D data calculations. We also demonstrate a quick self-affinity test for the data of interest using our program.

2. Methods and results

Conventional algorithm for the Hurst exponent calculation runs as follows: Original time series of length N are divided into the sets of shorter partial series $\{X_i\}$ with length $n = N$, $N/2$, $N/4, \ldots, 4, 3$, and 2 points. The upper, $n = N$, and lower, $n = 2$, cutoff limits differ from study to study and depend on data availability and studied phenomena. For each set with particular n value, and for every partial series $\{X_i\}$ within this set, we calculate the mean-adjusted series $\{Y_i\}$ derived from $\{X_i\}$ by subtracting from each element the series' mean value m

$$Y_t = X_t - m;\ t = 1, 2, \ldots, n;\ m = 1/n \sum_{i=1}^{n} X_i \tag{1}$$

We define the rescaled range $R(n)$ for a given subseries $\{Y_i\}$ as follows

$$R(n) = max(Z_1, Z_1, Z_n) - min(Z_1, Z_1, Z_n) \tag{2}$$

where the cumulative deviate series Z_n are given by the following equation

$$Z_i = \sum_{1=1}^{t} Y_i;\ t = 1, 2, , n. \tag{3}$$

Now, we can built the tabulated power law function $E(n)$ for all possible values of n such as

$$E\left[\frac{R(n)}{S(n)}\right] = Cn^H \tag{4}$$

where $S(n) = \{1/n \sum_{i=1}^{n} (X_i - m)^2\}^{1/2}$ is a standard deviation. The value of H then could be estimated from the slope of the line $log(E) = log(C) + H \cdot log(n)$ [7].

Fractal dimension D is closely related to the Hurst exponent and could be calculated, according to Higuchi cornerstone paper [3], by constructing the following sets of subseries X_k^m

$$X(m), X(m+k), X(m+2k), \ldots, X(m + \left[\frac{N-m}{k}\right] k);$$
$$m = 1, 2, \ldots, k;\ k = 1, 2, \ldots, \left[2^{(j-1)/4}\right];\ j = 11, 12, 13, \ldots \tag{5}$$

where the square brackets are used to denote the closest integer after rounding the fraction to zero.

Next, we calculate the normalized lengths $L_m(k)$ of the constructed subseries

$$L_m(k) = \left\{ \frac{N-1}{\left[\frac{N-m}{k}\right] k} \sum_{i=1}^{\left[\frac{N-m}{k}\right]} |X(m+ik) - X(m+(i-1)k)| \right\} k^{-1} \tag{6}$$

These are expected to follow the similar power law in the form of $\langle L_m(k) \rangle \propto k^{-D}$ after averaging within all sets of k values for different $m = 1, 2, \ldots, k$.

Table 1. Curves slope estimates in D and H calculations.

Time series, name	Time series length, N	Fractal dimension, D	Hurst exponent, H
Higuchi series	2^{17}	-1.5345	0.9875
Gaussian noise	$1000 + 2^{17}$	-1.9999	0.5400
NM data, 1st channel	14703	-1.8128	0.9810
2nd channel	14703	-1.9015	0.9332
3rd channel	14703	-1.9363	0.9184
4th channel	14703	-1.9583	0.8734
5th channel	14703	-1.9064	0.9333
6th channel	14703	-1.9102	0.9425

To address a persistent need for the effective parallel algorithms in data processing we designed our piece of `C/C++` code compatible with available for all latest `Debian` operating system distribution of the Message Passing Interface (`OpenMPI`) parallel computations environment. We have configured our virtual cluster using the `Oracle VM VirtualBox`. The host is `64 bit Windows 8.1` operating system running on Intel Core i3-3220 CPU with 4 Gb of RAM. The guest operating systems are the three Linux machines with `64 bit Debian Wheezy 7.4` with 512 Mb of RAM per each server and two nodes. The number of processes was equal to the number of virtual machines and it was arranged to be three. All scales calculations were distributed between them sequentially in a circular fashion, see Figure 2.

The original neutron monitor data were retrieved from the underground monitor of the Tien Shan High-Altitude Station (3340 m above sea level) located in Almaty, Kazakhstan. The monitor has several sections with multiple detectors located under a ground layer 20 m of water equivalent thick [8]. Neutron counts were acquired at one minute interval from January the 1st, 2015 till January the 11th, 2015. All data are pressure corrected. Neutron monitor time series data additionally underwent simple exponential smoothing filtration procedure as described in [9].

Our H estimate for the first out of six registration channels of the nuclear monitor data file is shown on Figure 1. Each channel data contain week and a half long data on neutrons counts, taken with one minute resolution. Operator makes the choice of registration channel for analysis within the program. Calculated H and D values for all of the channels are given in Table 1.

To test the validity of the written program we reproduced the exact same time series $Y(i)$ (we using the original $Y(i)$ and $Z(j)$ notations from his paper) which were used by Higuchi in his derivations [3] and plotted the results on the same figure for reference, see Figure 1. Here $Y(i) = \sum_{j=1}^{1000+i} Z(j)$, where $Z(j)$ is a Gaussian noise with mean zero and standard deviation equals to 1. As we said, the set of $\langle L_m(k) \rangle$ values is expected to represent the power law function $\langle L_m(k) \rangle \propto k^{-D}$ and the value of D could be extracted within the same piece of code we wrote previously for the Hurst exponent.

Our calculations give us D value equals -1.5345 and $H = 0.9875$ for the Higuchi's time series. A separate test run of the code for the Gaussian time series $Z(j)$ produced another set of values H and D equal -1.9999 and 0.5400 correspondingly, see Table 1. These values are typical for the highly uncorrelated times series. Both sets of scales n and k are taking the same values ranging from 6 to 6888. The $Y(i)$, $Z(j)$ original series and neutron monitor data are plotted as the offsets on the same Figure 1.

The $\pm\Delta y_i$ error bars from the least squares fitting procedures were plotted on the Figure 1 around each experimental data points (x_i, y_i) using `polyfit`, `polyval` and `errobar` functions available in `GNU Octave` software in such a way that then $y_i \pm \Delta y_i$ contains at least 50% of the predictions of future observations at x_i, see `Octave` help notes on the selected functions.

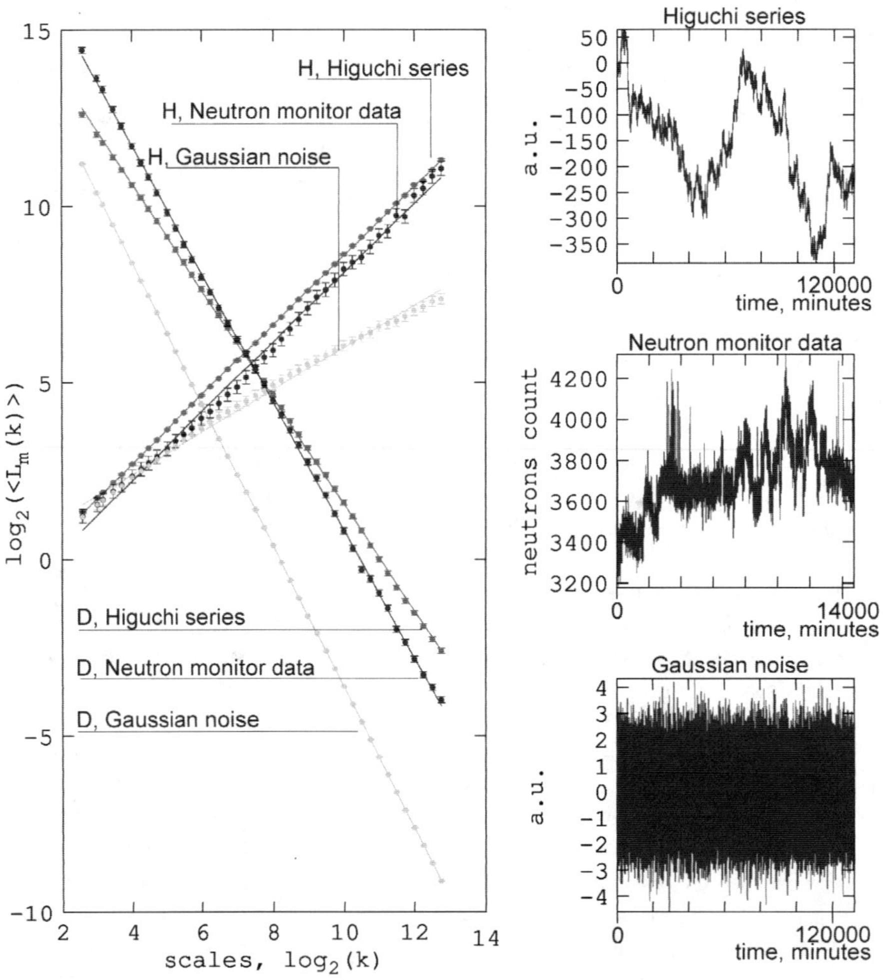

Figure 1. Calculations of H and D values for the Higuchi time series, the neutron monitor data and the Gaussian noise. Offsets on the right side contain the plots of analyzed time series.

As we mentioned before, the n-length of subseries for the Hurst exponent calculations, and k-value for the fractal dimension calculation, have been treated equally, as a single variable $k = n$, in our code. That is two separate arrays of the same size k were allocated for both types of concurrent calculations of the H and D values. In case of the Hurst exponent calculations the array's values have been filled sequentially from the data file readout. For the fractal dimension calculations, each array's element contained the single value $\langle L_m(k)\rangle$ was updated through the same data readout according to the scheme described in equations (5) and (6).

Calculated slopes for all three sets of data, where neutron monitor data contain information on all six cahnnels, are given in the Table 1. These are the Higuchi time series, Gaussian noise mpiuser@debian-lenovo-1: ~

File Edit View Search Terminal Help

scale	Lmk	<E>	id	hostname	wall time, sec	CPU time, sec
6	265472.18	2.33	0	debian-lenovo-1	0.5470	0.5200
9	121224.58	3.24	2	debian-lenovo-3	0.5611	0.5600
8	152193.36	2.94	1	debian-lenovo-2	0.5989	0.5300
11	82536.81	3.80	0	debian-lenovo-1	1.0630	1.0100
16	40731.49	5.08	2	debian-lenovo-3	1.0730	1.0500
13	59978.23	4.34	1	debian-lenovo-2	1.1194	1.0400
26	16499.70	7.32	2	debian-lenovo-3	1.5227	1.4900
19	29348.67	5.79	0	debian-lenovo-1	1.5434	1.4900
22	22270.44	6.45	1	debian-lenovo-2	1.6483	1.5800
45	6083.83	10.73	2	debian-lenovo-3	1.9563	1.9200
32	11264.44	8.50	0	debian-lenovo-1	2.0376	1.9700
38	8245.03	9.53	1	debian-lenovo-2	2.2169	2.1200
76	2474.98	15.45	2	debian-lenovo-3	2.4012	2.3600
53	4575.39	12.06	0	debian-lenovo-1	2.5082	2.4300
64	3302.50	13.73	1	debian-lenovo-2	2.7577	2.6500
128	1057.82	22.45	2	debian-lenovo-3	2.8852	2.8400
90	1872.18	17.32	0	debian-lenovo-1	2.9855	2.9100
107	1413.30	19.75	1	debian-lenovo-2	3.2591	3.1600
215	454.47	34.55	2	debian-lenovo-3	3.3674	3.3200
152	802.71	25.62	0	debian-lenovo-1	3.4643	3.3600
181	599.26	29.55	1	debian-lenovo-2	3.8289	3.7200
362	195.77	54.98	2	debian-lenovo-3	3.8630	3.8200

Figure 2. Typical workload distribution between the three processes on the virtual cluster with timing output.

and neutron monitor data H and D values. With such an approach, the H and D values may be calculated retrospectively, for the whole history of data acquisition, as well as at the present moment, within a fixed time window moving forward with the natural time flow and incorporated with the acquisition hardware.

3. Discussion

H and D values for the Higuchi time series in our calculations were found to be equal to the D and H values from [3], that is H value is close to 1 (see Table 1) and indicates long term positive autocorrelation, as expected from the $Y(i)$ series. For the neutron monitor series, Hurst exponent estimates are matching the majority of the previously obtained results for geomagnetic indices [10] where H value is well above 0.5. Drastic changes in H and D are observed if rigorous filtering is applied to the raw data. This is usually the case with intrinsically very noisy neutron monitor data. Between all six channels, the H values are within 0.9333-0.9810 range, suggesting that at the chosen series duration and quality, our time series do have some scalable order. However, possibility of a long-term positive autocorrelation requires further studies and clarification.

The Gaussian noise data $Z(j)$ in its turn produces values of D=-1.9999 and H=0.5400 as expected from the highly uncorrelated data. As we can see, for all sets of data the processes are not completely self-affine in the sense of $D + H = n + 1$ relationship, and the dependence between D and H is not of linear nature.

It is interesting to observe how the curve's slope changes according to the changes in data structure. We can see this transition when we come down from Higuchi time series to the chaos in the Gaussian noise. All lines for the Hurst exponent originate from the single point. That is rotations for all H data happen to be around (2,0) point due to the loss of differences between time series at the small scales. Through such a graphic representation and comparative, real-time analysis of our data with the purpose of data quality check and search for a particular fractal dimensional phenomena is possible.

4. Conclusions

Using `C/C++` programming language we implemented an algorithm for the simultaneous parallel calculations of the Hurst exponent H and the fractal dimension D for the different time series. Parallel programming environment was provided by an open source freeware such as `OpenMPI` package installed on three machines networked to the virtual cluster using the `Oracle VM VirtualBox` and operated by a `64 bit Debian Wheeze 7.4` operating systems. Additional data processing and graphic representation were made on `GNU Octave` freeware.

We used our program to perform a comparative analysis of the week and a half long, one minute resolution, six channels data file from a neutron monitor. To verify the validity of the written code, we compared these results with a similar data analysis of the random Gaussian noise signal and with time series which have manually introduced self-affinity features. Both reference time series have the known values of H and D. All results are in good correlation with each other and supported by the modern theories on signal processing, thus confirming the validity of the implemented algorithms.

Our algorithm has multiple applications such as quick data self-affinity [11] test and have great potentials for the future development. The code could be used as a tool to study time series of different nature and origin. Timing and optimization in this paper are the subjects for the further studies.

Acknowledgments

This study was supported by grant #2532/GF3 provided by the Science Committee at the Ministry of Science and Education of Republic of Kazakhstan to the principal investigator at the National Nanotechnology Open Laboratory. We also would like to thank Dr. Yelena White, from East Georgia State College, Swainsboro, GA, USA, for editing the manuscript.

References

[1] Alvarez-Ramirez J, Alvarez J, Rodriguez E and Fernandez-Anaya G 2008 *Physica A* **387** 6159-69
[2] Hurst J, Morandi O, Manfredi G and Hervieux P 2014 *EPJ D* **68** 176-179
[3] Higuchi T 1988 *Physica D* **31(2)** 277-283
[4] Patra S N, Ghosh K and Panja S C 2013 *IJEER* **3(2)** 237-246
[5] Flynn M N, Pereira W R L S 2013 *Biomatematica* **23** 1-14
[6] Mandelbrot B B, Passoja D E and Paullay A J 1984 *Nature* **308** 721-724
[7] Górski A Z, Drożdż S and Speth J 2002 *Physica A* **316** 496-510
[8] Shepetov Yu V 2014 *The Tien-Shan mountain station's database* (Institute of Terrestrial Magnetism, Ionosphere and Radio Wave Propagation named after Nikolay Pushkov of the Russian Academy of Sciences) http://cr29.izmiran.ru
[9] Brown R G 1956 *Exponential smoothing for predicting demand* (Cambridge, Massachusetts: Arthur D. Little Inc) p 15
[10] Pesnell W D 2012 *Sol. Phys* **281** 507532
[11] Gneiting T and Schlather M 2004 *SIAM Rev.* **46(2)** 269-282

Pseudo Magic Squares

Giuliano G. La Guardia, Ana Lucia Pereira Baccon

Department of Mathematics and Statistics, State University of Ponta Grossa, 84030-900, Ponta Grossa - PR, Brazil

E-mail: gguardia@uepg.br

Abstract. A magic square of order n is an $n \times n$ square (matrix) whose entries are distinct nonnegative integers such that the sum of the numbers of any row and column is the same number, the magic constant. In this paper we introduce the concept of pseudo magic squares, *i.e.*, magic squares defined over the ring of integers, without the restriction of distinct numbers. Additionally, we generalize this new concept by introducing a group (ring) structure over it. This new approach can provide useful tools in order to find new non-isomorphic pseudo magic squares.

1. Introduction

The concept of magic squares is well known in literature [5, 3, 2]. The Loh-Shu magic square

$$\begin{bmatrix} 4 & 9 & 2 \\ 3 & 5 & 7 \\ 8 & 1 & 6 \end{bmatrix}$$

is the oldest known magic square and its invention is attributed to Fuh-Hi (2858-2738 b.C.) [5]. There exist interesting papers available in the literature dealing with constructions of magic squares [3, 2, 4, 1]. Recently, Xin [7] have constructed all magic squares of order three. In most works available in literature, the approach adopted is to apply combinatorial methods to compute more classes of new (non-isomorphic) magic squares.

In this paper we introduce the concept of pseudo magic square (PMS). We show that a PMS have a natural group structure. Additionally, we generalize this new concept in order to obtain a generic magic square (GMS), which is derived from a arbitrary group (ring). The structure of a group (ring) induces a group (ring) structure in the set of GMS's. Based on these facts, one can see that our approach is quite different of the ones available in literature.

The paper is organized as follows. In Section 2, we fix the notation and also define the concept of pseudo magic squares. In Section 3, we present the contributions of this paper: some properties of PMS are shown as well as the introduction of the concept of generic magic squares with corresponding properties are exhibited. In Section 4, a brief summary of this paper is given.

2. Preliminaries

Notation. Throughout this paper, $\mathbb{N}$ denotes the set of nonnegative integers, $\mathbb{N} - \{0\} = \mathbb{N}^*$, $\mathbb{Z}$ is the set of integers. The set of square matrices of order n with entries in $\mathbb{Z}$ is denoted by $\mathbb{M}_n(\mathbb{Z})$; the cardinality of a set S is denoted by $\mid S \mid$.

Definition **2.1** *Let* $n \in \mathbb{N}^*$. *A pseudo magic square of order* n*, denoted by* $A_n^\square$*, is an element of* $\mathbb{M}_n(\mathbb{Z})$ *such that the sum of the numbers of any row and column is the same number* $c_{A_n^\square}$*, the pseudo magic constant (constant, for short). The set of PMS of order* n *is denoted by* $\mathcal{P}_n^\square$.

Remark **2.1** *To avoid stress of notation, we do not distinguish between an* $n \times n$ *matrix or an* $n \times n$ *square.*

Example **2.1** *As an example, we have a PMS of order* 4 *given by*

$$\begin{bmatrix} -5 & -5 & -5 & -5 \\ -5 & -5 & -5 & -5 \\ -5 & -5 & -5 & -5 \\ -5 & -5 & -5 & -5 \end{bmatrix}.$$

3. The Results

In this section we present the results of this work. We start by showing that the set of PMS endowed with the sum of matrices is a group:

Theorem **3.1** *The ordered pair* $(\mathcal{P}_n^\square, +)$ *is an abelian group, where the operation* $+$ *means the addition of matrices.*

Proof: If $A_n^\square$ and $B_n^\square$ are PMS of constants $c_{A_n^\square}$ and $c_{B_n^\square}$, respectively, then $A_n^\square + B_n^\square$ is a PMS with constant $c_{A_n^\square} + c_{B_n^\square}$, so $\mathcal{P}_n^\square$ is closed. It is clear that the null PMS of order n is the identity element. The associativity follows trivially of the associativity of the (additive) group $(\mathbb{Z}, +)$. The same applies to commutativity. It is clear that the inverse of a PMS $A_n^\square$ of constant $c_{A_n^\square}$ is the PMS $[-A]_n^\square$ of constant $-c_{A_n^\square}$. □

Corollary **3.2** $(\mathcal{P}_n^\square, +)$ *is a subgroup of* $(\mathbb{M}_n(\mathbb{Z}), +)$.

Proof: Straightforward. □

It is possible to construct new PMS's from old ones. In fact, the first method to construct new PMS's is by means of the direct sum structure (see Theorem 3.3). It is clear that several structures can be defined over PMS's as well, but we present such results in other paper.

Theorem **3.3** *Let* $A_n^\square$ *and* $B_n^\square$ *be two PMS's. Define the direct sum* $A_n^\square \oplus B_n^\square$ *as*

$$\begin{bmatrix} A_n^\square & B_n^\square \\ B_n^\square & A_n^\square \end{bmatrix} \in \mathbb{M}_{2n}(\mathbb{Z}).$$

Then the set $\mathcal{P}_n^\square \oplus \mathcal{P}_n^\square$ *of direct sums endowed with addition of matrices is an abelian group.*

Proof: The direct sum is well-defined. Let $A_n^\square$ and $B_n^\square$ be two PMS's of constants $c_{A_n^\square}$ and $c_{B_n^\square}$, respectively. Then $A_n^\square \oplus B_n^\square$ is a PMS of order $2n$ and constant $c_{A_n^\square} + c_{B_n^\square}$. If $C_n^\square$ and $D_n^\square$ are also PMS's of constants $c_{C_n^\square}$ and $c_{D_n^\square}$, respectively, then its sum $[A_n^\square \oplus B_n^\square] + [C_n^\square \oplus D_n^\square]$ is also a PMS of order $2n$ and constant $c_{A_n^\square} + c_{B_n^\square} + c_{C_n^\square} + c_{D_n^\square}$. Then $\mathcal{P}_n^\square \oplus \mathcal{P}_n^\square$ is closed. The remaining properties follow similarly as in the proof of Theorem 3.1. □

Corollary **3.4** $(\mathcal{P}_n^\square \oplus \mathcal{P}_n^\square, +)$ *is a subgroup of* $(\mathbb{M}_{2n}(\mathbb{Z}), +)$.

Journal of Physics: Conference Series **633** (2015) 012073 doi:10.1088/1742-6596/633/1/012073

Proof: Straightforward. □

Let $A_n^{\square}$ be a PMS of order n. We can multiply each entry of $A_n^{\square}$ by an integer k. Then the resulting square is also a PMS square. Similarly, by adding an integer to each entry we obtain again a PMS. Moreover, one can define the tensor product of PMS's. Note that such ideas can be explored in order to generate several properties as well as new structures for PMS's. These facts will be described in other paper.

Next, we introduce the concept of generic magic squares:

Definition **3.1** *Let $(G, \star)$ be an abelian group. A generic magic square, of order n, derived from $(G, \star)$, is an $n \times n$ matrix with entries in G such that for any row and any column, the result of the operation among the row(column) elements corresponds to the same element, the generic magic constant (constant, for short).*

We denote the set of GMS's of order n over G by $[\mathcal{G}g]_{G_n}^{\square}$.

Remark **3.5** *It is clear that $[\mathcal{G}g]_{G_n}^{\square} \neq \emptyset$. In fact, if e_G is the identity of G, then* $[E]_{G_n}^{\square} = \begin{bmatrix} e_G & e_G & \cdots & e_G \\ e_G & e_G & \cdots & e_G \\ \vdots & \vdots & \cdots & \vdots \\ e_G & e_G & \cdots & e_G \end{bmatrix} \in \mathbb{M}_n(G)$ *is a GMS over G.*

Theorem **3.6** *The ordered pair $([\mathcal{G}g]_{G_n}^{\square}, \star_p)$, where $\star_p$ denotes the group operation component-wise in $\mathbb{M}_n(G)$, is an abelian group.*

Proof: It is clear that the set is closed. The identity is the GMS $[E]_{G_n}^{\square}$. The remaining properties follows similarly as in the proof of Theorem 3.1. The detailed proof will be provided in other paper. □

By lack of space, we will not investigate the structures of this interesting group, *i.e.*, homomorphisms and isomorphisms, group actions, properties, and so on.

Consider now a commutative ring with unit $(R, +, \cdot)$. Similarly as was done with the group structure, the ring $(R, +, \cdot)$ induces a natural ring structure in the set os GMS as follows:

Definition **3.2** *Let $(R, +, \cdot)$ be an commutative ring with unit. A generic magic square of order n, derived from $(R, +, \cdot)$, is an $n \times n$ matrix with entries in R such that:*
(1) For any row and any column, the result of the operation + among the row(column) elements corresponds to the same element, the generic additive magic constant (additive constant, for short);
(2) For any row and any column, the result of the operation $\cdot$ among the row(column) elements corresponds to the same element, the generic multiplicative magic constant (multiplicative constant, for short).

We denote by $[\mathcal{G}r]_{R_n}^{\square}$ the set of GMS's of order n over R. Note that the additive constant can be different from the multiplicative constant. Analogously to groups, the set $[\mathcal{G}r]_{R_n}^{\square}$ is not empty.

Theorem 3.7, given in the following, states that the set $[\mathcal{G}r]_{R_n}^{\square}$ endowed with two component-wise operations form a commutative ring with unit. We give only a sketch of the proof.

Theorem **3.7** *The ordered triple $([\mathcal{G}r]_{R_n}^{\square}, +_p, \cdot_p)$, where $+_p$ and $\cdot_p$ denote the ring operations component-wise in $\mathbb{M}_n(R)$, is a commutative ring with unit.*

Proof: It is easy to see that the set $[\mathcal{G}r]_{R_n}^{\square}$ is closed under the operations $+_p$ and $\cdot_p$. The commutativity, associativity and distributivity of such operations follow from the commutativity, associativity and distributivity, respectively, of the operations + and $\cdot$ of the ring. The identity and the unit of $[\mathcal{G}r]^{\square}_{R_n}$ are obvious from the context. □

Given a GMS $([\mathcal{G}r]^{\square}_{R_n}, +_p, \cdot_p)$, we can operate (the operation considered here is $\cdot$ of the ring) a fixed element of the ring $r \in R$ with each entry in $[\mathcal{G}r]^{\square}_{R_n}$. It is easy to see that the resulting matrix is also a GMS. In this context, one can derive more algebraic structures over $([\mathcal{G}r]^{\square}_{R_n}, +_p, \cdot_p)$.

Definition **3.3** *A matroid M is an ordered pair $(S, \mathcal{I})$ consisting of a finite set S and a collection $\mathcal{I}$ of subsets of S satisfying the following three conditions:*
(I.1) $\emptyset \in \mathcal{I}$;
(I.2) If $I \in \mathcal{I}$ and $I' \subset I$, then $I' \in \mathcal{I}$;
(I.3) If I_1, $I_2 \in \mathcal{I}$ and $\mid I_1 \mid < \mid I_2 \mid$, then there exists an element $e \in I_2 - I_1$ such that $I_1 \cup \{e\} \in \mathcal{I}$.

It is well known that the concept of matroid generalizes several algebraic structures. In this context, we will describe in other manuscript, how to induce a matroid structure on the set $[\mathcal{G}r]^{\square}_{R_n}$. In particular, if we consider the class of vectorial matroid (see [6]), interesting results can be proved.

4. Summary

In this paper, we have introduced the concept of pseudo magic squares and generic magic squares. Moreover, we have shown that the set of generic (pseudo) magic squares has a group (ring) structure. Additionally, some properties of PMS and GMS as well as constructions of new PMS's from old ones have also been presented. From the content exhibited in this paper, it can be observed that there exist much work with respect to our new approach to be done. Other interesting possibility is the introduction of a topological structure in GMS's. Therefore, convergence, completeness, compactness in GMS's can also be investigated in future works.

Acknowledgment

This research was partially supported by the Brazilian Agencies CAPES and CNPq.

[1] Ahmed M, Algebraic combinatorics of magic squares, Ph.D. Thesis, University of California, Davis, Preprint math.CO/0405476.
[2] Ahmed M, Loera J D and Hemmecke R 2003 Discrete and Computational Geometry, Algorithms Combin. **25** pp.25-41.
[3] Andrews G E, Paule P, A. Riese A and Strehl V 2001 Algebraic Combinatorics and Applications Springer Berlin pp.1-39.
[4] Beck M, Cohen M, Cuomo J, Gribelyuk P 2003 Amer. Math. Monthly **110-8** pp.707-717.
[5] Pickover C 2002 *The Zen of Magic Squares, Circles, and Stars: An Exhibition of Surprising Structures Across Dimensions* (Princeton University Press).
[6] Whitney H 1935 Amer. J. Math. **57** pp.509–533.
[7] Xin G 2008, Discrete Math. **308** pp.3393–3398.

Ultrashort Pulse Interaction with Intersubband Transitions of Semiconductor Quantum Wells

Ioannis Katsantonis[1], Elias Stathatos[2] and Emmanuel Paspalakis[3]

[1]Physics Department, School of Natural Sciences, University of Patras, Patras 265 04, Greece
[2]Electrical Engineering Department, Technological and Educational Institute of Western Greece, Megalou Alexandrou 1, Patras 263 34, Greece
[3]Materials Science Department, School of Natural Sciences, University of Patras, Patras 265 04, Greece

E-mail: i.katsadonis@gmail.com, paspalak@upatras.gr

Abstract. We study coherent ultrashort pulse propagation in a two-subband system in a symmetric semiconductor quantum well structure, performing calculations beyond the rotating wave approximation and the slowly varying envelope approximation and taking into account the effects of electron-electron interactions. The interaction of the quantum well structure with the electromagnetic fields is studied with modified, nonlinear, Bloch equations. These equations are combined with the full-wave Maxwell equations for the study of pulse propagation. We present results for the pulse propagation and the population inversion dynamics in the quantum well structure for different electron sheet densities.

1. Introduction

In recent years, Rabi flopping and coherent propagation of ultrashort electromagnetic pulses have been studied both theoretically and experimentally when intersubband transitions of semiconductor quantum wells interact with electromagnetic fields [1, 2, 3, 4, 5, 6, 7, 8, 9]. In some of these studies the Rabi flopping [2, 3, 8] and the coherent propagation of ultrashort far-infrared electromagnetic pulses [4, 5, 7] that interact with a two-subband system in a symmetric semiconductor quantum well structure, taking into account the effects of electron-electron interactions, have been presented. All these studies have shown that at certain electron densities the electron-electron interactions make the intersubband transitions to behave quite differently from atomic transitions.

Here, we revisit the problem of coherent ultrashort pulse propagation in a two-subband system in a symmetric semiconductor quantum well structure, performing calculations beyond the rotating wave approximation (RWA) and the slowly varying envelope approximation (SVEA) and taking into account the effects of electron-electron interactions. The interaction of the quantum well structure with the electromagnetic fields is studied with modified, nonlinear, Bloch equations [10]. These equations are combined with the full-wave Maxwell equations for the study of pulse propagation, so the coupled Maxwell-nonlinear Bloch equations are solved computationally beyond the RWA and SVEA. In order to analyze the impact of propagation on self-induced transparency effects [11, 6], we present results for a hyperbolic secant envelope electromagnetic pulses with initial 2π pulse area.

Journal of Physics: Conference Series **633** (2015) 012074 doi:10.1088/1742-6596/633/1/012074

2. Coupled Maxwell-Nonlinear Bloch Equations

The system under study is a symmetric double semiconductor quantum well and is presented in Fig. 1. We assume that only the two lower energy subbands, $n = 0$ for the lowest subband and $n = 1$ for the excited subband, contribute to the system dynamics. The Fermi level is below the $n = 1$ subband minimum, so the excited subband is initially empty. This is succeeded by a proper choice of the electron sheet density. The two subbands are coupled by a time- and spatially-dependent electric field polarized along the x-direction $E_x(z,t)$, which is taken at the entrance of the medium as $E_x(z = 0, t) = E\text{sech}[1.76(t - t_0)/t_p]\cos[\omega(t - t_0)]$, where E is the maximum electric field amplitude, ω is the central carrier frequency, t_p is the full-width at half-maximum of the pulse intensity envelope of the electromagnetic pulse, and t_0 is the pulse delay. We consider the propagation properties of an ultrashort pulse $E_x(z,t)$ along the z-axis in multiple double symmetric quantum wells. The propagation of the electric field $E_x(z,t)$ is determined by the Maxwell-wave equations

$$\frac{\partial E_x}{\partial t} = -\frac{1}{\varepsilon}\frac{\partial H_y}{\partial z} - \frac{1}{\varepsilon}\frac{\partial P_x}{\partial t}, \tag{1}$$

$$\frac{\partial H_y}{\partial t} = -\frac{1}{\mu_r}\frac{\partial E_x}{\partial z}, \tag{2}$$

where $H_y(z,t)$ is the magnetic field along the y-direction, ε is the relative dielectric constant and μ_r is the magnetic permeability in the medium.

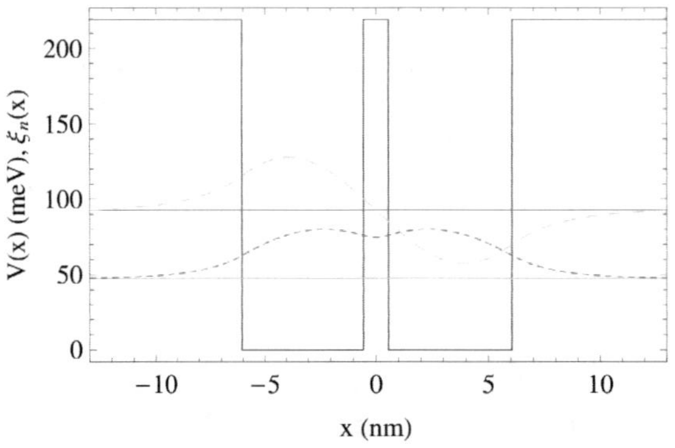

Figure 1. The confinement potential of the quantum well structure under study (blue solid curve) and the energies of the lower (green lower line) and upper states (red upper line). The envelope wavefunctions for the ground (dotted curve) and first excited (dashed curve) subbands.

Olaya-Castro *et al.* [10] showed that the quantum well dynamics is described by the following effective nonlinear Bloch equations

$$\frac{\partial S_1}{\partial t} = [\omega_{10} - \gamma S_3]S_2 - \frac{S_1}{T_2}, \tag{3}$$

$$\frac{\partial S_2}{\partial t} = -[\omega_{10} - \gamma S_3]S_1 + 2[\frac{\mu E_x}{\hbar} - \beta S_1]S_3 - \frac{S_2}{T_2}, \tag{4}$$

$$\frac{\partial S_3}{\partial t} = -2[\frac{\mu E_x}{\hbar} - \beta S_1]S_2 - \frac{S_3 + 1}{T_1}. \tag{5}$$

Here, $S_3(z,t)$ is the mean population inversion per electron (difference of the occupation probabilities in the upper and lower subbands) and the macroscopic polarization of the medium is given by $P_x(z,t) = -N_v\mu S_1(z,t)$, where N_v is the electron volume density of the quantum well. Also, $\mu = ex_{01}$ is the electric dipole matrix element between the two subbands and the parameters $\omega_{10}, \beta, \gamma$ are given by

$$\omega_{10} = \frac{E_1 - E_0}{\hbar} + \frac{\pi e^2}{\hbar\varepsilon} N \frac{L_{1111} - L_{0000}}{2}, \quad (6)$$

$$\gamma = \frac{\pi e^2}{\hbar\varepsilon} N \left(L_{1001} - \frac{L_{1111} + L_{0000}}{2} \right), \quad (7)$$

$$\beta = \frac{\pi e^2}{\hbar\varepsilon} N L_{1100} \,. \quad (8)$$

Here, N is the electron sheet density, e is the electron charge, E_0, E_1 are the eigenvalues of energy for the ground and excited states in the well, respectively, and $L_{ijkl} = \int\int dx dx' \xi_i(x)\xi_j(x')|x - x'|\xi_k(x')\xi_l(x)$, with $i, j, k, l = 0, 1$. Also $\xi_i(x)$, is the envelope wavefunction for the i-th subband along the growth direction. Finally, T_1 describes the population decay time and T_2 the dephasing time.

3. Numerical Results

We consider a GaAs/AlGaAs double quantum well. The structure consists of two GaAs symmetric square wells of width 5.5 nm and height 219 meV. The wells are separated by a AlGaAs barrier of width 1.1 nm. This system has been studied in several previous works [3, 4, 5, 7, 8, 10]. The electron sheet density takes values between 10^9 cm^{-2} and 5×10^{11} cm^{-2}. These values ensure that the system is initially in the lowest subband, so the initial conditions can be taken $S_1(z, t = 0) = S_2(z, t = 0) = 0$, $S_3(z, t = 0) = -1$.

The relevant parameters are calculated to be $E_1 - E_0 = 44.955$ meV and $x_{01} = -3.29$ nm. Also, for electron sheet density $N = 2 \times 10^{11}$ cm^{-2}, we obtain $\pi e^2 N(L_{1111} - L_{0000})/2\varepsilon = 0.412$ meV, $\hbar\gamma = 0.095$ meV and $\hbar\beta = -1.56$ meV. In all calculations we include the population decay and dephasing rates with values $T_1 = 100$ ps and $T_2 = 10$ ps. Also, in all calculations the angular frequency of the field is at exact resonance with the modified frequency ω_{10}, i.e. $\omega = \omega_{10}$.

We consider a structure with 40 quantum wells, each one equally separated by the other by 20 nm. We also take $N_v = N/L$, where $L = 11$ nm. In the propagation homogenization of the structure is considered, similar to what was done in previous studies [4, 5, 7]. The solution of the coupled Maxwell-nonlinear Bloch equations is performed computationally, without applying the SVEA and the RWA, using a combination of the finite-difference time-domain method for the Maxwell equations and a predictor-corrector scheme for the nonlinear Bloch equations, similar to the approach proposed initially by Ziolkowski *et al.* [12].

Typical results are presented in Fig. 2, where the pulse propagation and the population inversion for a pulse with area 2π at the entrance of the medium are shown for three different electron sheet densities. We find that for short propagation distances the pulse changes and leads to strong population inversion, that is in contrast to the behavior expected for an ultrashort 2π pulse [8] which, if no propagation is applied it leads to transient excitation and final de-excitation of the quantum well. Then, for longer propagation distances, near the middle of the medium, the pulse changes and the expected behavior of excitation and de-excitation is found. Finally, for even longer distances, near the end of the medium, the pulse is absorbed and leads to partial population inversion. It is interesting that similar behavior is found for both weak electron sheet densities and for large sheet densities. Qualitatively similar results have been found for other pulse durations, e.g. $t_p = 0.15$ ps and $t_p = 0.05$ ps. In addition, similar behavior is found for an initially 4π pulse.

4. Summary

In this work, we have studied coherent ultrashort pulse propagation in a two-subband system in a symmetric semiconductor quantum well structure. We have performed numerical calculations

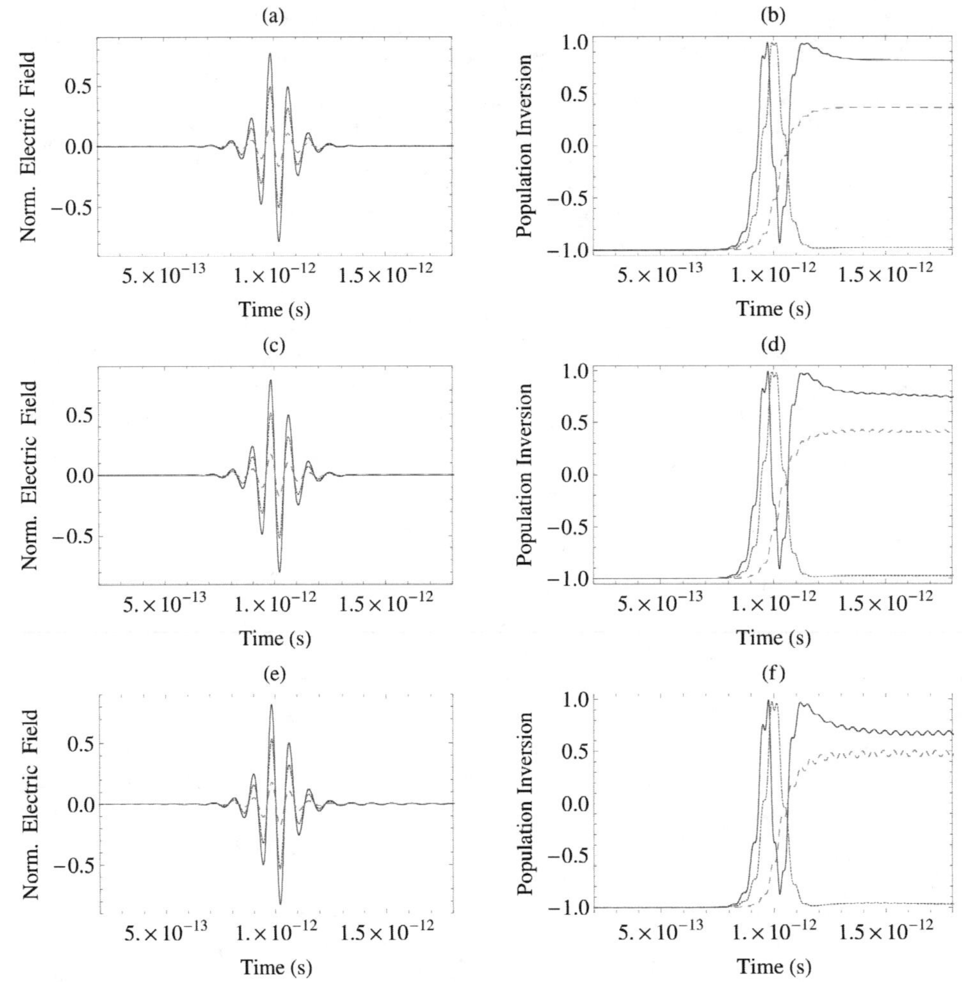

Figure 2. In (a), (c) and (e) we present the normalized electric field for an initially 2π pulse for $t_p = 0.1$ ps. In (b), (d) and (f) we present the respective population inversion. In (a), (b) $N = 10^9$ cm^{-2}, in (c), (d) $N = 2 \times 10^{11}$ cm^{-2} and in (e), (f) $N = 5 \times 10^{11}$ cm^{-2}. The solid curves are for $z = 20$ nm, the dotted curves are for $z = 320$ nm and the dashed curves are for $z = 640$ nm.

beyond the RWA and the SVEA, taking into account the effects of electron-electron interactions. The interaction of the quantum well structure with the electromagnetic fields was studied with nonlinear Bloch equations. These equations were combined with the full-wave Maxwell equations for the study of pulse propagation, and we presented results for the pulse propagation and the population inversion dynamics in the quantum well structure for different electron sheet densities.

Acknowledgments
This research has been co-financed by the European Union (European Social Fund - ESF) and Greek national funds through the Operational Program "Education and Lifelong Learning" of the National Strategic Reference Framework (NSRF) - Research Funding Program: Archimedes III. We thank Professors A. F. Terzis and M. M. Sigalas for useful discussions and help.

References

[1] Luo C W, Reimann K, Woerner M, Elsaesser T, Hey R and Ploog K H 2004 *Phys. Rev. Lett.* **92** 047402
[2] Batista A A and Citrin D S 2004 *Phys. Rev. Lett.* **92** 127404
[3] Paspalakis E, Tsaousidou M and Terzis A F 2006 *Phys. Rev. B* **73** 125344
[4] Cui N, Niu Y-P, Sun H and Gong S-Q 2008 *Phys. Rev. B* **78** 075323
[5] Xu X-S, Cui N, Zhu J and Gong S-Q 2008 *Chin. Opt. Lett.* **6** 689
[6] Choi H, Gkortsas V-M, Diehl L, Bour D, Corzine S, Zhu J, Höfler G, Capasso F, Kärtner F X and Norris T B 2010 *Nature Photonics* **4** 706
[7] Yao H-F, Niu Y-P, Peng Y and Gong S-Q 2011 *Opt. Commun.* **284** 4059
[8] Paspalakis E and Boviatsis J 2012 *Nanosc. Res. Lett.* **7** 478
[9] Dietze D, Darmo J and Unterrainer K 2013 *New J. Phys.* **15** 065014
[10] Olaya-Castro A, Korkusinski M, Hawrylak P and Ivanov M Yu 2003 *Phys. Rev. B* **68** 155305
[11] McCall S L and Hahn E L 1969 *Phys. Rev.* **183** 457
[12] Ziolkowski R W, Arnold J M and Gogny D M 1995 *Phys. Rev. A* **52** 3082

A Mathematical Model for the Exhaust Gas Temperature Profile of a Diesel Engine

C H G Brito[1], C B Maia[2] and J R Sodré[3]
Department of Mechanical Engineering, Pontifical Catholic University of Minas Gerais, Av. Dom José Gaspar, 500, Belo Horizonte, MG, 30531-910, Brazil

E-mail: cristianohbrito@globo.com[1]; cristiana@pucminas.br[2]; ricardo@pucminas.br[3]

Abstract. This work presents a heat transfer model for the exhaust gas of a diesel power generator to determine the gas temperature profile in the exhaust pipe. The numerical methodology to solve the mathematical model was developed using a finite difference method approach for energy equation resolution and determination of temperature profiles considering turbulent fluid flow and variable fluid properties. The simulation was carried out for engine operation under loads from 0 kW to 40 kW. The model was compared with results obtained using the multidimensional Ansys CFX software, which was applied to solve the governor equations of turbulent fluid flow. The results for the temperature profiles in the exhaust pipe show a good proximity between the mathematical model developed and the multidimensional software.

1. Introduction

There is a growing need to produce more efficient internal combustion engines, since the laws that regulate the level of pollutant emissions and fuel consumption by automobiles are increasingly demanding lower limits. Together with regulatory laws, government incentive laws also encourage the manufacturers to continuously develop new technologies that allow automobiles to reach and even exceed the consumption limits and emissions. Heat transfer is an important requirement in the analysis, development and design of internal combustion engine, being directly related to engine efficiency and emissions levels [1]. Within the various lines of research and approaches for heat transfer analysis at internal combustion engines, a special focus is given to the exhaust system.

The equations governing the fluid flow and determining the temperature profiles, when analyzed in the plane or space, results in complex calculations and hard analytical solution. This problem can be mitigated with the adoption of numerical methods, such as finite difference method (FDM) and finite volume method (FVM). FDM is a numerical method generally used to study heat transfer and fluid flow dynamics [2]. In the analysis of exhaust systems, the FDM has been employed to verify the energy availability of exhaust gases [3-7]. The exhaust gas fluid flow and heat transfer was simulated via FDM to design optimization of the exhaust system [8].

The main objective of this work is to evaluate the consistency of heat transfer solution from exhaust pipe modeling on FDM from comparison with results obtained in computational fluid dynamics (CFD) modeling using Ansys CFX software. The FDM modeling was adopted for energy equation resolution and determination of the exhaust gas temperature profiles, considering turbulent fluid flow and variable properties. The numerical simulation was carried out for engine operation under loads from 0 kW to 40 kW.

2. Problem formulation

The exhaust system is responsible by about 30% of the energy loss from an internal combustion engine [9]. The energy carried by the exhaust gas can be evaluated from the analysis of the internal fluid flow. The exhaust pipe taken as the physical domain of this work is shown in Fig. 1. The heat transfer and fluid flow is analyzed taking the duct as a straight pipe, as shown by Fig. 2.

Figure 1. Exhaust pipe

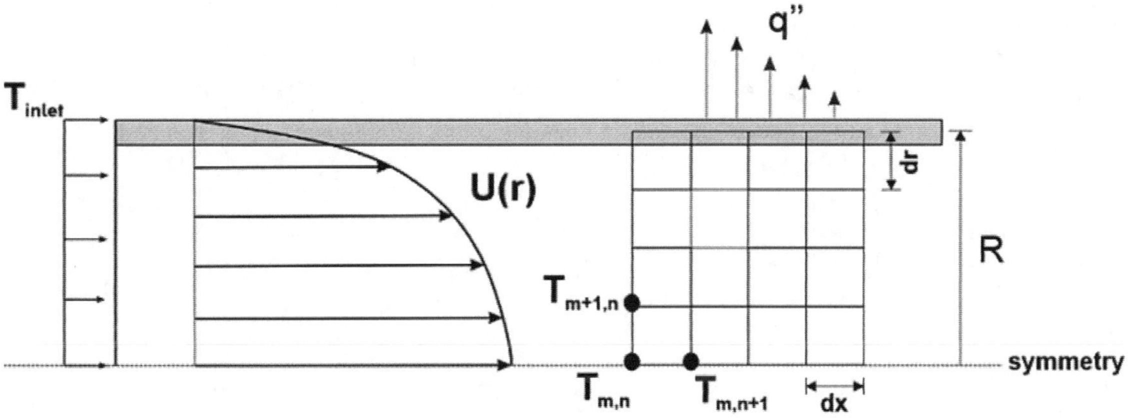

Figure 2. Nodal mesh

3. Mathematical model

In order to determine the temperature profiles it has been proposed a model based on FDM. The implemented model solve the energy equation from fundamental assumptions: (1) steady state; (2) fully developed turbulent flow; (3) turbulent velocity profile 1/7; (4) variable properties; (5) fluid modelled as air.

The results of this model are compared with results obtained from the Ansys CFX® software, in which the governing equations of turbulent flow are resolved. The governing equations are the continuity equation (Eq. 1), momentum equation (Eq. 2) and energy equation (Eq. 3) [10]:

$$\frac{\partial}{\partial t}\rho + \nabla \bullet \left(\rho \overline{\mathbf{U}}\right) = 0 \qquad (1)$$

$$\frac{D(\rho \overline{\mathbf{U}})}{Dt} = \rho \mathbf{g} - \nabla \overline{p} + \nabla \bullet \tau_{ij} \qquad (2)$$

$$\rho c_p \frac{D\overline{T}}{Dt} = -\frac{\partial}{\partial x_i} q_i + \overline{\Phi} \qquad (3)$$

The K-ω SST model [11] is used to treat the turbulence regime in the present study in the simulation with Ansys CFX®.

4. Numerical methodology

The use of numerical techniques to solve engineering problems arises from the difficulty to achieve analytical solutions to highly complex problems, which involve partial differential equations and implied equations, among others. The currently available computational speed enables the effectiveness of the solutions for such problems [2]. In this research two numerical methods were used to solve the governing equations of the exhaust gas flow. The FDM was used to solve the energy equation and FVM was used to solve the seven governing equations of the exhaust gas flow.

4.1. FDM Model

The discretization of the energy equation, based on FDM developed using MATLAB® software (Fig. 3), was done by the bidimensional nodal mesh shown by Fig. 2. The boundary conditions are [12]:

$$\frac{d\theta}{dr} = \begin{cases} 0 & \rightarrow r = 0 \\ -\mathrm{Bi}\cdot\theta & \rightarrow r = R \end{cases} \quad \text{where} \quad \theta = \frac{T_\infty - T(r,x)}{T_\infty - T_m(x)} \tag{4}$$

For turbulent, fully developed flow the 1/7 velocity profile is able to satisfactorily approximate the behaviour of the real flow velocity profiles [10].

To determine the equations of the fluid intensive properties as a function of the exhaust gas temperature, modelled as air, it was used the Engineering Equation Solver (EES®) software.

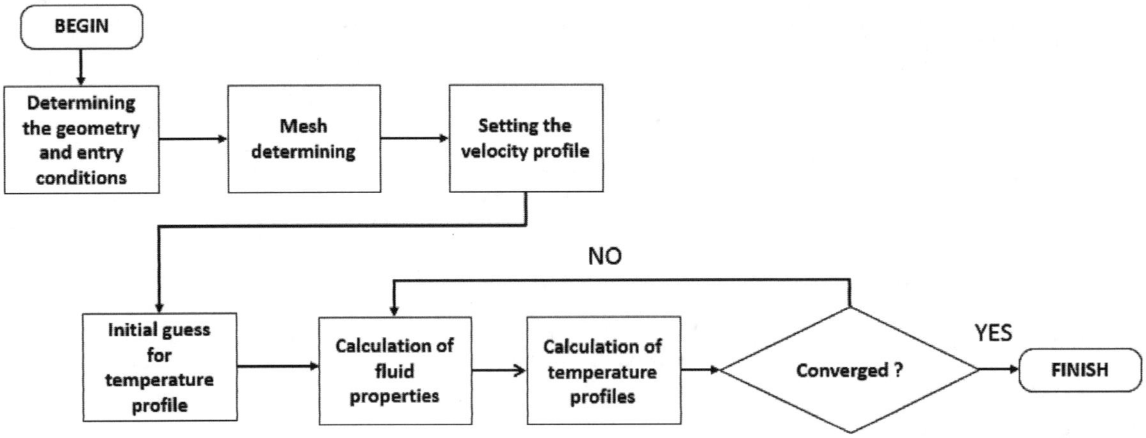

Figure 3. Algorithm flowchart of FDM implemented using MATLAB® software.

4.2. FVM Model

The FVM model was solved using Ansys CFX® 14.5 software, adopting a straight pipe geometry and considering symmetry and 3-dimensional fluid dynamics and heat transfer. This model was solved to compare with the FDM model. The fundamental hypotheses adopted are: (1) air at 25°C, (2) turbulent flow, (3) steady state condition, and (4) incompressible flow. The entry conditions are: (1) pipe outlet with relative static pressure of 0 Pa, (2) non-sliding walls conditions, (3) known experimental mass flow rate and flow temperature, (4) prescribed heat transfer coefficient in the pipe wall, and (5) ambient temperature 27°C.

4.3. Entry conditions of numerical models

In the numerical simulations it was adopted input and boundary conditions obtained experimentally [13] and described in Table 1. These data were collected from a diesel power generator operating under loads of 0, 10, 20 and 40 kW.

Table 1. Data entry conditions used in the numerical simulations.

Nominal power	kW	0	10	20	40
Ambient temperature	°C	27	27	27	27
Exhaust gas inlet temperature	°C	149.4	239.37	332.12	598.27
Exhaust gas mass flow rate	kg/h	148.4	144.71	142.72	132.95

5. Results and discussion

In the algorithm implemented in Matlab software it was adopted square mesh to ensure compliance with the proposed mesh stability criterion. The convergence criterion adopted the value of 1% [2]. Mesh with 0.17 mm element size was adopted for the simulation. The FVM model in Ansys CFX® 14.5 software was simulated with 11.6 million elements. The numerical solutions proposed to the problem of exhaust gas flow are here evaluated in a comparative way. For such it is evaluated the temperature profiles obtained in the section 150 mm before the pipe outlet. The FDM model was named Matlab-FDM and FVM model was named Ansys-SP. The temperature profiles in Fig. 4 and Fig. 5 show that the Matlab-MDF model depicts a behaviour similar to Ansys-DR model and the temperature gradients, increasing as the engine load power is increased.

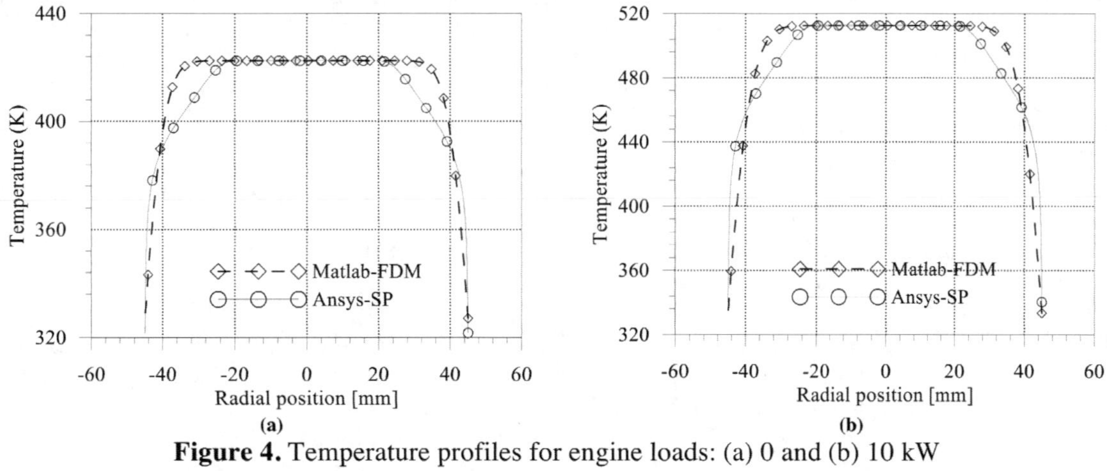

Figure 4. Temperature profiles for engine loads: (a) 0 and (b) 10 kW

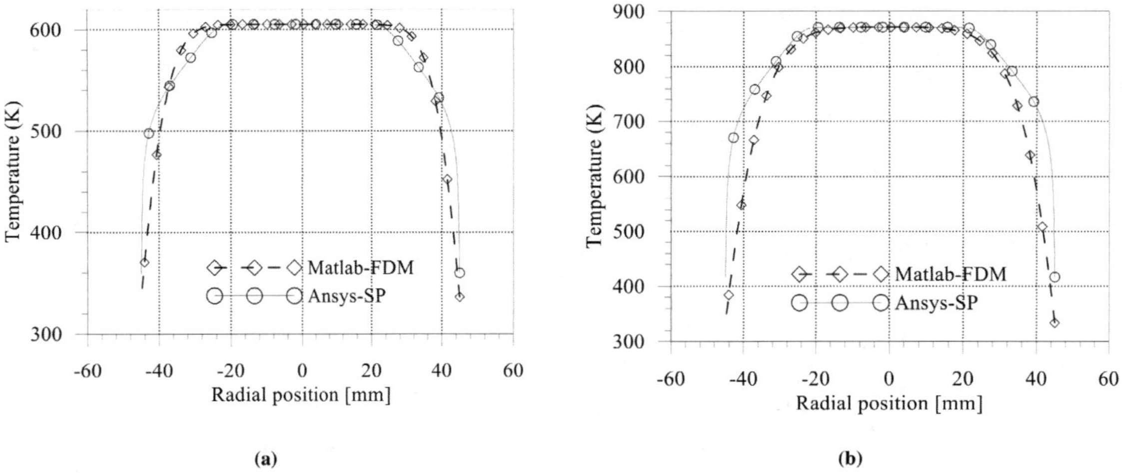

Figure 5. Temperature profiles for engine loads: (a) 20 and (b) 40 kW

6. Conclusions

With the numerical methodology used, it was possible to determine the temperature profiles in the exhaust pipe of a stationary engine operating under different loads. The results show a good correlation between the proposed Matlab-FDM model and the Ansys-SP model. Temperature profiles obtained in both cases show that, as the load applied to the engine is increased, the temperature gradient along the pipe length is also increased as a result of increased exhaust gas temperature and reduced mass flow rate.

Acknowledgments

The authors thank CAPES, CNPq and FAPEMIG for the financial support to this work.

References

[1] Drenth A C, Olsen D B, Cabot P E and Johnson J J 2014 Compression ignition engine performance and emission evaluation of industrial oilseed feedstock camelina, carinata and pennycress across three fuel pathways *Fuel* **136** 143-155

[2] Özisik N (1994) *Finite Difference Methods in Heat Transfer* (Florida : CRC Press)

[3] Rakopoulos C, Andritsakis E, Hountalas D 1995 The influence of exhaust system unsteady gas flow and insulation on the performance of turbocharged diesel *Heat Recovery Systems & CHP* **15** 51-72

[4] Baker C, Vuppuluri P, Shi L and Hall M 2012 Model of Heat Exchanger for Waste Heat Recovery from Diesel Engine Exhaust for Thermoelectric *Power Generation Journal of Electronic Materials* **41** 1290-1297

[5] Moazami N, Wyszynski M L and Mahmoudi H 2013 Modeling of catalytic monolith reactor for reforming of hexadecane with exhaust gas *International Journal of Hydrogen Energy* **38** 11826-11839

[6] Galindo J, Serrano J R, Piqueras P, García-Afonso O 2012 Heat transfer modelling in honeycomb wall-flow diesel particulate filters *Energy* **43** 201-213

[7] Brito C H G, Andrade G F and Sodré J R 2014 Análise do Perfil de Temperaturas no Duto de Exaustão de um Motor pelo Método das Diferenças Finitas *Proceeding Series of the Brazilian Society of Computational and Applied Mathematics* **2** 10511-10516

[8] Kesgin U 2005 Study on the design of inlet and exhaust system of a stationary internal combustion engine *Energy Conversion and Management* **46** 2258-2287

[9] Pulkrabek W W 1997 *Engineering Fundamentals of the Internal Combustion Engine* (New Jersey: Pratice Hall) pp 312

[10] White F (2007) *Fluid Mechanics* 6. Ed. (New York : McGraw-Hill)

[11] Versteeg H K and Malalasekera W (2007) *An Introduction to Computational Fluid Dynamics: The Finite Volume Method* 2. Ed. (London: Pearson Education)

[12] Bejan A (2012) *Convective Heat Transfer* 2. Ed. (New Jersey: John Wiley & Sons)

[13] Morais A et al 2013 Hydrogen impacts on performance and CO2 emissions from a diesel power generator *Internal Journal Hydrogen Energy* **38** 6857-6864

Journal of Physics: Conference Series **633** (2015) 012076 doi:10.1088/1742-6596/633/1/012076

New implementation of the first-order perturbation theory for calculation of interatomic vibrational amplitudes and corrections in gas electron diffraction

Yury V. Vishnevskiy[1] and Yuriy A. Zhabanov[2]

[1] Universität Bielefeld, Lehrstuhl für Anorganische Chemie und Strukturchemie, Universitätsstrasse 25, 33615, Bielefeld, Germany

[2] Ivanovo State University of Chemistry and Technology, Sheremetev av. 7, 153000 Ivanovo, Russian Federation

E-mail: yury.vishnevskiy@uni-bielefeld.de

Abstract. The problem of computation of vibrational molecular parameters in models of gas electron diffraction is discussed. Some peculiarities of the standard for these purposes Shrink program are analysed. An alternative code has been written to implement the well established Sipachev's first-order perturbation theory. It has been used for calculation of vibrational parameters of several molecules. Comparison of obtained results shows that the new program outperforms the available program Shrink in many aspects.

1. Introduction

A typical application of the gas electron diffraction (GED) method is to study geometrical structure and dynamics of free molecules.[1] Models of molecular part of electron diffraction in GED are usually expressed in terms of thermally averaged interatomic distances r_a or r_g, mean amplitudes of their variations l and asymmetry constants κ.[2, 3, 4] Successful interpretation of experimental electron diffraction patterns requires, excepting very simple cases, preliminary calculation of these geometrical and vibrational parameters. In structural analysis they are used for construction of initial model and as fixed constraints[5] or flexible restraints[6] to make the inverse problem stable. In addition, it is well known that the thermally average r_a and r_g distances are not geometrically consistent, which leads to a so-called "shrinkage effect".[7, 8] Therefore, in current practice GED structural analyses are internally done in terms of geometrically consistent r_{h0}, r_{h1} or r_e parameters.[9] This, in turn, requires corresponding corrections to interatomic distances $(r_{h0} - r_a)$, $(r_{h1} - r_a)$ or $(r_e - r_a)$, which are in most cases computed on the basis of theoretically optimized structures and calculated force constants. The usage of corrections to equilibrium structure $(r_e - r_a)$ is most preferable since they have clear physical meaning and provide common basis for combination of experimental data from different sources and for comparison of structures obtained by different methods.

For calculation of vibrational amplitudes l and corrections for interatomic distances most of the modern GED structural studies utilize the first-order perturbation theory of Sipachev implemented in program Shrink using classical formalism.[10] The current implementation of the corresponding algorithm has, however, several limitations restricting studies of large molecules by GED method.

In this contribution we are extending the current Sipachev's algorithm for the case of large molecules. This work has been done in scope of the UNEX project[11], a software platform for computational studies, simulations and inverse problems. Its second version UNEX2 provides a computer language and its implementation with standard library, which includes a toolkit for GED investigations. A new program VibModule has been written for calculations of vibrational amplitudes and corrections using standard Sipachev's theory. As a next step this standalone program will be converted to a corresponding module for the standard UNEX2 library.

2. Implementation of the first-order perturbation theory

As already stated before the vast majority of modern GED structural studies use the Sipachev's formulation of the first-order vibrational perturbation theory implemented in program Shrink [12, 13, 14] for calculations of vibrational amplitudes and distance corrections. Its algorithm uses a so-called "curvilinear approach" by solving the vibrational problem in redundant internal coordinates and requires at least Cartesian coordinates and a harmonic force field of a molecule as the input information. Additionally, a cubic force field is required in general case for calculation of anharmonic distance corrections to equilibrium geometry. Altogether $r_e - r_a$ corrections are calculated as:

$$r_e - r_a = \frac{l^2}{r_e} - \langle \Delta r \rangle_{har} - \langle \Delta r \rangle_{loc} - \langle \Delta r \rangle_{rot} - \langle \Delta r \rangle_{anh}, \quad (1)$$

where $\langle \Delta r \rangle_{har}$ is the shrinkage correction calculated in the curvilinear harmonic approximation [12], $\langle \Delta r \rangle_{loc}$ is the local centrifugal distortion caused by intramolecular vibrational motions [14], $\langle \Delta r \rangle_{rot}$ is the centrifugal distortion caused by rotational motion of a molecule [15] and $\langle \Delta r \rangle_{anh}$ is the anharmonic part of the correction.[13] A comparison of such contributions for different interatomic distances of ethane molecule is given in Table 1. Amplitudes of interatomic vibrations l are calculated in the similar manner [12] and are also collected in Table 1. The theory performs well for relatively stiff molecules without large amplitude motions. Calculations for molecules with lowest frequency below 85 cm^{-1} can be unreliable.[10] Current implementation of the program Shrink has limitations, which complicate GED studies of large molecules. Firstly, numbers of internal coordinates, of atoms and their pairs are limited to 600 (empirical observation), and secondly, this program is too slow. It is possible that these disadvantages can be fixed in future. However, to our knowledge this code is not supported anymore and in this work it has been decided to write a new program, which implements the aforementioned theory. The new program VibModule is coded in C computer language and uses the UNEX mathematical library (UML) from the UNEX2 project.[11] It utilizes algorithms similar to those described earlier and used in program Shrink.[12, 13, 14, 15] VibModule has a number of advantages:

- virtually no limits on numbers of molecular parameters mentioned earlier;
- higher computational performance;
- automatic generation of internal coordinates;
- variety of input possibilities: Shrink input files, output files of Gaussian [16] and Firefly [17] quantum chemical packages;
- output data are ready to use in UNEX [18] and KCED [19] programs for GED structural analysis.

In order to test the accuracy of results obtained with VibModule several molecules have been processed and corresponding parameters have been compared with values calculated by Shrink. Results for two of these objects, relatively small ethane (C_2H_6) [20] and large [30]trithia-2,3,5,10,12,13,15,20,22,23,25,30-dodecaazahexaphyrin ($C_{30}N_{15}H_{15}S_3$) [21] molecules, are summarized in Table 2. The small values of the root-mean-square deviations (RMSD) demonstrate the consistency of the results of both programs.

An another test has been done in order to measure the improvement in computational speed of VibModule. Magnesium octa(m-trifluoromethylphenyl)porphyrazine ($MgC_{72}H_{32}N_8F_{24}$) has been chosen as a test object, which is probably the largest molecule ever studied by GED method.[22] Calculations have been performed on a computer running under 64-bit Windows 7 operating system. The running times for VibModule and Shrink were 46.5 and 316.1 minutes, respectively. It should be noted that, in contrast to VibModule, Shrink was not able to process the full set of interatomic pairs for the test molecule due to the limitations mentioned earlier. A calculation of all required parameters would require 23 runs. Thus, the real effective timings were 46.5 and 7270.3 minutes for VibModule and Shrink, respectively.

3. Conclusions

A new program has been developed for calculations of interatomic vibrational amplitudes and corrections using the first-order perturbation theory. This code together with the well-known Shrink program has been used for processing of data for several test molecules of different sizes. Comparisons of the obtained parameter values have shown that both programs produce consistent results. The new code has, however, significant technical advantages including much higher computational speed and possibility of processing data for large molecules.

4. Acknowledgements

This work has been supported by Deutsche Forschungsgemeinschaft (DFG, project VI 713/1-1). Dr. Yu. A. Zhabanov thanks the Ministry of Education and Science of the Russian Federation (Projects 11.9166.2014 and N4.1385.2014/K) and German academic exchange service DAAD (project A/13/75239) for financial support.

Table 1. Amplitudes of vibrations and corrections for different pairs of atoms in C_2H_6 and $C_{30}N_{15}H_{15}S_3$ molecules calculated using VibModule program, [Å].

term	r_e	l	$\langle\Delta r\rangle_{har}$	$\langle\Delta r\rangle_{loc}$	$\langle\Delta r\rangle_{rot}$	$\langle\Delta r\rangle_{anh}$	$r_e - r_a$
C_2H_6							
C—H	1.095	0.077	0.0000	0.0037	0.0002	0.0176	-0.0161
C—C	1.530	0.051	0.0000	0.0000	0.0010	0.0115	-0.0107
H...H	1.766	0.123	-0.0063	0.0053	0.0003	0.0271	-0.0178
C...H	2.182	0.107	-0.0062	0.0032	0.0009	0.0250	-0.0176
H...H	2.542	0.182	-0.0055	0.0055	0.0008	0.0321	-0.0199
H...H	3.095	0.125	-0.0259	0.0079	0.0009	0.0417	-0.0195
$C_{30}N_{15}H_{15}S_3$							
C—H	1.083	0.070	0.0000	0.0038	0.0000	0.0125	-0.0118
N...N	2.462	0.064	-0.0069	0.0050	0.0001	0.0142	-0.0107
C...N	13.165	0.246	-0.2352	0.0173	0.0008	0.0822	0.1395
C...S	13.648	0.254	-0.2658	0.0137	0.0010	0.0937	0.1622
H...H	17.553	0.260	-0.3469	0.0194	0.0009	0.1268	0.2036

Table 2. Root-mean-square deviations between results obtained with VibModule and Shrink programs for C_2H_6 and $C_{30}N_{15}H_{15}S_3$ molecules, [Å].

Molecule	RMSD(l)	RMSD($r_e - r_a$)
C_2H_6	$4.08 \cdot 10^{-5}$	$4.08 \cdot 10^{-5}$
$C_{30}N_{15}H_{15}S_3$	$1.06 \cdot 10^{-3}$	$4.69 \cdot 10^{-4}$

References

[1] Hargittai I 1988 *Stereochemical Applications of Gas Phase Electron Diffraction, Part A: The Electron Diffraction Technique* ed Hargittai I and Hargittai M (VCH Publishers, Inc., New York)
[2] Wann D A, Less R J, Rataboul F, McCaffrey P D, Reilly A M, Robertson H E, Lickiss P D and Rankin D W H 2008 *Organomet.* **27** 4183–4187
[3] Wann D A, Zakharov A V, Reilly A M, McCaffrey P D and Rankin D W H 2009 *J. Phys. Chem. A* **113** 9511–9520
[4] Kochikov I V, Tarasov Y I, Kuramshina G M, Spiridonov V P, Yagola A G and Strand T G 1998 *J. Mol. Struct.* **445** 243 – 258
[5] Chiu N S, Ewbank J D, Askari M and Schäfer L 1979 *J. Mol. Struct.* **54** 185 – 195
[6] Blake A J, Brain P T, McNab H, Miller J, Morrison C A, Parsons S, Rankin D W H, Robertson H E and Smart B A 1996 *J. Phys. Chem.* **100** 12280–12287
[7] Almenningen A, Arnesen S P, Bastiansen O, Seip H M and Seip R 1968 *Chem. Phys. Lett.* **1** 569 – 574
[8] Hinchley S L, Wann D A and Rankin D W H 2005 *Int. J. Quant. Chem.* **101** 878–884
[9] Rankin D W H, Mitzel N W and Morrison C A 2013 *Structural methods in molecular inorganic chemistry* Inorganic chemistry : a Wiley textbook series (John Wiley & Sons, Chichester)
[10] Sipachev V A 2011 *Adv. Phys. Chem.* **2011** 1–14
[11] Vishnevskiy Y V 2015 Unex version 2, www http://unexprog.org
[12] Sipachev V A 1985 *J. Mol. Struct: THEOCHEM* **121** 143 – 151
[13] Sipachev V A 2000 *Struct. Chem.* **11**(2) 167–172
[14] Sipachev V A 2001 *J. Mol. Struct.* **567-568** 67 – 72
[15] Iwasaki M and Hedberg K 1962 *J. Chem. Phys.* **36** 2961–2963
[16] Frisch M J, Trucks G W, Schlegel H B, Scuseria G E, Robb M A, Cheeseman J R, Scalmani G, Barone V, Mennucci B, Petersson G A, Nakatsuji H, Caricato M, Li X, Hratchian H P, Izmaylov A F, Bloino J, Zheng G, Sonnenberg J L, Hada M, Ehara M, Toyota K, Fukuda R, Hasegawa J, Ishida M, Nakajima T, Honda Y, Kitao O, Nakai H, Vreven T, Montgomery J A J, Peralta J E, Ogliaro F, Bearpark M, Heyd J J, Brothers E, Kudin K N, Staroverov V N, Kobayashi R, Normand J, Raghavachari K, Rendell A, Burant J C, Iyengar S S, Tomasi J, Cossi M, Rega N, Millam J M, Klene M, Knox J E, Cross J B, Bakken V, Adamo C, Jaramillo J, Gomperts R, Stratmann R E, Yazyev O, Austin A J, Cammi R, Pomelli C, Ochterski J W, Martin R L, Morokuma K, Zakrzewski V G, Voth G A, Salvador P, Dannenberg J J, Dapprich S, Daniels A D, Farkas O, Foresman J B, Ortiz J V, Cioslowski J and Fox D J Gaussian 09, Revision D.01 Gaussian Inc. Wallingford CT 2009
[17] Granovsky A A Firefly version 8, www http://classic.chem.msu.su/gran/firefly/index.html
[18] Vishnevskiy Y V 2007 Unex version 1.5, www http://unexprog.org
[19] Anderson B, Seip H, Strand T and Stolevik R 1969 *Acta Chem. Scand.* **23** 3224–3234
[20] Bartell L S and Higginbotham H K 1965 *J. Chem. Phys.* **42** 851–856
[21] Zhabanov Y A, Zakharov A V, Shlykov S A, Trukhina O N, Danilova E A, Koifman O I and Islyaikin M K 2013 *J. Porphyr. Phthalocya.* **17** 220–228
[22] Zhabanov Y A, Zakharov A V, Giricheva N I, Shlykov S A, Koifman O I and Girichev G V 2015 *J. Mol. Struct.* **1092** 104–112

Journal of Physics: Conference Series **633** (2015) 012077 doi:10.1088/1742-6596/633/1/012077

Classification method for heterogeneity in monoclonal cell population

S Aburatani[1], K Tashiro[2] and S Kuhara[2]

[1] Senior Research Scientist, BRIDD, National Institute of AIST, AIST Tokyo Waterfront Bio-IT Research Building, 2-4-7 Aomi, Koto-ku, Tokyo, 135-0064, Japan
[2] Grad. Sch. of Biores. Bioenv. Sci, Department of Systems Life Sciences, Kyushu Univ., 6-10-1, Hakozaki, Higashi-ku, Fukuoka-city, Fukuoka, 812-8581, Japan

E-mail: s.aburatani@aist.go.jp

Abstract. Monoclonal cell populations are known to be composed of heterogeneous sub-populations, thus complicating the data analysis. To gain clear insights into the mechanisms of cellular systems, biological data from a homogeneous cell population should be obtained. In this study, we developed a method based on Latent Profile Analysis (LPA) combined with Confirmatory Factor Analysis (CFA) to divide mixed data into classes, depending on their heterogeneity. In general cluster analysis, the number of measured points is a constraint, and thereby the data must be classified into fewer groups than the number of samples. By our newly developed method, the measured data can be divided into groups depending on their latent effects, without constraints. Our method is useful to clarify all types of omics data, including transcriptome, proteome and metabolic information.

1. Introduction

According to progress in science and technology, we can obtain information about the intracellular behaviors of cell components as omics data, such as expression profiles, proteome data and metabolic data [1,2,3]. To analyze these data, it was assumed that the cell populations were basically homogeneous. However, recent studies have shown that cell populations, even monoclonal cell populations, are heterogeneous [4,5]. Furthermore, the differences in cell kinetics dependent on the heterogeneity affect organogenesis and cell differentiation controls in pluripotent stem cells [6], but the mechanisms have remained unclear. To clarify the mechanisms in living cells, homogeneous data should be analyzed.

In order to obtain homogeneous data from mixed heterogeneous data, we developed a method for dividing heterogeneous data into subclasses, depending on their heterogeneity. In our previous investigations, we utilized an improved Structural Equation Modelling (SEM) approach to infer the latent effect for gene regulatory networks [7], and latent variable models were employed to infer the unobserved effects for perturbations of cellular components and metabolism.

[1] Sachiyo Aburatani, s.aburatani@aist.go.jp.

In this study, we combined Latent Profile Analysis (LPA) with Confirmatory Factor Analysis (CFA) to estimate the effects of unknown sub-populations. According to the estimated effects, the measured data can be divided into subclasses. We then reapplied our developed method to each data subclass. Although the clustering method is restricted by the number of samples in general, our method allows classification into more subclasses than the measured data sample number.

2. Methodology

2.1. Latent profile analysis

In the latent mixture model, a population is composed of unobserved subclasses that depend on an independent distribution. The measured data in a monoclonal cell population are considered to be composed of different data populations. To detect the heterogeneity within the data, we applied Latent Profile Analysis (LPA). Latent profile analysis is a cluster analysis technique based on latent variable models, and is applied when the latent variables are categorical data and the measured data are numerical data [8,9]. The measured variables in LPA are assumed to be normal, and their conditional distribution given the latent variables is assumed to be normal [9]. The advantages of latent profile analysis over cluster analysis are that they are model-based, and thus generate probabilities for group membership.

In this analysis, the LPA model for a single item can be written as follows:

$$x_{vi} = \sum_{g=1}^{G} \pi_g f_{ig} \ . \tag{1}$$

Here, x_{vi} denotes the standardized data for a randomly selected individual sample v obtained on item i (i=1, …, I), and π_g is the class size parameter, which indicates the unconditional probability of belonging to latent class g (g=1, …, G). In the LPA model, each individual v belongs to one latent class. Thus, the sum of all class size parameters equals 1:

$$\sum_{g=1}^{G} \pi_g = 1 \, . \tag{2}$$

The parameter f_{ig} is the effect score on item i given membership in class g, and f_{ig} is calculated for each x_{vi}. This x_{vi} is often referred to as a conditional response, and it depends on both the latent class prevalence π_g and the class-specific effect.

2.2. Confirmatory factor analysis

To assess the effect of each latent variable, we applied Confirmatory Factor Analysis (CFA). Factor analysis is a statistical method for describing the variability among observed variables by assumed latent variables. The number of latent variables is usually smaller than the number of observed samples [10]. In the CFA model, it is assumed that each observed sample may be related to all latent variables. Let us assume that there are p latent variables $F_1, F_2, \dots, F_p$ and q observed samples $x_1, x_2, \dots, x_q$. Since all of the observed variables were standardized, the mean of each observed variable is 0, and can expressed as linear combinations of latent variables, as follows:

$$x_v = \alpha_{v1} F_1 + \alpha_{v2} F_2 + \cdots \alpha_{vp} F_p + \varepsilon_v \tag{3}$$

where x_v is the I-dimensional vector of the sample v, α_{vk} is the partial regression weight of the latent variable F_k, and ε_v is an independently distributed error term with zero mean and finite variance. In matrix form, equation (3) is expressed as

$$X = \Lambda F + E \, . \tag{2}$$

In this case, the item number is I, and then X and E are the (q×I) matrix composed of the observed data and error terms, respectively. The partial regression coefficients of each latent variable are indicated as elements of Λ, the (q×p) latent interaction matrix. In the matrix Λ, each column corresponds to a factor and each row corresponds to an observed sample, and thus each element of Λ

indicates the strength of the regulation from each latent variable to each sample. The matrix F is the latent variable vector.

The variance-covariance matrix between the observed variables is structurized by parameters, as follows:

$$Var[X] = E[(X - U)(X - U)^t] = \Sigma = \Lambda\Phi\Lambda^t + \Psi^2. \tag{3}$$

Here, Ψ^2 is the covariance matrix of error terms, Λ is the factor loading matrix of latent variables, and Φ is the covariance matrix among factors. From this structurized matrix, the values of the partial regression weight matrix Λ and the variances of the "errors" are estimated.

2.3. Estimation of subclass number

To estimate the optimal subclass number for dividing the data, the LPA and CFA results were compared. In LPA, the number of divided classes should be set as the class-parameter, and the prevalence of each class is calculated depending on the set parameter. Thus, we applied LPA with all possible numbers as the class-parameter.

In CFA, the number of latent variables should be assumed in the models, and all possible models were tested in this study. Basically, the number of latent variables is smaller than the number of samples, and thus the minimum number of latent variables is two and the maximum number is (I-1). By CFA, the partial regression weights of the latent variables are calculated for each sample. Although the weights of all assumed latent variables are estimated, an observed sample can be divided into some groups depending on the largest absolute values of the partial regression weight.

The ratio of samples that are classified into groups in CFA is considered to have the same meaning as the class prevalence in LPA. Thus, a comparison of the class prevalence in LPA and the classification ratio in CFA is useful to estimate a suitable number of subclasses.

2.4. Calculation of estimated subclass data

In LPA, the individual data x_{vi} is assumed to depend on both the latent class prevalence π_g and the effect score f_{ig}, which is the product of the latent class-specific effect $f_{.g}$ and the item-specific effect $f_{i.}$. By CFA, the latent class g-specific effect of the sample v is estimated as the partial regression weight α_{vg}. According to the class prevalence π_g and the partial regression weight α_{vg}, equation (1) is expressed as

$$x_{vi} = \kappa_{vi} \sum_{g=1}^{G} \pi_g \alpha_{vg} \tag{4}$$

where κ_{vi} is the individual weight for each x_{vi}. The data of each subclass can be calculated as follows:

$$X_{vi}^{g} = \kappa_{vi} \pi_g \alpha_{vg} . \tag{5}$$

Here, X^g_{vi} is the element of subclass data matrix X^g with the class prevalence π_g. By LPA and CFA, the number of subclasses is estimated as G, and thus the data will be divided into G subclasses. In this model, the sum of X^g_{vi} is equal to the former data, which were equal to x_{vi}.

2.5. Iteration algorithm

Since the classified subclasses still have heterogeneity, an iteration algorithm to classify the subclass data into smaller subclasses was developed, as follows:

Step 1: Application of LPA with all possible numbers of classes

In LPA, the range of the class-parameter number is from two to the number of samples. Thus, we applied LPA with all possible numbers as class-parameters. The Structural Equation Modelling (SEM) software Mplus is utilized to apply LPA. The quality of classification by LPA is estimated as entropy static in Mplus, and thus we evaluated the set class-parameter numbers by the entropy static values. Values of entropy static close to 1 indicate high classification accuracy, whereas values close to 0 indicate low classification certainty. In this study, we only used the entropy static value equal to 1.

Step 2: Application of CFA with all possible numbers of latent variables

In CFA, the number of latent variables should be smaller than the number of samples. We employed the CFA models with the combinations of all possible numbers of latent variables, all possible factoring methods, and all possible rotation methods.

Step 3: Comparison of the results of LPA and the classified variables in CFA

By comparing the absolute values of the factor loadings in CFA, each sample can be classified into one specific latent group. The ratio of classified samples is compared with the class prevalence in LPA for the same number of subclasses. When the classified ratio in CFA and the class prevalence in LPA are similar, the subclass number is considered to be suitable. If the results of CFA and LPA are not similar in any cases, then we decide that the data could not be separated.

Step 4: Calculation of subclass data

According to section 2.4, each data subclass is calculated as a $v \times I$ matrix. In this step, the individual weight κ_{vi} of the observed data is calculated from the individual data x_{vi}, the class prevalence π_g, and the partial regression α_{vg}. After the individual weight κ_{vi} is calculated, the individual data x_v is divided into G different numerical data by multiplication for partial regression weight and class prevalence.

Step 5: Return to step 1 for classification of subclass data

From Step 1 to Step 4, the original data can be divided into some smaller subclasses, but not sufficiently. To classify the data depending on their homogeneity, these classification steps are iterated for each subclass data matrix, which is calculated in Step 4. The stopping criterion in this iteration algorithm is that the LPA results could not fit the CFA results.

3. Results & Discussion

We applied our developed method to simulated data and some omics data compiled from the GEO database (http://www.ncbi.nlm.nih.gov/geo/). To evaluate the classification ability of our developed method, we created 4 sample datasets composed of 5 different Gaussian distributions. In this simulation, the individual effect κ_{vi} was not set for each data point, but the group effect κ_v was set for each sample. The 4 created samples were divided into 6 different subclasses, which is a larger number than the set of different types, in advance.

The omics data multiply measured under the same conditions were compiled for our evaluation. In the real data, the samples were classified into more subclasses than the measured points. Interestingly, the LPA results tended to be similar to the CFA results with no rotation. For both data types, our method facilitated division into groups depending on the latent effects, without a sample number constraint. Our method is useful to clarify all types of omics data, including transcriptome, proteome and metabolic information.

4. References

[1] Chu Y and Corey D R 2012 *Nucleic Acid Ther.* **22(4)** 271
[2] Mirza S P and Olivier M 2008 *Physiol. Genomics* **33(1)** 3
[3] Griffin J L and Vidal-Puig A 2008 *Physiol. Genomics* **34(1)** 1–5
[4] Altschuler S J and Wu L F 2010 *Cell* **141(4)** 559
[5] Newman J R, Ghaemmaghami S, Ihmels J, Breslow D K, Noble M, DeRisi J L and Weissman J S 2006 *Nature* **441(7095)** 840
[6] Torres-Padilla M E and Chambers I 2014 *Development* **141(11)** 2173
[7] Aburatani S 2012 Bioinformation 8(14) 652
[8] Bartholomew D J, Steel F, Moustaki I and Galbraith J I 2002 *The Analysis and Interpretation of Multivariate Data for Social Scientists* 2nd ed. (London: Chapman and Hall/CRC Press)
[9] Geiser C 2010 *DATA ANALYSIS WITH MPLUS* (New York: The Guilford Press)
[10] Spirtes P, Glymour C and Scheines R 2001 *Causation, Prediction, and Search* 2nd ed. Cambridge: The MIT Press

Simulation of frequency dependence of dielectric susceptibility for description of director reorientation dynamics of nematic LC

Vakulin D.A., Ivanov A.V.

ITMO University, Kronverkskii pr. 49A, 197101, Saint Petersburg, Russia

vakulin.dmitry@gmail.com

Abstract. Simulation of the liquid crystal behavior in an external electric field allows to determine the characteristics of the LC device for specific control signal. The reorientation of director tilt angles is described by the Frank-Oseen elasticity theory, by the Leslie-Ericksen hydrodynamics theory and also by Maxwell system of electromagnetic equations. Due to the fact that it is difficult to obtain analytical solution of dependencies polar and azimuthal angles on time it is necessary to use numerical methods. The numerical solution allows to determine the characteristics of the LC device for specific control signal, but it does not allow to solve the inverse problem. It is determining the necessary parameters of the control signal. To solve this problem, we obtained an approximate analytical solution of the dynamics equations of the director reorientation under the electric field for arbitrary values of the elastic constants and boundary conditions. The results of numerical simulation of the director reorientation under the electric field in hybrid-oriented structure of a dual-frequency nematic liquid crystal were obtained. It is shown that results of analytical solution are in good agreement with the results of computer simulation and experiment in case of small deformations of the LC layer and used approximations.

1. Introduction

Widespread use of nematic liquid crystals (NLC) in the optical components of high-speed telecommunication systems due to the presence of specific properties of these materials: stability, fast response, low loss operation over a wide temperature range management via low voltage, low power consumption. With LC technology to produce fiber-optic systems, optical switches, modulators, filters, attenuators, equalizers, polarization controllers and other devices.

One of the important characteristics of the electrically controlled optical elements based on the LC that affect the performance of telecommunication networks is the time to switch between operating states. To speed up the switch to LC devices use hybrid-oriented (HAN - hybrid aligned nematic) cell with a dual frequency (DF) NLC [1, 2]. Have DF addressing HAN devices allows you to control the process of switching from the «OFF» → «ON» → «OFF» by changing the frequency applied to the cell of an external electric signal.

Changes angles director of NLC by an electric field described by the theory of elasticity Franca Oseen [3,4], the theory of hydrodynamic Leslie-Ericksen for nematic phase [5], as well as the system of Maxwell's electromagnetic equations. Given the complexity of the resulting system of equations to determine the spatial and temporal dependence of the polar and azimuthal angles of the NLC director usually use numerical methods [6-8]. The numerical calculation allows you to get listed according to the given parameters of the control signal: field strength, frequency, pulse duration and shape. However, the simulation LC devices often need to solve the inverse problem - to define the parameters of the control signal to produce a predetermined angle distribution director. To achieve this goal it is necessary to find an analytical solution of the equations describing the dynamics of the director reorientation that can only be done using some approximations.

The purpose of this paper is to obtain approximate analytical solutions of equations describing the dynamics of the reorientation of the molecules in the NLC DF HAN structure under the influence of an electric signal of arbitrary shape for modeling the optical response.

2. Simulation optical response

2.1. Equations of dynamics

A characteristic feature is the presence of DF NLC frequency dispersion of the dielectric constant. The values of the components of the dielectric constant of the NLC electronic transitions are determined by the external field and the corresponding polarizabilities. LC full polarization of the medium can be represented as the sum of the contributions of the electronic polarization of individual atoms and molecules and the orientation of the polarization of permanent molecular dipoles $\mathbf{P} = \mathbf{P}^{e} + \mathbf{P}^{o}$. The anisotropy of the NLC molecules is determined by various polarizability with respect to the molecular axis (the director), hence the need to introduce as the sum of the polarization parallel and perpendicular to the axis of the component $\mathbf{P} = \mathbf{P}_{||} + \mathbf{P}_{\perp}$.

As in the present study carried out modeling of the optical response of the NLC to the millisecond time range, used for the theoretical description of the Debye model of dielectric relaxation. In this model it is assumed that the forced response of the orientation of the polarization component parallel to the director obeys Langevin type [7]:

$$\tau \frac{dP_{||}^{o}}{dt} + P_{||}^{o} = \alpha E_{||}, \tag{1}$$

We write the equations for calculating the dynamics of change in the angle of inclination of the director in the structure of the HAN in the approximation when the moment of inertia of the molecules and streaming effects can be neglected. Then we obtain the following system of equations:

$$\gamma_1 \frac{\partial \theta_{z,t}}{\partial t} = \left(K_{33} - K_{11}\right) \sin\theta_{z,t} \cos\theta_{z,t} \left(\frac{\partial \theta_{z,t}}{\partial z}\right)^2 + f_{z,t} \frac{\partial^2 \theta_{z,t}}{\partial z^2} + P_{z,t}^{-} E_{z,t},$$
$$\tau \frac{dP_{||}^{o}}{dt} + P_{||}^{o} = \chi_{||l}^{o} \varepsilon_0 E_{z,t} \sin\theta_{z,t}, \quad f_{z,t} = K_{11} \cos^2\theta_{z,t} + K_{33} \sin^2\theta_{z,t}, \tag{2}$$

$$P_{z,t}^{\pm} = \left(P_{||}^{e} + P_{||}^{o}\right) \cos\theta_{z,t} \pm \left(P_{\perp}^{e} + P_{\perp}^{o}\right) \sin\theta_{z,t},$$
$$P_{\perp}^{e,o} = \chi_{\perp}^{e,o} \varepsilon_0 E_{z,t} \cos\theta_{z,t}, \quad P_{||}^{e} = \chi_{||}^{e} \varepsilon_0 E_{z,t} \sin\theta_{z,t}, \tag{3}$$

$$E_{z,t} = \frac{V_t}{d_0} - \frac{P_{z,t}^{+}}{\varepsilon_0} + \frac{1}{d_0 \varepsilon_0} \int_0^d P_{z,t}^{+} dz, \tag{4}$$

where γ_1 - rotational viscosity LC, ε_0 - dielectric constant, K_{11} and K_{33} - elastic coefficients Frank, $\chi_i^{e(o)}$ - electronic (orientation) dielectric susceptibility for the i-component of polarization, $\chi_{||l}^{o}$ - orientation dielectric susceptibility in the low-frequency limit, V_t - potential difference between two electrodes, d_0 - the thickness of the LC layer. The expression for the electric field in the z direction obtained by the Maxwell equations. To describe the dynamics of the director reorientation under the influence of the electric field required to find a self-consistent solution of the above recorded for initial angles director. The system of equations (2) - (4) and a numerical method of its calculation model polar structure of DF bistable nematic liquid crystal has been proposed in [7], and the computer simulation of the dynamics of the director reorientation for a more general case of a twist structure is made in [8].

2.2. Approximate analytic solution

As previously noted, the analytical solutions for the resulting system of equations is required to use additional approximation [8, 9]. So in the case of large amplitude of the control voltage can be assumed that the forces caused by an external electric field is much larger than the effect of the elastic forces, and neglect terms depending on the elasticity coefficients Frank. We make one more assumption, often used when considering the impact of a control signal to the liquid crystal layer, we assume that the electric field in the LC layer is constant and equal to $E_t = V_t/d_0$ [10]. Then, equations (2) - (4), which will now determine the change in tilt of the director in time to a certain point z_0, takes the following form:

$$\gamma_1 \frac{\partial \theta_{z_0,t}}{\partial t} = \varepsilon_0 \Delta\chi E^2_{z_0,t} \sin\theta_{z_0,t} \cos\theta_{z_0,t} + P^{\mathrm{o}}_{\parallel} E_{z_0,t} \cos\theta_{z_0,t},$$

$$\tau \frac{dP^{\mathrm{o}}_{\parallel}}{dt} + P^{\mathrm{o}}_{\parallel} = \varepsilon_0 \chi^{\mathrm{o}}_{\parallel l} E_{z_0,t} \sin\theta_{z_0,t}, \quad \Delta\chi = \chi^{\mathrm{e}}_{\parallel} - \chi^{\mathrm{e}}_{\perp} - \chi^{\mathrm{o}}_{\perp}. \tag{5}$$

By setting the initial spatial distribution of the director tilt LC layer thickness and calculating the time dependence for each point of this distribution, you can get a full dependence $\theta_{z,t}$.

Consider the dynamics of the director reorientation in time under the influence of the periodic electrical signal of arbitrary shape, presenting it as a Fourier series:

$$V_t = \frac{a_0}{2} + \sum_{n=1}^{\infty} [a_n \cos(nft) + b_n \sin(nft)], \tag{6}$$

where f - frequency electric field, a_n and b_n - Fourier coefficients. A solution of the equations for the director parallel component orientation polarization $P_{\parallel}^{\mathrm{o}}$ will be sought in the form of a product of two functions - a slowly varying amplitude $\chi(t)$ and a periodic function with the period of the external field $g(t)$:

$$P^{\mathrm{o}}_{\parallel}(t) = \chi(t) g(t), \quad g(t) = \frac{\alpha_0}{2} + \sum_{n=1}^{\infty} [\alpha_n \cos(nft) + \beta_n \sin(nft)]. \tag{7}$$

Periodic function $g(t)$ in (7) is represented as a Fourier series with coefficients α_n and β_n. Using approximate equation for the derivative with respect to time $P'_t = \chi'_t g + \chi g'_t \approx \chi g'_t$ in the case of the slowly varying amplitude, the second equation (5) can be rewritten as follows:

$$\tau \frac{dg(t)}{dt} + g(t) = \frac{\varepsilon_0 \chi^{o}_{\parallel l}}{d_0} \frac{\sin\theta(t) V(t)}{\chi(t)}. \tag{8}$$

Further, in the equation (8) perform the replacement of periodic functions of their Fourier series (6) and (7) and express the expansion coefficients α_n and β_n of the coefficients a_n and b_n. Then the expression (7) for the orientation of the polarization $P_{\parallel}^{\mathrm{o}}$ takes the following form:

$$P^{\mathrm{o}}_{\parallel}(t) = \frac{\varepsilon_0 \chi^{o}_{\parallel l}}{d_0} \sin\theta(t) P(t), \quad P(t) \equiv \frac{a_0}{2} + \sum_{n=1}^{\infty} \left[\frac{a_n - nf\tau b_n}{1 + (nf\tau)^2} \cos(nft) + \frac{b_n + nf\tau a_n}{1 + (nf\tau)^2} \sin(nft) \right]. \tag{9}$$

Substituting (9) in the first equation (5) which will carry out the separation of variables in the differential equation and find a solution to the director tilt. Finally, we obtain the expression

$$\ln(tg\theta)\,|_{\theta_1}^{\theta}=\frac{\varepsilon_0}{\gamma_1 d_0^2}\left(\Delta\chi\int_0^t V^2(t)dt+\chi_{\|l}^{o}\int_0^t P(t)V(t)dt\right), \tag{10}$$

where θ_1 - the initial value of the angle of the director and the integrands Expressions containing Fourier series (6) and (9) that can be easily integrated (due to the bulkiness of these expressions, they write in this paper not shown). Equation (10) can be used to determine the time dependence of the tilt of the director, and for determining the parameters of the control signal under the influence of which the angle will move to a predetermined position θ_2. Of particular note is the fact that the frequency dependence of the anisotropy of the dielectric constant is clearly taken into account in the resulting formula (10).

In the case of a sinusoidal waveform $V(t) = V_0 \sin(f\,t)$ ($a_0 = 0$, $a_n = 0$, $b_1 = V_0$, $b_n = 0$) solution takes the form:

$$\ln\left(\frac{tg\theta_2}{tg\theta_1}\right)=\frac{\varepsilon_0 V_0^2}{2\gamma_1 d_0^2}\left[\left(\Delta\chi+\frac{\chi_{\|l}^{o}}{1+(f\tau)^2}\right)\left(t_i-\frac{\sin(2ft_i)}{2f}\right)+\frac{\chi_{\|l}^{o}}{1+(f\tau)^2}\frac{\tau\cos(2ft_i)}{2}\right], \tag{11}$$

where t_i - duration pulse V_0 - the amplitude of the electric field.

3. Conclusion

For HAN structure DF NLC an approximate analytical solution of equations describing the dynamics of the director reorientation under the influence of the periodic electrical signal of arbitrary shape. Calculation formulas to determine the size of the control signal to produce a symmetrical optical response of the structure HAN DF NLC. Comparison of experimental data with the results of numerical and analytical calculations showed that when modeling the optical response of the structure HAN DF NLC analytical solution is in good agreement with the experimental results in the case of small deformations of the LC layer.

References

[1] Jewell S.A., Taphouse T.S., Sambles J.R. // Appl. Phys. Lett. 2005. V. 87. P. 021106(3).
[2] Lu Y.-Q., Liang X., Wu Y.-H., Du F., Wua S.-T. // Appl. Phys. Lett. 2004. V. 85. No 16. P. 3354-3356.
[3] Oseen C.W. // Trans. Faraday Soc. 1933. V. 29. P. 883-899.
[4] Frank F.C. // Discuss. Faraday Soc. 1958. V. 25. P. 19–28.
[5] Leslie F.M. // Arch. Ration. Mech. Anal. 1968. V. 28. No 4. P. 265-283.
[6] Chen C.-J., Lien A., Nathan M.I. // J. Appl. Phys. 1997. V. 81. No 1. P. 70-73.
[7] Mottram N.J., Brown C.V. // Physical Review E. 2006. V. 74. No 3. P. 031703 (7).
[8] Ivanov A. V., Vakulin D. A., Konshina E. A. // J. Opt. Technol. 2014. V. 81. No 3. P. 23-29.
[9] Dayton D., Browne S., Gonglewski J., Restaino S. // Appl. Opt. 2001. Vol. 40. No 15. P. 2345-2355.
[10] Stewart I.W. The static and dynamic continuum theory of liquid crystals. London. Taylor & Francis. 2004. 360 p.

Design of a decision support system, trained on GPU, for assisting melanoma diagnosis in dermatoscopy images

Dimitris Glotsos[1], Spiros Kostopoulos[1], Stella Lalissidou[1], Konstantinos Sidiropoulos[2], Pantelis Asvestas[1], Christos Konstandinou[3], George Xenogiannopoulos[1], Eirini Konstantina Nikolatou[4], Konstantinos Perakis[5], Thanassis Bouras[5] and Dionisis Cavouras[1]

[1]Department of Biomedical Engineering, Technological Educational Institute of Athens, Greece
[2]European Bioinformatics Institute (EMBL-EBI), European Molecular Biology Laboratory, Welcome Trust Genome Campus, Hinxton, Cambridge, UK
[3]Department of Medical Physics, University of Patras, 26504, Rio, Patras, Greece
[4]Department of Economic Sciences, University of Patras, Patras, Greece
[5]UBITECH Research Department, UBITECH Ltd., Athens, Greece

E-mail: dimglo@teiath.gr

Abstract. The purpose of this study was to design a decision support system for assisting the diagnosis of melanoma in dermatoscopy images. Clinical material comprised images of 44 dysplastic (clark's nevi) and 44 malignant melanoma lesions, obtained from the dermatology database Dermnet. Initially, images were processed for hair removal and background correction using the Dull Razor algorithm. Processed images were segmented to isolate moles from surrounding background, using a combination of level sets and an automated thresholding approach. Morphological (area, size, shape) and textural features (first and second order) were calculated from each one of the segmented moles. Extracted features were fed to a pattern recognition system assembled with the Probabilistic Neural Network Classifier, which was trained to distinguish between benign and malignant cases, using the exhaustive search and the leave one out method. The system was designed on the GPU card (GeForce 580GTX) using CUDA programming framework and C++ programming language. Results showed that the designed system discriminated benign from malignant moles with 88.6 % accuracy employing morphological and textural features. The proposed system could be used for analysing moles depicted on smart phone images after appropriate training with smartphone images cases. This could assist towards early detection of melanoma cases, if suspicious moles were to be captured on smartphone by patients and be transferred to the physician together with an assessment of the mole's nature.

1. Introduction

Melanoma comprises one of the most lethal and difficult to treat forms of cancer with more than 100 000 cases worldwide each year [1]. The annual incidence rate of the disease has an increasing tendency, which is attributed to modern lifestyle and long exposure to ultraviolet radiation that has been identified as among the most influential factors for triggering the disease. Early stage detection of melanoma has been shown as the most crucial step for treating the disease, since at early phases melanoma seems to be more vulnerable to available treatments [2, 3].

Diagnosis of melanoma is done using special observation instruments called dermoscopes. Dermoscopes use polarized light in order to enhance the visibility of pigmented lesions [4]. The accuracy in melanoma detection using dermoscopes has been shown significantly higher (>90%) as compared to conventional visual clinical examination (approximately 65%) [5], rendering dermoscopy the standard every day clinical tool of dermatologists. The evaluation of suspected skin lesions follows various guidelines, among which the ABCD rule is probably the most popular [6]. According to the ABCD guidelines the diagnostic conclusion is derived based on estimations regarding the suspected lesion's asymmetry (A), border irregularity (B), color (C) and diameter (D). Although dermoscopy has improved melanoma detection rates, it requires training and experience for seizing the benefits that this technology may offer to the dermatologist [6]. Moreover, even with experienced dermatologists, difficulties may arise due to differential diagnosis issues, since melanoma may be mixed with other skin lesion types, such as seborrheic keratosis [7, 8]. The above factors may comprise potential sources of diagnostic misleads that might guide inappropriate treatments with controversial efficacy [9-11]. The purpose of this study was to design a decision support system to be used as a second opinion tool available to the dermatologist in order to safeguard from potential sources diagnostic misleads in melanoma detection.

2. Material and Methods

Clinical material comprised images of 44 dysplastic (clark's nevi) and 44 malignant melanoma lesions, obtained from the dermatology database Dermnet [12]. Initially, images were processed for hair removal and background correction using the Dull Razor algorithm [13]. This method consists of three phases: hair are identified using an operation of closure over each of the color segments, hair pixels are then replaced by interpolation with neighbourhooding tissue pixels and finally, a median filter with a 7×7 kernel is applied in order to eliminate the impulse noise. Processed images were segmented to isolate moles from surrounding background, using a combination of level sets/gradient flow vector snakes and an automated thresholding approach [14].

Morphological and textural features were calculated from each segmented mole. Morphological features comprised measurements of the area, perimeter, roundness and eccentricity [15] describing the size, the shape and asymmetry of segmented moles. Textural features consisted of measurements regarding the mole's gray-level histogram by means of the mean value, standard deviation, skeweness and kurtosis. Moreover, the spatial distribution of the mole's illumination was investigated with the aid of the co-occurrence matrix and run-length matrixes [16, 17]. The co-occurrence matrix encodes the frequency of appearance of the same pairs of pixel intensities, whereas the run-length matrix describes the frequency of appearance of a set of consecutive pixel- intensities having the same grey value. In total, twenty (20) textural features were computed (4 histogram-based, 11 co-occurrence matrix-based and 5 run-length matrix-based). Thus, each case was represented by 24 features, 4 morphological and 20 textural. Extracted features were fed into a pattern recognition system, assembled with the Probabilistic Neural Network (PNN) classifier [18], which was trained to distinguish between benign and malignant cases, using the exhaustive search and the leave one out method.

The exhaustive search generates all possible combinations of available features. Each feature combination is then evaluated by the classifier using the leave-one-out method that constructs the model using all cases but one. The left-one-out case is considered as an unknown case. The classifier is then asked to classify the left-one-out case to either the benign or malignant class. If the classifier decides that the left-one-out case belongs, for example, to the benign class and the left-one-out case comes from the benign class, then the classifier has a successful/correct classification result, else an unsuccessful/error classification result. By measuring the correct and error classifications then it is possible to determine the prediction accuracy of the model for each feature combination. Then, the feature combination, which provides the highest classification result, is selected as the most informative one [19]. The system was designed on the GPU card (GeForce 580GTX) using CUDA programming framework and C++ programming language.

3. Results and Discussion

Figure 1 illustrates an example of the application of the proposed image processing strategy for segmentation of melanoma mole. Figure 1(a) is the original color dermoscopy image. Figure 1(b) is image following the application of the DullRazor hari removal algorithm,. It is clear that the effect of darker hairs covering the mole has been greatly reduced. Figure 1(c) is the gray scale transformed image with corrected background. Figure 1(d) is the thresholded image that separated brighter from darker pixels based on automated threshold calculation based on the Otsu's method [20]. Figure 1(e) illustrates the elimination of stray pixels outside and inside the mole's region. Figure 1(f) depicts the result following the application of morphological opening and closing filters for smoothing of boundaries. Figure 1(g) shows the border of the mole, which was obtained following the application of an edge-detection Roberts filter. The result of Figure 1(g) was considered as the initialization for the level set method that converged to the final boundaries of the mole using the gradient flow vector approach (Figure 1(h)).

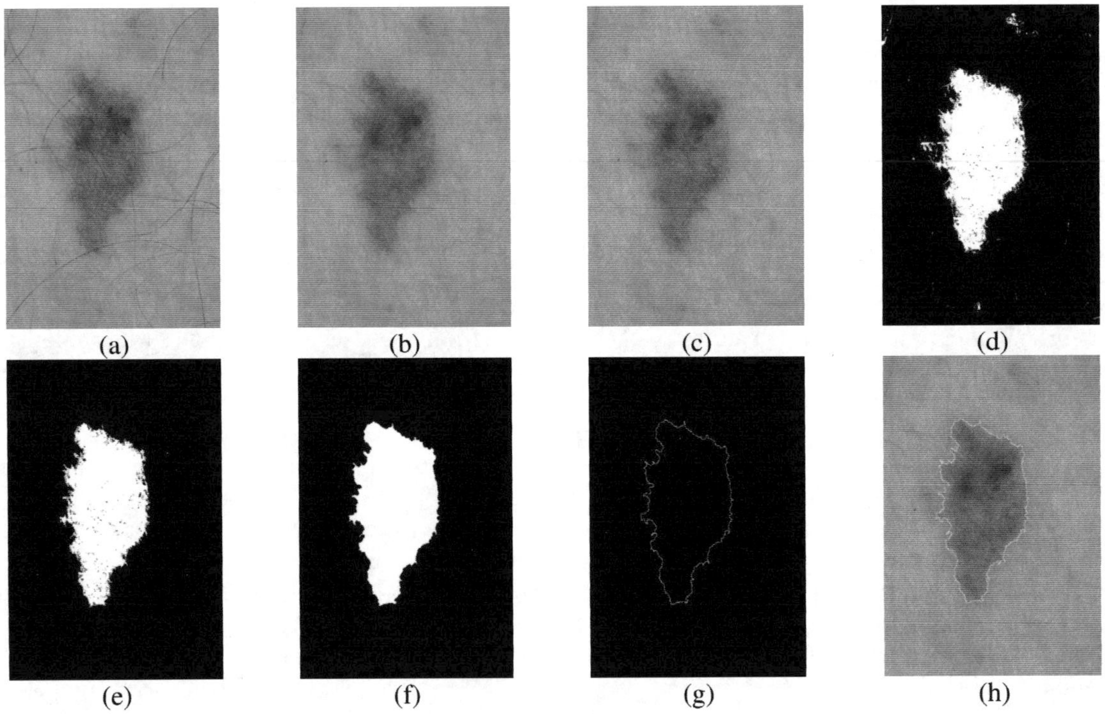

Figure 1. (a) Original Image, (b) Application of the DullRazor algorithm for hair removal, (c) Background corrected grey scale image, (d) Thesholded Image, (e) Elimination of stray objects outside and inside the mole's region, (f) Morphological closing and opening, (g) retaining of the mole's boundary using Roberts edge detection filter, (h) Final delineation of mole's boundary following the application of the gradient flow vector algorithm

Results showed that the designed system discriminated benign from malignant moles with accuracy higher than 88.6% employing three (3) morphological features (perimeter, roundness and eccentricity) and six (6) textural features extracted from the co-occurrence matrix according to the leave-one-out method. This particular feature combination was also tested on an external cross validation scheme to investigate the generalization of the model to new cases [21]. According to this set-up, the performance of the system was estimated 84%±5%. The proposed system could be used for analysing moles depicted on smart phone images after appropriate training with smartphone images cases. This could assist towards early detection of melanoma cases, if suspicious moles were to be captured on smartphone by patients and be transferred to the physician together with an assessment of the mole's nature.

Acknowledgments
The research activities that led to this work, were co-financed by National Funds and by the European Regional Development Fund (ERDF) under the Hellenic National Strategic Reference Framework (NSRF) 2007-2013, concerning the project "MARK1" with ref. number (ISR_3233) within the Bilateral Cooperation between Greece & Israel action.

4. References

[1] Holterhues C, Hollestein LM, Nijsten T, Koomen ER, Nusselder W, de Vries E. 2013 *The British journal of dermatology* **169** 389-97.
[2] Ruiz D, Berenguer V, Soriano A, Sánchez B. 2011 *Expert Sys Appl* **38** 15217-23.
[3] Belli F, Santinami M, Baldini MT, Testori A, Maio A, Cascinelli N. 1989 *International journal of radiation applications and instrumentation Part B, Nuclear medicine and biology* **16** 621-4.
[4] Tenenhaus A, Nkengne A, Horn JF, Serruys C, Giron A, Fertil B. 2010 *Skin research and technology : official journal of International Society for Bioengineering and the Skin* **16** 85-97.
[5] Schein O, Westreich M, Shalom A. 2009 *Harefuah* **148** 820-3, 55.
[6] Abbasi NR, Shaw HM, Rigel DS, Friedman RJ, McCarthy WH, Osman I, Kopf AW, Polsky D. 2004 *Jama* **292** 2771-6.
[7] Sahin MT, Ozturkcan S, Ermertcan AT, Gunes AT. 2004 *The Journal of dermatology* **31** 884-9.
[8] Deshabhoina SV, Umbaugh SE, Stoecker WV, Moss RH, Srinivasan SK. 2003 *Skin research and technology : official journal of International Society for Bioengineering and the Skin* **9** 348-56.
[9] Veierod MB, Parr CL, Lund E, Hjartaker A. 2009 *Melanoma research* **19** 61.
[10] Pfahlberg AB, Gefeller O. 2008 *Melanoma research* **18** 300-1.
[11] Lorentzen HF, Weismann K, Larsen FG. 2001 *Melanoma research* **11** 495-501.
[12] Dermnet-Skin-Disease-Atlas, www.dermnet.com.
[13] Lee T, Ng V, Gallagher R, Coldman A, McLean D. 1997 *Computers in biology and medicine* **27** 533-43.
[14] Erkol B, Moss RH, Stanley RJ, Stoecker WV, Hvatum E. 2005 *Skin research and technology : official journal of International Society for Bioengineering and the Skin* **11** 17-26.
[15] Wolberg WH, Street WN, Mangasarian OL. 1999 *Clinical cancer research : an official journal of the American Association for Cancer Research* **5** 3542-8.
[16] Haralick RM, Shanmugam K, Dinstein I. 1973 *IEEE Trans Systems Man and Cybernetics* **SMC-3** 610-21.
[17] Galloway MM. 1975 *Computer Graphics and Image Processing* **4** 172-9.
[18] Specht DF. 1990 *IEEE transactions on neural networks / a publication of the IEEE Neural Networks Council* **1** 111-21.
[19] Jain AK, Duin RPW, Mao JC. 2000 *IEEE Transactions on Pattern Analysis and Machine Intelligence* **22** 4-37.
[20] Nyma A, Kang M, Kwon YK, Kim CH, Kim JM. 2012 *Journal of biomedicine & biotechnology* **2012** 830252.
[21] Ambroise C, McLachlan GJ. 2002 *Proceedings of the National Academy of Sciences of the United States of America* **99** 6562-6.

Mathematical Model for Absolute Magnetic Measuring Systems in Industrial Applications

Armin Fügenschuh[1], Marzena Fügenschuh[2], Marina Ludszuweit[1], Aleksandar Mojsic[1] and Joanna Sokół[1]

[1]Helmut Schmidt University University of the Federal Armed Forces Hamburg, Department of Mechanical Engineering, Holstenhofweg 85, 22043 Hamburg, Germany

[2]Beuth Hochschule für Technik Berlin, Luxemburger Straße 10, 13353 Berlin, Germany

E-mail: marina.ludszuweit@hsu-hh.de

Abstract. Scales for measuring systems are either based on incremental or absolute measuring methods. Incremental scales need to initialize a measurement cycle at a reference point. From there, the position is computed by counting increments of a periodic graduation. Absolute methods do not need reference points, since the position can be read directly from the scale. The positions on the complete scales are encoded using two incremental tracks with different graduation. We present a new method for absolute measuring using only one track for position encoding up to micrometre range. Instead of the common perpendicular magnetic areas, we use a pattern of trapezoidal magnetic areas, to store more complex information. For positioning, we use the magnetic field where every position is characterized by a set of values measured by a hall sensor array. We implement a method for reconstruction of absolute positions from the set of unique measured values. We compare two patterns with respect to uniqueness, accuracy, stability and robustness of positioning. We discuss how stability and robustness are influenced by different errors during the measurement in real applications and how those errors can be compensated.

1. Introduction

In many areas of industrial processing, length measurement systems are important. To determine the distance between two objects along a linear distance, these systems consist of a sensor device, a measurement scale and a processing unit. Possible applications are positioning control of the milling head in CNC machines or position control of industrial robots.

In length measurement two methods are common. In incremental measuring, the sensor reads increments of periodic graduation from the scale. The measured distance is computed by counting the read increments, so only relative distances are measurable. For absolute measuring, the position is computed directly from a uniquely encoded scale. Both measurement methods are used in several modalities of systems, e.g., magnetic and optical systems.

The unique encoding in absolute positioning is realized using properties such as patterns of light wavelength for optical scales or magnetic field information for magnetic scales [1, 2]. To ensure uniqueness, more than one pattern track may be necessary. The different magnetic poles induce magnetic fields according to the pattern. This allows reading the information using hall sensor devices, which convert magnetic field information into signals. Measuring values with a

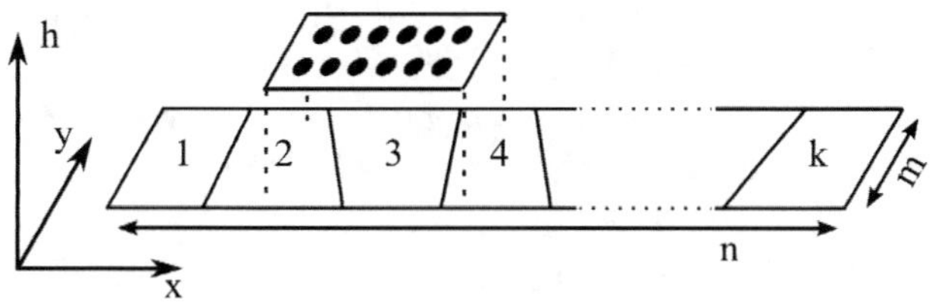

Figure 1. Illustration of the measurement setup.

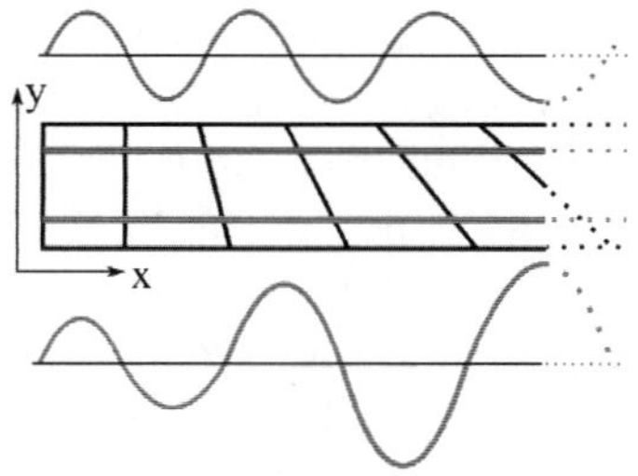

Figure 2. Schematic of a pattern with related magnetic signals.

hall cell along a magnetic pattern results in a function of magnetic signals with zeros at the boundaries between two poles [3].

We present a method to be used in magnetic systems which requires only one magnetic track for unique encoding and resolution up to micrometer range. This saves overall production costs and allows for smaller, less energy consuming systems to be used in applications where size is important.

2. Experimental setup

We define a three-dimensional, right-handed coordinate system with the x-axis parallel and the y-axis orthogonal to the measuring direction of the magnetic scale and the h-axis pointing away from the scale. See also Figure 1.

The sensor containing the hall cells is placed above the scale and moved in x-direction. For our experiments, we consider a sensor with a 2×6 hall cell placement (see Figure 1). This hall cell array allows for reading the magnetic signal on two parallel tracks at once. The distance between two adjacent cells is 1.5mm in x- and 5mm in y-direction.

The measurement stripe with its scale is described as an area of width $m \in \mathbb{R}^+$ and length $n \in \mathbb{R}^+$ divided into $k \in \mathbb{N}$ consecutive trapezoidal magnetic poles. The initial value $n = 0$ is equal to the left pole boundary of the first magnet. Considering the distance between two hall sensors in y-direction, $m \geq 5$mm is required for our setup; usually 10mm in practice.

In order to encode for an absolute position, each measurement on the scale must be described by a set of unique values of orthogonal magnetic field components as measured by the hall cell array. Following, the magnetic field information from the hall sensor is denoted as magnetic signal. The measured values of magnetic signals are influenced by the specific length l_i of magnetic pole i, $i = 1, \ldots, k$ and the corresponding magnetic field. Constructing a magnetic pattern with trapezoidal magnetic poles, therefore allows the sensor to read different pole lengths on each track. Hence, we introduce l_i^{lower} for the lower and l_i^{upper} for the upper reading track.

For computations, we used two different magnetic patterns A and B. For pattern A, the pole boundaries in the middle of the reading lines have a constant length d_i^A and continuously increased angles α_i^A from α_{min}^A to α_{max}^A in steps of α_{step}^A. For Pattern B we define a set of lengths d_i^B from $d_{min}^B \leq d_i^B \leq d_{max}^B$ with steps of d_{step}^B and a limitation d_ε^B for total lengths for available trapezoids. The construction of pattern B is based on a mixed-integer linear method programming model, where the constraints of the pattern construction were formulated by algebraic inequalities. The resulting optimization problem is solved by a linear programming based branch-and-bound procedure. A schematic example of a pattern and its magnetic signals is presented in Figure 2. Both patterns were produced by our project partner Bogen Electronic GmbH, Berlin

3. Method

In order to apply absolute positioning, the actual scale needs to be known. Hence we measure the function f of the magnetic signal from the manufactured tape. Computing the absolute position is done by finding the best fit between the set of measured values and functions $f_i : [0, l_i] \to \mathbb{R}$ of approximatives to the magnetic signals on each reading track. Here, the f_i are piecewise functions of f. For each pole i, $i = 1, \dots k$ with length l_i we use a linear combination $g_i : [0, 1] \to \mathbb{R}$, $g_i(x) = \lambda_{i,1}p_1(x) + \lambda_{i,2}p_2(x) + \ldots + \lambda_{i,5}p_5(x)$ of 5 polynomials $p_1, p_2, \ldots, p_5$ of 5th degree and coefficients $\lambda_{i,1}, \lambda_{i,2}, \ldots, \lambda_{i,5}$ to construct $f_i(x) \approx g_i\left({}^{x}/_{l_i}\right)$. For every pole i, we save the 5 coefficients of the linear combination on the processing unit. Due to memory limitations on the processing unit, we did not implement higher polynomial approximation methods.

To find the best fit of measured data on f, we use a two-step algorithm. The first step evaluates the sequence of signs of values measured to find all sets of possible subsequences of magnets from the scale. In the second step the set of measured values is fitted to f for all found subsequences from the first step by calculating the best fit according to the smallest residual $\varepsilon_f = min\left(\|f_i - \mathbf{v}\|\right)$ for the measured values $\mathbf{v} = v_1, \ldots, v_6$ from each reading line where ε_f is assumed to give the correct position on the scale. Our method of reconstructing the correct position from measured data was implemented on a DSP and on a FPGA by our project partners at the Technical University Berlin.

We evaluate our results with respect to accuracy and stability. Let the error of a measurement $\varepsilon_p = |P_a - P_r|$ be the distance between actual position $P_a \in [0, n]$ and reconstructed position $P_r \in [0, n]$. As a score for the accuracy we define the length l_a of the longest interval with an error below a threshold t_ε. Considering also additional positions with a residual close to ε_f, we get additional potential positions for every reconstructed one. For small differences in the residuals between the potential positions and the position reconstructed from ε_f, it becomes more likely, that the position chosen by our algorithm may be wrong. Hence, we measure stability as the mean distance l_s between the residuals of the two best candidates. The larger l_s, the better the scale, because it is more robust against small measuring errors.

4. Results

For testing we constructed scales for patterns A and B with lengths of 675mm(A) and 672mm (B). For Pattern A we used $d_i^A = 2$mm and increased the angles from $\alpha_{min}^A = 0°$ to $\alpha_{max}^A = 32°$ in steps of $\alpha_{step}^A = 0.095°$. In pattern B we used $d_{min}^B = 2$mm and $d_{max}^B = 6.5$mm with stepsize $d_{step}^B = 0.35$mm and $d_\varepsilon^B = 3.4$mm.

We recorded f with 1.5kHz sampling rate at a movement speed of 20mm/s using a 2×6 hall cell array device. The polynomials and coefficients for f are stored on the processing unit. The same scale is measured again with same sampling rate and speed and to recover the positions for each set of measured values from the approximated functions stored on the processing unit.

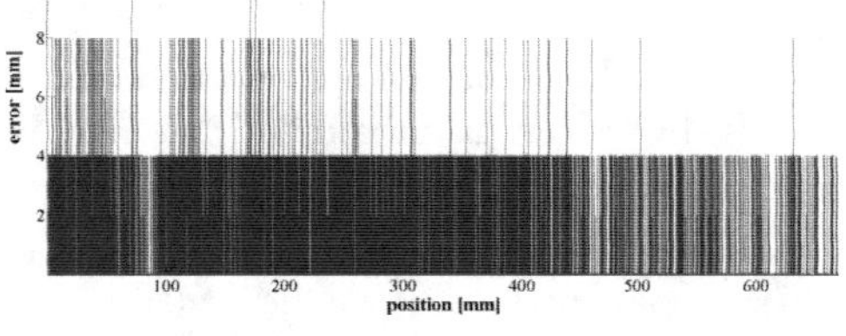

Figure 3. Error ε_p of computed positions for magnetic pattern A.

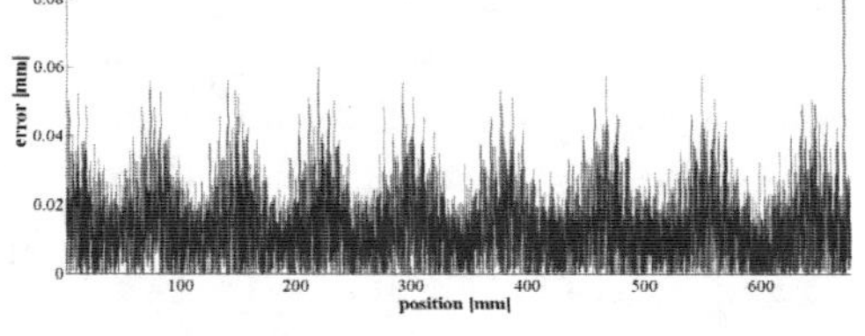

Figure 4. Error ε_p of computed positions for magnetic pattern B.

Table 1. Accuracy l_a of A and B.

t_ε [mm]	0.01	0.04	0.06	4.00	4.02	8.01	12.01
A [mm]	2.02	6.26	6.26	8.67	130.20	436.70	675.00
B [mm]	1.89	64.10	671,10	671,10	...	...	671,10

The results in figure 3 and 4 show the error for each position. A value of less than 0.1mm is acceptable. For our scales we measured stabilities $l_s = 14$ for A and $l_s = 305$ for B.

5. Discussion

Our results show huge differences in both accuracy and stability as defined between both scales, with scale B delivering far better results than scale A. Good accuracy for scale A is only achievable at very small intervals with the error being in the range of several centimeters for most reconstructions while a value of 0.1mm would be acceptable. The main reason for this is the high regularity in the construction of the pattern leading to overall many candidates after the first step of our algorithm. Additionally, the small changes in the angles between the magnetic poles result in locally very similar magnetic signals and pole lengths. This corresponds to the low stability of the scale. Hence, for most of the positions the candidates for reconstruction have similar high errors and therefore only small changes in the signal (e.g., due to noise) can lead to different results in the reconstruction. Such effects do not occur regularly on scale B, leading to a far better accuracy which also corresponds to the better stability. Still due to similar magnetic areas that occur in the construction of B, we observed some spikes which are also the result of too small changes in pole lengths.

6. Conclusions

We were able to show that for appropriate scales our algorithm delivers precise results, sufficient for practical applications. Still, local errors arise and should be corrected by adjusting the scale or introducing additional rules in construction to avoid these patterns.

For real world industrial applications, we have to consider additional errors like waves on the stripe, shaking of the machines or shifts in the track. Additionally the distance between sensor and stripe, either in h- or y-direction influences the amplitude. Hence, the stability decreases. We are currently working on data from different sensor-stripe shifts to improve stability. Further placement errors of the sensor like twists and turns need to be addressed in the future, too.

References

[1] K. Engelhardt and P. Seitz: *High-resolution optical position encoder with large mounting tolerances*, Applied Optics Vol. 36 (1997) 2912-2916

[2] F. Perez-Quintin, A. Lutenberg and M.A. Rebollo: *Linear displacement measurement with a grating and speckle pattern illumination*, Applied Optics Vol. 45 (2006) 4821-4825

[3] J. Hoyer and T. Becker 2013 *Messvorrichtung und Verfahren zum Messen von Körpern*, European Patent no. EP2846126 (2013)

Acknowledgments

This work is part of a joint project funded by the Central Innovation Program SME supported by the German Federal Ministry for Economic Affairs and Energy. The results were achieved in cooperation with BOGEN Electronic GmbH and the Technical University of Berlin, chair of Electronics and Medical Signal Processing.

Validating Knowledge-Based Framework through Mission-Oriented Sensors Array and Smart Sensor Protocol

Douglas Rodrigues, Rayner M Pires and Kalinka R L J C Branco

Institute of Mathematics and Computer Science, University of São Paulo, São Carlos, Brazil

E-mail: {douglasr, rayner, kalinka}@icmc.usp.br

Abstract. This paper addresses the problem of using Service-Oriented Architecture (SOA) in critical embedded systems, mainly in Unmanned Aerial Vehicles (UAVs). We present the use of a SOA approach to provide the integration of the payload in the UAV. The integration is provided by a plug and play protocol named Smart Sensor Protocol (SSP) that validates the SOA approach.

1. Introduction

Although the use of Service-Orient Architecture (SOA) in the field of applications related to business and IT (Information Technology) is well established, several aspects should be considered in the field of critical embedded systems, especially for Unmanned Aerial Vehicles (UAV). For instance, the availability of sufficient resources, such as processing power and memory capacity, is one of some issues to be considered. This class of systems can range from small to large, complex and multi-core systems. Therefore, in many of them, there are enough resources to be able to use Web services.

Then, we proposed the KBF (Knowledge-Based Framework for Dynamically Changing Applications) [1]. The KBF allows dynamic definition of Web services, depending on the current system configuration, the availability of Web services component, and a database with predefined context information. Specifically, the KBF uses both context and monitoring information to compose or generate services with different levels of reliability, security and performance, in a dynamic and intelligent way.

The MOSA (Mission-Oriented Sensor Array) [2], in turn, is a concept of sensors arrangement capable of performing a specific mission. The sensors to be used are determined according to the mission characteristics. Once these characteristics have been established, MOSA is then made representing a new specific sensor to that mission.

As described in [3], the SSP (Smart Sensor Protocol) is an application level protocol that was designed to rule the interaction between MOSA mission execution system and UAV system.

Thus, the architecture presented in this paper is designed to facilitate the communication between the MOSA and the KBF, which occurs using the SSP. The KBF is able to support the MOSA to perform missions autonomously, once KBF selects the service that best fits its system usage scenario, so the mission is performed in a more intelligent way.

2. Related works

Different implementations of UAVs are found in the literature and most of these implementations and developments typically uses traditional approaches [4] [5]. The USAF roadmap [6] discusses and suggests the use of SOA in UAVs. This is a new approach, once these kind of systems have special requirements, like real-time performance.

There are also approaches that discuss the use of SOA in embedded systems [7] [8] [9]. Unlike those, the approach here presented discusses the use of SOA in a critical embedded system to achieve reusability, maintainability, interoperability, flexibility, and security using KBF and MOSA.

3. Communication between the KBF and the MOSA using the SSP

The operation of the KBF can be summarized as the enumerated steps illustrated in Figure 1. In this case, notice that the MOSA acts as the client.

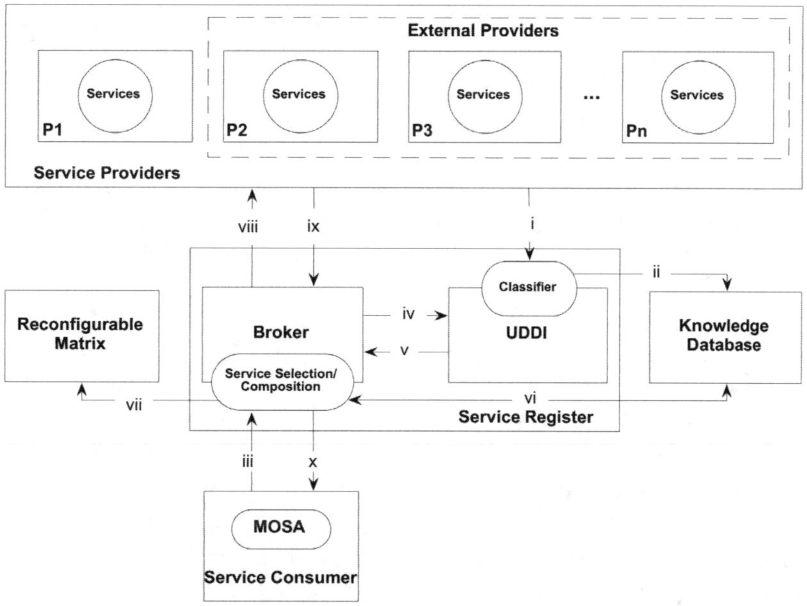

Figure 1. Knowledge Based Framework for Dynamically Changing Applications (KBF).

(i) Initially, a service provider develops its service, describes its interface (e.g., WSDL) and publishes in a service registry (e.g., UDDI). After the publication, a classifier automatically classifies the service according to defined selection criteria (reliability, security and performance). This operation will be repeated to every new service published.

(ii) The resulting information from the classification are sent and added to the database.

(iii) A MOSA is connected to the UAV and sends a request to the broker, specifying necessary functionality features for defined mission.

(iv) The broker performs a search in the service registry, looking for an appropriate service, based on the features requested by the MOSA.

(v) Following the search, the broker gets a set of service interfaces with same functionality, changing only their selection criteria. i.e., among the services with same functionality, one of them may be more secure, another may have better performance, another may have greater reliability and so on.

(vi) With the obtained information from these interfaces, now the service selection and composition module queries the database to verify these services selection criteria, choosing the one that is the best for the situation, according to the defined mission.

(vii) Then the chosen service, its functionality and its selection criteria are related to the mission and sent to the reconfigurable matrix. As each mission is defined by a MOSA, the next time that same MOSA is connected, the service selection and composition module will perform a search on the reconfigurable matrix and return the service previously used, removing the need to repeat the search.

(viii) After performing this procedure, the broker finally sends the request to the chosen service.

(ix) After processing the request, the response is returned to the broker.

(x) Finally, the response is forwarded to the MOSA that initially requested the service.

In this case, MOSA is a client connected to the UAV, and every time that a MOSA board is connected to an UAV, it starts a series of requests to the aircraft system in order to decide about its qualification to accomplish a predefined mission stored in that MOSA (item iii illustrated in Figure 1).

Queries for the UAV characteristics are done in the first step of the SSP protocol, named Negotiation, when the `REQ` primitive requests a predefined list of features. The default list of queries requests information related to the aircraft physical capabilities.

Once the MOSA is compatible with the KBF, it must try to discover KBF services in the first step. The second primitive, named `SEND`, was designed in [3] for this purpose. It is used to encapsulate data exchanged between client and provider of Web services. This exchange will happen not only in the Negotiation step, but also during the whole period of mission execution.

A software architecture describes the system components and the way they interact at a high level. In this application, the components are MOSA and KBF, and they interact via SSP messages, as illustrated in Figure 2.

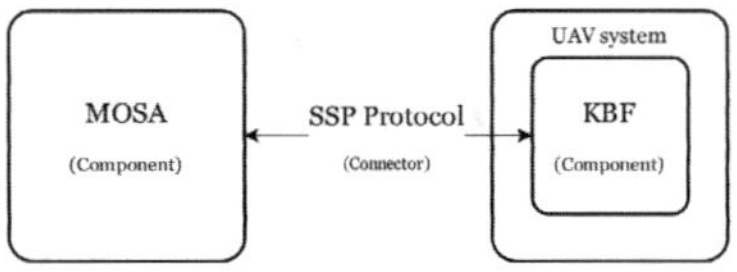

Figure 2. The MOSA-SSP-KBF software architecture.

SOA is a design pattern in which application components provide services to other components via a communication protocol and, in this scenario, SSP is the appropriate protocol. Through primitives `SEND` and `ACK`, both MOSA and KBF will exchange data for the SOA Find-Bind-Execute Paradigm.

In the Negotiation step, that primitive peer will be firstly used to detect KBF presence on UAV side. Once KBF announces its presence, both primitives `SEND` and `ACK` will encapsulate all messages for services discovery on KBF registry.

A non-exhausted example of services enumeration that is pertinent to this scenario and is in the KBF registry is listed in Table 1.

While the mission is being executed, KBF will eventually serve MOSA with its services. All data exchanged will be encapsulated over SSP messages again, by means of that last peer of primitives. The SSP protocol specification defines the format of `SEND` and `ACK` messages, when they are triggered, and how much data can be put in the message payload.

We simulated a mission using a MOSA board plugged in a computer (as an UAV) and exchanging information about the mission. The mission was implemented using a specific grammar [3] and was based on a real flight. This simulation using the MOSA board and the KBF is illustrated in Figure 3.

Table 1. Example of services that can be consumed by MOSA application.

Static services	Dynamic services		
	Displacement	*Autonomy*	*Status*
Inform cruising altitude	Inform/define current height	Inform remaining autonomy	Inform current action
Inform minimum stall speed	Inform/define current altitude	Inform payload weight	
Inform maximum weight capacity	Inform/define current speed		
Inform cruising speed	Define destination coordinates		
Inform maximum speed	Inform current GPS position		
Inform autonomy with maximum load			

Figure 3. MOSA board (right), connected to an Arduino (left) that makes the interface with the computer.

4. Conclusions

This paper presents the validation of SOA in a specific critical embedded system (UAV). The communication between the UAV and MOSA was described. The results show that is possible to fly an UAV using SOA and a MOSA for a specific mission.

As future work, we expect to show a complete mission implemented using SOA and MOSA. The takeoff and landing and all the services necessary to complete a entire mission using SOA is the next step of this work.

Acknowledgments

The authors acknowledge the support granted by FAPESP, processes 2011/06086-9 and 2011/044161.

References

[1] Rodrigues D, Pires R M, Estrella J C, Vieira M, Correa M, Camargo J B, Branco K R L J C and Trindade O 2011 *Communications in Computer and Information Science* **206** 345–354

[2] Rodrigues D, Pires R M, Estrella J C, Marconato E A, Trindade O and Branco K R L J C 2011 *Proceedings of the 2011 IEEE International Conferences on Internet of Things (iThings), and Cyber, Physical and Social Computing (CPSCom)* pp 733–738

[3] Pires R, Chaves A and Branco K 2014 *ICUAS '14: International Conference on Unmanned Aircraft Systems* pp 1300–1310

[4] DoD 2007 Unmanned systems roadmap 2007-2032 U.S. Department of Defense. Office of the Secretary of Defense

[5] DoD 2009 Unmanned systems integrated roadmap FY2009-2034 U.S. Department of Defense. Office of the Secretary of Defense

[6] USAF 2009 United states air force unmanned aircraft systems flight plan 2009-2047 Headquarters, United States Air Force

[7] Thramboulidis K C, Doukas G and Koumoutsos G 2008 *EURASIP Journal on Embedded Systems - Embedded System Design in Intelligent Industrial Automation* **2008** 1–15 ISSN 1687-3955

[8] Lee M H, Yoo C J and Jang O B 2008 *International Journal of Advanced Science and Technology* **1** 55–64 ISSN 2005-4238

[9] Moritz G, Prüter S, Timmermann D and Golatowski F 2008 *ETFA '08: IEEE International Conference on Emerging Technologies and Factory Automation* pp 432–435

4th International Conference on Mathematical Modeling in Physical Sciences (IC-MSquare2015) IOP Publishing
Journal of Physics: Conference Series **633** (2015) 012082 doi:10.1088/1742-6596/633/1/012082

A random field approach to the Lagrangian modeling of turbulent transport in vegetated canopies

Rita Cesari
Istituto di Scienze dell'Atmosfera e del Clima (ISAC-CNR), Str. Prov. Lecce-Monteroni, Km 1.2, 73100 Lecce, Italy
E-mail: r.cesari@isac.cnr.it

Paolo Paradisi
Istituto di Scienza e Tecnologie dell'Informazione "A. Faedo" (ISTI-CNR), Via G. Moruzzi 1, 56124 Pisa, Italy
E-mail: paolo.paradisi@cnr.it

Abstract. We present an application of a Lagrangian Stochastic Model (LSM) to turbulent dispersion over complex terrain, where turbulent coherent structures are known to play a crucial role. We investigate the case of a vegetated canopy by using semi-empirical parameterizations of turbulence profiles in the region inside and above a canopy layer. The LSM is based on a 4-dimensional Fokker-Planck (4DFP) equation, which extends the standard Thomson87 Lagrangian approach. The 4DFP model is derived by means of a Random Field description of the turbulent velocity field. The main advantage of this approach is that not only the experimental Eulerian one-point statistics, but also the Eulerian two-point two-time covariance structure can be included explicitly in the LSM. At variance with the standard Thomson87 approach, the 4DFP model allows to consider explicit parameterizations of the turbulent coherent structures as it explicitly includes both spatial and temporal correlation functions. In order to investigate the effect of the turbulent geometrical structure on a scalar concentration profile, we performed numerical simulations with two different covariance parameterizations, the first one isotropic and the second anisotropic. We show that the accumulation of scalars near the ground is due to the anisotropic geometrical properties of the turbulent boundary layer.

1. Introduction
Lagrangian Stochastic Models (LSMs) provide a useful tool to describe turbulent dispersion. They are based on a (Markovian) Langevin Equation for the Lagrangian particle velocity [3], whose drift and noise terms are related through the *well-mixedness criterion* (WMC) stated in the 1987 milestone paper by Thomson (Th87) [2]. In the Th87 approach the 1-point 1-time Probability Density Function (PDF) of the Eulerian velocity is experimentally known. The noise term is described by the 2nd order structure function, whose scaling is linear when the time increments are limited to the inertial subrange scales: $\langle [u(t+\Delta t) - u(t)]^2 \rangle \simeq C_0 \epsilon \Delta t$, being ϵ the turbulent energy dissipation rate and C_0 a constant [1]. Applying the WMC, the drift is

[1] This is assumed to be a constant in the original 1941 Kolmogorov similarity (K41) theory, but it is nowadays well-known that this assumption is not exactly true due to intermittency of the velocity field at the inertial scales and small higher-order corrections, related to the velocity statistics, are considered.

Journal of Physics: Conference Series **633** (2015) 012082 doi:10.1088/1742-6596/633/1/012082

expressed as a function of ϵ and of the Eulerian PDF. Apart from the homogeneous isotropic Gaussian turbulence, the drift is not determined uniquely, but only up to a velocity curl [2, 9, 11]. The standard Th87 approach, based on the WMC, is a powerful technique extensively used in Lagrangian transport modeling. However, the spatio-temporal geometrical structure of the velocity field is given by the universal K41 theory, thus inserting a strong modeling constrain that does not allow to take into account the experimental knowledge about space-time correlations of the velocity field, but only that about the Eulerian PDF. Then, Th87 shows some limitations in highly structured velocity fields (e.g., wall turbulence [4]) and, in general, in conditions where the geometrical features of the velocity field play a crucial role. From the the first paper by Thomson [2], Th87 approach has since then been applied in attempts to overcome the limitations related to the geometrical structure and to other aspects such as the non-uniqueness problem (see, e.g., [10, 8, 7] for recent developments).
A different approach, with respect to the Th87, was proposed in Ref. [1]. These authors developed a mathematical formulation which explicitly relates a Random Velocity Field (RVF), describing the Eulerian statistics, to the associated Lagrangian Stochastic Model (LSM), describing the trajectory dynamics of a passive scalar. In this approach, the application of the Ito's lemma [3] to a 3-Dimensional RVF depending on a 4-Dimensional (4D) space-time vector allows to derive a sort of 4D Fokker-Planck (4DFP) equation for the PDF of the Eulerian velocity. Given a semi-empirical expression for the Eulerian PDF with experimentally known parameters, the WMC [2] is directly applied to the 4DFP equation describing the RVF and not at the level of the LSM. At variance with Th87, the noise term is not defined from the K41 theory. This freedom is exploited to explicitly include the 2-point 2-time velocity statistics by means of the 2-points 2-time covariance tensor, which describes the space-time RVF correlations. When the coefficients of the 4DFP equation are known from the Eulerian statistics (1-point Eulerian PDF and 2-points covariance tensor), the ordinary Fokker-Planck and Langevin equations describing the LSM are simply derived by applying the dynamical equation of the passive tracers: $\dot{\mathbf{x}} = \mathbf{v}$. The main advantage of 4DFP is that not only the experimental Eulerian one-point PDF, but also the space-time RVF correlation structure, can be explicitly included in the LSM, thus allowing to consider explicit parameterizations of the turbulent coherent structures.
In this work, a particular version of the 4DFP-LSM, introduced in Ref. [1], is derived and applied to the case of the turbulent transport of passive scalars in vegetated canopies, thus allowing to get some information about the role of coherent structures. In particular, two different modeling choices, one with isotropic and the other with anisotropic motion structures, are compared.
The paper is organized as follows. In Section 2 we introduce the the 4DFP-LSM, in Section 3 we discuss the RVF statistics, both the Eulerian PDF and the covariance tensor, in Section 4 we show the numerical simulations and, finally, in Section 5 we draw some conclusions. In Appendix A the expression for the noise covariance tensor is introduced.

2. The 4DFP Lagrangian model

We recall here the main definitions and equations of the 4DFP Lagrangian model (see Ref. [1] for details). A RVF is formally defined through the assignment of n-point statistics for every n. The theoretical treatment of the general case is much involved and, actually, it is not really interesting for many applications, as n-point statistics is hardly obtained with the currently available instrumental technology. Then, the Lagrangian properties of the RVF are defined only in terms of 1-point 1-time Eulerian PDF and of 2-point 2-time covariance tensor, this being related to the 2nd order space-time structure function.
Now, we can define a zero-mean, incompressible RVF $\mathbf{u}(\mathbf{x}, t)$, with 2nd order space-time structure function scaling linearly in the increments $|\mathrm{d}\mathbf{x}|$ and $\mathrm{d}t$ at small space-time separations. We introduce 4-vector notation: $x^\mu = \{x^0, x^i\} \equiv \{t, \mathbf{x}\}, \partial_\mu \equiv \frac{\partial}{\partial x^\mu}$ and we stick rigorously to the Einstein convention of summation over covariant-contro- variant repeated indices. We have the

following equation for the velocity increment:

$$\mathrm{d}u^i \equiv \mathrm{d}x^\mu \partial_\mu u^i = A_\mu{}^i(\mathbf{u}, x^\mu)\mathrm{d}x^\mu + \mathrm{d}w^i \ ; \quad A_\mu{}^i(\mathbf{u}, x^\mu) = \langle \partial_\mu u^i(\mathbf{x}, t)|\mathbf{u}(\mathbf{x}, t)\rangle \tag{1}$$

and $\langle \mathrm{d}\mathbf{w}|\mathbf{u}\rangle = 0$. A possible expression for the noise correlation structure $\langle \mathrm{d}w^i \mathrm{d}w^j|\mathbf{u}(\mathbf{x}, t)\rangle$ is given in Appendix A. In the case of Gaussian Eulerian PDF, we have:

$$\rho_E(\mathbf{u}, x^\mu) \equiv \rho_G(\mathbf{u}, x^\mu) = (8\pi^3||R||)^{-\frac{1}{2}} \exp(-\frac{1}{2}S_{ij}u^i u^j) \tag{2}$$

where $S_{ij} = (R^{-1})_{ij}$ is the inverse of the Reynolds covariance tensor $R_{ij} = \langle u_i' u_j' \rangle$, being u_i' the components of the fluctuating velocity . In this case, the coefficients in Eq. (1) are given by the following expressions:

$$A_\mu^i = \bar{A}_\mu^i + \frac{1}{\rho_E}\Phi_\mu{}^i + \frac{1}{\rho_E}\Psi_\mu{}^i \ ; \quad \bar{A}_\mu{}^i \mathrm{d}x^\mu = -\frac{1}{2}\langle \mathrm{d}w^i \mathrm{d}w^j\rangle S_{jk}u^k \tag{3}$$

$$\Phi_\mu{}^i = \frac{1}{2}(\partial_\mu R^{ik})S_{kl}u^l \rho_E \ ; \quad \Psi_j{}^i = \frac{1}{4}[-\delta_j^i(\partial_l R^{lk})S_{km} + (\partial_l R^{li})S_{jm}]u^m \rho_E \tag{4}$$

The Langevin and Fokker-Planck equations associated with 4DFP-LSM are derived by substituting the dynamical equations of the passive tracer $dx^\mu = \{dt, \mathbf{u}dt\}$:

$$\mathrm{d}u^i \equiv \dot{u}^i \mathrm{d}t = (\mathbf{u} \cdot \mathbf{A}^i + A_0{}^i)\mathrm{d}t + \mathrm{d}w^i \tag{5}$$

$$(\partial_t + \mathbf{u} \cdot \nabla)\rho_L + \partial_{u^i}[(\mathbf{u} \cdot \mathbf{A}^i + A_0{}^i)\rho_L] = \frac{1}{2}\partial_{u^i}\partial_{u^j}(\mathcal{B}^{ij}\rho_L) \tag{6}$$

where $\mathcal{B}^{ij}(x^\mu, \mathbf{u}) = \frac{\mathrm{d}}{\mathrm{d}\Delta^0}\langle \Delta w^i \Delta w^j\rangle = \frac{u_T^2}{\tau_E}[B_t^{ij}(1) + B^{ij}(\mathbf{u})]$ [see Eqn. (A.1)]. We used the notation $A_\mu^i = \{A_0^i, \mathbf{A}^i\}$.

3. RVF statistics in a vegetated canopy

We limit ourselves to consider a 2D version of the 4DFP-LSM and the Gaussian Eulerian PDF, Eq. (2), as a 2D PDF. We consider only the streamwise and vertical directions $x_1 \equiv x$ and $x_3 \equiv z$ and the corresponding velocity components $u_1 \equiv u$ and $u_3 \equiv w$. In order to define the 1-point statistics in Eq. (2), the Reynolds covariance tensor and the mean velocity must be given. Firstly, it results $R_{12} = R_{22} = R_{23} = 0$. Then, the mean vertical velocity W is zero, as such as the spanwise component V. The vertical profiles of the 1-point statistics (mean wind $U(x_3)$, Reynolds covariance tensor and Eulerian time) are taken from [5]. Let us indicate the mean canopy height with h and the friction velocity with u_*. Then, we have:

$$U(z) = \begin{cases} \frac{u_*}{k}\log\left(\frac{z-d}{z_0}\right); & z \geq h \\ U(h)\exp\left(a_1(1 - \frac{z}{h})\right); & z < h \end{cases} \qquad R_{11} = \langle uu\rangle = \begin{cases} (a_4 u_*)^2\,; & z \geq h \\ (a_4 u_*)^2 \exp\left(2a_5(1 - \frac{z}{h})\right); & z < h \end{cases} \tag{7}$$

$$R_{33} = \langle ww\rangle = \begin{cases} (a_6 u_*)^2\,; & z \geq h \\ (a_6 u_*)^2\left[0.5\,(a_7 + a_6) + 0.5\,(a_7 - a_6)\cos\left(\pi z/h\right)\right]^2; \ z < h \end{cases} \tag{8}$$

$$R_{13} = \langle uw\rangle = \begin{cases} u_*^2; \ z \geq h \\ u_*^2 \max\left(a_3 - a_2 z/h; a_3 - 0.45a_2\right); z < h \end{cases} \tag{9}$$

$$\tau_E = \max\left(a_8 h/u_*; k\,(z - d)\,u_*/R_{33}\right) \tag{10}$$

where $k = 0.4$ is the Von Karman constant, d the displacement height and z_0 the aerodynamic roughness. These formulations are partially derived from Boundary Layer similarity theory and from turbulence measurements. More specifically, the parameters used in Ref. [5] for the turbulent transport of CO_2 are the following:

$$a_1 = -3; a_2 = 1.79; a_3 = 0.79; a_4 = 2.0; a_5 = -2; a_6 = 1.2; a_7 = 0.07; a_8 = 0.3\ . \tag{11}$$

Furthermore, following [5], we set $\frac{d}{h} = \frac{2}{3}$ and $z_0 = 0.1$.
Regarding the 2-point covariance tensor, we restrict ourselves to coherent structures with axial simmetry around the vertical axis. As shown in Appendix A, only three coefficients need to be estimated. In order to do this we should know the behavior of at least three components of the tensor β_l^{ij} defined in Eq. (A.10). Alternatively, we need to know only two components as we can estimate the isotropic coefficient a in Eqs. (A.11) as a rate between the eddy life-time (given approximately by τ_E) and the eddy rotation time. This last one is estimated by means of the rate between a sort of isotropic scale L_{iso} and u_T (see Appendix A). Then, we have:

$$a = C\,\frac{u_T \tau_E}{L_{iso}} \quad ; \qquad \frac{L_i so}{h} = \max\left(h - d; k(z - d)\right) \tag{12}$$

where C is a free adimensional model parameter, which is set $C = 1$ in our simulations. Then, b^{33} and c^{33} are given by Eq. (A.12). The normalized structure function can be evaluated in the following way:

$$\beta_l^{ij} = \frac{\tau_E R^{ij}}{u_T L_l^{ij}} \tag{13}$$

where L_l^{ij} is the correlation lenght of the velocity components u^i and u^j along the direction x_l.
In Finnigan [6] it is possible to find some experimental results about the behavior of L_1^{11} and L_1^{33}. According to these data, we can propose the following interpolation formulas:

$$\frac{L_1^{11}}{h} = \begin{cases} f_1\left(\frac{z}{h}\right)^3 + f_2\left(\frac{z}{h}\right)^2 + f_3\ ;\ z \le z_1 \\ f_4\left(\frac{z}{h}\right)^2 + f_5\left(\frac{z}{h}\right) + f_6\ ;\ z_1 \le z \le z_2 \\ k\left(\frac{z-d}{h}\right)\ ;\ z > z_2 \end{cases} \quad ; \quad \frac{L_1^{33}}{h} = \begin{cases} f_7\left(\frac{z-z_3}{h}\right)^2 + f_8\ ;\ z \le z_3 \\ f_9\left(\frac{z-z_3}{h}\right)^3 + f_{10}\left(\frac{z-z_3}{h}\right)^2 + f_8\ ;\ z_3 \le z \le z_4 \\ k\left(\frac{z-d}{h}\right) \qquad z > z_4 \end{cases} \tag{14}$$

where: $f_1 = -0.3$; $f_2 = 0.9$; $f_3 = 2$; $f_4 = 0.033136$; $f_5 = -0.32817$; $f_6 = 3.7337$; $f_7 = -0.03$; $f_8 = 0.85$; $f_9 = -0.22083$; $f_{10} = 0.7625$; $z_1 = 2.1h$; $z_2 = 10.987h$; $z_3 = 4h$; $z_4 = 6h$.
Note that far from the canopy the correlation lengths are required to satisfy a isotropy condition. For this reason they are equal to each other and equal to L_{iso} (see Appendix A).

4. Numerical simulations

Simulations have been performed for CO_2 dispersion in a Neutral Boundary Layer. The source is placed at the bottom ($z = 0$), while the top boundary condition at $z/h = 20$ is an absorbing one. Two different versions of 4DFP model have been considered. The first one is a isotropic version, with the coefficients b^{33} and c^{33} set to zero, while the second one is the version with axy-symmetric coherent structures described in the previous section. In this way, in the isotropic version the role of anisotropy (related to coherent structures) is neglected. The comparison between the isotropic and anisotropic versions of the model are reported in Fig. 4. It is evident that the inclusion of coherent structures in the anisotropic version substantially modify the concentration profiles and, in particular, it determines an accumulation of tracers at a given height. It follows that the role of coherent structures cannot be neglected in 4DFP model and, in particular, the accumulation is related to the presence of anisotropic motion structures. This result is in qualitative agreement with experimental and modeling studies on the subject [8].

Journal of Physics: Conference Series **633** (2015) 012082 doi:10.1088/1742-6596/633/1/012082

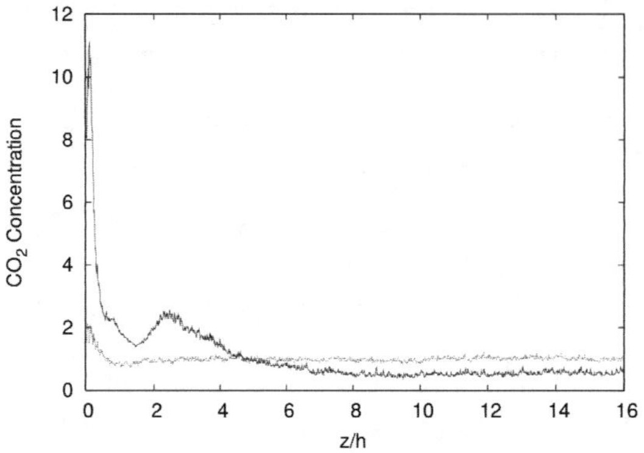

Figure 1. Numerical results for the CO_2 vertical profile. The isotropic (dotted line) and anisotropic (continuous line) models are compared.

5. Conclusions

We have shown that 4DFP-LSM can esplicitly include the geometrical properties of the turbulent structures. Numerical simulations have been performed in the case of vegetated canopies, comparing a isotropic version of the model with a version including coherent structures with rotational simmetry around vertical axis. Far from being a complete treatment of turbulent transport in vegetated canopies, the results suggest that the role of geometrical structure should be explicitly included. Even if limited to time scales comparable to or larger than the Eulerian/Lagrangian time scales, 4DFP model can explicitly include experimental information about space-time correlation structure of the velocity field. This is a crucial aspect in many situations such as wall turbulence and the motion of solid particles on time and space scales where the trajectory crossing effect is important [1]. In particular, the application to turbulent transport in vegetated canopies prove that accumulation features are associated with anisotropic motion structures in the fluid flow.

Appendix A. The model for the noise covariance tensor

A general form for the noise tensor, satisfying the incompressibility condition $\partial\langle\Delta w^i \Delta w^j\rangle/\partial\Delta^i = 0$, is a superposition of terms in the form:

$$\langle\Delta w^i \Delta w^j\rangle = \frac{2u_T^2}{\tau_E}[B_t^{ij}(\Delta^0) + B^{ij}(\mathbf{\Delta} - \bar{\mathbf{u}}\Delta^0))]; \qquad \partial_i B^{ij}(\mathbf{\Delta}) = 0. \tag{A.1}$$

where $\Delta^0 = \Delta t$ and $\mathbf{\Delta} = \Delta\mathbf{x}$, τ_E is the time scale of fluctuations, given by the Eulerian correlation time in the Gaussian case, $u_T^2 = \frac{1}{3}R_i^i$, and $B_t^{ij} = |\Delta^0|\delta^{ij} + \hat{B}_t^{ij}$, with $\hat{B}_t^{ij}$ symmetric and traceless[2].

In moderately anisotropic situations, it may be expedient to expand the space component B^{ij} in spherical tensors, following the SO(3) decomposition technique [12]: $B^{ij}(\mathbf{\Delta}) = \sum_{J=0} B_J^{ij}(\mathbf{\Delta})$, where B_J^{ij} indicates a combination of J-th order spherical tensors. A symmetric two-index tensor function can be decomposed in spherical tensors in the form

$$\delta^{ij}Y_J(\mathbf{x}), \qquad \partial^i\partial^j Y_J(\mathbf{x}), \qquad x^i x^j Y_J(\mathbf{x}), \qquad (x^i\partial^j + x^j\partial^i)Y_J(\mathbf{x}),$$

$$x_n(\epsilon^{jnm}x^i + \epsilon^{inm}x^j)\partial_m Y_J(\mathbf{x}) \quad \text{and} \quad x_n(\epsilon^{jnm}\partial^i + \epsilon^{inm}\partial^j)\partial_m Y_J(\mathbf{x}) \tag{A.2}$$

[2] For lighter notation we leave in this section the dependence from the space-time position unindicated.

Journal of Physics: Conference Series **633** (2015) 012082 doi:10.1088/1742-6596/633/1/012082

where $Y_J(\mathbf{x})$ is a J-order polynomial in the components x_i:

$$Y_J(\mathbf{x}) = y^{i_1 i_2 \dots i_J} x_{i_1} x_{i_2} \dots x_{i_J} \tag{A.3}$$

and $y^{i_1 i_2 \dots i_J}$ is traceless with respect to any pair of indices. In consequence of this, the spherical tensors in Eq. (A.3) will be polynomials of order $L = J,\ J-2,\ J+2,\ J,\ J+1$ and $J-1$ respectively. In the case of the noise tensor $\langle \Delta w^i \Delta w^j \rangle$, we have the additional symmetry with respect to spatial inversion, which imposes the condition that L be even. This implies J even for the first four spherical tensors and J odd for the last two. Limiting the analysis to $J \leq 2$, we notice immediately that the last spherical tensor in Eq. (A.2) disappears. Similarly the $J = 1$ contribution from $x_n(\epsilon^{jnm}x^i + \epsilon^{inm}x^j)\partial_m Y_J(\mathbf{x})$ is absent due to incompressibility. Writing $Y_1(\mathbf{x}) = y_m x^m$ it results:

$$\partial_i x_n(\epsilon^{jnm}x^i + \epsilon^{inm}x^j)\partial_m Y_1(\mathbf{x}) = 5\epsilon^{jnm}x_n y_m = 0 \ ,$$

which imposes $y_m = 0$. We are thus left only with the $J = 0$ and $J = 2$ contributions.
From the SO(3) decomposition introduced above, the $J = 0$ and $J = 2$ contributions to B^{ij} will have the form:

$$u_T B_0^{ij}(\mathbf{x}) = a|\mathbf{x}|\delta^{ij} + \hat{a}\frac{x^i x^j}{|\mathbf{x}|} \tag{A.4}$$

$$u_t B_2^{ij}(\mathbf{x}) = \frac{4b^{lm}}{|\mathbf{x}|}x_l x_m \delta^{ij} + 2c^{lm}|\mathbf{x}|\partial_i\partial_j x_l x_m + \frac{d^{lm}}{|\mathbf{x}|^3}x_l x_m x^i x^j$$
$$+\frac{e^{lm}}{2|\mathbf{x}|}(x^j\partial^i + x^i\partial^j)x_l x_m \tag{A.5}$$

Applying the incompressibility condition $\partial_i B^{ij} = 0$ leads to the equations:

$$\begin{cases} \hat{a} = -\frac{a}{3} \\ 8b^{lm} + 4c^{lm} + 8e^{lm} = 0 \\ 3d^{lm} - 4b^{lm} - e^{lm} = 0 \end{cases} \tag{A.6}$$

Substituting the solution to Eq. (A.6) into Eqs. (A.4-A.5) leads to:

$$B^{ij}(\mathbf{\Delta}) = \frac{|\mathbf{\Delta}|}{u_T}[(a + 4b^{lm}\hat{\Delta}_l\hat{\Delta}_m)\delta^{ij} + \frac{1}{3}(-a + (2b^{lm} - c^{lm})\hat{\Delta}_l\hat{\Delta}_m)\hat{\Delta}^i\hat{\Delta}^j$$
$$-\hat{\Delta}_l[(2b^{li} + c^{li})\hat{\Delta}^j + (2b^{lj} + c^{lj})\hat{\Delta}^i] + 4c^{ij}] \tag{A.7}$$

where a gives the $J = 0$ part, while the tensors b^{ij} and c^{ij}, which are symmetric and traceless, account for the the $J = 2$ part.
We consider next some relevant limit cases. In the isotropic case, the $J = 2$ contribution disappears and we are left with the simple expression:

$$\langle \Delta w^i \Delta w^j \rangle = \frac{2u_T^2}{\tau_E}[|\Delta^0|\delta^{ij} + \frac{a|\mathbf{\Delta}|}{u_T}(\delta^{ij} - \frac{1}{3}\hat{\Delta}^i\hat{\Delta}^j)] \tag{A.8}$$

The parameter a identifies a length-scale $l_u = u_T\tau_E/a$ for the fluctuations and has therefore the meaning of a ratio between the eddy life-time τ_E and the eddy rotation time l_u/u_T.

In the axisymmetric case, taking the symmetry axis along x^3, we have that only spherical harmonics with azimuthal quantum number $m = 0$ can contribute, i.e. Y_{00} and Y_{20}, the last one depending only on $(\hat{\Delta}_3)^2$. This means that all coefficients b^{ij} and c^{ij} with $i \neq j$ are zero and we have, remembering also the zero trace condition $b^i_i = c^i_i = 0$:

$$b^{11} = b^{22} = -\frac{1}{2}b^{33}, \qquad c^{11} = c^{22} = -\frac{1}{2}c^{33}. \tag{A.9}$$

So, in this case, there are only three unknown: a, b^{33} and c^{33}. Let us denote with $\boldsymbol{\Delta}_l$ the spatial increment in the direction x_l. Then, we define:

$$\beta^{ij}_l = \frac{u_T B^{ij}(\boldsymbol{\Delta}_l)}{|\boldsymbol{\Delta}_l|} \tag{A.10}$$

as the normalized structure function relative to the velocity u^i and u^j along the direction x_l. If we know at least three components of the tensor β^{ij}_l, we can take the corresponding equations given in (A.7) and derive a solution for the coefficients a, b^{33} and c^{33}. If only two components are known, then it is possible to close the linear system by considering that a still represents the ratio of the eddy lifetime to the eddy rotation time.

For example, let us suppose that only information along x_1 is known and consider Eqn. (A.7) for the velocity components u_1 and u_3 in the direction x_1:

$$\begin{cases} \frac{1}{3}\left(2a - b^{33} - \frac{5}{2}c^{33}\right) = \beta^{11}_1, \\ a + 4b^{33} - 2c^{33} = \beta^{33}_1, \end{cases} \tag{A.11}$$

Taking the coefficient a as given and solving with respect to b^{33} and c^{33}, we find immediately the result:

$$\begin{cases} b^{33} = \frac{1}{8}a - \frac{1}{2}\beta^{11}_1 + \frac{5}{24}\beta^{33}_1 \\ c^{33} = \frac{3}{4}a - \beta^{11}_1 - \frac{1}{12}\beta^{33}_1 \end{cases} \tag{A.12}$$

References

[1] P. Olla and P. Paradisi, Phys. Rev. E, **70**, 046305 (2004)
[2] D.J. Thomson, J. Fluid Mech. **180**, 529-556 (1987).
[3] C.W. Gardiner, *Handbook of Stochastic Methods* (Springer, Berlin 1985)
[4] J. Kim, P. Moin and R. Moser, J. Fluid Mech. **177**, 3990 (1987)
[5] C.I. Hsieh, M. Siqueira, G. Katul, C.R. Chu, Boundary Layer Meteorology **109**, 113-138 (2003)
[6] J. Finnigan: Ann. Rev. Fluid Mech., **32**, 519-571 (2000)
[7] T. Duman, G.G. Katul, M.B. Siqueira, M. Cassiani, Boundary-Layer Meteorol. **152**:118 (2014).
[8] A.M. Reynolds, PhytoPathology **102**(11), 1026-33, 2012.
[9] S.B. Pope, Turbulent Flows (Cambridge University Press, Cambridge, 2000).
[10] J.J.H. Brouwers, Phys. Rev. E **86**, 066309 (2012).
[11] M.S. Borgas and B.L. Sawford, J. Fluid Mech. **279**, 69 (1994); B. Sawford, Boundary-Layer Meteorol. **93**, 411-424 (1999); T.K. Flesch and J.D. Wilson, Boundary Layer Meteorology **61**, 349 (1992); J.D. Wilson and B.L. Sawford, Boundary Layer Meteorology **78**, 191-210 (1996); B.L. Sawford and P.K. Yeung, Phys. Fluids 13, 2627 (2001).
[12] I. Arad, V.S. L'vov and I.Procaccia, Phys. Rev. E, **59**, 6753 (1999)

Comparison of serial and parallel simulations of a corridor fire using FDS

L Valasek[1]
Slovak Academy of Sciences/Institute of Informatics, Dubravska cesta 9, 845 07 Bratislava, Slovakia

E-mail: lukas.valasek@savba.sk

Abstract Current fire simulators allow to model the course of fire in large areas and its impact on structure and equipment. This paper deals with a comparison of serial and parallel calculations of simulation of a corridor fire by the FDS (Fire Dynamics Simulator) system. In parallel case, the whole computational domain is divided into several computational meshes, the computation on each mesh is considered as a single MPI (Message Passing Interface) process realised on one computational core and communication between MPI processes is provided by MPI. The aim of this paper is to determine the size of error caused by parallelization of computation, which occurs at touches of computational meshes.

1. Introduction

Current fire simulation systems are capable to utilize a rapid progress of computational power of contemporary computers and model fires and their effects also in large areas. However, parallelization can introduce inaccuracies into calculation or even waste simulation results [1-5]. In this paper, error of parallel simulation of fire in a short corridor by FDS is investigated. The FDS (Fire Dynamics Simulator) system [6, 7] is an advanced CFD-based fire simulator intended for simulation of fire in various environments. It numerically solves a form of Navier-Stokes equations for low-speed fire induced flows with emphasis on transport of smoke and heat from fire. The system is capable to utilize various types of parallelization of calculation. In this paper, we focus on MPI (Message Passing Interface) version of parallelization, which is the most efficient [1] and most commonly used. We use the 64-bit MPI version 5.5.3 of FDS for MS Windows installed on a fast multi-core PC. In the MPI approach, the whole computational domain is divided into several computational meshes, the computation on each mesh is considered as a single MPI process assigned to one computational core and communication between MPI processes is provided by MPI. The aim of this paper is to determine the size of error caused by parallelization which occurs at touches of computational meshes.

2. Fire scenario

Let us consider a simple rectangular parallelepiped corridor with the dimensions of 24 m x 4 m x 3 m (length x width x height) with 10 cm thick concrete walls and the OPEN boundary condition on both exits of the corridor. The 20 °C ambient temperature is considered inside and outside the corridor. We do not assume any additional flammable materials inside the corridor, nor any airflow at the beginning of simulation. We consider a 10 MW fire source represented by a 0.4 m x 0.4 m burning surface

[1] To whom any correspondence should be addressed.

placed at the 21.8 m distance from the left corridor exit producing the 62500 kW/m2 HRRPUA (heat release rate per unit area) from the beginning until the end of simulation. In order to study the impact of parallelization on accuracy of simulation, we placed 239 thermocouples in the 50 cm height below the ceiling in the middle of the corridor. We consider simulation of the 5-minute fire.

3. Simulation results and discussion of errors caused by parallelization

For the purposes of this paper, we realised one sequential simulation 1M using one computational mesh, and three parallel simulations 3M, 6M and 12M using 3, 6 and 12 computational meshes, respectively. In all cases, we use the same 10 cm mesh density. According to the mesh sensitivity study [7] performed, such meshes can be considered as "fine" for all considered calculations already since the 1st s of fire. In the case of 1M, the calculation is realised as a single MPI process assigned to one computer core. In parallel simulation 3M, 6M and 12M, the calculation is realised as 3, 6 and 12 MPI processes assigned to 3, 6 and 12 computer cores, respectively.

The course of fire was as follows. At first, a hot gas induced by the fire began to spread from the fire source towards the ceiling. After the smoke hit the ceiling at the 1st s of fire, it spread towards both exits. The right and left exit was achieved at the 2nd and 8th s, respectively. At the end of simulation, a relatively steady flow can be observed (see figure 1). In figure 1, turbulent flows above the fire and at the front of hot layer below the ceiling, moving fast towards the left exit are highlighted.

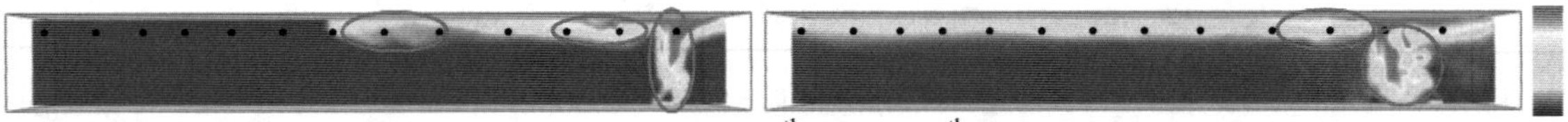

Figure 1. Vertical gas temperature slices at the 6th and 300th s of fire with selected thermocouples highlighted: temperature values vary from 0 °C (blue) to 1000 °C (red).

For the purposes of analysis of errors introduced into simulation by parallelization, we calculate the mean values of temperatures on 239 thermocouples obtained by calculations 1M, 3M, 6M and 12M, and compare the values obtained by parallel calculations with the values obtained by the sequential calculation. Figure 2 shows the differences between the mean values obtained by the parallel calculations and the sequential calculation as well as the values of their absolute values. For individual parallel calculations, these curves represent errors at selected thermocouples caused by parallelization. For simplicity, they are denoted by $e_{1,j}$ and $|e_{1,j}|$, $j = 3, 6, 12$. In figure 3, graphs of relative errors $r_{1,j}$ and $|r_{1,j}|$ corresponding to the errors $e_{1,j}$ and $|e_{1,j}|$, $j = 3, 6, 12$ are illustrated.

It follows from analysis of the curves in figure 2, that the considered thermocouples can be divided into three groups according to behaviour of errors of the calculation parallelization, depending on their distance from the fire source. *The 1st group* consists of the thermocouples placed in the distance smaller than 16 m from the left exit. This corresponds to the first 2, 4, and 8 meshes in the 3M, 6M, and 12M calculation, respectively, which are numbered from left to right from the left exit. The errors caused by parallelization at this group of thermocouples are relatively low and have a relatively very similar course (see e.g. the course of the errors over the <8 m, 14 m> interval in figure 2), whereby an increase of fluctuations towards the fire source can be observed. *The 2nd group* consists of the thermocouples located in the distance from 16 m to 19.2 m from the left exit, whereby the last thermocouple corresponds to the point from which the tendency of the $e_{1,3}$ curve changes from ascending to descending (see figure 2). It can be also observed that the course of all three analysed errors is qualitatively very similar. *The 3rd group* consists of the thermocouples laying on the right from the last thermocouple of the 2nd group. The course of errors at these thermocouples is characterized by high fluctuations, which partially change a relatively very similar qualitative behaviour of individual errors, in spite of high fluctuations of their values (see figure 2). These fluctuations are caused by high fluctuations of temperatures in the area, where the thermocouples are located (relatively close to the fire source).

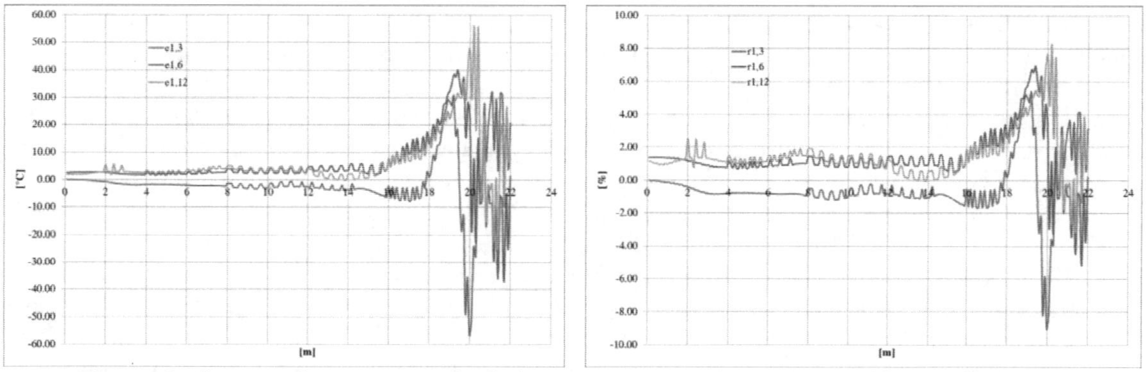

Figure 2. Course of errors $e_{1,j}$ and relative errors $r_{1,j}$, $j = 3, 6, 12$ caused by parallelization

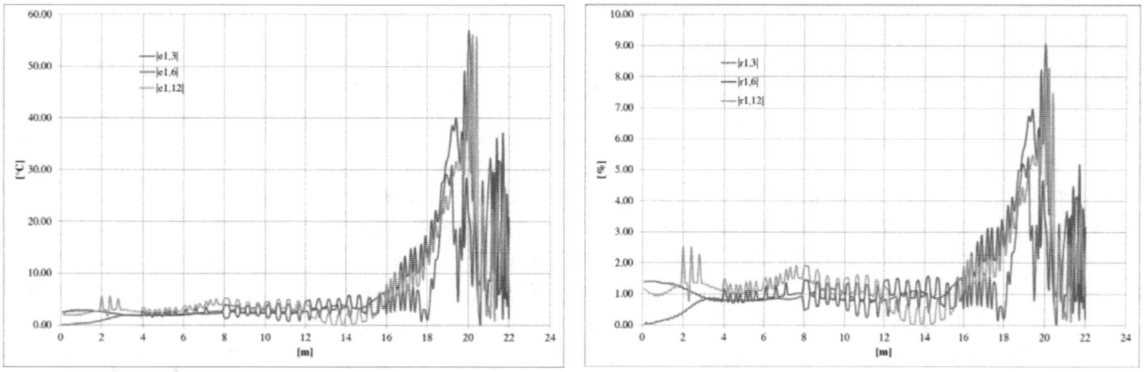

Figure 3. Course of absolute errors $|e_{1,j}|$ and absolute values of relative errors $|r_{1,j}|$, $j = 3, 6, 12$

It follows from figure 2 and figure 3 that relatively small errors of the mean values of temperatures caused by the calculation parallelization occur on the first two groups of thermocouples reaching values not exceeding 2% or 5% at the most of the thermocouples. This behavior corresponds to the places where relatively small turbulent flows occur. At the rest of thermocouples, i.e., in a neighborhood of the fire source, the error of mean values is increasing, but does not exceed the value of 9%. This error increase is related with stronger turbulent flows in places, where fast moving hot and cold gases mix each other. In the figures, it can be also seen that fluctuations of the errors of temperatures mean values as well as the corresponding relative errors above the fire source are relatively not very high, which is a consequence of the fact that the space above the fire source is already heated with only a small amount of cold air. The figures also show that the relative errors above the fire source rarely exceed 5%.

From a thorough analysis of the $e_{1,3}$ error behavior, it can be seen that in the 1st mesh, the error relatively slightly decreases and then a rapid increase of the error between the thermocouples located in the last cell of the 1st mesh and the 1st cell of the next mesh occurs. In the 2nd mesh, the course of the error is also decreasing with small fluctuations and then a rapid increase of the error between the thermocouples located in the last cell of the 2nd mesh and the 1st cell of the next mesh can be observed. In the 3rd mesh, the error behavior is strongly affected by turbulences of temperature caused by proximity to the fire source and by mixing of fast moving hot gases and cold air. Similarly, the course of the error $e_{1,6}$ is relatively slightly decreasing in the 1st mesh and then a rapid increase of the error follows between to thermocouples passing through the touch of meshes. In the 2nd, 3rd and 4th meshes, the error slightly increases with moderate fluctuations, which grow with decreasing the distance from the fire source. At the touches of meshes, increase of the error can be observed similarly as in the previous case. The error behavior in other meshes is markedly influenced by turbulent flow from the fire source. The error $e_{1,12}$ has a very similar course to the error $e_{1,6}$. The course of error $e_{1,12}$ in the first two meshes is probably partially affected by the boundary OPEN condition at the computational domain boundary which causes error fluctuations at the right from the touch of meshes. In seven subsequent meshes, the course of errors $e_{1,12}$ and $e_{1,6}$ is qualitatively very similar. The $e_{1,12}$ error behavior in other meshes can be interpreted similarly as in the previous cases; it is strongly affected by increasing the error fluctuations caused by temperatures fluctuations caused by mixing of fast moving hot gases and cold air in the vicinity of the fire source. The course of the relative errors $r_{1,3}$, $r_{1,6}$ and $r_{1,12}$ qualitatively corresponds to the course of the errors $e_{1,3}$, $e_{1,6}$ and $e_{1,12}$, respectively (see figure 2 and figure 3). The size of fluctuations in individual meshes is related to the distance from the fire source (as it was mentioned above), where the greatest fluctuations/values of the absolute error are concentrated in the 3rd mesh. The course of $|r_{1,3}|$, $|r_{1,6}|$ and $|r_{1,12}|$ is qualitatively the same as the course of the absolute errors $|e_{1,3}|$, $|e_{1,6}|$ and $|e_{1,12}|$, respectively. It follows from figure 3 that the values of the relative errors caused by parallelization do not exceed 5% in the upper corridor part, where turbulent mixing of fast moving hot gases inducted by fire with cold air does not occur.

4. Conclusion

In this paper, we realised three parallel simulations of corridor fire using 3, 6 and 12 computer cores, and sequential calculation with the same density of meshes. In each calculation, the mean values of temperature recorded on thermocouples placed in the 50 cm height below the ceiling in the middle of the corridor were calculated. The size of error caused by the parallelization of calculation, which was obtained by comparison of the mean temperature values calculated by the parallel simulations in regard of the sequential simulation, is described. From thorough analysis of the course of errors, it follows that in the upper part of the corridor, where the flow of hot gases induced by fire is concentrated, the values of relative errors at the thermocouples considered do not exceed 5% in the area, in which turbulent mixing of hot and cold gases does not occur. The error at the left from the fire source is affected by significant fluctuations and large turbulences caused by mixing of fast moving hot gases and cold air. The analysis of simulation results also shows that the size of error does not grow so significantly in places with turbulent flows without mixing of gases with high differences of temperature (above the fire source). Moreover, the analysis also indicates that, although the values of the errors caused by parallelization are not significantly high for relatively low numbers of processors, the problem of impact of parallelization on accuracy of calculation should be carefully investigated, especially in cases, where it is necessary to use more robust parallelization and numbers of processors.

Acknowledgement

This paper was supported by the Slovak Science Foundation VEGA (project No. 2/0184/14).

5. References

[1] Weisenpacher P, Glasa J, Halada L, Valasek L and Sipkova V 2014 Comput. Inform. 33 6 1237-1268
[2] Glasa J, Valasek L, Halada L and Weisenpacher P 2014 J. Phys.: Conf. Ser. 490 012067 DOI: 10.1088/1742-6596/490/1/012067
[3] Halada L, Weisenpacher P and Glasa J 2012 Computer modeling of automobile fires (chapter 9) In Advances in modeling of fluid dynamics InTech Publ. 203-228
[4] Weisenpacher P, Halada L, Glasa J and Astalos J 2014 Influence of parked cars on smoke propagation during car park fire Proc. of the 26th European Modeling & Simulation Symposium Bordeaux 384-391
[5] Zhao Z, Yao H, Liang D and Hu Z 2011 Procedia Eng. 11 723-729 DOI:10.1016/j.proeng.2011.04.719
[6] McGrattan K, McDermott R, Floyd J, Hostikka S, Forney G and Baum H 2012 Int. J. of Comput. Fluid Dynamics 26 (6-8) 349-361
[7] McGrattan K, Klein B, Hostikka S and Floyd J 2009 Fire Dynamics Simulator (Version 5): User's Guide, NIST Special Publication 1019-5, NIST, Washington

Journal of Physics: Conference Series **633** (2015) 012084 doi:10.1088/1742-6596/633/1/012084

A parallel approach of COFFEE objective function to multiple sequence alignment

G F D Zafalon[1], J M V Visotaky[1], A R Amorim[1], C R Valêncio[1], L A Neves[1], R C G de Souza[1], J M Machado[1]

[1] Department of Computer Science and Statistics (DCCE), São Paulo State University (UNESP), São José do Rio Preto, Brazil.

E-mail: zafalon@sjrp.unesp.br

Abstract. The computational tools to assist genomic analyzes show even more necessary due to fast increasing of data amount available. With high computational costs of deterministic algorithms for sequence alignments, many works concentrate their efforts in the development of heuristic approaches to multiple sequence alignments. However, the selection of an approach, which offers solutions with good biological significance and feasible execution time, is a great challenge. Thus, this work aims to show the parallelization of the processing steps of MSA-GA tool using multithread paradigm in the execution of COFFEE objective function. The standard objective function implemented in the tool is the Weighted Sum of Pairs (WSP), which produces some distortions in the final alignments when sequences sets with low similarity are aligned. Then, in studies previously performed we implemented the COFFEE objective function in the tool to smooth these distortions. Although the nature of COFFEE objective function implies in the increasing of execution time, this approach presents points, which can be executed in parallel. With the improvements implemented in this work, we can verify the execution time of new approach is 24% faster than the sequential approach with COFFEE. Moreover, the COFFEE multithreaded approach is more efficient than WSP, because besides it is slightly fast, its biological results are better.

1. Introduction

The biology has always been a study of the human object, trying to understand the functioning of living organisms and their possible relationships. So with the technological advances brought in recent decades, several studies have enabled a better understanding of living beings of all kinds, from the simplest to the most complex [1].

In this context, it is necessary to conduct experiments to prove the studies, which produce a large amount of information that need to be organized and arranged to become intelligible. This organization is not feasible using conventional means, ie a manual organization, emerging as important field of genetic computing resources, which led to Bioinformatics [2].

In this work, we proposed the implementation of parallel version of COFFEE objective function using multithread paradigm in the multiple sequence alignment tool MSA-GA. Thus, it is possible to reach improvements in execution time of the objective function without lost the biological significance of the alignments.

This work is organized as follows: in the section 2 a brief review about parallel sequence alignment is provided. In the section 3 are described the materials and methods with the special attention to the implementation of the objective function in the MSA-GA tool. Some analysis and results are presented in the section 4. Finally, in the section 5, the conclusions and future perspectives are presented.

2. Parallel Multiple Sequence Alignments

Bioinformatics has as primary goal to provide computational solutions to problems that are important to biology, assisting in the analysis of experimental data in search of biological identities between sequences [1]. However, the difficulty lies in the fact that, given the huge amount of scheduled sequences, using deterministic techniques to treat them is highly unfeasible due to the computational cost [2]. Thus, as an alternative, we started to use heuristic techniques, especially for multiple sequence alignment.

It is interesting the development of tools that can mitigate the high cost in terms of execution time in bioinformatics problems, using the processing provided by parallelism [3]. As an example, we can cite parallel versions of CLUSTALW [4], the MAFFT [5], T-COFFEE [6], among others.

3. Materials and Methods

The COFFEE is an objective function that works on a pairwise alignment reference library. The alignment of the evaluation is performed by positions. For each estimated alignment position, a scoring matrix is built, which is filled with the weights assigned to each pairwise alignment available in the library. In position (column) analyzed, each cell of the matrix corresponds to the alignment between two residues of the position. If the alignment between the two residues are found in the library, then is allocated to the cell the weight of the library, otherwise it is assigned the value 0. The score of the position is given by the sum of all values in the scoring matrix divided by the sum of weights of the alignments involved. Then the total score of alignment is calculated by the sum of the scores of all columns divided by the alignment length [7]. In the Figure 1 is presented the schematic calculation of COFFEE score.

To achieve greater gains in terms of execution time and use any idle computing resources in tournaments phase of the genetic algorithm, the implementation of COFFEE objective function was slightly altered to allow its calculations are performed simultaneously by the processing elements processing that are idle or incomplete recovery. As shown in Figure 1, the function is able to calculate multiple column scores simultaneously for alignment being evaluated, rather than just one at a time as it was in the original version. All obtained column scores are used to calculate the final score of the alignment.

4. Tests and Results

The tests were performed in a computer with Intel Xeon X3430@2.4 GHz Quad-Core processor, 16 GB of RAM memory and Windows 7 Home Premium 64-bits. Due the stochastic approach of genetic algorithms, we have executed each test case five times, in order to compare the execution time of the new approach with multithreaded COFFEE and the version with sequential COFFEE objective function. The result used in each case test, for comparison purposes, was the average of the five alignments execution times.

We have used the well-known test cases from BAliBase[1]. The sets we have chosen are divided into two main categories: the less similar ones (< 25% of identity and 20% ~ 40% of identity) and the more similar ones (> 35% of identity). Thus, in the Table 1 are shown the obtained execution times, given in seconds, of COFFEE Multithread (COFFEE-M) comparing with the sequential COFFEE (COFFEE-S) approach.

Thus, it can be noticed that multithread COFFEE approach was able to improve the execution time of alignments of the less similar test cases, when compared with the sequential COFFEE. The results obtained in the present work for these cases have achieved at least 7% of improvements and 28% at most, as showed in Figure 2 (a) and Figure 2 (b). Moreover, it can be noticed that the improvements here presented were also obtained for alignments of more similar sequences sets, as can be seen in Figure 2 (c). In these cases, the multithread COFFEE was at least 5% faster than sequential COFFEE and 29% at most.

We have calculated the maximum, minimum and the average standard deviation for COFFEE multithreaded and COFFEE sequential. For the multithreaded approach, the maximum, the minimum and the average were 1.61, 0.01 and 0.40, respectively.

[1] http://www.lbgi.fr/balibase/.

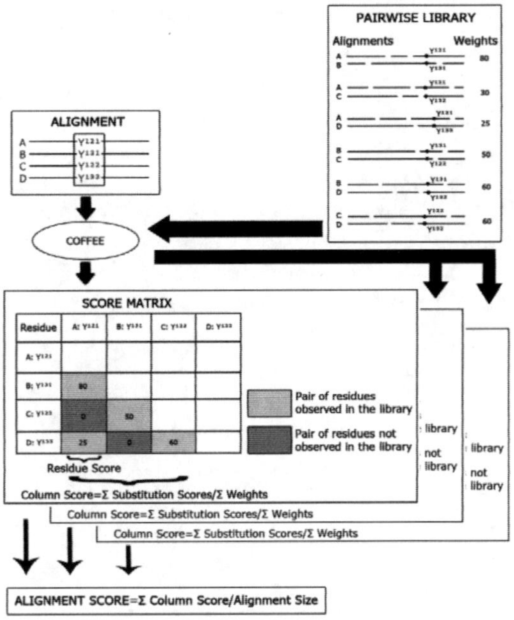

Figure 1. Operation mode of COFFEE objective function.

Table 1. Execution time of each test case.

Sequence Set	COFFEE-M (s)	COFFEE-S (s)	Sequence Set	COFFEE-M (s)	COFFEE-S (s)	Sequence Set	COFFEE-M (s)	COFFEE-S (s)
<25% identity			**20% ~ 40% identity**			**>35% identity**		
1idy	008,7	010,0	1ycc	012,1	013,2	1amk	034,4	046,8
1tvxA	008,7	009,4	1ad2	022,0	030,8	1aho	009,7	010,2
Kinase	052,4	067,6	1aym3	026,3	033,8	1csp	009,8	011,0
1r69	008,7	010,0	1fieA	091,4	121,4	1ar5A	019,8	023,6
1ubi	010,4	012,0	1ldg	034,6	044,4	1ad3	065,4	090,0
1wit	014,5	017,6	1sesA	093,7	125,4	1gpb	201,4	282,2
1ped	031,1	039,8	3cyr	011,9	013,2	1krn	011,0	013,2
2myr	070,4	089,2						

For the sequential approach, the maximum, the minimum and the average were 5.35, 0.44 and 1.29, respectively. This indicates that the variation of multithreaded approach is smaller than sequential one, which shows the stability of the multithread strategy. Finally, when comparing the total execution time of all test cases, it can be noticed that the overall multithread COFFEE execution time was 24% faster than the sequential COFFEE approach. Moreover, the quality of the results was kept without lost the biological significance of the final alignments.

5. Conclusion

Through extensive evaluation of the final version of the multithreaded COFFEE objective function developed in this work, we proved to be able the reduction of execution time of COFFEE objective function, improving the efficiency in relation to the initial version and using all avaliable computational resources. Moreover, we did not depreciate the quality of final alignments. We expect to implement the stratategy using Graphics Processing Units (GPUs).

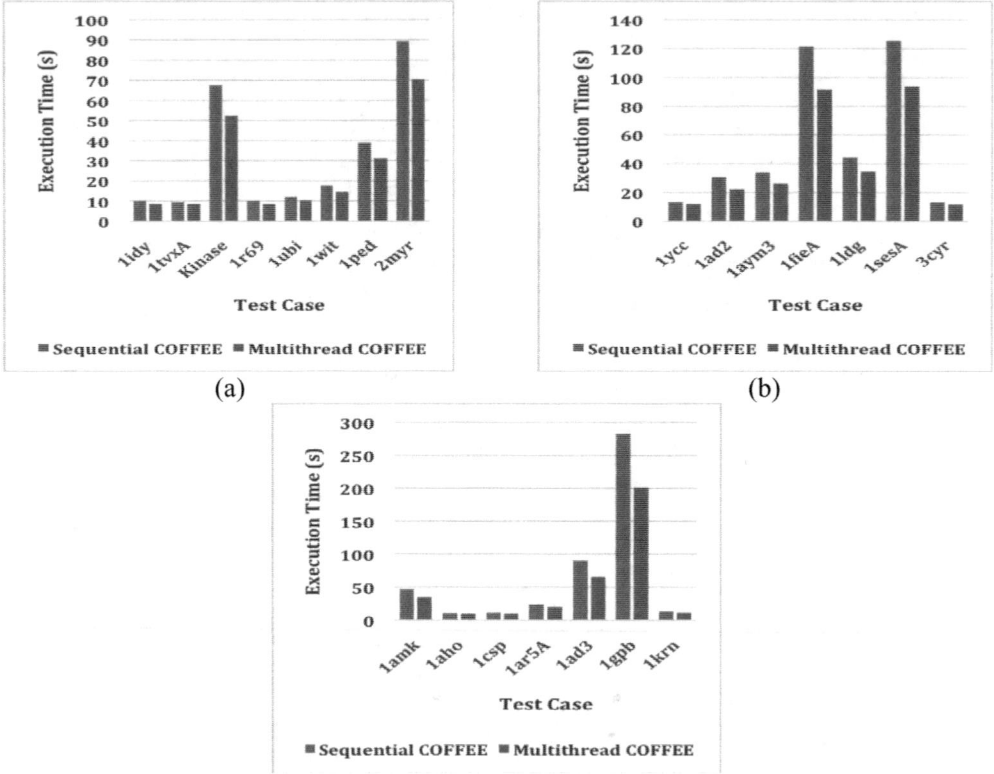

(c)

Figure 2. Execution time given in seconds: (a) sets with < 25% of similarity, (b) sets with similarity between 20% and 40% and (c) sets with > 35% of similarity.

6. Acknowledgments

This work was partially supported by São Paulo Research Foundation (FAPESP), under grant 2013/08289-0 and São José do Rio Preto Extension and Research Foundation (FAPERP).

References

[1] Khuri S 2008 A bioinformatics track in computer science. *ACM SIGCSE Bulletin* **40** n. 1 p. 508–512

[2] Amorim A R et al. 2015 Improvements in the sensibility of MSA-GA tool using COFFEE objective function *Journal of Physics: Conference Series* **574** n. 1 p. 12104

[3] Marucci A M et al. 2014 An Efficient Parallel Algorithm for Multiple Sequence Similarities Calculation Using a Low Complexity Method. *Biomed Research International* **2014** p. 1-6

[4] Li, K-B 2003 ClustalW-MPI: ClustalW analysis using distributed and parallel computing *Bioinformatics* **19** n. 12 p. 1585–1586

[5] Katoh, K and Toh H 2010 Parallelization of the MAFFT multiple sequence alignment program *Bioinformatics* **26** n. 15 p. 1899–900

[6] Zola J et al. 2007 PARALLEL-TCOFFEE: A parallel multiple sequence aligner. *ISCA PDCS* p. 248–253

[7] Notredame C and Holm L and Higgins D 1998 COFFEE: an objective function for multiple sequence alignments *Bioinformatics* **14** n. 5 p. 407–422

Journal of Physics: Conference Series **633** (2015) 012085 doi:10.1088/1742-6596/633/1/012085

Large deviation estimates for a Non-Markovian Lévy generator of big order

Rémi Léandre

Laboratoire de mathématiques. Université de Franche-Comté. route de Gray. 25030. Besançon. FRANCE

E-mail: `remi.leandre@univ-fcomte.fr`

Abstract. We give large deviation estimates for a non-markovian convolution semi-group with a non-local generator of Lévy type of big order and with the standard normalisation of semi-classical analysis. No stochastic process is associated to this semi-group.

1. Introduction

There are much more semi-groups than semi-groups which are represented by stochastic processes. On the other hand, there are a lot of formulas in stochastic analysis which are natural. The theory of pseudodifferential operators [1–3] allow to understand a lot of partial differential equations, including parabolic equations. On the other hand we have imported in the theory of non-markovian semi-groups a lot of tools of stochastic analysis [4–17]. Stochastic analysis formulas are valid for the whole process. Their interpretation for non-markovian semi-groups work only for the semi-group.

In [16] and [17], we have done with the classical normalization of semi-classical analysis [18] Wentzel-Freidlin estimates [19] for four order differential operators. Here we extend the method of [16] to the case of an integro-differential operator of big order which generates a non-markovian convolution semi-group. Normalisation are of Maslov type [18].

2. Statement of the theorems

Let $C_b^\infty(R)$ the set of smooth functions on R with bounded derivatives at each order endowed with its natural topology. $C_b(R)$ is the space of bounded contiunous functions endowed with the uniform norm. L^2 is the space of square integrable norms for the Lebesgue measure. This is an Hilbert space endowed with its natural scalar product $<,>$.

Let h be a smooth positive function on R with compact support such that $h(y) = h(-y)$ and such that $h(y) = 1$ is equal to 1 on $[-\beta, \beta]$.

Let be $\alpha \in [0, 1]$. We introduce the Levy generator acting on $C_b^\infty(R)$:

$$Lf(x) = (-1)^{d+1} \int_R (f(x+y) - f(x) - \sum_{i=1}^{d} \frac{y^{2i}}{2i!} f^{(2i)}(x)) \frac{h(y)}{|y|^{2d+\alpha}} dy \tag{1}$$

$\frac{h(y)}{|y|^{2d+\alpha}} dy$ is called the Lévy measure.

Theorem 1:: *L is symmetric positive on L^2. It has therefore a natural essentially-self adjoint extension which generates a semi-group of contraction P_t* on L^2.

We consider the Hamiltonian

$$H(\xi) = \int_R (\exp[\xi y] - 1 - \sum_{i=1}^{d} \frac{(\xi y)^{2i}}{2i!}) \frac{h(y)}{|y|^{2d+\alpha}} dy \tag{2}$$

Theorem 2: $H(\xi)$ i *is a smooth convex function equals to* 1 *in* 0.

Associate to it, we consider its Legendre transform:

$$L(p) = \sup_{\xi \in R}(\xi p - H(\xi)) \tag{3}$$

If ϕ is a finite energy function in R, we consider the action functional

$$S(\phi) = \int_0^1 L(\frac{d\phi}{dt})dt \tag{4}$$

Let us recall some basis of the pseudodifferential calculus. $\hat{f}$ is the Fourier transform of f. Let L_1 be an operator acting on $C_b^\infty(R^d)$ by

$$L_1 f(x) = \int_R a(x,\xi)\hat{f}(\xi)\exp[2\sqrt{-1}\pi\xi x]d\xi \tag{5}$$

We say that $a(.,.)$ is its symbol. If

$$|\frac{\partial^n}{\partial x^n}\frac{\partial^m}{\partial \xi^m}a(x,\xi)| \leq C|\xi|^{r-m} \tag{6}$$

and if for $|\xi| > C_0$

$$|a(x,\xi)| \geq C|\xi|^r \tag{7}$$

we say that L_1 is an elliptic operator of order r. Let us recall that our thesis underline the relationship between pseudodifferential operators and Poisson processes [20].

Theorem 3: *L is an elliptic pseudodifferential operator.*

By elliptic theory, it generates a semi-group on $C_b(R)$.

According the theory of semi-classical analysis [19], we consider the symbol L_1^ϵ associated to the symbol $\epsilon^{-1}a(x,\epsilon\xi)$. This leads to the operator

$$L^\epsilon f(x) = (-1)^{d+1}\frac{1}{\epsilon}\int_R (f(x+\epsilon y) - f(x) - \sum_{i=1}^{d}\frac{(\epsilon y)^{2i}}{2i!}f^{(2i)}(x))\frac{h(y)}{|y|^{2d+\alpha}}dy \tag{8}$$

By elliptic theory L^ϵ generates a semi-group on L^2 and even on $C_b(R)$ P_t^ϵ. We consider its absolute value $|P_t^\epsilon|$. We have

Theorem 4 (Wentzel-Freidlin estimates): *Let O be the complement in R of the interval $[x-\delta, x+\delta]$. We have when $\epsilon \to 0$*

$$\overline{Lim}\epsilon Log|P_1^\epsilon|[1_O](x) \leq -\inf_{\phi(1)\in O} S(\phi) \tag{9}$$

if $d+1$ is even.

The proof is very similar to the proof of [16], the only difference being in the algebraic treatment of Davies method [21].

3. Proofs of Theorem 1, Theorem 2 and Theorem 3

Proof of theorem 1: Let us show that L is symmetric. Let f anf g be smooth with compact supports:

$$(-1)^{d+1} < Lf, g > = \int_{R^2} g(x)((f(x+y) - f(x) - \sum_{i=1}^{d} \frac{y^{2i}}{2i!} f^{(2i)}(x)) \frac{h(y)}{|y|^{2d+\alpha}}) dx dy \quad (10)$$

The symmetry holds by integrating by parts and since $h(-y) = h(y)$.

Let us show that L is positive. We have if $y > 0$

$$f(x+y) - f(x) = \sum_{i=1}^{2d-1} \frac{y^i}{i!} f^{(i)}(x) + \int_{0<s_1<..<s_{2d}<y} f^{(2d)}(x+s_1) ds_1..ds_{2d} \quad (11)$$

Due to the parity of h, we have only to look at

$$\int_{R\times R^+} f(x) \int_{0<s_1<..<s_{2d}<y} f^{(2d)}(x+s_1) ds_1..ds_{2d} \frac{h(y)}{|y|^{2d+\alpha}} dx dy \quad (12)$$

By integrating by parts, it is equal to:

$$(-1)^d \int_{R\times R^+} f^{(d)}(x) \int_{0<s_1<..<s_{2d}<y} f^{(d)}(x+s_1) ds_1..ds_{2d} \frac{h(y)}{|y|^{2d+\alpha}} dx dy \quad (13)$$

By Cauchy -Schwartz inequality,

$$\int_R f^{(d)}(x) f^{(d)}(x+s_1) dx \leq \int_R (f^{(d)}(x))^2 dx \quad (14)$$

Therefore

$$(-1)^{d+1} \int_{R\times R^+} f(x)$$

$$\int_{0<s_1<..<s_{2d}<y} (f^{(2d)}(x+s_1) - f^{(2d)}(x)) ds_1..ds_{2d} \frac{h(y)}{|y|^{2d+\alpha}} dx dy \geq 0 \quad (15)$$

This shows the result. The fact that the operator has a natural self-adjoint extension which is essentially self-adjoint holds by standard results.◊

Proof of theorem 2: $H(\xi)$ is smooth. We have clearly

$$H^{(1)}(\xi) = \int_R y(\exp[\xi y] - \sum_{i=1}^{d} \frac{(\xi y)^{2i-1}}{(2i-1)!}) \frac{h(y)}{|y|^{2d+\alpha}} dy \quad (16)$$

$$H^{(2)}(\xi) = \int_R y^2(\exp[\xi y] - 1 - \sum_{i=1}^{d-1} \frac{(\xi y)^{2i}}{2i!}) \frac{h(y)}{|y|^{2d+\alpha}} dy \quad (17)$$

Due to the fact that $h(y) = h(-y)$, the result holds from the fact by induction on l that

$$\exp[y\xi] + \exp[-\xi y] - \sum_{i=0}^{l} \frac{\xi^{2i} y^{2i}}{2i!} \quad (18)$$

Journal of Physics: Conference Series **633** (2015) 012085 doi:10.1088/1742-6596/633/1/012085

is positive convex in ξ.$\diamondsuit$

Proof of theorem 3: Let us compute the symbol of L.

$$f(x) = C\int_R \hat{f}(\xi)\exp[\sqrt{-1}x\xi]d\xi \tag{19}$$

Therefore

$$Lf(x) + \int_R \frac{h(y)}{|y|^{2d+\alpha}}dy\int_R(\exp[\sqrt{-1}(x+y)\xi] - \sum_{i=0}^{2d}\frac{(\sqrt{-1}y\xi)^{2i}}{2i!}\exp[\sqrt{-1}x\xi])\hat{f}(\xi)d\xi =$$
$$\int_R \hat{f}(\xi)\exp[\sqrt{-1}x\xi]d\xi\int_R(\exp[\sqrt{-1}\xi y] - 1 - \sum_{i=1}^{d}\frac{(\sqrt{-1}\xi y)^{2i}}{2i!})\frac{h(y)}{|y|^{2d+\alpha}}dy \tag{20}$$

Therefore the symbol is given by

$$a(\xi) = H(\sqrt{-1}\xi) \tag{21}$$

By puting $y\xi = z$ if $\xi > 0$ we get that

$$a(\xi) = \xi^{(2d-1+\alpha)}\int_R h(\frac{z}{\xi})(cos[z] - \sum_{i=0}^{2d}\frac{(-1)^i z^{2i}}{(2i)!})dz \tag{22}$$

In (22), we consider a smooth h_1 function which is equal to zero near 0 and which is equal to 1 in a neighborhood of the infinity and which takes its values in $[0, 1]$ and we write

$$a(\xi) = a_1(\xi) + a_2(\xi) \tag{23}$$

$$a_1(\xi) = \xi^{(2d-1+\alpha)}\int_R h_1(\frac{z}{\xi})h(\frac{z}{\xi})(cos[z] - \sum_{i=0}^{2d}\frac{(-1)^i z^{2i}}{(2i)!})dz \tag{24}$$

By integrating by parts succesively, $|a_1(\xi)| \leq C\xi^{-n}$ for all n. On the other if the support of $1 - h_1$ is small enough, we have

$$|a_2(\xi)| \geq C\xi^{(2d-1+\alpha)} \tag{25}$$

for some positive C. The result arise by symmetry for $\xi < 0$. $\diamondsuit$

4. Proof of the Wentzel-Freidlin estimates

Let us begin by some elementary remarks. We remark that

$$\hat{Lf} = H(\sqrt{-1}.)\hat{f} \tag{26}$$

such that

$$\hat{P_t f} = \exp[-tH(\sqrt{-1}.)]\hat{f} \tag{27}$$

These elementary remarks (which are true a lot of convolution semi-groups) will allow us to adapt the proof of [16].

Lemma 5: *For all $\delta > 0$, all C there exist t_δ such that if $t < t_\delta$*

$$|P_t^\epsilon|(|1_{[x-\delta,x+\delta]^c})(x) \leq \exp[-\frac{C}{\epsilon}] \tag{28}$$

Proof: We consider the semi-group

$$\exp[-\frac{x\xi}{\epsilon}]P_t^\epsilon[\exp[\frac{x'\xi}{\epsilon}]f(x')](x) \tag{29}$$

Journal of Physics: Conference Series **633** (2015) 012085 doi:10.1088/1742-6596/633/1/012085

The symbol of its generator is

$$F_\xi^\epsilon(\xi') = \frac{1}{\epsilon}H(\epsilon\sqrt{-1}\xi' + \xi) \tag{30}$$

This is the symbol of an elliptic operator which is positive if $|\xi'|$ is big. It generates therefore a semi-group on $C_b(R)$ $Q_t^{\epsilon,\xi}$. We get the expansion

$$F_\xi^\epsilon(\xi') = \frac{H(\xi)}{\epsilon} + H^{(1)}(\xi)\sqrt{-1}\xi' + \epsilon\int_{0<s_1<s_2<1}(\xi')^2 H^{(2)}(\epsilon s_1\sqrt{-1}\xi' + \xi)ds_1 ds_2 +$$
$$\frac{H(\xi)}{\epsilon} + H^{(1)}\sqrt{-1}\xi' + R_\xi^\epsilon(\xi') \tag{31}$$

Therefore We get

$$\widehat{Q_t^{\epsilon,\xi} f} = \exp[-\frac{tH(\xi)}{\epsilon}]\exp[-t(+H^{(1)}\sqrt{-1}\xi' + R_\xi^\epsilon(\xi'))]\hat{f} \tag{32}$$

The uniform norm of $\exp[-t(+H^{(1)}\sqrt{-1}\xi' + R_\xi^\epsilon(\xi'))$ is bounded and the uniform norm of its derivative is bounded by $\exp[C|\xi|]/\epsilon$. Therefore the norm on $C_b(R)$ of $Q_t^{\epsilon,\xi}$ is bounded by $\exp[-\frac{CtH(\xi)}{\epsilon}]\exp[C|\xi|]$. Therefore

$$|P_t^\epsilon|(|1_{[x-\delta,x+\delta]^c})(x) \leq \exp[-\frac{CtH(\xi)}{\epsilon}]\exp[\frac{\delta\xi}{\epsilon}]\exp[C\xi] \tag{33}$$

But $H(\xi) \geq C|\xi|$ if $|\xi| > K(C)$ for all $C.\diamondsuit$

Remark: This inequality where the classical Davies gauge transform plays a fundamental role [21] replace the role of exponential martingales of [19].

When we have proved this lemma, the estimates follow closely the lines of [16] and [19].

We cut the time interval $[0,1]$ is small intervals of length $[t_i, t_{i+1}]$. By the semi group property we use that

$$|P_1^\epsilon|[1_{[x-\delta,x+\delta]^c}](x) \leq |P_{t_1}^\epsilon|..|P_{1-t_n}^\epsilon|[1_{[x-\delta,x+\delta]^c}](x) \tag{34}$$

In $P_{t_{i+1}-t_i}^\epsilon$, we distinguish if $x_{t_{i-1}}$ and x_{t_i} are far or not. If they are we use the previous lemma. If they are close, we deduce a positive measures $|W_\epsilon|$ on polygonal paths ϕ_t which joins x_{t_i} to $x_{t_{i+1}}$. By the previous lemma, it remains to estimate $|W_\epsilon|[|1_{[x-\delta,x+\delta]^c}(\phi_1)]$. But $|W_\epsilon|$ is a positive measure, we have

$$|W_\epsilon|[|1_{[x-\delta,x+\delta]^c}](\phi_1)] \leq |W_\epsilon|[\exp[\frac{S(\phi)}{\epsilon}]1_{[x-\delta,x+\delta]^c}(\phi_1)]\exp[-\inf_{\phi_1\in[x-\delta,x+\delta]^c}\frac{S(\phi)}{\epsilon}] \tag{35}$$

Therefore we have only to estimate $|W_\epsilon|[\exp[\frac{S(\phi)}{\epsilon}]1_{[x-\delta,x+\delta]^c}(\phi_1)]$. The sequel follows [19] p 152 ([16]. We can choose some p_i in finite numberrs such that if we put

$$L'(p) = \sup_i(L(p_i) + \frac{\partial}{\partial p}L(p_i)(p - p_i)) \tag{36}$$

we have for all polygonal paths considereded for a small χ

$$L(\frac{d\phi_t}{dt}) - L'(\frac{d\phi_t}{dt}) \leq \chi \tag{37}$$

Let us put

$$S'(\phi) = \int_0^1 L'(\frac{d\phi_t}{dt})dt \tag{38}$$

Since $|W_\epsilon|$ is a positive measure, we have only to estimate the quantity

$$|W_\epsilon|[\exp[\frac{S'(\phi)}{\epsilon}]1_{[x-\delta,x+\delta]^c}(\phi_1)] \tag{39}$$

We remark that

$$\exp[\sup a_i] \leq \sum \exp[a_i] \tag{40}$$

Moreover

$$L'(p) = \sup(\xi_i p - H(\xi_i)) \tag{41}$$

where $\xi_i = \frac{\partial}{\partial p}L(p_i)$. Therefore it is enough to show that

$$\sup_{x,|\xi|<C} |P^\epsilon_{t_\delta}|[\exp[\frac{\xi}{\epsilon}(x'-x) - t_\delta H(\xi)]](x) \tag{42}$$

has a small blowing up when $\epsilon \to 0$. We do as in the previous lemma. We consider the generator of the semi group

$$f \to P^\epsilon_t[\exp[\frac{\xi}{\epsilon}(x'-x) - tH(\xi)]f](x) \tag{43}$$

Its symbol is

$$\frac{1}{\epsilon}H(\epsilon\sqrt{-1}\xi' + \xi) - \frac{1}{\epsilon}H(\xi) \tag{44}$$

Its asymptotic expansion in ϵ is

$$((H^{(1)}\sqrt{-1}\xi' + R^\epsilon_\xi(\xi') \tag{45}$$

The result follows as in the lemma.$\diamondsuit$

5. Conclusion

We have adapted the standard proof of large deviation estimates of jump processes of [19] (with the standard normalisation of semi-classical analysis [18]) to the case of a non-markov Lévy generator of big order. The main difference with [16] is that the classical gauge transform of Davies [21] induces a simple transformation on the symbol of the Lévy generator [20]

References

[1] Dieudonné J 1977 *Eléments d'analyse VII* (Paris: Gauthiers-Villars).
[2] Hoermander L 1984 *The analysis of linear partial operators III* (Berlin: Springer).
[3] Hoermander L *The analysis of linear partial operators IV* (Berlin: Springer).
[4] Léandre R 2012 A path-integral approach to the Cameron-Martin-Maruyama-Girsanov formula associated to a Bilaplacian *Integral and differential systems in function spaces. Journal of function spaces and applications.*, 458738. ed Ti-Jun Xiao
[5] Léandre R 2011 *2011 World Congress Engineering Technology*,ed Victor Jin C.D. (Los Alamitos: I.E.E.E.) p 68
[6] Léandre R Stochastic analysis for a non-markovian generator: an introduction Preprint to appear *Festchrift in memory S. Belavkin.Rus. Journal. Math. Phys.*, ed J. Gough
[7] Léandre R 2013 A Girsanov formula associated to a big order pseudodifferential operator *Festchrift in honour of G. N'Guérékata. Cubo: a Mathematical Journal.* **15** 113, ed B. Anglade
[8] Léandre R 2008 *Non-Euclidean geometry and its applications. Acta. Physica Debrecin.* **42** 133. ed P. Nagy.
[9] Léandre R 2009. Itô-Stratonovitch for the Schroedinger equation associated to a big order operator on a torus,. *Fractional Order differentiation. Physica Scripta,* **136** 014028. ed G. Zaslavsky.

[10] Léandre R. 2010*Computations of stochastic systems* (Trans. Comp. Sciences. VII. L.N.C.S. Vol 5890) ed M A El Tawil .(Berlin: Springer) p 68.
[11] Léandre R 2010 *ISAAC 2009* ed M Ruzhansky and J Wirth (Singapore: World Scientific) p.225.
[12] Léandre R 2010 *Journal Pseudo Differential Operators and Applications.* **1** 389
[13] Léandre R 2011 Long time behaviour on a path group of the heat semi-group associated to a bilaplacian *Symmetry measures on complex networks. Symmetry.* **3** 72 ed. A. Garrido
[14] Léandre R 2012 An Itô formula for an accretive operator *Axioms: feature papers Axioms.* **1** 4. ed A. Garrido
[15] Léandre R 2013 *Geometry of science information 2013*, (L.N.C.S. Vol 8085) ed F. Nielsen (Berlin: Springer) p 497.
[16] Léandre R 2014 *2014 Int. Conf. Comput Science and Comput Intel.* Ed B. Akhgar (Los Alamitos: IEEE computer Society.) p 360
[17] Léandre R 2014 The Itô-Stratonovitch formula for an operator of order four and some applications to large deviation theory *Preprint.*
[18] Maslov V.P. and Fedoriuk M.V 1981 *Semiclassical approximation in quantum mechanics* (Dordrecht: Reidel).
[19] Wentzel A.D and Freidlin M.J 1984 *Random perturbations of dynamical systems*(Berlin: Springer).
[20] Léandre R 1984 *Extension du théoreme de Hoermander a divers processus de sauts.* PHD thesis (Université de Besançon).
[21] Davies E.B 1997 *Bulletin London Mathematical Society* **29** 513.

Analysis of the chaotic maps generating different statistical distributions

M Lawnik

Faculty of Applied Mathematics, Silesian University of Technology, ul. Kaszubska 23, 44-100 Gliwice, Poland

E-Mail: marcin.lawnik@polsl.pl

Abstract. The analysis of the chaotic maps, enabling the derivation of numbers from given statistical distributions was presented. The analyzed chaotic maps are in the form $x_{k+1} = F^{-1}(U(F(x_k)))$, where F is the cumulative distribution function, U is the skew tent map and F^{-1} is the inverse function of F. The analysis was presented on the example of chaotic map with the standard normal distribution in view of his computational efficiency and accuracy. On the grounds of the conducted analysis, it should be indicated that the method not always allows to generate the values from the given distribution.

1. Introduction

Pseudo-random numbers are very significant in many disciplines of science, including cryptography and simulations. In cryptography, they are essential elements of ciphers, whereas in simulations they enabling the modelling of many important physical, economic and other related processes.

Thus, many methods have been developed to make it possible to generate pseudo-random numbers from given distributions. The most popular one is the inverse cumulative distribution function method based on a specific distribution [1]. The method involves the generation of random variable $U \in [0,1]$ derived from the uniform distribution and in the next step, transformed by means of the following dependence:

$$X = F^{-1}(U), \tag{1}$$

where F is the cumulative distribution function (CDF) and F^{-1} denotes the inverse function of F.

It is then that the determined random variable has a distribution determined by F. The disadvantage of this method is the fact that for some distributions, for example, normal distribution, the cumulative distribution function cannot be expressed by means of elementary functions. In such cases, approximations should be used by means of other functions that are invertible.

Apart from (1), to derive numbers from the required distribution, transformations may be applied [1], which makes it possible to transform one distribution (in most cases uniform distribution) to another one. Another very popular methods for generating pseudorandom numbers from given distribution are the so called rejection and decomposition methods [1].

In [2] Lai and Chen presented an interesting way of generating numbers from given distributions, involving the construction of chaotic maps in the following form:

$$x_{k+1} = F^{-1}(U(F(x_k))), \tag{2}$$

where F is the cumulative distribution function of the required distribution and U is the skew tent map. The successive iterations of (2) generate numerical values from the distribution designated by F.

Skew tent map U used in (2) is designated by the equation:

$$x_{k+1} = f(x_k) = \begin{cases} \dfrac{x_k}{p}, & 0 \le x_k \le p \\ \dfrac{1-x_k}{1-p}, & p < x_k \le 1 \end{cases}. \quad (3)$$

For each value of parameter $p \in (0,1)$ equation (3) is chaotic and its distribution is uniform. Lyapunov exponent for such chaotic map is calculated in accordance with:

$$\lambda = \lim_{n\to\infty} \frac{1}{n} \sum_{k=0}^{n-1} \ln\left|f'(x_k)\right|, \quad (4)$$

is equal to [3]

$$\lambda = -p \ln p - (1-p)\ln(1-p). \quad (5)$$

Furthermore equation (5) describes also the Lyapunov exponent of recursion (2) [2]. The graph of Lyapunov exponent (5) is shown in figure 1.

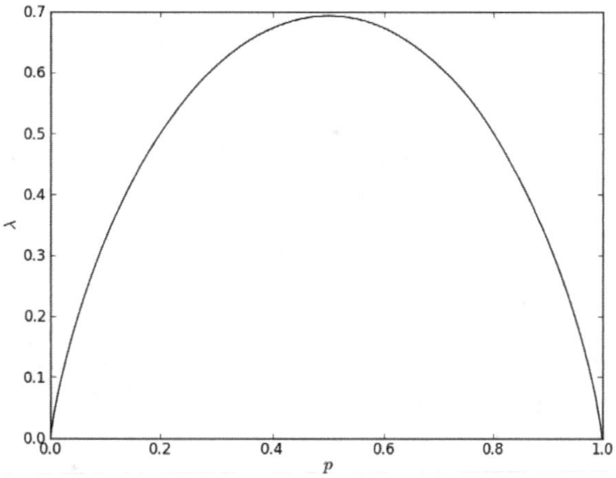

Figure 1. Lyapunov exponent of skew tent map (3).

The scope of the paper is to analyze the chaotic maps in the form of (2) in view of their computational accuracy and efficiency on the example of the constructed recursive map with the normal distribution.

1.1. Normal distribution

The normal distribution is designated by density function $\rho(x)$ expressed by the following equation [1]:

$$\rho(x) = \frac{1}{\sigma\sqrt{2\pi}} \exp\left(\frac{-(x-\mu)^2}{2\sigma^2}\right), \quad (6)$$

where μ denotes the arithmetic mean value, σ is the standard deviation.

The cumulative distribution function of the above equation cannot be expressed by means of elementary functions. Therefore, to take advantage of (1) or (2) it should be approximated by another function, for which an inverse function can be designated. An example of such approximation is [4]:

$$F(x) = \frac{e^{2kx}}{1+e^{2kx}} \qquad \left(k = \left(\frac{2}{\pi}\right)^{1/2}\right). \quad (7)$$

List of another functions that approximate the CDF of the normal distribution and methods that enables the derivation of variables from the normal distribution can be found in [5-8].

Journal of Physics: Conference Series **633** (2015) 012086 doi:10.1088/1742-6596/633/1/012086

2. Analysis

On the grounds of (2) and (7) a chaotic map was derived. Its distribution for different values of parameter p is shown in figure 2. As seen in the graph, the derived chaotic map does not well approximate the standard normal distribution for some values of p.

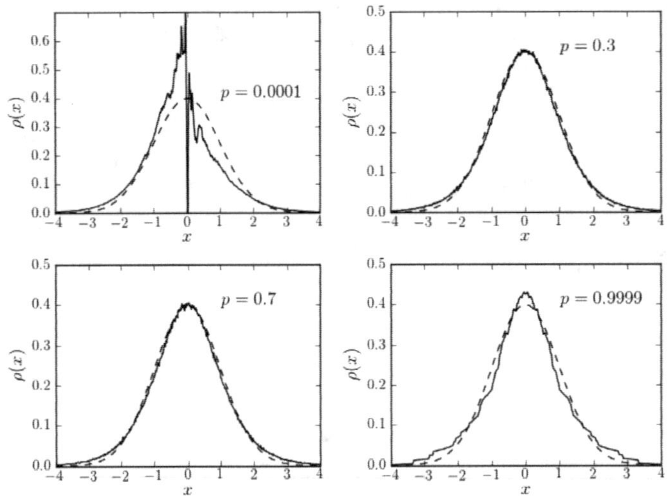

Figure 2. Distributions of constructed map for different values of parameter p (continuous line); the dotted line indicates the normal distribution.

Furthermore as seen in figure 3, the distribution numerically generated from part $U(F(x))$ of the constructed map in form (2), is not a uniform distribution. This is a consequence of a low value of Lyapunov exponent of (2) which is equal to (5). This means that by the application of chaotic map (2) extreme values of parameter p should not be selected. It follows that recursion in the form (2) in conjunction with (7) is not appropriate for generating numbers from standard normal distributions. Such conclusion may be generalized to each recursive map in the form of (2) with any distribution, where the value of p is close to 0 or 1.

To eliminate the above discussed disadvantage of recursion (2), the skew tent map (3) should be replaced by another chaotic map, with uniform distribution, but, at the same time, with Lyapunov exponent of a stable and positive value. Such maps can be found in [9].

Furthermore, in table 1, the mean time required for generating the numbers from the normal distribution was indicated, in comparison with (1). The results indicate that constructed map takes much more time than method (1). This is an outcome of the form of (2), which additionally requires the use of CDF (7) in the calculations.

Table 1. Results obtained for the analyzed function generating the normal distribution for 1000000 iterations.

Method		Time in [s]
method (1) with (7)		1.862
constructed map	$p = 0.0001$	3.333
	$p = 0.3$	3.158
	$p = 0.7$	3.086
	$p = 0.9999$	3.051

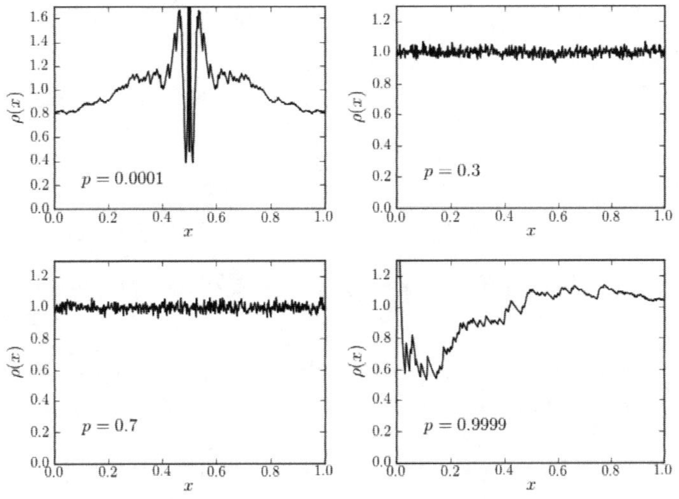

Figure 3. Distributions derived from part $U(F(x))$ of the analyzed recursion for different values of parameter p.

3. Conclusions

The scope of the paper was the presentation of the analysis of a chaotic map that generates numerical values from the standard normal distribution, in the form of (2). On the grounds of the conducted analysis, it should be indicated that for some values of parameter p in the skew tent map, the recurrence in the form of (2) does not generate the values from the given distribution. This is due to the fact that part $U(F(x))$ of the recursion expressed by equation (2) does not render the uniform distribution. This, in turn, is a consequence of a low value of Lyapunov exponent for such values of p. Furthermore chaotic maps in the form of (2) takes more time to generate numbers from the given distribution in comparison with method (1).

References

[1] Devroye L 1986 *Non-Uniform Random Variate Generation* (New York: Springer)

[2] Lai D and Chen G 2000 Generating different statistical distributions by the chaotic skew tent map *International Journal of Bifurcation and Chaos* **10(6)** 1509–12

[3] Hasler M and Maistrenko Y L 1997 An introduction to the synchronization of chaotic systems: coupled skew tent maps *IEEE Transactions on Circuits and Systems—I: Fundamental theory and Applications* **44(10)** 856–66

[4] Tocher KD 1963 *The Art of Simulation* (London: English University Press)

[5] Aludaat K M and Alodat M T 2008 A note on approximating the normal distribution function *Applied Mathematical Sciences* **2(9)** 425–29

[6] Bowling S R Khasawneh M T Kaewkuekool S and Cho B T 2009 A logistic approximation to the cumulative normal distribution *Journal of Industrial Engineering and Management* **2(1)** 114–27

[7] Thomas D B Luk W Leong P H W and Villasenor J D 2007 Gaussian random number generators *ACM Computing Surveys* **39(4)** 1–38

[8] Lawnik M 2014 The approximation of the normal distribution by means of chaotic expression *Journal of Physics: Conference Series* **490** 012072

[9] Anikin V M Arkadaksky S S Kuptsov S N Remizov A S and Vasilenko L P 2008 Lyapunov exponent for chaotic 1D maps with uniform invariant distribution *Bulletin of the Russian Academy of Sciences: Physics* **72(12)** 1684–88

Parameters analysis of a porous medium model for treatment with hyperthermia using OpenMP

Ruy Freitas Reis[1], **Felipe dos Santos Loureiro**[2], **Marcelo Lobosco**[3]
[1,2,3]Programa de Pós-Graduação em Modelagem Computacional, Universidade Federal de Juiz de Fora, Brasil

E-mail: [1]`ruyfreis@gmail.com`, [2]`felipe.loureiro@ufjf.edu.br`, [3]`marcelo.lobosco@ufjf.edu.br`

Abstract. Cancer is the second cause of death in the world so treatments have been developed trying to work around this world health problem. Hyperthermia is not a new technique, but its use in cancer treatment is still at early stage of development. This treatment is based on overheat the target area to a threshold temperature that causes cancerous cell necrosis and apoptosis. To simulate this phenomenon using magnetic nanoparticles in an under skin cancer treatment, a three-dimensional porous medium model was adopted. This study presents a sensibility analysis of the model parameters such as the porosity and blood velocity. To ensure a second-order solution approach, a 7-points centered finite difference method was used for space discretization while a predictor-corrector method was used to time evolution. Due to the massive computations required to find the solution of a three-dimensional model, this paper also presents a first attempt to improve performance using OpenMP, a parallel programming API.

1. Introduction

According to the World Health Organization (WHO), cancer is the second cause of death in the world. There are several kinds of treatments trying to mitigate this world health problem, like radiotherapy, chemotherapy or even surgical extraction. An attempt to minimize trauma due to invasive treatments is the use of local hyperthermia, which is a minimally non-invasive therapy based on overheat the target area up to a threshold temperature that causes necrosis and apoptosis of the cancerous tissue. One important point is how to heat the tumor: according to [1], magnetic nonoparticles are a good choice. Due to the biocompatibility with the human body, iron oxides magnetite (Fe_3O_4) and maghemite ($\gamma - Fe_2O_3$) are the most used nonoparticles.

In this work, hyperthermia is modeled as follows. The bioheat transfer is modeled as a porous medium[2], but using the external heat as shown in [3]. The heat generation due to the nanoparticles is modeled as suggested by [1]. Then a sensibility analysis of the blood velocity and porosity is carried out. Finally, this paper presents a parallelization strategy applied to the numerical method in order to speedup the simulation time since its solution in the three-dimensional domain demands a huge computational time. The parallelization was implemented using OpenMP, a parallel programming API.

2. Bioheat Transfer Model

The transient bioheat model adopted in this work is based on the theory of porous media in biological tissues adapted from [2]. This model, in a well-posed problem, can be written as:

Journal of Physics: Conference Series **633** (2015) 012087 doi:10.1088/1742-6596/633/1/012087

$$\begin{cases} [\rho c_p(1-\varepsilon)+\rho_b c_{p_b}\varepsilon]\dfrac{\partial T}{\partial t} = \nabla\cdot\kappa\nabla T + \beta\dfrac{\rho_b(u_b)_{AVG}\varepsilon c_{p_b}}{\delta}(T_b - T) + \\ \quad (1-\varepsilon)(Q_m+Q_r), & \text{in } \Omega\times I \\ \alpha T + \gamma\nabla T\cdot\vec{n} = f, & \text{on } \Gamma\times I. \\ T(\cdot,0) = T_0, & \text{in } \Omega \end{cases} \tag{1}$$

where $\kappa = (1-\varepsilon)k_t + \varepsilon k_b$; ρ, c_p and k_t are density, specific heat and thermal conductivity of the tissue, respectively; ρ_b, c_{p_b}, k_b and $(u_b)_{AVG}$ are density, specific heat, thermal conductivity and velocity of the blood, respectively; ε, δ and β are porosity of the medium, average distance between the transverse blood vessels and a correction factor, respectively; T_b is the blood temperature; Q_m is the metabolic heat source; Q_r external spatial heating.

3. Numerical Scheme and Parallel Strategy

The finite difference method in both time and space is employed to solve Eq. (1). Briefly, the domain $\Omega\cup\Gamma\subset\mathbb{R}^3$ is discretized into a set of equally spaced points defined by $\boldsymbol{S} = \{(x_i,y_j,z_k);i=0,\dots,I_x;j=0,\dots,I_y;k=0,\dots,I_z\}$ with I_x, I_y and I_z being the number of intervals of length Δx, Δy and Δz, respectively in each direction. To ensure a second order convergence, the 7-points centered finite difference scheme [4] was applied for the spatial discretization. To march in time, the time domain I was partitioned into I_t equal time intervals of length Δt, and an explicit predictor-multicorrector algorithm originated from the generalized alpha-family of time integration method [5] was adopted into Eq. (1) as shown by Alg. 1. According to [5], the use of $r_f = 2$ and $\alpha = 0.5$ in this scheme ensures a second order convergence when applied to linear equations.

Algorithm 1: Parallel Predictor-Multicorrector Algorithm

begin
  # pragma omp parallel
  foreach $n\in I_t$ **do**
    for $r\leftarrow 0$ *to* r_f **do**
      # pragma omp for
      foreach $T\in\Omega\cup\Gamma$ **do**
        if *($r=0$)* **then**
          $T^{n+1,r}_{i,j,k} = T^n_{i,j,k} + (1-\alpha)\Delta t\dfrac{\partial T^n_{i,j,k}}{\partial t}$
        end if
        $\dfrac{\partial T^{n+1,r}_{i,j,k}}{\partial t} = \left[\dfrac{1}{\rho c_p(1-\varepsilon)+\rho_b c_{p_b}\varepsilon}\right]_{i,j,k}\Phi(T^{n+1,r}_{i,j,k},T^{n+1,r}_{i\pm1,j,k},T^{n+1,r}_{i,j\pm1,k},T^{n+1,r}_{i,j,k\pm1}) +$
        $\left[\beta\dfrac{\rho_b(u_b)_{AVG}\varepsilon c_{p_b}}{\delta}\right]_{i,j,k}(T_b - T_{i,j,k}) + [(Q_m+Q_r)(1-\varepsilon)]^n_{i,j,k}$
        $T^{n+1,r+1}_{i,j,k} = T^{n+1,0}_{i,j,k} + \alpha\Delta t\dfrac{\partial T^{n+1,r+1}_{i,j,k}}{\partial t}$
      end foreach
    end for
  end foreach
end

In Alg. 1, $T^n_{i,j,k}\approx T(x_i,y_j,z_k,t_n)$, and $\Phi(T^{n+1}_{i,j,k},T^{n+1}_{i\pm1,j,k},T^{n+1}_{i,j\pm1,k},T^{n+1}_{i,j,k\pm1}) = q^{n+1}_{x,i,j,k} + q^{n+1}_{y,i,j,k} + q^{n+1}_{z,i,j,k}$, where $q^{n+1}_{x,i,j,k} = \frac{1}{\Delta x^2}\left[\kappa_{i+\frac{1}{2},j,k}\left(T^{n+1}_{i+1,j,k} - T^{n+1}_{i,j,k}\right) - \kappa_{i-\frac{1}{2},j,k}\left(T^{n+1}_{i,j,k} - T^{n+1}_{i-1,j,k}\right)\right]$ (all the other fluxes are approximated in a similar manner). Due to the fact that κ is a piecewise continuous function, the harmonic mean $k_{i\pm1/2,j}\approx\frac{2k_{i,j}k_{i\pm1,j}}{k_{i,j}+k_{i\pm1,j}}$ was adopted to ensure the flux continuity.

4. Simulation Results

The aforementioned mathematical model was used to simulate the hyperthermia in a cubic domain with lengths equal to 0.1 m in each direction, discretized using $\Delta x = \Delta y = \Delta z = \Delta =$ 0.001 m and over 3000 s of simulation time with $\Delta t = 1$ s, in addition to a tumor formed by two intercepted spheres seated at (0.5, 0.5, 0.5) m and (0.4, 0.4, 0.4) m with radius of 0.015 m and 0.01 m, respectively. All the parameters used to perform the simulation are shown in Tab. 1 and are in accordance with the literature [6, 7]. In addition, the heat generated by the nanoparticles is mathematically modeled as a function $Q_r(\vec{x}) = \sum_{i=1}^{n} A_i e^{-r(\vec{x}_i^2)/r_{0,i}^2}$ with one ($n = 1$) injection point seated at (0.05,0.05,0.05) m; and considering $A = 1.3\text{x}10^6$ W/m^3 and $r_0 = 3.1\text{x}10^{-3}$ m.

Table 1. Model parameters

Sign	Unit	Muscle	Tumor
c	(J/Kg^oC)	4000.0	4000.0
c_b	(J/Kg^oC)	4000.0	4000.0
k_t	(W/m^oC)	0.51	0.56
k_b	(W/m^oC)	0.64	0.64
ρ	(Kg/m^3)	1000.0	1000.0
ρ_b	(Kg/m^3)	1000.0	1000.0

Sign	Unit	Muscle	Tumor
u_b	(mm/s)	2.0	$[1.5, 4.5]$
δ	(mm)	6.0	4.0
ε	−	0.02	$[0.001, 0.1]$
β	−	0.2	0.2
Q_m	W/m^3	420.0	4200.0

To evaluate the sensibility analysis, two parameters were chosen: blood velocity and porosity. For the porosity analysis, the velocity was fixed in 1.6 mm/s with the porosity values ranging from 0.001 to 0.1 in increments of 0.02 at the tumor tissue, whereas for the blood velocity analysis the porosity was fixed in 0.06 with the blood velocity values ranging from 1.5 mm/s to 4.5 mm/s in increments of 0.6 mm/s, according to the experimental data shown in [6], at the tumor tissue. These parameters were chosen mainly because both are not constant at the body. As shown at [6] the blood velocity depends on the cancer stage, and the porosity can vary as well, it depends on, for instance, the shape of the cell and the interstitial fluid [8].

First of all, it is important to observe that these temperature curves represent an hyperthermia treatment since they are qualitatively in accordance with experimental data in [1]. Furthermore, according to Fig. 1(a) it is possible to observe that increasing the value of the porosity parameter in 100 times affects the steady state temperature in an hyperthermia treatment about 4°C; moreover, Fig. 1(b) shows that increasing the value of the blood velocity in 3 times results in a difference of 1.4^{o}C, i.e. this model is more sensitive to the velocity of blood than to the medium porosity. Finally, it is worthwhile to note that the hyperthermia treatment starts having effects at 43^oC, and in some cases it is necessary to use more injection points [1]

All the simulations were performed in a SMP Linux (3.9.2-200) computer with two Intel Xeon CPUs E5620 with 2.40GHz of clock frequency, the Hyper-Threading option turned off and 12 GB of RAM. Each CPU has four cores, so a total of 8 cores were available. The speedup was computed using the average of three executions. The standard deviation for these executions were below 2.73%. According to Fig. 2, the best speedup result, 7.12, was obtained using the dynamic scheduling with chunk size equals to 1. The speedup results are quite encouraging since the serial version was executed in 11.5 min and the best parallel version executed in 1.62 min using 8 cores.

5. Conclusions

This work presented a parameter analysis of the blood velocity and the medium porosity using a bioheat model in porous media. It has been observed that the mathematical model is more sensitive to blood velocity parameter when compared to porosity. Furthermore, a finite difference

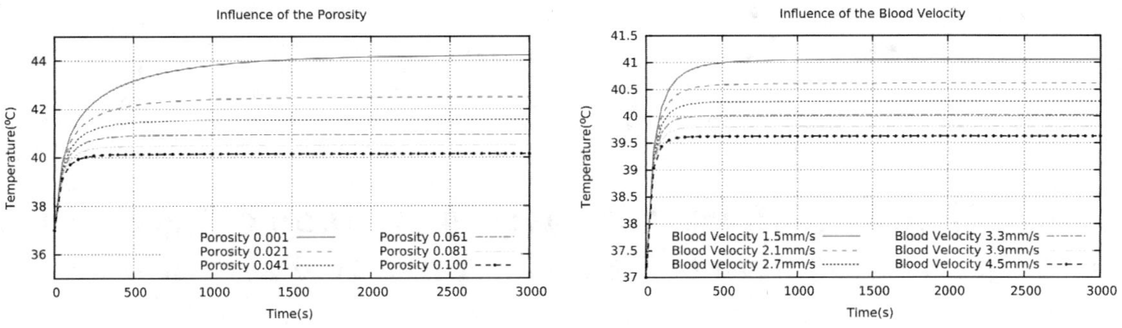

(a) (b)

Figure 1. Influence of porosity (a) and blood velocity (b) at the point (0.048, 0.048, 0.048) m inside the tumor.

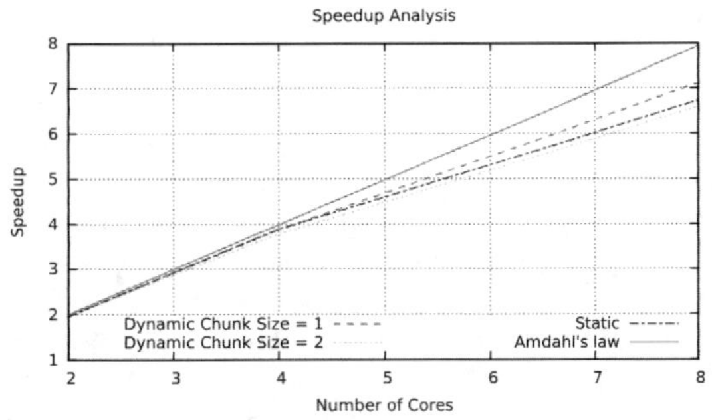

Figure 2. Speedup analysis

strategy was employed to solve the equations and a parallel implementation was used to speedup the simulation up to 7.1 times in a 8 core processor. As future work, we plan to handle with more complex porous media bioheat models, e.g. two phases, and try a parallelization strategy using GPUs as shown in [9].

Acknowledgements

The financial support by CNPq, CAPES, UFJF and FAPEMIG is greatly acknowledged.

References

[1] Salloum M, Ma R and Zhu L 2008 *International Journal of Hyperthermia* **24** 589–601
[2] Khaled A R and Vafai K 2003 *International Journal of Heat and Mass Transfer* **46** 4989 – 5003
[3] Minkowycz W J and Sparrow E M 2009 *Advances in numerical heat transfer* 1st ed vol 3 (CRC press)
[4] LeVeque R J 2007 *Finite Difference Methods for Ordinary and Partial Differential Equations, Steady State and Time Dependent Problems* 1st ed (SIAMI)
[5] Hughes T J 2012 *The finite element method: linear static and dynamic finite element analysis* (Courier Dover Publications)
[6] Ishida H, Hachiga T, Andoh T and Akiguchi S 2012 *Journal of Applied Physics* **112** 104703
[7] Dombrovsky L A, Timchenko V and Jackson M 2012 *International Journal of Heat and Mass Transfer* **55** 4688 – 4700
[8] Karch R, Neumann F, Ullrich R, Heinze G, Neumuller J, Podesser B and Neumann M 2010 *Annals of Biomedical Engineering* **38** 308–318 ISSN 0090-6964
[9] Reis R F, Loureiro F S and Lobosco M 2015 *Journal of Computational and Applied Mathematics* – ISSN 0377-0427

Oxygen induced effects on avascular tumour growth: a preliminary simulation using an adaptive grid algorithm

Antonino Amoddeo

Department of Civil, Energy, Environmental and Materials Engineering, Università 'Mediterranea' di Reggio Calabria,Via Graziella 1, Feo di Vito, I-89122 Reggio Calabria, Italy

E-mail: antonino.amoddeo@unirc.it

Abstract. Cancer cells oxygenation from surrounding healthy tissue influences tumour growth in the avascular phase. The effects induced by oxygen on cancer cells dynamics during their interaction with the urokinase plasminogen activator system, are simulated mimicking hypoxic conditions in the early stage of human tumour proliferation, using oxygen supply parameters determined from *in vivo* experiments. The system of six coupled partial differential equations, arising from the problem modelling, is solved over a one-dimensional domain implementing a moving mesh numerical technique, using the finite element method. Our preliminary computations show that oxygen concentration at hypoxic conditions cause cancer cells to build inhomogeneous proliferation pattern, similarly to what happens in absence of oxygen.

1. Introduction

Cancer cells dynamics is the result of the interaction among several biological mechanisms, covering phenomena occurring over different length scales [1]. On one hand it is relatively easy to obtain data on avascular tumour growth from *in vitro* experiments performed on multicellular spheroids, on the other cannot be ignored that during *in vivo* proliferation, microenvironment mechanisms occur, not present in *in vitro* experiments [2]. The presence of oxygen in cancer tissue, coming from the surrounding healthy tissue, can influence the sensitivity of the cells to, for example, nonsurgical treatments as radiotherapy, and most of malignancies are characterized by the heterogeneous presence of hypoxic tissue areas [3]. The deadly stage of tumour proliferation begins when the sprout of new blood vessels allows cancer cells to reach region of healthy tissue, or organs, located far from the initial tumour mass, sowing with metastasis [4]. Reviews about modelling of cancer cell growth and proliferation can be found, among others, in references [5-7].

Models of tumour growth inside biological tissue are expressed as systems of partial differential equations (PDEs) [8], considering an initial cluster of cancer cells seeded in the extracellular matrix (ECM), a nutrient-rich aqueous environment allowing the cells to grow and proliferate [1]. A mechanism recognized as promoting cancer cell growth and proliferation is their interaction with the urokinase plasminogen activator (uPA) system: in short, the uPA serine protease degrades an ECM macromolecule, vitronectin (VN), while the ECM enzyme plasmin activates uPA; the plasminogen activator inhibitor type-1 (PAI-1) secreted by healthy cells regulates excess of proteolysis, see [9] for details about biological foundations and descriptions.

Journal of Physics: Conference Series **633** (2015) 012088 doi:10.1088/1742-6596/633/1/012088

The above mechanism has been recently simulated over a one-dimensional domain, and the PDEs system arising from the problem modelling has been solved implementing the moving mesh partial differential equation (MMPDE) numerical technique [10], using the finite element method (FEM) [11]. As elsewhere shown and discussed [12-15], in such technique the discretization of the integration domain, typical of the FEM procedure, is controlled by the gradient of a parameter calculated during the solution procedure, in our case the gradient of the cancer cells concentration [10], using the equidistribution principle [16]: it results a non-uniform node distribution, but nodal connectivity and number of mesh points inside the integration domain are kept constant, while the mesh points are moved in regions where more detail is required.

We present a preliminary simulation of the tumour progression in a one-dimensional portion of biological tissue, in the early stage of the avascular phase under hypoxic conditions. The PDEs system which arises modelling the interaction of cancer cells dynamics with both the uPA system and a model for oxygen supply is solved using the MMPDE numerical technique. The method results appropriate in presence of rapidly varying parameters, as in shocks or defects, leading to sharp solution features [12-15], allowing to capture finer spatial details related to malignancy.

2. Theory and method

We assume that cell migration inside a given biological domain originates from random diffusion, chemotaxis and haptotaxis: in particular, chemotaxis is the cell movement towards chemicals dissolved in the ECM, while haptotaxis is the attraction of the cells by chemicals adhered to some ECM component [10].

In a fixed domain $\Omega \subset \mathbb{R}^3$, we consider explicitly the concentration evolution of cancer cell, VN protein, uPA serine protease, PAI-1 inhibitor, plasmin degrading enzyme and oxygen concentration, denoted, respectively, by $c(\mathbf{x},t)$, $v(\mathbf{x},t)$, $u(\mathbf{x},t)$, $p(\mathbf{x},t)$, $m(\mathbf{x},t)$ and $n(\mathbf{x},t)$, according to the following system of coupled PDEs:

$$\frac{\partial c}{\partial t} = \nabla \cdot \left[D_c \nabla c - c\left(\chi_u \nabla u + \chi_p \nabla p + \chi_v \nabla v \right) \right] + \mu_1 c\left(1-c\right) + k_2 nc \tag{1}$$

$$\frac{\partial v}{\partial t} = -\delta vm + \phi_{21} up - \phi_{22} vp + \mu_2 v\left(1-v\right) \tag{2}$$

$$\frac{\partial u}{\partial t} = \nabla \cdot \left(D_u \nabla u \right) - \phi_{31} pu - \phi_{33} cu + \alpha_{31} c \tag{3}$$

$$\frac{\partial p}{\partial t} = \nabla \cdot \left(D_p \nabla p \right) - \phi_{41} pu - \phi_{42} pv + \alpha_{41} m \tag{4}$$

$$\frac{\partial m}{\partial t} = \nabla \cdot \left(D_m \nabla m \right) + \phi_{52} pv + \phi_{53} cu - \phi_{54} m \tag{5}$$

$$\frac{\partial n}{\partial t} = \nabla \cdot \left(D_n \nabla n \right) + k_1 n_0 (1-c) - k_1 n - k_2 nc \tag{6}$$

In equation (1), the migration of cancer cells is accounted for by random motility, chemotaxis due to the presence of both uPA and PAI-1 molecules, and haptotaxis due to VN, via, respectively, D_c, χ_u, χ_p, and χ_v; cells production term includes contributions from logistic growth at a rate μ_1, and from oxygen interaction through k_2. Vitronectin cannot diffuse: it is degraded by plasmin at a rate δ, indirectly produced from the interaction of PAI-1 with uPA at a rate ϕ_{21}, neutralized by PAI-1 inhibitor at a rate ϕ_{22}, and produced according to cell proliferation at a rate μ_2, see equation (2). Concerning equation (3), uPA diffuses at a rate D_u and is inhibited by PAI-1 at a rate ϕ_{31}; cancer cells cause its degradation at a rate ϕ_{33} through their mutual interaction, and its production at a rate α_{31}. In equation (4), PAI-1 inhibitor diffuses at a rate D_p, the interactions with uPA and VN degrade it at a rate of, respectively, ϕ_{41} and ϕ_{42}, while it is produced by plasmin at a rate α_{41}. Considering plasmin, equation (5), it moves according to D_m, is produced by both PAI-1 binding to VN and uPA binding to cancer cells at a rate, respectively, ϕ_{52} and ϕ_{53}, and degrades at a rate ϕ_{54}. Finally, let n_0 be the initial concentration of oxygen in absence of cancer cells, assuming its utilization from cancer cells only, it diffuses according to D_n, degrades at a rate k_1, while it is utilized by cancer cells at a rate k_2, equation (6).

The coupled equations (1) - (6) constitute a PDEs system, which solution starts from the discretization of the one dimensional integration domain, equidistributing in each subinterval a monitor function depending on the gradient of the unknown solution. Denoting as $\Omega_p = [0,1]$ and $\Omega_c = [0,1]$, respectively, physical and computational domains, $u(z,t)$ a solution of a PDEs system over Ω_p with z physical coordinate, and ξ a computational coordinate of a fixed computational domain, a one-to-one mapping from physical space $\Omega_p \times (0,T]$ to computational space $\Omega_c \times (0,T]$ is defined as $\xi = \xi(z,t)$, $z \in \Omega_p$, $t \in (0,T]$, while $z = z(\xi,t)$, $\xi \in \Omega_c$, $t \in (0,T]$ is from computational space $\Omega_c \times (0,T]$ to physical space $\Omega_p \times (0,T]$. From the equidistribution of a monitor function $M(u(z,t))$ over our one-dimensional domain we obtain the mesh equation

$$M(z(\xi,t))\frac{\partial}{\partial\xi}z(\xi,t) = \int_0^1 M(u(s,t))ds = C(t), \tag{7}$$

and the choice of the monitor function tested by Beckett et al. [17], ensures a good quality control of the node distribution and final convergence of the FEM solution, with the additional advantages of being naturally smoothed, and not demanding external interventions, then

$$M(u(z,t)) = \int_0^1 \left(\left|\frac{\partial u(z,t)}{\partial z}\right|\right)^{1/2} \mathrm{d}z + \left(\left|\frac{\partial u(z,t)}{\partial z}\right|\right)^{1/2}. \tag{8}$$

The adaptive solution involves an iterative process in which, at each time step, (i) a mesh is generated using the equidistribution principle based on the numerical solution at the current time step and (ii) the PDEs is solved on a new generated grid and the solution is updated in time.The one-dimensional domain is initially discretized via the method of lines, using an implicit Euler method for time discretization. See [12-15] for details about the MMPDE procedure. We used dimensionless quantities obtained by scaling all variables and parameters by reference quantities: distance x is scaled with the maximum distance for cancer cells $L= 10^{-1}$ cm; let $D = 1 \times 10^{-6}$ cm^{2}s^{-1} be a representative diffusion coefficient, time t is scaled according to $\tau=L^2D^{-1}$. A reference cancer cell concentration is $c_0=6.7 \times 10^7$ cell cm^{-3}, while uPA, VN, PAI-1 and plasmin, are scaled according to appropriate reference concentration values, u_0=1 nM, v_0=1 nM, p_0=1 nM, m_0=0.1 nM, respectively [9]. Furthermore, $D_c=3.5 \times 10^{-4}$, $D_u=2.5 \times 10^{-3}$, $D_p=3.5 \times 10^{-3}$, $D_m=4.91 \times 10^{-3}$, $\chi_u=3.05 \text{ x } 10^{-2}$, $\chi_p=3.75 \text{ x } 10^{-2}$, $\chi_v=2.85 \text{ x } 10^{-2}$, $\mu_1=0.25$, $\mu_2=0.15$, $\alpha_{31}=0.215$, $\alpha_{41}=0.5$, $\delta=8.15$, $\phi_{21}=0.75$, $\phi_{22}=0.55$, $\phi_{31}=0.75$, $\phi_{33}=0.3$, $\phi_{41}=0.75$, $\phi_{42}=0.55$, $\phi_{52}=0.11$, $\phi_{53}=0.75$, $\phi_{54}=0.5$.

Concerning oxygen, its diffusion coefficient has been estimated in various biological environment in the 1.3×10^{-5} cm^{2}s^{-1} ÷ 3.1×10^{-5} cm^{2}s^{-1} range [18], hence, we put D_n=25; the others model parameters come from *in vivo* measurements on human tumours, and to mimic hypoxic conditions typical of malignant proliferations [3] we put k_2=0.3, n_0=1.37 μM, serving also as reference concentration, while we assume $k_1=0.5\times k_2=0.15$ [8]. We impose that at $t = x = 0$, a cluster of cancer cells is seeded in the domain, while the remaining portion of tissue is filled by ECM; moreover, the initial concentrations of uPA and PAI-1 are proportional to the cancer cells concentration, plasmin is not yet produced, and the oxygen concentration is at the initial concentration. Therefore, $c(\mathbf{x},0)=\exp(-|\mathbf{x}|^2\varepsilon^{-1})$ with $\varepsilon=0.01$, $v(\mathbf{x},0)=1-0.5c(\mathbf{x},0)$, $u(\mathbf{x},0)=0.5c(\mathbf{x},0)$, $p(\mathbf{x},0)=0.05c(\mathbf{x},0)$, $m(\mathbf{x},0)=0$ [13,14] and $n(\mathbf{x},0)=n_0$. The components cannot break-out from the simulated domain, then zero-flux boundary conditions have been imposed on our PDEs system, apart from the ordinary differential equation (2). We show the numerical simulation over a one dimensional portion of biological tissue 2 mm thick,

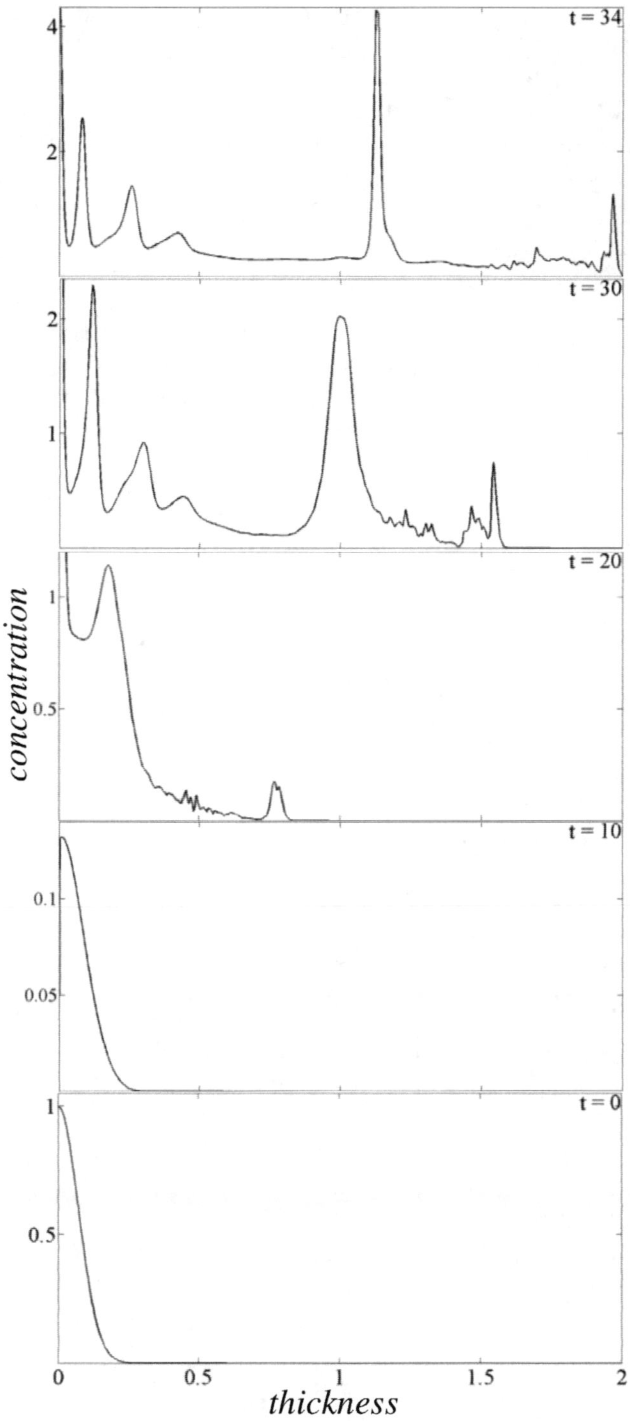

Figure 1. Concentration evolution of cancer cells *c* along the tissue thickness. Each panel corresponds to a snapshot taken during the time evolution.

with 200 grid points, for $x \in [0,2]$, $t \in [0,100]$ and steps δt=0.5. Both the dimensional thickness and timescale are recovered multiplying, respectively, x by L and t by τ. At each time step our PDEs system has been solved replacing the exact solution $u(x,t)$ in equation (8), with $c(x,t)$.

3. Results and discussion

In figure 1 we show snapshots of the cancer cells density profile taken at selected times during the proliferation process: horizontal and vertical axes report, respectively, the tissue thickness in the $0 \le x \le 2$ range and the cancer cell concentration. At t=0 the density profile reflects the initial conditions, and until t=10 the cells remains clustered around x=0. At t=20 it appears radically changed: the invasion front moves near x=1; a sharp onset marks the cancer/healthy cells interface, coming with an irregular and jagged fine structure. For t=30 the invading front is beyond x=1.5, while, at t=34, it reaches x=2.

In a continuum frame, the whole set of density profiles builds the spatio-temporal proliferation pattern shown in figure 2, where the cancer cells density is linearly mapped in a grey level scale between black and white colours, respectively zero and maximum density. On the vertical axis the time scale of the solution evolution is in the $0 \le t \le 100$ interval. The interaction between uPA system and cancer cells, in presence of oxygen in the micromolar range, gives rise to marked inhomogeneity of the spatio-temporal proliferation: such behaviour is consistent to what recently observed in absence of oxygen [10], and evidence of malignancy [5].

4. Conclusions

We have presented a preliminary numerical simulation of the effects induced by oxygen during the interaction of uPA system with cancer cell dynamics. The results indicate malignant proliferation through heterogeneous spatio-temporal patterns, in continuity with our previous modelling performed in

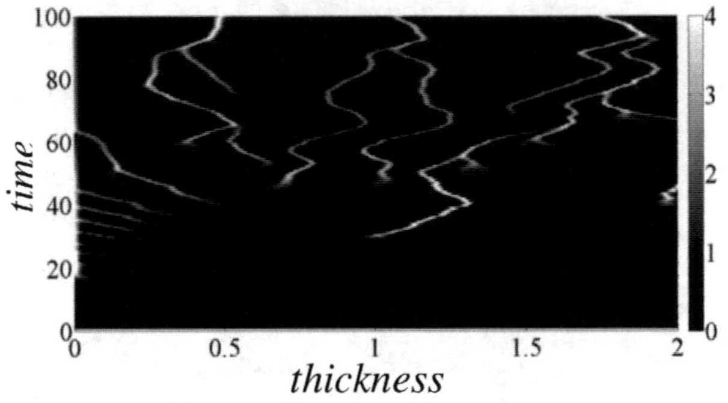

Figure 2. Surface contour plots of the cancer cells evolution in the $0 \leq x \leq 2$ spatial domain and in the $0 \leq t \leq 100$ time interval. The cell concentration is linearly mapped in a grey levels scale between the black (zero concentration) and white (maximum concentration) colours.

absence of oxygen. The peculiarity of the MMPDE technique relies on the cell density-driven mechanism used to discretize the integration domain, letting the grid points to cluster in region where ∇c is growing, and allowing to resolve the fine structure coming with the irregular invading front at the boundary between cancer ad healthy cells. It results, with respect to techniques based on uniform discretizations, an improved resolution of features related to the heterogeneous spatial distribution of malignant cells.

References

[1] Preziosi L and Tosin AR 2009 *J. Math. Biol.* **58** 625-56
[2] Sherratt JA and Chaplain MAJ 2001 J. *Math. Biol.* **43** 291-312
[3] Vaupel P, Kallinowski F and Okunieff P 1989 *Cancer Research* **49** 6449-65
[4] Chaplain MAJ and Anderson ARA 2003 Mathematical modelling of tissue invasion *Cancer Modelling and Simulation* ed L Preziosi (Boca Raton: Chapman & Hall/CRC Press) p. 269
[5] Araujo RP and McElwain DLS 2004 *Bull. Math. Biol.* **66** 1039-91
[6] Roose T, Chapman SJ and Maini PK 2007 *SIAM Rev.* **49** 179-208
[7] Lowengrub JS, Frieboes HB, Jin F, Chuang YL, Li X, Macklin P, Wise SM and Cristini V 2010 *Nonlinearity* **23** R1-R91
[8] Hubbard ME and Byrne HM 2013 *J. Theor. Biol.* **316** 70-89
[9] Andasari V, Gerisch A, Lolas G, South AP and Chaplain MAJ 2011 *J. Math. Biol.* **63** 141-71
[10] Amoddeo A 2015 *Comput. Math. Appl.* (in press).
[11] Zienkiewicz OC and Taylor RL 2002 *The Finite Element Method* (Oxford: Butterworth–Heinemann)
[12] Amoddeo A, Barberi R and Lombardo G 2010 *Comput. Math. Appl.* **60** 2239-52
[13] Amoddeo A, Barberi R and Lombardo G 2011 *Liq. Cryst.* **38** 93-103
[14] Amoddeo A, Barberi R and Lombardo G 2012 *Phys. Rev.* E **85** 061705-10
[15] Amoddeo A, Barberi R and Lombardo G 2013 *Liq Cryst.* **40** 799-809
[16] de Boor C 1974 Good approximation by splines with variable knots II *Lecture Notes in Mathematics* vol 263, ed J Morris (Berlin: Springer-Verlag) pp 12-20
[17] Beckett G, Mackenzie JA, Ramage A and Sloan DM 2001 *J. Comput. Phys.* **167** 372-92
[18] Buchwald P 2009 *Theor. Biol. Med. Model.* **6** 5

Stochastic effects in a discrete RT model with critical behaviour

D. Rodriguez-Perez[1], **O. Sotolongo-Grau**[2], **M.M. Desco**[1], **J.C. Antoranz**[1]

[1]Dept. Fisica Matematica y de Fluidos, UNED. Senda del Rey, 9 — 28040 Madrid, Spain
[2]Fundacio ACE, Institut Catala de Neurociencies Aplicades — 08029 Barcelona, Spain

E-mail: jcantoranz@dfmf.uned.es

Abstract. The effects of radiation on a tissue (being it healthy or cancerous) are well described by current linear-quadratic (LQ) radiobiological model for low absorbed doses around the 2 Gy often used in clinical fractionation. However, experimental data show a disagreement between the predicted and the observed effect of large doses. The Sotolongo et al. (2011) radiobiological (SRB) model, derived from Tsallis nonextensive entropy, has shown a good agreement with experiments for high absorbed doses, where LQ overestimates the dose required for a required effect. Other studies have reported a crossover in LQ model where its effects are underestimated for large doses. In this paper we develop a mechanistic version of the SRB model and show that it can reproduce both behaviors with a minimum set of assumptions. We compare the results of our simulations with some data reported in the literature. We also trivially adapt this model to fractionated radiotherapy and, in particular, to hypofractionation for which we draw some conclusions.

1. Introduction

In [1] the Sotolongo radiobiological model (SRB) was introduced to explain the observed deviations of the linear-quadratic model (LQ) observed at high doses, mainly in *in vitro* experiments. The LQ model was introduced by Kellerer & Rossi as a good approximation to more involved theoretical mechanistic models which included responses of cells to radiation [2]; despite being an approximation to these models it has been used as a sort of constitutive relation in which to include new correcting factors that fit better the observations [3]. That is, what was once introduced as an *ab initio* model, has become just a practical one.

SRB model was introduced as the result of Tsallis entropy, S_q, maximization [4, 5]. The problem of the probability distribution of the dose D required to kill a cell then becomes

$$S_q = \frac{1 - \int_D p^q(x)dx}{q-1} = \max, \qquad \int_D p(x)dx = 1$$

requiring a finite q-mean [5]

$$\int_D p^q(x)x dx = \langle D \rangle_q < \infty$$

The success of the prediction required a working hypothesis to fix the support of the D distribution (its valid interval): the existence of a critical dose D_0 which could obliterate the radiated tissue. For D ranging from 0 to D_0 the resulting survival fraction F_s (the probability of killing a cell with a dose lower or equal to D) becomes

$$F_s(D) = \left(1 - \frac{D}{D_0}\right)^{-\gamma} \tag{1}$$
$$\gamma = (2-q)/(1-q)$$

According to [1], γ only depends on tissue and type of radiation used in the therapy, whereas D_0 may include these and other irradiation conditions (e.g. dose rate). At first glance, the LQ model can be obtained as a second order approximation of the SRB for the logarithm of the survival fraction around $D = 0$, such that$\alpha = \gamma/D_0, \quad \beta = \gamma/2D_0^2$. For doses close to that critical value, however, both models are expected to disagree. In the case of multifractionation based on the SRB, it was shown in [6] that mathematical consistency of the nonextensive operations requires that the survival fraction of two successive dose fractions d_1 and d_2 takes the form

$$F_s(d_1 \oplus d_2) = \left(1 - \frac{d_1}{D_0}\right)^{-\gamma} \left[1 - \frac{d_2}{D_0 - \epsilon d_1}\right]^{-\gamma} \tag{2}$$

the session-coupling coefficient ϵ takes values from 0 (independent doses) to 1 (continuous irradiation) and thus describes the tissue radiosensitivity recovery between successive fractions.

The objective of this paper is to study the consequences of the finite size effects on the results predicted by SRB radiobiological model [1]. These effects become important close and above the critical dose D_0 and will become important to assess the results of large doses for radioresistive (low γ) tissues.

2. Methods

Let the radiation may affect randomly a number of N finite discrete units in each cell. Let these units represent N alternative pathways in the function of a cell that chooses one randomly at a given moment among those available, such that just one of them is enough to keep the cell alive, but if the chosen pathway is suddenly not available the cell will die. Then if at a given moment the cell has n of its units damaged, the probability of being killed by a damage of δn of the remaining $N-n$ units will be $\delta n/(N-n)$. We will, however, consider a probability proportional to this, with a proportionality factor s. If we write this killing probability as $p = s\delta n/(N-n)$ and identify it with a decrease in the survival fraction $p = -\delta F_s/F_s$, we arrive at

$$\frac{\delta F_s}{F_s} = -s\frac{\delta n}{N-n}$$

It is simple to relate the absorbed dose with the (expected) number of units damaged, $D \sim n$, and the critical dose with the total number of units, $D_0 \sim N$, to get to

$$\frac{\delta F_s}{F_s} = -s\frac{\delta D}{D_0 - D}$$

The SRB model can be seen as a mean field approach to this discrete model of cell damage and survival, just identifying the s parameter with the γ exponent by taking the first derivative in eq. 1.

Let us start after a radiation dose that has damaged n_1 units. A fraction r of the damaged units will not recover, thus in the next session we will start from only $N - rn_1$ intact units. That is, if no recovery happens between fractions, the number of units available at the start of the second fraction will be $N - n_1$; $r = 1$, no time to recover, corresponding to continuous

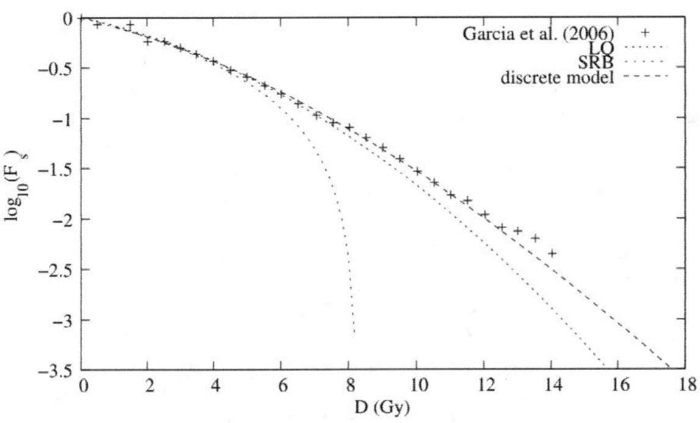

Figure 1. Comparison among LQ ($\alpha = 0.157$ Gy^{-1}, $\beta = 0.0227$ Gy^{-2}), SRB and discrete model ($\gamma = 1.6$, $D_0 = 8.27$ Gy and$N = 8$) and experimental data from [8].

irradiation. If all the n_1 damaged units get recovered between fractions, the second fraction will affect the N original functional units: $r = 0$, complete recovery, thus independent doses. This means that the survival fraction after the next radiation dose would be

$$\frac{\delta F_s}{F_s} = -s\frac{\delta n}{(N - rn_1) - n}$$

As before, this recovery fraction can be identified with the ϵ parameter of the SRB model, eq. 2.

3. Results and discussion

Figure 1 compares experimental data from reference [8] with different models: LQ, SRB and discrete model. As can be seen in this figure for low doses (and high survival fraction) all models work. The SRB model, that was fitted for doses lower than 6 Gy (before the onset of the linear exponential behavior) overestimates the effect of radiation dose. LQ model follows the experimental data up to 8 Gy, but detaches from it from this point. Only the discrete model follows the experimental data up to 13 Gy (around a 60% further in the dose). Other models exist [9, 10] that fit the experimental data more or less up to the same extreme value, but all of them either need more hypotheses or are ad hoc mathematical constructions, which provide no hints as to the radiobiological behavior.

The results of different simulations covering a wide range of parameter values (γ and N) can be seen in figures 2. Each simulation includes the statistics of an initial number of 10^8 cells. Figure 2a shows how, for a fixed γ the survival curves extend to higher doses when the number of units decreases, i.e., the SRB corresponds to a very large number of units, whereas the exponential behavior (shown as linear in the semilog plot) appears for small values of N. On the other hand, figure 2b shows how, for a fixed number of units, as γ decreases a crossover separating two well defined behaviors appears; for a given difference between the theoretical SRB model and the simulation, it is reached at lower doses (relative to the critical one, D_0) as γ increases.

What does N mean, radiobiologically? As the γ parameter, introduced in [SRb], describes the radiosensitivity of a tissue for low doses, the N describes the change of radiosensitivity at high doses. That is, as shown in figure 2a, N determines (for a given cell line) the effectivity of radiation. Counterintuitively, a larger number of units does not imply less radiosensitivity: that is because the cost in dose of removing one unit is also lower, as D_0 (the critical dose) remains the same. In other words, as N decreases, we need a larger ammount of radiation to destroy the same number of cells.

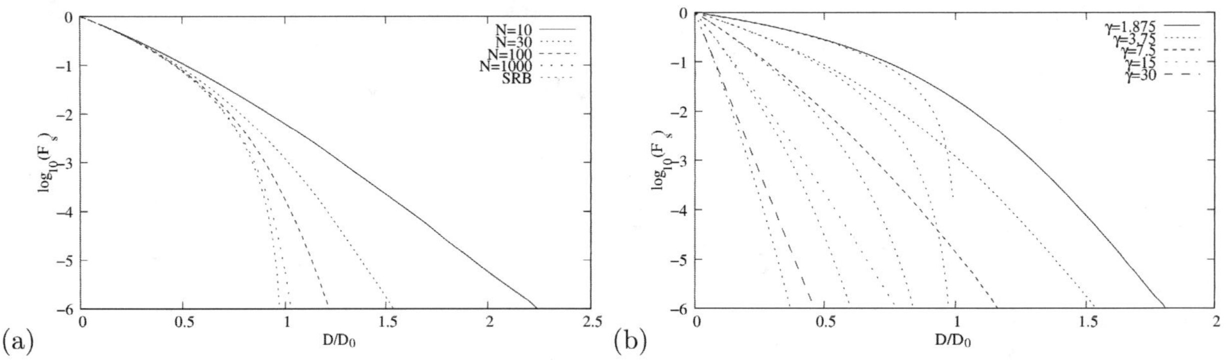

Figure 2. (a) Dimensionless results of simulations for a fixed $\gamma = 3.75$ and different number of units: $N = 10$, 30, 100, 1000. (b) Dimensionless results of simulations for different $\gamma = 1.875\times2^k$ $(k = 1, \ldots, 5)$ and fixed number of units, $N = 30$.

4. Conclusions

The discrete model provides a natural continuation to describe the linear exponential behavior observed at high doses in radiotherapy. As the SRB model accounted for observed changes in the usual α and β parameters of the LQ model (dependent on the range of doses used to fit the model to the data), the new model describes the change of regime at high doses for low radiosensitivity tissues. As we mentioned before, other models either need more radiobiological hypotheses or are ad hoc mathematical constructions; ours provides a new simple parameter that rules this behavior.

Acknowledgements
Authors acknowledge the financial support from the Spanish Ministerio de Economa y Competitividad under project FIS2012-37408.

References

[1] O. Sotolongo-Grau, D. Rodriguez-Perez, J.C. Antoranz, O. Sotolongo-Costa (2010). Tissue radiation response with maximum Tsallis entropy. Physical Review Letters, 105(15):158105
[2] A.M. Kellerer, H.H. Rossi (1974). The theory of dual radiation action. Current Topics in Radation Research. 8:85-158.
[3] D. J. Brenner, L. R. Hlatky, P. J. Hahnfeldt, Y. Huang and R. K. Sachs (1998). The Linear-Quadratic Model and Most Other Common Radiobiological Models Result in Similar Predictions of Time-Dose Relationships. Radiation Research 150:83-91.
[4] C. Tsallis (1988). Possible generalization of Boltzmann-Gibbs statistics. J. Stat. Phys. 52: 479-487.
[5] A.R.P.A. Plastino, A. R. Plastino (1999). Tsallis entropy and Jaynes' information theory formalism. Brazilian Journal of Physics, 29:50-60.
[6] O. Sotolongo-Grau, D. Rodrguez-Prez, O. Sotolongo-Costa, J.C. Antoranz (2013). Tsallis entropy approach to radiotherapy treatments. Physica A, 392:2007-2015
[7] D. Rodriguez-Perez, O. Sotolongo-Grau, O. Sotolongo-Costa, J. C. Antoranz (2014). Radiotherapy treatments by using Tsallis entropy statistical approach. Journal of Physics: Conference Series, 490:012132.
[8] L.M. Garcia, J. Leblanc, D. Wilkins and G.P. Raaphorst (2006). Fitting the linear-quadratic model to detailed data sets for different dose ranges Phys. Med. Biol. 51:2813-2823
[9] C. Park, L. Papiez, S. Zhang, M. Story, and R.D. Timmerman (2008). Universal Survival Curve and Single Fraction Equivalent Dose: Useful Tools in Understanding Potency of Ablative Radiotherapy. Int. J. Radiation Oncology Biol. Phys., 70(3), 847-852.
[10] K.E. Ekstrand (2010). The Hug-Kellerer equation as the universal cell survival curve. Phys. Med. Biol. 55:N267-N273

Thermal stress analysis for laser cutting corner with a fluctuant cutting speed in steel plate

H.B. Xu[1], Jun Hu[1], Hong Sheng[1], Z.C Du[1]
[1] School of Mechanical Engineering, Shanghai Jiao Tong University, China

E-mail: xuhebing@sjtu.edu.cn

Abstract. The temperature and thermal stress field are important factors for controlling laser cutting quality. In the present study, thermal stress field analysis for laser cutting corner with a fluctuant cutting speed is carried out. Firstly, a three-dimensional (3-D) finite element model is established to simulate the temperature and thermal stress field. Then the factors of influencing the cutting speed fluctuation such as corner angle and cutting speed are studied in laser cutting corner. It is found that the constrained cutting speed at the corner introduces the phenomenon of corner burning and uneven residual stress distribution. The predicted temperature, thermal stress results agree well with the experiment data through the laser cutting corner experiment of 3 mm steel plate.

1. Introduction

Laser cutting is a kind of non-contact thermal cutting process and has many advantages such as good cutting quality, high cutting speed, small heat-affected zone, little pollution and so on. And it has been widely used in the industrial production. However, there is a certain decrease of cutting speed in order to avoiding cutting machine shock at the cutting corner [1, 2]. And this results in bad cutting quality such as corner burning, heat concentration and uneven residual stress distribution. So it is in dire need of solving this problem to improve laser cutting efficiency and extend application.

The heat transfer during laser cutting has been analysed by many scholars. Sheng and Joshi [3] analysed the heat affected zone formation for laser cutting of stainless steel. The energy balance at the cutting front was used to derive the steady-state kerf shape, and the extent of the heat affected zone (HAZ) can be determined from the area where properties of the workpiece material were significantly affected by the resulting temperature field. Prusa, Venkitachalam [4] presented a mathematical model to estimate the heat conduction losses during laser cutting. They predicted the thermal field in the heat affected zone (HAZ). The influence of gas jet velocity in laser heating with a moving workpiece was studied by Shuja and Yilbas [5]. They found that the effect of assisting gas jet velocity was more pronounced in the cooling cycle than in the heating cycle of the laser heating process. The temperature field during laser forming of plate metal was analysed by Ji and Wu [6]. An increase in the maximum values of temperature was found with an increase in laser output power, while increasing the cutting speed and workpiece thickness has an inverse impact on the temperature. Thermal stress of the workpiece in laser cutting is always the result of the thermal load which is determined by the coupled temperature field. Yilbas, Arif [7] established a three-dimensional finite model to predict thermal stress field for different materials. The thermal stress formation during cutting in different shapes such as circle, sharp edges has been studied [8-11]. In their studies, a constant temperature heat source at the melting temperature of material for the laser beam was introduced in the simulation. Comparatively, Nyon, Nyeoh [12] used the same method to simulate kerf width formation and thermal stress during the laser cutting process. And a Guassian-distributed laser heat source was adopted to model the heat input during the laser irradiation.

However, in above studies few research studies have been conducted on the factor of cutting speed fluctuation at corner. So in this paper, the influence of cutting speed fluctuation for thermal stress field is carefully analysed, and a Guassian-distributed heat source is also adopted as laser heat source in the 3-D finite simulation model. Meanwhile, the process of material melting is simulated by the method of element death and birth in ANSYS that the temperature of finite element over the melt temperature will be deactivated. A deactivated element remains in the model but contributes a near-zero conductivity value to the overall matrix.

2. Model of cutting speed fluctuation at corner

According to reference [13], the corner size and maximum speed for the machine tool and the distance of acceleration and deceleration must be considered to calculate the cutting speed at corner. The workpiece size and B, C, D locations are shown in Fig.1.

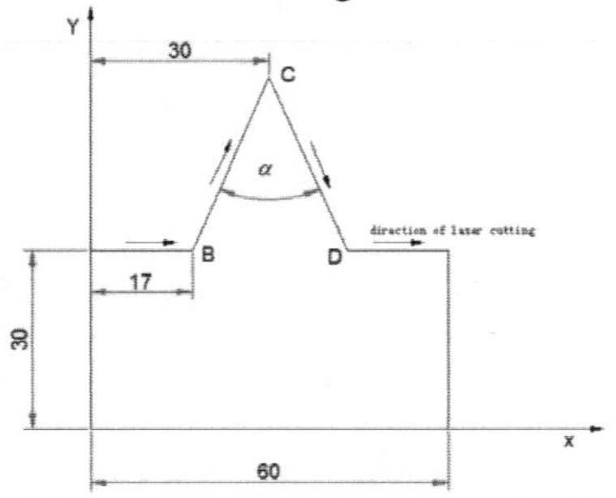

Fig.1 Workpiece size and B, C, D locations (unit: mm)

Considering the corner size, then

$$v_i \leq \frac{Ta_{max}}{2sin\frac{\alpha}{2}} \quad (7)$$

where T is interpolation period, is maximum acceleration, is corner size.

Considering the maximum speed v_{max} of the cutting machine, then

$$v_i \leq v_{max} \quad (8)$$

Considering the distance of acceleration and deceleration, then

$$v_i^2 \leq v_{i-1}^2 + a_m L_i \quad (9)$$
$$v_i^2 \leq v_{i+1}^2 + 2a_m L_{i+1} \quad (10)$$

where v_i is the initial velocity at acceleration distance L_i to the corner, v_{i+1} is the end velocity at deceleration distance L_i away from the corner. a_m is the maximum acceleration of cutting machine.

Combining Eqs.(7)-(10), the cutting speed at corner can be calculated as

$$v_i \leq \frac{Ta_{max}}{2sin\frac{\alpha}{2}} \quad (11)$$

In this study, specified cutting speed v_0 is 50mm/s, and maximum acceleration a_{max} is 8000mm/s2, and the interpolation period is 5ms. The cutting speed at location B, C, D can be calculated by speed fluctuation model as: $v_B = 36.8mm/s$, v_C=21.92mm/s, v_D=36.8mm/s. It is obvious that the speed at location C decreases most. The cutting speed variation along the laser scanning path for location B, C, D is shown in Fig.2.

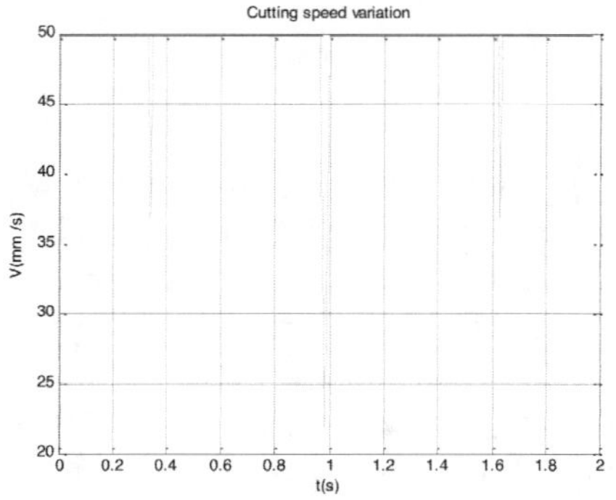

Fig.2 The curve of cutting speed variation

3. Finite element simulation

The cutting parameters used in the simulation are listed in Table1 which are the same with that in the experiment for comparison. The process of laser beam moving along the corner is realized by program as shown in Fig.3, and only half part of the cutting kerf is modeled due to its symmetry. The temperature distribution and the resulted thermal stress in the workpiece are performed by the FEM software ANSYS using ANSYS parametric Language (APDL). The mesh in finite element analysis is shown as Fig.3. The mesh is more intensive near the cutting edge. The mesh comprises about 70000 elements.

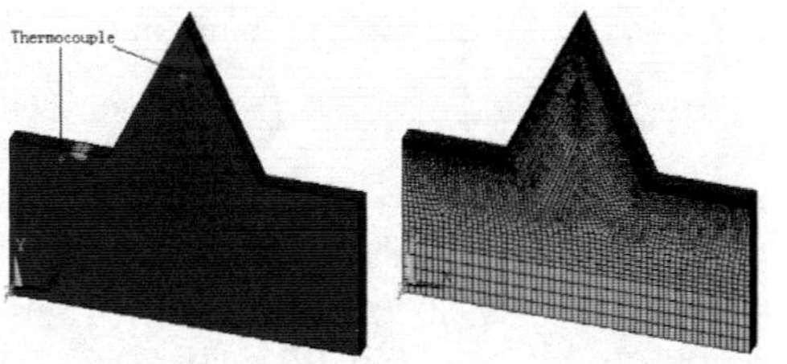

Fig.3 Model and meshing

The moving distance of laser source in each time step is the length of an element. The cutting parameters and material mechanical properties in simulation model are shown in Table 1. Temperature dependent thermal properties of plate are shown in Table 2.

Table 1. Cutting parameters and material properties used in the simulation

Cutting Speed (mm/s)	Laser power (W)	Plate thickness (mm)	thermal expansion coefficient α(1/K)	Poisson ratio ν	elasticity modulus E(GPa)	ρ(kg/m^3)
50	3000	3	63 x10-6	0.29	210	7854

Table 2. Thermal properties parameters of plate

Temp(K)	300	400	600	800	1000
K(W/mgK)	60.5	56.7	48	39.2	30
Cp(J/kgK)	434	487	559	685	1169

4. Results and discussion

4.1. Effects of corner size

Firstly, the corner size declines from 48.3° to 38.5°. The acceleration is 5000mm/s^2, and the cutting speeds for location B, C, D are calculated as: B=34.45mm/s, C=21.16mm/s, D=34.45mm/s. It can be found that the cutting speed for location C declines because of the smaller corner size. Temperature distribution contrast along Y-axis at location C for two kinds of corner size is separately shown in Fig.4 (a) and (b).The temperature distribution along Y-axis for two kinds of corner size is almost the same when the laser beam is at location C in Fig.4 (a). However, the temperature for smaller corner size at cut edge is higher when laser cutting finishes, as shown in Fig.4 (b). The reason for this is that smaller corner size leads to smaller heat diffusion area. Meanwhile, smaller corner size results in bigger decline of cutting speed and the corner attains more energy. Consequently, the corner burning is more serious.

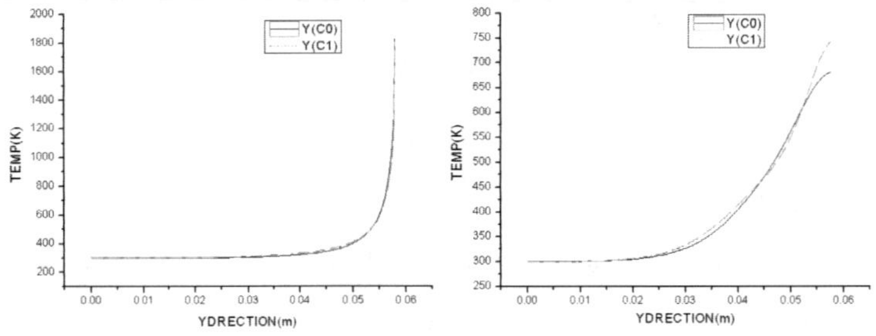

(a) (b)

Fig.4 Temperature distribution contrast along Y-axis at location C $\alpha_i = 48.3^{\circ}$(C0), $\alpha_i = 38.5^{\circ}$(C1) (a) Laser beam at location C (b) after cutting

Von Mises stress distribuion along Y-axis contrast with different corner size for location C is shown in Fig.5. Time T2 is when the laser beam is located at point C, and time T4 is when the workpiece is completely cool. The Von Mises stress has always rose from distance 0 to 0.056m at time T2 for different corner size. But the von mises stress has declined from distance 0.056m to cutting edge because of modulus of elasticity declining obviously near the melting temperature. It also should be noted that the Von Mises stress is smaller for smaller corner size because of more serious heat concentretionand and smaller temperature gradient. The residual stress has the same variation trend as Von Mises stress at time T4. The reason for this is that residual stress results from the Von Mises stress. Meanwhile, the corner with smaller size also has smaller residual stress. The maximum residual stress of different corner size at different locactions is shown in Table 3. It can be found that the residual stress reach the maximum at location B, D. Meanwhile, residual stress has increased when the corner size declines.

Table 3 residual stress at locations B, C, D

	B	C	D
C=48.3rad	275MPa	130MPa	270MPa
C=38.05rad	290MPa	120MPa	280MPa

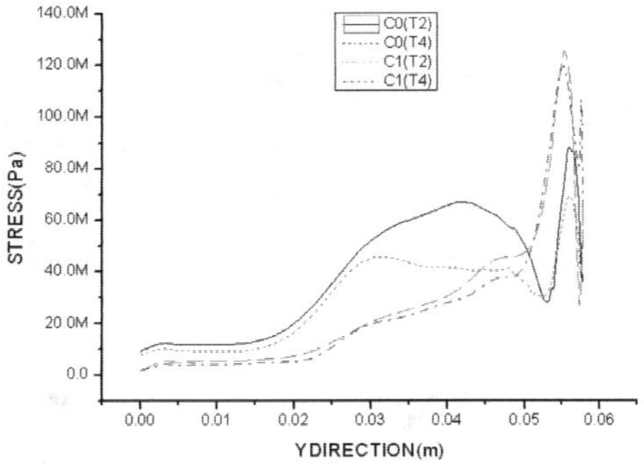

Fig.5 Von Mises stress distribution contrast along Y-axis for location C

4.2. Effects of cutting speed

The cutting speed declines from 50mm/s to 30mm/s. The acceleration is 5000mm/s^2. Then the cutting speed for location B, C, D is changed to 30mm/s, 21.92mm/s, 30mm/s. Temperature distribution contrast along Y-axis at location C for two kinds of cutting speeds is shown in Fig.6. The temperature distribution along Y-axis for two cutting speeds near cutting edge is almost the same when the laser beam is at location C in Fig.6 (a). It should be noted that the temperature is higher for smaller cutting speed at distance from 0.036mm to 0.055mm. The reason for this is that smaller cutting speed leads to the material attaining more energy. The temperature for two kinds of cutting speed at cut edge is the same when laser cutting finished in Fig.6 (b). The reason for this is that the cutting speed at corner declines almost to the same value. And this corner burnt phenomenon is the same for two cutting speeds. However, the temperature is higher for smaller cutting speed at distance 0.02mm to 0.051mm for the same reason that smaller cutting speed leads to more absorbed energy.

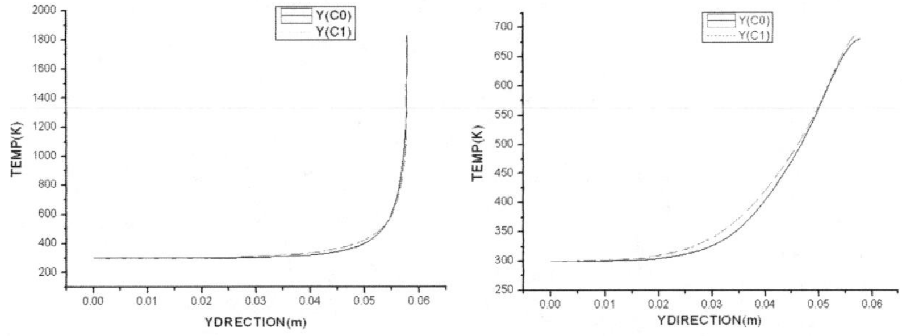

(a) (b)

Fig.6 Temperature distribution along Y-axis contrast at location C, 50mm/s(C0), (a) Laser beam at location C (b) after cutting

The Von Mises stress distribuion along Y-axis contrast with different cutting speed for location C is shown in Fig.7. Time T2 is when the laser beam is at location C, and time T4 is when the workpiece is completely cool. It can be found that the Von Mises stress is smaller for smaller cutting speed because of more serious heat concentration. Meanwhile, the corner with smaller cutting speed also has a smaller residual stress.The maximum residual stress of different cutting speed at different locactions is shown in Table 4. Residual stress has declined when the cutting speed declined.

Table 4 residual stress at locations B, C, D

	B	C	D
V=50mm/s	275MPa	130MPa	270MPa
V=30mm/s	265MPa	120MPa	265MPa

Journal of Physics: Conference Series **633** (2015) 012090 doi:10.1088/1742-6596/633/1/012090

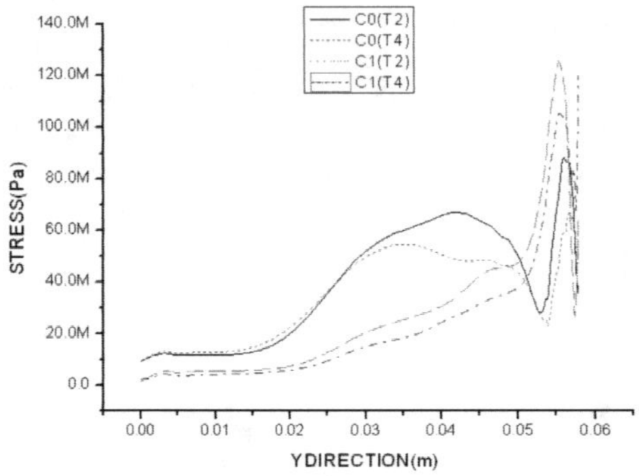

Fig.7 Von Mises stress distribution contrast along the y-axis at location C

5. Experiment

The laser used in the experiment is continuous CO2 laser and cutting parameters are the same with that used in the simulation. The temperature data is measured by thermocouple. And the residual stress is obtained by XRD. Then the surface quality of corner can be gained by optical microscope.

Fig.8 shows the temperature variation curve of location A for simulation and experiment. It can be found that the simulation has good agreement with the experiment. The residual stress at different locations for simulation and experiment is shown in Table 5. It also can be found that the simulation has good agreement with the experiment. Fig.9 shows the surface quality of corner. The surface has a poor quality in Fig.9. At the same time, it can be found that there exist serious corner burning and large adhering slag at bottom.

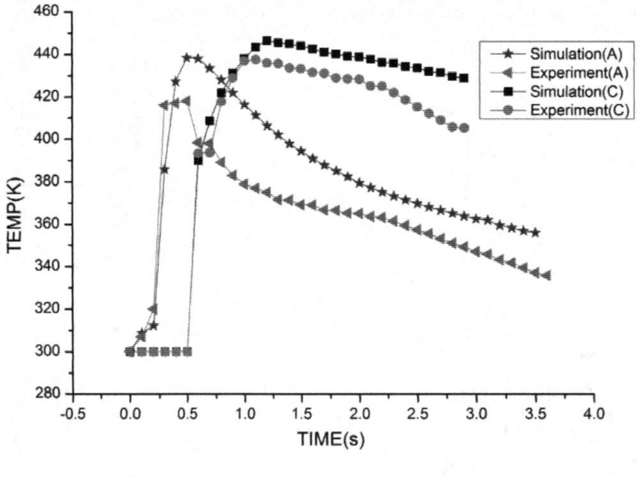

Fig.8 Temperature variation curve of location A, C

Table 5 residual stress (Mpa) at locations B, C, D

	B	C	D
Experiment(X-axis)	83.32 ± 8.39	74.31 ± 6.4	59.87 ± 8.19
Model predict(X-axis)	72	81	70
Experiment(Y-axis)	127.20 ± 7.13	-21.06 ± 8.89	115.13 ± 14.35
Model predict(Y-axis)	123	-35	130

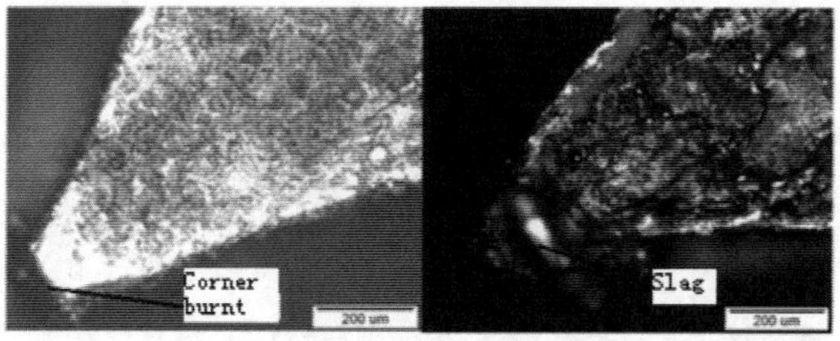

Fig.9 Surface quality at the corner

6. Conclusion

A 3D model of laser cutting corner with a fluctuant cutting speed is established, and the cutting process of laser heat source moving along the corner trajectory is simulated. Then the temperature and thermal stress field are analysed, and also the effects of changing cutting parameters such as corner size and cutting speed are studied.

1) Based on heat transfer, thermodynamics theory and speed fluctuation model, the modeling approach for laser cutting corner with a fluctuant speed is analyzed. The phenomenon of slower heat diffusion at the corner and corner burnt are been found in the 3D simulation model. At the same time, the influence of fluctuant speed on laser cutting quality is confirmed.
2) The influence of laser cutting parameters in the 3D simulation model is carried out. It is found that the temperature changes more slowly at the corner and the phenomenon of corner burning happens. Smaller thermal stress is obtained when the corner size decreases. And if the cutting speed decreases, the same phenomenon can also been found.

7. Acknowledgements

This work was supported by Research Project of State Key Laboratory of Mechanical System and Vibration (MSV201204).

8. References

[1] Farin, G.E., J. Hoschek, and M.-S. Kim 2002 *Handbook of computer aided geometric design* (North Holland)
[2] Tsai, M.-S., H.-W. Nien, and H.-T. Yau 2008 *Comput.-Aided Des.* **40** 554-566.
[3] Sheng, P.S. and V.S. Joshi 1995 *Journal of materials processing technology* **53** 879-892.
[4] Prusa, J.M., G. Venkitachalam, and P.A. Molian 1999 *International Journal of Machine Tools and Manufacture* **39** 431-458.
[5] Shuja, S. and B. Yilbas 2000 *Proceedings of the Institution of Mechanical Engineers, Part C: Journal of Mechanical Engineering Science* **214** 1059-1078.
[6] Ji, Z. and S. Wu 1998 *Journal of Materials Processing Technology* **74** 89-95.
[7] Yilbas, B., A. Arif, and B. Aleem 2009 *Optics and Lasers in Engineering* **47** 909-916.
[8] Yilbas, B. and S. Akhtar 2012 *Optics and Lasers in Engineering* **50** 204-209.
[9] Yilbas, B., S. Akhtar, and C. Karatas 2011 *Journal of Materials Processing Technology* **211** 1296-1304.
[10] Yilbas, B., A. Arif, and B. Abdul Aleem 2010 Optics and Lasers in Engineering **48** 10-19.
[11] Yilbas, B., A. Arif, and B.A. Aleem 2010 *Journal of materials engineering and performance* **19** 177-184.
[12] Nyon, K., et al. 2012 *The International Journal of Advanced Manufacturing Technology* **60** 995-1007.
[13] Hu, J., et al. 2006 *The International Journal of Advanced Manufacturing Technology* **28** 930-935.

4th International Conference on Mathematical Modeling in Physical Sciences (IC-MSquare2015) IOP Publishing
Journal of Physics: Conference Series **633** (2015) 012091 doi:10.1088/1742-6596/633/1/012091

Graph Model of Intradisciplinary Connections In Example of General Physics Course

Tatyana Gnitetskaya
The School of Natural Sciences, Far Eastern Federal University,
8 Sukhanova St., Vladivostok, 690950, Russia

E-mail: gnitetskaya.tn@dvfu.ru

Abstract. The model of an intradisciplinary connections was elaborated on the base of the theory of graphs. Every connection which appears in training content may be presented as oriented marked graphs. Each graph is a tree. In this paper we presented definition and model of intradisciplinary connections for example of physics course. The quantitative parameters of model are described in this paper. Quantitative method based on this model could be help to optimize a content of physics course. Furthermore using this model we can distinguish fundamental notions, laws and other elements of knowledge to separate hierarchical groups. This group is very important during the process of creating training course. Method of semantic structure attached to content of physics' lections and physical problems was described.

1. Introduction

The significance of modeling intradisciplinary connections in optimization of training course content is obvious. Optimizing the content of physics academic course is necessary due to several reasons including availability of different ways of describing natural phenomena in physics. The content of physics courses compiled by different authors can have different arrangement of sections and materials and hence, different interconnections. These connections depend on the level of "integrity" of intradisciplinary connections in the content of the course. If the degree of integrity (force) of every connection could be assessed, then the sum of forces would be equal to the integral characteristics of the course interconnectedness, i.e. its integrity. The course of physics with a higher integrity has a more interconnected content and is obviously more preferable for studying physics. Based on the course of general physics, the author of the article presents a graph model of an intradisciplinary connection whose quantitative features allows assessing the integrity of the course.

2. Graph Model of intradisciplinary connection

2.1. Definition of intradisciplinary connection

Many research works have been devoted to interdisciplinary connections; however, insufficient attention has been paid to intradisciplinary connections so far, a reason for that being a widespread opinion about the single nature of interdisciplinary and intradisciplinary connections. For example, V.A. Dalinger – "Internal connections between the elements of course structure can be established either within one subject or within different subjects, thus distinguishing between interdisciplinary and intradisciplinary connections" [1].

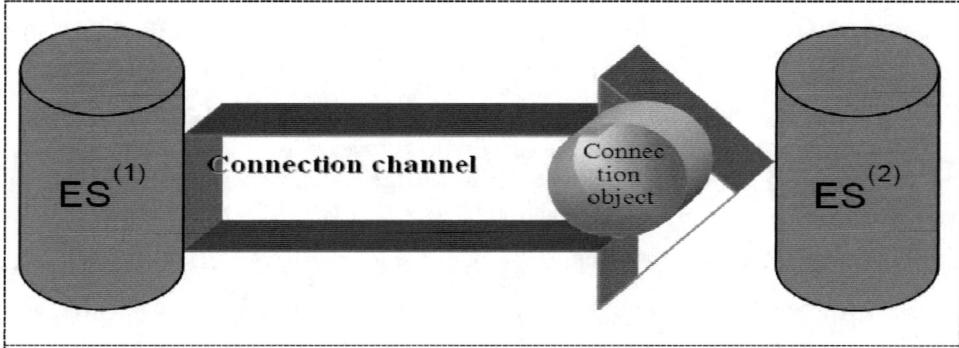

Figure 1. Illustration to intradisciplinary connection definition

Different interpretations of disciplinary connections suggested by other authors are similar in one area – pointing out that the nature of intra- and interdisciplinary connections is the same. In fact, connections appear in the moment of transmitting academic information. Therefore, in order to define the notion of intradisciplinary connections we will use the definition of interdisciplinary connections made in our previous research works [4, 6, 7, 8]. We define intradisciplinary connections as a construction of elements within a pedagogical system [3], which connects the structural elements of the educational intradisciplinary content and consists of (see fig.1) the following: object of connection – any element of knowledge, abilities and skills of the discipline – used at least in two elements of its structure; channel of connection – one or two elements of educational technologies coherent to the discipline where the connection is formed. The direction of intradisciplinary connections is defined by the sequence of studying the discipline and is set by the direction of transmitting academic information – from a structural element where the object initially appears to a structural element the connection is formed with. The definition of intradisciplinary connections determines that they all have the same direction.

2.2. Model

Definition of graph characteristics of intradisciplinary connections will be shown on the example of the traditional course of general physics (see, for example, [5]); we are going to use the approach defined in the works.

The structure of intradisciplinary content of the physics training course $\{ES^i\}$ can be presented as a set of twenty structural elements shown in table 1. The **objects of connection** are interpreted as elements of knowledge (principles, models, theories, laws and notions), skills and abilities used in the course of general physics.

Structuring the general physics course makes it possible to use the same elements of knowledge (principles, models, theories, laws and notions), the skills and abilities in studying various physical phenomena within an element of structure ES^i. We can unite the principles, models, theories, laws, notions and skills into the $\{EG_\mu^\nu\}$ sets or groups: $\{EG_\mu^1\}$ – group of notions; $\{EG_\mu^2\}$ – group of laws; $\{EG_\mu^3\}$ – group of theories; $\{EG_\mu^4\}$ – group of models; $\{EG_\mu^5\}$ – group of principles; $\{EG_\mu^6\}$ – group of skills and $\{EG_\mu^7\}$ – group of abilities. We use the following symbols: $\nu = 1,2,....,7$ – index number of the group, $\mu = 1,2,....,\mu_\nu$ - index number of the element, μ_ν - number of elements in ν-group. For example, element $\{EG_8^2\}$ corresponds to the Lorentz force, and in the fragment of the group of laws shown in Table 1 it is $\mu_2 = 25$.

Intersection of $\{EG_\mu^\nu\}$ with a set of elements of the general physics structure $\{ES^{(i)}\}$ creates many nodes $\{J_i(EG_1^\nu), J_i(EG_2^\nu), ...J_i(EG_\mu^\nu)\}$. This set falls into subsets of tagged nodes (nodes of Intradisciplinary connections) $\{J_i^*(EG_1^\nu), J_i^*(EG_2^\nu), \ldots J_i^*(EG_\mu^\nu)\}$.; each of them gets tagged only if the element EG_μ^ν is used in the i-element of the $ES^{(i)}$ structure. In Table 1, these sets are tagged with "diamonds". For example, the subset of $\{J_i^*(EG_5^4)\}$ includes the nodes tagged by the fact of applying the law EG_8^2 ("Lorentz force"), in the following structural elements of general physics course: $ES^{(2)}$ ("Dynamics"), $ES^{(13)}$ («Magnetic field in the vacuum»), $ES^{(15)}$ («Quasi-stationary electromagnetic field»), $ES^{(18)}$ ("Electromagnetic waves in matter"), $ES^{(19)}$ ("Atomic physics"). Since students acquire intradisciplinary knowledge, skills and abilities in the process of studying, the intradisciplinary connections and corresponding tagged nodes are formed in the course of time. Therefore we will supply the subsets of nodes of intradisciplinary connections $\{J_i(EG_1^\nu), J_i(EG_2^\nu), \ldots J_i(EG_\mu^\nu)\}$ with the subsets of nodes $\{J_{k\mu}^*(EG_\mu^\nu)\}$, tagged in case the element EG_μ^ν appears and is formed in the course of general physics for the first time. Here the k_μ index determines the index number of the structural element of general physics course where $\{EG_\mu^\nu\}$ appears first. For example, the Lorentz force $\{EG_8^2\}$ (see Table 1) first (time moment $\tau_{1\mu}$) is introduced in the second structural element -$ES^{(2)}$ («Dynamics»), whereas the well-grounded and most general concepts of the Lorentz force are formed much later (time moment $\tau_{0\mu}$)), in $ES^{(13)}$ («Stationary magnetic field in the vacuum»). This example illustrates time-distributed process of forming EG_μ^ν and hence, the intradisciplinary connection.

Thus, like interdisciplinary connections, the intradisciplinary connections realized through the EG_μ^ν object of connection can be described as orientable tagged graphs $\tilde{G}_\iota(EG_\mu^\nu)$ (see Table 1). Each $\tilde{G}_\iota(EG_\mu^\nu)$ graph of intradisciplinary connections is a tree, which starts from a tagged $J_{k\mu}^*(EG_\mu^\nu)$ node – the same for every intradisciplinary connection realized through EG_μ^ν. This node is tagged by the fact that the property of element EG_μ^ν is manifested sooner in it than in other tagged nodes. The graphs finish in the nodes of $\{J_i^*(EG_\mu^\nu)\}$ subset, which are tagged if EG_μ^ν is used in the ***i*** –element of $ES^{(i)}$ structure. Like in the graphs of interdisciplinary connections [4, 7], the $J_{k\mu}^*(EG_\mu^\nu\ )$ node is a root a single circle at the root of the tree of the abovementioned orientable tagged graph where there are no cycles. This peculiarity manifests itself in the fact that the leaf node of the graph of intradisciplinary connections $\tilde{G}_\iota(EG_\mu^\nu)$ is adjacent to the leaf node of the subgraph $\tilde{G}_{i-1}(EG_\mu^\nu)$ (or overgraph $\tilde{G}_{i+1}(EG_\mu^\nu)$, if the latter exists. Therefore, the process of forming intradisciplinary connections can be interpreted as a transfer from the leaf node of one intradisciplinary connection of graph $\tilde{G}_\iota(EG_\mu^\nu)$ to the leaf node of a different intradisciplinary connection, corresponding to the overgraph $\tilde{G}_{i+1}(EG_\mu^\nu)$. According to the sequence of studying the general physics course, this transfer geos from one structural element of general physics course to another, or through the nodes $J_k(EG_\mu^\nu)$, which reflects the dependency of the described process on the time. In this dynamic interpretation, the capacity of the intradisciplinary connection is a function of time or index number of the nod:

$$p_{im}^{(in)}(t, EG_\mu^\nu) = p_{im}^{(in)}(J_k(EG_\mu^\nu). \tag{1}$$

It is a quantitative feature of the connection considering the non-equivalence of connection channels, which appears due to the differences, for example, between the elements of educational technologies used in studying the same EG_μ^ν in different structural elements of general physics course.

This model of intradisciplinary connections allows introducing their quantitative features. If take that T_k – is a length of studying k – element of structure $ES^{(k)}$, then the time of creating the node $J_i^*(EG_\mu^\nu)$, calculated relative to the beginning of studying the general physics course is equal to:

$$t_i(J_\mu^\nu(EG_\mu^\nu)) = \sum_{i=0}^{n} T_k \tag{2}$$

Let's consider the homogenous model, constructed in the approximation of the time homogeneity of the course structure which says that the length of studying every element of structure $ES^{(i)}$ is the same and equals to the unit of *(T_i=1)*. Then, the time (2) of forming the node $J_i^*(EG_\mu^\nu)$ calculated relative to the beginning of studying the general physics course is equal to the index number of the element of structure $ES^{(i)}$. $t_i(J_\mu^*(EG_\mu^\nu)) = i$, and the length of connection is:

$$t_{k_\mu i}\left(EG_\mu^\nu\right) = i - k_\mu. \quad (3)$$

In the taken approximation of structural homogeneity of the course by time ($T_i = 1$), using (3), and capacity (1) of the connection $\left(p_{im}^{(ex)}\left(J_k\left(EG_\mu^\nu\right)\right) = 1\right)$, the expressions for effective length

$$L_{k_\mu i}^{(in)}\left(EG_\mu^\nu\right) = i - k_\mu \quad (4)$$

and effective continuous length have been simplified

$$\tilde{L}_{k_\mu i}^{(in)}\left(EG_\mu^\nu\right) = i - k_\mu - R_{k_\mu i}\left(EG_\mu^\nu\right), \quad (5)$$

where the length of gaps $R_{k_\mu i}\left(EG_\mu^\nu\right)$ equals to the number of non-tagged nodes between root $J_{k_\mu}^*\left(EG_\mu^\nu\right)$ and leaf $J_i^*\left(EG_\mu^\nu\right)$ nodes of the graph of intradisciplinary connections $\tilde{G}\ \left(EG_\mu^\nu\right)$. Relative maximal length of the intradisciplinary connection $L^{(in)}\left(EG_\mu^\nu\right)$ can be calculated by:

$$L^{(in)}\left(EG_\mu^\nu\right) = \frac{i_{max} - k_\mu}{N-1} \quad (6)$$

Existence of gaps, i.e. the structural elements of general physics course where EG_μ^ν is not used weakens the intradisciplinary connection. Therefore, on the base (4) and (5) the force of intradisciplinary connection is implemented $f_{k_\mu i}^{(in)} EG_\mu^\nu$:

$$f_{k_\mu i}^{(in)} EG_\mu^\nu = \frac{\tilde{L}_{k_\mu i}^{(in)}\left(EG_\mu^\nu\right)}{L_{k_\mu i}^{(in)}\left(EG_\mu^\nu\right)} \quad (7)$$

Then, integrity of the course can be interpreted as a sum of branching of the graphs of intradisciplinary connections within the total of knowledge elements for the group of knowledge elements

$$F_\nu^{(in)} = \sum_{\mu=1}^{\mu_\nu} \sum_i f_{k_\mu i}^{(in)} EG_\mu^\nu \quad (8)$$

or for all groups

$$F^{(in)} = \sum_\nu \sum_{\mu=1}^{\mu_\nu} \sum_i f_{k_\mu i}^{(in)} EG_\mu^\nu \quad (9)$$

3. Findings

Table 1 shows a fragment of intradisciplinary space of the general physics course. It is generated by intersection of multiple laws of physics with the set (N=20) of structural elements of general physics course. The cell is tagged with a diamond if a law is used in the material of a corresponding structural element of the course. The connection is established through every law between the structural element where this law is introduced and the one where it is applied. The table shows that the Lorentz force helps establish four connections. The existing chronological classification of intradisciplinary connections is well illustrated in this model. The objects of interdisciplinary connection EG_μ^ν can be used in the course of general physics during or after they have been formed, which corresponds to the current and subsequent intradisciplinary connections. For example, the intradisciplinary connections established through the object of connection EG_8^2 («Lorentz force»), are classified as follows. If we consider that the process of formation EG_8^2 ends in the 13th element of the structure of general physics course $ES^{(m_5=13)}$, then the intradisciplinary connection between $ES^{(2)}$ and $ES^{(13)}$ is current, and the connections between $ES^{(2)}$ и $ES^{(15)}$, $ES^{(2)}$ and $ES^{(18)}$, $ES^{(2)}$ and $ES^{(19)}$ – are subsequent. Calculation of the features of intradisciplinary connections by correlations (6) and (7) helped establishing the laws applied in this course of physics into a hierarchical sequence. First the laws are grouped by values of relative maximal length of connection $L^{(in)}\left(EG_\mu^\nu\right)$, use the formula (6), then they are ranked by the value of force of the intradisciplinary connection $f_{k_\mu i}^{(in)}\left(EG_\mu^\nu\right)$ – formula (7). Thus, the degree of significance of the element EG_μ^ν is defined by its position (determined by index μ) in the ranked set $\left\{EG_\mu^\nu\right\}$. Table 1 shows that the Newton's first law ranked 1st in the hierarchy; this law is a fundamental one by right. Among other fundamental laws there are the laws, which are used at least once in every section of the course. For example, in the list of fundamental laws, the Law of energy conservation ranks second, the

Law of conservation of momentum – fourth, and the Law of charge conservation – seventh, since, with the equal values of relative maximal length (equals to 0.89 for each law), the forces of intradisciplinary connections are different: 0.59, 0.24, and 0.12 correspondingly.

Table 1 Fragment of distributing the laws in the general physics course with the graphs of intradisciplinary connections established through the Lorentz force.

	Physics' Course Structure	*Mechanics*						*Molecular Physics and Thermodynamics*				*Electricity and Magnetism*					*Waves*			*Atomic and Nuclear*		Relative maximal length	Connection relative strength
												Stationary electric field		Stationary magnetic field				ElectromagneticWaves					
	Group of Laws	Kinematics	Dynamics	Conservation laws	Dynamics of rigid body	Vibration	Gravitation field	Thermodynamics	Statistic laws	Gases, Liquids	Solids	Vacuum	Matter	Vacuum	Matter	Electromagnetic Field	Elastic waves	Vacuum	Matter	Atom	Nuclear		
	an object for study	1	2	3	4	5	6	7	8	9	10	11	12	13	14	15	16	17	18	19	20		
1	1-st Newton Law		◆	◆	◆	◆	◆	◆	◆		◆	◆	◆	◆	◆	◆	◆	◆	◆	◆		0,89	0,79
2	Conservation of energy			◆	◆	◆	◆	◆	◆	◆	◆					◆	◆	◆	◆	◆	◆	0.89	0.59
…																							
4	Conservation of impulse			◆	◆	◆		◆		◆								◆		◆	◆	0.89	0.24
…																						…	…
7	Conservation of charge law			◆						◆		◆	◆	◆				◆		◆	◆	0.89	0.12
8	Lorentz force		◆											◆		◆			◆	◆			
…	……																						
25	Ampere Law													◆									

4. Conclusions

Thus, we have elaborated the model of intradisciplinary connections, which interprets the connection as a tree-structured orientable graph. The model involves qualitative characteristics of intradisciplinary connections, which help optimize the structure of general physics course, conduct a comparative analysis of the course content, define the level of their integrity - formulas (8) and (9) and the hierarchical order, the calculation of which – formulas (6), (7) we noted above, of the notions, laws, models, theories and principles within the course, as well as define the fundamental core of the course. For example, the fundamental laws of impulse conservation, impulse momentum and Coulomb's law are studied not in every section in the described course. In comparison with the law of energy conservation, their significance is lower, since the traditional course of general physics does not contain the material, which involves using these laws; therefore, the connection is weakened. We can suggest, for example, supplying the section of Electromagnetism with the tasks of motion of charged particles in electric and magnetic fields that can be solved based on the laws of conservation of energy, impulse, and impulse momentum. The table shows the necessity of other revisions, which can strengthen the intradisciplinary connections and contribute to the formation of more comprehensive knowledge of general physics with the students.

References

[1] Dalinger V.A. Perfection of mathematics educating process on the basis of results-oriented realizing of intrasubject connections, Omsk: OmIATE, 1993.- 323 p.

[2] Bakmaev Sh.A. About realizing of intrasubject connections while studying of trigonometric expressions transformations // Ways of formal description precaution in student's knowledge while studying mathematics: Methodological Recommendations under ed. of E.I. Liaschenko, Z.I.Novoseltseva-L.: Publication of Leningrad pedagogics institute, 1989.-p.45-53.

[3] Bespalko V.P. Components of educational technology.-M.:Pedagogics, 1989.-192 p.

[4] Gnitetskaya T.N.2006. Information models of intra- and interdisciplinary connections as a basis of training physics technology. Dr. dissert., (Moscow: MPGU Press).

[5] Saveliev I.V. Course of General physics. In 3 books. Study guide: 3-rd publ. ed. M.: Science, 1982

[6] Gnitetskaya T.N., Ivanova E.B. Interdisciplinary Physical Cluster for Environmental Engineers Training //Advanced Materials Research Vols. 753-755 (2013) pp 3153-3156 © (2013) Trans Tech Publications, Switzerland doi:10.4028/www.scientific.net/AMR.753-755.3153

[7] Gnitetskaya T.N. Modeling of Interdisciplinary Connections in Science Courses Journal of Physics: Conference Series 490 (2014) 012068 doi:10.1088/1742-6596/490/1/012068

[8] Gnitetskaya Tatyana N., Ivanova Elena B. Interdisciplinary Conceptual Cluster in Mathematics and Physics on the Basis of a Graph Model of Interdisciplinary Links. Advanced Materials Research Vols. 889-890 (2014) pp 1704-1707 © (2014) Trans Tech Publications, Switzerland doi:10.4028/www.scientific.net/AMR.889-890.1704-1707

Journal of Physics: Conference Series **633** (2015) 012092 doi:10.1088/1742-6596/633/1/012092

Effects of molecular complexity and reservoir conditions on the discharge coefficient of adapted planar nozzles

Alberto Guardone
Department of Aerospace Science and Technology
Politecnico di Milano
Via La Masa 34, 2016 Milano, Italy

E-mail: alberto.guardone@polimi.it

Abstract. The transonic flow at throat section of a convergent-divergent nozzle is studied in adapted conditions to assess the influence of the fluid molecular complexity and total thermodynamic state on the discharge coefficient. The standard Sauer method is applied to solve the transonic perturbation potential equation in the vicinity of the nozzle throat. An analytic expression is derived that allows one to compute the discharge coefficient in terms of the nozzle curvature at the throat section and of the value of the fundamental derivative of gasdynamics at sonic conditions, which depends on the fluid molecular complexity and on the thermodynamic state in the reservoir. A linear dependence of the discharge coefficient on the sonic value of the fundamental derivative of gasdynamics is exposed.

1. Introduction

The flow expansion through a nozzle from a reservoir into a constant pressure ambient is a fundamental problem of fluid mechanics. In the case where the exit conditions are supersonic, namely, if the discharge velocity is supersonic, a convergent-divergent nozzle geometry is implemented. This is the case for example of rocket exhaust nozzle, supersonic wind tunnel and supersonic blade turbine passages. In most relevant conditions, the fluid thermodynamic properties can be computed according to the ideal dilute gas model. However, for operating conditions in the close proximity of the liquid-vapor saturation curve, see figure 1, so-called non-ideal compressible-fluid effects are observed and non-ideal thermodynamics is to be accounted for [1, 2, 3].

In the present work, the well known transonic flow solution of Sauer [4] is applied to the case of a planar nozzle operating in adapted conditions in the non-ideal compressible-fluid regime to evaluate the influence of the fluid molecular complexity and of the reservoir thermodynamic state on the discharge coefficient, namely, on the maximum mass flow that can be processed by the nozzle itself [5].

The paper structure is as follows. In section 2, the Sauer theory is extended to non-ideal compressible fluids and an explicit expression of the perturbation velocity is derived. In section 3, a Taylor-series expansion of the density is derived to compute the discharge coefficient in transonic conditions. In section 4, final remarks and comments are given.

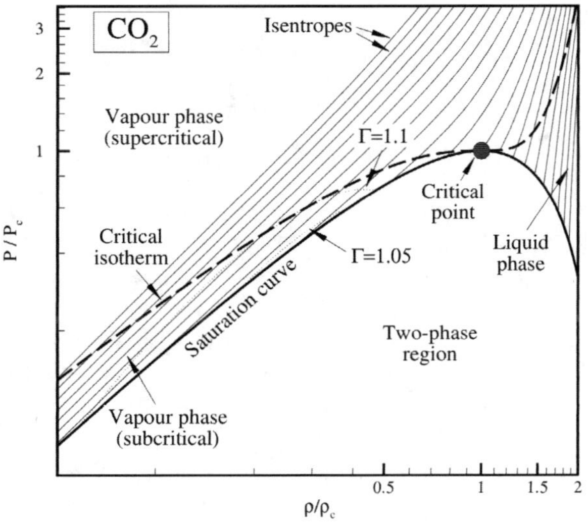

Figure 1. Liquid-vapor saturation curve in the density-pressure thermodynamic plane according to the van der Waals model of carbon dioxide. The critical point and the critical isotherms are also shown, together with a number of isentropes crossing the non-ideal compressible-fluid region. The iso-Γ lines $\Gamma = 1.1$ and $\Gamma = 1.05$ are depicted.

2. Transonic pertubation-potential flows at the throat section

The governing equations describing planar nozzle flows of interest here are now specialized to the case of non-ideal compressible fluids. Two-dimensional flow fields with uniform total enthalpy $h^{\rm t}$ and entropy s per unit mass, namely, $h^{\rm t} \equiv h_{\rm r}$ and $s \equiv c_{\rm r}$, with the subscript r indicating reservoir conditions, are irrotational by virtue of the well-known Crocco theorem [6]. Therefore, a (scalar) velocity potential ϕ, with $\boldsymbol{v} = (u, v)^{\rm T} = \boldsymbol{\nabla}\phi$ velocity vector, can be introduced to obtain the well-known velocity potential equation [7]

$$\left(\phi_x^2 - c^2\right)\phi_{xx} + 2\,\phi_x\phi_y\phi_{xy} + \left(\phi_y^2 - c^2\right)\phi_{yy} = 0, \tag{1}$$

where x and y are the spatial coordinates and where the subscripts indicate partial derivatives and where c is the speed of sound, $c^2(s,\nu) = -\nu^2(\partial P/\partial \nu)_s$, with ν the specific volume, P the pressure and where, according to standard thermodynamic nomenclature, the subscript s indicates that the pressure derivative is computed at constant entropy per unit mass s.

It is remarkable that the thermodynamic description of the fluid enters Eq. (1) only via the definition of the speed of sound function. Choosing the entropy s and the enthalpy $h = h(s,\nu)$ per unit mass as independent thermodynamic variables, $c^2 = c^2(s,h)$ and one immediately obtains $h(h_{\rm r}, w^2) = h_{\rm r} - \frac{1}{2}w^2$, with $w^2 = u^2 + v^2 = |\boldsymbol{\nabla}\phi|^2$. Therefore, the speed of sound in Eq. (1)—and all thermodynamic quantities—is a function of the velocity module only, or $c^2 = c^2(w^2)$. For example, for an ideal gas with constant isochoric specific heat, $c^2(w^2) = (\gamma - 1)(h_{\rm r} - \frac{1}{2}w^2)$, where γ is a constant equal to the ratio of the isobaric to the isochoric specific heat.

A perturbation velocity potential $\tilde{\phi}$, $\phi = \bar{w}\,L(x + \tilde{\phi})$, with $\bar{w}$ velocity module at the reference state, where the flow is parallel to the x axis, and L reference length, is now introduced to study the small-perturbation flow occurring in the vicinity of the nozzle throat. In the following, the symbols $\bar{}$ and $\tilde{}$ indicate quantities evaluated at the (constant) reference state and perturbation quantities, respectively. Under the assumption of small perturbation velocities, i.e. $||\tilde{\boldsymbol{v}}|| \ll 1$, with $\tilde{\boldsymbol{v}} = \boldsymbol{\nabla}\tilde{\phi}$ perturbation velocity vector, the equation for the perturbation velocity potential in the transonic, i.e. $\bar{M} \sim 1$, regime is then obtained from (1) as

$$\left[\bar{M}^2 - 1 + 2\bar{\Gamma}\bar{M}^2\,\tilde{\phi}_x\right]\tilde{\phi}_{xx} - \tilde{\phi}_{yy} = 0, \tag{2}$$

where the thermodynamic quantity Γ is the so-called fundamental derivative of gasdynamics introduced by Thompson [8]

$$\Gamma(s, v) \equiv 1 - \frac{\nu}{c}\left(\frac{\partial c}{\partial \nu}\right)_s. \tag{3}$$

In deriving (2), $\bar{\Gamma}$ has been assumed to be $\mathcal{O}(1)$. This assumption, that is valid for most typical fluids, see [9], leads to a simpler equation for the perturbation velocity potential with respect to those derived by [10, 11, 12, 13], which are valid also for the $\Gamma < 0$ region of so-called BZT fluids, see e.g. [10]. For an ideal polytropic gas, Eq. (2) assumes the familiar form $\left[\bar{M}^2 - 1 + (\gamma+1)\bar{M}^2\,\tilde{\phi}_x\right]\tilde{\phi}_{xx} - \tilde{\phi}_{yy}$, see e.g.[7].

The non-ideal fluid equation (2) differs from the ideal gas one considered by Sauer [4] only for the constant multiplying the first term, whose value is of the same order of the ideal-gas constant $\gamma + 1$, since $\bar{\Gamma} \sim 1$ in the region of interest. The Sauer technique can therefore be applied without modifications to obtain the perturbation potential as [7]

$$\tilde{\phi}(\hat{x}, y) = \frac{1}{16}\alpha\left[8(\hat{x}^2 + \alpha\Gamma\hat{x}y^2) + (\alpha\Gamma y^2)^2\right] \tag{4}$$

where $\alpha = \sqrt{1/(\Gamma\, r_{\rm t} y_{\rm t})}$, with $r_{\rm t}$ and $y_{\rm t}$ the throat radius of curvature and the height of the wall at $x = x_{\rm t}$, namely, at the throat section. In the expression above, $\hat{x} = x - x_{\rm t} - \frac{1}{4}\Gamma\alpha y_{\rm t}^2$. The x and y component of the perturbation velocity field are then computed from the perturbation potential (4) as

$$\tilde{u}(\hat{x}, y) = \tilde{\phi}_{\hat{x}} = \alpha\hat{x} + \bar{\Gamma}(\alpha y)^2 \qquad \text{and} \qquad \tilde{v}(\hat{x}, y) = \tilde{\phi}_y = 2\bar{\Gamma}\alpha^2 y\left[\hat{x} + \tfrac{1}{3}\bar{\Gamma}\alpha y^2\right] \tag{5}$$

3. Discharge coefficient

The discharge coefficient of a convergent-divergent planar nozzle in chocked conditions is now computed by integrating the momentum density at the nozzle section, using the Sauer solution of the perturbation potential (4) for non-ideal compressible-fluid flows.

The mass flux per unit height across the throat section ($x_{\rm t} = 0$) is defined as

$$\dot{m} = \int_{-\frac{y_{\rm t}}{2}}^{\frac{y_{\rm t}}{2}} \rho(0, y)u(0, y)\,\frac{{\rm d}y}{y_{\rm t}} \tag{6}$$

The momentum along the x direction is $\rho u = \bar{\rho}\bar{u}\,(1 + \tilde{\rho})\,(1 + \tilde{u}) = \bar{\rho}\bar{u}\,(1 + \tilde{\rho} + \tilde{u} + \tilde{\rho}\tilde{u})$ and therefore the discharge coefficient per unit height is computed from (6) as

$$\frac{\dot{m}}{\dot{m}_{\rm 1D}} - 1 = \int_{-\frac{y_{\rm t}}{2}}^{\frac{y_{\rm t}}{2}} \frac{\rho(0, y)u(0, y)}{\bar{\rho}\bar{u}}\,\frac{{\rm d}y}{y_{\rm t}} - 1 = \int_{-\frac{y_{\rm t}}{2}}^{\frac{y_{\rm t}}{2}} \left[\tilde{\rho}(0, y) + \tilde{u}(0, y) + \tilde{\rho}(0, y)\tilde{u}(0, y)\right]\frac{{\rm d}y}{y_{\rm t}} \tag{7}$$

where $\dot{m}_{\rm 1D} = \bar{\rho}\bar{u}$ is the mass flux per unit height in one-dimensional conditions.

To compute the integral (7), an approximate expression of the function $\tilde{\rho} = \tilde{\rho}(w)$ is now derived. Expansions in the vicinity of the reference state gives

$$\rho = \bar{\rho} + \left.\frac{{\rm d}\rho}{{\rm d}w^2}\right|_{\bar{w}}\left[w^2 - \bar{w}^2\right] + \frac{1}{2}\left.\frac{{\rm d}^2\rho}{{\rm d}^2 w^2}\right|_{\bar{w}}\left[w^2 - \bar{w}^2\right]^2 + \mathcal{O}([\Delta w^2]^3) \tag{8}$$

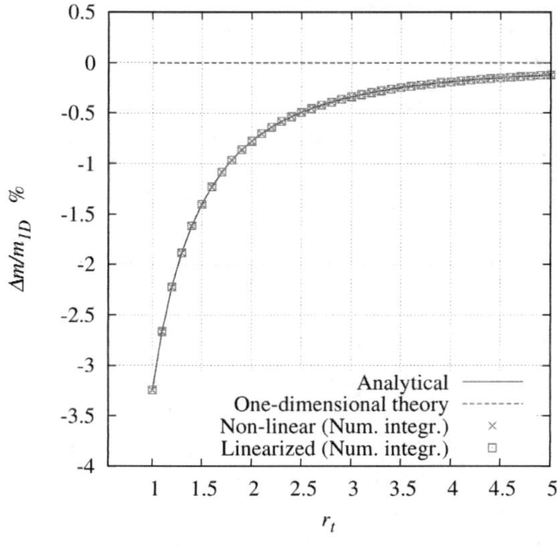

(a)

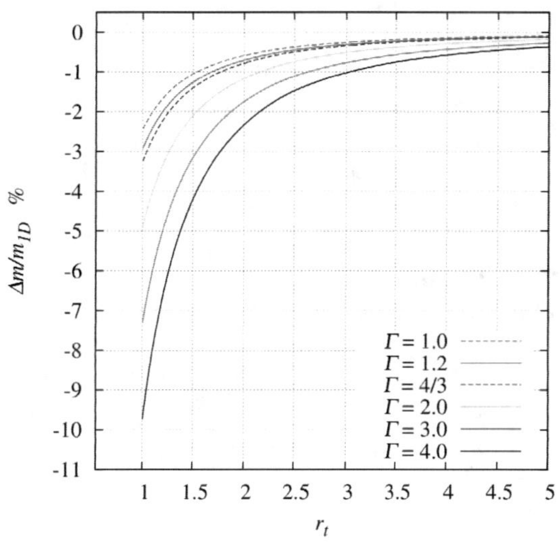

(b)

Figure 2. a) Discharge coefficient computed using the Sauer [4] method for non-ideal compressible-fluid flows. a) Comparison of the results from the analytic expression (12) to numerical integration of the Sauer solution including all non-linear terms and of the linearized form (11). b) Mass flow defect from the analytic expression (12) for different values of Γ.

where the first and second order derivative of the density function reads

$$\frac{\mathrm{d}\,\rho}{\mathrm{d}\,w^2} = \left(\frac{\partial \rho}{\partial h}\right)_s \frac{\mathrm{d}\,h}{\mathrm{d}\,w^2} = -\frac{1}{2}\frac{\rho}{c^2} \qquad \text{and} \qquad \frac{\mathrm{d}^2\,\rho}{\mathrm{d}^2\,w^2} = \frac{3-2\Gamma}{4}\frac{\rho}{c^4}. \tag{9}$$

By retaining first and second order terms of the velocity potential, the perturbation density is finally obtained as follows

$$\tilde{\rho} = \frac{\rho}{\bar{\rho}} - 1 = -\frac{\bar{M}^2}{2}(2\tilde{u} + \tilde{u}^2 + \tilde{v}^2) + \frac{\bar{M}^4}{4}(3 - 2\bar{\Gamma})\tilde{u}^2 + \mathcal{O}(\tilde{v}^3, \tilde{v}^3). \tag{10}$$

The momentum defect in the x direction for $\bar{M} = 1$ therefore reads

$$\frac{\rho u}{\bar{\rho}\bar{u}} - 1 = \tilde{\rho} + \tilde{u} + \tilde{\rho}\tilde{u} \simeq -\bar{\Gamma}\tilde{u}^2 - \frac{1}{2}\tilde{v}^2, \tag{11}$$

an expression that exposes the second-order nature of the mass flux defect correction within the potential perturbation theory. Substituting the expressions of the perturbation velocity components (5) evaluated at the nozzle throat ($x = 0$, i.e., $\hat{x} = \epsilon$), one finally obtains

$$\frac{\dot{m}}{\dot{m}_{1\mathrm{D}}} - 1 = \frac{1}{45}\frac{\bar{\Gamma}}{r_{\mathrm{t}}^3}\left(r_{\mathrm{t}} + \frac{2}{21}\right). \tag{12}$$

Retaining higher-order terms in the expansion (10) was found not to improve the accuracy of expression if compared to the numerical integration of (7), as it can be appreciated from figure

2a, where the results from the analytic expression (12) are compared to numerical integration of the Sauer solution including all non-linear terms and of the linearized form (11). If figure 2b, the mass flow defect from the analytic expression (12) is reported for different values of $\Gamma \geq 1$. Value of Γ in between 1 and 4/3 are attainable by ideal gases with constant specific heats: $\Gamma = 1$ corresponds to the limiting case of an infinite value of the isochoric specific heat, $\Gamma = 1.2$ to standard air, $\Gamma = 4/3$ to a monoatomic gas. Values of Γ outside this range can be attained only in the non-ideal compressible-fluid regime. In particular, large values of Γ are expected in the liquid phase and in the close proximity of the critical point, where $\Gamma \to \infty$ [14].

4. Conclusions

The Sauer method for solving the perturbation potential equation in the transonic regime was applied to non-ideal compressible-fluid flows in the close proximity of the throat of convergent-divergent planar nozzles. The analytic expression of the perturbation velocity was derived as a function of the curvature of the nozzle geometry at the throat section and of the sonic value of the fundamental derivative of gasdynamics Γ, which embeds non-ideal fluid behavior and its dependence on the reservoir conditions. An analytic expression of the discharge coefficient was devised for the first time which exposes its linear dependence on Γ. For large values of Γ, which are expected to be observed in the liquid phase and in the close proximity of the liquid-vapor saturation curve, the discharge coefficient is as low as 90% (for $\Gamma = 4$), to be confronted to a minimum value of 97% corresponding to mono-atomic ideal gases. The discharge coefficient depends also on the inverse of the second and third powers of the curvature radius at the throat section. Experimental and numerical activities are planned to verify the present findings.

Acknowledgments

The research is partially funded by the European Research Council under Grant ERC Consolidator 2013, project NSHOCK 617603.

References

[1] Harinck J, Colonna P, Guardone A and Rebay S 2010 *Journal of turbomachinery* **132** 011001
[2] Guardone A 2011 *8th International Conference on Flow Dynamics* (Sendai)
[3] Guardone A, Spinelli A and Dossena V 2013 *ASME Journal of Engineering for Gas Turbines and Power* **135** 042307
[4] Sauer R 1947 General characteristics of the flow through nozzles at near critical speeds TM 1147 NACA
[5] Nakao S i 2007 **122** 730–734
[6] Crocco L 1937 *Z. Angew. Math. Mech.* **17** 1–7
[7] Zucrow M H and Hoffman J D 1976 *Gas dynamics* vol 1 (Wiley, Jhon & Son)
[8] Thompson P A 1971 *Phys. Fluids* **14** 1843–1849
[9] Thompson P A 1988 *Compressilbe Fluid Dynamics* (McGraw-Hill)
[10] Kluwick A 1993 *J. Fluid Mech.* **247** 661–688
[11] Schnerr G H and Molokov S 1994 *Phys. Fluids* **6** 3465
[12] Schnerr G H and Molokov S 1995 *Phys. Fluids* **7** 2867
[13] Rusak Z and Wang C W C W 1997 *Journal of Fluid Mechanics* **346** 1–21
[14] Nannan N R, Guardone A and Colonna P 2012 *Fluid Phase Equilib.* **337** 259–273

Optimum filter selection for Dual Energy X-ray Applications through Analytical Modeling

V Koukou[1], N Martini[1], C Michail[2], P Sotiropoulou[1], N Kalyvas[2], I Kandarakis[2], G Nikiforidis[1] and G Fountos[2]

[1] Department of Medical Physics, Medical School, University of Patras, 265 00 Patras, Greece
[2] Radiation Physics, Materials Technology and Biomedical Imaging Laboratory, Department of Biomedical Engineering, Technological Educational Institute of Athens, Egaleo, 122 10 Athens, Greece

E-mail: gfoun@teiath.gr

Abstract. In this simulation study, an analytical model was used in order to determine the optimal acquisition parameters for a dual energy breast imaging system. The modeled detector system, consisted of a 33.91mg/cm^2 Gd_2O_2S:Tb scintillator screen, placed in direct contact with a high resolution CMOS sensor. Tungsten anode X-ray spectra, filtered with various filter materials and filter thicknesses were examined for both the low- and high-energy beams, resulting in 3375 combinations. The selection of these filters was based on their K absorption edge (K-edge filtering). The calcification signal-to-noise ratio (SNR_{tc}) and the mean glandular dose (MGD) were calculated. The total mean glandular dose was constrained to be within acceptable levels. Optimization was based on the maximization of the SNR_{tc}/MGD ratio. The results showed that the optimum spectral combination was 40kVp with added beam filtration of 100 μm Ag and 70kVp Cu filtered spectrum of 1000 μm for the low- and high-energy, respectively. The minimum detectable calcification size was 150 μm. Simulations demonstrate that this dual energy X-ray technique could enhance breast calcification detection.

1. Introduction

Breast cancer is the most common cancer in women both in the developed and the developing world. According to World Health Organization (WHO), early detection in order to improve breast cancer outcome and survival remains the key of breast cancer control [1]. Dual energy (DE) imaging technique can enhance the visualization of microcalcifications (μC), which are the principal indicator of breast cancer. Previous dual energy numerical calculations showed that the minimum detectable calcification size was 250 μm to 300 μm using polyenergetic X-rays and flat panel detectors [2,3]. The signal to noise ratio (SNR) in the subtracted (dual energy) image depends on various parameters such as: (i) the low- and high-energy X-ray spectra, (ii) the elemental composition of the breast, the microcalcifications, and (iii) the efficiency of the image receptor [2].

A wide range of X-ray pairs can be used in dual energy breast applications, resulting in the same SNR for calcification detection. In general, optimization can be performed with respect to patient entrance dose, mean glandular dose, image acquisition time, or X-ray tube heat loading [4].

In this simulation study, an analytical model was developed in order to determine the optimal acquisition parameters for a dual energy breast imaging system. The system modeled, consisted of a tungsten anode and a high resolution CMOS sensor. The calcification signal-to-noise ratio (SNR_{tc}) and the mean glandular dose (MGD) were calculated. Optimization was based on the maximization of the SNR_{tc}/MGD ratio.

2. Materials and Methods

The calcification SNR was determined considering a three-component system: adipose tissue of thickness t_a, glandular tissue of thickness t_g, and cubic calcification of thickness t_c. Two images are used, obtained with exposures R_{low}, R_{high} and polyenergetic X-ray spectra at different kVp $\Phi_{low}(E_i)$, $\Phi_{high}(E_i)$ at the entrance of the breast. The digital detector system has a pixel size d, spectral matching factor A_s, and quantum detection efficiency per unit energy $Q(E_i)$ .The mean signals in the low- and high-energy images, S_{low} and S_{high}, can be calculated by [3]:

$$S_j = \int_{E_{\min}}^{E_{\max}} dE \cdot R_j \cdot d^2 \cdot \Phi_j(E_i) \cdot e^{-\frac{\mu}{\rho}a(E_i)\rho_a t_a - g_r\left(\frac{\mu}{\rho}g(E_i)\rho_g - \frac{\mu}{\rho}a(E_i)\rho_a\right)T - \frac{\mu}{\rho}c(E_i)\rho_c t_c} \cdot A_s \cdot Q(E_i),\ \ j = low, high \tag{1}$$

where μ/ρ_a, μ/ρ_g, μ/ρ_c are the mass attenuation coefficients and ρ_a, ρ_g, ρ_c are the densities of the glandular, adipose tissue and μC, respectively. The calcification SNR can be expressed as [3]:

$$SNR_{tc} = \frac{t_c}{\sqrt{\sigma_{tc}^2}} \tag{2}$$

where σ_{tc}^2 represents the noise level in the subtracted image.

The elemental compositions of adipose and glandular breast tissue were obtained from a previous study [5]. Hydroxyapatite, which is a calcium-phosphate mineral form, was used to simulate microcalcifications. The mass attenuation coefficients were calculated by NISTIR published data using the computer program XMuDat [6]. The SNR_{tc} has been evaluated for the subtracted digital images of 4cm thick compressed breast, with composition of 50% glandular and 50% adipose tissue. The microcalcification sizes tested were 100 μm to 500 μm, in 50 μm increments.

The low- and high-energy unfiltered spectra were obtained from Boone *et al* for a tungsten anode [7]. The kVp examined were 40kVp for the low-energy (LE) and 70kVp for the high-energy (HE). The filter materials tested were silver (Ag), cadmium (Cd), palladium (Pd) and copper (Cu), europium (Eu), neodymium (Nd), samarium (Sm), holmium (Ho) for the low- and high-energy respectively [8]. The filter thicknesses varied from 10-150 μm and 100-1500 μm for the low- and high-energy respectively.

The assumed digital detector consisted of a 33.91mg/cm^2 terbium-doped gadolinium oxysulfide (Gd_2O_2S:Tb) placed in direct contact with the RadEye HR CMOS sensor. This scintillator was selected due to its efficiency and imaging properties, which are superior in comparison to other detectors [9]. The sensor pitch is 22.5 μm. The matching factor A_s and the quantum detection efficiency per unit energy $Q(E_i)$ were calculated according to Michail *et al* [10,11].

The mean glandular dose (MGD) was calculated for all the filtered X-ray spectra according to the following equation [12]:

$$MGD = DgN \cdot K_a \tag{3}$$

where *DgN* is a numerical factor [13], and K_a is the entrance dose. MGD was calculated for the low- and the high-energy exposures and then summed to obtain the total glandular dose. *DgN* data for a breast thickness of 4cm, with 0% and 100% glandularity, were obtained from published data [13]. For 50% glandular tissue the mean MGD value was used.

Optimization was based on the maximization of the SNR_{tc}/MGD ratio. SNR_{tc}/MGD was calculated for every possible filtered spectrum resulting in 3375 combinations. The SNR_{tc} threshold was set to be 3 ($SNR_{tc} \geq 3$) [3]. The MGD was limited to be below 2mGy, which lies within the EUREF acceptable limit [14].

3. Results and Discussion

In the following figures, the numbers of y-axis correspond to the examined thicknesses of the low energy (Ag:1-15, Cd:16-30, Pd:31-45). Respectively, the numbers of x-axis correspond to the examined thicknesses of the high energy (Cu:1-15, Eu:16-30, Nd:31-45, Sm:46-60, Ho:61-75).

Figure 1 shows SNR_{tc}/MGD values as a function of all low- and high-energy filter materials and thicknesses for 150 μm calcification size. Holmium with thicknesses in the range of 500-1000 μm gave values between 1.75 and 2.23. Copper combined with all low-energy filters resulted in the highest values ($SNR_{tc}/MGD \geq 2.5$).

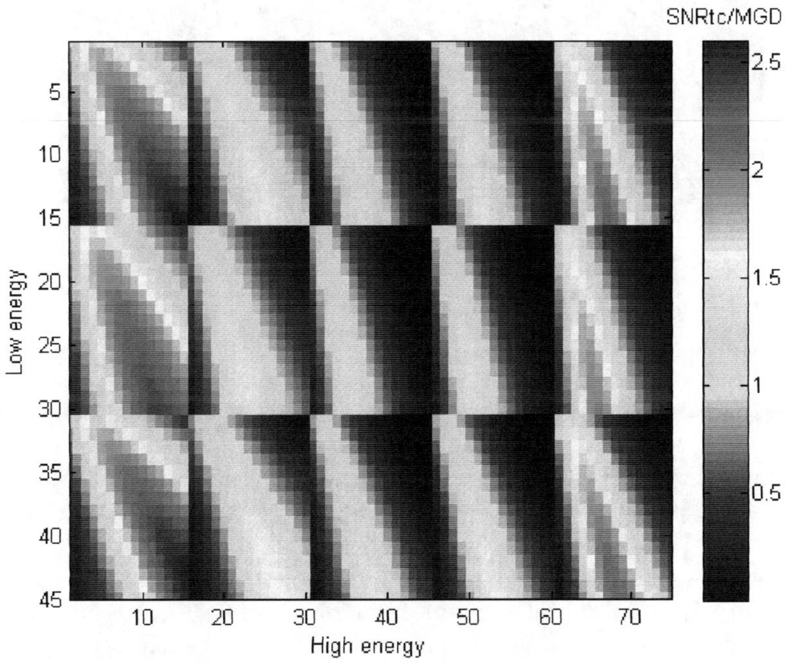

Figure 1. SNR_{tc}/MGD values as a function of low- and high-energy filter materials and thicknesses.

Figure 2 shows SNR_{tc}/MGD values as a function of all low- and high-energy filter materials and thicknesses for a μC size of 150 μm, after applying the limitations ($SNR_{tc} \geq 3$ and MGD<2mGy). It is obvious that maximum values were obtained from Cu filter (1000-1200 μm thick) combined with Ag filter (90-110 μm thick) or Cd filter (100-130 μm thick). These differences in the thicknesses of LE filters can be explained by considering the density of the materials. However, the combination of Cd-100 μm filtered spectrum (LE) and Cu-1000 μm filtered spectrum (HE) led to maximization of the optimization parameter (SNR_{tc}/MGD=2.34). Particularly, for 150 μC size, the SNR_{tc} was 3.61 while the corresponding MGD was 1.54mGy.

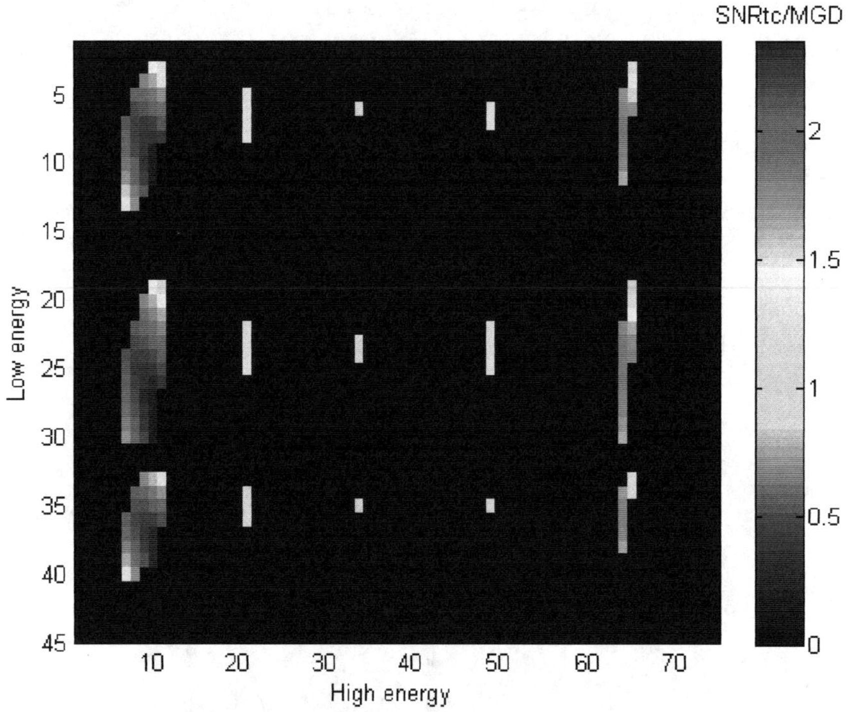

Figure 2. SNR_{tc}/MGD values as a function of low- and high-energy filter materials and thicknesses after applying the limitations.

4. Conclusion

In this simulation study, an analytical model was developed for optimal X-ray spectra determination using a dual energy breast imaging system. The modelled system consisted of a CMOS based imaging detector combined with different X-ray tungsten spectra. Optimization was based on the maximization of SNR_{tc}/MGD. Simulation results show that a SNR_{tc} value of 3.61 can be achieved for a calcification size of 150 μm using 40kVp with added beam filtration of 100 μm Cd and 70kVp Cu filtered spectrum of 1000 μm for the low- and high-energy, respectively. Compared with previous studies, this method can improve detectability of microcalcifications, while keeping mean glandular dose within acceptable levels.

Acknowledgement
This research has been co-funded by the European Union (European Social Fund) and Greek national resources under the framework of the "Archimedes III: Funding of Research Groups in TEI of Athens" project of the "Education & Lifelong Learning" Operational Programme.

5. References

[1] World Health Organization 2009 Available via: http://www.who.int/gender/women_health_report/full_report_20091104_en.pdf
[2] Brandan M and Ramirez V 2006 *Phys. Med. Biol.* **51** 2307
[3] Lemacks M, Kappadath S, Shaw C, Liu X, Whitman G 2002 *Med. Phys.* **29** 1739
[4] Johns P, Yaffe M 1985 *Med. Phys.* **12** 289
[5] Hammerstein R, Miller D, White D, Masterson M, Woodward H and Laughlin J 1979 *Radiology* **130** 485
[6] Hubbell J and Seltzer S 1995 US Department of commerce NISTIR 5632
[7] Boone J and Seibert A 1997 *Med. Phys.* **24** 1661
[8] Martini N, Koukou V, Michail C, Sotiropoulou P, Kalyvas N, Kandarakis I, Nikiforidis G and Fountos G *Journal of Spectroscopy* **2015** Article ID 563763
[9] Seferis I, Michail C, Valais I, Fountos G, Kalyvas N, Stromatia F, Oikonomou G, Kandarakis I, Panayiotakis G. 2013 *Nucl. Instrum. Meth. Phys. Res. A.***729** 307
[10] Michail C, Spyropoulou V, Fountos G, Kalyvas N, Valais I, Kandarakis I and Panayiotakis G 2011 *IEEE Trans. Nucl. Sci.* **58** 314
[11] Michail C, Fountos G, Liaparinos P, Kalyvas N, Valais I, Kandarakis I and Panayiotakis G. 2010 *Med. Phys.* **37** 3694
[12] ACR 1999 Mammography, Quality Control Manual (Reston, VA: Americal Collage of Radioliogy)
[13] Boone J 1999 *Radiol.* **213** 23
[14] EUREF 2006 European guidelines for quality assurance in breast cancer screening and diagnosis 4th ed European Commission

Modeling of the Calcium/Phosphorus Mass ratio for Breast Imaging

N Martini[1], V Koukou[1], C Michail[2], P Sotiropoulou[1], N Kalyvas[2], I Kandarakis[2], G Nikiforidis[1] and G Fountos[2]

[1] Department of Medical Physics, Medical School, University of Patras, 265 00 Patras, Greece
[2] Radiation Physics, Materials Technology and Biomedical Imaging Laboratory, Department of Biomedical Engineering, Technological Educational Institute of Athens, Egaleo, 122 10 Athens, Greece

E-mail: gfoun@teiath.gr

Abstract. Breast microcalcifications are mainly composed of calcite ($CaCO_3$), calcium oxalate (CaC_2O_4) and apatite (a calcium-phosphate mineral form). Any pathologic alteration (carcinogenesis) of the breast may produce apatite. In the present simulation study, an analytical model was implemented in order to distinguish malignant and non-malignant lesions. The Calcium/Phosphorus (Ca/P) mass ratio and the standard deviation (SD) of the calcifications were calculated. The size of the calcifications ranged from 100 to 1000 μm, in 50 μm increments. The simulation was performed for hydroxyapatite, calcite and calcium oxalate calcifications. The optimum pair of energies for all calcifications was 22keV and 50keV. Hydroxyapatite and calcite calcifications were sufficiently characterized through their distinct confidence interval (99.7%, 3SD) values for calcifications sizes above 500 μm, while the corresponding sizes for hydroxyapatite and calcium oxalate characterization were found above 250 μm. Initial computer simulation results indicate that the proposed method can be used in breast cancer diagnosis, reducing the need for invasive methods, such as biopsies.

1. Introduction

The most common cause of cancer deaths is the breast cancer, responsible for approximately 16% of cancer deaths in adult women [1]. Calcifications are a diagnostic key for the localization of malignancy. Type I calcifications are composed of calcium oxalate (CaC_2O_4), and are associated with benign lesions of the breast [2,3]. Type II calcifications are composed of calcium phosphate, mainly hydroxyapatite (HAp), and calcium carbonate ($CaCO_3$) associated with malignant and benign breast conditions, respectively [4,5]. Where suspected calcifications are detected, traditionally a needle biopsy follows in order to determine whether the associated lesion is malignant or benign [6]. This could be avoided if a rapid and safe noninvasive method was available [7].

In this simulation study, a dual energy method was used for the calculation of the Ca/P mass ratio and the standard deviation (SD) of the calcifications. The study was performed for hydroxyapatite calcifications, indicating malignancy, as well as for calcite and calcium oxalate calcifications, indicating benign lesions. The optimum pair of energies for all calcifications was 22keV and 50keV. Hydroxyapatite and calcite calcifications thicker than 500 μm were sufficiently characterized through their distinct confidence interval (99.7%, 3SD) values, while the corresponding sizes for hydroxyapatite and calcium oxalate characterization were found above 250 μm. Initial simulation results indicate that the proposed method can be used in breast cancer diagnosis, reducing the need for invasive methods, such as biopsies.

2. Materials and Methods

An analytical model was developed in order to determine the calcifications' Ca/P mass ratio and the Coefficient of Variation (CV), assuming only Poisson distribution. The breast is considered to be composed of a mixture of 50% adipose and 50% glandular tissue, of a thickness t_m and a calcification with thickness t_c. Malignant calcifications were simulated with hydroxyapatite ($Ca_{10}(PO_4)_6(OH)_2$) having a density of 3.18g/cm^3 [8]. For the simulation of benign calcifications calcium carbonate ($CaCO_3$) and calcium oxalate (CaC_2O_4) with densities of 2.93g/cm^3 and 2.20g/cm^3 [9], respectively, were used.

A 4cm compressed breast [10] was irradiated twice. In the first irradiation, for the low- and high-energy, the intensities attenuate after passing through the mixture only, while in the second irradiation the intensities attenuate after passing through the mixture and calcification. The examined calcification thicknesses were 100 μm to 1000 μm (microcalcifications), and 550 to 1000 μm (macrocalcifications) in 50 μm increments respectively.

In order to determine the Ca/P mass ratio, the above attenuated intensities were calculated according to equations (1) & (2) considering monoenergetic X-rays ranging from 15 to 70keV, in 1keV increments. The attenuated intensities, for the low- and high-energy, after passing through a mixture of 50% adipose and 50% glandular tissue can be expressed as:

$$I_{m,E_i} = I_{o,E_i} \cdot e^{-\left(\frac{\mu}{\rho_{a(E_i)}} \cdot \rho_a + \frac{\mu}{\rho_{g(E_i)}} \cdot \rho_g\right) \cdot 0.5 \cdot T} \quad , i = l,h \tag{1}$$

where, I_{o,E_l} and I_{o,E_h} are the unattenuated low- and high-energy photon intensity, per unit energy $\left(photons/cm^2 keV\right)$, at the detector input. The energy-dependent mass attenuation coefficients $\left(cm^2/g\right)$, as well as the density $\left(g/cm^3\right)$ for adipose and glandular are given by $\mu/\rho_{a(E_i)}$, $\mu/\rho_{g(E_i)}$, respectively.

The corresponding attenuated intensities by both the mixture and the calcification were calculated as follows:

$$I_{c,E_i} = I_{o,E_i} \cdot e^{-\frac{\mu}{\rho_{Ca(E_i)}} \cdot \rho_{Ca} \cdot t_{Ca} - \frac{\mu}{\rho_{j(E_i)}} \cdot \rho_j \cdot t_j - \left(\frac{\mu}{\rho_{a(E_i)}} \cdot \rho_a + \frac{\mu}{\rho_{g(E_i)}} \cdot \rho_g\right) \cdot 0.5 \cdot t_m} \quad , \ i = l,h, \ j = PO_4, CO_3, C_2O_4 \tag{2}$$

where, the energy-dependent mass attenuation coefficients $\left(cm^2/g\right)$, as well as the density $\left(g/cm^3\right)$ for adipose, glandular, calcium, phosphate, carbonate and oxalate are given by $\mu/\rho_{a(E_i)}$, $\mu/\rho_{g(E_i)}$, $\mu/\rho_{Ca(E_i)}$, $\mu/\rho_{PO_4(E_i)}$, $\mu/\rho_{CO_3(E_i)}$, $\mu/\rho_{C_2O_4(E_i)}$ and ρ_a, ρ_g, ρ_{Ca}, ρ_{PO_4}, ρ_{CO_3}, $\rho_{C_2O_4}$, respectively.

Data for the elemental compositions of glandular and adipose tissues in a human breast were used for the mass attenuation coefficients determination [11]. The densities used were 0.93 g/cm^3 for adipose tissue and 1.04 g/cm^3 for glandular tissue [11]. Mass attenuation coefficients of the materials used in this work were computed by XMuDat software from NIST [12].

Thus, the Ca/P mass ratio can be determined by [13,14]:

$$\frac{Ca}{P}=\frac{\ln\frac{I_{m,E_l}}{I_{c,E_l}}\cdot\left(\frac{\mu}{\rho_{PO_4,E_h}}\cdot\rho_{HAp}-\left(\frac{\mu}{\rho_{a,E_h}}\cdot\rho_a+\frac{\mu}{\rho_{g,E_h}}\cdot\rho_g\right)\cdot 0.5\right)-\ln\frac{I_{m,E_h}}{I_{c,E_h}}\cdot\left(\frac{\mu}{\rho_{PO_4,E_l}}\cdot\rho_{HAp}-\left(\frac{\mu}{\rho_{a,E_l}}\cdot\rho_a+\frac{\mu}{\rho_{g,E_l}}\cdot\rho_g\right)\cdot 0.5\right)}{-\ln\frac{I_{m,E_l}}{I_{c,E_l}}\cdot\left(\frac{\mu}{\rho_{Ca,E_h}}\cdot\rho_{HAp}-\left(\frac{\mu}{\rho_{a,E_h}}\cdot\rho_a+\frac{\mu}{\rho_{g,E_h}}\cdot\rho_g\right)\cdot 0.5\right)+\ln\frac{I_{m,E_h}}{I_{c,E_h}}\cdot\left(\frac{\mu}{\rho_{Ca,E_l}}\cdot\rho_{HAp}-\left(\frac{\mu}{\rho_{a,E_l}}\cdot\rho_a+\frac{\mu}{\rho_{g,E_l}}\cdot\rho_g\right)\cdot 0.5\right)}\cdot 3.0679$$

(3)

The simulation was performed for hydroxyapatite, calcite and calcium oxalate indicating malignant (hydroxyapatite) and benign (calcite and calcium oxalate) calcifications. When malignant calcifications are present, the model estimates the Ca/P mass ratio of hydroxyapatite where Ca and P are both elements in this molecule. On the contrary, P is not present in benign calcifications. In this case, the model calculates an effective-Ca/P mass ratio, which differs compared to that calculated for hydroxyapatite. The differences between these Ca/P mass ratios were determined.

3. Results and Discussion

For all the calcification sizes examined, the optimum pair of energies was 22keV and 50keV for the low- and high-energy, respectively.

Figure 1 shows the Ca/P mass ratios of hydroxyapatite and calcite for both microcalcifications and macrocalcifications. In all figures error bars correspond to 3SD. Calcifications thicker than 500 μm can be discriminated, since the confidence interval (99.7%, 3SD) for the estimated Ca/P mass ratios, between calcite and hydroxyapatite, can be clearly distinguished. This allows characterization between malignant and benign calcifications. The corresponding Ca/P mass ratios and SDs were 1.2782 and 0.0862 for calcite, while for hydroxyapatite were 1.9713 and 0.1366, respectively.

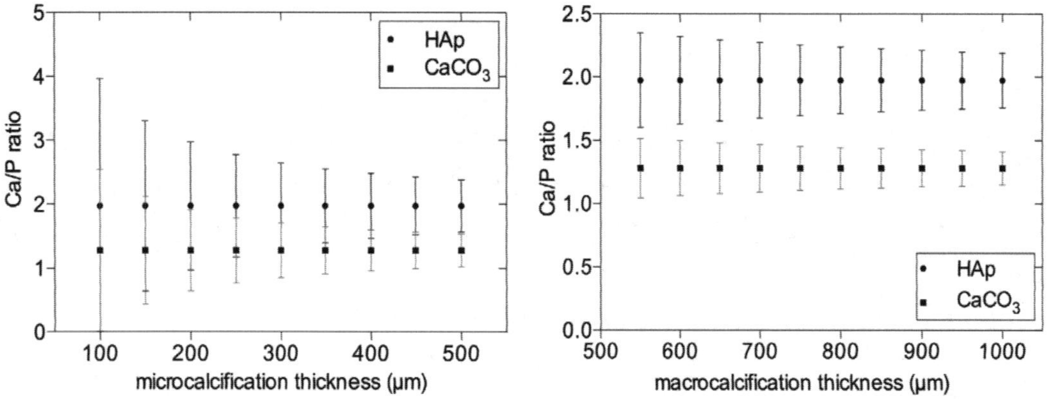

Figure 1. Ca/P_{HAp} and Ca/P_{CaCO3} as a function of microcalcifications (left) and macrocalcifications (right) sizes.

Figure 2 shows the Ca/P mass ratios of hydroxyapatite and calcium oxalate for both microcalcifications and macrocalcifications. Discrimination between malignant (hydroxyapatite) and benign (calcium oxalate) microcalcifications, with sizes above 250 μm, can be observed. The corresponding Ca/P mass ratios and SDs were 0.7198 and 0.1484 for calcium oxalate, and for hydroxyapatite were 1.9713 and 0.2679, respectively.

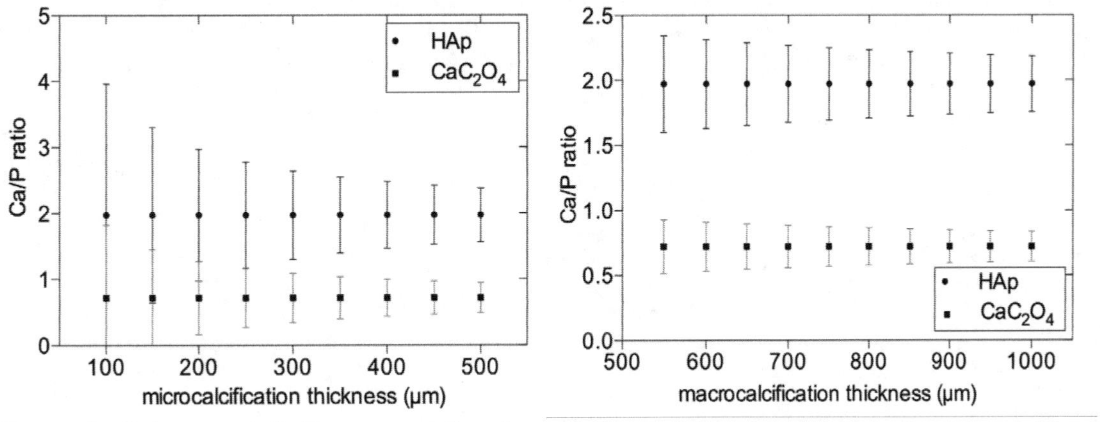

Figure 2. Ca/P_{HAp} and Ca/P_{CaCO3} as a function of microcalcifications (left) and macrocalcifications (right) sizes.

4. Conclusion

In this study characterization of breast calcifications through the determination of the Calcium/Phosphorus (Ca/P) mass ratio was performed, by implementing a dual energy method through analytical modeling. Sufficient characterization resulted between malignant and non-malignant calcifications for sizes above 500 μm. The corresponding size between hydroxyapatite and calcite or calcium oxalate was 250 μm, respectively. The method implemented in this work could be used to discriminate between malignant and benign lesions, reducing the need for biopsies.

Acknowledgement

This research has been co-funded by the European Union (European Social Fund) and Greek national resources under the framework of the "Archimedes III: Funding of Research Groups in TEI of Athens" project of the "Education & Lifelong Learning" Operational Programme.

5. References

[1] World Health Organization 2008 World health statistics 2008
[2] Frappart L, Remy I, Lin HC, Bremond A, Raudrant D, Grousson B, and Vauzelle JL 1986 *Virchows Arch A* **410** 179
[3] Busing CM, Keppler U, and Menges V 1981 *Virchows Arch [Pathol Anat]* **393** 307
[4] Haka AS, Shafer-Peltier KE, Fitzmaurice M, Crowe J, Dasari RR, and Feld MS 2002 *Cancer Res.* **62** 5375
[5] Koukou V, Martini N, Fountos G, Sotiropoulou P, Michail C, Valais I, Kandarakis I and Nikiforidis G 2014 *IFMBE Proceedings* **41** 459
[6] Arkidis N, Skiadopoulos S, Sakellaropoulos F, Karahaliou A, Vassiou K, Kazantzi A, and ostaridou L 2014 *Phys Medica* **30** e75
[7] Matousek P and Stone N 2013 *J Biophoton* **6** 7
[8] Gong JK, Arnold JS, and Cohn SH 1964 *Anat. Rec.* **149** 319
[9] Brandan M and Ramirez V 2006 *Phys. Med. Biol.* **51** 2307
[10] Samei E and Saunders RSJr 2011 *Phys. Med. Biol.* **56** 6359
[11] Hammerstein R, Miller D, White D, Masterson M, Woodward H and Laughlin J 1979 *Radiology* **130** 485

[12] Hubbell J and Seltzer S 1995 US Department of commerce NISTIR 5632
[13] Fountos G, Yahumura S, and Glaros D 1997 *Med. Phys.* **24** 1303
[14] Sotiropoulou P, Fountos G, Martini N, Koukou V, C Michail C, Valais I, Kandarakis I and Nikiforidis G 2015 *Phys Medica* doi:10.1016/j.ejmp.2015.01.019

4th International Conference on Mathematical Modeling in Physical Sciences (IC-MSquare2015) IOP Publishing
Journal of Physics: Conference Series **633** (2015) 012095 doi:10.1088/1742-6596/633/1/012095

A theoretical investigation of spectra utilization for a CMOS based indirect detector for dual energy applications

N Kalyvas[1], N Martini[2], V Koukou[2], C Michail[1], P Sotiropoulou[2], I Valais[1], I Kandarakis[1] and G Fountos[1]

[1] Radiation Physics, Materials Technology and Biomedical Imaging Laboratory, Department of Biomedical Engineering, Technological Educational Institute of Athens, Egaleo, 122 10 Athens, Greece

[2] Department of Medical Physics, Medical School, University of Patras, 265 00 Patras, Greece

E-mail: gfoun@teiath.gr

Abstract. Dual Energy imaging is a promising method for visualizing masses and microcalcifications in digital mammography. Currently commercially available detectors may be suitable for dual energy mammographic applications. The scope of this work was to theoretically examine the performance of the Radeye CMOS digital indirect detector under three low- and high-energy spectral pairs. The detector was modeled through the linear system theory. The pixel size was equal to 22.5μm and the phosphor material of the detector was a 33.9 mg/cm^2 Gd_2O_2S:Tb phosphor screen. The examined spectral pairs were (i) a 40kV W/Ag (0.01cm) and a 70kV W/Cu (0.1cm) target/filter combinations, (ii) a 40kV W/Cd (0.013cm) and a 70kV W/Cu (0.1cm) target/filter combinations and (iii) a 40kV W/Pd (0.008cm) and a 70kV W/Cu (0.1cm) target/filter combinations. For each combination the Detective Quantum Efficiency (DQE), showing the signal to noise ratio transfer, the detector optical gain (DOG), showing the sensitivity of the detector and the coefficient of variation (CV) of the detector output signal were calculated. The second combination exhibited slightly higher DOG (326 photons per X-ray) and lower CV (0.755%) values. In terms of electron output from the RadEye CMOS, the first two combinations demonstrated comparable DQE values; however the second combination provided an increase of 6.5% in the electron output.

1. Introduction

Breast cancer, which is a common cause of death among female population, may manifest as microcalcifications. Modern breast examination techniques, includes irradiation with dual energy spectra [1-5]. Dual-energy subtraction imaging techniques offer an alternative approach to the detection and visualization of microcalcifications. With this technique, high- and low-energy images are separately acquired and "subtracted" from each other in a weighted fashion to cancel out the cluttered tissue structure so as to decrease the obscurity from overlapping tissue structures [6]. Although this technique reduces the contrast to noise ratio of the final image, it makes microcalcifications better visualized [1-5]. Digital mammography utilizes direct or indirect detection methods. The latter uses scintillators coupled to amorphous Silicon (a-Si) sensors [7-8]. Detector modelling has been carried out to determine the detector design and the incident X-ray spectra for an optimum detector performance. One method to determine the optimum detector parameters is linear cascaded systems theory (LCS). This theory calculates the output of a detector as a series of cascaded stages [7-10]. These stages describe the statistics of signal carrier interactions and are divided into gain stages and blur stages. In this study, the aforementioned theory was used in order to investigate the performance of three low energy (LE) and high energy (HE) X-ray spectra combinations incident at a commercially available CMOS detector. The performance was evaluated through image quality metrics like the detective quantum efficiency (DQE), showing the signal-to-noise transfer, the detector optical gain (DOG) showing the detector output per incident X-ray and the coefficient of variation (CV) of the output signal [7-10].

Journal of Physics: Conference Series **633** (2015) 012095 doi:10.1088/1742-6596/633/1/012095

2. Materials and Methods

In this study the LCS theory was used. This theory calculates the output of a detector as a series of cascaded stages. Every stage has a frequency domain input $S_{in}(u)$, where *u* is the spatial frequency, a mean input value $\bar{x}_{in}$, a frequency domain output $S_{out}(u)$ and a mean output $\bar{x}_{out}$. Every gain stage is characterized by a statistical mean value $\bar{q}$ and variance σ_q^2, while every blur stage is characterized by a Modulation Transfer Function $MTF(u)$. The blur stages are either stochastic or deterministic. The frequency depended output and the mean output signal for each stage can be calculated from the following relationships: (i) $S_{out}(u) = \bar{q}^2 S_{in}(u) + \bar{x}_{in}\sigma_q^2$ and $\bar{x}_{out} = \bar{q}\bar{x}_{in}$, for gain stages, (ii) $S_{out}(u) = (S_{in}(u) - \bar{x}_{in})MTF^2(u) + \bar{x}_{in}$ and $\bar{x}_{out} = \bar{x}_{in}$ for stochastic blur stages and (iii) $S_{out}(u) = S_{in}(u)MTF^2(u)$ and $\bar{x}_{out} = \bar{x}_{in}$ for deterministic blur stages [7-10]. In this work the following stages were considered: the X-ray absorption in the phosphor material, the optical photon production per absorbed X-ray, the optical photon escape and spread to the output, the impingement of the optical photons at the CMOS surface and the production of electrons at the CMOS output. A more extensive analysis of the above stages can be found in current literature. Through these stages the total Noise Power Spectrum (NPS(u)) was calculated, were u is the spatial frequency. In addition the total signal output in electrons M was determined. DOG was calculated as [10]:

$$DOG = \frac{M}{\sum_E f(E)} \tag{1}$$

where *f(E)* is the incident X-ray photons of energy E,

$$CV = \frac{\sqrt{\sum_u NPS(u)}}{M} \tag{2}$$

and [11]:

$$DQE(u) = \frac{[M \cdot MTF(u)]^2}{\sum_E f(E) \cdot NPS(u)} \tag{3}$$

where *MTF(u)* is the modulation transfer function.

The X-ray spectra combinations tested were obtained by considering polyenergetic X-rays filtered with various filter materials and thicknesses [12]. An analytical model was developed for the calculation of the calcification signal to noise ratio (SNR_{tc}) and the mean glandular dose (MGD) for various LE and HE filter combinations. The filters selection was based on the maximization of the SNR_{tc}/MGD ratio. This work is presented in an accompanied paper [13].
The data used for calculating the equations were obtained from literature [7,8,10,11]. The model was applied to a commercially available indirect CMOS detector (RadEye CMOS), incorporating a 34mg/cm^2 Gd_2O_2S:Tb phosphor screen in close contact with a 22.5μm pixel size photodetector array [14-15].

3. Results and Discussion

The low and high energy spectra combinations are presented in Figure 1. It may be observed from the spectra that there is a small spectral overlap in the range between 35 keV and 40 keV.
In table 1 the DOG and CV values for the presented spectra is shown.

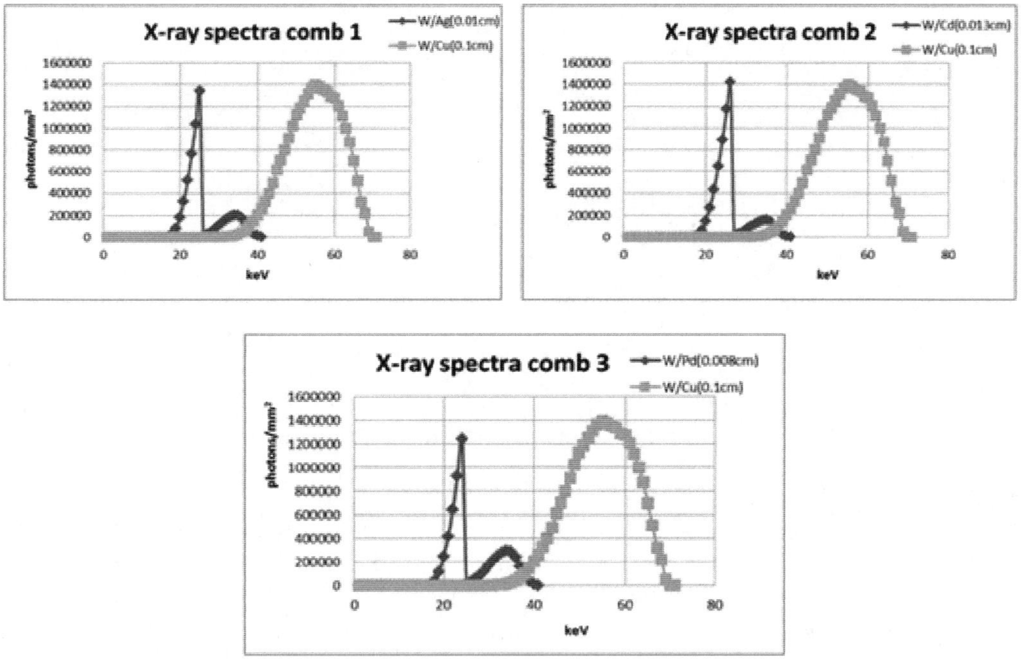

Figure 1. The used low and high energy spectra combinations.

Table 1. The DOG and CV values for the X-ray spectra under consideration.

X-ray spectrum (target/filter)	DOG (electrons/X-ray photon)	CV (%)
70 kVp W / Cu (0.1cm)	0.080	0.50
40 kVp W/ Ag (0.01cm)	0.060	0.78
40 kVp W/ Cd (0.013 cm)	0.060	0.76
40 kVp W/Pd (0.008cm)	0.058	0.79

It can be observed from table 1 that for the low energy spectra the larger CV and the lowest DOG values have been calculated for the W/Pd (0.008 cm) target/filter combination. From the other two LE combinations the 40 kVp W/Cd (0.013 cm) is slightly better than the 40 kVp W/Ag (0.01cm), due to its lower calculated CV value. In contrast the DOG value is higher for the HE spectrum. Although the thin Gd_2O_2S:Tb phosphor screen (34 mg/cm^2) of the RadEye CMOS sensor provides better absorption characteristics the phosphor intrinsic gain (optical photons produced per absorbed X-ray) is higher for higher X-ray energies deposited.

Figure 2 present the DQE for the low and the high energy spectra for the detector under consideration. It can be observed from Figure 2 that the W/Cd (0.013 cm) LE target/filter exhibit slightly better DQE values per spatial frequency than the other LE combinations. In addition the lowest DQE values are that of the HE combination due to the reduced X-ray absorption of the 34mg.cm^2 Gd_2O_2S:Tb phosphor screen.

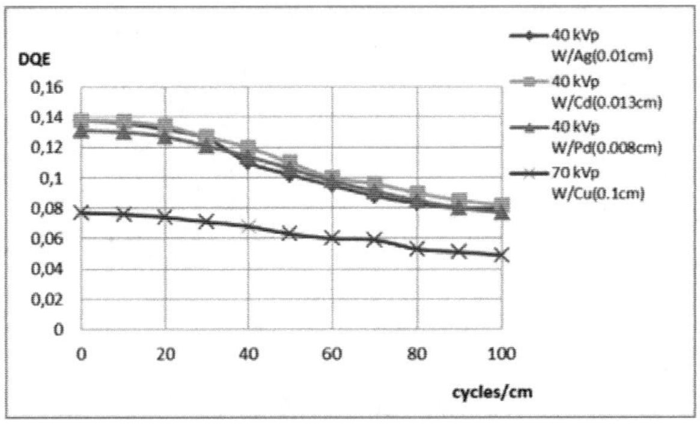

Figure 2. The DQE of the spectral components used for the dual energy spectra combinations.

4. Conclusions

In this work, the applicability of three target/filter combinations impinging on a commercially available RadEye CMOS detector was evaluated in terms of DOG, CV and DQE. It was found that the 40kVp W/ Cd (0.013 cm) provided the best low energy component. The low DQE and DOG values are mainly attributed (i) to the thin mammographic screen, which is designed for smaller X-ray energies.

Acknowledgement

This research has been co-funded by the European Union (European Social Fund) and Greek national resources under the framework of the "Archimedes III: Funding of Research Groups in TEI of Athens" project of the "Education & Lifelong Learning" Operational Programme.

5. References

[1] Johns P, Drost D, Yaffe M and Fenster A 1985 *Med. Phys.* 12 297.
[2] Boone M, Shaber G and Tecotzky M 1990 *Med. Phys.* 17 665.
[3] Boone M 1991 *Invest. Radiol.* 26 521.
[4] Lemacks M, Kappadath S, Shaw C, Liu X and Whitman G 2002 *Med. Phys.* 29 1739.
[5] Jong R, Yaffe M, Skarpatiotakis M, Shumak R, Danjoux N, Gunesekara A and Plewes D 2003 *Rad.* 228 842.
[6] Sotiropoulou P, Fountos G, Martini N, Koukou V, C Michail C, Valais I, Kandarakis I and Nikiforidis G 2015 *Phys Medica* doi:10.1016/j.ejmp.2015.01.019.
[7] Liaparinos P, Kalyvas N, Kandarakis I and Cavouras D 2013 *Nucl. Instrum. and Meth. A* 697 87.
[8] Michail C, Spyropoulou V, Fountos G, Kalyvas N, Valais I, Kandarakis I and G Panayiotakis 2011 *IEEE Trans. Nucl. Sci.* 58 314.
[9] Meter R and Rabbani M 1990 *Med. Phys.* 17 65.
[10] Martini N, Koukou V, Kalyvas N, Sotiropoulou P, Michail C, Valais I, Kandarakis I, Nikiforidis G and Fountos G. 2014 *J. Phys.: Conf. Ser.* 574 012075.
[11] Nishikawa R and Yaffe M 1990 *Med. Phys.* 17 894.
[12] Martini N, Koukou V, Michail C, Sotiropoulou P, Kalyvas N, Kandarakis I, Nikiforidis G and Fountos G *Journal of Spectroscopy* 2015 Article ID 563763
[13] Koukou V, Martini N, Michail C, Sotiropoulou P, Kalyvas N, Kandarakis I, Nikiforidis G, and Fountos G, Optimum filter selection for Dual Energy X-ray Applications through analytical modelling, accepted in ICMSQUARE 2015.
[14] http://www.rad-icon.com/products-software.php
[15] Rad-icon Imaging Corp. Remote Radeye data sheet.

Medical Imaging Image Quality Assessment with Monte Carlo Methods

C. M. Michail[1] G. E. Karpetas[2], G. P. Fountos[1], N. I. Kalyvas[1], Niki Martini[2], Vaia Koukou[2], I. G. Valais[1] and I. S. Kandarakis[1]

[1] Radiation Physics, Materials Technology and Biomedical Imaging Laboratory, Department of Biomedical Engineering, Technological Educational Institute of Athens, Egaleo, 122 10 Athens, Greece

[2] Department of Medical Physics, Faculty of Medicine, University of Patras, 265 00 Patras, Greece

e-mail: gfoun@teiath.gr

Abstract. The aim of the present study was to assess image quality of PET scanners through a thin layer chromatography (TLC) plane source. The source was simulated using a previously validated Monte Carlo model. The model was developed by using the GATE MC package and reconstructed images obtained with the STIR software for tomographic image reconstruction, with cluster computing. The PET scanner simulated in this study was the GE DiscoveryST. A plane source consisted of a TLC plate, was simulated by a layer of silica gel on aluminum (Al) foil substrates, immersed in 18F-FDG bath solution (1MBq). Image quality was assessed in terms of the Modulation Transfer Function (MTF). MTF curves were estimated from transverse reconstructed images of the plane source. Images were reconstructed by the maximum likelihood estimation (MLE)-OSMAPOSL algorithm. OSMAPOSL reconstruction was assessed by using various subsets (3 to 21) and iterations (1 to 20), as well as by using various beta (hyper) parameter values. MTF values were found to increase up to the 12th iteration whereas remain almost constant thereafter. MTF improves by using lower beta values. The simulated PET evaluation method based on the TLC plane source can be also useful in research for the further development of PET and SPECT scanners though GATE simulations.

Keywords: PET; MTF; image quality; iterative reconstruction; Monte Carlo; GATE; STIR

1. Introduction

The most common image quality metrics in order to assess spatial resolution in Positron Emission Tomography (PET) scanners are the Point Spread Function (PSF) and the corresponding Full Width at Half Maximum (FWHM) obtained from point sources [1]. Only in a few studies spatial resolution was assessed in terms of the Modulation Transfer Function (MTF), calculated from PSF data [2]. The same method was employed in studies concerning combined micro PET/CT scanners, in which MTF was estimated in the CT detector of the combined system [3,4]. MTF can be also assessed from a line source through the estimation of the line spread function (LSF). The use of the LSF method for determining the MTF in tomographic imagers was initially introduced by Boone [5] who applied this method for CT scanners evaluation. Fountos *et al.* [6,7] recently introduced a similar method for SPECT scanners by immersing an Agfa MammoRay HDR Medical X-ray film in a solution of dithiothreitol (DTT)/Tc-99m(III)-DMSA to obtain the MTF through the LSF method. The purpose of the present study was to extend a previous validated Monte Carlo model [8-10] in order to obtain PET image resolution through the estimation of the MTF. The model could be useful in implementing experimental methods for systems evaluation in clinical practice as well as in research for further improvement of the image quality performance in PET scanners. Within the context of the model, MTF was estimated by the simulation of a thin layer chromatography (TLC) plane source filled with 18F-FDG using a software model based on the GATE Monte Carlo package. The GATE package, which was developed by the Open-GATE collaboration, is an open-source extension of the Geant4 Monte Carlo toolkit [11]. GATE was used in combination with the STIR image reconstruction software [12] to obtain the plane source reconstructed images. The influence of different number of iterations (corresponding to a complete cycle) with various subsets in the iterative image reconstruction was also investigated. The simulation of this plane source phantom provides an accurate model that is useful to fully characterize the performance of nuclear medicine imaging systems.

2. Materials And Methods

2.1 Geometry of the modeled PET scanner

In this study the Discovery ST PET/CT system was modeled. The system incorporates bismuth germinate oxide (BGO) [13] crystals with dimensions of 6.3x6.3x30 mm in the tangential, axial and radial directions, respectively. The detector ring is finally comprised of 35 modules, i.e. 280 crystal blocks, or 24 rings of 420 crystals (for a total of 10080 BGO crystals). The dimensions of the rings are 88.6 cm diameter with a 15.7 cm axial and 70cm transaxial fields of view (FOV).

2.2. Software for simulation

The simulation software employed in the present study was built around the GATE toolkit, which in turn is based on the Geant4 platform. In every BGO crystal of the detector block, a Gaussian energy blur is applied with an average energy resolution of 17%, referenced at 511 keV. Afterwards, a 300 ns dead time value was applied on the single events in the BGO crystal [14] by a paralysable Deadtime module. The Quantum Detection Efficiency (QDE) [15,16] of the crystals was found 0.94 at 511 keV. The sinogram output file (.ima) was used by STIR as input file for the reconstruction of the simulated plane source image. The coincidence time window was set to 11.7 ns.

2.3. Preparation of the MTF test object

The MTF test object, which was simulated in this study, was a planar source similar to those used in previous works of Boone [5] for CT and Fountos *et al.* [6,7] for SPECT systems. The idea for the particular 18F-FDG plane source was based on the excellent binding of 18F-FDG with TLC plates [10]. The plate was considered to consist of a layer of silica gel on Al foil substrates (Al density 2.7 g/cm^3). The dimensions of the TLC plate were 5x10 cm^2 and it was assumed to be immersed in an 18F-FDG bath solution (1 MBq). The MTF test object (plane source, i.e. the radioactive plate) was simulated between two semi-cylindrical polyethylene blocks with 20 cm diameter and 70 cm length, in the horizontal and vertical direction. The importance of the hyper parameter (beta value) (which controls the weight of the penalty term and the importance of the prior) was also examined in order to investigate the corresponding influence on the MTF [17].

2.4. Modulation Transfer Function (MTF)

MTF was obtained, by using the LSF method. In this method, the thin plane source was simulated at a slight angle of 3° with respect to the horizontal or vertical axis, following the technique proposed by Fujita to avoid aliasing effects [18]. The final LSF was obtained by averaging all line LSF profiles after angle correction. The angle correction was performed following the procedure described in Fountos *et al.* [6].

3. Results and | Discussion

3.1. Iterative Image Reconstruction

Figure 1a shows transverse image slices, used for the MTF calculation through the LSF method [6], of the plane source reconstructed with the OSMAPOSL algorithm with subsets ranging from 3 to 21 and iterations from 2 to 20. Figure 1b shows transverse plane source reconstructed images with the OSMAPOSL using 2, 5 and 20 iterations with fixed subsets (15) using various values of the beta (hyper) parameter (0.01, 1 and 3). These data were obtained in order to investigate the influence of the beta parameter on image resolution.

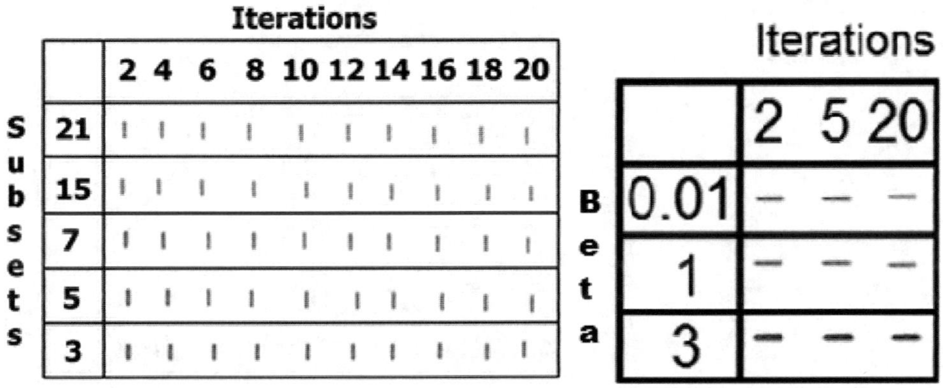

Figure 1. Transverse slices, from the plane source reconstructed images with the OSMAPOSL a) (3, 5, 7 and 21 subsets, various iterations) and b) (15 subsets, 2, 5 and 20 iterations) using various values of the beta (hyper) parameter.

3.2. Modulation Transfer Function (MTF)

Figure 2 shows MTF curves obtained from iterative STIR reconstructed LSF images. The curves correspond to transverse reconstructed slices in which the number of subsets was kept fixed and the number of iterations was increased with a step of 2. In every case MTF was found to increase up to 10 iterations and remain almost stable thereafter. However, the range of this increase in the MTF is limited as the number of subsets increases. In Fig.2 (3 subsets) the MTF values at 0.04 cycles/mm range from 0.208 (2 iterations) to 0.507 (20 iterations). MTF reaches a plateau at a value of 0.492 (12 iterations). With 21 subsets, the MTF values at 0.04 cycles/mm range from 0.432 (2 iterations) to 0.495 (20 iterations). The MTF value where the increase stops is 0.489 (12 iterations). Furthermore in figure 2 the influence of the number of subsets on the reconstructed image is shown. In order to obtain these values, the number of iterations was kept constant (2 and 10 iterations) and the number of subsets was allowed to vary from 3 to 21. The obtained MTF values show that as the number of iterations increases, the impact on image resolution is restricted.

Figure 3 shows the resulted MTF curves when the hyper parameter is allowed to vary from a value close to zero up to 3, in order to investigate its impact on PET image resolution through the MTF. As it can be seen, spatial resolution is improved by using lower beta values [17]. Larger beta values result in smoother images, while lower values decrease the importance of the penalty term [17,19]. By increasing the number of iterations, the influence of the beta parameter on image resolution tends to decrease showing smaller deviations. For example, in the 15 subsets/2 iterations image, deviations of the order of 60.2% are shown between the 0.01 and 3 beta value, whereas in the 15 subsets and 20 iterations image the corresponding difference decrease to 31.3%, at the spatial frequency of 0.05 cycles/mm.

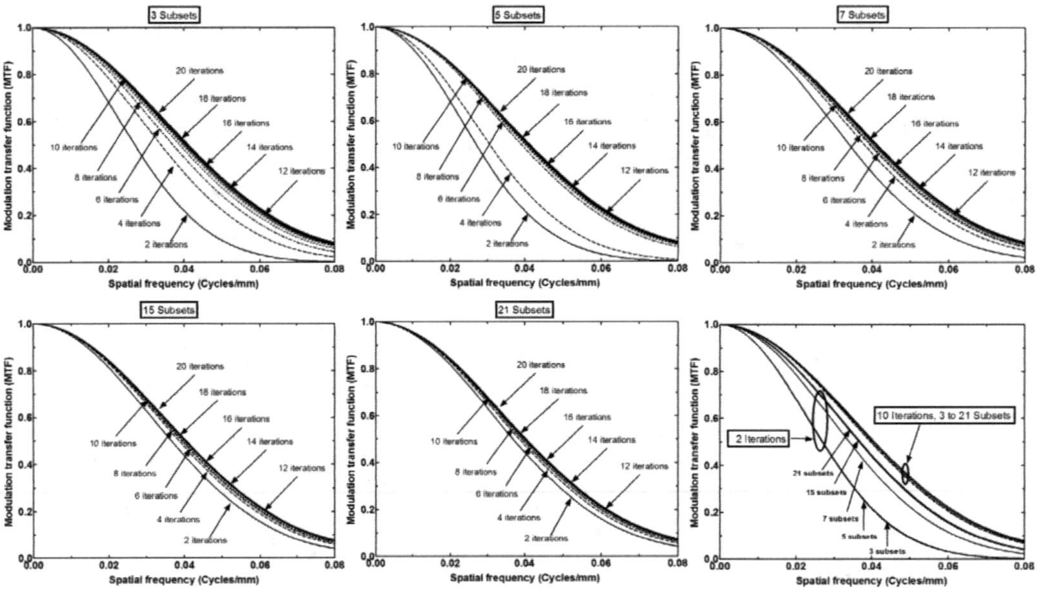

Figure 2. Comparison between the MTFs obtained with the LSF method, from the plane source reconstructed image with the OSMAPOSL (3 to 21 subsets, various iterations and various subsets, 2 and 10 iterations).

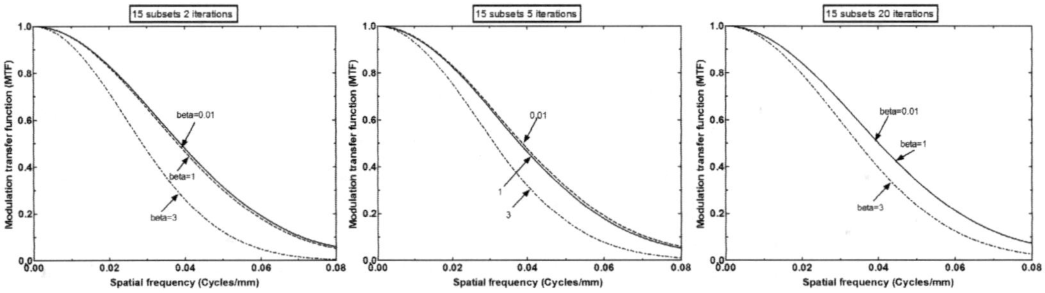

Figure 3. Comparison between the MTFs obtained from the OSMAPOSL (15 subsets, 2, 5 and 20 iterations) reconstructed images using various values of the beta (hyper) parameter.

4. Conclusions

Image quality characterisation of a PET scanner was achieved through the estimation of the MTF of thin 18F-FDG plane source reconstructed images, simulated by a Monte Carlo model. The influence of iterative image reconstruction on the plane source simulated images was investigated by using various subsets (3 to 21) and iterations (1 to 20). Image quality was found to improve by increasing the number of iterations (number of image updates) (up to 12). Spatial resolution is also improved by using lower beta values. Larger beta values have as a result smoother images, while lower values decreases the importance of the penalty term. The method was used for the image quality assessment and optimisation, but it can be also useful in research for the further development of PET and SPECT scanners though GATE simulations.

Acknowledgments
This research has been co-funded by the European Union (European Social Fund) and Greek national resources under the framework of the "Archimedes III: Funding of Research Groups in TEI of Athens" project of the "Education & Lifelong Learning" Operational Programme.
STIR is the successor of the PARAPET software library which was the result of a (European Union funded) collaboration between 6 different partners http://stir.sourceforge.net.

5. References

[1] NEMA Standards Publication NU 2, 2007.
[2] Defrise M, Geissbuhler A and Townsend D W 1994 *Phys. Med. Biol.* **39** 305.
[3] Liang H, Yang Y, Yang K, Wu Y, Boone J M and Cherry S R 2007 *Phys. Med. Biol.* **52** 3881.
[4] Valais I, Michail C, David S, Liaparinos P, Fountos G, Paschalis T, Kandarakis I, Panayiotakis G 2010 *IEEE Trans. Nucl. Sci.* **57(1)** 3.
[5] Boone J M 2001 *Med. Phys.* **28** 356.
[6] Fountos G P, Michail C M, Zanglis A, Samartzis A, Martini N, Koukou V, Kalatzis I and Kandarakis I S 2012 *Med. Phys.* **39(3)** 1561.
[7] Fountos G, Zanglis A, Michail C, Kalatzis I, Cavouras D, Samartzis A, Kounadi E, Valsamaki P, Gerali S, Nikiforidis G and Kandarakis I, IFMBE Proceedings 25/II, pp. 802-805, 2009.
[8] Karpetas G E, Michail C M, Fountos G P, Valsamaki P N, Kandarakis I S and Panayiotakis G S 2013 *Hell J. Nucl. Med.* **16(2)** 111.
[9] Karpetas G E, Michail C M, Fountos G P, Kalyvas N I, Valais I G, Kandarakis I S, Panayiotakis G S 2014 *J. Phys.: Conf. Ser.* **490** 012139.
[10] Karpetas G, Michail C, Fountos G, Kandarakis I and Panayiotakis G 2014 *Nucl. Med. Com.* **35(8)** 967.
[11] OpenGATE Collaboration, Users Guide V6.1: http://www.opengatecollaboration.org.
[12] Thielemans K, Tsoumpas C, Mustafovic S, Beisel T, Aguiar P, Dikaios N and Jacobson M W 2012 *Phys. Med. Biol.* **57** 867.
[13] Valais I, Michail C, David S, Costaridou L, Nomicos C D, Panayiotakis G S, Kandarakis I 2008 *Physica Medica* **24** 122.
[14] Valais I G, Michail C M, David S L, Konstantinidis A, Cavouras D A, Kandarakis I S and Panayiotakis G S 2008 *IEEE Trans. Nucl. Sci.* **55(2)** 785.
[15] Michail C M, Fountos G P, Liaparinos P F, Kalyvas N E, Valais I, Kandarakis I S and Panayiotakis G S 2010 *Med. Phys.* **37(7)** 3694.
[16] Michail C, Kalyvas N, Valais I, David S, Seferis I, Toutountzis A, Karabotsos A, Liaparinos P, Fountos G and Kandarakis I 2013 *J. Lumin.* **144** 45.
[17] Cheng J-C, Shoghi K and Laforest R 2012 *Med. Phys.* **39(2)** 1029.
[18] Fujita H, Tsia D, Itoh T, Doi K, Morishita J, Ueda K, and Ohtsuka A 1992 *IEEE Trans. Med. Imaging* **11** 34.
[19] Shepp L and Vardi Y 1982 *IEEE Trans. Med. Imaging* **1(2)** 113.

New analytical approach for transition to slow 3-D turbulence

J Foukzon[1], **E Men'kova**[2], **A Potapov**[3]
[1]Department of mathematics, Israel Institute of Technology,
Haifa, Israel

E-mail: jaykovfoukzon@list.ru

[2]All-Russian Research Institute for Optical and Physical Measurements,
Moscow, Russia

E-mail: E_Menkova@mail.ru

[3]Kotel'nikov Institute of Radioengineering and Electronics of the Russian Academy of Sciences, Moscow, Russia

E-mail: potapov@cplire.ru

Abstract. Analytical non-perturbative study of the three-dimensional nonlinear stochastic partial differential equation with additive thermal noise analogous to that proposed by Nikolaevskii V.N. to describe longitudinal seismic waves is presented. The equation has a threshold of short-wave instability and symmetry providing long wave dynamics. New mechanism of quantum chaos generating in nonlinear dynamical systems with infinite number of degrees of freedom is proposed. The hypothesis says that physical turbulence could be identified with quantum chaos of considered type. It is shown that the additive thermal noise destabilizes dramatically the ground state of the Nikolaevskii system causing it to make a direct transition from a spatially uniform to a turbulent state.

1. Introduction

In the presented work a non-perturbative analytical approach to the studying of problem of quantum chaos in dynamical systems with infinite number of degrees of freedom is proposed. Statistical descriptions of dynamical chaos and investigations of noise effects on chaotic regimes are studied. Proposed approach also allows estimate the influence of additive (thermal) fluctuations on the formation processes of developed turbulence modes in essentially nonlinear processes like electro-convection and others. A principal role of the influence of thermal fluctuations on the dynamics of some types of dissipative systems in the approximate environs of derivation rapid of a short-wave instability was ascertained. Important physical results following from Theorem 2.1 are numerically illustrated by example of 1D stochastic model system:

---

[1] To whom any correspondence should be addressed.

$$\frac{\partial u_\eta(x,t,\varepsilon)}{\partial t} + \Delta[\varepsilon - (1+\Delta)^2]u_\eta(x,t,\varepsilon) + \delta\frac{\partial u_\eta(x,t,\varepsilon)}{\partial x_1}u_\eta(x,t,\varepsilon) + f(x,t) - \sqrt{\eta}w(x,t) = 0,$$
$$x \in \mathbb{R}, u_\eta(x,0,\varepsilon) = 0, w(x,t) = \frac{\partial^2 W(x,t)}{\partial x \partial t}, \eta \ll 1, 0 < \delta \quad (1.1)$$

Systematic study of a different type of chaos at onset ''soft-mode turbulence'' based on numerical integration of the simplest 1D Nikolaevskii model has been executed and analyzed by many authors [1]-[3].

2. Main Theoretical Results

We study the stochastic partial differential equation (1.1) in the sense of Colombeau generalized functions [4].

Theorem 2.1. [5]-[6]. (**Strong Large Deviation Principle for Colombeau-Ito's SPDE**)

(**I**) Let $(u_\epsilon(x,t,\varepsilon,\eta,\omega))_\epsilon, \epsilon \in (0,1]$ be a solution of the Colombeau-Ito's SPDE [5]:

$$\frac{\partial(u_\epsilon(x,t,\varepsilon,\eta,\omega))_\epsilon}{\partial t} + \Delta[\varepsilon - (1+\Delta)^2](u_\epsilon(x,t,\varepsilon,\eta,\omega))_\epsilon$$
$$+\left(\left(F_\epsilon(u_\epsilon(x,t,\varepsilon,\eta,\omega))\right)_\epsilon\right)\sum_{i=1}^{r}\delta_i\left(F_\epsilon\left(\frac{\partial u_\epsilon(x,t,\varepsilon,\eta,\omega)}{\partial x_i}\right)\right)_\epsilon + (f_\epsilon(x,t))_\epsilon$$
$$-\sqrt{\eta}(w_\epsilon(x,t,\omega))_\epsilon = 0, \quad (2.1)$$
$$x \in \mathbb{R}^r, u_\epsilon(x,t,\varepsilon,\eta,\omega) \equiv 0, \delta_j > 0, j = 1,\dots,r. \quad (2.2)$$

Here: (1)$(F_\epsilon(z))_\epsilon \in \mathcal{G}(\mathbb{R}), \mathcal{G}(\mathbb{R})$ is the Colombeau algebra of Colombeau generalized functions and $F_0(z) = z$.

(2) $(f_\epsilon(x,t))_\epsilon \in \mathcal{G}(\mathbb{R}^{r+1})$.

(3)$w_\epsilon(x,t)$ is a smoothed with respect to $\mathbb{R}^r$ white noise.

(**II**) Assume that Colombeau-Ito's SDE (2.1)-(2.2) is a strongly dissipative [5].

(**III**) Let $\Re(x,t,\varepsilon,\lambda)$ be the solutions of the linear PDE:

$$\frac{\partial\Re(x,t,\varepsilon,\lambda)}{\partial t} + \Delta[\varepsilon - (1+\Delta)^2]\Re(x,t,\varepsilon,\lambda) + \lambda\sum_{i=1}^{r}\delta_i\frac{\partial\Re(x,t,\varepsilon,\lambda)}{\partial x_i} - f(x,t) = 0, \lambda \in \mathbb{R}, \quad (2.3)$$
$$\Re(x,0,\varepsilon,\lambda) = -\lambda. \quad (2.4)$$

Then

$$\liminf_{\epsilon\to 0}\mathbf{E}[|u_\epsilon(x,t,\varepsilon,\eta,\omega) - \lambda|^2] \leq \Re(x,t,\varepsilon,\lambda).$$

Definition 2.1. (**Differential Master Equation**) The linear PDE [5]-[6]

$$\frac{\partial\Re(x,t,\varepsilon,\lambda)}{\partial t} + \Delta[\varepsilon - (1+\Delta)^2]\Re(x,t,\varepsilon,\lambda) + \lambda\sum_{i=1}^{r}\delta_i\frac{\partial\Re(x,t,\varepsilon,\lambda)}{\partial x_i} - f(x,t) = 0, \lambda \in \mathbb{R}, \quad (2.5)$$

$$\Re(x,0,\varepsilon,\lambda) = -\lambda \quad (2.6)$$

Definition 2.2. (**Transcendental Master Equation**) The transcendental equation [5]-[6]

$$\Re\big(x,t,\varepsilon,\lambda(x,t,\varepsilon)\big) = 0, \tag{2.7}$$

we will call as the transcendental master equation.
We note that concrete structure of the Nikolaevskii chaos is determined by the solution $\lambda(x,t,\varepsilon)$ variety by transcendental master equation (2.7). Master equation (2.7) is only determined by the way of some many-valued function $\lambda(x,t,\varepsilon)$ which is the main constructive object, determining the characteristics of quantum chaos in the corresponding model of Euclidian quantum field theory.

3. Criterion of the existence quantum chaos in Euclidian quantum N-model

Definition 3.1. Let $u_\eta(x,t,\varepsilon,\omega)$ be the solution of the equation (2.1). Assume that for almost all points $(x,t) \in \mathbb{R}^r \times \mathbb{R}_+$ (in the sense of Lebesgue measure on $\mathbb{R}^r \times \mathbb{R}_+$), there exist a function $u(x,t)$ such that $\lim_{\eta\to 0}\mathbf{E}\left[\big(u_\eta(x,t,\varepsilon,\omega) - u(x,t)\big)^2\right] = 0$.Then we will say that a function $u(x,t)$ is a quasi-determined solution (QD-solution of the equation (2.1). Assume that there exist a set $\mathfrak{H} \subset \mathbb{R}^r \times \mathbb{R}_+$ that is positive Lebesgue measure, i.e. $\mu_L(\mathfrak{H}) > 0$ and $\forall(x,t)\{(x,t) \in \mathfrak{H} \to \neg\exists\lim_{\eta\to 0}\mathbf{E}[u_\eta^2(x,t,\varepsilon,\omega)]\}$,i.e., $(x,t) \in \mathfrak{H}$ imply that the limit: $\lim_{\eta\to 0}\mathbf{E}[u_\eta^2(x,t,\varepsilon,\omega)]$ does not exist. Then we will say that Euclidian quantum N-model has the quasi-determined Euclidian quantum chaos (**QD**-quantum chaos). For each point $(x,t) \in \mathbb{R}^r \times \mathbb{R}_+$ we define a set $\{\widetilde{\Re}(x,t,\varepsilon)\} \subset \mathbb{R}$ by the condition:

$$\forall\lambda\big[\lambda \in \{\widetilde{\Re}(x,t,\varepsilon)\} \Leftrightarrow \Re(x,t,\varepsilon,\lambda) = 0\,\big].$$

Definition 3.2. Assume that Euclidian quantum N-model (2.1) has the Euclidian **QD**-quantum chaos. For each point $(x,t) \in \mathbb{R}^r \times \mathbb{R}_+$ we define a set-valued function $\widetilde{\Re}(x,t)\colon \mathbb{R}^r \times \mathbb{R}_+ \to 2^{\mathbb{R}}$ by the condition: $\widetilde{\Re}(x,t,\varepsilon) = \{\widetilde{\Re}(x,t,\varepsilon)\}$.We will say that the set-valued function $\widetilde{\Re}(x,t,\varepsilon)$ is a quasi-determined chaotic solution (**QD**-chaotic solution) of the quantum N-model.

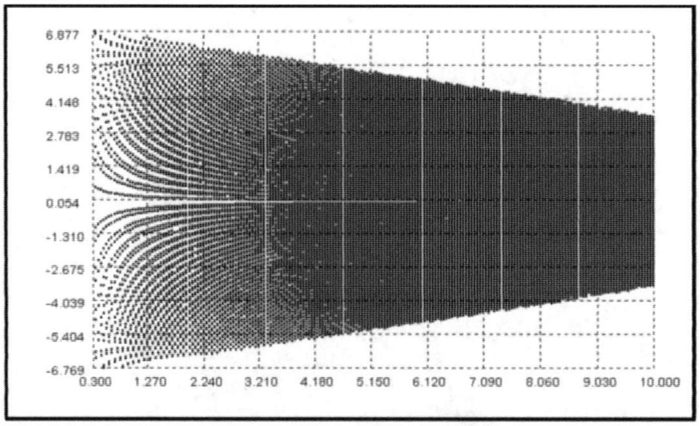

Figure 1. Evolution of **QD**-chaotic solution $\widetilde{\Re}(x,t,\varepsilon)$ in time $t \in [0,10]$ at point $x = 3$.
$t \in [0,10], \varepsilon = -10^{-2}, \sigma = 10^3, p = 1.1.$

Theorem 3.1. Assume that $f(x,t) = \sigma \sin(p \cdot x)$. Then for all values of parameters $r, \varepsilon, \sigma, \delta_j, j = 1, \dots, r$ such that $r \in \mathbb{N}, \delta_j \in \mathbb{R}_+, j = 1, \dots, r, \varepsilon \in [-1,1], p \in \mathbb{R}^r, \sigma \neq 0$, quantum N-model (2.1) has the **QD**-chaotic solutions.

Definition 3.3. For each point $(x,t) \in \mathbb{R}^r \times \mathbb{R}_+$ we define the functions such that: (i)$u_+(x,t,\varepsilon) = \limsup_{\eta \to 0} \mathbf{E}[u_\eta(x,t,\varepsilon,\omega)]$, (ii)$u_-(x,t,\varepsilon) = \liminf_{\eta \to 0} \mathbf{E}[u_\eta(x,t,\varepsilon,\omega)]$, (iii)$u_w(x,t,\varepsilon) = u_+(x,t,\varepsilon,\omega) - u_-(x,t,\varepsilon,\omega)$. (iv)Function $u_+(x,t,\varepsilon)$ is called upper bound of the **QD**-quantum chaos at point (x,t), (v) function $u_-(x,t,\varepsilon)$ is called lower bound of the **QD**-quantum chaos at point (x,t), (vi) function $u_w(x,t,\varepsilon)$ is called width of the **QD**-quantum chaos at point (x,t).

Definition 3.4. For each point $(x,t) \in \mathbb{R}^r \times \mathbb{R}_+$ we define the functions such that:

$$\text{(i)}\widetilde{\mathfrak{R}}_+(x,t,\varepsilon) = \sup\{\widetilde{\mathfrak{R}}(x,t,\varepsilon)\}, \text{(ii)}\widetilde{\mathfrak{R}}_-(x,t,\varepsilon) = \inf\{\widetilde{\mathfrak{R}}(x,t,\varepsilon)\},$$
$$\text{(iii)}\widetilde{\mathfrak{R}}_w(x,t,\varepsilon) = \widetilde{\mathfrak{R}}_+(x,t,\varepsilon) - \widetilde{\mathfrak{R}}_-(x,t,\varepsilon).$$

Theorem 3.2. (Criterion of QD-quantum chaos in Euclidian quantum N-model) Assume that $\mu_L\{(x,t)|\widetilde{\mathfrak{R}}_w(x,t,\varepsilon) > 0\} > 0$. Then Euclidian quantum N-model has **QD**-quantum chaos.

4. Quasi-determined quantum chaos and physical turbulence nature

The physical nature of quasi-determined chaos is simple and mathematically is associated with discontinuously of the trajectories of the stochastic process $u_\eta(x,t,\varepsilon,\omega)$ on parameter η. In order to obtain the characteristics of this turbulence we have used some appropriate functions [6].

Definition 4.1. The normalized local auto-correlation function is defined by the formula

$$\Phi_n(x,\tau) = \frac{\Phi(x,\tau)}{\Phi(x,0)}. \tag{4.1}$$

Now let us consider 1D Euclidian quantum N-model corresponding to the classical dynamics. Corresponding differential master equations (2.5)-(2.6) are

$$\frac{\partial \mathfrak{R}(x,t,\varepsilon,\lambda)}{\partial t} + \Delta[\varepsilon - (1+\Delta)^2]\mathfrak{R}(x,t,\varepsilon,\lambda) + \lambda\delta \frac{\partial \mathfrak{R}(x,t,\varepsilon,\lambda)}{\partial x} - \sigma \sin(px) = 0, \tag{4.2}$$

$$\mathfrak{R}(x,0,\varepsilon,\lambda) = -\lambda. \tag{4.3}$$

Corresponding transcendental master equations (2.7) are

$$\frac{\{\cos(p\cdot x) - \exp[t\cdot\chi(p)]\cos[p(x-\lambda\cdot\delta\cdot t)]\}\cdot\lambda\cdot\delta\cdot p}{\chi^2(p)+\lambda^2\cdot\delta^2\cdot p^2} + \frac{\{\sin(p\cdot x) - \exp[t\cdot\chi(p)]\sin[p(x-\lambda\cdot\delta\cdot t)]\}\cdot\chi(p)}{\chi^2(p)+\lambda^2\cdot\delta^2\cdot p^2} + \frac{\lambda}{\sigma} = 0, \tag{4.4}$$

$$\chi(p) = p^2[\varepsilon - (p^2-1)^2]. \tag{4.5}$$

We assume now that $\chi(p) = 0$. The result of calculation of the corresponding function $\widetilde{\mathfrak{R}}(x,t,\varepsilon)$ using master equation (4.5) is presented in figure 1. Let us calculate now the corresponding normalized local auto-correlation function $\Phi_n(x,\tau)$. The result of calculation is presented in figure 2.

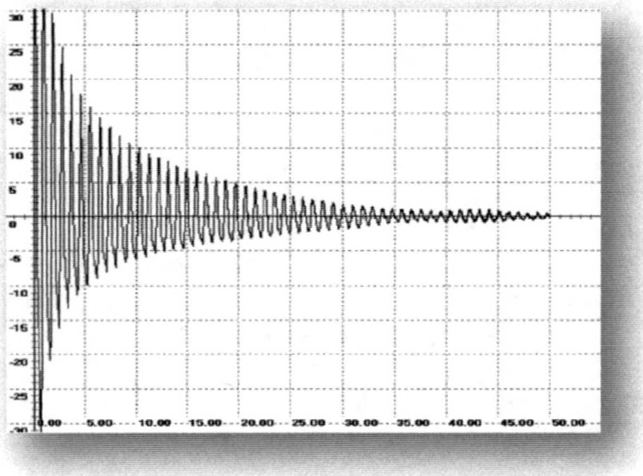

Figure 2. Normalized local auto-correlation function $\Phi_{\mathrm{n}}(1,\tau), \tau \in [0,50], \varepsilon = 10^{-7}, \sigma = 10^{2}, \delta = 1, p = 1.$

5. Conclusion

A non-perturbative analytical approach to the studying of problem of quantum chaos in dynamical systems with infinite number of degrees of freedom is proposed and developed successfully. It is shown that the additive thermal noise dramatically destabilizes the ground state of the system thus causing it to make a direct transition from a spatially uniform to a turbulent state.

6. References

[1] Tribelsky M I, Tsuboi K 1996 New scenario to transition to slow turbulence *Phys. Rev. Lett.* **76** 1631

[2] Toral R, Xiong J D, Gunton and Xi H W 2003 Wavelet description of the Nikolaevskii model *Journ. Phys. A* **36** 1323

[3] Haowen Xi, Toral R, Gunton D, Tribelsky M I 2003 Extensive chaos in the Nikolaevskii model *Phys. Rev. E* **61** R17

[4] Oberguggenberger M, 2001 Generalized functions in nonlinear models-a survey *Nonlinear Analysis: Theory, Methods & Applications* Proceedings of the Third World Congress of Nonlinear Analysts **47**, Issue 8, 5029–5040

[5] Foukzon J 2014 *Communications in Applied Sciences* **2, No 2** 230-363

[6] Foukzon J 2015 New scenario for transition to slow 3-D turbulence *Journal of Applied Mathematics and Physics* **3** 371-389. doi: 10.4236/jamp.2015.33048

Journal of Physics: Conference Series **633** (2015) 012098 doi:10.1088/1742-6596/633/1/012098

The effect of thermal losses on traveling waves for in-situ combustion in porous medium

G. Chapiro[1] **and D. Marchesin**[2]

[1] Universidade Federal de Juiz de Fora, Juiz de Fora, MG 36036-900, Brazil
[2] Instituto de Matemática Pura e Aplicada, Rio de Janeiro, RJ 22460-320, Brazil.

E-mail: grigori@ice.ufjf.br

Abstract. We study a model for the injection of air into an underground porous medium that contains a solid fuel. In our previous works the model was simplified and all wave sequences for the Riemann problem solution were obtained without taking into account thermal losses to the surrounding rock. In this work the first step is made to understand the effect of heat losses, which are important especially in laboratory experiments. In order to simplify the proof of the existence and uniqueness of the traveling wave solution, we disregard diffusion effects and the dependence of gas density on temperature.

1. Introduction

Air injection with in-situ combustion offers several potential technical and economic advantages that may include faster oil production, reduced operational costs and increased CO_2 content with decreased oil viscosity. Despite other difficulties related to engineering and chemical modeling, solving the equations for such models is a challenge.

This paper is part of long term research project the purpose of which is to identify waves that arise in one-dimensional models of combustion in porous media, and to understand how the waves fit together in solutions of Riemann problems; see [1, 2, 3, 4, 5, 6], and references therein. The paper is motivated by a model for the injection of air into a porous medium that contains oil so viscous that it can be considered a solid fuel. The model was proposed in [7] and further studied in [1]. This model was simplified in [8] in order to (i) reproduce the variety of phenomena observed when air is injected into a porous medium containing a solid fuel, yet (ii) to be simple enough to permit a rigorous investigation.

In this paper we prove the existence of a traveling wave solution corresponding to the combustion wave for a simple in-situ combustion model. This model is more general then the previously considered one [8] in three aspects. First it considers more physically realistic thermal capacity of the medium, leading to distinct thermal and gas velocities. Second, we utilize a more correct form of Arrhenius law, allowing chemical reaction to happen at any temperatures, see also [9]. Third, we take into account the effect of thermal losses, which is important for laboratory experiments. We solve possible Riemann problems and classify the resulting solutions depending on the presence of slow, fast or resonance combustion waves. We emphasize that the existence and uniqueness proofs for the results presented in this paper are technically simple and should be accessible to students. For other approaches addressing combustion with heat losses see [10].

A model for combustion is presented in Section 2. It consists of three balance laws for energy, oxygen, and fuel. We use a reaction rate described by Arrhenius law combined with Law of Mass Action. The combustion waves studied in this paper have been called "reaction-trailing smolder waves" [11] and "coflow (or forward) filtration combustion waves" [12] in the context of more realistic models of air injection into a porous medium. We formulate the main results of the paper in Section 3. Due to lack of space we present only schematically the proofs of the existence and uniqueness of the combustion traveling waves in Section 4. The complete proof will be published soon in a companion paper with some numerical examples showing the phase portrait of the traveling wave.

2. Combustion model

We consider one-dimensional flow due to air injection into a porous medium. We use notation and assumptions from [1]. We neglect changes of porosity during the reaction and gas expansibility under temperature increase. We assume that the temperature of solid and gas is the same (local thermal equilibrium). This work is concerned with heat losses, which we consider to depend linearly on the temperature difference with the prevailing temperature, see e.g., [7]. We also assume that pressure variations are small compared to the prevailing pressure. In order to prove the existence of traveling waves minimizing technical difficulties we neglect thermal diffusion effects. In dimensionless form the governing system of equations is written as follows [8]:

$$\frac{\partial \theta}{\partial t} + v_\theta \frac{\partial \theta}{\partial x} = -\beta\theta + \rho Y \Phi, \tag{1}$$

$$\frac{\partial Y}{\partial t} + v_Y \frac{\partial Y}{\partial x} = -\mu_Y \rho Y \Phi, \tag{2}$$

$$\frac{\partial \rho}{\partial t} = -\rho Y \Phi, \tag{3}$$

$$\Phi = \exp\left(\frac{-\mathcal{E}}{\theta + \theta_0}\right), \tag{4}$$

where the dependent variables are temperature θ, oxygen fraction Y and fuel ρ. Here v_θ and v_Y are dimensionless thermal and oxygen wave speeds; β is the constant thermal loss coefficient; $\bar{\lambda}$ represents the dimensionless thermal diffusion coefficient; μ_Y represents the dimensionless quantity of oxygen consumed during the reaction; $\mathcal{E}$ is the scaled activation energy and θ_0 is the scaled reservoir temperature. The oxygen Y is a component of the gas moving with velocity $v_Y > 0$. The heat θ is transported with velocity v_θ. We are of course interested in solutions with $\rho \geq 0$ and $Y \geq 0$ everywhere. We consider (1)–(3) on $0 < x < \infty$, $t \geq 0$, with the (constant) boundary conditions

$$(\theta, \rho, Y)(0) = (\theta^L, \rho^L, Y^L), \qquad (\theta, \rho, Y)(\infty) = (\theta^R, \rho^R, Y^R). \tag{5}$$

We assume that the reaction does not occur at the boundaries, i.e., the reaction terms in (1)–(3) vanish. Differently from [6, 8], here we consider the correct Arrhenius law (4) and thus there are only two reasons for the reaction terms to vanish: (i) fuel control (FC) – the reaction ceases due to lack of fuel, $\rho = 0$; (ii) oxygen control (OC) – the reaction ceases due to lack of oxygen, $Y = 0$. In the next section we study the solution of the Riemann problem for system (1)–(4).

3. Wave sequences

In this section we follow [6, 8] and denote by $(\theta^-, \rho^-, Y^-) \xrightarrow{v} (\theta^+, \rho^+, Y^+)$ a wave of velocity v that connects (θ^-, ρ^-, Y^-) at the left to (θ^+, ρ^+, Y^+) at the right. At the end states of the wave, the reaction terms in (1)–(3) vanish. States at which the reaction terms vanish can be classified as FC or OC. The type of the state indicates exactly which conditions hold at that state. For simplicity we consider that $\rho = 1$ in OC state and $Y = 1$ in FC state.

3.1. Contact waves

In the non-combustion waves supported by system (1)–(3) the source terms vanish. The characteristic eigenvalues and corresponding eigenfunctions of the resulting hyperbolic system are (see [8] for details) $\lambda_\theta = v_\theta$ and $(1,0,0)^T$, $\lambda_Y = v_Y$ and $(0,1,0)^T$, $\lambda_f = 0$ and $(0,0,1)^T$. We can see that the Riemann problem possesses three non-combustion waves. As the characteristic speeds are constant the waves correspond to contact discontinuities. The latter separate moving spatial intervals in which the reaction does not occur (since (θ, ρ, Y) is constant).

3.2. Combustion waves

As in [8], system (1)–(4) possesses a traveling combustion wave. We formulate the main result below and give a sketch of the proof in Section 4.

Theorem 3.1. *The system* (1)–(3) *possesses a unique combustion wave in the following cases*

(i) If $v_Y < (\mu_Y + 1)v_\theta$ *then a unique slow combustion wave with speed* $v < v_\theta$ *exists.*

(ii) If $v_Y > (\mu_Y + 1)v_\theta$ *then a unique fast combustion wave with speed* $v > v_\theta$ *exists.*

(iii) If $v_Y = (\mu_Y + 1)v_\theta$ *then there exists a unique combustion wave if and only if: either* $\mathcal{E} < 4\theta_0$ *or there are exactly three values of* $\theta > 0$ *satisfying Eq.* $4\beta\theta \exp(\mathcal{E}/(\theta + \theta_0)) = 1$.

In all cases the combustion wave is of type $FC \xrightarrow{v} OC$, *where its speed is given by* v *in Eq.* (7).

3.3. Solutions of the Riemann problem

For the existence of a wave sequence describing Riemann solution obviously it is has to start as one equilibrium state (FC or OC) and finish at another equilibrium state (FC or OC). The waves speeds in the sequence appear in increasing order from left to right. This fact together with the results concerning contact and combustion waves described above lead to three possibilities for wave sequences corresponding respectively to three cases described in Theorem 3.1.

(i) If the sequence contains a slow combustion wave $FC \xrightarrow{v} OC \xrightarrow{v_\theta} OC$.

(ii) If the sequence contains a fast combustion wave $FC \xrightarrow{v_\theta} FC \xrightarrow{v} OC$.

(iii) If the sequence is composed of a single resonance combustion wave $FC \xrightarrow{v} OC$.

In Fig. 1 we plot the wave sequences separated by constant states for the cases (i) (left figure), (ii) (center figure) and (iii) right figure.

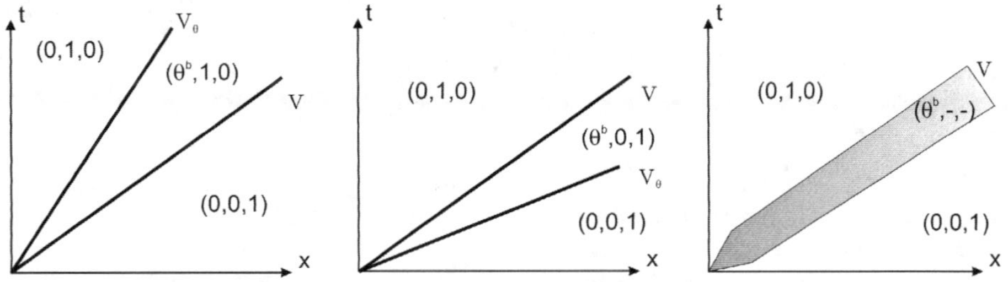

Figure 1. Wave sequences in Riemann solution separated by constant states: (i) (left figure), (ii) (center figure) and (iii) right figure. θ^b is the temperature of the combustion wave.

4. Existence and uniqueness of the combustion traveling waves

We rewrite the system (1)–(4) in traveling coordinates $(x, t) \to (\xi = x - vt, t)$, where v is positive velocity of the traveling wave. We integrate the resulting system of ODEs once and substitute the boundary conditions (5), obtaining

$$d_\xi \rho = \rho(1-\rho)\frac{\Phi}{v}, \qquad d_\xi \theta = \frac{\rho(1-\rho)\Phi - \beta\theta}{v_\theta - v}. \tag{6}$$

After some manipulations, similar to those done in [8], we obtain the combustion wave speed v and the oxygen concentration as function of fuel concentration

$$v = v_Y/(\mu_Y + 1), \quad Y = 1 - \rho. \tag{7}$$

Following [8], waves with velocities $v > v_\theta$ and $v < v_\theta$ are called *fast* and *slow combustion waves* respectively. The case when $v_\theta = v$ is known as resonance condition for the combustion wave, see [2, 12, 13] and references therein.

The proof of Theorem 3.1 is extensive and can not be presented here due to limitation in space, it will be published later in a companion paper. The existence and uniqueness of the traveling wave solution for the system (1)–(4) is equivalent to the existence and uniqueness of an heteroclinic orbit of system (6) connecting two equilibria, which are hyperbolic. The local analysis based on the Stable Manifold Theorem is performed to understand the behavior of stable and unstable manifolds in the vicinity of these equilibria and verify the necessary conditions for the existence of such an heteroclinic orbit. The rest of the proof is based on the geometrical analysis of the flux defined by Eqs. (6).

In the model (1)–(3) without thermal losses the traveling wave equations under the resonance condition $v = v_\theta$ yields in a degenerate traveling wave without combustion. The same situation happens for the model studied in [8]. Taking into account thermal losses allows the appearance of a different type of solution, which is interesting from the physical point of view.

Acknowledgments

The authors would like to thank Prof. S. Schecter (NCSU), L. Furtado (U. Columbia), F. Ozbag (NCSU) for preliminary work and Prof. J. Bruining for enlightening discussions.

G.C. was supported in part by CNPq, FAPEMIG and CAPES.

D.M. was supported in part by: ANP–PRH32, grant 731948/2010; Petrobras–PRH32, grant 6000.0069459.11.4; CAPES Nuffic, grant 024/2011; CNPq, grants 402299/2012-4, 301564/2009-4, 470635/2012-6; FAPERJ, grants E-26/210.738/2014, E-26/201.210/2014, E-26/110.658/2012, E-26/111.369/2012, E-26/110.114/2013, E-26/010.002762/2014.

References

[1] Chapiro G, Mailybaev A A, Souza A, Marchesin D and Bruining J 2012 *Comput. Geosciences* **16** 799–808
[2] Mailybaev A, Marchesin D and Bruining J 2011 *SIAM Journal on Mathematical Analysis* **43** 2230–2252
[3] Marchesin D and Schecter S 2003 *Zeitschrift für Angewandte Mathematik und Physik (ZAMP)* **54** 48–83
[4] Mota J and Schecter S 2006 *Journal of Dynamics and Differential Equations* **18** 615–665
[5] Schecter S and Marchesin D 2001 *Bulletin of the Brazilian Mathematical Society* **32** 237–270
[6] Chapiro G, Furtado L, Marchesin D and Schecter S 2015 *Accepted in DCDS*
[7] Akkutlu I and Yortsos Y 2003 *J. of Combustion and Flame* **134** 229–247
[8] Chapiro G, Marchesin D and Schecter S 2014 *Journal of Hyperbolic Differential Equations* **11** 295–328
[9] Chapiro G and de Souza A J 2015 *Applicable Analysis* 1–15
[10] Ghazaryan A, Schecter S and Simon P 2013 *SIAM J. on App. Math.* **73** 1303–1326
[11] Schult D, Matkowsky B, Volpert V and Fernandez-Pello A 1996 *Combustion and Flame* **104** 1–26
[12] Aldushin A, Rumanov I and Matkowsky B 1999 *J. of Combustion and Flame* **118** 76–90
[13] Chapiro G and Bruining J 2015 *Journal of Petroleum Science & Engineering* **127** 179–189

A viscoelastic model to simulate soft tissue materials

J. S. Espinoza Ortiz[1], **R. E. Lagos**[2]

[1]Departamento de Física, IFQC, Universidade Federal de Goiás, Catalão, Goiás 75704/020, Brasil.
[2]Departamento de Física, IGCE, UNESP, Rio Claro, São Paulo 13500/970, Brasil.

E-mail: `jsespino@physics.org`

Abstract. Continuum mechanic theories are frequently used to simulate the mechanical behavior of elastic and viscous materials, specifically soft tissues typically showing incompressibility, nonlinear deformation under stress, fading memory and insensitivity to the strain-rate. The time dependence of a viscoelastic material could be better understood by considering it as composed by an elastic solid and a viscous fluid. Different types of mechanical devices can be constructed provided a particular configuration of elastic springs and dashpots. In this work our aim is to probe many of the soft tissue mechanical behavior, by considering a Kelvin's device coupled to a set of *in parallel* Maxwell's devices. Then, the resulting model composed of a long series of modified Kelvin bodies must span a broad range of characteristic times resulting in a suitable model for soft tissue simulation. Under driving static and dynamic stress applied to a *2-Dim* system, its time dependence strain response is computed. We obtain a set of coupled Volterra integral equations solved via the extended trapezoidal rule scheme, and the Newton-Raphson method to solve nonlinear coupled equations.

1. Introduction

Modeling the mechanical behavior of soft tissue are relevant for applications in surgical simulations in real time and fast precise calculations of tissue mechanical deformations [1, 2, 3, 4]. Soft tissue material response to stress achieves large deformations at the beginning of a relatively stress low level and subsequently stiffening at higher stress level. Their structural composition of collagen fibers distribution leads to a pronounced anisotropy [5, 6, 7, 8]. Almost all the biological soft tissues are mechanically less sensitive under different strain rates. Their stress-strain curve exhibit a hysteresis loop showing a nonlinear stress-strain relationship, and their hysteresis loop-area does not depend on the strain rate [9, 10, 11].

Viscoelastic materials time behavior are well understood wether considered as an elastic solid and a viscous fluid. Different types of mechanical devices are constructed provided a particular arrangement of elastic springs and dashpots [9]. Maxwell and Voigt models supply a logarithmic functional type relationship between strain-rate and frequency (inverse of the characteristic time), describing a decreasing curve for the former; while for the latter an increasing curve is observed. Kelvin's model strain-rate vs logarithm of the frequency describes a bell-shaped curve. None of these models are able to present a typical flat strain-rate vs frequency curve of living tissues. In this work our aim is to probe many of the features of the soft-tissue mechanical behavior, considering a Kelvin's device coupled to a set of in parallel Maxwell devices in *2 Dim*. The resulting model composed of a long series of modified Kelvin bodies will span a broad range of characteristic times, becoming a suitable model for soft tissue simulation[9]. In section 2 we present a mathematical formalism applicable to systems showing fading memory,

Figure 1. Left side: A single modified Kelvin device consisting of $N+1$ springs and N dashpots. In the middle: An structured network consisting of a set of modified Kelvin's devices. Right side: Neighbors connectivity graph for the $(i\,j)$ network link, where nx is the number of arbitrary elements in a row.

we include an alternative formulation suitable to treat nonlinear elastic materials. The physical formalism applied to an ensemble of modified Kelvin devices *N-Dim* network system is presented in section 3. Results considering a bi-dimensional network are shown in section 4. Conclusions and perspectives are presented in section 5.

2. Systems with fading memory

In this section we reproduce an established one dimensional model, with a more convenient notation for our purposes [9]. Consider a single mechanical device consisting of $(N+1)$ springs and N dashpots set up in parallel (left side of Figure 1). A force is supposed to cause an uniaxial deformation on springs identified as stress and an strain (proportional to the spring constant K) and denoted by σ and u; respectively. Meanwhile into a dashpot, at any instant, it causes an strain rate $\dot{\alpha}$ (proportional to the viscosity η). Denoting the spring constants as K_o and K_n, the dashpot viscosity by $\eta_n\,(n=1,2,\ldots,N)$, its inelastic mechanical response is an internal variable denoted by $\dot{\alpha}_n(t)$. For the viscous stress we have $\sigma^v(t)=\sum_{n=1}^N \eta_n\,\dot{\alpha}_n(t)$. To complete the constitutive hypotheses is assumed a linear, elastic, stress-strain spring response. Since $u(t)$ is the strain on the spring, $K_o u(t)=\sigma(t)-\sigma^v(t)$, thus $\sigma(t)=\tilde{K}u(t)-\sum_{n=1}^N K_n\alpha_n(t)$ where $\tilde{K}=K_o+\sum_{n=1}^N K_n$ is the initial modulus, and the internal variables must satisfy the detailed balance evolution equation, $u(t)=\tau_n\dot{\alpha}_n(t)+\alpha_n(t)$, with $\alpha_n(0)=0\,;\,n=1,2,\ldots N$ and $\tau_n=\eta_n/K_n$ being the relaxation times . Then, $\alpha_n(t)=u(t)-\int_0^t \exp[-(t-s)/\tau_n]\,\dot{u}(s)ds$, and we find

$$\sigma(t)=\int_0^t G(t-s)\,\dot{u}(s)ds \quad , \quad G(t)=K_o+\sum_{n=1}^N K_n\,\exp(-t/\tau_n)\,. \tag{1}$$

As a straightforward extension including nonlinear elastic response, we replace the viscous strain by a stress-like set of internal variables $Q_n(t)=K_n\alpha_n(t)\,;\,n=1,\ldots,N$, so the stress response $\sigma(t)=\tilde{K}u(t)-\sum_{n=1}^N Q_n(t)$. By defining the nondimensional relative moduli becomes $\gamma_o=K_o/\tilde{K}$, $\gamma_n=K_n/\tilde{K}\,;\,n=1,\ldots,N$, $(\gamma_o+\sum_{n=1}^N\gamma_n=1)$ and the initial store energy function

Journal of Physics: Conference Series **633** (2015) 012099 doi:10.1088/1742-6596/633/1/012099

as $\mathcal{U}(u) = 1/2\, u\, \tilde{K}\, u$, the model is recasted in the following form:

$$\sigma(t) \; = \; \tfrac{\partial}{\partial u}\mathcal{U}(u) \; - \; \textstyle\sum_{n=1}^{N} Q_n \quad , \quad \tfrac{\partial}{\partial u}\mathcal{U}(u) \; = \; \tfrac{\tau_n}{\gamma_n}\dot{Q}_n + \tfrac{1}{\gamma_n} Q_n \, . \tag{2}$$

chemical energy rate canceling [9], the equation of motion is given by

$$m\ddot{q}(t) + \frac{\partial \mathcal{U}}{\partial\, q} \; - \; F^e \; = 0 \, , \tag{3}$$

where F^e is the external force on the device. From the results on section 2, this equation of motion yields a typical equation with fading memory [9],

$$m\ddot{q}(t) + \int_0^t ds\, G(t-s)\dot{q} \; - \; F^{(e)} = 0 \, . \tag{4}$$

Now we propose the following model for an ensemble of l connected modified Kelvin's devices, by generalizing the storage energy as,

$$\mathcal{U} = 1/2 \sum_{(i,j)=1}^{l} K^o_{ij}\, q_i q_j + 1/2 \sum_{(ij)=1}^{l} \sum_{n=1}^{N} K^n_{ij}(q_i - \alpha_{i,n})(q_j - \alpha_{j,n}) \, .$$

The equation of motion is straightforward, for $(i,\, j) \; = \; (1, 2, \ldots,\, l)$,

$$m_i \ddot{q}_i(t) + \textstyle\sum_{j=1}^{l} \int_0^t ds\, G_{ij}(t-s)\dot{q}_j \; - \; F_i^{(e)} = 0 \quad , \quad G_{ij}(t) = \tilde{K}_{ij}\left[\gamma_o^{ij} + \sum_{n=1}^{N} \gamma_n^{ij} \exp\left(-t/\tau_n\right)\right] \, , \tag{5}$$

with $\gamma_o^{ij} = K^o_{ij}/\tilde{K}_{ij}$ and $\gamma_n^{ij} = K^n_{ij}/\tilde{K}_{ij}$. Although $\mathcal{U}(q)$ may be a nonlinear function of strain, the relaxation process is linear (a quasi-linear viscoelastic theory).

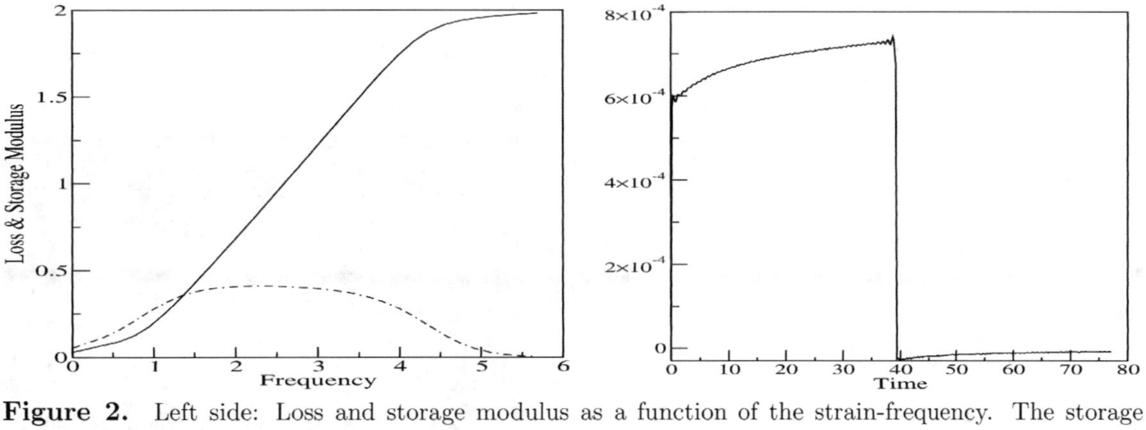

Figure 2. Left side: Loss and storage modulus as a function of the strain-frequency. The storage modulus is represented by continuous lines, the loss modulus is in long dotted line showing a flat curve reproducing soft tissue insensitivity to the strain-rate. Right side: A typical creep viscoelastic behavior for the network system.

4. Numerical results

For a two dimensional network of size $nx \times ny$ structured as in the middle of Figure 1, where each link $(i\,j)$ being one modified Kelvin device with preset relaxation times τ_n, a single relative moduli γ_o^{ij} and $\gamma_n^{ij} = (1 - \gamma_o^{ij})/N$; for $n = 1, 2, \ldots, N$ (independent of i and j). We consider the connectivity between neighboring elements (six), as shown in the right side of Figure 1. In the overdamped regime ($m_i\ddot{q}_i \approx 0$), and by integrat by parts Eqn. 5, we obtain the following expression for the i-th particle: $\sum_{j=1}^{6}\left[G_{ij}(0)\,\delta r_{ij} - \int_0^t \tilde{K}_{ij} \sum_{n=1}^{N} \frac{\gamma_n^{ij}}{\tau_n}\, e^{-(t-s)/\tau_n} \delta r_{ij}\, ds\right] = \vec{F}_i^e$. The summation is carried over all six neighbors: $j_1 = i - nx - 1$, $j_2 = i - nx$, $j_3 = i - 1$, $j_4 = i + 1$, $j_5 = i + nx$, $j_6 = i + nx + 1$, with $i = 1, \ldots, ny$; δr_{ij} are the relative displacements (chosen to be quadratic [9]) between neighbors $(i\,j)$ and we set $\tilde{K}_{ij}$ to be unity . The resulting coupled system of Volterra integral equations are nonlinear. These are solved via extended trapezoidal rule scheme, and the Newton-Raphson method to solve nonlinear equations [13]. We consider $N = 6$ dashpots with different $\tau_n = 3^n$ and $\gamma_o = 0.25$, for a network of 20×20 links.

We compute: *a)* the response of the system to a dynamic oscillatory strain $F_i^e = \cos\omega t$, *b)* the energy stored due to the applied strain (storage modulus) and *c)* the dissipation energy as determined by the strain-rate (loss modulus). In Figure 2, left side, it is shown both the loss and storage modulus behavior vs the inverse of the relaxation time, in the range $(0.0, 6.0)\,Hz$. The loss modulus is represented by the curve in long dotted lines showing a plateau that resembles insensitivity to the strain-frequency. The storage modulus is represented by and increasing continuous curve until it reaches a constant value; delimiting the viscous region. Next we computed the systems response to a cyclic suddenly applied stress (step function). The procedure results in an instantaneous elastic deformation followed by a delayed time-dependent deformation; i.e. the creeping effect. This is shown in Figure 2, right side, where the applied stress is hold for $\approx 80\,sec$, while the strain response is pre-set to be linear. Then it is observed an instantaneous elastic deformation $\approx 6 \times 10^{-4}$ units, followed by a delayed time-deformation in the range $(0, 40)\,sec$ and a delayed time-recovery in a range of $(40, 80)\,sec$. Finally, we considered the response of the system to a cyclic stress linearly increasing and decreasing with slope $\pm\,1/2\pi$ during successive time intervals and considered pre-set quadratic deformations. In Figure 3, left side, is plotted the deformation history curve for the elastic viscous deformation. Next, in Figure 3, right side, it is presented a typical intensity color stress deformation for the present network. The plot shown that the relative strain distributed on the network increases from the blue to the red color.

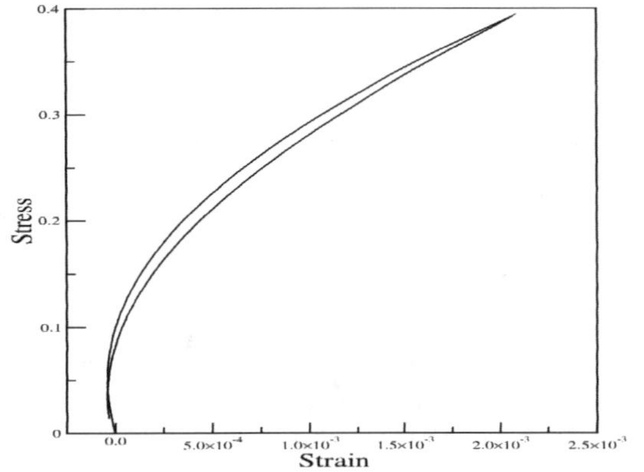

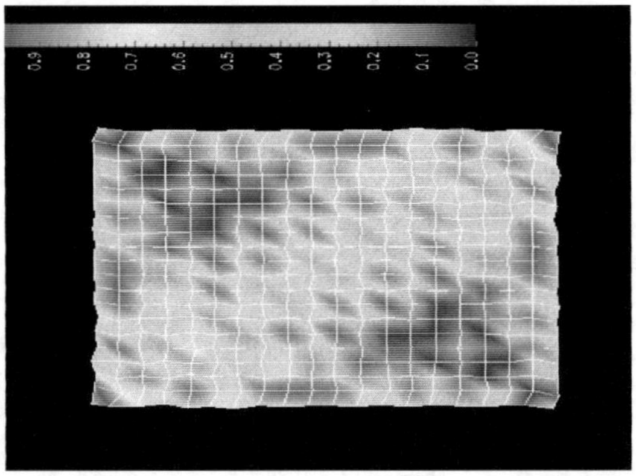

Figure 3. Left side: A typical hysteresis curve for *2-Dim* network system. Right side: A color intensity stress-deformation for a 20 × 20 square network.

5. Conclusions and perspectives

This model reproduces soft tissue mechanical behavior as it is numerically verified by simulated mechanical test. Storage-loss modulus, creeping and hysteresis are common features of viscoelastic mechanical behavior. It is remarkable that the model is capable to compute efficiently nonlinear elastic deformations in real time. It seems that methods based on continuum mechanics are more realistic than their spring-mass viscoelastic counterpart. However, the latter can be faster than the former, therefore more suitable for real time applications. It is curious that quasi-linear viscoelastic formulation is based on the assumption that the viscous stress is related to linearly superposed strain rates. Similar constitutive relations are also useful in a continuum mechanics formulation. There is not an exhaustive comparison between this two alternative approaches, when attempted in applications relying on advantages and precision [14]. The model presented here is appropriated to describe the viscoelastic behavior of bubbles foams, useful when studying microfluidics [15].

Acknowledgments

The authors would like to thank the support of the Goiás Research Foundation - FAPEG.

References

[1] Bro-Nielsen M (1998) Finite element modeling in surgery simulation. Proc. of the IEEE: Special issue on virtual and augmented reality in medicine **86**(3) 524-30.

[2] Mosegaard J (2003) LR-spring-mass model for cardiac surgical simulation. Medicine meets virtual reality, **12** 256-58.

[3] Teschrer M, Girod S, Girod B (2000) Direct computation of nonlinear soft-tissue deformation. Vision modeling and visualization VMV'000, Saarbucken, Germany, November 22-24.

[4] Keeve E, Girod S, Pfeifel P, Girod B (1996) Anatomy-based facial tissue modeling using the finite element method. Proc. IEEE visualization, San Francisco, CA. USA.

[5] Yamamoto E, Hayashi K and Yamamoto N (1999) Mechanical properties of collagen fascicles from the rabbit pattelar tendom. ASME J. Biomech. Eng. **121** 124-31.

[6] Hasegawa M, and Azuna T (1974) Wall structure and static viscoelasticities of large veins. J. Jap. College Angiol. **14** 87-92.

[7] Chuong C J, and Fung Y C (1984) Compressibility and constitutive equation of arterial wall in radial compression experiments. J. Biomech, **17** 35-40.

[8] Woo S L Y, Lubock P, Gomez M A, Jemmott G F, Kuei S C and Akeson W H (1979) Large deformation nonhomogeneus and directional properties of articular cartilage. J. Biomech. **12** 437-46.

[9] Fung Y C (1994) A first course in continuum mechanics. Prentice Hall, Englewood Cliff, New Jersey.

[10] Fung Y C, Fronek K and Patitucci P (1979) Pseudoelasticity of arteries and the choice of its mathematical expression. Am. J. Physiol. **237** H620-631.

[11] Tong P and Fung Y C (1976) The stress-strain relationship for the skin. J. Biomech. **9** 649-57.

[12] The stability of the system is guaranteed if all the principal minors, including the determinant of the full matrices are positives.

[13] Press W H, Teukolsky S A, Vetterling W T, and Flannery B P (2002) Numerical Recipes in C++ (The art of Scientific Computing), Cambridge University Press.

[14] De Pascalis R, Abrahams I D, Parnell W J (2014) On nonlinear viscoelastic deformations: a reappraisal of Fungs quasi-linear viscoelastic model. Proc. R. Soc. A **470** 20140058.

[15] Espinoza Ortiz J S, Belich H, Orlando M T D, Lagos R E (2014) Topological rearrangements and flow simulation of dry ordered foams. Int. J. Model. Simul. Sci. Comput. **05**, 1441007.

Magnetization reversal modes in fourfold Co nano-wire systems

T Blachowicz[1,3] and A Ehrmann[2,3]
[1] Institute of Physics – Center for Science and Education, Silesian University of Technology, 44-100 Gliwice, Poland
[2] Faculty of Engineering Sciences and Mathematics, Bielefeld University of Applied Sciences, 33609 Bielefeld, Germany
[3] VIARAM – Virtual Institute for Applied Research on Advanced Materials

E-mail: tomasz.blachowicz@polsl.pol

Abstract. Magnetic nano-wire systems are, as well as other patterned magnetic structures, of special interest for novel applications, such as magnetic storage media. In these systems, the coupling between neighbouring magnetic units is most important for the magnetization reversal process of the complete system, leading to a variety of magnetization reversal mechanisms. This article examines the influence of the magnetic material on hysteresis loop shape, coercive field, and magnetization reversal modes. While iron nano-wire systems exhibit flat or one-step hysteresis loops, systems consisting of cobalt nano-wires show hysteresis loops with several longitudinal steps and transverse peaks, correlated to a rich spectrum of magnetization reversal mechanisms. We show that changing the material parameters while the system geometry stays identical can lead to completely different hysteresis loops and reversal modes. Thus, especially for finding magnetic nano-systems which can be used as quaternary or even higher-order storage devices, it is rational to test several materials for the planned systems. Apparently, new materials may lead to novel and unexpected behaviour – and can thus result in novel functionalities.

1. Introduction

Magnetic nano-structures are possible candidates for research aiming at increasing the data density of magnetic storage devices. Especially magnetic particles exhibiting more than two magnetization states at vanishing external magnetic field are of great interest for this purpose. A quaternary storage medium, showing four distinguishable states (zero, one, two, three) at zero magnetic field, would allow for storing two binary bits, thus increasing the storage density by a factor of two without changing the storage position dimension.

This idea has stimulated research by several groups, leading to reports about magnetic systems with three, four, or even eight different magnetization states in diverse nano-structures [1-5]. The detection of these states, however, is normally not easy, since magnetization has to be measured at different spatial positions to differentiate between similar states, a procedure which is not compatible with contemporary hard disk drive technology.

Thus, our previous work has concentrated on fourfold magnetic nano-wire systems produced from iron (Fe), showing four stable intermediate states at remanence which can easily be detected by an overall magnetization measurement, as it is used in hard disk write/read heads [6-8]. However, Fe wire systems seem to be limited to the aforementioned four states in square nano-structures. In Fe particles of higher symmetry, such as six-fold wire systems, more intermediates states can be found; however, these states are not stable enough to be used as reliable storage possibilities [9]. Thus, other materials have been chosen to be tested. While permalloy (Py) or FeCu show smaller anisotropies and mostly flat hysteresis loops without any additional steps indicating possible stable intermediate states, simulations of cobalt (Co) nano-wire systems exhibit a rich spectrum of different magnetization reversal mechanisms, resulting in several steps in the hysteresis loops for some angular orientations. This article gives an overview of the different reversal mechanisms and hysteresis loop forms found in fourfold Co nano-wire systems.

2. Mathematical Model

To allow for modelling the round wire shapes in the best possible way, the micromagnetic simulation program Magpar is used [10]. Since it is based on the finite element method, Magpar approximates spherical systems better than programs using finite differences, such as OOMMF [11].

Magpar is based on dynamically integrating the Landau-Lifshitz-Gilbert equation of motion. Finite elements are meshed as tetrahedral elements with dimensions of max. 3 nm; the element sizes were significantly decreased along the edges to include the influence of demagnetizing fields more exactly. The exchange constant was chosen as $A = 1.3 \cdot 10^{-11}$ J/m, the magnetic polarization at saturation J = 1.76 T, and the Gilbert damping constant $\alpha = 0.1$. Both values A and J are smaller than in Fe, but larger than in Py [12].

The simulation was carried out starting at zero external field, sweeping to + 600 kA/m (with the external field in the sample plane), reverse the magnetic field to – 600 kA/m and close the hysteresis loop by sweeping to + 600 kA/m again. The field sweeping speed was 10 kA/(m ns), comparable to typical values in MRAM applications [13].

3. Results

The simulated coercive fields of a Co particle consisting of four wires with length 70 nm and diameter 10 nm are depicted in Fig. 1; the inset shows the meshed particle used for modelling. The angular orientation of 0° is defined by one of the wires being oriented parallel to the external magnetic field. For comparison, the angle-dependent coercive fields of the same particle modelled for Fe are added.

While the coercivities of Co show values quantitatively similar (about 50 % higher) to those simulated for Fe, the angle-dependence has a different form, with broader maxima and sharper minima in the Co nano-particle.

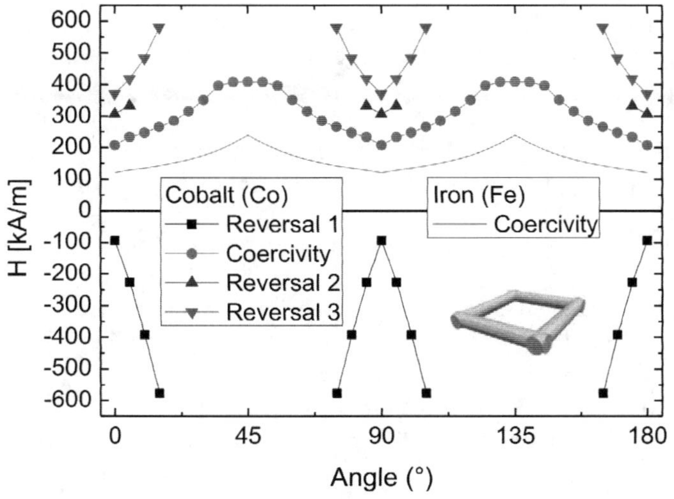

Figure 1. Coercive field and reversal fields simulated for a fourfold Co nano-structure (see inset). Coercivities of Fe are added for better comparability. From [14], modified.

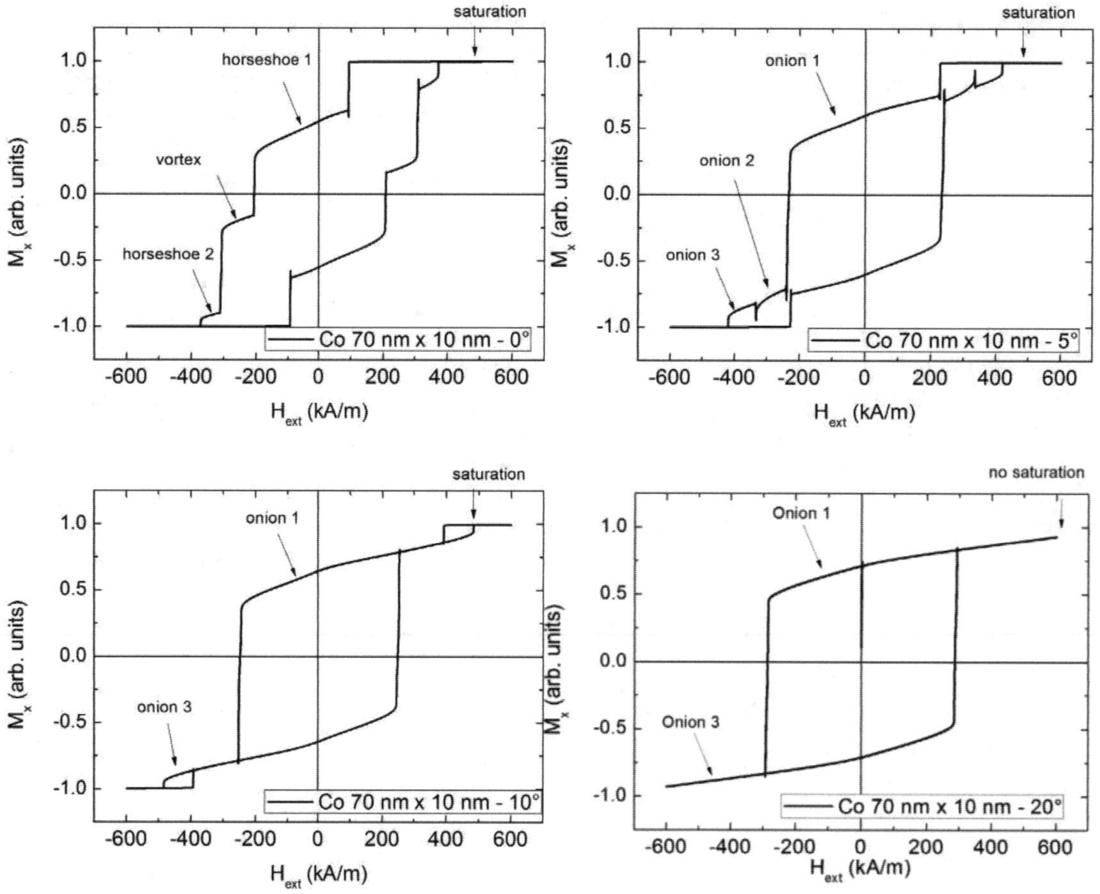

Figure 2. Simulated hysteresis curves for the Co sample depicted in Fig. 1 (Inset), exemplarily shown for orientations relative to the external magnetic field of 0° (i.e. parallel to one pair of wires), 5°, 10°, and 20°. From [14], modified.

The most intriguing feature, however, are the additional reversal fields which can be attributed to steps in the respective hysteresis loops and are found in Fe only for small angular regions [6,8]. In the fourfold Co wire particle, however, there seem to be additional reversal fields for all angles – 3 relatively small fields around 0°, 90° etc., and 2 significantly larger reversal fields around 45°, 135° etc.

To examine these reversal fields further, Fig. 2 shows hysteresis loops taken at different angles between the wire system and the external magnetic field. For 0° sample orientation, starting at positive saturation, the magnetization reverses into a horseshoe state before the external magnetic field vanishes. This behaviour has not been recognized in fourfold Fe wire samples, but is well-known from magnetic nano-dots [15,16]. Magnetization reversal is continued via a vortex state and a second horseshoe state to negative saturation.

Rotating the sample to 5° changes this magnetization reversal mechanism. Here, magnetization reversal from positive to negative saturation occurs via 3 different onion states. Rotating the sample further to 15°, the second onion state vanishes.

For 20° and larger angles around 45°, saturation is not reached. However, since there are no energy barriers between positive saturation and the first onion state, this finding does not influence the magnetization reversal processes in the simulation.

Figure 3. Magnetization reversal of the Co wire system from positive to negative saturation for sample angles of 0° (top row), 5° (2nd row), 10° (3rd row), and 45° (bottom row). The colors depict a magnetization orientation in positive (red) or negative (blue) x-direction; the external field direction is marked in each row. From [14], modified.

Fig. 3 shows snapshots of the magnetization reversal process for four different angular orientations of the magnetic wire system. The different magnetization states are defined as follows: In saturation, the magnetization is aligned along the external magnetic field in all four wires. In a vortex state, magnetization orientation follows the wires steadily clockwise or counter-clockwise (e.g. 1st row, 3rd panel). In an onion state, magnetization "splits" into two different ways to be aligned from one position of the system to an opposite position (e.g. 2nd row, 2nd panel). A horseshoe state, finally, is defined by the magnetization following a horseshoe-like form, which means that magnetization is aligned parallel in one pair of wires and antiparallel in the other pair (e.g. 1st row, 2nd panel).

For 0° (upper row), positive saturation is followed by a first horseshoe state, a vortex state and the second horseshoe state, before negative saturation is reached (now shown here).

At a sample orientation of 5° (second row), three different onion states are visible after positive saturation, finally leading to negative saturation.

For a sample orientation of 10° (third row), the second onion state is no longer visible in the hysteresis loop; however, as the snapshots of magnetization reversal show, a new non-diagonal onion state

occurs. This new state is not correlated with additional steps in the hysteresis loops, which is typical for domain wall processes.
Finally, the last row shows magnetization reversal for 45° (identical to 20° and the angles between). It is clearly visible that saturation is not completely reached for the maximum external fields used in the simulation. The first onion state is transformed into an "extreme onion state" (2nd panel), as it is already known from fourfold Fe wire samples [8]. Magnetization reversal into the opposite onion state (4th panel) happens via an asymmetric domain wall state (3rd panel), before negative saturation is approached.

4. Conclusion

Concluding, we have shown that different materials used to model identical nano-structure geometries can lead to completely different hysteresis loops and reversal modes. Especially for finding magnetic nano-systems which can be used as higher-order storage devices to enhance data density limits, testing diverse materials by simulations is necessary, before the most promising systems can be realized using lithographically produced samples, to allow for experimental examination. As depicted for the example of Co, in comparison with recently tested Fe systems, new materials may lead to novel and unexpected behavior – and can thus result in new functionalities.

References

[1] Wang R-H, Jiang J-S, Hu M 2009 *Mater. Res. Bull.* **44** 1468
[2] Huang L, Schofield M A and Zhu Y 2010 *Adv. Mater.* **22** 492
[3] Thevenard L, Zeng H T, Petit D and Cowburn D 2010 *J. Magn. Magn. Mater.* **322** 2152
[4] Zhang W and Haas S 2010 *Phys. Rev. B* **81** 064433
[5] Moritz J, Vinai G, Auffret S and Dieny B 2011 *J. Appl. Phys.* **109** 083902
[6] Blachowicz T, Ehrmann A 2011 *J. Appl. Phys.* **110** 073911
[7] Blachowicz T, Ehrmann A, Steblinski P, Palka J 2013 *J. Appl. Phys.* **113** 013901
[8] Blachowicz T and Ehrmann A 2013 *Sci World J* **2013** 472597
[9] Blachowicz T, Ehrmann A 2013 *J. Magn. Magn. Mat.* **331** 21-23
[10] Scholz W, Fidler J, Schrefl T, Suess D, Dittrich R, Forster H and Tsiantos V 2003 *Comp. Mat. Sci.* **28** 366
[11] Donahue M J, Porter D G 1999 OOMMF User's Guide, Version 1.0. Interagency Report NISTIR 6376, National Institute of Standards and Technology, Gaithersburg, MD
[12] Smith N, Markham D and la Tourette D 1989 *J. Appl. Phys.* **65** 4362
[13] Tehrani S, Engel B, Slaughter J M, Chen E, DeHerrera M, Durlam M, Naji P, Whig R, Janesky J and Calder J. 2000 *IEEE Trans. Magn.* **36** 2752
[14] Ehrmann A, *Examination and simulation of new magnetic materials for the possible application in memory cells*, Logos Verlag, Berlin / Germany 2014, ISBN 978-3-8325-3772-2
[15] He K, Smith D J and McCartney M R 2010 *J. Appl. Phys.* **107** 09D307
[16] Blachowicz T, Ehrmann A 2015 *Journal of Physics: Conference Series* accepted

Reliability of statistic evaluation of microscopic pictures taken from knitted fabrics

A Ehrmann[1], T Blachowicz[2], H Zghidi[3] and M O Weber[4]
[1] Faculty of Engineering Sciences and Mathematics, Bielefeld University of Applied Sciences, 33609 Bielefeld, Germany
[2] Institute of Physics – Center for Science and Education, Silesian University of Technology, 44-100 Gliwice, Poland
[3] Faculty of Automatic Control, Electronics and Computer Science, Silesian University of Technology, 44-100 Gliwice (Poland)
[4] Faculty of Textile and Clothing Technology, Niederrhein University of Applied Sciences, 41065 Mönchengladbach, Germany

E-mail: andrea-ehrmann@gmx.de

Abstract. One of the techniques which can be used to quantitatively evaluate images statistically is the so-called random-walk approach. The resulting Hurst exponent is a measure of the complexity of the picture. Especially long, fine elements in the image, such as fibres, influence the Hurst exponent significantly. Thus, determination of the Hurst exponent has been suggested as new method to measure the hairiness of yarns or knitted fabrics, since existing hairiness measurement instruments are based on different measurement principles which are not comparable. While the principal usability of this method for hairiness detection has been shown in former projects, the absolute value of the calculated Hurst exponents depends on the technique to take the photographic image of a sample, to transfer it into a monochrome picture, and on possible image processing steps. This article gives an overview of edge detection filters, possible definitions of the threshold value between black and white for the transformation into a monochrome image, etc. It shows how these parameters should be chosen in case of typical textile samples and correlates the challenges of this novel method with well-known problems of common techniques to measure yarn and fabric hairiness.

1. Introduction

The description of textile fabrics is often related to their structure and to yarn parameters such as material, fineness, hairiness etc. While the structure, e.g. of a woven or a knitted fabric, can be defined well, the yarn parameters can often be evaluated only by extracting the yarn from the fabric, resulting in a destruction of the original textile sample.

One of the parameters which can hardly be detected in a textile fabric by a non-destructive method is the yarn hairiness, being defined as number of hairs per yarn length which have a defined minimum length. Since yarn hairiness can be measured by different devices, based on different definitions which are not always compatible, with different results for different yarns [1-6], a novel approach to measure yarn hairiness optically was suggested [7]. While this approach, based on the statistical evaluation of microscopic pictures, was successfully tested for single yarns and knitted fabrics [8,9], several parameters of microscopy and image processing have shown to significantly influence the results.

This article aims at depicting the influence of image processing – i.e. reduction of a coloured microscopic picture to a monochrome image by different methods – on the statistical evaluation process.

2. Methods

Single jersey fabrics were knitted on a Stoll flat knitting machine CMS 302TC with machine gauge E8, using a blended staple fibre yarn with 70 % polyacrylonitrile (PAN) and 30 % new wool (WV) (Nm 30/2) and a filament yarn 100 % viscose (CV) (2 x 330 dtex).

Photographic images with a nominal magnification of 20 x were taken using a digital optical microscope VHX-600D by Keyence with an objective VH-Z20R. Fig. 1 (1st row) shows two photographs of knitted fabrics after 10 washing cycles, leading to an enhanced hairiness compared with the untreated fabric.

These pictures were transformed into 1-bit monochrome images by two different methods. Using the first method, they were imported into CorelDRAW® X5 and transformed into a monochrome line graphic with different threshold values of 70, 90, 110, and 130, with a value of 0 resulting in a nearly completely white and a value of 255 in an almost completely black picture. For the following statistical evaluation, the pictures were inverted afterwards. The results for the extreme values of the threshold, 70 and 130, are depicted in Fig. 1 (2nd and 3rd row, respectively). Using the second, automatic method, the pictures were transferred into black-and-white images by a self-prepared algorithm which denoises the image by means of median filtering, unsharp masking and a special filter based on a feature vector and a given noise sample. Next the algorithm segments the image and extracts fibres by subtracting high pass filtered image from the denoised part and then skeletonizes it to make length measurement possible. The results of this process are depicted in Fig. 1 (last row). Here, structures inside the yarns become visible, while the fine fibres outside the threads vanish.

Afterwards, a so-called random walking experiment is performed on the monochromic pictures. Starting from a random point on the textile (i.e. on the black areas), a defined number of steps $t = 1, 2, \dots 1000$ is carried out, with the same probabilities for all four possible directions (up, down, left, right). The starting and ending point of this random walk differ by distance $R(t)$. By plotting log $R(t)$ vs. log t, a linear correlation is visible. The slope of the linear fit is the so-called Hurst exponent H. This process is repeated 1000 times to get a statistical distribution of Hurst exponents. A more detailed description of the method can, e.g., be found in [10].

While a completely black area results in $H = 0.5$, all white areas – which are forbidden for the random walking process, thus restricting the possible movements – lead to a reduction of this value. Especially long, tiny structures – e.g. single fibres – result in a decreased Hurst exponent. Thus, this method can be used to give a quantitative measure of the fabric structure, with smaller values of H for finer yarn and higher hairiness.

3. Results

Fig. 2 shows Hurst distributions calculated for the monochrome pictures shown in Fig. 1. Obviously, the pictures with large black areas (Fig. 2, 1st row) result in Hurst distributions with large peaks near $H = 0.5$, as expected due to a black square having $H = 0.5$. The large disadvantage of this method is that the Hurst distributions of clearly different pictures look quite similar.

On the other hand, in pictures with significantly more white areas (Fig. 2, 2nd row) the fine structures are mostly lost – amongst them the single fibres or hairs in which we are interested.

The automatic algorithm, however, results in the peak near $H = 0.5$ being completely deleted, since the picture is broken up in single stitches without connections. This concentration on smaller features, suppressing the influence of long-range effects in the whole fabric, may support the differentiation between different fabrics, since the average value is now dominated by the short-range effects. To test this idea, the average Hurst exponents were calculated using the previously described methods of creating a monochrome image from the same coloured microscopic picture.

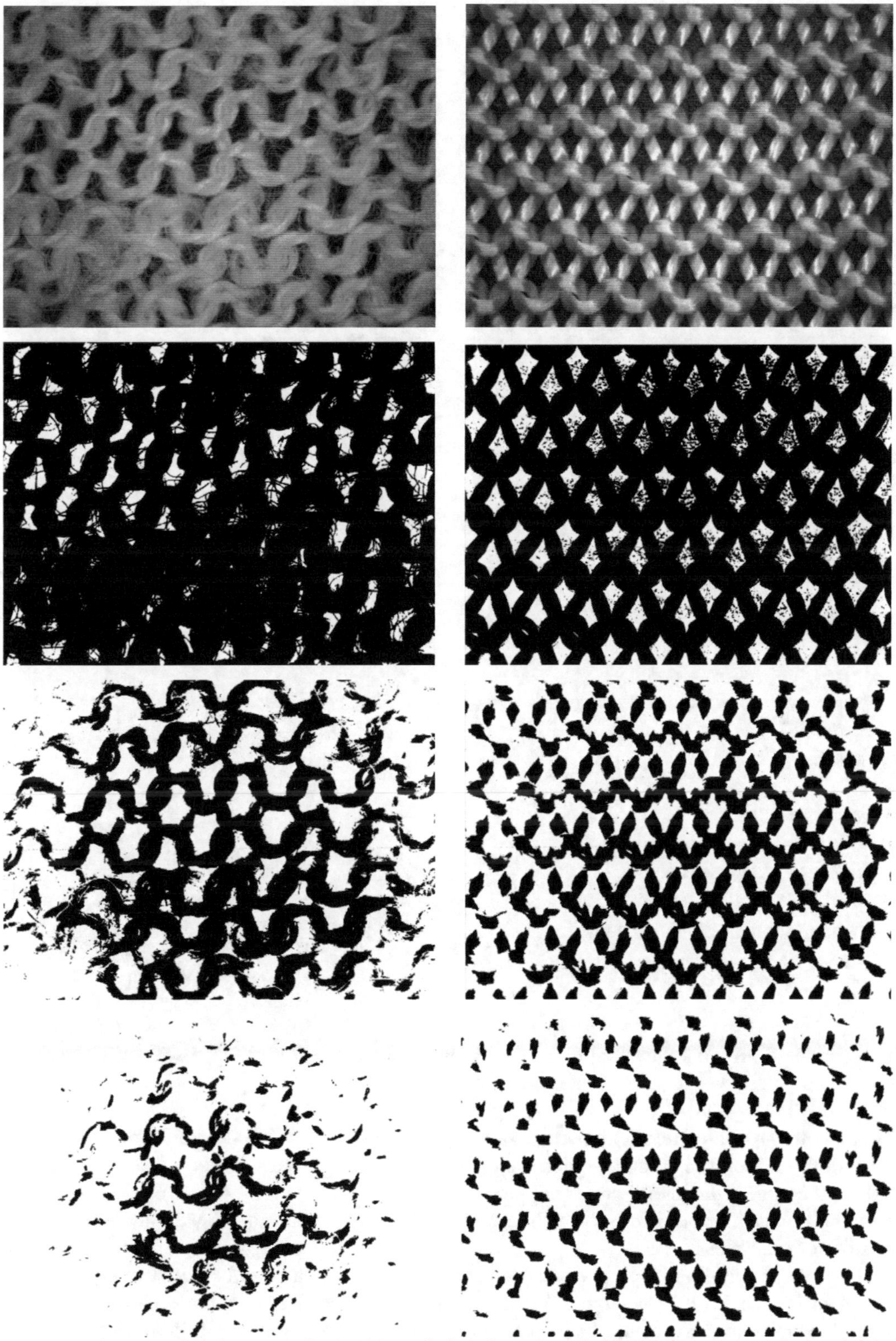

Figure 1. Microscopic pictures of PAN/WV (left column) and CV (right column) knitted fabrics (1st row), inverted monochrome images with threshold values of 70 (2nd row) and 130 (3rd row), automatic reduction to black-and-white images (last row).

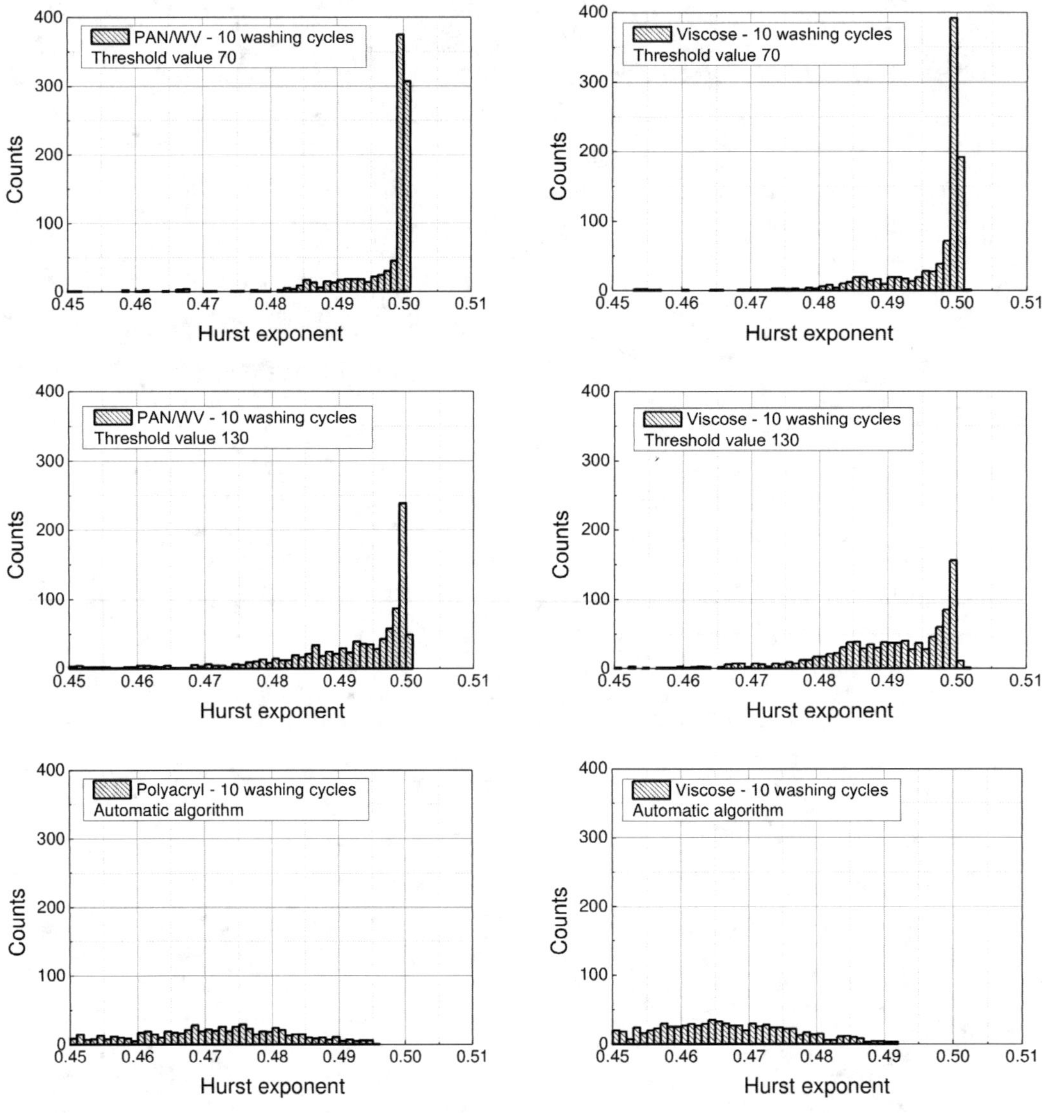

Figure 2. Hurst distributions, calculated for PAN/WV (left column) and viscose (right column) knitted fabrics from monochrome pictures with threshold values of 70 (1st row) and 130 (2nd row) as well as automatic reduction to black-and-white images (last row).

Fig. 3 depicts the average Hurst exponents, calculated for each set of 1000 single measurements. For small threshold values (i.e. mostly black pictures), the average H is larger for PAN/WV, what is changed for the largest threshold value (i.e. pictures with more white regions). This finding is based on the different colour uniformities of the materials under examination as well as on the different hairiness of both samples.

Based on the structure, the viscose sample is more open, thus a *smaller* Hurst exponent would be expected than for the PAN/WV sample. On the other hand, the viscose sample shows less hairiness, thus a *larger* Hurst exponent would be expected. Apparently, the pictures with dominating black areas – resulting in a smaller H for viscose – probe mostly the structure, while the brightest pictures probe mostly the hairiness.

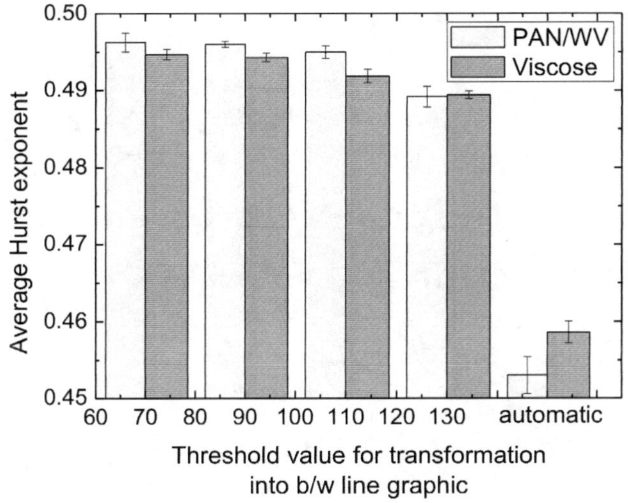

Figure 3. Average Hurst exponent, calculated for Hurst distributions (cf. Fig. 2) of monochrome pictures with different threshold values as well as automatic reduction to black-and-white images.

The automatic algorithm, however, pursues this trend even further. While both average Hurst exponents are, as expected from Fig. 2, significantly smaller than those calculated from the other pictures, the random walk on the automatically created monochrome image shows a significant difference between the PAN/WV and the CV sample, obviously probing the small-scale structures.

Future examinations will show whether different automatic algorithms can be used to probe the hairiness and the stitch structure independently, which would allow for detecting two important parameters of textile fabrics unambiguously from a single microscopic picture.

4. Conclusion

To conclude, we have shown the significant influence of different image processing methods and parameters on the calculation of Hurst exponents from pictures of knitted fabrics.

Depending on the transformation process of the original colour picture into a monochrome image, different parameters of the pictures are probed. While this effect has to be taken into account to avoid evaluation errors, it can also be utilized to gain more information from one picture by choosing different image processing parameters.

References

[1] Basu A 1999 *Indian Journal of Fibre & Textile Research* **24** 86-92
[2] Canoglu S, Yukseloglu S M 2008 *Fibres & Textiles in Eastern Europe* **16** 34-38
[3] Tyagi G K 2004 *Indian Journal of Fibre & Textile Research* **29** 35
[4] Alay S, Göktepe F 2008 *Tekstil ve Konfeksiyon* **18** 28-34
[5] Kilic M, Okur A 2014 *Indian Journal of Fibre & Textile Research* **39** 49-54
[6] Standard Guide for Measuring Hairiness of Yarns by the Photo-Electric Apparatus, *ASTM D5647-2007*
[7] Blachowicz T, Aumann S, Pruß B, Reiners P, Ehrmann A, Weber M O, Weide T 2014 *Proceedings of Aachen-Dresden International Textile Conference*, Dresden / Germany, November 27-28, 2014
[8] Blachowicz T, Ehrmann A, Zieliński M, Weber M O 2015 Quantitative estimation of textile structural complexity using random walking approach, *submitted*
[9] Blachowicz T, Ehrmann A, Domino K, Weber M O 2015 Examination of washing relaxation of knitted fabrics by random walking approach, *submitted*
[10] Schepers H E, van Beek J H G M, Bassingthwaighte J B 1992 *IEEE Engineering in Medicine and Biology* **11** 57-71

Journal of Physics: Conference Series **633** (2015) 012102 doi:10.1088/1742-6596/633/1/012102

An accurate solution of elastodynamic problems by numerical local Green's functions

F S Loureiro[1,2*], J E A Silva[2] and W J Mansur[3]

[1]Department of Thermal and Fluid Sciences, Federal University of São João del-Rei, Brazil
[2]Postgraduate Program in Computational Modeling, Federal University of Juiz de Fora, Brazil
[3]Department of Civil Engineering, COPPE/Federal University of Rio de Janeiro, Brazil

E-mail: felipe.loureiro@ufjf.edu.br

Abstract. Green's function based methodologies for elastodynamics in both time and frequency domains, which can be either numerical or analytical, appear in many branches of physics and engineering. Thus, the development of exact expressions for Green's functions is of great importance. Unfortunately, such expressions are known only for relatively few kinds of geometry, medium and boundary conditions. In this way, due to the difficulty in finding exact Green's functions, specially in the time domain, the present paper presents a solution of the transient elastodynamic equations by a time-stepping technique based on the Explicit Green's Approach method written in terms of the Green's and Step response functions, both being computed numerically by the finite element method. The major feature is the computation of these functions separately by the central difference time integration scheme and locally owing to the principle of causality. More precisely, Green's functions are computed only at $t = \Delta t$ adopting two time substeps while Step response functions are computed directly without substeps. The proposed time-stepping method shows to be quite accurate with distinct numerical properties not presented in the standard central difference scheme as addressed in the numerical example.

1. Introduction

The elastodynamic equations appear in a great deal of applications such as geophysics, soil dynamics, structural dynamics, earthquake engineering, etc [1]. The elastodynamic equations for a general problem in the time domain can be written, in index notation, as follows:

$$\rho \ddot{u}_i - \sigma_{ij,j} = b_i \text{ in } \Omega \times (0, t_f] \tag{1}$$

where u_i, $\sigma_{ij,j}$ stand for the displacement and Cauchy stress tensor, respectively and ρ is the mass density while b_i is the given body force. In addition to the above equation, Dirichlet and Neumann type boundary conditions as well as initial conditions are given by:

$$u_i = \bar{u}_i \text{ on } \Gamma_{D_i} \times (0, t_f]; \quad \sigma_{ij} n_j = \bar{t}_i \text{ on } \Gamma_{N_i} \times (0, t_f]; \quad u_i = u_{0i}; \quad \dot{u}_i = v_{0i} \text{ in } \Omega \text{ for } t = 0 \tag{2}$$

where $\bar{u}_i$, $\bar{t}_i$, u_{0i} and v_{0i} denote prescribed functions with n_j being the normal vector components.

Here Eqs. (1)-(2), within the context of elastic wave problems limited to the linear case (i.e., $\sigma_{ij} = C_{ijkl} u_{k,l}$), are solved numerically by means of the Explicit Green's approach (ExGA) method as described in what follows.

2. Time domain solution

In the ExGA method, the problem domain Ω is partitioned into non-overlapping element domains, i.e., $\bar{\Omega} = \bar{\Omega}^h = \cup_e \bar{\Omega}_e$. Then an approximation of unknown fields (i.e., Green's functions, displacements and velocities), in a similar way to that of the semi-discrete FEM formulation where the space is driscretized independently of the time domain, is employed [2, 3]. Finally, the ExGA displacement and velocity time integral expressions written in a step-by-step manner where the time domain is split into N equal time intervals of length $\Delta t = t_{k+1} - t_k = t_f/N$ read (convolution integrals over the time range $[0, \Delta t]$ have been approximated by the trapezoidal rule) [4]:

$$\mathbf{U}^{k+1} = \mathbf{H}(\Delta t)\,\mathbf{M}\mathbf{U}^k + \mathbf{G}(\Delta t)\,\bar{\mathbf{U}}^k; \quad \dot{\mathbf{U}}^{k+1} = \dot{\mathbf{H}}(\Delta t)\,\mathbf{M}\mathbf{U}^k + \dot{\mathbf{G}}(\Delta t)\,\bar{\mathbf{U}}^k + \mathbf{M}^{-1}\mathbf{F}^{k+1}\Delta t/2 \quad (3)$$

where $\bar{\mathbf{U}}^k = \mathbf{M}\dot{\mathbf{U}}^k + \mathbf{F}^k\Delta t/2$. In addition, $\mathbf{G}(\Delta t)$ and $\mathbf{H}(\Delta t)$ stand, respectively, for the Green's and the Step response matrices for the discrete system with entries representing their nodal values. Moreover, $\mathbf{M}$ and $\mathbf{F}$ denote, respectively, the standard FEM mass matrix (here assumed to be lumped) and load vector.

The continuum equations that govern the Green's functions and also the Step response functions are those of the problem under consideration (Eqs. (1)-(2)) considering homogeneous (null) boundary conditions. Their computations for a given nodal source point, say $\mathbf{y} = \mathbf{y}_p$, along an $\mathbf{e}_l$ direction by the semi-discrete FEM are carried out by the following discrete weak form:

$$\int_{\Omega_s^h} \rho v_i^h \ddot{w}_i^h + v_{i,j}^h C_{ijkl} w_{k,l}^h d\Omega = 0; \quad \int_{\Omega_s^h} \rho \left(v_i^h w_i^h - v_i^h w_{0i} \right) d\Omega = 0; \quad \int_{\Omega_s^h} \rho \left(v_i^h \dot{w}_i^h - v_i^h \tilde{w}_{0i} \right) d\Omega = 0 \quad (4)$$

where w_i^h can be either an approximation for the Green's function ($w_i^h(\mathbf{x}, t) \equiv G_{il}^h(\mathbf{x}, \mathbf{y}_p, t)$) or for the Step response function ($w_i^h(\mathbf{x}, t) \equiv H_{il}^h(\mathbf{x}, \mathbf{y}_p, t)$) components that when evaluated at the nodal points gives the entries of a column of the Green's (or Step response) matrix [3]. Physically, the Green's function can be interpreted as the response of the system due to an initial velocity applied along an $\mathbf{e}_l$ direction defined as $\tilde{w}_{0i} = \delta(\mathbf{x} - \mathbf{y}_p)\,\delta_{il}/\rho$, whereas the Step response function is the response of the system due to an initial displacement defined as $w_{0i} = \delta(\mathbf{x} - \mathbf{y}_p)\,\delta_{il}/\rho$. Due to the causality, the response is confined to a region defined as $\Omega_s = \left\{\mathbf{x}, \mathbf{y} \in \mathbb{R}^2 : \|\mathbf{x} - \mathbf{y}\| \leq c_d \Delta t\right\} \subset \Omega \subset \mathbb{R}^2$ called local subdomain that may be very small once the time step length Δt is also small which is the case here. Notice that this upper bound for the local subdomain is defined in terms of the largest wave velocity which is the dilatational (or primary) one (c_d); besides, the discrete local subdomain Ω_s^h constructed around the nodal source point $\mathbf{y} = \mathbf{y}_p$ is slightly larger than the subdomain inferred by the principle of causality because of the semi-discretization procedure [3].

The semi-discretization procedure of Eq. (4) gives rise to a system of second-order ordinary differential equation in the time domain. In this sense its solution must be performed adopting, for instance, time integration schemes. In this work, the well-known Central Difference time integration scheme [5] is employed to compute Green's and Step response functions only at time $t = \Delta t$ and in a independent way. The novel feature of the proposed formulation is as follows:

- Green's functions are computed from $t = 0$ to $t = \Delta t/2$ and then from $t = \Delta t/2$ to $t = \Delta t$ or, in other words, two time substeps are adopted until reach the desired solution at the end of the first time interval;
- Step response functions are computed directly from $t = 0$ to $t = \Delta t$ without substeps.

After computing the Green's and Step response functions locally taking into account all the nodal source points, the Green's and Step response matrices, which are very sparse, are

Journal of Physics: Conference Series **633** (2015) 012102 doi:10.1088/1742-6596/633/1/012102

readily constructed, and the time marching process can be initialized by Eq. (3). The final task is to show that such time-stepping technique is convergent yielding meaningful results. To do so, a convergence analysis following the standard literature guideline [5] is performed and the conclusions are summarized in Fig. 1 as a function of the circular frequency. From this figure we can conclude that the proposed ExGA method has some distinct features over the standard Central Difference scheme, namely: (i) the stability constraint obtained from the spectral radius is not decreased and remains the same as that of the CD scheme; (ii) less numerical dispersion; and (iii) numerical dissipation is introduced (spectral radius less than one). It is worthy mentioning that the numerical dissipation in spite of leading to a non-conservative scheme may be beneficial to the solution of some practical problems to dissipate the inaccurate higher frequencies as presented later in the example.

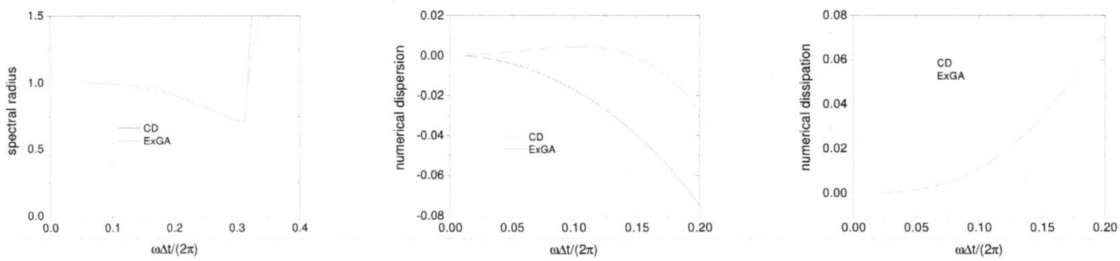

Figure 1. Convergence analysis for the ExGA and standard Central Difference (CD) methods.

3. Results and discussion

In this section, an example consisting of a homogeneous plane strain two-dimensional soil strip model is considered and the results are analyzed. A sketch of the FE model is depicted in Fig. 2(left); the time-dependent load and the physical properties are in Fig. 2(right) and Table 1, respectively. An element length $l_e = 2.5$ m and time interval $\Delta t = 5.5 \times 10^{-3}$ s are adopted.

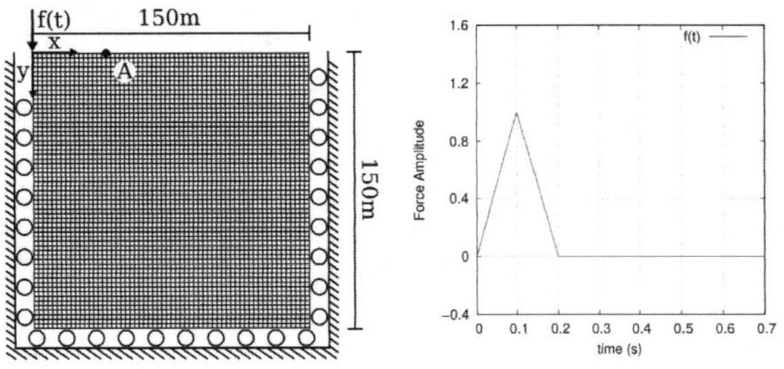

Figure 2. Soil strip FE model (left) and time variation of the load (right).

ν	$\rho(\mathrm{kg/m^3})$	$E(\mathrm{N/m^2})$
0.25	2000	2.5×10^8
c_d(m/s)	c_s(m/s)	c_r(m/s)
388	224	205.54

Table 1. Physical properties of the model.

The time-histories of the vertical displacement and velocity at point A=(40,0) are plotted, respectively, in Fig. 3(left) and Fig. 3(middle), considering both the CD and the ExGA methods. Stress σ_y time-histories at the Gauss integration point B=(41.25,1.125), close to point A, are computed, and the results are presented in Fig. 3(right). Analyzing the figures, it can be observed the presence of high spurious oscillations into the CD results, which can lead to a misinterpretation. On the other hand, due to the capability of filtering spurious oscillations by means of the numerical dissipation, the results concerning the ExGA method are free of oscillations, indicating that accurate and reliable results are achieved. This is better visualized in Fig. 4 where the Von-Mises stress snapshots of the CD and ExGA methods are compared.

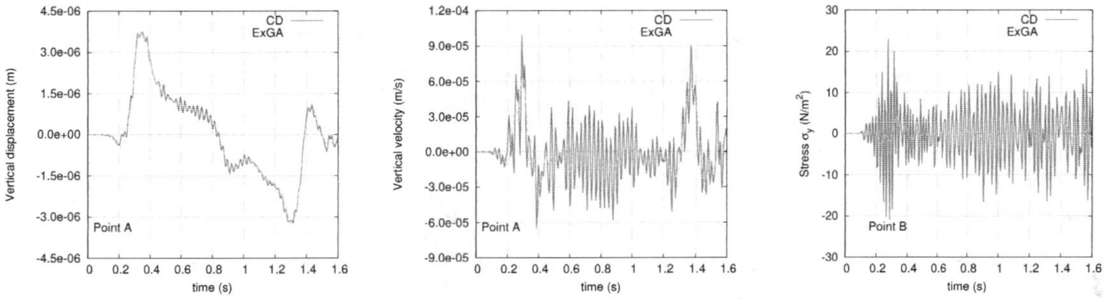

Figure 3. Comparison of the numerical results: vertical displacement (left), vertical velocity (middle); and stress σ_y (right).

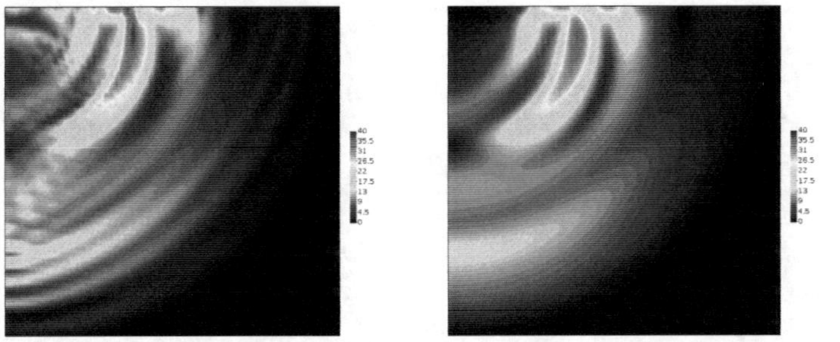

Figure 4. Snapshots of the Von-Mises stress at $t = 0.385$ s: CD(left) and ExGA(right).

4. Conclusions

An improved Explicit Green's Approach (ExGA) method based on concepts of numerical local Green's and Step response functions has been presented in this paper for solving the elastodynamic equations. The use of time substeps to compute only the Green's functions in the ExGA method provided some important numerical properties leading to a very accurate time-stepping scheme. One such property is the numerical dissipation responsible for filtering out undesirable spurious oscillations that may arise in some simulations. It is worth pointing out that the time substeps are applied only once at $t = \Delta t$ and both the Green's and Step response functions are computed locally due to the causality, resulting at a low computational cost. The numerical example analyzed here has confirmed the effectiveness and robustness of the proposed ExGA method. Further research with the goal of incorporating the so-called fundamental solution (analytical expression) by modifying the standard ExGA formulation [2] is in progress.

5. Acknowledgements

The financial support by CNPq, UFJF and FAPEMIG is greatly acknowledged.

References

[1] Morse P M and Feshbach H 1953 *Methods of Theoretical Physics* (New York: McGraw-Hill)

[2] Loureiro F S and Mansur W J 2010 *Comput Methods Appl Mech Eng* **199** 1502–1512

[3] Loureiro F S, Silva J E A and Mansur W J 2015 *Eng Anal Bound Elem* **50** 381–394

[4] Mansur W J, Loureiro F S, Soares Jr D and Dors C 2007 *J Comp Phys* **227** 851–870

[5] Hughes T J 2000 *The finite element method: Linear Static and Dynamic Finite Element Analysis* (Dover Publications)

Aggregate driver model to enable predictable behaviour

A Chowdhury[1], T Chakravarty[1], T. Banerjee[1] and P. Balamuralidhar[2]
[1]TCS Innovation Labs, BIPL, Salt Lake, Sector-V, Kolkata, India
[2]TCS Innovation Labs, Abhilash, Plot-96, EPIP, Whitefield, Bangalore

E-mail: tapas.chakravarty@tcs.com

Abstract. The categorization of driving styles, particularly in terms of aggressiveness and skill is an emerging area of interest under the broader theme of intelligent transportation. There are two possible discriminatory techniques that can be applied for such categorization; a micro-scale (event based) model and a macro-scale (aggregate) model. It is believed that an aggregate model will reveal many interesting aspects of human-machine interaction; for example, we may be able to understand the propensities of individuals to carry out a given task over longer periods of time. A useful driver model may include the adaptive capability of the human driver, aggregated as the individual propensity to control speed/ acceleration. Towards that objective, we carried out experiments by deploying smartphone based application to be used for data collection by a group of drivers. Data is primarily being collected from GPS measurements including position & speed on a second-by-second basis, for a number of trips over a two months period. Analysing the data set, aggregate models for individual drivers were created and their natural aggressiveness were deduced. In this paper, we present the initial results for 12 drivers. It is shown that the higher order moments of the acceleration profile is an important parameter and identifier of journey quality. It is also observed that the Kurtosis of the acceleration profiles stores major information about the driving styles. Such an observation leads to two different ranking systems based on acceleration data. Such driving behaviour models can be integrated with vehicle and road model and used to generate behavioural model for real traffic scenario.

1. Introduction

Modern smartphones have sensors like accelerometer, GPS, gyroscope, magnetometer etc. Numbers of applications are being built that uses smartphone as data collection sensors. Some of these sensors can simply be used to obtain the speed/ acceleration of a moving vehicle. Thus, smartphone based driving style recognition and classification is an important field of ongoing study [1]. Authors [1] have identified that the said work needs to be "extended to differentiate how individual drivers vary in their style of driving from day-to-day". Smartphone based sensing and modelling aggressive driving is also studied through machine learning technique [2]. Some important factors associated with driving pattern are identified in Näätanen & Summala [3]; these factors are important indicators of propensity of accident. Further, driver classification based on extracted features is outlined in [4]. This ve-dyna model allows driver classification and can be used to simulate "behaviour for common driving tasks". It is believed that a useful driver model may include the adaptive capability of the human driver, aggregated as the individual propensity to control speed/ acceleration. It can also be hypothesized that

[1] To whom any correspondence should be addressed.

an aggregate model describing natural driving styles may reveal many more interesting aspects of human-machine interaction. Our investigations for similar purpose also use smartphone as a sensor. We take a statistical approach towards identification of aggression/ skill. The main objective of our study is the parametric identification of driving behaviour from the GPS data where different parameters identifying different driving styles are extracted. This is done in two ways: One is to identify significant parameters which influence signatures relevant to different journeys of a given driver and secondly to identify significant parameters that indicate different driving styles when a number of drivers are compared. The experimentation method and preliminary analysis based on statistical approach is presented in following sections.

2. Experiment and observation

A mobile application created by Tata Consultancy Services Ltd (TCS) was used to collect appropriate data from a number of registered TCS associates. The application residing in each user's smartphone is used to capture GPS data (location & speed) at 1 second interval, when the person is driving. The user assortment consists of both Android & iOS platforms. The data collection pilot was carried out for two months where the registered drivers went about their usual routine. For the preliminary analysis, we utilized 39 observations (trips) for each of the 12- randomly selected drivers. It is understood that the data represents their individual driving behaviour when faced with real life scenario and not a simulated set-up.

2.1 Data pre-processing

For purpose of analysis, we have selected trips of typical duration of approximately 80-90 minutes. The forward/longitudinal acceleration is calculated from the respective GPS speed measurements by numerical differentiation. While analyzing driver behavior, the tail portion of acceleration pdf (representing one completed trip) is considered important since these represent high acceleration/deceleration zones. An acceleration profile with heavy tail i.e. high value of kurtosis is considered as a signature of relatively high risk driving. Hence, kurtosis is an important parameter for these distributions. The kurtosis of forward acceleration is computed for each journey. After this 1st level – trip level of aggregation, we collate kurtosis values for all the journeys completed by the said driver. The analysis is focused on relationship of kurtosis with each driver. Towards that objective, we make an ANOVA (Analysis of Variance) based Design of Experiment to check whether the kurtosis values have dependence on individual driving styles or not. Essentially, we follow a completely randomized single-factor experiment where each driver constitutes *treatment* and the observed kurtosis value for each completed trip is considered a *replicate*.

2.3 Test for kurtosis

Table 1 shows the results of ANOVA test with null hypothesis H_0: Driver is not relevant for kurtosis value of acceleration". ANOVA shows that drivers impact kurtosis of forward acceleration.

Table 1: ANOVA results for kurtosis of 12 drivers

Cause of the variation	df (Degree of freedom)	SS (Sum of Squares)	MS (Mean Square)	F_0
driver	11	60949	5541	2.425
Residuals	456	1034850	2284	

Table 1 shows the result of ANOVA test for 12 drivers' kurtosis data with null hypothesis H_0. We know that the critical value of F-distribution at 1% probability level and degrees of freedom of 11 & 456 respectively, $F_{0.01,11,456} = 2.2867$. Since $F_0 > F_{0.01,11,456}$, we reject H_0 and conclude that the individual driving style significantly influences kurtosis value of the acceleration profile. The analysis is yet to be performed for a larger sized set of drivers. From the above, one can state that the kurtosis of forward acceleration profile is an important indicator of driving behaviour and can potentially serve as aggression/ skill indicator. Thus, 'risk-propensity ranking' based on these values (collected for a group of drivers in similar spatiotemporal environment) will be an important statistical classification. Figure 1 gives the box plot for these 12 drivers' kurtosis data. Variation of kurtosis values across the drivers is clearly visible. Now to evaluate the performance of drivers, we compute different performance metrics. Based on those metrics we decide the rankings of these drivers. As described earlier, each driver has distinct nature of observed kurtosis values of forward acceleration. Every driver has his/her own propensity defined by his/her of kurtosis data.

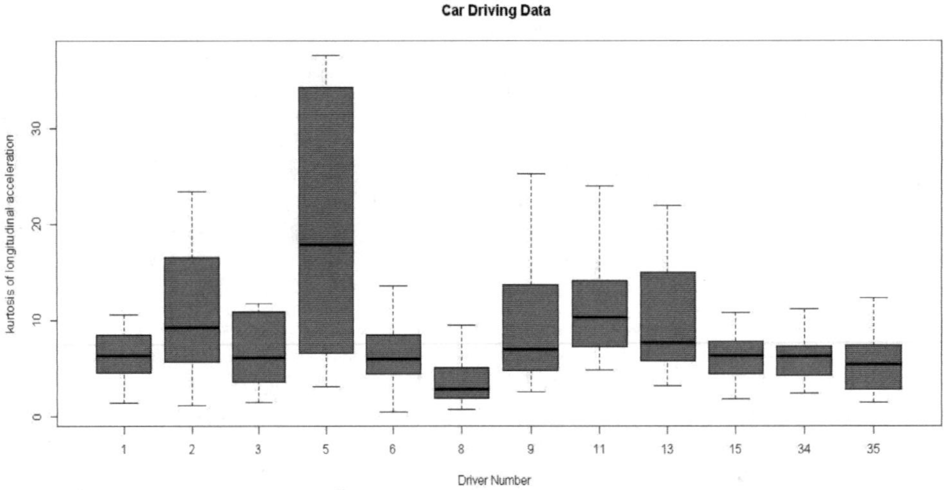

Figure 1: Box plot of kurtosis of longitudinal acceleration for different drivers (with outlier)

3. Results

We compute kurtosis for each completed trip performed by a driver. This is a trip level identifier of the driver behaviour, in terms of aggression and skill; this is a trip level aggregation. Further, for each driver, all the kurtosis values for all the trips completed in 2 month time-window are stored. For our nonparametric descriptive statistical approach, we choose median of all the kurtosis data corresponding to a driver to be representative of that driver's normal driving behaviour. Higher value of kurtosis means heavy tail. This indicates incidents of harsh acceleration or hard brake which, in turn, is usually considered as a signature of aggression. Hence, having higher median will mean more aggressive driving behaviour. Simultaneously, the kurtosis profile for each driver is unique in terms of the kurtosis data range seen for the set of completed journeys. Since each driver is facing a very similar contextual environment (on a daily basis) within the test cycle, one can logically infer that a widely varying acceleration profile signifies lack of driving skill. To quantify such, we implement IQR (Interquartile range) as a robust sample estimator of skill. Thus, low median of kurtosis samples correspond to caution and low IQR for the same corresponds to unskilled driving. Next, we rank the drivers on these two bases and attempt to classify each driver on aggression-skill plane. It is understood that the two said measures, for an individual, do not form orthogonal bases. The final summary of results containing rank is given in table 2.

Table 2: Ranking (Aggression Rank and Skill Rank) based on kurtosis data for 12 drivers

Driver ID	D001	D002	D003	D005	D006	D008	D009	D011	D013	D015	D034	D035
Aggression Rank (median)	6	10	4	12	3	1	8	11	9	7	5	2

Journal of Physics: Conference Series **633** (2015) 012103 doi:10.1088/1742-6596/633/1/012103

Skill Rank (IQR)	4	11	7	12	5	2	9	8	10	3	1	6
Spearman's rank correlation 0.727												

Table 2 shows that for this set of 12 drivers', there exists a positive strong correlation between the two respective rankings. We further use the two ranks (normalized) for classifying the drivers in the aggression-skill plane. This is shown in figure 2. The 12 drivers are placed in one of the four quadrants.

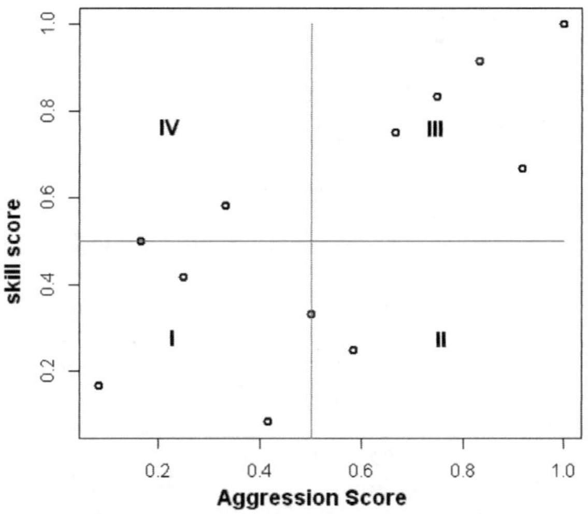

Figure 2: Driver classification: location of the 12 drivers in skill-aggression plane (I: Cautious; II: skilled & aggressive; III: Unskilled & aggressive; IV. Anxious)

Due to the fact that the aforesaid classification is not dependent on specific parametric threshold value, this classification is considered relative and pertains to this group only.

5. Concluding remarks

In this paper, the methods of analysing the statistical parameters are used to obtain aggregated driver ranking, amongst peer group of drivers. Such rankings signify driver's natural aggression and skill at an aggregated level. Since kurtosis is estimated from entire acceleration profile, handfuls of anomalous events have less influence and the overall driving pattern plays a greater role. These rankings can be used for performance comparison; for cases of demographic & contextual similarity. Also in terms of parametric representation, such analysis will eventually lead to general driver behaviour model based on statistical estimation.

Reference

[1] Van Ly M., Martin S., Trivedi M. M. "Driver classification and driving style recognition using inertial sensors". In Intelligent Vehicles Symposium (IV), June 2013 IEEE, pp. 1040-1045. IEEE.
[2] Hong, J. H., Margines, B., & Dey, A. K. (2014, April). "A smartphone-based sensing platform to model aggressive driving behaviors". In Proceedings of the 32nd annual ACM conference on Human factors in computing systems (pp. 4047-4056). ACM.
[3] Näätänen R. and Summala H. "A model for the role of motivational factors in drivers' decision-making." Accident Analysis & Prevention 6.3 (1974): pp. 243-261.
[4] Irmscher M., and Ehmann M. "Driver classification using ve-dyna advanced driver". No. 2004-01-0451. SAE Technical Paper, 2004.

4th International Conference on Mathematical Modeling in Physical Sciences (IC-MSquare2015) IOP Publishing
Journal of Physics: Conference Series **633** (2015) 012104 doi:10.1088/1742-6596/633/1/012104

Modifier mass transfer kinetic effect in the performance of solvent gradient simulated moving bed (SG-SMB) process

L. D. T. Câmara[1]

Polytechnic Institute of State University of Rio de Janeiro (IPRJ-UERJ), Dep. of Mechanical Engineering and Energy, DEMEC, Nova Friburgo-RJ, Brazil

Abstract. The solvent-gradient simulated moving bed process (SG-SMB) is the new tendency in the performance improvement if compared to the traditional isocratic solvent conditions. In such SG-SMB separation process the modulation of the solvent strength leads to significant increase in the purities and productivity followed by reduction in the solvent consumption. A stepwise modelling approach was utilized in the representation of the interconnected chromatographic columns of the system combined with lumped mass transfer models between the solid and liquid phase. The influence of the solvent modifier was considered applying the Abel model which takes into account the effect of modifier volume fraction over the partition coefficient. The modelling and simulations were carried out and compared to the experimental SG-SMB separation of the amino acids phenylalanine and tryptophan. A lumped mass transfer kinetic model was applied for both the modifier (ethanol) as well as the solutes. The simulation results showed that such simple and global mass transfer models are enough to represent all the mass transfer effect between the solid adsorbent and the liquid phase. The separation performance can be improved reducing the interaction or the mass transfer kinetic effect between the solid adsorbent phase and the modifier. The simulations showed great agreement fitting the experimental data of the amino acids concentrations both at the extract as well as at the raffinate.

1. Introduction

The simulated moving bed (SMB) process has been implemented successfully in the pharmaceutical industry to perform difficult separation of molecules. The simulated moving bed (SMB) is a continuous process of interconnected high-performance liquid chromatographic (HPLC) columns which can be used to separate enantiomeric mixtures with high purity and low solvent consumption. The process consists in the simulated countercurrent movement of solid adsorbent phase by switching the positions of the two inlet (Feed and Desorbent) and the two outlet (Extract and Raffinate) streams in the clockwise direction. The more and less adsorbable molecules of enantiomers are collected in the extract and raffinate stream, respectively.

The solvent-gradient simulated moving bed process (SG-SMB) is the new tendency in the performance improvement if compared to the traditional isocratic solvent conditions [1-2]. In such SG-SMB process the modulation of the solvent strength leads to significant increase in the purities and productivity followed by reduction in the solvent. The solvent strength in the SG-SMB process is altered through the volume fraction of the modifier (ϕ) as can be seen in Fig. 1. Nam *et al.* [1] carried out SG-SMB experiments in the performance evaluation of phenylalanine and tryptophan separation

1 diogenescamara@gmail.com, dcamara@iprj.uerj.br

which was compared to isocratic solvent conditions. Its operating conditions were determined from the SG-SMB optimization tool based on genetic algorithm.

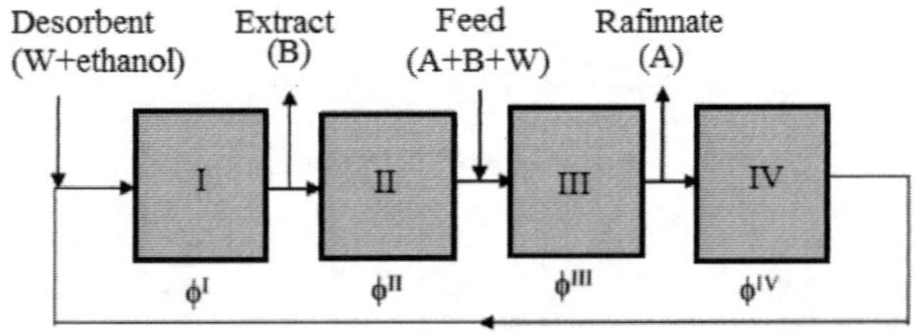

Figure 1 – Schematic representation of SG-SMB

The utilization of modifier (ethanol) to modulate the strength of the solvent in SG-SMB process brings the question about the influence of modifier adsorption on the adsorbent phase over the separation performance. So this work focus on the study of modifier adsorption over the separation performance of SG-SMB applying a linear mass transfer kinetic model for the modifier combined with stepwise modelling approach for the chromatographic column representation.

1. Modelling and optimization approaches

The chromatographic columns of the SG-SMB process are modeled as a discrete representation of N mixed cells in series combined with lumped mass transfer solid-liquid model,

$$\frac{dC_i^p}{dt} + \frac{dq_i^p}{dt} = (C_{i0}^p - C_i^p)\sigma_i^p \tag{1}$$

in which C_i^p, C_{i0}^p, q_i^p and t represents the concentration of compound i in the liquid phase, the concentration of compound i in the liquid phase at the entrance of the mixed cell, the concentration of compound i in the solid adsorbent phase and the time, respectively, all at column p. In Eq. 1 $\sigma_i^p = F/\epsilon V$ with F, ϵ and V representing the volumetric flow rate, the total porosity and the volume of the mixed cell, respectively. It was utilized the Abel's model for the determination of the partition coefficient (K)

$$K = \frac{p_1}{(1+p_2\phi)^{p_3}} \tag{2}$$

where p_1, p_2 and p_3 are the model parameters of either tryptophan or phenylalanine.

The following linear mass transfer kinetic model is applied both for the solutes as well as for the modifier (ethanol)

$$\frac{dq_i^p}{dt} = \tau_i\, C_i^p - \sigma_i q_i^p \tag{3}$$

in which τ_i and σ_i correspond to the mass transfer kinetic parameter in the adsorption and desorption, respectively. The residence times of the two kinds of molecules (phenylalanine and tryptophan) were utilized in the columns characterization for the determination of the mass transfer parameters of the lumped mass transfer model (Eq. 3). For each aminoacid in each condition of the volume fraction of modifier (ϕ=0, 0.05 and 0.2) it was determined the mass transfer parameters of adsorption and desorption that led to the same experimental value of residence time [3]. For the modifier the mass transfer parameters were get through the Lagmuir isotherm equation obtained in the literature [4]

3. Results and discussions

The Fig. 2 shows the effect of mass transfer adsorption parameter of modifier (ethanol), τ_i, over the separation performance of the aminoacids phenylalanine and tryptophan. The results of Fig. 2 are in terms of concentration along the chromatographic columns. It can be noted from Fig. 2A to 2C that the mass transfer adsorption parameter of modifier, τ_i, was increased keeping the equilibrium constant (K_{eq}) leading to an increase in the adsorption of modifier on the solid adsorbent phase. Such procedure altered the concentration of modifier in the liquid phase which becomes more distributed along the columns as it can be seen in Fig. 2C (vide dotted line). The distribution of modifier along the columns altered the concentration profiles of the aminoacids along the columns which become wider reducing the purity in the extract as well as raffinate (see Table 1).

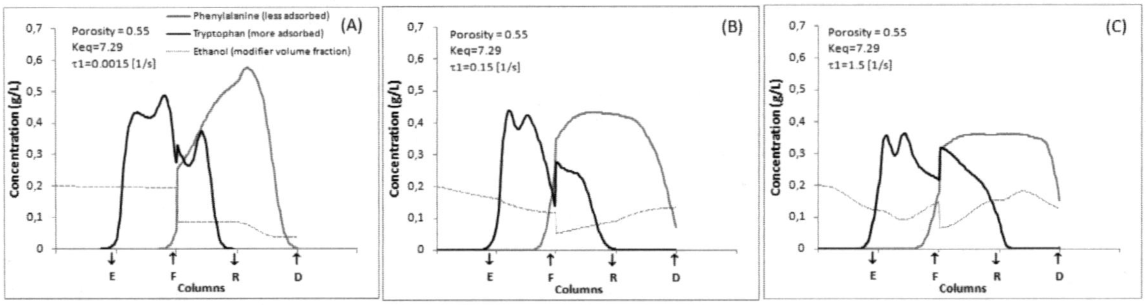

Figure 2 – Amioacids concentration along time according to different values of mass transfer kinetic of modifier; $\phi_0 = 0.2$

The Table 1 presents the values of purity and productivity according to the mass transfer adsorption parameter values of modifier. As can be seen in Table 1 the reduction in the extract purity is more significant than in the raffinate both contribution to the reduction in the productivity of the SG-SMB process.

Table 1 – Separation performance of SG-SMB according to τ_i

Figure	(Fig. 2A)	(Fig. 2B)	(Fig. 2C)
Adsorption Parameter, τ_i [1/s]	0.0015	0.15	1.5
Raffinate purity (%)	99.99	99.98	97.69
Extract purity (%)	99.36	95.74	96.20
Productivity [g/h.L]	0.581	0.548	0.508

The Fig. 3 presents the concentration profiles of the phenylalanie and tryptophan along time in the raffinate and extract, respectively. The dotted lines correspond to the cyclic simulations obtained as the streams change the positions along time in the process. The full continuous lines are the average simulations in which the black results correspond to the conditions obtained assuming mass transfer of modifier (ethanol) between the liquid and solid adosorbent phase. The points corresponds to the experimental data obtained by Nam et al. [1]. As can be seen in Fig. 3 there is a better agreement between the simulation results and the experiments for the cases considering the modifier mass transfer effect. The "roll-up" phenomenon (described by Nam et al.[1]) or the peak in the phenylalanine concentration in Fig. 3A was observed in the simulation results which is more evident in the cyclic simulations (dotted lines) of Fig. 3A.

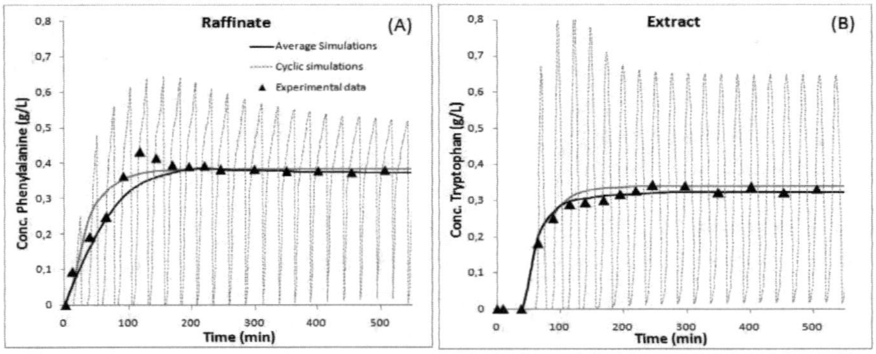

Figure 3 – Correlation between the simulation (lines) and the experimental data (points)

The Fig. 4 presents the concentration of phenylalanine along the columns showing in detail the concentration in the extract along time. The "roll-up" phenomenon can be also observed in the extract as described in detail in Fig. 4A and 4B. In the transient period of the SG-SMB process there is an increase in the concentration of modifier along time (dotted lines) which reduces the adsorption affinity of phenylalanine releasing such aminoacid from the adsorbent phase to the liquid phase. Therefore there is an increase in the phenylalanine concentration (black line) followed by a decrease (thick gray line) as the molecules migrate forward through the liquid phase. The arrows in Fig. 4 indicate the position the extract stream is collected in the process. In Fig. 4A it was utilized an hypotetical value of mass transfer kinetic only to show clearly the "roll-up" phenomenon. The experimental data of Nam et al. [1] shows the "roll-up" in the extract in detail in Fig. 4B. The values of modifier mass transfer kinetic in Fig. 4B are the same utilized in Fig. 3. It should be noted that the "roll-up" phenomenon in the extract is very small due to the low concentration of phenylalanine in such stream.

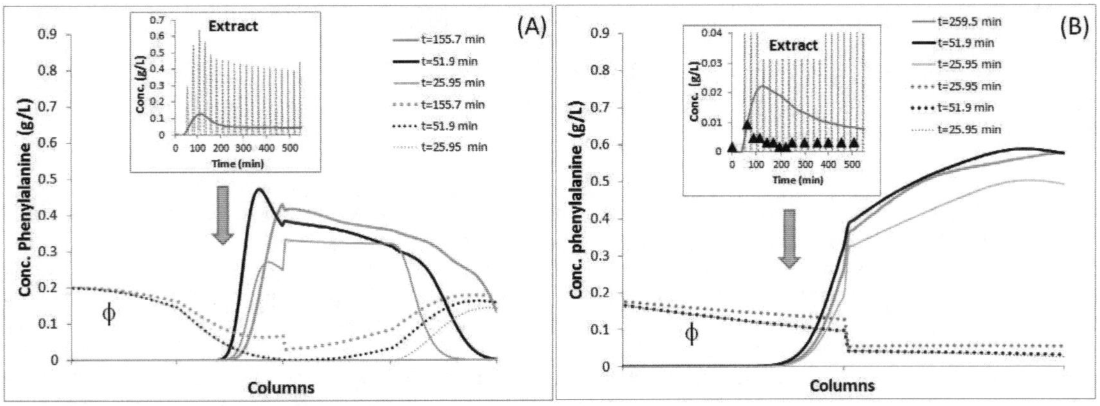

Figure 4 – The "roll-up" phenomenon in the extract

3. Conclusions

The proposed modeling approach based on modifier mass transfer kinetic was able to represent the main phenomenon aspects of the separation of the aminoacids phenylalanine and tryptophan in SG-SMB processes. The modifier mass transfer effect is related to the experimental observation of "roll-up" phenomenon of phenylalanine both in the extract as well as in the raffinate. The separation performance of the SG-SMB process can be improved reducing the interaction or the mass transfer kinetic effect between the solid adsorbent phase and the modifier.

Acknowledgements

The author thanks to UERJ, Faperj, CNPq and Capes for the financial support.

References

[1] Nam, H-G., Jo, S-H., Park, C., Mun, S., *Process Biochemistry*, 47 (2012) 401-409.
[2] Antos, D. Seidel-Morgenstern, A., *Chemical Engineering Process*, 56 (2001) 6667-6682.
[3] Nam, H-G., Kim, T-H., Mun, S. , *J. Chem. Eng. Data*, 55 (2010) 3327-3333.
[4] Mun, S. *J. Liq. Chromatogr. Relat. Technol.*, 34 (2011) 1518.

Short-range correlations in modified planar rotator model

M Žukovič[1] and D T Hristopulos[2]

[1] Institute of Physics, Faculty of Science, P.J. Šafárik University, Park Angelinum 9, 041 54 Košice, Slovakia

[2] Geostatistics Laboratory, School of Mineral Resources Engineering, Technical University of Crete, Chania 73100, Greece

E-mail: milan.zukovic@upjs.sk

Abstract. We introduce a model inspired from statistical physics that is shown to display flexible short-range spatial correlations which are potentially useful in geostatistical modeling. In particular, we consider a suitably modified planar rotator or XY model, traditionally used for modeling continuous spin systems in magnetism, and we demonstrate that it can capture spatial correlations typically present in geostatistical data. The empirical study of the spin configurations produced by Monte Carlo simulations at various temperatures and stages in the nonequilibrium regime shows that their spatial variability can be modeled by the flexible class of Matérn covariance functions. The correlation range and the smoothness of these functions vary significantly in the parameter space that consists of the temperature and the simulation time. We briefly discuss the potential of the model for efficient and automatic prediction of spatial data with short-range correlations, such as commonly encountered in geophysical and environmental applications.

1. Introduction

Modern remote sensing techniques allow recording enormous amounts of spatial data, including geographical, natural resources, land use, and environmental remote sensing images, which need efficient (preferably real-time) processing [1, 2]. Such processing involves the reconstruction of missing data that often occur due to different reasons, such as equipment malfunctions and gaps in the coverage of the targeted area that appear as a result of restricted satellite paths or bad weather conditions [3, 4, 5, 6, 7]. However, such massive data sets can not be efficiently handled by standard geostatistical methods, such as kriging [8]. The main drawbacks of such methods are high computational complexity, difficulties to automatize the algorithms to work without subjective user inputs (selection of variogram and search radius for interpolation) and often the necessity of some data pre-processing (e.g., lognormal transformation) if the data does not comply with the Gaussianity requirement [9]. Thus, there is a need to develop new spatial prediction techniques that overcome these shortcomings [10, 11]. Recently, we have proposed efficient spatial classification methods based on models inspired from statistical physics, in particular discrete spin Ising, Potts, and clock models, employing a heuristic "energy matching" principle [12, 13].

In the present study we extend the idea of using spin models for spatial prediction purposes and introduce a new method that is based on a continuous spin planar rotator model. Even though the above mentioned nonparametric discrete-spin-based classification models have been shown to be rather competitive, their performance was based on the assumption of the existence of relevant correlations at some unknown parameters and the prediction results were discrete values even for continuous data. As we empirically demonstrate, the modified planar rotator model *inherently* displays flexible short-range spatial correlations that vary significantly over the model's parameter space and could be used for spatial interpolation of continuous geostatistical data.

2. Model and simulations

2.1. Standard planar rotator model

The Hamiltonian of the standard two-dimensional planar rotator model with nearest-neighbor interactions on a square lattice is defined as

$$H = -J\sum_{\langle i,j\rangle} \boldsymbol{s}_i\boldsymbol{s}_j = -J\sum_{\langle i,j\rangle}\cos(\phi_i - \phi_j), \tag{1}$$

where $\boldsymbol{s}_i = (\cos\phi_i, \sin\phi_i)$ is a continuous spin on i-th lattice site, represented by a two-dimensional unit vector, $\phi_i \in [0, 2\pi]$ is an angle associated with the spin $\boldsymbol{s}_i$, J is an exchange interaction parameter and $\langle i,j\rangle$ denotes the sum over nearest neighbors. The Mermin-Wagner theorem [14] prevents any long-range ordering at finite temperatures in such a model, which has, however, been intensively studied in connection with quasi long-range ordering (the so called Kosterlitz-Thouless phase) that appears at low temperatures [15, 16]. The phase transition is produced by the unbinding of vortex-antivortex pairs at the Kosterlitz-Thouless critical temperature T_{KT}, below which all spins are almost aligned even though true long-range order is destroyed by spin fluctuations. The quasi long-range ordering for $T < T_{\mathrm{KT}}$ is characterized by the power-law decaying correlation function

$$C(h;\eta) = \langle \boldsymbol{s}(h)\boldsymbol{s}(0)\rangle = \left(\frac{h}{L}\right)^{-\eta(T)}, \tag{2}$$

where h is the spin separation distance, L is the linear lattice size, and $\eta(T) = T/2\pi J$ is the temperature-dependent exponent.

2.2. Modified planar rotator model

Even though the algebraic correlations with the tunable exponent $\eta(T)$ could be relevant for modeling of some spatial processes, there are some deficiencies that prevent the straightforward use of the standard planar rotator model for such purposes. First of all, one needs to define a proper mapping between the spin values and the geostatistical values. Geostatistical data are often positively correlated reflecting a degree of spatial continuity, which means that neighbors with similar values are more likely (i.e., have lower energy) than those with different values. Apparently, this condition is not fulfilled in the standard planar rotator model, described by the Hamiltonian (1); the latter, for example, assigns equal energies to a pair of neighbors whose angles differ by θ as to a pair whose angles differ by $2\pi + \theta$. This degeneracy could be fixed by defining the energy functional so that it monotonically increases with the turn angle between the neighboring spins on the entire interval $[0, 2\pi]$. The latter can be achieved by modifying the Hamiltonian (1) to take the following form of the modified planar rotator (MPR) model

$$H' = -J\sum_{\langle i,j\rangle}\cos[q(\phi_i - \phi_j)], \tag{3}$$

where $0 < q \leq 1/2$ is a modification factor. In the following we will use a fixed value of $q = 1/2$. The difference between the nature of spatial correlations in the standard and modified models

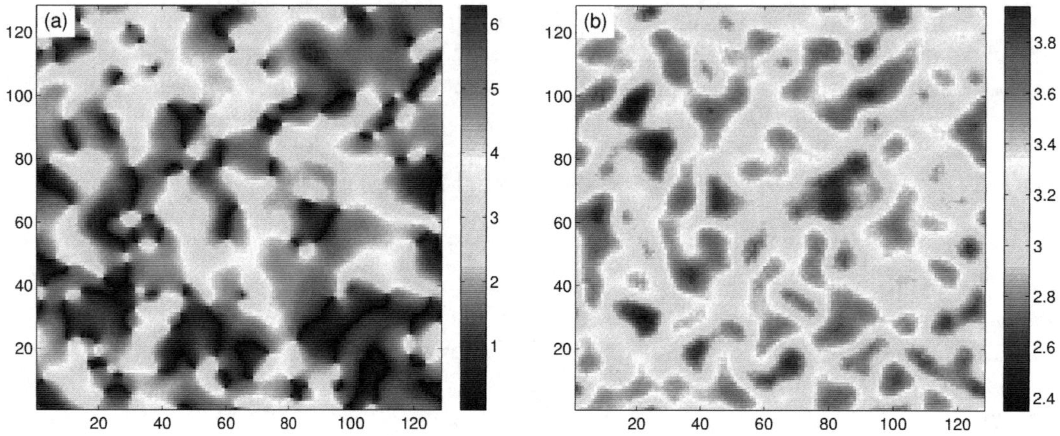

Figure 1. Snapshots of spin configurations in the low-temperature regime ($T = 10^{-4}$) for (a) standard and (b) modified planar rotator model, with periodic boundary conditions.

is apparent from snapshots of spin configurations (turn angles) in the low-temperature region, shown in Fig. 1. While the presence of the vortex-antivortex pairs shows up in Fig. 1(a) in the form of sharp boundaries between the domains of similarly oriented spins, the variation of the spin values in the MPR model, shown in Fig. 1(b) is rather smooth, as it is typical for geostatistical data.

2.3. Monte Carlo simulations

We use Monte Carlo (MC) simulations with the Metropolis update rule and vectorized checkerboard algorithm. The results are presented for a square lattice with the size $L \times L$. We chose $L = 128$ as a compromise value between the computational resources needed for MC simulations and calculations of the empirical variograms (see below) on one side and the effort to secure ergodic conditions on the other side. In typical simulations of magnetic systems one would like to suppress boundary effects by employing periodic boundary conditions (see the snapshots in Fig. 1). However, these are not appropriate for simulation of geostatistical data, since normally there is no reason to assume the data on the opposite boundaries are correlated. Free boundary conditions are not appropriate either since the spatial process is not necessarily confined within the domain boundaries. Instead, we consider it reasonable to apply boundary conditions that assume the smooth continuation of the spatial process beyond the borders. This assumption is implemented by inserting additional nodes that decorate the lattice and requiring that these additional points have the same values as their nearest neighbors inside the lattice. Therefore, if $\boldsymbol{s}_{i,j}$ is a spin in the i-th row and j-th column of the lattice, $i, j = 1, \ldots L$, then $\boldsymbol{s}_{i,L+1} = \boldsymbol{s}_{i,L}$, $\boldsymbol{s}_{L+1,j} = \boldsymbol{s}_{L,j}$, $\boldsymbol{s}_{i,0} = \boldsymbol{s}_{i,1}$ and $\boldsymbol{s}_{0,j} = \boldsymbol{s}_{1,j}$, where the rows and columns with indices 0 and $L + 1$ respectively are formed by auxiliary nodes around the lattice added to deal with the boundary effects.

2.4. Modeling spatial variability

The MC simulations produce spatially correlated spin realizations $\boldsymbol{s}_i$, which can be represented by their turn angles ϕ_i. It turns out that there is great flexibility in the spatial correlations as we move in the temperature and simulation time parameter space. Therefore, in order to model the spatial variability (the local variogram) of the simulated spin values we use a very flexible Matérn covariance model. In addition to the trivial parameters of the standard models (Gaussian, exponential, spherical, etc.), controlling the variance, σ^2, and the characteristic covariance length, ξ, the Matérn model involves one more parameter, ν, controlling the smoothness of the spatial process. Therefore, it can be considered appropriate for those geostatistical applications in which controlling the smoothness is important. Furthermore, the Matérn model includes the Gaussian and the exponential models as special cases for $\nu \to \infty$ and $\nu = 0.5$, respectively, as well as several other bounded models [17]. Due to its flexibility, it has been used in various areas including pedology [17, 18, 19], hydrology [20, 21], topography [22], health modeling [23], meteorology [24] and environmental modeling [25]. The Matérn correlation function has the general form

$$C(h;\boldsymbol{\theta}') = \frac{2^{1-\nu}}{\Gamma(\nu)}\Big(\frac{h}{\xi}\Big)^{\nu} K_{\nu}\Big(\frac{h}{\xi}\Big), \tag{4}$$

where $\boldsymbol{\theta}' = (\xi, \nu)$ and K_ν is the modified Bessel function of order ν. Then the corresponding variogram function $\gamma(h;\boldsymbol{\theta})$, which is typically used in geostatistics, is related to the correlation function as

$$\gamma(h;\boldsymbol{\theta}) = \sigma_n^2 + \sigma^2 \left[1 - C(h;\boldsymbol{\theta}')\right], \tag{5}$$

where σ^2 is the random field variance, σ_n^2 is the nugget variance that corresponds to uncorrelated fluctuations and $\boldsymbol{\theta} = (\sigma_n, \sigma, \xi, \nu)$ is a vector of the complete model parameters.

Having generated realizations of spin values from MC simulations, we can assess their spatial variability by calculating the experimental (i.e., sample-based) variogram as

$$\hat{\gamma}(h) = \frac{1}{2N(h)} \sum_{i=1}^{N(h)} \left[\phi(x_i) - \phi(x_i + h)\right]^2, \tag{6}$$

where $N(h)$ is the number of pairs of spins separated by the distance h.

There are several methods of fitting the model to the experimental variogram. In the present study we used the weighted least squares estimator (hereafter WLS) that was reported to give the best overall results [27]. The WLS estimator is based on minimizing the weighted sum of squared residuals (objective function), which is defined as follows:

$$O(\boldsymbol{\theta}) = \sum_{k=1}^{k_{\max}} \frac{N(h_k)}{[\gamma(h_k;\boldsymbol{\theta})]^2}[\hat{\gamma}(h_k) - \gamma(h_k;\boldsymbol{\theta})]^2, \tag{7}$$

where $\hat{\gamma}(h_k)$ is the experimental and $\gamma(h_k;\boldsymbol{\theta})$ the model variogram. The summation runs over the $k = 1, \ldots, k_{\max}$ lags containing $N(h_k)$ pairs of points, where $h(k_{\max})$ is less than one half of the maximum lag in the sample.

3. Results and discussion

In order to empirically study spatial correlations in the realizations of spin values generated by the modified planar rotator model we run MC simulations at various temperatures T (in units J/k_B, where k_B is Boltzmann constant). Each simulation starts from an initial configuration of spatially uncorrelated random spin angles uniformly distributed in the interval $[0, 2\pi]$ and the snapshots are collected after τ MC sweeps. Then, using Eq. (6) the experimental variogram is calculated and subsequently fitted to the appropriate model by the WLS method defined by Eq. (7). The results obtained at relatively higher temperatures are demonstrated in Fig. 2, for $T = 10^{-2}$ and $\tau = 10$, 100, 1000 and 10000. Soon after the start of the simulation one can observe the development of spatial correlations both in the snapshots in the form of nucleation of small spin domains with similar values as well as in the small-scale behavior of the experimental variograms. The fitted parameters $\hat{\xi} = 1.62$ and $\hat{\nu} = 0.62$ for $\tau = 10$ in Fig. 2(a) indicate that the realization is quite rough with a rather small characteristic length. As the relaxation proceeds

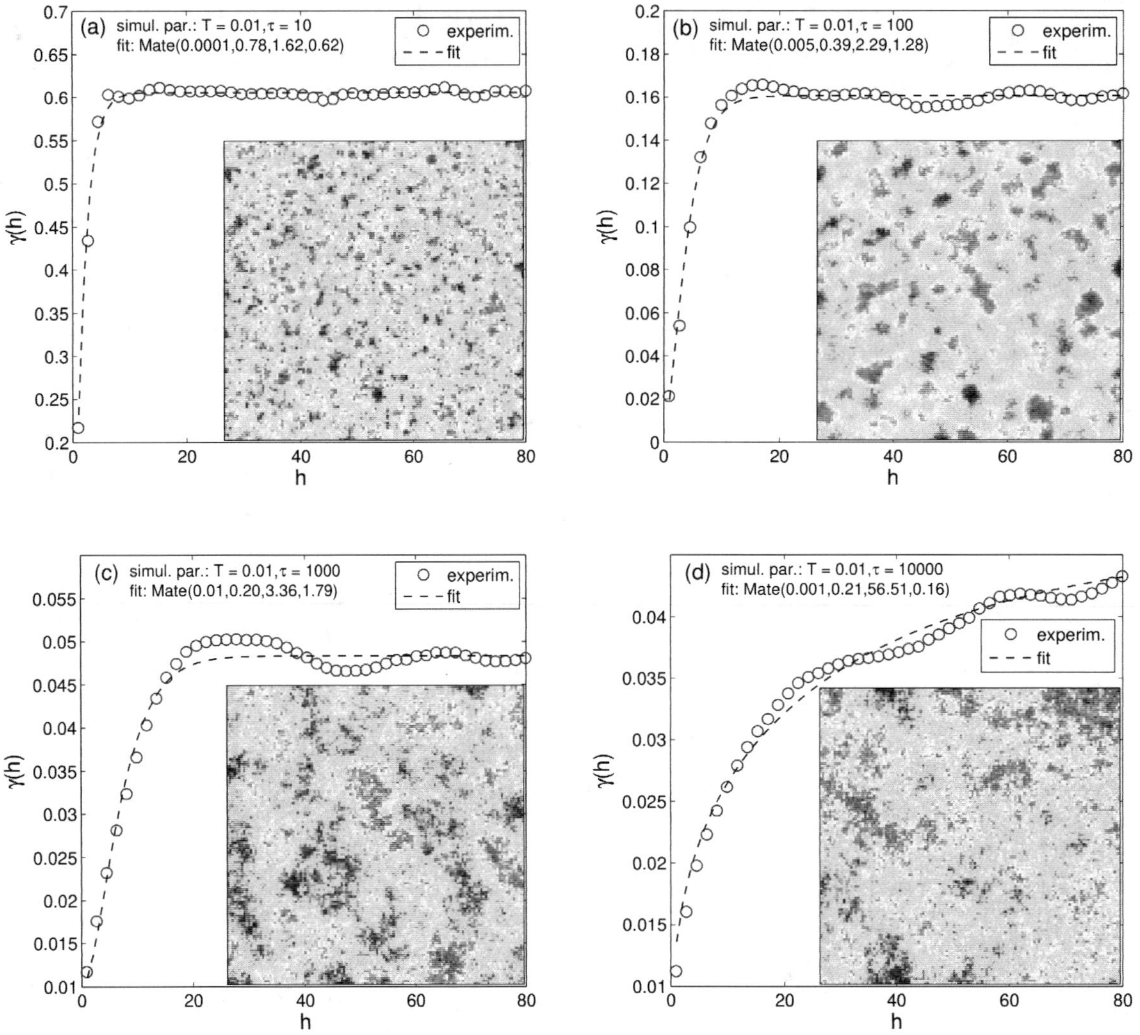

Figure 2. Experimental and fitted variograms of spin realizations —shown in the inset— obtained at temperature $T = 10^{-2}$ and at various simulation times τ equal to (a) 10, (b) 100, (c) 10^3 and (d) 10^4 MC sweeps. Mate($\hat{\boldsymbol{\theta}}$) denotes the Matérn model for the inferred parameter values $\hat{\boldsymbol{\theta}} = (\hat{\sigma}_n, \hat{\sigma}, \hat{\xi}, \hat{\nu})$.

the variance decreases and both the characteristic length and the smoothness parameter increase, as shown in Figs. 2(b) and 2(c), for $\tau = 100$ and 1000, respectively. Nevertheless, as equilibrium is approached the realizations become rougher again, whereas the characteristic length further increases and saturates only in equilibrium to a temperature-dependent value. The estimated characteristic length $\hat{\xi} = 56.51$ for the equilibrium configuration at $\tau = 10000$ in Fig. 2(d) implies that the samples on the 128×128 lattice suffer from lack of ergodicity, which is also reflected in the trend displayed by the corresponding experimental variogram.

As the temperature is decreased the trend of building up correlations in the equilibration process is further enhanced. Since thermal fluctuations are gradually suppressed the realizations become smoother. This tendency is apparent by visual comparison of the snapshots obtained at

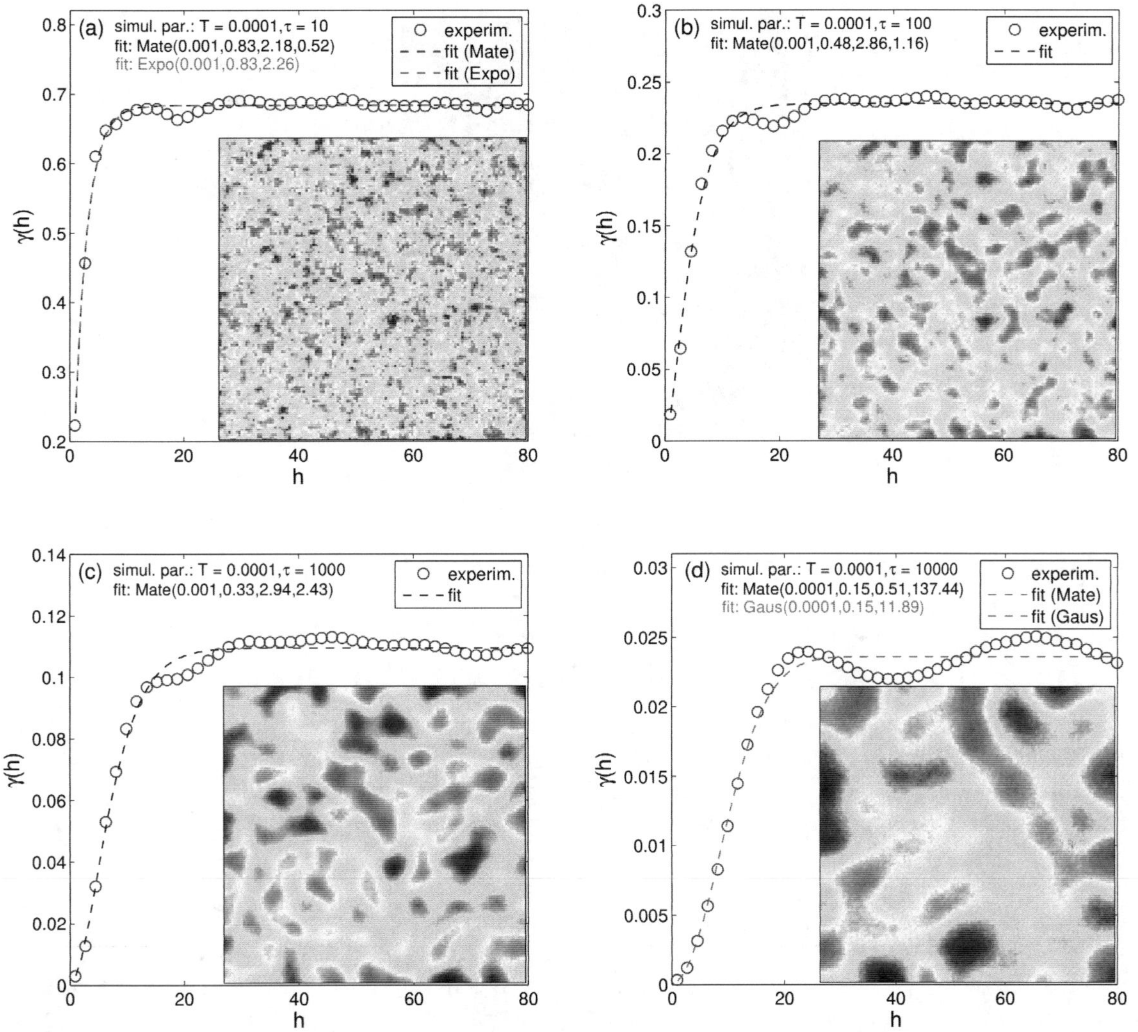

Figure 3. The same as in Fig. 2 for $T = 10^{-4}$. Expo($\hat{\boldsymbol{\vartheta}}$) and Gaus($\hat{\boldsymbol{\vartheta}}$) denote respectively the exponential and Gaussian models for the inferred parameter values $\hat{\boldsymbol{\vartheta}} = (\hat{\sigma}_n, \hat{\sigma}, \hat{\xi})$. Note that in Fig. 3(a) the fitted Matérn and exponential models coincide as do in Fig. 3(d) the Matérn and the Gaussian models.

$T = 10^{-4}$, shown in Fig. 3 with those generated in the same stages of the relaxation at $T = 10^{-2}$, shown in Fig. 2. The difference between the inferred values of the smoothness parameter for the respective cases becomes evident particularly at later stages ($\tau = 10^3$ and 10^4). As mentioned above, the exponential covariance model with quite rough (non-differentiable) realizations and the Gaussian model with very smooth (infinitely differentiable) realizations can be obtained from the Matérn model as special cases for $\nu = 0.5$ and $\nu \to \infty$, respectively. To emphasize the flexibility of the correlations developed in the relaxation process we also fit the relatively rough data for $\tau = 10$ (Fig. 3(a)) to the exponential and the very smooth data for $\tau = 10^4$ (Fig. 3(d)) to the Gaussian models. One can see an excellent collapse of both curves with the best fits to the Matérn model. In spite of the very small value of $\hat{\xi}$ inferred in the Matérn model for $\tau = 10^4$

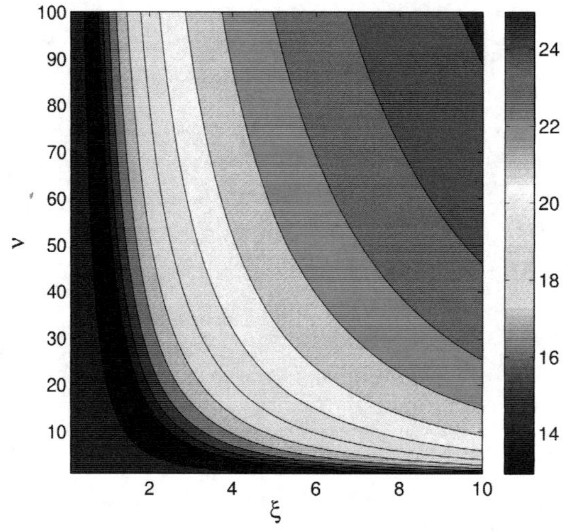

Figure 4. The objective function O given by (7) over the $\xi - \nu$ parameter space. O represents the fit of the data obtained at $T = 10^{-4}$ and $\tau = 10000$ (see Fig. 3(d)) to the Matérn model with $\hat{\sigma}_n = 0.001$ and $\hat{\sigma} = 0.15$.

(see the explanation below), the characteristic length is expected to increase with decreasing temperature and increasing simulation time. This expectation can be justified by the fact that the ground state of the MPR model (the equilibrium state at $T = 0$ corresponding to the lowest energy), is the ferromagnetic state with all the spins aligned in the same direction. Thus, for $T \to 0$ one can expect that the correlation (characteristic) length will span the entire lattice and eventually go to infinity for $L \to \infty$.

Next, we comment on the big discrepancy between the characteristic length parameters $\hat{\xi}$, estimated for the Matérn and Gaussian models, for $T = 10^{-4}$ and $\tau = 10^4$. In fact, in the Matérn model the parameter ξ has been found to be highly negatively correlated with the smoothness parameter ν [18, 26, 28]. This correlation is also apparent by looking at the objective function of the WLS fit, defined by Eq. (7), which is shown in Fig. 4 in the $\xi - \nu$ parameter space, for the data presented in Fig. 3(d). It demonstrates that the samples with small $\hat{\xi}$ and large $\hat{\nu}$ values are quite "close" to the samples with large $\hat{\xi}$ and small $\hat{\nu}$ values. Put differently, a small difference between samples from the same population (particularly in non-ergodic samples) can lead to completely different parameter estimates. Thus, one should be careful when performing parameter inference from the empirical variogram. Recently, we have presented a model-independent definition of the correlation length borrowed from statistical field theory and proposed to use a so-called ergodicity index to compare coarse-grained measures corresponding to both trivial (standard) and non-trivial, e.g. Matérn, covariance models with different parameters [28].

4. Summary and outlook

We have modified the standard planar rotator spin model from statistical physics to display a flexible type of short-range spatial correlations relevant in geostatistical modeling. In particular, the empirical study of spin configurations produced by Monte Carlo simulations in the nonequilibrium regime at various temperatures and stages shows that the smoothness and correlation range vary greatly versus the temperature and the simulation time. This behavior implies that the model has good potential for the simulation and prediction of spatial processes in geophysical and environmental applications. In particular, one can use the model to perform conditional simulations of gridded data, such as remote sensing images. Conditional simulations honor the sample values and reconstruct the variability of missing data, based on simulation parameters inferred from the available samples. Owing to the fact that the model does not show undesirable critical slowing down, the relaxation process is rather fast; furthermore, the short-range nature of the interactions between spin variables allows vectorization of the algorithm. Consequently, the proposed method is significantly more efficient than the conventional geostatistical approaches, and thus applicable to huge datasets, such as satellite and radar images. The implementation details are now under investigation.

Acknowledgments

This work was supported by the Scientific Grant Agency of Ministry of Education of Slovak Republic (Grant No. 1/0331/15). We also acknowledge support for a short visit by Prof. Žukovič at TUC from the Hellenic Ministry of Education - Department of Inter-University Relations, the State Scholarships Foundation of Greece and the Slovak Republic's Ministry of Education through the Bilateral Programme of Educational Exchanges between Greece and Slovakia.

References

[1] Rossi R E, Dungan J L and Beck L R 1994 *Remote Sens. Environ.* **49** 32
[2] Foster M P and Evans A N 2008 *IEEE T. Geosci. Remote* **46** 2153
[3] Jun M and Stein M L 2004 *Atmos. Environ.* **38** 4427
[4] Lehman J, Swinton K, Bortnick S, Hamilton C, Baldridge E, Eder B and Cox B 2004 *Atmos. Environ.* **38** 4357
[5] Albert M F M A, Schaap M, Manders A M M, Scannell C, O'Dowd C D and de Leeuw G 2012 *Atmos. Environ.* **57** 289
[6] Bechle M J, Millet D B and Marshall J D 2013 *Atmos. Environ.* **69** 345
[7] Emili E, Popp C, Wunderle S, Zebisch M and Petitta M 2011 *Atmos. Environ.* **45** 4344
[8] Wackernagel H 2003 *Multivariate Geostatistics* (New York: Springer)
[9] Diggle P J and Ribeiro Jr P J 2007 *Model-based Geostatistics. Series: Springer series in statistics* (New York: Springer)
[10] Cressie N 2008 *J. Roy. Stat. Soc. B* **70** 209
[11] Hartman L and Hössjer O 2008 *Comput. Stat. Data An.* **52** 2331
[12] Žukovič M and Hristopulos D T 2009 *J. Stat. Mech.-Theory E.* **2009** P02023
[13] Žukovič M and Hristopulos D T 2009 *Phys. Rev. E* **80** 011116
[14] Mermin N D and Wagner H 1966 *Phys. Rev. Lett.* **17** 1133
[15] Kosterlitz J M and Thouless D J 1973 *J. Phys. C* **6** 1181
[16] Kosterlitz J M 1974 *J. Phys. C* **7** 1046
[17] Minasny B and McBratney A B 2005 *Geoderma* **128** 192
[18] Minasny B and McBratney A B 2007 *Geoderma* **140** 324
[19] Marchant B P and Lark R M 2007 *Geoderma* **140** 337
[20] Rodríguez-Iturbe I and Mejía J M 1974 *Water Resour. Res.* **10** 713
[21] Pardo-Iguzquiza E, Mardia K V and Chica-Olmo M 2009 *Comput. Geosci.* **35** 1139
[22] Handcock M S and Stein M L 1993 *Technometrics* **35** 403
[23] Kammann E E and Wand M P 2003 *Appl. Statist.* **52** 1
[24] Handcock M S and Wallis J R 1994 *J. Am. Stat. Assoc.* **89** 368
[25] Fuentes M 2002 *Biometrika* **89** 197
[26] Žukovič M and Hristopulos D T 2009 *Technometrics* **51** 173
[27] Zimmerman D L and Zimmerman M B 1991 *Technometrics* **33** 77
[28] Hristopulos D T and Žukovič M 2011 *Stoch. Environ. Res. Risk. Assess.* **25** 11

Higher order Peregrine breathers solutions to the NLS equation.

Pierre Gaillard Université de Bourgogne Dijon France

Abstract. The solutions to the one dimensional focusing nonlinear Schrödinger equation (NLS) can be written as a product of an exponential depending on t by a quotient of two polynomials of degree $N(N+1)$ in x and t. These solutions depend on $2N-2$ parameters : when all these parameters are equal to 0, we obtain the famous Peregrine breathers which we call P_N breathers. Between all quasi-rational solutions of rank N fixed by the condition that its absolute value tends to 1 at infinity and its highest maximum is located at point $(x=0, t=0)$, the P_N breather is distinguished by the fact that $P_N(0,0) = 2N+1$.
We construct Peregrine breathers of the rank N explicitly for $N \leq 11$. We give figures of these P_N breathers in the $(x;t)$ plane; plots of the solutions $P_N(0;t)$, $P_N(x;0)$, never given for $6 \leq N \leq 11$ are constructed in this work. It is the first time that the Peregrine breather of order 11 is explicitly constructed.

PACS numbers :
33Q55, 37K10, 47.10A-, 47.35.Fg, 47.54.Bd

1. Introduction

After the first results concerning the NLS equation obtained in 1972 by Zakharov and Shabat who solved it using the inverse scattering method [1, 2] a lot of studies have been carried out. The first quasi rational solutions to NLS equation were constructed in 1983 by Peregrine [3]. Akhmediev, Eleonski and Kulagin obtained in 1986 in particular the first higher order analogue of the Peregrine breather [4, 5]. Other analogues of the Peregrine breathers of order 3 and 4 were constructed in a series of articles by Akhmediev et al. [6, 7] using Darboux transformations.
Rational solutions to the NLS equation were written in 2010 as a quotient of two wronskians in [8]. Another representation of the solutions to the NLS equation in terms of a ratio of two wronskians of even order $2N$ composed of elementary functions using truncated Riemann theta functions in 2011 has been given in [9]. In 2013 rational solutions in terms of determinants which do not involve limits were given in [10].
We recall the representation of the solutions to the NLS equation in terms quasi rational solutions depending a priori on $2N-2$ parameters at order N as a ratio of two polynomials of degree $N(N+1)$ of x and t multiplied by an exponential depending on t. The present paper presents Peregrine breathers as particular case of multi-parametric families of quasi rational solutions to NLS of order N depending on $2N-2$ real parameters : Peregrine breather P_N of order N are obtained when all the parameters are equal to 0.

2. Families of solutions to NLS equation depending on $2N-2$ parameters and P_N breathers

We consider the focusing NLS equation

$$iv_t + v_{xx} + 2|v|^2 v = 0. \tag{1}$$

Then we get the following result [9, 10] :

Theorem 2.1 *The function v defined by*

$$v(x,t) = \exp(2it - i\varphi) \times \frac{\det((n_{jk})_{j,k\in[1,2N]})}{\det((d_{jk})_{j,k\in[1,2N]})} \tag{2}$$

is a quasi-rational solution to the NLS equation (1)

$$iv_t + v_{xx} + 2|v|^2 v = 0,$$

quotient of two polynomials $N(x,t)$ and $D(x,t)$ depending on $2N-2$ real parameters $\tilde{a_j}$ and $\tilde{b_j}$, $1 \leq j \leq N-1$.
N and D are polynomials of degrees $N(N+1)$ in x and t, where

$$\begin{array}{l}
n_{j1} = \varphi_{j,1}(x,t,0),\ 1 \leq j \leq 2N \quad n_{jk} = \frac{\partial^{2k-2}\varphi_{j,1}}{\partial\epsilon^{2k-2}}(x,t,0),\\
n_{jN+1} = \varphi_{j,N+1}(x,t,0),\ 1 \leq j \leq 2N \quad n_{jN+k} = \frac{\partial^{2k-2}\varphi_{j,N+1}}{\partial\epsilon^{2k-2}}(x,t,0),\\
d_{j1} = \psi_{j,1}(x,t,0),\ 1 \leq j \leq 2N \quad d_{jk} = \frac{\partial^{2k-2}\psi_{j,1}}{\partial\epsilon^{2k-2}}(x,t,0),\\
d_{jN+1} = \psi_{j,N+1}(x,t,0),\ 1 \leq j \leq 2N \quad d_{jN+k} = \frac{\partial^{2k-2}\psi_{j,N+1}}{\partial\epsilon^{2k-2}}(x,t,0),\\
2 \leq k \leq N,\ 1 \leq j \leq 2N
\end{array}$$

The functions φ and ψ are defined in (3),(4), (5), (6).

$$\begin{array}{ll}
\varphi_{4j+1,k} = \gamma_k^{4j-1}\sin X_k, & \varphi_{4j+2,k} = \gamma_k^{4j}\cos X_k,\\
\varphi_{4j+3,k} = -\gamma_k^{4j+1}\sin X_k, & \varphi_{4j+4,k} = -\gamma_k^{4j+2}\cos X_k,
\end{array} \tag{3}$$

for $1 \leq k \leq N$, and

$$\begin{array}{ll}
\varphi_{4j+1,N+k} = \gamma_k^{2N-4j-2}\cos X_{N+k}, & \varphi_{4j+2,N+k} = -\gamma_k^{2N-4j-3}\sin X_{N+k},\\
\varphi_{4j+3,N+k} = -\gamma_k^{2N-4j-4}\cos X_{N+k}, & \varphi_{4j+4,N+k} = \gamma_k^{2N-4j-5}\sin X_{N+k},
\end{array} \tag{4}$$

for $1 \leq k \leq N$.

$$\begin{array}{ll}
\psi_{4j+1,k} = \gamma_k^{4j-1}\sin Y_k, & \psi_{4j+2,k} = \gamma_k^{4j}\cos Y_k,\\
\psi_{4j+3,k} = -\gamma_k^{4j+1}\sin Y_k, & \psi_{4j+4,k} = -\gamma_k^{4j+2}\cos Y_k,
\end{array} \tag{5}$$

for $1 \leq k \leq N$, and

$$\begin{array}{ll}
\psi_{4j+1,N+k} = \gamma_k^{2N-4j-2}\cos Y_{N+k}, & \psi_{4j+2,N+k} = -\gamma_k^{2N-4j-3}\sin Y_{N+k},\\
\psi_{4j+3,N+k} = -\gamma_k^{2N-4j-4}\cos Y_{N+k}, & \psi_{4j+4,N+k} = \gamma_k^{2N-4j-5}\sin Y_{N+k},
\end{array} \tag{6}$$

for $1 \leq k \leq N$.
Arguments Y_k and Y_k are defined by

$$\begin{array}{l}
X_\nu = \kappa_\nu x/2 + i\delta_\nu t - ix_{3,\nu}/2 - ie_\nu/2,\\
Y_\nu = \kappa_\nu x/2 + i\delta_\nu t - ix_{1,\nu}/2 - ie_\nu/2,
\end{array}$$

for $1 \leq \nu \leq 2N$.
These terms are defined by means of λ_ν such that $-1 < \lambda_\nu < 1$, $\nu = 1, \dots, 2N$,

$$\begin{array}{l} -1 < \lambda_{N+1} < \lambda_{N+2} < \dots < \lambda_{2N} < 0 < \lambda_N < \lambda_{N-1} < \dots < \lambda_1 < 1 \\ \lambda_{N+j} = -\lambda_j, \quad j = 1, \dots, N. \end{array} \tag{7}$$

κ_ν, δ_ν and γ_ν are defined by

$$\begin{array}{llll} \kappa_j = 2\sqrt{1-\lambda_j^2}, & \delta_j = \kappa_j \lambda_j, & \gamma_j = \sqrt{\frac{1-\lambda_j}{1+\lambda_j}}, & \\ \kappa_{N+j} = \kappa_j, & \delta_{N+j} = -\delta_j, & \gamma_{N+j} = 1/\gamma_j, & j = 1 \dots N. \end{array} \tag{8}$$

Parameters a_j and b_j in the form

$$a_j = \sum_{k=1}^{N-1} \tilde{a_k} j^{2k+1} \epsilon^{2k+1}, \quad b_j = \sum_{k=1}^{N-1} \tilde{b_k} j^{2k+1} \epsilon^{2k+1}, \quad 1 \leq j \leq N. \tag{9}$$

Complex numbers e_ν $1 \leq \nu \leq 2N$ are defined by

$$e_j = ia_j - b_j, \quad e_{N+j} = ia_j + b_j, \quad 1 \leq j \leq N, \quad a, b \in \mathbf{R}. \tag{10}$$

The terms $x_{r,\nu}$ $(r = 3, 1)$ are defined by

$$x_{r,\nu} = (r-1) \ln \frac{\gamma_\nu - i}{\gamma_\nu + i}, \quad 1 \leq j \leq 2N. \tag{11}$$

In particular we have the following structure for the P_N breathers

Theorem 2.2 *The function v_0 defined by*

$$v_0(x,t) = \exp(2it - i\varphi) \times \left(\frac{\det((n_{jk})_{j,k \in [1,2N]})}{\det((d_{jk})_{j,k \in [1,2N]})} \right)_{(\tilde{a_j} = \tilde{b_j} = 0,\, 1 \leq j \leq N-1)} \tag{12}$$

is the Peregrine breather of order N solution to the NLS equation (1) whose highest amplitude in module is equal to $2N+1$.

3. Peregrine breathers P_N of order N

We have already constructed in [9, 11, 12, 13, 14, 15, 16, 17] solutions for the cases $N = 1$ until $N = 10$.
Because of the length of the expressions of polynomials N and D of the solutions v to the NLS equation defined by

$$v(x,t) = \frac{N(x,t)}{D(x,t)} \exp(2it - i\varphi),$$

we cannot give it in this paper. We only give figure in the $(x;t)$ plane and plots of the solutions $v(0;t)$, $v(x;0)$.

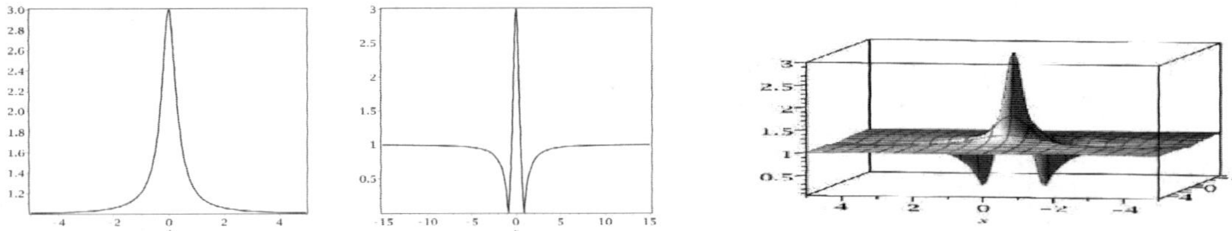

Figure 1. Solution to NLS, N=1, on the left $v(x,0)$; in the center $v(0,t)$; on the right $v(x,t)$.

Journal of Physics: Conference Series **633** (2015) 012106 doi:10.1088/1742-6596/633/1/012106

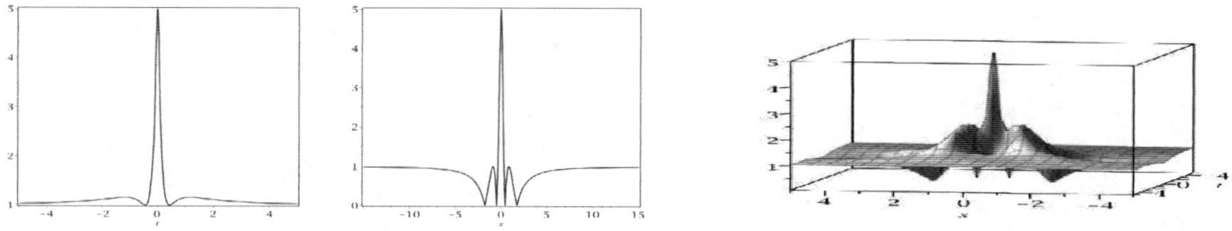

Figure 2. Solution to NLS, N=2, on the left $v(x,0)$; in the center $v(0,t)$; on the right $v(x,t)$.

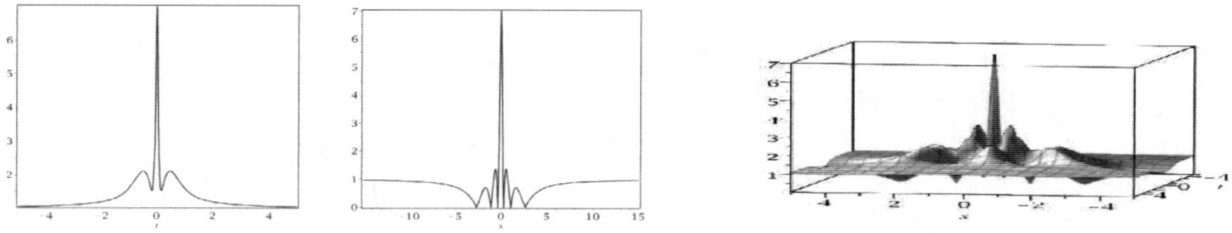

Figure 3. Solution to NLS, N=3, on the left $v(x,0)$; in the center $v(0,t)$; on the right $v(x,t)$.

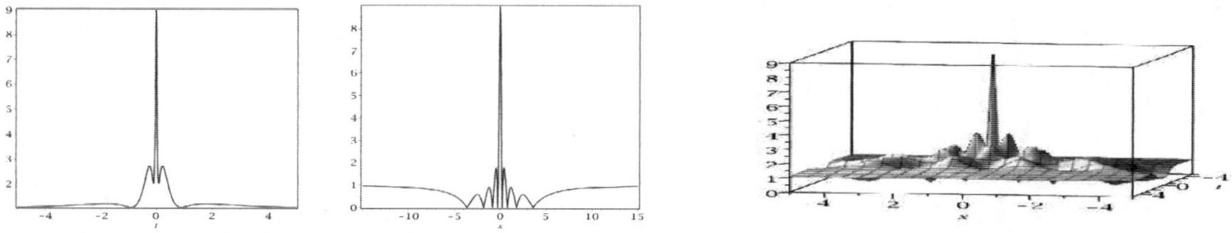

Figure 4. Solution to NLS, N=4, on the left $v(x,0)$; in the center $v(0,t)$; on the right $v(x,t)$.

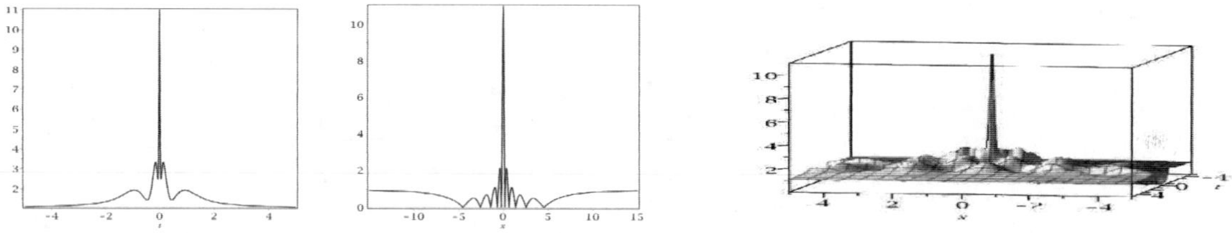

Figure 5. Solution to NLS, N=5, on the left $v(x,0)$; in the center $v(0,t)$; on the right $v(x,t)$.

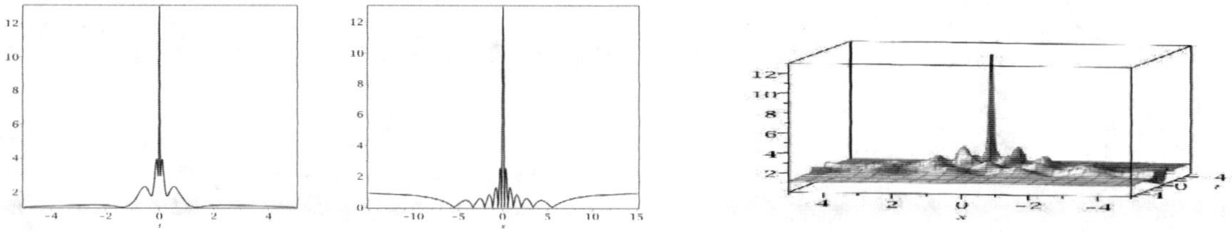

Figure 6. Solution to NLS, N=6, on the left $v(x,0)$; in the center $v(0,t)$; on the right $v(x,t)$.

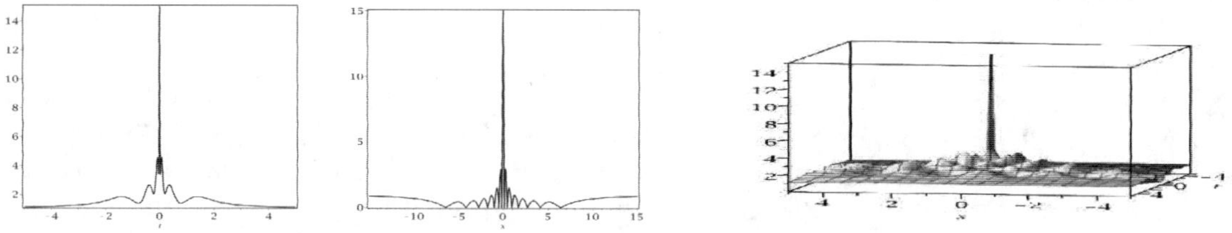

Figure 7. Solution to NLS, N=7, on the left $v(x,0)$; in the center $v(0,t)$; on the right $v(x,t)$.

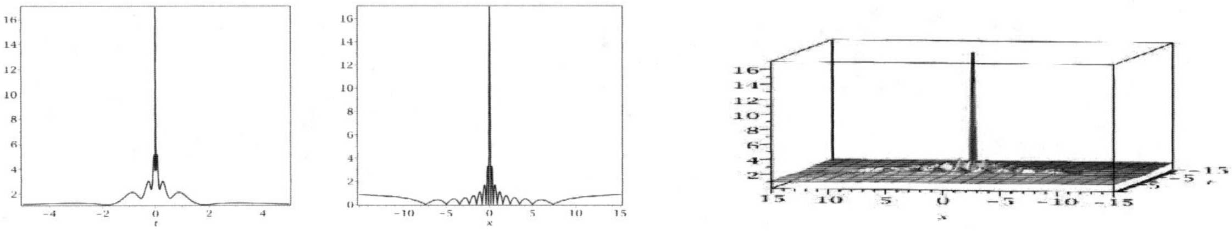

Figure 8. Solution to NLS, N=8, on the left $v(x, 0)$; in the center $v(0, t)$; on the right $v(x, t)$.

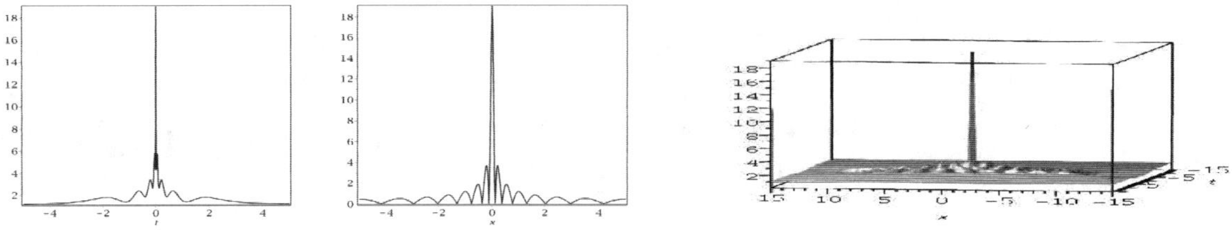

Figure 9. Solution to NLS, N=9, on the left $v(x, 0)$; in the center $v(0, t)$; on the right $v(x, t)$.

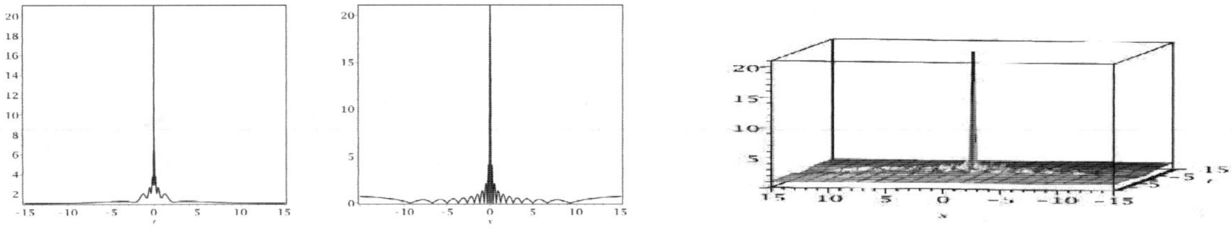

Figure 10. Solution on NLS, N=10, to the left $v(x, 0)$; in the center $v(0, t)$; on the right $v(x, t)$.

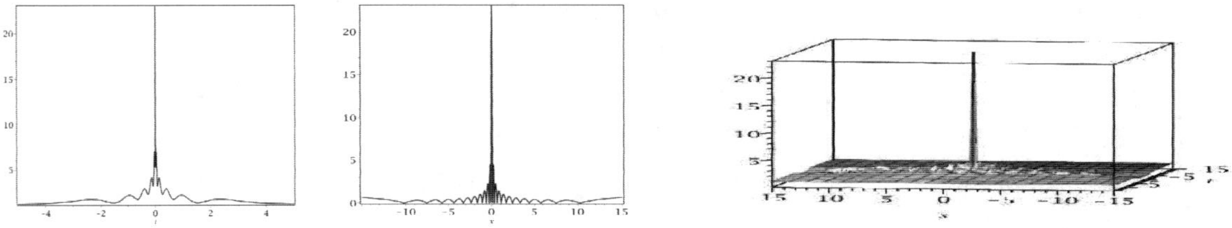

Figure 11. Solution on NLS, N=11, to the left $v(x, 0)$; in the center $v(0, t)$; on the right $v(x, t)$.

It is important to say that, contrary to the P_1 breather, all higher ranks P_N breathers can be deformed thus generating the multi-rogue-waves (MRW) solutions.
Actually $N = 11$ is a greatest rank for which this work is completed and this is one of the results of this article. We postpone to give a more precise study of this eleventh Peregrine breather to another publication.

4. Conclusion

The method described in the present paper provides a powerful tool to get explicit solutions to the NLS equation and to understand the behavior of rogue waves. It is the first time that the Peregrine breather order 11 is explicitly constructed; the complete expression as a quotient of polynomials of order 132 in x and t is too long to be presented here. Moreover, the studies for the initial conditions $x = 0$ and $t = 0$, in the cases $N = 6$ until $N = 10$ are completely new, as they had never been presented.

We hope that several promising applications may be a result of this theoretical study. Recently, the notion of the rogue wave has been transferred into the realm of nonlinear optics. Experimental studies have shown that continuous-wave laser radiation in optical fibers splits into separate pulses and those pulses can reach very high amplitudes [18]. This kind of solution has been observed in plasma [19], Bose-Einstein condensates [20], fiber optics [21] or on a water surface [22, 23]. In the case of nonlinear optics and hydrodynamics, the results were checked up to order 5.

This study leads to a better understanding of the phenomenon of rogue waves, and it would be relevant to go on with more higher orders.

References

[1] Zakharov V E 1968 J. Appl. Tech. Phys **9** 86
[2] Zakharov V E Shabat A B Sov. 1972 Sov. Phys. JETP **34** 62
[3] Peregrine H. 1983 J. Austral. Math. Soc. Ser. B **25** 16
[4] Akhmediev N Eleonski V Kulagin N 1985 Sov. Phys. J.E.T.P. **62** 894
[5] Akhmediev N Eleonski V Kulagin N 1987 Th. Math. Phys. **72** 183
[6] Akhmediev N Ankiewicz A Soto-Crespo J M 2009 Physical Review E **80** 026601-1
[7] Akhmediev N Ankiewicz A Clarkson P A 2010 J. Phys. A : Math. Theor. **43** 122002-1
[8] Dubard P Gaillard P Klein C Matveev V B 2010 Eur. Phys. J. Special Topics **185** 247
[9] Gaillard P 2011 J. Phys. A : Meth. Theor. **44** 1
[10] Gaillard P 2013 Jour. Of Math. Phys. **54** 013504-1
[11] Gaillard P 2013 Phys. Rev. E **88** 042903-1
[12] Gaillard P 2014 J. Math. Phys. **54** 073519-1
[13] Gaillard P 2014 Commun. Theor. Phys. **61** 365
[14] Gaillard P 2014 Physica Scripta V. **89** 015004-1
[15] Gaillard P 2014 Jour. Of Phys. : conferences Series **482** 012016-1
[16] Gaillard P 2014 Jour. Of Math. Phys. **5** 093506-1
[17] Gaillard P 2013 Jour. Of Math. **2013** 1-111
[18] Solli DR Ropers C Koonath P Jalali B 2007 Nature **450** 1054
[19] Bailung H Sharma SK Nakamura Y 2011 Phys. Review Letters **107** 255005-1
[20] Bludov YV Konotop VV Akhmediev N 2009 Phys. Rev. A **80** 033610-1
[21] Kibler B Fatome J Finot C Millot G Dias F Genty G Akhmediev N Dudley J M 2010 Nature Physics **6** 790
[22] Chabchoub A Hoffmann H Onorato M Akhmediev N 2012 Phys. Review X **2** 1
[23] Chabchoub A Hoffmann N Akhmediev N 2011 Phys. Rev. Lett. **106** 204502-1

Improvement of rotated comb decimation filter magnitude characteristic using sharpening technique

Gordana Jovanovic Dolecek

Department of Electronics, INAOE, E. Erro 1, 72740 Puebla , Mexico

E-mail: gordana@ieee.org

Decreasing of sampling rate in a digital form is called decimation. This process may introduce aliasing, i.e. non desired replicas of the main spectrum of the decimated signal. That is why we need the filter before decimation. This filter is called the decimation or antialiasing filter. The most popular decimation filter is comb filter which is usually used in the first stage of decimation. However it has a high passband droop and low attenuations in the folding bands. Different methods have been proposed to improve comb magnitude characteristic. The methods based on simple zero rotations have been proposed recently. In this paper we use the sharpening technique to improve both the passband and the stopband of comb filter with the rotated zeros. The method is compared with methods based on the rotation of zeros of comb filter.

1. Introduction

The decreasing of the sampling rate in digital form is called decimation. This process may introduce the aliasing effects that deteriorate the signal and have to be eliminated by the decimation filter, as shown in Figure 1, where M is the decimation factor [1].

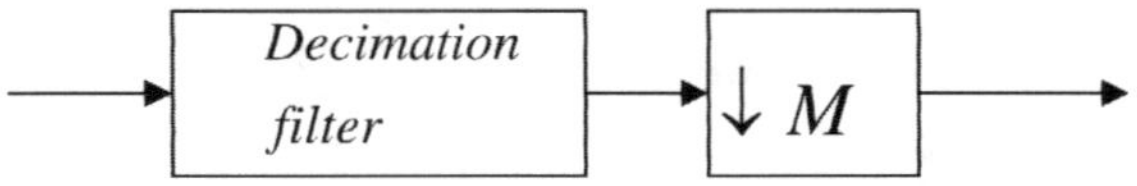

Figure 1. Decimation.

The decimation is usually accomplished using a cascade of two or more stages. Because the filter in the first stage must be very simple, it is usually used comb filter, which does not require the multipliers. The magnitude response of comb filter must be flat in the passband to avoid the distortion of the decimated signal. Additionally, the comb magnitude response must have high attenuations in the so called folding bands, (bands around the zeros of comb filter). However, the comb magnitude response exhibits a high passband droop and the folding bands are narrow and have low attenuations. Different methods have been proposed to improve comb magnitude characteristic in the passband [2-7], in the folding bands [8-9] and in both, the passband and the folding bands [10-11]. Recently, the simple zero-rotation of comb filter has been introduced in [12] to improve the alias rejections in the folding bands, for even decimation factors. However, by introducing the zero rotation, the passband droop is increased. The main goal of this work is to apply the sharpening technique to decrease comb passband droop and further improve the alias rejection in comb filters with the simple zero rotation.

The rest of the paper is organized in the following way. Next section briefly describes the simple zero rotation of comb filters. Section 3 presents the sharpening technique. The proposed method is described in Section 4 and illustrated with examples. Section 5 describes a two-stage structure, where the sharpening is applied only at the second stage.

2. Simple zero-rotation of comb filters [12]

The system function of a comb filter can be expressed, either in the recursive form,

$$H(z) = \left[\frac{1}{M}\frac{1-z^{-M}}{1-z^{-1}}\right]^{K}, \tag{1}$$

or in recursive form,

$$H^{K}(z) = \left[\frac{1}{M}\sum_{k=0}^{M-1} z^{-k}\right]^{K}, \tag{2}$$

where M is a decimation factor and K is order of the filter and presents the number of the cascaded comb filters.

The magnitude response of the filter is in the sinx/x form :

$$\left|H(e^{j\omega})\right|^{K} = \left|\frac{\sin(\omega M/2)}{M\sin(\omega/2)}\right|^{K}. \tag{3}$$

The rotation in [12] is obtained by subtracting a positive constant a at the point of symmetry of the comb impulse response. Taking into account that we consider here comb filters with the even length, in order to obtain odd length of the comb filter we apply the convolution of two even length comb filters. The resulting point of symmetry is at the point M-1. By omitting the normalisation constant, the rotation term is given as:

$$H_r(z) = \left[\sum_{k=0}^{M-1} z^{-k}\right]^{2} - az^{-(M-1)},. \tag{4}$$

The Rouche's theorem is used to prove that the rotated zeros are on unit circle. The system function of the rotated comb is the cascade of K_1 combs and the rotated term:

$$H_R(z) = H^{K_1}(z)H_r(z). \tag{5}$$

Example 1:
Consider M=10 and a=1/2. The rotated term is given as:

$$H_r(z) = \left[\sum_{k=0}^{9} z^{-k}\right]^{2} - 0.5z^{-9}. \tag{6}$$

The system function of the corresponding rotated comb is:

$$H_R(z) = \left[\frac{1}{10}\frac{1-z^{-10}}{1-z^{-1}}\right]^{K_1}\left[\left[\sum_{k=0}^{9} z^{-k}\right]^{2} - 0.5z^{-9}\right]. \tag{7}$$

Figure 2a presents the z-plane plot of the corresponding comb with K=2. Similarly, Figure 2b shows the corresponding z-plane plot for the rotated term (6). Note, that the zeros of the rotated term, are rotated from their original positions, shown in Fig.2a. As a result, the widths of the folding bands are increased. The overall magnitude responses and the passband zooms of the rotated comb (5) for K_1=2 and the corresponding comb with K=4, are shown in Figure 3.

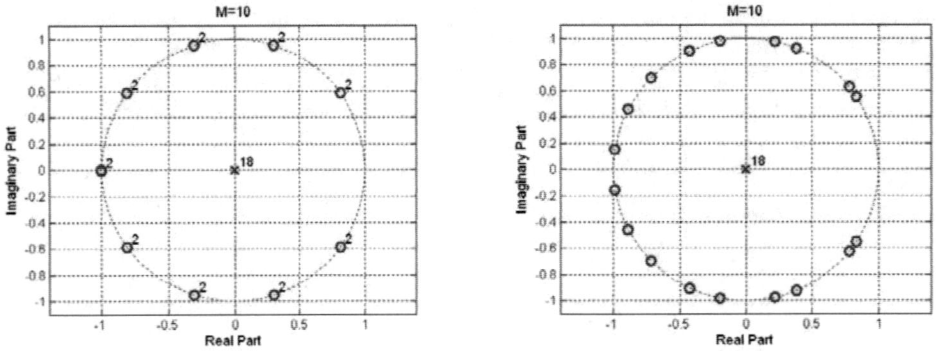

a. Comb. b. Rotated term.

Figure 2. Zeros of comb and rotated term.

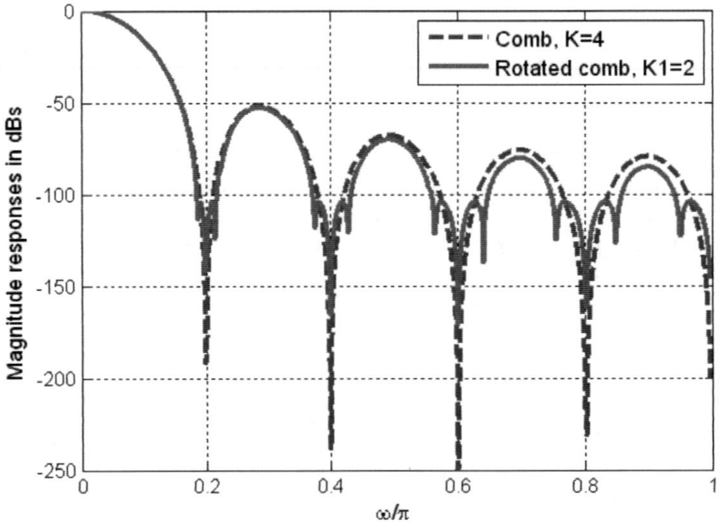

Figure 3. Magnitude responses of rotated comb and corresponding comb.

3. Sharpening technique

The sharpening technique [13] simultaneously improves the passband and the stopband of symmetrical prototype filter *H*, using a polynomial relationship of the form $H_{sh} = f(H)$ between the amplitudes of the sharpened H_{sh} and the prototype filters *H*. The improvement in the magnitude response near the passband edge *H*=1, or near the stopband edge *H*=0, depends on the order of tangencies m and n of the polynomial relationship at *H*=1, or at *H*=0, respectively.

The polynomial relationship in terms of m and n is given [13]

$$H_{sh} = H^{n+1}\sum_{s=0}^{m}\frac{(n+s)!}{n!s!}(1-H)^{s} = H^{n+1}\sum_{s=0}^{m}C(n+s,s)(1-H)^{s}, \tag{8}$$

where $C(n+s,s)$ is the binomial coefficient.

The most simple polynomial for the passband improvement, (m=1, n=0) is in the form $H_{sh1}=2H-H^2$.

Similarly, the most simple polynomial for improvement of passband and stopband is given as $H_{sh2}=3H^2-2H^3$.

Example 2:

We apply both sharpening polynomials to comb filter with $M=10$ and $K=2$. Figure 4 shows the corresponding magnitude responses.

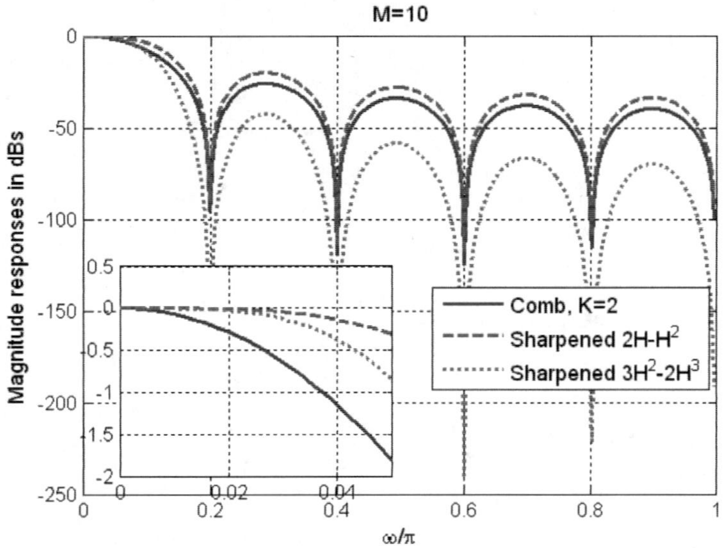

Figure 4. Magnitude responses of sharpened combs and the prototype comb.

4. Sharpened rotated comb

We propose to use simple sharpening polynomials $H_{sh1}=2H-H^2$, and $H_{sh2}=3H^2-2H3$, to improve the magnitude response of the rotated comb.

The sharpened rotated combs are defined by,

$$H_{sh1}(z) = 2H_R(z) - H_R^2(z), \tag{9}$$

$$H_{sh2}(z) = 3H_R(z) - 2H_R^3(z). \tag{10}$$

where $H_R(z)$ is given in (5).

4.1. Comparison with comb

Example 3:

In this example we first compare magnitude responses of the sharpened rotated comb from Example 1, using the sharpening polynomial (9) with the magnitude response of comb with $K=4$. Figure 5a shows the corresponding magnitude responses along with the passband zooms. Similarly, Figure 5b compares the magnitude responses of sharpened rotated comb (10) and comb filter.

4.2. Comparison with rotated comb

Example 4:

We compare sharpened rotated combs with the original rotated combs. To this end, Figures 6a and 6b show magnitude responses of the sharpened rotated combs and the original combs taking the sharpening polynomials (9) and (10).

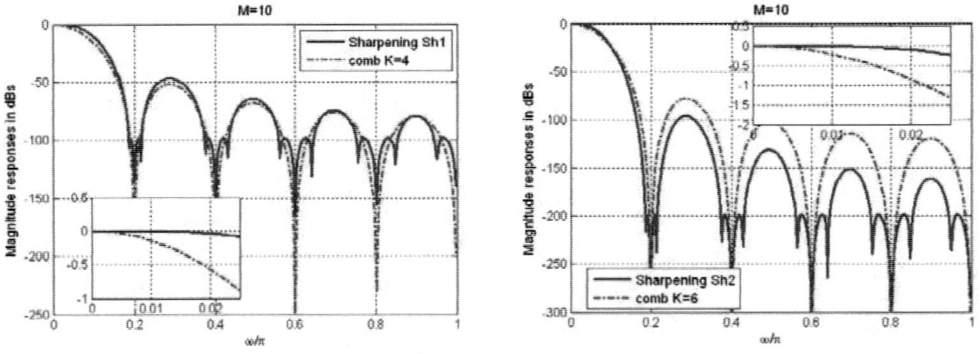

a. Sharpening polynomial (9). b. Sharpening polynomial (10).

Figure 5. Magnitude responses of comb, and sharpened rotated combs.

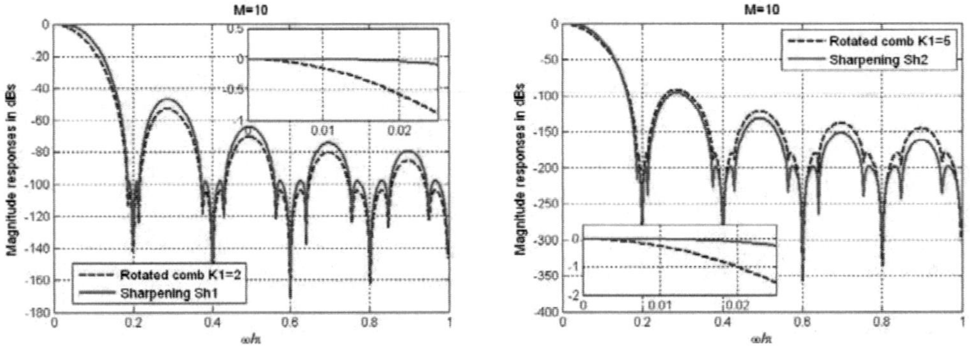

a. Sharpening polynomial (9). b. Sharpening polynomial (10).

Figure 6. Magnitude responses of rotated comb, and sharpened rotated comb.

5. Two-stage sharpened compensated filter

In order to decrease the complexity, consider that the decimation factor can be factorized in the form $M=M_1M_2$, where M_2 is even number. In this case we can split decimation into two stages, where in the first stage is the comb filter,

$$H_1(z) = \left[\frac{1}{M_1} \frac{1-z^{-M_1}}{1-z^{-1}} \right]^{K_{11}}, \tag{11}$$

decimated by M_1, while in the second stage is the comb filter,

$$H_2(z) = \left[\frac{1}{M_2} \frac{1-z^{-M_2}}{1-z^{-1}} \right]^{K_{22}}, \tag{12}$$

decimated by M_2.

In that way, we can apply the rotation only to the second stage, and thus to apply sharpening only to the rotated comb at that stage:

$$H_{2sh1}(z) = 2H_{2R}(z) - H_{2R}^{2}(z) \tag{13}$$

$$H_{2sh2}(z) = 3H_{2R}(z) - 2H_{2R}^{3}(z), \tag{14}$$

where

$$H_{2R}(z) = H_2^{K_{22}}(z)H_{2_r}(z) \,, \tag{15}$$

$$H_{2_r}(z) = \left[\sum_{k=0}^{M_2-1} z^{-k}\right]^2 - az^{-(M_2-1)} \,. \tag{16}$$

The corresponding structure is given in Figure 7.

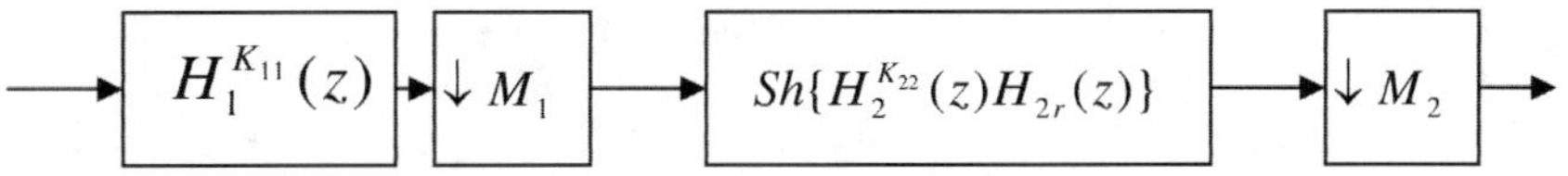

Figure 7. Two-stage sharpening structure.

Next example illustrates the characteristics of two-stage structure.
Example 6
We consider a two-stage structure with M_1=3 and M_2=4. Figure 8 shows the magnitude response of the two stage sharpening structure, and the corresponding comb with K=6.In the first stage is the comb filter with the parameter K_{11}=6, while the second sharpening polynomial is applied to the rotated comb $H_2(z)$, with K_{22}=2. Note that the two-stage structure provides more attenuations in the folding bands, and less passband droop.

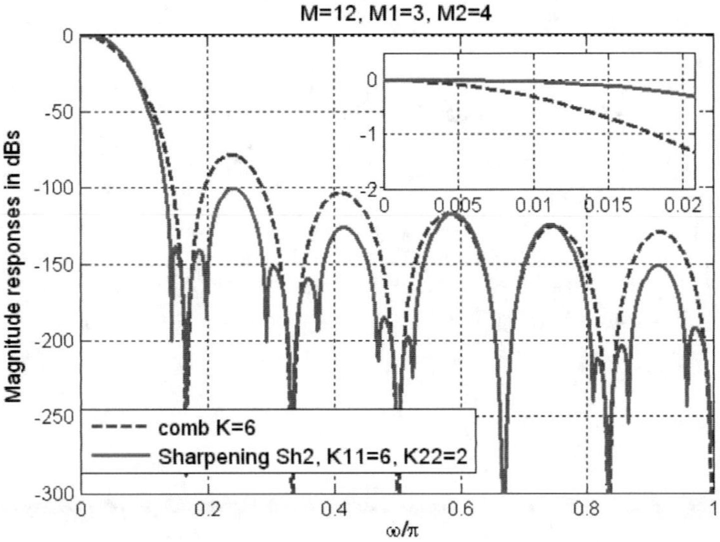

Figure 8. Magnitude responses of two-stage sharpened structure and the comb.

6. Concluding remarks

This paper presents the application of sharpening technique for the improvement of magnitude response of a simple rotated comb. Two simple sharpening polynomials have been considered. In that way the improvements in both, passband and stopband, in comparisons with the original and rotated combs, have been achieved. In order to decrease the complexity, introduced by the sharpening technique, a novel two-stage structure is also proposed, in which the zero rotation and the corresponding sharpening are applied only at the second stage.

Acknowledgement
This work is supported by Conacyt Mexico under grant No.179587.

References

[1] Jovanovic Dolecek G 2002 *Multirate Signal Processing: Design and Application* (Hershey: IGP).

[2] Kim S Lee W C Alm S Choi 2006 *Digital Signal Processing* Design of CIC roll-off compensation filter in a W-CDMA digital receiver **16**, 846-854.

[3] Jovanovic Dolecek G Mitra S K 2008 *Electronics Letters* Simple method for compensation of CIC decimation filter **44** 1162-1163.

[4] Fernandez-Vazquez A Jovanovic Dolecek G 2009 *IEEE Transactions on Circuits and Systems II: Express Brief* A general method to design GCF compensation filter **56** 409-413.

[5] Jovanovic Dolecek G 2009 *Electronics Letters* Simple wideband CIC compensator **45**, 1270-1272.

[6] Molnar G Vucic M 2011 *IEEE Transactions on Circuits and Systems II: Express Brief* Closed-form design of CIC compensators based on maximally flat error criterion **58** 926-30.

[7] Jovanovic Dolecek G Fernandez-Vazquez 2014 *International Journal for Electronics and Communications* (*AEUE*) Trigonometrical Approach to Design a Simple Wideband Comb Compensator **68** 437–441.

[8] Presti L 2000 *IEEE Trans on Circ. and Syst.-II* Efficient modified-sinc filters for sigma-delta A/D converters **47** 1204–1213.

[9] Laddomada M, 2007 *IEEE Trans on Circ. and Syst.-I* Generalized comb decimation filters for sigma-delta A/D converters **54**, 994–1005.

[10] Kwentus A Willson, Jr.A 1997 *IEEE Trans. on Signal Processing* Application of filter sharpening to cascaded integrator-comb decimation filters **45**, 457-467.

[11] Jovanovic Dolecek G Feranandez-Vazquez A *International Journal for Electronics and Communication (AEUE)* Novel droop-compensated comb decimation filter with improved alias rejections **67**, 387-396.

[12] Jovanovic Dolecek G Baez R 2015 Very Simple Method to Improve Alias Rejection of Comb Decimation Filter for Even Decimation Factors (submitted).

[13] Kaiser J F Hamming R W 1977 *IEEE Transactions on Acoustics, Speech, and Signal Processing,* Sharpening the response of a symmetric non-recursive filter **ASSP-25** 415-422.

Magnetization of layered superconductors with ferromagnetic nanorods

V A Kashurnikov, A N Maksimova and I A Rudnev
National Research Nuclear University, Moscow, Kashirskoe shosse, 31, 115409, Russia

e-mail: nastymaksimova@yandex.ru

Abstract. The magnetization of layered high-temperature superconductors (HTSC) with ferromagnetic nanorods as bulk pinning centers is studied in the 2D model of layered HTSC by using Monte Carlo method. Magnetic part of the interaction energy between a ferromagnetic cylinder of arbitrary radius and constant magnetization and an Abrikosov vortex was calculated in London approximation. The magnetization at different radius of magnetic defects was investigated. The results of calculations were compared with the results for extended nonmagnetic defects.

1. Introduction

The transport properties of type-II superconductors strongly depend on the dynamics of vortex system and the pinning properties of the defects structure. The experimental and theoretical study of the interplay between superconductivity and ferromagnetism has suggested that ferromagnetism can actually enhance superconductivity in artificial superconductor-ferromagnet hybrid structures under certain conditions. In general, artificial ferromagnet-superconductor hybrid structures are characterized by a superconducting film on the top or underneath a textured ferromagnetic layer. The ferromagnetic layer can be either solid (superconductor-ferromagnet bilayer, [1-4]) or made of magnetic dots or dipoles [5,6]. In [5,6] the dynamics of ac-driven vortices and antivortices in a superconducting film with an array of magnetic dipoles was investigated by hybrid molecular dynamic – Monte Carlo simulations. The spontaneous appearance and stabilization of vortex-antivortex (v-av) pairs was observed, the *E-J* curves were obtained. The periodic sequence of production and annihilation v-av pairs results in the peaks of the *E-J* curve. In [7] the superconductor with ferromagnetic nanoparticles as bulk pinning centers was studied experimentally, the magnetization curves were obtained. The interaction of a vortex with ferromagnetic particle of arbitrary radius was theoretically analyzed. In [8] the enhancement of a single vortex pinning by a magnetic cylinder was calculated, it was shown that a magnetic inclusion can reduce the Lorenz force on a vortex yielding an enhanced critical current density.

The Monte Carlo method has demonstrated the high efficiency in the calculation of the transport properties and magnetization of layered HTSC. The calculations in presence of nonmagnetic defects were performed in [9-11], the effect of ferromagnetic impurities in the case of bulk ferromagnetic defects were analyzed in [12]. But the model [12] suggests the size of ferromagnetic particles much smaller than superconducting magnetic penetration depth and does not work for magnetic inclusions of arbitrary radius and shape. The aim of our work is to investigate the pinning properties of the array of ferromagnetic rods in the bulk of the superconductor and to investigate the magnetization and the transport properties at different concentration and size of ferromagnetic defects.

2. Model and the calculation method

The calculations were performed within the two-dimensional model of a layered HTSC by using Monte Carlo algorithm [9-11]. The model is a limiting case of the realistic three-dimensional model involving various types of in-plane interactions and the inter-plane interaction. In this model, a vortex line in the bulk of the superconductor is represented in the form of a set of interacting planar vortices (pancakes). Early calculations in this approximation as well as comparison with the experimental data confirm that such an approach provides an adequate description of the situation. Thus, the Gibbs thermodynamic potential of a two-dimensional system with variable number of interacting pancakes (in the absence of inter-plane interaction) takes a form:

$$G = N\varepsilon + \sum_{i<j} U_{in-plane}(r_{ij}) + \sum_{i,j} U_p(r_{ij}) + \sum_{i,j} U_{surf}(r_{ij}^{(im)}),$$

where $\varepsilon = \varepsilon_0(\ln[\lambda(T)/\xi(T)] + 0.52)$ is a self-energy of a vortex, N is a number of pancakes in the layer under consideration, the second term describes the pair interaction of vortices, the third term is the interaction of vortices with pinning centers and the fourth term corresponds to the interaction of vortices with the surface and Meissner current. $\varepsilon_0 = \Phi_0^2 s/(4\pi\lambda)^2$, $\Phi_0 = \pi\hbar c/e$ is the quantum of magnetic flux (see [9-11] for details).

In this work, we introduce the ferromagnetic rods of arbitrary radius as bulk pinning centers. For calculations, the interaction energy between the vortex line and ferromagnetic rod must be obtained. Let us consider an infinite ferromagnetic cylinder of radius R and magnetization $\mathbf{M}$, embedded into an infinite type-II superconductor containing straight vortex lines. Similarly to [7], we solve in cylindric coordinates the London equation for the vector-potential $\mathbf{A}$ in a superconductor and the Maxwell equation inside the ferromagnetic particle and obtain the supercurrent induced around the rod:

$$j_\varphi = -\frac{cM}{\lambda^2} \frac{K_1\left(\frac{r}{\lambda}\right)}{\frac{2}{R}K_1\left(\frac{R}{\lambda}\right) + \frac{1}{\lambda}K_0\left(\frac{R}{\lambda}\right)},$$

where K_0 and K_1 are the McDonald functions, λ is a superconductive penetration depth. The energy of interaction between ferromagnetic rod and vortex line is calculated as an inverse work of Lorenz force when the vortex moves from infinity to the position r at the distance r from the centre of the rod:

$$U = -\int_\infty^r F dx = -\delta \frac{\Phi_0 M}{\frac{2\lambda}{R}K_1\left(\frac{R}{\lambda}\right) + K_0\left(\frac{R}{\lambda}\right)} K_0\left(\frac{r}{\lambda}\right) \qquad (1)$$

The coefficient δ denotes a thickness of a superconductive layer. It is worth noting that at R<<λ this expression turns to the energy of point magnetic dipole in the field of the vortex. The calculations were performed for temperatures of 1-10 K and for typical parameters of $Bi_2Sr_2CaCu_2O_{8-\delta}$: λ(0)=80 nm, ξ(0)=2 nm, T_C=84 K. Ferromagnetic particles have a size about $0.1\lambda - \lambda$, the magnetization of the particles is about 10^2 - 10^3 Gs, the size of superconducting region under consideration is 6×3 μm.

3. Results

Let us consider the superconductive slab with periodic lattice of ferromagnetic rods. The magnetic moment of the defects is perpendicular to the superconductive layers and does not change during the magnetization process. We analyze now the magnetization curves at different size of ferromagnetic defects. Fig.1 represents the two typical magnetization loops at different radius, magnetization and concentration of the rods (a) and the dependence of the square of the loop (which is proportional to the value of hysteretic losses) for the configurations from (a) on the amplitude of external magnetizing field (b). According to [11], the smaller the square of the loop, the greater the critical current (in the regime of hysteretic losses). A zero residual magnetization at the second half of the magnetization cycle (when the external field drops from maximal negative value to zero; see the curves in fig. 1a) is a special feature of magnetic defects with constant magnetization and takes place because the ferromagnetic particles do not pin the vortices of opposite polarity. The calculations were done for both periodic and non-periodic lattices of defects and it was shown that the results in both cases qualitatively coincide.

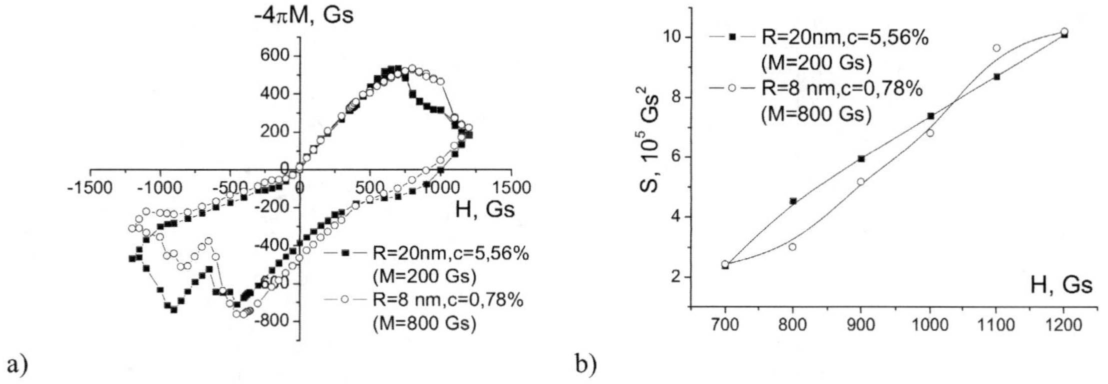

Fig. 1. The magnetization loops at different size of ferromagnetic rods (a) and the square of the magnetization loop as a function of the amplitude of the external field (b). R is a radius of the ferromagnetic defect, c is a volume concentration of a ferromagnetic in the superconductor.

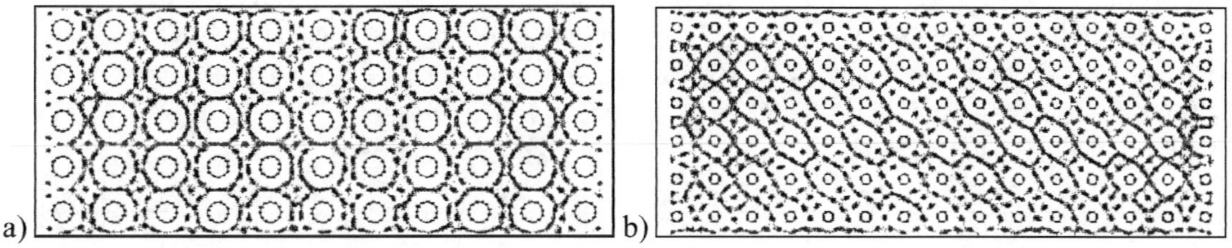

Fig. 2. Different configurations of vortices arising during magnetization process: a) c=31%, R=180 nm, b) c=13%, R=90 nm. The opened circles denote the magnetic defects.

Note that at some sizes and concentrations of magnetic defects the vortices form superlattices like shown in fig.2 during magnetization.

Let us analyze now the transport properties of the superconductor with extended magnetic defects. We fix the volume concentration of the ferromagnetic and then change the size of individual rod. The smaller the radius of the defect, the higher the number of defects per unit square and vice versa. We then obtain the current-voltage characteristics and estimate the critical current j_c as a current at which the electric field strength in the superconductor reaches 0.1 μ V/cm. The results are shown in fig. 3. It is readily seen that at each concentration there is some optimal combination of R, N at which j_c reaches a peak and the peak shifts left at smaller concentrations.

The computational model introduced above is applicable for cylindric ferromagnetic defects but does not include the case of defects of arbitrary shape. In order to describe this case we may consider the extended defect as an ensemble of point defects (of the radius ξ). The potential produced by extended defect is therefore the sum of point defects potential. The simulations show a satisfactory agreement between two potentials, especially far from the defect's surface.

Journal of Physics: Conference Series **633** (2015) 012108 doi:10.1088/1742-6596/633/1/012108

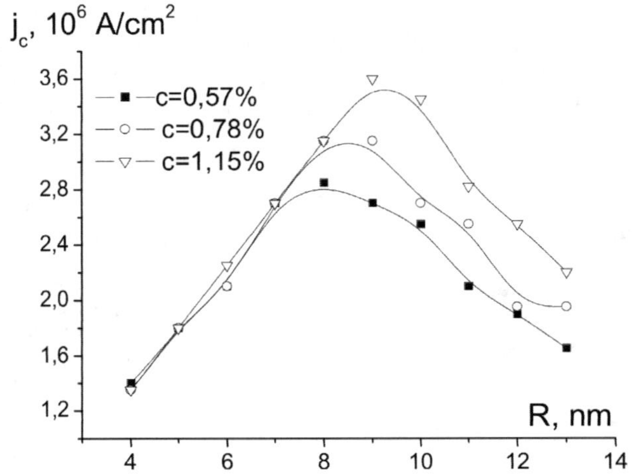

Fig. 3. The critical current as a function of the radius of the rod at different concentrations of the ferromagnetic. The magnetization of the ferromagnetic M=800 Gs and is the same for all curves

4. Conclusion

The description of the interaction between straight vortex line and ferromagnetic rod of arbitrary radius in the bulk of the superconductor was introduced in the 2D – model of layered high-temperature superconductor. The magnetization curves of the superconductor were obtained at the different size of magnetic defects. The transport properties at different configurations of ferromagnetic defects were investigated. The model can be used to investigate the magnetization and transport properties of the layered HTSC with bulk ferromagnetic defects of arbitrary size and configuration.

Acknowledgements

Research was done with the financial support of RSF under grant № 14-22-00098.

References

[1] L.S. Uspenskaya, S.V.Egorov, Physica B (2013), http://dx.doi.org/10.1016/j.physb.2013.09.040
[2] T. Tamegai, Y. Nakao, Y. Tsuchiya, Y. Nakajima, Physica C 468 (2008) 1308–1312
[3] Dong Ho Kim, T.J. Hwang, Physica C 455 (2007) 58–62
[4] Yoshio Nakao, Yasuyuki Nakajima, Tsuyoshi Tamegai, Physica C 470 (2010) S801–S803
[5] Clessio L.S. Lima, Clecio C. de Souza Silva, J. Albino Aguiar, Physica C 479 (2012) 147–150
[6] Cléssio L. S. Lima and Clécio C. de Souza Silva, Phys. Rev. B 80, 054514 (2009)
[7] A. Snezhko, T. Prozorov, R. Prozorov, Phys. Rev. B 71 024527 (2005)
[8] M G Blamire, R B Dinner, S C Wimbush and J L MacManus-Driscoll, Supercond. Sci. Technol. 22 (2009) 025017
[9] D.S. Odintsov, I.A. Rudnev, V.A. Kashurnikov, JETP 103 (2006), 66
[10] I.A. Rudnev, D.S. Odintsov, V.A. Kashurnikov, Phys. Lett. A 372 (2008) 3934-3936
[11] D.S. Odintsov, I.A. Rudnev, V.A. Kashurnikov, JETP 105 (2007), 253
[12] V.A. Kashurnikov, A.N. Maksimova, I.A. Rudnev, Physics of the Solid State 56 (5) (2014) 894-911

Band structure and density of states in FeAs-based superconductors

V A Kashurnikov, A V Krasavin

National Research Nuclear University MEPhI (Moscow Engineering Physics Institute), 31 Kashirskoye shosse, 115409, Moscow, Russia

avkrasavin@gmail.com

Abstract. The generalized quantum Monte Carlo algorithm was used to obtain one-particle excitation spectrum and electron density of states for two-dimensional FeAs-clusters modeling iron-based superconductors within the limits of the full two-orbital model. The calculations were performed for clusters with sizes up to 10×10 FeAs-cells. The excitation spectra were reconstructed from Matsubara Green's function. The spectral density of states and the total density of states near the Fermi level were obtained. The data are in accordance with known experimental results. The influence of the cluster size, temperature, and the interaction strength on the density of states was analysed.

1. Introduction

Iron-based high-temperature superconductors (HTSC) [1–3] are among the most interesting objects of research in the physics of superconductivity nowadays. As well as copper-based HTSC, these compounds have a layered structure. However, due to the features of the multi-gap band structure they have a complex phase diagram which includes antiferromagnetic, structural, and superconducting phase transitions [4–9].

Iron-based HTSC due to their pronounced anisotropy can be successfully described within the limits of two-dimensional tight-binding models. One of the most appropriate models for this purpose is the two-orbital model [10, 11] which is a typical multi-band generalized Hubbard model. The lack of expansion parameters for analytical approximations, and the presence of strong Coulomb correlations assume the use of exact numerical quantum methods for the study of iron-based HTSC.

The present work continues the study of electronic correlations in big FeAs-clusters and focuses on the calculation of the spectrum of elementary excitations which determines the Fermi surface and the density of states. The quantum continuous time world line Monte Carlo (MC) algorithm (CTWL-algorithm) [12] was used to simulate the full two-orbital model [13, 14].

According to experimental data and calculations based on DMFT and LDA techniques [4, 6, 15-17], it is the quasiparticle spectrum which forms a complex multi-gap band structure. Furthermore, the density of states near the Fermi level is almost entirely determined by d-states of iron atoms, which gives reason to believe that all the phenomena associated with conducting and superconducting properties of these compounds are played on the FeAs-plane.

The quantum algorithm developed in [13] allows one to calculate the Matsubara Green's function and to obtain the quasiparticle spectrum and its dependence on the system size and interaction parameters. The first data are presented on the excitation spectrum in the framework of the full two-orbital model for clusters with sizes up to 10×10 FeAs cells.

2. Model

The minimal electron model for FeAs-based HTSC is defined by the crystal structure of these compounds [1] and band structure calculations, that show that the largest contribution to the electronic density of states near the Fermi level give 3d-states of iron atoms [11, 17]. These reasons give the following Hamiltonian of the two-orbital model [10, 11]:

$$H = H_{int} + H_{kin},$$

$$H_{int} = U\sum_{i,\alpha} n_{i,\alpha,\uparrow}n_{i,\alpha,\downarrow} + V\sum_{i} n_{i,x}n_{i,y} - \mu\sum_{i} n_i - J\sum_{i}(n_{i,x,\uparrow}n_{i,y,\uparrow} + n_{i,x,\downarrow}n_{i,y,\downarrow})$$

$$-J\sum_{i}(a^+_{i,x,\downarrow}a_{i,x,\uparrow}a^+_{i,y,\uparrow}a_{i,y,\downarrow} + a^+_{i,x,\uparrow}a_{i,x,\downarrow}a^+_{i,y,\downarrow}a_{i,y,\uparrow} + a^+_{i,x,\uparrow}a_{i,y,\downarrow}a^+_{i,x,\downarrow}a_{i,y,\uparrow} + a^+_{i,y,\uparrow}a_{i,x,\downarrow}a^+_{i,y,\downarrow}a_{i,x,\uparrow}),$$

$$H_{kin} = -t_1\sum_{i,\sigma}(a^+_{i,x,\sigma}a_{i+x,x,\sigma} + a^+_{i,y,\sigma}a_{i+y,y,\sigma} + h.c.) - t_2\sum_{i,\sigma}(a^+_{i,y,\sigma}a_{i+x,y,\sigma} + a^+_{i,x,\sigma}a_{i+y,x,\sigma} + h.c.)$$

$$-t_3\sum_{i,\sigma}(a^+_{i,x,\sigma}a_{i+x+y,x,\sigma} + a^+_{i,x,\sigma}a_{i+x-y,x,\sigma} + a^+_{i,y,\sigma}a_{i+x+y,y,\sigma} + a^+_{i,y,\sigma}a_{i+x-y,y,\sigma} + h.c.)$$

$$-t_4\sum_{i,\sigma}(a^+_{i,x,\sigma}a_{i+x+y,y,\sigma} + a^+_{i,y,\sigma}a_{i+x+y,x,\sigma} + a^+_{i,x,\sigma}a_{i+x-y,y,\sigma} + a^+_{i,y,\sigma}a_{i+x-y,x,\sigma} + h.c.) \quad (1)$$

where operator $a^+_{i,x(y),\sigma}(a_{i,x(y),\sigma})$ creates (annihilates) an electron with spin σ on site $\boldsymbol{i}$ and orbital $x(y)$; t_i, $i = 1, \ldots, 4$ are the hopping amplitudes between d_{xz} and d_{yz} orbitals defined as x and y, respectively; parameters U and V describe the Coulomb interaction within and between orbitals, respectively; J is the exchange integral; μ is the chemical potential.

The two-orbital model is quite complex to be studied by CTWL-algorithm; the encoding of the basis states, and the features of calculation of pair correlations are presented in detail in [14, 18, 19]. The set of parameters of H_{kin} in (1) was taken from [11]:

$$t_1 = 0.058; t_2 = 0.22; t_3 = -0.0208; t_4 = -0.079. \quad (2)$$

The relationship between the parameters of H_{int} was determined as

$$V = 0.5U; J = 0.25U. \quad (3)$$

All energy quantities are measured in 0.2 eV units.

3. Results

The Matsubara Green's function

$$\langle T_\tau a_{i\alpha\sigma}(\tau)a^+_{j\beta\sigma'}(0)\rangle, \quad (4)$$

where i, j are the coordinates of Fe atoms; α, β are the orbitals; σ, σ' are the spin projections, was calculated for clusters of sizes from 4×4 to 10×10 FeAs-sells in the temperature range $\beta = 5 \div 20$, and the range of the interaction parameter $U = 2 \div 16$ while maintaining the relation (3). The data are presented for the half-filling; the typical number of MC steps for the convergence was $\sim 10^{10}$.

Figure 1a shows typical Matsubara Green's functions in logarithmic scale for various values of the interaction parameter.

Assuming that near the forbidden band the quasiparticle spectrum should be well resolved, the asymptotic behavior of the Matsubara Green's function has the following form:

$$G(k,\tau) \sim e^{-E(k)\tau}. \quad (5)$$

This dependence allows to derive the quasiparticle dispersion $E(k)$. Figure 1b shows a typical excitation spectrum along the main crystallographic directions for the 10×10 cluster. According to our data, the temperature dependence is weak at $\beta \geq 10$. This is the sufficient condition for realization of the asymptote of Green's function (5) and for the correct analysis of the dispersion. Furthermore, the comparison of the calculations for clusters of size 4×4, 6×6, 8×8, and 10×10 indicates that the size of the system 8×8 is big enough for obtaining results which are free of size effect. Similar dispersion curves were obtained under the simplified S_4-model [15, 20].

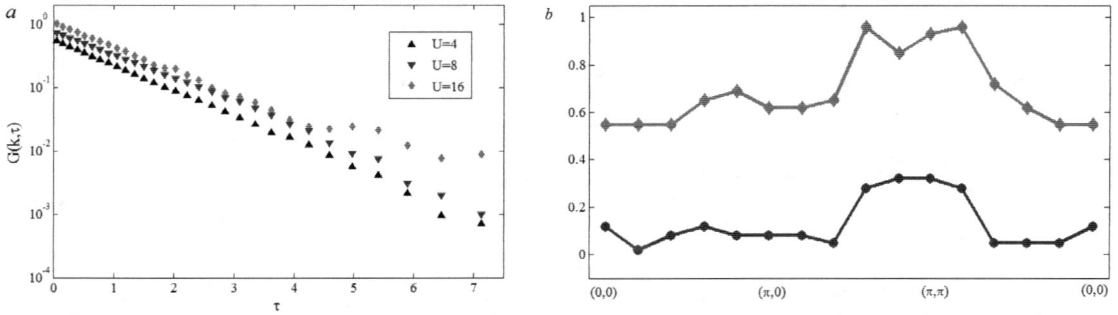

Figure 1. (a) Fourier component of the Matsubara Green's function $G(k,\tau) = \sum G_{ij}(\tau)e^{\boldsymbol{kr}_{ij}}$, $G_{ij}(\tau) = \langle T_\tau a_{ix\uparrow}(0)a^{+}_{jx\uparrow}(\tau)\rangle$, $k_x, k_y = 0$; (b) two branches of the excitation spectrum, $U = 4$. Cluster 10×10, $\beta = 15$.

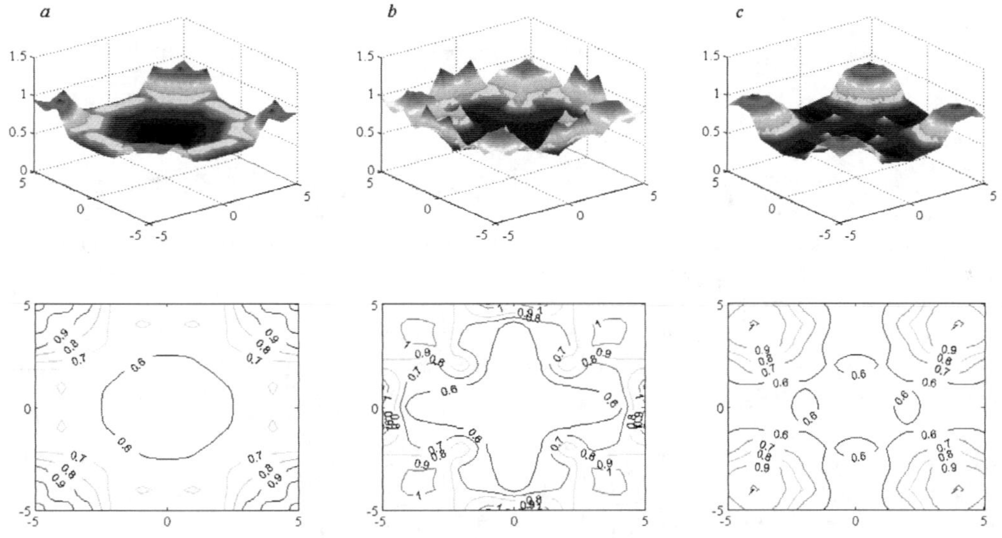

Figure 2. Dispersion and constant-energy curves for 10×10 cluster. (a) $U = 4$; (b) $U = 8$; (c) $U = 16$. $\beta = 15$.

Figure 2 shows the excitation spectrum and Fermi surfaces for the upper mode at different interaction parameters for the entire Brillouin zone. It can be seen that the Coulomb interaction strongly changes the picture of the constant-energy curves.

With increasing the interaction strength peculiar hole pockets appear in the vicinity of regions $k_x, k_y = \pm\pi/2$; such behavior was also observed in calculations [4, 7, 8, 17] and experimentally [9].

Figure 3 presents the results for the density of states $N(\omega) = \sum_{p\sigma} A_\sigma(p,\omega)$, where the spectral density $A_\sigma(p,\omega)$ was restored from the integral relation

$$G_\sigma(p,\tau) = -\int \frac{A_\sigma(p,\omega)e^{-\tau\omega}}{1 + e^{-\beta\omega}} d\omega. \quad (6)$$

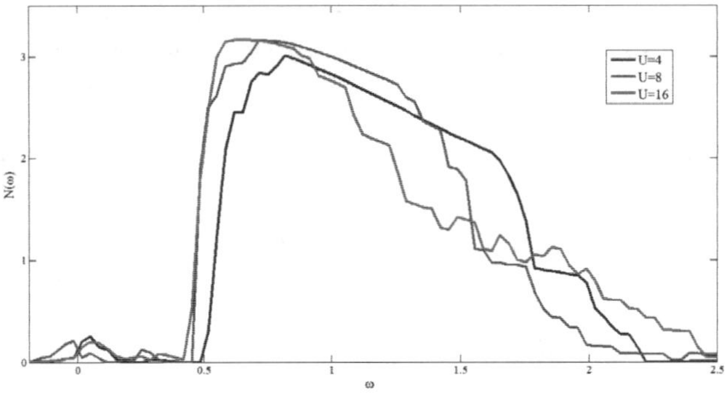

Figure 3. The density of states $N(\omega)$. The Fermi level corresponds to $\omega = 0$.

The calculation of the kernel in (6) was carried out by direct iteration [21]. Problems with the convergence of the method for this typical ill-posed problem allowed us to introduce only preliminary data for the density of states, reliable mostly in the region of unoccupied states above the Fermi level. It can be seen that the total density of states is weakly dependent on the interaction parameter, and there is a gap in the spectrum in the range $(0 - 0.5)$ on the energy scale. The weak dependence of the density of states on the interaction parameters was also mentioned in several works [4, 6, 15, 16].

It should also be noted that even these preliminary results indicate a definite, though weak, peak at the Fermi level, confirming the experimental data on the metallic nature of the undoped FeAs systems [2, 4, 9, 22].

4. Conclusions

Calculations of FeAs clusters of size up to 10×10 within the limits of the full two-orbital model were performed using quantum Monte Carlo algorithm at half-filling and different values of temperature and the interaction parameter. The data obtained on the excitation spectrum and the density of states indicate the possibility of a strong correlation between the features of the spectrum and the value of the interaction, which is consistent with the known experimental data and calculations.

The preliminary data on the total density of states show the weak dependence on the parameters of the interaction. The gap in the excitation spectrum and the finite density of states at the Fermi level are observed, which confirms the supposition of the metallic state of undoped FeAs systems.

Acknowledgements
The work was supported by Russian Science Foundation (project # 14-22-00098) and Russian Found for Basic Research (projects # 14-08-00509 and # 15-02-02764).

References

[1] Y.Kamihara, T. Watanabe, M. Hirano, and H. Hosono, J. Am. Chem. Soc., **130** (2008) 3296.
[2] Yu. Isyumov, E. Kurmaev. High-Tc Superconductors Based on FeAs Compounds. Springer, 2010.
[3] E. Dagotto, Rev. Mod. Phys. **85** (2013) 849.
[4] M.V. Sadovskii, Physics-Uspekhi **51** (2008) 1201.
[5] K. Ishida, Y. Nakai, and H. Hosono, J. Phys. Soc. Jpn. **78** (2009) 062001.
[6] M.V. Sadovskii, E.Z. Kuchinskii, I.A. Nekrasov, Journal of Magnetism and Magnetic Materials **324** (2012) 3481.
[7] M.V. Medvedev, I.A. Nekrasov, M.V. Sadovskii, JETP Lett. **95** (2012) 37.
[8] I.A. Nekrasov, M.V. Sadovskii, JETP Lett. **99** (2014) 598.
[9] P.D. Johnson, G. Xu, W.-G. Yin. Iron-Based Superconductivity. Springer, 2015.
[10] S. Raghu, Xiao-Liang Qi, Chao-Xing Liu et al., Phys. Rev. B **77** (2008) 220503(R).
[11] A. Moreo, M. Daghofer, J.A. Riera, and E. Dagotto, Phys. Rev. B **79** (2009) 134502.
[12] N.V. Prokof'ev, B.V. Svistunov, I.S. Tupitsyn, J. Exp. Theor. Phys. **114** (1998) 570.

[13] V.A. Kashurnikov, A.V. Krasavin, J. Exp. Theor. Phys. **111** (2010) 180.
[14] V.A. Kashurnikov, A.V. Krasavin, JETP Lett. **97** (2013) 333.
[15] Yang Wu, Guangkin Liu and Tianxing Ma, EPL **104** (2013) 27013.
[16] I.A. Nekrasov, Z.V. Pchelkina, M.V. Sadovskii, JETP Lett. **88** (2008) 621; **88** (2008) 777.
[17] S. Liang, G. Alvarez, C. Sen, A. Moreo, and E. Dagotto, Phys. Rev. Lett. **109** (2012) 047001.
[18] V.A. Kashurnikov, A.V. Krasavin, JETP Lett. **100** (2014) 16.
[19] V.A. Kashurnikov, A.V. Krasavin, Phys. Lett. A **378** (2014) 2894.
[20] Tianxing Ma, Hai-Qing Lin, Jiangping Hu, Phys. Rev. Lett. **110** (2013) 107002.
[21] S.R. White, D.I. Scalapino, R.L. Sugar, and N.E. Bickers, Phys. Rev. Lett. **63** (1989) 1523.
[22] S. Liang, A. Moreo, and E. Dagotto, Phys. Rev. Lett. **111** (2013) 047004.

Journal of Physics: Conference Series **633** (2015) 012110 doi:10.1088/1742-6596/633/1/012110

Active Flow Control on Laminar flow over a Backward facing step

Aditya Mushyam*, Josep M Bergada
Fluid Mechanics Department
Technical University of Catalunya
Terrassa, Barcelona, Spain-08222

*mushyam.aditya@gmail.com

Abstract. In the present study, two dimensional flow over a backward-facing step in laminar flow regime with application of active flow control (AFC) technique is analysed. The aim of the present work is to gauge the effectiveness of implementing AFC to reduce drag and study its effects on flow characteristics. In order to analyse the influence of AFC on the boundary layer and the downstream vortex shedding, two different kinds of AFC techniques have been used in this study namely zero net mass flow actuators and fluidic actuators. A parametric non dimensional analysis has been carried out by varying the frequency from 0.025 to 0.1 and jet amplitude from 0.05 and 1. Four different positions of the groove were simulated; groove was respectively located at 0.024a, 0.047a, 0.072a and 0.097a, measured upstream from the right side upper edge. Three different non dimensional groove widths 0.023a, 0.048a and 0.073a were also evaluated, where a is the step height. The idea behind this study was to determine an optimal configuration to reduce the drag on the step and to suppress the vortex dissipation in the wake of the step. It was observed that when using an AFC frequency ± 10% of the vortex shedding one, was causing the maximum drag reduction. When comparing the effects of zero net mass flow actuators with the fluidic actuators, it was observed that zero net mass flow actuators were more effective.

Nomenclature:

a	Height of the inclined step	(m)
A	Amplitude of the AFC jet	
C_D	Drag coefficient	
C_L	Lift coefficient	
f	Frequency of AFC	
h	Height of the non-orthogonal physical domain	
l	Length of the non-orthogonal physical domain	
p	Non dimensional Pressure	
Re	Reynolds number	
t	Non-dimensional Time	
ul	Upstream Length of the physical domain	
u	Non-dimensional Velocity Y direction	
U	Free stream velocity in X direction at the inlet	(m/sec)
v	Non-dimensional Velocity Y direction	
w	Width of the groove	

x	Non-Dimensional Eulerian coordinates in horizontal direction
y	Non-Dimensional Eulerian coordinates in vertical direction
ϕ	Additional default amplitude of the AFC jet

1. Introduction:

Among the previous research undertaken on AFC on backwards facing step, study carried out by Dahan et al [1] can be noted. They performed a 2D and 3D numerical study using Reynolds number, Re=280 and 1500 respectively with both open and closed loop control. Two different locations of the zero net mass flow actuators were considered, namely near the separation border and close to the step foot. They observed that perturbing the flow with a frequency close to the shedding frequency, increased the turbulent kinetic energy in the near separation region and decreased the base pressure. Dejoan et al [2] numerically studied the effect of a 2D ZNMF actuator located just before the separation corner with a jet inclination of 45°, the same inclination also used by Dahan et al [1]. Marrot et al [3] used a loudspeaker located downstream of the step to activate the boundary layer, that maximum effect was observed when the loudspeaker frequency was equal to the boundary layer instability mode. At these conditions a small reduction of the recirculation length was obtained. The pre exited flow was controlled using two rows of synthetic jets located on the vertical side of the step. A similar method, consisting of two arrays of jets located on the vertical side of the step of which the upper row steadily absorbs and the lower row steadily injects, was analysed numerically in 2D by Creusé et al [4], in fact different AFC techniques including pulsating flow were evaluated. As in Dahan et al [1], and Marrot et al [3], open loop and closed loop control were studied. For open loop control, the optimum actuation frequency, as already found by many other researchers, coincided with the fundamental frequency, which was St_h =0.14, the optimum amplitude being A=0.2, this was the same amplitude used by Dahan et al [1]. According to the work presented in Hassan [5], in laminar flow conditions, the shear layer has two instability modes defined by a Strouhal number, $S_{r\theta} = f\,\theta/U$, where θ is the momentum thickness of the boundary layer on the step edge, f is the natural roll up frequency of the shear layer and U is the free stream velocity. The step mode based on the step height h, is defined by Strouhal number, $S_{rh} = f\,h/U$. He observed these values to be around $S_{r\theta} = 0.012$ and $S_{rh} = 0.185$ and these values match with the values obtained lately in Marrot et al [3], St_h =0.14 and Dejoan et al [2], St_h =0.2. It was reported that a slight waviness of energy in the boundary layer results in a periodic vortex train, as defined by Vukasinovic et al [6]. The location, frequency, amplitude and of course the overall amount of energy of the AFC in each particular application is nowadays an interesting field of research.

2. Problem Statement

Figure 1 presents the physical domain of the inclined step. Flow over a Backward facing step with upstream length, ul=2a and downstream length dl=20a, was simulated at Reynolds number Re 500. No slip boundary condition was used at all solid boundaries. Flow was simulated in two dimensions and under laminar regime. The AFC grove was placed at four different locations, measured upstream of the upstream surface edge. Figure 1 also presents the boundary conditions used for the simulation. Structured grid has been generated using the stretching transformation proposed by Roberts [7], which refines the grid in the vicinity of the solid walls. Two different types of active flow control techniques were studied namely zero net mass flow and fluidic actuators and a brief explanation and formulation follows.

Zero-net mass flow control technique, (ZNMF):

In this technique, the flow is injected into the flow upstream of the step such that the net mass flow from the groove is zero, in a cycle. A sinusoidal form of wave equation was used to induce the flow control technique as presented in equation 1.

$$u = 0$$
$$v = A\,U \sin(2\,\pi\,f\,t) \tag{1}$$

Where u is the horizontal x-component and v is the vertical y-component of jet velocity respectively, f is the frequency of the active flow control jet and t is the time. A is the amplitude of the jet with respect to free stream velocity U.

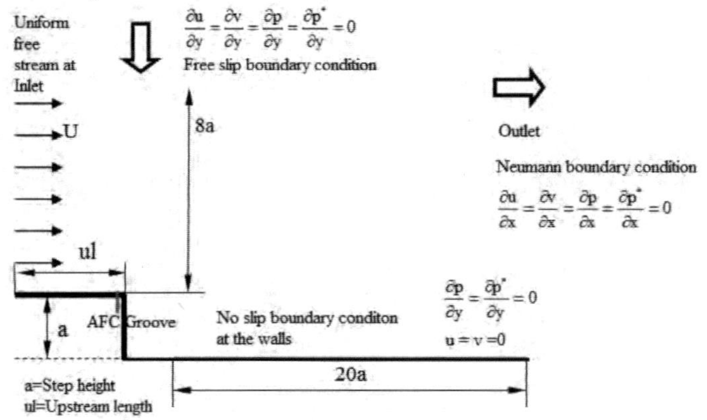

Figure 1: Physical Domain of the backwards facing Step

Fluidic Actuator flow control technique, (FA):
In fluidic actuators, the flow control is induced into the flow upstream of the step, in such a way that there is net outgoing mass flow from the groove. This the primary difference between (ZNMF) and (FA) techniques. This net outwards mass flow is applied by inducing a default constant amplitude Φ, in addition to the sinusoidal wave equation used in zero net mass flow technique, as presented in equation 2. In the the present study for all the cases of FA, the amplitude Φ was considered equal to the amplitude A.

$$u = 0$$
$$v = \Phi + A\,U \sin(2\,\pi\,f\,t) \tag{2}$$

For both configurations (ZNMF) and (FA), the non-dimensional frequency was varied in a range from f = 0.025 to 0.1 and the jet amplitude, A between 0.05 and 1.

The effect of groove implementing AFC, was studied for four different positions of the groove at 0.024a, 0.047a, 0.072a and 0.097a, measured on the upstream surface edge of the step. Also three different non dimensional groove widths were considered, 0.023a, 0.048a and 0.073a, where a, is the step height. The main aim of this study is to determine an optimum configuration which results in the suppression or minimization of drag on the step.

2.1 Governing Equations:
Simulations have been done with non-dimensional form of Navier-Stokes equations. Flow was considered as isothermal. The height of the step, a, was taken as the characteristic length, upstream velocity U was used as characteristic velocity. The characteristic pressure was defined as ϱU^2 and the characteristic time was defined as the characteristic length divided by the characteristic velocity a/U.
Reynolds number was defined as $R_e = \frac{\rho\, U a}{\mu}$; Fluid density ϱ and viscosity μ, were taken as constant.

2.2 Numerical Strategy and Boundary Conditions:

A second-order Adams Bashforth-Crank Nicholson scheme for temporal discretization was applied to Navier Stokes equations in finite volume formulation. The boundary conditions used in the simulation were depicted in figure 1. MAC (Marker and Cell) method with velocity and pressure coupling was applied using a predictor-corrector strategy. In the present study, the respective pressure correction factor in all the neighboring cells were also considered, notice that this is a further improvement with respect to the original MAC method and minimizes the error involved in the calculation.

3. Code Validation:

The code was validated by simulating the flow over a backward step and the results were compared with the results presented in the literature. Figure 2 presents the streamline plots for the backward facing step case. Figure 2 (a, b, c), present the results obtained from the code developed and the graphs presented in Figure 2 (d, e, f), were obtained by Chowdhary and Dhiman [8]. Three Reynolds numbers of 100, 150 and 200 were studied in both cases. The comparison was in complete agreement in all cases studied, the downstream vortexes were steady in all cases and their respective length was the same in all comparisons undertaken.

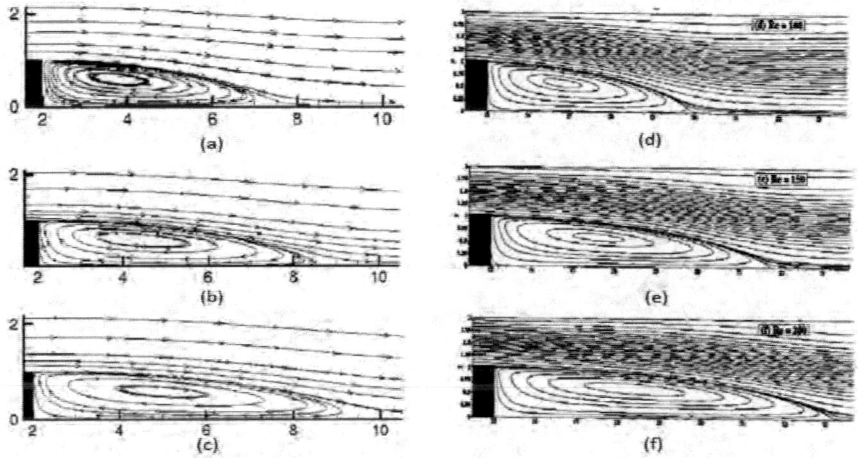

Figure 2: Instantaneous Streamline plots for Reynolds Numbers: (a), (d) Re=100, (b), (e) Re=150 and (c), (f) Re=200. Left hand side figures were generated by the present code, right hand side figures were created by Chowdhary and Dhiman [8].

4. Results and Discussion:

From the global flow characteristics, it can be said that the use of ZNMF and FA produce a very similar effect on the downstream wake patterns. Figure 3 presents the equally spaced temporal streamline and vorticity plots in a vortex shedding cycle. It was observed that, at frequency f=0.055, the vortex shedding was subdued and the drag reduction was found to be maximum. Jet velocity amplitude, A was 10% of the free stream velocity. It was already observed in previous studies, Creusé et al [4], Dahan et al [1], Hassan [5], among others, that exciting the flow at the vortex shedding natural frequency diminishes downstream vorticity for an appropriate amplitude of the injected jet. Figure 4 presents the vortex shedding when AFC frequency is f= 0.055 and the velocity amplitude is A=1.

For a jet velocity amplitude A=0.1 and the optimum frequency f=0.055, the drag on the step was minimum, as presented in figure 5b. As the velocity amplitude of the AFC jet increases above 0.1, the drag acting on the body also increases exponentially, figure 5b, and the same trend was observed in both zero net mass flow and fluidic actuator techniques. The increase in drag with jet velocity amplitude appeared to be higher in the case of fluidic actuators than in the case of zero net mass flow, proving that the technique involving absorption and injection is more effective than just periodic blowing. At high jet velocity amplitudes the energy of the dissipating vortex increases, leading to higher tendency of separation, see figures 4 (a, g) and 4 (d, j). Shear stresses on the upstream surface also increase causing the drag acting on the body to increase exponentially.

On figure 5a, the variation of Drag coefficient with active flow control frequency is presented. It can be observed that for a jet frequency around f=0.05, the drag coefficient oscillates. The minimum drag values appear at f=0.045 and f=0.055, which happens to be ±10% of the natural vortex shedding frequency when no flow control is applied. In the present case +10% i.e. f=0.055 is found to be slightly more effective than -10% in drag reduction, as seen in figure 5a, hence flow patterns in figures 3 and 4 are presented at this frequency. It is interesting to observe that between the two effective frequencies 0.045 and 0.055 and in both control techniques, the drag seems to increase and decrease following a zigzag cycle. This could be attributed to the resonance effect on the boundary layer in combination of the AFC frequency. Therefore a slight deviation from this frequency leads to instabilities until reaching an effective frequency of f=0.05±10%. The frequencies f=0.055 and f=0.045 seem to be the limits of oscillations in the drag coefficient, then the frequencies above and below them, leads to an increase of the drag coefficient.

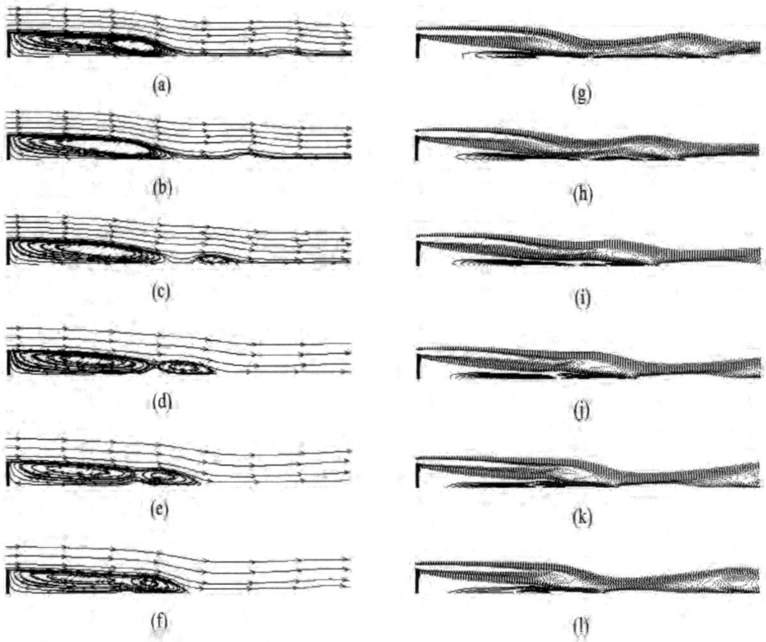

Figure 3: Streamline and corresponding Vorticity plots of vortex shedding cycle in the wake of Backward Step at Re=500, with upstream zero net mass flow control frequency f=0.055, A=0.1 and w=0.023a. Groove position 0.047a. (a, g) T/6; (b, h) T/3; (c, i) T/2; (d, j) 2T/3; (e, k) 5T/6; (f, l) T.

An interesting phenomena which can be observed from figures 5(a,b) and 6(a,b), is that with the variation of the parameters, the two flow control techniques exhibit similar behavior, although the zero net mass flow technique was found to be more effective than the fluidic actuator technique, regarding drag reduction. In ZNMF actuators, the flow is both absorbed and injected, whereas in FA the flow is only injected, causing a smaller absorption of the dissipating vortex energy in the wake.

Four different groove locations were as well evaluated, 0.024a, 0.047a, 0.072a and 0.097a, measured upstream surface edge. The groove optimum position for both actuating techniques, with respect to maximum drag reduction was 0.047a, as presented in figure 6a. At this position, flow control seems to influence the boundary layer in an efficient manner, to delay the separation and thereby reducing the drag on the body. When the groove is moved further upstream, it leads to boundary layer earlier separation and the drag on the body increases. When the groove is very close to the right edge, the boundary layers seems to become more unstable and separate earlier than in the case of optimum position, thereby enhancing the vortex dissipation and causing the drag to increase, as seen in figure 6a.

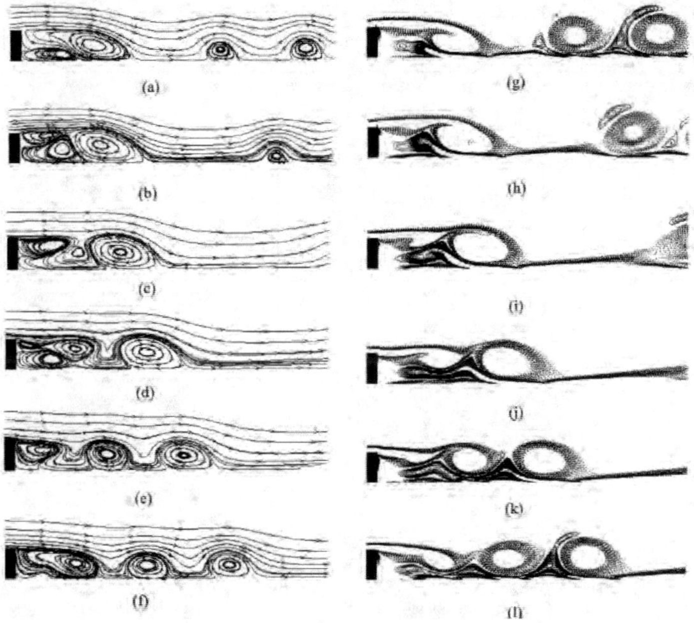

Figure 4: Streamline and corresponding Vorticity plots of vortex shedding cycle in the wake of Backward Step at Re=500, with upstream zero net mass flow control frequency f=0.055, A =1 and w=0.023a. Groove position 0.047a. (a, g) T/6; (b, h) T/3; (c, i) T/2; (d, j) 2T/3; (e, k) 5T/6; (f, l) T.

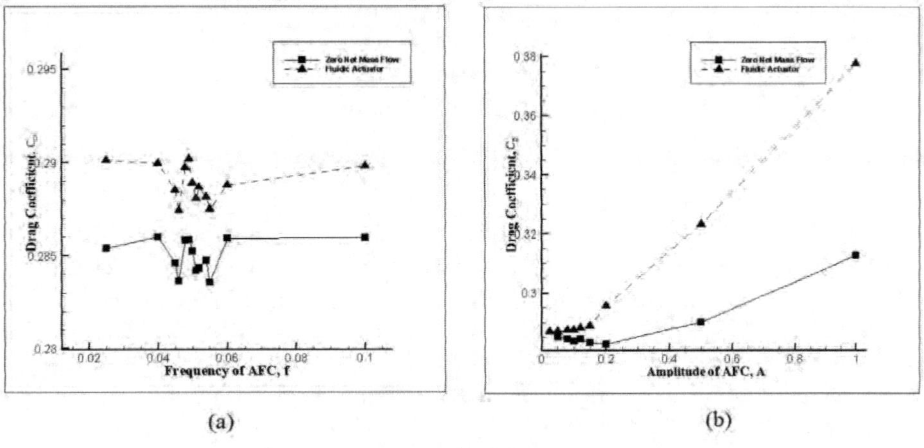

Figure 5: (a)Averaged Drag coefficient versus frequency, f of active flow control (AFC) (b) Averaged-Drag coefficient versus amplitude of AFC, A at Reynolds number, Re=500. Groove position 0.047a.

Groove width also plays a very important role regarding the body drag reduction, in the case of fluidic actuators, the drag acting on the body is directly proportional to the groove width, figure 6b. This is caused due to only injection of the fluid into the boundary layer. As the width increase, the volumetric flow into the boundary layer also increases, leading to more energy being induced into the boundary layer which causes it to separate quickly, increasing the drag acting on the body. In the case of zero net mass flow the drag decreases when the groove width is increased from 0.023a to 0.048a and increases when the width is further increased to 0.074a. The critical width at which the most effective amount of fluid being injected and absorbed into the boundary layer, which reduces the boundary layer instability, is around 5% of the step height. Whereas in the case of fluidic actuators, the most efficient width, was found to be around 2% of the step height, as seen in figure 6b.

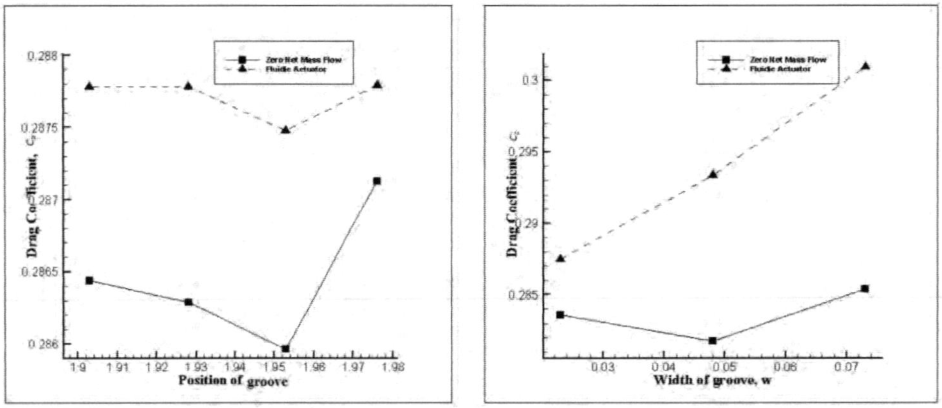

(a) (b)

Figure 6: (a)Averaged Drag coefficient versus position of groove (b) Averaged Drag coefficient versus width of the groove, w at Reynolds number, Re=500

5. Conclusions:

- It was found that ZNMF and FA exhibit a similar global behavior.
- The predominant factor affecting the flow is the AFC frequency.
- The optimum frequency was found to be ± 10% of the flow vortex shedding frequency.
- Velocity amplitude of 10% of the upstream velocity was found to be effective in both AFC techniques.
- The most effective position of the groove was 0.047a.
- Optimum groove width for ZNMF was about 5% of the step height, while for FA the optimum groove width was 2% of the step height.

References

[1] Dahan Jeremy A., Morgans A.S., Lardeau S. "Feedback control for form-drag reduction on a bluff body with a blunt trailing edge". J. Fluid Mech. 704; 360-387. (2012).

[2] Dejoan A., Jang Y.-J., Leschziner M.A. "Comparative LES and Unsteady rANS Computations for a Periodically-Perturbed Separated Flow Over a Backward-Facing Step". ASME-Journal of Fluids Engineering. 127; 872-878. (2005).

[3] Marrot F., Gajan P., Pauzin S., Simon F. "Experimental Application of an Active Control Loop on Backward-Facing Step Flow". AIAA Journal, 43:6; 1176-1186. (2005).

[4] Creusé E., Giovannini A., Mortazavi I. "Vortex simulation of active control strategies for transitional backward-facing step flows". Conputers & Fluids 38; 1348-1360. (2009).

[5] Hasan M. A. Z. "The flow over a backward-facing step under controlled perturbation: laminar separation". J Fluid Mech. 238; 73-96. (1992).

[6] Vukasinovic B., Rusak Z., Glezer A. "Dissipative small-scale actuation of turbulent shear layer". J Fluid Mech. 656; 51-81. (2010).

[7] Roberts GO. "Computational meshes for boundary layer problems". Proceedings of the Second International Conference on Numerical Methods and Fluid Dynamics, Lecture Notes in Physics. Springer-Verlag: New York. 8;171–177. (1971).

[8] Chowdhary H., Dhiman A. Two-Dimensional Laminar Fluid Flow and Heat Transfer over a Backward-Facing Step: Effects of Reynolds and Prandtl Numbers. Heat transfer research journal. 42: 4; 379-402 (2011).

Journal of Physics: Conference Series **633** (2015) 012111 doi:10.1088/1742-6596/633/1/012111

On the exit probability of the extended Sznajd model and the Kirkwood approximation

Andr M Timpanaro[1,2] **and Serge Galam**[2]

[1] Instituto de Fsica, Universidade de So Paulo, So Paulo, Brazil [2] CEVIPOF, SciencesPo, Paris, France

E-mail: timpa@if.usp.br

Abstract. The Sznajd model is a spin model, inspired by sociophysics, where 2 aligned spins influence a neighbouring spin to change its orientation. A bit of controversy arised in the literature recently about the properties of its one dimensional case, namely the exit probability, the probability that the system ends up in the "all spins up" state as a function of the proportion of spins pointing up in the initial condition. The Kirkwood approximation, which is a type of mean-field treatment, gives a surprisingly simple expression for this probability that has been verified for simulations with up to 10^7 sites. Nevertheless some aspects of the Kirkwood approximation cast doubt upon its validity. In this work, we use an extended version of the Sznajd model due to G. Kondrat, that allows for any interaction between a pair of spins and a neighbouring spin that respects the symmetry between up and down spins. Looking at the Kirkwood approximation for this case we obtain an expression for the exit probability in the cases where the "all up" state is absorbing and compare this probability with simulations, finding cases where the results are indeed not valid.

1. Introduction

The aplication of tools from physics (mainly statistical mechanics and dynamical systems) to sociology problems has gathered interest in the last decade, under the label of sociophysics. By taking these tools away from their usual domain of validity new questions arise that are themselves interesting to physical theories. This has happened recently with the study of one dimensional spin models inspired by social interactions.

The Sznajd model [1], which is the model in question has the following simple definition

- Consider a linear chain with periodic boundary conditions and sites that can be in one of 2 states: $+$ or $-$.
- At each time step choose a pair of neighbouring sites, i and $i+1$.
- If these 2 sites are in different states, then nothing happens.
- However, if they have the same state, them we choose at random one of their neighbours ($i-1$ or $i+2$) and change the state of this neighbour to match the state of the pair.
- We continue to follow these rules until all the sites in the linear chain have the same state.

It is trivial that this system possesses 2 absorbing states that be reached: "all spins up" and "all spins down". The question then arises about what is the probability that we end up in each of the absorbing states. This is given by the so called exit probability, more precisely, the exit

Journal of Physics: Conference Series **633** (2015) 012111 doi:10.1088/1742-6596/633/1/012111

probability is the probability $E(\rho)$ that we end up in the " all spins up" state when we start with an initial condition having no spatial correlations and a proportion ρ of up spins.

In recent papers [2, 3], the exit probability of this model was measured in simulations and a remarkably simple expression was found:

$$E(\rho) = \frac{\rho^2}{\rho^2 + (1-\rho)^2}.$$

This expression can be found analytically through a mean field taking into account pair correlations, called Kirkwood approximation in the literature. However a simple change in the way correlations are truncated in this approximation leads to nonsensical results, which cast doubt upon the validity of the approximation.

2. The Kondrat model and a test for the validity of the Kirkwood approximation

We turn now to the extension of the Sznajd model due to G. Kondrat [4]:

- Consider a linear chain with periodic boundary conditions and sites that can be in one of 2 states: $+$ or $-$.
- At each time step choose $s = \pm 1$ randomly (50% chance for each) and a site i. Let $j = i + s$ and $k = i + 2s$.
- Flip the state of site k with probability $p(\sigma_i, \sigma_j, \sigma_k)$, where σ_X denotes the state of site X.

The probabilities p are the most general possible while still keeping the symmetry between states $+$ and $-$. These probabilities (together with the social interpretation of the interactions related to them) are:

- $p(\sigma_i, \sigma_j, \sigma_k) = p(-\sigma_i, -\sigma_j, -\sigma_k)$
- $p(+,+,+) = p_1$ (anticonformity)
- $p(+,+,-) = p_2$ (Sznajd-like conformity)
- $p(-,+,-) = p_3$ (voter-like conformity)
- $p(+,-,-) = p_4$ (disagreement propagation)

The Kirkwood approximation can be used in this generalized model and the results concerning the states "all up" and "all down" are as follows:

- If $p_1 \neq 0$ then neither "all up" nor "all down" are absorbing states and so we can't talk about an exit probability.
- If $p_1 = 0$ but $p_2 \leq p_4$ the approximation predicts that the time to reach one of these states diverges, meaning that neither are reached in the thermodynamical limit (in a finite lattice we should see a 50-50 split between them, with very little dependance on the initial condition and extremelly long transients).
- If $p_1 = 0$ and $p_2 > p_4$ then the approximation predicts the following exit probability:

$$E(\rho) = \frac{\rho(1-\rho)p_3 + \rho^2(p_2 - p_4)}{2\rho(1-\rho)p_3 + (\rho^2 + (1-\rho)^2)(p_2 - p_4)}$$

which is a mixture of sorts of the Sznajd and voter exit probabilities.

We now show simulations for 2 parameter combinations that don't correspond to any model where the approximation has been tested yet. Figure 1 shows a case where the exit probability agrees with the Kirkwood approximation while figure 2 shows a case with disagreement.

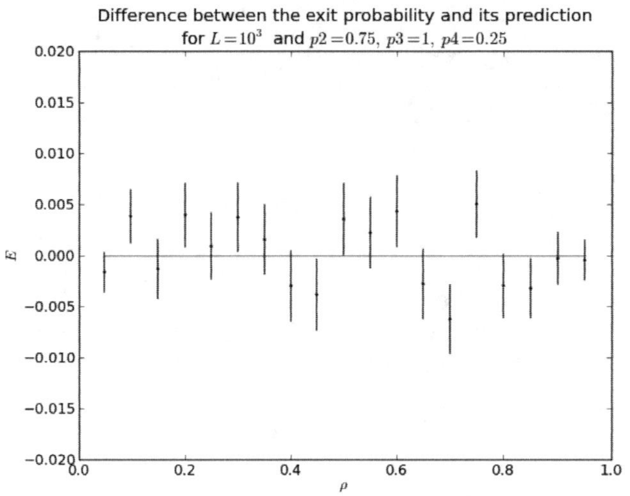

Figure 1. Simulations agreeing with the Kirkwood approximation

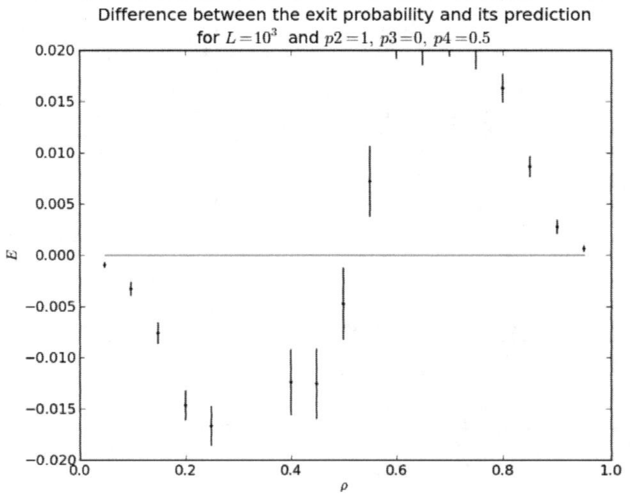

Figure 2. Simulations disagreeing with the Kirkwood approximation

Acknowledgments
A. M. Timpanaro would like to thank FAPESP for financial support.

References
[1] Katarzyna Sznajd-Weron, Józef Sznajd. Opinion evolution in closed community. *International Journal of Modern Physics C*, 11:1157, 2000.
[2] František Slanina, Katarzyna Sznajd-Weron, and Piotr Przybyła. Some new results on one-dimensional outflow dynamics. *Europhysics Letters*, 82:18006, 2008.
[3] Renaud Lambiotte and Sidney Redner. Dynamics of non-conservative voters. *Europhysics Letters*, 82:18007, 2008.
[4] Grzegorz Kondrat. How to introduce temperature to the 1D Sznajd model. *Physica A*, 390:2087, 2011.

Journal of Physics: Conference Series **633** (2015) 012112 doi:10.1088/1742-6596/633/1/012112

Using UAVs and digital image processing to quantify areas of soil and vegetation

A.A. Chaves[1], R.A. La Scalea[2], A.B. Colturato[1], C.L.O. Kawabata[2], E.L. Furtado[3], K.R.L.J. Castelo Branco[1]

[1] Institute of Mathematical and Computer Sciences, University of São Paulo - São Carlos, Brazil

[2] Federal Institute of Education, Science and Technology of São Paulo - São Carlos, Brazil

[3] Faculty of Agronomical Sciences, State University of São Paulo - Botucatu, Brazil

E-mail: arthurac@usp.br, la.scalea@me.com, adimara@gmail.com, celiak@gmail.com, elfurtado@fca.unesp.br, kalinka@icmc.usp.br

Abstract. Unmanned aerial vehicles (UAVs) are becoming a very popular tool for remote sensing and crop monitoring. They are more easily deployed, cheaper and can obtain images with higher spatial-resolution than satellites. Some small, commercial UAVs can obtain images with spatial-resolution as low as 1.5cm per pixel. This opens up the range of possible remote sensing and monitoring applications. Moreover, they can cover large areas in very little time, such as 50 ha in about 20min, which makes UAVs the ideal tool for monitoring large farms and plantations. On the other hand, it is important to know precisely the area covered by farms in order to avoid invasion of other properties or preserved areas, and also to detect flaws in the plantation area. However, it is difficult to measure planted areas in some cases, such as Eucalyptus crops. Therefore, this paper aims to evaluate the use of UAV imagery for precise area measurement in Eucalyptus crops. We developed an image-processing algorithm to segment regions of soil, low biomass and high biomass and tested it on a Eucalyptus plantation in the city of Lenis Paulista -SP, Brazil. Results show that the area quantification is very accurate especially for bare soil regions and this method can be used to estimate areas in other scenarios.

1. Introduction

The production of eucalyptus presents an important role in the international economy. With rapid growth, high yield, wide variety of species, great capacity of adaptation and applications in different industrial processes, the eucalyptus is the raw material for a wide range of products such as wooden panels, cellulose and paper and charcoal [1].

Precision agriculture [2], combined with unmanned aerial vehicles (UAVs) and remote sensing, can be used to increase productivity by monitoring crop status, estimating yield, counting plants, detecting flaws in the crop, assessing water stress and detecting plagues and diseases.

Remote sensing consists in the combined use of sensors, airships, satellites, and other equipments in order to study the Earth environment by recording and analysing the interactions between electromagnetic radiation and the materials that make up the planet Earth [3]. The advantages of the remote sensing in agriculture are the capacity of mapping vast areas in little time and also the possibility of an aerial vision of the crop, which can reveal details that otherwise couldnt be seen from ground level.

UAVs have been used in remote sensing because they have a smaller cost of acquiring high spatial resolution data when compared to manned airplanes or satellites [4]. Moreover, images obtained by UAVs tend to be more accurate and can be obtained anytime, including more than once per day, as opposed to satellites that only obtain images from the same area every few days.

Vegetation indices are values obtained from measures of spectral reflectance and can be used for several ends, including estimation of biomass, vegetation cover and production assessment. Among these indices, the Normalized Difference Vegetation Index (NDVI) [5] stand out as a tool for agricultural monitoring. It is calculated by the difference between Near-Infrared (NIR) reflectance and red reflectance as shown in Equation 1.

$$NDVI = \frac{NIR - Red}{NIR + Red} \tag{1}$$

2. Methodology

The proposed methodology can be divided into two stages: (i) image acquisition and (ii) image processing.

The first stage was performed in a commercial plantation of eucalyptus in the city of Lençóis Paulista in Brazil. The images were obtained using a Canon ELPH 100 HS, a camera that captures the green (G), red (R) and near-infrared (NIR) channels, aboard the eBee, a small sized UAV made by Sensefly[1]. Spatial resolution is 4 centimeters per pixel, flight altitude was about 130 meters and radiometric resolution is 2^8 bits.

After image acquisition, we can begin the second stage, which is the processing of the images obtained using C++ with the OpenCV library.

The first step is to calculate the value of the NDVI for each pixel and then create a pseudo color image with the HSV (Hue, Saturation, Value) model. In this model, the H value of a given pixel is proportional to the NDVI value of that same pixel. That is, if a pixel has a high NDVI value (close to 1), the pixel will be colored with a high H value (such as blue or magenta).

The use of this color scheme makes it easier to differentiate the elements in the image and therefore, facilitates the segmentation of the objects. In that way, it is possible to isolate the soil by selecting pixels with values between red and yellow (very low NDVI). Moreover, the same can be made for vegetation with low and high biomass by selecting, respectively, pixels between green and cyan (low to medium NDVI), and between blue and magenta (high NDVI).

This threshold generates six binary images (pixels with value either 0 or 255), one for each primary color of the HSV scheme, which are then merged into three images, one for each area of interest (soil, low biomass and high biomass). This method, however, generates some noise in the low and high biomass images that can be removed with a median filter of size 7x7, making the segmented areas more homogeneous and continuous.

Finally, with the regions already split, the process of quantifying the areas becomes very straightforward: simply count the number of pixels in each image and multiply by the area of each pixel and multiply by the spatial resolution. Therefore, we can estimate the areas of soil, and of vegetation with low and high biomass. Figure 1 shows every step in the algorithm described.

3. Results

The image of the area observed obtained from the eBee is shown in Figure 2a. Figures 2b and 2c show the results of the NDVI in gray scale and in pseudo colors using the HSV model, respectively. Figure 3 shows the segmentation results for (a) soil, (b) low biomass and (c) high biomass. Table 1 shows the size of the three types of areas considered.

[1] www.sensefly.com

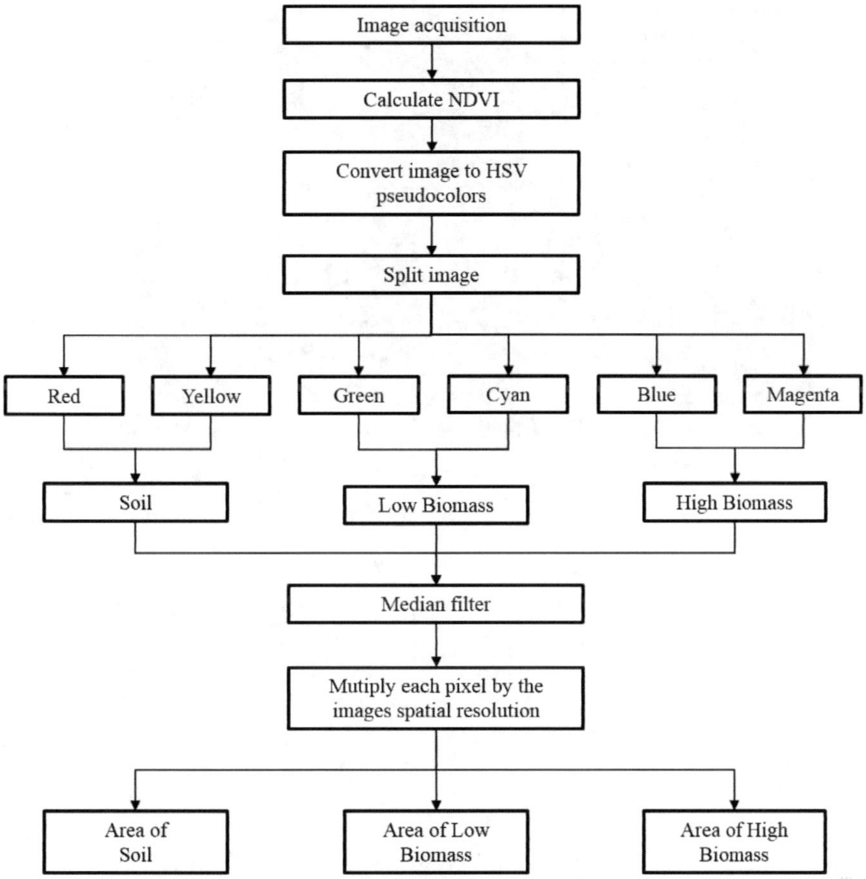

Figure 1. Algorithm for area quantification

Figure 2. (a) Area of study, (b) Grayscale NDVI, (c) Pseudo colors NDVI

The results of the segmentation can be seen in Figures 3a, 3b and 3c, respectively.

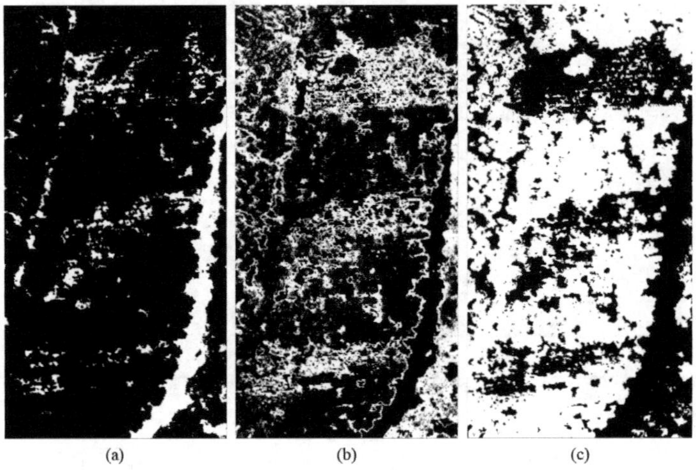

Figure 3. Segmentation results for (a) soil, (b) low biomass, (c) high biomass

Table 1. Areas of soil, low and high biomass present in the region shown in Figure 2a

	Soil	Low biomass	High Biomass	Total
Pixels	799559	1920186	3773618	6493363
m^2	1279.29	3072.3	6037.79	10389.38
ha	0.12	0.3	0.6	1.02

4. Conclusions

The goal of this study was to use aerial near-infrared images obtained from an UAV to quantify areas of soil and of vegetation with low and high biomass. For that, we developed an algorithm to process the images, segment the regions and estimate the given areas.

The algorithm for area quantification achieved very good estimatives, specially with the area of soil. However, it obtained a few errors in the classification of areas of low and high biomass because the values of the NDVI can be similar in some cases. Therefore, future works should aim to enhance the accuracy of this method in order to obtain more precise information about planted areas or areas with flaws in the plantation.

5. Acknowledgments

The authors acknowledge the support granted by CAPES, CNPq and FAPESP to the INCT-SEC (National Institute of Science and Technology - safety-critical Embedded Systems - Brazil), processes 573963/2008-9 and 08/57870-9 and FAPESP process number 2012/08498-5, 2014/12134-4.

References

[1] ABRAF 2013 Anurio estatstico abraf 2013: Ano base 2012 Tech. rep. Associao Brasileira Florestal dos Produtores de Florestas Plantadas Braslia, DF

[2] Mantovani E C, Queiroz D M and Dias G P 1998 *Mecanizao e agricultura de preciso* ed Silva F M d (Poos de Caldas: UFLA/SBEA) pp 109–157

[3] Novo E d M 1989 *Sensoriamento remoto: principios e aplicaes* (So Paulo: Edgard Blucher) 308p.

[4] Jensen J R 2009 *Sensoriamento Remoto do ambiente: Uma perspectiva sobre recursos terrestres* (So Jos dos Campos, SP: Ed. Parntese)

[5] Rouse J W, Haas R H, Schell J A and Deering D W 1973 *Third ETRS Symposium, NASA SP353* vol 1 (Washington, DC) pp 309–317

Journal of Physics: Conference Series **633** (2015) 012113 doi:10.1088/1742-6596/633/1/012113

Analysis of Stomata Distribution Patterns for Quantification of the Foliar Plasticity of *Tradescantia Zebrina*

Joao Batista Florindo, Gabriel Landini, Humberto Almeida Filho, Odemir Martinez Bruno

University of Sao Paulo, Brazil / University of Birmingham, UK

E-mail: jbflorindo@gmail.com

Abstract. Here we propose a method for the analysis of the stomata distribution patterns on the surface of plant leaves. We also investigate how light exposure during growth can affect stomata distribution and the plasticity of leaves. Understanding foliar plasticity (the ability of leaves to modify their structural organization to adapt to changing environmental resources) is a fundamental problem in Agricultural and Environmental Sciences. Most published work on quantification of stomata has concentrated on descriptions of their density per unit of leaf area, however density alone does not provide a complete description of the problem and leaves several unanswered questions (e.g. whether the stomata patterns change across various areas of the leaf, or how the patterns change under varying observational scales). We used two approaches here, to know, multiscale fractal dimension and complex networks, as a means to provide a description of the complexity of these distributions. In the experiments, we used 18 samples from the plant *Tradescantia Zebrina* grown under three different conditions (4 hours of artificial light each day, 24 hours of artificial light each day, and sunlight) for a total of 69 days. The network descriptors were capable of correctly discriminating the different conditions in 88% of cases, while the fractal descriptors discriminated 83% of the samples. This is a significant improvement over the correct classification rates achieved when using only stomata density (56% of the samples).

1. Introduction

Phenotypic plasticity is the ability of an organism to change its behaviour or structural organization to cope with the surrounding environment [1]. This is an important characteristic of plants, considering their immobility, in that it facilitates the survival in inhospitable environments such as those with inadequate temperatures or amount of light and nutrients.

Understanding where and how plasticity works in plants is an important challenge in the context of the climate changes observed in recent times. Plasticity can modulate organism structure faster than natural evolution and hence play a fundamental role in the adaptation process. Furthermore changes in plant behaviour may reflect the nature of environmental changes (e.e. pollution levels), or pathologies, where an understanding of plasticity might provide clues for disease diagnosis and prognosis.

Leaves are important interfaces between the plant and the environment and are very susceptible to plasticity. This may be expressed by external changes such as size, shape, thickness or by changes in the organization of their structure. Among the latter, the relation between the stomata density and the plasticity is well known. For instance, plants from the same species grown under different amounts of light tend to vary in the density of stomata on the leaf surface. However, their distribution is not fully described only by the density. In this context, we propose a mathematical-physical approach to describe such distribution in a more complete and precise way. The proposed analysis was performed over samples from species *Tradescantia Zebrina*, as the stomata can be easily identified in leaves from that species.

2. Method

Two state-of-the-art approaches in shape analysis were investigated: multiscale fractal dimension and complex networks.

2.1. Multiscale fractal dimension

Fractal descriptors provide multiple scale features of an image or point set considered to be an approximation of a fractal set. This modelling is suitable to analyse complex biological morphologies, including the stomata distribution considered here, that exhibit high degrees of complexity.

We propose using a method called 'multiscale fractal dimension' (MFD), originally applied to shape analysis in [2], but generalized to sets of disjoint points in the current application. Essentially, this consists of simultaneously dilating all points in the set with circular structural elements of radius r and then computing the total area of the dilated structure $A(r)$. Finally, the descriptors are computed from the derivative of the function $\log(r) \times \log(A(r))$ when r is ranged within an interval of values. Here r ranges from 1 to 100 pixels.

2.2. Complex networks

Complex networks is an area of study combining Graph Theory with analytical tools from Statistical Physics [3]. The essential idea is to model a problem structure in terms of a graph to then extract measures from that graph, which then can be used as features to represent the modelled system.

In the case when a set of points is to be analysed, the strategy adopted in [4] can be adapted by considering each set point as a graph vertex and the Euclidean distance between points as a graph edge weight. Likewise in [4] we follow an approach named dynamic diffusion, where starting from a complete graph (with all vertices are connected) a threshold function $\phi(G, t)$ is applied to remove all edges with weights smaller than t.

When a range of values of t is used, one network is generated for each threshold and measures can be computed for each of those networks and used as features characterising the original set of points. Here a range of Euclidean distances between 1 and 100 was applied and the average degree (number of vertices connected to each node) was computed for each thresholded network.

3. Results

Fractal and complex network descriptors were extracted from the distribution of stomata over the leaf surface from the plant *Tradescantia Zebrina* under different illumination conditions: sunlight, 4 hours of artificial light every day and 24 hours of artificial light every day, during 69 days in a greenhouse with controlled conditions. The images were collected by using a Zeiss discovery V.20 stereo microscope coupled to an Axiocam ICc1 camera. Figure 1 shows images of three leaves (one for each light condition) and the stomata manually identified as points in a binary image while figure 2 shows the average descriptors (complex networks and fractal) for each light condition. To discriminate between the conditions, both descriptor types were submitted to a 2-Nearest-Neighbour classifier using χ^2 distance and following a leave-one-out scheme [5].

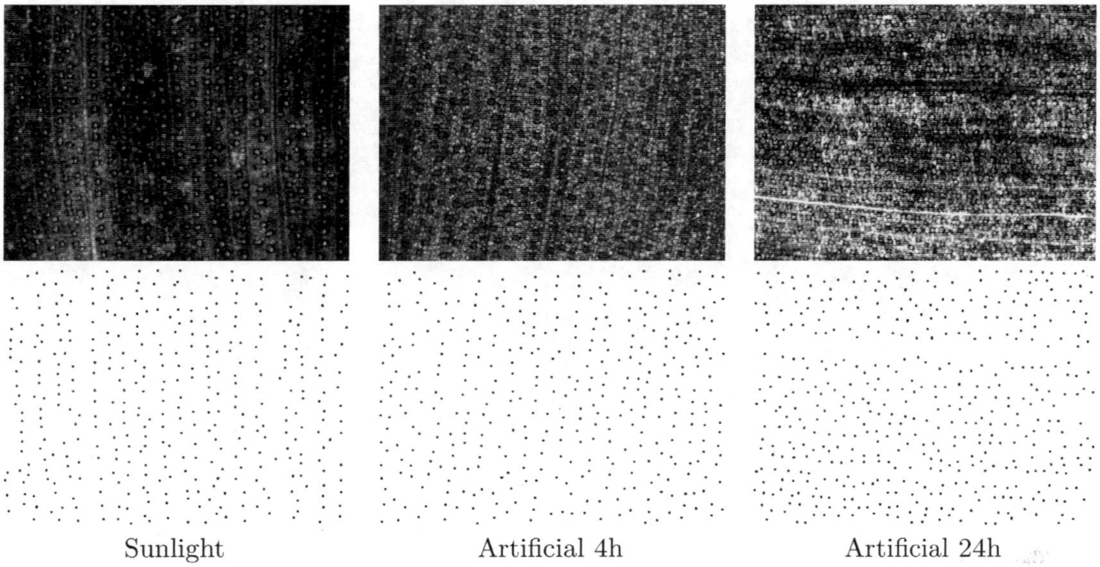

Figure 1. Examples of *Tradescantia Zebrina* leaf samples and respective stomata distribution.

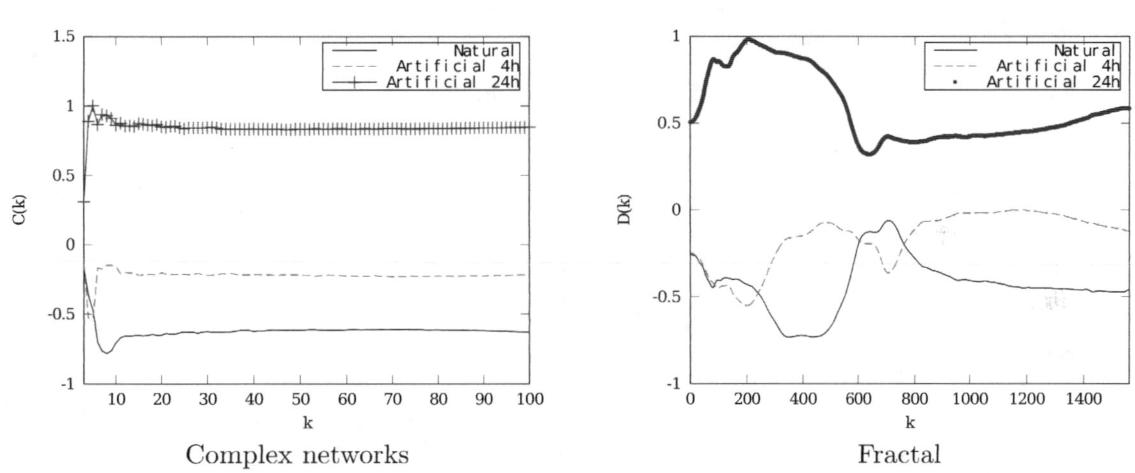

Figure 2. Average Descriptors (z-score normalized) of stomata distribution in each light condition.

Figure 3 shows the rates of correct classification when using varying numbers of descriptors in both fractal and network modelling. Twenty-one fractal descriptors were sufficient to correctly discriminate 83.3% of the samples (15 samples), whereas when 62 network descriptors were used, the discrimination increased to 88.9% (16 samples). This result significantly improves the classification results obtained when considering only stomata density (55.6% correct). Table 1 shows the respective confusion matrices. We observe that it is possible to separate samples of plants grown under sunlight from those under artificial light. However, when only the exposition hours to artificial light is taken into account, the difference in stomata organization appears to be more subtle. Both approaches reveal that stomata patterns on leaves can be successfully quantified and that detectable differences appear under the various experimental conditions investigated here. These preliminary results suggest that it might be possible to reveal the effects of other environmental factors such as pollution or diseases affecting forests or crops, hence providing more quantitative tools to allow precise planning in the detection and solution of such problems.

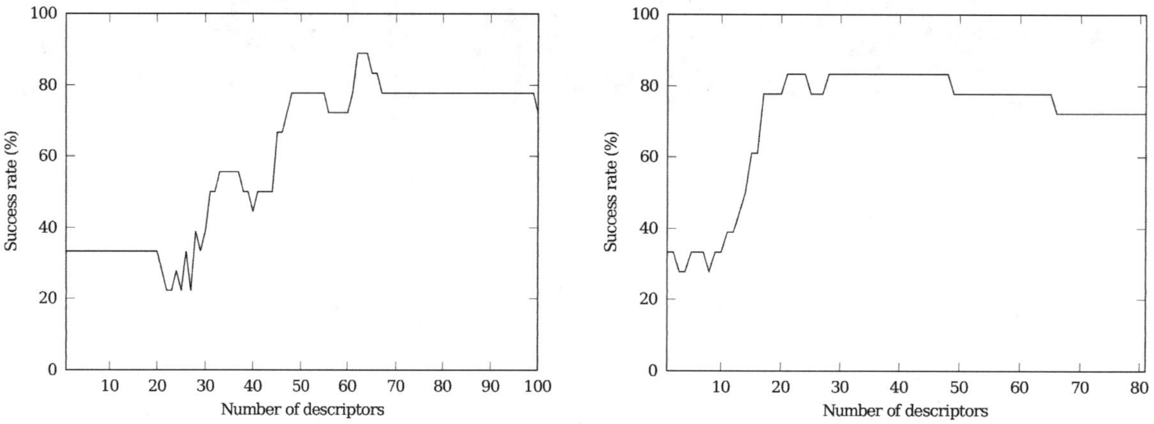

Complex networks Fractal

Figure 3. Correctness rates.

Density

		Predicted		
		S	A4h	A24h
Actual	S	6	0	0
	A4h	3	2	1
	A24h	1	3	2

Complex networks

		Predicted		
		S	A4h	A24h
Actual	S	6	0	0
	A4h	1	5	0
	A24h	1	0	5

Fractal

		Predicted		
		S	A4h	A24h
Actual	S	6	0	0
	A4h	2	4	0
	A24h	1	0	5

Table 1. Confusion matrices. S = Sunlight. A4h = Artificial light for 4 hours. A24h = Artificial light for 24 hours.

4. Conclusions

We showed a systematic representation and quantification of the stomata distribution over the leaf surfaces of (*Tradescantia Zebrina*) by means of two state-of-the-art shape analysis methods, namely the multiscale fractal descriptors and complex networks. The results showed that both approaches were capable of discriminating between three different experimental lighting conditions used during the plant growth.

The results confirm that the amount and type of exposition light influences the stomata arrangements in the leaves. Furthermore, although the stomata density could identify the samples grown under sunlight, our analysis was able to discriminate between samples grown under 4 and 24 hours of artificial light a day. These results exemplify how imaging methods can help in quantify and elucidating the relationships between plants and the environment.

ACKNOWLEDGEMENTS

Joao Batista Florindo acknowledges support from FAPESP (The State of São Paulo Research Foundation) (Grants No. 2013/22205-3 and 2012/19143-3). Odemir Martinez Bruno acknowledges support from CNPq (National Council for Scientific and Technological Development, Brazil) (Grant Nos. 484312/2013 and 308449/2010).

References

[1] Molina-Montenegro M A and Naya D E 2012 *PLoS ONE* **7** e47620
[2] Florindo J B, Backes A R, de Castro M and Bruno O M 2012 *Pattern Recognition Letters* **33** 798–806
[3] Costa L F, Rodrigues F A, Travieso G and Villas Boas P R 2006 *Advances in Physics* **56** 167–242
[4] Backes A R, Casanova D and Bruno O M 2009 *Pattern Recognition* **42** 54–67
[5] Duda R O and Hart P E 1973 *Pattern Classification and Scene Analysis* (New York: Wiley)

Journal of Physics: Conference Series **633** (2015) 012114 doi:10.1088/1742-6596/633/1/012114

Movement of a vortex filament near oscillating pinning centers in the hard superconductor

D Apushkinskaya[1], E Apushkinskiy[2] and M Astrov[3]

[1] Department of Mathematics, Saarland University, Saarbrücken, Germany
[2] St. Petersburg Polytechnical University, St. Petersburg, Russia
[3] D.V. Efremov Research Institute of Electrophysical Apparatus, St. Petersburg, Russia

E-mail: darya@math.uni-sb.de

Abstract. A problem on a vortex behavior near oscillating pinning centers in the type III superconductors (SCs) is considered. The movement of a vortex filament is described by the parabolic free boundary problem, where an a priori unknown free boundary is the set of points, at which a transition from superconducting to normal state and vice versa occurs. A number of statements on the qualitative properties of a phase boundary superconductor - nonsuperconductor are considered.

1. Introduction

As is known, the real type II superconductors (the so called hard or type III SCs) involve different defects of a crystal lattice, and the magnetic field penetrates into a superconductor in the form of separate vortices interacting with these defects. The arising forces (known as pinning forces) form a state of vortex matter with special electro-acoustic properties. The vortices have different lengths provided by the intricate configuration of the superconducting grains. The vortices may be pinned to a crystal lattice in cites of its defect. These are regions with broken superconductivity due to the changes in the carrier-lattice interaction, that results in the deterioration of conditions for superconducting pair production (low density of superconducting pairs at defects). Since the crystal defects are distributed inside a grain in a complicated manner and the magnetic field penetrates into the sample regions without superconducting pairs only, the configuration of each magnetic filament (vortex) is also quite complicated (Fig. 1).

Determination of the vortex interaction with a concrete pinning center (elementary pinning force) is a hard problem because of the complicated microstructure of a real sample. Direct experimental investigations of the elementary pinning forces require very complicated instrumentation as well as experimenter skills [1]. Theoretical investigations of the elementary pinning forces are made, mainly, in the linear approximation and far from all the defect types [2]. It is well-known that hard SCs are a strongly nonlinear medium. So, the works, where elementary pinning force analysis is carried out without substantial limitations related with a kind of interaction between vortex and crystal lattice defects, are of a great interest.

2. Theory

We consider the vortex oscillations excited by an external ac field action [3] in the absence of any vortex jumps between the pinning centers. The surface exposed to the vortex magnetic moment

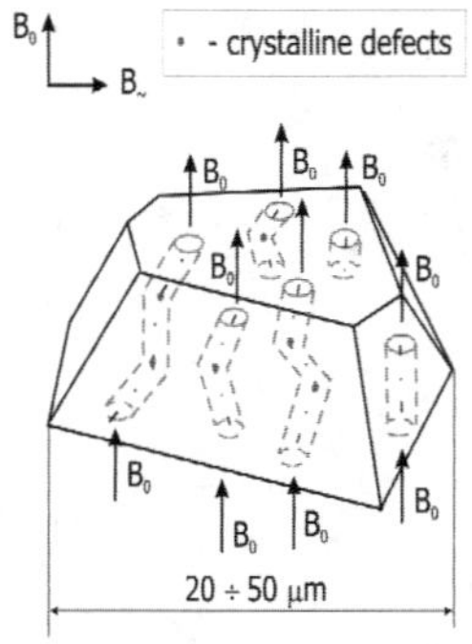

Figure 1. A superconductor powder grain in the external dc magnetic field $\mathbf{B}_0$

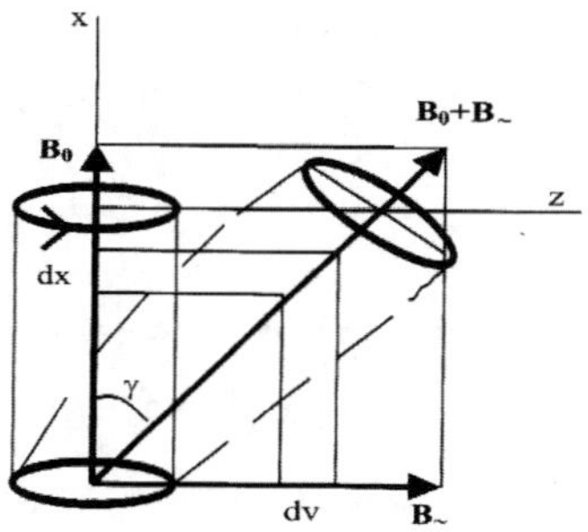

Figure 2. Vortex oscillation caused by ac field $\mathbf{B}_\sim$

should oscillate following the total external magnetic field (Fig. 2). Simultaneously, the vortex, which forms it, should also oscillate around the equilibrium position. The vortex axis deviates from the equilibrium position by $v(x,t)$. In geometry terms, $\tan\gamma = v_x = B_0^{-1}B_\sim(t)$ (Fig. 2) that determines a formula for the vortex deviation at boundaries.

Let us write the equation of motion for a vortex displacement v for volume element ΔV

$$-G_d\mathbf{v}_t - \mathbf{F}_{rest} + [\mathbf{j}_{RF}\mathbf{B}_0]\,\Delta V + \mathbf{F}_T = m_f\mathbf{v}_{tt}, \tag{1}$$

similar to that presented in [4]. The forces entering in (1) are:

- $G_d\mathbf{v}_t$ is the force of viscous decay of the vortex motion, which manifests itself only at vortex filament edges due to radiation and at the pinning sites;
- $\mathbf{F}_{rest}$ is the retrieving force conditioned by the interaction with other fluxoids and the voltage induced by the randomly distributed separate pinning centers;
- $[\mathbf{j}_{RF}\mathbf{B}_0]\,\Delta V$ is the Ampere force; The induction $B_\sim(x)$ at sample boundary is determined by the external action and it decreases while penetrating deep into the sample. Hence, $v_{xx} = B_0^{-1}(B_\sim)_x$ and $[\mathbf{j}_{RF}\mathbf{B}_0] = \mathbf{k}\mu_0^{-1}B_0^2 v_{xx} = \mathbf{k}C_{44}v_{xx}$, where $C_{44} = \mu_0^{-1}B_0^2$ is introduced for the vortex lattice modulus of elastisity.
- $\mathbf{F}_T$ is the random force of thermal deviation, which is considered to be zero;
- $m_f\mathbf{v}_{tt}$ is the force of inertia, which is usually small near the grain boundary in the external ac magnetic field, because the vortex inertia mass m_f of its core is very small.

If the external action is quite strong, one can neglect m_f in Eq. (1). This assumption is valid only at boundary of the type II SC or in the vicinity of the pinning centers inside SC. $\mathbf{F}_{rest}$ is zero in the equilibrium state; for the nonequilibrium state it can be written as

$$F_{rest} = \alpha_L v dV + f(x,t)\chi_{\{v>0\}}. \tag{2}$$

Here α_L is the Labusch parameter, while the values of $f(x,t)$ are determined by the pinning center distribution, the entanglement of the vortices, the anisotropic properties of SC as well as by the transport current if the latter exists. The function $\chi_{\{v>0\}}$ denotes the characteristic function of a set $\{v>0\}$, i.e., $\chi_{\{v>0\}}(x,t) = 1$ if $v(x,t) > 0$ and $\chi_{\{v>0\}}(x,t) = 0$ otherwise.

Considering the aforesaid, Eq. (2) at the sample boundaries and in the vicinity of pinning centers under the strong external action takes the form

$$C_{44}dV v_{xx} - G_d v_t - \alpha_L v dV = f(x,t)\chi_{\{v>0\}}. \tag{3}$$

Since the vortex deviations from the equilibrium state in the positive and negative directions are absolutely symmetric, we restrict our analysis by a case $v \geq 0$.

The model equation (3) can be treated as the parabolic obstacle problem

$$a(x,t)v_{xx} + b(x,t)v_x + c(x,t)v - v_t = f(x,t)\chi_{\{v>0\}}, \quad (x,t) \in Q_R, \tag{4}$$

$$v(x,t) \geq 0 \quad \text{in} \quad Q_R := \left\{(x,t) \in \mathbb{R}^2 : |x| < R, k|t| < R^2\right\}. \tag{5}$$

For coefficients a, b, c and function f we assume that

(i) a and f are nondegenerate in Q_R, i.e., there is $\delta_0 > 0$ such that $a(x,t) > \delta_0$ and $f(x,t) > \delta_0$ for any $(x,t) \in Q_R$;

(ii) a, b, c and f belong to $C^{\alpha,\alpha/2}(Q_R)$ for some $\alpha \in (0,1)$.

Note,$\{v > 0\}$ is a priori unknown subset of Q_R. We denote by Γ the intersection of Q_R with the boundary of the set $\{v = 0\}$. Observe that an a priori unknown interface Γ describes a set of points in the plane (x,t) where transition from superconducting to normal state or vice versa is realized. Indeed, suppose that a vortex is initially fixed and its core has a stationary boundary separating the superconducting area outside a vortex and normal region inside it. After beginning of the external action on the vortex, some deviation v of the core boundary from its stationary state has place. The separation point starts moving from the sample boundary or from the pinning centers. So, on the plane (x,t) the trajectory of separation point is given by $x = g(t)$ determining the interface between superconducting and normal areas.

3. Discussion

3.1. Qualitative properties of v

(S1) The problem (4)-(5) has a unique solution v under suitable initial and boundary data [5].

(S2) Functions v and v_x are continuous and the set $\{v = 0\}$ is closed [6, 7].

(S3) Locally, v has bounded second derivatives in space and first derivative in time [8].

(S4) v_{xx} is discontinuous across Γ.

(S5) v_t is continuous for almost every t. More precisely, $v_t = 0$ in the interior of the set $\{v = 0\}$. If $v(x^0,t^0) > 0$, the standard parabolic estimates imply that v_t is continuous near (x^0,t^0). So, the only difficulty is the regularity of v_t on Γ. We emphasize that the continuity of v_t cannot be obtained everywhere in t. The relevant counterexample can be found in [8].

(S6) If, in addition, $v_t \geq 0$, then v_t is continuous for all t and satisfies $v_t = 0$ on Γ. The assumption, that v_t is nonnegative, can be established in some special cases.

(S7) v is nondegenerate on the closure of $\{v > 0\}$, i.e., there is an absolute constant c such that

$$\sup_{Q_\rho^-(x^0,t^0)} v \geq c\rho^2, \qquad c > 0, \tag{6}$$

where $Q_\rho^-(x^0,t^0) := \left\{(x,t) \in \mathbb{R}^2 : |x - x^0| < \rho,\ t^0 - t < \rho^2\right\}$ and ρ is small enough [8]. Ineq. (6) prevents Γ from being flat with the set $\{v = 0\}$ below (such case is shown on Fig. 3).

3.2. Properties of the phase boundary Γ

(FB1) Ineq. (6) guarantees that Γ cannot appear or disappear suddenly, or is not "blurred".

(FB2) There is a set $\Gamma^* \subset \Gamma$ such that the Lebesgue measure of the set $\Gamma \setminus \Gamma^*$ is equal to zero. The set Γ^* is locally contained in a $C^{1/2}$-graph as a function of t [9].

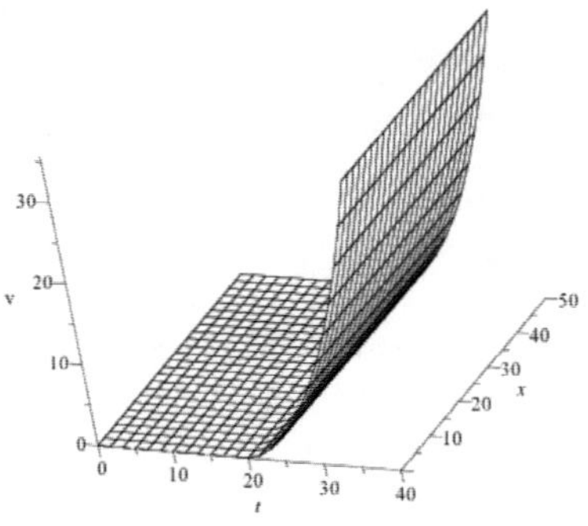

Figure 3. Impossible case of Γ dynamics

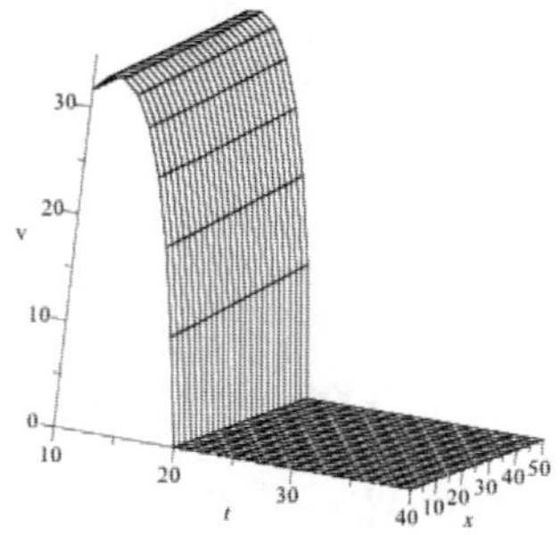

Figure 4. Possible case of Γ dynamics

3.3. *Interpretation*

- Properties (S2) and (S3) guarantees that the propagation of superconductivity along a sample proceeds with the speed bounded for almost all t. The latter means that the Lebesgue measure of the set t, where v_t is unbounded, is equal to zero. Formation or destruction of superconducting carriers occur precisely in such moments.
- It follows from (S7) that the formation of superconducting carriers cannot occur simultaneously on the spatial interval. The opposite case, i.e., simultaneous destruction of carriers on the spatial interval, is allowed (Fig. 4).
- Due (S2), a vortex remains elastic at all x and t. It was verified experimentally [1].
- Γ describes the set of points, where transition from SC to normal state or vice versa is realized. In view of (FB2), for almost every t the set Γ can be locally defined by the equation $x = g(t)$ with bounded continuous function g satisfying the inequality $|g(t)| \leq C\sqrt{t}$ with an absolute constant $C > 0$. Thus, in the local coordinate system for almost all t we have a growth estimate for the function describing Γ.

4. Conclusions

Based on analysis of the model we prove finiteness of propagation speed for the boundary between normal and superconducting (SC) states and show that

- The force of decay of the oscillating vortex is continuous for almost all t;
- A transition from normal to SC state is impossible on the spatial interval. The opposite case, i.e., a transition from SC to normal state on the spatial interval is allowed;
- The boundary between normal and SC states cannot appear or disappear suddenly.

References

[1] Rai M, Luc T, Masamitsu H, Bazaliy Y B, at all 2008 *Nature Physics* **4** 368-72
[2] Kramer E J 1973 *J. App. Phys.* **44** 1360-70
[3] Petrov M P, Pleshakov I V, Paugurt A P, et al. 1991 *Solid State Com.* **78** 893-95
[4] Van der Beek C J, Geshkenbein V B and Vinokur V M 1993-I *Phys. Rev.* B **48** 3393-403
[5] Friedman 1993 *Variational Principles and Free-Boundary Problems* (Malabar. FL: R E Krieger Publ. Co.)
[6] Ladyzhenskaja O A, Solonnikov V A and Uralceva N N 1967 *Linear and Quasilinear Equations of Parabolic Type* (Providence RI: AMS)
[7] Lieberman G M 1996 *Second Order Parabolic Differential Equations* (River Edge. NJ: World Sc. Publ. Co.)
[8] Blanchet A, Dolbeault J and Monneau R 2006 *J. Math. Pures Appl.* **85** 371-414
[9] Blanchet A 2006 *Nonlinear Anal.* **65** 1362-78

Analysis of paramagnetic centers for threevalent iron in aluminosilicates

D Apushkinskaya[1], E Apushkinskiy[2], B Popov[2], V N Romanov[2], V Saveliev[2] and V Sobolevskiy[2]

[1] Department of Mathematics, Saarland University, Saarbrücken, Germany
[2] St. Petersburg Polytechnical University, St. Petersburg, Russia

E-mail: darya@math.uni-sb.de, apushkinsky@hotmail.com

Abstract. We present the results of investigation of the defects in fluorine aluminosilicates from the Volyn-field $Al_2[SiO_4][F,OH]_2$ by the Electron Paramagnetic Resonance (EPR) method. The studies were carried out on the spectrometer Bruker ER 220D. Three types of EPR spectra of single centers were obtained. Their angular dependence was also investigated. The obtained EPR spectra correspond to the paramagnetic ion Fe^{3+} in the high-spin state $S = 5/2$. Three types of paramagnetic centers were found: one with cubic-symmetry and two with orthorhombic-symmetries.

1. Introduction

The processes of minerals formation in geological environments are still not quite understood because of their complexity and diversity. A complex multi-stage evolution of natural materials causes to formation of various defects in the crystal lattice, i.e., to deviations from the ideal crystal structure.

Studying of local defects in the minerals-silicates formed in the process of their growth is of great interest. Recent investigations [1] show that the local defects arising during the minerals growth are stable. Such defects define the optical, electrical and magnetic properties of the silicates [1,2].

On the other hand, there are so called radiation defects of the minerals. They arise after irradiation of the mineral that upset the crystalline structure energy balance. In this case, opening of separate chemical bonds, redistribution of free electrons between the lattice ions and impurities as well as formation of short-lived isotopes of chemical elements are possible. This type of defects is unstable. It manifests itself by degradation of mechanical, electrical and other properties of silicates that limits probable application of such materials.

Unfortunately, there are no reliable criteria to identify the defects of natural or radiation origin. In the main, this is due to our insufficient knowledge about the nature of point defects in aluminosilicates.

Embedding of the iron ions in the state ^6S $(3d^5)$ into the crystalline lattice of a mineral can provide data accessing about its structure [2]. As a brilliant representative of aluminosilicates one can point to topaz. Its chemical formula is $Al_2[SiO_4][F,OH]_2$, and the crystal structure is built with silicon-oxygen tetrahedra (SiO_4) and aluminum octahedra surrounded by four oxygen ions and two fluoride ions. Defects formed by the substitutional ions are of most practical interest. In the topazes there are two types of defects: octahedral and tetrahedral. The type of defect depends on the substitutional ions location. If silicon in silicon-oxygen tetrahedra (SiO_4) is replaced by aluminum ($Al^{3+} \rightarrow Si^{4+}$), as well as by impurity ions such as chromium (Cr^{3+}) or iron (Fe^{3+}), we observe luminescence in different spectral regions. For example, luminescence in aluminosilicates can be caused by tetrahedrally coordinated impurity ions Cr^{3+}, Fe^{3+} and Mn^{2+} which isomorphically replaced Si^{4+} in silicon-oxygen tetrahedra [1].

2. Experiment / Calculation / Theory

The intrinsic defects in fluorine aluminosilicates (topazes) were investigated by the EPR method. The studies were carried out with spectrometer "Bruker" ER 220D. We have obtained three types of EPR spectra of single centers (Fig. 1) and studied their angular dependence (Fig. 2).

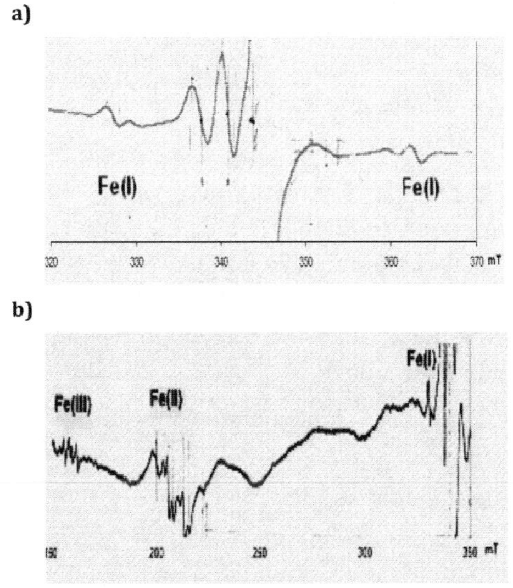

Figure 1. Examples of EPR spectrum in topaz.
a) Fe^{3+} in an octahedral encirclement;
b) Fe^{3+} in an tetrahedral encirclement.

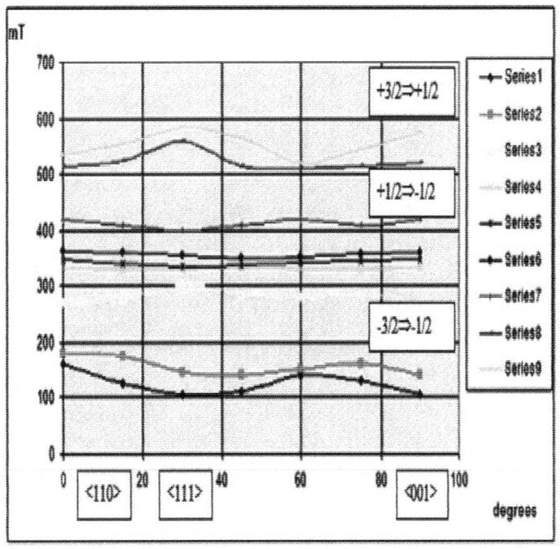

Figure 2. The angular dependence of EPR spectrum (Fe^{3+}) in topaz. Series 1, 2, 4, 5, 8, 9 - tetrahedral encirclement; Series 3, 6, 7 - octahedral encirclement.

Three types of spectra of iron centers with different symmetry were observed. Each spectrum consists of 5 lines of the fine structure and corresponds to high-spin state $S = 5/2$. However, the ratios of the intensities of the individual transitions are different. The ratios of intensities in the spectrum of Fe (I) equal $8 : 5 : 9 : 5 : 8$. The intensity ratios for similar transitions for Fe (II) and Fe (III) are $5 : 8 : 9 : 8 : 5$.

Thus, it can be assumed that the center of Fe (I) corresponds to the octahedral coordination of the oxygen, while the centers of Fe (II) and Fe (III) correspond to the tetrahedral one. In order to prove the assumption about the symmetry of location centers we made numerical calculations and simulation of the EPR spectrum of ion Fe^{3+} using the model of the spin Hamiltonian. This

Hamiltonian for the ground state ^{6}S has the form [3]:

$$\mathbf{H} = g\beta HS + \frac{a}{6}\left\{\left[Sx^4 + Sy^4 + Sz^4\right] - \frac{1}{5}\left[S(S+1)(3S^2+3S+1)\right]\right\} \tag{1}$$

where g is the spectroscopic splitting factor, b is the Bohr magneton, H is the strength of the constant magnetic field, and a is the constant of the level splitting for the ground state by a crystal field of cubic symmetry. Energy levels determining the position of the EPR lines are calculated in the second-order perturbation theory in a strong magnetic field with direction cosines l, m and n with respect to the axis of the fourth order:

$$\begin{aligned}
E_{\pm 5/2} &= \pm\frac{5}{2}g\beta H + \frac{1}{2}pa \pm \left(212 - 24p - 113p^2\right)\frac{a^2}{240g\beta H} \\
E_{\pm 3/2} &= \pm\frac{3}{2}g\beta H - \frac{3}{2}pa \pm \left(12 + 8p - 15p^2\right)\frac{a^2}{16g\beta H} \\
E_{\pm 1/2} &= \pm\frac{1}{2}g\beta H + pa \pm \left(-2 - 3p + p^2\right)\frac{a^2}{3g\beta H}
\end{aligned} \tag{2}$$

where $p = l - 5(l^2m^2 + m^2n^2 + l^2n^2)$.

Extreme values of p are equal to +1 for the direction $< 001 >$, -2/3 for the direction $< 111 >$ and -1/4 for the direction $< 110 >$. Consequently, the angular dependence let us to determine the direction of chemical bonds in the complex. All the measurements were carried out in high magnetic fields ($g\beta H \gg a$). The values of the second order with respect to a were neglected. In this case, the EPR spectrum consists of 5 lines of the fine structure, for those the magnetic quantum number m is changed to 1. The latter corresponds to transitions:

$$\begin{aligned}
\Delta E_{(5/2 \Rightarrow 3/2)} &= g\beta H + 2pa \\
\Delta E_{(3/2 \Rightarrow 1/2)} &= g\beta H - \frac{5}{2}pa \\
\Delta E_{(1/2 \Rightarrow -1/2)} &= g\beta H \\
\Delta E_{(-1/2 \Rightarrow -3/2)} &= g\beta H + \frac{5}{2}pa \\
\Delta E_{(-3/2 \Rightarrow -5/2)} &= g\beta H - 2pa
\end{aligned} \tag{3}$$

The following values of parameters of the spin Hamiltonian for Fe (I) in an octahedral encirclement of oxygen were obtained: $g = 2.004 \pm 0.003$; $a = 120 \cdot 10^{-4}$ cm^{-1}. In addition, the central transition width measured between the peaks of the first derivative points, $\Delta H = (11 \pm 1)$ Oe is established as well. The parameters of EPR spectra for Fe (II) and Fe (III) in tetrahedral symmetry are shown in Table 1.

Table 1. The parameters of the EPR spectra.

g-factor	ΔH [Oe]	a [cm^{-1}]
2.004	11	$120 \cdot 10^{-4}$
2.022	14	$620 \cdot 10^{-4}$
2.040	17	$700 \cdot 10^{-4}$

3. Discussion

Thus, it can be concluded that three types of paramagnetic iron centers in the S-state are formed in aluminosilicates, the total spin of centers equals 5/2, but the fine structure parameters as well as the symmetry of centers are different. The analysis of the angular dependence shows that centers such as Fe (I) are formed by iron ions in the state $3d^5$ replacing aluminum in a cubic lattice surrounded by four oxygen ions and two fluoride ions. This is also confirmed by the presence the lines of the superhyperfine structure of low intensity of the nuclear magnetic moments of fluorine ($I = 1/2$) in the EPR spectrum of Fe (II) center. The centers of Fe (II) and Fe (III) occur in tetrahedras of SiO_4 with one (center Fe (II)) or two (center Fe (III)) oxygen vacancies by substituting $Fe^{3+} \rightarrow Si^{4+}$. The fragment of the aluminosilicate lattice with a tetrahedral iron center and one oxygen vacancy is shown in Fig. 3.

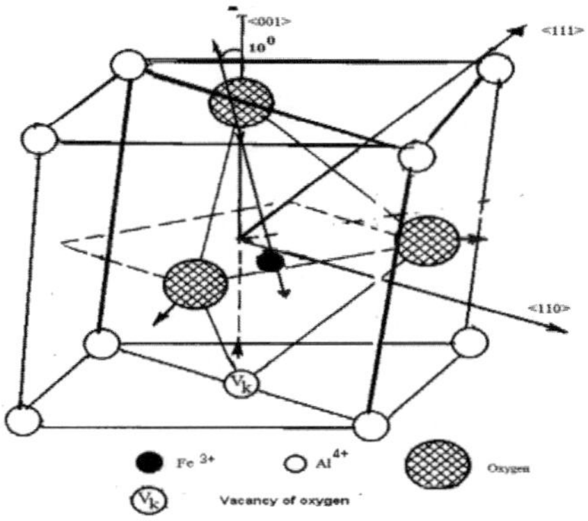

Figure 3. Fragment of the center Fe^{3+} with oxygen vacancies at the angle $\varphi = 10°$.

It is important to note that the presence of vacancies implies a significant distortion the symmetry of the center [4]. Substituting ion is displaced from its equilibrium position at a distance $d = 0.544L \tan\varphi$. The displacement for the center Fe (II) is equal 0.17 Å, while the center of Fe (III) is shifted to 0.27 Å from the center of the tetrahedron.

4. Conclusion

The above-described centers were observed in non-irradiated topaz crystals. So, their structure and electron states are stable. In addition, the study of the temperature dependence of EPR spectra in the temperature range 77-300 K showed that the intensity of the EPR signal is well described by the Curie-Weiss law.

References

[1] Lazarenko E K, Pavlishin V I, Lett V T and Sorokin Y G 1973 *Mineralogy and Genesis Chamber Pegmatites of Volyn* (Lviv)
[2] Pleshakov I V, Klekhta N S and Kuzmin Yu I 2012 *Technical Phys. Letters* **38** 60-66
[3] Abragam A and Bleaney B 1970 *Electron Paramagnetic Resonance of Transition Ions* (Oxford)
[4] Vasiukov V N, Chabanenko V V, Kochkanyan R O et all 2004 *Low Temp. Phys.* **30** 1272-78

Development of a Support Vector Machine – Based Image Analysis System for Focal Liver Lesions Classification in Magnetic Resonance Images

I Gatos[1], S Tsantis[1], M Karamesini[2], A Skouroliakou[3], G Kagadis[1]
[1]Department of Medical Physics, School of Medicine, University of Patras, Rion, GR 26504, Greece,
[2]Department of Radiology, Olympion General Clinic, Patra, Greece
[3]Department of Energy Technology Engineering, Technological Education Institute of Athens, 12210, Greece

E-mail: stsantis@teiath.gr

Abstract. Purpose: The design and implementation of a computer-based image analysis system employing the support vector machine (SVM) classifier system for the classification of Focal Liver Lesions (FLLs) on routine non-enhanced, T2-weighted Magnetic Resonance (MR) images. Materials and Methods: The study comprised 92 patients; each one of them has undergone MRI performed on a Magnetom Concerto (Siemens). Typical signs on dynamic contrast-enhanced MRI and biopsies were employed towards a three class categorization of the 92 cases: 40-benign FLLs, 25-Hepatocellular Carcinomas (HCC) within Cirrhotic liver parenchyma and 27-liver metastases from Non-Cirrhotic liver. Prior to FLLs classification an automated lesion segmentation algorithm based on Marcov Random Fields was employed in order to acquire each FLL Region of Interest. 42 texture features derived from the gray-level histogram, co-occurrence and run-length matrices and 12 morphological features were obtained from each lesion. Stepwise multi-linear regression analysis was utilized to avoid feature redundancy leading to a feature subset that fed the multiclass SVM classifier designed for lesion classification. SVM System evaluation was performed by means of leave-one-out method and ROC analysis. Results: Maximum accuracy for all three classes (90.0%) was obtained by means of the Radial Basis Kernel Function and three textural features (Inverse-Different-Moment, Sum-Variance and Long-Run-Emphasis) that describe lesion's contrast, variability and shape complexity. Sensitivity values for the three classes were 92.5%, 81.5% and 96.2% respectively, whereas specificity values were 94.2%, 95.3% and 95.5%. The AUC value achieved for the selected subset was 0.89 with 0.81 – 0.94 confidence interval. Conclusion: The proposed SVM system exhibit promising results that could be utilized as a second opinion tool to the radiologist in order to decrease the time/cost of diagnosis and the need for patients to undergo invasive examination.

1. Introduction

Focal liver lesions (FLL) have been considered as very difficult task faced by gastroenterologists and hepatologists. The increasing and widespread use of contemporary imaging modalities has led to an increase in incidental FLL detection. Thus, the identification of not only malignant liver lesions, but benign solid and cystic liver lesions (such as hemangioma, focal nodular hyperplasia, hepatocellular adenoma, and hepatic cysts) as well is crucial in differential diagnosis [1]. The majority of FLLs present in non-cirrhotic livers are considered benign. Hemangiomas, focal nodular hyperplasias (FNH), and adenomas (HCA) are the most commonly encountered solid benign lesions. On the other hand, the most commonly encountered malignant lesions in noncirrhotic livers are metastases. Hepatocellular carcinoma (HCC) and intrahepatic cholangiocarcinoma (ICC) also occur in cases of chronic liver disease [2].

Specificity and accuracy optimization of cross-sectional imaging in the context of these incidental liver lesions is of importance in avoiding unnecessary biopsies, which may portend a post-procedural morbidity of 2.0% to 4.8% and mortality of 0.05%. Ultrasound, computed tomography (CT), and magnetic resonance imaging (MRI) are the main liver imaging modalities. A meta-analysis comparing contrast-enhanced ultrasound, CT, and MRI in evaluating incidental FLLs demonstrated similar diagnostic performance with specificities ranging from 82%-89% and no significant differences in the receiver operating characteristic analysis between modalities. Given the lack of ionizing radiation and relative non availability of ultrasound contrast in the U.S., Magnetic resonance imaging (MRI) is considered a sensitive diagnostic method towards characterization of hepatic cirrhosis that competes or outperforms the diagnostic utility of liver biopsies. However, hepatic MRI has some shortcomings in detection and classification of focal liver lesions without the need for administration of contrast media, despite its superior tissue contrast and the combination of different pulse sequences [3]. A comprehensive liver protocol evaluates the parenchyma, vasculature, and biliary system. This in MRI is accomplished by way of a combination of single-shot T2-weighted fast spin-echo, gradient echo T1-weighted in- and opposed-phase, fat suppressed T2-weighted, dynamic pre- and post-contrast T1-weighted imaging and potentially subtraction of pre from post-contrast image sets [3].

Computer-aided diagnostic (CAD) systems that include quantitative lesion description and classification can provide radiologists or physicians an alternative second opinion towards optimization of the diagnostic procedure. The need to strengthen the diagnostic value of non-enhanced T1 and T2 sequences, Mayerhoefer et al. has tried to apply texture analysis for an automatic classification of liver cyst and hemangiomas showing promising results [4]. A study with more cases including more types of lesions is needed though to establish this method as a reliable non-enhanced tool for differential diagnosis. In this study, the feasibility of Computer Aided Diagnosis system for the classification of Focal Liver Lesions (FLLs) on routine non-enhanced, T2-weighted Magnetic Resonance images is evaluated.

The design and implementation of a computer-based image analysis system employing the support vector machine (SVM) classifier system for the classification of Focal Liver Lesions (FLLs) on routine non-enhanced, T2-weighted Magnetic Resonance (MR) is presented in this study.

2. Materials and Methods

2.1. Clinical Data

Clinical material includes 30 – benign FLLs, 19 – Hepatocellular Carcinomas (HCC) within Cirrhotic liver parenchyma and 22 – liver metastases from Non-Cirrhotic liver. All three classes' diagnosis was established by means of typical signs on dynamic contrast-enhanced MRI and biopsies.

2.2. FLLs Detection

The lesion detection – segmentation procedure comprise an initialization step that combines the edge information derived from Dyadic Wavelet Transform (DWT) and the clustering properties of unsupervised Fuzzy C-means (FCM) clustering algorithm followed by Marcov Random Fields (MRF) segmentation for final lesion extraction.

Wavelet transform is a multiresolution analysis technique that has been developed and applied in various fields, such as astronomy, finance, quantum physics, signal processing, video compression, and image processing. Throughout this study, the 2D redundant DWT was employed for multiresolution analysis [5]. The Dyadic Wavelet Transform (DWT) of a function $f(x,y) \in L^2R^2$ is the set of functions $W_{2^j}^1 f(x,y), W_{2^j}^2 f(x,y)$ which are respectively the partial derivative along the horizontal and vertical orientation of the convolution of $f(x,y)$ by the smoothing function $\theta(x,y)$, dilated along a dyadic sequence $(2^j)_{j\in Z}$ and is given by:

$$\begin{pmatrix} W_{2^j}^1 f(x,y) \\ W_{2^j}^2 f(x,y) \end{pmatrix} = f\cdot\begin{pmatrix} \psi_{2^j}^1 \\ \psi_{2^j}^2 \end{pmatrix} = 2^j\begin{pmatrix} \frac{\partial}{\partial x}(f\cdot\theta_{2^j})(x,y) \\ \frac{\partial}{\partial y}(f\cdot\theta_{2^j})(x,y) \end{pmatrix} \tag{1}$$

where $\psi_{2^j}^1$ and $\psi_{2^j}^2$ are the analyzing wavelets and j the dyadic scale. We performed the wavelet analysis with the DWT using Mallat's filters [5].

The FCM algorithm is an iterative clustering algorithm in which each data point is assigned to a cluster to a degree specified by a fuzzy membership grade. The procedure for assigning each cluster is considered an iterative optimization procedure that minimizes a cost function when pixels close to the centroid of their clusters are assigned to high membership values [6].

Let $V = \{v_1, \ldots, v_n\}$ be the set of n image pixels and let C be the number of clusters. The cost or objective function for FCM is described as follows:

$$Q_{fcm} = \sum_{j=1}^{n}\sum_{i=1}^{C} u_{ij}^{m}\left\|v_j - \mu_i\right\|^2 \qquad 1 \leq m < \infty \tag{2}$$

subject to:

$$\sum_{i=1}^{C} u_{ij} = 1, \forall\ j = \{1,\ldots,n\} \tag{3}$$

where, $m \in (1, \infty)$ controls the fuzziness of the resulting partition, u_{ij} denotes the membership of data pixel v_j to fuzzy cluster i whose value is between $[0,1]$; μ_i is the cluster center of fuzzy cluster i, and $\left\|v_j - \mu_i\right\|^2$ represents the Euclidian distance between the pixel v_i and the cluster center μ_i.

The FCM cost function is minimized when high membership values are assigned to pixels whose intensities are close to the centroid of their clusters and low membership values are assigned when the point is far from the centroid. The membership degrees and the cluster centers are updated according to the following formulas:

$$u_{ij}^{m} = \frac{1}{\sum_{k=1}^{C}\left(\frac{\left\|v_j - \mu_i\right\|}{\left\|v_j - \mu_k\right\|}\right)^{\frac{2}{m-1}}} \tag{4}$$

$$\mu_i = \frac{\sum u_{ij}^{m} v_i}{\sum_{i=1}^{n} u_{ij}^{m}} \tag{5}$$

Starting with an initial guess for each cluster center, the FCM converges to a solution for μ_i representing the local minimum or a saddle point of the cost function. Convergence can be detected by comparing the changes in the membership function or the cluster center at two successive iteration steps.

MRF modelling combines conditional (local intensity distribution) with contextual (intensity similarity within small neighborhoods) information under the Bayesian framework in order to estimate the true intensities of the image rather than those based only on the conditional information [7]. It assumes that the class probability of a pixel is only dependent on class membership of its spatial neighbors (also called lattice) which in turn reduces the possible influence of noise and overlapping structures. The model assumption that the conditional distribution depends on the pixels in the near neighborhood is subject to the Bayesian framework which states that the decision rule for labelling an image pixel combines the conditional intensity distribution of an individual region with prior knowledge regarding that region.

Given the fact that the observed image y is a realization of a random field *Y*, x^* is the true unknown label of the observed pixel, and $\hat{x}$ indicates the estimate of x^*, the main objective of the MRF segmentation model is to find $\hat{x}$ given the observed image y. Let's assume that *P(X)* is our prior knowledge and *P(Y|X)* is the probability of realizing the observed image given the regions distribution in the image. Then, in accordance to Bayes theorem:

$$P(X \mid Y) = \frac{P(Y \mid X)P(X)}{P(Y)} \tag{6}$$

where, *P(X|Y)* is our posterior probability. The most widely used conditional intensity distribution is the Gaussian distribution, whose function, given the class x_s is given by:

$$P(Y = y \mid X = x_s) = \frac{1}{\sqrt{2\pi\sigma_s^2}} \exp\left(-\frac{(y - \mu_s)}{\sigma_s^2} \right) \tag{7}$$

Where, μ_s and σ_s are the distribution parameters of class x_s.

Then, $\hat{x}$ can be obtained by taking the posterior's probability natural logarithm and minimizing its negative resultant:

$$\hat{x} = \arg\min_{x}(-\log(P(X \mid Y))) \tag{8}$$

In our case this optimization task is solved within the deterministic approach (iterated conditional mode – ICM). The ICM solves the minimization problem by sequentially updating (i.e. raster scanning the image) labels by minimizing the following equation at each pixel s:

$$\hat{x} = \arg\min_{x_s \in L} = \left\{ \frac{(y - \mu_s)}{\sigma_s^2} + \frac{1}{2}\log(2\pi\sigma_s^2) + \beta U(x_s) \right\} \tag{9}$$

Where, $U(x_s)$ is the number of pixels in the neighborhood that have color x_s. and β is a positive constant that controls the interaction between the pixels within the neighborhood.

An Edge Indicator Function (EIF) is computed from the wavelet image in order to locate the edge positions that correspond to local maxima in the wavelet domain (Figure 1c). The mean intensity value of the area between consecutive edges from the MR images is then calculated and fed as input to the FCM algorithm to acquire the initialization image (Figure 1d). The resulted image is applied as initial image to the Marcov – Random – Field (MRF) Model towards final lesion segmentation (Figure 1e).

2.3. Feature Extraction, Selection & Classification

For each lesion extracted by the aforementioned segmentation procedure, a single Region-of-Interest (ROI) was constructed on the image section depicting the maximum lesion diameter. 42 texture features derived from the gray-level histogram, co-occurrence and run-length matrices and 12 morphological features were obtained from each lesion [8-9]. Stepwise multi-linear regression analysis was utilized to avoid feature redundancy leading to a feature subset that fed the multiclass SVM classifier designed for lesion classification.

An SVM based classifier [10] is designed to work for two class problems and can be applied to linearly or nonlinearly separable data, with or without class data overlap. In the most difficult case of nonlinearly separable and overlapped data, which is often the case, data are first transformed from the input space to a higher dimensionality feature space, where classes are linearly separable. Then two parallel hyperplanes are determined with maximum distance between them and at the same time with minimum number of training points in the area between them (also called the margin). Finally, a third hyperplane through the middle of the margin is defined, which is the decision boundary of the two classes. SVM System evaluation was performed by means of leave-one-out method.

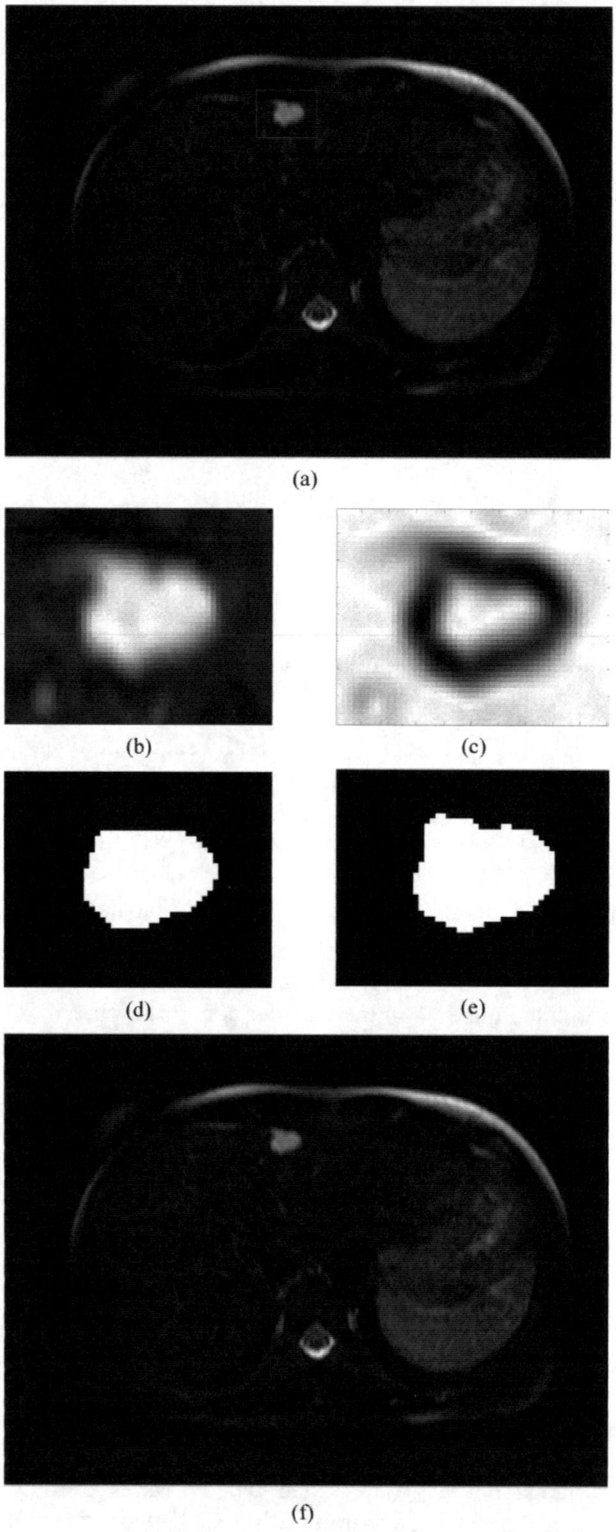

Figure 1. MRF segmentation with edge driven FCM initialization. (a) MR image with lesion selected, (b) Lesion cropped, (c) Edge Indicator Function, (d) FCM initialization, (e) MRF segmentation, (f) Lesion outlined.

3. Results

Maximum accuracy for all three classes (90.2%) was obtained by means of the Radial Basis Kernel Function and three textural features (Inverse-Different-Moment, Sum-Variance and Long-Run-Emphasis) that describe lesion's contrast, variability and shape complexity. Sensitivity values for the three classes were 92.5%, 81.4% and 96.0% respectively, whereas specificity values were 94.2%, 95.3% and 95.5%. The AUC value achieved for the selected subset was 0.89 with 0.81 – 0.94 confidence interval (Table 1).

Table 1 Confusion matrix of the SVM classifier employing the Inverse-Different-Moment, Sum-Variance and Long-Run-Emphasis best feature combination.

	Multi-Class SVM classification				
Verified FLLs	Benign	HCC	Metastases	Sensitivity/ Specificity	LOO precision
Benign	37	2	1	92.5 % / 94.2 %	92.5 %
HCC	3	22	2	81.5 % / 95.3 %	88.0 %
Metastases	0	1	24	96.2 % / 95.5 %	88.8 %
Overall accuracy					90.0 %

4. Discussion & Conclusion

The proposed SVM system exhibit promising results that could be utilized as a second opinion tool to the radiologist in order to decrease the time/cost of diagnosis and the need for patients to undergo invasive examination.

5. References

[1] Marrero J, Ahn J, and Rajender K 2014 *Am J Gastroenterol.* 109(9) 1328
[2] Fowler K, Brown J, and Narra V 2011 *Hepatology*, 54(6), 2227
[3] Elsayes K, Narra V, Yin Y, Mukundan G, Lammle M, Brown J 2005 *RadioGraphics*. 25 1299
[4] Mayerhoefer ME, Schima W, Trattnig S, Pinker K, Kulemann VB and BaSsalamah J 2010 *Magn. Res. Imag*. 32 352
[5] Mallat S and Hwang WL 1992 *IEEE Trans. Inform.* 2 617
[6] Bezdek JC, Ehrlich R and Full W 1984 *Comput. Geosci.* 10 191
[7] Parzen E 1962 *Ann. Math. Stat.* 33 1065.
[8] Haralick M, Shanmugam K and Dinstein I 1973 *IEEE Trans. Syst., Man* 610
[9] Galloway MM 1975 *Comput. Graph. Imag. Proc*. 4 172
[10] Kecman V, 2001 *MIT Press, Cambridge*, 121

Acknowledgments

This research has been co-financed by the European Union (European Social Fund – ESF) and Greek national funds through the Operational Program "Education and Lifelong Learning" of the National Strategic Reference Framework (NSRF) – Research Funding Program: ARCHIMEDES III. Investing in knowledge society through the European Social Fund

Wave modelling for the German Bight coastal-ocean predicting system

J Staneva[1], **A Behrens and K Wahle**
Institute for Coastal Research, HZG, Max-Planck-str. 1, 21502 Geesthacht, Germany

E-mail: joanna.staneva@hzg.de

Abstract. Ocean wave modeling has shown impressive developments, both in theoretical aspects as in the quality of the results available to users. The recent advances in development of the WAM wave model for forecasts applications at operational services and climate assessments for the North Sea and the German Bight are presented here. Ocean waves control the exchange of energy, momentum, heat, moisture, gas, etc. between the ocean and the atmosphere in the earth system. The impact of waves on currents and water levels in coastal areas is demonstrated. This is an important step towards a fully coupled atmosphere-wave-ocean modelling system. The synergy between wave observations and models for the North Sea and German Bight is increased on the road to improving the ocean state estimate and predictions in the coastal areas and generating up-to-date information, products and knowledge.

1. Introduction

Ocean wave modeling has shown impressive developments during the last decades, both in theoretical aspects as in the quality of the results available to users. The state-of the art WAve Model (WAM) for forecasts applications at operational services and for hindcasts and climate assessments is presented with focus on the new advances in the wave model development and applications for the North Sea and German Bight areas.

During the last decade the north European coasts has been affected by severe storms which caused serious damages on the North Sea coastal areas. Additionally, the human activities, e.g. offshore wind power, offshore oil industry, coastal recreations urges information about the sea state with a high detail (resolution) in the coastal environment. Predictions of extreme events like storm surges and flooding caused by storms are very important in order to avoid or at least minimize losses and human and material damages. Therefore, reliable wave forecasts and long term statistics of extreme wave conditions are needed for the coastal areas where various human activities are carried out, e.g. coastal securities, harbor activities, offshore wind energy, search and rescue, etc.

For the North Sea and German Bight regions the past and future wave conditions cannot be fully assessed from the analyses of observational data only, which, as it is well known, are sparse in both time and space. Numerical wave model systems have become the most common tool for producing high quality forecasts and long term hindcast wave data and to analyze trends and capability in severe extreme events [10]. The climate change can influence the multi-decadal wave conditions in the North Sea and thus may lead to the intensification of wave extremes in the future which will increase the risk

[1] To whom any correspondence should be addressed.

in the coastal area [4]; and the capability of the wave model to predict extreme events as severe winter storms for the North Sea and Baltic Sea has been evaluated in [1].

The ocean waves control the exchange of energy, momentum, heat, moisture, gas, etc. between the ocean and the atmosphere in the earth system. Understanding these processes is of utmost importance towards fully integrating of the atmosphere-wave-ocean models and their further coupling with biological, morphological, hydrographical systems. This topic reflects the increased interest in operational oceanography on order to reduce prediction errors of state estimates at coastal scales. The uncertainties in most of the presently used models result from the nonlinear feedback between strong tidal currents and wind-waves, which can no longer be ignored, in particular in the coastal zone where its role seems to be dominant. A nested-grid modelling system is used to producing reliable now- and short-term forecasts of ocean state variables; including wind waves and hydrodynamics. Analyses of observations, as well as the results of numerical simulations are presented in [8].

The structure of the paper is as follows. The WAM is described in Section 2. Section 3 shows the results from wave model forecasts for the North Sea and German Bight. Section 4 covers ocean-current interactions followed finally by concluding remarks.

2. Model Description WAM

WAM is a third generation wave model which solves the wave transport equation explicitly without any presumptions on the shape of the wave spectrum. It represents the physics of the wave evolution in accordance with our knowledge today for the full set of degrees of freedom of a 2D wave spectrum. WAM computes the 2D wave variance spectrum through integration of the transport equation in spherical coordinates:

$$\frac{\partial F}{\partial t} + (\cos\phi)^{-1}\frac{\partial}{\partial\phi}(\dot{\phi}\cos\phi F) + \frac{\partial}{\partial\lambda}(\dot{\lambda}F) + \sigma\frac{\partial}{\partial\sigma}\left(\dot{\sigma}\frac{F}{\sigma}\right) + \frac{\partial}{\partial\theta}(\dot{\theta}F) = S$$

with:

$$\dot{\phi} = (c_g\cos\theta + u_{North})/R$$
$$\dot{\lambda} = (c_g\sin\theta + u_{East})/(R\cos\phi)$$
$$\dot{\theta} = c_g\sin\theta\tan\phi/R + \dot{\theta}_D + \dot{\theta}_C$$
$$\dot{\sigma} = \dot{\sigma}_C$$

$F(\lambda,\phi,\sigma,\theta,t)$ wave energy density spectrum
(λ, ϕ) Longitude, Latitude
(σ, θ) intrinsic frequency, wave direction

The source functions on the right of the transport equation comprise the contributions of wind input (S_{in}), nonlinear interaction (S_{nl}), dissipation (S_{dis}), bottom friction (S_{bf}) and wave breaking (S_{br}):
$S = S_{in} + S_{nl} + S_{ds} + S_{bf} + S_{br}$

The last release of the third generation wave model WAM Cycle 4.5.4 is an update of the WAM Cycle 4 wave model, which is described in [7] and [5]. The basic physics and numeric are kept in the new release. The source function integration scheme made by [6] and the up-dates model [2] are incorporated. A big advantage of the new state-of-the-art version WAM Cycle 4.5.4 is its high-grade modular composition which allows an easy replacement of individual parts of the code. The wave model code is freely available under http://mywave.github.io/WAM/.

3. Short-term pre-operational wave model for the North Sea and German Bight

Within the framework of Coastal Observing System for Northern and Arctic Seas (COSYNA), a pre-operational wave forecast system has successfully been implemented at HZG and is running continuously since December 2009. It provides 24-hour wave forecasts twice a day and makes the results available in the web under http://www.coastlab.org. The system includes a regional WAM model for the North Sea (spatial resolution: $\Delta\varphi*\Delta\lambda = 0.05°*0.08333°\sim5$ km) and a finer meshed local model for the German Bight ($\Delta\varphi*\Delta\lambda = 0.00928°* 0.015534°\sim900$ m). The driving wind fields for both areas are provided by the German Met Service (DWD: Deutscher Wetterdienst).

The required boundary information used at the open boundaries of the North Sea model is derived from the regional wave model EWAM for Europe that is running twice a day in the operational wave forecast routine of the DWD. The depth distribution in the model grid for EWAM is given in figure 1 (left panel). The local model for the German Bight receives its boundary values from the North Sea wave model. The model grids and the depth distributions for the North Sea and German Bight COSYNA model areas are given on figure 1, (in the middle and right panels correspondingly).

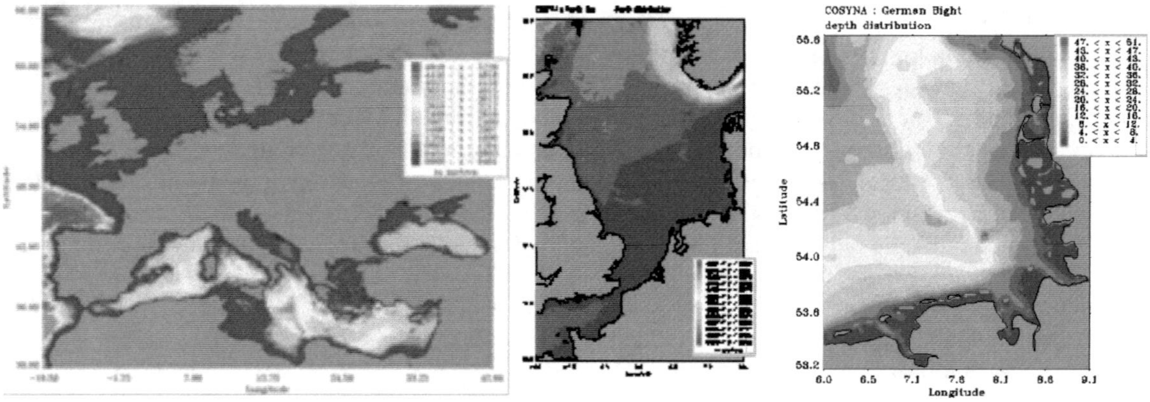

Figure 1. Setup of the pre-operational COSYNA wave forecast system for North Sea and German Bight. Boundary values by the regional European wave model EWAM (left: EWAM depth distribution). Depth distribution of the model for the North Sea (middle) and for the German Bight (right).

The wave models run in shallow water mode including depth refraction and wave breaking and calculate the two dimensional energy density spectrum at the active model grid points in the frequency-/direction space. The solution of the WAM transport equation is provided for 24 directional bands at 15° each with the first direction being 7.5° measured clockwise with respect to true north and 30 frequencies logarithmically spaced from 0.042 Hz to 0.66 Hz at intervals of $\Delta f/f = 0{,}1$. Figure 2 shows an example of the horizontal distribution of the significant wave height in the North Sea and in the nest for the German Bight on the 15th of February 2012 at 06 UTC with significant wave heights up to 6.8 m.

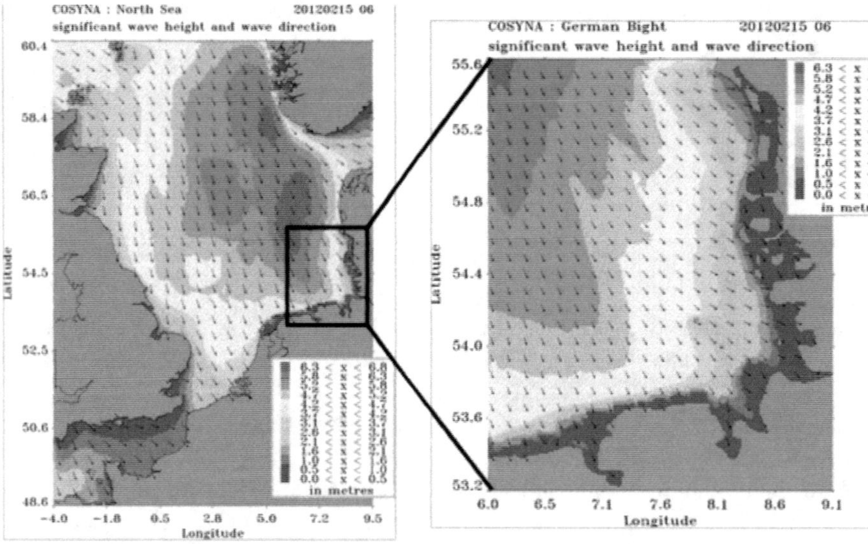

Figure 2. COSYNA wave forecast system for North Sea and German Bight (right)

The wave model simulations of the fine resolution German Bight set-up have been validated with measurements recorded by the buoys within the German Bight area (figure 3 indicates the location of the buoys in the German Bight where wave measurements are available). As representative examples for the validation of the wave model results in the German Bight some comparisons with measurements will be discussed for October 2013. In the end of that month the severe storm Christian afflicted the German coast with high wind speeds above 30 m/s and significant wave heights of about 8 m. Time series of wind and wave heights at FINO station are given on figure 3. At 28th of October during storm Christian the wind speed increases rapidly to 30 m/s causing brake down of several buoys and making impossible to provide measurements for this extreme event. The agreement between measured and modelled values is very good. Here we have to note that the measurements are compared with wave model forecasts: the first 12 hours of each forecast have been used, respectively.

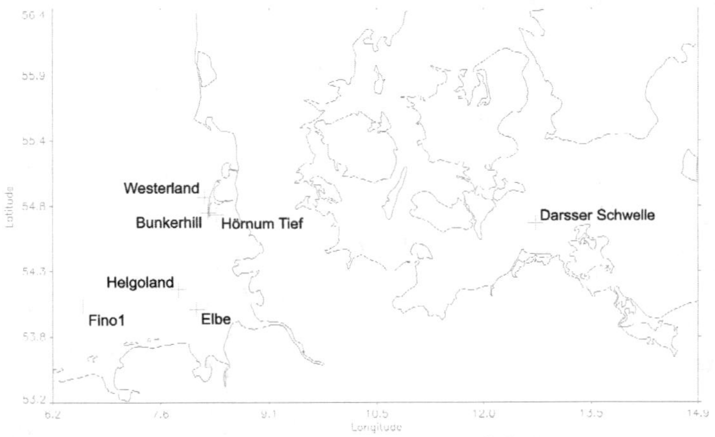

Figure 3. Buoy locations with measurements in the German Bight

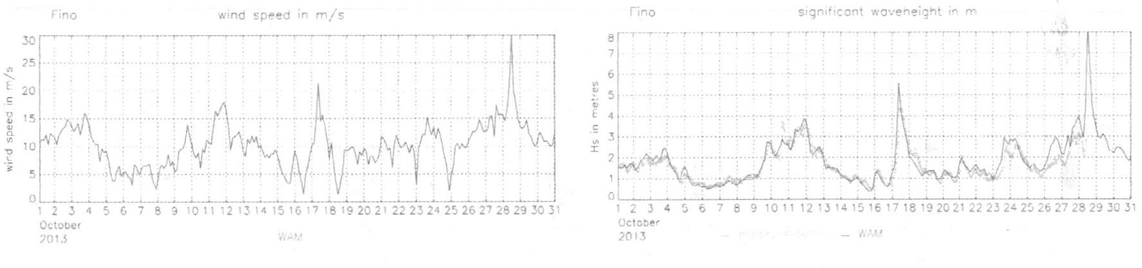

Figure 4. Time series of wind and wave heights at FINO station (storm Christian).

At Elbe station (see figure 4 for their coordinates) the wave heights were lower than the ones at FINO station during storm Christian and the buoy continuously recoded the wave parameters during the storm. Figure 5 includes the corresponding comparisons for significant wave heights, Tm_2/T_z-periods and total wave directions at the location Elbe. The agreement between measured and modeled wave parameters indicates that the model is capable in reliable forecasting wind waves. The peak on the 28th of October (3 pm UTC) in H_s of about 6 m and in Tm_2 of about 8 s is well predicted by the wave model. The statistical analysis of the comparisons (see table 1) supports the good quality of the wave forecasts for the German Bight area.

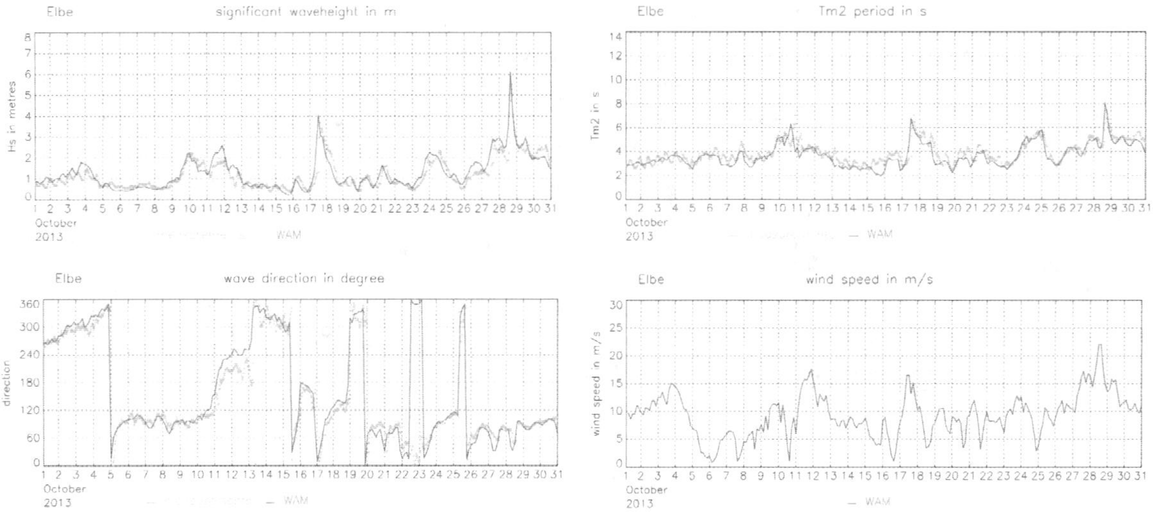

Figure 5. Time series of measured and computed wave parameters at the location Elbe

Table 1. H_s statistics for October 2013 at buoys located in the German Bight

Buoy	**Number of comparisons**	**Mean of measurements**	**Bias**	**Root mean square error**	**Skill**	**Scatter index**
$\mathbf{H_s}$	-	(m)	(m)	(m)	-	(%)
Fino	218	1.59	0.11	0.33	0.86	19
Elbe	247	1.23	0.08	0.31	0.84	25
Westerland	247	1.17	0.14	0.28	0.88	21
Helgoland	247	1.45	-0.03	0.30	0.90	20

skill : reduction of variance, scatter index : standard deviation*100/mean of the measurements

4. Wave-current interaction

The role of the coupling of wave and circulation models on improving the ocean forecast is demonstrated for the German Bight region. The German Bight (southern North Sea) is characterized by wind-waves and strong tidal currents. As a result, processes like nonlinear feedback between currents and waves play an important role in this area. The coupling between the wave model (WAM) and hydrodynamical model , which is the General Estuarine Transport Model (GETM, [3]) improves the estimates of ocean state variables, especially in coastal areas like the Wadden Sea and estuaries (for more details about the model configuration see [9]. The coupling takes into consideration both: the effect of currents on waves and the effects of waves on upper ocean dynamics, in particular on mixing and drift currents. In WAM the depth and/or current fields can be non-stationary, grid points can fall dry and refraction due to spatially varying current and depth is accounted for in the quasi-stationary approach. GETM was modified to account for wave effects by introducing the depth dependent radiation stresses and Stokes drift. The gradient of the radiation stresses serves as an additional explicit wave forcing term in the momentum equations for the horizontal velocity components. The transfer of momentum by waves becomes important for the mean water level setup and for the alongshore currents generated by waves in the surf zone.

We demonstrate the role on coupling by analyzing the impact of waves on extreme events (storm on 06.12.2013, see figure 7). The radiation stress increases the average water levels, which is much pronounced in the coastal area. During normal conditions the differences of the sea level due to the coupling with wave model maximum 10-15 cm in the Elbe area. However, during the storm Xavier, the differences of simulated sea level when considering waves are about 30-40 cm along the whole German coast. Therefore the uncertainties in most of the presently used models result from the nonlinear feedback between strong tidal currents and wind-waves, which can no longer be ignored in the operational oceanography, in particular in the coastal zone where its role seems to be dominant.

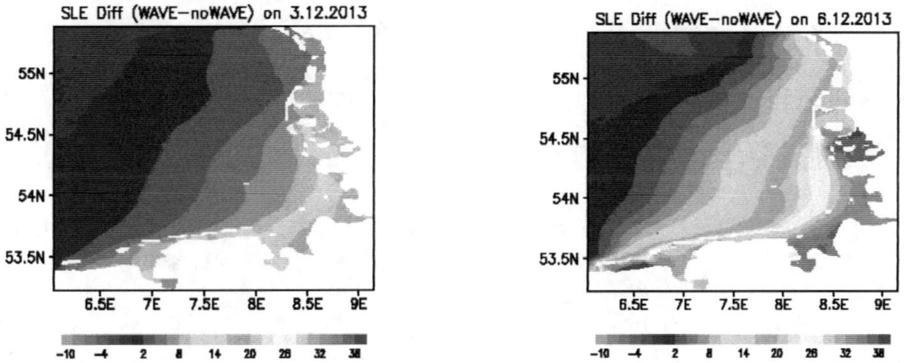

Figure 7. Sea surface elevation (SLE) difference between coupled wave-circulation model (WAM-GETM) and only circulation model (GETM) for the German Bight at 03.12.2013 (left) and during the storm Xavier at 06.12.2013.

5. Conclusions

Wave hindcasts and forecasts for the North Sea and German Bight are of great importance for the management of coastal zones, ship navigation, off-shore wind energy, naval operations etc. Storms and wind waves which they generate have direct impacts on the on the coastal and marine environment. The population living in the coastal areas is recently concerned with the impacts of erosion and flooding, and activities of what can be done to predict and further to minimize them. Important driving forces that cause serous damages on coastal environment are the wave conditions. Latter can be determined by using as a tool coastal numerical wave model systems. In this paper we summarized the recent advances in the field of wave modelling for the North Sea and German Bight regions. The state-of the art development of the WAM wave model for forecasts applications at operational services and for hindcasts and climate assessments for the North Sea and the German Bight in HZG is demonstrated. The synergy between observations and models for the North Sea and German Bight is increased on the road to improving the ocean state estimate and predictions in the coastal areas and generating up-to-date information, products and knowledge. The very good agreement between observations and model simulations is being demonstrated for both the long term wave hindcasts and short term wave forecasts for the North Sea and German Bight area. It enables to provide reliable predictions as well as to analyze long term changes of wave conditions, including extreme events. The performance of the forecasting system is illustrated for the cases of several extreme events. Effects of ocean waves on coastal circulation are investigated during extreme events, as well. The improved skill resulting from the recent wave model developments, in particular during storms, justifies further enhancements of the both forecasts applications at operational services and long-term hindcasts and climate for the North Sea and the German Bight. Short-term wave forecasts, sea state reconstructions and climate scenarios computations with the WAM model have created a huge interest to use the data in industrial applications. The pre-operational COSYNA wave forecast system for the North Sea and the German Bight provides wave forecasts twice a day delivering a number of wave parameters such as wave height, period and direction and is a very good example of how wave modelling products can support coastal management in the context of climate change and human activities. Data from wave forecasts and hindcasts form an essential part of the COSYNA data base that are being actively used by partners from industry, administration and research.

6. References

[1] Behrens A and Günther, H 2009 *Natural Hazards* **49 2** 387–399 DOI 10.1007/s11069-008-9298-3

[2] Bidlot J, Janssen P and Abdalla S 2005 *MEMORANDUM RESEARCH DEPARTMENT of ECMWF*, April 2005 R60.9/JB/0516

[3] Burchard, H and Bolding K GETM 2002 A General Estuarine Transport Model *No EUR 20253 EN*, European Comission 2002

[4] Groll N, Grabemann I and Gaslikova 2014 *Ocean Dynamics* **64**1–12 DOI 10.1007/s10236-013-0666-5

[5] Günther, H, Hasselmann H and Janssen P 1992 The WAM Model Cycle 4.0 User Manual Technical Report No. 4, Deutsches Klimarechenzentrum, Hamburg, Germany. 102 pages.

[6] Hersbach H and Janssen P 1999 *J. Atmos. Ooceanic Techn.* **16** 884-892

[7] Komen G, Cavaleri L, DonelanM , Hasselmann K, Hasselmann S and Janssen P 1994 Dynamics and modelling of ocean waves. *Cambridge University Press* **560** pages

[8] Stanev E, Schulz-Stellenfleth J, Staneva J, Grayek S, Seemann J and Petersen W 2011 Coastal observing and forecasting system for the German Bight – estimates of hydrophysical states *Ocean Sci.* 7 569-583

[9] Staneva J, Stanev E, Wolff J-O, Badewien T, Reuter R, Flemming B, Bartholomae A and Bolding K 2009 *Cont. Shelf Res.* **29** 302-319

[10] Weisse R, Günther H 2007 *Ocean Dynamics* **57**:161–172 DOI 10.1007/s10236-006-0094-x

Journal of Physics: Conference Series **633** (2015) 012118 doi:10.1088/1742-6596/633/1/012118

The Approximate Bayesian Computation methods in the localization of the atmospheric contamination source

P Kopka[1], **A Wawrzynczak**[1,2], **M Borysiewicz**[1]

[1] National Centre for Nuclear Research, Andrzeja Soltana 7, 05-400 Otwock,Poland
[2] Institute of Computer Sciences, Siedlce University, Poland

E-mail: piotr.kopka@ncbj.gov.pl

Abstract. In many areas of application, a central problem is a solution to the inverse problem, especially estimation of the unknown model parameters to model the underlying dynamics of a physical system precisely. In this situation, the Bayesian inference is a powerful tool to combine observed data with prior knowledge to gain the probability distribution of searched parameters. We have applied the modern methodology named Sequential Approximate Bayesian Computation (S-ABC) to the problem of tracing the atmospheric contaminant source. The ABC is technique commonly used in the Bayesian analysis of complex models and dynamic system. Sequential methods can significantly increase the efficiency of the ABC. In the presented algorithm, the input data are the on-line arriving concentrations of released substance registered by distributed sensor network from OVER-LAND ATMOSPHERIC DISPERSION (OLAD) experiment. The algorithm output are the probability distributions of a contamination source parameters i.e. its particular location, release rate, speed and direction of the movement, start time and duration.The stochastic approach presented in this paper is completely general and can be used in other fields where the parameters of the model bet fitted to the observable data should be found.

1. Introduction and problem setup

Accidental atmospheric releases of hazardous material pose significant risks to human health and the environment. In this context, it is valuable to develop the emergency techniques able to identify the probable location and characteristics of the release source using only the concentrations of released substance by sensors ([1] ,[2]). Given concentration measurements and knowledge of the wind field and other atmospheric air parameters, finding the location of the source and its parameters is ambiguous ([3] ,[4]). In this paper, we present the stochastic dynamic data-driven event reconstruction model that through Sequential Approximate Bayesian Computation (S-ABC) obtains a solution to the inverse problem. This algorithm based on the successively arriving concentrations of given substance finds the most probable values of contamination source parameters. The developed algorithm is tested using the data from the Over-Land Alongwind Dispersion (OLAD) experiment [5]. The OLAD Experiment was conducted during 8-25 September 1997 at the U.S. Army Dugway Proving Ground. The experimental objective was to acquire information on alongwind dispersion of passive contaminants over distances from 2 to 20 km for evaluation of atmospheric dispersion models.

Meteorological data, including near-surface winds and temperatures at locations across the test site, vertical wind and thermodynamic profiles, were collected to supplement the tracer concentration data. In the reconstruction procedure, we used second experiment data ($Trial252-2$). The domain area is 25 km $\times$ 25 km (see fig. 1); contamination source moves uniformly in a certain direction (see fig. 1 black arrow); the gas samplers have registered the tracer concentration in 45 one-minute time intervals since 06:45 up to 07:30; six meteorological towers registered the wind speed and its direction (see fig. 2 and fig. 1 green points).

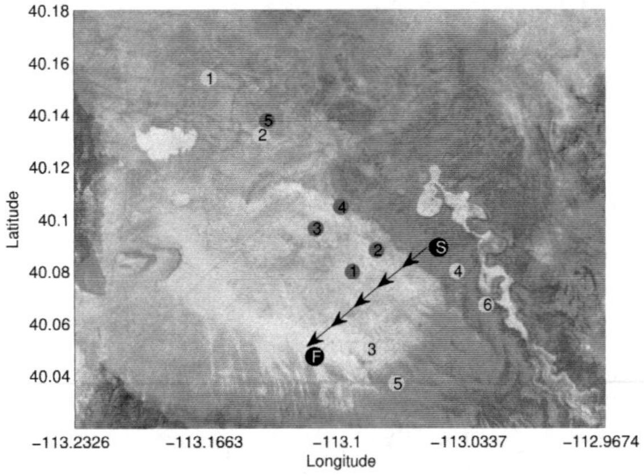

Figure 1: Domain of the OLAD Trial-252-2 experiment. The S and F denotes the starting and final point of the release, respectively. The locations of the sampling points are marked in red while the meteorological towers in green.

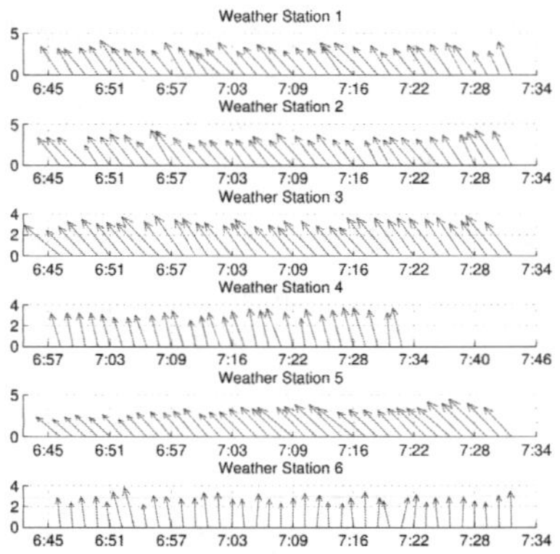

Figure 2: The wind speed and direction measured on 9 September since 6:45 up to 7:30. The weather stations numbers corresponds to the meteorological towers marked in Fig. 1 by the green dots.

2. Sequential Approximate Bayesian Computation

Let θ be a parameter vector, given the prior distribution $\pi(\theta)$. The goal of Bayesian inference is to approximate the posterior distribution, $\pi(\theta|x) \propto \pi(x|\theta)\pi(\theta)$, where $\pi(x|\theta)$ is the likelihood of θ given the data x. The main idea of the Approximate Bayesian Computation (ABC) methods is to accept θ as an approximate posterior draw if its associate data x are close enough to the observed one x_{obs}. Accepted parameters are sampled from $\pi(\theta|\rho(x, x_{obs}) < \epsilon)$, where the $\rho(x, x_{obs})$ is the chosen measure of discrepancy, and ϵ is a threshold defining 'closeness margin'.

In ABC methods, Sequential Monte Carlo (SMC) [2] is used to automatically sequently 'clean' approximation of the posterior distribution used to generate proposals for further stages. In $ABC - SMC$ methods, the set of samples with weights, called particles sampled from the prior distribution $\pi(\theta)$ are propagated through a sequence of intermediate posterior distributions $\pi(\theta|\rho(x, x_{obs}) < \epsilon_t)$, $t = 1, ..., T$. Process is repeated until the set represents a sample from the target distribution, $\pi(\theta|\rho(x, x_{obs}) < \epsilon_T)$. In [6] authors proposes strategies called ABC SMC with Adaptive Weights. This methods includes a stage where the weights are modified according to the respective values of x. The algorithm 1 presents the $ABC - SMC - AW$ procedure [6].

After initialization of the threshold schedule, first N samples are simulated based on the predefined priori $\pi(\theta)$ and the corresponding acceptance condition $\rho(x, x_{obs}) < \epsilon_1$. In time step $t = 2$ simple uniform weights are changed based on additional kernel $K_{x,t}(x_{obs}|x_i^{t-1})$ proposed in [6]. Samples, denoted by a tilde are drawn from the previous generation with probabilities v_i^{t-1}. Using perturbation kernel $K_{\theta,t}(\theta_i^t|\tilde{\theta}_i)$ new 'fresh' samples θ_i^t are obtained, with the veracity of the condition $\rho(x, x_{obs}) < \epsilon_t$. The weights are calculated according to the formula in stage (11); in stage (12) weights are normalized and the time step is increased $t = t + 1$. Procedure is

Algorithm 1 Approximate Bayesian Computation-Sequential Monte Carlo-Adaptive Weights

1. Initialize threshold schedule $\epsilon_1 < \epsilon_2 < ... < \epsilon_T$
2. Set $t = 1$

for $i = 1$ to N **do**
- 3. Simulate $\theta_i^t \sim \pi(\theta)$ and $x \sim \pi(x|\theta_i^t)$
- 4. Until $\rho(x, x_{obs}) < \epsilon_t$
- 5. Set $w_i^t = \frac{1}{N}$

end for

for $i = 2$ to T **do**
- 6. Compute new weights $v_i^{t-1} \propto w_i^{t-1} K_{x,t}(x_{obs}|x_i^{t-1})$
- 7. Normalize weights v_i^{t-1}
- **for** $i = 1$ to N **do**
  - 8. Pick $\tilde{\theta}_i$ from the θ_j^{t-1} set with probabilities v_i^{t-1}
  - 9. Draw $\theta_i^t \sim K_{\theta,t}(\theta_i^t|\tilde{\theta}_i)$ and $x \sim \pi(x|\theta_i^t)$
  - 10. Until $\rho(x, x_{obs}) < \epsilon_t$
  - 11. Compute new weights as $w_i^t \propto \frac{\pi(\theta_i^t)}{\sum_j v_j^{t-1} K_{\theta,t}(\theta_i^t|\theta_j^{(t-1)})}$
  - 12. Normalize weights w_i^t for $i = 1 : N$
- **end for**

end for

repeated until $t \leq T$. In the presented reconstruction we assume that the substance concentration registered by the sensors $(S1, S2, S3, S4, S5)$ (Fig. 1) arrive subsequently in time intervals, hereafter called time steps. It is important to know that for time step t only data $\hat{C}_1^{Sj}\hat{C}_2^{Sj}\ldots\hat{C}_t^{Sj}$ are available. Finally we have forty-five time steps ($T = 45$). To get the predicted concentration a Scipuff dispersion model [7] model was run; it refers to the procedure $x \sim \pi(x|\theta_i^t)$ in algorithm 1. To run dispersion model and obtain data x we use source parameter vector θ_i^t and hiherto known weather data I. The simulated concentrations C_t^{Sj} for time step t correspond to five sensors locations $(S1, S3, S3, S4, S5)$.

The choice of distance measure or summary statistics is crucial in ABC. For number of sensors greater than one, overall measure of $\rho(.,.)$ will be simple mean of distances from all sensors:

$$\rho(x, x_{obs}) = \frac{1}{5}\sum_{j=1}^{5}(\frac{1}{t}\sum_{i=1}^{t}\frac{|C_i^{Sj} - \hat{C}_i^{Sj}|}{C_i^{Sj} + \hat{C}_i^{Sj}}), \tag{1}$$

under the assumption, that when $C_i^{Sj} = \hat{C}_i^{Sj} = 0$ then $\frac{|C_i^{Sj} - \hat{C}_i^{Sj}|}{C_i^{Sj} + \hat{C}_i^{Sj}} = 0$. Adaptive scheme for the threshold is based on the quantile of the empirical distribution of the distances between the simulated data and observations from the previous population (see [8]). We selected transition kernel $\sim K_{\theta,t}(\cdot|\cdot)$ to be a multivariate normal distribution.Unfortunately, in the type of inverse problems presented here the parameters often are highly correlated, and multimodality is very common. In [9] one the proposed methods is to use the multivariate normal kernel based on the M neighbours. Based on some pre-processed experiment we determined the optimal number of neighbors $M = 30$.

3. Results

The reconstruction algorithm starts to search for the starting source location (x, y), speed and direction of movement (s, d), release rate (q), duration of release and start time (u, t) after 10 minutes from the first sensors measurements (east->west direction line correspond to 0° and $t = 0$ refers to 06:45) . Consequently, the algorithm is run with obtaining the substance concentrations and continues working, as long as sensors report a new data. The scanned parameters space is $\theta \equiv \{x, y, d, s, q, u, t\}$. Fig. 3 presents the posterior distributions of all parameters estimated during the reconstruction employing $S-ABC$ method after all time steps.

The corresponding most probable values (MPV) of parameters are listed in table 1. The most probable release source initial coordinates were estimated as: $P(x = 16775m \pm 65) = 0.058$ and $P(y = 12565m \pm 65) = 0.033$, which is very neighbouring to the real location (RV) i.e. $x = 16369m$ and $y = 12052m$. In fig. 3 and table 1 we can see that all parameters were estimated very close to its actual values.
This results confirm that the presented $S - ABC$ method can be successfully applied in the problem of the localization of the atmospheric contamination source.

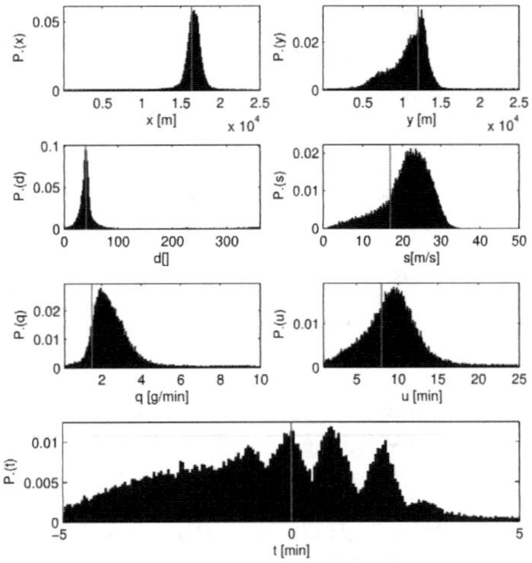

Figure 3: The posterior distributions of searched parameters x, y, d, s, q, u, t obtained by the reconstruction algorithm for the OLAD experiment. The red vertical lines represent the target value of the parameter.

Parameter	RV	MPV	Prob.
x [m]	16369	16775 ± 65	0.058
y [m]	12052	12565 ± 65	0.033
d [°]	40.96	40.75 ± 0.9	0.096
s [m/s]	16.8	22.23 ± 0.13	0.021
q [g/min]	1.51	1.90 ± 0.03	0.028
u [min]	8	8.52 ± 0.06	0.018
t [min]	0	1 ± 0.03	0.011

Table 1: The real values (RV) and estimated during the reconstruction most probable values (MPV) of searched parameters and its probabilities.

4. Acknowledgments and References

This work was supported by the Welcome Programme of the Foundation for Polish Science operated within the European Union Innovative Economy Operational Programme 2007-2013 and by the EU and MSHE grant nr POIG.02.03.00-00-013/09.

[1] Borysiewicz M, Wawrzynczak A and Kopka P 2012 Stochastic algorithm for estimation of the model's unknown parameters via Bayesian inference *Proceedings of the Federated Conference on Computer Science and Information Systems* 501–508
[2] Wawrzynczak A, Borysiewicz M and Kopka P. 2014 Sequential Monte Carlo in Bayesian assessment of contaminant source localization based on the distributed sensors measurements *Lecture Notes in Computer Sciences* **8385**
[3] Wade D and Senocak I 2013 Stochastic reconstruction of multiple source atmospheric contaminant dispersion events *Atmospheric Environment* **74** 45-51
[4] Senocak I, Hengartner N.W, Short M. B., et al. 2008 Stochastic event reconstruction of atmospheric contaminant dispersion using Bayesian inference, *Atmospheric Environment* **42** 7718–7727
[5] Biltoft C A, Turley S D et.al. Final test report of the over-land alongwind dispersion field trials *Report No. WDTC/JCP-00/004*
[6] Bonassi F V, and West M 2014 Sequential Monte Carlo with Adaptive Weights for Approximate Bayesian Computation *Bayesian Analysis* **10(1)** 71–187
[7] Sykes R I et.al. 1998 PC-SCIPUFF Version 1.2PD Technical Documentation *ARAP Report* **718**
[8] Del Moral P, Doucet A, and Jasra A 2012 An adaptive sequential Monte Carlo method for approximate Bayesian computation *Statistics and Computing* **22(5)** 1009–1020
[9] Filippi S, Barnes C P, Cornebise J and Stumpf M P 2013 On optimality of kernels for approximate Bayesian computation using sequential Monte Carlo *Statistical applications in genetics and molecular biology* **12(1)** 87–107

Fractal fragmentation and small-angle scattering

E M Anitas[1,2]

[1]Joint Institute for Nuclear Research, Dubna 141980, Moscow region, Russia
[2]Horia Hulubei National Institute of Physics and Nuclear Engineering, RO-077125 Bucharest, Romania

E-mail: anitas<at>theor.jinr.ru

Abstract. The small-angle scattering form factor of a three-dimensional idealized fragmentation model based on the concept of renormalization is calculated. The system consists of randomly oriented microscopic fractal objects whose positions are uncorrelated. It is shown that in the fractal region, the monodisperse form factor is characterized by a succession of maxima and minima superimposed on a simple power-law decay, and whose scattering exponent coincide with the fractal dimension of the scatterer. The results could be used to obtain additional structural information about systems obtained through fragmentation processes at microscale.

1. Introduction

Fragmentation processes such as those produced by earth's crust, rocks weathering or explosions, usually lead to to a fractal [1–3] distribution of number of fragments as a function of their sizes over a wide range of scales, and a quantification of these processes using the renormalization group approach has been suggested in [4, 5]. However, an important issue concerns the distribution of fragments at microscales, since they are responsible for the physical properties, such as hydraulic conductivity or moisture characteristics in soils [6].

An important technique for investigating the microstructure of various types of systems and which addresses the issue of size distribution, including the smallest and largest components, is small-angle scattering (SAS) [7, 8] which yields the differential elastic cross section per unit solid angle as a function of the momentum transfer. This technique has been successfully used in studying the property of self-similarity across nano- and microscales [2], such as various types of elastomeric membranes [9–12], cements [13], semiconductors [14], magnetic [15, 16] or biological structures [17–19], and therefore the concept of fractal geometry coupled with SAS technique can give new insights regarding the structural characteristics of such fractal systems [20–27]. One of the main parameters which can be obtained is the fractal dimension D [1]. For a mass fractal it is given by the scattering exponent of the power-law SAS intensity $I(q) \propto q^{-D}$ where $0 < \mathrm{D} < 3$.

In this paper, we develop an idealized theoretical model based on renormalization group approach which could describe fragmentation processes, and calculate the mono- and polydisperse form factor. We show how to obtain the fractal dimension, and how to estimate the smallest and largest radius of the fractal from SAS data.

2. Model

We start with a cube of edge length l_0 ($m = 0$; initiator) and in the first iteration ($m = 1$) the number of cubes which are kept is related to the probability f that the initiator will be fragmented into 27 cubes of

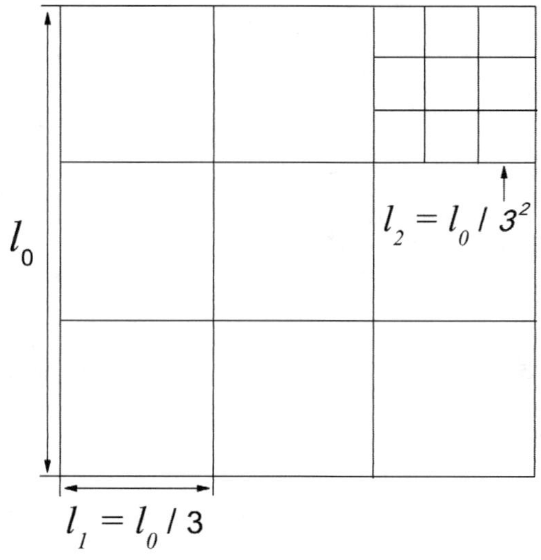

Figure 1. A two-dimensional projection of the fragmentation model. The probability that the cube with edge length l_0 ($m = 0$; initiator) will be fragmented into 27 cubes of edge length l_1 is f.

edge length $l_1 = l_0/3$. By calculating the number of fragmented elements and their linear sizes, it can be shown that the fractal dimension is given by [4]:

$$D = 3\frac{\log(27f)}{\log 27}, \tag{1}$$

with $1/27 < f < 1$, and thus $0 < D < 3$. Fig. 1 illustrates the construction of a generic model of fragmentation, where the sizes of remaining cubes at m-th iteration is given by $l_m = l_0/3^m$.

3. Theoretical background on SAS

Here we restrict ourselves to two-phase systems, which are composed from homogeneous units of mass density ρ_{m}. The units are immersed into a solid matrix of pore density ρ_{p}. Then [7, 8] we can consider the system as if the units were frozen in a vacuum and had the density $\Delta\rho = \rho_{\mathrm{m}} - \rho_{\mathrm{p}}$. The density $\Delta\rho$ is called scattering contrast, and thus, the scattering intensity is given by

$$I(q) = n|\Delta\rho|^2 V^2 \left\langle |F(\boldsymbol{q})|^2 \right\rangle, \tag{2}$$

where n is the fractal concentration, V the volume of each fractal, and $F(\boldsymbol{q})$ the normalized form factor $F(\boldsymbol{q}) = (1/V)\int_V e^{-i\boldsymbol{q}\cdot\boldsymbol{r}} \mathrm{d}\boldsymbol{r}$, obeying $F(0) = 1$. The brackets $\langle\cdots\rangle$ stand for the ensemble averaging over all orientations of $\boldsymbol{q}$. Once a deterministic fractal is composed of the same objects, say, N cubes of edge length l, then $F(\boldsymbol{q}) = \rho_{\boldsymbol{q}} F_0(ql)/N$, with $\rho_{\boldsymbol{q}} = \sum_j e^{-i\boldsymbol{q}\cdot\boldsymbol{r}_j}$ the Fourier component of the density of the cubes, and $\boldsymbol{r}_j$ are the center-of-mass positions of cubes. Here the cube form factor of unit edge length is given by [7] $F_0(\boldsymbol{t}) = (8/t_x t_y t_z)\sin(t_x/2)\sin(t_y/2)\sin(t_z/2)$. Therefore, the scattering intensity becomes $I(q) = I(0)S(q)|F_0(ql)|^2/N$, with $I(0) = n|\Delta\rho|^2 V^2$ and the structure factor is defined by $S(q) \equiv \langle \rho_{\boldsymbol{q}}\rho_{-\boldsymbol{q}}\rangle / N$.

4. Results and Discussion

The monodisperse form factor at m-th iteration can be written as [23,25] $F_m(\boldsymbol{q}) = F_0(l_m q)\prod_{i=1}^{m} G_i(\boldsymbol{q})$ with $m = 1, 2, \cdots$ is the generative function of the fractal and it specifies the positions of the scattering

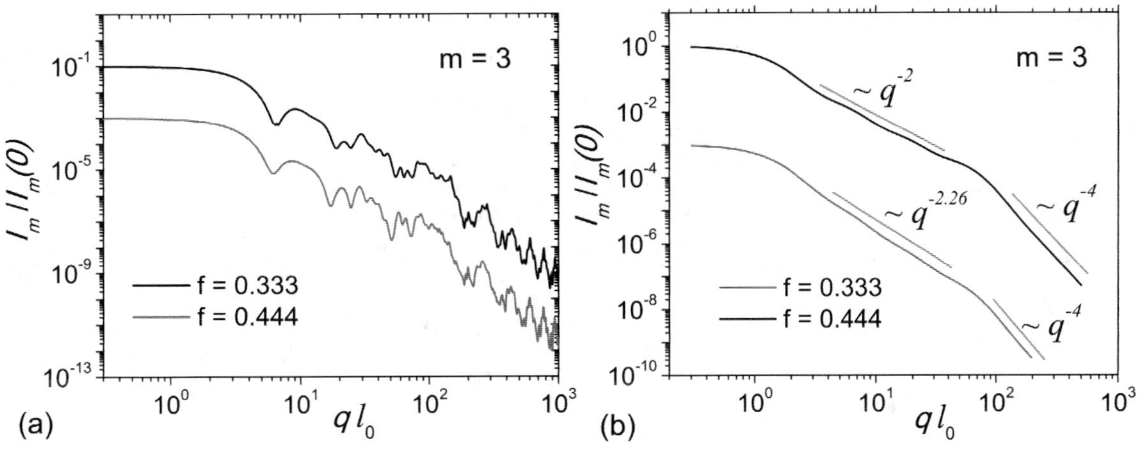

Figure 2. SAS from an idealized fragmentation model for two values of the probability f. a) Monodisperse intensity; b) Polydisperse intensity ($\sigma_r = 0.4$). The curves are shifted vertically for clarity.

cubes inside the fractal. For well-known systems such as Cantor sets or Vicsek fractals $G_i(\boldsymbol{q})$ has known analytical expressions [23, 25]. Thus, the scattering intensity becomes [25]

$$I_m(q)/I_m(0) = \langle |F_m(\boldsymbol{q})|^2 \rangle . \quad (3)$$

The monodisperse SAS intensity for various values of f is shown in Fig. (2a), where it can be seen that the intensity is characterized by three main regions: a plateau (Guinier region) at low q ($q \lesssim 1/l_0$), an intermediate (fractal) region at $1/l_0 \lesssim q \lesssim 1/l_m$, and a decay proportional to q^{-4} (Porod region) at high q ($1/l_m \lesssim q$). These regions allows us to obtain the size of the generator (from the end of Guinier region), and the smallest size of the scatterers found in the fractal (from the end of the fractal region). In addition, from the Porod region we can extract information about the specific surface of the fractal.

The value of the fractal dimension can be obtained if we take into account the polydispersity of the scatterers. Thus, we consider a distribution function $D_N(l)$ of sizes in such a way that $D_N(l)\mathrm{d}l$ gives the probability of finding a fractal whose size falls within $(l, l + \mathrm{d}l)$. We consider a log-normal distribution, and therefore, the polydisperse SAS intensity is given by [25]

$$I_m(q) = n|\Delta\rho|^2 \int_0^\infty \langle |F_m(\boldsymbol{q})|^2 \rangle V_m^2 D_N(l)\mathrm{d}l, \quad (4)$$

where V_m is the total volume at the m-th iteration.

Fig. (2b) shows the polydisperse SAS intensity obtained from Fig. (2a) together with Eq. (4). The results clearly show a smoothing of the scattering curves, as expected, and with the scattering exponent in the fractal region, being given by Eq. (1).

5. Conclusion

We have developed a model for describing fragmentation processes based on $3D$ fractal and have used the SAS technique to characterize the microstructural properties of such fragments.

We have calculated the monodisperse and polydisperse SAS intensities, and have shown how to obtain the smallest and largest sizes inside the fractal. We have shown that the slope of SAS intensity coincide with the fractal dimension of the fractal, as given by Eq. (1).

The suggested model can be used in understanding various physical properties such as hydraulic conductivity or moisture characteristics for systems obtained through fragmentation.

References

[1] Mandelbrot B B 1983 *The Fractal Geometry of Nature* (W.H. Freeman)
[2] Gouyet J F 1996 *Physics and Fractal Structures* (Springer)
[3] Vicsek T 1989 *Fractal Growth Phenomena* (Singapore: World Scientific)
[4] Turcotte D L 1986 *Journal of Geophysical Research* **91** 1921–1926
[5] Turcotte D L 1997 *Fractals and Chaos in Geology and Geophysics* (Cambridge University Press)
[6] Bittelli M, Campbell G S and Flury M 1999 *Soil Sci. Soc. Am. J.* **63** 782–788
[7] Glatter O and Kratky O 1982 *Small-angle X-ray Scattering* (London: Academic Press)
[8] Feigin L A and Svergun D I 1987 *Structure Analysis by Small-Angle X-Ray and Neutron Scattering* (NY: Plenum press)
[9] Balasoiu M, Anitas E M, Bica I, Erhan R, Osipov V A, Orelovich O L, Savu D, Savu S and Kuklin A I et al 2008 *Optoelectronics and Advanced Materials - Rapid Communications* **2** 730–734
[10] Balasoiu M, Craus M L, Kuklin A I, Plestil J, Garamus V, Islamov A K, Erhan R, Anitas E M, Lozovan M, Tripadus V, Petrescu C, Savu D, Savu S and Bica I 2008 *Journal of Optoelectronics and Advanced Materials* **10(11)** 2932–2936
[11] Anitas E M, Balasoiu M, Bica I, Osipov V A and Kuklin A I 2009 *Optoelectronics and Advanced Materials - Rapid Communications* **3** 621–624
[12] Balasoiu M, Craus M L, Anitas E M, Bica I, Plestil J and Kuklin A I 2010 *Physics of the Solid State* **52(5)** 917–921
[13] Das A, Mazumder S, Sen D, Yalmali V, Shah J G, Ghosh A, Sahu A K and Wattal P K 2014 *J. Appl. Cryst.* **47** 421–429
[14] Cho K, Biswas P and Fraundorf P 2014 *J. Ind. Eng. Chem.* **20** 558
[15] Craus M L, Islamov A K, Anitas E M, Cornei N and Luca D 2014 *Journal of Alloys and Compounds* **592** 121–126
[16] Naito T et al 2014 *Eur. Phys. J. B* **86** 410
[17] Yadav I, Kumar S, Aswal V K and Kohlbrecher J 2014 *Phys. Rev. E* **89** 032304–1
[18] Gebhardt R 2014 *J. Appl. Cryst.* **47** 29–34
[19] Maji J, Bhattacharjee S M, Seno F and Trovato A 2014 *Phys. Rev. E* **89** 012121
[20] Schmidt P W and Dacai X 1986 *Phys. Rev. A* **33** 560–566
[21] Martin J E and Hurd A J 1987 *J. of Appl. Cryst.* **20** 61–78
[22] Schmidt P W 1991 *J. of Appl. Cryst.* **24** 414–435
[23] Cherny A Y, Anitas E M, Kuklin A I, Balasoiu M and Osipov V A 2010 *J. Appl. Cryst.* **43** 790–797
[24] Cherny A Y, Anitas E M, Kuklin A I, Balasoiu M and Osipov V A 2010 *J. Surf. Invest.* **4** 903–907
[25] Cherny A Y, Anitas E M, Kuklin A I and Osipov V A 2011 *Phys. Rev. E* **84** 036203–1–036203–11
[26] Cherny A Y, Anitas E M, Osipov V A and Kuklin A I 2014 *J. Appl. Cryst.* **47** 198–206
[27] Anitas E M 2014 *Eur. Phys. J. B* **87** 139

Early breast cancer detection method based on a simulation study of single-channel passive microwave radiometry imaging

Spiros A. Kostopoulos[1], Andonis D. Savva[1,2], Pantelis A. Asvestas[1], Christos D. Nikolopoulos[2], Christos N. Capsalis[2], and Dionisis A. Cavouras[1]
[1] Department of Medical Instruments Technology, Technological Education Institute of Athens, Greece
[2] School of School of Electrical and Computer Engineering, National Technical University of Athens, Greece

E-mail: skostopoulos@teiath.gr

Abstract. The aim of the present study is to provide a methodology for detecting temperature alterations in human breast, based on single channel microwave radiometer imaging. Radiometer measurements were simulated by modelling the human breast, the temperature distribution, and the antenna characteristics. Moreover, a simulated lesion of variable size and position in the breast was employed to provide for slight temperature changes in the breast. To detect the presence of a lesion, the temperature distribution in the breast was reconstructed. This was accomplished by assuming that temperature distribution is the mixture of distributions with unknown parameters, which were determined by means of the least squares and the singular value decomposition methods. The proposed method was validated in a variety of scenarios by altering the lesion size and location and radiometer position. The method proved capable in identifying temperature alterations caused by lesions, at different locations in the breast.

1. Introduction

Microwave Radiometry (MWR) has been explored as complementary to mammography and clinical examination in breast cancer detection [1]. It has been shown that MWR imaging can be used to detect differences in temperature distribution in the human breast when a lesion is present [2]. This is because breast tumours have electrical properties at microwave frequencies that are significantly different from those of healthy breast tissues [3], especially at early lesion development [4]. Two methods for MWR imaging have been studied; the so called active and the passive methods. Active methods involve transmission of microwave signals to the breast area and measurement of corresponding reflections. Passive methods incorporate radiometers to detect lesions based on temperature differences from normal tissues [5].

MWR involves measurement of the power in the microwave region of the natural thermal radiation from body tissues to obtain the so-called brightness temperature of the tissue under observation. The relationship between real temperature value $T(x, y, z)$ at a location $[x, y, z]$ and brightness temperature value T_{b_j} is given by [6]:

$$T_{b_j} = \iiint_{FOV} W_j(x, y, z) \cdot T(x, y, z) \cdot dxdydz \qquad (1)$$

where FOV is the antenna's field of view, $W_j(x, y, z)$ is a weighting coefficient function, which depends on the antenna properties and tissue characteristics, and j indicates the number of different tuned frequencies. The calculation of $T(x, y, z)$ cannot be made explicitly, thus, other numerical techniques must be applied. In the present study, a methodology is proposed for estimating $T(x, y, z)$, where the temperature distribution can be approximated by a mixture model of distributions with unknown parameters. The proposed method assumes a passive single-channel microwave radiometer, it simulates sets of radiometer measurements and attempts to find an estimate of the original temperature distribution, in order to detect and visualize alterations, if any, in internal breast temperature.

2. Material and Methods

2.1. Simulation of single-channel radiometer measurements

Initially, the human female breast was simulated as an hemisphere of radius R and with three basic homogeneous layers, the thoracic muscle, the connective tissue and the skin layer [7]. Next, a single channel microwave radiometer was considered, with an antenna tuned at frequencies

$$f_j = 1 + j * 0.5GHz$$

with $j =$ 0,1,2,3,4 which accounted for electrical field measurements at different depths (11.60, 0.80, 0.62, 0.48, 0.38 cm) in the human body. The weighting coefficient function $\mathrm{W_j(x, y, z)}$ was estimated based on a recent work of Nikolopoulos *et al* [8]. Also, a spherical lesion of varying diameter and location was considered for inducing temperature variations. The internal temperature T at point (x, y, z) was modelled by means of a Butterworth function. The average temperature was approximated to 37°C inside the human body, about 36°C at skin depth, and that the lesion induced 0.7°C divergence. Subsequently, the brightness temperature $\mathrm{T_{b_j}}$, was calculated as the weighted average of the real temperature values in the radiometer's FOV by employing equation (1). Figure 1 shows the simulated human breast, along with the different tissues, the lesion (as "hot spot"), and the simulated temperature distribution.

2.2. Temperature reconstruction

In general, temperature reconstruction is a process that approximates the original temperature distribution for a given set of sampled values T_{b_j}. This enables the extraction of further information to facilitate the detection and/or visualization of the unknown temperature distribution. The proposed method assumes that the temperature distribution can be approximated by a Mixture Model (MM) of known distributions [9], as shown in equation (2).

$$\tilde{T}(x, y, z) = a_1 D_1(x, y, z, \theta_1) + \cdots + a_K D_K(x, y, z, \theta_K) \tag{2}$$

where $D_i(x, y, z, \theta_i)$ indicate the distributions with parameters θ_i and a_i $(i = 1, \ldots, K)$ are the coefficients that express the influence of each distribution to the final temperature approximation. The calculation of each weight a_i is derived from the solution of a system of linear equations [10]. The least squares and singular value decomposition methods were employed for finding the unknown coefficients. The Gaussian distribution was used for the modeling of the real temperature distribution $T(x, y, z)$, that has two parameters, the mean value μ and the covariance matrix Σ and is given by the following equation:

$$G(\vec{r}) = \frac{1}{\sqrt{\det(\Sigma)\,(2\pi)^3}} e^{-\frac{1}{2}(\vec{r}-\mu)\Sigma^{-1}(\vec{r}-\mu)^T} \tag{3}$$

where $\mu = (\mu_x \quad \mu_y \quad \mu_z)^T$ is the mean value vector of grid points inside the antenna's FOV.

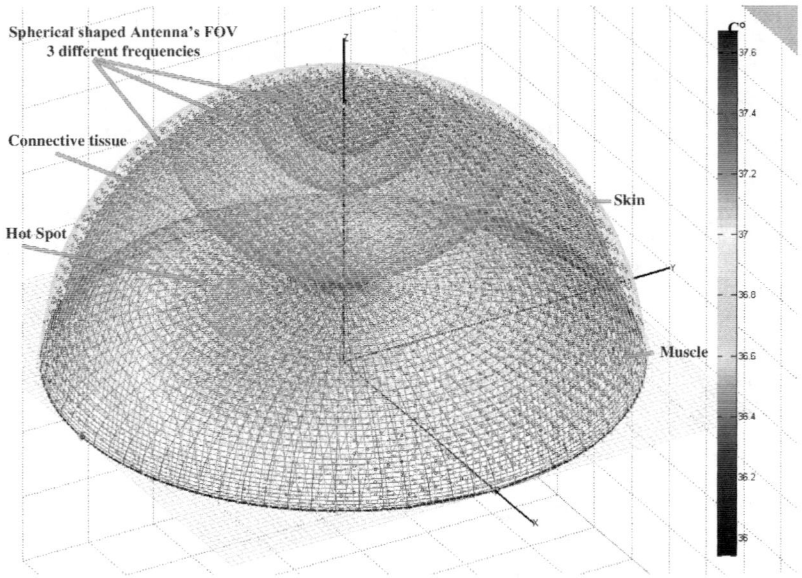

Figure 1. Human breast modelled as hemisphere with various layers, a lesion (red sphere), the approximated field of view of the antenna for three frequencies (purple), and the temperature distribution based on Butterworth function.

2.3. Testing scenarios

For simulating the concept of an early warning system in detecting a possible breast malignancy, several testing scenarios were considered at different examination periods and for lesions of increasing size. The aim was to detect differences in temperature distribution in the breast, due to lesion presence.

Table 1. Radiometer positions in spherical coordinates

Position	φ (°)	θ (°)		**Antenna at positions**
1[a]	0	0	Set 1	1
2	90	0	Set 2	1,2,3
3	90	90	Set 3	1,2,3,4,5
4	90	180	Set 4	1,2,3,…9
5	90	270		
6	45	45		
7	45	135		
8	45	225		
9	45	315		

[a] First position is considered the one on top of the hemisphere

Eleven different lesion sizes (between 0.01 and 1cm in diameter at times t_0 and t_{10} respectively) were considered, at eight (8) different locations. Table 1 presents the various antenna positions around the breast that were tested. Temperature differences at different examination periods were monitored in terms of L_1, L_2 norms, as shown in equations (4).

$$L_1 = \|t_1 - t_m\|_1 = \sum_{fov} |t_1 - t_m|$$
$$L_2 = \|t_1 - t_m\|_2 = \left(\sum_{fov} (t_1 - t_m)^2 \right)^{1/2} \tag{4}$$

3. Results and Discussion

Figures 2a and 2b show the mean temperature differences in terms of L_1 and L_2 with varying lesion diameter and radiometer sets of measurements at different locations. All temperatures were normalized to [0-1] interval. Each curve corresponds to the mean temperature difference at different lesion sites. As it can be observed from figure 2a, mean temperature differences (L_1) increased non-linearly with increasing lesion diameter. Similar behavior was observed regarding L_2.temperature differences (see figure 2b). This suggests that an early warning system may be plausible for lesion diameters larger than about 0.4 cm in diameter.

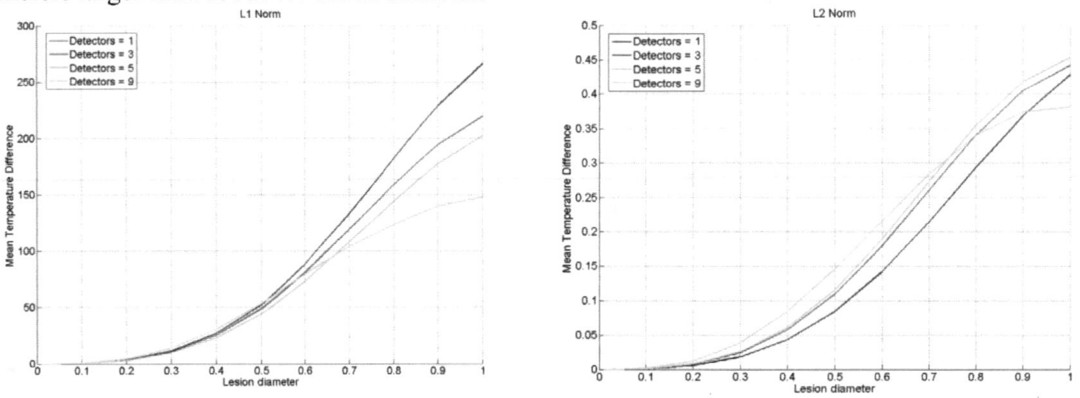

Figure 2a,b. Variation of mean temperature differences (L_1 and L_2) with increasing lesion diameter.

Acknowledgments

The research activities that led to these work, were co-financed by National Funds and by the European Regional Development Fund (ERDF) under the Hellenic National Strategic Reference Framework (NSRF) 2007-2013, concerning the project "MMRIBTM" with ref. number (12CHN181) within the Bilateral Cooperation between Greece & China action.

4. References

[1] Ng EYK, et al. 2001 *Proceedings of the Institution of Mechanical Engineers, Part H: Journal of Engineering in Medicine* **215** 25-38.
[2] Scheeler R, et al. 2014 *IEEE Transactions on Antennas and Propagation* **62** 1293-303.
[3] Fear EC, et al. 2000 *IEEE Transactions on Microwave Theory and Techniques* **48** 1854-63.
[4] Bond EJ, et al. 2003 *IEEE Transactions on Antennas and Propagation* **51** 1690-705.
[5] Bardati F, et al. 2008 *IEEE Transactions on Biomedical Engineering* **55** 214-21.
[6] Sugiura T, et al. 2011 *Proc. IFMBE Proceedings* 332-5.
[7] Sudharsan NM, et al. 1999 *Computer Methods in Biomechanics and Biomedical Engineering* **2** 187-99.
[8] Nikolopoulos CD, et al. 2014 *Progress In Electromagnetics Research C* **49** 19-29.
[9] Guoshen Y, et al. 2012 *IEEE Transactions on Image Processing* **21** 2481-99.
[10] Hand JW, et al. 2001 *Physics in Medicine and Biology* **46** 1885-903.

Emergence of periodic behaviours from randomness

J N Pickton, K I Hopcraft and E Jakeman

School of Mathematical Sciences, University of Nottingham, Nottingham, NG7 2RD, UK

E-mail: pmxjp5@nottingham.ac.uk

Abstract. This paper discusses how periodic behaviours can arise in discrete systems where the underlying dynamics are purely random. We consider non-interacting particles moving randomly on a network of nodes forming a closed loop. The population dynamics describing the number of particles at a node is a stochastic birth-death process, augmented by particles migrating randomly to adjacent nodes. This can result in the emergence of periodic behaviours occurring because of the interaction between the dynamics of the particles and the spatial structure through which they move. The conditions for this requires the network to comprise of three or more nodes and the migration to have a preferred direction. Moreover there are three classes of equilibria for the populations at nodes that depend on the relative values of the migration and birth rates.

1. Introduction

Periodic effects can be observed in collections of discrete objects, be they fire-flies signalling to attract mates [1], synapses firing in the brain [2] or photons emerging from a cavity [3]. The identification and origin of wave-like properties becomes difficult to interpret without the conceptual aid of a field. In this paper we demonstrate how periodic behaviours may emerge in such systems when the individual components' dynamics are random. In section 2 we describe how a system of non-interacting particles moving spontaneously on a network of nodes forming a loop can display oscillatory behaviours provided two conditions are satisfied – that the network comprise at least three nodes and that there be a preferred direction for the particles spontaneous movement. The oscillations so produced are damped and lead to the particles being uniformly distributed around the network. However, in section 3 we show that the disturbance can be made to persist if the particles' dynamics are augmented by a birth-death process operating at each of the nodes [4]. Then three distinct dynamic equilibria emerge for the distribution of particles on the network whose adoption is dependent on the relative values of the birth and migration rates – an incoherent uniform distribution, a coherent propagating wave-packet and a collapse into a non-propagating state located at a single node.

2. Model

Consider a population of non-interacting particles that are constrained to move around a network of N nodes forming a closed loop. Such a network can be viewed as a line with periodic boundary conditions. We impose that the whole system cannot hold any information other than its current state. This implies that each particle moves through random migration to an adjacent node at constant rate r, the time between these spontaneous events being independent and exponentially distributed. These dynamics and the network's structure can be represented through an $N \times N$

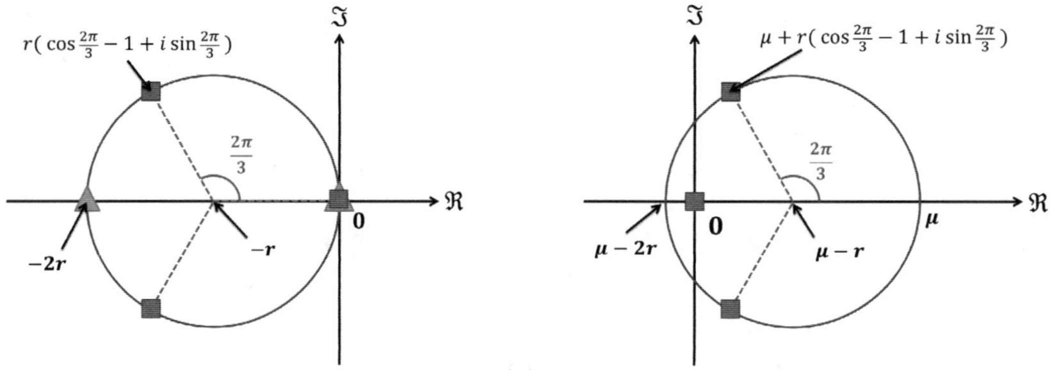

(a) (b)

Figure 1. (a) Argand diagram of the eigenvalues without a birth and death process. Green triangles and blue squares denote the eigenvalues for when $N = 2, 3$ respectively. (b) Corresponding eigenvalues with the birth-death process.

rate-matrix $\mathbf{Q}$ with elements Q_{ij} representing the rate at which a particle jumps from state i to state j and diagonal entries Q_{ii} defined such that the rows sum to zero

$$Q_{ij} = \begin{cases} -r & \text{if } i = j, \\ r & \text{if } j \equiv i + 1 \pmod{N}, \\ 0 & \text{otherwise.} \end{cases} \tag{1}$$

Representing the distribution of particles at time t by the N-dimensional row vector $\mathbf{n}(t)$ we can write down a linear differential equation for the expectation of this distribution $\langle \mathbf{n}(t) \rangle$

$$\frac{d}{dt}\langle \mathbf{n}(t) \rangle = \langle \mathbf{n}(t) \rangle \mathbf{Q} \quad \Longrightarrow \quad \langle \mathbf{n}(t) \rangle = \mathbf{n}(0) \exp(t\mathbf{Q}) \,. \tag{2}$$

The behaviour of the system is therefore dependent on the eigenvalues ω_k of the rate-matrix

$$\omega_k = r\left(e^{i\frac{2\pi}{N}(k-1)} - 1\right), \qquad k = \{1, 2, \ldots, N\}. \tag{3}$$

The eigenvalue ω_1 is zero and corresponds to the equilibrium distribution of the particles, which is uniform. For $k > 1$, the eigenvalues are complex provided that $N > 2$. The imaginary part of ω_k gives rise to oscillations, but because the real part of ω_k is negative these are damped. Thus the system exhibits damped oscillations towards a uniform equilibrium. The complex spectrum of eigenvalues can be represented graphically as in Figure 1(a) which shows these for a network comprising 2 and 3 nodes. Eigenvalues lie on a circle of radius r with centre located on the real axis at $-r$. If $N = 2$, then the eigenvalues $(0, -2r)$ are real and no oscillatory behaviour can occur. If $N = 3$, two eigenvalues form a complex conjugate pair and can exhibit oscillations, damped due to the negative real part. If the particles were permitted to migrate in both directions around the loop the imaginary part of ω_k would tend towards the real axis, eventually becoming zero when the probabilities of moving in both directions are equal. Hence for oscillations to occur there must be a preferred direction to the migration.

3. Birth and death processes

The damped oscillations can be sustained by augmenting the migration dynamics with a birth-death process acting at each node. The births occur at rate μ for each particle and thus at a

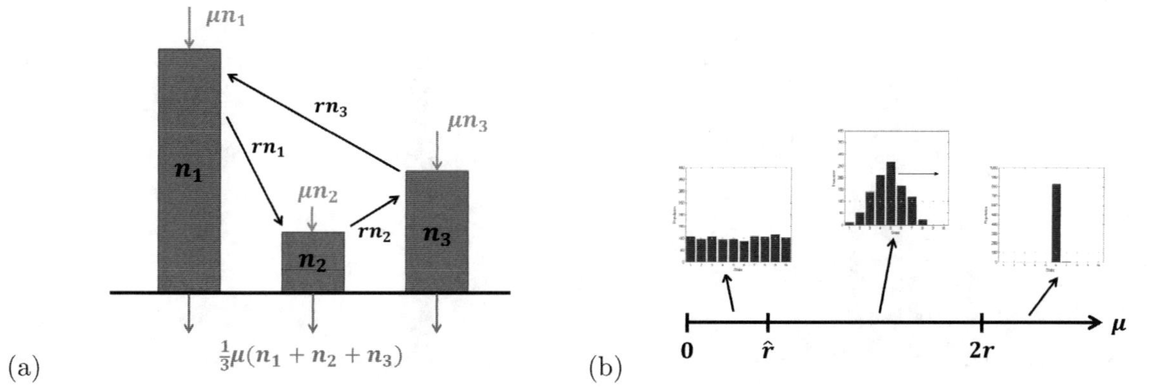

Figure 2. (a) The dynamics for $N = 3$. (b) The rate regimes and corresponding equilibrium distributions.

combined rate μn_i for a node containing a population of n_i particles. In contrast the combined death rate occurring at one of the $N^+(\leq N)$ nodes hosting particles is $\mu n/N^+$, proportional to the *total* instantaneous population size n on the entire network, with the effect that information about the global state of the network is introduced to a node and not just local information via interactions with itself and its nearest neighbours. Thus although the births at a node occur independently of other nodes, the deaths do not. This non-locality leads to the emergence of coherent behaviours. A schematic of these dynamics is illustrated in Figure 2(a) for a three node network. When there exist particles at every node ($N^+ = N$) the dynamics given by equation (2) are modified according to

$$\frac{d}{dt}\langle \mathbf{n}(t)\rangle = \langle \mathbf{n}(t)\rangle \left(\mathbf{Q} + \mu\mathbf{I} - \mu\frac{1}{N}\mathbf{1}_{N\times N}\right) \equiv \langle \mathbf{n}(t)\rangle\mathbf{A} \quad \Longrightarrow \quad \langle \mathbf{n}(t)\rangle = \mathbf{n}(0)\exp(t\mathbf{A})\,. \tag{4}$$

where $\mathbf{I}$ is an $N \times N$ identity matrix and $\mathbf{1}_{N\times N}$ is a matrix with all elements unity. Comparing with equation (2) we have two additional terms on the right-hand side corresponding to the effect of births and deaths respectively. Equation (4) is adjusted when the population at some of the nodes is zero ($N^+ < N$) to account for the changing death rates. These adjustments ensure the populations cannot become negative.
The eigenvalues λ_k of $\mathbf{A}$ can be expressed in terms of the eigenvalues (3) of the rate-matrix

$$\lambda_1 = \omega_1 = 0, \quad \lambda_k = \omega_k + \mu, \qquad k = \{2, 3, \ldots, N\}. \tag{5}$$

All non-zero eigenvalues have been shifted by the birth rate and now lie on a circle with centre $\mu - r$, see Figure 1(b). This shift dramatically changes how the system behaves compared to the incoherent uniform state that results when $\mu = 0$. A diagram of the values of μ corresponding to three different classes of behaviour is given by Figure 2(b). By increasing the birth and death rates we find that a bifurcation occurs when the eigenvalue λ_2 and its conjugate λ_N cross the imaginary axis and the uniform state becomes unstable. This happens when

$$\mu = \hat{r} \equiv r - r\cos\left(\frac{2\pi}{N}\right). \tag{6}$$

Increasing the number of nodes in the network causes $\hat{r}$ to decrease towards zero as N^{-2}, meaning the change is more likely to occur in larger sized networks. Eigenvalues with positive real part allow oscillations to emerge from small perturbations from the stationary distribution, growing

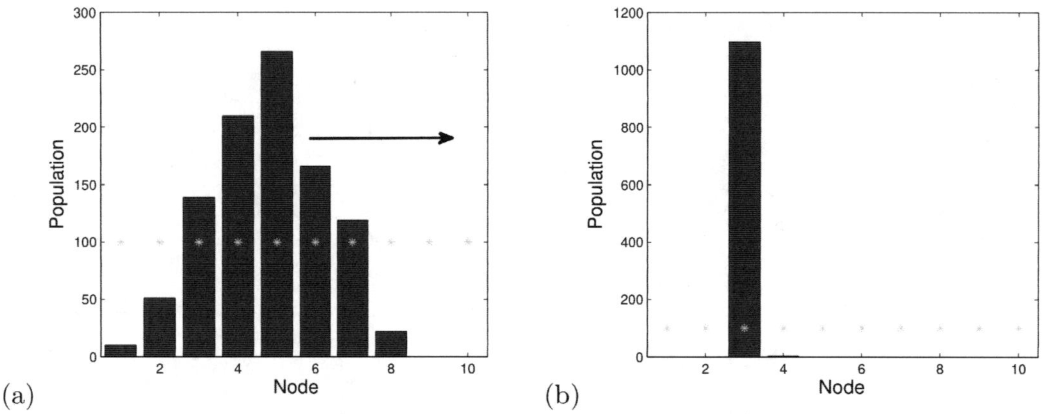

Figure 3. Simulation results of an initial population of 1000 particles in a network of $N = 10$ nodes in a loop. Both cases started in the stationary distribution with 100 particles at each node, indicated by green asterisks. Corresponding parameter values are (a) $\mu = 0.3r$ and (b) $\mu = 3r$.

in amplitude until the growth is halted by the restriction of non-negative populations. The resulting equilibrium behaviour of the system is a wave of particles that propagates around the loop at a constant speed determined by the imaginary parts of the eigenvalues. Simulation results demonstrating this behaviour are shown in Figure 3(a).
As the rate μ is increased further, the effect of the boundaries becomes more prominent causing the wave-packet to propagate at a slower speed. This effect continues until $\mu = 2r$ at which point a second bifurcation occurs and the equilibrium behaviour of system is characterised by a collapse into a non-propagating state locked at a single node, say node 1. Simulation results are shown in Figure 3(b). The second bifurcation point is explained by considering what happens to the small number of particles that migrate to the neighbouring node 2. The equation for the expected population $\langle n_2 \rangle$ at this node is then

$$\frac{d}{dt}\langle n_2 \rangle = r\langle n_1 \rangle - r\langle n_2 \rangle + \mu\langle n_2 \rangle - \frac{1}{2}\mu\left(\langle n_1 \rangle + \langle n_2 \rangle\right) = \left(r - \frac{1}{2}\mu\right)\left(\langle n_1 \rangle - \langle n_2 \rangle\right). \tag{7}$$

When $\mu > 2r$ the right-hand side is negative and the population at node 2 is driven downwards, the rate of birth and deaths dominate the rate of the migrations.

4. Conclusion

We have described how periodic behaviours are capable of emerging from discrete systems with purely random dynamics, specifically particles migrating around a network can exhibit wave-like properties when a birth-death process is present at each node. Key to the emergent oscillations was arresting the attenuation with a non-local birth-death process at each node. In order for wave-like behaviour to emerge the birth-death rates are required to be large enough to amplify oscillations but small enough to allow particles to migrate.

References

[1] Buck J 1988 *Quart. Rev. Biol.* **63** 265-89
[2] Wang XJ 2010 *Physiol. Rev.* **90** 1195-268
[3] Dagenais M, Mandel L 1978 *Phys. Rev.* A **18** 5 2217-28
[4] Hopcraft KI, Jakeman E and Ridley KD 2014 *The Dynamics of Discrete Populations and Series of Events* (CRC Press) ed D Raine

MaNIAC-UAV - a methodology for automatic pavement defects detection using images obtained by Unmanned Aerial Vehicles

Luiz Henrique Castelo Branco[1] and Paulo César Lima Segantine[2]

[1]Instituto Federal de Educação, Ciência e Tecnologia de São Paulo – IFSP,
Rua Dr Aldo Benedito Pierri 250, Jd. Manacás, cep 14.801-600, Araraquara, SP, BR
[2]Departamento de Transportes – Escola de Engenharia de São Carlos – USP,
Av. Trabalhador São Carlense 400, Centro, cep 13566-590, São Carlos, SP, BR

E-mail: luiz.branco@ifsp.edu.br[1], seganta@sc.usp.br[2]

Abstract. Intelligent Transportation Systems - ITS is a set of integrated technologies (Remote Sensing, Image Processing, Communications Systems and others) that aim to offer services and advanced traffic management for the several transportation modes (road, air and rail). Collect data on the characteristics and conditions of the road surface and keep them update is an important and difficult task that needs to be currently managed in order to reduce accidents and vehicle maintenance costs. Nowadays several roads and highways are paved, but usually there is insufficient updated data about current condition and status. There are different types of pavement defects on the roads and to keep them in good condition they should be constantly monitored and maintained according to pavement management strategy. This paper presents a methodology to obtain, automatically, information about the conditions of the highway asphalt pavement. Data collection was done through remote sensing using an UAV (Unmanned Aerial Vehicle) and the image processing and pattern recognition techniques through Geographic Information System.

1. Introduction

The highways preservation and monitoring are something of great importance especially when we are talking about countries witch a large geographical area, like Brazil. Roads are increasingly presenting emergency maintenance situations, especially in areas of difficult access and in remote locations.

From the user point of view, the state of the pavement surface is the most important because the defects or irregularities are perceived as affecting both the comfort as raises the related operating costs to vehicle maintenance parts, the fuel consumption, tires and others. Furthermore, the precarious highway conditions may result in accidents, thus endangering the human beings lives.

The use of traditional techniques for fieldwork, or even ground vehicles for detecting the quality of the roads is something that is very time consuming and costly. Thus, the use of modern technologies that allow monitoring the conditions of these roads and act more effectively in the recovery of these highways is an important task.

Starting from this current reality we aim with this research provide a more efficient and cost-effective method for the highways preservation using Unmanned Aerial Vehicles (UAVs).

In this paper we present MaNIAC-UAV (Methodology for Asphalt Automatic Characterization – using Unmanned Aerial Vehicles) a low cost and automatic methodology for the detection of defects in the asphalt surface using UAVs to support the Pavement Management Systems.

2. Related Work

In this regard, some works are highlighted. Through the development of algorithms, [1] extracted points of interest on various scales of hyperspectral images obtained by UAVs. In [2], the author presents georeferencing video accurately and in near real time in a small UAV developed for fire monitoring. In [3], determines the best ways of measuring textures associated with the scale of analysis obtained by UAV images that present the highest spatial resolution in pursuance of monitoring vegetation.

The study presented in [4] discusses the use of spectrometry combined to hyperspectral remote sensing to explore the potential road network conditions. The pavements aging and degradation are represented by distinct spectral characteristics. This study used image indices and the spatial variance measures relate to the remote sensing signal to the pavement condition index. Similarly, the same authors continue the study presented in [5-7]. In these studies the authors point out that the drones are an emerging technology in the context of transport. In this way, MaNIAC-UAV uses UAVs and provides an automatic pavement defects detection.

3. MaNIAC-UAV definition

The first step to use a set of high-resolution images to assist in automatic detection of defects in pavements is the definition of well-defined steps. Thus, we propose MaNIAC-UAV (Methodology for Asphalt Automatic Characterization – using Unmanned Aerial Vehicles) that is illustrated in Figure 1a.

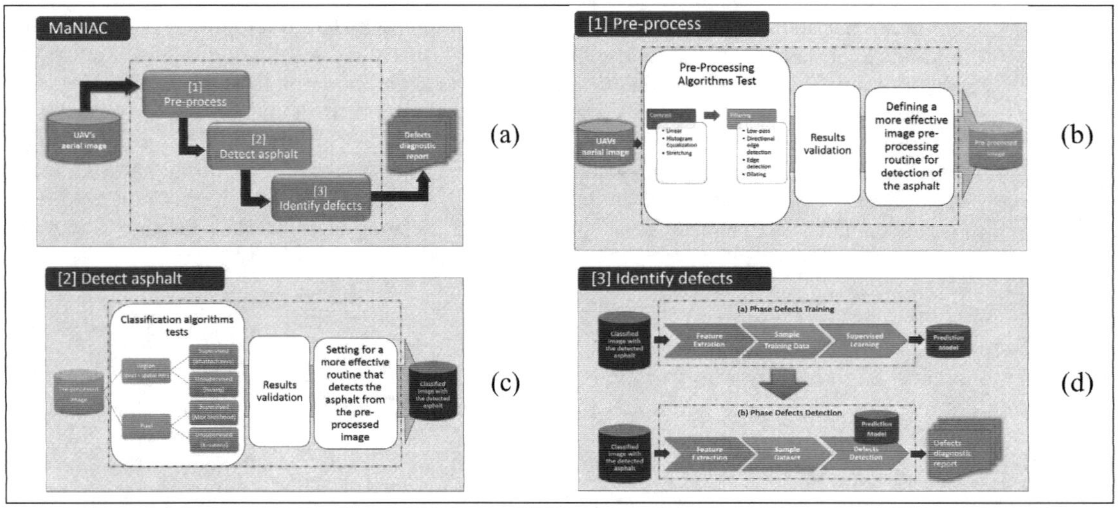

Figure 1. (a) MaNIAC - UAV. (b) Pre-process detail. (c) Detect Asphalt detail. (d) Identify Defects

The methodology is divided into three processes: Pre-Process, Detect Asphalt and Identify Defects. As input we have the high spatial resolution images, like 4cm/px, obtained by UAVs and as outputs the defects reports found under the Standard DNIT 005/2003 - B [8].

Initially, the Pre-Process (Figure 1b) refers to the initial processing of the raw data that allows the radiometric calibration image, correction of geometrical distortions and noise removal. The objective is to improve image quality, allowing objects in the image are differentiated.

The second process, Detect Asphalt (Figure 1c), aims to the classification of the images obtained and is made by SPRING (Sistema de PRocessamento de INformações Georeferenciadas) [9], a GIS (Geographic Information System) open-source software with image processing features.

The output result of this process is the classified images. These images are divided to be automatically identified using Machine Learning for the defects identification. Thus, Identify Defects (Figure 1d) makes use of two data sets: one is used as a training set, and the results obtained from other images as sample dataset. From then obtained the report with the defects identified in the reported images.

4. Results and Discussions

The images were obtained in São Carlos in Area II of the USP Campus and its surroundings. The camera resolution was 16 megapixel while the flight altitude was 130m with a spatial resolution of 4.0cm/pixel.

The flight adjustments are made directly on the UAV software. The missions were planned by polygons to delimit the areas to be imaged. The flight altitude and the images overlap that will be obtained were defined during the programming of the mission.

It is possible see that in areas with higher population density the pavement identification presents a challenge when compared to areas with lower density. This is because of the presence not only of vegetation but also constructs.

The first step was performed Pre-Process. The Histogram Equalization, the Dilation and the Edges Detection methods were used, illustrated in Figure 2a, 2b and 2c respectively.

For the second step, the Detect Asphalt two-classification process was used: a pixel supervised and unsupervised. In the unsupervised the image was loaded as monochromatic and the classification obtained using linear operation. In a supervised were extracted samples of the pavement surface and the MAXVER (Maximum Likelihood), that analysis each class based in its value. Figures 3a and 3b, shows a comparison of supervised and unsupervised process.

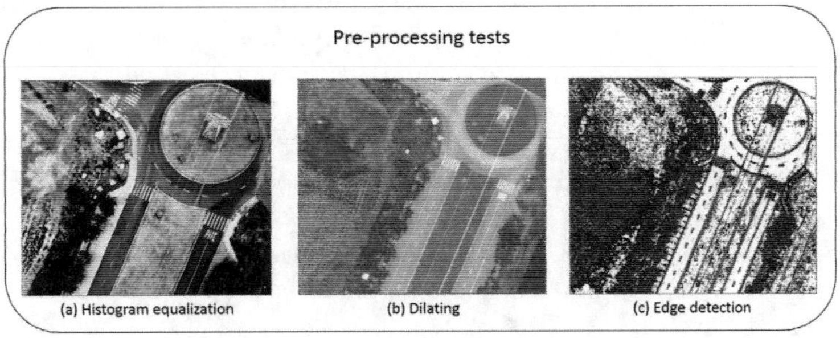

Figure 2. Pre-Process – selection areas. (a) Histogram Equalization; (b) Morphologic Dilatation; (c) Edge Detection.

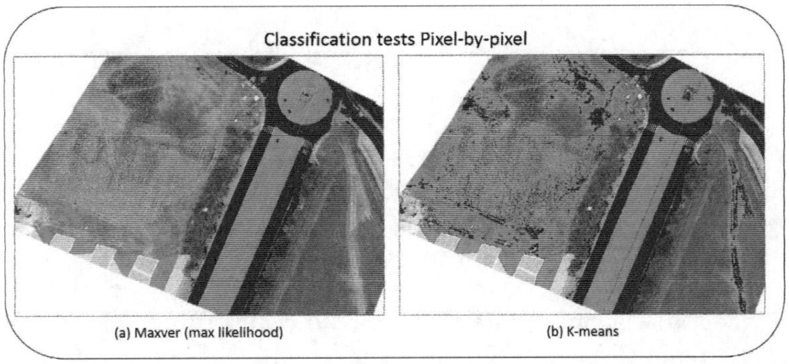

Figure 3. Classification process comparison – (a) Maximum Likelihood – (b) k-means

The last step was Identify Defects. The last step was obtained in this fist moment using a specific algorithm to detect one class of defect that is illustrated in Figure 4.

Figure 4. Asphalt defect characterized.

5. Conclusion

The main objective of this paper was present MaNIAC-UAV a methodology to automatic pavement defects detection. The MaNIAC-UAV was presented and some stages were validated by images taken, pre-processed and classified.

The results obtained from the use of image processing and geographic information systems allowed the proof of concept needed to thereafter perform the classification of defects types. These defects will be presented in a future work that will use Machine Learning to provide an automatic detection of different asphalt defects classes.

6. References

[1] MUKHERJEE A, VELEZ-REYES M and ROYSAM B 2009 Interest points for hyperspectral image data *In: IEEE Transactions on Geoscience and Remote Sensing* vol 47 n 3.

[2] ZHOU G 2009 Near real-time orthorectification and mosaic of small UAV video flow for time-critical even response *In: IEEE Transactions on Geoscience and Remote Sensing* vol 47 n 3.

[3] LALIBERTE A S and RANGO Albert 2009 Texture and scale in object-based analysis of subdecimeter resolution Unmanned Aerial Vehicle (UAV) imagery *In: IEEE Transactions on Geoscience and Remote Sensing* v 47 n 3.

[4] Herold M and Roberts D A 2005 Mapping asphalt road conditions with hyperspectral remote sensing *Proceedings of URS2005 conference* (Phoenix AZ) March.

[5] Herold M and Roberts D A 2005 Spectral characteristics of asphalt road aging and deterioration: Implications for remote sensing applications *Applied Optics* 44, 20, 4327-4334.

[6] Herold M, Roberts D A, Smadi O and Noronha V 2004 Road Condition Mapping Using Hyperspectral Remote Sensing *Proceedings of the 2004 AVIRIS workshop* March 31 April 2 (Pasadena CA).

[7] Herold M, Roberts D A, Noronha V and Omar S 2007 Imaging spectrometry and asphalt road surveys *Transport. Res. Part C.*

[8] DNIT – Departamento Nacional de Estruturas de Transportes 2003 *DNIT 009-PRO: avaliação subjetiva da superfície de pavimentos flexíveis e semi-rígidos: procedimento* (Brazil: Rio de Janeiro).

[9] INPE - Instituto Nacional de Pesquisas Espaciais 2004 *SPRING: Sistema de Processamento de Informações Georreferenciadas: Manual de Ajuda* (Brazil: São José dos Campos).

Acknowledgment

The authors acknowledge the support granted by CNPq (Conselho Nacional de Desenvolvimento Científico e Tecnológico - Brasil) and IFSP (Instituto Federal de Educação Ciência e Tecnologia de São Paulo.

Use of genetic algorithm for the selection of EEG features

P Asvestas[1], A Korda[2], S Kostopoulos[1], I Karanasiou[3], A Ouzounoglou[2], K Sidiropoulos[1], E Ventouras[1] and G Matsopoulos[2]
[1] Department of Biomedical Engineering, Technological Educational Institute of Athens, Greece
[2] School of Electrical and Computer Engineering, National Technical University of Athens, Greece
[3] Institute of Communications and Computer Systems, Athens, Greece

E-mail: pasv@teiath.gr

Abstract. Genetic Algorithm (GA) is a popular optimization technique that can detect the global optimum of a multivariable function containing several local optima. GA has been widely used in the field of biomedical informatics, especially in the context of designing decision support systems that classify biomedical signals or images into classes of interest. The aim of this paper is to present a methodology, based on GA, for the selection of the optimal subset of features that can be used for the efficient classification of Event Related Potentials (ERPs), which are recorded during the observation of correct or incorrect actions. In our experiment, ERP recordings were acquired from sixteen (16) healthy volunteers who observed correct or incorrect actions of other subjects. The brain electrical activity was recorded at 47 locations on the scalp. The GA was formulated as a combinatorial optimizer for the selection of the combination of electrodes that maximizes the performance of the Fuzzy C Means (FCM) classification algorithm. In particular, during the evolution of the GA, for each candidate combination of electrodes, the well-known (Σ, Φ, Ω) features were calculated and were evaluated by means of the FCM method. The proposed methodology provided a combination of 8 electrodes, with classification accuracy 93.8%. Thus, GA can be the basis for the selection of features that discriminate ERP recordings of observations of correct or incorrect actions.

1. Introduction

A major task of medical signal analysis is to develop algorithms (also known as classifiers) that will be capable to discriminate between signals from various classes of interest (for example, normal vs. pathological). A significant step towards this direction is to select quantitative indices (features) that: a/ provide a compact representation of the signals under consideration and b/ present different values for signals that do not belong to the same class. Thus, instead of "feeding" the classifier with the whole signal (which can contain a large number of samples), a much smaller number of feature values is used. The process of feature selection helps in understanding data, reducing computation requirement, and improving the performance of the classifier [1].

Usually, an initial set of candidate features is formed and an optimization method is applied in order to select the optimum subset of features. The optimization process involves the definition of an objective function that evaluates a subset of features, a generator of candidate subsets of features and a termination condition. The subset of features that provides the best value of the objective function is selected. One of the most popular global optimization methods is Genetic Algorithm (GA), which is a heuristic algorithm that mimics the process of natural selection [2]. GA have been widely used for feature selection [3], [4], wherein the chromosome bits represent whether the feature is selected or not.

The aim of this paper is to present a methodology, based on GA, for the selection of the optimal subset of features that can be used for the efficient classification of Event Related Potentials (ERPs), which are recorded during the observation of correct or incorrect actions. ERPs are a special category of electroencephalographic (EEG) signals, which are recorded from various locations on a subject's scalp when the subject is presented with external stimuli or events. ERPs provide non-invasive measurements of the electrical activity of the brain and describe the specific cognitive processes that are responsible for processing the stimuli or the events [5]. A significant area of application of ERPs is for generating models that represent the brain activity when a subject is committing errors or observing other people's errors. The reliable detection of correct/incorrect actions is the basis for the implementation of brain computer interface (BCI) systems that decode brain electrical activity into actions controlling devices that will assist the users of the system.

2. Material and Methods

The ERP data used in the present study were recorded in the study described in [6]. The data were acquired from sixteen (16) healthy volunteers (observers), who observed correct or incorrect actions of subjects (actors) performing. The actors were seated in front of a table facing an experimenter, having in front of them, on the table, two joystick devices positioned to the left and right of a LED stimulus device. The actors were asked to respond to the direction of a centre arrowhead surrounded by distracting flankers pointing either in the same direction as the centre arrow, or in opposite direction (figure 1). The brain electrical activity of the observers was recorded from 47 Ag/AgCl electrodes as well as vertical and horizontal electro-oculograms and was sampled with sampling rate 250 Hz. Electrodes were mounted in an elastic cap (Easy cap, Montage 10) configured for equal arrangement of the electrodes over the scalp (figure 2).

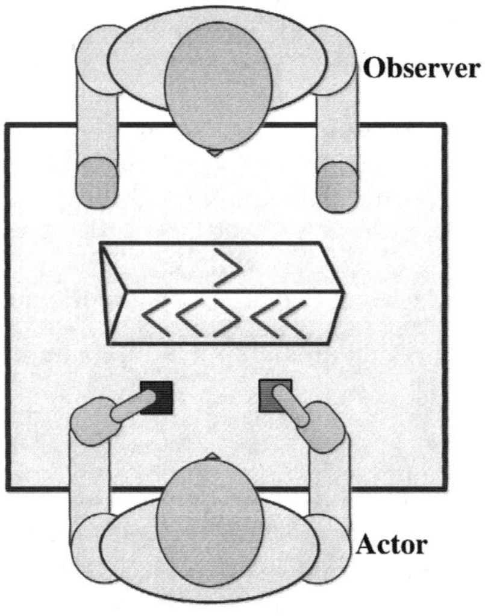

Figure 1. Experimental setup.

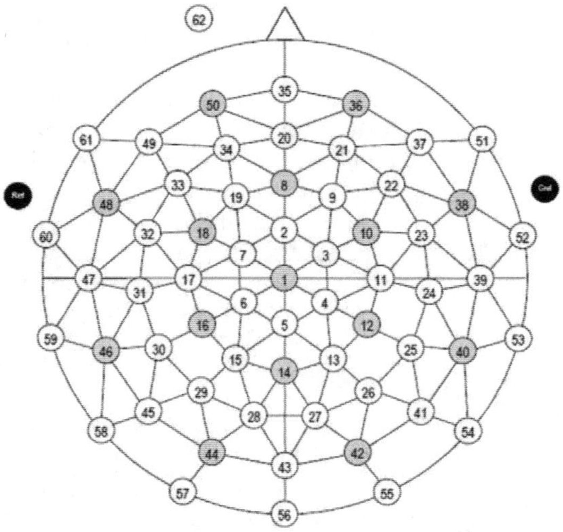

Figure 2. Electrode placement according to Easy Cap, Montage 10.

The electrode common was placed on the sternum. Ocular artifacts were corrected using the method described in [7]. The experimental session involved 8 runs of 100 trials of the task and the observations of correct and incorrect responses were averaged over a 800ms epoch (baseline [-100 , 0] ms before response) (figure 3). This procedure is necessary in order to discriminate the ERP signal from noise (brain activity that is not relevant to the task).

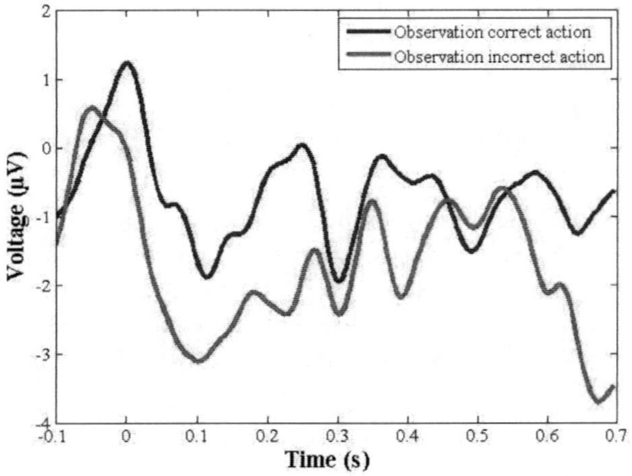

Figure 3. Representative ERP signals for observation of correct and incorrect actions

A time window, starting at -6ms and ending at 700ms (corresponding to 176 samples) after the response, was selected for analysis. A total of $32\times47=1504$ ERP recordings were available for analysis. From the available recordings, 16×47=752 recordings corresponded to observations of correct actions and the rest 16×47=752 recordings corresponded to observations of incorrect actions.

The quantitative characterisation of multichannel EEG recordings can be performed by means of the (Σ, Φ, Ω) features [8]. The feature Σ provides a measure of global signal strength. The feature Φ provides a measure of global frequency of signal changes. Finally, the feature Ω is a measure of spatial complexity. Let $K=47$ be the number of electrodes, $N=176$ be the number of samples per recording and $\Delta t=\frac{1}{250}s$ be the sampling period. If $u_k(n)$, $k=1,2,\ldots,K$ and $n=0,1,\ldots,N-1$ is the recording from the electrode site k, then K-dimensional vectors can be defined as follows:

$$\mathbf{u}(n)=\left[u_1(n),\mathrm{u}_2(n),\ldots,u_K(n)\right]^T \tag{1}$$

Assuming that the vectors are detrended such that $\sum_{n=0}^{N-1}\mathbf{u}(n)=\mathbf{0}$ and $\sum_{i=1}^{K}u_i(n)=0$ for each n, then the Σ feature is defined by the following equation [8]:

$$\Sigma=\sqrt{\frac{m_0}{K}} \tag{2}$$

where $m_0=\frac{1}{N}\sum_{n=0}^{N-1}\left\|\mathbf{u}(n)\right\|^2$. If $m_1=\frac{1}{N}\sum_{n=0}^{N-1}\left\|\frac{\Delta\mathbf{u}(n)}{\Delta t}\right\|^2$, where $\Delta\mathbf{u}(n)=\mathbf{u}(n)-\mathbf{u}(n-1)$, then the feature Φ is defined as follows [8]:

$$\Phi=\frac{1}{2\pi}\sqrt{\frac{m_1}{m_0}} \tag{3}$$

Furthermore, let $\mathbf{C}=\frac{1}{N}\sum_{n=0}^{N-1}\mathbf{u}(n)\mathbf{u}^T(n)$ be the $K\times K$ covariance matrix and $\lambda_1,\ldots,\lambda_K$ be its eigenvalues. Then, the feature Ω is given by the following equation:

$$\Omega=-\sum_{k=1}^{K}\lambda_k'\log\lambda_k' \tag{4}$$

where $\lambda'_k = \frac{\lambda_k}{\sum_{k=1}^{K} \lambda_k}$.

The next step is to find the best combination of electrodes which will be used for the calculation of the values of the (Σ, Φ, Ω) features. The best combination of the electrodes can be obtained by optimizing the clustering performance (which will be defined below) of the fuzzy c-means (FCM) algorithm [9]. The FCM algorithm is an unsupervised clustering algorithm, which allows one feature vector to belong to two or more clusters according to the value of a membership function, which represents the fuzzy behaviour of this algorithm. Specifically, since 32 sets of recordings are available, for each set of recordings a vector of (Σ, Φ, Ω) values can be computed resulting in 32 vectors. The FCM algorithm will group the 32 vectors in two clusters. Ideally, each cluster should contain vectors from one class only (observation of correct actions or observation of incorrect actions). In order to evaluate the clustering performance, CP, the following metric can be defined:

$$CP = \max\{cp_{11} + cp_{22}, cp_{12} + cp_{21}\} \tag{5}$$

where $cp_{ij} = \sum_{j=1}^{32} w_{ij}$ and $w_{ij} \in [0,1]$ is the degree of membership of feature vector j ($j = 1,2,\ldots,32$) in cluster i ($i = 1,2$). Thus, the objective is to find the combination of electrodes that will provide the best value of the clustering performance. This can be accomplished using GA optimization. Each individual of the GA population is a vector with 47 components. The components are either 0 (electrode not selected) or 1 (electrode selected). For each individual, the (Σ, Φ, Ω) values of the 32 sets of recordings for the selected electrodes are calculated and the clustering performance is evaluated using (5). The GA will provide the individual with the best clustering performance.
The parameters of the GA algorithm are given in table1.

Table 1. Parameters of the GA.

Parameter	Value
Population size	200
Number of generations	1000
Selection process	Tournament
Mutation process	Uniform
Crossover process	Scattered

The algorithm terminates if the predefined number of generations is completed or the weighted average relative change in the best value of the objective function over 150 generations is less than 10^{-6}.

3. Results

The proposed methodology was developed in Matlab and applied on the available data resulting in the following electrodes (see figure 2): 8, 10, 11, 13, 14, 20, 23 and 48. The best value of the objective function was -27.08 (the problem was formulated as a minimisation one). Sixteen (16) feature vectors corresponding to class 1 (observation of correct actions) were grouped into cluster 1 and 14 out of 16 feature vectors from class 2 (observation of incorrect actions) were grouped into cluster 2. Thus, the accuracy was 100% for cluster 1, 87.5% for cluster 2 and 93.8% in total.

Figure 4 presents the evolution of the best value and the average value of the objective function with respect to generation number. As can be seen, the best value is achieved quite early (generation 34).

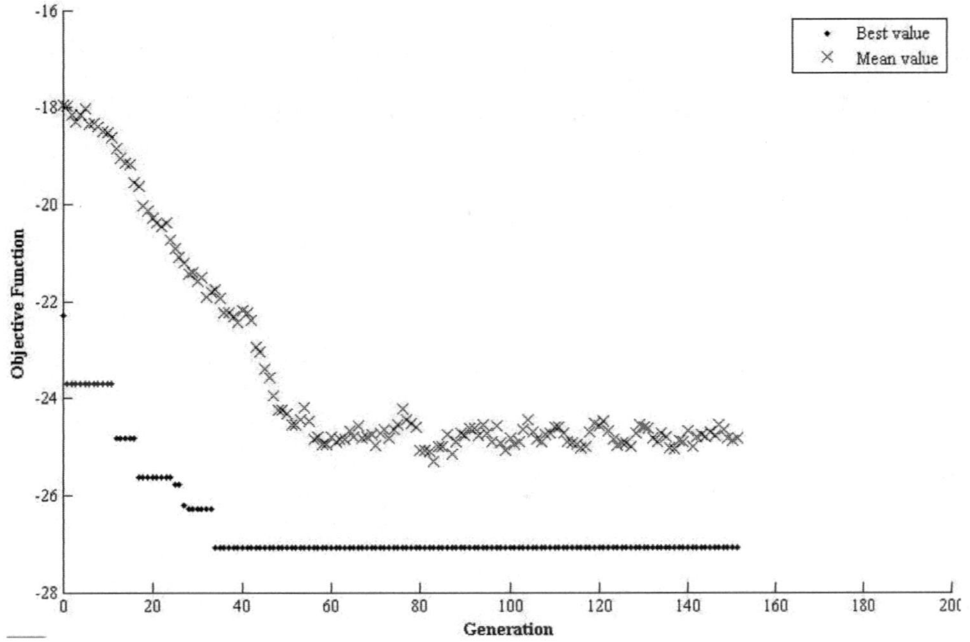

Figure 4. Evolution of the best value (dots) and the average value (stars) of the objective function with respect to generation number.

4. Conclusion

In this paper, a GA based feature selection method for the characterization of ERP data that are collected during the observation of correct or incorrect actions was presented. The method finds the combination of electrodes that are used to extract the well-known (Σ, Φ, Ω) features. Each combination of electrodes is evaluated using the clustering performance of the FCM algorithm. A total of eight electrodes were finally selected resulting in at total accuracy of correct classification 93.8%.

Acknowledgements

The authors would like to thank Hein van Schie and Ellen de Bruijn from the Nijmegen Institute for Cognition and Information (NICI), The Netherlands, for kindly providing the data of their experiments and for their contribution to initial stages of the research.

This research has been co-funded by the European Union (European Social Fund) and Greek national resources under the framework of the "Archimedes III: Funding of Research Groups in TEI of Athens" project of the "Education & Lifelong Learning" Operational Programme.

European Union
European Social Fund
OPERATIONAL PROGRAMME
EDUCATION AND LIFELONG LEARNING
investing in knowledge society
MINISTRY OF EDUCATION & RELIGIOUS AFFAIRS, CULTURE & SPORTS
MANAGING AUTHORITY
Co-financed by Greece and the European Union
NSRF
2007-2013
EUROPEAN SOCIAL FUND

References

[1] Chandrashekar G and Sahin F 2014 *Computers and Electrical Engineering* **40** 16-28

[2] Goldberg D 1989 *Genetic Algorithms in Search, Optimization and Machine Learning* (Boston, MA: Addison-Wesley)

[3] Sun Y, Babbs C and Delp E 2005 *Conf. Proc. IEEE Eng. Med. Biol.* Soc. **6** 6532-5

[4] Yang J and Honavar V 1998 *IEEE Intell. Syst. Appl.* **13** 44–9
[5] Fabiani M, Gratton G and Coles M 2000 *Event-Related Potentials: Methods, Theory, and Applications* (New York: Cambridge University Press)
[6] van Schie H, Mars R B, Coles M G H and Bekkering H 2004 *Nat. Neurosci.* **7** 549-54
[7] Gratton G, Coles M G H and Donchin E 1983 *Electroencephalogr. Clin. Neurophysiol.* **55** 468-84
[8] Wackermann J 1999 *Int. J. of Psychophysiol.* **34** 65-80
[9] Bezdek J-C 1981 *Pattern Recognition with Fuzzy Objective Function Algorithms* (Norwell, MA: Plenum Press)

A model for anomaly classification in intrusion detection systems

V O Ferreira, V V Galhardi, L B L Gonçalves, R C Silva, A M Cansian

Department of Computer Science and Statistics (DCCE), São Paulo State University (UNESP), São José do Rio Preto, Brazil.

E-mail: viniciusoliveira@acmesecurity.org

Abstract. Intrusion Detection Systems (IDS) are traditionally divided into two types according to the detection methods they employ, namely (i) misuse detection and (ii) anomaly detection. Anomaly detection has been widely used and its main advantage is the ability to detect new attacks. However, the analysis of anomalies generated can become expensive, since they often have no clear information about the malicious events they represent. In this context, this paper presents a model for automated classification of alerts generated by an anomaly based IDS. The main goal is either the classification of the detected anomalies in well-defined taxonomies of attacks or to identify whether it is a false positive misclassified by the IDS. Some common attacks to computer networks were considered and we achieved important results that can equip security analysts with best resources for their analyses.

1. Introduction

Intrusion Detection Systems (IDS) are aimed at monitoring the computer network traffic in some environment. They can be divided in two groups according to its detection method: anomaly detection and misuse detection [1]. In abuse detection, signatures representing previously known attacks are confronted with the current traffic in the monitored network to detect intrusions. In the other hand, anomaly detection tries to learn the normal pattern of the monitored traffic and detect any unusual behavior or anomaly. Although anomaly based methodologies appears quite efficient because they do not need signatures and have the capability to detect unknown attacks, they have some downsides. Anomaly based IDS suffer with the generation of high amounts of false positives and, also, the detected anomalies often have no clear information about the malicious events they represent, causing the analysis process difficult to take place.

In this context, this paper presents a model for automatic classification of anomalies detected by an anomaly based IDS in well-known taxonomies of attacks. Our main goal is to help security analysts in their analysis so that they can take the adequate counter measures in timely fashion. Besides, our methodology helps to identify the false positives, by classifying them into different classes of those that represent real attacks. We built a model based on the algorithm Autoclass and showed its capability to anomaly classification as well as the identification of false positives.

This paper is organized as follows: in section 2 we provide a brief discussion about anomaly classification field. In section 3 are detailed the Autoclass, the features we use for classification and other details about our model. In section 4 some discussion and results are presented. Finally, conclusions are shown in the section 5.

2. Anomaly Classification

Machine learning algorithms are widely used to classify data. Among machine learning algorithms, there are supervised and unsupervised methods. The main difference between these methods is that the supervised methods require previous labeled data while unsupervised methods work with unlabeled data. Nguyen and Armitage, in [2], do an extensive review about the use of machine learning algorithm for network traffic classification. They showed the efficiency of these algorithms compared to more traditional methods (port-based and payload based). Autoclass is an unsupervised algorithm which has shown good results in network classification as can be observed in [3], [4] and [5].

Although Autoclass has been used to classify legitimate network traffic, as far as we know, it had not been used yet to classify anomalous network traffic. Anomalous traffic classification is quite unexplored [6], with just a few works related in the literature, e.g. [6] and [7]. Therefore, our work presents a model which uses the unsupervised method Autoclass to classify common anomalies in computer networks.

3. Materials and Methods

The main component of our model is the machine learning algorithm we used for the classification. Autoclass is a partitional clustering algorithm and a fuzzy classificator. It uses an unsupervised approach which allows automatic discovery of the clusters inherent in a training dataset. This dataset contains unlabelled instances which are grouped according to its similar features, then, the clusters generated can be used in later steps to classify new unseen instances (i.e. anomalies, in the context of this work). To accomplish this, Autoclass combines Expectation Maximization algorithm with Bayesian theory to build a probabilistic model with distinct probability distributions that governs the classification in each cluster. As a fuzzy methodology, Autoclass allows the instances to be partly classified in more than one cluster. For further details about Autoclass we strongly encourage the reader to read [8].

Table 1. Autoclass classification features.

Name	Description
total_con	Number of flows/connections.
total_src_pkts	Number of packets.
total_src_bytes	Number of bytes.
synrate	Rate of flows with the TCP SYN flag set.
nullrate	Rate of flows without answer.
rstrate	Rate of flows with the TCP RST flag set.
distinct_dstports	Rate of distinct destination ports.
distinct_dstwkps	Rate of distinct well know destination ports.
distinct_srcwkps	Rate of distinct well know source ports.
distinct_dstaddrs	Rate of distinct destinations IPs accessed.

In our model, the clusters represent the possible attack classes into which the anomalies can be classified, in the case an anomaly is classified in more than one cluster we choose the classification with the highest probability as the anomaly's attack class. Autoclass classify the anomalies according to some pre-selected features of them, which must be grouped by each source IP address that access the monitored network. In this work we selected the features shown in Table 1, they are able to represent the traffic behaviour and can be easily extracted from protocols for traffic summarization such as Netflow, a widely deployed protocol in operational environments [6].

We designed our model to serve as generic approach for anomaly classification for different anomaly based IDS. In order to provide this interoperability, our model implements the IDMEF (Intrusion Detection Message Exchange Format) protocol, described by RFC 4765. Therefore, any IDS that exports its alerts in IDMEF standard, including the information from Table 1, can be served by the anomaly classification model proposed in this paper.

Journal of Physics: Conference Series **633** (2015) 012124 doi:10.1088/1742-6596/633/1/012124

4. Tests and Results

In order to evaluate our model, we established a data collection environment from both: 1) a simulated network, with virtual machines, where real attacks were performed against target machines and 2) the internal network from ACME! Cybersecurity Research, the laboratory where this research was conducted. In the former network we collected anomalous traffic from the five attack classes our model is able to classify: Denial of Service (DoS), Dictionary Attack, Port Scan, Network Scan and Web Vulnerability Scan (Webscan). From the latter network we collected the legitimate traffic data generated from the laboratory users using the network in business days. This legitimate traffic includes HTTP traffic, chatting, remote access and other common operations performed by legitimate users.

The IDS we choose for the tests is described in [9]. We configured it to analyse the traffic generated in the data collection environment for two days. The IDS generated 7411 alerts, in the IDMEF standard, of which 5333 are anomalies caused by the attacks and 2078 are false positives caused by the legitimate traffic.

In order to use our model for the first time, one has to train the Autoclass so that it can learn about the different attack classes and the normal traffic. We trained the Autoclass with one day data from the data collection environment. At the end, Autoclass found 35 clusters of which 30 were mapped to legitimate traffic and the other five were mapped one for each attack. This mapping was possible due the attacks have been performed with specific source IP address.

After the training step we submitted the alerts to our model, a correct classification is when an alert of an unknown anomaly is correctly classified in the cluster mapped to the attack that generates such anomaly. Besides classify the anomalies, we also innovate in classifying the false positives detected by the IDS. False positives represent the misclassification of legitimate traffic as anomaly. Once we can classify the false positives, from the IDS, as legitimate traffic it is possible to save the time security analysts spend analysing these false alerts. Our model considers all clusters mapped to legitimate traffic as one comprehensive cluster able to classify the false positives detected by the IDS. In order to measure the classification performance we used three standard metrics [6] showed in equations 1, 2 and 3. Respectively, the first one measures the global performance and the last ones evaluate the performance for each cluster.

$$overall\_accuracy = \frac{TruePositives}{TruePositives + FalsePositives} \tag{1}$$

$$recall(x) = \frac{TruePositives(x)}{TruePositives(x) + FalseNegatives(x)} \tag{2}$$

$$precision(x) = \frac{TruePositives(x)}{TruePositives(x) + FalsePositives(x)} \tag{3}$$

The recall and precision for each cluster as well as the overall classification accuracy is show in Figure 1. Our model had an overall accuracy of 87.17% which is in conformity with other works, presented in [3], [4] and [5], that used Autoclass for legitimate traffic classification. We achieved moderate results with some classes, in the Dictionary Attack cluster we achieved a recall of 71.83% because some instances were misclassified as false positives, the traffic generated for this attack was not so disparate from the legitimate traffic generated by the users. In the Webscan cluster we achieved a precision of 78.89% because some instances of DoS alerts were misclassified in this class, we believed it happened because these attacks are similar, since both perform several requests to a certain target.

Despite that, we achieved a recall of 99.9% in the false positives classification which means that almost all false positives detected by the IDS were correctly identified.

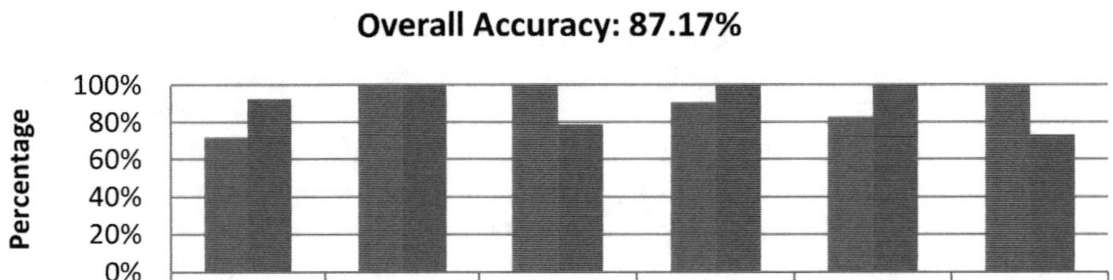

	Dictionary Attack	Portscan	WebScan	DoS	Networks can	False Positives (IDS)
Recall	71.83%	100.00%	99.53%	90.52%	82.86%	99.90%
Precision	92.47%	100.00%	78.89%	100.00%	100.00%	73.43%

Figure 1. Autoclass classification accuracy.

5. Conclusion

In this work we present a new model for anomaly traffic classification using the algorithm Autoclass. We achieved an overall accuracy of 87.17%, a value in conformity with related work that used Autoclass for legitimate traffic classification. Therefore, we achieved our goal to show the Autoclass potentiality to also classify previously unknown anomalies in well-defined taxonomies of attacks. Moreover, we showed another benefit of anomaly classification, since our model could identify 99.9% of false positives detected by the IDS, it could be used as an additional feature to reduce false alerts in Intrusion Detection Systems.

6. Acknowledgments

The authors thank FAPESP, CAPES and FAPERP (proc. 68/2015) for supporting this research.

References

[1] Wu S X, Banzhaf W 2010 The use of computational intelligence in intrusion detection systems: A review *Appl Soft Comput.* **10** 1–35

[2] Nguyen T, Armitage G 2008 A survey of techniques for internet traffic classification using machine learning *IEEE Commun Surv Tutorials.* **10** 56–76

[3] Zander S, Nguyen T, Armitage G 2005 Automated traffic classification and application identification using machine learning. *IEEE Conf Local Comput Networks 30th Anniv.* 250–2577

[4] Erman J, Arlitt M 2006 Traffic classification using clustering algorithms *SIGCOMM work on min net data* 281–286

[5] Erman J, Mahanti A, Arlitt M 2006 QRP05-4: Internet Traffic Identification using Machine Learning *IEEE Globecom* 1–6.

[6] Paredes-Oliva I, Castell-Uroz I, Barlet-Ros P, Dimitropoulos X, Sole-Pareta J 2012 Practical anomaly detection based on classifying frequent traffic patterns *Comput Commun Work (INFOCOM WKSHPS) IEEE Conf* 49–54

[7] Tellenbach B, Burkhart M, Schatzmann D, Gugelmann D, Sornette D 2011 Accurate network anomaly classification with generalized entropy metrics *Comput Networks.* **55** 3485-3502.

[8] Cheeseman P and Stutz J 1996 Bayesian classification (AutoClass): theory and results *In Advances in knowledge discovery and data mining.* American Association for Artificial Intelligence, Menlo Park, CA, USA 153-180.

[9] Batista M L and Cansian A M 2011 Detecção de eventos em redes de computadores utilizando detecção de novidade *IADIS Ibero Americana WWW/Internet.* Rio de Janeiro, RJ, Brazil.

AVALON: definition and modeling of a vertical takeoff and landing UAV

N B F Silva, E A Marconato, K R L J C Branco
Universidade de São Paulo, São Carlos, SP, Brasil
E-mail: ays@icmc.usp.br, emerson@icmc.usp.br, kalinka@icmc.usp.br

Abstract. Unmanned Aerial Vehicles (UAVs) have been used in numerous applications, like remote sensing, precision agriculture and atmospheric data monitoring. Vertical takeoff and landing (VTOL) is a modality of these aircrafts, which are capable of taking off and landing vertically, like a helicopter. This paper presents the definition and modeling of a fixed-wing VTOL, named AVALON (Autonomous VerticAL takeOff and laNding), which has the advantages of traditional aircrafts with improved performance and can take off and land in small areas. The principles of small UAVs development were followed to achieve a better design and to increase the range of applications for this VTOL. Therefore, we present the design model of AVALON validated in a flight simulator and the results show its validity as a physical option for an UAV platform.

1. Introduction

Unmanned Aerial Vehicles (UAVs) are aircrafts that can fly autonomously or controlled remotely by a human pilot. These vehicles can be used in several applications, like remote sensing, precision agriculture and atmospheric data monitoring. Different configurations of aircrafts were created to supply the requirements of all these applications. One example is the class of Vertical takeoff and landing (VTOL) UAVs, which can take off and land in small areas and whose more representative archetype is the helicopter. However, aircrafts based on rotary wings suffer penalty performance regarding speed of operation, which reduces the area covered in a mission.

In this paper, we present a VTOL definition and design with a configuration called tailsitter, which has the benefits of VTOL and fixed-wing aircrafts. The tailsitter configuration is characterized by a fixed-wing aircraft, capable of reaching higher velocities, and that can also take off and land vertically in small areas without a runway. Our tailsitter UAV, named AVALON (Autonomous VerticAL takeOff and laNding), is a small electric aircraft with only one propeller and a structure to keep it in the vertical position during the takeoff and landing. The AVALON definition and modeling is demonstrated here with simulation results that show its physical appropriateness as an UAV platform.

2. Related work

Different configurations of VTOLs can be found in the literature, however they are usually based on complex mechanical structures like tilt rotors [1, 2], thrust vectoring [3] and ducted-fans [4, 5]. Other examples of VTOL adopt rotary wings, but can only operate at low velocities

[6, 7]. Tailsitters with simpler designs were also explored before as an alternative more suitable for small UAVs, such as in the T-Wing [8] and C-Plane II [9]. However, these aircrafts have contra-rotating propellers to avoid the torque effect from the thrust, which increases the design and cost of the platform. Differently, AVALON design rests on the premises of simplicity to achieve a low-cost and easy to maintain vehicle, which are desirable characteristics for small UAVs.

3. AVALON definition

The operation of a tailsitter UAV is divided in five stages: vertical takeoff, transition from vertical to horizontal position, horizontal flight, transition from horizontal to vertical position and vertical landing. The platform definition must be able to encompass all stages, with hover for short periods and cruise flight for a longer range. Hence, our tailsitter UAV was designed following these criteria.

AVALON is a small electric UAV with fixed-wings and a structure to keep it in the vertical position, as seen in Figure 1 in different perspectives. It has just one propeller, two ailerons, two elevators and one rudder, as a typical fixed-wing aircraft. The design is symmetrical in the xy plane with a wingspan of 1.74 m, height of 1.7 m and weight of 2.5 kg. The airfoil of the main components is the flat plate, again chosen due to simplicity.

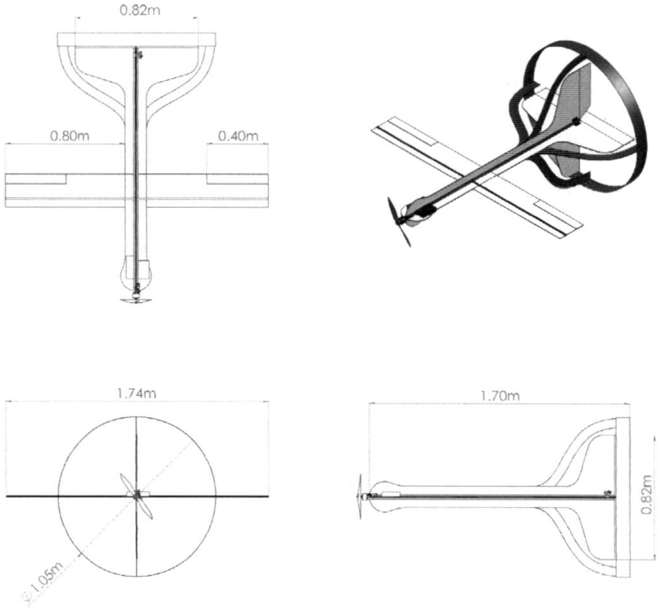

Figure 1. AVALON design in different perspectives: isometric view in the top right, top view in the top left, front view in the bottom left and right view in the bottom right.

4. AVALON modeling

To validate the aircraft design, we used the flight simulator X-Plane, which is considered the flight simulator with the higher physical and functional fidelity between simulator engines [10]. The AVALON model was constructed using Plane Maker, a software that is part of X-Plane, and each part was designed separately, resulting in the model seen in Figure 2. During the simulation, X-Plane calculates the dynamic and kinematic effects according to blade element analysis, which is influenced by aerodynamic coefficients related to the components airfoil. AVALON flat plate airfoil coefficients for lift, C_L, drag, C_D, and moment, C_M, for different angles of attack can be seen in Figures 3, 4 and 5, respectively.

Figure 2. AVALON model in the flight simulator X-Plane.

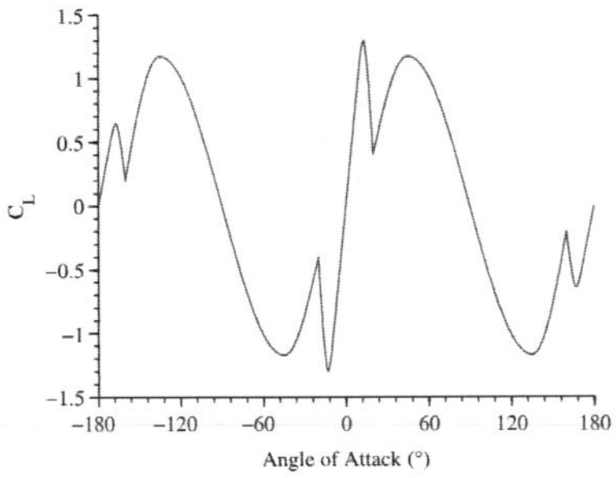

Figure 3. Flat plate airfoil coefficient for lift, C_L, versus angle of attack.

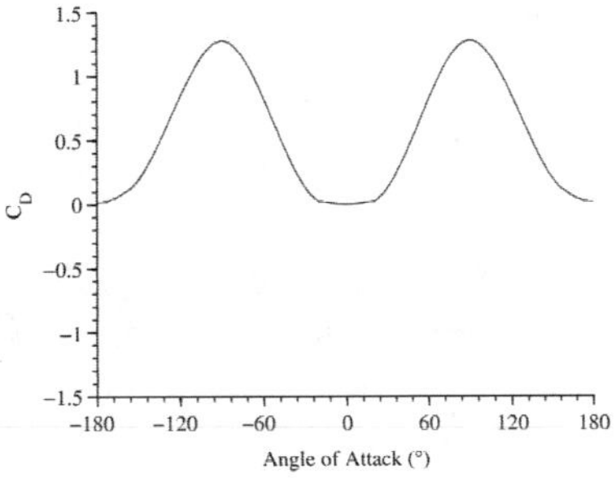

Figure 4. Flat plate airfoil coefficient for drag, C_D, versus angle of attack.

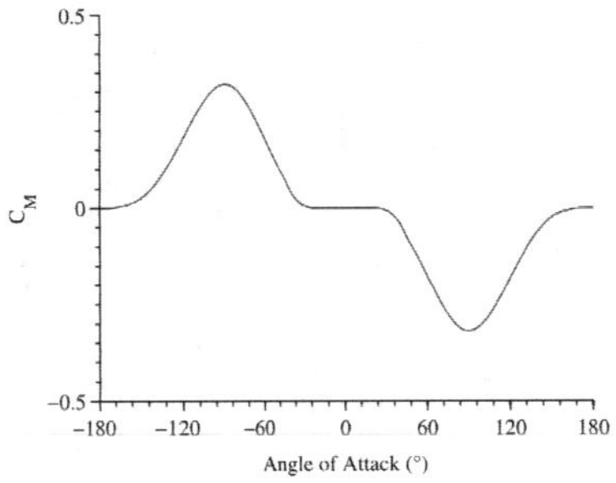

Figure 5. Flat plate airfoil coefficient for moment, C_M, versus angle of attack.

5. Simulation results

The desired trajectory was simulated with a vertical takeoff, then a transition between vertical position and horizontal position, followed by a new transition to the vertical position and an attempt to land. The latitude, longitude and altitude were retrieved from X-Plane and converted to the navigation frame represented as North-East-Up using MATLAB. The resulting trajectory of AVALON test can be seen in Figure 6.

The results show that AVALON design allows it to behave as a manoeuvrable VTOL during the vertical position and the transitions. They also illustrate the possibility of using a flight simulator as a tool to the development of UAV platforms and its capabilities to represent the dynamic and kinematic effects of UAV models.

6. Conclusions

This paper presents AVALON, a tailsitter UAV that can take off and land in small areas and does not suffer performance penalties during horizontal flight. The results in a simulation environment show that AVALON design enables the platform to act as an UAV during all operation stages.

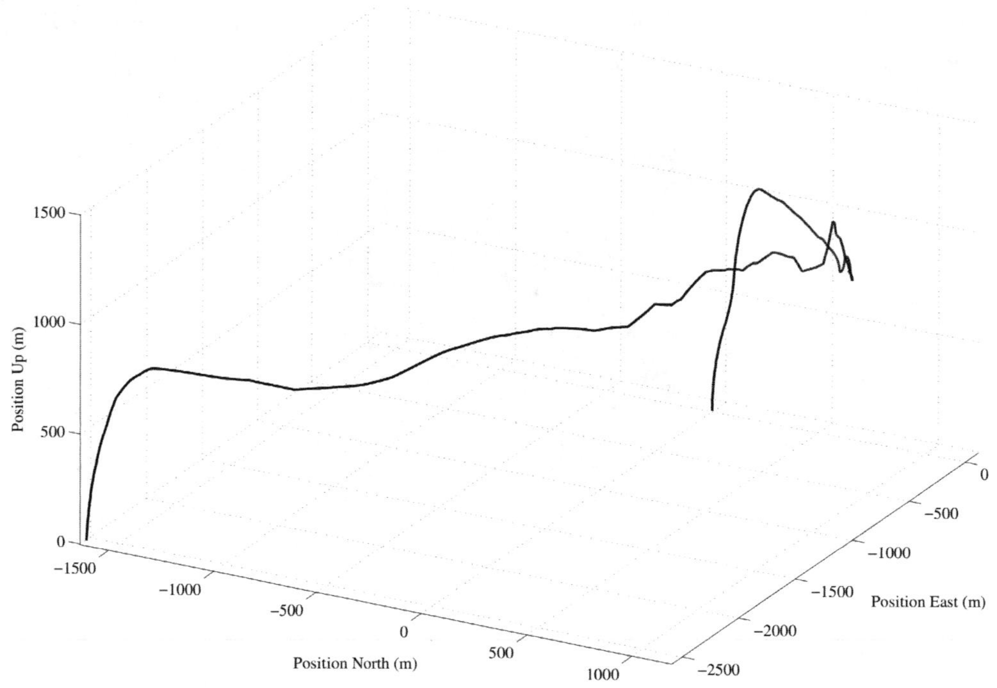

Figure 6. Resulting trajectory represented in the navigation frame North-East-Up.

As future work, we expect to use the AVALON modeling to perform autopilot tests and validate control laws for each flight operation. Afterwards, we plan to use the simulation process to perform tests of multiple platforms and their communications in a mission scenario.

Acknowledgment

The authors acknowledge the support granted by FAPESP through processes 2012/13641-1, 2014/12134-4 and 2013/18562-5.

References

[1] Papachristos C, Alexis K and Tzes A 2013 *2013 21st Mediterranean Conference on Control Automation (MED)* pp 369–374

[2] Ta D A, Fantoni I and Lozano R 2012 *American Control Conference (ACC), 2012* pp 131–136 ISSN 0743-1619

[3] Hang Y, Jihong Z, Xiaming Y and Chao Z 2013 *2013 10th IEEE International Conference on Control and Automation (ICCA)* pp 735–738 ISSN 1948-3449

[4] Naldi R and Marconi L 2013 *Automatica* **49** 1693–1704 ISSN 00051098

[5] Argyle M E, Beard R W and Morris S 2013 *American Control Conference (ACC), 2013* pp 806–811 ISSN 0743-1619

[6] Oosedo A, Abiko S, Konno A, Koizumi T, Furui T and Uchiyama M 2013 *2013 IEEE International Conference on Robotics and Automation (ICRA)* pp 317–322 ISSN 1050-4729

[7] Sinha P, Esden-Tempski P, Forrette C A, Gibboney J K and Horn G M 2012 *2012 IEEE Aerospace Conference* pp 1–17 ISSN 1095-323X

[8] Stone R b c, Anderson P b, Hutchison C b, Tsai A b, Gibbens P b and Wong K b 2008 *Journal of Aircraft* **45** 673–685 ISSN 00218669

[9] Garcia O, Castillo P, Wong K and Lozano R b 2012 *Journal of Intelligent and Robotic Systems: Theory and Applications* **65** 123–136 ISSN 09210296

[10] Craighead J, Murphy R, Burke J and Goldiez B 2007 *2007 IEEE International Conference on Robotics and Automation* pp 852–857

Estimation of bone Calcium-to-Phosphorous mass ratio using dual-energy nonlinear polynomial functions

P Sotiropoulou[1], V Koukou[1], N Martini[1], C Michail[2], E Kounadi[3], I Kandarakis[2], G Nikiforidis[1] and G Fountos[2]
[1]Department of Medical Physics, Medical School University of Patras, 265 00 Patras, Greece
[2]Department of Medical Instruments Technology, Technological Educational Institution of Athens, 122 10 Athens, Greece

E-mail: gfoun@teiath.gr

Abstract: In this study an analytical approximation of dual-energy inverse functions is presented for the estimation of the calcium-to-phosphorous (Ca/P) mass ratio, which is a crucial parameter in bone health. Bone quality could be examined by the X-ray dual-energy method (XDEM), in terms of bone tissue material properties. Low- and high-energy, log-intensity measurements were combined by using a nonlinear function, to cancel out the soft tissue structures and generate the dual energy bone Ca/P mass ratio. The dual-energy simulated data were obtained using variable Ca and PO_4 thicknesses on a fixed total tissue thickness. The XDEM simulations were based on a bone phantom. Inverse fitting functions with least-squares estimation were used to obtain the fitting coefficients and to calculate the thickness of each material. The examined inverse mapping functions were linear, quadratic, and cubic. For every thickness, the nonlinear quadratic function provided the optimal fitting accuracy while requiring relative few terms. The dual-energy method, simulated in this work could be used to quantify bone Ca/P mass ratio with photon-counting detectors.

1. Introduction

Crystal structure and chemical composition of biological apatite reflect their physiological function as a mechanical support, which is very important to the skeletal system, storage of mineral, and storage of chemical energy. Alterations in the bone quality can be attributable to disease-related structural and chemical changes. Due to the complexity of bone, various methods have been used to measure bone tissue properties [1]. The relative content of calcium (Ca) and phosphorus (P) regulates the mineral homeostasis and bone metabolism [2]. Since variations in the concentrations of either Ca or P are not necessarily in mutual association, changes of the Ca/P ratio are crucial for the valuation of bone health. The Ca/P bone ratio in cases of diseases such as osteogenesis imperfecta or osteoporosis is systematically lower than that derived from normal bone samples [3,4]. Consequently, the estimation of the Ca/P ratio as a bone quality index may add valuable information regarding bone mineralization state and bone diseases [5,6,7,8,9].

In this study an analytical approximation of dual energy inverse functions is presented for the estimation of Calcium-to-Phosphorous (Ca/P) mass ratio. An X-ray dual energy (XRDE) method is used to provide the tissue materials attenuation data in order to estimate the bone quality index. The functional forms of inverse functions that have been investigated are the linear, quadratic, and cubic. Inverse fitting functions with the least-squares estimation were used to obtain fitting coefficients and calculate the thickness of each material.

2. Materials and Methods

2.1. Theory

The theory, upon which the *in vivo* estimation of skeletal Ca/P mass ratio was based, described firstly by Fountos et al [5]. The Ca/P mass ratio is obtained assuming a three-component system: Ca, PO_4 and water. Using the dual energy formalism, the water (soft tissue) is cancel out and the two unknown parameters are the calcium thickness (t_{Ca}) and the phosphate thickness (t_{PO_4}). The log-intensity function, $Y(t_{Ca}, t_{PO_4})$, defined as the logarithm of the ratio of the reference intensity (I_w) to the bone intensity (I_b), can be written as [7]

$$Y_i(t_{Ca}, t_{PO_4}) = \ln\left(I_{w,E_i} / I_{b,E_i}\right),\ i = \ell, h \quad (1)$$

where, I_{o,E_ℓ} and I_{o,E_h} are the unattenuated low- and high-energy intensity per unit energy $\left(photons/cm^2 keV\right)$ at the detector input.

The low- and high-energy log-intensity values are made independently by using X-ray beams of different spectra. The goal of the presented dual energy Ca/P determination is to invert the obtained log-intensity functions, $Y_i(t_{Ca}, t_{PO_4})$, into Ca thickness, $t_{Ca}(Y_\ell, Y_h)$ and PO_4 thickness, $t_{PO_4}(Y_\ell, Y_h)$. The inverse functions are nonlinear due to the polyenergetic spectra and can be approximated by surfaces of second or third order, in three dimensions [10,11]. The functional forms of inverse functions that have been investigated [10,11] for each thickness are linear, quadratic and cubic functions.

2.2. Incident spectra

Experimental measurements of incident spectra were carried out using an XRDE system [7] incorporating a Cadmium Telluride (CdTe) photon counting detector [12,13,14,15]. The 100kVp energy spectrum from the Tungsten anode, after 0.88 g/cm^2 Ce filtration resulted in two energy bands with mean energies of 38keV (low-energy spectrum, LE) and 86keV (high-energy spectrum, HE).

2.3. Simulated calibration data

Simulations of scale phantoms were performed for investigation of the inverse functions and the determination of polynomial's coefficients. Software phantoms consisted of water and hydroxyapatite were used, to simulate a finger of 2.8cm total thickness. The simulated calibration data were obtained from seven thicknesses of each component, ranging from 0.0210-0.0438cm and 0.1260-0.1488cm for Ca and PO_4 thicknesses, respectively. Thus 49 calibration points were acquired. The reference intensity (I_{w,E_i}) and the unattenuated intensity level (I_{o,E_i}) were measured considering Poisson noise and source stability. The attenuated intensities, I_{b,E_i}, were estimated from the software phantoms. To estimate the fitting accuracy of the different inverse functions the residuals of the fit were determined.

3. Results and Discussion

The calculated low-energy and high-energy log-intensity values are plotted in Figure 1. The range of the log-intensity values are 0.28 to 0.46 and 0.07 to 0.10 for the low- and high-energy, respectively. The maximum log-intensity values are attainable in highest thicknesses where there is the maximum attenuation. In LE the modulation of log-intensity values is greater than in HE, due to higher attenuation of the photons in low energies.

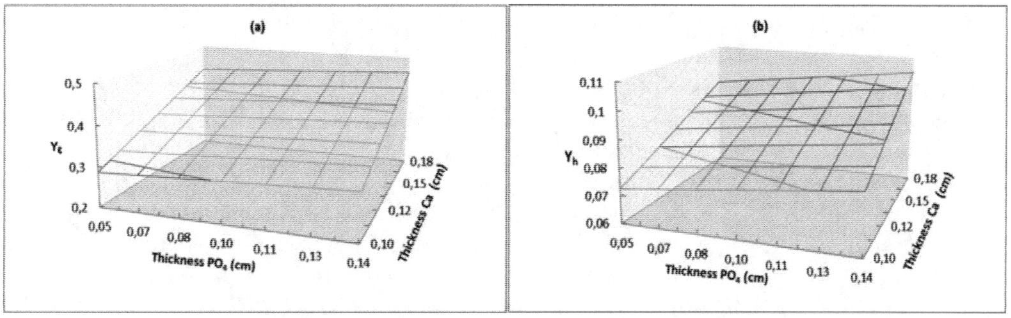

Figure 1. The calibration (a) LE and (b) HE log-intensity values, as a function of the *Ca* and PO_4 thickness, used to compute the various inverse functions.

The thickness of each material was fitted separately using a nonlinear least squares analysis. The median, δ_{med}, the standard deviation, δ_{rms}, and the maximum absolute deviation, $\delta_{\max}$ of the residual values for each of the inverse function are presented in Table 1.

Table 1. The median, standard deviation and maximum absolute deviation of the residual values from the fit to Ca and PO_4 thicknesses of the simulated data.

Functional models	Ca thickness (cm)			PO_4 thickness (cm)		
	δ_{med}	δ_{rms}	$\delta_{\max}$	δ_{med}	δ_{rms}	$\delta_{\max}$
Linear	$-6.6\cdot10^{-6}$	$2.4\cdot10^{-5}$	$4.9\cdot10^{-5}$	$-3.1\cdot10^{-5}$	$8.0\cdot10^{-5}$	$1.8\cdot10^{-4}$
Quadratic	$-3.4\cdot10^{-9}$	$1.5\cdot10^{-8}$	$2.7\cdot10^{-8}$	$4.0\cdot10^{-8}$	$1.3\cdot10^{-6}$	$3.1\cdot10^{-6}$
Cubic	$1.8\cdot10^{-10}$	$3.9\cdot10^{-9}$	$8.7\cdot10^{-9}$	$-1.2\cdot10^{-9}$	$2.3\cdot10^{-8}$	$5.1\cdot10^{-8}$

As it can be seen from Table 1, the linear function inadequately modeled both thicknesses. The accuracy of the cubic function, with δ_{med} of $1.8\cdot10^{-10}$cm and δ_{rms} of $3.9\cdot10^{-9}$ cm, for Ca thickness and with δ_{med} of $-1.2\cdot10^{-9}$cm and δ_{rms} of $2.3\cdot10^{-8}$cm for PO_4 thickness, is slightly better than the accuracy of the quadratic function. However, the quadratic function requires a smaller number of terms compared to the cubic function.

4. Conclusions

A nonlinear quadratic function demonstrated high fitting accuracy, for each thickness, while requiring relative few terms. The proposed method showed that it is possible to quantify bone Ca/P mass ratio by using dual energy techniques and photon-counting detectors.

Acknowledgement

This research has been co-funded by the European Union (European Social Fund) and Greek national resources under the framework of the “Archimedes III: Funding of Research Groups in TEI of Athens” project of the “Education & Lifelong Learning” Operational Programme.

5. References

[1] Malluche H, Porter D and David Pienkowski 2013 *Nat. Rev. Nephro.* **11** 671
[2] Masuyama R, Nakaya Y, Katsumata S, Kajita Y, Uehara M, Tanaka S, Sakai A, Kato S, Nakamura T and Suzuki K 2003 *J. Bone Miner. Res.* **18** 1217
[3] Tzaphlidou M 2008 *J. Biol. Phys.* **34** 39

[4] Kounadi E, Fountos G and Tzaphlidou M 1998 *Connect. Tissue Res.* **37** 69
[5] Fountos G, Yasumura S and Glaros D 1997 *Med. Phys.* **24** 1303
[6] Fountos G, Tzaphlidou M, Kounadi E and Glaros D 1999 *Appl. Radiat. Isot.* **51** 273
[7] Sotiropoulou P, Fountos G, Martini N, Koukou V, Michail C, Kandarakis I and Nikiforidis G 2015 *Phys. Medica* doi:10.1016/j.ejmp.2015.01.019
[8] Sotiropoulou P, Fountos G, Martini N, Koukou V, C Michail C, Valais I, Kandarakis I and Nikiforidis G 2014 X-ray spectra for bone quality assessment using energy dispersive counting and imaging detectors with dual energy method *XIII Mediterranean Conf. on Medical and Biological Engineering and Computing 2013 (Seville, Spain, 25–28 September 2013) (IFMBE Proceedings Vol 41)* ed LM Roa Romero (Springer Int.) pp. 463
[9] Zoehrer R, Perilli E, Kuliwaba JS, Shapter NL, Fazzalari NL and Voelcker NH 2012 *Osteoporos. Int.* **23** 1297
[10] Cardinal HN and Fenster A 1991 *Med. Phys.* **18** 867
[11] Lemacks M, Kappadath S, Shaw C, Liu X and Whitman G 2002 *Med. Phys.* **29** 1739
[12] Martini N, Koukou V, Michail C, Sotiropoulou P, Kalyvas N, Kandarakis I, Nikiforidis G and Fountos G *Journal of Spectroscopy* **2015** Article ID 563763
[13] Michail C M, Spyropoulou V A, Fountos G P, Kalyvas N E, Valais I G, Kandarakis I S and Panayiotakis G S 2011*IEEETrans. Nucl. Sci.* **58(1)** 314
[14] Michail CM, Fountos G P, Valais I G, Kalyvas N, Liaparinos P, Kandarakis I S, Panayiotakis G S 2011 *IEEETrans. Nucl. Sci.* **58(5)** 2503
[15] Seferis I E, Michail C M, Valais I G, Fountos G P, Kalyvas N I, Stromatia F, Oikonomou G, Kandarakis I S, Panayiotakis G S 2013 *Nucl. Instrum. Meth. Phys. Res.* A. **729** 307

Basic model for traffic interweave

Ding-wei Huang
Department of Physics, Chung Yuan Christian University, Chung-li, Taiwan
E-mail: dwhuang@cycu.edu.tw

Abstract. We propose a three-parameter traffic model. The system consists of a loop with two junctions. The three parameters control the inflow, the outflow (from the junctions,) and the interweave (in the loop.) The dynamics is deterministic. The boundary conditions are stochastic. We present preliminary results for a complete phase diagram and all possible phase transitions. We observe four distinct traffic phases: free flow, congestion, bottleneck, and gridlock. The proposed model is able to present economically a clear perspective to these four different phases. Free flow and congestion are caused by the traffic conditions in the junctions. Both bottleneck and gridlock are caused by the traffic interweave in the loop. Instead of directly related to conventional congestion, gridlock can be taken as an extreme limit of bottleneck. This model can be useful to clarify the characteristics of traffic phases. This model can also be extended for practical applications.

1. Introduction

Practical applications of traffic dynamics are well appreciated because of their importance to modern society. Due to the complexity of real traffic, a workable model often has numerous parameters. A lot of research has focused to explore the fundamentals of complex dynamics [1]. A basic model is indispensable in understanding the underlying dynamics. Previously, we proposed a cellular automaton model for traffic dynamics at roundabout, which has twenty four parameters [2]. In this work, we aim to reduce the number of parameters and propose a basic model for the traffic interweave.

On a simple roadway without traffic interweave, traffic dynamics is a competition between inflow and outflow which results in two phases: free flow and conventional congestion. On the complex roadway with traffic interweave, there are two more types of congestion: bottleneck and gridlock. Together, there are three types of congestion. We propose a three-parameter model to present economically a clear perspective to these distinct phases. The three parameters can each be associated to a different characteristics of traffic phases: free flow, congestion, and bottleneck. Conventional wisdom would take the gridlock as an extreme limit of congestion. We show that it can be more appropriate to consider the gridlock as an extreme limit of bottleneck. This model can be useful to clarify the characteristics of traffic phases. This model can also be extended for practical applications.

2. Model

The model system consists of a rotary with two junctions as shown in Fig. 1(a). All the roadways are single-lane and unidirectional. Vehicles move into the system through one of the straight roadway, go around the loop a few times, and leave the system from the other straight roadway.

Roadway is divided into discrete cells, on which vehicles hop with cellular automaton rules. The dynamics is deterministic and governed by the Asymmetric Simple Exclusion Processes (ASEP) [3]. If an empty cell in front is available, vehicle moves forward in the next time step. The system configuration is updated in parallel. In contrast to the deterministic dynamics, the boundary conditions are stochastic. The inflow and outflow are controlled by parameters α and β, respectively. When the first cell of the incoming roadway is empty, a new vehicle is added stochastically with a probability α in the next time step. When a vehicle reached the last cell of the outgoing roadway, that vehicle is removed stochastically with a probability β in the next time step. Upon entering the system, each vehicle is assigned an integer n which specifies the number of turns $(1+n)$ for that vehicle to move along the loop. The parameter n is taken from a Poisson process, i.e. with a probability $(\gamma^n/n!)\ \exp(-\gamma)$. The average number of turns in the loop is $(1+\gamma)$. In summary, the model has three parameters: $\alpha \in (0,1)$ controls the inflow, $\beta \in (0,1)$ controls the outflow, and $\gamma \in (0,\infty)$ controls the interweave.

The model configuration can also be taken as a T-shaped intersection imposed with periodic boundary condition [4], as shown in Fig. 1(b). The related traffic rules can be applied directly. On all cells, except the shaded one, a vehicle has only one direction to move. The shaded cell indicates the exit junction, on which a vehicle has two choices: to exit or to go another round. The interweave implies a traffic conflict on the cell next to the shaded one, i.e. the entry junction. The model prescribes the incoming vehicles to yield to the traffic in the loop, as required by most traffic regulations. Our results show that the gridlock cannot be avoided even if all vehicles perfectly follow the well-intended traffic regulations.

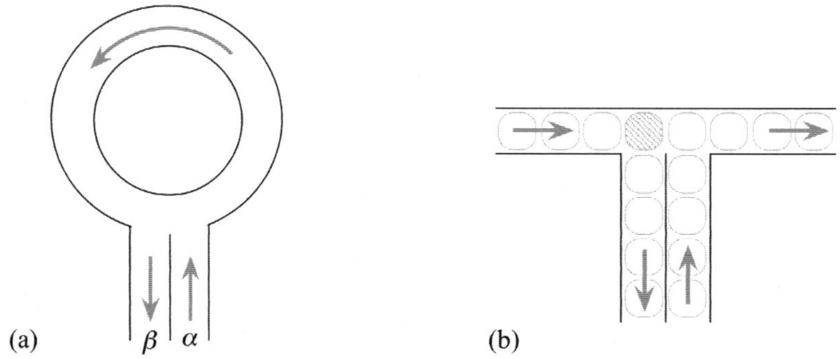

Figure 1. (a) System configuration, where the arrow indicates the traffic direction. **(b)** T-shaped intersection, where the shaded cell indicates the exit junction.

3. Results

If all vehicles go around the loop just one time, i.e. $\gamma = 0$, there is no traffic interweave. The model reduces to ASEP on a straight roadway. The simulation results for the density ρ on the loop are shown in Fig. 2(a). When $\alpha < \beta$, traffic is free-flowing and ρ depends on α only. When $\alpha > \beta$, traffic is congested and ρ depends on β only. An abrupt transition along $\alpha = \beta$ separates the free flow phase and the congestion phase. When vehicles go around the loop more than one times, traffic in the loop will interweave with the incoming traffic. Two new traffic phases can be observed. At $\gamma = 0.5$, vehicles go around the loop one and a half times on average, where 61% of vehicles go around the loop one time, 30% two times, 8% three times, and 1% four times. The results are shown in Fig. 2(b). Four different phases can be observed. Besides the free flow and congestion, the plateau indicates a bottleneck phase and the narrow stripe with a maximum density $\rho = 1$ indicates a gridlock phase. When both α and β are larger than the critical value, the plateau emerges and ρ is independent of both α and β. When β becomes very small, the severe congestion leads to the gridlock. The density reaches the maximum $\rho = 1$ on the loop and also on the incoming roadway; while the outgoing roadway becomes empty. When the traffic interweave further increases, the gridlock phase becomes dominant. At $\gamma = 1$, vehicles go around the loop two times on average. The ratio of vehicles travelling around the loop one time decreases to 37% and the ratios of multi-time traveller increase accordingly. The results are shown in Fig. 3(c), where we observe only free flow and gridlock. The free flow phase is limited to $\alpha < \beta$ and small α.

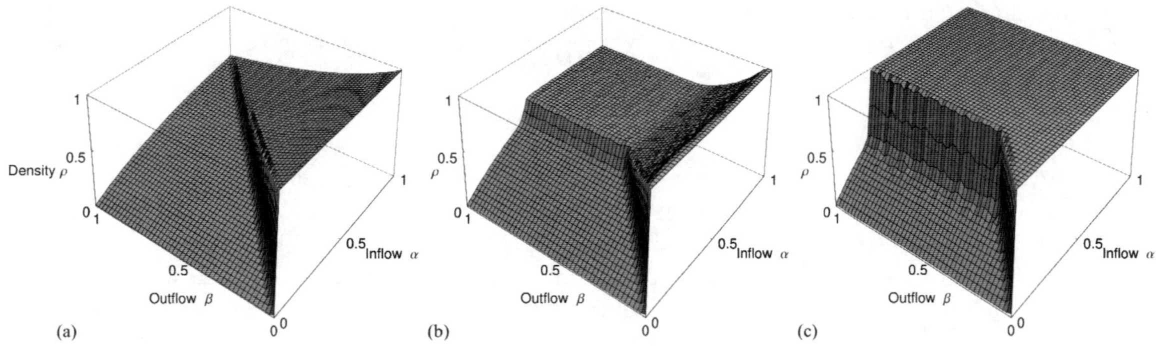

Figure 2. Density on the loop as a function of inflow and outflow $\rho(\alpha, \beta)$: **(a)** $\gamma = 0$; **(b)** $\gamma = 0.5$; **(c)** $\gamma = 1$.

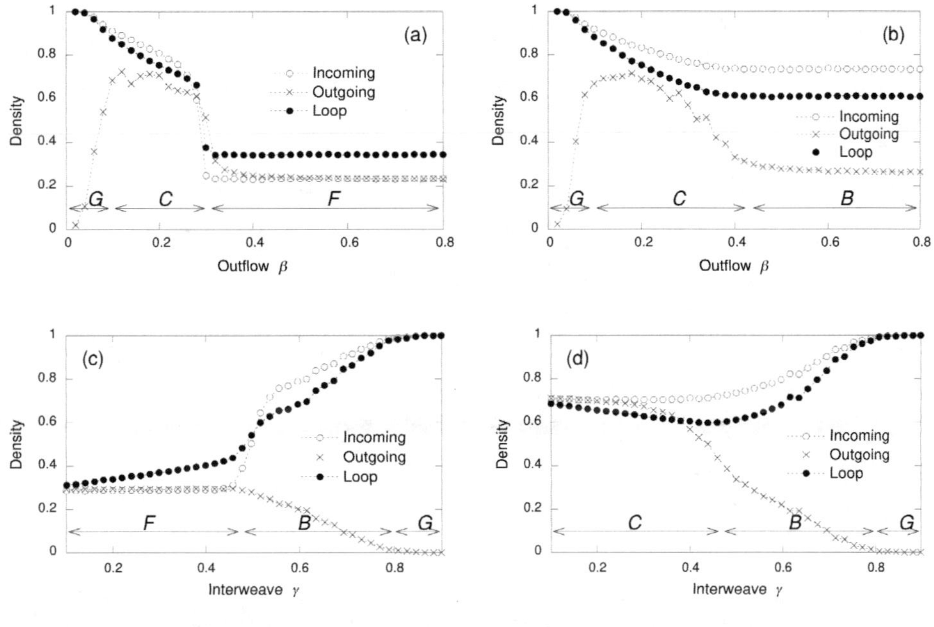

Figure 3. Density evolution in the consecutive transitions: **(a)** $\rho(\beta)$ with fixed $\alpha = 0.3$ and $\gamma = 0.5$; **(b)** $\rho(\beta)$ with fixed $\alpha = 0.6$ and $\gamma = 0.5$; **(c)** $\rho(\gamma)$ with fixed $\alpha = 0.4$ and $\beta = 0.6$; **(d)** $\rho(\gamma)$ with fixed $\alpha = 0.6$ and $\beta = 0.4$;

This model presents four steady phases: Free flow (F), Congestion (C), Bottleneck (B), and Gridlock (G). Simulation results in Fig. 2 indicate various kind of phase transitions as the three control parameters (α, β, γ) vary. Basically the traffic conditions become worse and the free flow can no longer be maintained when either α increases, β decreases, or γ increases. First we consider the simple transition, which involves only two phases. As α increases, we observe three abrupt transitions ($F \to C$, $F \to B$, $F \to G$). As β decreases, we observe two abrupt transitions ($F \to C$, $F \to G$) and one smooth transition ($B \to G$). As γ increases, we observe one smooth transition ($C \to G$). Then we consider the consecutive transition, which involves three phases. As α increases, there is no consecutive transition. As β decreases, we observe two consecutive transitions ($F \to C \to G$, $B \to C \to G$). As γ increases, we observe two consecutive transitions ($F \to B \to G$, $C \to B \to G$). Typical results are shown in Fig. 3. As α increases, all three types of congestion can be the final phase. As β decreases or γ increases, gridlock becomes the dominate final phase. Bottleneck cannot be the final phase. Congestion can be the final phase only if there is no interweave ($\gamma = 0$).

4. Discussions

This model is able to differentiate the four traffic phases. The free flow is characterized by the inflow (α). The conventional congestion is caused by a direct comparison between inflow (α) and outflow (β), i.e. $\alpha > \beta$, and characterized by the outflow (β). The bottleneck is caused mainly by the interweave (γ), i.e. independent of both α and β. The model also presents the gridlock, which often occurs in real traffic but is totally absent in ASEP. We observe two kinds of gridlock: one as a narrow stripe and the other as a plateau in Fig. 2. The stripe phase is a gridlock transited from the conventional congestion as shown in Figs. 3(a) and 3(b); while the plateau phase is a gridlock transited from the bottleneck as shown in Figs. 3(c) and 3(d). The stripe phase appears only at the boundary of the parameter space. In the limit $\beta \to 0$, the conventional congestion is the final phase provided there is no interweave $\gamma = 0$. With interweave $\gamma \neq 0$, the conventional congestion becomes unstable. The gridlock can be triggered when the density on the loop is high. As a result, the high density in the outgoing roadway takes a drastic turn to reduce to zero in the limit $\beta \to 0$, as shown in Figs. 3(a) and 3(b). In contrast, the plateau phase occupies a finite region of the parameter space. All densities evolve gradually from bottleneck to gridlock, as shown in Figs. 3(c) and 3(d). Unlike the other three traffic phases, the gridlock implies the loss of mobility which cannot be restored. Thus the observation of gridlock will depend on the time scale. Typical results are shown in Figs. 4(a) and 4(b).

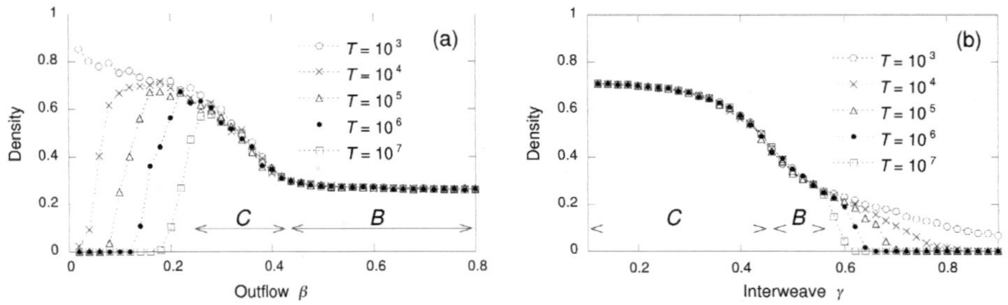

Figure 4. Density on the loop at various time scales T, where the simulation runs a period of T before taking the average: **(a)** the same as Fig. 3(b); **(b)** the same as Fig. 3(d).

5. References

[1] Chowdhury D, Santen L and Schadschneider A 2000 *Phys. Rep.* **329** 199
[2] Huang D W 2015 *Comput. Phys. Comm.* **189** 72
[3] Derrida B 1998 *Phys. Rep.* **301** 65
[4] Huang D W 2010 *Internat. J. Modern Phys.* **C21** 189

Numerical solution of viscous and viscoelastic fluids flow through the branching channel by finite volume scheme

Radka Keslerová, David Trdlička

Czech Technical University in Prague, Faculty of Mechanical Engineering, Dep. of Technical Mathematics, Czech Republic

E-mail: Radka.Keslerova@fs.cvut.cz, David.Trdlicka@fs.cvut.cz

Abstract. This work deals with the numerical modelling of steady flows of incompressible viscous and viscoelastic fluids through the three dimensional channel with T-junction. The fundamental system of equations is the system of generalized Navier-Stokes equations for incompressible fluids. This system is based on the system of balance laws of mass and momentum for incompressible fluids. Two different mathematical models for the stress tensor are used for simulation of Newtonian and Oldroyd-B fluids flow. Numerical solution of the described models is based on cetral finite volume method using explicit Runge-Kutta time integration.

1. Introduction

Branching of pipes occurs in many technical or biological applications. The use of proper mathematical model is important to get the correct results. In the practice, the numerical simulations are crucial for the solution of a large number of technical problems.

In this work the incompressible laminar viscous and viscoelastic fluids flow are numerically simulated. The mathematical model is based on the system of balance laws of mass and momentum for incompressible fluids. For the different choice of fluids model the different model of the stress tensor is used. For viscous flows the simple *Newtonian model* is considered. In the case of viscoelastic fluids, the simplest viscoelastic model can be used. This model is denoted as *Maxwell model.* By combination of these two models (Newtonian and Maxwell models) the behaviour of mixture of viscous and viscoelastic fluids can be described. This model is called *Oldroyd-B model.* Both models (Newtonian and Oldroyd-B) could be generalized. Viscosity function is defined according to the cross model.

In prewious work we studied the numerical simulation of generalized Newtonian and Oldroyd-B fluids flow in two dimensional branching channel, [4]. In this work we extended our problem to the numerical modelling of tested fluids flow in three dimensional branching channel with T-junction.

2. Mathematical model

The governing system of equations is the system of generalized Navier-Stokes equations, see [6]. This system consists of the continuity equation and the momentum equation

$$\mathrm{div}\ \boldsymbol{u} = 0 \tag{1}$$

Journal of Physics: Conference Series **633** (2015) 012128 doi:10.1088/1742-6596/633/1/012128

$$\rho \frac{\partial \boldsymbol{u}}{\partial t} + \rho(\boldsymbol{u}.\nabla)\boldsymbol{u} = -\nabla P + \text{div } \mathbf{T}, \tag{2}$$

where P is the pressure, ρ is the constant density, $\boldsymbol{u}$ is the velocity vector. The symbol $\mathbf{T}$ represents the stress tensor.

The stress tensor $\mathbf{T}$ in the system can be decomposed to the viscous $\mathbf{T}_s$ and viscoelastic part $\mathbf{T}_e$ ($\mathbf{T} = \mathbf{T}_s + \mathbf{T}_e$). Both tensors are defined by corresponding rheological models, Newtonian and Maxwell, see e.g. [1], [2], [4]

$$\mathbf{T}_s = 2\mu_s \mathbf{D}, \qquad \mathbf{T}_e + \lambda_1 \frac{\delta \mathbf{T}_e}{\delta t} = 2\mu_e \mathbf{D}, \tag{3}$$

where λ_1 is the relaxation time. By combination of these two mathematical models the behaviour of mixture of viscous and viscoelastic fluids can be described. This model is called Oldroyd-B and it's used together with Newtonian model for numerical simulation of tested fluids. The symbol $\frac{\delta}{\delta t}$ used in the viscoelastic part of the stress tensor denotes the upper convected derivative [2], [4].

For the numerical modelling of the generalized Newtonian and Oldroyd-B fluids flow it is necessary to generalize the mathematical models. In this case the viscosity function is defined by cross model [7]

$$\mu(\dot{\gamma}) = \mu_\infty + \frac{\mu_0 - \mu_\infty}{(1 + (\lambda\dot{\gamma})^b)^a}, \quad \dot{\gamma} = 2\sqrt{\frac{1}{2}\text{tr } \mathbf{D}^2}, \tag{4}$$

with special parameters $\mu_0 = 1.6 \cdot 10^{-1} Pa.s, \mu_\infty = 3.6 \cdot 10^{-3} Pa.s, a = 1.23, b = 0.64, \lambda = 8.2s$. For Newtonian and Oldroyd-B flow, the viscosity is kept constant and equal to μ_∞.

3. Numerical solution

The mathematical models described above are solved numericaly the artificial compressibility approach combined with the finite-volume discretization. The artificial compressibility method [3], [4], [6] is used to obtain equation for pressure. It means that the continuity equation is completed by a pressure time derivative term $\frac{\partial p}{\beta^2 \partial t}$, where β is positive parameter, making the inviscid part of the system of equations hyperbolic. The parameter β in this work is chosen equal to the maximum inlet velocity. This value ensures good convergence to steady state but is not large enough to make the transient solution accurate in time. Therefore it is suitable for steady flows only. The discretization is done by a cell-centered finite-volume method with hexahedral finite volumes. The system including the modified continuity equation and the momentum equations can be written

$$\tilde{R}_\beta W_t + F^c_x + G^c_y + H^c_z = F^v_x + G^v_y + H^v_z + S, \quad \tilde{R}_\beta = \text{diag}(\frac{1}{\beta^2}, 1, \cdots, 1), \tag{5}$$

where W is vector of unknowns, $W = (p, u, v, w, t_{e1}, \ldots, t_{e6})$, by superscripts c and v the inviscid and the viscous fluxes are denoted. The symbol S denotes the source term. Eq. (5) is discretized in space by the finite volume method and the arising system of ODEs is integrated in time by the explicit multistage Runge–Kutta scheme ([4], [5]).

The flow is modelled in a bounded computational domain where a boundary is divided into three mutually disjoint parts: a solid wall, an outlet and an inlet. At the inlet Dirichlet boundary condition for velocity vector and for the stress tensor is used and for the pressure homogeneous Neumann boundary condition is used. At the outlet parts the pressure value is prescribed and for the velocity vector and the stress tensor homogeneous Neumann boundary condition is used. The no-slip boundary condition for the velocity vector is used on the wall. For the pressure and stress tensor homogeneous Neumann boundary condition is considered.

4. Numerical Results

This section deals with the comparison of the numerical results of generalized Newtonian and generalized Oldroyd-B fluids flow. Numerical tests are performed in an idealized branched channel with the square cross-section. Fig. 1 (left) shows the shape of the tested domain. The computational domain is discretized using a structured, wall fitted mesh with hexahedral cells. The domain is divided to 4 blocks with 153 000 cells.

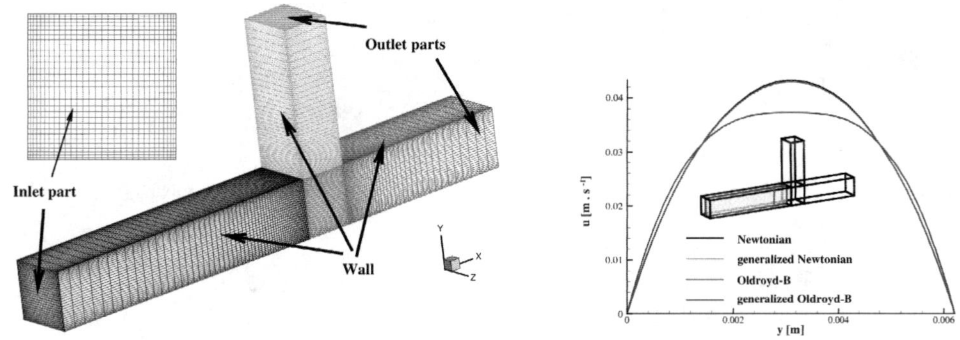

(a) structure of the domain (b) axial velocity profile

Figure 1. Structure of the tested domain (left) and axial velocity profile of tested fluids (right).

As initial condition the following model parameters are used: $L_0 = 0.0031m$, $\mu_e = 0.0004Pa.s$, $\mu_s = 0.0036Pa.s$, $\lambda_1 = 0.06s$, $U_0 = 0.0615m.s^{-1}$, $\rho = 1050kg.m^{-3}$. At the inlet

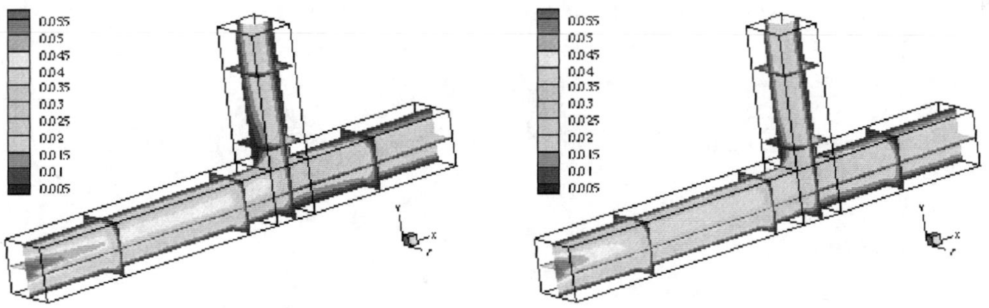

(a) Newtonian (b) generalized Newtonian

Figure 2. Velocity isolines of steady flows for generalized Newtonian fluids.

the Dirichlet boundary conditions for velocity are used, the parabolic profile with maximum velocity value U_0. At the outlet the constant pressure values are prescribed (0.0005 Pa (main channel) and 0.00025 Pa (branch)). In Fig. 1 the axial velocity profile for tested types of fluids close to the branching is shown. The lines for Newtonian and Oldroyd-B fluids are similar to the parabolic line, as was assumed. From this velocity profile is clear that the shear thinning fluids attain lower maximum velocity in the central part of the channel (close to the axis of symmetry) which is compensated by the increase of local velocity in the boundary layer close to the wall.

In Figs. 2 and 3 the velocity isolines and the cuts through the main channel and the small branch are shown.

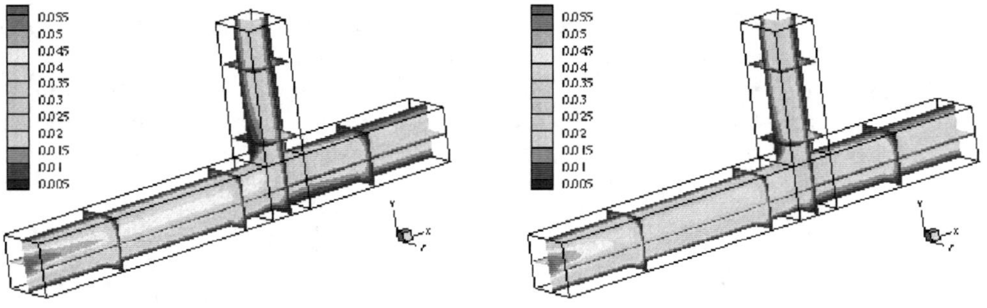

(a) Oldroyd-B (b) generalized Oldroyd-B

Figure 3. Velocity isolines of steady flows for generalized Oldroyd-B fluids.

It can be observed from these that the size of separation region for generalized Newtonian and generalized Oldroyd-B fluids is smaller than for Newtonian and Oldroyd-B fluids.

5. Conclusion

In this paper a finite volume solver for incompressible laminar viscous and viscoelastic flows in the branching channel with T-junction and circle cross section was described. Newtonian and Oldroyd-B fluids models were generalized by the cross model for numerical solution of generalized Newtonian and Oldroyd-B fluids flow. The explicit Runge-Kutta method was considered for time integrating. The numerical results obtained by this method were presented.

Numerical results were compared. It has been shown that in this type of the channel the numerical results for Newtonian and Oldroyd-B fluids are similar. From the presented velocity profile is clear that the shear thinning fluids (generalized Newtonian and Oldroyd-B fluids) attain lower maximum velocity in the central part of the channel (close to the axis of symmetry) which is compensated by the increase of local velocity in the boundary layer close to the wall.

The future work will be occupy with the numerical simulation of viscous and viscoelastic fluids flow in the branching channel with circle cross-section and with the comparison numerical results of both type of channels.

Acknowledgments

This work was supported by the grant SGS13/174/OHK2/3T/12.

References

[1] Anand M, Kwack J and Masud A 2013 *International Journal of Engineering Science* **72** 78
[2] Bodnar T and Sequeira A 2010 *Advances in Mathematical Fluid Mechanics* 83
[3] Chorin AJ 1967 *Journal of Computational Physics* **135** 118
[4] Keslerová R and Kozel K 2012 *Proc. 16th Sem. on Programs and Algorithms of Num. Math.* (Inst. of Math. Ac. of Sc. of CR) 100
[5] LeVeque R 2004 *Finite-Volume Methods for Hyperbolic Problems* (Cambridge University Press)
[6] Louda P, Kozel K, Příhoda J, Beneš L and Kopáček T 2011 *Journal of Computers & Fluids* **46** 318
[7] Rabby MG, Razzak A and Molla MM 2013 *Procedia Engineering* **56** 225

A Green-Naghdi approach for thermo-electroelasticity

Montanaro A
University of Padua, Italy
E-mail: adriano.montanaro@unipd.it

Abstract. The constitutive relations of piezoelectric ceramics are essentially nonlinear since the so-called piezoelectric moduli depend on the induced strains. Pioneering papers in these topics dealt mainly with the isothermal case. In view of applications, however, thermal effects have to be taken into account in connection with thermo-electric behaviors. Here we briefly compare continuum theories for nonlinear thermoelettroelasticity. In particular we describe an extension of Green-Naghdi thermoelasticity theory for an electrically polarizable and finitely deformable heat conducting elastic continuumn, which interacts with the electric field. In this theory, unlike other, thermal waves propagate at a finite speed.

1. Introduction

The materials exhibiting couplings between elastic, electric, magnetic and thermal fields have attracted great attention in the last decades, in connection with increasing wide use in sensing and actuation. The interest of many researchers turned to mathematical theories of such materials, in order to give certainty to experimental results and applications. Many applications have their mathematical formulation within a linear framework, and the theoretical study began from this context. Foundamental is Nowacki's paper [1], where a uniqueness theorem for the solutions of the initial boundary value problems is proved in linear thermopiezoelectricity referred to a natural state, i.e., with no biasing (or initial) fields. Hence Nowacki [2] also deduced the generalized Hamilton principle and a theorem of reciprocity of work.

For nonlinear continuous media, in order to find the thermodynamic restrictions on the constitutive relations for an electrically polarizable and finitely deformable heat conducting elastic continuum which interacts with the electric field, some authors (e.g. Tiersten [4]) use a theory based on the Clausius-Duhem entropy inequality, following the standard Coleman-Noll procedure [3]. Such theories, which we call here theories of type (A), predict an infinite speed of heat propagation. To avoid such physical contradiction, several papers were written to present continuum theories capable of predicting thermal waves propagating at finite speeds in various media, referred as *second sound*; such theories are often referred as generalized thermoelasticity.

In addition, Rybalko [9] in 2004 showed by experiments that a second-sound (thermal) wave is accompanied by the appearance of electric induction. This fact demonstrates the need for second sound theories of an electrically polarizable and finitely deformable heat conducting elastic continuum which interacts with the electric field.

A group of such theories, referred here as theories of type (B), assumes the Clausius-Duhem inequality together with a generalized Cattaneo equation [5].

Another theoretical approach of continuum thermodynamics is proposed by Green-Naghdi theories [6], [7]. It is based on an integral thermodynamic equality rather than an entropy inequality and uses the notion of thermal displacement α associated with empirical temperature T. Such theories have attracted considerable interest and have been applied in many physical situations, where heat propagation is coupled e.g. with elasticity (see [8]). We refer to them as theories of the type (C).

In [13] there is an extension of the thermodynamic theories [6], [7] to an electrically polarizable and finitely deformable heat conducting elastic continuumn. Here we briefly present it and show the thermodynamic restrictions on response mappings within each type $(A)-(C)$ of theory.

2. Notations and preliminary definitions

Let $\mathcal{E}$ denote a three-dimensional Euclidean point space. We consider a body B whose particles are identified with the positions $\boldsymbol{X} \in \mathcal{E}$ they occupy in a given reference configuration $\boldsymbol{B}$. The material filling B is characterized by a given process class $I\!P(B)$ of B as a set of ordered 10−tuples of functions on $\boldsymbol{B} \times I\!R$

$$p = \Big(\boldsymbol{x}(.),\, \theta(.),\, \varphi(.),\, \varepsilon(.),\, \eta(.),\, \boldsymbol{\tau}(.),\, \boldsymbol{P}(.),\, \boldsymbol{q}(.),\, \boldsymbol{b}(.),\, r(.)\Big) \in I\!P(B) \tag{1}$$

defined with respect to $\boldsymbol{B}$, satisfying the balance laws of linear momentum, moment of momentum, energy, an entropy inequality or equality and the field equations of electrostatics, where

- $\boldsymbol{x} = \boldsymbol{x}(\boldsymbol{X},\, t)$ is the *motion*, $\quad \boldsymbol{v} = \dot{\boldsymbol{x}}(\boldsymbol{X},\, t)$ is the *velocity*,
- $\boldsymbol{F} = \frac{\partial \boldsymbol{x}}{\partial \boldsymbol{X}}$ is the *deformaton gradient*,
- $\theta = \theta(\boldsymbol{X},\, t) > 0$ is the *absolute temperature*, $\quad \boldsymbol{g} = \frac{\partial \theta}{\partial \boldsymbol{x}}$ is the *temperature gradient*,
- $\varphi = \varphi(\boldsymbol{X},\, t)$ is the *electric potential*,
- $\boldsymbol{E}^M = -\nabla_{\boldsymbol{x}}\varphi$ is the *(Maxwellian) spatial electric vector*
- $\boldsymbol{P} = \boldsymbol{P}(\boldsymbol{X},\, t)$ is the *polarization vector*, $\boldsymbol{\pi} = \boldsymbol{P}/\rho$,
- $\boldsymbol{D} = \varepsilon_0 \boldsymbol{E}^M + \boldsymbol{P}$ is the *electric displacement* ($\varepsilon_0 =$ *vacuum electric permittivity*),
- $\varepsilon = \varepsilon(\boldsymbol{X},\, t)$ is the specific *internal energy* per unit mass,
- $\eta = \eta(\boldsymbol{X},\, t)$ is the specific *entropy* per unit mass,
- $\boldsymbol{\tau} = \boldsymbol{\tau}(\boldsymbol{X},\, t) \quad \Big(\boldsymbol{S} = \boldsymbol{S}(\boldsymbol{X},\, t)\Big)$ is the Cauchy stress tensor,
- $\boldsymbol{q} = \boldsymbol{q}(\boldsymbol{X},\, t)$ is the spatial *heat flux vector*,
- $\boldsymbol{b} = \boldsymbol{b}(\boldsymbol{X},\, t)$ is the external specific *body force* per unit mass,
- $r = r(\boldsymbol{X},\, t)$ is the *radiating heating* per unit mass

The *free energy function* is then defined by

$$\psi = \varepsilon - \theta\eta - \boldsymbol{E}^M \cdot \boldsymbol{\pi} \tag{2}$$

3. Theories of type (A)

3.1. Local balance laws in spatial form

Under suitable assumptions of regularity the usual integral forms of the balance laws of linear momentum, moment of momentum, energy, the field equations of electrostatics, and the entropy inequality are equivalent to the spatial field equations

$$\rho\dot{\boldsymbol{v}} = div\boldsymbol{\tau} + \boldsymbol{P}\cdot\nabla_x \boldsymbol{E}^M + \rho\boldsymbol{b}\,, \tag{3}$$

$$\rho\dot{\varepsilon} = \boldsymbol{\tau}\cdot\nabla\boldsymbol{v} - div\boldsymbol{q} + \boldsymbol{E}^M\cdot\rho\dot{\boldsymbol{\pi}} + \rho r\,, \tag{4}$$

$$\boldsymbol{E}^M = -\nabla_{\boldsymbol{x}}\varphi\,, \qquad div\boldsymbol{D} = \boldsymbol{0}\,, \tag{5}$$

$$\rho\dot{\eta} \geq \rho(r/\theta) - div(\boldsymbol{q}/\theta)\,. \tag{6}$$

3.2. Constitutive Assumptions in Spatial Form

Let $\mathcal{D}$ be an open, simply connected domain consisting of 4−tuples $(\boldsymbol{F},\, \theta, \boldsymbol{E}^M,\, \boldsymbol{g})$, and assume that if $(\boldsymbol{F},\, \theta, \boldsymbol{E}^M,\, \boldsymbol{g}) \in \mathcal{D}$, then $(\boldsymbol{F},\, \theta, \boldsymbol{E}^M,\, \boldsymbol{0}) \in \mathcal{D}$.

For every $p \in I\!\!P(B)$ the specific free energy $\psi(\boldsymbol{X},\, t)$, the specific entropy $\eta(\boldsymbol{X},\, t)$, the Cauchy stress tensor $\boldsymbol{T}(\boldsymbol{X},\, t)$, the specific polarization vector $\boldsymbol{P}(\boldsymbol{X},\, t)$, and the heat flux $\boldsymbol{q}(\boldsymbol{X},\, t)$ are given by continuously differentiable functions on $\mathcal{D}$ such that

$$\psi = \overline{\psi}(\boldsymbol{F},\, \theta, \boldsymbol{E}^M,\, \boldsymbol{g})\,, \tag{7}$$

$$\eta = \overline{\eta}(\boldsymbol{F},\, \theta, \boldsymbol{E}^M,\, \boldsymbol{g})\,, \tag{8}$$

$$\boldsymbol{\tau} = \overline{\boldsymbol{\tau}}(\boldsymbol{F},\, \theta, \boldsymbol{E}^M,\, \boldsymbol{g})\,, \tag{9}$$

$$\boldsymbol{P} = \overline{\boldsymbol{P}}(\boldsymbol{F},\, \theta, \boldsymbol{E}^M,\, \boldsymbol{g})\,, \tag{10}$$

$$\boldsymbol{q} = \overline{\boldsymbol{q}}(\boldsymbol{F},\, \theta, \boldsymbol{E}^M,\, \boldsymbol{g})\,. \tag{11}$$

Note that the dependence upon $\boldsymbol{X}$ is not written only for brevity; when the body is not materially homogeneous it becomes active.

3.3. Dissipation Principle

For any given motion, temperature field and electric potential field, the process p constructed from the constitutive equations (7)-(11) belongs to the process class (1) of B. Therefore the constitutive functions (7)-(11) are compatible with the second law of thermodynamics in the sense that they satisfy the dissipation inequality (6).

3.4. Constitutive restrictions implied by the entropy inequality

The following proposition holds (e.g. see [12]).

The Dissipation Principle is satisfied if and only if the conditions $(i - ii)$ below hold:

(i) The free energy response function $\overline{\psi}(\boldsymbol{F},\, \theta, \boldsymbol{E}^M,\, \boldsymbol{g})$ is independent of $\boldsymbol{g}$ and determines the response functions for entropy, first Piola-Kirchhoff stress, and polarization vector through the relations

$$\overline{\eta}(\boldsymbol{F},\, \theta, \boldsymbol{E}^M) = -\partial_\theta \overline{\psi}(\boldsymbol{F},\, \theta, \boldsymbol{E}^M) \tag{12}$$

$$\overline{\boldsymbol{\tau}}(\boldsymbol{F},\, \theta, \boldsymbol{E}^M) = \rho \boldsymbol{F} \partial_{\boldsymbol{F}} \overline{\psi}(\boldsymbol{F},\, \theta, \boldsymbol{E}^M) \tag{13}$$

$$\overline{\boldsymbol{\pi}}(\boldsymbol{F},\, \theta, \boldsymbol{E}^M) = -\partial_{\boldsymbol{E}^M} \overline{\psi}(\boldsymbol{F},\, \theta, \boldsymbol{E}^M) \tag{14}$$

(ii) The reduced dissipation inequality (Fourier inequality)

$$\boldsymbol{q} \cdot \boldsymbol{g} \leq 0 \tag{15}$$

is satisfied along any process.

4. Theories of type (B)

4.1. Local balance laws in spatial form

They are just (3)-(6) in Section 3.1.

4.2. Spatial Constitutive Assumptions

Let $\mathcal{D}$ be an open and simply connected domain consisting of 5−tuples $(\boldsymbol{F},\, \theta, \boldsymbol{E}^M,\, \boldsymbol{q},\, \boldsymbol{g})$ and assume that *if* $(\boldsymbol{F},\, \theta, \boldsymbol{E}^M,\, \boldsymbol{q},\, \boldsymbol{g}) \in \mathcal{D}$, *then* $(\boldsymbol{F},\, \theta, \boldsymbol{E}^M,\, \boldsymbol{0},\, \boldsymbol{0}) \in \mathcal{D}$.

Below we use the time derivative for the heat flux vector

$$\overset{\circ}{\boldsymbol{q}} = \dot{\boldsymbol{q}} - \boldsymbol{L}\boldsymbol{q} + (tr\boldsymbol{L})\boldsymbol{q}\,, \qquad \boldsymbol{L} = grad\boldsymbol{v}\,. \tag{16}$$

The equality $\dot{\boldsymbol{Q}} = J\boldsymbol{F}^{-1}\overset{\circ}{\boldsymbol{q}}$, where $\boldsymbol{Q}$ is the material heat flux vector, shows that the spatial counterpart of the material derivative $\dot{\boldsymbol{Q}}$ is represented by $\overset{\circ}{\boldsymbol{q}}$ rather than by $\dot{\boldsymbol{q}}$.

For every $p \in I\!\!P(B)$ the specific free energy $\psi(\boldsymbol{X},\, t)$, the specific entropy $\eta(\boldsymbol{X},\, t)$, the Cauchy stress tensor $\boldsymbol{\tau}(\boldsymbol{X},\, t)$, the polarization vector $\boldsymbol{P}(\boldsymbol{X},\, t)$, and the time rate of the heat flux $\overset{\circ}{\boldsymbol{q}}(\boldsymbol{X},\, t)$ are given by continuously differentiable functions on $\mathcal{D}$ such that

$$\psi = \overline{\psi}(\boldsymbol{F},\, \theta, \boldsymbol{E}^M,\, \boldsymbol{q},\, \boldsymbol{g})\,, \tag{17}$$

$$\eta = \overline{\eta}(\boldsymbol{F},\, \theta, \boldsymbol{E}^M,\, \boldsymbol{q},\, \boldsymbol{g})\,, \tag{18}$$

$$\boldsymbol{\tau} = \overline{\boldsymbol{\tau}}(\boldsymbol{F},\, \theta, \boldsymbol{E}^M,\, \boldsymbol{q},\, \boldsymbol{g})\,, \tag{19}$$

$$\boldsymbol{P} = \overline{\boldsymbol{P}}(\boldsymbol{F},\, \theta, \boldsymbol{E}^M,\, \boldsymbol{q},\, \boldsymbol{g})\,, \tag{20}$$

$$\overset{\circ}{\boldsymbol{q}} = \boldsymbol{h}(\boldsymbol{F},\, \theta, \boldsymbol{E}^M,\, \boldsymbol{q},\, \boldsymbol{g})\,. \tag{21}$$

Further, the tensors $\partial_{\boldsymbol{q}}\boldsymbol{h}(.)$ and $\partial_{\boldsymbol{g}}\boldsymbol{h}(.)$ are non-singular.

4.3. Dissipation Principle

Just as in Section 3.3.

4.4. Constitutive restrictions implied by the entropy inequality

The following proposition holds (see [5]). *The Dissipation Principle is satisfied if and only if the conditions* $(i - ii)$ *below hold:*

(i) *The free energy response function* $\overline{\psi}(\boldsymbol{F},\, \theta, \boldsymbol{E}^M,\, \boldsymbol{q},\, \boldsymbol{g})$ *is independent of the temperature gradient* $\boldsymbol{g}$ *and determines the entropy, the Cauchy stress tensor, and the polarization vector through the relations*

$$\overline{\eta}(\boldsymbol{F},\, \theta, \boldsymbol{E}^M,\, \boldsymbol{q}) = -\frac{\partial\overline{\psi}}{\partial\theta}(\boldsymbol{F},\, \theta, \boldsymbol{E}^M,\, \boldsymbol{q})\,, \tag{22}$$

$$\overline{\boldsymbol{\tau}}(\boldsymbol{F},\, \theta, \boldsymbol{E}^M,\, \boldsymbol{q}) = \rho\boldsymbol{F}\frac{\partial\overline{\psi}}{\partial\boldsymbol{F}}(\boldsymbol{F},\, \theta, \boldsymbol{E}^M,\, \boldsymbol{q})\,, \tag{23}$$

$$\overline{\boldsymbol{\pi}}(\boldsymbol{F},\, \theta, \boldsymbol{E}^M,\, \boldsymbol{q}) = -\frac{\partial\overline{\psi}}{\partial\boldsymbol{E}^M}(\boldsymbol{F},\, \theta, \boldsymbol{E}^M,\, \boldsymbol{q})\,; \tag{24}$$

(ii) *The reduced dissipation inequality*

$$\rho\theta\frac{\partial\overline{\psi}}{\partial\boldsymbol{q}}(\boldsymbol{F},\, \theta, \boldsymbol{E}^M,\, \boldsymbol{q})\cdot\dot{\boldsymbol{q}} + \boldsymbol{q}\cdot\boldsymbol{g} \leq 0\,, \tag{25}$$

where

$$\dot{\boldsymbol{q}} = \boldsymbol{h}(\boldsymbol{F},\, \theta, \boldsymbol{E}^M,\, \boldsymbol{q},\, \boldsymbol{g}) + [\boldsymbol{L} - (tr\boldsymbol{L})\boldsymbol{I}]\,\boldsymbol{q}\,, \tag{26}$$

is satisfied.

5. Theories of type (C)

Now we present the extension [13] of Green-Naghdi theories [6], [7] to thermoelectroelasticity, which is based on an integral thermodynamic equality rather than an entropy inequality and use the notion of thermal displacement α associated with empirical temperature T. The following relations hold:

$$\alpha = \alpha(\boldsymbol{x}, t) = \int_0^t T(\boldsymbol{x}, \tau)\, d\tau + \alpha_0(\boldsymbol{x}), \qquad t > 0\,, \tag{27}$$

$$T = \dot{\alpha}\,, \qquad \boldsymbol{\beta} = \frac{\partial \alpha}{\partial \boldsymbol{X}}\,, \qquad \boldsymbol{\gamma} = \frac{\partial T}{\partial \boldsymbol{x}}\,, \qquad \boldsymbol{g} = \frac{\partial \theta}{\partial \boldsymbol{x}} = \frac{\partial \theta}{\partial T}\boldsymbol{\gamma}\,. \tag{28}$$

5.1. Local balance laws in spatial form

$$\begin{aligned} \dot{\rho} + \rho\, div\boldsymbol{v} &= 0\,, \\ \rho\dot{\boldsymbol{v}} &= div\boldsymbol{\tau} + \boldsymbol{P}\cdot\nabla_x \boldsymbol{E}^M + \rho\boldsymbol{f}\,, \\ skw\boldsymbol{\tau} + skw\boldsymbol{T}^E &= \boldsymbol{0}\,, \\ \rho\dot{\eta} &= \rho(s+\xi) - div\boldsymbol{p}\,, \\ \rho\dot{e} &= \boldsymbol{\tau}\cdot\nabla\boldsymbol{v} - div\boldsymbol{q} + \boldsymbol{E}^M\cdot\rho\dot{\boldsymbol{\pi}} + \rho r\,, \\ \nabla_x\cdot\boldsymbol{D} = 0\,, \qquad \boldsymbol{E}^M &= -\nabla_{\boldsymbol{x}}\phi\,, \end{aligned} \tag{29}$$

where $s = r/\theta$ is the *external rate of supply of entropy per unit mass* and ξ is the *internal rate of supply of entropy per unit mass.* Eliminating r between equations $(29)_4$, $(29)_5$ and using the free energy (2) yields the *reduced energy equation*

$$\rho(\dot{\psi} + \dot{\theta}\eta) + \boldsymbol{p}\cdot\boldsymbol{g} + \rho\theta\xi - \boldsymbol{\tau}\cdot\nabla\boldsymbol{v} + \dot{\boldsymbol{E}}^M\cdot\boldsymbol{P} = 0\,, \tag{30}$$

where $\boldsymbol{p} = \boldsymbol{q}/\theta$ is the entropy flux vector per unit area.

5.2. Constitutive equations and Dissipation Principle

Constitutive equations are assumed that have the form

$$\psi = \hat{\psi}(T,\, \boldsymbol{\beta},\, \boldsymbol{\gamma},\, \boldsymbol{F},\, \boldsymbol{E}^M) \tag{31}$$

$$\eta = \hat{\eta}(T,\, \boldsymbol{\beta},\, \boldsymbol{\gamma},\, \boldsymbol{F},\, \boldsymbol{E}^M) \tag{32}$$

$$\theta = \hat{\theta}(T,\, \boldsymbol{\beta},\, \boldsymbol{\gamma},\, \boldsymbol{F},\, \boldsymbol{E}^M) \tag{33}$$

$$\xi = \hat{\xi}(T,\, \boldsymbol{\beta},\, \boldsymbol{\gamma},\, \boldsymbol{F},\, \boldsymbol{E}^M) \tag{34}$$

$$\boldsymbol{p} = \hat{\boldsymbol{p}}(T,\, \boldsymbol{\beta},\, \boldsymbol{\gamma},\, \boldsymbol{F},\, \boldsymbol{E}^M) \tag{35}$$

$$\boldsymbol{\tau} = \hat{\boldsymbol{\tau}}(T,\, \boldsymbol{\beta},\, \boldsymbol{\gamma},\, \boldsymbol{F},\, \boldsymbol{E}^M) \tag{36}$$

$$\boldsymbol{P} = \hat{\boldsymbol{P}}(T,\, \boldsymbol{\beta},\, \boldsymbol{\gamma},\, \boldsymbol{F},\, \boldsymbol{E}^M) \tag{37}$$

The dissipation inequality

$$\xi \geq 0 \tag{38}$$

holds along any process of $\mathcal{B}$ ([10], [11]).

5.3. Constitutive restrictions implied by the reduced energy equality

The following proposition holds (e.g. see [13]).

Assume constitutive equations of the forms (31), *with*

$$\boldsymbol{q} = \theta \boldsymbol{p}\,, \tag{39}$$

$$\frac{\partial \hat{\theta}}{\partial T} > 0\,, \tag{40}$$

and let the internal energy response function be defined by (2). *Then the validity of the reduced energy equation* (30) *along any smooth enough process implies the following conditions for the response functions*

$$\psi = \hat{\psi}(T,\, \boldsymbol{\beta},\, \boldsymbol{F},\, \boldsymbol{E}^M)\,, \qquad \theta = \hat{\theta}(T)\,, \tag{41}$$

$$\hat{\boldsymbol{\tau}} = \rho \boldsymbol{F} \frac{\partial \hat{\psi}}{\partial \boldsymbol{F}}\,, \qquad \hat{\boldsymbol{P}} = -\rho \frac{\partial \hat{\psi}}{\partial \boldsymbol{E}^M}\,, \qquad \hat{\eta} = -\frac{\partial \hat{\psi}}{\partial \theta}\,. \tag{42}$$

$$\rho \frac{\partial \hat{\psi}}{\partial \boldsymbol{\beta}}(T,\, \boldsymbol{\beta},\, \boldsymbol{F},\, \boldsymbol{E}^M) \cdot \boldsymbol{F}^T \boldsymbol{\gamma} + \rho \hat{\theta}(T) \hat{\xi}(T,\, \boldsymbol{\beta},\, \boldsymbol{\gamma},\, \boldsymbol{F},\, \boldsymbol{E}^M) + \hat{\boldsymbol{p}}(T,\, \boldsymbol{\beta},\, \boldsymbol{\gamma},\, \boldsymbol{F},\, \boldsymbol{E}^M) \cdot \boldsymbol{g} = 0\,. \tag{43}$$

References

[1] Nowacki W 1965 *Proc. Vibrations Probl.* **6** 3
[2] Nowacki W 1978 *J. Thermal Stresses* **1** 171
[3] B.D. Coleman BD and Noll W 1963 *Arch. Rat. Mech. Anal.* **13** 167
[4] Tiersten HF 1971 *Int. J. Engng Sci.* **9** 587
[5] Montanaro A 2011 *Arch. Mech.* **63** 225
[6] Green A E and Naghdi P M 1991 *Proc. of the Royal Society London* **432** 171
[7] Green A E and Naghdi P M 1992 *J. Thermal Stresses* **15** 253
[8] Straughan B 2011 *Heat Waves* (Applied Mathematical Sciences **177**) (Berlin:Springer)
[9] Rybalko A S 2004 *J. Low Temp. Phys.* **30** 994
[10] Fabrizio M and Giorgi C 1986 *Boll. Un. Mat. Ital.* **5B** 441
[11] Amendola G, Fabrizio M and Golden JM 2012 Thermodynamics of Materialswith Memory. Theory and Applications, Springer, 2012
[12] Montanaro A 2009 *Report n.102 October 2009 DMMMSA Univ.Padua arXiv:0910.1344v3 [math-ph]* ,
[13] Giorgi C and Montanaro A 2015 On Constitutive Equations in Green-Naghdi Type III Thermo-Electroelasticity *Preprint* gr-qc/0401010

Journal of Physics: Conference Series **633** (2015) 012130 doi:10.1088/1742-6596/633/1/012130

Human-computer interfaces applied to numerical solution of the Plateau problem

Antonio Elias Fabris[1], **Ivana Soares Bandeira**[1], **Valério Ramos Batista**[2]

[1] *IME-USP, r. do Matão 1010, 05508-090 São Paulo-SP, Brazil*
[2] *CMCC-UFABC, av. dos Estados 5001, 09210-580 St André-SP, Brazil*

E-mail: aefabris@gmail.com

Abstract. In this work we present a code in Matlab to solve the Problem of Plateau numerically, and the code will include human-computer interface. The Problem of Plateau has applications in areas of knowledge like, for instance, Computer Graphics. The solution method will be the same one of the Surface Evolver, but the difference will be a complete graphical interface with the user. This will enable us to implement other kinds of interface like ocular mouse, voice, touch, etc. To date, Evolver does not include any graphical interface, which restricts its use by the scientific community. Specially, its use is practically impossible for most of the Physically Challenged People.

1. Introduction

In 1989 K. Brakke introduced the first version of a general-purpose program to perform physical experiments in a virtual way. His program is devoted to studying the behaviour of surfaces when submitted to surface tension and other energies. For this reason it is named *Surface Evolver*, and its latest version is 2.70 [1]. Even some earlier versions were already complete and broad enough for several applications in many Areas of Knowledge like Physics, Chemistry, Mathematics and Engineering [2–5]. Most recently we have [6, 7] and [8] in Medicine.

The source-code of *Evolver* is open and freely available. It is all written in standard C-language, and consists of circa 100 files with a total of nearly 180,000 lines of source-code. A curious fact is that such a complex and complete program results from the dedication of a work team that consists of a sole member, namely Prof Kenneth Brakke. Studying the Evolver code is a challenging task, not to mention the maintenance of this program in a distant future.

One of the reasons for which Evolver's development was so rapid is that K. Brakke gave priority to the *technical* part of his program. Consequently, in Evolver data input is only possible by means of datafiles and command lines. Namely, there is no Graphical User Interface (GUI) like clicking a mouse arrow on a virtual button. This seems to be just a minor detail but nowadays academia has been using human-computer interfaces increasingly. Voice recognition, touchscreen and head mouses are resources that greatly simplify the daily work of researchers, professors and students.

In this sense our present work is also devoted to *Physically Challenged People* (PCP) that cannot use the hands when working on a computer. Their participation in academia has been growing steadily and their access to programs like Evolver is highly desirable.

Already in the version 1.0 Evolver included the *Numerical Solution of the Plateau Problem* (NSPP). It consists of finding the least area surface that spans a given 3D curve in space. The solution is called *minimal surface*, which also has the least overall surface tension among all spanners of the given curve. Physically they are soap films obtained by dipping a closed wire curve into soap solution.

Our approach is to implement Evolver's technique to NSPP in one of the easiest programming languages, namely Matlab [9]. This language can be used through voice recognition and touch pen. Moreover, it is highly compact and we estimate that it can reduce the C-code of Evolver's NSPP by almost 70%.

This is because our Matlab version will *not* be a mere translation between programming languages. Rather, our NSPP has been implemented directly from the Technical References of Evolver Manual (Chapter 16), and already in accordance with Matlab's short-programming philosophy: toolboxes, syntactical and semantical conciseness, GUI, etc. This task is not trivial because it requires a perfect understanding of Evolver's NSPP technique, which includes deep knowledge of Discrete Differential Geometry and also Optimization Algorithms.

We are still at the first stage of a year long research. Our Matlab code will be available in future and it will enable researchers to include other capabilities of Evolver. Additionally, our GUI will be implemented to work with the *Eye Tribe Tracker*, which is one of the most comfortable eye tracking devices to date. See Section 2 for details. In this way we hope to contribute to PCP like many other researchers' works in the literature [10–13].

2. Methodology

The *Eye Tribe* is a Danish enterprise devoted to eye tracking technology. One of its products is a case box with photosensitive cells. The case box can be attached to the bottom of a screen, as depicted in Figure 1. The photosensitive cells detect eye movements, which can be used as a mouse.

Evolver requires the description of initial surfaces within a datafile. These are very rough polyhedral surfaces that are later refined and smoothened by Evolver commands. In order to describe an initial surface one has to write a list of its vertices and edges within the datafile. Our Matlab version does it graphically with a mouse (see Figure 2).

Figure 1: The Eye Tribe Tracker.[1]

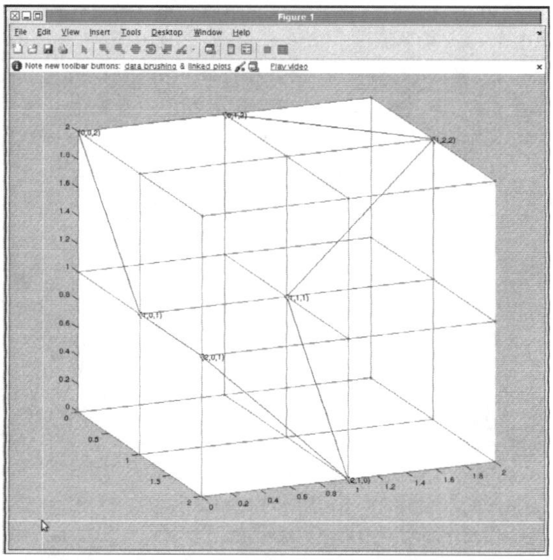

Figure 2: Plotting a 3D polygonal curve.

[1] Taken from `https://theeyetribe.com/products`

Regarding the NSPP, we invoke Evolver built-in commands that refine the initial triangulation, iterate the surface towards minimum energy, make triangles uniform, add or remove them either locally or globally, save current surface, etc. These commands can be repeated and invoked in any order. For the time being we have implemented just two of them: surface refinement and iteration. See Figures 3-5.

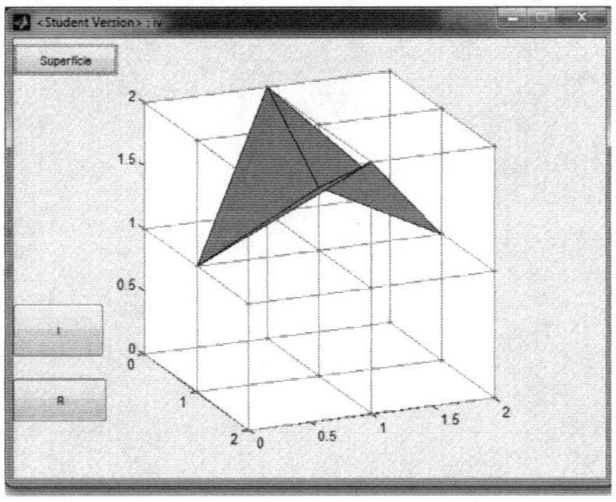

Figure 3: Buttons I and R.

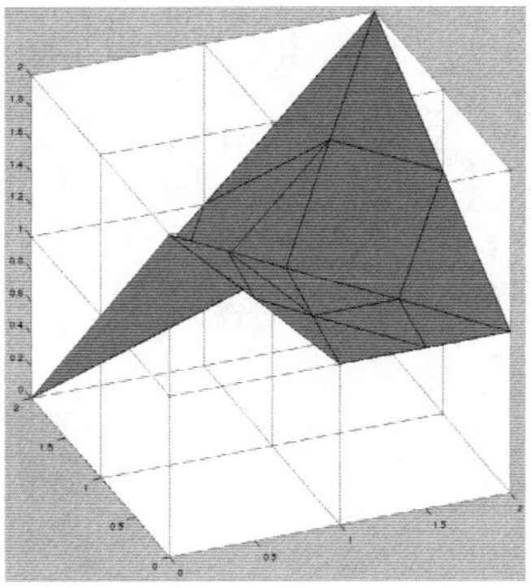

Figure 4: Refining.

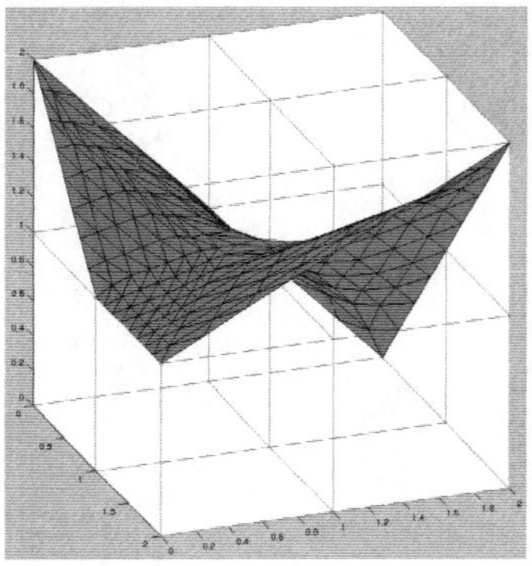

Figure 5: Minimal surface.

Our commands are activated by mouse click on virtual buttons, as shown in Figure 3. In future we shall implement the Eye Tribe Tracker in Matlab to work as a mouse in our program.

As mentioned in the Introduction, our code is being programmed according to the Technical References of the Evolver Manual. Therefore, details about how to manipulate the triangulated surface are omitted here. However, the 2nd author is preparing a PhD-Thesis that will present the theory behind NSPP, together with pseudo-codes, block diagrams and illustrations that explain how the whole program works. In this way we strive for accessibility in a much broader sense, for the code will be comprehensible even to undergraduate students.

3. Results

The main result is our program, devoted both to users and developers. Its source code will be freely available in future, together with a complete technical documentation.

4. Conclusions

We have just presented the achievements of a year long research. Admittedly, both the involved theory and programming strategies are highly non-trivial. However, we are focusing only on the NSPP. Because of that, a short code that is also easy to handle and to understand is a totally feasible task, even considering that it will include the programming of human-computer interfaces. In our case it is the Eye Tribe Tracker as a mouse to control some of Evolver's capabilities.

5. Acknowledgements

This research is supported by FAPESP proc.No.14/20219-0.

References

[1] Brakke K 2013 *Susquehanna University* URL `http://www.susqu.edu/brakke/evolver/evolver.html`
[2] Tegart J 1991 *AIAA, SAE, ASME, and ASEE, 27th Joint Propulsion Conference* vol 1
[3] Michalet X and Bensimon D 1995 *Science* **269** 666–668
[4] Francis G, Sullivan J M, Kusner R B, Brakke K A, Hartman C and Chappell G 1997 *Visualization and Mathematics* (Springer) pp 3–20

[5] Racz L, Szekely J and Brakke K 1993 *ISIJ international* **33** 328–335
[6] Nascimento M Z and Ramos Batista V 2014 *Journal of Physics: Conference Series* vol 490 (IOP Publishing) p 012146
[7] Nascimento M Z and Ramos Batista V 2015 *Journal of Physics: Conference Series* vol 574 (IOP Publishing) p 012072
[8] Nascimento M Z and Ramos Batista V 2015 *Recent Patents on Medical Imaging* **to appear**
[9] MatLab M 2012 *The MathWorks, Inc.* URL `http://www.mathworks.com`
[10] Hornof A, Cavender A and Hoselton R 2004 *ASSETS '04 ACM SIGACCESS Conference on Computers and Accessibility* (ACM Press) pp 86–93
[11] Perini E, Soria S, Prati A and Cucchiara R 2006 *Proceedings of the 2006 International Conference on Computer Vision in Human-Computer Interaction* ECCV'06 (Berlin, Heidelberg: Springer-Verlag) pp 99–108
[12] Chau T, Memarian N, Leung B, Treherne D, Hobbs D, Worthington-Eyre B, Lamont A and Pla-Mobarak M 2012 *Handbook of Ambient Assisted Living* pp 581–597
[13] Jian Y and Jin J 2012 *Proceedings of the 2012 International Conference on Systems and Informatics (ICSAI)* (IEEE) pp 505–509

4th International Conference on Mathematical Modeling in Physical Sciences (IC-MSquare2015) IOP Publishing
Journal of Physics: Conference Series **633** (2015) 012131 doi:10.1088/1742-6596/633/1/012131

Algorithmic improvements to an exact region-filling technique

Antonio Elias Fabris[1], **Valério Ramos Batista**[2]

[1]*IME-USP, r. do Matão 1010, 05508-090 São Paulo-SP, Brazil*

[2]*CMCC-UFABC, av. dos Estados 5001, 09210-580 St André-SP, Brazil*

E-mail: aefabris@gmail.com

Abstract. We present many algorithmic improvements in our early region filling technique, which in a previous publication was already proved to be correct for all connected digital pictures. Ours is an integer-only method that also finds all interior points of any given digital picture by displaying and storing them in a locating matrix. Our filling/locating program is applicable both in computer graphics and image processing.

1. Introduction

Until recently, two classical algorithms predominated as definite choices to implement either point location within a given digital picture, or the filling-up of its inside. The first is called *Seed Fill* [1, 2] and the second *Scan Conversion* [3, 4]. Their strategies are illustrated in Figures 1 and 2, respectively.

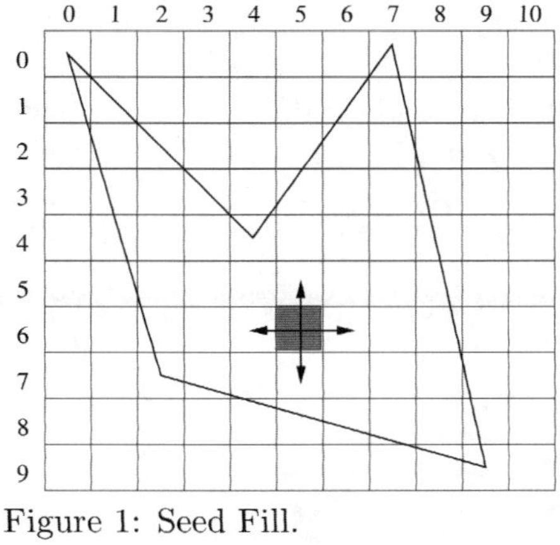

Figure 1: Seed Fill.

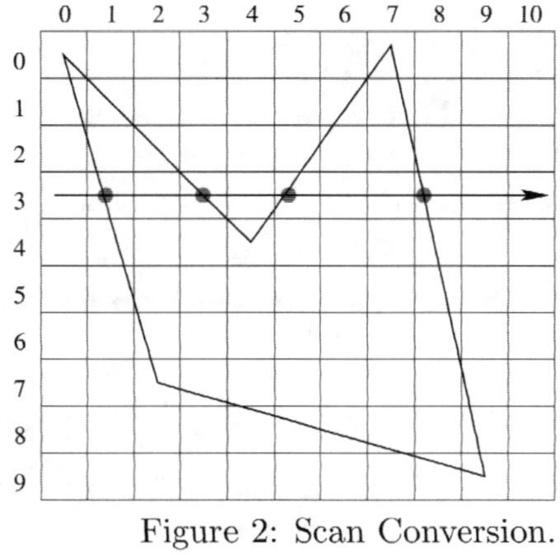

Figure 2: Scan Conversion.

Seed Fill starts from painting a given pixel called *seed* and then all of its subsequent neighbour pixels until meeting boundary conditions. Scan Conversion considers each scan line and determines its overlap intervals within the region by means of pixel parity. The weak and strong points of these algorithms were already discussed in Section 2 of [5] and therefore will be omitted here.

But none of them is free of failures, and these are caused by the following characteristic they have in common: the use of *continuous* Euclidean Geometry constructions within the finite *discrete* space of the computer. Of course, at implementing these classical algorithms one can add several strategies to reduce failures, but challenging regions like nearly degenerate polygons will always prove that they never achieve perfection for all cases. Needless to test complicate regions, in [6] the author presents many *simple* polygons that represent German States, together with a list of failures of the most known implementations of both Seed Fill and Scan Conversion.

Some minor failures can be often dismissed in many circumstances, but not always. For instance, microchips are produced in millions out of a single mould. Just one flaw in the mould will cause a great loss in all senses: time, money, etc. In [5] we illustrate this example and some others in which perfect filling is crucial. There we introduced an integer-only algorithm called CoTRA&FUA that finds all interior points of any given digital picture *exactly*. They are displayed and stored in a locating matrix, hence CoTRA&FUA is *correct* as both filling and locating procedure.

It works for absolutely *any* connected digital picture, no matter how complex or intricate it might be. It is applicable in Computer Graphics and Image Processing even without penalizing speed, for its complexity is quasi-linear.

In this present paper we show many improvements in the CoTRA&FUA algorithm. Among other achievements, we rewrote its implemented source code to have just 193 lines, which represents an approximate 20% reduction compared with the original version of [5]. Moreover, CoTRA&FUA now count on internal procedures that gave much more clarity to the original program. Details are given in the next section, but for the readers who already want to test our program it is available in the link *Codes* of

`https://sites.google.com/site/aefabris`

in Matlab p-code. In Section 7 of [5] the reader will find instructions to run it, together with some examples.

2. Methodology

First of all we need to explain that the input of our program is just an image file in either TIF or JPG format. It should contain a rectangular *canvas* of only black and white pixels. It could have been scanned or created by a mathematical model, but what originated it is unimportant.

Then we *construct* a mathematical model that results in the given canvas, perform the region filling and finally generate the filling matrix. This matrix consists of three different entries: points of the interior, exterior and of the picture itself. In our works they are always graphically painted in yellow, white and black, respectively.

Since no mathematical model is given beforehand, then we disambiguate situations like the ones illustrated in Figure 3 by means of a a precise definition of *exterior* of a digital picture.

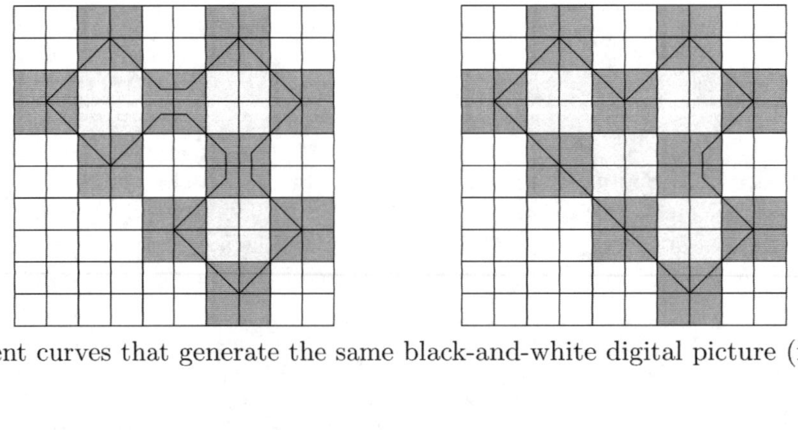

Figure 3: Different curves that generate the same black-and-white digital picture (from [5, Fig. 3])

Journal of Physics: Conference Series **633** (2015) 012131 doi:10.1088/1742-6596/633/1/012131

Technical definitions are given in [5]. Now, CoTRA&FUA are both abbreviations of a double-barreled algorithm. The first stands for *Connectivity and Thickness Reduction Algorithm.* It obtains the so-called *Lego curve* L out of the digital picture. The second stands for *Filling Up Algorithm*, which is always exact for any Lego curve, and its interior coincides with the one of the digital picture. See details in [5].

Now, the main improvements we have got in this present work are the following: **1.** Addition of the internal procedures *Enlarge* and *Tighten* to CoTRA; **2.** Reduction of the code by working only with *thin* instead of *locally thin* curves (Definitions 3.6 and 3.7 of [5], respectively); **3.** Simplification of the part of the code devoted to the *stalk problem* (not explained in [5]);

Regarding **1**, the procedures *Enlarge* and *Tighten* simplified CoTRA because some of its parts had to be repeated in order to replace false *L-pixels* with true ones. Figure 4 illustrates this problem. It shows a Lego curve in blue and magenta, this latter to enhance its L-pixels (Definition 5.4 of [5]).

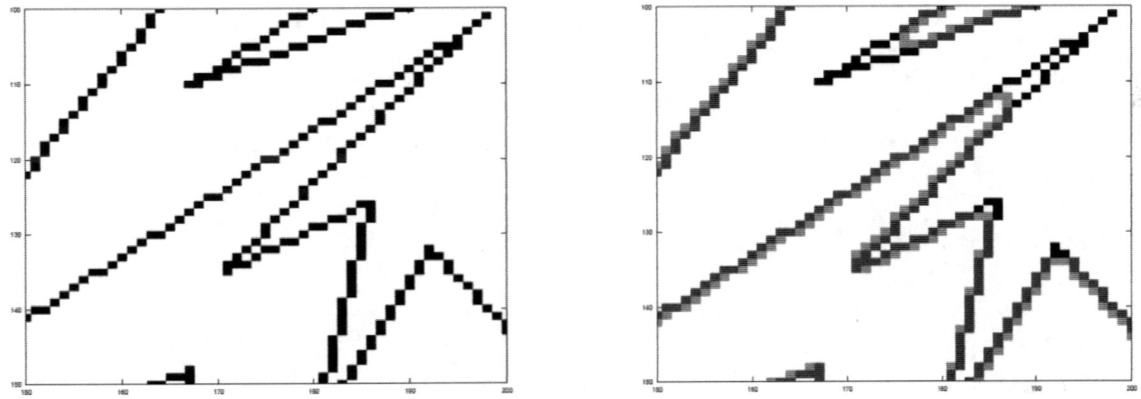

Figure 4: Zoomed input curve (left) and its Lego curve with false L-pixels (right, [5, Fig. 8]).

Roughly saying, *Enlarge* replaces each input pixel p by a 4×3 rectangle of pixels of the same colour of p (black or white), and *Tighten* reverts this process. On the one hand, this introduces an "a priori" multiplication of the computational time by 12. On the other hand, an enlarged input curve always corresponds to a thin Lego curve. Consequently, it is already free of false L-pixels and the problem depicted in Figure 4 does not occur. Then CoTRA runs much much faster, and this explains the improvement **2**.

Regarding **3**, first we need to explain what the *stalk problem* is. For example, consider that our digital picture D has a pixel p with a *unique* neighbour $q \in D$, and q is a *diagonal* neighbour of p. After enlarging D, p will become a black 4×3 rectangle that we call "apple", depicted at the left bottom of Figure 5. The stalk is its connection with a diagonal pixel originated from q, this one also expanded to a 4×3 matrix of black pixels. Now CoTRA marks the Lego curve L as in Figure 6, and there we see a 2×2-square S of diagonal blue and magenta pixels.

Now L is *not* thin because of S, and then FUA will mark internal pixels incorrectly after passing through S. There are some other circumstances in which the *stalk problem* can happen, and all of them were already known in [5]. But there we got round this problem by means of too complicated a strategy, which was then not explained in that work.

Our present version follows a new and much easier strategy: it just makes two copies of the enlarged D with its Lego curve, then shifts right one copy and left the other, and finally overlaps them in order to get an thin Lego curve, as shown in Figure 7.

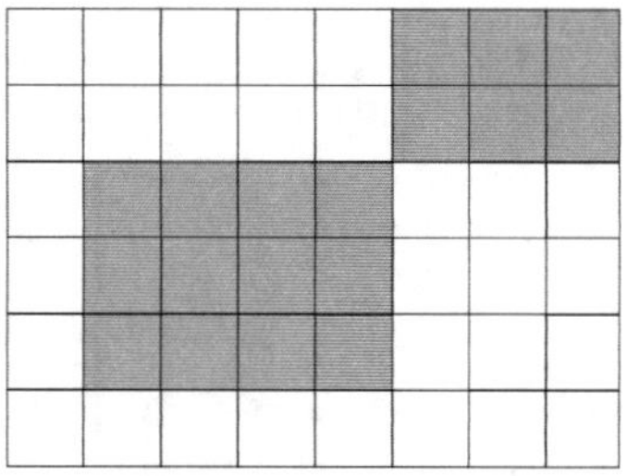

Figure 5: Apple and stalk.

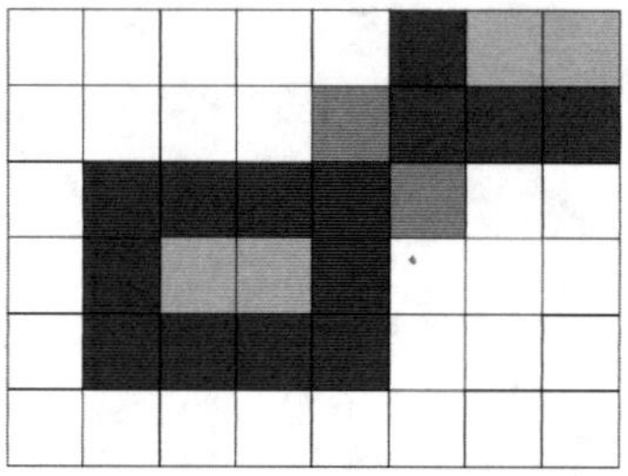

Figure 6: Coloured Lego curve.

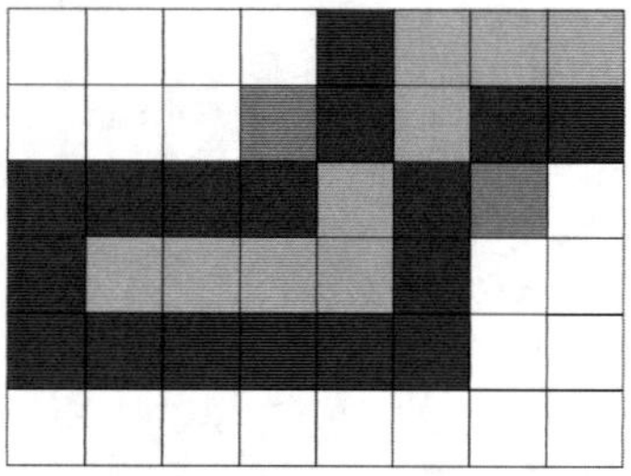

Figure 7: New strategy.

This new strategy also works faster than our previous version in [5]. Together with the fact that we now work only with thin Lego curves, computational time has in fact remained almost the same. Table 1 of [5] summarizes the computational time of 8 tests performed in both Matlab and Octave. It also applies to our new version since there have not been significant changes in time performance.

3. Results

We now have a much simpler and shorter filling/locating program compared with our previous version of [5]. The program is correct for all connected digital pictures, and one of the main reasons for that is our integer-only approach.

4. Conclusions

Our new filling/locating program requires more memory because dimensions are now multiplied by 12. However, modern computers do not have space complexity as a limitation any longer. Hence, increasing space complexity to the detriment of the algorithm is in fact harmless. Time complexity remained the same, but we have gained in clarity and conciseness.

References

[1] Fishkin K P and Barsky B A 1984 *ACM SIGGRAPH Computer Graphics* **18** 235–244
[2] Tsai Y H and Chung K L 2000 *Computer & Graphics* **24** 529–537
[3] Codrea M C and Nevalainen O S 2005 *Computer & Graphics* **29** 441–450
[4] Haines E 1994 ed Heckbert P S (San Diego, CA, USA: Academic Press Professional, Inc.) chap Point in Polygon Strategies, pp 24–46 ISBN 0-12-336155-9 URL `http://dl.acm.org/citation.cfm?id=180895.180899`
[5] Fabris A E and Ramos Batista V 2014 *Aditi International Journal of Computational Mathematics* **3** 1–23
[6] Schirra S 2008 *Algorithms-ESA 2008* (Springer) pp 744–755

A software tool based on the Surface Evolver for precise location of tumours as a preoperative procedure to partial mastectomy

Antonio Elias Fabris[1], **Marcelo Zanchetta do Nascimento**[2], **Valério Ramos Batista**[3]

[1] *IME-USP, r. do Matão 1010, 05508-090 São Paulo-SP, Brazil*
[2] *FACOM-UFU, Av. João Naves de Ávila 2121, Bl.A, 38400-902 Uberlândia-MG, Brazil*
[3] *CMCC-UFABC, R. Sta. Adélia 166, Bl.B, 09210-170 St. André-SP, Brazil*

E-mail: aefabris@gmail.com

Abstract. We present a fast and reliable program that gives precise location of breast tumours for a partial mastectomy. Our program is fully implemented in the Surface Evolver, which is a general-purpose simulator of physical experiments. By starting from the mammograms that show a tumour one takes its 2D coordinates in each view (CC and MLO). These coordinates, together with some measurements of the patient's breast, are given as input to our simulator. From this point on the simulator reproduces all main steps of taking mammography with a virtual transparent breast that matches the patient's. The virtual mammography procedure is graphically displayed on the computer screen, so that users can track the virtual tumour inside the breast. As output we have the coordinates of the tumour position when the woman lies on the operating table for the surgery. With these coordinates the surgeon can make a small incision into the breast and reach the tumour for its removal. The whole structure of the breast is preserved after a simple plastic correction.

The 3rd author dedicates this work to his wife Clarice.

1. Introduction

Mammographies are X-ray images of the interior of the breast. There are two different views: Craniocaudal (CC) and Mediolateral Oblique (MLO). They are obtained by compressions with the mammographer. Together they increase the chances of detecting cancer but the breast is shown in strongly deformed shapes. Because of that cancer location is highly uncertain for the surgery. However, by means of a manual examination the surgeon can estimate a quadrant of the breast that contains the tumour and then proceed with a partial mastectomy called *quadrantectomy*.

This technique was introduced in 1981 by a famous oncologist called Umberto Veronesi. Along the decades there have been efforts to reduce the quadrant to smaller regions. In order to accomplish this task one can resort to softwares that simulate the mammography procedure. Herewith we cite [1–7]. Each of them focus on a different approach. In [8–10] we have already discussed their weak and strong points, hence they will be omitted here. Anyway, to the best of our knowledge, until now, not one of them has been officially approved by a Medical Council as a reliable nodule locator permitting it to become part of surgical preparations.

We present here a new software tool that differs from previous ones by three main characteristics: 1) our method utilizes the Surface Evolver; 2) it includes the study of nodule displacements in transparent breast phantoms; 3) it achieves a fast, reliable and relatively simple algorithm thanks to the application of many geometrical and physical properties.

Our software is the final result of a long-term research. Earlier achievements of this research can be found in [8–10]. For the time being we have released only the first part of our simulator. It is available in the public link

`https://www.copy.com/s/4fdQnwMb2IRqkPOo/ubuntu11.10.ova`

and instructions to use it are in the link *Softwares* of our webpage

`http://www.facom.ufu.br/~nascimento`

in the PDF-file `icmmps`. This file gives some general information about the first version of our simulator and includes a user manual. In the next sections we shall explain how the complete software works.

2. Methodology

Our virtual mammography simulator starts from the CC and MLO views that show a tumour. One takes its 2D coordinates in each view, together with some measurements of the patient's breast. These data are the input of our simulator. As output we have the coordinates of the tumour position when the woman lies on the operating table for the surgery. The simulator reproduces all main steps of taking mammography. They are named SRG (*surgery*), STU (*stand-up*), LAT (*lay-on-table*), CRC (*cranio-caudal*), LET (*lean-on-table*) and MLO (*medio-lateral-oblique*). Since we can use a *table* to take measurements of the patient's breast we prefer that term instead of *plate*. This one is only used when the mammographer is really necessary.

Now we give a practical example. Figure 1 was taken from the *Digital Database for Screening Mammography* [11]. Figure 2 shows how to mark our coordinate system on the image with colour felt-tip pens. Intuitively, the woman places her breast on the lower plate so that most of it will lie thereon. Figure 3 shows the point of contact: it is at the origin of Oxz. The edge is tangent to the thorax where the red vector N makes an angle θ with Ox, $\theta = \arccos((x_d/x_t)^{3/2})$, $N = (z_t \cos^{\frac{4}{3}} \theta,\, 0,\, x_t \sin^{\frac{4}{3}} \theta)$. See [9] for details about formulas and parameters like x_d, z_r, x_t, etc.

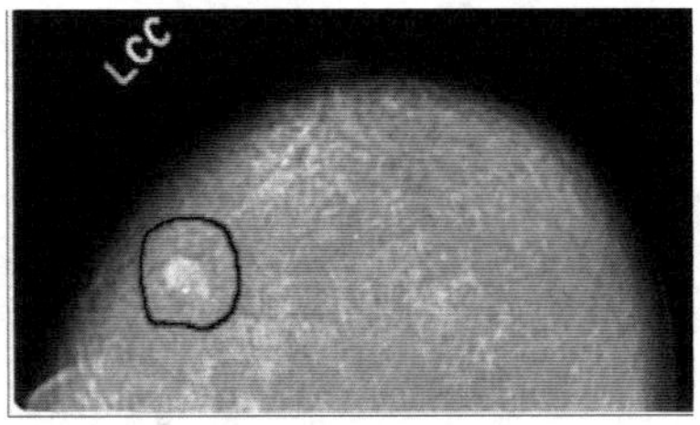

Figure 1: Case 0023-1.

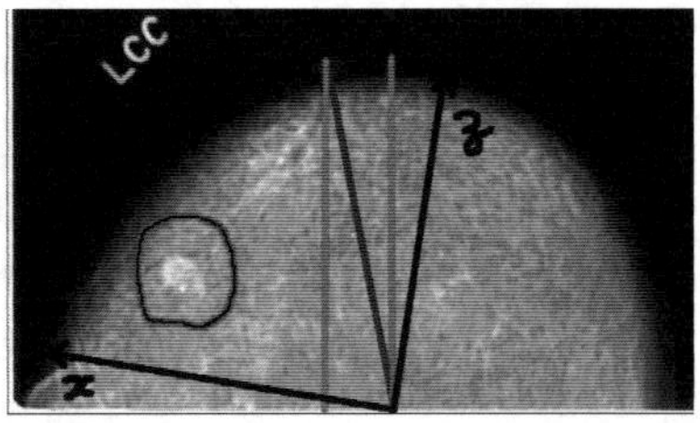

Figure 2: Tracing Oxz.

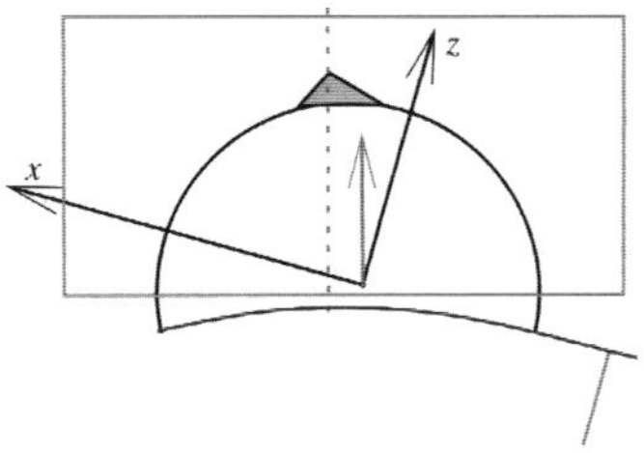

Figure 3: Nipple and plate for CC.

The nodule in Figure 2 is located at $(x_c, z_c) = (6.83, 3.20)$, where the subscript c stands for "coordinate". Figure 4 was obtained by following the first instructions of our manual in `icmmps.pdf`, followed by the commands `coors` and `mk`. The former prints x_c, z_c and H_c, which is where both Ox and Oz cross the breast contour. It is quite close to a half-circumference at CC. Right after invoking `coors` the user can change x_c, z_c and H_c according to another patient's. Now `mk` deduces the nodule position at SRG as $(x_n, z_n) = (x_r x_c, z_r z_c)/H_c = (4.07, 2.13)$, where the subscript n stands for "nodule". By invoking `mk` the user sees that both L and R are marked on the virtual breast, as depicted in Figure 4, where $\overline{LR}$ is the segment that contains the nodule.

The simulation proceeds with the commands `stu`, `let` and `mlo`, which finally gives Figure 5. However, it does not match Figure 6 very well. This is because at MLO the patient must lift her elbow and hold a handle of the mammographer. This projects part of the breast forwards. Because of that one can see the image of the pectoralis major muscle at the lower left corner of Figure 6.

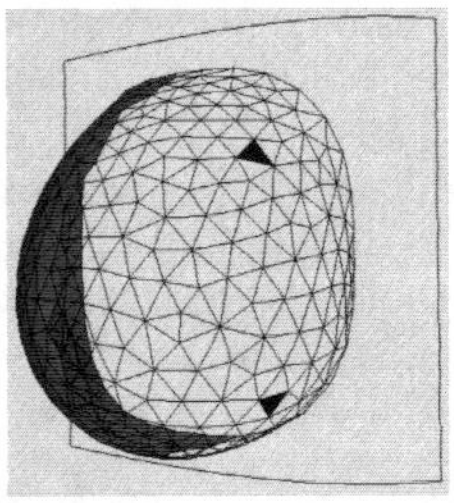
Figure 4: Marking L and R.

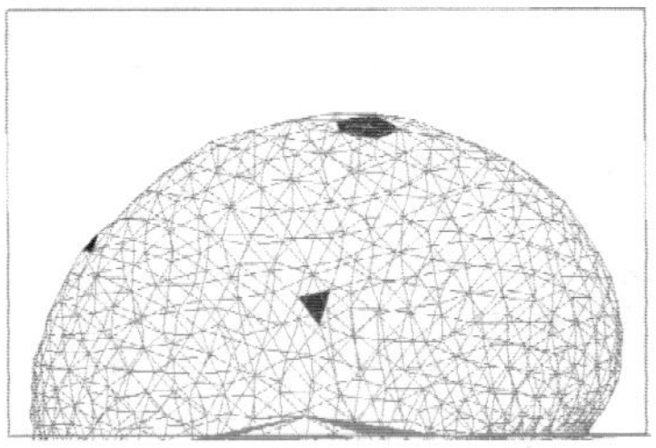
Figure 5: Comparing virtual with real MLO.

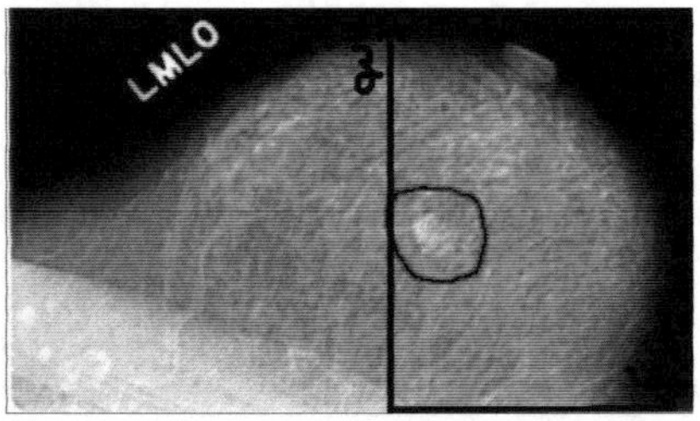

Figure 6: Marking Ozw on MLO.

Such a movement is hard to implement computationally. For our purposes one can simply take for granted that

$$f(\zeta) := (\zeta - bH\frac{b-i}{1-ib})/(\frac{1+ib}{1-ib} - \frac{ib\zeta}{H}) \tag{1}$$

Journal of Physics: Conference Series **633** (2015) 012132 doi:10.1088/1742-6596/633/1/012132

maps Figure 7 to Figure 8. The former is the upper half of $D_H := \{\zeta = u + iv \in \mathbb{C} \,|\, u^2 + v^2 \leq H\}$ whereas the latter is determined by the fixed point $f(H) = H$ and any chosen $b \in]0,1[$ for which $f(0) = ibH$.

We have $b = b_c/H$, where b_c is the height at which Oz transposes the pectoralis major muscle in Figure 6. In our example $b_c = 0.95$ and $H = H_c = 10.5$, thus $b = 9\% = 0.09$ and these are the values in (1) that generated Figure 8. Notice that its lower blue arc is concave because it represents the contour of the breast base, *not* the convex pectoralis major muscle.

The tumour is located at $(p_w, p_z) = (1.2, 4.7)$ in Figure 6. Hence its position in Figure 5 is given by $\mathcal{P} = f^{-1}(1.2 + 4.7i) = 0.57 + 4.1i$. The command `frho` prints the coordinates of the black triangles in Figure 5: $\mathcal{L} = -7.47 + 6.22i$ and $\mathcal{R} = -0.332 + 4.25i$. Technically speaking the tumour coincides with the rightmost of these triangles. In Figure 4 it corresponds to R (the lower triangle), and the tumour is almost on the skin.

For the sake of generality let us take another (p_w, p_z). The default in our simulator is $(p_w, p_z) = (-5.2, 6.1)$. Then $\mathcal{P} = f^{-1}(-5.2 + 6.1i) \simeq -6.6 + 4.1i$, and $|\widehat{\mathcal{LP}}|/|\widehat{\mathcal{LR}}| \stackrel{\text{def}}{=} \rho = 0.317$. Figure 5 was obtained with the command `mlo` of our simulator. Now the command `coors` indicate example values that the user can change according to another patient's. These are `bc`, `pw` and `pz`. Now the data (x_n, z_n) and ρ will give a unique position of the tumour inside the breast since

$$y_n = (1 - 2\rho)\sqrt{a^2 - x_n^2 - z_n^2}, \tag{2}$$

where $a = (x_r + y_r + z_r)/3$. At this point our simulator gives the polar coordinates (r, p, d) that will guide the surgeon to reach the nodule, as illustrated in Figures 9 and 10. The triple (r, p, d) is computed via

$$p = \arccos\left(\frac{x_n}{\sqrt{x_n^2 + y_n^2}}\right); \; r = a \arccos\left(\frac{z_n}{\sqrt{x_n^2 + y_n^2 + z_n^2}}\right); \; d = a - \sqrt{x_n^2 + y_n^2 + z_n^2}. \tag{3}$$

As a matter of fact (3) are just an "a priori" estimate. As explained in [9] we use a layer-approach in order to track the nodule inside the virtual breast. After obtaining Figure 5 the user invokes `frho`. This command prints (p, r, d), ρ and `lf`, which is a mnemonic for "layer-factor". In our example $(p, r, d) = \big(7.5921, 23.977, 1.8125\big)$, $\rho = 0.317$ and `lf` $= 73\%$. We proceed with the simulation, which now includes the layer and the virtual nodule, as depicted in Figure 11. Eventually it will end up with the virtual CRC depicted in Figure 12. But for this one the black triangle has coordinates

$$(x_c, z_c) = (4.94, 2.82), \tag{4}$$

which are considerably different from the original $(6.83, 3.20)$. Since we have been using approximations the user can try another layer factor, for instance 81%. This will give another picture similar to Figure 12.

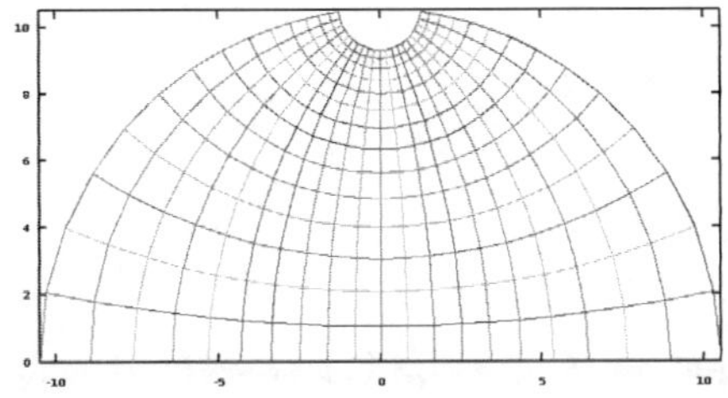

Figure 7: Upper half of D_H.

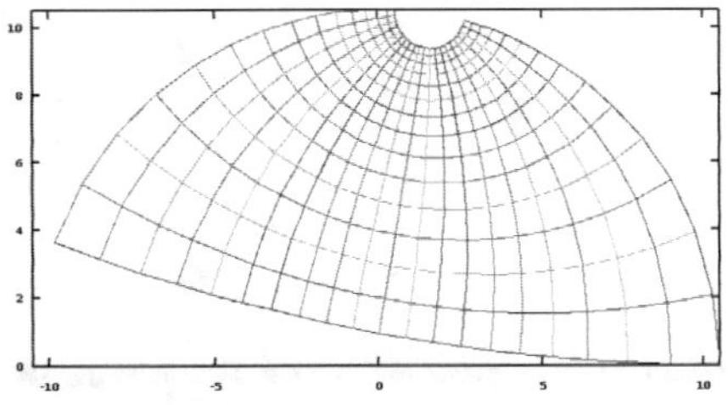

Figure 8: Its image under f.

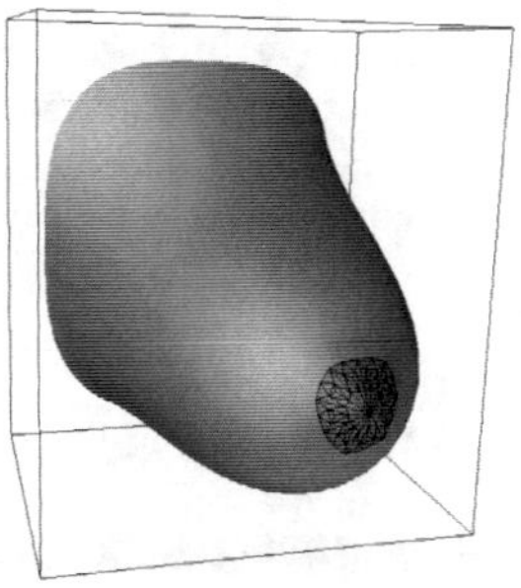
Figure 9: Patient's breast.

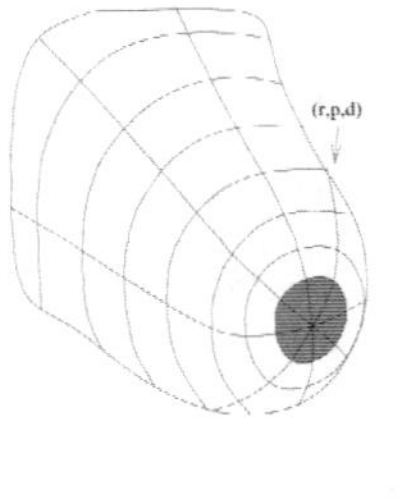

Figure 10: Polar coordinates.

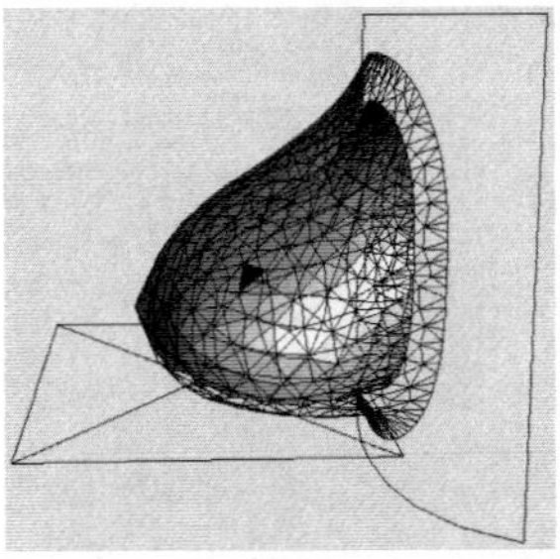
Figure 11: Nodule at LAT.

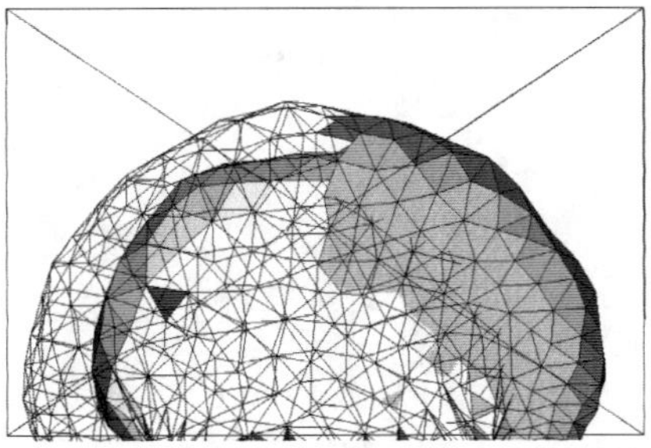
Figure 12: Checking `lf` at 73%.

For `lf` $= 0.81$ our simulator will print $(x_c, z_c) = (5.95, 3.26)$, which is pretty closer to the original $(6.83, 3.20)$ than (4). Moreover, the new $\rho = 0.2$ now gives

$$(r, p, d) = \Big(7.9024, 36.096, 1.2812\Big). \quad (5)$$

Of course, now we must simulate MLO. The virtual mammographer will give a picture very similar to Figure 12 and will print

$$(p_w, p_z) = (-5.83, 4.79), \quad (6)$$

which is distant from the original $(p_w, p_z) = (-5.2, 6.1)$ by less than 1.5cm, as in the CC-view. However, we recall that our original (p_w, p_z) is hypothetical because Figure 5 shows a nodule that technically gives $\rho = 1$. Since this is too extreme for a general example then we *guessed* $(-5.2, 6.1)$.

In order to improve accuracy one should start a convergence process: replace the initial guess with another point closer to (6); simulate again to find new values of `rho` and `lf`; apply these new values in the simulator; get a new picture like Figure 5; change `lf` to approach the original $(x_c, z_c) = (6.83, 3.20)$ at CRC, as depicted in Figure 12; simulate MLO and get new values at (6). Repeat this whole process until obtaining the desired precision.

However, we consider that a margin of error of 1.5cm is acceptable. The surgeon will then try (5) as illustrated by Figure 10. Once the point $(r, p) \simeq (7.9, 36^\circ)$ is marked on the breast the scalpel will cut till depth $d \simeq 1.3$cm.

3. Results

The main result is our software, of which the first version is available in the aforementioned virtual machine.

4. Conclusions

This work is the outcome of a long-term research, which can be evaluated in previous publications [8–10]. As explained at the Introduction, along the decades there has been an effort to improve Breast-Conserving Surgery (BCS). For this purpose one of the crucial tasks is to minimize the size of breast portion for removal. Therefore, one seeks the exact position of the tumour inside the breast.

The results presented herein are a valuable help for BCS. In order to test them one should start with real patients for whom a mastectomy was prescribed. Section 2 will guide the medical staff as part of the surgical preparation. Finally, if the tumour is really found this will validate our method, in spite of the simplifications that we have been using.

5. Acknowledgements

We thank Dr Ana Cláudia Veronesi for details about gynaecological surgery, and her husband Alejandro Montepeloso for translating some technical terms. We are grateful to *Ribeirão Pires Hospital (SP - Brazil)* for their assistance and for the concession to use their mammographer in our compression experiments with the phantom. This research was supported by FAPESP proc.No.12/16244-3.

References

[1] Bliznakova K, Bliznakov Z, Bravou V, Kolitsi Z and Pallikarakis N 2003 *Physics in Medicine and Biology* **48** 3699–3721

[2] Bhatti S N and Sridhar-Keralapura M 2012 *Medical Physics* **39** 1748–1768

[3] Hsu C M, Palmeri M L, Segars W P, Veress A I and Dobbins III J T 2011 *Medical Physics* **38** 5756–5770

[4] Kuhlmann M, Fear E, Ramirez-Serrano A and Federico S 2013 *Medical Engineering & Physics* **35** 470–478

[5] Nishikawa R M 2007 *Computerized Medical Imaging and Graphics* **31** 224–235

[6] Pathmanathan P, Gavaghan D J, Whiteley J P, Chapman S J and Brady J M 2008 *Biomedical Engineering, IEEE Transactions on* **55** 2471–2480

[7] Tang J, Rangayyan R M, Xu J, El Naqa I and Yang Y 2009 *Information Technology in Biomedicine, IEEE Transactions on* **13** 236–251

[8] Nascimento M Z and Ramos Batista V 2014 *Journal of Physics: Conference Series* vol 490 (IOP Publishing) p 012146

[9] Nascimento M Z and Ramos Batista V 2015 *Journal of Physics: Conference Series* vol 574 (IOP Publishing) p 012072

[10] Nascimento M Z and Ramos Batista V 2015 *Recent Patents on Medical Imaging* **to appear**

[11] Heath M, Bowyer K, Kopans D, Moore R and Kegelmeyer P 2001 *Proceedings of the Fifth International Workshop on Digital Mammography* (Medical Physics Publishing) pp 212–218 URL `http://marathon.csee.usf.edu/Mammography/Database.html`

Effects of magnetic field and Hall current to the blood velocity and LDL transfer

I Abdullah[1]**, N Naser, A H Talib and S Mahali**
School of Informatics and Applied Mathematics, Universiti Malaysia Terengganu, 21030 Kuala Terengganu, Terengganu, MALAYSIA

E-mail: [1]ilyani@umt.edu.my

Abstract. The magnetic field and Hall current effects have been considered on blood velocity and concentration of low-density lipoprotein (LDL). It is important to observe those effects to the flowing blood in a stenosed artery. The analysis from the obtained results may be useful to some clinical procedures, such as MRI, where the radiologists may have more information in the investigations before cardiac operations could be done. In this study, the uniform magnetic field and Hall current are applied to the Newtonian blood flow through an artery having a cosine-shaped stenosis. The governing equations are coupled with mass transfer and solved employing a finite difference Marker and Cell (MAC) method with an appropriate initial and boundary conditions. The graphical results of velocity profiles and LDL concentration are presented in this paper and the results show that the velocity increases and concentration decreases as Hall parameter increased.

1. Introduction

Cardiovascular disease which is also known as heart disease is proved as a leading cause of death globally from the study conducted by World Health Organization [1]. The heart failure is a condition in which the heart unable to pump enough blood to meet the body's needs [2]. This situation can be caused by aortic stenosis, a narrowing of the aortic orifice of the heart or the aorta near the valve. The stenosis is caused by atherosclerosis or the accumulation of macromolecules such as low-density lipoprotein (LDL) in the arterial wall which is known as plaque. The development of atherosclerosis is strongly related to the characteristics of the blood flow in the arteries [3].

Chakravarty et al. [4-6] investigated the flow behavior of blood in a different shape of stenotic artery. The blood is considered as a Newtonian fluid. Further studies of unsteady blood flow coupled with mass transfer through arterial stenosis have been reported by [7-9]. Mass transport is a movement of atherogenic molecules such as LDL and oxygen within the blood flow and arterial wall. This movement has been proved to contribute in the formation of stenosis.

Dynamics of biological fluid in the presence of magnetic field has developed great interest among researchers due to its importance in various implications in the bioengineering and medical technology. The development of magnetic devices for cell separation, targeted transport of magnetic particles as cancer tumor treatment, magnetic wound or drug carriers causing magnetic hyperthermia, reduction of

[1] To whom any correspondence should be addressed.

bleeding during surgeries and the development of magnetic tracers, are well-known applications in this domain of research [10-11]. The most characteristic biological fluid is blood, due to the complex interaction in which behaves as a magnetic fluid. If a magnetic field is applied to a moving and electrically conducting fluid, it will induce electric and magnetic field and the interaction of these fields produces Lorentz force. By Lenz's law, the Lorentz force opposes the motion of conducting fluid and since blood is an electrically conducting fluid, the magnetohydrodynamic (MHD) principles may be used to decelerate the flow of blood in human arterial system and thereby it is useful in the treatment of certain cardiovascular disorders.

The current trend for MHD flow is towards a strong magnetic field, so that the influence by electromagnetic force is noticeable. When the applied magnetic field strength is large and the conducting fluid is an ionized gas where the density is low, the conductivity normal to the magnetic field is reduced due to the spiraling of electrons and ions about the magnetic lines of force before collisions take place and a current is induced in a direction normal to both the electric and magnetic fields. This phenomenon is known as the Hall effect [12-14]. Fluid flow characteristics of blood flow in stenosed arteries with the presence of magnetic field are investigated by [15-16]. Recent study by [17] investigated the influence of magnetic field with Hall currents on blood flow through a stenotic artery. The results obtained show that the axial velocity increases as the Hall parameter increases.

Therefore, the aim of this present paper is to investigate the distribution of the velocity and mass concentration with the presence of magnetic field and Hall current. It is assumed that the arterial segment to be a cylindrical tube with a cosine-shaped stenosis and the blood flow through it to be a Newtonian fluid. The corresponding governing equations consists of continuity, axial and radial momentum and also mass transfer equations are solved using finite-difference scheme known as MAC method. This study will explore the effects of Hall parameter, Hartmann, Reynolds and Schmidt numbers on axial and radial velocity as well as mass transfer concentration.

2. Mathematical formulation

2.1. *The geometry of stenosis*

The geometry of stenosis, $R(z,t)$ for this study is described mathematically as

$$R(z,t)=\begin{cases} a(t)\left\{1-\dfrac{\delta}{2R_0}\left[1+\cos\left\{\dfrac{\pi(z-z_1)}{z_0}\right\}\right]\right\}, & z_1-z_0\le z\le z_1+z_0 \\ a(t), & otherwise \end{cases} \tag{1}$$

where R_0 is a constant radius of normal artery in the non-stenotic region, z is axial direction of stenosis, z_0 is half-length of stenosis, z_1 is centre of stenosis and δ is critical height of the stenosis. Time-variant parameter in (1), $a(t)$ is given by

$$a(t)=1+k_R\cos(\omega t-\varphi)$$

where k_R is a constant, t is time, $\omega = 2\pi f$ is the angular frequency with f is the pulse frequency and φ is the radius phase angle. The diagram of the geometry of stenosis is shown in figure 1.

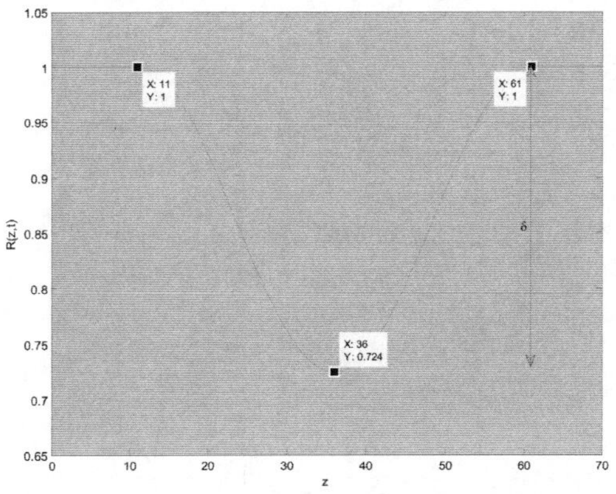

Figure 1. The schematic diagram of cosine-shaped stenosis.

2.2. *Governing equations*

Let us consider (r, z, θ) to be the coordinates of the material point in the cylindrical coordinates system where the z-axis is taken along the axis of the artery while r and θ are taken along the radial and circumferential directions, respectively. In this study, blood flow in the constricted artery segment is considered to be two-dimensional, unsteady, incompressible and the fluid is assumed to be Newtonian fluid. The governing equations corresponding with this study consist of continuity, momentum and mass transfer equations. The dimensionless conservative governing equations can be written as:

$$r\frac{\partial w}{\partial z}+\frac{\partial (ur)}{\partial r}=0, \tag{2}$$

$$\frac{\partial w}{\partial t}+\frac{\partial (wu)}{\partial r}+\frac{\partial w^2}{\partial z}+\frac{(wu)}{r}=-\frac{\partial p}{\partial z}+\frac{1}{\mathrm{Re}}\left(\frac{\partial^2 w}{\partial r^2}+\frac{1}{r}\frac{\partial w}{\partial r}+\frac{\partial^2 w}{\partial z^2}\right)+\frac{M^2}{\mathrm{Re}}\frac{(mu-w)}{(1+m^2)}, \tag{3}$$

$$\frac{\partial u}{\partial t}+\frac{\partial u^2}{\partial r}+\frac{\partial (wu)}{\partial z}+\frac{u^2}{r}=-\frac{\partial p}{\partial r}+\frac{1}{\mathrm{Re}}\left(\frac{\partial^2 u}{\partial r^2}+\frac{1}{r}\frac{\partial u}{\partial r}+\frac{\partial^2 u}{\partial z^2}-\frac{u}{r^2}\right)-\frac{M^2}{\mathrm{Re}}\frac{(u+mw)}{(1+m^2)}, \tag{4}$$

$$\frac{\partial C}{\partial t}+u\frac{\partial C}{\partial r}+w\frac{\partial C}{\partial z}=\frac{1}{\mathrm{Re}\cdot \mathrm{Sc}}\left(\frac{\partial^2 C}{\partial r^2}+\frac{1}{r}\frac{\partial C}{\partial r}+\frac{\partial^2 C}{\partial z^2}\right) \tag{5}$$

where u and w are the radial and axial velocity components along the r- and z-axes respectively, p is pressure, r is radial axis, C is the mass concentration, $\mathrm{Re}=U_0 r_0 \rho/\mu$ is Reynolds number, $M = B_0 r_0 (\sigma/\mu)^{\frac{1}{2}}$ is the Hartmann number and used as the dimensionless parameter of the magnetic field, m is Hall parameter and $Sc=\mu/\rho D$ is the Schmidt number with ρ is density of blood, μ is dynamic viscosity, D is the coefficient of diffusion, U_0 is mean velocity and r_0 is the normal radius in the arterial segment.

2.3. *Boundary and initial conditions*

Along the symmetry axis of the artery, the radial flow is zero. Therefore, the normal component of the radial velocity and the axial velocity gradient and the mass concentration gradient of the blood along the axis are vanished which means that the shear stress does not exist. These may be stated mathematically as

$$\text{at } r=0, \quad \frac{\partial w(r,z,t)}{\partial r}=u(r,z,t)=\frac{\partial C(r,z,t)}{\partial r}=0. \tag{6}$$

The radial velocity at arterial wall may be assumed to be equal to the changes of radius of the arterial segment in the constricted artery. Hence the velocity boundary conditions on the arterial wall obey the no-slip condition and can be written as

$$\text{at } r = R,\quad w(r,z,t) = C(r,z,t) = 0\,,\quad u(r,z,t) = \frac{\partial R}{\partial t}\,. \tag{7}$$

The inlet velocity (when $z = 0$) conditions are assumed to be fully developed with parabolic velocity profile corresponding to Hagen-Poiseuille flow through a long circular tube. The mass concentration is assumed to be constant. It is provided with evidence by study of Mustapha et al. [11] as

$$\text{at } z = 0, \text{ for } M = 0\,,\quad w(r,z,t) = 2(1-x^2),\, u(r,z,t) = 0\,,\, C(r,z,t) = 1\,, \tag{8}$$

$$\text{at } z = 0, \text{ for } M \neq 0\,,\quad w(r,z,t) = 2\left(\frac{I_0(MR)}{I_0(MR)-1}\right)\left(1-\left(\frac{I_0(Mr)}{I_0(MR)}\right)\right),\, u(r,z,t) = 0\,,\, C(r,z,t) = 1\,. \tag{9}$$

At the outlet (when $z = L$), where L is the finite length of the arterial segment, the velocity gradients and mass concentration gradient are taken to be traction-free conditions and can be stated as

$$\text{at } z = L,\quad \frac{\partial w(r,z,t)}{\partial z} = \frac{\partial u(r,z,t)}{\partial z} = \frac{\partial C(r,z,t)}{\partial z} = 0\,. \tag{10}$$

It is also assumed that no flow takes place when the system is at rest ($t = 0$) except at the inlet,

$$\text{at } t = 0, \text{ for } z > 0,\quad w(r,z,0) = u(r,z,0) = p(r,z,0) = C(r,z,0) = 0\,. \tag{11}$$

3. Marker and Cell (MAC) method

The governing equations (2-5) together with initial and boundary conditions (6-11) are solved numerically using the finite difference scheme known as MAC method. This method is based on the central difference approximations for the uniform spatial derivatives and the forward difference formula for time derivatives. The discretized form for the continuity equation is

$$x_{lj} R_{li}^k \left(\frac{w_{i,j}^k - w_{i-1,j}^k}{\Delta z}\right) - \left(x_{lj}\right)^2 \left(\frac{\partial R}{\partial z}\right)_{li}^k \left(\frac{w_{at} - w_{ab}}{\Delta x}\right) + \frac{x_j u_{i,j}^k - x_{j-1} u_{i,j-1}^k}{\Delta x} = 0 \tag{12}$$

with z_{lj} and x_{lj} are represents the respective coordinates of the cell centre while z_i and x_j represents the cell at the top right corner of the $(i, j)^{\text{th}}$ control volume.

Meanwhile, the discretized form of the axial momentum equation can be written into the form

$$\frac{w_{i,j}^{k+1} - w_{i,j}^k}{\Delta t} = \left(\frac{p_{i,j}^k - p_{i+1,j}^k}{\Delta z}\right) + \frac{x_{lj}}{R_i^k}\left(\frac{\partial R}{\partial z}\right)_i^k \left(\frac{p_t - p_b}{\Delta x}\right) + \left(w_{me}\right)_{i,j}^k \tag{13}$$

Finally, the discretized form of the radial momentum equation can be put in the form

$$\frac{u_{i,j}^{k+1} - u_{i,j}^k}{\Delta t} = \frac{1}{R_{li}^k}\left(\frac{p_{i,j}^k - p_{i,j+1}^k}{\Delta x}\right) + \left(u_{me}\right)_{i,j}^k \tag{14}$$

where all the terms are given in [9], in details.

On the other hand, the discretization of the convection-diffusion equation for mass transfer can be written as

$$\frac{C_{i,j}^{k+1}-C_{i,j}^{k}}{\Delta t}=\frac{x_j}{R_i^k}\left(\frac{\partial R}{\partial t}\right)_i^k\left(\frac{\partial C}{\partial x}\right)_{i,j}^k-\frac{u_{i,j}^k}{R_i^k}\left(\frac{\partial C}{\partial x}\right)_{i,j}^k-w_{i,j}^k\left(\left(\frac{\partial C}{\partial z}\right)_{i,j}^k-\frac{x_{ij}}{R_i^k}\left(\frac{\partial R}{\partial z}\right)_i^k\left(\frac{\partial C}{\partial x}\right)_{i,j}^k\right)$$
$$-\frac{1}{\mathrm{Re}\cdot Sc\cdot R_i^{k^2}}\left[\left\{1+\left(x_j\left(\frac{\partial R}{\partial z}\right)_i^k\right)^2\right\}\left(\frac{\partial^2 C}{\partial x^2}\right)_{i.j}^k\right.$$
$$\left.+\left\{\frac{1}{x_j}+2x_j\left(\left(\frac{\partial R}{\partial z}\right)_i^k\right)^2-x_jR_i^k\left(\frac{\partial^2 R}{\partial z^2}\right)_{i,j}^k\right\}\left(\frac{\partial C}{\partial x}\right)_{i,j}^k+\left(R_i^k\right)^2\left(\frac{\partial^2 C}{\partial z^2}\right)_{i,j}^k\right] \quad (15)$$

with

$$\frac{\partial C}{\partial x}=\frac{C_{i,j+1}^k-C_{i,j-1}^k}{2\Delta x},\quad \frac{\partial C}{\partial z}=\frac{C_{i+1,j}^k-C_{i-1,j}^k}{2\Delta z},\quad \frac{\partial^2 C}{\partial x^2}=\frac{C_{i,j+1}^k-2C_{i,j}^k+C_{i,j-1}^k}{(\Delta x)^2},\quad \frac{\partial^2 C}{\partial z^2}=\frac{C_{i+1,j}^k-2C_{i,j}^k+C_{i-1,j}^k}{(\Delta z)^2}$$

4. Results and discussion

The governing equations (12-15) corresponding with the boundary and initial conditions have been solved numerically using finite difference method which is known as Marker-and-Cell (MAC) method. Results for present paper are obtained for Re = 450, Sc = 5, M = 1, 2, 5, 10, 20, and m = 1, 2, 5, 10. The geometry of stenosis considered in this study is cosine-shaped stenosis. From figure 1, it is shown that there are two regions represent the non-stenotic region or normal artery and stenotic region. For z = 0 until 1 and z = 3 until 5 are known as non-stenotic regions while the stenotic region located at z = 1 until 2 where the critical height of the stenosis occur at z = 0.7241 with δ = 0.276.

Figure 2 illustrates the behavior of the axial velocity profiles for different magnetic field at z = 3.5 and Re = 450. It can be seen that the shape of the velocity profiles varies depending on the intensity of the magnetic field applied. As magnetic field increases, the axial velocity becomes flatter and the velocity gradient becomes steeper towards the wall surface of the stenosed artery. Meanwhile, the magnitude of the maximum velocity decreases as the magnetic field increases. This is because the magnetic field leads to a rising drag-like force known as Lorentz force. This force has tendency to reduce the fluid velocity in the boundary layer.

On the other hand, the dimensionless axial velocity for different Hall parameter is illustrated as in figure 3. It is shown that as Hall parameter increases, the axial velocity will increase. Theoretically, the influence of the magnetic field may cause the positive and negative ions that bounce back and forth between the sides of the vessel and induce a Hall current. The combination of the electromotive force, altered ionic pattern, and currents may potentially cause the blood vessel dilation which corresponds to the increase in blood flow.

Figure 4 shows the dimensionless mass concentration of LDL at different z for a fixed Schmidt number, Sc = 5. The results illustrate that at normal artery (at z = 1), the mass concentration is higher than at the critical height of the stenosis (at z = 3.5). This is because the arterial wall is thicker at the stenotic region, thus the resistance to LDL is higher than on the non-stenotic region. Hence, the mass concentration decreases toward the stenosed artery region.

Figure 5 shows the behavior of dimensionless mass concentration for different Hall parameter. It is shown that as the value of Hall parameter increases, the mass concentration will decrease. We can also see that the mass concentration fluctuates in most radial part of the artery and it becomes flatter near the wall surface of the artery. As for various values of magnetic field, M, there is insignificant result for dimensionless mass concentration.

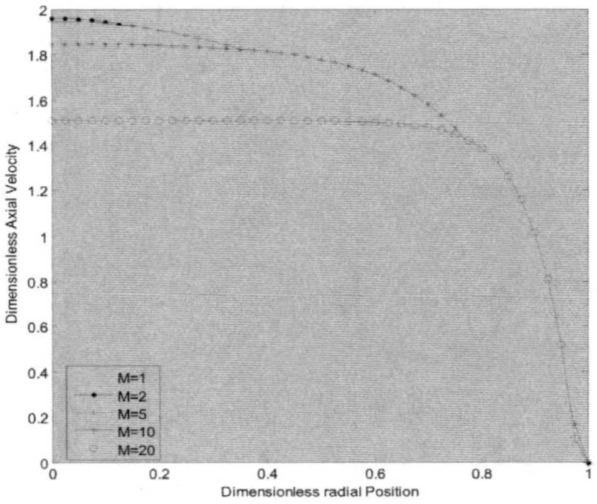

Figure 2. Dimensionless axial velocity for different magnetic field, M.

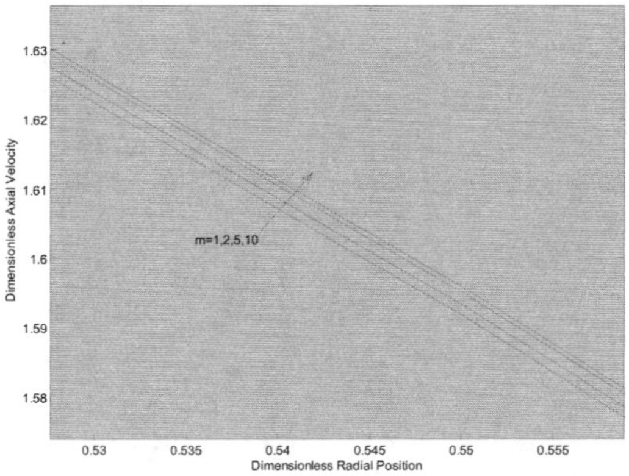

Figure 3. Dimensionless axial velocity for different Hall parameter, *m*.

5. Conclusions

The problem of blood flow coupled with mass transfer through an arterial stenosis with Hall current effect has been studied numerically in this paper. The effects of magnetic field, M and Hall parameter, *m*, on the axial velocity and mass concentration has been examined in details in this paper. As conclusion, the present study shows that how the magnetic field plays a role that reduces the velocity of blood but at the same time, the presence of Hall current increases the velocity. However, the effect from magnetic field is much more apparent compared to the Hall effect. With the large enough intensity of applied magnetic field, it is believed that it can streamlining the flow of blood around the body. Thus, it still needs further investigation to conclude the adequate of the intensity applied magnetic field in order to have any such effects.

Furthermore, it is evidence from the results that the LDL concentration is lower in the stenotic region. This finding shows that the possibility of the LDL adheres at the upstream of stenosis, thus contributes to the enlargement of stenosis. As for different Hall parameter, the concentration having a fluctuate profile near the axis, and slowly approaches zero at the wall.

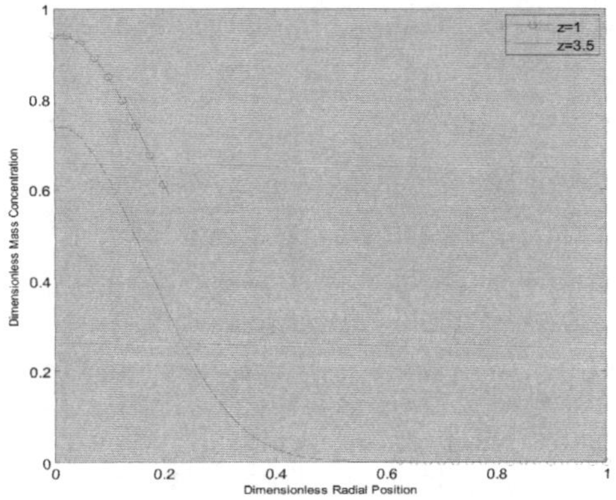

Figure 4. Dimensionless mass concentration at different *z*.

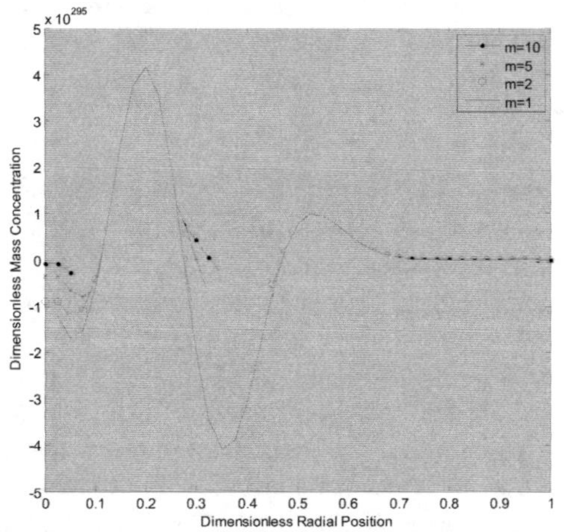

Figure 5. Dimensionless mass concentration for different Hall parameter, *m*.

Acknowledgments
Authors wishing to acknowledge Ministry of Higher Education (MOHE) and Research Management Centre UMT for the financial support through vote number 56020 (RACE) for this research.

References

[1] World Health Organization (WHO) 2011 *Global Atlas on Cardiovascular Disease Prevention and Control* ed S Mendis, P Puska and B Norrving (Geneva: WHO, World Heart Federation, and World Stroke Organization)
[2] Tan Y B and Mustapha N 2013 Numerical modeling of blood flow in irregular stenosed artery with the effects of gravity *Jurnal Teknologi* **62** 57-64
[3] Abdullah I, Amin N and Hayat T 2011 Magnetohydrodynamic effects on blood flow through an irregular stenosis *Int. J. Num. Meth. in Fluids* **67** 1624-36
[4] Chakravarty S, Datta A and Mandal P K 1996 Effect of body acceleration on unsteady flow of blood past a time-dependent arterial stenosis *Mathl. Comput. Modelling* **24** 57-74
[5] Chakravarty S and Sannigrahi A K 1999 A nonlinear mathematical model of blood flow in a constricted artery experiencing body acceleration *Mathl. Comput. Modelling* **29** 9-25
[6] Chakravarty S, Mandal P K and Mandal A 2000 Mathematical model of pulsatile blood flow in a distensible aortic bifurcation subject to body acceleration *Int. J. Engineering Sci.* **38** 215-38
[7] Kaazempur-Mofrad M R, Wada S, Myers J G and Ethier C R 2005 Mass transport and fluid flow in stenotic arteries: axisymmetric and asymmetric models *Int. J. Heat and Mass Transfer* **48** 4510–17
[8] Sun N, Wood N B, Hughes A D, Thom S A M and Xu X Y 2006 Fluid-wall modelling of mass transfer in an axisymmetric stenosis: effects of shear-dependent transport properties *Annals of Biomedical Engineering* **34** 1119–28
[9] Abdullah I, Talib A H and Sabri I 2013 Numerical simulation on convection-diffusion in an overlapping stenosed artery *Malaysian J. Fundamental and Applied Sciences* **9** 156-60
[10] Hayat T, Awais M and Alsaedi A 2012 Newtonian heating and magnetohydrodynamic effects in flow of a Jeffery fluid over a radially stretching surface *Int. J. Physical Sciences* **7** 2838-44
[11] Mustapha N, Amin N, Chakravarty S and Mandal P K 2009 Unsteady magnetohydrodynamic blood flow through irregular multi-stenosed arteries *Comp. in Biology and Medicine* **39** 896-906
[12] Abo-Eldahab E M, El-Aziz M A, Salem A M and Jaber K K 2007 Hall current effect on MHD mixed convection flow from an inclined continuosly stretching surface with blowing/suction and internal heat generation/absorption *Applied Mathematical Modeling* **31** 1829-46
[13] Shit G C 2010 Hall effects on MHD free convective flow and mass transfer over a stretching sheet *Int. J. Applied Mathematics and Mechanics* **5** 22-38
[14] Ahmed N, Kalits H and Barua D P 2010 Unsteady MHD free convective flow pass a vertical porous plate immersed in a porous medium with Hall current, thermal diffusion and heat source *Int. J. Engineering Science and Technology* **2** 55-74
[15] Tashtoush B and Magableh A 2008 Magnetic field effect on heat transfer and fluid flow characteristics of blood flow in multi-stenosis arteries *Int. J. Heat and Mass Transfer* **44** 297–304
[16] Prakash O, Singh S P, Kumar D and Dwivedi Y K 2011 A study of effects of heat source on MHD blood flow through bifurcated arteries *AIP Advances* **1** 042128-1-7
[17] Mekheimer K S and Kot M A E 2008 Influence of magnetic field and Hall currents on blood flow through a stenotic artery *Appl. Math. Mech.* **29** 1093–104

Modelling of poro-visco-elastic biological systems

Y Bilotsky and M Gasik

Department of Material Science and Engineering, Aalto University Foundation School of Chemical Technology, P.O. Box 16200, FIN-00076 AALTO, Finland

E-mail: yevgen.bilotsky@aalto.fi

Abstract. The research of mechanical properties of poro-visco-elastic biomaterials is an important task, especially for tailoring the best conditions for in-growth and healing of implants. In this work we analysed the behaviour of biomaterials under different static and dynamic loading regimes, in "dry" and "wet" conditions. Retrieved data revealed nonlinear relations between applied force and resulting deformation, with time and frequency dependence. These features were described by a nonlinear model, which reasonably fits mentioned peculiarities. The simplified model was validated with numerical simulations using COMSOL software. Upon validation it allows incorporation of the experimental data obtained by biomechanical spectroscopy towards prediction of biomaterials behaviour in "in vitro" conditions, with the purpose to extrapolate to clinically-relevant environment.

1. Introduction

Investigation of mechanical properties of biological systems is very complicated objective and requires specific knowledge of every part of this system and creation a general model is a problematic task [1,2]. Even in the case of biomechanical behavior of the natural (bone, cartilage) or artificial (relevant implants), without inclusion of cellular signals, biochemical pathways and external environment, it is very difficult to obtain an engineering model capable of prediction of the poro-visco-elastic behavior of such materials, needed for their proper design.

It is known that biomechanical properties are critical for biological processes, and these properties are regulating the signals sent to the cells [3]. Due to variety of parameters involved and the obvious limitations of *in vivo* studies, computational mechanobiology is often applied to determine the quantitative rules that govern the effects of mechanical loading on tissue differentiation, growth, adaptation and maintenance by trial-and-error. From a mechanics point of view, the task is considered a 'boundary value problem', whereby the boundary loads of a domain are translated into local mechanical variables within the domain, depending on the geometry and mechanical properties of its materials [1-3]. Such problems are usually solved with finite element analysis (FEA).

Whereas numerous studies were performed in this area and their results were sometimes correlated with *in vivo* experiments, the feedback did not allow deployment of a reverse design scheme, which would be capable to tell the desired properties of the scaffold or implant for specific application and patient. In the most cases, this is limited to purely geometrical constrains (whether implant fits the site and how well) or mechanical criteria (maximal expected stress vs. body weight, expected local stresses concentrators or fatigue life).

In this work we attempt to derive simplified yet practical model of characterization of porous scaffolds intended for articular cartilage (AC) correction surgery. The structure, functions and behavior of AC are very complex and highly anisotropic [4]. Articular cartilage is a living material composed of a relatively small number of cells known as chondrocytes surrounded by a multicomponent matrix. Mechanically, AC (Figure 1) is a composite with widely differing properties, of which weight 70 to 85% is water. The remainder of the tissue is composed primarily of proteoglycans (protein core to which glycosaminoglycans are attached to form a bottlebrush-like structure) and collagen II [4,5]. These proteoglycans can bind or aggregate to a backbone of hyaluronic acid (HA) to form a macromolecule with a weight up to 200 million [5]. Approximately 30% of the dry AC weight is composed of proteoglycans, which concentration and water content vary through the depth of the tissue.

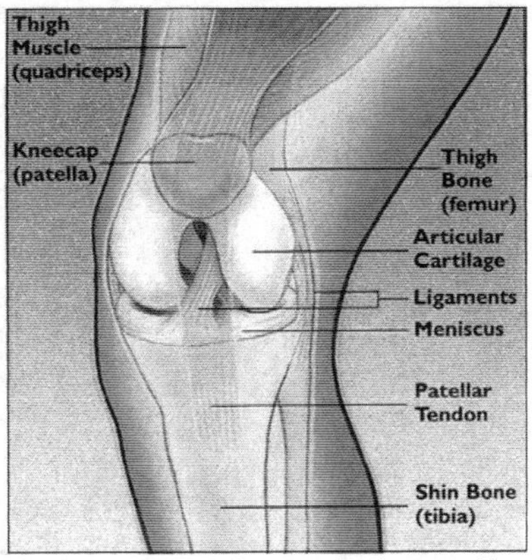

Figure 1. The structure of the knee joint. Articular cartilage is shown in light color.

Synthetic materials with fibrous origin are often used for AC correction applications. There 75-85% porous scaffolds are usually attached to the bony side of the cartilage, leaving other side to exposure to synovial fluid with HA. In this work the target was to find the proper simple model which can describe the poro-visco-elastic biomaterials for easier use in clinical practice, such as to find answers which scaffold design would be the most appropriate for specific patient conditions.

2. Models for porous scaffolds.

There are many simple and more complex expressions for porous materials intended for AC applications. Mechanically the description of these materials might be divided into two categories, called here respectively "dry" and "wet" conditions. In the first case, the material is analyzed without liquid, assuming complete drained conditions and neglecting the gas (air) viscosity and compressibility in the pores. This is justified for low loading frequencies, but as most of the physiological frequencies fit within 1 Hz region, dry conditions might be seen as relevant for the porous skeleton analysis. In the wet case, the material is fully saturated with a fluid (can be Newtonian or non-Newtonian) and the analysis can be performed for two categories: drained (free fluid in and out flow) and undrained (fluid flow constrained). Here we are limiting analysis only for drained conditions, as supported by our experiments.

Besides mechanical characterization (e.g. simple compression tests), the model sought is required to provide consistent answers for two important modes, namely creep at constant load and frequency response under oscillating dynamic load. These modes are combined in nature as the AC is constantly under the mean stress, whereas dynamic load changes for different activities [1-3]. Upon analysis of published data and the models limitations, we may observe that purely elastic (Hookean) and viscous (Newtonian) models have to be properly combined to correspond to experimental data. Usually this is being done by application of several Kelvin and Maxwell spring-dashpot models, of which Zener (also known as standard linear solid) and Burgers models are the most comprehensive. Based on the experimental analysis with BEST method [6], we selected Burgers model as the best compromise, which does not require assumption of relaxation times spectra and has a minimal set of fitting parameters. The Burger model consists of a linear spring and a linear dashpot in parallel (Kelvin element) and another linear spring and linear dashpot in series (Maxwell element, [7]). In this model response equation reading as

$$\sigma + p_1\dot{\sigma} + p_2\ddot{\sigma} = q_1\dot{\varepsilon} + q_2\ddot{\varepsilon} \tag{1.1}$$

where σ is stress, ε is strain for one-dimensional loading. Using E for the moduli of elasticity (spring elements) and η_i for the viscosities (dashpot elements), we can write

$$p_1 = \tau_1\left[1+\frac{E_1}{E_2}\right]+\tau_2,\ p_2 = \tau_1\tau_2,\ q_1 = \tau_1 E_1,\ q_2 = \tau_2\eta_2,\ \tau_1 = \frac{\eta_1}{E_1},\ \tau_2 = \frac{\eta_2}{E_2}, \tag{1.2}$$

where τ_i are the relaxation times. The creep function connects the strain and stress by the functional relation

$$\varepsilon = cr\cdot\sigma \tag{1.3}$$

After solving the equations (1.1), (1.3) comes as

$$cr = \frac{1}{E_1}\left(1+\frac{t}{\tau_1}\right)+\frac{1}{E_2}\left[1-\exp\left(-\frac{t}{\tau_2}\right)\right]. \tag{1.4}$$

This result is well-known in materials engineering especially for creep of solids with substantial viscosity component. Applying this equation to porous fibrous scaffold under creep conditions, relevant to the implant use [6], we found an excellent fit of the data in both wet and dry cases, Figure 2. Despite the parameters used in (1.2) are different for these two cases, the model predicts well behavior of the specimens, which is also reproducible in the experimental setup. It was observed that parameters with the model equations (1.2 and 1.4) are not exactly the materials constants, but they depend on the applied load (stress), and in this sense this Burgers model becomes non-linear.

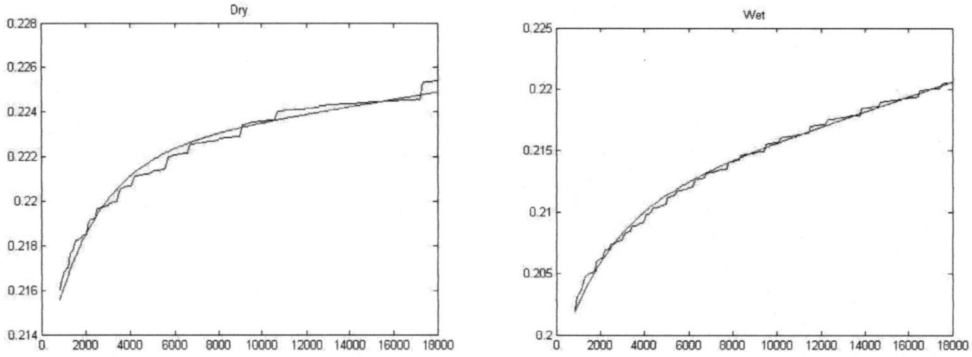

Figure 2. Experimental data (creep function vs. time in seconds) for porous scaffolds under 0.8 N load and the Burgers model fit for dry (left) and wet (right) conditions.

3. Numerical simulation and the experimental validation

The evidence of reasonable Burgers model fit for creep response allows calculations of the dynamic response of the scaffold under prescribed loading, which might simulate realistic knee joint operations. Here COMSOL software was used for such analysis, although other FEA algorithms can be also applied. The scaffolds were subjected to dynamical mechanical analysis (DMA) under 1 Hz oscillating load with a contacts part which is responsible for creep behavior for about 5 h experiment. Obtained axial strain values were compared with the FEA simulations along the axial symmetry axis for different models. Figure 3 show experiment data with linear (red) and non-linear Burgers model predictions, which clearly shows that non-linearity effects result in better prediction of materials behavior especially for long time scale.

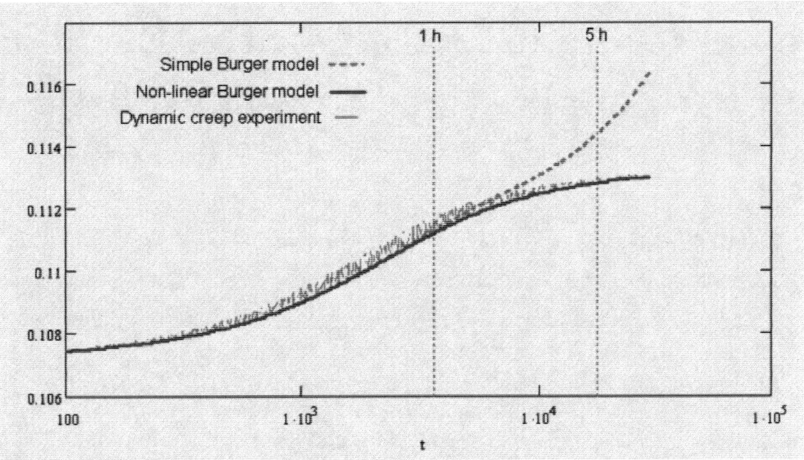

Figure 3. DMA of the scaffold disk specimen and modeling predictions for axial strain vs. time (s).

4. Discussion

Characterization of the visco-elastic behavior of the biomaterials can be achieved with different models, which basic elements could be combined with high complexity. However, whereas large number of virtual springs and dashpots could be calculated as in Weichert model [8], this does not allow reverse design of the scaffold with minimal number of parameters, which could be determined in a simple creep experiment and later expand to dynamic loading conditions.

In this work we have demonstrated that adding a simple non-linearity into Burgers model allows a very good description of the material behavior. This results in two important practical outputs. First, biomechanical conditions of the patient allow formulation of scaffold properties in terms of parameters (1.2) as functions of expected loads. Second, biomaterial manufacturers could make a fast screening of potential candidates to determine these parameters and thus predict how well particular structure and properties of the material would fit clinical demands.

Acknowledgement. Financial support from the Finnish National Agency for Innovations (Tekes) under project No. 211493 is gratefully acknowledged.

References

[1] S. Scheiner, P. Pivonka and C. Hellmich: Comput. Methods Appl. Mech. Engr. 254 (2013) p. 181
[2] F.M. René, B. Rietbergen, K. Ito, P.A.J. Hilbers and R. Huiskes: Biomech. Model. Mechanobiol. 87 (2011), p. 965.
[3] M.C.H. van der Meulen and R. Huiskes: J. Biomech. 35 (2002), p. 401.
[4] W. Wilson, C. C. van Donkelaar and J. M. Huyghe: Trans. ASME 127 (2005), p. 258.
[5] V.C. Mow and W.M. Lai: SIAM Rev 22 (1980), p. 275.
[6] M. Gasik: ALTEX Proc. 3 (2014), p. 33.
[7] F. Irgens: Continuum Mechanics (Berlin), 2008.
[8] M.L. Williams: AIAA Journal 5 (1964), p. 785.

Journal of Physics: Conference Series **633** (2015) 012135 doi:10.1088/1742-6596/633/1/012135

A functional technique based on the Euclidean algorithm with applications to 2-D acoustic diffractal diffusers

Luis Cortés-Vega

Mathematics Department, Antofagasta University, Casilla 170, Antofagasta, Chile

E-mail: `luis.cortes@uantof.cl, luisvega@vtr.net`

Abstract. We built, based on the Euclidean algorithm, a functional technique, which allows to discover a direct proof of Chinese Remainder Theorem. Afterwards, by using this functional approach, we present some applications to 2-D acoustic diffractal diffusers. The novelty of the method is their functional algorithmic character, which improves ideas, as well as, other results of the author and his collaborators in a previous work [2].

1. Introduction

In a historical context, the Chinese remainder theorem, CRT, was proposed to solve some astronomical problems. Now, is one of the main theorems of number theory [10]. Over the years, CRT has been played a prominent role due to its applicability in diverse fields of science and engineering, as photo radar [6], cryptography and code theory [1,4], circuits on systems [7], technology of diffusers [3] and matrix theory [8], among others.

The aim of this paper is twofold: 1) The first one, is to exploit some functionals expressions based on the Euclidean algorithm in order to describe a direct proof of CRT. It should be stressed that the origins of this approach appear in [2], however, there no analysis of the CRT has been considered, so far. In this context, the results of this paper are new and they complement previous theoretical results given in [2]. 2) Second one, we are interested in to characterize functionally the depths $d_{h,k}$ of a 2-D acoustic diffractal diffuser, which have the following form

$$d_{h,k} = \left\{ \left\lfloor \frac{h}{N} \right\rfloor^2 + \left\lfloor \frac{k}{N} \right\rfloor^2 \right\}_{mod\, M} \times \frac{\lambda_M}{2M} + \left\{ (h_{mod\, M})^2_{mod\, N} + (k_{mod\, M})^2_{mod\, N} \right\} \times \frac{\lambda_N}{2N}. \quad (1)$$

Here, $M \in \mathbb{N}^*$ and $N \in \mathbb{N}^*$ are prime numbers associated with low and high frequency; λ_M and λ_N are the wavelengths of the low frequency diffusser and high frequency diffuser, respectively. Also in the same formula, the expression $\lfloor \bullet \rfloor$ denotes the integer floor function, and the expression $(a)_{mod\, N}$ is the usual modular representation of a. It is important to emphasize that the $d_{h,k}$ depths are associated to the $(h,k)^{\text{th}}$ position of a series of wells separated by thin fins, all inside of a 2-D acoustic diffractal diffuser. We address in this part of paper to answer the next two questions: 1) Which is the minimum amount of information needed in order to accelerate the computation of the $d_{h,k}$ numbers ?. - This is certainly an interesting question, since it plays a role in many facets associated to the building of diffractal for industrial use, see [3]-, and 2) It plays some role, in the answer to this first question the CRT?.

In all of our discussion, $\mathbb{N}$ will designate the set of natural numbers, $a \in \mathbb{N}^*$ if $a \in \mathbb{N} \cup \{0\}$. Furthermore, it will be used the simbol $gcd(b, d)$ to denote the greatest common divisor between b and d (not both zero). In this notation, if $gcd(a, b) = 1$, we say that a and b are relatively prime.

2. Preliminaries

We present first given some basic facts about the representation of natural number. Indeed, the Euclidean algorithm implies that

$$a = r_0 + r_1 b + r_2 b^2 + \ldots + r_{n-1} b^{n-1} + r_n b^n, \ 0 \leq r_i \leq b-1, \text{for all } i = 0, 1, 2 \ldots, n. \qquad (2)$$

Here, $a, b \in \mathbb{N}^*$ and $b \geq 2$, additional details can be found in [10]. From here, we can define on $\mathbb{N}^*$ the following functions

$$\phi_b : \mathbb{N}^* \to \mathbb{N}^*, \ \phi_b(a) = \begin{cases} a, & \text{if } 0 \leq a \leq b-1, \\ r_0, & \text{if } \ a \geq b. \end{cases} \quad \text{and} \quad C_b : \mathbb{N}^* \to \mathbb{N}^*, \ C_b(a) = \frac{a - \phi_b(a)}{b}.$$

Definition 1 *Let $b, d \in \mathbb{N}^*$ with $b, d \geq 2$. We define the composition of C_d with C_b, and of ϕ_d with ϕ_b by a usual way, that is: $C_d \circ C_b : \mathbb{N}^* \to \mathbb{N}^*$, such that $(C_d \circ C_b)(a) \stackrel{def}{=} C_d(C_b(a))$ and $\phi_d \circ \phi_b : \mathbb{N}^* \to \mathbb{N}^*$ such that $(\phi_d \circ \phi_b)(a) \stackrel{def}{=} \phi_d(\phi_b(a))$, for all $a \in \mathbb{N}^*$.*

We begin by establishing the following

Theorem 2 *Let $a, b, d \in \mathbb{N}^*$, with $b \geq 2$. Then the following properties are valid:*

(ϕ1) $\phi_b(0) = 0$,
(ϕ2) $\phi_b(d \times b) = 0$, *for all* $d \in \mathbb{N}^*$,
(ϕ3) $\phi_b(a) = \phi_b(\phi_b(a))$,
(ϕ4) $\phi_b(a + d) = \phi_b(\phi_b(a) + \phi_b(d)) = \phi_b(a + \phi_b(d)) = \phi_b(\phi_b(a) + d)$,
(ϕ5) $\phi_b(a \times d) = \phi_b(\phi_b(a) \times \phi_b(d)) = \phi_b(a \times \phi_b(d)) = \phi_b(\phi_b(a) \times d)$,
(ϕ6) $\phi_b(a + b) = \phi_b(a)$ *("periodicity" of ϕ_b).*
(c1) $C_b(0) = 0$,
(c2) $C_b(b \times a) = a$, *in particular* $C_b(b) = 1$,
(c3) $C_b(b^n) = b^{n-1}$,
(c4) $C_b(\phi_b(a)) = 0$, *(C_b is a "annihilator" of ϕ_b),*
(c5) $C_b(a + d) = C_b(a) + C_b(d) + C_b(\phi_b(a) + \phi_b(d))$,
(c6) $C_b(a \times d) = C_b(a) \times d + \phi_b(a) \times C_b(d) + C_b(\phi_b(a) \times \phi_b(d))$,
(c7) $C_b(a + b) = C_b(a) + 1$ *(C_b is "quasi-periodic"),*
(e1) $\phi_b(a) + b \times C_b(a) = a$,
(e2) $a < b$ *if and only if* $C_b(a) = 0$,
(e3) $(C_d \circ C_b)(a) = C_{d \times b}(a)$.

Proof. In order to prove the Theorem 2, we have that (ϕ1), (ϕ2), (c1), (c2), (c3) and (e1) are direct. The proof of (ϕ3), (ϕ4), (ϕ5) and (ϕ6) can be found in [2]. In the same lines of reasoning as in [2], one can also check that the rest of the items are straightforward from the definition of ϕ_b and C_b, respectively. Note that in particular, the identity (c5) follows from definition of $C_b(a)$ and the property (ϕ4). Also, (c6) follows direct from the extension of the property (c5) for "d" summands, (e2) follows of definitions of C_b and ϕ_b. Finally, the proof of property (e3) follows by using all previous properties. This last property will be key in the proof of CRT. ∎

3. A direct proof of CRT via a functional approach

As a first step, we start with the following

Definition 3 *In our approach, if a and b are relatively prime, hencenfort the simbol $i_b(a)$ will be denoted the moduli multiplicative inverse of a. That is, $i_b(a) \in \{0, 1, 2, \ldots, b-1\}$ and*

$$\phi_b\left(a \times i_b(a))\right) = 1.$$

Note that if a and b are relatively prime, the number $i_b(a)$ always exists and verifies the following aditional identity:

$$i_b(a) = i_b(\phi_b(a)). \tag{3}$$

Now, we will center on the following Theorem:

Theorem 4 *Assume $b, d \in \mathbb{N}^*$, with $b, d \geq 2$. Then, for all $a \in \mathbb{N}^*$ we have*

$$\phi_{d\times b}(a) = \phi_d(a) + d \times \phi_b(C_d(a)). \tag{4}$$

Proof. Is direct, by applyIng C_d to the numerical value of $C_b(a)$, together with the definitions of C_d and $C_{d\times b}$ and the properties (e1) and (e3) of Theorem 2, respectively. ■

Remark 5 *Our first observation is that, once established the expression (4), one can proof the validity of the expression (5), given below:*

$$\phi_{d\times b}(a) = \phi_b(a) + b \times \phi_d(C_b(a)), \tag{5}$$

for all $a \in \mathbb{N}^*$. With the folowing Theorem, we reached our first goal.

Theorem 6 *Let b_1 and b_2 relatively prime in $\mathbb{N}$, with $b_1, b_2 \geq 2$. Let γ and β be two arbitaries numbers such that $\gamma \in \{0, 1, \ldots, (b_1 - 1)\}$ and $\beta \in \{0, 1, \ldots, (b_2 - 1)\}$, respectively. Then we can find in the set $\{0, 1, 2 \ldots, (b_1 \times b_2 - 1)\}$ one unique element $\widehat{a}$ that sastifies the system:*

$$\begin{cases} \phi_{b_1}(a) = \gamma, \\ \\ \phi_{b_2}(a) = \beta. \end{cases} \tag{6}$$

Proof. We shall prove that the expression $\hat{a} = \phi_{b_1\times b_2}(a)$, obtained from the expression (4)-(5), is the unique solution of system (6) in $\{0, 1, 2 \ldots, (b_1 \times b_2 - 1)\}$. In fact, we first note that $\hat{a} = \phi_{b_1\times b_2}(a) < b_1 \times b_2$, follows direct by the definition of $\phi_{b_1\times b_2}(a)$. By applying ϕ_{b_1} to $\hat{a}$, plus the aid of Theorem 4, we get $\phi_{b_1}(\hat{a}) = \phi_{b_1}(\phi_{b_1\times b_2}(a)) = \phi_{b_1}(\phi_{b_1}(a) + b_1 \times \phi_{b_2}(C_{b_1}(a)))$.

Then by property (ϕ4) of Thoerem 2 we find that

$$\phi_{b_1}(\hat{a}) = \phi_{b_1}(\,\phi_{b_1}(\phi_{b_1}(a)) + \underbrace{\phi_{b_1}(b_1 \times \phi_{b_2}(C_{b_1}(a)))}_{0}\,).$$

So, by the property (ϕ2), from this equality we get: $\phi_{b_1}(\hat{a}) = \phi_{b_1}(\phi_{b_1}(\phi_{b_1}(a))))$.

Finally, the property (ϕ3) yields to $\phi_{b_1}(\hat{a}) = \phi_{b_1}(\phi_{b_1}(a)) = \phi_{b_1}(a) = \gamma$. Now, by a similar argument, and the aid of the Remark 5, we get the other identity, that is, $\phi_{b_2}(\hat{a}) = \phi_{b_2}(\phi_{b_1\times b_2}(a)) = \phi_{b_2}(a) = \beta$. The uniqueness is direct. Now, it is important to count in applications with an explicit version of the solution. Indeed, if the min $\{b_1, b_2\} = b_1$ the solution $\widehat{a}$ of (6) can be expressed by the formula:

$$\widehat{a} = \beta + b_2 \times \phi_{b_1}\left\{i_{b_1}(\phi_{b_1}(b_2)) \times [\gamma + \beta \times (b_1 - 1)]\right\}. \tag{7}$$

If on the contrary, $\min\{b_1, b_2\} = b_2$ the the form explicit to consider, it must be

$$\widehat{a} = \gamma + b_1 \times \phi_{b_2}\left\{i_{b_2}\left(\phi_{b_2}(b_1)\right) \times [\beta + \gamma \times (b_2 - 1)]\right\}. \tag{8}$$

Note that in both expressions (7) and (8), it is important the fact that both b_1 and b_2 are relatively prime in $\mathbb{N}$. ■

Finally, we shall given an example to ilustrate an application of (7) and (8). In fact, for the system

$$\begin{cases} \phi_7(a) = 3, \\ \\ \phi_9(a) = 7. \end{cases} \tag{9}$$

all the conditions of Theorem 6 can be verified. Now, $\min\{7, 9\} = 7$. As we have already mentioned, the expression (7) yields to: $\widehat{a} = 7 + 9 \times \phi_7\left\{i_7\left(\phi_7(9)\right) \times [3 + 7 \times (7-1)]\right\}$. Hence,

$$\widehat{a} = 7 + 9 \times \phi_7\left\{i_7(2) \times [3 + 7 \times 6]\right\}.$$

As, $i_7(2) = 4$, we get $\widehat{a} = 7 + 9 \times \phi_7\{4 \times [3 + 7 \times 6]\}$. Now, by the properties ($\phi 5$), ($\phi 4$) and ($\phi 2$) of Theorem 2, we get $\phi_7\{4 \times [3 + 7 \times 6]\} = 5$. Therefore, $\widehat{a} = 52 < 63$.

4. About Diffractal diffusers

A diffuser is a technology to obtain a pleasant sound atmosphere for one or more receivers. To appreciate the importance of the study of this type of thecnology, see e.g., the book by Cox and D'Antonio [3]. Here, the mathematical expression vital in the building of 2-D acoustic diffractal diffuser is:

$$d_{h,k} = \left\{\left\lfloor\frac{h}{N}\right\rfloor^2 + \left\lfloor\frac{k}{N}\right\rfloor^2\right\}_{mod\,M} \times \frac{\lambda_M}{2M} + \left\{\left(h_{mod\,M}\right)^2_{mod\,N} + \left(k_{mod\,M}\right)^2_{mod\,N}\right\} \times \frac{\lambda_N}{2N}, \tag{10}$$

where $d_{h,k}$ denotes the depth of the $(h,k)^{th}$ well of 2-D diffractal, also $N \in \mathbb{N}^*$ and $M \in \mathbb{N}^*$ are two prime numbers, λ_N and λ_M denote the wavelengths of the low frequency diffusser and high frequency diffuser, respectively. In the next Theorem we give one response to the two questions posed in the introduction. Before, we need some preliminaries Lemmas.

Lemma 7 *Assume $b \in \mathbb{N}^*$, with $b \geq 2$. Then for all $a \in \mathbb{N}^*$ we have that:*

$$(I)\;\; \phi_b(a) = \begin{cases} a, & if\; 0 \leq a \leq b-1, \\ \\ (a)_{mod\,b}, & if\;\; a \geq b \end{cases} \quad and \quad (II)\;\; C_b(a) = \begin{cases} 0, & if\; 0 \leq a \leq b-1, \\ \\ \lfloor\frac{a}{b}\rfloor, & if\;\; a \geq b. \end{cases}$$

Proof. In the last section of [2], we prove that under these hypotheses the identity (I) above is valid. On the other hand, the identity (II) follows direct of the property (e2) on Theorem 2, togheter with the identity (I), the definition of $C_b(a)$ and the equality $a = (a)_{mod\,b} + b \times \lfloor\frac{a}{b}\rfloor$, as shown in [5]. ■

Of this form, by using the Lemma 7, we can rewrite the $d_{h,k}$ numbers in a transparent functional way. More precisely we have

$$d_{h,k} = \phi_M\left(C_N^2(h) + C_N^2(k)\right) \times \frac{\lambda_M}{2M} + \left\{\phi_N\left(\phi_M^2(h)\right) + \phi_N\left(\phi_M^2(k)\right)\right\} \times \frac{\lambda_N}{2N}. \tag{11}$$

Remark 8 *In (11), we have that $h, k \in \mathbb{N}^*$ are arbitraries, $C_N^2(h) = C_N(h) \times C_N(h)$, $\phi_M^2(k) = \phi_M(k) \times \phi_M(k)$, and the other variables have the same Physical significance than in (10).*

Lemma 9 *Assume $b,\, d \in \mathbb{N}^*$, with $b,\, d \geq 2$. Then for all $a \in \mathbb{N}^*$ we have*

$$\phi_d\left(\phi_b(a)\right) = \phi_d\left[\phi_b(\phi_{b\times d}(a))\right] \tag{12}$$

and

$$\phi_d\left(C_b(a)\right) = C_b\left(\phi_{b\times d}(a)\right). \tag{13}$$

Proof. The proof of (12) and (13) follows direct by of the expressions (4) and (5), and properties of Theorem 2. ∎

Theorem 10 *Assume $M,\, N \in \mathbb{N}^*$, with $M,\, N \geq 2$ are primes numbers. Then for all $h,\, k \in \mathbb{N}$ we have*

$$d_{h,k} = F_{h,k} \times \frac{\lambda_M}{2M} + G_{h,k} \times \frac{\lambda_N}{2N}, \tag{14}$$

where

$$F_{h,k} = \phi_M\left[C_N^2(\phi_{M\times N}(h)) + C_N^2(\phi_{M\times N}(k))\right]$$

and

$$G_{h,k} = \left\{\phi_N\left(\phi_M^2(\phi_{M\times N}(h))\right) + \phi_N(\phi_M^2(\phi_{M\times N}(k))\right\}.$$

Proof. The proof follows direct by of (11) and the equalities given in Lemma 9 given above, together with the properties of Theorem 2. ∎

Theorem 10, from a computational point of view, it is exactly what we need, since rather than using natural numbers h and k on direct way, the arguments of functions C_N and ϕ_M in formula (11) can be used their residual representations on $M \times N$, and so to make a fast computation on a residual number system. This is transparent, since the functional expressions $\phi_{M\times N}(h)$ and $\phi_{M\times N}(k)$ are intrisic to the $d_{h,k}$ depths. It is interesting to note that here, the CRT plays a role important to convert the computation of $\phi_N(h)$, $\phi_M(h)$, $\phi_N(k)$ and $\phi_M(k)$ into $\phi_{M\times N}(h)$ and $\phi_{M\times N}(k)$, respectively. Theorem 10 then, answers the two questions posed at the beginning of this section. Numerical simulations will be presented in a future work.

5. Conclusions

We considered a technique based on the Euclidean algorithm, where the novel scenario is their functional algorithmic character. We obtained a direct proof of Chinese Remainder Theorem, CRT, and then it is shown that the functional expressions $\phi_{M\times N}(h)$ and $\phi_{M\times N}(k)$ are intrisic to the $d_{h,k}$ depths, this is in order to accelerate their computation. Here it is a key to note that the diffuser diffractal combines the property of self-similarity of fractals [9] with uniform scattering exhibiting Schroeder diffuser, and producing an an extended bandwidth of sound. Results, related with the theoretical part of this paper, are a complement of those given in [2], which, with appropriate modifications, can be extended to general Euclidean domains. Finally, with our exposition, we seek to motivate and to familiarize the reader with this type of approach and with the technology of diffusers.

Acknowledgments

I am especially indebted to professor Dr. Pablo Martín de Julian, academic of Department of Physics at the University of Antofagasta, who went through the whole paper and recommended several valuable modifications and corrections.

Work supported by Decanatura de la Facultad de Ciencias Básicas, Universidad de Antofagasta, and Grant FONDECYT N° 1121103, Chile.

6. References

[1] Asmuth, C. A. and Bloom J., A modular approach to key safeguarding, IEEE Transactions on Information Theory IT-**29**, 208-210, (1983).

[2] Cortés–Vega L. A., Rojas-Castro D. E., Santiago–Ayala Y. S. and Rojas–Romero S. C., "A technique based on the Euclidean algorithm and its applications to Cryptography and Nonlinear Diophantine Equations"., Proyecciones. J. Math., **26**, 309–333 (2007).

[3] Cox T. J. and D'Antonio P.,"Acoustic Absorbers and Diffusers: Theory, Design and Application", 2nd ed. Abington, Oxfordshire: Taylor & Francis, (2009).

[4] Ding C., Pei D. and Salomaa A.,"Chinese remainder theorem: applications in computing, coding, cryptography, World Scientific Publishing Co., Inc., (1996).

[5] Knuth, D. E.,"The Art of Computer Programming", Addison-Wesley, (1972).

[6] Sullivan R. J.,"Microwave Radar Imaging and Advanced Concepts", 2nd ed. Scitech Pub Inc, (2004).

[7] Wang Y., Residue to binary converters based on net Chinese remainder theorems. IEEE Trans Circuits Syst II, **47**, 197-204. (2000).

[8] Fiol M.A., Congruences in Z^n, finite Abelian groups and the Chinese remainder theorem. Discrete Math. **67**,101-105 (1987) .

[9] Mandelbrot B. B.,"The fractal Geometry of Nature", Freeman, San Fransisco (1983).

[10] Niven I., Zuckerman S. H. and Montgomery H., "An introduction to the theory of numbers", 5nd ed. Jhon Wiley & Sons, Inc., (1991).

CFD Modeling of LNG Spill: Humidity Effect on Vapor Dispersion

S G Giannissi[1,2], A G Venetsanos[1] and N Markatos[2]
[1] Environmental Research Laboratory, National Center for Scientific Research Demokritos, Agia Paraskevi 15310, Greece
[2] School of Chemical Engineering, National Technical University of Athens, Heroon Polytechniou 9, Zografou, 15780, Greece

E-mail:sgiannissi@ipta.demokritos.gr

Abstract. The risks entailed by an accidental spill of Liquefied Natural Gas (LNG) should be indentified and evaluated, in order to design measures for prevention and mitigation in LNG terminals. For this purpose, simulations are considered a useful tool to study LNG spills and to understand the mechanisms that influence the vapor dispersion. In the present study, the ADREA-HF CFD code is employed to simulate the TEEX1 experiment. The experiment was carried out at the Brayton Fire Training Field, which is affiliated with the Texas A&M University system and involves LNG release and dispersion over water surface in open-obstructed environment. In the simulation the source was modeled as a two-phase jet enabling the prediction of both the vapor dispersion and the liquid pool spreading. The conservation equations for the mixture are solved along with the mass fraction for natural gas. Due to the low prevailing temperatures during the spill ambient humidity condenses and this might affect the vapor dispersion. This effect was examined in this work by solving an additional conservation equation for the water mass fraction. Two different models were tested: the hydrodynamic equilibrium model which assumes kinetic equilibrium between the phases and the non hydrodynamic equilibrium model, in order to assess the effect of slip velocity on the prediction. The slip velocity is defined as the difference between the liquid phase and the vapor phase and is calculated using the algebraic slip model. Constant droplet diameter of three different sizes and a lognormal distribution of the droplet diameter were applied and the results are discussed and compared with the measurements.

1. Introduction

The use of natural gas as fuel has increased over the past few years and its flammable nature generates the necessity of performing safety assessment, in order to indentify the risks and design measures for prevention and mitigation in case of an accident. A common practice for handling, storage and transportation of natural gas is its liquefaction at low temperatures. Therefore, it is significant to investigate the dispersion of liquefied natural gas (LNG).

A useful tool to investigate the behavior and dispersion of LNG are the CFD codes. Prior to the use of the CFD codes a validation of the code should be carried out, in order to asses its performance. The validation should be performed against several experiments.

In the present study the ADREA-HF CFD code [1] is used to simulate the TEEX1 trial of the BFTF experimental series. TEEX1 involves the release and dispersion of LNG over water surface in a confined area [2]. The ADREA-HF code has been validated against other LNG experiments [3], and against the dispersion of others gases [1] and cryogenic fluids [4] and has shown satisfactory agreement with the experiments.

During a LNG release the ambient humidity condenses due to the low prevailing temperatures. The liquid droplets that are formed might develop different velocity than the vapor phase and fall faster to the ground. The difference between the velocity of the liquid and the velocity of the vapor phase is called slip velocity. A non-hydrodynamic equilibrium model which takes into account the slip velocity is employed in the present study. Three different constant droplet diameters (50, 100 and 200 μm) and a lognormal diameter distribution with 50μm mean diameter were considered for the sensitivity study.

The computational results showed that the results with the slip model assuming a 200μm constant diameter or a lognormal distribution with mean diameter 50 μm are in better agreement with the experiment, especially at the higher sensors.

2. Experimental description

Six small and medium tests were performed at the Brayton Fire Training Field (BFTF), Texas A&M University between 2005 and 2009. During these experiments LNG was spilled above water surface with 45-degree angle in a confined area. TEEX1 test (07LNG01) is simulated here. The experimental site is depicted in Figure 1. The pit was filled with water, and 1.3m high wooden boards were erected around the pit as obstacles. 40 gas detector poles were located outside the fence with south east direction to monitor the methane concentration at several heights.

Figure 1. The experimental site [2].

During the TEEX1 the wind direction was varied from south east to south and the average velocity at 2.3 m was 1.2 m. Table 1 summarizes the release and weather conditions for TEEX1.

Table 1. A summary of the release and weather conditions.

Parameter	Value	Parameter	Value
Source diameter (m)	0.063	Pond temperature (K)	293
Spill rate (m^3/min)	0.75	Wind speed (m/s) #1	1.2
Spill duration (sec)	619	Measur. height for wind speed (m) #1	2.3
Fence length (m)	10	Wind speed (m/s) #2	1.9
Fence width (m)	6.4	Measur. height for wind speed (m) #2	10
Fence height (m) - above water surface	1.3	Wind direction (deg)	160
Relative humidity (%)	32.6	Measur. height for wind direction (m)	2.3

Ambient temperature #1 (K)	289.05	Roughness length z0 for water (m)	0.0004
Measur. height for temperature #1 (m)	2.3	Roughness length z0 (m)	0.00085
Ambient temperature #2 (K)	289.05	Friction velocity u-star (m/s)	0.053
Measur. height for temperature #2 (m)	10	Monin-Obukhov length (m)	8.3

3. Modeling set-up

3.1. Mathematical equations

The transient three dimensional mass, momentum and enthalpy conservation equations for the mixture and the mass fraction conservation equation of each component (methane and water) are solved,

$$\frac{\partial \rho}{\partial t}+\frac{\partial \rho u_i}{\partial x_i}=0 \tag{1}$$

$$\frac{\partial \rho u_i}{\partial t}+\frac{\partial \rho u_j u_i}{\partial x_j}=-\frac{\partial p}{\partial x_i}+\frac{\partial}{\partial x_j}\left((\mu+\mu_t)\left(\frac{\partial u_i}{\partial x_j}+\frac{\partial u_j}{\partial x_i}\right)\right)+\rho g_i-\delta_{3i}\frac{\partial}{\partial z}\left(\rho q_\ell w_{sl}(1-q_\ell)w_{sl}\right) \tag{2}$$

$$\frac{\partial \rho q_k}{\partial t}+\frac{\partial \rho u_j q_k}{\partial x_j}=\frac{\partial}{\partial x_j}\left(\left(\rho d+\frac{\mu_t}{Sc_t}\right)\frac{\partial q_k}{\partial x_j}\right)-\frac{\partial \rho w_{sl}\left(q_{\ell_k}-q_k q_\ell\right)}{\partial z} \tag{3}$$

$$\frac{\partial \rho H}{\partial t}+\frac{\partial \rho u_j H}{\partial x_j}=\frac{\partial}{\partial x_j}\left(\frac{\mu_t}{Pr_t}\frac{\partial H}{\partial x_j}\right)+\frac{Dp}{Dt}+\frac{\partial}{\partial z}\left[\rho q_\ell w_{sl}(H-H_\ell)\right]-q_\ell w_{sl}\left(1-\frac{\rho}{\rho_\ell}\right)\frac{\partial p}{\partial z} \tag{4}$$

For the turbulence modeling the standard k-ε model with extra buoyancy terms [1] is used. The last terms in the right side of equations (2), (3) and (4) are activated when the non-hydrodynamic equilibrium model is employed. In the hydrodynamic equilibrium model these terms are zero.

In the above equations ρ is the mixture density (kg/m^3), u is the velocity vector (m/s), P is the pressure (Pa), g is the gravity vector (m/s^2), μ and μ_t are the laminar and turbulent viscosity, respectively (kg/m/s), Pr_t is the dimensionless turbulent Prandtl number, d is the molecular diffusivity of methane to air (m^2/s), Sc_t is the dimensionless turbulent Schmidt number, H is the enthalpy, and q is the mass fraction. Prandtl number and Schmidt number are set equal to 0.72. The subscripts i, j, k denote the Cartesian i, j coordinates and the component k, respectively, the indexes ℓ, v stand for the liquid and vapor phase, respectively, and index sl stands for the slip velocity.

3.2. Initial and inflow weather conditions

Initial and inflow wind profile should be set during the simulation. To derive the velocity profile of the wind and the k, ε values the atmospheric conditions should be taken into account. The velocity, the k and the ε values can be estimated based on the Monin-Obukhov similarity theory. In the present study the Monin-obukhov length is positive and corresponds to stable conditions. The value of Monin-Obukhov length as indicated in Table 1 was used, in order to calculate the wind profile according to the following equations,

$$u(z)=\frac{u_*}{\kappa}\left[\ln\left(\frac{z}{z_0}\right)-\psi_m\left(\frac{z}{L}\right)\right] \tag{6}$$

$$\text{where } \psi_m\left(\frac{z}{L}\right) = -5\frac{z}{L} \tag{7}$$

$$k = 6u_*^2 \tag{8}$$

$$\varepsilon = \frac{u_*}{\kappa z}\left(1.24 + 4.3\frac{z}{L}\right) \tag{9}$$

In [2] it is mentioned that even though a south wind was observed at the early stage of the release as the test proceeded the wind shifted and according to the experimental concentration contours the south east direction seems to be dominant during the experiment. Therefore, a south east direction was assumed during the entire simulation.

3.3. Computational grid

The zero reference point was set on the right, leeward corner (see Figure 2). The computational domain was extended in the y-axis 9.6 m from the windward fence wall and 40m from the leeward fence wall. A symmetry plane was assumed along the x-axis and the computational domain is extended 50 m from the right wall. The domain in the z-axis was extended 6 m. A grid with total number of cells equal to 70576 was used. Refinement was applied near the release point, the fences and the ground.

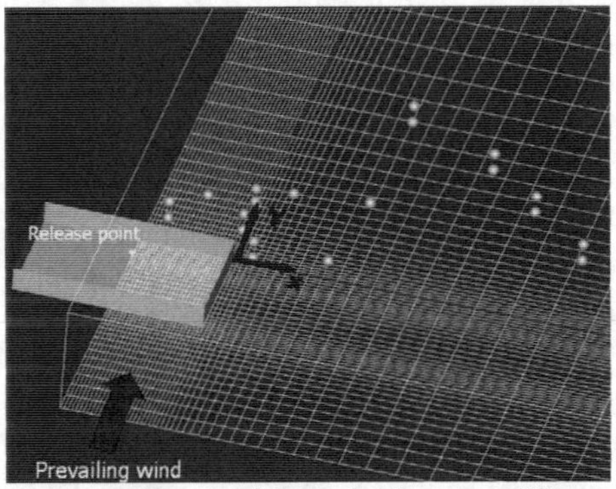

Figure 2. The test site and the computational grid on the bottom. The orange spheres represent the sensors that deployed in the simulation.

3.4. Boundary conditions

Symmetry boundary conditions were applied on symmetry plane. In the outlet boundaries a constant pressure boundary conditions was imposed, except for the north plane were zero gradient boundary condition was applied. For the components' concentration and the temperature either a zero gradient boundary condition was applied if outflow occurs or a given value boundary condition if inflow occurs at all outlet boundaries. At the bottom boundary zero gradient boundary conditions was set for all variable except for temperature. The boundary condition for the temperature inside the confined area (water substrate) was constant value equal to 293 K. The surface roughness for water was set as indicated in Table 1. At bottom boundary outside the confined area a concrete substrate was set with properties $\rho = 2371\ \mathrm{kg/m^3}$, $\lambda = 1.13$ J/m/s/K and $c_p = 880\ \mathrm{J/kg/K}$ and a roughness length equal to $8.5 \cdot 10^{-4}$. A 1-D transient temperature equation is solved inside the underground. The initial and boundary condition at the bottom boundary of the underground was set equal to 289 K.

3.5. Methane inlet conditions

The LNG was released into the pit with 45-degree angle. However, the exact elevation of the pipeline exit is not available. Based on the photograph of the experimental site (Figure 1) a height of approximately 30cm was assumed. Pure methane was assumed as natural gas and the source was modeled as a two-phase jet. A small fraction of the LNG was flashed vaporized in the pipeline, therefore, a small vapor fraction of 5% v/v was assumed at the source. Due to lack of detailed information this fraction is not certain. However, a sensitivity study with 10 and 15 % v/v flash fractions showed little effect on the results.
The measured volumetric flow rate was equal to 0.75 m^3/min. Using the liquid density the mass flow rate is estimated. The velocity at the source was estimated using the mass flow rate and the mixture density, which was calculated with the help of the flashed vapor volume fractions.

3.6. Slip model
In the presence of humidity in the atmosphere the water condenses and freezes since the prevailing temperatures (near the release point the temperature is 111.5 K) are below the water boiling point. The humidity phase change liberates heat and the cloud becomes more buoyant. The droplets that are formed can develop different velocities and might fall faster to the ground. The difference between the phases' velocity is called slip velocity.
The slip velocity is derived by,

$$w_{sl} = \frac{1}{18\mu f_{drag}}(\rho_\ell - \rho) g D^2 \tag{10}$$

To model this phenomenon a non hydrodynamic equilibrium model should be employed, which allows the phases to obtain different velocities.
According to the diameter size the flow regime can be laminar, transition or turbulent. The drag function, f_{drag}, defines the flow regime, and its value is given by [5]:

$$f_{drag} = \begin{cases} 1+0.15\mathrm{Re}_p^{0.687}, & \mathrm{Re}_p \le 1000 \\ 0.01833\mathrm{Re}_p, & \mathrm{Re}_p > 1000 \end{cases} \tag{11}$$

As humidity condenses a cloud droplet of small size, approximately 20μm is formed. The cloud droplets are so small that they are considered to follow the mean flow. Then, as more water condenses, two or more droplets collide and coalescence to form larger droplets. The larger droplets may develop different velocities due the gravitational acceleration. The droplet diameter is an unknown parameter; therefore, a sensitivity study was performed assuming a constant diameter of 50, 100 and 200μm. Furthermore, a model that assumes a lognormal droplet diameter distribution was developed and used in the present study,

$$n(D) = \frac{n_0}{D\tilde{\sigma}\sqrt{2\pi}} \exp\left(-\frac{\left(\ln D/\tilde{D}\right)^2}{2\tilde{\sigma}^2} \right) \quad \text{για } D>0 \tag{12}$$

The unknown parameters are the parameters $\tilde{D}$ and $\tilde{\sigma}$ and the total number of droplets, n_0. The dimensionless parameter $\tilde{\sigma}$ is given by,

$$\tilde{\sigma} = \ln(\sigma), \quad \sigma \in (1.5, 2.5) \tag{13}$$

A value of σ=2.0 was used here and a mean diameter of 50μm was assumed. The total number of droplets can be estimated by the mean diameter and the liquid mass fraction using the following relationship,

$$n_0 = \frac{6\rho q_\ell}{\rho_\ell \pi \bar{D}^3 \exp(3\tilde{\sigma}^2)} \tag{14}$$

Finally, in the case with the lognormal diameter distribution the slip velocity is estimated by solving numerically the following integral in order to account for all the flow regimes based on the diameter size,

Journal of Physics: Conference Series **633** (2015) 012136 doi:10.1088/1742-6596/633/1/012136

$$\rho q_{\ell} w_{sl} = \int_{0}^{\infty} N(D) \rho_{\ell} \frac{\pi}{6} D^3 w_{sl} dD \tag{15}$$

4. Results and discussion

Figure 3 compares the predictions using the hydrodynamic equilibrium model (no slip model) and the non-hydrodynamic equilibrium model (slip model) with the different droplet sizes with the experiment. Each figure corresponds to a different sensor.

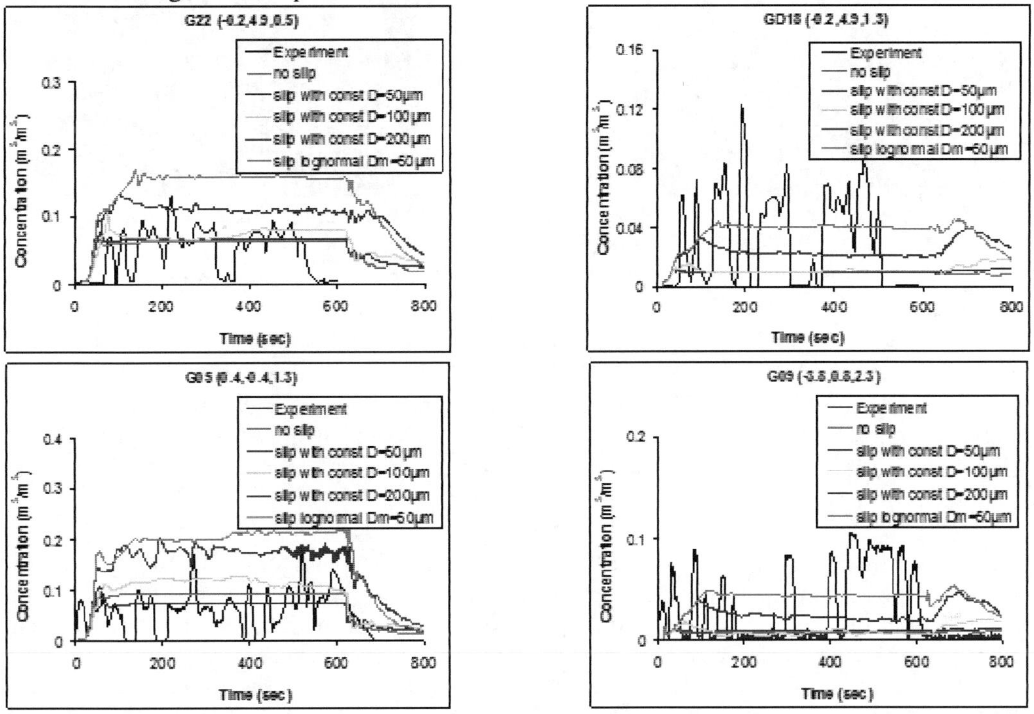

Figure 3. Methane concentration (by volume) time series compared against the experimental data.

The concentration with the hydrodynamic equilibrium model and with the slip model assuming 50μm or 100μm constant droplet diameter is under-predicted at most of the sensors. Assuming diameter of 200μm or lognormal diameter distribution (with mean diameter of 50μm) the predictions are improved and are in better agreement with the experiment at most of the sensors. At the higher sensors the predicted concentration is increased and it is closer to the measurements. This is attributed to the fact that with the slip model the heavy droplets are allowed to obtain different velocity and to fall faster to the ground, leaving a lighter cloud with positive buoyancy. In general, the predictions with the slip model assuming diameter of 200μm or assuming diameter that follows lognormal distribution fall into the range of concentration fluctuations and give less under-prediction of the peak values compared to the no slip model or the slip model assuming small droplet diameters at the higher sensors. The fluctuations in the measurements are due to the turbulence in the wind and they are not reproduced by the simulation. At the low height (0.5m) the simulation results with the slip model and the large diameter (or the lognormal diameter distribution) over-predict reasonably the concentration. This behavior is desirable for determination of the safety zones.

Figure 4 illustrates the predicted concentration contours at 0.5m height using no slip model and the slip model with lognormal diameter distribution. According to the figure the slip model predicts a narrower cloud and a less extended Lower Flammability Limit (LFL) distance. This is attributed to the fact that a more buoyant cloud is predicted and the dense gas behavior is limited.

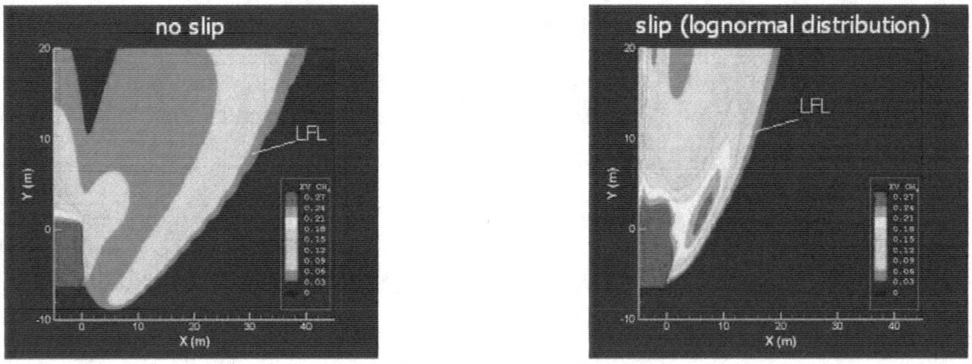

Figure 4. The predicted methane concentration contours at 0.5m height without slip model (left) and with slip model and assuming lognormal diameter distribution (right).

5. Conclusions

In the present study, the spill of liquefied natural gas (LNG) in open-obstructed environment was modeled using the ADREA-HF CFD code. The effect of ambient humidity and its condensation on the vapor dispersion was studied. Humidity condenses and liberates heat. The droplets that are formed by the humidity condensation develop different velocity from the vapor phase, fall faster to the ground and leave a lighter, more buoyant cloud. To model this phenomenon a non-hydrodynamic equilibrium model (slip model) should be employed.

A droplet size parameterization was performed assuming constant diameter of 50, 100 and 200 μm and a lognormal distribution with mean diameter 50 μm. It is shown that the results with a constant diameter of 200μm or assuming lognormal distribution are closer to the measurements compared to the smaller diameters assumption and the no slip model prediction, especially at the higher heights. They tend to slightly over-predict the concentration at the lower heights, which is desirable for determination of the safety zones. However, they still under-predict the peak concentrations at the higher sensors. The predictions assuming smaller diameter and the no slip model underestimate the peak concentrations at all sensors.

Finally, according to the contours figures the simulation without slip model predicts a wider cloud influenced by the dense gas behavior and larger LFL distance crosswind the relase.

6. References

[1] Venetsanos A G, Papanikolaou E and Bartzis J G 2010 *Int. J. Hydrogen Energy* **35** pp. 3908

[2] Cormier B. R., Qi R, Yun G, Zhang Y and Mannan M S 2009 *J. Loss Prev. Process Ind.* **22** 332

[3] Giannissi S G, Venetsanos A G, Markatos N and Bartzis j G 2013 *J. Loss Prev. Process Ind.* **26** 245

[4] Giannissi S G, Venetsanos A G, Markatos N and Bartzis j G 2014 *Int. J. Hydrogen Energy* **39** 15851

[5] L.Schiller and Z.Naumann, 1935 *Z. Ver. Deutsch. Ing.* 77:318

Modelling of the Influence of Vegetative Barrier on Concentration of PM10 from Highway

Hynek Řezníček, Luděk Beneš

Department of Technical Mathematics, FME - Czech Technical University in Prague, Karlovo nám. 293/13, 121 35 Praha 2, Czech Republic, CR

E-mail: hynek.reznicek@fs.cvut.cz, benes@marian.fsik.cvut.cz

Abstract. The influence of different types of the vegetative barrier near a highway on dustiness was studied. Transport and dispersion of pollutants PM10 emitted from the highway were numerically simulated. Mathematical model was based on the Navier-Stokes equations for turbulent fluid flow in Boussinesq approximation. The AUSM-MUSCL scheme in finite volume formulation on structured orthogonal grid was used.

1. Introduction

The pollution in the Atmospheric Boundary Layer (ABL) is increasing these days. The numerical modelling can help us to predict the transmission of the pollution and to understand its behavior ([1] and [2]). The core of this problem lies in a accurate fluid flow modelling.

The increasing car traffic on highways is one of the major problems. The concentration of pollution produced (or swirled) by the traffic is increasing as well. Therefore areas around the highways become contaminated and dangerous for human activity. Instalation of the various types of obstacles is the most widely used method for solving of this problem. Influence of the vegetative barrier near the highway on the dustiness was numerically studied in this paper. Effect of the selected parameters, especially shape and type of vegetation on the concentration of PM10 was investigated.

2. Mathematical model

The air flow in ABL is described by the 2D system of RANS (Reynolds Averaged Navier-Stokes) equations for the viscous, incompressible, turbulent and stratified flow with variable density (in general), simplified by the Boussinesq approximation.

$$
\begin{aligned}
\frac{\partial u_j}{\partial x_j} &= 0, \\
\frac{\partial \rho}{\partial t} + \frac{\partial u_j \rho}{\partial x_j} &= u_2 \frac{\partial \rho_0}{\partial x_2}, \\
\frac{\partial u_i}{\partial t} + \frac{\partial u_j u_i}{\partial x_j} + \frac{1}{\rho_0}\frac{\partial p}{\partial x_i} &= \frac{\partial}{\partial x_j}\nu\left(\frac{\partial u_i}{\partial x_j}\right) - \frac{\rho}{\rho_0} g \delta_{i2} + T_i, \\
\frac{\partial \rho C}{\partial t} + \frac{\partial (\rho u_j C)}{\partial x_j} &= \frac{\partial}{\partial x_j} D \left(\frac{\partial C}{\partial x_j}\right) + Z
\end{aligned}
\tag{1}
$$

where u_i ($i \in \{1, 2\}$) are velocity components, ρ resp. p are density and pressure perturbations, C is pollution concentration, ρ_0 is the background density field and T_i is the aerodynamic resistance of the barrier, Z is source term placed on the highway. The modelled PM10 particles are assumed as a non-hygroscopic, primary emitted contaminant which can be considered as passive scalar in the flow field.

The viscosity was composed from the molecular kinematic viscosity and the turbulent viscosity $\nu = \nu_m + \nu_T$. The turbulent viscosity is computed according to [5], [4] from

$$\nu_T = l^2 \left| \frac{\partial u_1}{\partial x_2} \right|, \qquad l = \frac{\kappa(x_2 + y_0)}{1 + \kappa \frac{(x_2+z_0)}{l_\infty}}, \qquad l_\infty = \frac{27\,|V_g|\,10^{-5}}{f}, \tag{2}$$

where κ is von Karman constant, f marks Coriolis parameter $f = 1.1 \cdot 10^{-4}\,\mathrm{rad.s}^{-1}$, z_0 is surface roughness parameter (choose as tenth of barrier height h), l_∞ means mixing length for $x_2 \to \infty$ and V_g is geostrofic velocity on the top of boundary domain.

2.1. Vegetation model

The vegetative barrier is modelled by the volume force T in momentum equations, which simulates the aerodynamic resistance caused by the vegetation

$$T_i = r_h |U| u_i, \qquad r_h(z) = \begin{cases} r\,\frac{x_2/h}{0.75} & \forall (x_2/h);\quad 0 \le x_2/h \le 0.75 \\ r\,\frac{1-x_2/h}{1-0.75} & \forall (x_2/h);\; 0.75 \le x_2/h \le 1.0, \end{cases} \tag{3}$$

here $|U|$ is the velocity magnitude and r_h is an obstructing coefficient of the vegetative barrier. The vertical profile of this parameter has been set–up as rhombus which simulate the distribution of the tree mass and its ability to obstruct the flow where $r \in \langle 0; 1 \rangle$. Different values of these constants have been tested.

3. Numerical approximation

The AUSM scheme in finite volume formulation is used for space discretization of the inviscid fluxes. Velocities on the cell faces are computed using MUSCL reconstruction according to van Leer [6] with Hemker-Koren limiter. Since the pressure is approximated by the central way, the scheme has to be stabilized by the artificial pressure diffusion introduced in [7]. The viscous fluxes are discretized in the central way on dual (diamond type) mesh. This scheme is of the second order in space. For details see [3], [8]. After the space discretization the time derivative is approximated by the robust second order BDF formula. The time step Δt can be chosen according to the problem. The artificial compressibility method in an artificial (dual) time is used. The resulting system of ODEs is solved by an explicit 3-stage Runge-Kutta method.

The validation of the model was done on case study solved by Bodnar [5], for more information see [9].

4. The numerical experiment and setup

The computational domain was 300 $\times$ 150 m large, the highway is situated in position $x_1 \in \langle 20, 45 \rangle$ m. The vegetative barrier of height $h = 15$ m is located in position $x_1 \in \langle 50, 80 \rangle$ m. Dust source was situated to the center of the highway $x_1 = 32$ m. The source term was set to $Z/\rho_0 = 1$ for all time. Some other parameters are $g = -10\ \frac{\mathrm{m}}{\mathrm{s}^2}$, kinematic viscosity of air $\nu = 10^{-5}\ \frac{\mathrm{m}^2}{\mathrm{s}}$, $\frac{\partial \rho_0}{\partial x_2} = 0$.

The structured Cartesian grid of 120 $\times$ 300 cells is equidistantly spaced in x_1-direction. The vertical step till height 50 m is exponentially diluted, above 50 m is equidistant with 2 m step. The smallest step is 0.1464334293 m high in x_2-direction (designates in experiments for clarity y-direction).

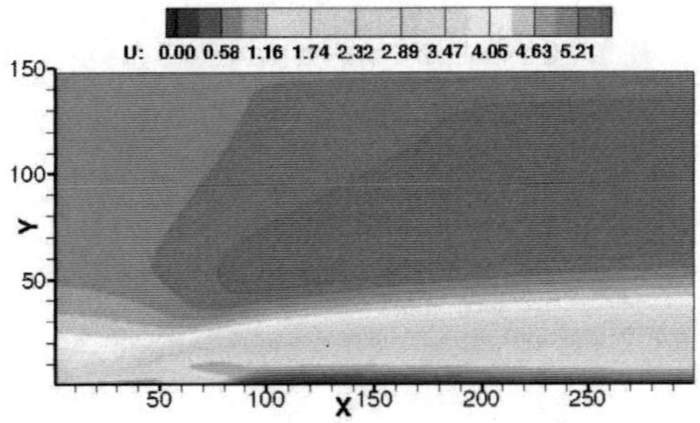

Figure 1. Vertical velocity u_1 for $r = 0.3$.

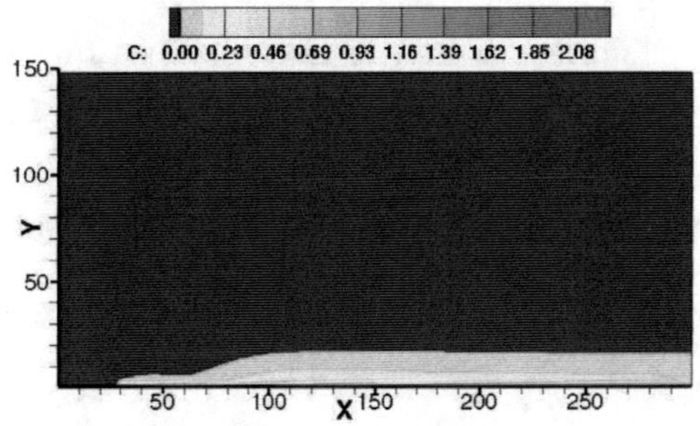

Figure 2. PM10 concentration field for $r = 0.3$.

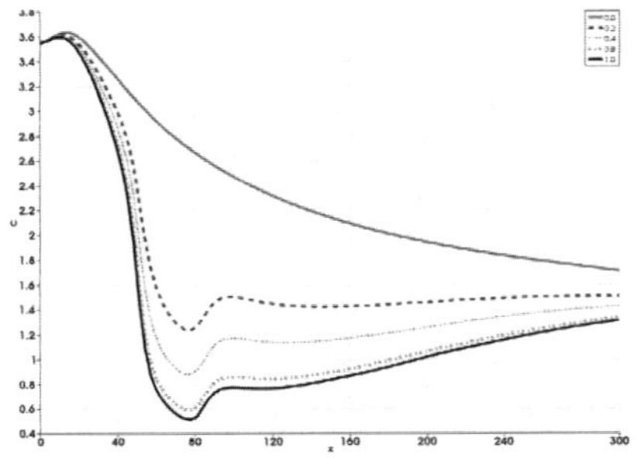

Figure 3. Vertical velocities on sectional line in 10 m height

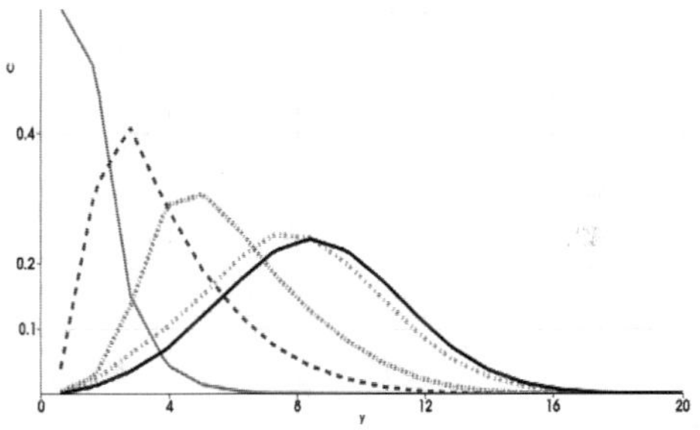

Figure 4. Vertical distribution of PM10 concentration for different r in $x = 150$ m

5. Results

The influence of the obstructing constant on the flow field and concentration was studied. Five different cases for $r = 0,\ 0.2,\ 0.4,\ 0.8,\ 1.0$ has been considered. The typical plot of u_1 velocity and PM10 concentration are shown in Figs. 1 and 2.

Dependance of the horizontal velocity component on parameter r in y=10 m is plotted in Fig. 3. The red line shows the situation, when $r = 0.0$ (without barrier). No breaking effect appeared for this line. However the velocity at the same high is decreasing. The explanation can be in boundary layer evolution, turbulent viscosity draw kinetic energy. The rapid acceleration of the flow for other r away from the barrier (behind $x = 90$ m) is caused probably by the faster stream of fluid close to the ground. These levels were not breaking so well, because the drag coefficient is lower here (see Eq. (3)).

The vertical distribution of concentration in point $x = 150$ m for different r is plotted in Fig. 4 and summarized in Tab. 1. The maxima of concentrations are decreasing with increasing of density of vegetation. The large part of the PM10 is transported in higher part of ABL due to flow field deformation caused by the vegetative barrier. So the maxima are shifted to the higher levels.

This tendency illustrate Fig. 6 where the distribution of concentration on parameter r in two heights ($y = 2$ m and $y = 10$ m) is shown. The concentration in 2 m level decrease more then 5 times till $r = 0.4$ and then remains approximately the same. On the other hand, concentration in 10m level uniformly increase.

Table 1. Values of concentrations for position $x = 150$ m depending on r and height y

r=	0.0	0.1	0.2	0.3	0.4	0.5	0.8	1.0
y=2 m	0.40	0.54	0.34	0.12	0.06	0.04	0.03	0.02
y=10 m	0.00	0.01	0.02	0.04	0.07	0.09	0.18	0.20

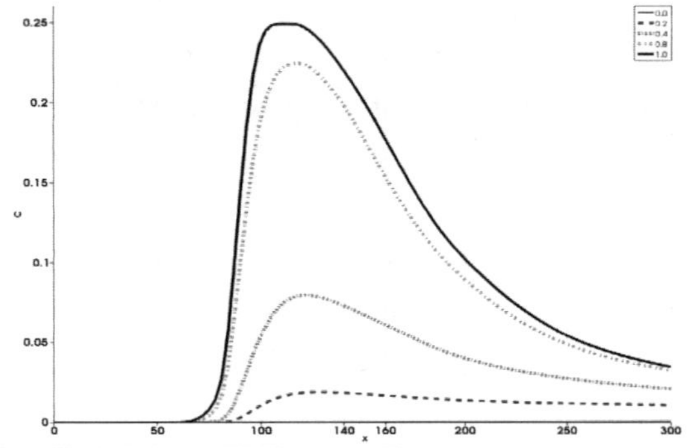

Figure 5. Horizontal distribution of the concentration for different r in $y = 10$ m

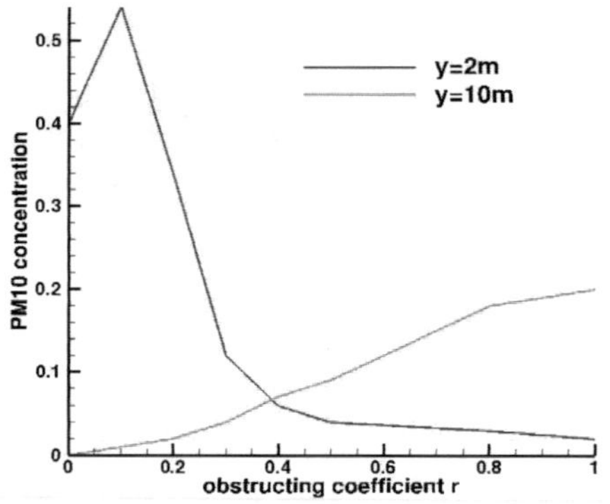

Figure 6. Dependence of the PM10 concentration on obstructing coefficient.

6. Conclusions

The numerical scheme for ABL flows with passive concentration has been developed and they has been used successfully for simulation of the flow over vegetative barrier. Several numerical results for different obstructing coefficient was obtained. The influence of the barrier density on the PM10 concentration was demonstrated.

Acknowledgments

This work was supported by grant SGS13/174/OHK2/3T/12 of the Czech Science Foundation.

References

[1] Matejiček, L., et al., 2008, "Spatio-temporal Modelling of Dust Transport over Surface Mining Areas and Neighbouring Residential Zones", *SENSORS*, Vol 8(6)
[2] Bodnár, T., Beneš, L., Kozel, K., 2008,"Numerical Simulation of Flow and Pollution Dispersion in the Area of Opencast Coal Mine", *NUMERICAL ANALYSIS AND APPLIED MATHEMATICS*, Vol: 1048
[3] Beneš, L., Fürst, J., Fraunié, Ph., 2011, "Comparison of two Numerical Methods for the Stratified Flow " , *Computers & Fluids*, Vol 46
[4] Blackadar, A.K., 1962, "The Vertical Distribution of Wind and Turbulent Exchange in a Neutral Atmosphere", *J. Geophys. Res.*, vol 67
[5] Bodnar, T., et. al, 2011, "Application of Compact Finite-Difference Schemes to Simulations of Stable Stratified Fluid Flows", *J of Applied Matematics and Computation*, Vol. 08, pp 58
[6] Van Leer, B., 1979, "Towards the Ultimate Conservative Difference Scheme. V. A Second-order Sequel to Godunov's Method",*J. of Computational Physics*, Vol. 32
[7] Dick, E., Vierendeels, J., Riemslagh, K., 1999, "A Multigrid Semi-implicit Line-method for Viscous Incompressible and Low-Mach-number Flows on High Aspects Ratio Grids, *J. of Comp. Physics*, Vol. 154
[8] Ferziger, J. H., Peric, M., 1997, *Computational Methods for Fluid Dynamics*, Springer (2nd edition)
[9] Řeznicek, H., Beneš, L., "Influence of Boundary Conditions on the Stratified Fluid Flow in Atmospheric Boundary Layer", proceeding of CMFF'15, in press.

A Fast Greedy Sparse Method of Current Sources Reconstruction for Ventricular Torsion Detection

Lu Bing[1], Shiqin Jiang[1*], Mengpei Chen[1], Chen Zhao[1], D Grönemeyer[2], B Hailer[3] and P Van Leeuwen[2]

[1] Department of Control Science and Engineering, School of Electronics and Information Engineering, Tongji University, Shanghai 201804, China
[2] Departments of Biomagnetism, Grönemeyer Institute for Microtherapy, Faculty of Health, University Witten/Herdecke, Bochum D-44799, Germany
[3] Department of Internal Medicine, Philippusstift, University Witten/Herdecke, Essen, Germany

E-mail: sqjiang@tongji.edu.cn

Abstract. A fast greedy sparse (FGS) method of cardiac equivalent current sources reconstruction is developed for non-invasive detection and quantitative analysis of individual left ventricular torsion. The cardiac magnetic field inverse problem is solved based on a distributed source model. The analysis of real 61-channel magnetocardiogram (MCG) data demonstrates that one or two dominant current source with larger strength can be identified efficiently by the FGS algorithm. Then, the left ventricle torsion during systole is examined on the basis of x, y and z coordination curves and angle change of reconstructed dominant current sources. The advantages of this method are non-invasive, visible, with higher sensitivity and resolution. It may enable the clinical detection of cardiac systolic and ejection dysfunction.

1. Introduction

Left ventricular torsion can be derived from the twisting of the heart along its long axis [1, 2]. It has been investigated by several measurement and imaging techniques, such as magnetic resonance imaging (MRI), tissue Doppler imaging by echocardiography, two-dimensional ultrasound speckle tracking imaging, speckle tracking imaging, and velocity vector imaging. There has been much interest in the analysis and description the function of the left ventricular torsion, torsion assessment of systolic and diastolic dysfunction regarding various cardiac diseases, and the knowledge for clinical diagnosis.

It is very important to evaluate the accuracy of left ventricular torsion detection. However, detection and quantitative analysis of left ventricular torsion are difficult, because of the physiological variability of ventricular torsion, e.g. the size and diameter of the left ventricular vary in a cardiac cycle, and the limitations of various measurement approaches that may show some conflicting results [2]. In previous research, most of the approaches focused on the angle and direction of left ventricular torsion [1]. At present, no gold standard of quantitative analysis has been established for the ventricular torsion detection.

In this paper, we propose a magnetic imaging approach to detect the left ventricular torsion. It is divided into three processes: acquisition of magnetocardiography (MCG) data using a multichannel superconducting quantum interference device (SQUID) system, reconstruction and imaging cardiac moving current sources through solving the cardiac magnetic field inverse problem, and detection of

* Corresponding author, e-mail: sqjiang@tongji.edu.cn.

the left ventricular torsion by means of the position coordination and angle change curves of reconstructed dominant current sources.

In general, current source reconstruction needs to solve a highly ill-posed inverse problem, in which the distributed source number is more than that of measurement points, so that an optimization method needs to be used. Since MCG maps were mainly dipolar, we assume the unknown sources are sparse. Thus, the underlying inverse problem can be expressed as finding a minimal numbers of basis vectors that represent the solution of interest, i.e., to find the l^0-based sparsest solution [3].

We developed a sparse solution of the cardiac source reconstruction based on the distributed current source model [4]. In addition, we investigated a fast greedy sparse (FGS) algorithm for solving the highly ill-posed problem of source reconstruction. Thereby, the sparse sources with larger strength can be identified efficiently. This has been demonstrated by a series of simulation experiments and the results of real MCG data.

We try to detect a twisting movement of the ventricle by means of the continuity analysis of the dominant current sources trajectories of x, y and z direction, i.e., the number of times of the position and the angle change of the dominant current sources occurring in the QRS complex of a cardiac cycle, based on a fixed coordinate system corresponding the measurement plane on the body surface.

2. Methods

The MCG inverse problem is expressed by the linear distributed source model as

$$\boldsymbol{b} = \boldsymbol{A}\boldsymbol{x} + \boldsymbol{v} \tag{1}$$

where $\boldsymbol{b}$ is a $M \times 1$ measurement signal vector, which acquired by $\boldsymbol{M}$ SQUID sensors on a measurement plane over the human thoracic surface. $\boldsymbol{x}$ is a $N \times 1$ source moment vector of Q distributed current sources. $\boldsymbol{A}$ is a $\boldsymbol{M} \times \boldsymbol{N}$ lead field matrix. $\boldsymbol{v}$ represents the measurement noise. Due to $\boldsymbol{M} \ll \boldsymbol{N}$, $\boldsymbol{A}$ is highly underdetermined, so that equation (1) has no unique solution. It is common to search for an optimal $\boldsymbol{x}$ by reducing the number of unknowns or constraining dipole orientations.

We developed a fast greedy sparse method of current source reconstruction and achieved a sparse solution on the basis of the sparse decomposition theory [5]:

$$\boldsymbol{x}^n_{\Gamma^n} = \boldsymbol{A}^{\dagger}_{\Gamma^n}\boldsymbol{b} \tag{2}$$

where x_{Γ^n} is a sub-vector of $\boldsymbol{x}, \boldsymbol{\Gamma}^n = \{i: \boldsymbol{x}_i \neq 0\}$. n is the iteration number. $\boldsymbol{A}_{\Gamma^n}$ is a sub-matrix of $\boldsymbol{A}$. $\boldsymbol{A}^{\dagger}_{\Gamma^n}$ is the Moore-Penrose inverse. The directional optimization of the source moment vector can be obtained [6]:

$$\boldsymbol{x}^n = \boldsymbol{x}^{n-1} + a^n \boldsymbol{d}_{\Gamma^n} \tag{3}$$

where a^n is the step-size that can be determined explicitly. $\boldsymbol{d}_{\Gamma^n}$ denotes an approximate conjugate gradient pursuit to optimize the source direction.

We adopted the strategy of selecting several elements per iteration, calculating the depth of distributed sources based on a hierarchical source space and simultaneously improving the time-consuming and accuracy of sparse source reconstruction based on a priori knowledge of the MCG map [5, 6], thereby significantly decreasing the computational complexity. In addition, we chose the reconstructed sparse sources whose strength ≥ 90% of the maximum at any instant as dominant equivalent current sources with the aim of reducing the number of sources under consideration, which may be caused by the measurement noise or the cases that can be neglected here.

3. Current source reconstruction

The fast greedy sparse (FGS) algorithm is tested by real MCG data for equivalent current source reconstruction. As shown in figure 1, MCG data were recorded by a 61 channel biomagnetometer (Magnes1300C, 4D Neuroimaging, San Diego, USA) inside a standard 3-layered magnetically shielded room (AK3b, Vacuumschmelze, Hanau, Germany). Over a circular area of 824 cm² with a sensor spacing of ca. 38 mm, the signals were measured for 3 minutes in each subject, at a sampling rate of 1 kHz with a 200 Hz low pass filter.

The sparse current sources were reconstructed by the FGS algorithm using MCG data during QRS complex. The x, y and z coordination and angle change curves of reconstructed dominant current sources from a healthy subject and a CAD patient are shown in figure 2, where the time interval of (t2-t1) and (t2-Qon) are shown by red and blue broken lines, respectively. The reconstructed source trajectories start at Q-wave onset and end at the Jpoint.

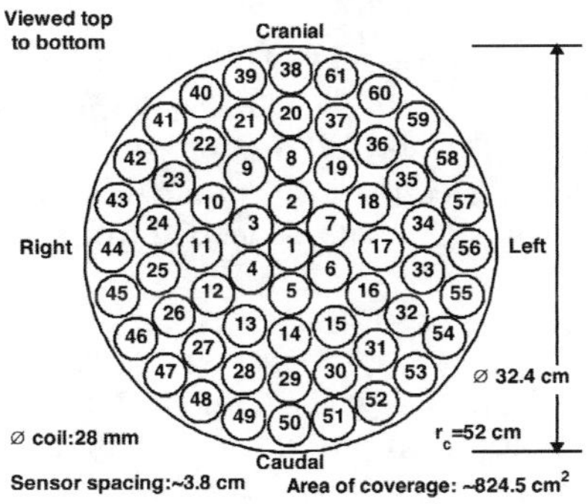

Figure 1. A circular measurement area with 61 sensors

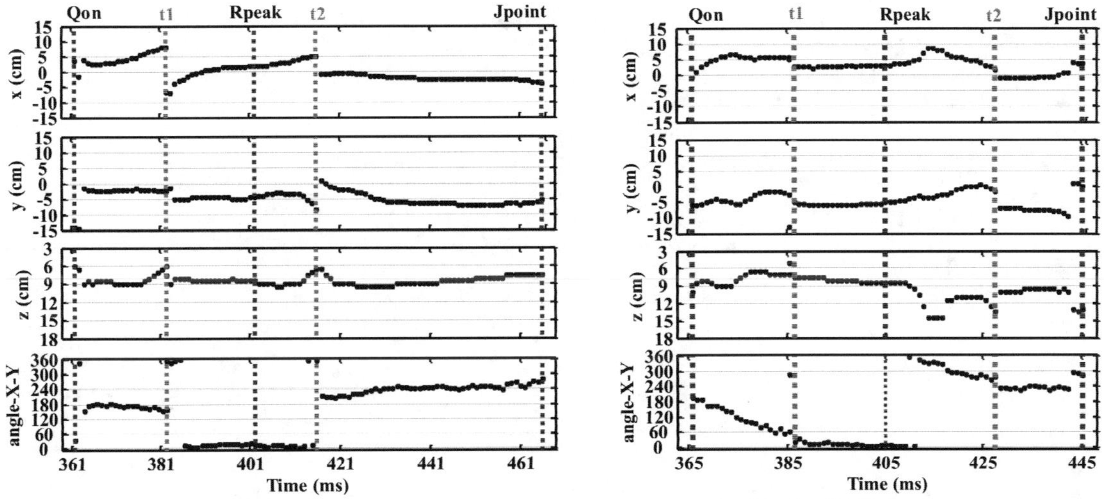

(a) the healthy subject (b) the CAD patient

Figure 2. The reconstructed spatial source trajectories during QRS complex from the healthy subject (a) and the CAD patient (b) The time interval of (t2-t1) and (t2-Qon) are shown by red and blue broken lines, respectively.

4. Detection of the left ventricular torsion

The trajectories permit the calculation of the time interval between two obvious position changes occurring before and after the R peak in a cardiac cycle indicated by (t2-t1), and a second time interval starting at Q-wave onset and the second position change (t2-Qon), which is the time, about 40-60 ms, that the mechanical motion falls behind electrical activities of the atrium and the ventricle of the heart.

Figure 2 shows the time intervals of t2-t1 and t2-Qon of the CAD patient are a little longer than that of the healthy subject. The results of detection the left ventricular torsion from 39 healthy subjects and 15 coronary artery disease (CAD) patients showed that source trajectories of 35 healthy subjects and 14 CAD patients displayed at least two position changes in x (or y) direction during the QRS complex, where the time interval between the two obvious position changes of the 33 healthy subjects was 12 to 35ms, and 10 CAD patients was 16 to 46ms. 26 healthy subjects had position changes between 41 and 61ms after the beginning of the QRS complex, and 10 CAD patients were 48 to 64ms.

Because of the given source space resolution in z-axis different from that in x and y-axes, the trajectory change in z direction also has a little different. We chose two better examples of reconstructed source trajectories to show in figure 2. Namely, in some results with poor reconstructions may be difficult to accurately detect and determine the time of twisting of the heart.

5. Conclusion

We developed a fast greedy sparse (FGS) method and it has been tested for non-invasive cardiac equivalent current source reconstruction. The left ventricle torsion during systole is detected and quantitatively analyzed on the basis of position curves and angle change of reconstructed dominant current sources. The advantages of this method are noninvasive, visible, with higher sensitivity and resolution. It may enable the clinical detection of cardiac systolic and ejection dysfunction. The statistical analysis with a number of real MCG data is needed to fully validate the effectiveness of this technique.

Acknowledgments

Research supported by the National Natural Science Foundation of China (60771030), the National High-Technology Research and Development Program of China (2008AA02Z308), the Shanghai Science and Technology Development Foundation (08JC1421800), the Open Project of State Key Laboratory of Function Materials for Information (Shanghai Institute of Microsystem and Information Technology, Chinese Academy of Sciences).

References

[1] Sengupta P P, Tajik A J, Chandrasekaran K and Khandheria B K 2008 Twist Mechanics of the Left Ventricle Principles and Application *JACC Cardiovasc. Imag.* **1** 366–376
[2] Shaw S M, Fox D J and Williams S G 2008 The development of left ventricular torsion and its clinical relevance *Int. J. Cardiol.* **130** 319–25
[3] Mohimani G H, Babaie-Zadeh and Christian J 2009 A fast approach for overcomplete sparse decomposition based on smoothed l^0 norm *IEEE T. Signal Proces.* **57** 289–301
[4] Bing L, Wang W-Y, Wang Y-L and Jiang S-Q 2013 MCG source reconstruction based on greedy sparse method *Acta Phys. Sin.* **62** 118703
[5] Rao B D and Kreutz-Delgado K 1999 An affine scaling methodology for best basis selection *IEEE T. Signal Proces.* **47** 187–200
[6] Blumensath T, Davis M E 2008 Gradient pursuits *IEEE T. Signal Proces.* **56** 2370–82

Size Effects in Elastic Thin Plates

R.J.Zhang

Institute of Applied Mechanics
Tongji University, Shanghai 200092, China

E-mail: zhangrj@tongji.edu.cn

Abstract. The couple stress theory is applied to describe the size effect in elastic thin plates. An extended equation is developed which is available for the bending problem of the plate whose thickness is close to the material length. A comparison with micropolar theory of plates is made.

1. Introduction

In elasticity, $\omega_{ij} = \left(u_{j,i} - u_{i,j}\right)/2$ and $\theta_k = e_{kij}\omega_{ij}/2$ are called, respectively, the rotation tensor and rotation vector. However, the two rotations do not play role in the classical elasticity actually. Mindlin [1] discussed the effect of the rotation gradient on deformation and indicated that the rotation gradient gives rise to bending and twisting of an element. Then, a concept of couple stress was introduced, which acts on the element and produce the bending and twisting. A linear relationship between the couple stress and the curvature is assumed. The proportional coefficient is called the modulus of curvature. Thus, a material constant is introduced into the equations of elasticity, which is a length equal to the square root of the ratio of the modulus of curvature and the modulus of shear. Since the presence of the material length as an intrinsic parameter in the equations of elasticity one can analyze the size effects in deformation problem.

In the classical theory of thin plates, the effect of the rotation gradient on deformation is also neglected. The contribution of the couple stresses on the moment resultant is not taken into account. Of course, the classical equation for thin plates can not be use to describe the size effects.

In the present paper an extended theory for thin plates is developed. In the theory, the bending and twisting of element are allowable, the curvatures of the bending and twisting are considered to be associated with couple stresses. Total moment resultants consist of two parts, one of which is the resultant due to stresses, and the other is that due to couple stresses. In extended equations the material length is explicitly included in the bending stiffness of plates as a size effect term. The presence of the size effect term indicates that the bending stiffness of plates significantly increases and is larger than the classical theory of plates suggests as the thickness of plates is close to the material length. This coincides with the observation in lots of experiments. As a special case, when the thickness of plates is greatly larger than the material length, the extended equation is reduced to the classical equation of plates.

2. Curvature and Couple Stress

The gradient of the rotations θ_x and θ_y results in the curvatures, which are measured by the curvatures tensor $\underset{\sim}{\chi}$ in the form

$$\chi_{xx} = \frac{\partial^2 w}{\partial x \partial y}, \quad \chi_{xy} = \frac{\partial^2 w}{\partial y^2}, \quad \chi_{yx} = -\frac{\partial^2 w}{\partial x^2}, \quad \chi_{yy} = -\frac{\partial^2 w}{\partial x \partial y} \tag{1}$$

The quantity energetically conjugated with curvature $\underset{\sim}{\chi}$ is couple-stress $\underset{\sim}{\mu}$. The isotropic constitutive equations between $\underset{\sim}{\mu}$ and $\underset{\sim}{\chi}$ are given by Eringen in [3] and Nowacki in [4] as

$$\mu_{ij} = \alpha \delta_{ij} \chi_{kk} + (\beta + \gamma) \chi_{ij} + (\beta - \gamma) \chi_{ji} \tag{2}$$

where the constants β and γ are the curvature modulus. Substituting (1) into (2) yields

$$\begin{aligned} \mu_{xx} &= 2\beta \frac{\partial^2 w}{\partial x \partial y}, \quad \mu_{xy} = (\beta + \gamma) \frac{\partial^2 w}{\partial y^2} - (\beta - \gamma) \frac{\partial^2 w}{\partial x^2} \\ \mu_{yx} &= -(\beta + \gamma) \frac{\partial^2 w}{\partial x^2} + (\beta - \gamma) \frac{\partial^2 w}{\partial y^2}, \quad \mu_{yy} = -2\beta \frac{\partial^2 w}{\partial x \partial y} \end{aligned} \tag{3}$$

3. Moment Resultant

The moment resultants produced by stresses are:

$$\begin{aligned} M_x^\sigma &= \int_{-h/2}^{h/2} \sigma_x z dz = -D\left(\frac{\partial^2 w}{\partial x^2} + \nu \frac{\partial^2 w}{\partial y^2} \right), \quad M_y^\sigma = \int_{-h/2}^{h/2} \sigma_y z dz = -D\left(\frac{\partial^2 w}{\partial y^2} + \nu \frac{\partial^2 w}{\partial x^2} \right) \\ M_{xy}^\sigma &= M_{yx}^\sigma = \int_{-h/2}^{h/2} \tau_S z dz = -D(1-\nu) \frac{\partial^2 w}{\partial x \partial y} \end{aligned} \tag{4}$$

where

$$\tau_S = \frac{1}{2}\left(\tau_{xy} + \tau_{yx} \right) = G\varepsilon_{xy} = -\frac{Ez}{1+\nu} \frac{\partial^2 w}{\partial x \partial y} \quad \text{and} \quad D = \frac{Eh^3}{12(1-\nu^2)} \tag{5}$$

in which D is the bending rigidity of plates, G is the modulus of shear, E is Young's modulus, ν is Poisson's ratio, respectively.

Additionally, the moment resultants produced by couple stresses are:

$$\begin{aligned} M_x^\mu &= \int_{-h/2}^{h/2} \mu_{yx} dz = h\left[-(\beta + \gamma) \frac{\partial^2 w}{\partial x^2} + (\beta - \gamma) \frac{\partial^2 w}{\partial y^2} \right] \\ M_y^\mu &= \int_{-h/2}^{h/2} \mu_{xy} dz = h\left[(\beta + \gamma) \frac{\partial^2 w}{\partial y^2} - (\beta - \gamma) \frac{\partial^2 w}{\partial x^2} \right] \\ M_{xy}^\mu &= \int_{-h/2}^{h/2} \mu_{xx} dz = 2\beta h \frac{\partial^2 w}{\partial x \partial y}, \quad M_{yx}^\mu = \int_{-h/2}^{h/2} \mu_{yy} dz = -2\beta h \frac{\partial^2 w}{\partial x \partial y} \end{aligned} \tag{6}$$

Then the total moment resultants are

$$M_x = M_x^\sigma + M_x^\mu, \quad M_y = M_y^\sigma - M_y^\mu, \quad M_{xy} = M_{xy}^\sigma - M_{xy}^\mu, \quad M_{yx} = M_{yx}^\sigma + M_{yx}^\mu \tag{7}$$

4. Strain energy and equilibrium equation

The strain energy of a plate can be written as

$$U = U^{\sigma} + U^{\mu} \tag{8}$$

where

$$U^{\sigma} = \frac{1}{2}\iiint\left(\sigma_x\varepsilon_x + \sigma_y\varepsilon_y + 2\tau_S\varepsilon_{xy}\right)dxdydz \tag{9}$$

and

$$U^{\mu} = \frac{1}{2}\iiint\left(\mu_{xx}\chi_{xx} + \mu_{yy}\chi_{yy} + \mu_{xy}\chi_{xy} + \mu_{yx}\chi_{yx}\right)dxdydz \tag{10}$$

By defining

$$\{\kappa\} = \left[-\frac{\partial^2 w}{\partial x^2}, -\frac{\partial^2 w}{\partial y^2}, -\frac{\partial^2 w}{\partial x \partial y}, -\frac{\partial^2 w}{\partial x \partial y}\right]^T \tag{11}$$

the strain energy (9) can then be written in the matrix form:

$$U^{\sigma} = \frac{1}{2}\iiint\{\varepsilon\}^T\{\sigma\}\,dxdydz = \frac{1}{2}\iint\{\kappa\}^T\left[D^{\sigma}\right]\{\kappa\}\,dxdy \tag{12}$$

where

$$\left[D^{\sigma}\right] = \frac{Eh^3}{12\left(1-\nu^2\right)}\begin{bmatrix}1 & \nu & 0 & 0\\ \nu & 1 & 0 & 0\\ 0 & 0 & 1-\nu & 0\\ 0 & 0 & 0 & 1-\nu\end{bmatrix} \tag{13}$$

By defining

$$\{\chi\} = \left[\chi_{yx}, \chi_{xy}, \chi_{xx}, \chi_{yy}\right]^T, \{\mu\} = \left[\mu_{yx}, \mu_{xy}, \mu_{xx}, \mu_{yy}\right]^T \tag{14}$$

the strain energy (10) can then be written in the matrix form:

$$U^{\mu} = \frac{1}{2}\iiint\left[\{\chi\}^T\{\mu\}\right]dxdydz = \frac{1}{2}\iint\{\kappa\}^T\left[D^{\mu}\right]\{\kappa\}\,dxdy \tag{15}$$

where

$$\left[D^{\mu}\right] = \frac{2h\beta}{1-\nu}\begin{bmatrix}1 & \nu & 0 & 0\\ \nu & 1 & 0 & 0\\ 0 & 0 & 1-\nu & 0\\ 0 & 0 & 0 & 1-\nu\end{bmatrix} \tag{16}$$

Here, the relation $\frac{\gamma-\beta}{\beta+\gamma} = \nu$ is used. Finally, the total stiffness matrix is

$$[D] = \left[D^{\sigma}\right] + \left[D^{\mu}\right] = \left[\frac{Eh^3}{12\left(1-\nu^2\right)} + \frac{2h\beta}{1-\nu}\right]\begin{bmatrix}1 & \nu & 0 & 0\\ \nu & 1 & 0 & 0\\ 0 & 0 & 1-\nu & 0\\ 0 & 0 & 0 & 1-\nu\end{bmatrix} \tag{17}$$

It may be seen that the ratio of the curvature modulus β and the shear modulus G has the dimensions of the square of a length such that

$$\beta = Gl^2 \tag{18}$$

where l is called as material length on which the influence of couple stresses depends strongly.

By substituting (18) into (17) the total stiffness matrix finally becomes

$$[D] = D\left[1+12\left(\frac{l}{h}\right)^2\right]\begin{bmatrix} 1 & \nu & 0 & 0 \\ \nu & 1 & 0 & 0 \\ 0 & 0 & 1-\nu & 0 \\ 0 & 0 & 0 & 1-\nu \end{bmatrix} \tag{19}$$

The total strain energy is written as

$$U = \frac{1}{2}\iint \{\kappa\}^T [D]\{\kappa\}\,dxdy \tag{20}$$

It may be seen that the bending stiffness of plates increases from D to $D\left[1+12\left(\frac{l}{h}\right)^2\right]$ if the local curvature of materials is taken into account. The term $12(l/h)^2$ describes the size effect.

Sequentially, the classical equilibrium equation is extended to the new form:

$$D\left[1+12\left(\frac{l}{h}\right)^2\right]\nabla^2\nabla^2 w = p(x,y) \tag{21}$$

It may be seen that the stiffness in the equilibrium equation (21) is identical with the stiffness in the strain energy (19).

5. Conclusions

It can be seen that taking couple stresses into account results in a term $12(l/h)^2$ arising in the stiffness of plates which expresses the size effect. Obviously, when the thickness of plates h is greatly larger than the material length l, equation (21) reduces to the classical equation of plates. Contrarily, the size effect term $12(l/h)^2$ even becomes a leading term in the stiffness of plates. It is well known that the material length of most metals is of the order of micrometers. This indicates that the proposed equation (21) in the present paper is applicable for the metal plates of micrometer scale.

Acknowledgements

Supporting by the China National Science Foundation (Grant No. 11072178 and 11172214) is gratefully acknowledged.

References

[1] Mindlin R D 1963 Influence of Couple-stresses on Stress Concentrations Experimental Mechanics Vol.3 pp 1-7
[2] Timoshenko S P, Woinowsky-Krieger S 1959 Theory of Plates and Shells 2d ed. (New York: McGraw-Hill)
[3] Eringen A C 1999 Microcontinuum Field Theory (Springer)
[4] Nowacki W 1986 Theory of asymmetric Elasticity (Pergamon press)
[5] Gauthler R D, Jahsman W E 1975 A Quest for micropolar elastic constants Journal of Applied Mechanics pp369-373
[6] Zhang R J 2010 Size effects in Kirchhoff flexible rods Physical Review E Vol.81 056601

4th International Conference on Mathematical Modeling in Physical Sciences (IC-MSquare2015) IOP Publishing
Journal of Physics: Conference Series **633** (2015) 012140 doi:10.1088/1742-6596/633/1/012140

Hierarchical Bayesian approach for estimating physical properties in spiral galaxies: Age Maps for M74

M. Carmen Sánchez Gil[1,3], Angel Berihuete[1,2], Emilio J. Alfaro[3], Enrique Pérez[3] and Luis M. Sarro[2,4]

[1] Dpt. Statistics and Operations Research, UCA, Avd. Universidad s/n, Jerez de la Frontera, Spain
[2] Spanish Virtual Observatory
[3] Instituto de Astrofísica de Andalucía (CSIC), E18008, Granada, Spain
[4] Dpt. Inteligencia Artificial, ETSI Informática, UNED, Juan del Rosal, 6, E-28040, Madrid, Spain

E-mail: mcarmen.sanchez@uca.es

Abstract. One of the fundamental goals of modern Astronomy is to estimate the physical parameters of galaxies from images in different spectral bands. We present a hierarchical Bayesian model for obtaining age maps from images in the Hα line (taken with Taurus Tunable Filter (TTF)), ultraviolet band (far UV or FUV, from GALEX) and infrared bands (24, 70 and 160 microns (μm), from Spitzer). As shown in [1], we present the burst ages for young stellar populations in the nearby and nearly face on galaxy M74.

As it is shown in the previous work, the Hα to FUV flux ratio gives a good relative indicator of very recent star formation history (SFH). As a nascent star-forming region evolves, the Hα line emission declines earlier than the UV continuum, leading to a decrease in the HαFUV ratio. Through a specific star-forming galaxy model (Starburst 99, SB99), we can obtain the corresponding theoretical ratio Hα / FUV to compare with our observed flux ratios, and thus to estimate the ages of the observed regions.

Due to the nature of the problem, it is necessary to propose a model of high complexity to take into account the mean uncertainties, and the interrelationship between parameters when the Hα / FUV flux ratio mentioned above is obtained. To address the complexity of the model, we propose a Bayesian hierarchical model, where a joint probability distribution is defined to determine the parameters (age, metallicity, IMF), from the observed data, in this case the observed flux ratios Hα / FUV. The joint distribution of the parameters is described through an i.i.d. (independent and identically distributed random variables), generated through MCMC (Markov Chain Monte Carlo) techniques.

1. Introduction

The study of the star formation history (SFH) and star formation rate (SFR) in galaxies provide vital information on the evolutionary properties of galaxies and the physical processes which drive that evolution. The variation in star formation (SF) activity across galaxies of different types is influenced by many factors, including gas content, mass and dynamical environment ([2]). But even galaxies with the same morphological type show a diversity in SF and enrichment histories (e.g. [3]). Some SFR measurements using Hα emission suggest episodic bursts of SF, rather than continuous SF ([4]). Trends in SFR and SFH across galaxy types allow predictions of galactic evolution with cosmic lookback time. In those galaxies with resolved stars it is possible to determine the SFH from analysis of the color-magnitude diagrams in regions with enough stars to reach sensible results. However individual stars are only resolved in the closest galaxies, so global SF properties of galaxies are mostly obtained by integrated light measurements. The most common diagnostic methods use measurements in the ultra-violet (UV), far-infrared (FIR), or nebular recombination lines ([2]). Of particular interest for this work are measurements in the UV and the Hα recombination line.

One way to deal with the study of the SFH in galaxies is to obtain the age map of the galaxy. In this paper we propose to study the spatial distribution of SFH in galaxies on a pixel by pixel basis, without any a priori hypothesis about the extent and spatial delineation of the star forming regions. Ages are obtained from comparison between Hα and UV emission, so this method is only sensitive to exploring the age map of the youngest stellar population, less than about 10 Myr. This age map should provide a global vision of the current star formation processes taking place in the galactic disk, the maximum scale of coherent star formation, and its relation with other large scale processes such as density waves.

This work is part of a larger study of a sample of nearby galaxies, where star forming regions are spatially resolved, in order to place the relationship between star formation, ultraviolet and Hα emission on a stronger empirical foundation ([1]; hereinafter Paper I).

The comparison of UV and Hα fluxes is used as a tracer of the recent star formation history. While the infrared images are used to correct for extinction via the TIR to FUV ratio. This flux ratio is defined as a robust and universal tracer of dust extinction ([5]; [6]; [7]). The measured Hα/FUV ratios are compared with the stellar population synthesis (SPS) models to study the dynamics of stellar populations of different ages within the galaxies; we use Starburst99, SB99 ([8]; [9])[1]. Model flux ratios are compared to the observed values for each pixel in the images, resulting in two dimensional spatial *age maps*.

We address the problem of deriving the galaxy age map from Hα/FUV flux ratio images by establishing a probabilistic framework which explain the relationship of the random variables involved in the problem.

This relationship will be formulated in terms of a joint probability distribution given the observations and their uncertainties. We define this joint probability distribution in terms of a hierarchical Bayesian model (HBM) and use Markov Change Monte Carlo (MCMC) to describe it (Nested Sampling)

In order to set up the HBM we need to generate a *model flux ratio* [2] from SB99 model and compare it with the observations by using a likelihood. Finally, we explore the parameters of interest (age) marginalizing nuisance parameters in the posterior distribution defined by HBM model.

2. Observations

The Hα image was taken with the Taurus Tunable Filter (TTF; [10]) on the William Herschel Telescope (WHT) on 1999 March 4–6. Conditions were photometric with stable seeing of 1.0 arcsec. TTF was tuned to a bandpass of width $\Delta\lambda = 20\,\text{Å}$ centred at $\lambda_c = 6570\,\text{Å}$. The intermediate-width R0 blocking filter ($\lambda_c/\Delta\lambda = 6680/210\,\text{Å}$) was used to remove transmissions from all but a single interference order. The pixel scale is 0.56 arcsec. More details on the data and data reduction can be found in Paper I.

[1] http://www.stsci.edu/science/starburst99/

[2] We prefer use the term *model* instead than *true* because its value depends on the assumption that SB99 is an appropriate model

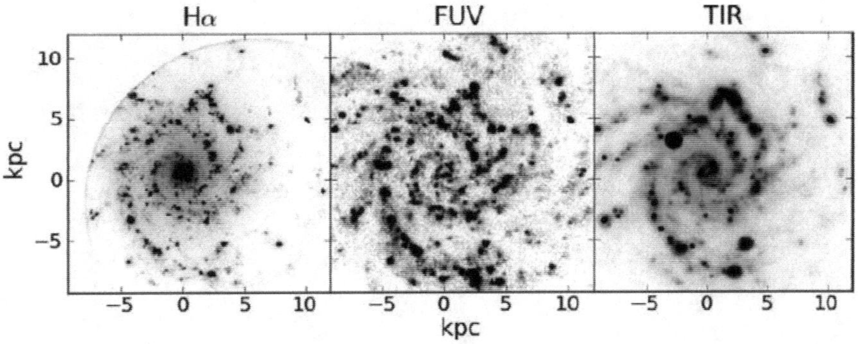

Figure 1. Examples of processed frames in Hα (left), far ultraviolet (FUV; centre), and total infrared (TIR; right) for the galaxy M74. The images have been resampled to have identical size, orientation, and pixel scale (1.5"/pix).

The UV image comes from the Nearby Galaxies Survey of the Galaxy Evolution Explorer mission (NGS survey, GALEX, [11]). This survey contains well-resolved imaging (1.5 arcsec pix) of 296 and 433 nearby galaxies for GR2/GR3 and GR4 releases, respectively, in two passbands: a narrower far-ultraviolet band (FUV; $\lambda_{\rm eff}/\Delta\lambda = 1516/268\,\text{Å}$), and a broader near-ultraviolet band (NUV; $\lambda_{\rm eff} = 2267/732\,\text{Å}$).

Archival *Spitzer*[3] images, at 24, 70, and 160μ, were used to provide additional estimates of extinction. They were resampled to a common 1.5 arcsec/px scale and combined into an image of TIR flux, $F_{TIR} = \zeta_1 \nu F_\nu(24\mu) + \zeta_2 \nu F_\nu(70\mu) + \zeta_3 \nu F_\nu(160\mu)$, with $[\zeta_1,\zeta_2,\zeta_2] = [1.559, 0.7686, 1.347]$ (Dale & Helou 2002).

The extinction within our galaxy is corrected using the [12][4] dust maps, which measure the Galactic extinction in all directions. The internal reddening is corrected using a straight relation between the A_{FUV} extinction factor and the F_{TIR}/F_{FUV} flux ratio: $A_{FUV} = -0.0333y^3 + 0.3522y^2 + 1.1960y + 0.4967$, where $y = log(F_{dust}/F_{FUV})$ ([13]). The F_{TIR}/F_{FUV} ratio is a robust and universal tracer of dust extinction, almost independent of the dust/stars geometry and of the dust properties, provided that the galaxies are actively forming stars ([5]; [6]; [7]). The A(Hα) extinction factor was derived through the relation $A_{FUV} = 1.4A(H\alpha)$ ([14]).

The resampled and aligned Hα, FUV and TIR images are shown in Fig.1, center panel.

3. Methodology

In this section, we address the problem of deriving the galaxy age map (GAM) from F_{H_α}/f_{FUV} flux ratio images by establishing a probabilistic framework which explain the relationship between random variables involved in the problem. This relationship will be formulated in terms of a joint probability distribution given the observations and their uncertainties using a hierarchical Bayesian model (HBM) [15]. Specifically, we want to describe the probability distribution $p(\text{age}|r_{obs})$, where r_{obs} is the observed Hα/FUV flux ratio.

Because the use of SB99 we extend the previous distribution to

$$p(\theta|r_{obs}), \quad \theta \in \mathbb{R}^3, \tag{1}$$

where the first component of θ is related to the age, the second is related to the initial mass function (IMF) and third component is related to metallicity. Furthermore, by marginalizing we can derive the age distribution according to

[3] http://irsa.ipac.caltech.edu/data/SPITZER/SINGS/summary.html
[4] http://www.astro.princeton.edu/~schlegel/dust/data/data.html

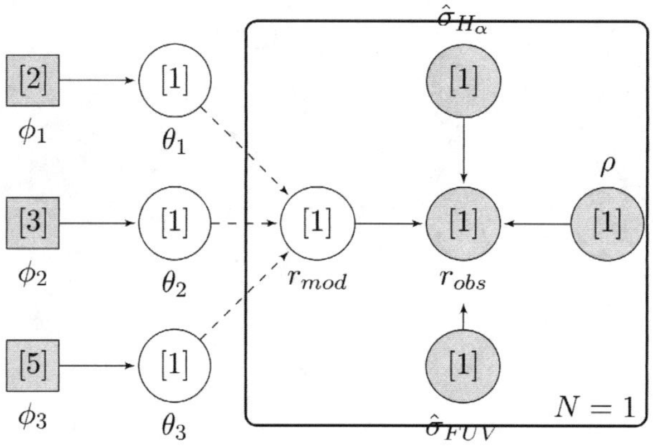

Figure 2. Hierarchical model using plate notation. Hyperparameters and parameters are written with ϕ_k and θ_k respectively. Observations and model ratios are inside the plate. The letter N notes the sample size per pixel, in our case, we only observed one pixel at time, then $N = 1$. Squares are fixed quantities and circles random variables. If the node is shaded then the variable is known (observed).

$$p(\text{age}|r_{obs}) = p(\theta_1|r_{obs}) = \iint p(\theta|r_{obs})\, d\theta_2 d\theta_3. \tag{2}$$

The joint probability distribution $p(\theta|r_{obs})$ can be rewritten by using the Bayes theorem according to

$$p(\theta|r_{obs}) = \frac{p(r_{obs}|\theta)p(\theta)}{p(r_{obs})} \propto p(r_{obs}|\theta)p(\theta), \tag{3}$$

where $p(\theta)$ is the prior distribution and $p(r_{obs}|\theta)$ is the likelihood function and $p(r_{obs})$ is a normalization constant according to

$$p(r_{obs}) = \int p(r_{obs}|\theta)p(\theta)\, d\theta \tag{4}$$

3.1. The prior distribution

The HBM is plot in figure 2 using plate notation. Prior distributions are represented with solid line linking hyperparameters ϕ_i and θ_i with $i \in \{1, 2, 3\}$. These hypermaramters are konwn and fixed in order to draw the parameters θ according to different prior distributions. Priors for θ were set as

$$\begin{aligned} \theta_1 &\sim \mathcal{U}(0, 10) \\ \theta_2 &\sim \text{Cat}(3) \\ \theta_3 &\sim \text{Cat}(5) \end{aligned} \tag{5}$$

where Cat(n) is a categorical distribution with n categories. For the IMF, the categories are: first, a Salpeter law with $\alpha = 2.35$ and $M_{up} = 100M_\odot$; second, a truncated Salpeter law with $\alpha = 2.35$ and $M_{up} = 30M_\odot$; third, and a Miller-Scalo law with $\alpha = 3.3$ and $M_{up} = 100M_\odot$. For metallicity, categories are $Z = 0.04$, 0.02 (solar, $Z_\odot$), 0.008, 0.004 and 0.001. Finally the prior probability distribution for age was set to the uniform distribution in the interval 0-10 Myr.

3.2. The likelihood

The likelihood function is represented by lines linking the variables θ_i with r_{obs} in Fig. 2. In order to obtain *model flux ratio* [5] $r_{mod} = F_{H_\alpha}/f_{FUV}$ from SB99 model, we need to set up SB99 with parameter $\theta = (\theta_1, \theta_2, \theta_3)$, but let us notice that the SB99 model only permits a finite set of values for θ_2 and θ_3, see Fig. 3. This situation means we cannot work with a continuous parameter space in three dimensions but a grid of parameters $\{\theta\}^{\text{grid}}$.

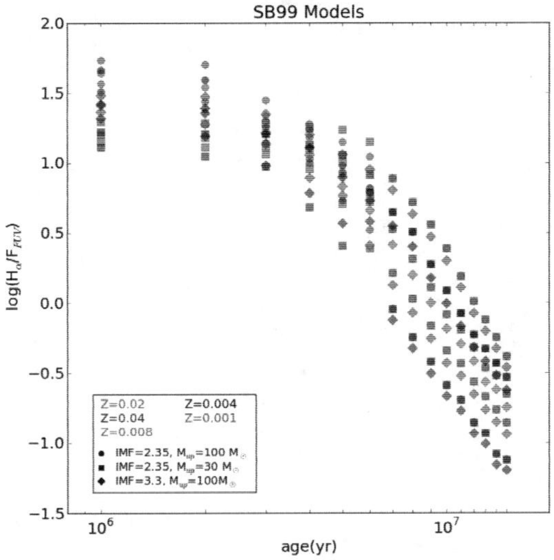

Figure 3. Model flux ratios for every combination of SB99 parameters.

Because this finite combination of parameters in SB99, we can only derive a finite number of model flux ratios from them. Therefore, in order to obtain more model flux ratios different from those on the grid, we use an Artificial Neural Network (ANN) which will interpolate the grid of parameters. A deep information about ANN can be found in [16]. Specifically the ANN is a *multilayer perceptron* with four layers: two hidden layers with 20 and 50 nodes, the input layer for the parameters θ and the output layer for the Hα and FUV flux. The ANN was trained with 70 % and validate with 30 % of the grid points. The uncertainties due this interpolation were considered negligible. Dashed lines in Fig. 2 represent the ANN.

3.2.1. Independence between pixels For explanation purposes this section assume independence between the fluxes of different pixels. Also the results are established for an unique pixel with $r_{obs} > 0$. Once the SB99 have generated the model fluxes we assume that the observed flux from each pixel has a normal distribution

$$F_{H_\alpha,obs} = F_{H_\alpha,mod} + \epsilon_{H_\alpha}, \qquad \epsilon_{H_\alpha} \sim \mathcal{N}(0, \hat{\sigma}_{H_\alpha}) \tag{6}$$

$$f_{FUV,obs} = f_{FUV,mod} + \epsilon_{FUV}, \quad \epsilon_{FUV} \sim \mathcal{N}(0, \hat{\sigma}_{FUV}), \tag{7}$$

where $F_{H_\alpha,obs}$, $f_{FUV,obs}$, $F_{H_\alpha,mod}$ and $f_{FUV,mod}$ are the observed and model Hα and FUV fluxes for the selected pixel respectively. The quantities $\hat{\sigma}_{H_\alpha}$ and $\hat{\sigma}_{H_\alpha,obs}$ are the flux uncertainties due

[5] We prefer use the term *model* instead than *true* because its value depends on the assumption that SB99 is an appropriate model

Journal of Physics: Conference Series **633** (2015) 012140 doi:10.1088/1742-6596/633/1/012140

to the instrumental, reduction and correction processes. Therefore, $r_{obs} = F_{H_\alpha,obs}/f_{FUV,obs}$ is the ratio of two correlated normal random variables, which exact distribution is given by ([17])

$$\psi(r) = \frac{b(r)d(r)}{\sqrt{2\pi}\hat{\sigma}_{H_\alpha}\hat{\sigma}_{FUV}}\left[\Phi\left(\frac{b(r)}{\sqrt{1-\rho^2}a(r)}\right) - \Phi\left(-\frac{b(r)}{\sqrt{1-\rho^2}a(r)}\right)\right] + \frac{\sqrt{1-\rho^2}}{\pi\hat{\sigma}_{H_\alpha}\sigma_{FUV}a^2(r)}\exp\left\{-\frac{c}{2(1-\rho^2)}\right\} \tag{8}$$

where,

$$a(r) = \left(\frac{r^2}{\hat{\sigma}^2_{H_\alpha}} - \frac{2\rho r}{\hat{\sigma}_{H_\alpha}\hat{\sigma}_{FUV}} + \frac{1}{\hat{\sigma}^2_{FUV}}\right)^{1/2} \tag{9}$$

$$b(r) = \frac{r^2}{\hat{\sigma}^2_{H\alpha}} - \frac{2\rho r}{\hat{\sigma}_{H\alpha}\hat{\sigma}_{FUV}} + \frac{1}{\hat{\sigma}^2_{FUV}} \tag{10}$$

$$c = \frac{F^2_{H_\alpha,mod}}{\hat{\sigma}^2_{H\alpha}} - \frac{2\rho F_{H_\alpha,mod}f_{FUV,mod}}{\hat{\sigma}_{H\alpha}\hat{\sigma}_{FUV}} + \frac{f^2_{FUV,mod}}{\hat{\sigma}^2_{FUV}} \tag{11}$$

$$d(r) = \exp\left\{\frac{b^2(r) - ca^2(r)}{2(1-\rho^2)a^2(r)}\right\} \tag{12}$$

where ρ is the correlation between Hα and FUV, and Φ is the cumulative density distribution of the standard normal,

$$\Phi(z) = \int_{-\infty}^{y} \phi(u)du, \quad \phi(u) = \frac{1}{\sqrt{2\pi}}e^{-\frac{1}{2}u^2} \tag{13}$$

The $F_{H\alpha}$ and f_{FUV} are correlated, in fact in normal star forming galaxies it is expected a constant flux ratio $\frac{F_{H\alpha}}{f_{FUV}} \simeq 11$ Å ([18], [19]).

In order to compare model and observed flux ratio, pixel-by-pixel, we consider the likelihood according to

$$p(r_{obs}|\theta) = \psi(r_{obs}|F_{H_\alpha,mod}, f_{FUV,mod}, \hat{\sigma}_{H\alpha}, \hat{\sigma}_{FUV}, \rho) \tag{14}$$

where $F_{H_\alpha,mod}$ and $f_{FUV,mod}$ are obtained from θ through the ANN. The theoretical correlation, ρ, between $F_{H\alpha}$ and f_{FUV} is estimated from all the pixels of the images .

3.3. *The posterior distribution*

We characterize the posterior distribution in equation 1 by extracting independent samples using the Nested Sampling Algorithm (NSA) [20]. The aim of the NSA is the estimation of $p(r_{obs})$ defined in 4, and as by-product an independent sample of the posterior distribution is obtained. The NSA is based on the relationship between the likelihood $p(r_{obs}|\theta)$ and the *prior volume* $X(\lambda)$ defined by

$$X(\lambda) = \int_{p(r_{obs}|\theta)>\lambda} p(\theta)\, d\theta, \tag{15}$$

which means the bulk of prior distribution contained within an iso-contour of the likelihood.

Finally we obtain the posterior distribution of our parameter of interest (age) by marginalizing nuisance parameters in the posterior distribution as we set in equation 2.

Figure 4 shows the sample of the marginalized the age distribution for a certain pixel of the image, as an example of the resulting posterior distribution.

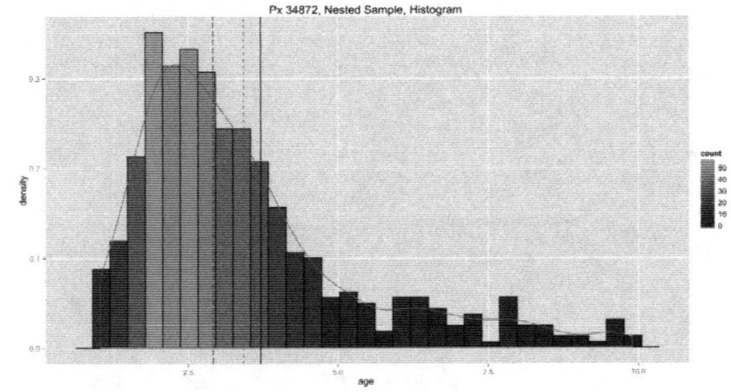

Figure 4. Sample of the marginal probability distribution for the age parameter, for the pixel #34872 of the image. The blue and red dashed lines represent the median and the mean respectively. The solid black line shows the age with the maximum likelihood.

4. Results

Figures 5 and 6 show some age maps for M74. In the Figure 5, we can compare between the age maps for M74 obtained in Paper I, by applying a robust, only based on astrophysical models. Whereas on the left, it is shown the age map when applying the Bayesian approach. The main difference, and more remarkable is that the first one is a discrete age map mode. However, with the Bayesian inference we can determine a single age to each pixel. Resulting therefore in more rich age patterns ans structures, which could not been observed with the latter.

To have an idea of the uncertainty in the age estimation, it is included the Figure 6, where it is represented the 50% central distribution of the age. On the left it is shown the Percentile 25, P_{25}, of the age distribution. The central age map corresponds to the median ages, P_{50}, and on the left the P_{75}. With this comparison we get as a confidence interval, checking out the robust of this methodology for age estimation.

Finally, with respect to the physical conclusions we can observe that these are basically the same that were obtained in Paper I. The age map shows an age gradient from the inner to the outer parts of the galaxy, from very recent to less recent episodes of star formation, in agreement with previous authors ([21], [22]).

Specifically, we find that the Hα luminosity decrease in radius is more pronounced in the inner 5 to 6 kpc, while the UV luminosity shows a shallower rate of change. Consequently, the Hα/FUV ratio decreases with radius indicating an age increases in the outward direction.

On more localised scales, the short arm that opens S-SW at 4 kpc shows a clear age gradient across it. However, the outer longer arm, that runs SE-S at $\sim$ 5 to 10 kpc, showed a less marked age gradient across its width in Paper I. But improving the "resolution" of our age maps with the bayesian technique, it is more remarkable the age gradient across the outer longer arm as well. If the age gradients across spiral arms are a direct product of the spiral density wave, then the dilution of the gradient in this southern arm may be related to the weakness of the density wave or the approach to corotation. As discussed by [23], the presence of a shock produced by the spiral density wave, (made visible by dust lanes along the spiral arms), is incompatible with the creation of star forming complexes, because of the absence of visible dust lanes in this arm, despite a chain of complexes along its length. The thickness of the longer arm is basically dominated by a single age range, and the youngest population located in the inner edge of the arm maps the location of the chain of complexes observed in this arm.

Upcoming work is applying this method to the rest of the sample of 6 galaxies from the original work, Paper I. To analyze with more detail now the age patterns. Next, we are studying also the more complex case and model of dependence between pixels of the image. To develop a parallelization of the algorithm, high perfomance computing to emcee, and / or

Journal of Physics: Conference Series **633** (2015) 012140 doi:10.1088/1742-6596/633/1/012140

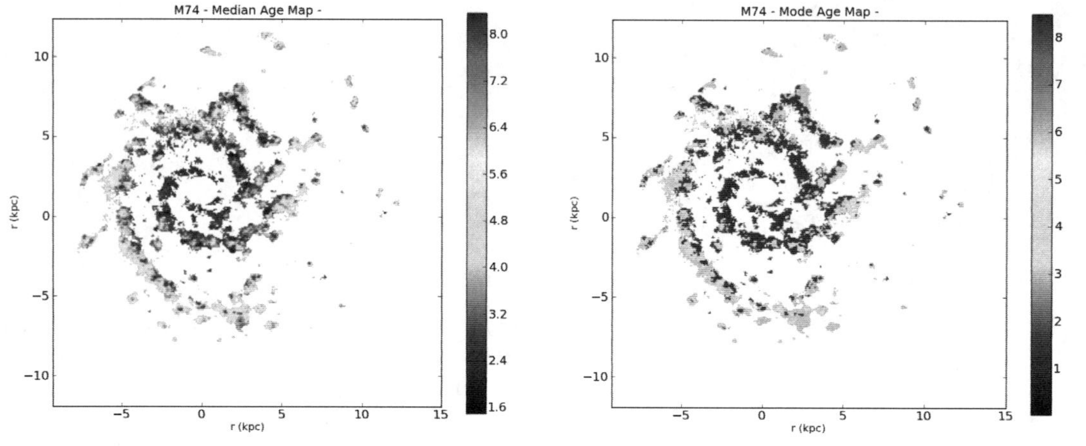

Figure 5. Comparison between the Agemaps for M74: on the right when applying the robust, astrophysical method in Paper I. And on the left, applying the Bayesian approach.

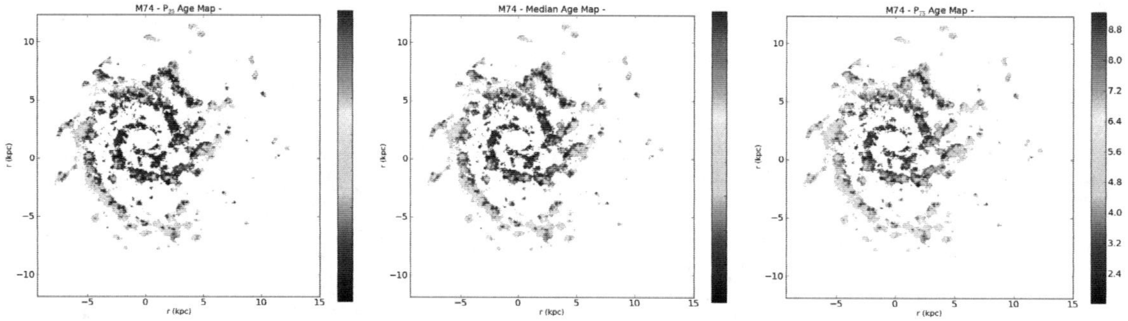

Figure 6. Agemaps for M74: On the left it is shown the Percentile 25, P_{25}, of the Age distribution. The central age map corresponds to the Median ages, and on the left the P_{75}. With this comparison we can check the robust of this methodology for age estimation.

HMC (programming in Python and R) to get better computing times. And finally, to get a wider, new sample of galaxies, to apply the method and determine not only their age maps, but also to study possible patterns/gradients in the metallicity maps.

Appendix A. Empirical Model: Starbust 99

Comparison of UV and Hα fluxes provides a good indication of the recent star formation history. As a young star forming region evolves, the Hα emission line declines before the UV emission does, leading to a decrease in $F_{H\alpha}/F_{FUV}$, which is very sensitive to the age of the youngest stellar population. Images at these wavelengths can be examined to track the recent star formation history and distribution, and to analyze how the SFH correlate with the galactic structure and the galactic environment.

Model Hα/FUV flux ratios were generated from Starburst99, as a function of age, to be compared with the measured flux ratios at each pixel in the image. The full models cover the age range 10^6 to 10^9 yr in steps of 1 Myr with spectral energy distributions (SEDs) spanning

100 Å to 1 μm in wavelength. However, we restricted the ages from 1 to 10 Myr as we are only interested in the youngest stellar populations responsible for the Hα and UV output.

The code can be run with two different star formation modes: an instantaneous burst or continuous. It has three alternatives in the stellar initial mass function, (IMF): a Salpeter law with $\alpha = 2.35$ and $M_{up} = 100 M_{\odot}$; a truncated Salpeter law with $\alpha = 2.35$ and $M_{up} = 30 M_{\odot}$; and a Miller-Scalo law with $\alpha = 3.3$ and $M_{up} = 100 M_{\odot}$. And five metallicities are available for each IMF: $Z = 0.04$, 0.02 (solar, $Z_{\odot}$), 0.008, 0.004 and 0.001.

We take as our reference model a Salpeter IMF, the interpolated metallicity value Z=0.02 (according to [24], M74 has 12+log(O/H)$\sim$8.15), and an instantaneous star formation law, since it is more sensitive to the age variation for younger regions (see paper I).

The age map is contoured in four age bins from the set 0 – 4 Myr, 4 – 6 Myr, 6 – 9 Myr and older than 9 Myr, (this binning corresponds to the internal precision of the flux ratio, Paper I) see right panel of Fig.1 . We find in the resulting age map that the youngest star forming population (0 – 4 Myr) surrounds an older one (>9 Myr) located at the centre of the complex, and a clear center to outer rim age gradient.

The $F_{H\alpha}/F_{FUV}$ depends on the model parameters, such as the IMF, the metallicity, the SFH, etc. To assess the suitability and robustness of our reference model we calculated the effect of changing various model inputs on the SB99 $F_{H\alpha}/F_{FUV}$ ratios (cf. Fig.2 of Paper I). The photometric uncertainty on the $F_{H\alpha}/F_{FUV}$ ratio also affects to the age calibration. Aside the model and photometric uncertainties, the age-dating technique is subject to a number of potential sources of systematic error as: (i) the lowest limit on cluster mass allowable for our assumptions on ionizing flux, (ii) the effect of changing the spatial bin size, and (iii) the effect of changes to the metallicity and IMF assumptions in the model. We analyze the reliability and robustness of our age-dating technique taking into account all these uncertainties. See Paper I for further details.

The total uncertainty of the extinction corrected Hα/FUV flux ratio is estimated, by error propagation, from the Hα , FUV and TIR flux uncertainties as: $\Delta log(F_{H\alpha}/F_{FUV})_0 = \Delta[0.4(A_{H\alpha} - A_{FUV})] + \Delta log(F_{H\alpha}/F_{FUV}) = \frac{1}{ln10}[f(IRX)(E_{TIR} + E_{FUV}) + (E_{H\alpha} + E_{FUV})]$. Where $f(IRX) = 4/35 \mid -0.0999 IRX^2 + 0.7044 IRX + 0.196 \mid$, and $IRX = log(F_{TIR}/F_{FUV})$ (from equation of [13]). These uncertainties are computed pixel by pixel in the image. An average of the relative errors of the respective images are $E_{H\alpha} \simeq 5\%$, $E_{FUV} \simeq 25\%$, and $E_{TIR} \simeq 6\%$, resulting in an overall uncertainty in the $F_{H\alpha}/F_{FUV}$ flux ratio of $\sim$28%.

Now, we calculate the underlying stellar mass per pixel and compare with the values of the lower mass limit M^{min}, given by [25]. The pixel stellar mass is a lower limit derived by comparing the linearly scaled extinction-corrected observed L_{FUV} with the highest expected value from SB99, for a stellar population mass of 10^6 $M_{\odot}$ at the youngest cluster age (1 Myr). In our map, these lowest mass values range log$M^{min} = 3.2 - 5.2$, with an average of 3.84 ± 0.33, around the M^{min} estimated by [25]. Binning the IC 2574 image by 3×3 puts the spatial sampling of IC 2574 well above the threshold of M^{min} (cf. Fig.5 of Paper I), to remove any cause of bias by IMF fluctuations.

References

[1] Sánchez-Gil M C, Jones D H, Pérez E, Bland-Hawthorn J, Alfaro E J and O'Byrne J 2011 *MNRAS* **415** 753–772
[2] Kennicutt Jr R C 1998 *ApJ* **498** 541–552 (*Preprint* astro-ph/9712213)
[3] Grebel E K 2000 *Star Formation from the Small to the Large Scale* (*ESA Special Publication* vol 445) ed Favata F, Kaas A and Wilson A p 87 (*Preprint* astro-ph/0005296)
[4] Glazebrook K, Blake C, Economou F, Lilly S and Colless M 1999 *MNRAS* **306** 843–856 (*Preprint* astro-ph/9808276)
[5] Buat V and Xu C 1996 *A&A* **306** 61
[6] Buat V, Donas J, Milliard B and Xu C 1999 *A&A* **352** 371–382 (*Preprint* astro-ph/9911081)

[7] Gordon K D, Clayton G C, Witt A N and Misselt K A 2000 *ApJ* **533** 236–244 (*Preprint* astro-ph/9912034)
[8] Leitherer C, Schaerer D, Goldader J D, Delgado R M G, Robert C, Kune D F, de Mello D F, Devost D and Heckman T M 1999 *ApJS* **123** 3–40 (*Preprint* astro-ph/9902334)
[9] Vazquez G A, Leitherer C and Geneva Group Collaboration 2005 *American Astronomical Society Meeting Abstracts #206* (*Bulletin of the American Astronomical Society* vol 37) p 464
[10] Bland-Hawthorn J and Jones D H 1997 *The Hubble Space Telescope and the High Redshift Universe* ed Tanvir N R, Aragon-Salamanca A and Wall J V p 121
[11] Martin D C and GALEX Science Team 2005 *American Astronomical Society Meeting Abstracts* (*Bulletin of the American Astronomical Society* vol 37) p 1235
[12] Schlegel D J, Finkbeiner D P and Davis M 1998 *ApJ* **500** 525–553 (*Preprint* astro-ph/9710327)
[13] Buat V, Iglesias-Páramo J, Seibert M, Burgarella D, Charlot S, Martin D C, Xu C K, Heckman T M, Boissier S, Boselli A, Barlow T, Bianchi L, Byun Y I, Donas J, Forster K, Friedman P G, Jelinski P, Lee Y W, Madore B F, Malina R, Milliard B, Morissey P, Neff S, Rich M, Schiminovitch D, Siegmund O, Small T, Szalay A S, Welsh B and Wyder T K 2005 *ApJL* **619** L51–L54 (*Preprint* astro-ph/0411343)
[14] Boissier S, Gil de Paz A, Madore B F, Boselli A, Buat V, Burgarella D, Friedman P G, Barlow T A, Bianchi L, Byun Y I, Donas J, Forster K, Heckman T M, Jelinsky P N, Lee Y W, Malina R F, Martin D C, Milliard B, Morrissey P, Neff S G, Rich R M, Schiminovich D, Siegmund O H W, Small T, Szalay A S, Welsh B Y and Wyder T K 2005 *ApJL* **619** L83–L86 (*Preprint* astro-ph/0411375)
[15] Gelman A, Carlin J B, Stern H S and Rubin D B 2003 *Bayesian Data Analysis* (Chapman and Hall/CRC) ISBN 158488388X
[16] Haykin S 1999 *Neural Networks, A Comprehensive Foundation* (Prentice-Hall) ISBN 0-13-273350-1
[17] Hinkley D V 1969 *Biometrika* **56** pp. 635–639 ISSN 00063444 URL http://www.jstor.org/stable/2334671
[18] Meurer G R, Wong O I, Kim J H, Hanish D J, Heckman T M, Werk J, Bland-Hawthorn J, Dopita M A, Zwaan M A, Koribalski B, Seibert M, Thilker D A, Ferguson H C, Webster R L, Putman M E, Knezek P M, Doyle M T, Drinkwater M J, Hoopes C G, Kilborn V A, Meyer M, Ryan-Weber E V, Smith R C and Staveley-Smith L 2009 *ApJ* **695** 765–780 (*Preprint* 0902.0384)
[19] Iglesias-Páramo J, Boselli A, Gavazzi G and Zaccardo A 2004 *A&A* **421** 887–897 (*Preprint* astro-ph/0403620)
[20] Skilling J 2006 *Bayesian Anal.* **1** 833–859 URL http://dx.doi.org/10.1214/06-BA127
[21] Cornett R H, O'Connell R W, Greason M R, Offenberg J D, Angione R J, Bohlin R C, Cheng K P, Roberts M S, Smith A M, Smith E P, Talbert F D and Stecher T P 1994 *ApJ* **426** 553–562
[22] Lelièvre M and Roy J R 2000 *AJ* **120** 1306–1315
[23] Efremov Y N 2009 *Astronomy Letters* **35** 507–517
[24] Miller B W and Hodge P 1996 *ApJ* **458** 467
[25] Cerviño M, Luridiana V, Pérez E, Vílchez J M and Valls-Gabaud D 2003 *A&A* **407** 177–190 (*Preprint* astro-ph/0304062)

Soltionic solutions generated by quantum groups

Alexander Zuevsky
Institute of Mathematics, Academy of Sciences of the Czech Republic, Prague
E-mail: zuevsky@yahoo.com

Abstract. We review the quantum group technique in construction of quantum soliton solutions to the affine Toda models. Various approaches are illustrated on the main example, the sine–Gordon equation.

1. Affine Toda systemsin the classical region

Let $\widehat{\mathcal{G}}$ be an affine Kac–Moody Lie algebra [18] of rank $r+1$, k. Denote by h_i and $x_{\pm i}$ Cartan and Chevalley generators of $\widehat{\mathcal{G}}$. In the principal grading, the elements of $\widehat{\mathcal{G}}_0$ and $\widehat{\mathcal{G}}_{\pm 1}$, respectively, satisfy the defining relations

$$[h_i, h_j] = 0, \; [h_i, x_{\pm j}] = \pm k_{ij} x_{\pm j}, \; [x_i, x_j] = \delta_{ij} h_i, \; 0 \leq i, j \leq r, \tag{1}$$

where k is Cartan matrix of $\widehat{\mathcal{G}}$, an affine (degenerated with the single zero eigenvalue) matrix [19], [18]. In the decomposition $\widehat{\mathcal{G}} = \oplus_{m \in \mathbb{Z}} \widehat{\mathcal{G}}_m$, the subspace $\widehat{\mathcal{G}}_0$ is abelian. Let $\widehat{G}$ be an infinite–dimensional Lie group corresponding to $\widehat{\mathcal{G}}$ of rank r endowed with the principal grading. Denote by $\widehat{G}_0$ and $\widehat{G}_\pm$ the subspaces corresponding to $\widehat{\mathcal{G}}_0$ and $\oplus_{m>1} \widehat{\mathcal{G}}_{\pm m}$ respectively. It is convenient to enlarge Cartan subalgebra of $\widehat{\mathcal{G}}$ by the derivative element d, such that $[d, h_i] = 0$, $[d, x_{\pm i}] = \pm x_{\pm i}$, and then completed Cartan subalgebra has dimension $r+2$. Positive integers m_i in (3) are defined as the lowest for which $\sum_i k_{ji} m_i = 0$. In terms of these integers the dual Coxeter numbers of $\widehat{\mathcal{G}}$ are written as $\tilde{h} = \sum_{i=0}^r m_i$, while the centre of $\widehat{\mathcal{G}}$ is $c = \sum_{i=0}^r m_i h_i$. Finally, $|\Lambda_i\rangle$ is the highest vector of the i-th fundamental representation of affine algebra $\widehat{\mathcal{G}}$. The subspaces $\widehat{\mathcal{G}}_{\pm 1}$ of an affine Lie algebra $\widehat{\mathcal{G}}$ endowed with the principal grading contain elements $\widehat{E}_{\pm 1} = \sum_{i=0}^r \sqrt{m_i}\, x_{\pm i}$ of the principal Heisenberg subalgebra of $\widehat{\mathcal{G}}$. These elements, thanks to the defining relations, satisfy a very important condition $[\widehat{E}_{+1}, \widehat{E}_{-1}] = c$. The whole principal Heisenberg subalgebra of $\widehat{\mathcal{G}}$ commuting with $\widehat{E}_{+1}$, modulo the central term c, is spanned by $\{\widehat{E}_m\}$ with m being exponents of $\widehat{\mathcal{G}}$. One can normalise the invariant bilinear form on $\widehat{\mathcal{G}}$ in such a way that $[\widehat{E}_m, \widehat{E}_{-n}] = \delta_{mn} mc$. The subspace of $\widehat{\mathcal{G}}$, $\widehat{\mathcal{G}}_m$, $m \neq 0$ contains, in addition to $\widehat{E}_m$, the elements $\widehat{F}^i_m$, $1 \leq i \leq r$, satisfying the relations $[\widehat{E}_m, \widehat{F}^i_n] = \chi_i \widehat{F}^i_{m+n}$, with some constants χ_i that belong to the root space of $\widehat{\mathcal{G}}$. Then the generating elements for $\widehat{F}^i_n$, $\widehat{F}^i := \widehat{F}^i(t) = \sum_{n \in \mathbb{Z}} t^{-n} \widehat{F}^i_n$ with a complex parameter t, is an eigenvector of $\widehat{E}_m$, $[\widehat{E}_m, \widehat{F}^i(t)] = \chi_i^{(m)} t^m \widehat{F}^i(t)$.

Let M be a manifold, $\mathbb{R}^2$ or $\mathbb{C}^1$, with the standard coordinates $z_\pm$. For $\mathbb{C}^1$ we assume that $z_- = z_+^*$. Affine Toda fields satisfy the equations

$$\partial_+ \partial_- \phi + \frac{4\eta^2}{\beta} \left(\sum_{i=1}^r m_i \frac{\alpha_i}{\alpha_i^2} e^{\beta \alpha_i \cdot \phi} - \frac{\psi}{\psi^2} e^{-\beta \psi \cdot \phi} \right) = 0. \tag{2}$$

Here $\partial_\pm$ stand for the partial derivatives with respect to $z^\pm$, η denotes a real inverse length scale, and β is an imaginary coupling constant. The formal general solution to (2) was found in [16]

$$e^{-\beta\lambda_j\cdot\phi} = e^{-\beta\lambda_j\cdot\phi^0}\langle\Lambda_j|\gamma_+^{-1}\mu_+^{-1}\mu_-\gamma_-|\Lambda_j\rangle\left(\langle\Lambda_0|\gamma_+^{-1}\mu_+^{-1}\mu_-\gamma_-|\Lambda_0\rangle\right)^{-m_j}, \tag{3}$$

$1 \le i \le r$. Here $\gamma_\pm(z_\pm)$ and $\mu_\pm(z_\pm)$ are holomorphic and antiholomorphic mappings $M \to \widehat{G}_0$, $M \to \widehat{G}_\pm$ to the Gauss decomposition subgroups $\widehat{G}_0$, $\widehat{G}_\pm$, respectively; $|\Lambda_i\rangle$ is the highest vector of the i-th fundamental representation of $\widehat{\mathcal{G}}$. Mappings $\mu_\pm(z_\pm)$ satisfy the initial value problem

$$\partial_\pm\mu_\pm = \mu_\pm\kappa_\pm, \quad \kappa_\pm(z_\pm) = \sum_{i=0}^{r}\phi_i^0\, x_{\pm i}, \tag{4}$$

where $\kappa_\pm$ are mappings $M \to \widehat{\mathcal{G}}_{\pm 1}$ that can be also represented as

$$\kappa_\pm(z_\pm) = \gamma_\pm^{-1}\widehat{E}_{\pm 1}\gamma_\pm, \quad \widehat{E}_{\pm 1} = \sum_{i=0}^{r} m_i x_{\pm i}, \tag{5}$$

with some nonzero constants m_i. The simplest example of affine Toda theories is the sine–Gordon equation

$$\partial_+\partial_-\phi + \frac{4\eta^2}{\beta}\left(\frac{\alpha_1}{\alpha_1^2}e^{\beta\alpha_1\cdot\phi} - \frac{\psi}{\psi^2}e^{-\beta\psi\cdot\phi}\right) = 0, . \tag{6}$$

In [16], [17] a remarkable specialisation of the general solution (3) which led to soliton solutions was suggested. Let $\widehat{G}$ be an infinite-dimensional group with the Lie algebra $\widehat{\mathcal{G}}$. One chooses mappings $\gamma_\pm$ in (5) to be unit elements of the subgroup $\widehat{G}_0$ of $\widehat{G}$. Therefore, because of (5), $\kappa_\pm = \widehat{E}_{\pm 1}$. Then it is easy to integrate the equations (4) and write $\mu_\pm = \mu_\pm(0)\, e^{\eta^\pm z_\pm \widehat{E}_{\pm 1}}$, with some constants $\eta^\pm$, and constant mapping $\mu(0) \equiv (\mu(0)_+)^{-1}\mu(0)_- : M \to \widehat{\mathcal{G}}$ independent of the coordinates $z_\pm$. The general solution (3) reads

$$e^{-\beta\lambda_j\cdot\phi} = \langle\Lambda_j|e^{-\eta^+\widehat{E}_1 z^+}\mu(0)e^{-\eta^-\widehat{E}_{-1}z^-}|\Lambda_j\rangle\left(\langle\Lambda_0|e^{-\eta^+\widehat{E}_1 z^+}\mu(0)e^{-\eta^-\widehat{E}_{-1}z^-}|\Lambda_0\rangle\right)^{-m_j}, \tag{7}$$

where $\mu(0) \equiv (\mu(0)_+)^{-1}\mu(0)_-$ is a constant mapping $\mathcal{M} \to \widehat{G}$ independent of the coordinates $z^\pm$. The elements $\widehat{E}_{\pm 1}$ can be removed from (3) and N–soliton solutions can be obtained by choosing the group element $\mu(0)$ as

$$\mu(0) = \prod_{n=1}^{N} e^{Q_n\widehat{F}^n(\zeta_n)}, \tag{8}$$

where $\zeta_n = 4\pi\theta_n/\gamma$, $\gamma = |\beta|^2/(1-|\beta|^2/8\pi)$, ($\theta_n$ are rapidities of solitons), and Q_n is the logarithm of the n-th soliton initial coordinate. In (8) $\widehat{F}^n(\zeta_n)$ are vertex operators in the principal grading [18], i.e., elements of the principal vertex operator construction. These operators are eigenvectors with respect to elements of the principal Heisenberg subalgebra $\widehat{E}_{\pm k}$. This fact helps us to eliminate $\widehat{E}_{\pm 1}$ from the solution (7) commuting them with $\mu(0)$. Note also that exponentiation series of $\widehat{F}^n(\zeta_n)$ operators terminate after the order which coincides with the level of the highest weight representation. In the homogeneous grading the mappings $\gamma_\pm$ can be parameterized as $\gamma_\pm = e^{d\phi_d}e^{c\phi_c}e^{\phi_0^\pm x_0^\pm}$, where d is the grading operator, c is the center of $\widehat{sl_2}$ and $x_k^\pm$ are generators of the subspaces $\widehat{\mathcal{G}}_k$ corresponding to the homogeneous grading. The mappings $\mu_\pm$ satisfy (4) where $\kappa_\pm(z^\pm) = a_{\pm 1} + \phi^\pm x_1^\mp$. In order to obtain a soliton solution we put $\phi^\pm = 0$, $\phi_0^\pm = 0$. Then the general solution reduces to

$$e^{-\beta\phi(z^+,z^-)} = \langle\Lambda_1|e^{a_{+1}z^+}\mu(0)e^{a_{-1}z^-}|\Lambda_1\rangle\left(\langle\Lambda_0|e^{a_{+1}z^+}\mu(0)e^{a_{-1}z^-}|\Lambda_0\rangle\right)^{-1}. \quad (9)$$

The following group element $\mu(0)$ in (9) $\mu(0) = e^{-\frac{\alpha}{2}N}\prod_{n=1}^{N}\left[exp\left((-1)^{\partial_\alpha+1}iQ_n\Phi(\zeta_n)\right)e^{\frac{\alpha}{2}}\zeta_n^{\frac{1}{2}\partial_\alpha}\right]$, generates an N–soliton solution. Here the action of the operators $\frac{1}{2}\partial_\alpha$ and $e^{\frac{\alpha}{2}}$ on the highest vectors $|\Lambda_n\rangle = |1\otimes e^{\frac{\alpha}{2}n}\rangle$, $n = 0,1$ is the same as in the case of $U_q'(\widehat{sl_2})$ [18] when $q = 1$. The operator $\Phi(\zeta)$ is given by $\Phi(\zeta) = exp\left(\sum_{n=1}^{\infty}\frac{a_{-n}}{n}\zeta^n\right)exp\left(-\sum_{n=1}^{\infty}\frac{a_{+n}}{n}\zeta^{-n}\right)$, and diagonalises the action of $a_{\pm k}$, $k\in\mathbb{N}$, i.e., $[a_{\pm k},\Phi(\zeta)] = \zeta^{\pm k}\Phi(\zeta)$. The product of two vertex operators can be normal ordered as $\Phi(\zeta_1)\Phi(\zeta_2) = X(x) : \Phi(\zeta_1)\Phi(\zeta_2) :$, where $X(x) = exp(-\sum_{n=1}^{\infty}x^{2n}/n) = \exp\left(\log(1-x^2)\right)$. When $x = 1$, $X(x)$ vanishes which results in $\Phi(\zeta)\cdot\Phi(\zeta) = 0$. Therefore the exponential of $\Phi(\zeta)$ operator terminates after the first order. In the limit $q \longrightarrow 1$ soliton–soliton, antisoliton–antisoliton and soliton–antisoliton scattering reduce to the classical case, i.e., $F^a(\zeta_1)F^b(\zeta_2) = \frac{1}{x}\frac{X(x)}{X(x^{-1})}F^b(\zeta_2)F^a(\zeta_1)$, where $x^2 = \zeta_2/\zeta_1$, a,b denote soliton (antisoliton), and the factor $1/x$ comes from the commutation of $e^{\frac{\alpha}{2}}\zeta_1^{\frac{1}{2}\partial_\alpha}$ and $e^{\frac{\alpha}{2}}\zeta_2^{\frac{1}{2}\partial_\alpha}$ operators. Therefore the vertex operator generating a classical soliton solution is $F(\zeta) = Q\ \Phi(\zeta)\ e^{\frac{\alpha}{2}}\ \zeta_2^{\frac{1}{2}\partial_\alpha}$. The antisoliton solution can be associated with the vertex operator

$$\bar{F}(\zeta) = -Q\ \Phi(\zeta)\ e^{\frac{\alpha}{2}}\ \zeta^{\frac{1}{2}\partial_\alpha}. \quad (10)$$

2. Solutions generated by quantum groups

There are a few ways to quantize the conformal affine Toda models [1], [2], [3], [4], [5], [6], [7]. The quantised affine Toda Heisenberg field operators (see [7] for the conformal Toda model) associated to an affine Lie algebra $\widehat{\mathcal{G}}$ satisfy the equations $\partial_+\partial_-\phi^{(q)} + \frac{4\eta^2}{\beta}\left(\sum_{i=1}^{r}m_i\frac{\alpha_i}{\alpha_i^2} : e^{\beta\alpha_i\cdot\phi^{(q)}} : -\frac{\psi}{\psi^2} : e^{-\beta\psi\cdot\phi^{(q)}} :\right) = 0$. One shows [7], [11], [12], [13] that the quantum Toda system possess the Heisenberg field operator solution of the form (dots denoting normal ordering) for $1\le i\le r$,

$$: e^{-\beta\lambda_j^{(q)}\cdot\phi^{(q)}} :=: e^{-\beta\lambda_j^{(q)}\cdot\phi_0^{(q)}} : {}_q\langle\Lambda_j|{}_q\gamma_+^{-1}\ {}_q\mu_+^{-1}\ {}_q\mu_-\ {}_q\gamma_-|\Lambda_j\rangle_q\left[{}_q\langle\Lambda_0|{}_q\gamma_+^{-1}\ {}_q\mu_+^{-1}\ {}_q\mu_-\ {}_q\gamma_-|\Lambda_0\rangle_q\right]^{-m_j}, \quad (11)$$

where $\phi_0^{(q)}$ are free field operators, and $\partial_{\pm q}\mu_\pm = {}_q\mu_\pm\ {}_q\kappa_\pm$, ${}_q\gamma_\pm$, ${}_q\mu_\pm$, and ${}_q\kappa_\pm$ are mappings $M\to U(\widehat{\mathcal{G}})$. Quantum solutions [2], [3] generated by quantum soliton operators [10] and corresponding to classical soliton solutions [17] can be obtained from the quantum formal solution (11). For that purpose [11] we put ${}_q\gamma_\pm = Id$, i.e., the unit mapping, and let ${}_q\kappa_\pm(z_\pm) = {}_q\gamma_\pm^{-1}\ {}_q\widehat{E}_{\pm 1}\ {}_q\gamma_\pm$, where ${}_q\widehat{E}_{\pm 1}$ are supposed to be generators of a Heisenberg subalgebra in $U_q(\widehat{\mathcal{G}})$. Then ${}_q\mu_\pm$ are of the form ${}_q\mu_\pm = {}_q\mu_\pm(0)\ e^{\eta^\pm z_\pm\ {}_q\widehat{E}_{\pm 1}}$. By letting ${}_q\mu(0) = ({}_q\mu(0)_+)^{-1}\ {}_q\mu(0)_- = e^{Q\mathcal{F}}$ for $Q\in\mathbb{R}$, we therefore obtain from (11)

$$: e^{-\beta\Lambda_i\cdot\phi^{(q)}} :=: e^{-\beta\lambda_j\cdot\phi_0^{(q)}} : {}_q\langle\Lambda_j|e^{-{}_q\widehat{E}_{+1}z_+}e^{Q\mathcal{F}}e^{{}_q\widehat{E}_{-1}z_-}|\Lambda_j\rangle_q\left[{}_q\langle\Lambda_0|e^{-{}_q\widehat{E}_{+1}z_+}e^{Q\mathcal{F}}e^{{}_q\widehat{E}_{-1}z_-}|\Lambda_0\rangle_q\right]^{-m_j}. \quad (12)$$

Here $|\Lambda_i\rangle_q$ denotes the highest vector in the i–th fundamental representation of $U_q(\widehat{\mathcal{G}})$, and $\mathcal{F}$ are eigenvectors with respect to ${}_q\widehat{E}_{\pm 1}$. We use the quantum Lie algebra constructions given in section 4 of Appendix to generate quantum group soliton solutions based on $(sl_2)_q^t$ [11]. Starting from (11), we put ${}_q^t\mu_\pm = {}_q^t\mu_\pm(0)\, e^{\eta^\pm z_\pm\, \widehat{E}_h^\pm}$, ${}_q^t\mu(0) = ({}_q^t\mu(0)_+)^{-1}\, {}_q^t\mu(0)_- = e^{Q\mathcal{F}_q}$ for the group-like mappings associated with $(sl_2)_q^t$, and include the fundamental highest weight vectors $|\Lambda_j\rangle_q^t$, $j = 0, \ldots, r$. Then we obtain for $j = 0,\, 1$: $e^{-\beta\lambda_j\cdot\phi^{(q)}} :=: e^{-\beta\lambda_j\cdot\phi_0^{(q)}} : {}_q^t\langle\Lambda_j|e^{-\widehat{E}_h^+ z_+}e^{Q\mathcal{F}_q}e^{\widehat{E}_h^- z_-}|\Lambda_j\rangle_q^t \times \left[{}_q^t\langle\Lambda_0|e^{-\widehat{E}_h^+ z_+}e^{Q\mathcal{F}_q}e^{\widehat{E}_h^- z_-}|\Lambda_0\rangle_q^t\right]^{-m_j} =: e^{-\beta\lambda_j\cdot\phi_0^{(q)}} : {}_q^t\langle\Lambda_j|F|\Lambda_j\rangle_q^t \left[\langle\Lambda_0|F|\Lambda_0\rangle_q^t\right]^{-m_j}$, where we have denoted $F = exp\left(Qe^{-2z_+\zeta - 2\frac{1}{t}z_-\zeta^{-1}}\mathcal{F}_q\right)$. Thus, for $j = 0,1$ we have : $e^{-\beta\lambda_0\cdot\phi^{(q)}} :=: e^{-\beta\lambda_0\cdot\phi_0^{(q)}} :, :$ $e^{-\beta\lambda_1\cdot\phi^{(q)}} :=: e^{-\beta\lambda_1\cdot\phi_0^{(q)}} : exp\left(-Qe^{-2z_+\zeta-2\frac{1}{t}z_-\zeta^{-1}}\Xi(\zeta,t,q)\right)$, $\Xi(\zeta,t,q) \equiv \sum_{m=-\infty}^{+\infty} \zeta^{2m}t^m(q+q^{-1})^m$. The construction of a quantum Lie algebra $(\mathcal{G})_h$ finds its extensions to all semisimple Lie algebras $\mathcal{G}$ in [21].

3. Vertex operators for quantum solitons

In the paper [9] a remarkable dilogarithmic structure of the factorized S–matrix elements for quantum sine–Gordon model calculated in [8] was discovered. It turned out that the S–matrix elements can be expressed through a function $X_q(x)$ which is a ratio of two regularized quantum dilogarithms [24]. The usual interaction function $X(\theta)$ [9], [16], [17] comes from the normal ordering of two vertex operators generating soliton solutions in the classical region,

$$F(\zeta_1)F(\zeta_2) = X(\theta_{12}) : F(\zeta_1)F(\zeta_2) :, \tag{13}$$

where $\theta_{12} = \theta_2 - \theta_1$, ($\theta_i$, $i = 1, 2$ are rapidities of two solitons). It has been discovered earlier that the S–matrix for the sine–Gordon model coincides, up to a scalar function factor, with the R–matrix of $U_q(\widehat{sl_2})$, [25] (here we have used notations of [9]): $x = \exp(8\pi\theta/\gamma)$, $q = \exp(-i8\pi^2/\gamma)$. The quantum interaction function $X_q(x)$ can be expressed as the ratio

$$X_q(x) = S_{q^{-2}}^{reg}(e^{i\pi+2mi\pi}x^2q^2)/S_{q^{-2}}^{reg}(e^{i\pi+2mi\pi}x^2), \tag{14}$$

where $m \in \mathbb{Z}$ is an arbitrary integer and $S_{q^{-2}}^{reg}(z)$ is the regularized quantum dilogarithm [24].

The dilogarithmic structure of the quantum interaction function hints us the answer. In [10] the quantum vertex operator ${}_qF(\zeta)$ was proposed:

$${}_qF(\zeta) = {}_q\Phi(\zeta)e^{\frac{\alpha}{2}}\zeta^{\frac{1}{2}\partial_\alpha}. \tag{15}$$

Here ${}_q\Phi(\zeta)$ is the product

$${}_q\Phi(\zeta) = \Phi_a(\zeta)\cdot\Phi_b(\zeta), \tag{16}$$

$$\Phi_a(\zeta) = exp\left(\sum_{n=1}^{\infty}\frac{a_{-n}}{[2n]}q^{\theta^+ n}\zeta^n\right) exp\left(-\sum_{n=1}^{\infty}\frac{a_n}{[2n]}q^{-\theta^- n}\zeta^{-n}\right), \tag{17}$$

and similar expression for $\Phi_b(\widehat{\zeta})$ with deformed parameter $\widetilde{q}$, [10]. Here $\widehat{\zeta} = \zeta^{-\frac{\pi}{\mu}}$ and

$$q^{-2} = \exp(i\mu), \qquad \widetilde{q} = \exp(\pi^2/\mu^2). \tag{18}$$

The operator $\Phi_a(\zeta)$ (17) is the vertex operator introduced in [28]. $\Phi_a(\zeta)$ contains elements of the homogeneous quantum Heisenberg subalgebra of $U_q(\widehat{sl_2})$. The quantum vertex operator ${}_q\Phi(\zeta)$ consists of two commuting multipliers ${}_q\Phi(\zeta) = \Phi_a(\zeta)\cdot\Phi_b(\zeta)$. It easy to verify that the product of two operators (17) with different parameters ζ_1 and ζ_2 satisfies $\Phi_a(\zeta_1)\Phi_a(\zeta_2) = X_q^{unreg}(x) : \Phi_a(\zeta_1)\Phi_a(\zeta_2) :$, $\Phi_a(\zeta_1)\Phi_a(\zeta_2) = \frac{X_q^{unreg}(x)}{X_q^{unreg}(x^{-1})}\Phi_a(\zeta_2)\Phi_a(\zeta_1)$, where columns denote the normal ordering which means that all $\{a_{-n}, n \in \mathbb{N}\}$ generators are moved to the left with respect to $\{a_n, n \in \mathbb{N}\}$ generators. The function $X_q^{unreg}(x)$ (the superscript 'unreg' meaning will be explained later) is given by

$$X_q^{unreg}(x) = \exp\left(-\sum_{n=1}^{\infty} q^n x^{2n}[n]/n[2n]\right), \tag{19}$$

where $x^2 = \zeta_2/\zeta_1$, i.e., $\zeta_n = \exp(4\pi\theta_n/\gamma)$. Similar expressions exist for $\Phi_b(\zeta)$. The result of commutation of two operators ${}_q\Phi(\zeta_1)$ and ${}_q\Phi(\zeta_2)$ is

$${}_q\Phi(\zeta_1){}_q\Phi(\zeta_2) = \frac{X_q^{unreg}(x)}{X_q^{unreg}(x^{-1})}\frac{X_{\widetilde{q}}^{unreg}(\widehat{x})}{X_{\widetilde{q}}^{unreg}(\hat{x}^{-1})}{}_q\Phi(\zeta_2){}_q\Phi(\zeta_1). \tag{20}$$

Let us denote by

$$X_q^{reg}(x) = X_q^{unreg}(x) \cdot X_{\widetilde{q}}^{unreg}(\widehat{x}), \tag{21}$$

the regularized quantum interaction function. Then ${}_q\Phi(\zeta_1){}_q\Phi(\zeta_2) = \frac{X_q^{reg}(x)}{X_q^{reg}(x^{-1})}{}_q\Phi(\zeta_2){}_q\Phi(\zeta_1)$. It turns out that the function $X_q^{reg}(x)$ in (21) coincides with the quantum interaction function $X_q(x)$ (14) introduced in [9]. The regularized quantum dilogarithm splits as $S_{q^{-2}}^{reg}(x) = S_{q^{-2}}^{unreg}(x) \cdot S_{\widetilde{q^{-2}}}^{unreg}(\widehat{x})$. Then we obtain

$$X_q(x) = \frac{S_{q^{-2}}^{unreg}(e^{i\pi+2mi\pi}x^2q^2)}{S_{q^{-2}}^{unreg}(e^{i\pi+2mi\pi}x^2)}\frac{S_{\widetilde{q^{-2}}}^{unreg}(e^{\widehat{i\pi+2mi\pi}}x^2q^2)}{S_{\widetilde{q^{-2}}}^{unreg}(e^{\widehat{i\pi+2mi\pi}}x^2)}. \tag{22}$$

This is the reason why we call the interaction function $X_q^{unreg}(x)$ (19) and $X_q^{reg}(x)$ (21) unregularized and regularized respectively. We see that the commutation of two quantum vertex operators ${}_q\Phi(\zeta_n)$, $n = 1, 2$ (16) gives precisely that common factor in the right hand side of (20). The S–matrix element corresponding to the scattering of two solitons (or two antisolitons) is

$$S(\theta) = -q\frac{xq - x^{-1}q^{-1}}{1 - q^2x^2}\frac{X_q(x)}{X_q(x^{-1})} = \frac{1}{x}\frac{X_q(x)}{X_q(x^{-1})}. \tag{23}$$

Therefore the vertex operator (15) satisfy the algebra of quantum soliton operators in the sense of [8]. It is easy to see that ${}_q\Phi(\zeta)$ is an eigenvector with respect to the action of the generators $a_{\pm k}$ of $U_q(\widehat{sl_2})$ (recall also that $a_{\pm k}$ commute with $b_{\pm n}$ for every $k, n \in \{\mathbb{Z} - 0\}$), $[a_{\pm k}, {}_q\Phi(\zeta)] = q^{\theta^{\pm}}\frac{[k]}{k}\zeta^{\pm}{}_q\Phi(\zeta)$. Secondly, the operator ${}_q\Phi(\zeta)$ satisfies ${}_q\Phi(\zeta_1){}_q\Phi(\zeta_2) = X_q(x) : {}_q\Phi(\zeta_1){}_q\Phi(\zeta_2) :$, with (21) and

$$\begin{aligned} X_q^{unreg}(x) &= S_{q^{-2}}^{unreg}(e^{i\pi+2mi\pi}x^2q^2)/S_{q^{-2}}^{unreg}(e^{i\pi+2mi\pi}x^2), \\ S_{q^{-2}}^{unreg}(e^{i\pi+2mi\pi}x^2q^2) &= \prod_{k=0}^{\infty}\left(1 + q^{-2(2k+1)}(e^{i\pi+2mi\pi}x^2q^2)\right). \end{aligned} \tag{24}$$

Therefore when $\zeta_1 = \zeta_2$, i.e., $x = 1$, we infer that $X_q(1) = 0$ which means that ${}_q\Phi(\zeta) \cdot {}_q\Phi(\zeta) = 0$.

4. Appendix

In the spirit of [26], [27], the quantised enveloping algebra $U_q(sl_2)$ is an associative algebra generated by X^+, X^-, H with q-deformed commutation relations $X^+X^- - X^-X^+ = \left(q^H - q^{-H}\right)(q - q^{-1})^{-1}$, $HX^{\pm} - X^{\pm}H = \pm 2X^{\pm}$. It possesses a Hopf algebra structure with the deformed adjoint action $(ad_{X^{\pm}})_q a = X^{\pm} a q^{H/2} - q^{\mp 1} q^{H/2} a X^{\pm}$, $(ad_H)_q a = Ha - aH$, for all $a \in U_q(sl_2)$. Let us recall the second Drinfeld realization of the quantized universal enveloping algebra $U'_q(\widehat{sl_2})$, (i.e., $U_q(\widehat{sl_2})$ without grading operator) [26], [28], which is a natural quantum analogue of the algebra $\widehat{sl_2}$ in the loop realizations. $U'_q(\widehat{sl_2})$ is an associative algebra generated by $\{x_k^{\pm}, k \in \mathbb{Z}; a_n, n \in \{\mathbb{Z} - 0\}; \gamma^{\pm\frac{1}{2}}, K\}$, where $\gamma^{\pm\frac{1}{2}}$ belong to the center of the algebra, satisfying the commutation relations $[K, a_k] = 0$, $[a_k, a_l] = \delta_{k,-l} \frac{[2k]}{k} \frac{\gamma^k - \gamma^{-k}}{q - q^{-1}}$, $K x_k^{\pm} K^{-1} = q^{\pm 2} x_k^{\pm}$, $\left[a_n, x_k^{\pm}\right] = \pm \frac{[2n]}{n} \gamma^{\mp \frac{|n|}{2}} x_{n+k}^{\pm}$, $\left[x_k^+, x_n^-\right] = \frac{\gamma^{(k-n)/2} \psi_{k+n} - \gamma^{(n-k)/2} \phi_{k+n}}{q - q^{-1}}$, $x_{k+l}^{\pm} x_l^{\pm} - q^{\pm 2} x_l^{\pm} x_k^{\pm} = q^{\pm 2} x_k^{\pm} x_{l+1}^{\pm} - x_{l+1}^{\pm} x_k^{\pm}$. The generators ϕ_k and ψ_{-k}, $k \in \mathbb{Z}_+$ are related to a_k and a_{-k} by means of the expressions $\sum_{k=0}^{\infty} \psi_m z^{-m} = K exp\left((q - q^{-1}) \sum_{n=1}^{\infty} a_k z^{-k}\right)$, $\sum_{k=0}^{\infty} \phi_{-m} z^m = K^{-1} exp\left(-(q - q^{-1}) \sum_{n=1}^{\infty} a_{-k} z^k\right)$, i.e., $\psi_m = 0$, $m < 0$; $\phi_m = 0$, $m > 0$. Here $[k] = \frac{q^k - q^{-k}}{q - q^{-1}}$. The principal grading can be realized with the help of the operator $D_p x = \frac{1}{2} q K^{-1} \left(\frac{d}{dq}(K x K^{-1})\right) K + 2\lambda \frac{d}{d\lambda} x$, where $x \in U_q(\widehat{sl_2})$ and λ is an affinization parameter. The power of λ is denoted by the subscript of $U'_q(\widehat{sl_2})$ generators. Then the grading subspaces are ${}_q\widehat{\mathcal{G}}_0 = \{K, \gamma\}$, ${}_q\widehat{\mathcal{G}}_{2n+1} = \{x_n^+, x_{n+1}^-, n \in \mathbb{Z}\}$, ${}_q\widehat{\mathcal{G}}_{2n} = \{a_n, n \in \{\mathbb{Z} - 0\}\}$. The grading operator for the homogeneous grading is $D_h x = 2\lambda \frac{d}{d\lambda} x$, so that the grading subspaces are ${}_q\widehat{\mathcal{G}}_0 = \{K, \gamma, x_0^+, x_0^-\}$, ${}_q\widehat{\mathcal{G}}_n = \{x_n^+, x_n^-, a_n, n \in \{\mathbb{Z} - 0\}\}$. The level one irreducible integrable highest weight representation of $U'_q(\widehat{sl_2})$ can be constructed in the following way [28]. Let $P = \mathbb{Z}\frac{\alpha}{2}$, $Q = \mathbb{Z}\alpha$ be the weight/root lattice of sl_2. Consider the group algebras $F[P]$, $F[Q]$ of P and Q. The multiplicative basis of $F[P]$ is formed by $e^{\frac{\alpha}{2}n}$, $n \in \mathbb{Z}$. The $F[Q]$-module is split into $F[P] = F[P]_0 \oplus F[P]_1$ where $F[P]_n = F[Q] e^{\frac{\alpha}{2}n}$. The sl_2-module structure on the space $W = F[a_{-1}, a_{-2}, ...] \otimes F[P]$ is given by the action of the $a_k, k \in \{\mathbb{Z} - 0\}$ and e^{α}, $\partial_\alpha = a_0$ generators in accordance with the rules $a_k(f \otimes e^{\beta}) = (a_k f \otimes e^{\beta})$, $k < 0$, $a_k(f \otimes e^{\beta}) = ([a_k, f] \otimes e^{\beta})$, $k > 0$, $e^{\alpha}(f \otimes e^{\beta}) = (f \otimes e^{\alpha+\beta})$, $\partial_\alpha(f \otimes e^{\beta}) = (\alpha, \beta)(f \otimes e^{\beta})$, $K = 1 \otimes q^{\partial_\alpha}$, $\gamma = q \otimes id$. Then W is a $U'_q(\widehat{sl_2})$-module. Its submodules are isomorphic to irreducible highest weight modules $V(\Lambda_n)$ with the highest vectors $|\Lambda_n\rangle = |1 \otimes e^{\frac{\alpha n}{2}}\rangle$, $n = 0, 1$.

In [21] a new deformed basis for the generators in the quantised universal enveloping algebra $U_q(sl_2)$ was introduced

$$X_h^{\pm} = \sqrt{2(q + q^{-1})^{-1}}\, q^{-H/2} X^{\pm},\ H_h = 2(q + q^{-1})^{-1} \left(q X^+ X^- - q^{-1} X^- X^+\right), \tag{25}$$

which form a three dimensional subspace $(sl_2)_q$ generated by $\{X_h^+, X_h^-, H_h\}$ in $U_q(sl_2)$ closed under the quantum Lie bracket

$$[a, b]_h := (ad_a)_q b, \qquad a, b \in (sl_2)_q. \tag{26}$$

The generators (25) possess the following commutation relations: $[H_h, X_h^{\pm}]_h = \pm 2 q^{\pm 1} X_h^{\pm}$, $\left[X_h^{\pm}, H_h\right]_h = \mp 2 q^{\mp 1} X_h^{\pm}$, $[X_h^+, X_h^-]_h = H_h$, $\left[X_h^-, X_h^+\right]_h = -H_h$, $[H_h, H_h]_h = 2(q - q^{-1}) H_h$, $[X_h^{\pm}, X_h^{\pm}]_h = 0$. The algebra $(sl_2)_q$ is not a Lie algebra in the standard sense. The generators (25) do not satisfy Jacobi identity and deformed Lie bracket (26) is not skew-symmetric. Although the q-analogue of Jacobi identity for $(sl_2)_q$ is missing, nevertheless (26) is q-skew-symmetric in accordance with [21]. By q-skew-symmetry we mean a symmetry under q-conjugation (which we will denote with a tilde), the automophism of $(sl_2)_q$ defined by $q \mapsto 1/q$. Then for an element in $(sl_2)_q$ $(a X_h^+ + b X_h^- + c H_h)^{\sim} = \tilde{a} X_h^+ + \tilde{b} X_h^- + \tilde{c} H_h$, $a, b, c \in \mathbb{C}$, the q-deformed Lie bracket satisfies $\widetilde{[a, b]}_h = [\tilde{b}, \tilde{a}]_h$. Denote $\mathcal{G} = (sl_2)_q$. Let $\mathcal{L} = \mathbb{C}[t, t^{-1}]$ be the algebra of Laurent polynomials in t and $\mathcal{L}(\mathcal{G}) = \mathcal{L} \otimes_{\mathbb{C}} \mathcal{G}$. Introduce the complex vector space $(sl_2)_q^t$: $\tilde{\mathcal{L}}(\mathcal{G}) = \mathcal{L}(\mathcal{G}) \oplus \mathbb{C}c \oplus \mathbb{C}d$. This is a loop algebra $\mathcal{L}(\mathcal{G})$ completed with the derivation d (acting as $t\frac{d}{dt}$ in $\mathcal{L}$ and trivially on c) extended by one dimensional center c corresponding to $\mathbb{C}$–valued q-deformed 2-cocycle $\Psi_q(a,b) = (x|y)_h\Phi(P,Q)$, $\Phi(P,Q) = Res_t\frac{dP}{dt}Q$ on $\mathcal{L}(\mathcal{G})$. Here $(x|y)_h$ is a non-degenerate bilinear form on $(sl_2)_q$ and P, Q are polynomials in t. We define the q-deformed Lie bracket in this algebra as

$$[t^m\otimes x\oplus\omega c\oplus\nu d,\quad t^n\otimes y\oplus\omega_1 c\oplus\nu_1 d]_h = (t^{m+n}\otimes[x,y]_h+\nu n t^n\otimes y-n\nu_1 m t^m\otimes x)\oplus m\delta_{m+n,0}(x|y)_h c, \tag{27}$$

where $x,\ y\in\mathcal{G}$; $\nu,\ \omega,\ \nu_1,\ \omega_1\in\mathbb{C}$. Now we introduce generators that form the affinisation of the quantum algebra $(sl_2)_q$ $H_1 = 1\otimes H_h$, $H_0 = 1\otimes(c-H_h)$, $e_1 = 1\otimes X_h^+$, $e_0 = t\otimes X_h^-$, $f_1 = 1\otimes X_h^-$, $f_0 = t^{-1}\otimes X_h^+$. We then define the elements of the principal Heiseberg subalgebra $\widehat{E}_h^+ = 1\otimes X_h^+ + t\otimes X_h^-$, $\widehat{E}_h^- = 1\otimes X_h^- + t^{-1}\otimes X_h^+$, so that the following generating series for $\zeta\in\mathbb{C}$, $\mathcal{F}_q = \sum_{k=-\infty}^{+\infty}\zeta^k A_k$, $A_{2m} = (q+q^{-1})^m t^m\otimes(-H_h)$, $A_{2m+1} = (q+q^{-1})^m t^m\otimes q^{-1}X_h^+ - t^{m+1}\otimes qX_h^- + \left(X_h^-|X_h^+\right)_h$, $m\in\mathbb{Z}$, is an eigenvector of $\widehat{E}_h^{\pm}$ with respect to the bracket (27) with eigenvalies $\zeta^{\pm1}$. Using $\mathcal{F}_q$

5. References

[1] Fedoseev I. A. and Leznov A. N. *Phys. Lett.* **141B** (1984) 1–2 , 100–104.
[2] Hollowood T. *PUPT-1286* hep-th/9110010 (1991).
[3] Hollowood T. *Int. J. Mod. Phys.* **A8** (1993) 947–982.
[4] Hollowood T. and Mansfield P. *Nucl. Phys. B* **330** (1990) 2–3.
[5] Leznov A. N. and Fedoseev I. A. *Theor. Math. Phys.* **53** 3 (1982) 1175.
[6] Leznov A. N. and Mukhtarov M. A. *Theor. Math. Phys.* **71** 1 (1987)
[7] Mansfield P. *Nucl. Phys.* **B 222** (1983) 419–455.
[8] Zamolodchikov A.B., Zamolodchikov Al. B. *Ann. Phys.* **120** (1979) 253–291.
[9] Johnson P. R. *Nucl. Phys.* **B496** (1997) 505–550.
[10] Saveliev M. V., Zuevsky A. B. *Internat. J. Modern Phys.* **A 15** 24 (2000).
[11] Zuevsky A. *J. Phys. Conf. ser.*, 474 (1), 012036, (2013).
[12] Zuevsky A. Accepted to J Phys: Conf. Ser. (JPCS), (2015).
[13] Zuevsky A. Accepted to *Journal of Physics: Conference Series* (JPCS) (2013)
[14] Faddeev L. D. and Takhtajan L. A. *Hamiltonian methods in the theory of solitons*, Springer–Verlag, 594, (1987)
[15] Ferreira L. A., Gervais J.-L., Guillen J. S. and Saveliev M. V. *Nuclear Phys. B* **470** (1996) 1–2.
[16] Olive D. I., Turok N. and Underwood J. W. R. *Nucl. Phys.* **B401** (1993) 663–697.
[17] Olive D. I., Turok N. and Underwood J. W. R. *Nucl. Phys.* **B 409** (1993) 509–546.
[18] Kac V. G. *Infinite dimensional Lie algebras.* Third edition (Cambridge university Press, Cambridge) 400 (1990)
[19] Serre J.-P. *Lie Algebras and Lie Groups, 2nd ed* (Lecture Notes in Mathematics), Vol. 1500 168 (1992).
[20] Chari V., Pressley A. *A guide for quantum groups*, Cambridge University Press, 651 p. (1995)
[21] Delius G. W. and Hueffmann A. *J. Phys* **A29** (1996) 1703-1722
[22] Leznov A. N. and Saveliev M. V. *Group-Theoretical Methods for Integration of Non-Linear Dynamical Systems.* Progress in Physics Series, v. 15, (Birkhauser–Verlag, Basel) (1992).
[23] Razumov A. V., Saveliev M. V. and Zuevsky A. *In the Memorial volume dedicated to M. V. Saveliev, Dubna* (1999).
[24] Faddeev L.D. and Kashaev R. *Mod. Phys. Lett.* **A9** (1994) 423–434.
[25] Reshetikhin N., Smirnov F. *Comm. Math. Phys.* **131** (1990) 157–177.
[26] Drinfeld V. G. *Sov. Math. Dokl.* **32** (1985) 254.
[27] Jimbo M. *Lett. Math. Phys.* **10** (1995) 63.
[28] Jimbo M., Miki K., Miwa T. and Nakayashiki A. *Phys. Let. A.* **168** (1992) 256–263.
[29] Damaskinski E. V., Kulish P. P., Lyakhovsky V. V. and Sokolov M. A. 120–129 *Symmetry methods in physics, 1, Dubna, Joint Inst. Nuclear Res., Dubna* (1996)

Graphical imaging system for shifted spectrum in daughter nuclei of charged-current ν-nucleus reactions

J. Sinatkas[1], **T.S. Kosmas**[2], **V. Tsakstara**[2,3]

[1]Information Engineering Department, TEI of Western Macedonia, GR-52100 Kastoria, Greece
[2]Theoretical Physics Section, University of Ioannina, GR 45110 Ioannina, Greece
[3]School of Pedagogical and Technological Education, GR-45500 Ioannina, Greece

E-mail: `sinatkas@kastoria.teikoz.gr`

Abstract. In this work, we improve the automated algorithm derived previously for drawing the quantum energy levels, in order to be efficient for plotting the theoretical spectra of the daughter nuclei in charged-current neutrino-nucleus reactions. Such calculated spectra need to be shifted so as the charged-current transition energies to be measured from the ground state of the daughter nucleus. The improvement involves the development of a computational code and graphical imaging system able to treat appropriately and as accurate as possible the energy-shifting which is different for the various multipole sets of nuclear states. The method is applied in the case of antineutrino-nucleus reaction $^{66}_{30}Zn + \tilde{\nu}_\ell \rightarrow e^+ + ^{66}_{29}Cu$, which is of current experimental interest.

1. Introduction

In recent investigations of various physical applications, the development of appropriate computational tools as codes and graphical imaging system requires the simultaneous raw data insertion [5], theoretical calculations or simulated predictions and experimental data insertion [6, 7]. In our previous work, we have developed a graphical imaging system which first determines the appropriate scale to depict graphically as good as possible in the available space the data. This code operates locally and the results are displayed on the screen and can also be exported to a PostScript file. One of its main features is to arrange and visualize in the available space the energy levels of a quantized microscopic system taking care the existence in the final diagram the least auxiliary deviations. This system was applied to the initial nucleus of a charge preserving process. In such a process the final nuclear system is the same with the initial one. In the present work, we improve the aforementioned algorithm of automated plotting of energy levels in such a way that other requirements like those appeared in charge changing processes (see below) to be satisfied.

1.1. Motivation

The Charged-Current (CC) scattering of low- and intermediate-energy neutrinos and their antimatter counterpart (the anti-neutrinos), on the $^{66}_{30}Zn$ with natural abundance 27.3 % isotope are represented by the reactions

$$\nu_l + {}^{66}_{30}Zn \rightarrow l^- + {}^{66}_{31}Ga^* , \quad (1)$$

$$\widetilde{\nu}_l + {}^{66}_{30}Zn \rightarrow l^+ + {}^{66}_{29}Cu^* , \quad (2)$$

(where $\ell = e, \mu, \tau$ and Ga^*, Cu^* denote excited states of Ga and Cu, respectively). Note that, the above reactions of the Zn isotopes play a significant role in the astrophysical environment by affecting the electron fraction of Y_e of the matter ant its strong effect on the matter flow [8].

1.2. The new algorithm

The basic ingredients of the method involve raw data insertion, simulated and experimental data insertion etc [5] (alone or in a combination in methods). The system determines the appropriate scale to show graphically the best possible way within the available space. The main ideas used in the implementation of the developed code are: (i) Public abstract void draw(Shape s) and (ii) Public abstract void draw String(String str, int x,int y) which are described below.

1.2.1. Public abstract void draw(Shape s) Public abstract void draw(Shape s) strokes the outline of a Shape using the settings of the current Graphics2Dcontext.

Parameters: s - the Shape to be rendered. Most of the shapes inserted as a parameter in the draw operation can be defined by using an additional package which is java.awt.geom. In the following example we use java.awt.geom.Line2D which is part of java.awt.geom. In this specific example a line is drawn.

```
import java.awt.Graphics2D;
import java.awt.geom.Ellipse2D;
import java.awt.geom.Line2D;
import java.io.FileOutputStream;
import java.io.IOException;
import de.erichseifert.vectorgraphics2d.EPSGraphics2D;
EPSGraphics2D g = new EPSGraphics2D(0.0,0.0,210.0,297.0);
g.draw(new Line2D.Double(x, 0.0, x, 297.0));
```

1.2.2. public abstract void draw String(String str, int x,int y) Public abstract void draw String(String str, int x,int y) renders the text of the specified String, using the current text attribute state in the Graphics2D context. The baseline of the first character is at position (x, y) in the User Space.

Parameters:

str - the string to be rendered

x - the x coordinate of the location where the String should be rendered

y - the y coordinate of the location where the String should be rendered

VectorGraphics2D provides the ability to export the graphics in different formats of vector files.

More specifically in formats that are more suitable for incorporating them in research papers such as: (i) Encapsulated PostScript (EPS), (ii) Scalable Vector Graphics (SVG), (iii) Portable Document Format (PDF). Using VectorGraphics2D has some inherited restrictions. More specifically different text encodings are not supported as well as font embedding.

The program's features are to arrange and visualize, in the available space, the energy levels taking care however, to exist in the final diagram the least auxiliary deviations. Many programs of the existing are very cumbersome due to the complexity of the input data [3], or have not automated this feature and either manually the person concerned puts the deviation he desires in any case [4] either all deviations are only in one direction (usually upwardly). The present

Journal of Physics: Conference Series **633** (2015) 012142 doi:10.1088/1742-6596/633/1/012142

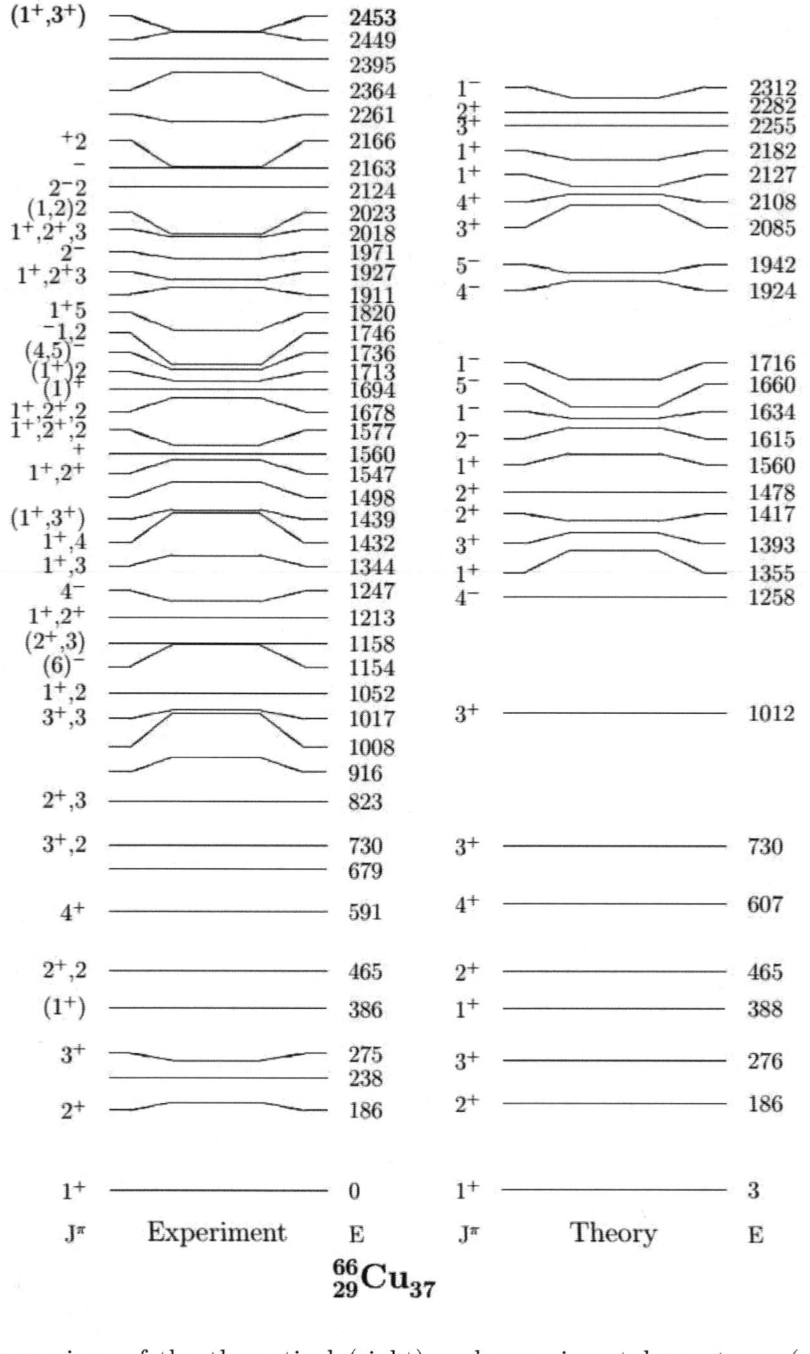

Figure 1. Comparison of the theoretical (right) and experimental spectrum (right)chart of calculated energy spectra in MeV generated in our own automated code for ^{66}Cu the daugter nucleus of the reaction $^{66}_{30}Zn + \tilde{\nu} \rightarrow e^{+} +^{66}_{29} Cu$ (left).

program is user-friendly and the user easily imports the data which are stored to a file. The resulted Figures are very well.

2. Concrete illustrated example

For the calculations of the charged-current (anti)neutrino-nucleus reaction cross sections for various interesting neutrino detectors in terrestrial experiments, one needs to calculate the excitation spectrum of the daughter nucleus. In the QRPA calculations of Ref.[10] for example, starting from the ground state of an even-even nucleus and the system goes to an excited state of the daughter nucleus. As an example, in the case of the COBRA experiment the detector materials adopted are the CdTe and CdZnTe semiconductors. In the present work, which is an extension of Refs [8] and [5], we choose the reaction (2), because the test of our QRPA method is done through checking the reproducibility of the experimental spectra for the daughter nuclear isotopes ${}^{66}_{31}Ga$ and ${}^{66}_{29}Cu$ one needs to compare theoretical and experimental spectra. The spectrum of the ${}^{66}_{29}Cu$ is designed by using the code constructed in this work. Because, however, the energy spectrum of the daughter nucleus must be shifted as discussed in Ref. [9], an improvement of the code is necessary as described below.

Code has been applied on several cases of older diagrams and both, ease of handling as well as the quality of results proved excellent.

3. Conclusions

In this work, e have improved the designing code constructed in Ref. [5] in such a way that the spectrum of the daughter nucleus of a charged current reaction of the type ${}^{A}_{Z}X(\nu_\ell, \ell^-)^{A}_{Z+1}Y^*$ or ${}^{A}_{Z}X(\tilde{\nu}_\ell, \ell^+)^{A}_{Z-1}Y^*$, is perfectly illustrated.

In such reactions the obtained theoretical spectrum, usually requires to be shifted so as the lower excitation of each multipole set of states to be well close to the corresponding experimental one. As a concrete example, in the present study we have designed the spectrum of the ${}^{66}Cu$ isotope, daughter nucleus of the charged-current reaction ${}^{66}(\tilde{\nu}_\ell, \ell^+)^{66}Cu^*$. The code is more advantageous than that of Ref. [9] for example, but a more refinement is needed for a faster and more efficient algorithm which is performing the matching of the shifting procedure optimum.

References

[1] VectorGraphic2D Library Home Page http://trac.erichseifert.de/vectorgraphics2d
[2] Graphics2D (Java Platform SE 7)-Oracle Documentation http://docs.oracle.com/javase/7/docs/api/java/awt/Graphics2D.html
[3] Charles L.Dunford, Robert R. Kinsey ENSDAT (Evaluated Nuclear Structure Drawings and Tables) Version 12.23 (2007) National Nuclear Data Center,http://www.nndc.bnl.gov/nndcscr/ensdf$_p gm$/utility/ensdat/
[4] M.A. Caprio LevelScheme: A level scheme drawing and scientific figure preparation system for Mathematica http://www.sciencedirect.com/science/article/pii/S0010465505002900 - item1-item1 Computer Physics Communications V171, 2, (2005) 107-118
[5] Stampoultzis M, Sinatkas J, Tsakstara, V, Kosmas T S, 2014 *J. Phys.: Conf. Ser.* **490** 012204.
[6] Tsakstara V and Kosmas T S, 2011 *Phys. Rev.* C **83** 054612.
[7] Tsakstara V and Kosmas T S, 2011 *Phys. Rev.* C **84** 064620.
[8] Tsakstara V, Kosmas T S, 2012 *Phys. Rev.* C **C 86** 044618.
[9] Giannaka P, Kosmas T S, 2015 *Advances in High Energy Physics* 398796.
[10] Tsakstara V and Kosmas T S, in preparation.

Journal of Physics: Conference Series **633** (2015) 012143 doi:10.1088/1742-6596/633/1/012143

First-principles study of migration mechanisms and diffusion of carbon in GaN

Alexandros Kyrtsos[1], Masahiko Matsubara[2] and Enrico Bellotti[1,2]

[1] Division of Materials Science and Engineering, Boston University, 15 St Mary's, Boston, USA
[2] Department of Electrical and Computer Engineering, Boston University, 8 St Mary's, Boston, USA

E-mail: akyrtsos@bu.edu

Abstract. Carbon related defects are readily incorporated in GaN due to its abundance during growth both with MBE and CVD techniques. Employing first-principles calculations we compute the migration barrier of neutral carbon interstitials in the wurtzite GaN crystal. The Minimum Energy Path (MEP) and the migration barriers of these defects are obtained using the Nudged Elastic Band (NEB) method with the climbing image modification (CI-NEB). In addition, the Dimer method is used to verify the results. The results yield a quantitative description of carbon diffusion in the crystal allowing for the determination of the most probable migration paths.

1 Introduction

Efficient LED lighting has developed in recent years thanks to III/nitride semiconductors such as GaN [1]. In addition to optoelectronic applications such as blue/green LEDs, GaN and its related alloys are used in power electronics [2] as well as photovoltaic applications [3].

Defects, both point [4] and extended ones [5, 6], affect the electronic properties of devices. Among them, carbon is a relevant and easily incorporated defect in GaN. Defect related levels in the band-gap may be the source of radiative recombination centers, leading to below gap emission compromising the performance of the device. A typical example of such a level is centered around $2.2 - 2.3\,\mathrm{eV}$ and is often referred to as the yellow luminescence (YL) [7–10]. This YL band can be present for both undoped samples [11] and samples containing carbon impurities [12–14]. Recent studies suggest that carbon related defects are responsible for the YL band [7, 15].

Carbon is a common impurity both in molecular beam epitaxy (MBE) [16, 17] and metal organic chemical vapor deposition (MOCVD) [18]. In the former case, carbon can contaminate the material during air exposure in standard substrate loading procedures or at the beginning of regrowth. In MOCVD, carbon is part of the metal organic compounds used as source material for gallium. In addition, carbon can be found as a contaminant in the source gases or it can be etched off the susceptor that transfers heat to the substrate. Even though carbon defects have been intensively investigated in the past, the current understanding of the migration mechanisms of these defects in GaN is still incomplete.

The migration paths and barriers of carbon interstitials in GaN were investigated using density-functional theory (DFT) [19, 20] in conjunction with the climbing image nudged elastic band method (CI-NEB) [21, 22]. Using the CI-NEB method we ensure that the calculated energy corresponds to the migration barrier for a given path. In addition, we get information about the atomic configuration in the saddle point.

2 Background

It is empirically observed that most diffusion processes in solids exhibit a temperature dependence described by an Arrhenius law

$$D = D_0 e^{-\frac{Q}{k_B T}}, \tag{1}$$

where D_0 is a temperature independent factor, Q is the activation energy for the atomic jump mechanism and $k_B T$ is the product of the Boltzmann constant with temperature.

Deviations from this equation may be caused by the existence of short circuiting diffusion paths such as dislocations and grain boundaries, multiple diffusion mechanisms and impurity effects [23].

The jump rate is characterized by an attempt frequency and a thermodynamic factor that dictates the probability of an attempt resulting in a successful jump.

$$\Gamma = \Gamma_0 e^{-\frac{\Delta G}{k_B T}}. \tag{2}$$

The attempt frequency, Γ_0, can be obtained within harmonic transition state theory via the Vineyard equation [24], but often the Einstein or Debye frequency is used instead. ΔG represents the isothermal work associated with the motion of the particle from the initial to the final state.

3 Computational Details

3.1 Computational method

The calculations were performed with the vienna *ab initio* simulation package (VASP) [25] using the projector augmented wave (PAW) method [26, 27]. The generalized gradient approximation (GGA) in the parametrization by Perdew, Burke and Ernzerhof (PBE) [28] was used. The investigation of the lowest energy paths was performed using the CI-NEB method [21, 22] as well as the Dimer method [29] as implemented in VASP through the VTST-Tools by Henkelman, Jónsson and others [30].

The convergence of the size of the supercell was investigated using supercells of 64, 72, 96, 128 and 144 atoms. The 64-atom supercell is found to be inadequate to contain the deformations induced by the migration of carbon. The 72-atom supercell might be sufficient for certain migrations but not all. Thus, we chose the 96-atom supercell to be the safest and most efficient option at hand. The plane-wave energy cutoff was set at 450 eV and a Γ-centered $2 \times 2 \times 2$ k-point mesh is used for the sampling of the Brillouin zone. The atomic configurations were relaxed until the maximum force per atom was less than 5×10^{-3} eV/Å. In the case of the NEB calculations, the images were relaxed until the maximum force per atom was no more than 10^{-2} eV/Å.

The crystallographic parameters of the w-GaN we obtained from our calculations are $a = 3.211$ Å, $c/a = 1.629$ and $u = 0.377$, which are in good agreement with the experimental values $a = 3.189$ Å, $c/a = 1.626$ and $u = 0.377$ and consistent with previous DFT calculations [31]. The bandgap at the Γ point is 1.757 eV which underestimates the experimental value of 3.4 eV as expected from GGA.

The main sources of error in DFT calculations of point defects are the electrostatic and elastic interactions of the defect between neighboring supercells and the underestimation of the bandgap. Lany and Zunger [32] describe a number of methods to overcome these shortcomings of DFT. However, migration barriers, the main focus of the present work, are calculated as

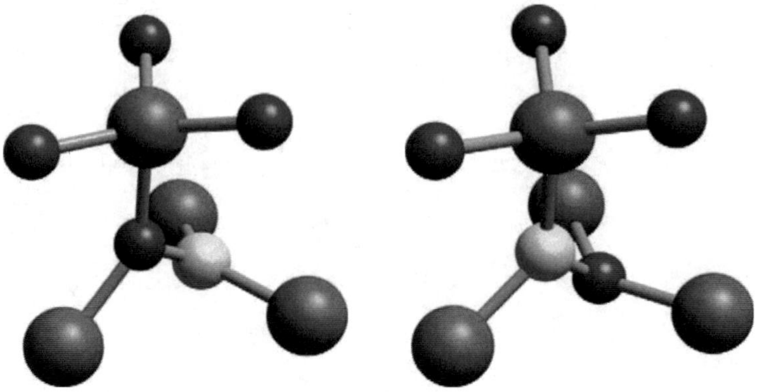

(a) C-N type-1 split interstitial (b) C-N type-2 split interstitial

Figure 1: Split interstitials of carbon (yellow) and nitrogen (blue) surrounded by four gallium atoms (red).

energy differences of electronically similar configurations. Thus, there is no need to apply any correction scheme.

A typical systematic error in the formation energies calculations is considered to be approximately 0.1 eV [33]. Performing identical migrations at different sites we verified that the uncertainty of the migration barrier also lies within 0.1 eV.

In the NEB method [34], a set of "images" of the system is used to represent the migration path from the initial to the final configuration. The images (i.e., atomic configurations along the migration path) are connected with springs to resemble a string (or band). An optimization algorithm is then applied to relax the string down towards the minimum energy path. In the climbing image modification the highest energy image is driven to the saddle point by neutralizing the forces along the band. Since the image converges to the saddle point, the exact migration barrier can be calculated. In addition, the Dimer method was used to verify the saddle point location and the migration barrier acquired via the CI-NEB.

3.2 Determination of diffusion paths

The two most stable configurations for the neutral carbon interstitial are the type-1 and type-2 C-N split complexes shown in Fig. 1. They differ by less than 0.1 eV [35]. Hence, both cases are investigated as potential initial and final configurations for the diffusion of carbon.

This small energy difference and the similarity of the two configurations allows for migrations accompanied by transformation from one type to the other. In this case, the migration barrier is defined as the energy difference between the saddle point and the lowest energy configuration. As a result, the barrier in this case could be smaller by ~ 0.1 eV following the opposite direction of the reaction.

The migration paths investigated in the present work have components both parallel and perpendicular to the c-axis of the wurtzite crystal. The endpoints of the migrations are either a type-1 or type-2 split interstitial. Fig. 2 presents the different migration paths considering jumps among the first and second nearest neighbors. The different combinations of paths and endpoint configurations result in nine different calculations.

Process A, shown in Fig. 2 refers to motion between two out of plane first nearest neighbors. A single gallium atom forms four bonds with nitrogens in the ideal crystal. Three of these nitrogens lie on a plane which is perpendicular to the [0001] direction. Process A describes the migration of a carbon atom forming a split with one of these three nitrogens to the fourth

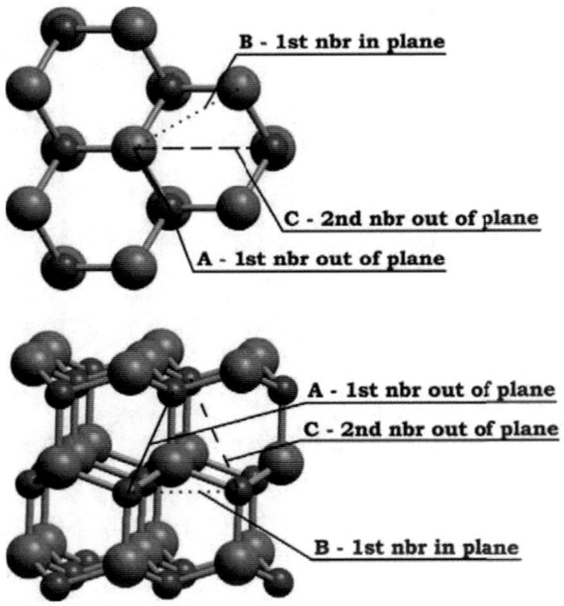

Figure 2: Diffusion paths available to carbon interstitial considering jumps among the first and second nearest neighbors.

nitrogen which is out of plane.

In the case of process B, migration occurs between two adjacent nitrogens lying on a plane perpendicular to the [0001] direction. This process is a first nearest neighbor migration process and together with process A is expected to exhibit the lowest migration barriers.

As an additional process, we consider process C in which migration occurs between the second nearest neighbors which lie in different planes with respect to the [0001] direction. Even though this process utilizes the hexagonal channel of the wurtzite structure, it is expected to exhibit the highest barriers since it is a second nearest neighbor migration.

4 Results

The difference of the formation energy between the two types of interstitials is 0.13 eV with the type-2 being the lowest in energy. The bond length between the carbon and the nitrogen is 1.29 Å and 1.31 Å for the type-1 and type-2 respectively.

The results obtained using the CI-NEB method are summarized in Table 1. The lowest migration barrier is observed in the case of process A from a type-2 interstitial to a type-1. This barrier becomes even smaller by approximately 0.1 eV if the reaction path is reversed. The diffusion potential energy along the reaction path is shown in Fig. 3.

The Dimer method verifies the results acquired by the CI-NEB for all the process A migrations. Both methods result in the same atomic configuration for the saddle point and consequently the same migration barrier. In the case of the lowest barrier migration, the saddle point lies closer to the type-1 interstitial. With respect to the straight line between the initial and final configurations, the saddle point forms an angle of $29°$ and $16°$ with the type-1 and type-2 ends respectively.

First nearest neighbor in plane migrations (process B) exhibit barriers close to 3 eV making them unfavorable compared to their out of plane counterparts. As expected, second nearest neighbor out of plane migrations require a much greater activation energy.

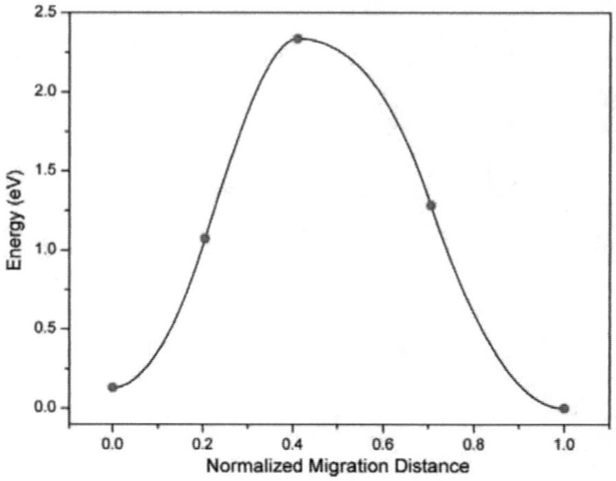

Figure 3: Diffusion energy barrier for a process A migration from a type-1 to type-2 split carbon interstitial.

Table 1: Energy barriers for carbon interstitial migration as obtained from CI-NEB calculations.

Migration Process	Initial	Final	Barrier (eV)
	s1	s1	2.6
A	s2	s2	2.7
	s1	s2	2.3
	s1	s1	2.9
B	s2	s2	3.1
	s1	s2	3.1
	s1	s1	4.0
C	s2	s2	–
	s1	s2	4.0

In order to estimate the temperature where carbon interstitials are activated we assume a jump rate of 1 Hz and a typical Debye frequency of 10 THz. Then, using eq. (2) and the calculated migration barriers, one can estimate the temperature where migration occurs. Hence, using $\Gamma = 1\,\text{Hz}$, $\Gamma_0 = 10^{13}\,\text{Hz}$ and $\Delta G = 2.3\,\text{eV}$ we estimate that carbon interstitials are activated at temperatures close to 890 K. This temperature is relevant during growth of GaN both with MBE and CVD techniques.

5 Conclusions

We employed the Climbing Image Nudged Elastic Band technique to explore migration barriers of carbon diffusion in GaN. The Dimer method was also used in order to verify the results obtained by the CI-NEB in the case of the three lowest energy barrier migrations. The two most stable interstitial types of carbon are investigated with respect to diffusion via first and second nearest neighbor mechanisms.

The first out of plane nearest neighbor mechanism exhibits the lowest migration barriers indicating a favorable direction of migration along the [0001] axis of the crystal. However, for growth temperatures, where carbon migration is relevant, the in plane migration is also competing. Migration via the second nearest neighbor mechanism exhibits high energy barriers minimizing the probability for this mechanism to contribute to the carbon diffusion.

Acknowledgments
The authors gratefully acknowledge financial support from the U.S. Army Research Laboratory through the Collaborative Research Alliance (CRA) for Multi-Scale multidisciplinary Modeling of Electronic Materials (MSME). The computational resources were provided by the 2014 ARO DURIP Award made to Dr. E. Bellotti.

References

[1] Pimputkar S, Speck J S, DenBaars S P and Nakamura S 2009 *Nature Photonics* **3** 180–182
[2] Mishra U K, Parikh P and Wu Y F 2002 *Proceedings of the IEEE* **90** 1022–1031
[3] Dahal R, Li J, Aryal K, Lin J Y and Jiang H X 2010 *Applied Physics Letters* **97** 1–4
[4] Limpijumnong S and Van de Walle C G 2004 *Phys. Rev.* B **69**(3) 035207
[5] Matsubara M, Godet J, Pizzagalli L and Bellotti E 2013 *Applied Physics Letters* **103**
[6] Matsubara M, Pizzagalli L and Bellotti E 2014 *physica status solidi (c)* **11** 521–524
[7] Demchenko D O, Diallo I C and Reshchikov M A 2013 *Physical Review Letters* **110** 1–5
[8] Suski T, Perlin P, Teisseyre H, Leszczynski M, Grzegory I, Jun J, Bockowski M, Porowski S and Moustakas T D 1995 *Applied Physics Letters* **67** 2188–2190
[9] Reshchikov M A and Morkoç H 2005 *Journal of Applied Physics* **97**
[10] Ogino T and Aoki M 1980 *Japanese Journal of Applied Physics* **19** 2395
[11] Glaser E, Kennedy T and Doverspike K 1995 *Physical Review* B **51** 326–336
[12] Kucheyev S O, Toth M, Phillips M R, Williams J S, Jagadish C and Li G 2002 *Journal of Applied Physics* **91** 5867–5874
[13] Seager C H, Wright A F, Yu J and Götz W 2002 *Journal of Applied Physics* **92** 6553–6560
[14] Armstrong A, Arehart A R, Green D, Mishra U K, Speck J S and Ringel S A 2005 *Journal of Applied Physics* **98**
[15] Lyons J L, Janotti A and Van De Walle C G 2010 *Applied Physics Letters* **97** 5–7
[16] Koblmüller G, Chu R M, Raman A, Mishra U K and Speck J S 2010 *Journal of Applied Physics* **107**
[17] Ber B Y, Kudriavtsev Y A, Merkulov A V, Novikov S V, Lacklison D E, Orton J W, Cheng T S and Foxon C T 1998 *Semiconductor Science and Technology* **13** 71
[18] Shah P B, Dedhia R H, Tompkins R P, Viveiros E A and Jones K A 2012 *Solid-State Electronics* **78** 121–126
[19] Hohenberg P and Kohn W 1964 *Phys. Rev.* B **136**(3B) 864–871
[20] Kohn W and Sham L J 1965 *Physical Review* A **140**
[21] Henkelman G, Uberuaga B P and Jónsson H 2000 *Journal of Chemical Physics* **113** 9901–9904
[22] Henkelman G and Jónsson H 2000 *The Journal of Chemical Physics* **113** 9978–9985

[23] Swalin R A 1973 *Atomic Diffusion in Semiconductors* ed Shaw D (Springer US) chap 2, pp 65–110

[24] Vineyard G H 1957 *Journal of Physics and Chemistry of Solids* **3** 121–127

[25] Kresse G and Furthmüller J 1996 *Phys. Rev.* B **54**(16) 11169–11186

[26] Blöchl P E 1994 *Phys. Rev.* B **50**(24) 17953–17979

[27] Kresse G and Joubert D 1999 *Phys. Rev.* B **59**(3) 1758–1775

[28] Perdew J P, Burke K and Ernzerhof M 1996 *Phys. Rev. Lett.* **77**(18) 3865–3868

[29] Henkelman G and Jónsson H 1999 *The Journal of Chemical Physics* **111** 7010–7022

[30] The Transition State Tools implementation for VASP can be obtained from http://theory.cm.utexas.edu/vtsttools

[31] Stampfl C and Van de Walle C G 1999 *Phys. Rev.* B **59**(8) 5521–5535

[32] Lany S and Zunger A 2008 *Phys. Rev.* B **78**(23) 235104

[33] Zhang S B, Wei S H and Zunger A 2001 *Phys. Rev.* B **63**(7)

[34] Jónsson H, Mills G and Jacobsen K W 1998 *Classical and quantum dynamics in condensed phase simulations* (World Scientific) chap 16, pp 385–404

[35] Wright A F 2002 *Journal of Applied Physics* **92** 2575–2585

Journal of Physics: Conference Series **633** (2015) 012144 doi:10.1088/1742-6596/633/1/012144

Relaxed elastic lines of second kind in semi-dual spaces

Gülşah Aydın Şekerci and A. Ceylan Çöken
Süleyman Demirel University, Isparta, Turkey
Akdeniz University, Antalya, Turkey
E-mail: gulsahaydin@sdu.edu.tr
E-mail: ceylancoken@akdeniz.edu.tr

Abstract. Theory of elasticity is a topic that keeps improving by using on many fields such as geometry, physics, chemistry and engineering. Energy density is given as some functions of curvature and torsion. If the curve of the Îś will be an external for the variation problem that minimizes the value of energy density; then this curve is called as relaxed elastic line. The relaxed elastic line on an oriented surface is considered as a model of DNA molecule. In this study, we worked on the second type relaxed elastic lines on the semi-dual spaces which has an important point on kinematic and Einstein's relativity theory. We also obtained boundary conditions for this type of curves. Moreover, the minimization problem of the energy which occurs with an applied force on an elastic line was discussed. Then, we researched the formed potential energy due to the applied force. Also, during the calculation of the potential energy on the elastic line, the amount of the potential energy for unit length of the elastic line was used. Afterwards, by integrating that amount, total potential energy calculated. So, we study to make a contribute both Einstein's relativity theory and kinematic.

1. Introduction

E^3 with the metric

$$\langle v, w\rangle = -v_1w_1 + v_2w_2 + v_3w_3, v = (v_1, v_2, v_3) \in E^3, w = (w_1, w_2, w_3) \in E^3$$

called Minkowski 3-space E_1^3. Scalar product in semi dual space is defined by

$$\langle x, y\rangle = \langle x^*, y^*\rangle + \varepsilon\left(\langle x^*, y^{**}\rangle + \langle x^{**}, y^*\rangle\right)$$

Let α be a curve with arc parameter s where

$$\begin{array}{rccl} \alpha: & I & \rightarrow & D_1^3 \\ & s & \rightarrow & \alpha(s) = \alpha^*(s) + \varepsilon\alpha^{**}(s) \end{array}$$

Then, unit tangent vector of $\alpha(s)$ is

$$\frac{d\alpha}{ds^*} = \frac{d\alpha}{ds}\frac{ds}{ds^*} = T$$

In here the curve α is parameterized with s, but s^* is a dual parameter. N is called principal normal vector and it is defined as $N = \frac{1}{\kappa}\frac{dT}{ds^*}$ where $\kappa = \kappa^* + \varepsilon\kappa^{**}$ is curvature. Morever, B is

Journal of Physics: Conference Series **633** (2015) 012144 doi:10.1088/1742-6596/633/1/012144

called binormal vector of $\alpha(s)$ if $B = T \times N$.

Definition. Let α is a curve with arc parameter. α is called second kind relaxed elastic line in D_1^3 dual space, if it is an extremal for variational problem which minimize value of H in family of all of curve with arc parameter and l length on S surface which have the same initial point and initial direction with α in D_1^3.

To understand elastic line, differential geometry of curves and surfaces is important. So, now we deal with relation between curve and surface used in the next section.

At $\alpha(s)$ the point of α, let T be unit tangent vector of α, n be unit normal vector of S and $n \times T = Q$. In that case, following equations is satisfied.

$$\begin{bmatrix} T' \\ Q' \\ n' \end{bmatrix} = \begin{bmatrix} 0 & \varepsilon_2 \kappa_g & \varepsilon_3 \kappa_n \\ -\varepsilon_1 \kappa_g & 0 & \varepsilon_3 \tau_g \\ -\varepsilon_1 \kappa_n & -\varepsilon_2 \tau_g & 0 \end{bmatrix} \begin{bmatrix} T \\ Q \\ n \end{bmatrix}$$

where κ_g is geodesic curvature, κ_n is normal curvature and τ_g is geodesic torsion.

2. Intrinsic equations for second kind relaxed elastic line in D_1^3

Suppose that the curve α lie on the surface S with $(u, v) \to x(u, v)$ where u and v are real parameter. Otherwise, if u and v are dual, S is hypersurface. Then, we take write as following when we take derivative according to u^* and v^* of x,

$$\begin{aligned} \frac{\partial x}{\partial u^*} &= \frac{\partial x}{\partial u}\frac{\partial u}{\partial u^*} = x_u \frac{\partial u}{\partial u^*} \\ \frac{\partial x}{\partial v^*} &= \frac{\partial x}{\partial v}\frac{\partial v}{\partial v^*} = x_v \frac{\partial v}{\partial v^*} \end{aligned}$$

So, we can state α the curve on S the surface.

$$\alpha(s) = x(u(s), v(s)), 0 \le s \le l$$

Also, T tangent vector of α the curve is found as following.

$$T(s) = \left(x_u \frac{\partial u}{\partial s} + x_v \frac{\partial v}{\partial s} \right) \frac{\partial s}{\partial s^*} = x_u \frac{\partial u}{\partial s^*} + x_v \frac{\partial v}{\partial s^*}$$

Let us take as $Q(s) = p(s) x_u + q(s) x_v$ where $p(s)$ and $q(s)$ are proper dual scalar functions. To obtain elastic line, let us extend to the arc $\underline{\alpha}$ which is defined by α for $0 \le s \le \underline{l}$ such that $\underline{l} > l$ to obtain variation arc with length $\underline{l}$ on S which has the same initial point and initial direction as α. But, $\underline{l}$ must be close enough to l so that both of them have the same coordinate neighbours.

For $0 \le s \le \underline{l}$, let $\mu(s)$ be a C^2 function which is not identically equal to zero. If we take as

$$\begin{aligned} \eta(s) &= \mu(s) \underline{p(s)} \\ \xi(s) &= \mu(s) \underline{q(s)} \end{aligned}$$

where $\underline{p(s)}$ and $\underline{q(s)}$ are coefficients of $Q(s)$ for a new obtained curve, then we can write that

$$\mu(s) Q(s) = \eta(s) x_u + \xi(s) x_v$$

and let $\mu(0) = 0 + \varepsilon 0$ and $\mu'(0) = 0 + \varepsilon 0$ be. For $0 \le \sigma \le \underline{l}$, we have

$$\beta(\sigma; t^*) = x(u(\sigma) + t^* \eta(\sigma), v(\sigma) + t^* \xi(\sigma))$$

where t^* is a dual parameter. For $0 \leq \sigma \leq \underline{l}$, let $\underline{L}(t^*)$ be a lenght of arc $\beta(\sigma; t^*)$. Then, we have

$$\underline{L}(t^*) = \int_0^l \sqrt{\left|\left\langle \frac{\partial \beta}{\partial \sigma^*}, \frac{\partial \beta}{\partial \sigma^*} \right\rangle\right|} d\sigma$$

Now, let us obtain equations which be used in the next. So, we have

$$\frac{\partial \beta}{\partial \sigma^*}|_{t^*=0} = \left(\frac{\partial x}{\partial u^*} \frac{du^*}{d\sigma^*} + \frac{\partial x^*}{\partial v^*} \frac{dv^*}{d\sigma^*} \right) = T(\sigma)$$

So, following equation is obtained by differentiated this equation

$$\frac{\partial^2 \beta}{\partial \sigma^{*2}}|_{t^*=0} = T'(\sigma) = \varepsilon_2 \kappa_g Q + \varepsilon_3 \kappa_n n$$

Also, the following equation is found.

$$\frac{\partial \beta}{\partial \sigma^*}|_{t^*=0} = \mu(\sigma) Q(\sigma)$$

If we differentiate this equation according to t*,

$$\frac{\partial^2 \beta}{\partial t^* \partial \sigma^*}|_{t^*=0} = \mu' Q - \varepsilon_1 \mu \kappa_g T + \varepsilon_3 \mu \tau_g n$$

Now, let us calculate the other equations.

$$\begin{aligned} \frac{\partial^3 \beta}{\partial t^* \partial \sigma^{*2}}|_{t^*=0} &= \left(-2\varepsilon_1 \mu' \kappa_g - \varepsilon_1 \mu \kappa_g' - \varepsilon_1 \varepsilon_3 \mu \tau_g \kappa_n\right) T \\ &+ \left(\mu'' - \varepsilon_1 \varepsilon_2 \mu \kappa_g^2 - \varepsilon_2 \varepsilon_3 \mu \tau_g^2\right) Q + (2\varepsilon_3 \mu' \tau_g \\ &- \varepsilon_1 \varepsilon_3 \mu \kappa_g \kappa_n + \varepsilon_3 \mu \tau_g') n \end{aligned}$$

$$\frac{d\lambda}{dt^*}|_{t^*=0} = \int_0^l \varepsilon_1 \mu \kappa_g ds$$

$$\begin{aligned} \frac{\partial^3 \beta}{\partial \sigma^{*3}}|_{t^*=0} &= \left(-\varepsilon_1 \varepsilon_2 \kappa_g^2 - \varepsilon_1 \varepsilon_3 \kappa_n^2\right) T + \left(\varepsilon_2 \kappa_g' - \varepsilon_2 \varepsilon_3 \tau_g \kappa_n\right) Q \\ &+ \left(\varepsilon_2 \varepsilon_3 \kappa_g \tau_g + \varepsilon_3 \kappa_n'\right) n \end{aligned}$$

$$\begin{aligned} \frac{\partial^4 \beta}{\partial t^* \partial \sigma^{*3}}|_{t^*=0} &= (-3\varepsilon_1 \mu'' \kappa_g - 3\varepsilon_1 \mu' \kappa_g' - \varepsilon_1 \mu \kappa_g'' - 3\varepsilon_1 \varepsilon_3 \mu' \tau_g \kappa_n \\ &- 2\varepsilon_1 \varepsilon_3 \mu \ \tau_g' \kappa_n - \varepsilon_1 \varepsilon_3 \mu \ \tau_g \kappa_n' + \varepsilon_2 \mu \ \kappa_g^3 \\ &+ \varepsilon_1 \varepsilon_2 \varepsilon_3 \mu \ \tau_g^2 \kappa_g + \varepsilon_3 \mu \ \kappa_g \kappa_n^2) T \end{aligned}$$

$$+(-3\varepsilon_1\varepsilon_2\mu^{'}\kappa_g^2 - 3\varepsilon_1\varepsilon_2\mu\kappa_g^{'}\kappa_g + \mu^{'''} - 3\varepsilon_2\varepsilon_3\mu^{'}\tau_g^2$$

$$-3\varepsilon_2\varepsilon_3\mu\tau_g\tau_g^{'})Q + (-3\varepsilon_1\varepsilon_3\mu^{'}\kappa_g\kappa_n - 2\varepsilon_1\varepsilon_3\mu\kappa_g^{'}\kappa_n$$

$$-\varepsilon_1\mu\tau_g\kappa_n^2 + 3\varepsilon_3\mu^{''}\tau_g - \varepsilon_2\mu\tau_g^3 + 3\varepsilon_3\mu^{'}\tau_g^{'}$$

$$-\varepsilon_1\varepsilon_3\mu\kappa_g\kappa_n^{'} + \varepsilon_3\mu\tau_g^{''} - \varepsilon_1\varepsilon_2\varepsilon_3\mu\tau_g\kappa_g^2)n$$

So, we can obtain equations which satisfy condition $\frac{dH}{dt^*}|_{t^*=0} = 0$ and so, α is must satisfy following boundary conditions for all of μ minimized α to H,

$$(i)\ \ (\kappa_n^3\tau_g\tau_g' + \varepsilon_3\kappa_n^2\kappa_g\tau_g'' - 3\varepsilon_1\varepsilon_3\kappa_g'\kappa_g\kappa_n^3 + 4\varepsilon_1\varepsilon_2\kappa_g^2\kappa_n') + (-7\varepsilon_1\varepsilon_2\kappa_g^4\kappa_n$$

$$-8\varepsilon_1\varepsilon_3\kappa_g^2\kappa_n^3 + \varepsilon_1\varepsilon_3\kappa_g^2\kappa_n'\kappa_n^2 - \tau_g^2\kappa_n'\kappa_n^2 + \varepsilon_3\kappa_g'\kappa_n^2\tau_g' + \varepsilon_1\varepsilon_2\kappa_g^4\kappa_n' + \varepsilon_2\varepsilon_3\kappa_g^2\tau_g^2\kappa_n'$$

$$-\varepsilon_2\kappa_g^2\tau_g'\kappa_g' + 2\varepsilon_2\varepsilon_3\tau_g'\kappa_n\tau_g\kappa_g^2 + \varepsilon_2\kappa_g^3\tau_g'' - 3\varepsilon_1\varepsilon_2\kappa_g^3\kappa_n\kappa_g') - [u\,(s)\,(-7\varepsilon_1\varepsilon_2\kappa_g^4\kappa_n$$

$$-8\varepsilon_1\varepsilon_3\kappa_g^2\kappa_n^3 - 2\varepsilon_2\kappa_g^2\kappa_g' - 3\varepsilon_2\varepsilon_3\kappa_g^2\tau_g^2\kappa_n + 3\varepsilon_2\kappa_g^3\tau_g' - \varepsilon_1\varepsilon_2\kappa_n^5 + 2\varepsilon_3\tau_g\kappa_n^2\kappa_g'$$

$$-3\tau_g^2\kappa_n^3 + 3\varepsilon_3\tau_g'\kappa_n^2\kappa_g]' + [u\,(s)\,(2\varepsilon_2\kappa_g^3\tau_g - \kappa_n'\kappa_g^2 + 2\varepsilon_3\kappa_g\kappa_n^2\tau_g + \varepsilon_2\varepsilon_3\kappa_n'\kappa_n^2)]''$$

$$-[u\,(s)\,(-\kappa_g^2\kappa_n - \varepsilon_2\varepsilon_3\kappa_n^3)]''' = 0$$

$$(ii)\ \ u\,(s)\,(-7\varepsilon_1\varepsilon_2\kappa_g^4\kappa_n - 8\varepsilon_1\varepsilon_3\kappa_g^2\kappa_n^3 - 2\varepsilon_2\kappa_g^2\kappa_g' - 3\varepsilon_2\varepsilon_3\kappa_g^2\kappa_n\tau_g^2 + 3\varepsilon_2\tau_g'\kappa_g^3)$$

$$-\varepsilon_1\varepsilon_2\kappa_n^5 + 2\varepsilon_3\kappa_n^2\kappa_g'\tau_g - 3\kappa_n^3\tau_g^2 + 3\varepsilon_3\kappa_g\kappa_n^2\tau_g') - [u\,(s)\,(2\varepsilon_2\kappa_g^3\tau_g - \kappa_g^2\kappa_n'$$

$$+2\varepsilon_3\kappa_g\kappa_n^2\tau_g + \varepsilon_2\varepsilon_3\kappa_n^2\kappa_n']' + [u\,(s)\,(-\kappa_g^2\kappa_n - \varepsilon_2\varepsilon_3\kappa_n^3]'' = 0$$

$$(iii)\ \ u\,(l)\,(-\kappa_g^2\kappa_n - \varepsilon_2\varepsilon_3\kappa_n^3)\,|_{s=l} = 0.\{But, dueto\}\kappa_g, \kappa_n \neq 0, u\,(l) = 0\{is.\}$$

$$(iv)\ \ u\,(0)\left(-\kappa_g^2\kappa_n\right) - \varepsilon_2\varepsilon_3\kappa_n^3)\,|0 = 0.\{But, dueto\}\kappa_g, \kappa_n \neq 0, u\,(0) = 0\{is.\}$$

$$(v)\ \ u\,(s)\,(2\varepsilon_2\kappa_g^3\tau_g - \kappa_g^2\kappa_n' + 2\varepsilon_3\kappa_g\kappa_n^2\tau_g + \varepsilon_2\varepsilon_3\kappa_n^2\kappa_n') - [u\,(s)\,(-\kappa_g^2\kappa_n - \varepsilon_2\varepsilon_3\kappa_n^3)]' = 0$$

3. Intrinsic equations for second kind relaxed elastic line deformed by external field in semi dual space

When the external field is applied, the following integration is used to find potential energy on all of elastic line, since the potential energy change according to intensity of external field. The variational problem is to minimize energy

$$H = \int_0^l \left(\frac{1}{2}b\tau^2 - \rho U\right) ds$$

Differential equation minimized the integral $\int_0^l U ds$ can be given as following.

$$\kappa_g \left(\varepsilon_1 U + U\left(l\right)\right) + \frac{\partial U}{\partial u^*} p_1 + \frac{\partial U}{\partial v^*} p_2 = 0$$

So, equations satisfied boundary conditions and minimized energy H can be obtain easily.

References

[1] Tutar A and Sarıoğlugil A 2006 *Applied Mathematics and Mechanics* B **27** 1481
[2] O'Neill B 1983 *Semi-Riemannian Geometry with Applications to Relativity* (New York: Academic Press)
[3] Manning G S 1987 *Quarterly of Applied Mathematics* **45** 515
[4] Ekici C and Görgülü A 2009 *Turk J. Math* **33** 397
[5] OYücesan A and Çöken A C 2008 *Science in China Series A: Mathematics* **51** 233
[6] Singer D 2008 *AIP Conf. Proc.* B **1002**